MyMathLab® = Your Resource for Success

In the lab, at home,...

- Access videos, PowerPoint® slides, and animations.
- Complete assigned homework and quizzes.
- Learn from your own personalized Study Plan.
- Print out MyMathGuide for additional practice.
- Explore even more tools for success.

...and on the go.

 Download the free MyDashBoard App to see instructor announcements and check your results on your Apple® or Android™ device. MyMathLab log-in is required.

 Download the free Pearson eText App to access the full eText on your Apple® or Android™ device. MyMathLab log-in is required.

 Use your Chapter Test as a study tool! Chapter Test Prep Videos show step-by-step solutions to all Chapter Test exercises. Access these videos in MyMathLab or by scanning the code.

Don't Miss Out! Log In Today.

MyMathLab delivers proven results in helping individual students succeed. It provides engaging experiences that personalize, stimulate, and measure learning for each student. And, it comes from a trusted partner with educational expertise and an eye on the future.

To learn more about how MyMathLab combines proven learning applications with powerful assessment, visit **www.mymathlab.com**

VIDEOS • POWERPOINT SLIDES • ANIMATIONS • HOMEWORK AND QUIZZES • PERSONALIZED STUDY PLAN • TOOLS FOR SUCCESS

NEW! Bittinger Integrated Video and *MyMathGuide* Program

The Bittinger video program and *MyMathGuide* are designed to work hand in hand. Each objective-based video corresponds exactly to the *MyMathGuide* workbook, and vice versa, so students can use both to achieve the conceptual understanding they need to succeed.

Video Program

- **NEW! To-the-Point Objective Videos**
  - **Concise:** Videos get right to the core of each concept.
  - **Objective based:** Simple navigation lets users view a whole section, choose an objective, or go straight to an example.
  - **Interactive: Your Turn Video Checks** let students pause to work problems and check answers.
  - **Integrated:** Designed for use with *MyMathGuide: Notes, Practice, and Video Path*.

- **Chapter Test Prep Videos**
  - **Step-by-step solutions** for every problem in the Chapter Tests
  - Available in MyMathLab,® on YouTube,™ and by scanning the QR code on the Chapter Test pages in the textbook

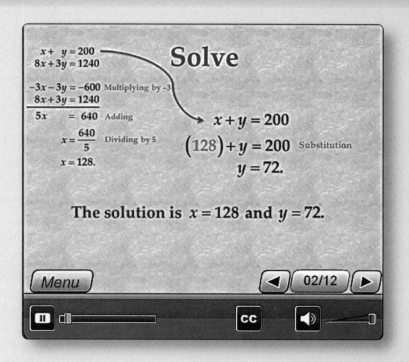

$$x + y = 200$$
$$8x + 3y = 1240$$

$$-3x - 3y = -600 \quad \text{Multiplying by } -3$$
$$8x + 3y = 1240$$

$$5x \qquad = 640 \quad \text{Adding}$$

$$x = \frac{640}{5} \quad \text{Dividing by 5}$$

$$x = 128.$$

Solve

$$x + y = 200$$
$$(128) + y = 200 \quad \text{Substitution}$$
$$y = 72.$$

The solution is $x = 128$ **and** $y = 72$.

Menu ◄ 02/12 ► CC

Section 5.3 | *Factoring Trinomials of the Type* $ax^2 + bx + c$ 7

Factoring with FOIL

ESSENTIALS

To factor $ax^2 + bx + c$ using FOIL:
1. Make certain that all common factors have been removed.
2. Find two First terms whose product is ax^2.
3. Find two Last terms whose product is c.
4. Check by multiplying to see if the sum of the Outer and Inner products is bx. If it is not, repeat steps (2) and (3) until the correct combination is found. If no correct combination exists, state that the polynomial is prime.

Example

- Factor $6x^2 + x - 12$.

  $6x^2 + x - 12 = (2x + 3)(3x - 4)$

GUIDED LEARNING:	📍 **Textbook**	📍 **Instructor**	📍 **Video**

EXAMPLE 1	YOUR TURN 1
Factor: $3x^2 + 13x - 10$.	Factor: $2x^2 + 5x - 12$.

1. There is no common factor other than 1 or −1.

2. $3x^2$ can be factored as $(3x)(\quad)$.

   The First terms can be $(3x \quad)(x \quad)$.

3. and 4. There are four pairs of factors of −10, and each can be listed in two ways.

Pairs of factors	Corresponding trial		Compare middle terms
−1, 10	$(3x-1)(x+10)$	$= 3x^2$	$\boxed{} -10$
1, −10	$(3x+1)(x-10)$	$= 3x^2$	$\boxed{} -10$
−2, $\boxed{}$	$(3x-2)(x+5)$	$= 3x^2$	$\boxed{} -10$
2, −5	$(3x+2)(x-5)$	$= 3x^2$	$\boxed{} -10$
10, $\boxed{}$	$(3x+10)(x-1)$	$= 3x^2$	$\boxed{} -10$
−10, 1	$(3x-10)(x+1)$	$= 3x^2$	$\boxed{} -10$
5, $\boxed{}$	$(3x+5)(x-2)$	$= 3x^2$	$\boxed{} -10$
−5, 2	$(3x-5)(x+2)$	$= 3x^2$	$\boxed{} -10$

$$3x^2 + 13x - 10 = (\boxed{})(\boxed{})$$

NEW! *MyMathGuide: Notes, Practice, and Video Path*

- **Objective-based, hands-on, guided learning:** students follow a textbook, instructor, or video path.

- Notes on **key concepts, skills, and definitions** for each learning objective

- **Vocabulary** practice and review

- **Examples** with guided solutions and Your Turn practice exercises

- **Space** to write questions and notes and to show work

- **Additional Practice Exercises** with Readiness Checks

- Designed to correlate exactly with the new **To-the-Point Objective Videos**

EDITION

9

Intermediate Algebra
Concepts and Applications

Marvin L. Bittinger
Indiana University Purdue University Indianapolis

David J. Ellenbogen
Community College of Vermont

Barbara L. Johnson
Indiana University Purdue University Indianapolis

PEARSON

Boston   Columbus   Indianapolis   New York   San Francisco   Upper Saddle River
Amsterdam   Cape Town   Dubai   London   Madrid   Milan   Munich   Paris   Montréal   Toronto
Delhi   Mexico City   São Paulo   Sydney   Hong Kong   Seoul   Singapore   Taipei   Tokyo

Editorial Director	Christine Hoag
Editor in Chief	Maureen O'Connor
Executive Editor	Cathy Cantin
Executive Content Editor	Kari Heen
Content Editor	Katherine Minton
Editorial Assistant	Kerianne Okie
Senior Managing Editor	Karen Wernholm
Senior Production Supervisor	Ron Hampton
Senior Author Support/ Technology Specialist	Joe Vetere
Composition	PreMediaGlobal
Production and Editorial Services	Martha K. Morong/Quadrata, Inc.
Art Editor and Photo Researcher	Geri Davis/The Davis Group, Inc.
Rights and Permissions Advisor	Sarah Smith/Creative Compliance
Associate Media Producer	Jonathan Wooding
Content Development Manager	Rebecca Williams (MXL)
Senior Content Developer	Mary Durnwald (TestGen)
Marketing Manager	Rachel Ross
Procurement Manager	Evelyn Beaton
Procurement Specialist	Debbie Rossi
Associate Design Director, USHE North and West	Andrea Nix
Senior Media Procurement Specialist	Ginny Michaud
Text Designer	Geri Davis/The Davis Group, Inc.
Cover Designer	Barbara T. Atkinson
Cover Photograph	Mountain Lake, Glacier National Park, Montana; © Doug Lemke/Shutterstock

Photo Credits
Photo credits appear on page A-66.

Library of Congress Cataloging-in-Publication Data
Bittinger, Marvin L.
Intermediate algebra : concepts and applications.
    Marvin L. Bittinger, David J. Ellenbogen, Barbara L. Johnson—9th ed.
        p. cm.
    Includes indexes.

1. Algebra—Textbooks.   I. Ellenbogen, David.   II. Title.
    QA154.3.B58 2013
    512. 9—dc23   2012033159

5 6 7 8 9 10—DORK—16 15 14

www.pearsonhighered.com

ISBN-13: 978-0-321-84828-4
ISBN-10: 0-321-84828-4

Contents

A KEY TO THE ICONS IN THE EXERCISE SETS

↪   Concept reinforcement exercises, indicated by blue exercise numbers, provide basic practice with the new concepts and vocabulary.

Aha!   Exercises labeled Aha! indicate the first time that a new insight can greatly simplify a problem and help students be alert to using that insight on following exercises. They are not more difficult.

   Calculator exercises are designed to be worked using either a scientific calculator or a graphing calculator.

⌗   Graphing calculator exercises are designed to be worked using a graphing calculator and often provide practice for concepts discussed in the Technology Connections.

📝   Writing exercises are designed to be answered using one or more complete sentences.

✔   A check mark in the annotated instructor's edition indicates Synthesis exercises that the authors consider particularly beneficial for students.

Preface

Welcome to the ninth edition of *Intermediate Algebra: Concepts and Applications*! As always, our goal is to present content that is easy to understand yet of enough depth to allow success in future courses. You will recognize many proven features, applications, and explanations; you will also find new material developed as a result of our experience in the classroom as well as from insights from faculty and students throughout North America.

This text is intended for students with a firm background in *elementary algebra*. It is one of three texts in an algebra series that also includes *Elementary Algebra: Concepts and Applications*, Ninth Edition, and *Elementary and Intermediate Algebra: Concepts and Applications*, Sixth Edition.

Approach Our goal is to help today's students learn and retain mathematical concepts. To achieve this, we feel that we must prepare developmental-mathematics students for the transition from "skills-oriented" elementary algebra courses to more "concept-oriented" college-level mathematics courses. This requires the development of critical thinking skills: to reason mathematically, to communicate mathematically, and to identify and solve mathematical problems. Following are aspects of our approach that we use to help meet the challenges we all face when teaching developmental mathematics.

Problem Solving We use problem solving and applications to motivate the students wherever possible, and we include real-life applications and problem-solving techniques throughout the text. Problem solving encourages students to think about how mathematics can be used, and it helps to prepare them for more advanced material in future courses.

In Chapter 1, we introduce our five-step process for solving problems: (1) Familiarize, (2) Translate, (3) Carry out, (4) Check, and (5) State the answer. Repeated use of this problem-solving strategy throughout the text provides students with a starting point for any type of problem they encounter, and frees them to focus on the unique aspects of the particular problem. We often use estimation and carefully checked guesses to help with the *Familiarize* and *Check* steps (see pp. 31 and 394).

What's New in the Ninth Edition? In addition to the following new features and other changes in this edition, including a new design, we have rewritten many key topics in response to user and reviewer feedback and have made significant improvements in design, art, pedagogy, and an expanded supplements package. Detailed information about the content changes outlined on p. ix is available in the form of a conversion guide. Please ask your Pearson representative for more information.

Applications Interesting applications of mathematics help motivate students and instructors. In the new edition, we have updated real-world data problems and examples to include subjects such as length of marriages (pp. 441–442), coral reefs (p. 60), and a satellite's escape velocity (pp. 412, 421, and 475). For a complete list of applications and the page numbers on which they can be found, please refer to the Index of Applications in MyMathLab or on the Instructor Resource Center.

Pedagogy

New! **Your Turn Exercises.** After every example, students are directed to work a similar exercise. This provides immediate reinforcement of concepts and skills. Answers to these exercises appear at the end of each exercise set. (See pp. 74 and 99.)

New! **Exploring the Concept.** Appearing once in every chapter, this feature encourages students to think about or visualize a concept. These activities lead into Active Learning Figures available in MyMathLab. Students can manipulate ALFs to further explore and solidify their understanding of the concepts. (See pp. 97, 335, and 480.)

Revised! **Connecting the Concepts.** Appearing at least once in every chapter, this feature summarizes concepts from several sections or chapters and illustrates connections between them. It includes a set of mixed exercises to help students make these connections. (See pp. 124–125 and 339.)

Algebraic–Graphical Connections.   Appearing occasionally throughout the text, this feature draws attention to visualizations of algebraic concepts. (See pp. 113, 154, and 698.)

Technology Connections.   These optional features in each chapter help students use a graphing calculator or a graphing calculator app to visualize concepts. Exercises are included with many of these features, and additional exercises in many exercise sets are marked with a graphing calculator icon to indicate more practice with this optional use of technology. (See pp. 77, 153, and 534.)

Student Notes.   Comments in the margin are addressed directly to students in a conversational tone. Ranging from suggestions for avoiding common mistakes to how to best read new notation, they give students extra explanation of the mathematics appearing on that page. (See pp. 75, 173, and 239.)

Study Skills.   We offer one study tip in every section. Ranging from time management to test preparation, these suggestions for successful study habits apply to any college course and any level of student. (See pp. 189, 327, and 383.)

New!   **Mid-Chapter Review.**   In the middle of every chapter, we offer a brief summary of the concepts covered in the first part of the chapter, two guided solutions to help students work step-by-step through solutions, and a set of mixed review exercises. (See pp. 118, 254, and 469.)

Exercise Sets

New!   **Vocabulary and Reading Check.**   These exercises begin every exercise set and are designed to encourage the student to read the section. Students who can complete these exercises should be prepared to work the remaining exercises in the exercise set. (See pp. 89, 286, and 404.)

Concept Reinforcement Exercises.   These true/false, matching, and fill-in-the-blank exercises appearing near the beginning of many exercise sets build students' confidence and comprehension. Answers to all concept reinforcement exercises appear in the answer section at the back of the book. (See pp. 60, 397, and 537.)

Aha! Exercises.   These exercises are not more difficult than their neighboring exercises; in fact, they can be solved more quickly, without lengthy computation, if the student has the proper insight. Designed to reward students who "look before they leap," the icon indicates the first time a new insight applies, and then it is up to the student to determine when to use that insight on subsequent exercises. (See pp. 126, 339, and 388.)

Revised!   **Skill Review Exercises.**   These exercises appear in every section beginning with Section 1.2. Taken together, each chapter's Skill Review exercises review all the major concepts covered in previous chapters in the text. Often these exercises focus on a single topic, such as solving equations, from multiple perspectives. (See pp. 263, 299, and 485.)

Synthesis Exercises.   Synthesis exercises appear in each exercise set following the Skill Review exercises. Students will need to use skills and concepts from earlier sections to solve these problems, and this will help them develop deeper insights into the current topic. The synthesis exercises are a real strength of the text, and in the annotated instructor's edition, the authors have placed a ✔ next to selected synthesis exercises that they suggest instructors "check out" and consider assigning. These exercises are not more difficult than the others, but they do use previously learned concepts and skills in ways that may be especially beneficial to students as they prepare for future topics. (See pp. 157, 235, and 342.)

Writing Exercises.   Two basic writing exercises appear just before the Skill Review exercises, and at least two more challenging exercises appear in the Synthesis exercises. Writing exercises aid student comprehension by requiring students to use critical thinking to explain concepts in one or more complete sentences. Because correct answers may vary, the only writing exercises for which answers appear at the back of the text are those in the chapter Review Exercises. (See pp. 104, 195, and 252.)

New!   **Quick Quiz.**   Beginning with the second section in each chapter, a five-question Quick Quiz appears near the end of each exercise set. Containing questions from sections already covered in the chapter, these quizzes provide a short but consistent review of the material in the chapter and help students prepare for a chapter test. (See pp. 166 and 564.)

New!   **Prepare to Move On.**   Beginning with Chapter 2, this short set of exercises appears at the end of every exercise set, and reviews concepts and skills previously covered in the text that will be used in the next section of the text. (See pp. 468 and 595.)

End of Chapter

Chapter Resources. These learning resources appear at the end of each chapter, making them easy to integrate into lessons at the most appropriate time. The mathematics necessary to use the resource has been presented by the end of the section indicated with each resource.

- **Translating for Success** and **Visualizing for Success.** These matching exercises help students learn to translate word problems to mathematical language and to graph equations and inequalities. (See pp. 63 and 213.)
- **Collaborative Activity.** Studies show that students who work in groups generally outperform those who do not, so these optional activities direct students to explore mathematics together. Additional collaborative activities and suggestions for directing collaborative learning appear in the *Instructor's Resources Manual with Tests and Mini Lectures.* (See pp. 272, 424, and 575.)
- New! **Decision Making: Connection.** Although many applications throughout the text can be applied in decision-making situations, this feature specifically applies the math of each chapter to a context in which students may be involved in decision making. (See pp. 214, 345, and 646.)

Revised! **Study Summary.** At the end of each chapter, this synopsis gives students a fast and effective review of key chapter terms and concepts. Concepts are paired with worked-out examples and practice exercises. (See pp. 346 and 576.)

Chapter Review and Test. A thorough chapter review and a practice test help prepare students for a test covering the concepts presented in each chapter. (See pp. 144 and 428.)

New! **Quick Response (QR) Codes.** These have been added to each chapter test, allowing students to link directly to chapter test prep videos with step-by-step solutions to all chapter test exercises. This effective tool allows students to receive help studying exactly when they need it at point of use.

Cumulative Review. This review appears after every chapter beginning with Chapter 2 to help students retain and apply their knowledge from previous chapters. (See pp. 148 and 502.)

New Design

While incorporating a new layout, a fresh palette of colors, and new features, we retain an open look and a typeface that is easy to read. In addition, we continue to pay close attention to the pedagogical use of color to ensure that it is used to present concepts in the clearest possible manner.

Content Changes

The exposition, examples, and exercises have been carefully reviewed and, as appropriate, revised or replaced. Some of the more significant changes are listed below.

- Section 2.5 now provides a comprehensive discussion of finding equations of lines.
- The explanation of simplifying powers of i in Section 7.8 is rewritten to provide students with a better understanding of the concept.
- Composition of functions in Section 9.1 is explained in greater detail to help students move from a concrete example to function notation.
- Examples and exercises that use real data are updated or replaced with current applications.
- The focus on students' conceptual understanding is enhanced. In addition to the Student Notes, Visualizing for Success, and Connecting the Concepts features present in earlier editions, an Exploring the Concept feature is now included in each chapter.
- Concept and skill review has been expanded. To aid students' review of the current chapter's material, Quick Quizzes have been added to the end of exercise sets, a Mid-Chapter Review has been added in each chapter, and the Study Summary has been expanded to include practice exercises. Also, cumulative review of material has been enhanced by a careful revision of the Skill Review exercises to provide an overall review of all material covered previously. In addition, Prepare to Move On exercises at the end of each exercise set review skills needed for the next section's material.
- The optional connection to technology has been revised to include new calculator operating systems and graphing calculator applications for mobile devices.

ANCILLARIES

The following ancillaries are available to help both instructors and students use this text more effectively.

STUDENT SUPPLEMENTS

NEW! Integrated Bittinger Video Program and MyMathGuide

Bittinger Video Program, available in MyMathLab, includes closed captioning and the following video types:

NEW! To-the-Point Objective Videos
- Concise, interactive, and objective based.
- View a whole section, choose an objective, or go straight to an example.
- Interactive Your Turn Video Check pauses for students to work problems and check answers.
- Seamlessly integrated with *MyMathGuide: Notes, Practice, and Video Path*.

Chapter Test Prep Videos
- Step-by-step solution for every problem in the Chapter Tests.
- Also available on YouTube, and by scanning the QR code on the Chapter Test pages in the text.

NEW! MyMathGuide: Notes, Practice, and Video Path
- Guided, hands-on learning for traditional, lab-based, hybrid, and online courses.
- Designed to correlate with *To-the-Point Objective Videos*.
- Highlights key concepts, skills, and definitions for each learning objective.
- Quick Review of key vocabulary terms and vocabulary practice problems.
- Examples with guided solutions and similar Your Turn exercises for practice.
- Space to write questions and notes and to show work.
- Additional Practice Exercises with Readiness Checks.
- Available in MyMathLab and printed.

ISBNs: 0-321-84832-2/978-0-321-84832-2

Student's Solutions Manual
by Christine Verity
- Contains completely worked-out solutions with step-by-step annotations for all the odd-numbered exercises in the text, with the exception of the writing exercises. Also contains all solutions to Chapter Review, Chapter Test, and Connecting the Concepts exercises.
- Available in MyMathLab and printed.

ISBNs: 0-321-84836-5/978-0-321-84836-9

INSTRUCTOR SUPPLEMENTS

Annotated Instructor's Edition
- Provides answers to all text exercises in color next to the corresponding problems.
- Includes Teaching Tips.
- Icons identify writing and graphing calculator exercises.

ISBNs: 0-321-84831-4/978-0-321-84831-4

Instructor's Solutions Manual (download only)
by Christine Verity
- Contains fully worked-out solutions to the odd-numbered exercises and brief solutions to the even-numbered exercises in the exercise sets.
- Available for download at www.pearsonhighered.com or in MyMathLab.

Instructor's Resource Manual (download only) with Tests and Mini Lectures
by Laurie Hurley
- Includes resources designed to help both new and adjunct faculty with course preparation and classroom management.
- Offers helpful teaching tips correlated to the sections of the text.
- Contains two multiple-choice tests per chapter, six free-response tests per chapter, and eight final exams.
- Available for download at www.pearsonhighered.com or in MyMathLab.

PowerPoint® Lecture Slides (download only)
- Present key concepts and definitions from the text.
- Available for download at www.pearsonhighered.com or in MyMathLab.

TestGen®
TestGen® (www.pearsoned.com/testgen) enables instructors to build, edit, print, and administer tests using a computerized bank of questions developed to cover all the objectives of the text. TestGen is algorithmically based, allowing instructors to create multiple equivalent versions of the same question or test with the click of a button. Instructors can also modify test bank questions or add new questions. The software and testbank are available for download from Pearson Education's online catalog.

AVAILABLE FOR STUDENTS AND INSTRUCTORS

MyMathLab® Online Course (access code required)
The Bittinger courses include all of MyMathLab's robust features, plus these additional highlights:

New! Two MyMathLab course options are now available:

- **Standard MyMathLab courses** allow instructors to build their course their way, offering maximum flexibility and control over all aspects of assignment creation.
New! - **Ready-to-Go courses** provide students with all the same great MyMathLab features, but make it easier for instructors to get started.

Both Bittinger course options include one pre-made pre-test and post-test for every chapter, section-lecture homework, and a chapter review quiz linked to personalized homework.

New! **Increased coverage** of skill-building, conceptual, and applications exercises, **plus new problem types,** provide even more flexibility when creating homework.

New! **Active Learning Figures**, available for key concepts, foster conceptual understanding for visual and tactile learners. Exploring the Concept in the text leads into ALFs available in MML. Students manipulate ALFs to explore concepts and solidify their understanding.

New! **Bittinger Video Program** includes all new To-the-Point Objective Videos and Chapter Test Prep Videos.

New! *MyMathGuide: Notes, Practice, and Video Path* can be viewed and printed.

MathXL® Online Course (access code required)
MathXL® is the homework and assessment engine that runs MyMathLab. (MyMathLab is MathXL plus a learning management system.)

With MathXL, instructors can

- Create, edit, and assign online homework and tests using algorithmically generated exercises correlated at the objective level to the textbook.
- Create and assign their own online exercises and import TestGen tests for added flexibility.
- Maintain records of all student work tracked in MathXL's online gradebook.

With MathXL, students can

- Take chapter tests in MathXL and receive personalized study plans and/or personalized homework assignments based on their test results.
- Use the study plan and/or the homework to link directly to tutorial exercises for the objectives they need to study.
- Access supplemental animations and video clips directly from selected exercises.

MathXL is available to qualified adopters. For more information, visit our website at www.mathxl.com, or contact your Pearson representative.

Acknowledgments

An outstanding team of professionals was involved in the production of this text. Judy Henn, Laurie Hurley, Helen Medley, Joanne Koratich, Monroe Street, and Holly Martinez carefully checked the book for accuracy and offered thoughtful suggestions. Michelle Lanosga and Daniel Johnson provided exceptional research support, and Christine Verity, Laurie Hurley, and Lisa Collette did remarkable work in preparing supplements. Special thanks are due Nelson Carter and Katherine Carter for their outstanding work on videos.

Martha Morong, of Quadrata, Inc., provided editorial and production services of the highest quality, and Geri Davis, of the Davis Group, Inc., performed superb work as designer, art editor, and photo researcher. Bill Melvin and Network Graphics provided the accurate and creative illustrations and graphs.

The team at Pearson deserves special thanks. Executive Editor Cathy Cantin, Executive Content Editor Kari Heen, Senior Production Supervisor Ron Hampton, Content Editor Katherine Minton, and Editorial Assistant Kerianne Okie provided many fine suggestions, coordinated tasks and schedules, and remained involved and accessible throughout the project. Marketing Manager Rachel Ross skillfully kept in touch with the needs of faculty. Editor in Chief Maureen O'Connor and Editorial Director Chris Hoag deserve credit for assembling this fine team.

We also thank the students at Indiana University Purdue University Indianapolis and the Community College of Vermont and the following professors for their thoughtful reviews and insightful comments.

Roberta Abarca, *Centralia College*; Darla J. Aguilar, *Pima Community College, Desert Vista Campus*; Bonnie Alcorn, *Waubonsee College*; Eugene Alderman, *South University*; Joseph Berland, *Chabot College*; Paul Blankenship, *Lexington Community College*; Susan Caldiero, *Cosumnes River College*; David Casey, *Citrus College*; Laura Crosby, *Bevill State Community College*; Steve Crowther, *Salt Lake Community College*; Emmett Dennis, *Southern Connecticut State University*; Jerrett Dumouchel, *Florida State College at Jacksonville*; Scott Eckert, *Cuyamaca College*; Nancy Eschen, *Florida State College at Jacksonville*; Henri Feiner, *Coastline University*; Gregory Fry, *El Camino College*; Joseph Gallegos, *Salt Lake Community College*; Gary Glaze, *Spokane Falls Community College*; Janet Hansen, *Dixie State College*; Nancy Hixson, *Columbia Southern University*; Linda Ho, *El Camino College*; Elizabeth Hodes, *Santa Barbara City College*; Weilin Jang, *Austin Community College*; Joanne Kawczenski, *Luzerne County Community College*; David Keller, *Kirkwood Community College*; Mojgan Khatoonabadi, *California State University–Fullerton*; Paulette Kirkpatrick, *Wharton County Junior College*; Susan Knights, *Boise State University*; Jeff Koleno, *Lorain County Community College*; Patricia Kopf, *Kellogg Community College*; Julianne Labbiento, *Lehigh Carbon Community College*; Kathryn Lavelle, *Westchester Community College*; Amy Marolt, *Northeastern Mississippi Community College*; Rogers Martin, *Louisiana State University, Shreveport*; Ben Mayo, *Yakima Valley Community College*; Laurie McManus, *St. Louis Community College–Meramac*; Carol Metz, *Westchester Community College*; Anne Marie Mosher, *St. Louis Community College–Florissant Valley*; Pedro Mota, *Austin Community College, South Austin Campus*; Brenda M. Norman, *Tidewater Community College*; Kristine Numrich, *El Camino College*; Kim Nunn, *Northeast State Technical College*; Michael Oppedisano, *Morrisville College, State University of New York*; Susan Quirarte, *Cosumnes River College*; Zaddock B. Reid, *San Bernardino Valley College*; Randall Robertson, *Solano Community College*; Nicole Saporito, *Luzerne County Community College*; Terri Seiver, *San Jacinto College–Central Campus*; Deborah Serdynski, *Florida State College at Jacksonville*; Kevin Shannon, *Orange Coast College*; Jean Shutters, *Harrisburg Area Community College*; Domingo Soria-Martin, *Cosumnes River College*; Richard Stevens, *Cuyamaca College*; Timothy Thompson, *Oregon Institute of Technology*; Diane Trimble, *Tulsa Community College–West Campus*; Jennifer Vanden Eynden, *Grossmont College*; Mariana Voicu, *Orange Coast College*; Beverly Vredevelt, *Spokane Falls Community College*; Michael Yarbrough, *Cosumnes River College*; Judy Wagner, *Blinn College–Bryan Campus*; Lisa Williams, *College of the Albemarle*

Finally, a special thank-you to all those who so generously agreed to discuss their professional use of mathematics in our chapter openers. These dedicated people all share a desire to make math more meaningful to students. We cannot imagine a finer set of role models.

M.L.B.
D.J.E.
B.L.J.

Algebra and Problem Solving

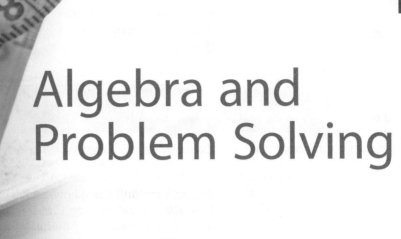

NAME	WEIGHT (in pounds)	HEIGHT (in inches)	BMI
Beyoncé	130	66	21
Tom Cruise	166	67	26.1
Serena Williams	150	69	22.2
Eli Manning	?	76	26.6

Source: www.health.com

Numbers Never Lie.

But numbers don't always give the whole picture. Take, for example, the **body mass index** (BMI), a figure that is often used to describe whether or not an individual is at a healthy weight. As the table above implies, BMI is a number calculated using weight and height; it does not take into account the percentage of weight that is lean muscle (rather than fat). For this reason, the BMI is best used in conjunction with other indicators when describing an individual's health. Later, we will use the formula for calculating BMI to estimate Eli Manning's weight. (*See Example 4 in Section 1.5*).

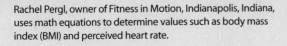

As a personal trainer, I find that math allows me to plan efficient workouts for my clients.

Rachel Pergl, owner of Fitness in Motion, Indianapolis, Indiana, uses math equations to determine values such as body mass index (BMI) and perceived heart rate.

The principal theme of this text is problem solving in algebra. In Chapter 1, we begin with a review of algebraic expressions and equations. The use of algebra as part of an overall strategy for solving problems is then presented in Section 1.4. Additional and increasing emphasis on problem solving appears throughout the book.

1.1 Some Basics of Algebra

Translating to Algebraic Expressions   ▪   Evaluating Algebraic Expressions   ▪   Sets of Numbers

The primary difference between algebra and arithmetic is the use of *variables*. A letter that can be any one of various numbers is called a **variable**. If a letter always represents a particular number that never changes, it is called a **constant**. If r represents the radius of the earth, in kilometers, then r is a constant. If a represents the age of a baby chick, in minutes, then a is a variable because a changes, or *varies*, as time passes. In this text, unless stated otherwise, we assume that all letters represent variables.

An **algebraic expression** consists of variables and/or numerals, often with operation signs and grouping symbols. Some examples of algebraic expressions are:

$t + 37;$ This contains the variable t, the constant 37, and the operation of addition.

$(s + t) \div 2.$ This contains the variables s and t, the constant 2, grouping symbols, and the operations addition and division.

Multiplication can be written in several ways. For example, "60 times n" can be written as $60 \cdot n$, $60 \times n$, $60(n)$, $60 * n$, or simply (and usually) $60n$. Division can also be represented by a fraction bar: $\frac{9}{7}$, or $9/7$, means $9 \div 7$.

When an equal sign is placed between two expressions, an **equation** is formed. We often **solve** equations.

For example, suppose that you collect $744 for group tickets to a concert. If you know that each ticket costs $12, an equation can be used to determine how many tickets were purchased.

One expression for total ticket sales is 744. Another expression for total ticket sales is $12x$, where x is the number of tickets purchased. Since these are equal expressions, we can write the equation

$$12x = 744.$$

To find a **solution**, we can divide both sides of the equation by 12:

$$x = 744 \div 12 = 62.$$

Thus, 62 tickets were purchased.

Using equations to solve problems like this is a major theme of algebra.

Translating to Algebraic Expressions

To translate phrases to expressions, we need to know which words correspond to which operations, as shown in the following table.

Key Words

Addition	Subtraction	Multiplication	Division
add	subtract	multiply	divide
sum of	difference of	product of	quotient of
plus	minus	times	divided by
increased by	decreased by	twice	ratio
more than	less than	of	per

When the value of a number is not given, we represent that number with a variable.

Phrase	Algebraic Expression
Five *more than* some number	$n + 5$
Half *of* a number	$\frac{1}{2}t$, or $\frac{t}{2}$
Five *more than* three *times* some number	$3p + 5$
The *difference* of two numbers	$x - y$
Six *less than* the *product of* two numbers	$rs - 6$
Seventy-six percent *of* some number	$0.76z$, or $\frac{76}{100}z$

EXAMPLE 1   Translate to an algebraic expression:

Five less than forty-three percent of the quotient of two numbers.

SOLUTION   We let r and s represent the two numbers.

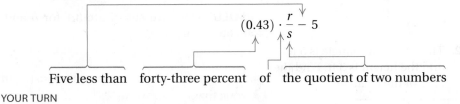

$$(0.43) \cdot \frac{r}{s} - 5$$

Five less than   forty-three percent   of   the quotient of two numbers

1. Translate to an algebraic expression: Half of the difference of two numbers.

YOUR TURN

Some algebraic expressions contain *exponential notation*. Many different kinds of numbers can be used as *exponents*. Here we establish the meaning of a^n when n is a counting number, $1, 2, 3, \ldots$.

EXPONENTIAL NOTATION

The expression a^n, in which n is a counting number, means

$$\underbrace{a \cdot a \cdot a \cdots a \cdot a}_{n \text{ factors.}}$$

In a^n, a is called the *base* and n is the *exponent*. When no exponent appears, the exponent is assumed to be 1. Thus, $a^1 = a$.

The expression a^n is read "a raised to the nth power" or simply "a to the nth." We read s^2 as "s-squared" and x^3 as "x-cubed." This terminology comes from the fact that the area of a square of side s is $s \cdot s = s^2$ and the volume of a cube of side x is $x \cdot x \cdot x = x^3$.

Area = s^2 s x Volume = x^3

s x x

Evaluating Algebraic Expressions

When we replace a variable with a number, we say that we are **substituting** for the variable. The calculation that follows the substitution is called **evaluating the expression**.

Geometric formulas are often evaluated. In the following example, we use the formula for the area of a triangle with a base of length b and a height of length h.

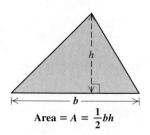

Area = $A = \frac{1}{2}bh$

EXAMPLE 2   The base of a triangular sail is 3.1 m and the height is 4 m. Find the area of the sail.

4 m

3.1 m

SOLUTION   We substitute 3.1 for b and 4 for h and multiply to evaluate the expression:

$$\tfrac{1}{2} \cdot b \cdot h = \tfrac{1}{2} \cdot 3.1 \cdot 4$$
$$= 6.2 \text{ square meters (sq m or m}^2).$$

2. The base of a triangle is 5 ft and the height is 3 ft. Find the area of the triangle.

YOUR TURN

Exponential notation tells us that 5^2 means $5 \cdot 5$, or 25, but what does $1 + 2 \cdot 5^2$ mean? If we add 1 and 2 and multiply by 25, we get 75. If we multiply 2 times 5^2 and add 1, we get 51. A third possibility is to square $2 \cdot 5$ to get 100 and then add 1. The following convention indicates that only the second of these approaches is correct: We square 5, then multiply, and then add.

Student Notes

Step (3) states that when division precedes multiplication, the division is performed first. Thus, $20 \div 5 \cdot 2$ represents $4 \cdot 2$, or 8. Similarly, $9 - 3 + 1$ represents $6 + 1$, or 7.

RULES FOR ORDER OF OPERATIONS

1. Simplify within any grouping symbols such as ( ), [ ], { }, working in the innermost symbols first.
2. Simplify all exponential expressions.
3. Perform all multiplication and division, working from left to right.
4. Perform all addition and subtraction, working from left to right.

EXAMPLE 3   Evaluate $5 + 2(a - 1)^2$ for $a = 4$.

SOLUTION

$$\begin{aligned} 5 + 2(a - 1)^2 &= 5 + 2(4 - 1)^2 && \text{Substituting} \\ &= 5 + 2(3)^2 && \text{Working within parentheses first} \\ &= 5 + 2(9) && \text{Simplifying } 3^2 \\ &= 5 + 18 && \text{Multiplying} \\ &= 23 && \text{Adding} \end{aligned}$$

3. Evaluate $2(x + 1)^2 - 10$ for $x = 5$.

YOUR TURN

Step (3) in the rules for order of operations tells us to divide before we multiply when division appears first, reading left to right. This means that an expression like $6 \div 2x$ means $(6 \div 2)x$.

> **CAUTION!**
>
> $6 \div 2x = (6 \div 2)x,$
>
> $6 \div (2x) = \dfrac{6}{2x}$
>
> $6 \div 2x$ *does not mean* $6 \div (2x)$.

EXAMPLE 4   Evaluate $9 - x^3 + 6 \div 2y^2$ for $x = 2$ and $y = 5$.

SOLUTION

$$\begin{aligned} 9 - x^3 + 6 \div 2y^2 &= 9 - 2^3 + 6 \div 2(5)^2 && \text{Substituting} \\ &= 9 - 8 + 6 \div 2 \cdot 25 && \text{Simplifying } 2^3 \text{ and } 5^2 \\ &= 9 - 8 + 3 \cdot 25 && \text{Dividing} \\ &= 9 - 8 + 75 && \text{Multiplying} \\ &= 1 + 75 && \text{Subtracting} \\ &= 76 && \text{Adding} \end{aligned}$$

4. Evaluate $8a^2 \div 5b - 4 + a$ for $a = 5$ and $b = 2$.

YOUR TURN

Sets of Numbers

When evaluating algebraic expressions, and in problem solving in general, we often must examine the *type* of numbers used. For example, if a formula is used to determine an optimal class size, fractions must be rounded up or down, since it is impossible to have a fractional part of a student. Three frequently used sets of numbers are listed below.

NATURAL NUMBERS, WHOLE NUMBERS, AND INTEGERS

Natural Numbers (or Counting Numbers)

Those numbers used for counting: $\{1, 2, 3, \ldots\}$

Whole Numbers

The set of natural numbers with 0 included: $\{0, 1, 2, 3, \ldots\}$

Integers

The set of all whole numbers and their opposites:

$$\{\ldots, -4, -3, -2, -1, 0, 1, 2, 3, 4, \ldots\}$$

The dots are called ellipses and indicate that the pattern continues without end.

Integers correspond to the points on the number line as follows:

The set containing the numbers -2, 1, and 3 can be written $\{-2, 1, 3\}$. This way of writing a set is known as **roster notation**. Roster notation was used for the three sets listed above. To describe sets containing the numbers between integers, we use a second type of set notation, **set-builder notation**, which specifies conditions under which a number is in the set. The following example of set-builder notation is read as shown:

$$\{x \mid x \text{ is a number between 1 and 5}\}$$

"The set of all x" — $\{x$

"such that" — $\mid$

"x is a number between 1 and 5"

Set-builder notation is generally used when it is difficult to list a set using roster notation.

EXAMPLE 5   Using both roster notation and set-builder notation, represent the set consisting of the first 15 even natural numbers.

SOLUTION

Using roster notation: $\{2, 4, 6, 8, 10, 12, 14, 16, 18, 20, 22, 24, 26, 28, 30\}$

Using set-builder notation: $\{n \mid n \text{ is an even number between 1 and 31}\}$

Note that other descriptions of the set are possible. For example, $\{2x \mid x \text{ is an integer } and\ 1 \leq x \leq 15\}$ is a common way of writing this set.

YOUR TURN

5. Using both roster notation and set-builder notation, represent the set of all multiples of 5 between 1 and 21.

The symbol $\in$ is used to indicate that an **element** or a **member** belongs to a set. Thus if $A = \{2, 4, 6, 8\}$, we can write $4 \in A$ to indicate that 4 *is an element of* A. We can also write $5 \notin A$ to indicate that 5 *is not an element of* A.

EXAMPLE 6   Classify the statement $8 \in \{x \mid x \text{ is an integer}\}$ as either true or false.

SOLUTION   Since 8 *is* an integer, the statement is true. In other words, since 8 is an integer, it belongs to the set of all integers.

YOUR TURN

6. Classify the statement $\frac{1}{2} \in \{x \mid x \text{ is a whole number}\}$ as either true or false.

Using set-builder notation, we can describe the set of all *rational numbers.*

RATIONAL NUMBERS

Numbers that can be expressed as an integer divided by a nonzero integer are called *rational numbers*:

$$\left\{ \frac{p}{q} \;\middle|\; p \text{ is an integer, } q \text{ is an integer, and } q \neq 0 \right\}.$$

Rational numbers can be written using fraction notation or decimal notation. *Fraction notation* uses symbolism like the following:

$$\frac{5}{8}, \quad \frac{12}{-7}, \quad \frac{-17}{15}, \quad -\frac{9}{7}, \quad \frac{39}{1}, \quad \frac{0}{6}.$$

In *decimal notation*, rational numbers either *terminate* (end) or *repeat* a block of digits.

For example, decimal notation for $\frac{5}{8}$ terminates, since $\frac{5}{8}$ means $5 \div 8$, and long division shows that $\frac{5}{8} = 0.625$, a decimal that ends, or terminates.

On the other hand, decimal notation for $\frac{6}{11}$ repeats, since $6 \div 11 = 0.5454\ldots$, a repeating decimal. Repeating decimal notation can be abbreviated by writing a bar over the repeating part—in this case, $0.\overline{54}$.

Many numbers, like π, $\sqrt{2}$, and $-\sqrt{15}$, are not rational numbers. For example, $\sqrt{2}$ is the number for which $\sqrt{2} \cdot \sqrt{2} = 2$. A calculator's representation of $\sqrt{2}$ as 1.414213562 is an approximation since $(1.414213562)^2$ is not exactly 2.

To see that $\sqrt{2}$ is a "real" point on the number line, we can show that when a right triangle has two legs of length 1, the remaining side has length $\sqrt{2}$. Thus we can "measure" $\sqrt{2}$ units and locate $\sqrt{2}$ on the number line.

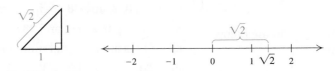

Numbers like π, $\sqrt{2}$, and $-\sqrt{15}$ are said to be **irrational**. Decimal notation for irrational numbers neither terminates nor repeats.

The set of all rational numbers, combined with the set of all irrational numbers, gives us the set of all **real numbers**.

REAL NUMBERS

Numbers that are either rational or irrational are called *real numbers*:

$$\{x \mid x \text{ is rational or } x \text{ is irrational}\}.$$

Every point on the number line represents some real number, and every real number is represented by some point on the number line.

The following figure shows the relationships among various kinds of numbers, along with examples of how real numbers can be sorted.

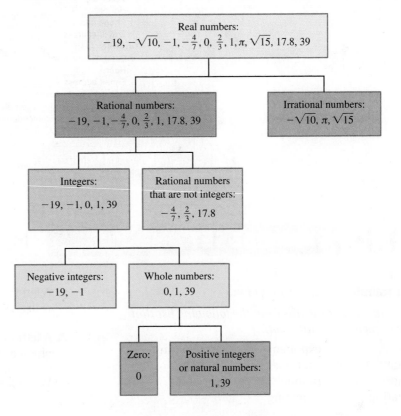

EXAMPLE 7 Which numbers in the following list are **(a)** whole numbers? **(b)** integers? **(c)** rational numbers? **(d)** irrational numbers? **(e)** real numbers?

$$-29, \quad -\tfrac{7}{4}, \quad 0, \quad 2, \quad 3.9, \quad \sqrt{42}, \quad 78$$

SOLUTION

a) 0, 2, and 78 are whole numbers.

b) -29, 0, 2, and 78 are integers.

7. Which numbers in the following list are integers?

$$-245, \quad 0, \quad 15, \quad \sqrt{11}, \quad \tfrac{2}{3}$$

c) -29, $-\tfrac{7}{4}$, 0, 2, 3.9, and 78 are rational numbers.

d) $\sqrt{42}$ is an irrational number.

e) -29, $-\tfrac{7}{4}$, 0, 2, 3.9, $\sqrt{42}$, and 78 are all real numbers.

YOUR TURN

When every member of one set is a member of a second set, the first set is a **subset** of the second set. Thus if $A = \{2, 4, 6\}$ and $B = \{1, 2, 4, 5, 6\}$, we write $A \subseteq B$ to indicate that *A is a subset of B*. Similarly, if $\mathbb{N}$ represents the set of all natural numbers and $\mathbb{Z}$ is the set of all integers, we can write $\mathbb{N} \subseteq \mathbb{Z}$. Additional statements can be made using other sets in the diagram above.

Study Skills

Get the Facts

Throughout this textbook, you will find a feature called *Study Skills*. These tips are intended to help improve your math study skills. On the first day of class, we recommend that you collect the course information shown here.

Instructor:
Name _____
Office hours and location _____
Phone number _____
E-mail address _____
Classmates:
1. Name _____
Phone number _____
E-mail address _____
2. Name _____
Phone number _____
E-mail address _____
Math lab on campus:
Location _____
Hours _____
Phone number _____
E-mail address _____
Tutoring:
Campus location _____
Hours _____
E-mail address _____
Important supplements:
(See the preface for a complete list of available supplements.)
Supplements recommended by the instructor.
_____
_____

1.1   EXERCISE SET

FOR EXTRA HELP

MyMathLab®   Math XL
PRACTICE   WATCH   READ   REVIEW

⤷ Vocabulary and Reading Check

Choose the word or words from the following list that best complete each statement.

base	exponent	terminating
constant	irrational	value
division	rational	variable
evaluating	repeating	

1. A letter that can be any one of a set of numbers is called a(n) _____.

2. A letter representing a specific number that never changes is called a(n) _____.

3. When $x = 10$, the _____ of the expression $4x$ is 40.

4. In a^b, the letter a is called the _____ and the letter b is called the _____.

5. When all variables in a variable expression are replaced by numbers and a result is calculated, we say that we are _____ the expression.

6. To calculate $4 + 12 \div 3 \cdot 2$, the first operation that we perform is _____.

7. A number that can be written in the form a/b, where a and b are integers (with $b \neq 0$), is said to be a(n) _____ number.

8. A real number that cannot be written as a quotient of two integers is an example of a(n) _____ number.

9. Division can be used to show that $\frac{7}{40}$ can be written as a(n) _____ decimal.

10. Division can be used to show that $\frac{13}{7}$ can be written as a(n) _____ decimal.

Translating to Algebraic Expressions

Use mathematical symbols to translate each phrase.

11. Five less than some number
12. Ten more than some number
13. Twice a number
14. Eight times a number
15. Twenty-nine percent of some number
16. Thirteen percent of some number
17. Six less than half of a number
18. Three more than twice a number
19. Seven more than ten percent of some number
20. Four less than six percent of some number
21. One less than the product of two numbers
22. One more than the difference of two numbers
23. Ninety miles per every four gallons of gas
24. One hundred words per every sixty seconds

Evaluating Algebraic Expressions

In Exercises 25–28, find the area of a square flower garden with the given length of a side. Use $A = s^2$.

25. Side = 6 ft
26. Side = 12 ft
27. Side = 0.5 m
28. Side = 2.5 m

In Exercises 29–32, find the area of a triangular fireplace with the given base and height. Use $A = \frac{1}{2}bh$.

29. Base = 5 ft, height = 7 ft
30. Base = 2.9 m, height = 2.1 m
31. Base = 7 ft, height = 3.2 ft
32. Base = 3.6 ft, height = 4 ft

To the student and the instructor: Throughout this text, selected exercises are marked with the icon Aha! *. Students who pause to inspect an Aha! exercise should find the answer more readily than those who proceed mechanically. This may involve looking at an earlier exercise or example, or performing calculations in a more efficient manner. Some Aha! exercises are left unmarked to encourage students to always pause before working a problem.*

Evaluate each expression using the values provided.

33. $3(x - 7) + 2$, for $x = 10$
34. $5 + (2x - 3)$, for $x = 8$
35. $12 + 3(n + 2)^2$, for $n = 1$
36. $(n - 10)^2 - 8$, for $n = 15$
37. $4x + y$, for $x = 2$ and $y = 3$
38. $8a - b$, for $a = 5$ and $b = 7$
39. $20 + r^2 - s$, for $r = 5$ and $s = 10$
40. $m^3 + 7 - n$, for $m = 2$ and $n = 8$
41. $2c \div 3b$, for $b = 2$ and $c = 6$
42. $3z \div 2y$, for $y = 1$ and $z = 6$

Aha! 43. $3n^2p - 3pn^2$, for $n = 5$ and $p = 9$
44. $2a^3b - 2b^2$, for $a = 3$ and $b = 7$
45. $5x \div (2 + x - y)$, for $x = 6$ and $y = 2$
46. $3(m + 2n) \div m$, for $m = 7$ and $n = 0$
47. $[10 - (a - b)]^2$, for $a = 7$ and $b = 2$
48. $[17 - (x + y)]^2$, for $x = 4$ and $y = 1$

49. $[5(r + s)]^2$, for $r = 1$ and $s = 2$

50. $[3(a - b)]^2$, for $a = 7$ and $b = 5$

51. $x^2 - [3(x - y)]^2$, for $x = 6$ and $y = 4$

52. $m^2 - [2(m - n)]^2$, for $m = 7$ and $n = 5$

53. $(m - 2n)^2 - 2(m + n)$, for $m = 8$ and $n = 1$

54. $(r - s)^2 - 3(2r - s)$, for $r = 11$ and $s = 3$

Sets of Numbers

Use roster notation to write each set.

55. The set of letters in the word "algebra"

56. The set of all days of the week

57. The set of all odd natural numbers

58. The set of all even natural numbers

59. The set of all natural numbers that are multiples of 10

60. The set of all natural numbers that are multiples of 5

Use set-builder notation to write each set.

61. The set of all even numbers between 9 and 99

62. The set of all multiples of 5 between 7 and 79

63. $\{0, 1, 2, 3, 4\}$

64. $\{-3, -2, -1, 0, 1, 2\}$

65. $\{11, 13, 15, 17, 19\}$

66. $\{24, 26, 28, 30, 32\}$

In Exercises 67–70, which numbers in the list provided are **(a)** *whole numbers?* **(b)** *integers?* **(c)** *rational numbers?* **(d)** *irrational numbers?* **(e)** *real numbers?*

67. $-8.7, -3, 0, \frac{2}{3}, \sqrt{7}, 6$

68. $-\frac{9}{2}, -4, -1.2, 0, \sqrt{5}, 3$

69. $-17, -0.01, 0, \frac{5}{4}, 8, \sqrt{77}$

70. $-6.08, -5, 0, 1, \sqrt{17}, \frac{99}{2}$

Classify each statement as either true or false. The following sets are used:

$\mathbb{N} =$ the set of natural numbers;

$\mathbb{W} =$ the set of whole numbers;

$\mathbb{Z} =$ the set of integers;

$\mathbb{Q} =$ the set of rational numbers;

$\mathbb{H} =$ the set of irrational numbers;

$\mathbb{R} =$ the set of real numbers.

71. $196 \in \mathbb{N}$

72. $\mathbb{N} \subseteq \mathbb{W}$

73. $\mathbb{W} \subseteq \mathbb{Z}$

74. $\sqrt{8} \in \mathbb{Q}$

75. $\frac{2}{3} \in \mathbb{Z}$

76. $\mathbb{H} \subseteq \mathbb{R}$

77. $\sqrt{10} \in \mathbb{R}$

78. $4.3 \notin \mathbb{Z}$

79. $\mathbb{Z} \not\subseteq \mathbb{N}$

80. $\mathbb{Q} \subseteq \mathbb{R}$

81. $\mathbb{Q} \subseteq \mathbb{Z}$

82. $\frac{8}{15} \in \mathbb{H}$

To the student and the instructor: *Writing exercises, denoted by* 📝*, are meant to be answered using English sentences. Because answers to many writing exercises will vary, solutions are not listed at the back of the book.*

83. What is the difference between rational numbers and integers?

84. Charlie insists that $15 - 4 + 1 \div 2 \cdot 3$ is 2. What error is he making?

Synthesis

To the student and the instructor: *Synthesis exercises are designed to challenge students to extend the concepts or skills studied in each section. Many synthesis exercises require the assimilation of skills and concepts from several sections.*

85. Is the following true or false, and why?

$$\{2, 4, 6\} \subseteq \{2, 4, 6\}$$

86. On a quiz, Mia answers $6 \in \mathbb{Z}$ while Giovanni writes $\{6\} \in \mathbb{Z}$. Giovanni's answer does not receive full credit while Mia's does. Why?

Translate to an algebraic expression.

87. The quotient of the sum of two numbers and their difference

88. Three times the sum of the cubes of two numbers

89. Half of the difference of the squares of two numbers

90. The product of the difference of two numbers and their sum

Use roster notation to write each set.

91. The set of all whole numbers that are not natural numbers

92. The set of all integers that are not whole numbers

93. $\{x \mid x = 5n, n \text{ is a natural number}\}$

94. $\{x \mid x = 3n, n \text{ is a natural number}\}$

95. $\{x \mid x = 2n + 1, n \text{ is a whole number}\}$

96. $\{x \mid x = 2n, n \text{ is an integer}\}$

97. Draw a right triangle that could be used to measure $\sqrt{13}$ units.

↪ **YOUR TURN ANSWERS: SECTION 1.1**

1. Let x and y represent the numbers: $\frac{1}{2}(x - y)$ **2.** 7.5 ft^2
3. 62   **4.** 81   **5.** $\{5, 10, 15, 20\}$,
$\{x \mid x \text{ is a multiple of 5 between 1 and 21}\}$ **6.** False
7. $-245, 0, 15$

1.2   Operations and Properties of Real Numbers

Absolute Value ▪ Inequalities ▪ Addition, Subtraction, and Opposites ▪ Multiplication, Division, and Reciprocals ▪ The Commutative, Associative, and Distributive Laws

In this section, we review addition, subtraction, multiplication, and division of real numbers. We also study important rules for manipulating algebraic expressions.

Absolute Value

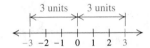

Both 3 and -3 are 3 units from 0 on the number line. Thus their distance from 0 is 3. We use *absolute-value* notation to represent a number's distance from 0. Note that distance is never negative.

> **ABSOLUTE VALUE**
>
> The notation $|a|$, read "the absolute value of a," represents the number of units that a is from 0 on the number line.

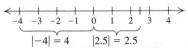

1. Find the absolute value: $|-237|$.

EXAMPLE 1   Find the absolute value: **(a)** $|-4|$; **(b)** $|2.5|$; **(c)** $|0|$.

SOLUTION

a) $|-4| = 4$ -4 is 4 units from 0.

b) $|2.5| = 2.5$ 2.5 is 2.5 units from 0.

c) $|0| = 0$ 0 is 0 units from itself.

YOUR TURN

Note that absolute value is never negative.

Inequalities

For any two numbers on the number line, the one to the left is said to be less than, or smaller than, the one to the right. The symbol $<$ means "is less than" and the symbol $>$ means "is greater than." The symbol $\leq$ means "is less than or equal to" and the symbol $\geq$ means "is greater than or equal to." These symbols are used to form **inequalities**.

As shown in the figure below, $-6 < -1$ (since -6 is to the left of -1) and $|-6| > |-1|$ (since 6 is to the right of 1).

EXAMPLE 2   Write out the meaning of each inequality and determine whether it is a true statement.

a) $-7 < -2$ b) $-3 \geq -2$

c) $5 \leq 6$ d) $6 \leq 6$

SOLUTION

	Inequality	*Meaning*
a)	$-7 < -2$	"-7 is less than -2" is *true* because -7 is to the left of -2.
b)	$-3 \geq -2$	"-3 is greater than or equal to -2" is *false* because -3 is to the left of -2.
c)	$5 \leq 6$	"5 is less than or equal to 6" is true if either $5 < 6$ or $5 = 6$. Since $5 < 6$ is true, $5 \leq 6$ is *true*.
d)	$6 \leq 6$	"6 is less than or equal to 6" is *true* because $6 = 6$ is true.

2. Write out the meaning of $-4 \leq -3$ and determine whether it is a true statement.

YOUR TURN

Addition, Subtraction, and Opposites

We are now ready to review addition of real numbers.

ADDITION OF TWO REAL NUMBERS

1. *Positive numbers*: Add the numbers. The result is positive.
2. *Negative numbers*: Add absolute values. Make the answer negative.
3. *A negative number and a positive number*: If the numbers have the same absolute value, the answer is 0. Otherwise, subtract the smaller absolute value from the larger one.

   a) If the positive number has the greater absolute value, the answer is positive.
   b) If the negative number has the greater absolute value, the answer is negative.

4. *One number is zero*: The sum is the other number.

EXAMPLE 3 Add: **(a)** $-9 + (-5)$; **(b)** $-3.24 + 8.7$; **(c)** $-\frac{3}{4} + \frac{1}{3}$.

SOLUTION

a) $-9 + (-5)$ We add the absolute values, getting 14. The answer is *negative*: $-9 + (-5) = -14$.

b) $-3.24 + 8.7$ The absolute values are 3.24 and 8.7. Subtract 3.24 from 8.7 to get 5.46. The positive number is further from 0, so the answer is *positive*: $-3.24 + 8.7 = 5.46$.

c) $-\frac{3}{4} + \frac{1}{3} = -\frac{9}{12} + \frac{4}{12}$ The absolute values are $\frac{9}{12}$ and $\frac{4}{12}$. Subtract to get $\frac{5}{12}$. The negative number is further from 0, so the answer is *negative*: $-\frac{3}{4} + \frac{1}{3} = -\frac{5}{12}$.

3. Add: $4.2 + (-12)$.

YOUR TURN

When numbers like 7 and -7 are added, the result is 0. Such numbers are called **opposites**, or **additive inverses**, of one another. The sum of two additive inverses is the **additive identity**, 0.

THE LAW OF OPPOSITES

For any two numbers a and $-a$,

$$a + (-a) = 0.$$

(The sum of opposites is 0.)

EXAMPLE 4   Find the opposite: **(a)** -17.5; **(b)** $\frac{4}{5}$; **(c)** 0.

SOLUTION

a) The opposite of -17.5 is 17.5 because $-17.5 + 17.5 = 0$.

b) The opposite of $\frac{4}{5}$ is $-\frac{4}{5}$ because $\frac{4}{5} + \left(-\frac{4}{5}\right) = 0$.

c) The opposite of 0 is 0 because $0 + 0 = 0$.

4. Find the opposite of -13.

YOUR TURN

To name the opposite, we use the symbol "$-$" and read the symbolism $-a$ as "the opposite of a."

> *CAUTION!* $-a$ does not necessarily represent a negative number. In particular, when a is *negative*, $-a$ is *positive*.

EXAMPLE 5   Find $-x$ for the following: **(a)** $x = -2$; **(b)** $x = \frac{3}{4}$.

SOLUTION

a) If $x = -2$, then $-x = -(-2) = 2$.   The opposite of -2 is 2.

b) If $x = \frac{3}{4}$, then $-x = -\frac{3}{4}$.   The opposite of $\frac{3}{4}$ is $-\frac{3}{4}$.

5. Find $-x$ for $x = -12$.

YOUR TURN

Using the notation of opposites, we can formally define absolute value.

> **ABSOLUTE VALUE**
>
> $$|x| = \begin{cases} x, & \text{if } x \geq 0, \\ -x, & \text{if } x < 0 \end{cases}$$
>
> (When x is nonnegative, the absolute value of x is x. When x is negative, the absolute value of x is the opposite of x. Thus, $|x|$ is never negative.)

A negative number is said to have a negative "sign" and a positive number a positive "sign." To subtract, we can add an opposite. This can be stated as: "Change the sign of the number being subtracted and then add."

EXAMPLE 6   Subtract: **(a)** $5 - 9$; **(b)** $-1.2 - (-3.7)$; **(c)** $-\frac{4}{5} - \frac{2}{3}$.

SOLUTION

a) $5 - 9 = 5 + (-9)$ Change the sign and add.

$\qquad\quad\ = -4$

b) $-1.2 - (-3.7) = -1.2 + 3.7$ Instead of *subtracting negative* 3.7, we *add positive* 3.7.

$\qquad\qquad\qquad\ = 2.5$

c) $-\frac{4}{5} - \frac{2}{3} = -\frac{4}{5} + \left(-\frac{2}{3}\right)$ Instead of *subtracting* $\frac{2}{3}$, we *add* the opposite, $-\frac{2}{3}$.

$\qquad\qquad = -\frac{12}{15} + \left(-\frac{10}{15}\right)$ Finding a common denominator

$\qquad\qquad = -\frac{22}{15}$

6. Subtract: $6 - (-13)$.

YOUR TURN

Technology Connection

Technology Connections highlight situations in which calculators (primarily graphing calculators) or computers can enrich learning. Most Technology Connections present information in a generic form—consult an outside reference for specific keystrokes.

Graphing calculators use different keys for subtracting and writing negatives. The key labeled ⊝ is used for a negative sign, whereas ⊖ is used for subtraction.

1. Use a graphing calculator to check Example 6.
2. Calculate: $-3.9 - (-4.87)$.

Multiplication, Division, and Reciprocals

Multiplication of real numbers can be regarded as repeated addition or as repeated subtraction that begins at 0. For example,

$$3 \cdot (-4) = 0 + (-4) + (-4) + (-4) = -12 \qquad \text{Adding } -4 \text{ three times}$$

and

$$(-2)(-5) = 0 - (-5) - (-5) = 0 + 5 + 5 = 10. \qquad \text{Subtracting } -5 \text{ twice}$$

When one factor is positive and one is negative, the product is negative. When both factors are positive or both are negative, the product is positive.

Division is defined in terms of multiplication. For example, $10 \div (-2) = -5$ because $(-5)(-2) = 10$. Thus the rules for division can be stated along with those for multiplication.

MULTIPLICATION OR DIVISION OF TWO REAL NUMBERS

1. To multiply or divide two numbers with *unlike signs*, multiply or divide their absolute values. The answer is *negative*.
2. To multiply or divide two numbers having the *same sign*, multiply or divide their absolute values. The answer is *positive*.

EXAMPLE 7 Multiply or divide as indicated.

a) $(-4)9$ **b)** $\left(-\frac{2}{3}\right)\left(-\frac{3}{8}\right)$

c) $20 \div (-4)$ **d)** $\frac{-45}{-15}$

SOLUTION

a) $(-4)9 = -36$ Multiply absolute values. The answer is negative.

b) $\left(-\frac{2}{3}\right)\left(-\frac{3}{8}\right) = \frac{6}{24} = \frac{1}{4}$ Multiply absolute values. The answer is positive.

c) $20 \div (-4) = -5$ Divide absolute values. The answer is negative.

d) $\frac{-45}{-15} = 3$ Divide absolute values. The answer is positive.

7. Multiply: $(-16)(-0.1)$.

↩ YOUR TURN

Note that since $\dfrac{-8}{2} = \dfrac{8}{-2} = -\dfrac{8}{2} = -4$, we have the following generalization.

Study Skills

Organize Your Work

When doing homework, consider using a spiral notebook or collecting your work in a three-ring binder. Because your course will probably include graphing, consider purchasing a notebook filled with graph paper. Write legibly, labeling each section and each exercise and showing all steps. Legible, well-organized work will make it easier for those who read your work to give you constructive feedback and will help you to review for a test.

THE SIGN OF A FRACTION

For any number a and any nonzero number b,

$$\frac{-a}{b} = \frac{a}{-b} = -\frac{a}{b}.$$

Recall that

$$\frac{a}{b} = \frac{a}{1} \cdot \frac{1}{b} = a \cdot \frac{1}{b}.$$

That is, rather than divide by b, we can multiply by $\frac{1}{b}$. The numbers b and $\frac{1}{b}$ are called **reciprocals**, or **multiplicative inverses**, of each other. Every real number except 0 has a reciprocal. The product of two multiplicative inverses is the **multiplicative identity**, 1.

THE LAW OF RECIPROCALS

For any two numbers a and $\frac{1}{a}$ ($a \neq 0$),

$$a \cdot \frac{1}{a} = 1.$$

(The product of reciprocals is 1.)

EXAMPLE 8 Find the reciprocal: **(a)** $\frac{7}{8}$; **(b)** $-\frac{3}{4}$; **(c)** -8.

SOLUTION

a) The reciprocal of $\frac{7}{8}$ is $\frac{8}{7}$ because $\frac{7}{8} \cdot \frac{8}{7} = 1$.

b) The reciprocal of $-\frac{3}{4}$ is $-\frac{4}{3}$.

c) The reciprocal of -8 is $\frac{1}{-8}$, or $-\frac{1}{8}$.

8. Find the reciprocal of $-\frac{1}{9}$.

YOUR TURN

To divide, we can multiply by the reciprocal of the divisor. We sometimes say that we "invert and multiply."

EXAMPLE 9 Divide: **(a)** $-\frac{1}{4} \div \frac{3}{5}$; **(b)** $-\frac{6}{7} \div (-10)$.

SOLUTION

$\frac{3}{5}$ is the divisor.

a) $-\frac{1}{4} \div \frac{3}{5} = -\frac{1}{4} \cdot \frac{5}{3}$ "Inverting" $\frac{3}{5}$ and changing division to multiplication

$= -\frac{5}{12}$

9. Divide: $12 \div \left(-\frac{2}{3}\right)$.

b) $-\frac{6}{7} \div (-10) = -\frac{6}{7} \cdot \left(-\frac{1}{10}\right) = \frac{6}{70}$, or $\frac{3}{35}$ Multiplying by the reciprocal of -10

YOUR TURN

There is a reason why we never divide by 0. Suppose 5 were divided by 0. The answer would have to be a number that, when multiplied by 0, gave 5. But any number times 0 is 0. Thus we cannot divide 5 or any other nonzero number by 0.

What if we divide 0 by 0? In this case, our solution would need to be some number that, when multiplied by 0, gave 0. But then *any* number would work as a solution to $0 \div 0$. This could lead to contradictions so we agree to exclude division of 0 by 0 also.

DIVISION BY ZERO

We never divide by 0. If asked to divide a nonzero number by 0, we say that the answer is *undefined*. If asked to divide 0 by 0, we say that the answer is *indeterminate*. Thus,

$\frac{7}{0}$ is *undefined*   and $\frac{0}{0}$ is *indeterminate*.

The rules for order of operations apply to all real numbers.

EXAMPLE 10 Simplify: **(a)** $(-5)^2$; **(b)** -5^2.

SOLUTION An exponent is always written immediately after the base. Thus in the expression $(-5)^2$, the base is (-5); in the expression -5^2, the base is 5.

a) $(-5)^2 = (-5)(-5) = 25$ Squaring -5

b) $-5^2 = -(5 \cdot 5) = -25$ Squaring 5 and then taking the opposite

Note that $(-5)^2 \neq -5^2$.

10. Simplify: -8^2.

YOUR TURN

EXAMPLE 11 Simplify: $7 - 5^2 + 6 \div 2(-5)^2$.

SOLUTION

$$7 - 5^2 + 6 \div 2(-5)^2 = 7 - 25 + 6 \div 2 \cdot 25 \quad \text{Simplifying } 5^2 \text{ and } (-5)^2$$
$$= 7 - 25 + 3 \cdot 25 \quad \text{Dividing}$$
$$= 7 - 25 + 75 \quad \text{Multiplying}$$
$$= -18 + 75 \quad \text{Subtracting}$$
$$= 57 \quad \text{Adding}$$

11. Simplify:

$24 \div (-3) \cdot (-2)^2 - 3(-6)$.

YOUR TURN

In addition to parentheses, brackets, and braces, groupings may be indicated by a fraction bar, an absolute-value symbol, or a radical sign ($\sqrt{}$).

EXAMPLE 12 Calculate: $\dfrac{12|7-9| + 4 \cdot 5}{(-3)^4 + 2^3}$.

SOLUTION We simplify the numerator and the denominator and divide the results:

$$\frac{12|7-9| + 4 \cdot 5}{(-3)^4 + 2^3} = \frac{12|-2| + 20}{81 + 8}$$
$$= \frac{12(2) + 20}{89}$$

12. Calculate:

$\dfrac{6 - 4 + 5 - 2^2}{2 - |35 - 6^2|}$.

$$= \frac{44}{89}. \quad \text{Multiplying and adding}$$

YOUR TURN

The Commutative, Associative, and Distributive Laws

When two real numbers are added or multiplied, the order in which the numbers are written does not affect the result.

THE COMMUTATIVE LAWS

For any real numbers a and b,

$a + b = b + a$; $a \cdot b = b \cdot a$.

(for Addition)   (for Multiplication)

The commutative laws provide one way of writing *equivalent expressions*.

> **EQUIVALENT EXPRESSIONS**
>
> Two expressions that have the same value for all possible replacements are called *equivalent expressions*.

EXAMPLE 13 Use a commutative law to write an expression equivalent to $7x + 9$.

SOLUTION Using the commutative law of addition, we have

$$7x + 9 = 9 + 7x.$$

We can also use the commutative law of multiplication to write

$$7 \cdot x + 9 = x \cdot 7 + 9.$$

The expressions $7x + 9$, $9 + 7x$, and $x \cdot 7 + 9$ are all equivalent. They name the same number for any replacement of x.

YOUR TURN

13. Use a commutative law to write an expression equivalent to $3 + mn$. Answers may vary.

The *associative laws* enable us to form equivalent expressions by changing grouping.

> **THE ASSOCIATIVE LAWS**
>
> For any real numbers a, b, and c,
>
> $$a + (b + c) = (a + b) + c; \qquad a \cdot (b \cdot c) = (a \cdot b) \cdot c.$$
> (for Addition)   (for Multiplication)

EXAMPLE 14 Write an expression equivalent to $(3x + 7y) + 9z$, using the associative law of addition.

SOLUTION We have

$$(3x + 7y) + 9z = 3x + (7y + 9z).$$

The expressions $(3x + 7y) + 9z$ and $3x + (7y + 9z)$ are equivalent. They name the same number for any replacements of x, y, and z.

14. Write an expression equivalent to $(2x)y$ using the associative law of multiplication.

YOUR TURN

The *distributive law* allows us to rewrite the *product* of a and $b + c$ as the *sum* of ab and ac.

Student Notes

The commutative, associative, and distributive laws are used so often in this course that it is worth the effort to memorize them.

> **THE DISTRIBUTIVE LAW**
>
> For any real numbers a, b, and c,
>
> $$a(b + c) = ab + ac.$$

EXAMPLE 15   Obtain an expression equivalent to $5x(y + 4)$ by multiplying.

SOLUTION   We use the distributive law to get

$$5x(y + 4) = 5x \cdot y + 5x \cdot 4 \qquad \text{Using the distributive law}$$
$$= 5xy + 5 \cdot 4 \cdot x \qquad \text{Using the commutative law of multiplication}$$
$$= 5xy + 20x. \qquad \text{Simplifying}$$

15. Obtain an expression equivalent to $-3(x + 7)$ by multiplying.

The expressions $5x(y + 4)$ and $5xy + 20x$ are equivalent. They name the same number for any replacements of x and y.

YOUR TURN

When we reverse what we did in Example 15, we say that we are **factoring** an expression. This allows us to rewrite a sum or a difference as a product.

EXAMPLE 16   Obtain an expression equivalent to $3x - 6$ by factoring.

16. Obtain an expression equivalent to $5x + 5y + 5$ by factoring.

SOLUTION   We use the distributive law to get

$$3x - 6 = 3 \cdot x - 3 \cdot 2 = 3(x - 2).$$

YOUR TURN

In Example 16, since the product of 3 and $x - 2$ is $3x - 6$, we say that 3 and $x - 2$ are **factors** of $3x - 6$. Thus the word "factor" can act as a noun or as a verb.

1.2   EXERCISE SET

FOR EXTRA HELP   MyMathLab® Math XP
PRACTICE WATCH READ REVIEW

Vocabulary and Reading Check

Classify each of the following statements as either true or false.

1. The sum of two negative numbers is always negative.

2. The product of two negative numbers is always negative.

3. The product of a negative number and a positive number is always negative.

4. The sum of a negative number and a positive number is always negative.

5. The sum of a negative number and a positive number is always positive.

6. If a and b are negative, with $a < b$, then $|a| > |b|$.

7. If a and b are positive, with $a < b$, then $|a| > |b|$.

8. The commutative law of addition states that for all real numbers a and b, $a + b$ and $b + a$ are equivalent.

9. The associative law of multiplication states that for all real numbers a, b, and c, $(ab)c$ is equivalent to $a(bc)$.

10. The distributive law states that the order in which two numbers are multiplied does not change the result.

Absolute Value

Find each absolute value.

11. $|-10|$ 12. $|-3|$ 13. $|7|$

14. $|13|$ 15. $|-46.8|$ 16. $|-36.9|$

17. $|0|$ 18. $|3\frac{3}{4}|$ 19. $|1\frac{7}{8}|$

20. $|7.24|$ 21. $|-4.21|$ 22. $|-5.309|$

Inequalities

Write the meaning of each inequality, and determine whether it is a true statement.

23. $-5 \le -4$ 24. $-2 \le -8$

25. $-9 > 1$ 26. $-9 < 1$

27. $0 \ge -5$ 28. $9 \le 9$

29. $-8 < -3$ 30. $7 \ge -8$

31. $-4 \ge -4$ 32. $2 < 2$

33. $-5 < -5$ 34. $-2 > -12$

Addition, Subtraction, and Opposites

Add.

35. $4 + 8$

36. $5 + 7$

37. $(-3) + (-9)$

38. $(-6) + (-8)$

39. $-5.3 + 2.8$

40. $9.3 + (-5.7)$

41. $\frac{2}{7} + \left(-\frac{3}{5}\right)$

42. $\frac{3}{8} + \left(-\frac{2}{5}\right)$

43. $-3.26 + (-5.8)$

44. $-2.1 + (-7.5)$

45. $-\frac{1}{9} + \frac{2}{3}$

46. $-\frac{1}{2} + \frac{4}{5}$

47. $-6.25 + 0$

48. $0 + (-3.69)$

49. $4.19 + (-4.19)$

50. $-8.35 + 8.35$

51. $-18.3 + 22.1$

52. $21.7 + (-28.3)$

Find the opposite, or additive inverse.

53. 2.37 **54.** 6.98 **55.** -56

56. -11 **57.** 0 **58.** $-2\frac{1}{3}$

Find $-x$ for each of the following.

59. $x = 8$

60. $x = 12$

61. $x = -\frac{1}{10}$

62. $x = -\frac{8}{3}$

63. $x = -4.67$

64. $x = 3.14$

65. $x = 0$

66. $x = -7$

Subtract.

67. $10 - 4$

68. $9 - 1$

69. $4 - 10$

70. $1 - 9$

71. $-5 - (-12)$

72. $-3 - (-7)$

73. $-5 - 14$

74. $-9 - 8$

75. $2.7 - 5.8$

76. $3.7 - 4.2$

77. $-\frac{3}{5} - \frac{1}{2}$

78. $-\frac{2}{3} - \frac{1}{5}$

Aha! **79.** $-31 - (-31)$

80. $-14 - (-14)$

81. $0 - (-5.37)$

82. $0 - 9.09$

Multiplication, Division, and Reciprocals

Multiply.

83. $(-3)8$

84. $(-5)9$

85. $(-2)(-11)$

86. $(-6)(-7)$

87. $(4.2)(-5)$

88. $(3.5)(-8)$

89. $\frac{3}{7}(-1)$

90. $-1 \cdot \frac{2}{5}$

91. $(-17.45) \cdot 0$

92. 15.2×0

93. $-\frac{2}{3}\left(\frac{3}{4}\right)$

94. $\frac{5}{6}\left(-\frac{3}{10}\right)$

Divide.

95. $\frac{-28}{-7}$ **96.** $\frac{-18}{-6}$ **97.** $\frac{-100}{25}$

98. $\frac{40}{-4}$ **99.** $\frac{73}{-1}$ **100.** $\frac{-62}{1}$

101. $\frac{0}{-7}$ **102.** $\frac{0}{-11}$

Find the reciprocal, or multiplicative inverse, if it exists.

103. 8 **104.** -7 **105.** $-\frac{5}{7}$

106. $\frac{4}{3}$ **107.** 0 **108.** $-\frac{9}{10}$

Divide.

109. $\frac{3}{5} \div \frac{6}{7}$ **110.** $\frac{2}{3} \div \frac{5}{6}$

111. $-\frac{3}{5} \div \frac{1}{2}$ **112.** $\left(-\frac{4}{7}\right) \div \frac{1}{3}$

113. $-\frac{2}{9} \div (-8)$ **114.** $\left(-\frac{2}{11}\right) \div (-6)$

Aha! **115.** $-\frac{12}{7} \div \left(-\frac{12}{7}\right)$ **116.** $\left(-\frac{2}{7}\right) \div (-1)$

Real-Number Operations

Calculate using the rules for order of operations. If an expression is undefined, state this.

117. -10^2 **118.** $(-10)^2$

119. $-(-3)^2$ **120.** $-(-2)^2$

121. $(2 - 5)^2$ **122.** $2^2 - 5^2$

123. $9 - (8 - 3 \cdot 2^3)$ **124.** $19 - (4 + 2 \cdot 3^2)$

125. $\frac{5 \cdot 2 - 4^2}{27 - 2^4}$ **126.** $\frac{7 \cdot 3 - 5^2}{9 + 4 \cdot 2}$

127. $\frac{3^4 - (5 - 3)^4}{8 - 2^3}$ **128.** $\frac{4^3 - (7 - 4)^2}{3^2 - 7}$

129. $\frac{(2 - 3)^3 - 5|2 - 4|}{7 - 2 \cdot 5^2}$

130. $\frac{8 \div 4 \cdot 6|4^2 - 5^2|}{9 - 4 + 11 - 4^2}$

131. $|2^2 - 7|^3 + 4$

132. $|-2 - 3| \cdot 4^2 - 3$

133. $32 - (-5)^2 + 15 \div (-3) \cdot 2$

134. $43 - (-9 + 2)^2 + 18 \div 6 \cdot (-2)$

The Commutative, Associative, and Distributive Laws

Write an equivalent expression using a commutative law. Answers may vary.

135. $6 + xy$ **136.** $4a + 7b$

137. $-9(ab)$ **138.** $(7x)y$

Write an equivalent expression using an associative law.

139. $(3x)y$

140. $-7(ab)$

141. $(3y + 4) + 10$

142. $x + (2y + 5)$

Write an equivalent expression using the distributive law.

143. $7(x + 1)$

144. $3(a + 5)$

145. $5(m - n)$

146. $6(s - t)$

147. $-5(2a + 3b)$

148. $-2(3c + 5d)$

149. $9a(b - c + d)$

150. $5x(y - z + w)$

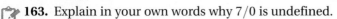
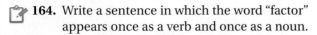

Find an equivalent expression by factoring.

151. $5x + 50$

152. $5d + 30$

153. $9p - 3$

154. $15x - 3$

155. $7x - 21y + 14z$

156. $6y - 9x - 3w$

157. $255 - 34b$

158. $13t - 143$

159. $xy + x$

160. $ab + b$

161. Describe in your own words a method for determining the sign of the sum of a positive number and a negative number.

162. Explain the difference between the expressions "five is less than x" and "five less than x."

Synthesis

163. Explain in your own words why $7/0$ is undefined.

164. Write a sentence in which the word "factor" appears once as a verb and once as a noun.

Insert one pair of parentheses to convert each of the following false statements into a true statement.

165. $8 - 5^3 + 9 = 36$

166. $2 \cdot 7 + 3^2 \cdot 5 = 104$

167. $5 \cdot 2^3 \div 3 - 4^4 = 40$

168. $2 - 7 \cdot 2^2 + 9 = -11$

Calculate using the rules for order of operations.

169. $17 - \sqrt{11 - (3 + 4)} \div [-5 - (-6)]^2$

170. $15 - 1 + \sqrt{5^2 - (3 + 1)^2}(-1)$

171. Find the greatest value of a for which $|a| \geq 6.2$ and $a < 0$.

172. Use the commutative, associative, and distributive laws to show that $5(a + bc)$ is equivalent to $c(b \cdot 5) + a \cdot 5$. Use only one law in each step of your work.

173. Are subtraction and division commutative? Why or why not?

174. Are subtraction and division associative? Why or why not?

175. Translate each of the following to an equation and then solve.

a) The temperature was $-16°$F at 6:00 P.M. and dropped $5°$ by midnight. What was the temperature at midnight?

b) Temperature drops about $3.5°$F for every 1000 ft increase in altitude. Ethan is flying a jet at 20,000 ft. If the ground temperature is $42°$F, what is the temperature outside Ethan's jet?

YOUR TURN ANSWERS: SECTION 1.2

1. 237   **2.** -4 is less than or equal to -3; true   **3.** -7.8
4. 13   **5.** 12   **6.** 19   **7.** 1.6   **8.** -9 **9.** -18
10. -64 **11.** -14 **12.** 3   **13.** $mn + 3; 3 + nm$
14. $2(xy)$ **15.** $-3x - 21$ **16.** $5(x + y + 1)$

QUICK QUIZ: SECTIONS 1.1–1.2

1. Translate to an algebraic expression: Eight less than twice a number.   [1.1]

2. Evaluate $8ac - a^2 \div 5c$ for $a = 10$ and $c = 2$.   [1.1]

3. Subtract: $-32 - (-40)$.   [1.2]

4. Multiply: $(-1.2)(5)$.   [1.2]

5. Simplify: $-1 - (4 - 10)^2 \div 2 \cdot (-3)$.   [1.2]

1.3   Solving Equations

Equivalent Equations ▪ The Addition and Multiplication Principles ▪ Combining Like Terms ▪ Types of Equations

Solving equations is an essential part of problem solving in algebra. In this section, we review and practice solving basic equations.

Equivalent Equations

Equation-solving principles in algebra are used to produce *equivalent equations* from which solutions are easily found.

> **EQUIVALENT EQUATIONS**
>
> Two equations are *equivalent* if they have the same solution(s).

EXAMPLE 1   Determine whether $4x = 12$ and $10x = 30$ are equivalent equations.

1. Determine whether $x + 1 = 5$ and $2x = 8$ are equivalent equations.

SOLUTION   The equation $4x = 12$ is true only when x is 3. Similarly, $10x = 30$ is true only when x is 3. Since both equations have the same solution, they are equivalent.

YOUR TURN

Note that the equation $x = 3$ is also equivalent to the equations in Example 1 and is the simplest equation for which 3 is the solution.

EXAMPLE 2   Determine whether $3x = 4x$ and $3/x = 4/x$ are equivalent equations.

2. Determine whether $5x = 10$ and $2x = 6$ are equivalent equations.

SOLUTION   Note that 0 is a solution of $3x = 4x$. Since neither $3/x$ nor $4/x$ is defined for $x = 0$, the equations $3x = 4x$ and $3/x = 4/x$ are *not* equivalent.

YOUR TURN

The Addition and Multiplication Principles

Suppose that a and b represent the same number and that some number c is added to a. If c is also added to b, we will get two equal sums, since a and b are the same number. The same is true if we multiply both a and b by c. In this manner, we can produce equivalent equations.

> **THE ADDITION AND MULTIPLICATION PRINCIPLES FOR EQUATIONS**
>
> For any real numbers a, b, and c:
>
> **a)** $a = b$ is equivalent to $a + c = b + c$;
> **b)** $a = b$ is equivalent to $a \cdot c = b \cdot c$, provided $c \neq 0$.

Either a or b usually represents a variable expression.

EXAMPLE 3   Solve: $y - 4.7 = 13.9$.

SOLUTION

$$y - 4.7 = 13.9$$
$$y - 4.7 + 4.7 = 13.9 + 4.7 \qquad \text{Using the addition principle;}$$
$$\text{adding 4.7 to both sides}$$
$$y + 0 = 13.9 + 4.7 \qquad \text{Using the law of opposites}$$
$$y = 18.6 \qquad \text{The solution of this equation is 18.6.}$$

Check:

$$\frac{y - 4.7 = 13.9}{18.6 - 4.7 \ \bigm| \ 13.9} \qquad \text{Substituting 18.6 for } y$$
$$13.9 \overset{?}{=} 13.9 \quad \text{\small TRUE}$$

3. Solve: $t - 9 = -2$.

The solution is 18.6.

YOUR TURN

In Example 3, we added 4.7 to both sides because 4.7 is the opposite of -4.7 and we wanted y alone on one side of the equation. Adding 4.7 gave us $y + 0$, or simply y, on the left side. This led to the equivalent equation $y = 18.6$.

Student Notes

The addition and multiplication principles can be used even when 0 is on one side of an equation. Thus to solve $\frac{2}{5}x = 0$, we would multiply both sides of the equation by $\frac{5}{2}$.

EXAMPLE 4   Solve: $\frac{2}{5}x = -\frac{9}{10}$.

SOLUTION   We have

$$\frac{2}{5}x = -\frac{9}{10}$$
$$\frac{5}{2} \cdot \frac{2}{5}x = \frac{5}{2} \cdot \left(-\frac{9}{10}\right) \qquad \text{Using the multiplication principle, we multiply by}$$
$$\frac{5}{2}, \text{ the reciprocal of } \frac{2}{5}, \text{ on both sides.}$$
$$1x = -\frac{45}{20} \qquad \text{Using the law of reciprocals}$$
$$x = -\frac{9}{4}. \qquad \text{Simplifying}$$

4. Solve: $\frac{1}{2}x = 7$.

The check is left to the student. The solution is $-\frac{9}{4}$.

YOUR TURN

In Example 4, we multiplied both sides by $\frac{5}{2}$ because $\frac{5}{2}$ is the reciprocal of $\frac{2}{5}$ and we wanted x alone on one side of the equation. Multiplying by $\frac{5}{2}$ gave us $1x$, or simply x, on the left side. This led to the equivalent equation $x = -\frac{9}{4}$.

There is no need for a subtraction principle or a division principle because subtraction can be regarded as adding opposites and division can be regarded as multiplying by reciprocals.

Combining Like Terms

In an expression like $8a^5 + 17 + 4/b + (-6a^3b)$, the parts that are separated by addition signs are called *terms*. A **term** is a number, a variable, a product of numbers and/or variables, or a quotient of numbers and/or variables. Thus, $8a^5$, 17, $4/b$, and $-6a^3b$ are terms in $8a^5 + 17 + 4/b + (-6a^3b)$. When terms have variable factors that are exactly the same, we refer to those terms as **like**, or **similar**, **terms**. Thus, $3x^2y$ and $-7x^2y$ are similar terms, but $3x^2y$ and $4xy^2$ are not. We can often simplify expressions by **combining**, or **collecting**, **like terms**.

EXAMPLE 5   Write an equivalent expression by combining like terms:

$$3a + 5a^2 - 7a + a^2.$$

SOLUTION

$$3a + 5a^2 - 7a + a^2 = 3a - 7a + 5a^2 + a^2 \quad \text{Using the commutative law}$$

$$= (3 - 7)a + (5 + 1)a^2 \quad \text{Using the distributive law. Note that } a^2 = 1a^2.$$

$$= -4a + 6a^2$$

5. Write an equivalent expression by combining like terms:

$$6x - 7 - x - 9.$$

_____ YOUR TURN

Sometimes we must use the distributive law to remove grouping symbols before combining like terms. Remember to remove the innermost grouping symbols first.

EXAMPLE 6   Simplify: $3x + 2[4 + 5(x - 2y)]$.

SOLUTION

$$3x + 2[4 + 5(x - 2y)] = 3x + 2[4 + 5x - 10y] \quad \text{Using the distributive law}$$

$$= 3x + 8 + 10x - 20y \quad \text{Using the distributive law (again)}$$

$$= 13x + 8 - 20y \quad \text{Combining like terms}$$

6. Simplify:

$$2(5a - 9) + 3(4 - 5a + 2).$$

_____ YOUR TURN

The product of a number and -1 is its opposite, or additive inverse. For example,

$$-1 \cdot 8 = -8 \quad \text{(the opposite of 8).}$$

In general, $-x = -1 \cdot x$. We can use this fact along with the distributive law when parentheses are preceded by a negative sign or subtraction.

EXAMPLE 7   Simplify $-(a - b)$ using multiplication by -1.

SOLUTION   We have

$$-(a - b) = -1 \cdot (a - b) \quad \text{Replacing } - \text{ with multiplication by } -1$$

$$= -1 \cdot a - (-1) \cdot b \quad \text{Using the distributive law}$$

$$= -a - (-b) \quad \text{Replacing } -1 \cdot a \text{ with } -a \text{ and } (-1) \cdot b \text{ with } -b$$

$$= -a + b, \text{ or } b - a. \quad \text{Try to go directly to this step.}$$

7. Simplify $-(5 - x)$ using multiplication by -1.

The expressions $-(a - b)$ and $b - a$ are equivalent. They represent the same number for all replacements of a and b.

_____ YOUR TURN

Example 7 illustrates a useful shortcut worth remembering:

The opposite of $a - b$ is $-a + b$, or $b - a$.

$$-(a - b) = b - a$$

EXAMPLE 8 Simplify: $9x - 5y - (5x + y - 7)$.

SOLUTION

$$9x - 5y - (5x + y - 7) = 9x - 5y - 5x - y + 7 \qquad \text{Using the distributive law}$$
$$= 4x - 6y + 7 \qquad \text{Combining like terms}$$

8. Simplify:

$6 - (3 - m - n) - 5n$.

 YOUR TURN

For many equations, before we use the addition and multiplication principles to *solve*, we must first *simplify* the expressions within the equation.

Study Skills

Do the Exercises

- When you have completed the odd-numbered exercises in your assignment, you can check your answers at the back of the book. If you miss any, closely examine your work and, if necessary, ask for help.

- Whether or not your instructor assigns the even-numbered exercises, try to do some on your own. Check your answers later with a friend or your instructor.

EXAMPLE 9 Solve: $5x - 2(x - 5) = 7x - 2$.

SOLUTION

$$5x - 2(x - 5) = 7x - 2$$
$$5x - 2x + 10 = 7x - 2 \qquad \text{Using the distributive law}$$
$$3x + 10 = 7x - 2 \qquad \text{Combining like terms}$$
$$3x + 10 - 3x = 7x - 2 - 3x \qquad \begin{array}{l}\text{Using the addition principle; adding} \\ -3x, \text{ the opposite of } 3x, \text{ to both sides}\end{array}$$
$$10 = 4x - 2 \qquad \text{Combining like terms}$$
$$10 + 2 = 4x - 2 + 2 \qquad \text{Using the addition principle}$$
$$12 = 4x \qquad \text{Simplifying}$$
$$\tfrac{1}{4} \cdot 12 = \tfrac{1}{4} \cdot 4x \qquad \begin{array}{l}\text{Using the multiplication principle;} \\ \text{multiplying both sides by } \tfrac{1}{4}, \text{ the} \\ \text{reciprocal of } 4\end{array}$$
$$3 = x \qquad \begin{array}{l}\text{Using the law of reciprocals;} \\ \text{simplifying}\end{array}$$

Check:

$$\begin{array}{c|c} 5x - 2(x - 5) & = 7x - 2 \\ \hline 5 \cdot 3 - 2(3 - 5) & 7 \cdot 3 - 2 \\ 15 - 2(-2) & 21 - 2 \\ 15 + 4 & 19 \\ 19 & \overset{?}{=} 19 \qquad \text{TRUE} \end{array}$$

9. Solve:

$t - 5 = 6 - 3(t - 7)$.

The solution is 3.

 YOUR TURN

Types of Equations

In Examples 3, 4, and 9, we solved *linear equations*. A **linear equation** in one variable—say, x—is an equation equivalent to one of the form $ax = b$ with a and b constants and $a \neq 0$. Since $x = x^1$, the variable in a linear equation is always raised to the first power.

We will often refer to the set of solutions, or the **solution set**, of a particular equation. The solution set for Example 9 is $\{3\}$. If an equation is true for all replacements, the solution set is $\mathbb{R}$, the set of all real numbers. If an equation is never true, the solution set is the **empty set**, denoted $\varnothing$, or { }. As its name suggests, the empty set is the set containing no elements.

Every equation is either an **identity**, a **contradiction**, or a **conditional equation**.

Type of Equation	Definition	Example	Solution Set
Identity	An equation that is true for all replacements	$x + 5 = 3 + x + 2$	$\mathbb{R}$
Contradiction	An equation that is never true	$n = n + 1$	$\varnothing$, or $\{\ \}$
Conditional equation	An equation that is sometimes true and sometimes false, depending on the replacement for the variable	$2x + 5 = 17$	$\{6\}$

EXAMPLE 10   Solve each equation. Then classify the equation as an identity, a contradiction, or a conditional equation.

a) $2x + 7 = 7(x + 1) - 5x$ **b)** $3x - 5 = 3(x - 2) + 4$

c) $3 - 8x = 5 - 7x$

SOLUTION

a) $2x + 7 = 7(x + 1) - 5x$

$2x + 7 = 7x + 7 - 5x$ Using the distributive law

$2x + 7 = 2x + 7$ Combining like terms

The equation $2x + 7 = 2x + 7$ is true regardless of what x is replaced with, so all real numbers are solutions. Note that $2x + 7 = 2x + 7$ is equivalent to $2x = 2x$, $7 = 7$, or $0 = 0$. The solution set is $\mathbb{R}$, and the equation is an identity.

b) $3x - 5 = 3(x - 2) + 4$

$3x - 5 = 3x - 6 + 4$ Using the distributive law

$3x - 5 = 3x - 2$ Combining like terms

$-3x + 3x - 5 = -3x + 3x - 2$ Using the addition principle

$-5 = -2$

Since the original equation is equivalent to $-5 = -2$, which is false regardless of the choice of x, the original equation has no solution. There is no solution of $3x - 5 = 3(x - 2) + 4$. The solution set is $\varnothing$, and the equation is a contradiction.

c) $3 - 8x = 5 - 7x$

$3 - 8x + 7x = 5 - 7x + 7x$ Using the addition principle

$3 - x = 5$ Simplifying

$-3 + 3 - x = -3 + 5$ Using the addition principle

$-x = 2$ Simplifying

$x = \dfrac{2}{-1}$, or -2 Dividing both sides by -1 or multiplying both sides by $\frac{1}{-1}$, or -1

There is one solution, -2. For other choices of x, the equation is false. The solution set is $\{-2\}$. This equation is conditional since it can be true or false, depending on the replacement for x.

10. Solve

$y - (2 - y) = 2(y - 1)$.

Then classify the equation as an identity, a contradiction, or a conditional equation.

 YOUR TURN

1.3 EXERCISE SET

FOR EXTRA HELP

MyMathLab® Math XL

PRACTICE   WATCH   READ   REVIEW

◆ Vocabulary and Reading Check

Choose the word from the following list that best completes each statement. Not every word will be used.

contradiction   like
equivalent   linear
identity   solution

1. Two equations are _____ if they have the same solutions.

2. An equation in x of the form $ax = b$ is a(n) _____ equation.

3. A(n) _____ is an equation that is never true.

4. A(n) _____ is an equation that is always true.

◆ Concept Reinforcement

Classify each of the following as either a pair of equivalent equations or a pair of equivalent expressions.

5. $2(x + 7)$, $2x + 14$

6. $2(x + 7) = 11$, $2x + 14 = 11$

7. $4x - 9 = 7$, $4x = 16$

8. $4x - 9$, $5x - 9 - x$

9. $8t + 5 - 2t + 1$, $6t + 6$

10. $5t - 2 + t = 8$, $6t = 10$

Equivalent Equations

Determine whether the two equations in each pair are equivalent.

11. $3t = 21$ and $t + 4 = 11$

12. $3t = 27$ and $t - 3 = 5$

13. $12 - x = 3$ and $2x = 20$

14. $3x - 4 = 8$ and $3x = 12$

15. $5x = 2x$ and $\dfrac{4}{x} = 0$

16. $6 = 2x$ and $5 = \dfrac{2}{3 - x}$

The Addition and Multiplication Principles

Solve. Be sure to check.

17. $x - 2.9 = 13.4$

18. $y + 4.3 = 11.2$

19. $8t = 72$

20. $9t = 63$

21. $\frac{2}{3}x = 30$

22. $\frac{5}{4}x = -80$

23. $4a + 25 = 9$

24. $5a - 11 = 24$

25. $2y - 8 = 9$

26. $3y + 4 = 2$

Combining Like Terms

Simplify to form an equivalent expression by combining like terms. Use the distributive law as needed.

27. $9t^2 + t^2$

28. $7a^2 + a^2$

29. $16a - a$

30. $11t - t$

31. $n - 8n$

32. $p - 3p$

33. $5x - 3x + 8x$

34. $3x - 11x + 2x$

35. $18p - 12 + 3p + 8$

36. $14y + 6 - 9y + 7$

37. $-7t^2 + 3t + 5t^3 - t^3 + 2t^2 - t$

38. $-9n + 8n^2 + n^3 - 2n^2 - 3n + 4n^3$

39. $2x + 3(5x - 7)$

40. $5x + 4(x + 11)$

41. $7a - (2a + 5)$

42. $x - (5x + 9)$

43. $m - (6m - 2)$

44. $5a - (4a - 3)$

45. $3d - 7 - (5 - 2d)$

46. $8x - 9 - (7 - 5x)$

47. $2(x - 3) + 4(7 - x)$

48. $3(y + 6) + 5(2 - 4y)$

49. $3p - 4 - 2(p + 6)$

50. $8c - 1 - 3(2c + 1)$

51. $-2(a - 5) - [7 - 3(2a - 5)]$

52. $-3(b + 2) - [9 - 5(8b - 1)]$

53. $5\{-2x + 3[2 - 4(5x + 1)]\}$

54. $7\{-7x + 8[5 - 3(4x + 6)]\}$

55. $8y - \{6[2(3y - 4) - (7y + 1)] + 12\}$

56. $2y + \{7[3(2y - 5) - (8y + 7)] + 9\}$

Solving Linear Equations

Solve. Be sure to check.

57. $4x + 5x = 63$

58. $3x - 7x = 60$

59. $\frac{1}{4}y - \frac{2}{3}y = 5$

60. $\frac{3}{5}t - \frac{1}{2}t = 3$

61. $4(t - 3) - t = 6$

62. $2(t + 5) + t = 4$

63. $3(x + 4) = 7x$

64. $3(y + 5) = 8y$

65. $70 = 10(3t - 2)$

66. $27 = 9(5y - 2)$

67. $1.8(2 - n) = 9$

68. $2.1(3 - x) = 8.4$

69. $5y - (2y - 10) = 25$

70. $8x - (3x - 5) = 40$

71. $\frac{9}{10}y - \frac{7}{10} = \frac{21}{5}$

72. $\frac{4}{5}t - \frac{3}{10} = \frac{2}{5}$

73. $7r - 2 + 5r = 6r + 6 - 4r$

74. $9m - 15 - 2m = 6m - 1 - m$

75. $\frac{2}{3}(x - 2) - 1 = \frac{1}{4}(x - 3)$

76. $\frac{1}{4}(6t + 48) - 20 = -\frac{1}{3}(4t - 72)$

77. $2(t - 5) - 3(2t - 7) = 12 - 5(3t + 1)$

78. $4t + 8 - 6(2t - 1) = 3(4t - 3) - 7(t - 2)$

79. $3[2 - 4(x - 1)] = 3 - 4(x + 2)$

80. $5 + 2(x - 3) = 2[5 - 4(x + 2)]$

Types of Equations

Find each solution set. Then classify each equation as a conditional equation, an identity, or a contradiction.

81. $2x + 2 = 2(x + 1)$

82. $x + 2 = x + 3$

83. $7x - 2 - 3x = 4x$

84. $3t + 5 + t = 5 + 4t$

85. $2 + 9x = 3(4x + 1) - 1$

86. $4 + 7x = 7(x + 1)$

87. $3x - (8 - x) = 6x - 2(x + 4)$

88. $\frac{1}{3}(2x - 7) = \frac{1}{2}(x + 3)$

Aha! **89.** $-9t + 2 = -9t - 7(6 \div 2(49) + 8)$

90. $-9t + 2 = 2 - 9t - 5(8 \div 4(1 + 3^4))$

91. $2\{9 - 3[-2x - 4]\} = 12x + 42$

92. $3\{7 - 2[7x - 4]\} = -40x + 45$

93. Explain the difference between the statements "The equation has no solution." and "The solution of the equation is zero."

94. As the first step in solving
$$2x + 5 = -3,$$
Pat multiplies both sides by $\frac{1}{2}$. Is this incorrect? Why or why not?

Synthesis

95. Explain how the distributive and commutative laws can be used to rewrite $3x + 6y + 4x + 2y$ as $7x + 8y$.

96. Explain the difference between equivalent expressions and equivalent equations.

Solve and check. The symbol indicates an exercise designed to be solved with a calculator.

 97. $-0.00458y + 1.7787 = 13.002y - 1.005$

98. $4.23x - 17.898 = -1.65x - 42.454$

99. $6x - \{5x - [7x - (4x - (3x + 1))]\} = 6x + 5$

100. $8x - \{3x - [2x - (5x - (7x - 1))]\} = 8x + 7$

101. $23 - 2\{4 + 3[x - 1]\} + 5\{x - 2(x + 3)\}$
$$= 7\{x - 2[5 - (2x + 3)]\}$$

102. $17 - 3\{5 + 2[x - 2]\} + 4\{x - 3(x + 7)\}$
$$= 9\{x + 3[2 + 3(4 - x)]\}$$

103. Create an equation for which it is preferable to use the multiplication principle *before* using the addition principle. Explain why it is best to solve the equation in this manner.

104. Jasmine is paid $500 per week plus 10% of all sales she makes.

a) Let x represent the amount of Jasmine's weekly sales, in dollars, and y her weekly paycheck. Write an equation expressing y in terms of x.

b) One week, Jasmine earned $900. What were her sales that week?

YOUR TURN ANSWERS: SECTION 1.3

1. Yes **2.** No **3.** 7 **4.** 14 **5.** $5x - 16$
6. $-5a$ **7.** $x - 5$ **8.** $m - 4n + 3$ **9.** 8
10. $\mathbb{R}$; identity

QUICK QUIZ: SECTIONS 1.1–1.3

1. Find the area of a triangle with height 5 m and base 2.4 m. [1.1]

2. Find $-x$ for $x = -6$. [1.2]

Solve. [1.3]

3. $4(y - 3) = 8 - y$ **4.** $\frac{2}{3}x - \frac{1}{4} = \frac{5}{6}$

5. Find an equivalent expression by factoring:
$5x - 10y + 20$. [1.2]

Mid-Chapter Review

It is important to distinguish between *equivalent expressions* and *equivalent equations*. We *simplify* an expression by writing equivalent expressions; we *solve* an equation by writing equivalent equations.

GUIDED SOLUTIONS

1. Simplify: $3x - 2(x - 1)$. [1.3]

Solution

$$3x - 2(x - 1) = 3x - \boxed{} + \boxed{}$$
$$= \boxed{} + \boxed{}$$

2. Solve: $3x - 2(x - 1) = 6x$. [1.3]

Solution

$$3x - 2(x - 1) = 6x$$
$$3x - \boxed{} + \boxed{} = 6x$$
$$\boxed{} + 2 = 6x$$
$$2 = \boxed{}$$
$$\boxed{} = x$$

MIXED REVIEW

3. Translate to an algebraic expression: Five less than three times a number. [1.1]

4. Evaluate $2a \div 3x - a + x$ for $a = 3$ and $x = 5$. [1.1]

5. Find the area of a triangle with base $\frac{1}{2}$ ft and height 3 ft. [1.1]

Perform the indicated operations. [1.2]

6. $\frac{1}{2} - \left(-\frac{1}{3}\right)$

7. $-32 \div (-0.8)$

8. $3.6 + (-1.08)$

9. $\left(\frac{3}{10}\right)\left(-\frac{2}{5}\right)$

10. Simplify: $8 - 2^3 \div 4 \cdot (-2) + 1 - 2$. [1.2]

11. Use an associative law to write an expression equivalent to $(x + 3) + y$. [1.2]

Simplify. [1.3]

12. $3x - 5 - x + 12$

13. $4t - (3t - 1)$

14. $8x + 2[x - (x - 1)]$

15. $-(p - 4) - [3 - (9 - 2p)] + p$

Solve. If the solution set is $\mathbb{R}$ or $\varnothing$, classify the equation as an identity or a contradiction. [1.3]

16. $2x - 6 = 3x + 5$

17. $5 - (t - 2) = 6$

18. $6(y - 1) - 2(y + 1) = 4(y - 2)$

19. $3(x - 1) - 2(2x + 1) = 5(x - 1)$

20. $\frac{1}{3}t - 2 = \frac{1}{6} + t$

1.4 Introduction to Problem Solving

The Five-Step Strategy ▪ Problem Solving

Study Skills

Seeking Help on Campus
Your college or university probably has resources to support you.

- A learning lab or tutoring center
- Study-skills workshops or group tutoring sessions
- A bulletin board or network for locating tutors
- Classmates interested in forming a study group
- Instructors available during office hours or via e-mail

We now begin to study and practice the "art" of problem solving. Although we are interested mainly in using algebra to solve problems, much of the strategy discussed applies to solving problems in all walks of life.

A problem is simply a question to which we wish to find an answer. Perhaps this can best be illustrated with some sample problems:

1. If I exercise twice a week and eat 2400 calories a day, will I lose weight?
2. Can I attend school full-time while working 20 hours a week?
3. My boat travels 12 km/h in still water. How long will it take me to cruise 25 km upstream if the river's current is 3 km/h?

Although these problems differ, there is a strategy that can be applied to all of them.

The Five-Step Strategy

Since you have already studied some algebra, you have some experience with problem solving. The following steps describe a strategy that you may have used already; they form a sound strategy for problem solving in general.

> **FIVE STEPS FOR PROBLEM SOLVING WITH ALGEBRA**
> 1. *Familiarize* yourself with the problem.
> 2. *Translate* to mathematical language.
> 3. *Carry out* some mathematical manipulation.
> 4. *Check* your possible answer in the original problem.
> 5. *State* the answer clearly.

Of the five steps, probably the most important is the first: becoming familiar with the problem situation. Here are some ways in which this can be done.

> **THE FAMILIARIZE STEP**
> 1. If the problem is written, read it carefully. Then read it again, perhaps aloud.
> 2. List the information given and restate the question being asked. Select a variable or variables to represent any unknown(s) and clearly state what each variable represents.
> 3. Find additional information. Look up formulas or definitions with which you are not familiar. Consult an expert in the field or a reference librarian.
> 4. Create a table, using variables, in which both known and unknown information are listed. Look for possible patterns.
> 5. Make and label a drawing.
> 6. Estimate an answer and check to see whether it is correct.

For example, consider Problem 1: "If I exercise twice a week and eat 2400 calories a day, will I lose weight?" To familiarize yourself with this situation, you might research the calorie deficit necessary to lose a pound. You might also visit a personal trainer to find out how many calories per day you burn without exercise and how many you burn doing various exercises.

As another example, consider Problem 3: "How long will it take the boat to cruise 25 km upsteam?" To familiarize yourself with this situation, you might read and even reread the problem carefully to understand what information is given and what information is required. We list the given information.

Given Information		*Information Required*	
Distance to be traveled:	25 km	Time required:	?
Speed of boat in still water:	12 km/h	Speed of boat upstream:	?
Speed of current:	3 km/h		

Since the problem asks for the time required, we let $t =$ the number of hours required for the boat to cruise 25 km upstream. You may need to consult outside references to determine any other relationships that exist among distance, speed, and time. The *distance formula* is a basic relationship of those three quantities:

Distance = Speed × Time.   It is important to remember this equation.

We also need to know that a boat's speed going upstream can be determined by subtracting the current's speed from the boat's speed in still water.

We're now ready to create a table and make a drawing.

Student Notes

It is extremely helpful to write down exactly what each variable represents before attempting to form an equation.

	25 km
	$12 - 3 = 9$ km/h
	t

Current

25 km
9 km/h
t h

Boat

Using the information in the table, we might try a guess. Suppose the boat traveled upstream for 2 hr. The boat would have then traveled

$$9 \, \frac{\text{km}}{\text{hr}} \times 2 \, \text{hr} = 18 \, \text{km}. \qquad \text{Note that } \frac{\text{km}}{\text{hr}} \cdot \text{hr} = \text{km}.$$

Speed × Time = Distance

Since $18 \neq 25$, our guess is wrong. Still, examining how we checked our guess sheds light on how to translate the problem to an equation.

The second step in problem solving is to translate the situation to mathematical language. In algebra, this often means forming an equation or an inequality. In the third step of our process, we work with the results of the first two steps.

THE TRANSLATE AND CARRY OUT STEPS

Translate the problem to mathematical language. This is sometimes done by writing an algebraic expression, but most often in this text it is done by translating to an equation or an inequality.

Carry out some mathematical manipulation. If you have translated to an equation, this means to solve the equation.

To complete the problem-solving process, we should always check our solution and then state the solution in a clear and precise manner.

THE CHECK AND STATE STEPS

Check your possible answer in the original problem. Make sure that the answer is reasonable and that all the conditions of the original problem have been satisfied.

State the answer clearly. Write a complete English sentence stating the solution.

Problem Solving

At this point, our study of algebra has just begun and our problems may seem simple; however, to gain practice with the problem-solving process, use all five steps. Later some steps may be shortened or combined.

EXAMPLE 1 *Purchasing.* Cheyenne pays $157.94 for a cordless headset. If the price paid includes a 6% sales tax, what is the price of the headset itself?

SOLUTION

1. **Familiarize.** First, we familiarize ourselves with the problem. Note that tax is calculated from, and then added to, the item's price. Let's guess that the headset's price is $140. To check the guess, we calculate the amount of tax, $(0.06)(\$140) = \$8.40$, and add it to $140:

$$\$140 + (0.06)(\$140) = \$140 + \$8.40$$
$$= \$148.40. \qquad \$148.40 \neq \$157.94$$

   Our guess was too low, but the manner in which we checked the guess will guide us in the next step. We let

   $p =$ the price of the headset, in dollars.

2. **Translate.** Our guess leads us to the following translation:

Rewording:	The price of the headset	plus	6% sales tax	is	the price with sales tax.
Translating:	p	$+$	$(0.06)p$	$=$	$157.94

3. **Carry out.** Next, we carry out some mathematical manipulation:

$$p + (0.06)p = 157.94$$
$$1.06p = 157.94 \qquad \text{Combining like terms:}$$
$$\qquad\qquad\qquad\qquad 1p + 0.06p = (1 + 0.06)p$$
$$\frac{1}{1.06} \cdot 1.06p = \frac{1}{1.06} \cdot 157.94 \qquad \text{Using the multiplication principle}$$
$$p = 149.$$

4. **Check.** To check the answer in the original problem, note that the tax on a headset costing $149 would be $(0.06)(\$149) = \$8.94$. When this is added to $149, we have

$$\$149 + \$8.94, \quad \text{or} \quad \$157.94.$$

   Thus, $149 checks in the original problem.

5. **State.** We clearly state the answer: The headset itself costs $149.

1. Service Printing offers a 5% discount for large-volume jobs. After the discount, Cesar's bill was $80.37. What was his bill before the discount?

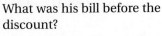 YOUR TURN

EXAMPLE 2 *Home Maintenance.* In an effort to make their home more energy-efficient, Jess and Drew purchased 200 in. of 3M Press-In-Place™ window glazing. This will be just enough to outline two square skylights. If the length of the sides of the larger skylight is $1\frac{1}{2}$ times the length of the sides of the smaller one, how should the glazing be cut?

SOLUTION

1. **Familiarize.** Note that the *perimeter* of (distance around) each square is four times the length of a side. Furthermore, if s represents the length of a side of the smaller square, then $\left(1\frac{1}{2}\right)s$ represents the length of a side of the larger square. We make a drawing and note that the two perimeters must add up to 200 in.

   Perimeter of a square $= 4 \cdot$ *length of a side*

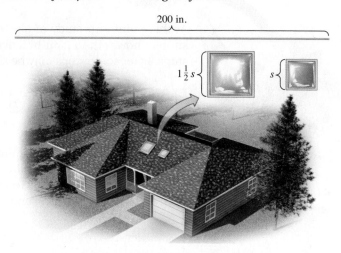

2. **Translate.** Rewording the problem can help us translate:

Rewording:	The perimeter of one square	plus	the perimeter of the other	is	200 in.
Translating:	$4s$	$+$	$4\left(1\frac{1}{2}s\right)$	$=$	200

3. **Carry out.** We solve the equation:

   $$4s + 4\left(1\frac{1}{2}s\right) = 200$$
   $$4s + 6s = 200 \qquad \text{Simplifying; } 4\left(1\frac{1}{2}s\right) = 4\left(\frac{3}{2}s\right) = 6s$$
   $$10s = 200 \qquad \text{Combining like terms}$$
   $$s = \tfrac{1}{10} \cdot 200 \qquad \text{Multiplying both sides by } \tfrac{1}{10}$$
   $$s = 20. \qquad \text{Simplifying}$$

4. **Check.** If the length of the smaller side is 20 in., then $\left(1\frac{1}{2}\right)(20\text{ in.}) = 30$ in. is the length of the larger side. The two perimeters would then be

   $$4 \cdot 20 \text{ in.} = 80 \text{ in.} \quad \text{and} \quad 4 \cdot 30 \text{ in.} = 120 \text{ in.}$$

   Since 80 in. $+ 120$ in. $= 200$ in., our answer checks.

5. **State.** The glazing should be cut into two pieces, one 80 in. long and the other 120 in. long.

2. Refer to Example 2. Jess and Drew purchased another 132 in. of window glazing to outline two other square skylights. For these skylights, the sides of the smaller skylight are five-sixths the length of the sides of the larger skylight. How should the glazing be cut?

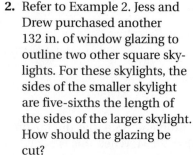 YOUR TURN

We cannot stress too greatly the importance of labeling the variables in your problem. In Example 2, solving for s is not enough: We need to find $4s$ and $4\left(1\frac{1}{2}s\right)$ in order to determine the numbers we are after.

EXAMPLE 3   Three numbers are such that the second is 6 less than three times the first and the third is 2 more than two-thirds the first. The sum of the three numbers is 150. Find the largest of the three numbers.

SOLUTION   We proceed according to the five-step process.

1. **Familiarize.**   We need to find the largest of three numbers. We list the information given in a table in which x represents the first number.

First Number	x
Second Number	6 less than 3 times the first
Third Number	2 more than $\frac{2}{3}$ the first

$$\text{First} + \text{Second} + \text{Third} = 150$$

Try to check a guess at this point. We will proceed to the next step.

2. **Translate.**   Because we wish to write an equation in just one variable, we need to express the second and third numbers using x ("in terms of x"). Using the table from the *Familiarize* step, we write the second number as $3x - 6$ and the third as $\frac{2}{3}x + 2$. We know that the sum is 150. Substituting, we obtain an equation:

$$\underbrace{\text{First}}_{\downarrow} + \underbrace{\text{second}}_{\downarrow} + \underbrace{\text{third}}_{\downarrow} \underset{\downarrow}{=} \underset{\downarrow}{150.}$$
$$x \quad + \quad (3x - 6) \quad + \quad \left(\tfrac{2}{3}x + 2\right) \quad = \quad 150$$

3. **Carry out.**   We solve the equation:

$$\begin{aligned}
x + 3x - 6 + \tfrac{2}{3}x + 2 &= 150 && \text{Leaving off unnecessary parentheses} \\
\left(4 + \tfrac{2}{3}\right)x - 4 &= 150 && \text{Combining like terms} \\
\tfrac{14}{3}x - 4 &= 150 && \text{Adding within parentheses; } 4\tfrac{2}{3} = \tfrac{14}{3} \\
\tfrac{14}{3}x &= 154 && \text{Adding 4 to both sides} \\
x &= \tfrac{3}{14} \cdot 154 && \text{Multiplying both sides by } \tfrac{3}{14} \\
x &= 33. && \text{Remember, } x \text{ represents the first number.}
\end{aligned}$$

Going back to the table, we can find the other two numbers:

Second: $3x - 6 = 3 \cdot 33 - 6 = 93$;
Third: $\frac{2}{3}x + 2 = \frac{2}{3} \cdot 33 + 2 = 24$.

Chapter Resources:
Translating for Success, p. 63;
Collaborative Activity, p. 64

4. **Check.**   We return to the original problem. There are three numbers: 33, 93, and 24. Is the second number 6 less than three times the first?

$$3 \times 33 - 6 = 99 - 6 = 93$$

The answer is *yes*. Is the third number 2 more than two-thirds the first?

$$\tfrac{2}{3} \times 33 + 2 = 22 + 2 = 24$$

The answer is *yes*. Is the sum of the three numbers 150?

$$33 + 93 + 24 = 150$$

The answer is *yes*. The numbers do check.

5. **State.**   The problem asks us to find the largest number, so the answer is: "The largest of the three numbers is 93."

3. Three numbers are such that the first is twice the second, and the third is 12 less than one-half the second. The sum of the three numbers is 247. Find the smallest of the three numbers.

YOUR TURN

> **CAUTION!** In Example 3, although the equation $x = 33$ enables us to find the largest number, 93, the number 33 is *not* the solution of the problem. By clearly labeling our variable in the first step, we can avoid thinking that the variable always represents the solution of the problem.

1.4 EXERCISE SET

FOR EXTRA HELP

MyMathLab® MathXL PRACTICE   WATCH   READ   REVIEW

Vocabulary and Reading Check

1. Write the five steps of the problem-solving process in the correct order.

   1. _____ Carry out.

   2. _____ Check.

   3. _____ State.

   4. _____ Familiarize.

   5. _____ Translate.

Each of the following corresponds to one of the five steps listed above. Write the name of the step during which each is done.

2. _____ Solve an equation.

3. _____ Give the answer clearly.

4. _____ Convert the wording into an equation.

5. _____ Read the problem carefully.

6. _____ Make certain that the question asked is answered.

The Five-Step Strategy

For each problem, familiarize yourself with the situation. Then translate to mathematical language. You need not actually solve the problem; just carry out the first two steps of the five-step strategy. You will be asked to complete some of the solutions as Exercises 35–42.

7. The sum of two numbers is 91. One of the numbers is 9 more than the other. What are the numbers?

8. The sum of two numbers is 88. One of the numbers is 6 more than the other. What are the numbers?

9. Many rowers from the Open Water Rowing Center in Sausalito, California, learn their skills in the Richardson Bay, a broad arm of San Francisco Bay. One suggested route for novice rowers is about 8 mi across Richardson Bay. In his single-person scull, Noah can maintain a speed of 4.6 mph in still water. If he is paddling into a 2.1-mph current, how long will it take him to complete the 8-mi Richardson Bay route?

   Source: Based on information from owrc.com

10. *Aviation.*   A Cessna airplane traveling 390 km/h in still air encounters a 65-km/h headwind. How long will it take the plane to travel 725 km into the wind?

11. *Angles in a Triangle.*   The degree measures of the angles in a triangle are three consecutive integers. Find the measures of the angles.

12. *Pricing.*   Becker Lumber gives contractors a 15% discount on all orders. After the discount, a contractor pays $272 for plywood. What was the original cost of the plywood?

13. *Escalators.*   A 205-ft long escalator at the CNN World Headquarters in Atlanta, Georgia, is the world's longest freestanding escalator. In a rush, Dominik walks up the escalator at a rate of 100 ft/min while the escalator is moving up at a rate of 105 ft/min. How long will it take him to reach the top of the escalator?

   Source: www.cnn.com/tour/atlanta/att.tour.home.html

100 ft/min

105 ft/min

14. *Moving Sidewalks.* A moving sidewalk in Pearson Airport, Ontario, is 912 ft long and moves at a rate of 6 ft/sec. If Alida walks at a rate of 4 ft/sec, how long will it take her to walk the length of the moving sidewalk?

    Source: *The Wall Street Journal, 8/16/07*

15. *Pricing.* Quick Storage prices flash drives by raising the wholesale price 50% and adding $1.50. What must a drive's wholesale price be if it is being sold for $22.50?

16. *Pricing.* Miller Oil offers a 5% discount to customers who pay promptly for an oil delivery. The Blancos promptly paid $142.50 for their December oil bill. What would the cost have been had they not paid promptly?

17. *Cruising Altitude.* The pilot of a Boeing 747 is instructed to climb from 8000 ft to a cruising altitude of 29,000 ft. If the plane ascends at a rate of 3500 ft/min, how long will it take to reach the cruising altitude?

18. *Angles in a Triangle.* One angle of a triangle is four times the measure of a second angle. The third angle measures 5° more than twice the second angle. Find the measures of the angles.

19. Find three consecutive odd integers such that the sum of the first, twice the second, and three times the third is 70.

20. Find two consecutive even integers such that two times the first plus three times the second is 76.

21. A steel rod 90 cm long is to be cut into two pieces, each to be bent to make an equilateral triangle. The length of a side of one triangle is to be twice the length of a side of the other. How should the rod be cut?

22. A piece of wire 10 m long is to be cut into two pieces, one of them two-thirds as long as the other. How should the wire be cut?

23. *Rescue Calls.* Rescue crews working for Stockton Rescue average 3 calls per shift. After his first four shifts, Cody had received 5, 2, 1, and 3 calls. How many calls will Cody need on his next shift if he is to average 3 calls per shift?

24. *Test scores.* Olivia's scores on five tests are 93, 89, 72, 80, and 96. What must the score be on her next test so that the average will be 88?

Problem Solving

Solve each problem. Use all five problem-solving steps.

25. *Pricing.* The price that Tess paid for her graphing calculator, $84, is less than what Tony paid by $13. How much did Tony pay for his graphing calculator?

26. *Class Size.* The number of students in Damonte's class, 35, is greater than the number in Rose's class by 12. How many students are in Rose's class?

27. Dan is moving from Greenville, South Carolina, to Charlotte, North Carolina. The average monthly rent of an apartment in Charlotte is $625. This is five-sixths of the average monthly rent of an apartment in Greenville. What is the average monthly apartment rent in Greenville?

    Source: Based on data from bankrate.com

28. Anne recently moved from Columbus, Ohio, to Boston, Massachusetts. She was told to expect to pay $44 for a haircut in Boston. This is four-thirds as much as she paid in Columbus. What did Anne pay for a haircut in Columbus?

    Source: Based on data from bankrate.com

29. *Nursing.* One Friday, Vance gave 11 more flu shots than Mike did. Together, they gave 53 flu shots. How many flu shots did Vance give?

30. *Sales.* In January, Meghan made 12 fewer sales calls than Paul did. Together, they made 256 calls. How many calls did Meghan make that month?

31. The length of a rectangular mirror is three times its width, and its perimeter is 120 cm. Find the length and the width of the mirror.

32. The length of a rectangular tile is twice its width, and its perimeter is 21 cm. Find the length and the width of the tile.

33. The width of a rectangular greenhouse is one-fourth its length, and its perimeter is 130 m. Find the length and the width of the greenhouse.

34. The width of a rectangular garden is one-third its length, and its perimeter is 32 m. Find the dimensions of the garden.

35. Solve the problem of Exercise 9.

36. Solve the problem of Exercise 13.

37. Solve the problem of Exercise 11.

38. Solve the problem of Exercise 18.

39. Solve the problem of Exercise 16.

40. Solve the problem of Exercise 12.

41. Solve the problem of Exercise 15.

42. Solve the problem of Exercise 20.

43. Write a problem for a classmate to solve for which fractions must be multiplied in order to get the answer.

44. Write a problem for a classmate to solve for which fractions must be divided in order to get the answer.

Synthesis

45. How can a guess or estimate help prepare you for the *Translate* step when solving problems?

46. Why is it important to check the solution from the *Carry out* step in the original wording of the problem being solved?

47. *Test Scores.* Tico's scores on four tests are 83, 91, 78, and 81. How many points above his current average must Tico score on the next test in order to raise his average 2 points?

48. *Geometry.* The height and sides of a triangle are four consecutive integers. The height is the first integer, and the base is the third integer. The perimeter of the triangle is 42 in. Find the area of the triangle.

49. *Self-Employment.* The number of self-employed Americans increased 3.8% from 2007 to 2008. It then decreased 6.1% from 2008 to 2009 and another 3.9% from 2009 to 2011. If there were 14.7 million self-employed Americans in 2011, how many were there in 2007?

Source: U.S. Bureau of Labor Statistics

50. *Adjusted Wages.* Emma's salary is reduced $n\%$ during a recession. By what number should her salary be multiplied in order to bring it back to where it was before the recession?

⤴ YOUR TURN ANSWERS: SECTION 1.4
1. $84.60   **2.** 72 in. and 60 in.   **3.** 25

QUICK QUIZ: SECTIONS 1.1–1.4

1. Use roster notation to write the set of letters in the word "college."   [1.1]

2. Divide and simplify: $\left(-\frac{3}{10}\right) \div \left(-\frac{2}{5}\right)$.   [1.2]

3. Use an associative law to write an expression equivalent to $8 + (y + 3)$.   [1.2]

4. Solve $2d - 3(5 - d) = 8 - (7 - 5d)$. If the solution set is $\varnothing$ or $\mathbb{R}$, classify the equation as a contradiction or as an identity.   [1.3]

5. Robbin took graduation pictures for 8 fewer seniors than Michelle did. Together, they photographed 40 seniors. How many seniors did Robbin photograph? [1.4]

1.5   Formulas, Models, and Geometry

Solving Formulas ▪ Formulas as Models

Study Skills

Avoiding Temptation

Choose a time and a place to study that will minimize distractions. For example, stay away from a coffee shop where friends may stop by. Once you begin studying, avoid answering the phone and do not check e-mail or text messages during your study session.

A **formula** is an equation that uses letters to represent a relationship between two or more quantities. Some important geometric formulas are $A = \pi r^2$ (for the area A of a circle of radius r), $C = \pi d$ (for the circumference C of a circle of diameter d), $A = bh$ (for the area A of a parallelogram of height h and base length b), and $A = \frac{1}{2}bh$ (for the area of a triangle of height h and base length b.)* A more complete list of geometric formulas appears at the very end of this text.

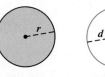

$A = \pi r^2$ $C = \pi d$

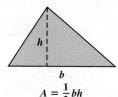

$A = bh$

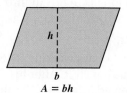

$A = \frac{1}{2}bh$

---

*The Greek letter π, read "pi," is *approximately* 3.14159265358979323846264. Often 3.14 or 22/7 is used to approximate π when a calculator with a π key is unavailable.

Solving Formulas

Suppose we know the floor area and the width of a rectangular room and want to find the length. To do so, we could "solve" the formula $A = l \cdot w$ (Area = Length · Width) for l, with the same principles that were used for solving equations.

$A = l \cdot w$

l

w

EXAMPLE 1 *Area of a Rectangle.* Solve the formula $A = l \cdot w$ for l.

SOLUTION

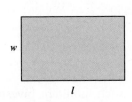

w

l

$$A = l \cdot w \qquad \text{We want this letter alone.}$$

$$\frac{A}{w} = \frac{l \cdot w}{w} \qquad \begin{array}{l} \text{Dividing both sides by } w \text{, or} \\ \text{multiplying both sides by } 1/w \end{array}$$

$$\left. \begin{array}{l} \dfrac{A}{w} = l \cdot \dfrac{w}{w} \\[2ex] \dfrac{A}{w} = l \end{array} \right\} \quad \text{Simplifying by removing a factor equal to 1: } \dfrac{w}{w} = 1$$

1. Solve $I = Prt$ for t.

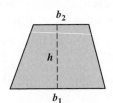

YOUR TURN

Thus to find the length of a rectangular room, we can divide the area of the floor by its width. Were we to do this calculation for a variety of rectangular rooms, the formula $l = A/w$ would be more convenient than repeatedly substituting into $A = l \cdot w$ and solving for l each time.

b_2

h

b_1

EXAMPLE 2 *Area of a Trapezoid.* A trapezoid is a geometric shape with four sides, exactly two of which, the bases, are parallel to each other. The formula for calculating the area A of a trapezoid with bases b_1 and b_2 (read "b sub one" and "b sub two") and height h is given by

$$A = \frac{h}{2}(b_1 + b_2), \qquad \begin{array}{l} \text{A derivation of this formula is outlined in} \\ \text{Exercise 79 of this section.} \end{array}$$

where the *subscripts* 1 and 2 distinguish one base from the other. Solve for b_1.

SOLUTION   There are several ways to "remove" the parentheses. We could distribute $h/2$, but an easier approach is to multiply both sides by the reciprocal of $h/2$.

$$A = \frac{h}{2}(b_1 + b_2)$$

$$\frac{2}{h} \cdot A = \frac{2}{h} \cdot \frac{h}{2}(b_1 + b_2)$$

Multiplying both sides by $\frac{2}{h}$ (or dividing by $\frac{h}{2}$)

$$\frac{2A}{h} = b_1 + b_2$$

Simplifying. The right side is "cleared" of fractions, since $(2h)/(h2) = 1$.

2. Solve $y = \frac{3x}{4}(a - b)$ for a.

$$\frac{2A}{h} - b_2 = b_1$$

Adding $-b_2$ on both sides

YOUR TURN

EXAMPLE 3   *Accumulated Simple Interest.*   The formula $A = P + Prt$ gives the amount A that a principal of P dollars will be worth in t years when invested at simple interest rate r. Solve the formula for P.

SOLUTION   We have

Student Notes

When solving for a variable that appears in more than one term, such as the P in Example 3, the key word to remember is *factor*. Once the variable is factored out, the next step in solving is often clear.

$$A = P + Prt$$ We want this letter alone.

$$A = P(1 + rt)$$ Factoring (using the distributive law) to write P just once, as a factor

$$\frac{A}{1 + rt} = \frac{P(1 + rt)}{1 + rt}$$ Dividing both sides by $1 + rt$, or multiplying both sides by $\frac{1}{1 + rt}$

$$\frac{A}{1 + rt} = P.$$ Simplifying

3. Solve $y = ax + cx$ for x.

This last equation can be used to determine how much should be invested at simple interest rate r in order to have A dollars t years later.

YOUR TURN

Note in Example 3 that factoring enabled us to write P once rather than twice. This is comparable to combining like terms when solving an equation like $16 = x + 7x$.

You may find the following summary useful.

TO SOLVE A FORMULA FOR A SPECIFIED LETTER

1. Get all terms with the specified letter on one side of the equation and all other terms on the other side, using the addition principle. To do this may require removing parentheses.

   • To remove parentheses, either divide both sides by the multiplier in front of the parentheses or use the distributive law.

2. When all terms with the specified letter are on the same side, factor (if necessary) so that the variable is written only once.
3. Solve for the letter in question by dividing both sides by the multiplier of that letter.

CONNECTING 🔗 THE CONCEPTS

Similarities between solving formulas and solving equations can be seen below. In (a), we solve as we did before; in (b), we do not carry out all calculations; and in (c), we cannot possibly carry out all calculations because the numbers are unknown. The same steps are used each time.

a)
$$9 = \frac{3}{2}(x + 5)$$
$$\frac{2}{3} \cdot 9 = \frac{2}{3} \cdot \frac{3}{2}(x + 5)$$
$$6 = x + 5$$
$$1 = x$$

b)
$$9 = \frac{3}{2}(x + 5)$$
$$\frac{2}{3} \cdot 9 = \frac{2}{3} \cdot \frac{3}{2}(x + 5)$$
$$\frac{2 \cdot 9}{3} = x + 5$$
$$\frac{2 \cdot 9}{3} - 5 = x$$

c)
$$A = \frac{h}{2}(b_1 + b_2)$$
$$\frac{2}{h} \cdot A = \frac{2}{h} \cdot \frac{h}{2}(b_1 + b_2)$$
$$\frac{2A}{h} = b_1 + b_2$$
$$\frac{2A}{h} - b_2 = b_1$$

EXERCISES

1. Solve: $2x - 3 = 7$.

2. Solve for x: $2x - c = h$.

3. Solve: $8 - 3(y - 7) = 2y$.

4. Solve for y: $8 - n(y - c) = ay$.

5. Solve: $\dfrac{a}{3} - \dfrac{1}{2} = 4a$.

6. Solve for a: $\dfrac{a}{3} - \dfrac{n}{2} = xa$.

Formulas as Models

A **mathematical model** can be a formula, or a set of formulas, developed to represent a real-world situation. In problem solving, a mathematical model is formed in the *Translate* step.

Name	Weight (in pounds)	Height (in inches)	BMI
Beyoncé	130	66	21
Tom Cruise	166	67	26.1
Serena Williams	150	69	22.2
Eli Manning	?	76	26.6

Source: www.health.com

EXAMPLE 4 *Body Mass Index.* *Body mass index*, or BMI, is based on height and weight and is often used as an indication of whether or not a person is at a healthy weight. A BMI between 18.5 and 24.9 is considered normal. Since this index does not take into consideration what percentage of a person's weight is lean muscle, it is not entirely accurate in evaluating an individual's weight. Some sample heights, weights, and BMIs are shown in the table at left.

Eli Manning, quarterback for the New York Giants, is 6 ft 4 in. tall and has a body mass index of approximately 26.6. What is his weight?

SOLUTION

1. Familiarize. From an outside source, we find that body mass index I depends on a person's height and weight and is found using the formula

$$I = \frac{704.5W}{H^2},$$

where W is the weight, in pounds, and H is the height, in inches.

Source: National Center for Health Statistics

2. Translate. Because we are interested in finding Manning's weight, we solve for W:

$$I = \frac{704.5W}{H^2}$$ We want this letter alone.

$$I \cdot H^2 = \frac{704.5W}{H^2} \cdot H^2$$ Multiplying both sides by H^2 to clear the fraction

$$IH^2 = 704.5W$$ Simplifying

$$\frac{IH^2}{704.5} = \frac{704.5W}{704.5}$$ Dividing by 704.5

$$\frac{IH^2}{704.5} = W.$$

3. Carry out. The model

$$W = \frac{IH^2}{704.5}$$

can be used to calculate the weight of someone whose body mass index and height are known. Using the information given, we have

$$W = \frac{26.6 \cdot 76^2}{704.5}$$ 6 ft 4 in. is 76 in.

$$\approx 218.$$ Using a calculator

4. Check. We could repeat the calculations or substitute in the original formula and then solve for W. The check is left to the student.

5. State. Eli Manning weighs about 218 lb.

YOUR TURN

4. The formula $E = w \cdot A$ is used to find the estimated blood volume E, in milliliters, of a patient with weight w, in kilograms, and average blood volume A, in milliliters per kilogram. Find the estimated blood volume of a toddler weighing 10.2 kg with an average blood volume of 80 mL/kg.

Source: www.manuelsweb.com

EXAMPLE 5 *Density.* A collector suspects that a silver coin is not solid silver. The density of silver is 10.5 grams per cubic centimeter (g/cm^3), and the coin is 0.2 cm thick with a radius of 2 cm. If the coin is really silver, how much should it weigh?

SOLUTION

1. Familiarize. From an outside reference, we find that density depends on mass and volume and that, in this setting, mass means weight. Since a coin is in the shape of a right circular cylinder, we also need the formula for the volume of such a cylinder. The applicable formulas are

$$D = \frac{m}{V} \quad and \quad V = \pi r^2 h,$$

where D is the density, m the mass, V the volume, r the length of the radius, and h the height of a right circular cylinder.

2. Translate. We need a model relating mass to the measurements of the coin, so we solve for m and then substitute for V:

$$D = \frac{m}{V}$$

$$V \cdot D = V \cdot \frac{m}{V}$$ Multiplying by V

$$V \cdot D = m$$ Simplifying

$$\pi r^2 h \cdot D = m.$$ Substituting

Chapter Resource:
Decision Making: Connection,
p. 64

5. The density of aluminum is 2.7 g/cm³. If the coin in Example 5 were made of aluminum instead of silver, how much would it weigh?

3. Carry out. The model $m = \pi r^2 hD$ can be used to find the mass of any right circular cylinder for which the dimensions and the density are known:

$$m = \pi r^2 hD$$
$$\qquad = \pi (2)^2 (0.2)(10.5) \qquad \text{Substituting}$$
$$\qquad \approx 26.3894. \qquad \text{Using a calculator with a } \pi \text{ key}$$

4. Check. To check, we could repeat the calculations. We can also check the model by examining the units:

$$r^2 h \cdot D = \text{cm}^2 \cdot \text{cm} \cdot \frac{\text{g}}{\text{cm}^3} = \text{cm}^3 \cdot \frac{\text{g}}{\text{cm}^3} = \text{g}.$$

Since g (grams) is the appropriate unit of mass, we have at least a partial check.

5. State. The coin, if it is indeed silver, should weigh about 26 g.

 YOUR TURN

1.5 EXERCISE SET

FOR EXTRA HELP   MyMathLab® MathXL
PRACTICE   WATCH   READ   REVIEW

Vocabulary and Reading Check

Complete each of the following statements.

1. A formula is a(n) _____ that uses letters to represent a relationship between two or more quantities.

2. The formula $A = \pi r^2$ is used to calculate the _____ of a circle.

3. The formula $C = \pi d$ is used to calculate the _____ of a circle.

4. The formula _____ is used to calculate the area of a triangle of height h and base length b.

5. The formula _____ is used to calculate the area of a parallelogram of height h and base length b.

6. The formula $l = A/w$ can be used to determine the _____ of a rectangle, given its area and width.

7. In the formula for the area of a trapezoid, $A = \dfrac{h}{2}(b_1 + b_2)$, the numbers 1 and 2 are referred to as _____.

8. When two or more terms on the same side of a formula contain the letter for which we are solving, we can _____ so that the letter is only written once.

Solving Formulas

Solve.

9. $E = wA$, for A (a nursing formula)

10. $F = ma$, for a (a physics formula)

11. $d = rt$, for r (a distance formula)

12. $P = EI$, for E (an electricity formula)

13. $V = lwh$, for h (a volume formula)

14. $I = Prt$, for r (a formula for interest)

15. $L = \dfrac{k}{d^2}$, for k (a formula for intensity of sound or light)

16. $F = \dfrac{mv^2}{r}$, for m (a physics formula)

17. $G = w + 150n$, for n (a formula for the gross weight of a bus)

18. $P = b + 1.5t$, for t (a formula for parking prices)

19. $2w + 2h + l = p$, for l (a formula used when shipping boxes)

20. $2w + 2h + l = p$, for w

21. $2x + 3y = 4$, for y

22. $3x - 7y = 2$, for y

23. $Ax + By = C$, for y (a formula for graphing lines)

24. $P = 2l + 2w$, for l (a perimeter formula)

25. $C = \frac{5}{9}(F - 32)$, for F (a temperature formula)

26. $T = \frac{3}{10}(I - 12{,}000)$, for I (a tax formula)

27. $V = \frac{4}{3}\pi r^3$, for r^3 (a formula for the volume of a sphere)

28. $V = \frac{4}{3}\pi r^3$, for π

29. $np + nm = t$, for n

30. $ab + ac = d$, for a

31. $uv + wv = x$, for v

32. $st + rt = n$, for t

33. $A = \dfrac{q_1 + q_2 + q_3}{n}$, for n (a formula for averaging)
(*Hint:* Multiply by n to "clear" fractions.)

34. $g = \dfrac{km_1m_2}{d^2}$, for d^2 (Newton's law of gravitation)

35. $v = \dfrac{d_2 - d_1}{t}$, for t (a physics formula)

36. $v = \dfrac{s_2 - s_1}{m}$, for m

37. $v = \dfrac{d_2 - d_1}{t}$, for d_1

38. $v = \dfrac{s_2 - s_1}{m}$, for s_1

39. $bd = c + ba$, for b

40. $st = n + sm$, for s

41. $v - w = uvw$, for w

42. $p - q = qrs$, for q

43. $n - mk = mt^2$, for m

44. $d - ct = ca^3$, for c

45. *Investing.* Eliana has \$2600 to invest for 6 months. If she needs the money to earn \$104 in that time, at what rate of simple interest must Eliana invest?

46. *Banking.* Chuma plans to buy a two-year certificate of deposit (CD) that earns 4% simple interest. If he needs the CD to earn \$150, how much should Chuma invest?

47. *Geometry.* The area of a parallelogram is 96 cm². The base of the figure is 6 cm. What is the height?

48. *Geometry.* The area of a parallelogram is 84 cm². The height of the figure is 7 cm. How long is the base?

Formulas as Models

For Exercises 49–56, make use of the formulas given in Examples 1–5.

49. *Body Mass Index.* Arnold Schwarzenegger, a former governor of California and a bodybuilder, is 6 ft 2 in. tall and has a body mass index of 30.8. How much does he weigh?

50. *Body Mass Index.* Actress Angelina Jolie has a body mass index of 17.9 and a height of 5 ft 8 in. What is her weight?

51. *Weight of Salt.* The density of salt is 2.16 g/cm³ (grams per cubic centimeter). An empty cardboard salt canister weighs 28 g, is 13.6 cm tall, and has a 4-cm radius. How much will a filled canister weigh?

52. *Weight of a Coin.* The density of gold is 19.3 g/cm³. If the coin in Example 5 were made of gold instead of silver, how much more would it weigh?

53. *Gardening.* A garden is constructed in the shape of a trapezoid, as shown in the following figure. The unknown dimension is to be such that the area of the garden is 90 ft². Find that unknown dimension.

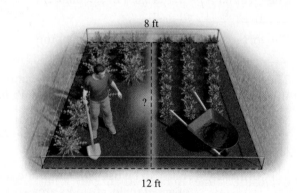

8 ft

?

12 ft

54. *Pet Care.* A rectangular kennel is being constructed, and 76 ft of fencing is available. The width of the kennel is to be 13 ft. What should the length be, in order to use just 76 ft of fence?

Aha! **55.** *Investing.* Do Xuan Nam is going to invest \$1000 at a simple interest rate of 4%. How long will it take for the investment to be worth \$1040?

56. Holli is going to invest \$950 at a simple interest rate of 3%. How long will it take for her investment to be worth \$1178?

57. *Baseball.*   Baseball analysts use the formula $r = 0.3b - 0.6c$ to estimate the number of runs r due to stolen bases for a runner who stole b bases and was caught stealing c times. In 2011, Ichiro Suzuki stole 40 bases and was credited with 7.8 stolen-base runs. How many times was he caught stealing?

58. *Nursing.*   The allowable blood loss L is the amount of blood that a patient can lose before a transfusion is necessary. This can be estimated by

$$L = \frac{E(H_i - H_f)}{H_i},$$

where E is the estimated blood volume of the patient, in milliliters, H_i is the initial hemoglobin level, and H_f is the lowest acceptable final hemoglobin level. What is the estimated blood volume of a patient with an allowable blood loss of 1470 mL, an initial hemoglobin of 13 g/dL, and a lowest final hemoglobin of 7 g/dL?

Source: Cecil B. Drain, *Perianesthesia Nursing: A Critical Care Approach.* Saunders, 2003.

Chess Ratings.   *The formula*

$$R = r + \frac{400(W - L)}{N}$$

is used to establish a chess player's rating R, after he or she has played N games, where W is the number of wins, L is the number of losses, and r is the average rating of the opponents.

Source: U.S. Chess Federation

59. Ulana's rating is 1305 after winning 5 games and losing 3 games in tournament play. What was the average rating of her opponents? (Assume there were no draws.)

60. Vladimir's rating fell to 1050 after winning twice and losing 5 times in tournament play. What was the average rating of his opponents? (Assume there were no draws.)

Female Caloric Needs.   *The number of calories K needed each day by a moderately active woman who weighs w pounds, is h inches tall, and is a years old can be estimated by the formula*

$$K = 917 + 6(w + h - a).$$

Source: M. Parker, *She Does Math.* Mathematical Association of America, p. 96

61. Julie is moderately active, weighs 120 lb, and is 23 years old. If Julie needs 1901 calories per day to maintain her weight, how tall is she?

62. Tawana is moderately active, 31 years old, and 5 ft 4 in. tall. If Tawana needs 1901 calories per day to maintain her weight, how much does she weigh?

Readability.   The reading difficulty of a text for elementary school grades 1–3 can be estimated by the Power–Sumner–Kearl Readability Formula,

$$g = 0.0778n + 4.55s - 2.2029,$$

where g is the grade level, n is the average number of words in a sentence, and s is the average number of syllables in a word.

Source: readabilityformulas.com

63. Elliot is writing a book for beginning third-graders (grade 3.0) using words with an average of 1.02 syllables per word. How many words, on average, should his sentences contain?

64. Autumn is writing a book for children near the end of third grade (grade 3.8). She uses, on average, 5 words per sentence. What should the average number of syllables per word be?

Energy-Efficient Lighting.   *The annual savings S realized from replacing a lighting fixture with a more efficient one is given by*

$$S = \frac{HR(W_i - W_n)}{1000},$$

where H is the number of burn hours per year, R is the cost of electricity per kilowatt-hour (kWh), W_i is the wattage of the existing lighting fixture, and W_n is the wattage of the replacement fixture.

65. Allison replaced a 100-watt fixture with a 15-watt fixture. She estimated that the fixture will burn 2000 hr per year and that the annual savings will be $20.40. What is the cost of her electricity per kWh?

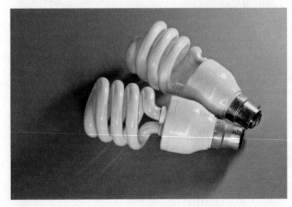

66. Connor calculated an annual savings of $42.90 when he replaced a 150-watt fixture. If the fixture will burn for 2600 hr per year and his electricity costs 15¢ per kWh, what was the wattage of the replacement fixture?

Blogging. A business owner's blog can be an effective marketing and advertising tool. The return on investment r of a blog can be estimated by

$$r = \frac{tmap}{hs},$$

where t is the average number of visits to the blog each day, m is the percentage of blog visitors who purchase merchandise, a is the average order size, p is the percentage of the average order that is profit, h is the number of hours spent blogging each day, and s is the hourly salary of the blogger. A return on investment less than 1 indicates that the blog is costing the company money.

Source: www.minethatdata.blogspot.com

67. Tomas earns $30 per hour writing a blog for his company. It takes him 4 hr per day to write the blog, and 5% of the blog visitors buy merchandise, with an average order size of $100 and a profit percentage of 15%. He calculates the return on investment to be 3.2. What is his average daily blog traffic?

68. Elyse earns $35 per hour writing a blog for her company. On average, 1200 people visit her blog daily, and 4% of them buy merchandise, with an average order size of $150 and a profit percentage of 14%. She calculates the return on investment to be 4.8. How long does it take her each day to write the blog?

Waiting Time. In an effort to minimize waiting time for patients at a doctor's office without increasing a physician's idle time, Michael Goiten of Massachusetts General Hospital has developed a model. Goiten suggests that the interval time I, in minutes, between scheduled appointments be related to the total number of minutes T that a physician spends with patients in a day and the number of scheduled appointments N according to the formula $I = 1.08(T/N)$.*

69. Dr. Cruz determines that she has a total of 8 hr per day to see patients. If she insists on an interval time of 15 min, according to Goiten's model, how many appointments should she make in one day?

70. A doctor insists on an interval time of 20 min and must be able to schedule 25 appointments per day. According to Goiten's model, how many hours per day should the doctor be prepared to spend with patients?

Projected Birth Weight. Ultrasonic images of 29-week-old fetuses can be used to predict weight. One model, developed by Thurnau,[†] is $P = 9.337da - 299$; a second model, developed by Weiner,[‡] is

---

* *New England Journal of Medicine*, 30 August 1990, pp. 604–608.

† Thurnau, G. R., R. K. Tamura, R. E. Sabbagha, et al., *Am. J. Obstet Gynecol* 1983; **145**: 557.

‡ Weiner, C. P., R. E. Sabbagha, N. Vaisrub, et al., *Obstet Gynecol* 1985; **65**: 812.

$P = 94.593c + 34.227a - 2134.616$. For both formulas, P is the estimated fetal weight, in grams; d is the diameter of the fetal head, in centimeters; c is the circumference of the fetal head, in centimeters; and a is the circumference of the fetal abdomen, in centimeters.

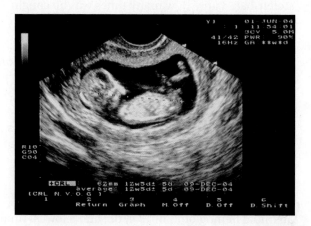

71. Solve Thurnau's model for d and use that equation to estimate the diameter of a fetus' head at 29 weeks when the estimated weight is 1614 g and the circumference of the fetal abdomen is 24.1 cm.

72. Solve Weiner's model for c and use that equation to estimate the circumference of a fetus' head at 29 weeks when the estimated weight is 1277 g and the circumference of the fetal abdomen is 23.4 cm.

73. Can the formula for the area of a parallelogram be used to find the area of a rectangle?

74. Predictions made using the models of Exercises 71 and 72 are often off by as much as 10%. Does this mean the models should be discarded? Why or why not?

Synthesis

75. Both of the models used in Exercises 71 and 72 have P alone on one side of the equation. Why?

76. See Exercises 59 and 60. Suppose Heidi plays in a chess tournament in which all of her opponents have the same rating. Under what circumstances will playing to a draw help or hurt her rating?

77. The density of platinum is 21.5 g/cm³. If the ring shown in the figure below is crafted out of platinum, how much will it weigh?

78. The density of a penny is 8.93 g/cm³. The mass of a roll of pennies is 177.6 g. If the diameter of a penny is 1.85 cm, how tall is a roll of pennies?

Solve.

79. To derive the formula for the area of a trapezoid, consider the area of two trapezoids, one of which is upside down, as shown below.

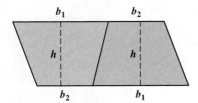

Explain why the total area of the two trapezoids is given by $h(b_1 + b_2)$. Then explain why the area of a trapezoid is given by $\frac{h}{2}(b_1 + b_2)$.

80. $A = 4lw + w^2$, for l

81. $s = v_i t + \frac{1}{2}at^2$, for a (a physics formula)

82. $\dfrac{P_1 V_1}{T_1} = \dfrac{P_2 V_2}{T_2}$, for T_2 (a chemistry formula)

83. $b = \dfrac{h + w + p}{a + w + p + f}$, for w (a baseball formula)

84. $m = \dfrac{(d/e)}{(e/f)}$, for d

85. $\dfrac{b}{a - b} = c$, for b

86. $\dfrac{a}{a + b} = c$, for a

Aha! **87.** $s + \dfrac{s + t}{s - t} = \dfrac{1}{t} + \dfrac{s + t}{s - t}$, for t

YOUR TURN ANSWERS: SECTION 1.5

1. $t = \dfrac{I}{Pr}$ **2.** $a = \dfrac{4y}{3x} + b$ **3.** $x = \dfrac{y}{a + c}$

4. 816 mL **5.** About 7 g

QUICK QUIZ: SECTIONS 1.1–1.5

1. Simplify: $\dfrac{8 \cdot 4 - 3(4 - 7 + 6)^2}{(-30) \div (-3)(-2)}$. [1.2]

Solve. If appropriate, classify the equation as a contradiction or as an identity. [1.3]

2. $3.3x - 1.5 = 4.1x$ **3.** $2 - (x - 7) = 5 + 3x$

4. Solve for p: $3p - 7 = bp$. [1.5]

5. The length of a rectangular table is twice its width, and its perimeter is 3 m. Find the dimensions of the table. [1.4]

1.6 Properties of Exponents

The Product Rule and the Quotient Rule ▪ The Zero Exponent ▪ Negative Integers as Exponents ▪ Simplifying $(a^m)^n$ ▪ Raising a Product or a Quotient to a Power ▪

Study Skills

Seeking Help Off Campus

Are you aware of all the supplements that exist for this textbook? See the preface for a description of each supplement. Many students find these learning aids invaluable when working on their own.

We now develop rules for manipulating exponents and determine what zero and negative integers will mean as exponents.

The Product Rule and the Quotient Rule

The expression $x^3 \cdot x^4$ can be rewritten as follows:

$$x^3 \cdot x^4 = \underbrace{x \cdot x \cdot x}_{3 \text{ factors}} \cdot \underbrace{x \cdot x \cdot x \cdot x}_{4 \text{ factors}}$$

$$= \underbrace{x \cdot x \cdot x \cdot x \cdot x \cdot x \cdot x}_{7 \text{ factors}}$$

$$= x^7.$$

The generalization of this result is the *product rule*.

> **MULTIPLYING WITH LIKE BASES: THE PRODUCT RULE**
>
> For any number a and any positive integers m and n,
>
> $$a^m \cdot a^n = a^{m+n}.$$
>
> (When multiplying, if the bases are the same, keep the base and add the exponents.)

Student Notes

Be careful to distinguish between how coefficients and exponents are handled. For instance, in Example 1(b), the product of the coefficients 5 and 3 is 15, whereas the product of b^3 and b^5 is b^8.

1. Multiply and simplify:

$$(-2x^2y)(7x^3y^6).$$

EXAMPLE 1 Multiply and simplify: **(a)** $m^5 \cdot m^7$; **(b)** $(5ab^3)(3a^4b^5)$.

SOLUTION

a) $m^5 \cdot m^7 = m^{5+7} = m^{12}$ Multiplying powers by adding exponents

b) $(5ab^3)(3a^4b^5) = 5 \cdot 3 \cdot a^1 \cdot a^4 \cdot b^3 \cdot b^5$ Using the associative and commutative laws; $a = a^1$

$= 15a^{1+4}b^{3+5}$ Multiplying coefficients; using the product rule

$= 15a^5b^8$

YOUR TURN

> **CAUTION!**
>
> $5^8 \cdot 5^6 = 5^{14}$ $\begin{cases} 5^8 \cdot 5^6 \neq 25^{14} & \text{Do not multiply the bases!} \\ 5^8 \cdot 5^6 \neq 5^{48} & \text{Do not multiply the exponents!} \end{cases}$

Next, we simplify a quotient:

$$\frac{x^8}{x^3} = \frac{x \cdot x \cdot x \cdot x \cdot x \cdot x \cdot x \cdot x}{x \cdot x \cdot x}$$ ⟵ 8 factors
    ⟵ 3 factors

$$= \frac{x \cdot x \cdot x}{x \cdot x \cdot x} \cdot x \cdot x \cdot x \cdot x \cdot x$$ Note that x^3/x^3 is 1.

$$= x \cdot x \cdot x \cdot x \cdot x$$ ⟵ 5 factors

$$= x^5.$$

The generalization of this result is the *quotient rule*.

> **DIVIDING WITH LIKE BASES: THE QUOTIENT RULE**
>
> For any nonzero number a and any positive integers m and n, $m > n$,
>
> $$\frac{a^m}{a^n} = a^{m-n}.$$
>
> (When dividing, if the bases are the same, keep the base and subtract the exponent of the denominator from the exponent of the numerator.)

EXAMPLE 2 Divide and simplify: **(a)** $\dfrac{r^9}{r^3}$; **(b)** $\dfrac{-10x^{11}y^5}{-2x^4y^3}$.

SOLUTION

a) $\dfrac{r^9}{r^3} = r^{9-3} = r^6$ Using the quotient rule

2. Divide and simplify:

$$\dfrac{32a^6c^8}{-4ac^7}.$$

b) $\dfrac{-10x^{11}y^5}{-2x^4y^3} = \dfrac{-10}{-2} \cdot x^{11-4} \cdot y^{5-3}$ Dividing coefficients; using the quotient rule

$$= 5x^7y^2$$

◀ YOUR TURN

CAUTION!

$$\dfrac{7^8}{7^2} = 7^6 \quad \begin{cases} \dfrac{7^8}{7^2} \neq 1^6 & \text{Do not divide the bases!} \\[2mm] \dfrac{7^8}{7^2} \neq 7^4 & \text{Do not divide the exponents!} \end{cases}$$

The Zero Exponent

Suppose now that the bases in the numerator and the denominator are identical and are both raised to the same power. On the one hand, any (nonzero) expression divided by itself is equal to 1. For example,

$$\dfrac{t^5}{t^5} = 1 \quad \text{and} \quad \dfrac{6^4}{6^4} = 1.$$

On the other hand, if we continue to subtract exponents when dividing powers with the same base, we have

$$\dfrac{t^5}{t^5} = t^{5-5} = t^0 \quad \text{and} \quad \dfrac{6^4}{6^4} = 6^{4-4} = 6^0.$$

This suggests that t^5/t^5 equals both 1 *and* t^0. It also suggests that $6^4/6^4$ equals both 1 *and* 6^0. This leads to the following definition.

THE ZERO EXPONENT

For any nonzero real number a,

$$a^0 = 1.$$

(Any nonzero number raised to the zero power is 1. The expression 0^0 is undefined.)

EXAMPLE 3 Evaluate each of the following for $x = 2.9$: **(a)** x^0; **(b)** $-x^0$; **(c)** $(-x)^0$.

SOLUTION

a) $x^0 = 2.9^0 = 1$ Using the definition of 0 as an exponent

b) $-x^0 = -2.9^0 = -1$ The exponent 0 pertains only to the 2.9.

3. Evaluate x^0 for $x = -16$.

c) $(-x)^0 = (-2.9)^0 = 1$ Because of the parentheses, the base here is -2.9.

◀ YOUR TURN

Parts (b) and (c) of Example 3 illustrate an important result:

$$-a^n \quad \text{means} \quad -1 \cdot a^n.$$

Thus, $-a^n$ and $(-a)^n$ are not equivalent expressions.*

Negative Integers as Exponents

We next develop a definition for negative integer exponents. We can simplify $5^3/5^7$ two ways. First we proceed as in arithmetic:

$$\frac{5^3}{5^7} = \frac{5 \cdot 5 \cdot 5}{5 \cdot 5 \cdot 5 \cdot 5 \cdot 5 \cdot 5 \cdot 5} = \frac{5 \cdot 5 \cdot 5 \cdot 1}{5 \cdot 5 \cdot 5 \cdot 5 \cdot 5 \cdot 5 \cdot 5}$$
$$= \frac{5 \cdot 5 \cdot 5}{5 \cdot 5 \cdot 5} \cdot \frac{1}{5 \cdot 5 \cdot 5 \cdot 5}$$
$$= \frac{1}{5^4}.$$

Were we to apply the quotient rule, we would have

$$\frac{5^3}{5^7} = 5^{3-7} = 5^{-4}.$$

These two expressions for $5^3/5^7$ suggest that

$$5^{-4} = \frac{1}{5^4}.$$

This leads to the definition of integer exponents, which includes negative exponents.

INTEGER EXPONENTS

For any nonzero real number a and any integer n,

$$a^{-n} = \frac{1}{a^n}.$$

(The numbers a^{-n} and a^n are reciprocals of each other.)

The definitions above preserve the following pattern:

$$4^3 = 4 \cdot 4 \cdot 4,$$
$$4^2 = 4 \cdot 4, \qquad \text{Dividing both sides by 4}$$
$$4^1 = 4, \qquad \text{Dividing both sides by 4}$$
$$4^0 = 1, \qquad \text{Dividing both sides by 4}$$
$$4^{-1} = \frac{1}{4}, \qquad \text{Dividing both sides by 4}$$
$$4^{-2} = \frac{1}{4 \cdot 4} = \frac{1}{4^2}. \qquad \text{Dividing both sides by 4}$$

---

*When n is odd, it is true that $-a^n = (-a)^n$. However, when n is even, we always have $-a^n \neq (-a)^n$, since $-a^n$ is always negative and $(-a)^n$ is always positive. We assume $a \neq 0$.

> **CAUTION!** A negative exponent does not, in itself, indicate that an expression is negative. For example,
>
> $$4^{-2} \neq 4(-2) \quad \text{and} \quad 4^{-2} \neq -4^2.$$

EXAMPLE 4   Express each of the following without negative exponents and, if possible, simplify: **(a)** 7^{-2}; **(b)** -7^{-2}; **(c)** $(-7)^{-2}$.

SOLUTION

a) $7^{-2} = \dfrac{1}{7^2} = \dfrac{1}{49}$ The base is 7. We use the definition of integer exponents.

b) $-7^{-2} = -\dfrac{1}{7^2} = -\dfrac{1}{49}$ The base is 7.

$$-7^{-2} = -1 \cdot 7^{-2} = -1 \cdot \frac{1}{7^2} = -1 \cdot \frac{1}{49} = -\frac{1}{49}.$$

c) $(-7)^{-2} = \dfrac{1}{(-7)^2} = \dfrac{1}{49}$ The base is -7. We use the definition of integer exponents.

Note that $-7^{-2} \neq (-7)^{-2}$.

4. Express -2^{-3} without negative exponents and simplify.

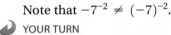 YOUR TURN

EXAMPLE 5   Express each of the following without negative exponents and, if possible, simplify: **(a)** $5x^{-4}y^3$; **(b)** $\dfrac{1}{6^{-2}}$.

SOLUTION

a) $5x^{-4}y^3 = 5\left(\dfrac{1}{x^4}\right)y^3 = \dfrac{5y^3}{x^4}$

5. Express $\dfrac{7}{n^{-1}}$ without negative exponents.

b) $\dfrac{1}{6^{-2}} = 6^{-(-2)} = 6^2$, or 36 $\dfrac{1}{a^n} = a^{-n}$; n can be a negative integer.

YOUR TURN

The result from Example 5(b) can be generalized.

FACTORS AND NEGATIVE EXPONENTS

For any nonzero real numbers a and b and any integers m and n,

$$\frac{a^{-n}}{b^{-m}} = \frac{b^m}{a^n}.$$

(A factor can be moved to the other side of the fraction bar if the sign of its exponent is changed.)

EXAMPLE 6   Write an equivalent expression without negative exponents:

$$\frac{vx^{-2}y^{-5}}{z^{-4}w^{-3}}.$$

6. Write an equivalent expression without negative exponents:

$$\frac{a^{-1}bc^2}{d^{-5}z}.$$

SOLUTION

$$\frac{vx^{-2}y^{-5}}{z^{-4}w^{-3}} = \frac{vz^4w^3}{x^2y^5}$$ Moving the factors with negative exponents to the other side of the fraction bar *and* changing the sign of those exponents

 YOUR TURN

Technology Connection

Most calculators have an exponentiation key, often labeled $\boxed{x^y}$ or $\boxed{\frown}$. To enter 4^7 on most scientific calculators, we press

On most graphing calculators, we press

1. List keystrokes that could be used to simplify 2^{-5} on a scientific or graphing calculator.
2. How could 2^{-5} be simplified on a calculator lacking an exponentiation key?

7. Simplify: $\dfrac{x^{-2}}{x^{-7}}$.

EXPLORING 🔍 THE CONCEPT

A factor can be moved to the other side of the fraction bar if the sign of its exponent is changed. Consider $\dfrac{a^3 b^{-6}}{c^{-9} d}$. Match each move described below with the resulting expression from the column on the right.

1. Move a^3 to the denominator.
2. Move b^{-6} to the denominator.
3. Move c^{-9} to the numerator.
4. Move d to the numerator.

a) $\dfrac{a^3}{b^6 c^{-9} d}$

b) $\dfrac{a^3 b^{-6} d^{-1}}{c^{-9}}$

c) $\dfrac{b^{-6}}{a^{-3} c^{-9} d}$

d) $\dfrac{a^3 b^{-6} c^9}{d}$

ANSWERS
1. (c) 2. (a) 3. (d) 4. (b)

The product rule and the quotient rule apply for all integer exponents.

EXAMPLE 7 Simplify: **(a)** $9^{-3} \cdot 9^8$; **(b)** $\dfrac{y^{-5}}{y^{-4}}$.

SOLUTION

a) $9^{-3} \cdot 9^8 = 9^{-3+8}$ — Using the product rule; adding exponents
$= 9^5$

b) $\dfrac{y^{-5}}{y^{-4}} = y^{-5-(-4)} = y^{-1}$ — Using the quotient rule; subtracting exponents
$= \dfrac{1}{y}$ Writing the answer without a negative exponent

YOUR TURN

Example 7(b) can also be simplified as follows:

$$\frac{y^{-5}}{y^{-4}} = \frac{y^4}{y^5} = y^{4-5} = y^{-1} = \frac{1}{y}.$$

Simplifying $(a^m)^n$

Next, consider an expression like $(3^4)^2$:

$$(3^4)^2 = (3^4)(3^4)$$ Raising 3^4 to the second power
$$= (3 \cdot 3 \cdot 3 \cdot 3)(3 \cdot 3 \cdot 3 \cdot 3)$$
$$= 3 \cdot 3 \cdot 3 \cdot 3 \cdot 3 \cdot 3 \cdot 3 \cdot 3$$ Using the associative law
$$= 3^8.$$

Note that in this case, we could have multiplied the exponents:

$$(3^4)^2 = 3^{4 \cdot 2} = 3^8.$$

The generalization of this result is the *power rule*.

THE POWER RULE

For any real number a and any integers m and n for which a^m and $(a^m)^n$ exist,

$$(a^m)^n = a^{mn}.$$

(To raise a power to a power, multiply the exponents.)

EXAMPLE 8   Simplify: **(a)** $(3^5)^4$; **(b)** $(y^{-5})^7$; **(c)** $(a^{-3})^{-7}$.

SOLUTION

a) $(3^5)^4 = 3^{5 \cdot 4} = 3^{20}$

b) $(y^{-5})^7 = y^{-5 \cdot 7} = y^{-35} = \dfrac{1}{y^{35}}$

c) $(a^{-3})^{-7} = a^{(-3)(-7)} = a^{21}$

8. Simplify: $(4^3)^{-9}$.

YOUR TURN

Raising a Product or a Quotient to a Power

When an expression inside parentheses is raised to a power, the inside expression is the base. Let's compare $2a^3$ and $(2a)^3$.

$$2a^3 = 2 \cdot a \cdot a \cdot a; \qquad (2a)^3 = (2a)(2a)(2a)$$
$$= 2 \cdot 2 \cdot 2 \cdot a \cdot a \cdot a$$
$$= 2^3 a^3 = 8a^3$$

We see that $2a^3$ and $(2a)^3$ are *not* equivalent. Note also that to simplify $(2a)^3$, we can raise each factor to the power 3. This leads to the following rule.

RAISING A PRODUCT TO A POWER

For any integer n and any real numbers a and b for which $(ab)^n$ exists,

$$(ab)^n = a^n b^n.$$

(To raise a product to a power, raise each factor to that power.)

EXAMPLE 9   Simplify: **(a)** $(-2x)^3$; **(b)** $(-3x^5 y^{-1})^{-4}$.

SOLUTION

a) $(-2x)^3 = (-2)^3 \cdot x^3$ Raising each factor to the third power

$\qquad\quad = -8x^3$ $\quad (-2)^3 = (-2)(-2)(-2) = -8$

b) $(-3x^5y^{-1})^{-4} = (-3)^{-4}(x^5)^{-4}(y^{-1})^{-4}$ Raising each factor to the negative fourth power

$$= \frac{1}{(-3)^4} \cdot x^{-20}y^4$$ Multiplying exponents; writing $(-3)^{-4}$ as $\frac{1}{(-3)^4}$

$$= \frac{1}{81} \cdot \frac{1}{x^{20}} \cdot y^4$$

$$= \frac{y^4}{81x^{20}}$$

9. Simplify: $(6a^{-1}b)^{-2}$.

YOUR TURN

There is a similar rule for raising a quotient to a power.

RAISING A QUOTIENT TO A POWER

For any integer n and any real numbers a and b for which a/b, a^n, and b^n exist,

$$\left(\frac{a}{b}\right)^n = \frac{a^n}{b^n}.$$

(To raise a quotient to a power, raise both the numerator and the denominator to that power.)

EXAMPLE 10 Simplify: **(a)** $\left(\frac{x^2}{2}\right)^4$; **(b)** $\left(\frac{y^2z^3}{5}\right)^{-3}$.

SOLUTION

a) $\left(\frac{x^2}{2}\right)^4 = \frac{(x^2)^4}{2^4} = \frac{x^8}{16}$ $\leftarrow 2 \cdot 4 = 8$ $\leftarrow 2^4 = 16$

b) $\left(\frac{y^2z^3}{5}\right)^{-3} = \frac{(y^2z^3)^{-3}}{5^{-3}}$

$$= \frac{5^3}{(y^2z^3)^3}$$ Moving factors to the other side of the fraction bar and changing the sign of those exponents

10. Simplify: $\left(\frac{2a^2x}{3c^{-4}}\right)^{-2}$.

$$= \frac{125}{y^6z^9}$$

YOUR TURN

The rule for raising a quotient to a power allows us to derive a useful result for manipulating negative exponents:

$$\left(\frac{a}{b}\right)^{-n} = \frac{a^{-n}}{b^{-n}} = \frac{b^n}{a^n} = \left(\frac{b}{a}\right)^n.$$

Using this result, we can simplify Example 10(b) as follows:

$$\left(\frac{y^2z^3}{5}\right)^{-3} = \left(\frac{5}{y^2z^3}\right)^3$$ Taking the reciprocal of the base and changing the exponent's sign

$$= \frac{5^3}{(y^2z^3)^3}$$

$$= \frac{125}{y^6z^9}.$$

DEFINITIONS AND PROPERTIES OF EXPONENTS

The following summary assumes that no denominators are 0 and that 0^0 is not considered and is true for any integers m and n.

1 as an exponent:	$a^1 = a$
0 as an exponent:	$a^0 = 1$
Negative exponents:	$a^{-n} = \dfrac{1}{a^n}$
	$\dfrac{a^{-n}}{b^{-m}} = \dfrac{b^m}{a^n}$
	$\left(\dfrac{a}{b}\right)^{-n} = \left(\dfrac{b}{a}\right)^n$
The Product Rule:	$a^m \cdot a^n = a^{m+n}$
The Quotient Rule:	$\dfrac{a^m}{a^n} = a^{m-n}$
The Power Rule:	$(a^m)^n = a^{mn}$
Raising a product to a power:	$(ab)^n = a^n b^n$
Raising a quotient to a power:	$\left(\dfrac{a}{b}\right)^n = \dfrac{a^n}{b^n}$

1.6 EXERCISE SET

FOR EXTRA HELP   MyMathLab® Math XL
PRACTICE   WATCH   READ   REVIEW

◥ Vocabulary and Reading Check

In each of Exercises 1–10, state whether the equation is an example of the product rule, the quotient rule, the power rule, raising a product to a power, or raising a quotient to a power.

1. $(a^6)^4 = a^{24}$

2. $\left(\dfrac{5}{7}\right)^4 = \dfrac{5^4}{7^4}$

3. $(5x)^7 = 5^7 x^7$

4. $\dfrac{m^9}{m^3} = m^6$

5. $m^6 \cdot m^4 = m^{10}$

6. $(5^2)^7 = 5^{14}$

7. $\left(\dfrac{a}{4}\right)^7 = \dfrac{a^7}{4^7}$

8. $(ab)^{10} = a^{10} b^{10}$

9. $\dfrac{x^{10}}{x^2} = x^8$

10. $r^5 \cdot r^7 = r^{12}$

The Product Rule and the Quotient Rule

Multiply and simplify. Leave the answer in exponential notation.

11. $6^4 \cdot 6^7$

12. $3^8 \cdot 3^9$

13. $m^0 \cdot m^8$

14. $t^6 \cdot t^0$

15. $5x^4 \cdot 4x^3$

16. $3a^5 \cdot 2a^4$

17. $(-3a^2)(-8a^6)$

18. $(-4m^7)(6m^2)$

19. $(m^5 n^2)(m^3 n p^0)$

20. $(x^6 y^3)(xy^4 z^0)$

Divide and simplify.

21. $\dfrac{t^8}{t^3}$

22. $\dfrac{a^{11}}{a^8}$

23. $\dfrac{15a^7}{3a^2}$

24. $\dfrac{24t^9}{8t^3}$

25. $\dfrac{m^7 n^9}{m^2 n^8}$

26. $\dfrac{m^6 n^9}{m^5 n^6}$

27. $\dfrac{32x^8 y^5}{8x^2 y}$

28. $\dfrac{35x^7 y^8}{7xy^2}$

29. $\dfrac{28x^{10} y^9 z^8}{-7x^2 y^3 z^2}$

30. $\dfrac{-20x^8 y^5 z^3}{-4x^2 y^2 z}$

The Zero Exponent

Evaluate each of the following for $x = -2$.

31. $-x^0$

32. $(-x)^0$

33. $(4x)^0$

34. $4x^0$

Negative Integers as Exponents

Write an equivalent expression without negative exponents and, if possible, simplify.

35. t^{-9}

36. m^{-2}

37. 6^{-2}

38. 5^{-3}

39. $(-3)^{-2}$

40. $(-2)^{-4}$

41. -3^{-2}

42. -2^{-4}

43. -1^{-10}

44. -10^{-2}

45. $\dfrac{1}{10^{-3}}$

46. $\dfrac{1}{2^{-4}}$

47. $6x^{-1}$

48 $9x^{-4}$

49. $3a^8b^{-6}$

50. $5a^{-7}b^4$

51. $\dfrac{2z^{-3}}{x^5}$

52. $\dfrac{5a^{-1}}{b}$

53. $\dfrac{3y^2}{z^{-4}}$

54. $\dfrac{t^{-6}}{7s^2}$

55. $\dfrac{ab^{-1}}{c^{-1}}$

56. $\dfrac{x^{-3}y^4}{z^{-5}}$

57. $\dfrac{pq^{-2}r^{-3}}{2u^5v^{-4}}$

58. $\dfrac{5a^{-3}bc^{-1}}{d^{-6}f^2}$

Write an equivalent expression with negative exponents.

59. $\dfrac{1}{x^3}$

60. $\dfrac{1}{n^4}$

61. $\dfrac{1}{(-10)^3}$

62. $\dfrac{1}{12^5}$

63. 8^{10}

64. $(-6)^4$

65. $4x^2$

66. $-4y^5$

67. $\dfrac{1}{(5y)^3}$

68. $\dfrac{1}{(5x)^5}$

69. $\dfrac{1}{3y^4}$

70. $\dfrac{1}{4b^3}$

Simplify. Should negative exponents appear in the answer, write a second answer using only positive exponents.

71. $6^{-3} \cdot 6^{-5}$

72. $4^{-2} \cdot 4^{-1}$

73. $a \cdot a^{-8}$

74. $b^5 \cdot b^{-2}$

75. $x^{-7} \cdot x^2 \cdot x^5$

76. $a^4 \cdot a^2 \cdot a^{-5}$

77. $(4mn^3)(-2m^3n^2)$

78. $(6x^6y^{-2})(-3x^2y^3)$

79. $(-7x^4y^{-5})(-5x^{-6}y^8)$

80. $(-4u^{-6}v^8)(-6u^{-4}v^{-2})$

81. $(5a^{-2}b^{-3})(2a^{-4}b)$

82. $(3a^{-5}b^{-7})(2ab^{-2})$

83. $\dfrac{10^{-3}}{10^6}$

84. $\dfrac{12^{-4}}{12^8}$

85. $\dfrac{2^{-7}}{2^{-5}}$

86. $\dfrac{9^{-4}}{9^{-6}}$

87. $\dfrac{y^4}{y^{-5}}$

88. $\dfrac{a^3}{a^{-2}}$

89. $\dfrac{24a^5b^3}{-8a^4b}$

90. $\dfrac{-12m^4}{-4mn^5}$

91. $\dfrac{15m^5n^3}{10m^{10}n^{-4}}$

92. $\dfrac{-24x^6y^7}{18x^{-3}y^9}$

93. $\dfrac{-6x^{-2}y^4z^8}{-24x^{-5}y^6z^{-3}}$

94. $\dfrac{8a^6b^{-4}c^8}{32a^{-4}b^5c^9}$

Simplifying $(a^m)^n$

95. $(x^4)^3$

96. $(a^3)^2$

97. $(9^3)^{-4}$

98. $(8^4)^{-3}$

99. $(t^{-8})^{-5}$

100. $(x^{-4})^{-3}$

Raising a Product or a Quotient to a Power

101. $(-5xy)^2$

102. $(-5ab)^3$

103. $(-2a^{-2}b)^{-3}$

104. $(-4x^6y^{-2})^{-2}$

105. $\left(\dfrac{m^2n^{-1}}{4}\right)^3$

106. $\left(\dfrac{3x^5}{y^{-4}}\right)^2$

107. $\dfrac{(2a^3)^3 4a^{-3}}{(a^2)^5}$

108. $\dfrac{(3x^2)^3 2x^{-4}}{(x^4)^2}$

Aha! **109.** $(8x^{-3}y^2)^{-4}(8x^{-3}y^2)^4$

110. $(2a^{-1}b^3)^{-2}(2a^{-1}b^3)^{-2}$

111. $\dfrac{(5a^3b)^2}{10a^2b}$

112. $\dfrac{(3x^3y^4)^3}{6xy^3}$

113. $\left(\dfrac{2x^3y^{-2}}{3y^{-3}}\right)^3$

114. $\left(\dfrac{-4x^4y^{-2}}{5x^{-1}y^4}\right)^{-4}$

Aha! **115.** $\left(\dfrac{21x^5y^{-7}}{14x^{-2}y^{-6}}\right)^0$

116. $\left(\dfrac{6a^{-2}b^6}{8a^{-4}b^0}\right)^{-2}$

117. $\left(\dfrac{5x^0y^{-7}}{2x^{-2}y^4}\right)^{-2}$

118. $\left(\dfrac{4a^3b^{-9}}{6a^{-2}b^5}\right)^0$

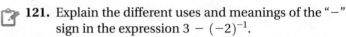 **119.** Explain why $(-1)^n = 1$ for any even number n.

120. Explain why $(-17)^{-8}$ is positive.

Synthesis

121. Explain the different uses and meanings of the "$-$" sign in the expression $3 - (-2)^{-1}$.

122. Is the following true or false, and why?
$$5^{-6} > 4^{-9}$$

Simplify. Assume all variables represent nonzero integers.

123. $\dfrac{8a^{x-2}}{2a^{2x+2}}$

124. $[7y(7-8)^{-4} - 8y(8-7)^{-2}]^{(-2)^2}$

125. $\{[(8^{-a})^{-2}]^b\}^{-c} \cdot [(8^0)^a]^c$

126. $(3^{a+2})^a$

127. $\dfrac{-28x^{b+5}y^{4+c}}{7x^{b-5}y^{c-4}}$

128. $\dfrac{4x^{2a+3}y^{2b-1}}{2x^{a+1}y^{b+1}}$

129. $\dfrac{3^{q+3} - 3^2(3^q)}{3(3^{q+4})}$

130. $\dfrac{25x^{a+b}y^{b-a}}{-5x^{a-b}y^{b+a}}$

131. $\left[\left(\dfrac{a^{-2c}}{b^{7c}}\right)^{-3}\left(\dfrac{a^{4c}}{b^{-3c}}\right)^2\right]^{-a}$

132. One cube has sides that are eight times as long as the sides of a second cube. How many times greater is the volume of the first cube than the volume of the second?

QUICK QUIZ: SECTIONS 1.1–1.6

1. Find the absolute value: $\left|\dfrac{3}{7}\right|$. [1.2]

2. Combine like terms: $5n^2 + 2n + 7n^3 - n^2$. [1.3]

3. Solve: $3(x-5) - 2(6-x) = 11 - x - 5$. [1.3]

4. The area of a triangle is 40 m². The base of the figure is 20 m. What is the height of the triangle? [1.5]

5. Simplify $\left(\dfrac{2w^4x^{-2}}{3wx^5}\right)^2$. Do not use negative exponents in the answer. [1.6]

1.7 Scientific Notation

Conversions • Multiplying, Dividing, and Significant Digits • Scientific Notation in Problem Solving

Study Skills

To Err Is Human

It is no coincidence that the students who experience the greatest success in this course work in pencil. We all make mistakes and by using pencil and eraser we are more willing to admit to ourselves that something needs to be rewritten. Please work with a pencil and eraser if you aren't doing so already.

We now study **scientific notation**, so named because of its usefulness in work with the very large and very small numbers that occur in science.

The following are examples of scientific notation:

7.2×10^5 means 720,000;

3.48×10^{-6} means 0.00000348.

The $\times$ in scientific notation is a multiplication symbol, not the variable x.

SCIENTIFIC NOTATION

Scientific notation for a number is an expression of the form $N \times 10^m$, where N is in decimal notation, $1 \le N < 10$, and m is an integer.

Conversions

Note that $10^b/10^b = 10^b \cdot 10^{-b} = 1$. To convert a number to scientific notation, we can multiply by 1, writing 1 in the form $10^b/10^b$ or $10^b \cdot 10^{-b}$.

EXAMPLE 1 *Computer Algorithms.* Scientists at the University of Alberta have proved that the computer program Chinook, designed to play the game of checkers, cannot ever lose. Checkers is the most complex game that has been solved with a computer program, with about 500,000,000,000,000,000,000 possible board positions. Write scientific notation for this number.

Source: sciencenetlinks.com

SOLUTION   To write 500,000,000,000,000,000,000 as 5×10^m for some integer m, we must move the decimal point in the number 20 places to the left. This can be accomplished by dividing and then multiplying by 10^{20}:

1. Write scientific notation for 32,100,000.

$$500{,}000{,}000{,}000{,}000{,}000{,}000 = \frac{500{,}000{,}000{,}000{,}000{,}000{,}000}{10^{20}} \times 10^{20} \qquad \text{Multiplying by 1: } \frac{10^{20}}{10^{20}} = 1$$

$$= 5 \times 10^{20}. \qquad \text{This is scientific notation.}$$

↪ YOUR TURN

EXAMPLE 2   Write scientific notation for the mass of a grain of sand:

0.0648 gram (g).

SOLUTION   To write 0.0648 as 6.48×10^m for some integer m, we must move the decimal point 2 places to the right. To do this, we multiply and then divide by 10^2:

$$0.0648 = \frac{0.0648 \times 10^2}{10^2} \qquad \text{Multiplying by 1: } \frac{10^2}{10^2} = 1$$

$$= 6.48 \times \frac{1}{10^2}$$

2. Write scientific notation for 0.0007.

$$= 6.48 \times 10^{-2}\,\text{g.} \qquad \text{Writing scientific notation}$$

↪ YOUR TURN

Try to make conversions to and from scientific notation mentally if possible. In doing so, remember that negative powers of 10 are used when representing small numbers and positive powers of 10 are used when representing large numbers.

EXAMPLE 3   Convert mentally to decimal notation.

a) 4.371×10^7 **b)** 1.73×10^{-5}

SOLUTION

3. Convert 7.04×10^{-3} to decimal notation.

a) $4.371 \times 10^7 = 43{,}710{,}000$ Moving the decimal point 7 places to the right

b) $1.73 \times 10^{-5} = 0.0000173$ Moving the decimal point 5 places to the left

↪ YOUR TURN

EXAMPLE 4   Convert mentally to scientific notation.

a) 82,500,000   **b)** 0.0000091

SOLUTION

a) $82{,}500{,}000 = 8.25 \times 10^7$ *Check*: Multiplying 8.25 by 10^7 moves the decimal point 7 places to the right.

4. Convert 3,401,000,000 to scientific notation.

b) $0.0000091 = 9.1 \times 10^{-6}$ *Check*: Multiplying 9.1 by 10^{-6} moves the decimal point 6 places to the left.

↪ YOUR TURN

Multiplying, Dividing, and Significant Digits

It is often important to know just how accurate a measurement is. For example, the measurement 5.72×10^4 km is more precise than the measurement 5.7×10^4 km. We say that 5.72×10^4 has three **significant digits** whereas 5.7×10^4 has only two significant digits. If 5.7×10^4, or 57,000, includes no rounding in the hundreds place, we would indicate that by writing 5.70×10^4.

When two or more measurements written in scientific notation are multiplied or divided, the result should be rounded so that it has the same number of significant digits as the measurement with the fewest significant digits. Rounding should be performed at the very end of the calculation.

Thus,

$$\underbrace{(3.1 \times 10^{-3}\,\text{mm})}_{\text{2 digits}}\underbrace{(2.45 \times 10^{-4}\,\text{mm})}_{\text{3 digits}} = 7.595 \times 10^{-7}\,\text{mm}^2$$

should be rounded to

$$\underbrace{7.6}_{\text{2 digits}} \times 10^{-7}\,\text{mm}^2.$$

EXAMPLE 5   Multiply and write scientific notation for the answer:

$$(7.2 \times 10^5)(4.3 \times 10^9).$$

SOLUTION   We have

$(7.2 \times 10^5)(4.3 \times 10^9) = (7.2 \times 4.3)(10^5 \times 10^9)$	Using the commutative and associative laws
$= 30.96 \times 10^{14}$	Adding exponents
$= (3.096 \times 10^1) \times 10^{14}$	Converting 30.96 to scientific notation
$= 3.096 \times 10^{15}$	Using the associative law
$\approx 3.1 \times 10^{15}.$	Rounding to 2 significant digits

 YOUR TURN

EXAMPLE 6   Divide and write scientific notation for the answer:

$$\frac{3.48 \times 10^{-7}}{4.64 \times 10^6}.$$

SOLUTION

$\dfrac{3.48 \times 10^{-7}}{4.64 \times 10^6} = \dfrac{3.48}{4.64} \times \dfrac{10^{-7}}{10^6}$	Separating factors. Our answer must have 3 significant digits.
$= 0.75 \times 10^{-13}$	Subtracting exponents; simplifying
$= (7.5 \times 10^{-1}) \times 10^{-13}$	Converting 0.75 to scientific notation
$= 7.50 \times 10^{-14}$	Adding exponents. We write 7.50 to indicate 3 significant digits.

YOUR TURN

Technology Connection

Both graphing and scientific calculators allow expressions to be entered using scientific notation. To do so, a key normally labeled (EE) or [EXP] is used. Often this is a secondary function and a key labeled (2ND) or [SHIFT] must be pressed first. To check Example 5, we press 7.2 (EE) 5 (×) 4.3 (EE) 9. When we then press (ENTER) or [=], the result 3.096E15 or 3.096 15 appears. We must interpret this result as 3.096×10^{15}.

5. Multiply and write scientific notation for the answer:
$$(3.9 \times 10^7)(4 \times 10^{15}).$$

6. Divide and write scientific notation for the answer:
$$\frac{(1.2 \times 10^{-1})}{(1.6 \times 10^{11})}.$$

Scientific Notation in Problem Solving

The following table lists common names and prefixes of powers of 10, in both decimal notation and scientific notation. These values are used in many applications.

One thousand	kilo-*	1000	1×10^3
One million	mega-	1,000,000	1×10^6
One billion	giga-	1,000,000,000	1×10^9
One trillion	tera-	1,000,000,000,000	1×10^{12}
One thousandth	milli-	0.001	1×10^{-3}
One millionth	micro-	0.000001	1×10^{-6}
One billionth	nano-	0.000000001	1×10^{-9}
One trillionth	pico-	0.000000000001	1×10^{-12}

EXAMPLE 7 *Social Networking.* In 2011, the 750 million active users of Facebook spent 700 billion min each month on Facebook. On average, how much time did each user spend on Facebook?

Source: Facebook

SOLUTION

1. **Familiarize.** In order to find the average time that each user spent on Facebook, we divide the total time spent by the number of users. We first write each number using scientific notation:

   $$700 \text{ billion} = 700 \times 10^9 = 7 \times 10^2 \times 10^9 = 7 \times 10^{11}, \quad \text{and}$$
   $$750 \text{ million} = 750 \times 10^6 = 7.5 \times 10^2 \times 10^6 = 7.5 \times 10^8.$$

   We also let $t =$ the average amount of time, in minutes, that each user spent on Facebook.

2. **Translate.** To find t, we divide:

   $$t = \frac{7 \times 10^{11}}{7.5 \times 10^8}.$$

3. **Carry out.** We calculate and write scientific notation for the result:

   $$t = \frac{7 \times 10^{11}}{7.5 \times 10^8}$$
   $$= \frac{7}{7.5} \times \frac{10^{11}}{10^8}$$
   $$\approx 0.9 \times 10^3 \qquad \text{Rounding to 1 significant digit}$$
   $$= (9 \times 10^{-1}) \times 10^3 \qquad \text{Writing scientific notation}$$
   $$= 9 \times 10^2.$$

---

*When these prefixes are used with -bytes, such as kilobytes and megabytes, the precise value is actually a power of 2. For example, 1 kilobyte is $2^{10} = 1024$ bytes. In practice, however, the powers of 10 listed in the table are often used as approximations.

7. In 2011, the 750 million active users of Facebook shared 30 billion pieces of content each month. On average, how many pieces of content did each user share each month on Facebook?

Source: Facebook

4. Check. To check, we multiply the average time spent by the number of users:

$$(9 \times 10^2 \text{ min/user})(7.5 \times 10^8 \text{ users}) \quad \text{We also check the units.}$$
$$= (9 \times 7.5)(10^2 \times 10^8) \text{ min/user} \cdot \text{users} \quad \text{Using the commutative and associative laws}$$
$$= 6.75 \times 10^{11} \text{ min.} \quad \text{This is approximately 700 billion. The unit, min, also checks.}$$

5. State. On average, each user spent 9×10^2, or 900, minutes per month on Facebook.

YOUR TURN

EXAMPLE 8 *Telecommunications.* A fiber-optic cable is to be used for 125 km of transmission line. The cable has a diameter of 0.6 cm. What is the volume of cable needed for the line?

SOLUTION

1. Familiarize. Making a drawing, we see that we have a cylinder (a very *long* one). Its length is 125 km and the base has a diameter of 0.6 cm.

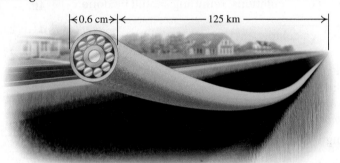

Recall that the formula for the volume of a cylinder is $V = \pi r^2 h$, where r is the radius and h is the height (in this case, the length of the cable).

2. Translate. Before we use the volume formula, we must make the units consistent. Let's express everything in meters:

Length: 125 km = 125,000 m, or 1.25×10^5 m;

Diameter: 0.6 cm = 0.006 m, or 6×10^{-3} m.

The radius, which we will need in the formula, is half the diameter:

Radius: 3×10^{-3} m.

We now substitute into the above formula:

$$V = \pi(3 \times 10^{-3} \text{ m})^2(1.25 \times 10^5 \text{ m}).$$

8. The diameter of the double helix of a DNA strand is 2×10^{-9} m. One such strand is 5 cm long. What is the volume, in cubic meters (m^3), of a cylinder with those dimensions?

3. Carry out. We do the calculations:

$$V = \pi \times (3 \times 10^{-3} \text{ m})^2(1.25 \times 10^5 \text{ m})$$
$$= \pi \times 3^2 \times 10^{-6} \text{ m}^2 \times 1.25 \times 10^5 \text{ m} \quad \text{Using } (ab)^n = a^n b^n$$
$$= (\pi \times 3^2 \times 1.25) \times (10^{-6} \times 10^5) \text{ m}^3$$
$$\approx 35.325 \times 10^{-1} \text{ m}^3 \quad \text{Using 3.14 for } \pi$$
$$\approx 3.5 \text{ m}^3. \quad \text{Rounding 3.5325 to 2 significant digits}$$

4. Check. We can recheck the translation and calculations. Note that m^3 is a unit of volume, as expected.

5. State. The volume of the cable is about 3.5 m^3 (cubic meters).

YOUR TURN

1.7 EXERCISE SET

FOR EXTRA HELP

MyMathLab® Math*XL*
PRACTICE   WATCH   READ   REVIEW

Vocabulary and Reading Check

Choose the word or phrase that best completes the statement from the choices listed below each blank.

1. The number 27×10^{16} _____ written in
    is/is not
   scientific notation.

2. Very small numbers are represented in scientific notation using _____ powers of 10.
    negative/positive

3. The number 4.587×10^5 has _____
    three/four/five
   significant digits.

4. In a series of calculations, rounding should be done _____.
   after each calculation/at the very end

Concept Reinforcement

State whether scientific notation for each of the following numbers would include either a positive power of 10 or a negative power of 10.

5. The length of an Olympic marathon, in centimeters

6. The thickness of a cat's whisker, in meters

7. The mass of a hydrogen atom, in grams

8. The mass of a pickup truck, in grams

9. The time between leap years, in seconds

10. The time between a bird's heartbeats, in hours

Conversions

Convert to scientific notation.

11. 64,000,000,000

12. 3,700,000

13. 0.0000013

14. 0.000078

15. 0.00009

16. 0.00000006

17. 803,000,000,000

18. 3,090,000,000,000

19. 0.000000904

20. 0.00000000802

21. 431,700,000,000

22. 953,400,000,000

Convert to decimal notation.

23. 4×10^5

24. 3×10^{-6}

25. 1.2×10^{-4}

26. 8.6×10^8

27. 3.76×10^{-9}

28. 4.27×10^{-2}

29. 8.056×10^{12}

30. 5.002×10^{10}

31. 7.001×10^{-5}

32. 2.049×10^{-3}

Multiplying, Dividing, and Significant Digits

Simplify and write scientific notation for the answer. Use the correct number of significant digits.

33. $(3.4 \times 10^{-8})(2.6 \times 10^{15})$

34. $(1.8 \times 10^{20})(4.7 \times 10^{-12})$

35. $(2.36 \times 10^6)(1.4 \times 10^{-11})$

36. $(4.26 \times 10^{-6})(8.2 \times 10^{-6})$

37. $(5.2 \times 10^6)(2.6 \times 10^4)$

38. $(6.11 \times 10^3)(1.01 \times 10^{13})$

39. $(7.01 \times 10^{-5})(6.5 \times 10^{-7})$

40. $(4.08 \times 10^{-10})(7.7 \times 10^5)$

Aha! 41. $(2.0 \times 10^6)(3.02 \times 10^{-6})$

42. $(7.04 \times 10^{-9})(9.01 \times 10^{-7})$

43. $\dfrac{6.5 \times 10^{15}}{2.6 \times 10^4}$

44. $\dfrac{8.5 \times 10^{18}}{3.4 \times 10^5}$

45. $\dfrac{9.4 \times 10^{-9}}{4.7 \times 10^{-2}}$

46. $\dfrac{4.0 \times 10^{-6}}{8.0 \times 10^{-3}}$

47. $\dfrac{3.2 \times 10^{-7}}{8.0 \times 10^8}$

48. $\dfrac{1.26 \times 10^9}{4.2 \times 10^{-3}}$

49. $\dfrac{9.36 \times 10^{-11}}{3.12 \times 10^{11}}$

50. $\dfrac{2.42 \times 10^5}{1.21 \times 10^{-5}}$

51. $\dfrac{6.12 \times 10^{19}}{3.06 \times 10^{-7}}$

52. $\dfrac{4.7 \times 10^{-9}}{2.0 \times 10^{-9}}$

Scientific Notation in Problem Solving

Solve.

53. *Coral Reefs.* There are 10 million bacteria per square centimeter of coral in a coral reef. The coral reefs near the Hawaiian Islands cover 14,000 km². How many bacteria are there in Hawaii's coral reef?

   Source: livescience.com; U.S. Geological Survey

54. *Agriculture.*   One of the shrimp ponds on Ana's farm contains 2.5×10^{11} mL of water. If the average number of suspended particles in the pond is 5.41×10^4 particles/mL, how many suspended particles are in the pond?

55. *High-Tech Fibers.*   A carbon nanotube is a thin cylinder of carbon atoms that, pound for pound, is stronger than steel. With a diameter of about 4.0×10^{-10} in., a fiber can be made 100 yd long. Find the volume of such a fiber.

Source: www.pa.msu.edu

56. *Home Maintenance.*   The thickness of a sheet of plastic is measured in *mils*, where 1 mil $= \frac{1}{1000}$ in. To help conserve heat, the foundation of a 24-ft by 32-ft rectangular home is covered with a 4-ft high sheet of 8-mil plastic. Find the volume of plastic used.

57. *Information Technology.*   International Data Corporation estimated that 988 exabytes of information was generated in 2010 by the worldwide population of 6.8 billion people. Given that 1 exabyte is 10^{12} megabytes, find the average number of megabytes of information generated per person in 2010.

Source: informationweek.com

58. *Computer Technology.*   Intel Corporation has developed silicon-based connections that use lasers to move data at a rate of 50 gigabytes per second. The printed collection of the U.S. Library of Congress contains 10 terabytes of information. How long would it take to copy the Library of Congress using these connections?

Sources: spie.org; newworldencyclopedia.org

59. *Office Supplies.*   A ream of copier paper weighs 2.25 kg. How much does a sheet of copier paper weigh?

60. *Printing and Engraving.*   A ton of five-dollar bills is worth $4,540,000. How many pounds does a five-dollar bill weigh?

For Exercises 61 and 62, use the fact that 1 light year = 5.88×10^{12} miles.

Aha! **61.** *Astronomy.*   The diameter of the Milky Way galaxy is approximately 5.88×10^{17} mi. How many light years is it from one end of the galaxy to the other?

62. *Astronomy.* The brightest star in the night sky, Sirius, is about 4.704×10^{13} mi from the earth. How many light years is it from the earth to Sirius?

Named in tribute to Anders Ångström, a Swedish physicist who measured light waves, 1 Å (read "one Angstrom") equals 10^{-10} meter. One parsec is about 3.26 light years, and one light year equals 9.46×10^{15} meters.

63. How many Angstroms are in one parsec?

64. How many kilometers are in one parsec?

For Exercises 65 and 66, use the approximate average distance from the earth to the sun of 1.50×10^{11} meters.

65. Determine the volume of a cylindrical sunbeam that is 3 Å in diameter.

66. Determine the volume of a cylindrical sunbeam that is 5 Å in diameter.

67. *Biology.*   An average of 4.55×10^{11} bacteria live in each pound of U.S. mud. There are 60.0 drops in one teaspoon and 6.0 teaspoons in an ounce. How many bacteria live in a drop of U.S. mud?

Source: Harper's Magazine, April 1996, p. 13

68. *Astronomy.*   If a star 5.9×10^{14} mi from the earth were to explode today, its light would not reach us for 100 years. How far does light travel in 13 weeks?

69. *Astronomy.*   The diameter of Jupiter is about 1.43×10^5 km. A day on Jupiter lasts about 10 hr. At what speed is Jupiter's equator spinning?

70. *Astronomy.*   The average distance of the earth from the sun is about 9.3×10^7 mi. About how far does the earth travel in a yearly orbit about the sun? (Assume a circular orbit.)

71. Write a problem for a classmate to solve. Design the problem so the solution is "The volume of the laser's light beam is 3.14×10^5 mm^3."

72. List two advantages of using scientific notation. Answers may vary.

Synthesis

73. A criminal claims to be carrying $5 million in twenty-dollar bills in a briefcase. Is this possible? Why or why not? (*Hint*: See Exercise 60.)

74. When a calculator indicates that $5^{17} = 7.629394531 \times 10^{11}$, an approximation is being made. How can you tell? (*Hint*: Examine the ones digit.)

75. *Density of the Earth.* The volume of the earth is approximately 1.08×10^{12} km^3, and the mass of the earth is about 5.976×10^{24} kg. What is the average density of the earth, in grams per cubic centimeter?

76. The Sartorius Microbalance Model 4108 can weigh objects to an accuracy of 3.5×10^{-10} oz. A chemical compound weighing 1.2×10^{-9} oz is split in half and weighed on the microbalance. Give a weight range for the actual weight of each half.

Source: Guinness Book of World Records

77. Given that the earth's average distance from the sun is 1.5×10^{11} m, determine the earth's orbital speed around the sun in miles per hour. Assume a circular orbit.

78. Write $\frac{4}{32}$ in decimal notation, in simplified fraction notation, and in scientific notation.

79. Compare $8 \cdot 10^{-90}$ and $9 \cdot 10^{-91}$. Which is the larger value? How much larger is it? Write scientific notation for the difference.

80. Write the reciprocal of 8.00×10^{-23} in scientific notation.

81. Evaluate: $(4096)^{0.05}(4096)^{0.2}$.

82. What is the ones digit in 513^{128}?

83. A grain of sand is placed on the first square of a chessboard, two grains on the second square, four grains on the third, eight on the fourth, and so on. Without a calculator, use scientific notation to approximate the number of grains of sand required for the 64th square. (*Hint*: Use the fact that $2^{10} \approx 10^3$.)

84. The Hubble-barn is the volume of a cylinder that has the cross-sectional area of an atom's nucleus (one barn) and the length of the radius of the universe (one Hubble). Although not used by scientists, this unit of volume illustrates the relative size of the units. A barn is 10^{-28} m^2 and one Hubble is about 10^{23} km. What is the size of a Hubble-barn, in gallons? (*Hint*: 1 m$^3 \approx 2.6417 \times 10^2$ gal.)

85. *Research.* Find the number of current active Facebook users and the total time spent each month on Facebook. Then estimate how much time the average user spends each month on Facebook. How does this compare with the average time spent per month in 2011? (See Example 7.)

YOUR TURN ANSWERS: SECTION 1.7

1. 3.21×10^7 **2.** 7×10^{-4} **3.** 0.00704 **4.** 3.401×10^9
5. 2×10^{23} **6.** 7.5×10^{-13}
7. 4.0×10^1, or 40, pieces of content per month
8. About 2×10^{-19} m^3

QUICK QUIZ: SECTIONS 1.1–1.7

1. Evaluate $x^2 \div yz - 3x$ for $x = 6, y = 2,$ and $z = 3$. [1.1]

2. Use a commutative law to write an expression equivalent to $6 + x$. [1.2]

3. Solve $d - (6 - d) = 2(d - 3)$. If appropriate, classify the equation as a contradiction or as an identity. [1.3]

4. Simplify $(2ac^{-3})(3a^{-6}c^2)$. Do not use negative exponents in the answers. [1.6]

5. Convert 0.000019 to scientific notation. [1.7]

1. *Consecutive Integers.* The sum of two consecutive even integers is 102. Find the integers.

2. *Dimensions of a Triangle.* One angle of a triangle is twice the measure of a second angle. The third angle measures 102° more than the second angle. Find the measures of the angles.

3. *Salary Increase.* After Susanna earned a 5% raise, her new salary was $25,750. What was her former salary?

4. *Dimensions of a Rectangle.* The length of a rectangle is 6 in. more than the width. The perimeter of the rectangle is 102 in. Find the length and the width.

5. *Population.* The population of Middletown is decreasing at a rate of 5% per year. The current population is 25,750. What was the population the previous year?

Translating for Success

Use after Section 1.4.

Translate each word problem to an equation or an inequality and select the most appropriate translation from A–O.

A. $0.05(25,750) = x$

B. $x + 2x = 102$

C. $2x + 2(x + 6) = 102$

D. $2x + x + x + 102 = 180$

E. $x - 0.05x = 25,750$

F. $x + (x + 2) = 102$

G. $6x - 102 = 180 - 5x$

H. $x + 5x = 150$

I. $x + 0.05x = 25,750$

J. $x + (2x + 6) = 102$

K. $x + (x + 1) = 102$

L. $102 + x = 180$

M. $0.05x = 25,750$

N. $x + 2x = x + 102$

O. $x + (x + 6) = 102$

Answers on page A–3

An additional, animated version of this activity appears in MyMathLab. *To use MyMathLab you need a course ID and a student access code. Contact your instructor for more information.*

6. *Numerical Relationship.* One number is 6 more than twice another. The sum of the numbers is 102. Find the numbers.

7. *DVD Collections.* Together, Ella and Ken have 102 DVDs. If Ken has 6 more DVDs than Ella, how many does each have?

8. *Sales Commissions.* Dakota earns a commission of 5% on his sales. One year, he earned commissions totaling $25,750. What were his total sales for the year?

9. *Fencing.* Brian has 102 ft of fencing that he plans to use to enclose dog runs at two houses. The perimeter of one run is to be twice the perimeter of the other. Into what lengths should the fencing be cut?

10. *Quiz Scores.* Lupe has a total of 102 points on the first 6 quizzes in her sociology class. How many total points must she earn on the 5 remaining quizzes in order to have 180 points for the semester?

Collaborative Activity   *Who Pays What?*

Focus: Problem solving
Use after: Section 1.4
Time: 15 minutes
Group size: 5

Suppose that two of the five members in each group are celebrating birthdays and the entire group goes out to lunch. Suppose further that each member whose birthday it is gets treated to his or her lunch by the other *four* members. Finally, suppose that all meals cost the same amount and that the total bill is $40.00.*

_____

* This activity was inspired by "The Birthday-Lunch Problem," *Mathematics Teaching in the Middle School*, vol. 2, no. 1, September–October 1996, pp. 40–42.

Activity

1. Determine, as a group, how much each group member should pay for the lunch described above. Then explain how this determination was made.

2. Compare the results and methods used for part (1) with those of the other groups in the class.

3. If the total bill is $65, how much should each group member pay? Again compare results with those of other groups.

4. If time permits, generalize the results of parts (1)–(3) for a total bill of x dollars.

Decision Making & Connection   (*Use after Section 1.5.*)

Grades. Estimating your current grade in a class can be complicated, especially when the grades for some assignments are weighted more heavily than others.

1. Ariel's syllabus for her general teaching methods class indicates that her grade will be based on the following.

Quizzes	100
Tests	500
Weekly Projects	800
Semester Project	500
Class Participation	100

So far, she has earned 75 of 80 possible points on quizzes, 360 of 400 possible points on tests, and 600 of 750 possible points on weekly projects. Estimate her grade in the class at this point.

a) First, calculate the *weight* that each of the five categories has toward the final grade. Adding, we see that there are 2000 possible points. Quizzes count for $\frac{100}{2000}$, or 5%, of the final grade. Calculate the weight of each of the other categories.

b) To calculate her course grade, Ariel will multiply the weight of each category from part(a) by the current grade in that category and add the results. Develop a formula that Ariel can use to calculate her grade.

c) Now calculate her current grade in each category, as a percent. For example, her quiz grade is $\frac{75}{80}$, or 93.75%. Calculate her current test grade and weekly project grade.

There are two categories, Semester Project and Class Participation, for which there is no grade. Ariel regularly attends class and participates in discussions, so she estimates that she will receive 100% in that category. She estimates her semester project grade to be the same as her weekly project grade.

d) Finally, calculate the current course grade using the formula developed in part (b).

2. Ariel (see Exercise 1) now has all of her scores except that for her semester project. She has earned 93 quiz points, 450 test points, 640 weekly project points, and 100 class participation points. She needs a 90% in order to receive an A in the course. Can she earn enough points on the semester project to receive an A?

3. *Research.* Find out how the grades are calculated in one or more of your classes. Choose at least one class in which assignments are weighted differently, and develop a formula for estimating your grade at any point during the term.

Study Summary

KEY TERMS AND CONCEPTS	EXAMPLES	PRACTICE EXERCISES

SECTION 1.1: *Some Basics of Algebra*

An **algebraic expression** consists of **variables**, numbers or **constants**, and operation signs.	*Phrase* *Translation* The difference of two numbers $x - y$ Twelve less than some number $n - 12$	**1.** Translate to an algebraic expression: Three times the sum of two numbers.
An algebraic expression can be **evaluated** by **substituting** specific numbers for the variables(s) and carrying out the calculations, following the rules for order of operations.	Evaluate $3 + 4x \div 6y^2$ for $x = 12$ and $y = -2$. $\begin{aligned} 3 + 4x \div 6y^2 &= 3 + 4(12) \div 6(-2)^2 & \text{Substituting} \\ &= 3 + 4(12) \div 6 \cdot 4 & \text{Squaring} \\ &= 3 + 48 \div 6 \cdot 4 & \text{Multiplying} \\ &= 3 + 8 \cdot 4 & \text{Dividing} \\ &= 3 + 32 & \text{Multiplying} \\ &= 35 & \text{Adding} \end{aligned}$	**2.** Evaluate $3 + 5a - b$ for $a = 6$ and $b = 10$.

SECTION 1.2: *Operations and Properties of Real Numbers*

Absolute Value $\|x\| = \begin{cases} x, & \text{if } x \geq 0 \\ -x, & \text{if } x < 0 \end{cases}$	$\|-15\| = 15;$ $\|4.8\| = 4.8;$ $\|0\| = 0$	**3.** Find the absolute value: $\|167\|$.
To **add** two real numbers, use the rules in Section 1.2.	$-8 + (-3) = -11;$ $-8 + 3 = -5;$ $8 + (-3) = 5;$ $-8 + 8 = 0$	**4.** Add: $-15 + (-10) + 20$.
To **subtract** two real numbers, change the sign of the number being subtracted and then add.	$8 - 14 = 8 + (-14) = -6;$ $8 - (-14) = 8 + 14 = 22$	**5.** Subtract: $7 - (-7)$.
Multiplication and Division of Real Numbers 1. Multiply or divide the absolute values of the numbers. 2. If the signs are different, the answer is negative. 3. If the signs are the same, the answer is positive.	$-3(-5) = 15;$ $10(-2) = -20;$ $-100 \div 25 = -4;$ $\left(-\frac{2}{5}\right) \div \left(-\frac{3}{10}\right) = \left(-\frac{2}{5}\right) \cdot \left(-\frac{10}{3}\right) = \frac{20}{15} = \frac{4}{3}$	**6.** Multiply: $-2(-15)$. **7.** Divide: $10 \div (-2.5)$.

The Commutative Laws

$a + b = b + a$;

$ab = ba$

$3 + (-5) = -5 + 3$;

$8(10) = 10(8)$

8. Use the commutative law of addition to write an expression equivalent to $6 + 10n$.

The Associative Laws

$a + (b + c) = (a + b) + c$;

$a \cdot (b \cdot c) = (a \cdot b) \cdot c$

$-5 + (5 + 6) = (-5 + 5) + 6$;

$2 \cdot (5 \cdot 9) = (2 \cdot 5) \cdot 9$

9. Use the associative law of multiplication to write an expression equivalent to $3(ab)$.

The Distributive Law

$a(b + c) = ab + ac$

Multiply: $3(2x + 5y)$.

$$3(2x + 5y) = 3 \cdot 2x + 3 \cdot 5y = 6x + 15y$$

Factor: $14x + 21y + 7$.

$$14x + 21y + 7 = 7(2x + 3y + 1)$$

10. Multiply: $10(5m + 9n + 1)$.

11. Factor: $26x + 13$.

SECTION 1.3: *Solving Equations*

Like terms have variable factors that are exactly the same. We can use the distributive law to **combine like terms.**

$$n - 9 - 4(n - 1) = n - 9 - 4n + 4$$
$$= n - 4n - 9 + 4$$
$$= -3n - 5$$

12. Combine like terms: $2(x - 3) - (3 - x)$.

The Addition and Multiplication Principles for Equations

$a = b$ is equivalent to $a + c = b + c$.

$a = b$ is equivalent to $a \cdot c = b \cdot c$, if $c \neq 0$.

Solve: $5t - 3(t - 3) = -t$.

$5t - 3(t - 3) = -t$

$5t - 3t + 9 = -t$	Using the distributive law
$2t + 9 = -t$	
$2t + 9 + t = -t + t$	Adding t to both sides
$3t + 9 = 0$	
$3t + 9 - 9 = 0 - 9$	Subtracting 9 from both sides
$3t = -9$	
$\frac{1}{3}(3t) = \frac{1}{3}(-9)$	Multiplying both sides by $\frac{1}{3}$
$t = -3$	

Check:

$$\begin{array}{c|c} 5t - 3(t - 3) = -t \\ \hline 5(-3) - 3(-3 - 3) & -(-3) \\ -15 - 3(-6) & 3 \\ -15 - (-18) & \\ 3 \stackrel{?}{=} 3 & \text{TRUE} \end{array}$$

The solution is -3.

13. Solve: $4(x - 3) - (x + 1) = 5$.

An **identity** is an equation that is true for all replacements.

A **contradiction** is an equation that is never true.

A **conditional equation** is true for some replacements and false for others.

$x + 2 = 2 + x$ is an identity. The solution set is $\mathbb{R}$, the set of all real numbers.

$x + 1 = x + 2$ is a contradiction. The solution set is $\varnothing$, the empty set.

$x + 2 = 5$ is a conditional equation. The solution set is $\{3\}$.

14. Solve: $3x - (2x - 7) = x + 7$. If the solution set is $\varnothing$ or $\mathbb{R}$, classify the equation as a contradiction or as an identity.

SECTION 1.4: *Introduction to Problem Solving*

Five-Step Strategy for Problem Solving in Algebra

1. *Familiarize* yourself with the problem.
2. *Translate* to mathematical language.
3. *Carry out* some mathematical manipulation.
4. *Check* your possible answer in the original problem.
5. *State* the answer clearly.

The perimeter of a rectangle is 70 cm. The width is 5 cm longer than half the length. Find the length and the width.

1. **Familiarize.** The formula for the perimeter of a rectangle is $P = 2l + 2w$. We can describe the width in terms of the length: $w = \frac{1}{2}l + 5$.

2. **Translate.**

$$2l + \quad 2w \quad = 70$$
$$2l + 2\left(\tfrac{1}{2}l + 5\right) = 70$$

3. **Carry out.** Solve the equation:

$$2l + 2\left(\tfrac{1}{2}l + 5\right) = 70$$
$$2l + l + 10 = 70 \qquad \text{Using the distributive law}$$
$$3l + 10 = 70 \qquad \text{Combining like terms}$$
$$3l = 60 \qquad \text{Subtracting 10 from both sides}$$
$$l = 20. \qquad \text{Dividing both sides by 3}$$

If $l = 20$, then $w = \frac{1}{2}l + 5 = \frac{1}{2} \cdot 20 + 5 = 15$.

4. **Check.**

$$w = \tfrac{1}{2}l + 5 = \tfrac{1}{2}(20) + 5 = 15;$$
$$2l + 2w = 2(20) + 2(15) = 70$$

The answer checks.

5. **State.** The length is 20 cm and the width is 15 cm.

15. Deborah rode a total of 120 mi in two bicycle tours. One tour was 25 mi longer than the other. How long was each tour?

SECTION 1.5: *Formulas, Models, and Geometry*

We can solve a **formula** for a specified letter using the same principles used to solve equations.

Solve $a - c = bc + d$ for c.
$$a - c = bc + d$$
$$a - d = c + bc$$
$$a - d = c(1 + b) \qquad \text{Factoring is a key step!}$$
$$\frac{a - d}{1 + b} = c$$

16. Solve $xy - 3y = w$ for y.

SECTION 1.6: *Properties of Exponents*

For $a, b \neq 0$ and n any integer:

$a^0 = 1;$

$a^{-n} = \dfrac{1}{a^n};$

$\dfrac{a^{-n}}{b^{-m}} = \dfrac{b^m}{a^n};$

$\left(\dfrac{a}{b}\right)^{-n} = \left(\dfrac{b}{a}\right)^n.$

$5^0 = 1$

$5^{-2} = \dfrac{1}{5^2} = \dfrac{1}{25}$

$\dfrac{x^{-4}}{5^{-7}} = \dfrac{5^7}{x^4}$

$\left(\dfrac{x^2}{6}\right)^{-3} = \left(\dfrac{6}{x^2}\right)^3$

17. Simplify: $(-6)^0$.

Write without negative exponents.

18. 10^{-1}

19. $\dfrac{x^{-1}}{y^{-3}}$

20. $\left(\dfrac{a}{b}\right)^{-1}$

The Product Rule $a^m \cdot a^n = a^{m+n}$	$2^5 \cdot 2^{10} = 2^{15}$	*Simplify.* **21.** $x^5 x^{11}$
The Quotient Rule $\dfrac{a^m}{a^n} = a^{m-n}$	$\dfrac{3^8}{3^7} = 3^1 = 3$	**22.** $\dfrac{8^{-9}}{8^{-11}}$
		23. $(t^{10})^{-2}$
The Power Rule $(a^m)^n = a^{mn}$	$(4^{-2})^{-5} = 4^{10}$	**24.** $(x^3 y)^{10}$
Raising a product to a power $(ab)^n = a^n b^n$	$(2y^3)^4 = 2^4 (y^3)^4 = 16y^{12}$	**25.** $\left(\dfrac{x^2}{7}\right)^5$
Raising a quotient to a power $\left(\dfrac{a}{b}\right)^n = \dfrac{a^n}{b^n}$	$\left(\dfrac{x^4}{5}\right)^2 = \dfrac{(x^4)^2}{5^2} = \dfrac{x^8}{25}$	

SECTION 1.7: *Scientific Notation*

Scientific Notation $N \times 10^m$, where N is in decimal notation, $1 \le N < 10$, and m is an integer	$1.2 \times 10^5 = 120{,}000$; $3.06 \times 10^{-4} = 0.000306$	**26.** Convert to scientific notation: 0.000904. **27.** Convert to decimal notation: 6.9×10^5.

Review Exercises: Chapter 1

The following review exercises are for practice. Answers are at the back of the book. If you need to, restudy the section indicated alongside the answer.

 Concept Reinforcement

In each of Exercises 1–10, match the expression or equation with an equivalent expression or equation from the column on the right.

1. ____ $2x - 1 = 9$ [1.3]   a) $2 + \frac{3}{4}x - 7$

2. ____ $2x - 1$ [1.3]   b) $2x + 14 = 6$

3. ____ $\frac{3}{4}x = 5$ [1.3]   c) $6x - 3$

4. ____ $\frac{3}{4}x - 5$ [1.3]   d) $2(3 + x)$

5. ____ $2(x + 7)$ [1.2]   e) $2x = 10$

6. ____ $2(x + 7) = 6$ [1.3]   f) $6x - 3 = 5$

7. ____ $4x - 3 + 2x = 5$ [1.3] g) $5x - 1 - 3x$

8. ____ $4x - 3 + 2x$ [1.3]   h) $3 = x$

9. ____ $6 + 2x$ [1.2]   i) $2x + 14$

10. ____ $6 = 2x$ [1.3]   j) $\frac{4}{3} \cdot \frac{3}{4}x = \frac{4}{3} \cdot 5$

11. Translate to an algebraic expression: Eight less than the quotient of two numbers.   [1.1]

12. Evaluate
$$7x^2 - 5y \div zx$$
for $x = -2$, $y = 3$, and $z = -5$.   [1.1], [1.2]

13. Name the set consisting of the first five odd natural numbers using both roster notation and set-builder notation.   [1.1]

14. Find the area of a triangular flag that has a base of 50 cm and a height of 70 cm.   [1.1]

Find the absolute value.   [1.2]

15. $|-19|$ 16. $|0|$ 17. $|6.08|$

Perform the indicated operation.   [1.2]

18. $-2.3 + (-8.7)$ 19. $-\frac{3}{4} - \left(-\frac{4}{5}\right)$

20. $10 + (-5.6)$ 21. $12.3 - 16.1$

22. $(-12)(-8)$ 23. $\left(-\frac{2}{3}\right)\left(\frac{5}{8}\right)$

24. $\frac{72.8}{-8}$ 25. $-7 \div \frac{4}{3}$

26. Find $-a$ if $a = -6.28$.   [1.2]

Use a commutative law to write an equivalent expression.   [1.2]

27. $12 + x$

28. $5x + y$

Use an associative law to write an equivalent expression.   [1.2]

29. $(4 + a) + b$

30. $x(yz)$

31. Obtain an expression that is equivalent to $12m + 4n - 2$ by factoring.   [1.2]

32. Combine like terms: $3x^3 - 6x^2 + x^3 + 5$.   [1.3]

33. Simplify: $7x - 4[2x + 3(5 - 4x)]$.   [1.3]

Solve. If the solution set is $\varnothing$ or $\mathbb{R}$, classify the equation as a contradiction or as an identity.   [1.3]

34. $3(t + 1) - t = 4$

35. $\frac{2}{3}n - \frac{5}{6} = \frac{8}{3}$

36. $-9x + 4(2x - 3) = 5(2x - 3) + 7$

37. $3(x - 4) + 2 = x + 2(x - 5)$

38. $5t - (7 - t) = 4t + 2(9 + t)$

39. Translate to an equation but do not solve: Fifteen more than twice a number is 21.   [1.4]

40. A number is 19 less than another number. The sum of the numbers is 115. Find the smaller number.   [1.4]

41. One angle of a triangle measures three times the second angle. The third angle measures twice the second angle. Find the measures of the angles. [1.4]

42. Solve for c: $x = \dfrac{bc}{t}$.   [1.5]

43. Solve for x: $c = mx - rx$.   [1.5]

44. The volume of a cylindrical candle is 538.51 cm³, and the radius of the candle is 3.5 cm. Determine the height of the candle. Use 3.14 for π.   [1.5]

45. Multiply and simplify: $(-4mn^8)(7m^3n^2)$.   [1.6]

46. Divide and simplify: $\dfrac{12x^3y^8}{3x^2y^2}$.   [1.6]

47. Evaluate a^0, a^2, and $-a^2$ for $a = -8$.   [1.6]

Simplify. Do not use negative exponents in the answer. [1.6]

48. $3^{-5} \cdot 3^7$

49. $(2t^4)^3$

50. $(-5a^{-3}b^2)^{-3}$

51. $\left(\dfrac{x^2y^3}{z^4}\right)^{-2}$

52. $\left(\dfrac{3m^{-5}n}{9m^2n^{-2}}\right)^4$

Simplify.   [1.2]

53. $\dfrac{4(9 - 2 \cdot 3) - 3^2}{4^2 - 3^2}$

54. $1 - (2 - 5)^2 + 5 \div 10 \cdot 4^2$

55. Convert 0.000307 to scientific notation.   [1.7]

56. One *parsec* (a unit that is used in astronomy) is 30,860,000,000,000 km. Write scientific notation for this number.   [1.7]

Simplify and write scientific notation for each answer. Use the correct number of significant digits.   [1.7]

57. $(8.7 \times 10^{-9}) \times (4.3 \times 10^{15})$

58. $\dfrac{1.2 \times 10^{-12}}{6.1 \times 10^{-7}}$

59. A sheet of plastic shrink wrap has a thickness of 0.00015 mm. The sheet is 1.2 m by 79 m. Use scientific notation to find the volume of the sheet.   [1.7]

Synthesis

 60. Describe a method that could be used to write equations that have no solution.   [1.3]

 61. Under what conditions is each of the following positive? **(a)** $-(-x)$; **(b)** $-x^2$; **(c)** $-x^3$; **(d)** $(-x)^2$; **(e)** x^{-2}. Explain.   [1.2], [1.6]

62. If the smell of gasoline is detectable at 3 parts per billion, what percent of the air is occupied by the gasoline?   [1.7]

63. Evaluate $a + b(c - a^2)^0 + (abc)^{-1}$ for $a = 3$, $b = -2$, and $c = -4$.   [1.1], [1.6]

64. What's a better deal: a 13-in. diameter pizza for $12 or a 17-in. diameter pizza for $15? Explain. [1.4], [1.5]

65. The surface area of a cube is 486 cm². Find the volume of the cube.   [1.5]

66. Solve for z: $m = \dfrac{x}{y - z}$.   [1.5]

67. Simplify: $\dfrac{(3^{-2})^a \cdot (3^b)^{-2a}}{(3^{-2})^b \cdot (9^{-b})^{-3a}}$.   [1.6]

68. Fill in the following blank so as to ensure that the equation is an identity.   [1.3]

$$5x - 7(x + 3) - 4 = 2(7 - x) + \underline{\quad}$$

69. Replace the blank with one term to ensure that the equation is a contradiction.   [1.3]

$$20 - 7[3(2x + 4) - 10] = 9 - 2(x - 5) + \underline{\quad}$$

70. Use the commutative law for addition once and the distributive law twice to show that

$$a \cdot 2 + cb + cd + ad = a(d + 2) + c(b + d).\ [1.2]$$

71. Find an irrational number between $\frac{1}{2}$ and $\frac{3}{4}$.   [1.1]

Test: Chapter 1

For step-by-step test solutions, access the Chapter Test Prep Videos in MyMathLab®, on YouTube® (search "Bittinger Inter Alg CA" and click on "Channels"), or by scanning the code.

1. Translate to an algebraic expression: Four less than the product of two numbers.

2. Evaluate $a^3 - 5b + b \div ac$ for $a = -2$, $b = 6$, and $c = 3$.

3. A triangular roof garden in Petach Tikva, Israel, has a base of length 7.8 m and a height of 46.5 m. Find its area.

Source: www.greenroofs.com

Perform the indicated operation.

4. $-15 + (-16)$

5. $-7.5 + 3.8$

6. $29.5 - 43.7$

7. $-6.4(5.3)$

8. $-\frac{7}{6} - \left(-\frac{5}{4}\right)$

9. $-\frac{2}{7}\left(-\frac{5}{14}\right)$

10. $\frac{-42.6}{-7.1}$

11. $\frac{2}{5} \div \left(-\frac{3}{10}\right)$

12. Simplify: $7 + (1 - 3)^2 - 9 \div 2^2 \cdot 6$.

13. Use a commutative law to write an expression equivalent to $3 + x$.

14. Combine like terms: $4y - 10 - 7y - 19$.

Solve. If the solution set is $\mathbb{R}$ or $\varnothing$, classify the equation as an identity or as a contradiction.

15. $10x - 7 = 38x + 49$

16. $13t - (5 - 2t) = 5(3t - 1)$

17. Solve for p: $2p = sp + t$.

18. Linda's scores on five tests are 84, 80, 76, 96, and 80. What must Linda score on the sixth test so that her average will be 85?

19. Find three consecutive odd integers such that the sum of four times the first, three times the second, and two times the third is 167.

Simplify. Do not use negative exponents in the answer.

20. $3x - 7 - (4 - 5x)$

21. $6b - [7 - 2(9b - 1)]$

22. $(7x^{-4}y^{-7})(-6x^{-6}y)$

23. -6^{-2}

24. $(-5x^{-1}y^3)^3$

25. $\left(\dfrac{2x^3y^{-6}}{-4y^{-2}}\right)^{-2}$

26. $(7x^3y)^0$

Simplify and write scientific notation for the answer. Use the correct number of significant digits.

27. $(9.05 \times 10^{-3})(2.22 \times 10^{-5})$

28. $\dfrac{1.8 \times 10^{-4}}{4.8 \times 10^{-7}}$

Solve.

29. The lightest known particle in the universe, a neutrino has a maximum mass of 1.8×10^{-36} kg. An alpha particle resulting from the decay of radon has a mass of 3.62×10^{-27} kg. How many neutrinos (with the maximum mass) would it take to equal the mass of one alpha particle?

Source: Guinness Book of World Records

Synthesis

Simplify. Do not use negative exponents in the answer.

30. $(2x^{3a}y^{b+1})^{3c}$

31. $\dfrac{-27a^{x+1}}{3a^{x-2}}$

32. Solve: $-\dfrac{5x + 2}{x + 10} = 1$.

Graphs, Functions, and Linear Equations

SEASON	GROSS REVENUE (in millions)
1999–2000	$ 603
2002–2003	721
2005–2006	862
2008–2009	943
2010–2011	1080

Source: The Broadway League

Give Our Regards to Broadway!

Even through years of recession, gross revenue from Broadway shows has grown steadily, as shown in the table above. The data are approximately linear, so we can fit a *linear function* to the data and use it to estimate gross revenue for seasons not shown here. (*See Example 9, Your Turn Exercise 9, and Exercise 114 in Section 2.5.*)

Theatre work is not all glamour; a great deal of preparation is put into every production. Math plays a vital role in bringing a play to life.

Tony Penna, a lighting designer from Clemson, South Carolina, uses math to determine how actors and scenery will appear, to balance electrical loads on equipment, and to optimize a show's budget.

Graphs help us to visualize information and allow us to see relationships. In this chapter, we will examine graphs of equations in two variables. A certain kind of relationship between two variables is known as a *function*. In this chapter, we explain what a function is as well as how it can be used in problem solving.

2.1 Graphs

Points and Ordered Pairs ▪ Quadrants and Scale ▪ Solutions of Equations ▪ Nonlinear Equations

Study Skills

Learn by Example

The examples in each section are designed to prepare you for success with the exercise set. Study the step-by-step solutions of the examples, noting that color is used to indicate substitutions and to call attention to the new steps in multistep examples. The time you spend studying the examples will save you valuable time when you do your assignment.

It has often been said that a picture is worth a thousand words. In mathematics, this is quite literally the case. Graphs are a compact means of displaying information and provide a visual approach to problem solving.

Points and Ordered Pairs

On the number line, each point corresponds to a number. On a plane, each point corresponds to an **ordered pair** of numbers. We use two perpendicular number lines, called **axes** (pronounced ak-sēz; singular, **axis**) to identify points in a plane. The point at which the axes intersect is called the **origin**. Arrows on the axes indicate the positive directions. The variable x is usually represented on the horizontal axis and the variable y on the vertical axis, so we often call such a plane an **x, y-coordinate system**.

To label a point on the x, y-coordinate system, we use a pair of numbers in the form (x, y). The numbers in the pair are called **coordinates**. In the ordered pair $(3, 2)$, the *first coordinate*, or *x-coordinate*, is 3 and the *second coordinate*, or *y-coordinate*,* is 2.

To plot, or graph, $(3, 2)$, we start at the origin, move horizontally 3 units, move up vertically 2 units, and then make a "dot." Thus, $(3, 2)$ is located above 3 on the first axis and to the right of 2 on the second axis. Note from the graph below that $(2, 3)$ and $(3, 2)$ are different points and that the origin has coordinates $(0, 0)$.

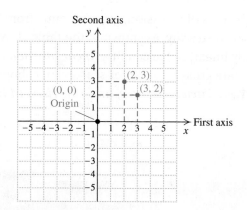

The idea of using axes to identify points in a plane is commonly attributed to the great French mathematician and philosopher René Descartes (1596–1650). In honor of Descartes, this representation is also called the **Cartesian coordinate system**.

---

*The first coordinate is sometimes called the **abscissa** and the second coordinate the **ordinate**.

EXAMPLE 1   Plot the points $(-4, 3)$, $(-5, -3)$, $(0, 4)$, $(4, -5)$, and $(2.5, 0)$.

SOLUTION   To plot $(-4, 3)$, note that the first coordinate, -4, tells us the distance in the first, or horizontal, direction. We go 4 units *left* of the origin. From that location, we go 3 units *up*. The point $(-4, 3)$ is then marked, or "plotted."

The points $(-5, -3)$, $(0, 4)$, $(4, -5)$, and $(2.5, 0)$ are also plotted below.

1. Plot the points $(-2, 5)$, $(3, -1)$, $(0, -1)$, $(-2, -4)$, and $(4, 0)$.

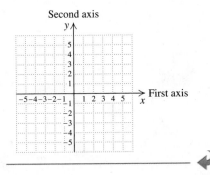

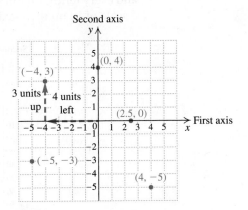

YOUR TURN

Quadrants and Scale

The horizontal axis and the vertical axis divide the plane into four regions, or **quadrants**, as indicated by Roman numerals in the figure below. Note that $(-4, 5)$ is in the second quadrant and $(5, -5)$ is in the fourth quadrant. The points $(3, 0)$ and $(0, 1)$ are on the axes and are not considered to be in any quadrant.

Second quadrant:
First coordinate negative, second coordinate positive:

$(-, +)$

Third quadrant:
Both coordinates negative:

$(-, -)$

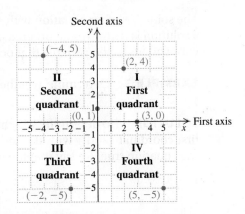

First quadrant:
Both coordinates positive:

$(+, +)$

Fourth quadrant:
First coordinate positive, second coordinate negative:

$(+, -)$

We draw only part of the plane when we create a graph. Although it is standard to show the origin and portions of all four quadrants, as in the graphs above, it may be more practical to show a different portion of the plane. Sometimes a different *scale* is selected for each axis.

EXAMPLE 2   Plot $(10, 44)$, $(14, 120)$, $(20, 130)$, and $(8, 15)$.

SOLUTION   The smallest first coordinate is 8 and the largest is 20. The smallest second coordinate is 15 and the largest is 130. We must show at least 20 units of the first axis and at least 130 units of the second axis. It is impractical to label the axes with all the natural numbers, so we label only every other unit on the x-axis and every tenth unit on the y-axis. We say that we use a *scale* of 2 on the x-axis and a scale of 10 on the y-axis. Only the first quadrant need be shown.

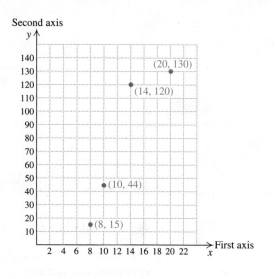

2. Plot $(2, 63)$, $(-15, 8)$, $(-5, 100)$, and $(8, 45)$.

YOUR TURN

Solutions of Equations

The solutions of an equation with two variables are pairs of numbers. When such a solution is written as an ordered pair, the first number listed in the pair generally corresponds to the variable that occurs first alphabetically.

EXAMPLE 3   Determine whether $(4, 2)$, $(-1, -4)$, and $(2, 5)$ are solutions of $y = 3x - 1$.

SOLUTION   To determine whether each pair is a solution, we replace x with the first coordinate and y with the second coordinate. When the replacements make the equation true, we say that the ordered pair is a solution.

$$
\begin{array}{c|c}
\multicolumn{2}{l}{y = 3x - 1} \\
\hline
2 & 3(4) - 1 \\
 & 12 - 1 \\
\hline
2 & \overset{?}{=} 11
\end{array}
\qquad
\begin{array}{c|c}
\multicolumn{2}{l}{y = 3x - 1} \\
\hline
-4 & 3(-1) - 1 \\
 & -3 - 1 \\
\hline
-4 & \overset{?}{=} -4
\end{array}
\qquad
\begin{array}{c|c}
\multicolumn{2}{l}{y = 3x - 1} \\
\hline
5 & 3(2) - 1 \\
 & 6 - 1 \\
\hline
5 & \overset{?}{=} 5
\end{array}
$$

Since $2 = 11$ is *false*, the pair $(4, 2)$ *is not* a solution.

Since $-4 = -4$ is *true*, the pair $(-1, -4)$ *is* a solution.

Since $5 = 5$ is *true*, the pair $(2, 5)$ *is* a solution.

3. Determine whether $(7, -1)$ is a solution of $x - y = 6$.

YOUR TURN

In fact, there is an infinite number of solutions of $y = 3x - 1$, and a graph provides a convenient way of representing them. To **graph** an equation means to make a drawing that represents all of its solutions.

EXAMPLE 4   Graph: $y = x$.

SOLUTION   We label the horizontal axis as the x-axis and the vertical axis as the y-axis, and find some ordered pairs that are solutions of the equation. In this case, since $y = x$, no calculations are necessary. Here are a few pairs that satisfy the equation $y = x$:

$$(0, 0), \quad (1, 1), \quad (5, 5), \quad (-1, -1), \quad (-6, -6).$$

Plotting these points, we see that if we were to plot many solutions, the dots would appear to form a line. Observing the pattern, we draw the line with a ruler. The line is the graph of $y = x$, so we label it $y = x$.

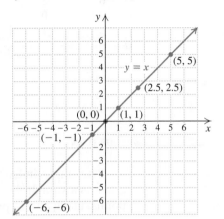

4. Graph: $y = x + 1$.

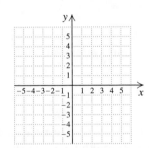

Note that the coordinates of *any* point on the line—for example, $(2.5, 2.5)$—satisfy the equation $y = x$. The line continues indefinitely in both directions, so we draw it to the edge of the grid and add arrows at both ends.

YOUR TURN

EXAMPLE 5   Graph: $y = 2x - 1$.

SOLUTION   We find some ordered pairs that are solutions. This time we list the pairs in a table. To find an ordered pair, we can choose *any* number for x and then determine y. For example, if we choose 3 for x, then

$$y = 2x - 1$$
$$y = 2(3) - 1 = 5.$$

We choose some negative values for x, as well as some positive ones (generally, we avoid selecting values beyond the edge of the graph paper). Next, we plot these points, draw the line with a ruler, and label it $y = 2x - 1$.

Student Notes

There is an infinite number of solutions of $y = 2x - 1$. When you choose a value for x and then compute y, you are determining one solution. Your choices for x may be different from those of a classmate. Although your plotted points differ, the graph of the line should be the same.

5. Graph: $y = 2x + 3$.

x	$y = 2x - 1$	(x, y)
0	-1	$(0, -1)$
1	1	$(1, 1)$
3	5	$(3, 5)$
-1	-3	$(-1, -3)$
-2	-5	$(-2, -5)$

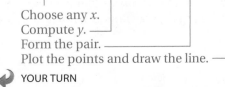

Choose any x.
Compute y.
Form the pair.
Plot the points and draw the line.

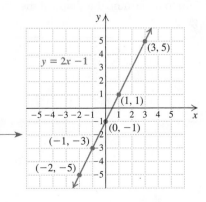

YOUR TURN

EXAMPLE 6   Graph: $y = -\frac{1}{2}x$.

SOLUTION   Since we can choose any number for x, let's select even integers in order to avoid fraction values for y. For example, if we choose 4 for x, we get $y = \left(-\frac{1}{2}\right)(4)$, or -2. When x is -6, we get $y = \left(-\frac{1}{2}\right)(-6)$, or 3. We find several ordered pairs, plot them, and draw the line.

x	$y = -\frac{1}{2}x$	(x, y)
4	-2	$(4, -2)$
-6	3	$(-6, 3)$
0	0	$(0, 0)$
2	-1	$(2, -1)$

↑ Choose any x.
↑ Compute y.
↑ Form the pair.
Plot the points and draw the line.

6. Graph: $y = -\frac{1}{3}x$.

_____

▶ YOUR TURN

The graphs in Examples 4–6 are straight lines. We refer to any equation whose graph is a straight line as a **linear equation**. To graph a line, we plot at least two points, using a third point as a check.

Nonlinear Equations

For many equations, the graph is not a straight line. Graphing these **nonlinear equations** often requires plotting many points in order to see the general shape of the graph.

EXAMPLE 7   Graph: $y = |x|$.

SOLUTION   We select numbers for x and find the corresponding values for y. For example, if we choose -1 for x, we get $y = |-1| = 1$. We list several ordered pairs and plot the points, noting that the absolute value of a positive number is the same as the absolute value of its opposite. Thus the x-values 3 and -3 both are paired with the y-value 3. The graph is V-shaped, as shown below.

Student Notes

If you *know* that an equation is linear, you can draw the graph using only two points. If you are not sure, or if you know that the equation is nonlinear, you must calculate and plot more than two points—as many as is necessary in order for you to determine the shape of the graph.

x	$y = |x|$	(x, y)
-3	3	$(-3, 3)$
-2	2	$(-2, 2)$
-1	1	$(-1, 1)$
0	0	$(0, 0)$
1	1	$(1, 1)$
2	2	$(2, 2)$
3	3	$(3, 3)$

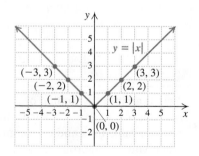

7. Graph: $y = |x| - 1$.

_____

▶ YOUR TURN

EXAMPLE 8 Graph: $y = x^2 - 5$.

SOLUTION We select numbers for x and find the corresponding values for y. For example, if we choose -2 for x, we get $y = (-2)^2 - 5 = 4 - 5 = -1$. The table lists several ordered pairs.

x	$y = x^2 - 5$	(x, y)
0	−5	(0, −5)
−1	−4	(−1, −4)
1	−4	(1, −4)
−2	−1	(−2, −1)
2	−1	(2, −1)
−3	4	(−3, 4)
3	4	(3, 4)

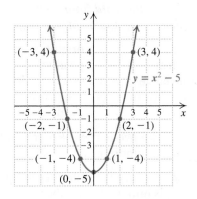

Next, we plot the points. The more points plotted, the clearer the shape of the graph becomes. Since the value of $x^2 - 5$ grows rapidly as x moves away from the origin, the graph rises steeply on either side of the y-axis.

8. Graph: $y = x^2 + 3$.

YOUR TURN

Technology Connection

The window of a graphing calculator is the rectangular portion of the screen in which a graph appears. Windows are described by four numbers of the form [L, R, B, T], representing the left and right endpoints of the x-axis and the bottom and top endpoints of the y-axis. If we enter an equation in the Y = screen and press (ZOOM) (6), the equation appears in the "standard" $[-10, 10, -10, 10]$ window. Below is the graph of $y = -4x + 3$ in the standard viewing window.

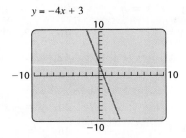

When (TRACE) is pressed, a cursor can be moved along the graph while its coordinates appear. To find the y-value that is paired with a particular x-value, we simply key in that x-value and press (ENTER).

Most graphing calculators can set up a table of pairs for any equation that is entered. By pressing (2ND) (TBLSET), we can control the smallest x-value listed using

TblStart and the difference between successive x-values using ΔTbl. Setting Indpnt and Depend both to Auto directs the calculator to complete a table automatically. To view the table, we press (2ND) (TABLE). For the table shown, we used $y_1 = -4x + 3$, with TblStart = 1.4 and ΔTbl = .1.

TblStart = 1.4 ΔTbl = .1

X	Y1
1.4	−2.6
1.5	−3
1.6	−3.4
1.7	−3.8
1.8	−4.2
1.9	−4.6
2	−5

X = 1.4

1. Graph $y = -4x + 3$ using a $[-10, 10, -10, 10]$ window. Then TRACE to find coordinates of several points, including the points with the x-values -1.5 and 1.

Graph each equation using a $[-10, 10, -10, 10]$ window. Then create a table of ordered pairs in which the x-values start at -1 and are 0.1 unit apart.

2. $y = 5x - 3$ **3.** $y = x^2 - 4x + 3$
4. $y = |x + 2|$

2.1   EXERCISE SET

Vocabulary and Reading Check

Choose the word from the following list that best completes each statement. Not every word will be used.

axes	origin
linear	positive
negative	second
nonlinear	solutions
ordered	third

1. The two perpendicular number lines that are used for graphing are called _____.

2. Because the order in which the numbers are listed is important, numbers listed in the form (x, y) are called _____ pairs.

3. In the _____ quadrant, both coordinates of a point are negative.

4. In the fourth quadrant, a point's first coordinate is positive and its second coordinate is _____.

5. To graph an equation means to make a drawing that represents all _____ of the equation.

6. An equation whose graph is a straight line is said to be a(n) _____ equation.

Points and Ordered Pairs

Give the coordinates of each point.

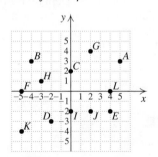

7. A, B, C, D, E, and F

8. G, H, I, J, K, and L

Plot the points. Label each point with the indicated letter.

9. $A(3, 0), B(4, 2), C(5, 4), D(6, 6), E(3, -4), F(3, -3),$ $G(3, -2), H(3, -1)$

10. $A(1, 1), B(2, 3), C(3, 5), D(4, 7), E(-2, 1), F(-2, 2),$ $G(-2, 3), H(-2, 4), J(-2, 5), K(-2, 6)$

11. Plot the points $M(2, 3), N(5, -3)$, and $P(-2, -3)$. Draw $\overline{MN}, \overline{NP}$, and $\overline{MP}$. ($\overline{MN}$ means the line segment from M to N.) What kind of geometric figure is formed? What is its area?

12. Plot the points $Q(-4, 3), R(5, 3), S(2, -1)$, and $T(-7, -1)$. Draw $\overline{QR}, \overline{RS}, \overline{ST}$, and $\overline{TQ}$. What kind of figure is formed? What is its area?

Quadrants and Scale

For Exercises 13–16, carefully choose a scale and plot the points. Scales may vary.

13. $(-75, 5), (-18, -2), (9, -4)$

14. $(-1, 83), (-5, -14), (5, 37)$

15. $(-100, -5), (350, 20), (800, 37)$

16. $(-83, 491), (-124, -95), (54, -238)$

In which quadrant or on which axis is each point located?

17. $(7, -2)$ 18. $(-1, -4)$ 19. $(-4, -3)$

20. $(1, -5)$ 21. $(0, -3)$ 22. $(6, 0)$

23. $(-4.9, 8.3)$ 24. $(7.5, 2.9)$ 25. $\left(-\frac{5}{2}, 0\right)$

26. $(0, 2.8)$ 27. $(160, 2)$ 28. $\left(-\frac{1}{2}, 2000\right)$

Solutions of Equations

Determine whether each ordered pair is a solution of the given equation. Remember to use alphabetical order for substitution.

29. $(2, -1); \ y = 3x - 7$ 30. $(1, 4); \ y = 5x - 1$

31. $(3, 2); \ 2x - y = 5$ 32. $(5, 5); \ 3x - y = 5$

33. $(3, -1); \ a - 5b = 8$

34. $(1, -4); \ 2u - v = -6$

35. $\left(\frac{2}{3}, 0\right); \ 6x + 8y = 4$ 36. $\left(0, \frac{3}{5}\right); \ 7a + 10b = 6$

37. $(6, -2); \ r - s = 4$ 38. $(4, -3); \ 2x - y = 11$

39. $(2, 1); \ y = 2x^2$ 40. $(-2, -1); \ r^2 - s = 5$

41. $(-2, 9); \ x^3 + y = 1$ 42. $(3, 2); \ y = x^3 - 5$

Graph.

43. $y = 3x$ 44. $y = -x$

45. $y = x + 4$ 46. $y = x + 3$

47. $y = x - 4$ 48. $y = x - 3$

49. $y = -2x + 3$ 50. $y = -3x + 1$

Aha! 51. $y + 2x = 3$ 52. $y + 3x = 1$

53. $y = -\frac{3}{2}x$ 54. $y = \frac{2}{3}x$

55. $y = \frac{3}{4}x - 1$ 56. $y = -\frac{3}{4}x - 1$

Nonlinear Equations

Graph.

57. $y = |x| + 2$ **58.** $y = |x| + 1$

59. $y = |x| - 2$ **60.** $y = |x| - 3$

61. $y = x^2 + 2$ **62.** $y = x^2 + 1$

63. $y = x^2 - 2$ **64.** $y = x^2 - 3$

Researchers at Yale University have suggested that the following graphs* may represent three different aspects of love.

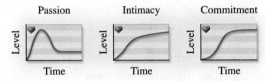

65. In what unit would you measure time if the horizontal length of each graph were ten units? Why?

66. Do you agree with the researchers that these graphs should be shaped as they are? Why or why not?

Skill Review

To the student and the instructor: Exercises included for Skill Review cover skills studied in earlier chapters of the text. The section(s) in which these types of exercises first appear is shown in brackets. Answers to all Skill Review exercises are at the back of the book.

Simplify. Do not use negative exponents in the answer.

67. $3 - 2(1 - 4)^2 \div 6 \cdot 2$ [1.2]

68. $3(x - 7) - 4(2 - 3x)$ [1.3]

69. $(2x^6y)^2$ [1.6]   **70.** $\dfrac{24a^{-1}b^{10}}{-14a^{11}b^{-16}}$ [1.6]

Synthesis

71. Without graphing, how can you tell that the graph of $y = x - 30$ passes through three quadrants?

72. At what point will the line passing through $(a, -1)$ and $(a, 5)$ intersect the line that passes through $(-3, b)$ and $(2, b)$? Why?

73. Match each sentence with the most appropriate of the four graphs shown below.

a) Austin worked part time until September, full time until December, and overtime until Christmas.

b) Marlo worked full time until September, half time until December, and full time until Christmas.

c) Liz worked overtime until September, full time until December, and overtime until Christmas.

From "A Triangular Theory of Love," by R. J. Sternberg, 1986, Psychological Review, **93(2), 119–135. Copyright 1986 by the American Psychological Association, Inc. Reprinted by permission.*

d) Roberto worked part time until September, half time until December, and full time until Christmas.

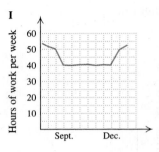

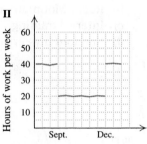

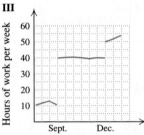

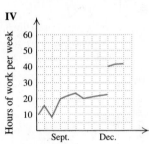

74. Match each sentence with the most appropriate of the four graphs shown below.

a) Carpooling to work, Jeremy spent 10 min on local streets, then 20 min cruising on the freeway, and then 5 min on local streets to his office.

b) For her commute to work, Chloe drove 10 min to the train station, rode the express for 20 min, and then walked for 5 min to her office.

c) For his commute to school, Theo walked 10 min to the bus stop, rode the express for 20 min, and then walked for 5 min to his class.

d) Coming home from school, Twyla waited 10 min for the school bus, rode the bus for 20 min, and then walked 5 min to her house.

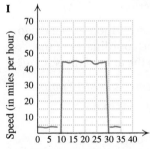

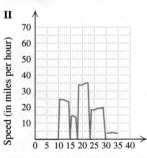

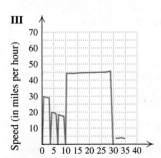

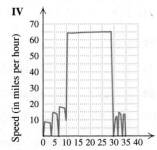

75. Match each program found on an exercise bike with the appropriate graph of speed shown below.

   a) Lakeshore loop
   b) Rocky Mountain monster hill
   c) Interval training
   d) Random mystery ride

I

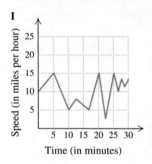

II

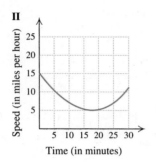

III

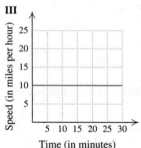

IV

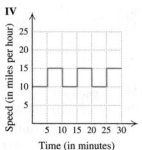

76. Indicate all of the following equations that have $\left(-\frac{1}{3}, \frac{1}{4}\right)$ as a solution.

   a) $-\frac{3}{2}x - 3y = -\frac{1}{4}$
   b) $8y - 15x = \frac{7}{2}$
   c) $0.16y = -0.09x + 0.1$
   d) $2(-y + 2) - \frac{1}{4}(3x - 1) = 4$

77. If $(-10, -2)$, $(-3, 4)$, and $(6, 4)$ are coordinates of three vertices (corners) of a parallelogram, determine the coordinates of three different points that could serve as the fourth vertex.

78. If $(2, -3)$ and $(-5, 4)$ are the endpoints of a diagonal of a square, what are the coordinates of the other two vertices? What is the area of the square?

📟 *Graph each equation after plotting at least 10 points.*

79. $y = 1/x^2$; use x-values from -4 to 4

80. $y = \dfrac{1}{x^2} + 3$; use x-values from -4 to 4

81. $y = \dfrac{1}{x} + 3$; use x-values from -4 to 4

82. $y = 1/x$; use x-values from -4 to 4

83. $y = \sqrt{x} + 1$; use x-values from 0 to 10

84. $y = \sqrt{x}$; use x-values from 0 to 10

85. $y = x^3$; use x-values from -2 to 2

86. $y = x^3 - 5$; use x-values from -2 to 2

Note: Throughout this text, the icon 📟 *indicates exercises designed for graphing calculators.*

📟 *In Exercises 87 and 88, use a graphing calculator to draw the graph of each equation. For each equation, select a window that shows the curvature of the graph and create a table of ordered pairs in which x-values extend, by tenths, from 0 to 0.6.*

87. a) $y = 0.375x^3$
   b) $y = -3.5x^2 + 6x - 8$
   c) $y = (x - 3.4)^3 + 5.6$

88. a) $y = 2.3x^4 + 3.4x^2 + 1.2x - 4$
   b) $y = -0.25x^2 + 3.7$
   c) $y = 3(x + 2.3)^2 + 2.3$

↘ YOUR TURN ANSWERS: SECTION 2.1

1.

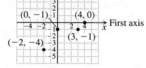

2.

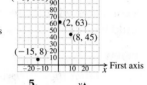

3. No

4.

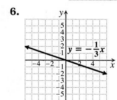

5.

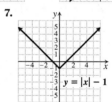

6.

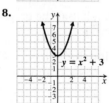

7.

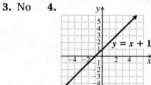

8.

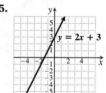

PREPARE TO MOVE ON

Evaluate.

1. $5t - 7$, for $t = 10$ [1.1]   **2.** $2r^2 - 7r$, for $r = -1$ [1.2]

3. $\dfrac{2x + 3}{x - 4}$, for $x = 0$ [1.2]   **4.** $\dfrac{4 - x}{3x + 1}$, for $x = 4$ [1.1]

Solve. [1.3]

5. $5 - x = 0$ **6.** $5x + 3 = 0$

2.2   **Functions**

Domain and Range   ▪   Functions and Graphs   ▪   Function Notation and Equations   ▪   Piecewise-Defined Functions

We now develop the idea of a *function*—one of the most important concepts in mathematics.

Domain and Range

A function is a special kind of correspondence between two sets. For example,

To each person in a class	there corresponds	a date of birth.
To each bar code in a store	there corresponds	a price.
To each real number	there corresponds	the cube of that number.

In each example, the first set is called the **domain**. The second set is called the **range**. For any member of the domain, there is *exactly one* member of the range to which it corresponds. This kind of correspondence is called a **function**.

Student Notes

Note that not all correspondences are functions.

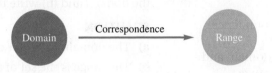

EXAMPLE 1   Determine whether each correspondence is a function.

a)

Domain	Range
4 ⟶ 2
1
−3 ⟶ 5

b)

Domain	Range
Ford ⟶ 200
Chrysler ⟶ Mustang
General Motors ⟶ Sonic
⟶ Volt

1. Determine whether the correspondence is a function.

Domain	Range
2 ⟶ 4
⟶ −4
3 ⟶ 9
⟶ −9

SOLUTION

a) The correspondence *is* a function because each member of the domain corresponds to *exactly one* member of the range.

b) The correspondence *is not* a function because a member of the domain (General Motors) corresponds to more than one member of the range.

 YOUR TURN

FUNCTION

A *function* is a correspondence between a first set, called the *domain*, and a second set, called the *range*, such that each member of the domain corresponds to *exactly one* member of the range.

EXAMPLE 2   Determine whether each correspondence is a function. Assume that the item mentioned first is in the domain of the correspondence.

a) The correspondence that assigns to a person his or her weight.

b) The correspondence that assigns to the numbers −2, 0, 1, and 2 each number's square

c) The correspondence that assigns to a best-selling author the titles of books written by that author

SOLUTION

a) For this correspondence, the domain is a set of people and the range is a set of positive numbers (the weights). We ask ourselves, "Does a person have *only one* weight?" Since the answer is Yes, this correspondence *is* a function.

b) The domain is $\{-2, 0, 1, 2\}$, and the range is $\{0, 1, 4\}$. We ask ourselves, "Does each number have *only one* square?" Since the answer is Yes, the correspondence *is* a function.

c) The domain is a set of authors, and the range is a set of book titles. We ask ourselves, "Has each author written *only one* book?" Since many authors have multiple titles published, the answer is No, the correspondence *is not* a function.

2. Determine whether the correspondence that assigns to a book the number of pages in the book is a function.

YOUR TURN

A set of ordered pairs is also a correspondence between two sets. The domain is the set of all first coordinates, and the range is the set of all second coordinates.

EXAMPLE 3   For the correspondence $\{(-6, 7), (1, 4), (-3, 4), (4, -5)\}$, **(a)** write the domain and **(b)** write the range.

SOLUTION

a) The domain is the set of all first coordinates: $\{-6, 1, -3, 4\}$.

b) The range is the set of all second coordinates: $\{7, 4, -5\}$.

3. For the correspondence $\{(0, -3), (4, 7), (5, -3)\}$, **(a)** write the domain and **(b)** write the range.

YOUR TURN

Functions and Graphs

The function in Example 1(a) can be written $\{(-3, 5), (1, 2), (4, 2)\}$ and the function in Example 2(b) $\{(-2, 4), (0, 0), (1, 1), (2, 4)\}$. We graph these functions in black as follows.

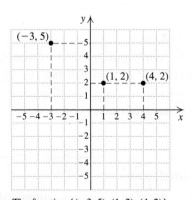

The function $\{(-3, 5), (1, 2), (4, 2)\}$
Domain is $\{-3, 1, 4\}$
Range is $\{5, 2\}$

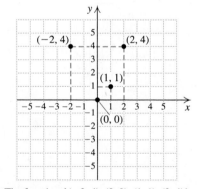

The function $\{(-2, 4), (0, 0), (1, 1), (2, 4)\}$
Domain is $\{-2, 0, 1, 2\}$
Range is $\{4, 0, 1\}$

We can find the domain and the range of a function directly from its graph. Note in the graphs above that if we move along the red dashed lines from the points to the horizontal axis, we find the members, or elements, of the domain. Similarly, if we move along the blue dashed lines from the points to the vertical axis, we find the elements of the range.

Functions are generally named using lowercase or uppercase letters. The function in the following example is named f.

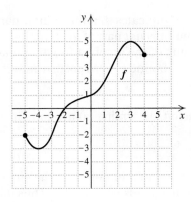

EXAMPLE 4 For the function f represented at left, determine each of the following.

a) The member of the range that is paired with 2

b) The domain of f

c) The member of the domain that is paired with -3

d) The range of f

SOLUTION

a) To determine what member of the range is paired with 2, we first note that we are considering 2 in the domain. Thus we locate 2 on the horizontal axis. Next, we find the point directly above 2 on the graph of f. From that point, we can look to the vertical axis to find the corresponding y-coordinate, 4. Thus, 4 is the member of the range that is paired with 2.

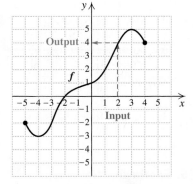

Student Notes

In mathematics, capitalization makes a difference! The variables r and R, for example, can be used in the same formula to represent two different quantities. Similarly, the function name f is different from the function name F.

b) The domain of the function is the set of all x-values that are used in the points of the curve. Because there are no breaks in the graph of f, these extend continuously from -5 to 4 and can be regarded as the curve's shadow, or *projection*, on the x-axis. This is illustrated by the shading on the x-axis. Thus the domain is $\{x \mid -5 \le x \le 4\}$.

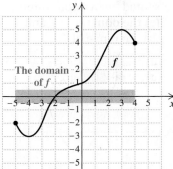

c) To determine what member of the domain is paired with -3, we note that we are considering -3 in the range. Thus we locate -3 on the vertical axis. From there, we look left and right to the graph of f to find any points for which -3 is the second coordinate. One such point exists, $(-4, -3)$. We observe that -4 is the only element of the domain paired with -3.

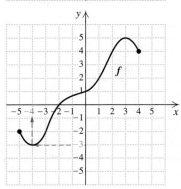

4. For the function g represented below, determine each of the following.

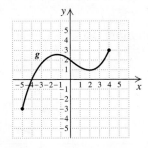

a) The member of the range that is paired with 2

b) The domain of g

c) The member of the domain that is paired with -3

d) The range of g

d) The range of the function is the set of all y-values that are in the graph. These extend continuously from -3 to 5, and can be regarded as the curve's projection on the y-axis. This is illustrated by the shading on the y-axis. Thus the range is $\{y \mid -3 \le y \le 5\}$.

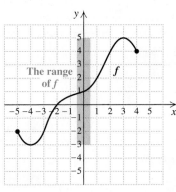

 YOUR TURN

A closed dot on a graph, such as in Example 4, indicates that the point is part of the function. An open dot indicates that the point is *not* part of the function.

The dots in Example 4 also indicate endpoints of the graph. A function may have a domain and/or a range that extends without bound toward positive infinity or negative infinity.

EXAMPLE 5 For the function g represented at left, determine **(a)** the domain of g and **(b)** the range of g.

SOLUTION

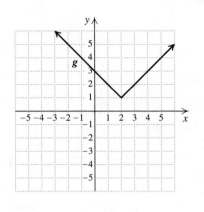

5. For the function f represented below, determine **(a)** the domain of f and **(b)** the range of f.

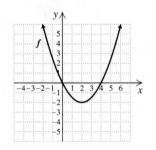

a) The domain of g is the set of all x-values that are used in points on the curve. Arrows on the ends of the graph indicate that it extends without end. Thus the shadow, or projection, of the graph on the x-axis is the entire x-axis, as shown on the left below. The domain is $\{x \mid x \text{ is a real number}\}$, or $\mathbb{R}$.

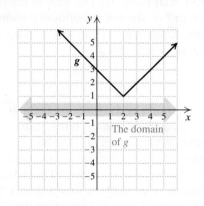
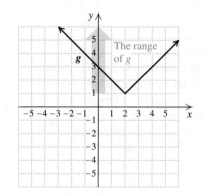

b) The range of g is the set of all y-values that are used in points on the curve. The graph indicates that every y-value greater than or equal to 1 is used at least once. Thus the projection of the graph on the y-axis is the portion of the y-axis greater than or equal to 1. (See the graph on the right above.) The range is $\{y \mid y \geq 1\}$.

YOUR TURN

Note that if a graph contains two or more points with the same first coordinate, that graph cannot represent a function (otherwise one member of the domain would correspond to more than one member of the range). This observation is the basis of the *vertical-line test*.

THE VERTICAL-LINE TEST

If it is possible for a vertical line to cross a graph more than once, then the graph is not the graph of a function.

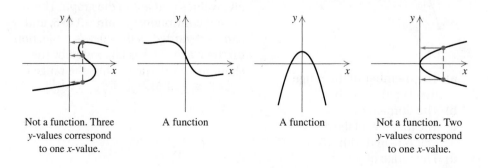

Not a function. Three y-values correspond to one x-value.   A function   A function   Not a function. Two y-values correspond to one x-value.

Although not all the graphs above represent functions, they all represent relations.

RELATION

A *relation* is a correspondence between a first set, called the *domain*, and a second set, called the *range*, such that each member of the domain corresponds to *at least one* member of the range.

Relations will appear throughout this book (indeed, every function is a special type of relation), but we will not focus on labeling them as such.

Function Notation and Equations

We often think of an element of the domain of a function as an **input** and its corresponding element of the range as an **output**. For example, consider the function

$$f = \{(-3, 1), (1, -2), (3, 0), (4, 5)\}.$$

Here, for an input of -3, the corresponding output is 1, and for an input of 3, the corresponding output is 0.

We use *function notation* to indicate what output corresponds to a given input. For the function f defined above, we write

$$f(-3) = 1, \quad f(1) = -2, \quad f(3) = 0, \quad \text{and} \quad f(4) = 5.$$

The notation $f(x)$ is read "f of x," "f at x," or "the value of f at x." If x is an input, then $f(x)$ is the corresponding output.

CAUTION! $f(x)$ *does not mean f times x.*

Most functions are described by equations. For example, $f(x) = 2x + 3$ describes the function that takes an input x, multiplies it by 2, and then adds 3.

$$f(x) \quad = \underset{\text{Double}}{2x} + \underset{\text{Add 3}}{3}$$

Input

To calculate the output $f(4)$, we take the input 4, double it, and add 3 to get 11. That is, we substitute 4 into the formula for $f(x)$:

$$f(4) = 2 \cdot 4 + 3$$
$$= 11. \longleftarrow \text{Output}$$

To understand function notation, it helps to imagine a "function machine." Think of putting an input into the machine. For $f(x) = 2x + 3$, the machine doubles each input and then adds 3. The result is the output.

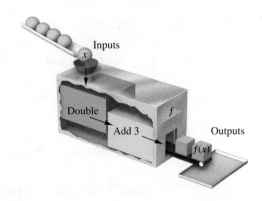

Inputs

Double

Add 3

f

Outputs

$f(x)$

Sometimes, in place of $f(x) = 2x + 3$, we write $y = 2x + 3$, where it is understood that the value of y, the *dependent variable*, depends on our choice of x, the *independent variable*. To understand why $f(x)$ notation is so useful, consider two equivalent statements:

a) Find the member of the range that is paired with 2.

b) Find $f(2)$.

Function notation is not only more concise; it also emphasizes that x is the independent variable.

EXAMPLE 6   Find each indicated function value.

a) $f(5)$, for $f(x) = 3x + 2$ **b)** $g(-2)$, for $g(r) = 5r^2 + 3r$

c) $h(4)$, for $h(x) = 7$ **d)** $F(a + 1)$, for $F(x) = 3x + 2$

e) $F(a) + 1$, for $F(x) = 3x + 2$

SOLUTION   Finding a function value is much like evaluating an algebraic expression.

a) $f(x) = 3x + 2$

$f(5) = 3 \cdot 5 + 2 = 17$ 5 is the input; 17 is the output.

b) $g(r) = 5r^2 + 3r$

Substitute.

$g(-2) = 5(-2)^2 + 3(-2)$

Evaluate.

$= 5 \cdot 4 - 6 = 14$

c) For the function given by $h(x) = 7$, every input has the same output, 7. Therefore, $h(4) = 7$. The function h is an example of a *constant function*.

Student Notes

In Example 6(d), it is important to note that the parentheses on the left are for function notation, whereas those on the right indicate multiplication.

d) $F(x) = 3x + 2$

$F(a + 1) = 3(a + 1) + 2$ The input is $a +1$.

$= 3a + 3 + 2 = 3a + 5$

e) $F(x) = 3x + 2$

$F(a) + 1 = [3(a) + 2] + 1$ The input is a.

$= [3a + 2] + 1 = 3a + 3$

6. Find $f(-1)$ for $f(t) = t^2 - 3$.

YOUR TURN

Note that whether we write $f(x) = 3x + 2$, or $f(t) = 3t + 2$, or $f(\ \blacksquare\ ) = 3 \cdot \blacksquare + 2$, we still have $f(5) = 17$. The variable in the parentheses (the independent variable) is the variable used in the algebraic expression.

When we find a function value, we determine an output that corresponds to a given input. To do this, we evaluate the expression for that input value.

Sometimes we want to determine an input that corresponds to a given output. To do this, we may need to solve an equation.

EXAMPLE 7 Let $f(x) = 3x - 7$.

a) What output corresponds to an input of 5?

b) What input corresponds to an output of 5?

SOLUTION

a) We ask ourselves, "$f(5) = \quad$?" Thus we find $f(5)$:

$$f(x) = 3x - 7$$
$$f(5) = 3(5) - 7 \qquad \text{The input is 5. We substitute 5 for } x.$$
$$= 15 - 7 = 8. \qquad \text{Carrying out the calculations}$$

The output 8 corresponds to the input 5; that is, $f(5) = 8$.

b) We ask ourselves, "$f(\quad) = 5$?" Thus we find the value of x for which $f(x) = 5$:

$$f(x) = 3x - 7$$
$$5 = 3x - 7 \qquad \text{The output is 5. We substitute 5 for } f(x).$$
$$12 = 3x$$
$$4 = x. \qquad \text{Solving for } x$$

The input 4 corresponds to the output 5; that is, $f(4) = 5$.

 YOUR TURN

7. Let $f(x) = \frac{1}{2}x$.

  a) What output corresponds to an input of 10?

  b) What input corresponds to an output of 10?

When a function is described by an equation, the domain is often unspecified. In such cases, the domain is the set of all numbers for which function values can be calculated.

EXAMPLE 8 For each equation, determine the domain of f.

a) $f(x) = |x|$

b) $f(x) = \dfrac{x}{2x - 6}$

SOLUTION

a) We ask ourselves, "Is there any number x for which we cannot compute $|x|$?" Since we can find the absolute value of *any* number, the answer is no. Thus the domain of f is $\mathbb{R}$, the set of all real numbers.

b) Is there any number x for which $\dfrac{x}{2x - 6}$ cannot be computed? Since $\dfrac{x}{2x - 6}$ cannot be computed when $2x - 6$ is 0, the answer is yes. To determine what x-value would cause $2x - 6$ to be 0, we set up and solve an equation:

$$2x - 6 = 0 \qquad \text{Setting the denominator equal to 0}$$
$$2x = 6 \qquad \text{Adding 6 to both sides}$$
$$x = 3. \qquad \text{Dividing both sides by 2}$$

> **CAUTION!** The denominator cannot be 0, but the numerator can be any number.

8. Determine the domain of the function given by
$$f(x) = \frac{x + 1}{x - 2}.$$

Thus, 3 is *not* in the domain of f, whereas all other real numbers are. The domain of f is $\{x \mid x \text{ is a real number } and \ x \neq 3\}$.

 YOUR TURN

Technology Connection

To visualize Example 8, note that the graph of $y_1 = |x|$ (which is entered $y_1 = \text{abs}(x)$, using the NUM option of the MATH menu) appears without interruption for any piece of the x-axis that we examine.

We show this graph on a graphing calculator app.

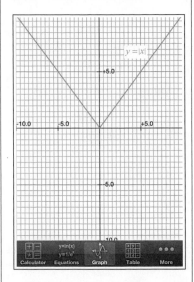

In contrast, the graph of $y_2 = \dfrac{x}{2x - 6}$ has a break at $x = 3$.

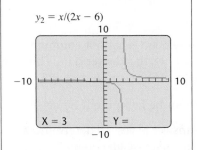

If the domain of a function is not specifically listed, it can be determined from a table, a graph, an equation, or an application.

DOMAIN OF A FUNCTION

The domain of a function f is the set of all inputs.

- If the correspondence is listed in a table or as a set of ordered pairs, the domain is the set of all first coordinates.
- If the function is described by a graph, the domain is the set of all first coordinates of the points on the graph.
- If the function is described by an equation, the domain is the set of all numbers for which the value of the function can be calculated.
- If the function is used in an application, the domain is the set of all numbers that make sense as inputs in the problem.

Piecewise-Defined Functions

Some functions are defined by different equations for various parts of their domains. Such functions are said to be **piecewise**-defined. For example, the function given by $f(x) = |x|$ can be described by

$$f(x) = \begin{cases} x, & \text{if } x \geq 0, \\ -x, & \text{if } x < 0. \end{cases}$$

To evaluate a piecewise-defined function for an input a, we determine what part of the domain a belongs to and use the appropriate formula for that part of the domain.

EXAMPLE 9   Find each function value for the function f given by

$$f(x) = \begin{cases} 2x, & \text{if } x < 3, \\ x + 1, & \text{if } x \geq 3. \end{cases}$$

a) $f(4)$ **b)** $f(-10)$

SOLUTION

a) The function f is defined using two different equations. To find $f(4)$, we must first determine whether to use the equation $f(x) = 2x$ or the equation $f(x) = x + 1$. To do this, we focus first on the two parts of the domain. It may help to visualize the domain on the number line, as shown below.

$$f(x) = \begin{cases} 2x, & \text{if } x < 3, \\ x + 1, & \text{if } x \geq 3 \end{cases} \quad \text{4 is in the second part of the domain.}$$

Since $4 \geq 3$, we use $f(x) = x + 1$. Thus, $f(4) = 4 + 1 = 5$.

b) To find $f(-10)$, we first note that $-10 < 3$, so we must use $f(x) = 2x$. Thus, $f(-10) = 2(-10) = -20$.

9. In Example 9, find $f(3)$.

YOUR TURN

2.2 EXERCISE SET

FOR EXTRA HELP   MyMathLab® Math XL
PRACTICE WATCH READ REVIEW

➤ Vocabulary and Reading Check

Choose the word from the following list that best completes each statement. Words may be used more than once.

correspondence   horizontal
domain   range
exactly   vertical
"*f* of 3"

1. A function is a special kind of _____ between two sets.

2. In any function, each member of the domain is paired with _____ one member of the range.

3. For any function, the set of all inputs, or first values, is called the _____.

4. For any function, the set of all outputs, or second values, is called the _____.

5. When a function is graphed, members of the domain are located on the _____ axis.

6. When a function is graphed, members of the range are located on the _____ axis.

7. The notation $f(3)$ can be read _____.

8. The _____-line test is used to determine whether or not a graph represents a function.

Domain and Range

Determine whether each correspondence is a function.

9. a → 2
   b →
   d → 3
   g → 4
   h →

10. 2 → a
    → b
    3 → d
    4 → g
    → h

11. *Girl's age (in months)* | *Average daily weight gain (in grams)*

2	21.8
9	11.7
16	8.5
23	7.0

Source: *American Family Physician,* December 1993, p. 1435

12. *Boy's age (in months)* | *Average daily weight gain (in grams)*

2	24.3
9	11.7
16	8.2
23	7.0

Source: *American Family Physician,* December 1993, p. 1435

13. *Actor/Actress* | *Movie*

Julia Roberts → My Best Friend's Wedding, Notting Hill, Pretty Woman
Denzel Washington → American Gangster, Safe House, Training Day

14. *Celebrity* | *Birthday*

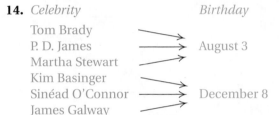

Tom Brady
P. D. James → August 3
Martha Stewart
Kim Basinger
Sinéad O'Connor → December 8
James Galway

Source: www.leannesbirthdays.com

15. *Predator* | *Prey*

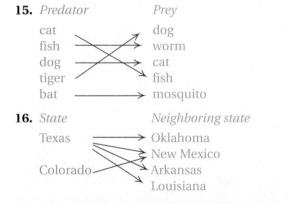

cat → dog
fish → worm
dog → cat
tiger → fish
bat → mosquito

16. *State* | *Neighboring state*

Texas → Oklahoma
→ New Mexico
Colorado → Arkansas
→ Louisiana

Determine whether each of the following is a function.

17. The correspondence that assigns to a USB flash drive its storage capacity

18. The correspondence that assigns to a member of a rock band the instrument the person can play

19. The correspondence that assigns to a player on a team that player's uniform number

20. The correspondence that assigns to a triangle its area

*For each correspondence, (**a**) write the domain, (**b**) write the range, and (**c**) determine whether the correspondence is a function.*

21. $\{(-3, 3), (-2, 5), (0, 9), (4, -10)\}$

22. $\{(0, -1), (1, 3), (2, -1), (5, 3)\}$

23. $\{(1, 1), (2, 1), (3, 1), (4, 1), (5, 1)\}$

24. $\{(1, 1), (1, 2), (1, 3), (1, 4), (1, 5)\}$

25. $\{(4, -2), (-2, 4), (3, -8), (4, 5)\}$

26. $\{(0, 7), (4, 8), (7, 0), (8, 4)\}$

Functions and Graphs

Determine the domain and the range of each function.

27.

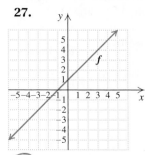

28.

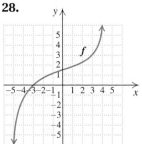

29.

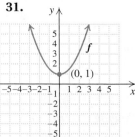

30.

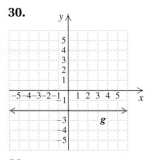

31.

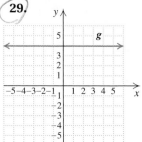

32.

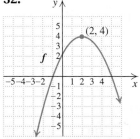

33.

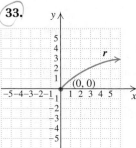

34.

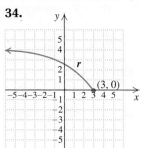

Determine whether each of the following is the graph of a function.

35.

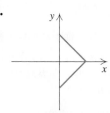

36.

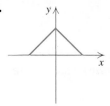

37.

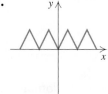

38.

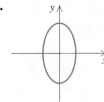

39.

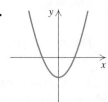

40.

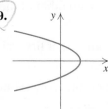

Function Notation and Equations

Find the function values.

41. $g(x) = 2x + 5$

    a) $g(0)$ **b)** $g(-4)$ **c)** $g(-7)$
    d) $g(8)$ **e)** $g(a + 2)$ **f)** $g(a) + 2$

42. $h(x) = 5x - 1$

    a) $h(4)$ **b)** $h(8)$ **c)** $h(-3)$
    d) $h(-4)$ **e)** $h(a - 1)$ **f)** $h(a) + 3$

43. $f(n) = 5n^2 + 4n$

    a) $f(0)$ **b)** $f(-1)$ **c)** $f(3)$
    d) $f(t)$ **e)** $f(2a)$ **f)** $f(3) - 9$

44. $g(n) = 3n^2 - 2n$

    a) $g(0)$ **b)** $g(-1)$ **c)** $g(3)$
    d) $g(t)$ **e)** $g(2a)$ **f)** $g(3) - 4$

45. $f(x) = \dfrac{x - 3}{2x - 5}$

    a) $f(0)$ **b)** $f(4)$ **c)** $f(-1)$
    d) $f(3)$ **e)** $f(x + 2)$ **f)** $f(a + h)$

46. $r(x) = \dfrac{3x - 4}{2x + 5}$

    a) $r(0)$ **b)** $r(2)$ **c)** $r\left(\frac{4}{3}\right)$
    d) $r(-1)$ **e)** $r(x + 3)$ **f)** $r(a + h)$

The function A described by $A(s) = \dfrac{\sqrt{3}}{4}s^2$ gives the area of an equilateral triangle with side s.

47. Find the area when a side measures 4 cm.

48. Find the area when a side measures 6 in.

The function V described by $V(r) = 4\pi r^2$ gives the surface area of a sphere with radius r.

49. Find the surface area when the radius is 3 in.

50. Find the surface area when the radius is 5 cm.

51. *Archaeology.* The function H described by

$$H(x) = 2.75x + 71.48$$

can be used to estimate the height, in centimeters, of a woman whose humerus (the bone from the elbow to the shoulder) is x cm long. Estimate the height of a woman whose humerus is 34 cm long.

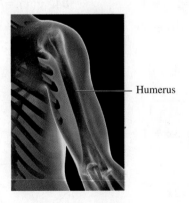

Humerus

52. *Chemistry.* The function F described by

$$F(C) = \tfrac{9}{5}C + 32$$

gives the Fahrenheit temperature corresponding to the Celsius temperature C. Find the Fahrenheit temperature equivalent to $-5°$ Celsius.

For each graph of a function, determine **(a)** $f(1)$; **(b)** the domain; **(c)** any x-values for which $f(x) = 2$; and **(d)** the range.

53.
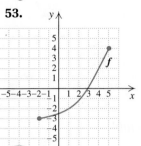

54.
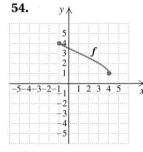

55.
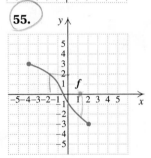

56.
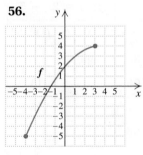

57.
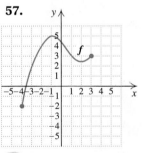

58.
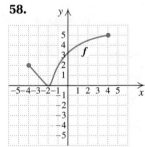

59.
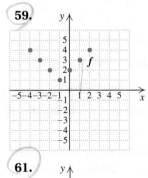

60.
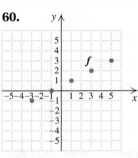

61.
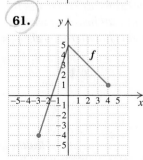

62.
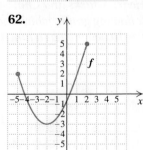

63.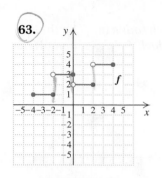

64.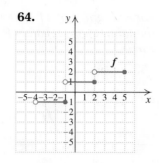

Fill in the missing values in each table.

$f(x) = 2x - 5$	
x	$f(x)$
65. 8	
66.	13
67.	-5
68. -4	

$f(x) = \frac{1}{3}x + 4$	
x	$f(x)$
69.	$\frac{1}{2}$
70.	$-\frac{1}{3}$
71. $\frac{1}{2}$	
72. $-\frac{1}{3}$	

73. If $f(x) = 4 - x$, for what input is the output 7?

74. If $f(x) = 5x + 1$, for what input is the output $\frac{1}{2}$?

75. If $f(x) = 0.1x - 0.5$, for what input is the output -3?

76. If $f(x) = 2.3 - 1.5x$, for what input is the output 10?

Heart Attacks and Cholesterol. For Exercises 77–80, use the following graph, which shows the annual heart attack rate per 10,000 men as a function of blood cholesterol level.*

* Copyright 1989, CSPI. Adapted from *Nutrition Action Health-letter* (1875 Connecticut Avenue, N.W., Suite 300, Washington, DC 20009-5728. \$20 for 10 issues).

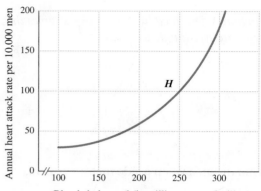

77. Approximate the annual heart attack rate for those men whose blood cholesterol level is 225 mg/dl. That is, find $H(225)$.

78. Approximate the annual heart attack rate for those men whose blood cholesterol level is 275 mg/dl. That is, find $H(275)$.

79. Approximate the blood cholesterol level for an annual heart attack rate of 100 attacks per 10,000 men. That is, find x for which $H(x) = 100$.

80. Approximate the blood cholesterol level for an annual heart attack rate of 50 attacks per 10,000 men. That is, find x for which $H(x) = 50$.

Find the domain of the function given by each equation.

81. $f(x) = \dfrac{5}{x - 3}$ **82.** $f(x) = \dfrac{7}{6 - x}$

83. $g(x) = 2x + 1$ **84.** $g(x) = x^2 + 3$

85. $h(x) = |6 - 7x|$ **86.** $h(x) = |3x - 4|$

87. $f(x) = \dfrac{3}{8 - 5x}$ **88.** $f(x) = \dfrac{5}{2x + 1}$

89. $h(x) = \dfrac{x}{x + 1}$ **90.** $h(x) = \dfrac{3x}{x + 7}$

91. $f(x) = \dfrac{3x + 1}{2}$ **92.** $f(x) = \dfrac{4x - 3}{5}$

93. $g(x) = \dfrac{1}{2x}$ **94.** $g(x) = \dfrac{1}{2}x$

Piecewise-Defined Functions

Find the indicated function values.

95. $f(x) = \begin{cases} x, & \text{if } x < 0, \\ 2x + 1, & \text{if } x \geq 0 \end{cases}$

a) $f(-5)$ **b)** $f(0)$ **c)** $f(10)$

96. $g(x) = \begin{cases} x - 5, & \text{if } x \le 5, \\ 3x, & \text{if } x > 5 \end{cases}$

   a) $g(0)$ **b)** $g(5)$ **c)** $g(6)$

97. $G(x) = \begin{cases} x - 5, & \text{if } x \le -1, \\ x, & \text{if } x > -1 \end{cases}$

   a) $G(-10)$ **b)** $G(0)$ **c)** $G(-1)$

98. $F(x) = \begin{cases} 2x, & \text{if } x < 3, \\ -5x, & \text{if } x \ge 3 \end{cases}$

   a) $F(-1)$ **b)** $F(3)$ **c)** $F(10)$

99. $f(x) = \begin{cases} x^2 - 10, & \text{if } x < -10, \\ x^2, & \text{if } -10 \le x \le 10, \\ x^2 + 10, & \text{if } x > 10 \end{cases}$

   a) $f(-10)$ **b)** $f(10)$ **c)** $f(11)$

100. $f(x) = \begin{cases} 2x^2 - 3, & \text{if } x \le 2, \\ x^2, & \text{if } 2 < x < 4, \\ 5x - 7, & \text{if } x \ge 4 \end{cases}$

   a) $f(0)$ **b)** $f(3)$ **c)** $f(6)$

101. Explain why the domain of the function given by $f(x) = \dfrac{x + 3}{2}$ is $\mathbb{R}$, but the domain of the function given by $g(x) = \dfrac{2}{x + 3}$ is not $\mathbb{R}$.

102. For the function given by $n(z) = ab + wz$, what is the independent variable? How can you tell?

Skill Review

103. Translate to an algebraic expression: Seven less than some number. [1.1]

104. Evaluate $9xy \div z^2x$ for $x = -2, y = 6$, and $z = 3$. [1.1], [1.2]

105. Use a commutative law to write an expression equivalent to $7 + t$. [1.2]

106. Convert 45,800,000 to scientific notation. [1.7]

Synthesis

107. Jaylan is asked to write a function relating the number of fish in an aquarium to the amount of food needed for the fish. Which quantity should he choose as the independent variable? Why?

108. Explain the difference between finding $f(0)$ and finding the input x for which $f(x) = 0$.

For Exercises 109 and 110, let $f(x) = 3x^2 - 1$ and $g(x) = 2x + 5$.

109. Find $f(g(-4))$ and $g(f(-4))$.

110. Find $f(g(-1))$ and $g(f(-1))$.

111. If f represents the function in Exercise 15, find $f(f(f(f(\text{tiger}))))$.

Determine the domain and the range of each function.

112.

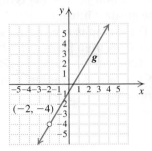

113.

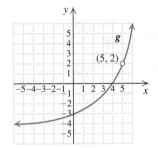

Pregnancy. For Exercises 114–117, use the following graph of a woman's "stress test." This graph shows the size of a pregnant woman's contractions as a function of time.

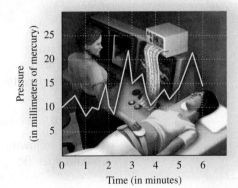

Stress test

114. How large is the largest contraction that occurred during the test?

115. At what time during the test did the largest contraction occur?

116. On the basis of the information provided, how large a contraction would you expect 60 seconds after the end of the test? Why?

117. What is the frequency of the largest contraction?

118. Suppose that a function g is such that $g(-1) = -7$ and $g(3) = 8$. Find a formula for g if $g(x)$ is of the form $g(x) = mx + b$, where m and b are constants.

119. The *greatest integer function* $f(x) = [\![x]\!]$ is defined as follows: $[\![x]\!]$ is the greatest integer that is less than or equal to x. For example, if $x = 3.74$, then $[\![x]\!] = 3$; and if $x = -0.98$, then $[\![x]\!] = -1$. Graph the greatest integer function for $-5 \le x \le 5$. (The notation $f(x) = \text{INT}(x)$ is used by many graphing calculators and computers.)

QUICK QUIZ: SECTIONS 2.1–2.2

1. In what quadrant or on what axis is $(9, 100)$ located? [2.1]

2. Determine whether $(-3, -5)$ is a solution of $y - x = 2$. [2.1]

3. Graph: $y - x = 2$. [2.1]

4. Find $h(20)$ when $h(x) = \frac{1}{2}x - 10$. [2.2]

5. Find the domain of the function given by $f(x) = \dfrac{7}{x}$. [2.2]

PREPARE TO MOVE ON

Simplify. [1.2]

1. $\dfrac{6 - 3}{-2 - 7}$

2. $\dfrac{-5 - (-5)}{3 - (-10)}$

3. $\dfrac{2 - (-3)}{-3 - 2}$

Solve for y. [1.5]

4. $2x - y = 8$

5. $5x + 5y = 10$

6. $5x - 4y = 8$

➤ **YOUR TURN ANSWERS: SECTION 2.2**

1. No   **2.** Yes   **3.** (a) $\{0, 4, 5\}$; (b) $\{-3, 7\}$ **4.** (a) 1;
(b) $\{x | -5 \le x \le 4\}$; (c) -5; (d) $\{y | -3 \le y \le 3\}$
5. (a) $\{x | x \text{ is a real number}\}$, or $\mathbb{R}$; (b) $\{y | y \ge -2\}$
6. -2 **7.** (a) 5; (b) 20   **8.** $\{x | x \text{ is a real number } and \ x \ne 2\}$
9. 4

2.3 Linear Functions: Slope, Graphs, and Models

Slope–Intercept Form ▪ Applications

The functions and real-life models that we examine in this section have graphs that are straight lines. Such functions and their graphs are called *linear* and can be written in the form $f(x) = mx + b$, where m and b are constants.

Slope–Intercept Form

The following two examples compare graphs of $f(x) = mx$ with $f(x) = mx + b$ for specific choices of m and b.

EXAMPLE 1   Graph $y = 2x$ and $y = 2x + 3$, using the same set of axes.

SOLUTION   We first make a table of solutions of both equations.

x	$y = 2x$	$y = 2x + 3$
0	0	3
1	2	5
-1	-2	1
2	4	7
-2	-4	-1
3	6	9

Study Skills

Strike While the Iron's Hot

Be sure to do your homework as soon as possible after each class. Make this part of your routine, choosing a time and a workspace where you can minimize distractions.

We then plot these points. Drawing a blue line for $y = 2x + 3$ and a red line for $y = 2x$, we observe that the graph of $y = 2x + 3$ is simply the graph of $y = 2x$ shifted, or *translated*, 3 units up. The lines are parallel.

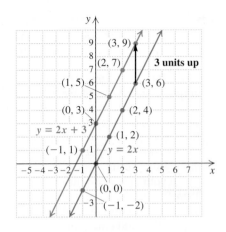

1. Graph $y = 3x$ and $y = 3x + 4$, using the same set of axes.

YOUR TURN

EXAMPLE 2 Graph $f(x) = \frac{1}{3}x$ and $g(x) = \frac{1}{3}x - 2$, using the same set of axes.

SOLUTION We first make a table of solutions of both equations. By choosing multiples of 3 for x, we can avoid fractions.

x	$f(x) = \frac{1}{3}x$	$g(x) = \frac{1}{3}x - 2$
0	0	-2
3	1	-1
-3	-1	-3
6	2	0

We then plot these points. Drawing a blue line for $g(x) = \frac{1}{3}x - 2$ and a red line for $f(x) = \frac{1}{3}x$, we see that the graph of $g(x) = \frac{1}{3}x - 2$ is simply the graph of $f(x) = \frac{1}{3}x$ shifted, or translated, 2 units down.

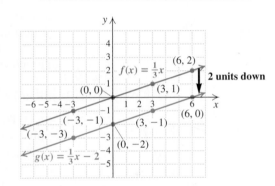

2. Graph $f(x) = \frac{1}{2}x$ and $g(x) = \frac{1}{2}x - 1$, using the same set of axes.

YOUR TURN

Note that the graph of $y = 2x + 3$ passes through the point $(0, 3)$ and the graph of $g(x) = \frac{1}{3}x - 2$ passes through the point $(0, -2)$. In general:

> **The graph of any line written in the form $y = mx + b$ passes through the point $(0, b)$. The point $(0, b)$ is called the y-intercept.**

For an equation $y = mx$, the value of b is 0 and $(0, 0)$ is the y-intercept.

> **The graph of $y = mx$ passes through the origin.**

EXAMPLE 3 For each equation, find the y-intercept.

a) $y = -5x + 7$

b) $f(x) = 5.3x - 12$

SOLUTION

a) The y-intercept is $(0, 7)$.

3. Find the y-intercept of the graph of $y = \frac{1}{3}x - 16$.

b) The y-intercept is $(0, -12)$.

YOUR TURN

A y-intercept $(0, b)$ is plotted by locating b on the y-axis. For this reason, we sometimes refer to the number b as the y-intercept.

In Examples 1 and 2, the graphs of $y = 2x$ and $y = 2x + 3$ are parallel and the graphs of $f(x) = \frac{1}{3}x$ and $g(x) = \frac{1}{3}x - 2$ are parallel.

It is the number m, in $y = mx + b$, that is responsible for the slant of the line. The following definition enables us to visualize this slant, or *slope*, as a ratio of two lengths.

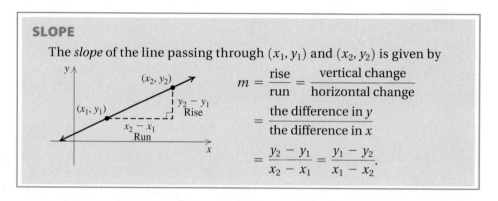

SLOPE

The *slope* of the line passing through (x_1, y_1) and (x_2, y_2) is given by

$$m = \frac{\text{rise}}{\text{run}} = \frac{\text{vertical change}}{\text{horizontal change}}$$

$$= \frac{\text{the difference in } y}{\text{the difference in } x}$$

$$= \frac{y_2 - y_1}{x_2 - x_1} = \frac{y_1 - y_2}{x_1 - x_2}.$$

In the definition above, (x_1, y_1) and (x_2, y_2)—read "x sub-one, y sub-one and x sub-two, y sub-two"—represent two different points on a line. It does not matter which point is considered (x_1, y_1) and which is considered (x_2, y_2) so long as coordinates are subtracted in the same order in both the numerator and the denominator.

The letter m is traditionally used for slope. This usage has its roots in the French verb *monter*, meaning "to climb."

EXAMPLE 4   Find the slope of the lines drawn in Example 1.

SOLUTION   To find the slope of a line, we can use the coordinates of any two points on that line. We use $(1, 5)$ and $(2, 7)$ to find the slope of the blue line in Example 1:

$$\text{Slope} = \frac{\text{rise}}{\text{run}} = \frac{\text{difference in } y}{\text{difference in } x}$$

$$= \frac{y_2 - y_1}{x_2 - x_1} = \frac{7 - 5}{2 - 1} = 2. \qquad \textit{Any pair of points on the line will give the same slope.}$$

To find the slope of the red line in Example 1, we use $(-1, -2)$ and $(3, 6)$:

$$\text{Slope} = \frac{\text{rise}}{\text{run}} = \frac{\text{difference in } y}{\text{difference in } x}$$

$$= \frac{6 - (-2)}{3 - (-1)} = \frac{8}{4} = 2. \qquad \textit{Any pair of points on the line will give the same slope.}$$

YOUR TURN

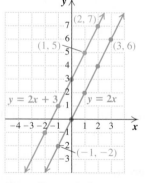

From Example 1

4. Find the slope of the lines drawn in Example 2.

In Example 4 and Your Turn Exercise 4, we see that the lines that are given by $y = 2x + 3$, $y = 2x$, $g(x) = \frac{1}{3}x - 2$, and $f(x) = \frac{1}{3}x$ have slopes 2, 2, $\frac{1}{3}$, and $\frac{1}{3}$, respectively. This supports (but does not prove) the following:

The slope of any line written in the form $y = mx + b$ is m.

A proof of this result is outlined in Exercise 113 of this section's exercise set.

━━━━━━━━━━ EXPLORING THE CONCEPT ━━━━━━━━━━

From each equation, determine the slope and the y-intercept of the graph. Then match each equation with its graph.

1. $y = \frac{2}{3}x + 1$ **2.** $y = \frac{2}{3}x - 5$

3. $y = \frac{1}{4}x + 1$ **4.** $y = \frac{1}{4}x - 5$

ANSWERS
1. A 2. C
3. B 4. D

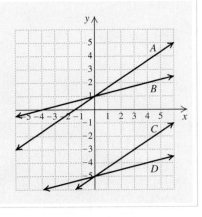

Technology Connection

To examine the effect of m when graphing $y = mx + b$, we can graph $y_1 = x + 1$, $y_2 = 2x + 1$, $y_3 = 3x + 1$, and $y_4 = \frac{3}{4}x + 1$ on the same set of axes.

   To see the effect of a negative value for m, we can graph $y_1 = x + 1$, $y_2 = -x + 1$, and $y_3 = -2x + 1$ on the same set of axes.

   If your calculator has a Transfrm application, use it to enter and graph $y_1 = Ax + B$. Choose a value for B and enter various positive and negative values for A. Uninstall Transfrm when you are finished.

   Describe how the sign of m affects the graph of $y = mx + b$.

---

5. Determine the slope and the y-intercept of the line given by $y = 6x - 7$.

---

6. Find a linear function whose graph has slope $-\frac{1}{2}$ and y-intercept $(0, 8)$.

SLOPE–INTERCEPT FORM

Any equation of the form $y = mx + b$ is said to be written in *slope–intercept* form and has a graph that is a straight line.

The slope of the line is m.

The y-intercept of the line is $(0, b)$.

EXAMPLE 5   Determine the slope and the y-intercept of the line given by $y = -\frac{1}{3}x + 2$.

SOLUTION   The equation $y = -\frac{1}{3}x + 2$ is in the form $y = mx + b$:

$$y = mx + b$$
$$ = -\tfrac{1}{3}x + 2.$$

Since $m = -\frac{1}{3}$, the slope is $-\frac{1}{3}$. Since $b = 2$, the y-intercept is $(0, 2)$.

YOUR TURN

EXAMPLE 6   Find a linear function whose graph has slope 3 and y-intercept $(0, -1)$.

SOLUTION   We use slope–intercept form, $f(x) = mx + b$:

$$f(x) = 3x + (-1) \qquad \text{Substituting 3 for } m \text{ and } -1 \text{ for } b$$
$$ = 3x - 1.$$

YOUR TURN

To graph $f(x) = 3x - 1$, we regard the slope of 3 as $\frac{3}{1}$. Then, beginning at the y-intercept $(0, -1)$, we count *up* 3 units (the *rise*) and *to the right* 1 unit (the *run*), as shown at right.

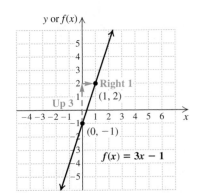

EXAMPLE 7 Determine the slope and the y-intercept of the line given by $f(x) = -\frac{1}{2}x + 5$. Then draw the graph.

SOLUTION The y-intercept is $(0, 5)$. The slope is $-\frac{1}{2}$, or $\frac{-1}{2}$. From the y-intercept, we go *down* 1 unit and *to the right* 2 units. That gives us the point $(2, 4)$. We can now draw the graph.

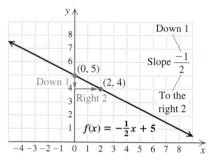

To check, we rewrite the slope in another form and find another point: $-\frac{1}{2} = \frac{1}{-2}$. Thus we can go *up* 1 unit and then *to the left* 2 units. This gives the point $(-2, 6)$. Since $(-2, 6)$ is on the line, we have a check.

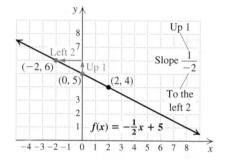

7. Determine the slope and the y-intercept of the line given by $f(x) = -\frac{2}{3}x + 1$. Then draw the graph.

↪ YOUR TURN

In Examples 1 and 2, the lines slant upward from left to right. In Example 7, the line slants downward from left to right.

Lines with positive slopes slant upward from left to right.

Lines with negative slopes slant downward from left to right.

Often the easiest way to graph an equation is to rewrite it in slope–intercept form and then proceed as in Example 7 above.

EXAMPLE 8 Determine the slope and the y-intercept for the equation $5x - 4y = 8$. Then graph.

SOLUTION We first write an equivalent equation in slope–intercept form:

$$5x - 4y = 8$$
$$-4y = -5x + 8 \qquad \text{Adding } -5x \text{ to both sides}$$
$$y = -\frac{1}{4}(-5x + 8) \qquad \text{Multiplying both sides by } -\frac{1}{4}$$
$$y = \frac{5}{4}x - 2. \qquad \text{Using the distributive law}$$

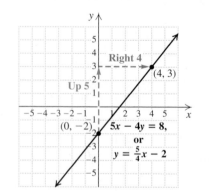

Because we now have the form $y = mx + b$, we know that the slope is $\frac{5}{4}$ and the y-intercept is $(0, -2)$. We plot $(0, -2)$, and from there go *up* 5 units, move *to the right* 4 units, and plot a second point at $(4, 3)$. We then draw the graph, as shown at left.

To check that the line is drawn correctly, we calculate the coordinates of another point on the line. For $x = 2$, we have

$$5 \cdot 2 - 4y = 8$$
$$10 - 4y = 8$$
$$-4y = -2$$
$$y = \frac{1}{2}.$$

8. Determine the slope and the y-intercept for the equation $3x - 2y = 6$. Then graph.

Thus, $\left(2, \frac{1}{2}\right)$ should appear on the graph. Since it *does* appear to be on the line, we have a check.

↪ YOUR TURN

Student Notes

Be careful to use proper units when writing your answers. When reading a rate of change from a graph, remember that the units from the vertical axis are used in the numerator and the units from the horizontal axis are used in the denominator.

9. The following graph shows projections for the number of Americans ages 65 and older. Use the graph to find the rate at which this number is expected to grow.

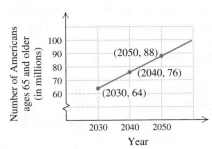

Source: Based on data from the U.S. Census Bureau

10. Between 2005 and 2008, the number of Freecycle groups grew from 2937 to 4224. At what rate was the number of Freecycle groups changing during these years?

Applications

Because slope is a ratio that indicates how a change in the vertical direction of a line corresponds to a change in the horizontal direction, it has many real-world applications. Foremost is the use of slope to represent a *rate of change*.

EXAMPLE 9   *Telephone Lines.*   As more people use cell phones as their primary phone line, the number of land lines in the world has been changing, as shown in the following graph. Use the graph to find the rate at which this number is changing. Note that the jagged "break" on the vertical axis is used to avoid including a large portion of unused grid.

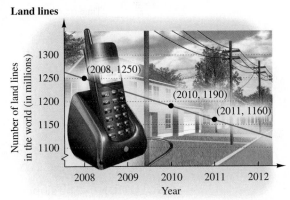

Source: International Telecommunication Union

SOLUTION   Since the graph is linear, we can use any pair of points to determine the rate of change. We choose (2008, 1250) and (2011, 1160), which gives us

$$\text{Rate of change} = \frac{1160 \text{ million} - 1250 \text{ million}}{2011 - 2008}$$

$$= \frac{-90 \text{ million}}{3 \text{ years}} = -30 \text{ million land lines per year.}$$

The number of land lines in the world is changing at a rate of -30 million land lines per year.

 YOUR TURN

EXAMPLE 10   *Recycling.*   The Freecycle Network was established in 2003 with the purpose of reducing waste by giving unwanted items to others who could use them. In January 2008, there were 4224 Freecycle recycling groups. By January 2012, this number had grown to 5018 groups. At what rate was the number of Freecycle groups changing?

Source: www.freecycle.org

SOLUTION   The rate at which the number of Freecycle groups changed is given by

$$\text{Rate of change} = \frac{\text{change in number of groups}}{\text{change in time}}$$

$$= \frac{5018 \text{ groups} - 4224 \text{ groups}}{2012 - 2008}$$

$$= \frac{794 \text{ groups}}{4 \text{ years}} = 198.5 \text{ groups per year.}$$

Between 2008 and 2012, the number of Freecycle groups grew at a rate of 198.5 groups per year.

YOUR TURN

EXAMPLE 11   *Running Speed.*   Stephanie runs 10 km during each workout. For the first 7 km, her pace is twice as fast as it is for the last 3 km. Which graph best describes her workout?

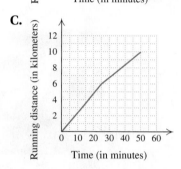

A.

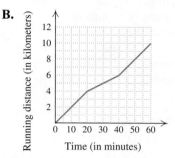

B.

C.

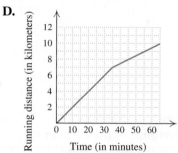

D.

SOLUTION   The slopes in graph A increase as we move to the right. This would indicate that Stephanie ran faster for the *last* part of her workout. Thus graph A is not the correct one.

The slopes in graph B indicate that Stephanie slowed down in the middle of her run and then resumed her original speed. Thus graph B does not correctly model the situation either.

11. Derek runs 10 km during each workout. For the last 4 km, his pace is two-thirds as fast as it is for the first 6 km. Which of the graphs in Example 11 best describes Derek's workout?

According to graph C, Stephanie slowed down not at the 7-km mark, but at the 6-km mark. Thus graph C is also incorrect.

Graph D indicates that Stephanie ran the first 7 km in 35 min, a rate of 0.2 km/min. It also indicates that she ran the final 3 km in 30 min, a rate of 0.1 km/min. This means that Stephanie's rate was twice as fast for the first 7 km, so graph D provides a correct description of her workout.

YOUR TURN

EXAMPLE 12   *Depreciation.*   Island Bike Rentals uses the function

$$S(t) = -125t + 750$$

to determine the value $S(t)$, in dollars, of a mountain bike t years after its purchase. Since this amount is decreasing over time, we say that the value of the bike is *depreciating*.

a) What do the numbers -125 and 750 signify?

b) How long will it take one of their mountain bikes to depreciate completely?

c) What is the domain of S?

SOLUTION   Drawing, or at least visualizing, a graph can be useful here.

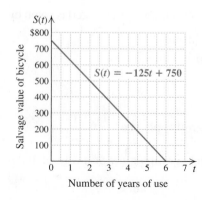

a) At time $t = 0$, we have $S(0) = -125 \cdot 0 + 750 = 750$. Thus the number 750 signifies the original cost of the mountain bike, in dollars.

This function is written in slope–intercept form. Since the output is measured in dollars and the input in years, the number -125 signifies that the value of the bike is decreasing at a rate of $125 per year.

12. Refer to Example 12. Island Bike Rentals uses the function $V(t) = -300t + 1200$ to determine the salvage value $V(t)$, in dollars, of a road bike t years after its purchase.

a) What do the numbers -300 and 1200 signify?

b) How long will it take a road bike to depreciate completely?

c) What is the domain of V?

b) The bike will have depreciated completely when its value drops to 0. To learn when this occurs, we determine when $S(t) = 0$:

$$S(t) = 0 \quad \text{Setting } S(t) \text{ equal to 0}$$
$$-125t + 750 = 0 \quad \text{Substituting } -125t + 750 \text{ for } S(t)$$
$$-125t = -750 \quad \text{Subtracting 750 from both sides}$$
$$t = 6. \quad \text{Dividing both sides by } -125$$

The bike will have depreciated completely in 6 years.

c) Neither the number of years of service nor the salvage value can be negative. In part (b), we found that after 6 years the salvage value will have dropped to 0. Thus the domain of S is $\{t \mid 0 \le t \le 6\}$.

 YOUR TURN

2.3 EXERCISE SET

FOR EXTRA HELP

MyMathLab® Math XL
PRACTICE   WATCH   READ   REVIEW

Vocabulary and Reading Check

In each of Exercises 1–6, match the word with the most appropriate choice from the column on the right.

1. ___ y-intercept

2. ___ Slope

3. ___ Rise

4. ___ Run

5. ___ Slope–intercept form

6. ___ Translated

a) $y = mx + b$
b) Shifted
c) $\dfrac{\text{Difference in } y}{\text{Difference in } x}$
d) Difference in x
e) Difference in y
f) $(0, b)$

Slope–Intercept Form

Graph.

7. $f(x) = 2x - 1$
8. $g(x) = 3x + 4$
9. $g(x) = -\frac{1}{3}x + 2$
10. $f(x) = -\frac{1}{2}x - 5$
11. $h(x) = \frac{2}{5}x - 4$
12. $h(x) = \frac{4}{5}x + 2$

Determine the y-intercept.

13. $y = 5x + 3$
14. $y = 2x - 11$
15. $g(x) = -x - 1$
16. $g(x) = -4x + 5$
17. $y = -\frac{3}{8}x - 4.5$
18. $y = \frac{15}{7}x + 2.2$
19. $f(x) = 1.3x - \frac{1}{4}$
20. $f(x) = -1.2x + \frac{1}{5}$
21. $y = 17x + 138$
22. $y = -52x - 260$

For each pair of points, find the slope of the line containing them.

23. $(10, 11)$ and $(8, 3)$
24. $(2, 9)$ and $(12, 4)$
25. $(-4, -5)$ and $(-8, 3)$
26. $(2, -3)$ and $(6, -2)$
27. $(13, 4)$ and $(-20, -7)$

28. $(-5, -11)$ and $(-8, -21)$
29. $\left(\frac{1}{2}, -\frac{2}{3}\right)$ and $\left(\frac{1}{6}, \frac{1}{6}\right)$
30. $\left(\frac{3}{4}, -\frac{2}{5}\right)$ and $\left(\frac{1}{3}, -\frac{1}{4}\right)$
31. $(-9.7, 43.6)$ and $(4.5, 43.6)$
32. $(-2.8, -3.1)$ and $(-1.8, -2.6)$

Determine the slope and the y-intercept.

33. $y = \frac{2}{3}x + 4$
34. $y = -x - 6$
35. $2x - y = 3$
36. $y = 4x + 9$
37. $y = x - 2$
38. $3x + y = 5$
39. $4x + 5y = 8$
40. $x + 6y = 1$

Find a linear function whose graph has the given slope and y-intercept.

41. Slope 2, y-intercept $(0, 5)$
42. Slope -4, y-intercept $(0, 1)$
43. Slope $-\frac{2}{3}$, y-intercept $(0, -2)$
44. Slope $-\frac{3}{4}$, y-intercept $(0, -5)$
45. Slope -7, y-intercept $\left(0, \frac{1}{3}\right)$
46. Slope 8, y-intercept $\left(0, -\frac{1}{4}\right)$

Determine the slope and the y-intercept. Then draw a graph. Be sure to check as in Example 7 or Example 8.

47. $y = \frac{5}{2}x - 3$
48. $y = \frac{2}{5}x - 4$
49. $f(x) = -\frac{5}{2}x + 2$
50. $f(x) = -\frac{2}{5}x + 3$
51. $F(x) = 2x + 1$
52. $g(x) = 3x - 2$
53. $4x + y = 3$
54. $4x - y = 1$
55. $6y + x = 6$
56. $4y + 20 = x$
57. $g(x) = -0.25x$ *Aha!*
58. $F(x) = 1.5x$

59. $4x - 5y = 10$

60. $5x + 4y = 4$

61. $2x + 3y = 6$

62. $3x - 2y = 8$

63. $5 - y = 3x$

64. $3 + y = 2x$

Aha! **65.** $g(x) = 4.5$

66. $g(x) = \frac{1}{2}$

Applications

For each graph, find the rate of change. Remember to use appropriate units. See Example 9.

67.

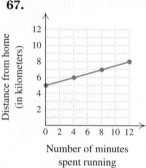

68.

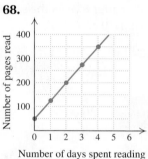

69.

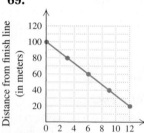

70.

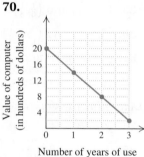

71.

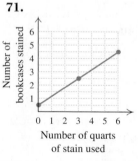

72.

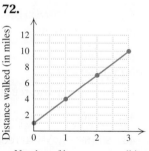

*73.

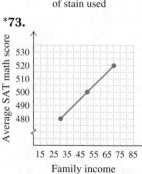

*74.

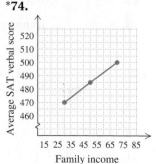

*Based on data from the College Board Online.

75. *Nursing.* Match each sentence with the most appropriate of the four graphs shown.

a) The rate at which fluids were given intravenously was doubled after 3 hr.

b) The rate at which fluids were given intravenously was gradually reduced to 0.

c) The rate at which fluids were given intravenously remained constant for 5 hr.

d) The rate at which fluids were given intravenously was gradually increased.

I

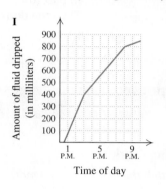

II

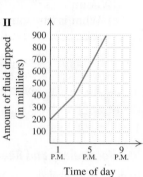

III

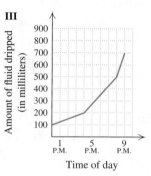

IV

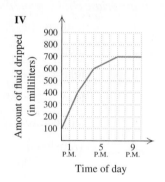

76. *Running Rate.* An ultra-marathoner passes the 15-mi point of a race after 2 hr and reaches the 22-mi point 56 min later. Assuming a constant rate, find the speed of the marathoner.

77. *Medicine.* The percentage of physicians who communicate electronically with patients has increased from 24% in 2005 to 39% in 2011. Determine the average rate of change of this percentage.

Source: Based on data from Manhattan Research

78. *Skiing Rate.*   A cross-country skier reaches the 3-km mark of a race in 15 min and the 12-km mark 45 min later. Assuming a constant rate, find the speed of the skier.

79. *Rate of Descent.*   A plane descends to sea level from 12,000 ft after being airborne for $1\frac{1}{2}$ hr. The entire flight time is 2 hr 10 min. Determine the average rate of descent of the plane.

80. *Work Rate.*   As a painter begins work, one-fourth of a house has already been painted. Eight hours later, the house is two-thirds done. Calculate the painter's work rate.

81. *Website Traffic.*   In early 2011, Starfarm.com had already received 80,000 hits at their website. By early 2013, that number had climbed to 430,000. Calculate the rate at which the number of hits is increasing.

82. *Market Research.*   Match each sentence with the most appropriate of the four graphs shown.

  a) After January 1, daily sales continued to rise, but at a slower rate.

  b) After January 1, sales decreased faster than they ever grew.

  c) The rate of growth in daily sales doubled after January 1.

  d) After January 1, daily sales decreased at half the rate that they grew in December.

In Exercises 83–92, each model is of the form $f(x) = mx + b$. *In each case, determine what m and b signify.*

83. *Cost of Renting a Truck.*   The cost, in dollars, of a one-day truck rental is given by $C(d) = 0.75d + 30$, where d is the number of miles driven.

84. *Weekly Pay.*   Each salesperson at Super Electronics is paid $P(x)$ dollars, where $P(x) = 0.05x + 200$ and x is the value of the salesperson's sales for the week.

85. *Hair Growth.*   After Lauren donated her hair to Locks of Love, the length $L(t)$ of her hair, in inches, was given by $L(t) = \frac{1}{2}t + 5$, where t is the number of months after she had the haircut.

86. *Trade Unions.*   The number of employed U.S. union members, in millions, can be estimated by $U(t) = \frac{9}{100}t + 17.5$, where t is the number of years after 1982.

Source: Based on data from unionstats.com

87. *Life Expectancy of American Women.*   The life expectancy of American women t years after 1970 is given by $A(t) = \frac{1}{7}t + 75.5$.

Source: Based on data from the National Center for Health Statistics

88. *Landscaping.*   After being cut, the length $G(t)$ of the lawn, in inches, at Harrington Community College is given by $G(t) = \frac{1}{8}t + 2$, where t is the number of days since the lawn was cut.

89. *Cost of a Sports Ticket.*   The average price $P(t)$, in dollars, of a major-league baseball ticket is given by $P(t) = 1.02t + 16.39$, where t is the number of years since 2000.

Source: Based on data from Team Marketing Report

90. *Cost of a Taxi Ride.*   The cost, in dollars, of a taxi ride in New York City is given by $C(d) = 2d + 2.5$,* where d is the number of miles traveled.

91. *Organic Food.*   The function given by $c(t) = 2.2t + 4.5$ can be used to estimate the U.S. sales of organic food, in billions of dollars, t years after 2000.

Source: Based on data from the Organic Trade Association

Rates are higher between 4 P.M. and 8 P.M. (Source: Based on data from New York City Taxi and Limousine Commission, 2012)

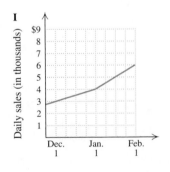

I

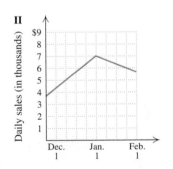

II

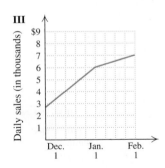

III

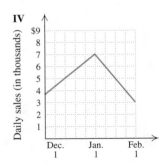

IV

92. *Catering.* When catering a party for x people, Chrissie's Catering uses the formula $C(x) = 25x + 75$, where $C(x)$ is the cost of the food, in dollars.

93. *Salvage Value.* Green Glass Recycling uses the function given by $F(t) = -5000t + 90,000$ to determine the salvage value $F(t)$, in dollars, of a waste removal truck t years after it has been put into use.

   a) What do the numbers -5000 and $90,000$ signify?

   b) How long will it take the truck to depreciate completely?

   c) What is the domain of F?

94. *Salvage Value.* Consolidated Shirt Works uses the function given by $V(t) = -2000t + 15,000$ to determine the salvage value $V(t)$, in dollars, of a color separator t years after it has been put into use.

   a) What do the numbers -2000 and $15,000$ signify?

   b) How long will it take the machine to depreciate completely?

   c) What is the domain of V?

95. *Trade-in Value.* The trade-in value of a Jamis Dakar mountain bike can be determined using the function given by $v(n) = -200n + 1800$. Here, $v(n)$ is the trade-in value, in dollars, after n years of use.

   a) What do the numbers -200 and 1800 signify?

   b) When will the trade-in value of the mountain bike be $600?

   c) What is the domain of v?

96. *Trade-in Value.* The trade-in value of a John Deere riding lawnmower can be determined using the function given by $T(x) = -300x + 2400$. Here, $T(x)$ is the trade-in value, in dollars, after x summers of use.

   a) What do the numbers -300 and 2400 signify?

   b) When will the value of the mower be $1200?

   c) What is the domain of T?

97. The number of meals served at the Knapp Memorial Soup Kitchen can be modeled by the function $K(t) = mt + 30$. Here, $K(t)$ is the average number of meals served each day t months after operations began. If you managed the soup kitchen, would you want the value of m to be positive or negative? Why?

98. Examine the function given in Exercise 97. What units of measure must be used for m? Why?

Skill Review

Perform the indicated operation. [1.2]

99. $-\frac{2}{3} - \frac{1}{2}$

100. $\left(\frac{4}{5}\right)\left(-\frac{10}{9}\right)$

101. $-12.9 \div (-3)$

102. $612 + (-1.7)$

Synthesis

103. The population of Valley Heights is decreasing at a rate of 10% per year. Can this be modeled using a linear function? Why or why not?

104. Janice claims that her firm's profits continue to go up, but the rate of increase is going down.

   a) Sketch a graph that might represent her firm's profits as a function of time.

   b) Explain why the graph can go up while the rate of increase goes down.

105. Match each sentence with the most appropriate of the four graphs shown.

   a) Ellie drove 2 mi to a lake, swam 1 mi, and then drove 3 mi to a store.

   b) During a preseason workout, Rico biked 2 mi, ran for 1 mi, and then walked 3 mi.

   c) Luis bicycled 2 mi to a park, hiked 1 mi over the notch, and then took a 3-mi bus ride back to the park.

   d) After hiking 2 mi, Marcy ran for 1 mi before catching a bus for the 3-mi ride into town.

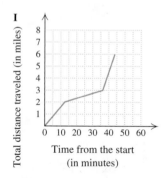

I

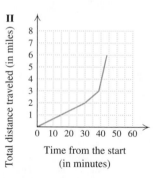

II

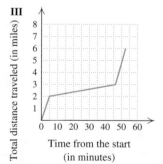

III

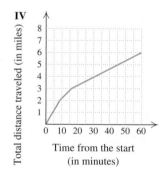

IV

The following graph shows the elevation of each section of a bicycle tour from Sienna to Florence in Italy. Use the graph for Exercises 106–110.

Source: greve-in-chianti.com

Bicycle route elevation

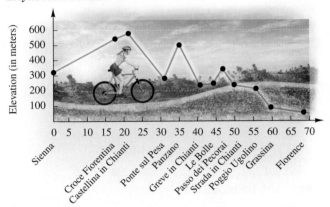

Distance from Sienna (in kilometers)

106. What part of the trip is the steepest?

107. What part of the trip has the longest uphill climb?

A road's grade *is the ratio, given as a percent, of the road's change in elevation to its change in horizontal distance and is, by convention, always positive.*

108. During one day's ride, Brittany rode uphill and then downhill at about the same grade. She then rode downhill at $\frac{1}{10}$ of the grade of the first two sections. Where did Brittany begin her ride?

109. During one day's ride, Scott biked two downhill sections and one uphill section, all at about the same grade. Where did Scott begin his ride?

110. Calculate the grade of the steepest section of the tour.

In Exercises 111 and 112, assume that r, p, and s are constants and that x and y are variables. Determine the slope and the y-intercept.

111. $rx + py = s - ry$

112. $rx + py = s$

113. Let (x_1, y_1) and (x_2, y_2) be two distinct points on the graph of $y = mx + b$. Use the fact that both pairs are solutions of the equation to prove that m is the slope of the line given by $y = mx + b$. (*Hint*: Use the slope formula.)

Given that $f(x) = mx + b$, classify each of the following as either true or false.

114. $f(cd) = f(c)f(d)$

115. $f(c + d) = f(c) + f(d)$

116. $f(c - d) = f(c) - f(d)$

117. $f(kx) = kf(x)$

118. Find k such that the line containing $(-3, k)$ and $(4, 8)$ is parallel to the line containing $(5, 3)$ and $(1, -6)$.

119. Find the slope of the line that contains the given pair of points.
   a) $(5b, -6c), (b, -c)$
   b) $(b, d), (b, d + e)$
   c) $(c + f, a + d), (c - f, -a - d)$

120. *Cost of a Speeding Ticket.* The penalty schedule shown below is used to determine the cost of a speeding ticket in certain states. Use this schedule to graph the cost of a speeding ticket as a function of the number of miles per hour over the limit that a driver is going.

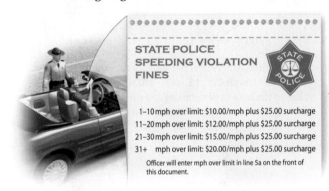

STATE POLICE
SPEEDING VIOLATION
FINES

1–10 mph over limit: $10.00/mph plus $25.00 surcharge
11–20 mph over limit: $12.00/mph plus $25.00 surcharge
21–30 mph over limit: $15.00/mph plus $25.00 surcharge
31+ mph over limit: $20.00/mph plus $25.00 surcharge

Officer will enter mph over limit in line 5a on the front of this document.

121. Graph the equations
$$y_1 = 1.4x + 2, \qquad y_2 = 0.6x + 2,$$
$$y_3 = 1.4x + 5, \quad \text{and} \quad y_4 = 0.6x + 5$$

using a graphing calculator. If possible, use the SIMULTANEOUS mode so that you cannot tell which equation is being graphed first. Then decide which line corresponds to each equation.

122. A student makes a mistake when using a graphing calculator to draw $4x + 5y = 12$ and the following screen appears. Use algebra to show that a mistake has been made. What do you think the mistake was?

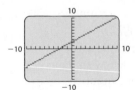

123. A student makes a mistake when using a graphing calculator to draw $5x - 2y = 3$ and the following screen appears. Use algebra to show that a mistake has been made. What do you think the mistake was?

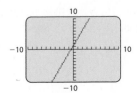

⤵ **YOUR TURN ANSWERS: SECTION 2.3**

1. 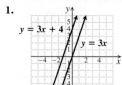 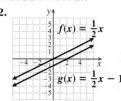   **2.**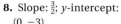

3. $(0, -16)$ **4.** Both slopes are $\frac{1}{3}$.   **5.** Slope: 6;
y-intercept: $(0, -7)$ **6.** $f(x) = -\frac{1}{2}x + 8$
7. Slope: $-\frac{2}{3}$; y-intercept: **8.** Slope: $\frac{3}{2}$; y-intercept:
     $(0, 1)$ $(0, -3)$

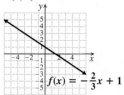 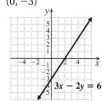

9. 1.2 million Americans per year   **10.** 429 groups per
year   **11.** Graph C   **12.** (a) -300 signifies that the
value of the bike is decreasing at a rate of $300 per year;
1200 signifies the original cost of the road bike, in dollars;
(b) 4 years; (c) $\{t \mid 0 \le t \le 4\}$

QUICK QUIZ: SECTIONS 2.1–2.3
Graph.

1. $y = 5x$ [2.1]   **2.** $f(x) = \frac{1}{2}x - 1$ [2.3]

3. Find the slope of the line containing the points $(3, 5)$
and $(4, 6)$.   [2.3]

*Determine the domain and the range of each
function.*   [2.2]

4. 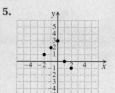   **5.**

PREPARE TO MOVE ON
Simplify.   [1.2]

1. $\dfrac{-8 - (-8)}{6 - (-6)}$ **2.** $\dfrac{-2 - 2}{-3 - (-3)}$

Solve.   [1.3]

3. $3 \cdot 0 - 2y = 9$ **4.** $4x - 7 \cdot 0 = 3$

5. If $f(x) = 2x - 7$, find $f(0)$.   [2.2]

6. If $f(x) = 2x - 7$, find any x-values for which $f(x) = 0$.
[2.2]

2.4   Another Look at Linear Graphs

Graphing Horizontal Lines and Vertical Lines   ▪   Parallel Lines and Perpendicular Lines   ▪   Graphing Using Intercepts   ▪
Solving Equations Graphically   ▪   Recognizing Linear Equations

Study Skills

Form a Study Group
Consider forming a study group
with some of your fellow students.
Exchange telephone numbers,
schedules, and e-mail addresses so
that you can coordinate study time
for homework and tests.

In this section, we graph lines that have a slope of 0 or that have an undefined
slope. We also learn to recognize whether the graphs of two linear equations are
parallel or perpendicular, as well as how to graph lines using x- and y-intercepts.

Graphing Horizontal Lines and Vertical Lines

To find the slope of a line, we use two points on
the line. For horizontal lines, those two points
have the same y-coordinate, and we can label
them (x_1, y_1) and (x_2, y_1). This gives us

$$m = \frac{y_1 - y_1}{x_2 - x_1} = \frac{0}{x_2 - x_1} = 0.$$

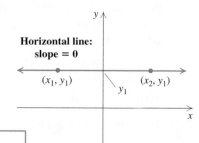

The slope of any horizontal line is 0.

EXAMPLE 1   Use slope–intercept form to graph $f(x) = 3$.

SOLUTION   A function of this type is called a *constant function*. Writing $f(x)$ in slope–intercept form,

$$f(x) = 0 \cdot x + 3,$$

we see that the y-intercept is $(0, 3)$ and the slope is 0. Thus we can graph f by plotting $(0, 3)$ and, from there, counting off a slope of 0. Because $0 = 0/2$ (any nonzero number could be used in place of 2), we can draw the graph by going up 0 units and to the right 2 units. As a check, we also find some ordered pairs. Note that for any choice of x-value, $f(x)$ must be 3.

x	$f(x)$
-1	3
0	3
2	3

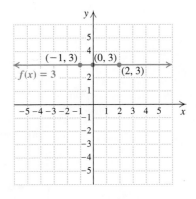

1. Graph: $f(x) = -1$.

YOUR TURN

HORIZONTAL LINES

The slope of any horizontal line is 0.

The graph of any function of the form $f(x) = b$ or $y = b$ is a horizontal line that crosses the y-axis at $(0, b)$.

Suppose two different points are on a vertical line. They then have the same first coordinate. In this case, when we calculate the slope, we have

$$m = \frac{y_2 - y_1}{x_1 - x_1} = \frac{y_2 - y_1}{0}.$$

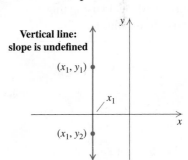

Since we cannot divide by 0, this slope is undefined. Note that when we say that $(y_2 - y_1)/0$ is undefined, it means that we have agreed to not attach any meaning to that expression.

The slope of any vertical line is undefined.

EXAMPLE 2   Graph: $x = -2$.

SOLUTION   The equation tells us that x is always -2. This will be true for any choice of y. The only way that an ordered pair can make $x = -2$ true is for the x-coordinate of that pair to be -2. Thus the pairs $(-2, 3)$, $(-2, 0)$, and $(-2, -4)$ all satisfy the equation. The graph is a line parallel to the y-axis. Note that this equation cannot be written in slope–intercept form, since it cannot be solved for y.

x	y
-2	3
-2	0
-2	-4

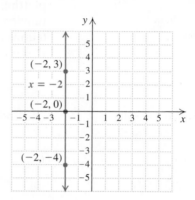

2. Graph: $x = 4$.

YOUR TURN

VERTICAL LINES

The slope of any vertical line is undefined.

The graph of any equation of the form $x = a$ is a vertical line that crosses the x-axis at $(a, 0)$.

EXAMPLE 3   Find the slope of each line. If the slope is undefined, state this.

a) $3y + 2 = 14$ **b)** $2x = 10$

SOLUTION

a) We solve for y:

$$3y + 2 = 14$$
$$3y = 12 \qquad \text{Subtracting 2 from both sides}$$
$$y = 4. \qquad \text{Dividing both sides by 3}$$

The graph of $y = 4$ is a horizontal line. Since $3y + 2 = 14$ is equivalent to $y = 4$, the slope of the line $3y + 2 = 14$ is 0.

b) When y does not appear, we solve for x:

$$2x = 10$$
$$x = 5. \qquad \text{Dividing both sides by 2}$$

The graph of $x = 5$ is a vertical line. Since $2x = 10$ is equivalent to $x = 5$, the slope of the line $2x = 10$ is undefined.

YOUR TURN

Student Notes

The slope of any horizontal line is 0, and the slope of any vertical line is undefined. Avoid using the ambiguous phrase "no slope."

3. Find the slope of $2y = y + 1$. If the slope is undefined, state this.

Parallel Lines and Perpendicular Lines

Two lines are parallel if they lie in the same plane and do not intersect no matter how far they are extended. If two lines are vertical, they are parallel. How can we tell if nonvertical lines are parallel? The answer is simple: We look at their slopes.

SLOPE AND PARALLEL LINES

Two lines are parallel if they have the same slope or if both lines are vertical.

EXAMPLE 4 Determine whether the line given by $f(x) = -3x + 4.2$ is parallel to the line given by $6x + 2y = 1$.

SOLUTION If the slopes of the lines are the same, the lines are parallel.
The slope of $f(x) = -3x + 4.2$ is -3.
To find the slope of $6x + 2y = 1$, we write the equation in slope–intercept form:

$$6x + 2y = 1$$
$$2y = -6x + 1 \qquad \text{Subtracting } 6x \text{ from both sides}$$
$$y = -3x + \tfrac{1}{2}. \qquad \text{Dividing both sides by 2}$$

4. Determine whether the line given by $8x + y = 2$ is parallel to the line given by $f(x) = 8x + 7$.

The slope of the second line is -3. Since the slopes are equal, the lines are parallel.

YOUR TURN

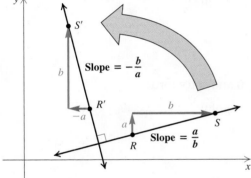

Two lines are perpendicular if they intersect at a right angle. If one line is vertical and another is horizontal, they are perpendicular. There are other instances in which two lines are perpendicular.

Consider a line $\overleftrightarrow{RS}$ as shown at left, with slope a/b. Then think of rotating the figure 90° to get a line $\overleftrightarrow{R'S'}$ perpendicular to $\overleftrightarrow{RS}$. For the new line, the rise and the run are interchanged, but the run is now negative. Thus the slope of the new line is $-b/a$. Let's multiply the slopes:

$$\frac{a}{b}\left(-\frac{b}{a}\right) = -1.$$

This can help us determine which lines are perpendicular.

SLOPE AND PERPENDICULAR LINES

Two lines are perpendicular if the product of their slopes is -1 or if one line is vertical and the other line is horizontal.

Thus, if one line has slope m ($m \neq 0$), the slope of any line perpendicular to it is $-1/m$. That is, we take the reciprocal of m ($m \neq 0$) and change the sign.

EXAMPLE 5 Determine whether the graphs of $2x + y = 8$ and $y = \tfrac{1}{2}x + 7$ are perpendicular.

SOLUTION The second equation is given in slope–intercept form:

$$y = \tfrac{1}{2}x + 7. \qquad \text{The slope is } \tfrac{1}{2}.$$

To find the slope of the other line, we solve for y:

$$2x + y = 8$$
$$y = -2x + 8. \qquad \text{Adding } -2x \text{ to both sides}$$
$$\text{The slope is } -2.$$

The lines are perpendicular if the product of their slopes is -1. Since

$$\tfrac{1}{2}(-2) = -1,$$

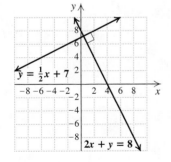

5. Determine whether the graphs of $x + y = 3$ and $x - y = 8$ are perpendicular.

the graphs are perpendicular. The graphs of both equations are shown at left, and they do appear to be perpendicular.

YOUR TURN

Technology Connection

To see if the graphs of two lines might be perpendicular, we use the ZSQUARE option of the ZOOM menu to create a "squared" window. This corrects for distortion that results from the default units on the axes being of different lengths.

1. Show that the graphs of

$$y = \tfrac{3}{4}x + 2$$

and

$$y = -\tfrac{4}{3}x - 1$$

appear to be perpendicular.

2. Show that the graphs of

$$y = -\tfrac{2}{5}x - 4$$

and

$$y = \tfrac{5}{2}x + 3$$

appear to be perpendicular.

3. To see that this type of check is not foolproof, graph

$$y = \tfrac{31}{40}x + 2$$

and

$$y = -\tfrac{40}{30}x - 1.$$

Are the lines perpendicular? Why or why not?

Graphing Using Intercepts

Any line that is neither horizontal nor vertical crosses both the x- and y-axes. We have already seen that the point at which a line crosses the y-axis is called the *y-intercept*. Similarly, the point at which a line crosses the x-axis is called the *x-intercept*.

> **TO DETERMINE INTERCEPTS**
>
> The x-intercept is $(a, 0)$. To find a, let $y = 0$ and solve for x.
> The y-intercept is $(0, b)$. To find b, let $x = 0$ and solve for y.

When the x- and y- intercepts are not both $(0, 0)$, the intercepts can be used to draw the graph of a line.

EXAMPLE 6 Graph $3x + 2y = 12$ by using intercepts.

SOLUTION To find the y-intercept, we let $x = 0$ and solve for y:

Let $x = 0$.	$3 \cdot 0 + 2y = 12$	For points on the y-axis, $x = 0$.
	$2y = 12$	
Solve for y.	$y = 6.$	

The y-intercept is $(0, 6)$.
To find the x-intercept, we let $y = 0$ and solve for x:

Let $y = 0$.	$3x + 2 \cdot 0 = 12$	For points on the x-axis, $y = 0$.
	$3x = 12$	
Solve for x.	$x = 4.$	

The x-intercept is $(4, 0)$.
We plot the two intercepts and draw the line. A third point could be calculated and used as a check.

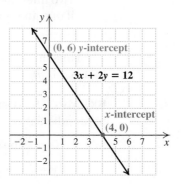

6. Graph $x - 5y = 5$ by using intercepts.

 YOUR TURN

7. Graph $f(x) = -\frac{1}{3}x + 1$ by using intercepts.

EXAMPLE 7   Graph $f(x) = 2x + 5$ by using intercepts.

SOLUTION   Because the function is in slope–intercept form, we know that the y-intercept is $(0, 5)$. To find the x-intercept, we replace $f(x)$ with 0 and solve for x:

$$0 = 2x + 5$$
$$-5 = 2x$$
$$-\frac{5}{2} = x.$$

The x-intercept is $\left(-\frac{5}{2}, 0\right)$.

We plot $(0, 5)$ and $\left(-\frac{5}{2}, 0\right)$ and draw the line. To check, we calculate the slope:

$$m = \frac{5 - 0}{0 - \left(-\frac{5}{2}\right)}$$

$$= \frac{5}{\frac{5}{2}}$$

$$= 5 \cdot \frac{2}{5}$$

$$= 2.$$

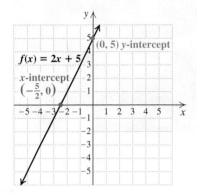

The slope is 2, as expected.

🔁 YOUR TURN

Solving Equations Graphically

In Example 7, $-\frac{5}{2}$ is the x-coordinate of the point at which the graphs of $f(x) = 2x + 5$ and $h(x) = 0$ intersect as well as the solution of $2x + 5 = 0$. Similarly, we can solve $2x + 5 = -3$ by finding the x-coordinate of the point at which the graphs of $f(x) = 2x + 5$ and $g(x) = -3$ intersect. From the graph shown at left, it appears that -4 is that x-value. To check, note that $f(-4) = 2(-4) + 5 = -3$.

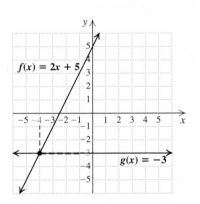

EXAMPLE 8   Solve graphically: $\frac{1}{2}x + 3 = 2$.

SOLUTION   To find the x-value for which $\frac{1}{2}x + 3$ will equal 2, we graph $f(x) = \frac{1}{2}x + 3$ and $g(x) = 2$ on the same set of axes. Since the intersection appears to be $(-2, 2)$, the solution is apparently -2.

Student Notes

Remember that it is only the first coordinate of the point of intersection that is the solution of the equation.

Check:

$$\begin{array}{c|c} \frac{1}{2}x + 3 = 2 & \\ \hline \frac{1}{2}(-2) + 3 & 2 \\ -1 + 3 & \\ 2 \overset{?}{=} 2 & \text{TRUE} \end{array}$$

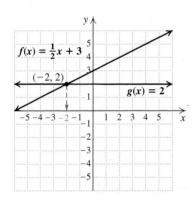

8. Solve graphically:

$$\frac{3}{2}x - 1 = 2.$$

The solution is -2.

🔁 YOUR TURN

EXAMPLE 9   *iPad Costs.*   In 2012, a 16-GB (gigabyte) Apple iPad cost $630. AT&T offered a plan allowing up to 250 MB (megabytes) of data for $15 per month. Write and graph a mathematical model for the total cost of an iPad 2 purchased in 2012 and put into use with a 250-MB data plan. Then use the model to estimate the number of months required for the total cost to reach $750.

Source: www.apple.com

SOLUTION

1. **Familiarize.** For this plan, a monthly fee is charged once the initial purchase is made. After 1 month of service, the total cost is $630 + $15 = $645. After 2 months, the total cost is $630 + $15 · 2 = $660. We can write a general model if we let $C(t)$ represent the total cost, in dollars, for t months of service.

2. **Translate.** We reword and translate as follows:

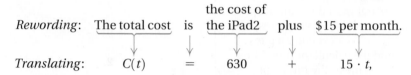

Rewording: The total cost is the cost of the iPad2 plus $15 per month.

Translating: $C(t)$ =   630   + $15 \cdot t$,

   with $t \geq 0$ (since there cannot be a negative number of months).

3. **Carry out.** Before graphing, we rewrite the model in slope–intercept form: $C(t) = 15t + 630$. We see that the vertical intercept is $(0, 630)$ and the slope, or rate, is $15 per month. Since we want to estimate the time required for the total cost to reach $750, we choose a scale for the vertical axis that includes 750.

   We plot $(0, 630)$ and, from there, count up $15 and to the right 1 month. This takes us to $(1, 645)$. We then draw a line passing through both points.

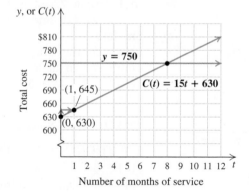

y, or C(t)

$810
780
750   y = 750
720
690   (1, 645)   C(t) = 15t + 630
660
630   (0, 630)
600

Total cost

1 2 3 4 5 6 7 8 9 10 11 12 t
Number of months of service

> **CAUTION!** When you are using a graph to solve an equation, it is important to use graph paper and to work as neatly as possible.

To estimate the time required for the total cost to reach $750, we are estimating the solution of

   $750 = 15t + 630.$ Replacing $C(t)$ with 750

We do so by graphing $y = 750$ and looking for the point of intersection. This appears to be $(8, 750)$. Thus we estimate that it takes 8 months for the total cost to reach $750.

4. **Check.** We evaluate:

   $C(8) = 15 \cdot 8 + 630 = 120 + 630 = 750.$

   Our estimate turns out to be precise.

5. **State.** It takes 8 months for the total cost to reach $750.

9. Use the model in Example 9 to estimate the number of months required for the total cost of the iPad to reach $690.

 YOUR TURN

Technology Connection

To solve $-\frac{3}{4}x + 6 = 2x - 1$, we can graph each side of the equation and then select the INTERSECT option of the CALC menu. Once this is done, we locate the cursor on each line and press **ENTER**. Finally, we enter a guess, and the calculator determines the coordinates of the intersection.

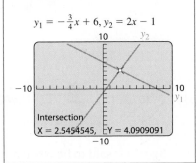

$y_1 = -\frac{3}{4}x + 6, \ y_2 = 2x - 1$

Intersection
X = 2.5454545,   Y = 4.0909091

Chapter Resource:
Visualizing for Success, p. 139

Algebraic & Graphical Connection

There are limitations to solving equations graphically. For example, on the left below we attempt to solve $-\frac{3}{4}x + 6 = 2x - 1$ graphically. It *appears* that the lines intersect at $(2.5, 4)$.

As the algebraic solution on the right indicates, however, the exact solution is $\frac{28}{11}$. This solution can be found graphically using a graphing calculator. (See the Technology Connection at left.)

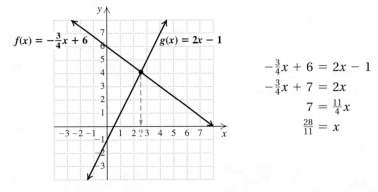

$$-\tfrac{3}{4}x + 6 = 2x - 1$$
$$-\tfrac{3}{4}x + 7 = 2x$$
$$7 = \tfrac{11}{4}x$$
$$\tfrac{28}{11} = x$$

Recognizing Linear Equations

One way to determine whether an equation is linear is to write it in the form $Ax + By = C$. We can show that every equation of this form is linear, so long as A and B are not both zero.

1. Suppose that $A = 0$. The equation becomes $By = C$, or $y = C/B$, which is the equation of a horizontal line.

2. Suppose that $B = 0$. The equation becomes $Ax = C$, or $x = C/A$, which is the equation of a vertical line.

3. Suppose that $A \neq 0$ and $B \neq 0$. Then if we solve for y, we have

$$Ax + By = C$$
$$By = -Ax + C$$
$$y = -\frac{A}{B}x + \frac{C}{B}.$$

This is the equation of a line with slope $-A/B$ and y-intercept $(0, C/B)$. We have now justified the following result.

STANDARD FORM OF A LINEAR EQUATION

Any equation of the form $Ax + By = C$, where A, B, and C are real numbers and A and B are not both 0, is a linear equation in *standard form* and has a graph that is a straight line.

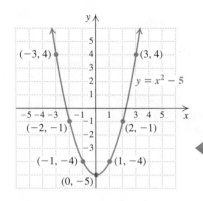

EXAMPLE 10 Determine whether the equation $y = x^2 - 5$ is linear.

SOLUTION We attempt to put the equation in standard form:

$$y = x^2 - 5$$
$$-x^2 + y = -5. \quad \text{Adding } -x^2 \text{ to both sides}$$

This last equation is not linear because it has an x^2-term. We can also see from the graph at left that $y = x^2 - 5$ is not linear.

YOUR TURN

Only linear equations have graphs that are straight lines. Also, only linear graphs have a constant slope. Were you to try to calculate the slope between several pairs of points in Example 10, you would find that the slopes vary.

10. Determine whether the equation $y = \dfrac{1}{x}$ is linear.

2.4 EXERCISE SET

FOR EXTRA HELP

MyMathLab®
PRACTICE   WATCH   READ   REVIEW

Vocabulary and Reading Check

Choose the word, number, or variable from the following list that best completes each statement. Choices may be used more than once or not at all.

0	undefined
horizontal	vertical
intersection	x
linear	x-axis
slope	y
standard	y-axis

1. Every _____ line has a slope of 0.

2. The graph of any function of the form $f(x) = b$ is a horizontal line that crosses the _____ at $(0, b)$.

3. The slope of a vertical line is _____.

4. The graph of any equation of the form $x = a$ is a(n) _____ line that crosses the x-axis at $(a, 0)$.

5. To find the x-intercept, we let $y =$ _____ and solve the original equation for _____.

6. To find the y-intercept, we let $x =$ _____ and solve the original equation for _____.

7. To solve $3x - 5 = 7$, we can graph $f(x) = 3x - 5$ and $g(x) = 7$ and find the x-value at the point of _____.

8. An equation like $4x + 3y = 8$ is said to be written in _____ form.

9. Only _____ equations have graphs that are straight lines.

10. Linear graphs have a constant _____.

Graphing Horizontal Lines and Vertical Lines

For each equation, find the slope. If the slope is undefined, state this.

11. $y - 2 = 6$ 12. $x + 3 = 11$

13. $8x = 6$ 14. $y - 3 = 5$

15. $3y = 28$ 16. $19 = -6y$

17. $5 - x = 12$ 18. $-5x = 13$

19. $2x - 4 = 3$ 20. $3 - 2y = 16$

21. $5y - 4 = 35$ 22. $2x - 17 = 3$

23. $4y - 3x = 9 - 3x$ 24. $x - 4y = 12 - 4y$

25. $5x - 2 = 2x - 7$ 26. $5y + 3 = y + 9$

Aha! 27. $y = -\frac{2}{3}x + 5$ 28. $y = -\frac{3}{2}x + 4$

Graph.

29. $y = 4$ 30. $x = -1$

31. $x = 3$ 32. $y = 2$

33. $f(x) = -2$ 34. $g(x) = -3$

35. $3x = -15$

36. $2x = 10$

37. $3 \cdot g(x) = 15$

38. $3 - f(x) = 2$

Parallel Lines and Perpendicular Lines

Without graphing, tell whether the graphs of each pair of equations are parallel.

39. $x + 2 = y,$
$y - x = -2$

40. $2x - 1 = y,$
$2y - 4x = 7$

41. $y + 9 = 3x,$
$3x - y = -2$

42. $y + 8 = -6x,$
$-2x + y = 5$

43. $f(x) = 3x + 9,$
$2y = 8x - 2$

44. $f(x) = -7x - 9,$
$-3y = 21x + 7$

Without graphing, tell whether the graphs of each pair of equations are perpendicular.

45. $x - 2y = 3,$
$4x + 2y = 1$

46. $2x - 5y = -3,$
$2x + 5y = 4$

47. $f(x) = 3x + 1,$
$6x + 2y = 5$

48. $y = -x + 7,$
$f(x) = x + 3$

Graphing Using Intercepts

Find the intercepts. Then graph by using the intercepts, if possible, and a third point as a check.

49. $x + y = 4$

50. $x + y = 5$

51. $f(x) = 2x - 6$

52. $f(x) = 3x + 12$

53. $3x + 5y = -15$

54. $5x - 4y = 20$

55. $2x - 3y = 18$

56. $3x + 2y = -18$

57. $3y = -12x$

58. $5y = 15x$

59. $f(x) = 3x - 7$

60. $g(x) = 2x - 9$

61. $5y - x = 5$

62. $y - 3x = 3$

63. $0.2y - 1.1x = 6.6$

64. $\frac{1}{3}x + \frac{1}{2}y = 1$

Solving Equations Graphically

Solve each equation graphically. Then check your answer by solving the same equation algebraically.

65. $x + 2 = 3$

66. $x - 1 = 2$

67. $2x + 5 = 1$

68. $3x + 7 = 4$

69. $\frac{1}{2}x + 3 = 5$

70. $\frac{1}{3}x - 2 = 1$

71. $x - 8 = 3x - 5$

72. $x + 3 = 5 - x$

73. $4x + 1 = -x + 11$

74. $x + 4 = 3x + 5$

Use a graph to estimate the solution in each of the following. Be sure to use graph paper and a straightedge.

75. *Fitness Centers.* Becoming a member at Keeping Fit Club costs $75 plus a monthly fee of $35. Estimate how many months Kerry has been a member if he has paid a total of $215.

76. *Seminars.* Efficiency Experts charges a $250 booking fee plus $150 per person for a one-day seminar. Estimate how many people attended a seminar if the charges totaled $1600.

77. *Painting.* To paint interior walls, Gavin charges 70¢ per square foot plus the cost of the paint. For a recent job, the paint cost $150 and the total bill was $710. Estimate the number of square feet that Gavin painted.

78. *Printing.* For large banners, Vistaprint charges $8 in upload and setup fees and $18.75 per banner in printing fees. Estimate the number of banners that can be printed for $120.50.

Source: vistaprint.com

79. *Healthcare.* Under one particular university's health-insurance plan, an employee pays the first $3500 of surgery expenses plus $\frac{1}{5}$ of all charges in excess of $3500. By approximately how much did Nancy's hospital bill exceed $3500 if a surgery cost her $6800?

Source: Based on the Ball State University Health Care Plan

80. *Cost of a FedEx Delivery.* In 2012, for 2nd day Zone 2 delivery of packages weighing from 10 to 50 lb, FedEx charged $8 plus 80¢ for each pound in excess of 10 lb. Estimate the weight of a package that cost $28 to ship.

Source: FedEx Service Guide

81. *Parking Fees.*   Cal's Parking charges $5.00 to park plus 50¢ for each 15-min unit of time. Estimate how long someone can park for $9.50.*

82. *Cost of a Road Call.*   Kay's Auto Village charges $50 for a road call plus $15 for each 15-min unit of time. Estimate the time required for a road call that cost $140.*

Recognizing Linear Equations

Determine whether each equation is linear. Find the slope of any nonvertical lines.

83. $5x - 3y = 15$

84. $3x + 5y + 15 = 0$

85. $8x + 40 = 0$

86. $2y - 30 = 0$

87. $4g(x) = 6x^2$

88. $2x + 4f(x) = 8$

89. $3y = 7(2x - 4)$

90. $y(3 - x) = 2$

91. $f(x) - \dfrac{5}{x} = 0$

92. $g(x) - x^3 = 0$

93. $\dfrac{y}{3} = x$

94. $\frac{1}{2}(x - 4) = y$

95. $xy = 10$

96. $y = \dfrac{10}{x}$

97. Explain why vertical lines are mentioned separately in the discussion of slope and parallel lines.

98. If line l has a positive slope, what is the sign of the slope of a line perpendicular to l? Explain your reasoning.

Skill Review

Solve. If appropriate, classify the equation as a contradiction or as an identity.   [1.3]

99. $2(x - 7) = 3 - x$

100. $\frac{1}{3}t - \frac{1}{4} = \frac{1}{2}t$

101. $n - (8 - n) = 2(n - 4)$

102. $9 - 6(x - 7) = 12$

103. $4(t - 6) = 7t - 15 - 3t$

104. $10 - x = 5(x + 2)$

Synthesis

105. Jim tries to avoid using fractions as often as possible. Under what conditions will graphing $Ax + By = C$ using intercepts allow him to avoid fractions? Why?

106. Under what condition(s) will the x- and y-intercepts of a line coincide? What would the equation for such a line look like?

107. Write an equation, in standard form, for the line whose x-intercept is 5 and whose y-intercept is -4.

108. Find the x-intercept of $y = mx + b$, assuming that $m \neq 0$.

In Exercises 109–112, assume that r, p, and s are nonzero constants and that x and y are variables. Determine whether each equation is linear.

109. $rx + 3y = p^2 - s$

110. $py = sx - r^2y - 9$

111. $r^2x = py + 5$

112. $\dfrac{x}{r} - py = 17$

113. Suppose that two linear equations have the same y-intercept but that equation A has an x-intercept that is half the x-intercept of equation B. How do the slopes compare?

Consider the linear equation
$$ax + 3y = 5x - by + 8.$$

114. Find a and b if the graph is horizontal and passes through $(0, 4)$.

115. Find a and b if the graph is vertical and passes through $(4, 0)$.

*More precise, nonlinear models for Exercises 81 and 82 appear in Exercises 117 and 116, respectively.

116. (Refer to Exercise 82.) A 32-min road call with Kay's costs the same as a 44-min road call. Thus a linear graph drawn for the solution of Exercise 82 is not a precise representation of the situation. Draw a graph with a series of "steps" that more accurately reflects the situation.

117. (Refer to Exercise 81.) It costs as much to park at Cal's for 16 min as it does for 29 min. Thus a linear graph drawn for the solution of Exercise 81 is not a precise representation of the situation. Draw a graph with a series of "steps" that more accurately reflects the situation.

Solve graphically and then check by solving algebraically.

118. $5x + 3 = 7 - 2x$

119. $4x - 1 = 3 - 2x$

120. $3x - 2 = 5x - 9$

121. $8 - 7x = -2x - 5$

Solve using a graphing calculator.

122. Weekly pay at The Furniture Gallery is $450 plus a 3.5% sales commission. If a salesperson's pay was $601.03, what did that salesperson's sales total?

123. It costs Gert's Shirts $38 plus $4.25 per shirt to print tee shirts for a day camp. Camp Weehawken paid Gert's $671.25 for shirts. How many shirts were printed?

YOUR TURN ANSWERS: SECTION 2.4

1. 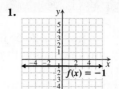 $f(x) = -1$

2. 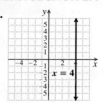 $x = 4$

3. 0   **4.** No   **5.** Yes

6. 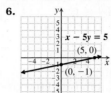 $x - 5y = 5$, $(5, 0)$, $(0, -1)$

7. 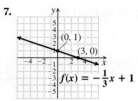 $(0, 1)$, $(3, 0)$, $f(x) = -\frac{1}{3}x + 1$

8. 2   **9.** 4 months   **10.** No

QUICK QUIZ: SECTIONS 2.1–2.4

Determine the slope of each line. If the slope is undefined, state this.

1. $2x - 4y = 5$ [2.3]   **2.** $5x = 7$ [2.4]

Graph.

3. $f(x) = 5$ [2.4]   **4.** $y = x^2$ [2.1]

5. $f(x) = 3 - x$ [2.3]

PREPARE TO MOVE ON

Simplify.

1. $-\frac{3}{10}\left(\frac{10}{3}\right)$ [1.2]   **2.** $2\left(-\frac{1}{2}\right)$ [1.2]

3. $-10[x - (-7)]$ [1.3]

4. $\frac{2}{3}\left[x - \left(-\frac{1}{2}\right)\right] - 1$ [1.3]   **5.** $-\frac{3}{2}\left(x - \frac{2}{5}\right) - 3$ [1.3]

Mid-Chapter Review

We can graph a line if we know any two points on the line. If we know an equation of a line, we can find two points by choosing values for one variable and calculating the corresponding values of the other variable. Depending on the form of the equation, it may be easiest to plot two intercepts and draw the line, or to plot the y-intercept and use the slope to find another point. The following is important to know.

- Slope–intercept form of a line: $y = mx + b$
- Standard form of a line: $Ax + By = C$
- Horizontal line: $y = b$
- Vertical line: $x = a$

GUIDED SOLUTIONS

1. Find the y-intercept and the x-intercept of the graph of $y - 3x = 6$.   [2.4]

Solution

y-intercept: $y - 3 \cdot \boxed{} = 6$

$y = \boxed{}$

The y-intercept is $(\boxed{}, \boxed{})$.

x-intercept: $\boxed{} - 3x = 6$

$-3x = 6$

$x = \boxed{}$

The x-intercept is $(\boxed{}, \boxed{})$.

2. Find the slope of the line containing the points $(1, 5)$ and $(3, -1)$.   [2.3]

Solution

$m = \dfrac{y_2 - y_1}{x_2 - x_1} = \dfrac{-1 - \boxed{}}{3 - \boxed{}}$

$= \dfrac{\boxed{}}{2}$

$= \boxed{}$

MIXED REVIEW

3. In which quadrant or on which axis is $\left(-16, \frac{1}{2}\right)$ located?   [2.1]

4. Find $f(6)$ for $f(x) = x - x^2$.   [2.2]

5. Find the domain of the function given by
$$g(x) = \frac{2x}{x - 7}.   [2.2]$$

Find the slope of the line containing the given pair of points. If the slope is undefined, state this.   [2.3]

6. $(-5, -2)$ and $(1, 8)$

7. $(0, 0)$ and $(0, -2)$

8. What is the slope of the line $y = 4$?   [2.4]

9. What is the slope of the line $x = -7$?   [2.4]

10. Determine the slope and the y-intercept of the line given by $x - 3y = 1$.   [2.3]

11. Find a linear function whose graph has slope -3 and y-intercept $(0, 7)$.   [2.3]

12. Tell whether the graphs of the equations are parallel, perpendicular, or neither.   [2.4]
$$f(x) = \tfrac{1}{4}x - 3,$$
$$4x + y = 8$$

13. Solve graphically: $\frac{1}{3}x - 2 = 2x + 3$.   [2.4]

14. Determine whether $\frac{1}{3}x = 6 - 5y$ is linear.   [2.4]

Graph.

15. $y = 2x - 1$ [2.3]

16. $3x + y = 6$ [2.4]

17. $y = |x| - 4$ [2.1]

18. $f(x) = 4$ [2.4]

19. $f(x) = -\frac{3}{4}x + 5$ [2.3]

20. $3x = 12$ [2.4]

2.5   Equations of Lines and Modeling

Point–Slope Form ■ Finding the Equation of a Line ■ Interpolation and Extrapolation ■ Linear Functions and Models

If we know the slope of a line and a point through which the line passes, then we can draw the line. With this information, we can also write an *equation* of the line.

Point–Slope Form

Suppose that a line of slope m passes through the point (x_1, y_1). For any other point (x, y) to lie on this line, we must have

$$\frac{y - y_1}{x - x_1} = m.$$

Note that if (x_1, y_1) itself replaces (x, y), the denominator is 0. To address this concern, we multiply both sides by $x - x_1$:

$$(x - x_1)\frac{y - y_1}{x - x_1} = m(x - x_1)$$

$$y - y_1 = m(x - x_1). \qquad \text{This equation } is \text{ true for } (x, y) = (x_1, y_1).$$

Every point on the line is a solution of this equation. This is the **point–slope form** of a linear equation.

> **POINT–SLOPE FORM**
>
> Any equation of the form $y - y_1 = m(x - x_1)$ is said to be written in *point–slope* form and has a graph that is a straight line.
>
> The slope of the line is m.
>
> The line passes through (x_1, y_1).

EXAMPLE 1   Graph: $(y + 4) = -\frac{1}{2}(x - 3)$.

SOLUTION   We first write the equation in point–slope form:

$$y - y_1 = m(x - x_1)$$
$$y - (-4) = -\tfrac{1}{2}(x - 3). \qquad y + 4 = y - (-4)$$

From the equation, we see that $m = -\frac{1}{2}$, $x_1 = 3$, and $y_1 = -4$. We plot $(3, -4)$, count off a slope of $-\frac{1}{2}$, and draw the line.

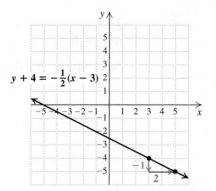

1. Graph: $y - 1 = 3(x + 2)$.

 YOUR TURN

We can use point–slope form to find an equation of the line.

EXAMPLE 2   Use point–slope form to find an equation of the line with slope 3 that passes through $(-7, 8)$.

SOLUTION   We substitute into the point–slope form:

$$y - y_1 = m(x - x_1)$$
$$y - 8 = 3(x - (-7)). \qquad \text{Substituting 3 for } m, -7 \text{ for } x_1, \text{ and 8 for } y_1$$

This is an equation of the line. If desired, we can solve for y to write it in slope–intercept form.

2. Use point–slope form to find an equation of the line with slope -5 that passes through $(1, -6)$.

YOUR TURN

The point–slope form can be used to find the equation of any line given the slope and a point. Other forms of linear equations can also be used and may be more convenient in some situations.

Finding the Equation of a Line

GIVEN THE SLOPE AND THE y-INTERCEPT

If we know the slope m and the y-intercept $(0, b)$ of a line, we can find an equation of the line by substituting into slope–intercept form, $y = mx + b$.

EXAMPLE 3   Find an equation for the line parallel to $8y = 7x - 24$ with y-intercept $(0, -6)$.

SOLUTION   We first find slope–intercept form of the given line:

$$8y = 7x - 24$$
$$y = \tfrac{7}{8}x - 3. \qquad \text{Multiplying both sides by } \tfrac{1}{8}$$
$$\text{The slope is } \tfrac{7}{8}.$$

The slope of any line parallel to the line given by $8y = 7x - 24$ is $\tfrac{7}{8}$. For a y-intercept of $(0, -6)$, we must have

3. Find an equation for the line parallel to $3y = 3x + 12$ with y-intercept $(0, 5)$.

$$y = mx + b$$
$$y = \tfrac{7}{8}x - 6. \qquad \text{Substituting } \tfrac{7}{8} \text{ for } m \text{ and } -6 \text{ for } b$$

YOUR TURN

Study Skills

Put It in Words

If you are finding it difficult to master a particular topic or concept, talk about it with a classmate. Verbalizing your questions about the material might help clarify it for you.

GIVEN THE SLOPE AND A POINT OR GIVEN TWO POINTS

When we know the slope m of a line and any point on the line, we can find the equation of the line either by using slope–intercept form, $y = mx + b$, and solving for b or by substituting directly into point–slope form, $y - y_1 = m(x - x_1)$.

EXAMPLE 4   Find an equation for the line perpendicular to $2x + y = 5$ that passes through $(1, -3)$.

SOLUTION   We first find slope–intercept form of the given line:

$$2x + y = 5$$
$$y = -2x + 5. \qquad \text{Subtracting } 2x \text{ from both sides}$$
$$\text{The slope is } -2.$$

4. Find an equation in point–slope form for the line perpendicular to $3x - 4y = 7$ that passes through $(8, 2)$.

The slope of any line perpendicular to the line given by $2x + y = 5$ is the opposite of the reciprocal of -2, or $\frac{1}{2}$. Substituting into the point–slope form, we have

$$y - y_1 = m(x - x_1)$$
$$y - (-3) = \tfrac{1}{2}(x - 1). \qquad \text{Substituting } \tfrac{1}{2} \text{ for } m, 1 \text{ for } x_1, \text{ and } -3 \text{ for } y_1$$

 YOUR TURN

EXAMPLE 5   Use slope–intercept form to find an equation of the line with slope 4 that passes through $(6, -5)$.

SOLUTION   Since the slope of the line is 4, we have

$$y = mx + b$$
$$y = 4x + b. \qquad \text{Substituting 4 for } m$$

To find b, we use the fact that if $(6, -5)$ is a point on the line, that ordered pair is a solution of the equation of the line.

$$y = 4x + b \qquad \text{We know that } m \text{ is 4.}$$
$$-5 = 4(6) + b \qquad \text{Substituting 6 for } x \text{ and } -5 \text{ for } y$$
$$-5 = 24 + b$$
$$-29 = b \qquad \text{Solving for } b$$

5. Use slope–intercept form to find an equation of the line with slope $\frac{1}{2}$ that passes through $(8, -3)$.

Now we know that $b = -29$, so the equation of the line is

$$y = 4x - 29. \qquad m = 4 \text{ and } b = -29$$

 YOUR TURN

We can also find the equation of a line if we know two points on the line.

EXAMPLE 6   Find a linear function that has a graph passing through $(-1, -5)$ and $(3, -2)$.

SOLUTION   We first determine the slope of the line and then write an equation in point–slope form. (We could also use slope–intercept form as in Example 5.) Note that

Find the slope.

$$m = \frac{-5 - (-2)}{-1 - 3} = \frac{-3}{-4} = \frac{3}{4}.$$

Since the line passes through $(3, -2)$, we have

Substitute the point and the slope in the point–slope form.

$$y - (-2) = \tfrac{3}{4}(x - 3) \qquad \text{Substituting into } y - y_1 = m(x - x_1)$$
$$y + 2 = \tfrac{3}{4}x - \tfrac{9}{4}. \qquad \text{Using the distributive law}$$

Before using function notation, we isolate y:

Write in slope–intercept form.

$$y = \tfrac{3}{4}x - \tfrac{9}{4} - 2 \qquad \text{Subtracting 2 from both sides}$$
$$y = \tfrac{3}{4}x - \tfrac{17}{4} \qquad -\tfrac{9}{4} - \tfrac{8}{4} = -\tfrac{17}{4}$$
$$f(x) = \tfrac{3}{4}x - \tfrac{17}{4}. \qquad \text{Using function notation}$$

6. Find a linear function that has a graph passing through $(6, -1)$ and $(-2, -3)$.

You can check that using $(-1, -5)$ as (x_1, y_1) in $y - y_1 = \tfrac{3}{4}(x - x_1)$ yields the same expression for $f(x)$.

 YOUR TURN

HORIZONTAL LINES OR VERTICAL LINES

If we know that a line is horizontal or vertical and we know one point on the line, we can find an equation for the line.

EXAMPLE 7   Find **(a)** the equation of the horizontal line that passes through $(1, -4)$, and **(b)** the equation of the vertical line that passes through $(1, -4)$.

SOLUTION

a) An equation of a horizontal line is of the form $y = b$. In order for $(1, -4)$ to be a solution of $y = b$, we must have $b = -4$. Thus the equation of the line is $y = -4$.

b) An equation of a vertical line is of the form $x = a$. In order for $(1, -4)$ to be a solution of $x = a$, we must have $a = 1$. Thus the equation of the line is $x = 1$.

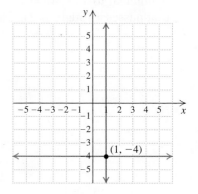

7. Find the equation of the vertical line that passes through $(2, 8)$.

 YOUR TURN

Interpolation and Extrapolation

When a function is given as a graph, we can use the graph to estimate an unknown function value. When we estimate the coordinates of an unknown point that lies *between* known points, the process is called **interpolation**. If the unknown point extends *beyond* the known points, the process is called **extrapolation**.

EXAMPLE 8   *Elementary School Math Proficiency.*   According to the National Assessment of Educational Progress (NAEP), the percentage of fourth-graders who are proficient in math has grown from 24% in 2000 to 39% in 2007 and 40% in 2011. Estimate the percentage of fourth-graders who showed proficiency in 2004 and predict the percentage who will demonstrate proficiency in 2015.

Source: nationsreportcard.gov

SOLUTION   The given information enables us to plot and connect three points. We let the horizontal axis represent the year and the vertical axis the percentage of fourth-graders demonstrating mathematical proficiency.

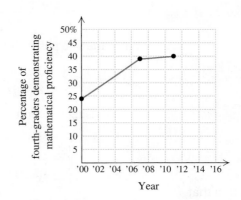

To estimate the percentage of fourth-graders showing mathematical proficiency in 2004, we locate the point directly above 2004. We then estimate its second coordinate by moving horizontally from that point to the *y*-axis. We see that about 33% of fourth-graders showed proficiency in math in 2004.

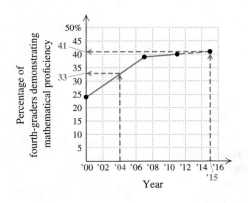

8. Use the graph in Example 8 to estimate the percentage of fourth-graders who showed proficiency in math in 2006.

To predict the percentage of fourth-graders showing proficiency in 2015, we extend the graph and extrapolate. It appears that about 41% of fourth-graders will show proficiency in math in 2015.

YOUR TURN

Season	Gross Revenue (in millions)
1999–2000	$ 603
2002–2003	721
2005–2006	862
2008–2009	943
2010–2011	1080

Source: The Broadway League

Linear Functions and Models

EXAMPLE 9 *Broadway Revenue.* Gross revenue from Broadway shows for several years is shown in the table at left. Use the data from 2005–2006 and from 2010–2011 to find a linear function that fits the data. Then use the function to estimate the gross revenue from Broadway shows in 2012–2013.

SOLUTION

1. **Familiarize.** We let t = the number of seasons after 1999–2000 and r = the gross revenue from Broadway shows, in millions of dollars, and form the pairs $(6, 862)$ and $(11, 1080)$. After choosing suitable scales on the two axes, we draw the graph.

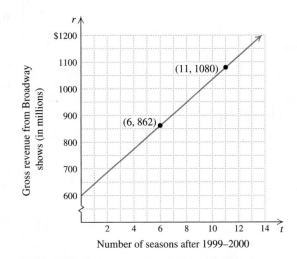

2. **Translate.** To find an equation relating r and t, we first find the slope of the line:

$$m = \frac{1080 - 862}{11 - 6} = \frac{218}{5} = 43.6.$$

The *growth rate* is $43.6 million per year.

Next, we write point–slope form and solve for r:

$$r - 862 = 43.6(t - 6) \qquad \text{Using (6, 862) to write point–slope form}$$
$$r - 862 = 43.6t - 261.6 \qquad \text{Using the distributive law}$$
$$r = 43.6t + 600.4.$$

3. **Carry out.** Using function notation, we have

$$r(t) = 43.6t + 600.4.$$

To predict the amount of income in 2012–2013, we find $r(13)$:

$$r(13) = 43.6(13) + 600.4 \qquad \text{2012–2013 is 13 seasons after 1999–2000.}$$
$$= 1167.2.$$

4. **Check.** To check, we can repeat our calculations. We could also extend the graph to see whether $(13, 1167.2)$ appears to be on the line. Since we are *extrapolating* from the data, our result is an approximation or estimate.

5. **State.** If we assume constant growth, gross revenue from Broadway shows in 2012–2013 will be about $1167.2 million.

 YOUR TURN

9. Refer to Example 9. Use the data from 2002–2003 and from 2008–2009 to find a linear function that fits the data. Use this function to estimate the gross revenue from Broadway shows in 2012–2013.

CONNECTING 🔗 THE CONCEPTS

Any line can be described by a variety of equivalent equations. Depending on the context, one form may be more useful than another.

Form of a Linear Equation	Example	Uses
Slope–intercept form: $y = mx + b$ or $f(x) = mx + b$	$f(x) = \frac{1}{2}x + 6$	Finding slope and y-intercept Graphing using slope and y-intercept Writing an equation given slope and y-intercept Writing linear functions
Standard form: $Ax + By = C$	$5x - 3y = 7$	Finding x- and y-intercepts Graphing using intercepts Future work with systems of equations
Point–slope form: $y - y_1 = m(x - x_1)$	$y - 2 = \frac{4}{5}(x - 1)$	Finding slope and a point on the line Graphing using slope and a point on the line Writing an equation given slope and a point on the line or given two points on the line Working with curves and tangents in calculus

EXERCISES

State whether each equation is in slope–intercept form, standard form, point–slope form, or none of these.

1. $2x + 5y = 8$

2. $y = \frac{2}{3}x - \frac{11}{3}$

3. $x - 13 = 5y$

4. $y - 2 = \frac{1}{3}(x - 6)$

5. $x - y = 1$

6. $y = -18x + 3.6$

Write each equation in standard form.

7. $y = \frac{2}{5}x + 1$

8. $y - 1 = -2(x - 6)$

Write each equation in slope–intercept form.

9. $3x - 5y = 10$

10. $y + 2 = \frac{1}{2}(x - 3)$

2.5 EXERCISE SET

FOR EXTRA HELP

MyMathLab® Math XL
PRACTICE   WATCH   READ   REVIEW

Vocabulary and Reading Check

Classify each of the following statements as either true or false.

1. The equation $y = -3x - 1$ is written in point–slope form.

2. The equation $y - 4 = -3(x - 1)$ is written in point–slope form.

3. Knowing the coordinates of just one point on a line is enough to write an equation of the line.

4. Knowing coordinates of just one point on a line and the slope of the line is enough to write an equation of the line.

5. Knowing the coordinates of just two points on a line is enough to write an equation of the line.

6. Point–slope form can be used with any point that is used to calculate the slope of that line.

Point–Slope Form

For each point–slope equation listed, state the slope and a point on the graph.

7. $y - 3 = \frac{1}{4}(x - 5)$

8. $y - 5 = 6(x - 1)$

9. $y + 1 = -7(x - 2)$

10. $y - 4 = -\frac{2}{3}(x + 8)$

11. $y - 6 = -\frac{10}{3}(x + 4)$

12. $y + 1 = -9(x - 7)$

Aha! **13.** $y = 5x$

14. $y = \frac{4}{5}x$

Graph.

15. $y - 2 = 3(x - 5)$

16. $y - 4 = 2(x - 3)$

17. $y - 2 = -4(x - 1)$

18. $y - 4 = -5(x - 1)$

19. $y + 4 = \frac{1}{2}(x + 2)$

20. $y + 7 = \frac{1}{3}(x + 5)$

21. $y = -(x - 8)$

22. $y = -3(x + 2)$

Finding the Equation of a Line

Find an equation for each line. Write your final answer in slope–intercept form.

23. Parallel to $y = 3x - 7$; y-intercept $(0, 4)$

24. Parallel to $y = \frac{1}{2}x + 6$; y-intercept $(0, -1)$

25. Perpendicular to $y = -\frac{3}{4}x + 1$; y-intercept $(0, -12)$

26. Perpendicular to $y = \frac{5}{8}x - 2$; y-intercept $(0, 9)$

27. Parallel to $2x - 3y = 4$; y-intercept $\left(0, \frac{1}{2}\right)$

28. Perpendicular to $4x + 7y = 1$; y-intercept $(0, -4.2)$

29. Perpendicular to $x + y = 18$; y-intercept $(0, -32)$

30. Parallel to $x - y = 6$; y-intercept $(0, 27)$

Find an equation in point–slope form for the line having the specified slope and containing the point indicated.

31. $m = 6, (7, 1)$

32. $m = 4, (3, 8)$

33. $m = -5, (3, 4)$

34. $m = -7, (1, 2)$

35. $m = \frac{1}{2}, (-2, -5)$

36. $m = 1, (-4, -6)$

37. $m = -1, (9, 0)$

38. $m = -\frac{2}{3}, (5, 0)$

Find an equation of the line having the specified slope and containing the indicated point. Write your final answer as a linear function in slope–intercept form. Then graph the line.

39. $m = 2, (1, -4)$

40. $m = -4, (-1, 5)$

41. $m = -\frac{3}{5}$, $(-4, 8)$ **42.** $m = -\frac{1}{5}$, $(-2, 1)$

43. $m = -0.6$, $(-3, -4)$ **44.** $m = 2.3$, $(4, -5)$

Aha! **45.** $m = \frac{2}{7}$, $(0, -6)$ **46.** $m = \frac{1}{4}$, $(0, 3)$

47. $m = \frac{3}{5}$, $(-4, 6)$ **48.** $m = -\frac{2}{7}$, $(6, -5)$

Write an equation of the line containing the specified point and parallel to the indicated line.

49. $(2, 5)$, $x - 2y = 3$

50. $(1, 4)$, $3x + y = 5$

51. $(-3, 2)$, $x + y = 7$

52. $(-1, -6)$, $x - 5y = 1$

53. $(-2, -3)$, $2x + 3y = -7$

54. $(3, -4)$, $5x - 6y = 4$

Aha! **55.** $(5, -4)$, $x = 2$

56. $(-3, 6)$, $y = 7$

Write an equation of the line containing the specified point and perpendicular to the indicated line.

57. $(3, 1)$, $2x - 3y = 4$

58. $(6, 0)$, $5x + 4y = 1$

59. $(-4, 2)$, $x + y = 6$

60. $(-2, -5)$, $x - 2y = 3$

61. $(1, -3)$, $3x - y = 2$

62. $(-5, 6)$, $4x - y = 3$

63. $(-4, -7)$, $3x - 5y = 6$

64. $(-4, 5)$, $7x - 2y = 1$

65. $(-3, 7)$, $y = 5$

66. $(4, -2)$, $x = 1$

Find an equation of the line containing each pair of points. Write your final answer as a linear function in slope–intercept form.

67. $(2, 3)$ and $(3, 7)$ **68.** $(3, 8)$ and $(1, 4)$

69. $(1.2, -4)$ and $(3.2, 5)$

70. $(-1, -2.5)$ and $(4, 8.5)$

Aha! **71.** $(2, -5)$ and $(0, -1)$ **72.** $(-2, 0)$ and $(0, -7)$

73. $(-6, -10)$ and $(-3, -5)$

74. $(-1, -3)$ and $(-4, -9)$

Find an equation of each line.

75. Horizontal line through $(2, -6)$

76. Horizontal line through $(-1, 8)$

77. Vertical line through $(-10, -9)$

78. Vertical line through $(4, 12)$

Interpolation and Extrapolation

Energy-Saving Lightbulbs.   *Compact fluorescent (CFL) bulbs are more efficient than incandescent lightbulbs. The following table lists incandescent wattage and the CFL wattage required to create the same amount of light.*

Source: U.S. Department of Energy

Input, Incandescent wattage	Output, Wattage of CFL equivalent
25	5
60	15
100	25

79. Use the data in the figure above to draw a graph. Estimate the wattage of a CFL bulb that creates light equivalent to a 75-watt incandescent bulb. Then predict the wattage of a CFL bulb that creates light equivalent to a 120-watt incandescent bulb.

80. Use the graph from Exercise 79 to estimate the wattage of a CFL bulb that creates light equivalent to a 40-watt incandescent bulb. Then predict the wattage of a CFL bulb that creates light equivalent to a 150-watt incandescent bulb.

Blood Alcohol Level.   *The data in the following table can be used to predict the number of drinks required for a person of a specified weight to be considered legally intoxicated (blood alcohol level of 0.08 or above). One 12-oz glass of beer, a 5-oz glass of wine, or a cocktail containing 1 oz of a distilled liquor all count as one drink. Assume that all drinks are consumed within one hour.*

12 oz   5 oz   1 oz

Input, Body Weight (in pounds)	Output, Number of Drinks
100	2.5
160	4
180	4.5
200	5

81. Use the data in the table above to draw a graph and to estimate the number of drinks that a 140-lb person would consume in order to be considered intoxicated. Then predict the number of drinks a 230-lb person would consume in order to be considered intoxicated.

82. Use the graph from Exercise 81 to estimate the number of drinks that a 120-lb person would consume in order to be considered intoxicated. Then predict the number of drinks a 250-lb person would consume in order to be considered intoxicated.

83. *Retailing.* Mountain View Gifts is experiencing constant growth. They recorded a total of $250,000 in sales in 2007, and $285,000 in 2012. Use a graph that displays the store's total sales as a function of time to estimate sales for 2008 and for 2015.

84. Use the graph in Exercise 83 to estimate sales for 2010 and for 2016.

Linear Functions and Models

85. *Recycling.* In 2005, Americans recycled 79.9 million tons of solid waste. In 2010, the figure grew to 85 million tons. Let $N(t)$ represent the number of tons recycled, in millions, and t the number of years since 2000.

Sources: U.S. EPA; Franklin Associates, Ltd.

a) Find a linear function that fits the data.
b) Use the function of part (a) to predict the amount recycled in 2016.

86. *National Park Land.* In 1990, the National Park system consisted of about 76.4 million acres. By 2009, the figure had grown to 80.4 million acres. Let $A(t)$ represent the amount of land in the National Park system, in millions of acres, t years after 1990.

Source: U.S. National Park Service

a) Find a linear function that fits the data.
b) Use the function of part (a) to predict the amount of land in the National Park system in 2015.

87. *Life Expectancy of Females in the United States.* In 2000, the life expectancy of females born in that year was 79.7 years. In 2010, it was 81.1 years. Let $E(t)$ represent life expectancy and t the number of years since 2000.

Source: National Center for Health Statistics

a) Find a linear function that fits the data.
Aha! b) Use the function of part (a) to predict the life expectancy of females in 2020.

88. *Life Expectancy of Males in the United States.* In 2000, the life expectancy of males born in that year was 74.3 years. In 2010, it was 76.2 years. Let $E(t)$ represent life expectancy and t the number of years since 2000.

Source: National Center for Health Statistics

a) Find a linear function that fits the data.
b) Use the function of part (a) to predict the life expectancy of males in 2020.

89. *History.* During the late 1600s, the capacity of ships in the English, French, and Dutch navies almost doubled, as shown in the graph below. Let $S(t)$ represent the average displacement of a ship, in tons, and t the number of years since 1650.

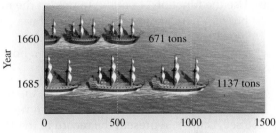

Naval history

Source: Harding, R., *The Evolution of the Sailing Navy, 1509–1815.* New York: St. Martin's Press, 1995

a) Find a linear function that fits the data.
b) Use the function of part (a) to estimate the average displacement of a ship in 1670.

90. *Consumer Demand.* Suppose that 6.5 million lb of coffee are sold when the price is $12 per pound, and 6.0 million lb are sold when it is $15 per pound.

a) Find a linear function that expresses the amount of coffee sold as a function of the price per pound.
b) Use the function of part (a) to predict how much consumers would be willing to buy at a price of $6 per pound.

91. *Pressure at Sea Depth.* The pressure 100 ft beneath the ocean's surface is approximately 4 atm (atmospheres), whereas at a depth of 200 ft, the pressure is about 7 atm.

a) Find a linear function that expresses pressure as a function of depth.
b) Use the function of part (a) to determine the pressure at a depth of 690 ft.

92. *Seller's Supply.* Suppose that suppliers are willing to sell 5.0 million lb of coffee at a price of $12 per pound and 7.0 million lb at $15 per pound.

a) Find a linear function that expresses the amount suppliers are willing to sell as a function of the price per pound.

b) Use the function of part (a) to predict how much suppliers would be willing to sell at a price of $6 per pound.

93. *Online Banking.* In 2000, about 39.4 million American households conducted at least some of their banking online. By 2010, that number had risen to about 72.5 million. Let $N(t)$ represent the number of American households using online banking, in millions, t years after 2000.

a) Find a linear function that fits the data.

b) Use the function of part (a) to predict the number of American households that will use online banking in 2015.

c) In what year will 157 million American households use online banking?

94. *Records in the 100-Meter Run.* In 1999, the record for the 100-m run was 9.79 sec. In 2012, it was 9.58 sec. Let $R(t)$ represent the record in the 100-m run and t the number of years since 1999.

Sources: International Association of Athletics Federation; *Guinness World Records*

a) Find a linear function that fits the data.

b) Use the function of part (a) to predict the record in 2015 and in 2030.

c) When will the record be 9.50 sec?

95. Suppose that you are given the coordinates of two points on a line, and one of those points is the y-intercept. What method would you use to find an equation for the line? Explain the reasoning behind your choice.

96. *Engineering.* Wind friction, or *air resistance*, increases with speed. Following are some measurements made in a wind tunnel. Plot the data and explain why a linear function does or does not give an approximate fit.

Velocity (in kilometers per hour)	Force of Resistance (in newtons)
10	3
21	4.2
34	6.2
40	7.1
45	15.1
52	29.0

Skill Review

Combine like terms. [1.3]

97. $-6x - x - 2x$

98. $ab^2 + 2a^2b - ab^2$

99. Solve for m: $x = \dfrac{mp}{c}$. [1.5]

100. Solve for y: $y + ax = dy$. [1.5]

Synthesis

101. Would an estimate found using interpolation be as reliable as one found using extrapolation? Why or why not?

102. On the basis of your answers to Exercises 87 and 88, would you predict that at some point in the future the life expectancy of males will exceed that of females? Why or why not?

For Exercises 103–107, assume that a linear equation models each situation.

103. *Temperature Conversion.* Water freezes at 32° Fahrenheit and at 0° Celsius. Water boils at 212°F and at 100°C. What Celsius temperature corresponds to a room temperature of 70°F?

104. *Depreciation of a Computer.* After 6 months of use, the value of Don's computer had dropped to $900. After 8 months, the value had gone down to $750. How much did the computer cost originally?

105. *Cell-Phone Charges.* The total cost of Tam's cell phone was $410 after 5 months of service and $690 after 9 months. What costs had Tam already incurred when her service just began? Assume that Tam's monthly charge is constant.

106. *Operating Expenses.* The total cost for operating Ming's Wings was $7500 after 4 months and $9250 after 7 months. Predict the total cost after 10 months.

107. Based on the information given in Exercises 90 and 92, at what price will the supply equal the demand?

108. Specify the domain of your answer to Exercise 90(a).

109. Specify the domain of your answer to Exercise 92(a).

110. For a linear function g, $g(3) = -5$ and $g(7) = -1$.

a) Find an equation for g.

b) Find $g(-2)$.

c) Find a such that $g(a) = 75$.

111. Find k so that the graph of $5y - kx = 7$ and the line containing $(7, -3)$ and $(-2, 5)$ are parallel.

112. Find k so that the graph of $7y - kx = 9$ and the line containing the points $(2, -1)$ and $(-4, 5)$ are perpendicular.

113. When several data points are available and they appear to be nearly collinear, a procedure known as *linear regression* can be used to find an equation for the line that best fits the data. Use a graphing calculator with a LINEAR REGRESSION option and the table that follows to find a linear function that predicts a woman's life expectancy as a function of the year in which she was born. Round coefficients to the nearest thousandth. Then use the function to predict the life expectancy in 2020 and compare this with the corresponding answer to Exercise 87 of this exercise set. Which answer seems more reliable? Why?

Life Expectancy of Women

Year, x	Life Expectancy, y (in years)
1930	61.6
1940	65.2
1950	71.1
1960	73.1
1970	74.7
1980	77.4
1990	78.8
2000	79.7
2010	81.1

Source: National Center for Health Statistics

114. Use linear regression (see Exercise 113) and the data accompanying Example 9 to find a linear function that predicts gross revenue from Broadway shows as a function of the number of seasons after 1999–2000. Round coefficients to the nearest thousandth. Then use the function to estimate gross revenue in 2012–2013 and compare this with the estimates found in Example 9 and Your Turn Exercise 9. Which answer seems most reliable? Why?

115. *Research.* Find the gross revenue from Broadway shows for the most recent year available. (See Example 9.) Use the functions developed in Example 9 and Exercise 114 to predict the gross revenue for that year. Did either function provide a close estimate? If not, what factors do you think caused a change in the rate of growth of the gross revenue?

116. Use a graphing calculator with a *squared* window to check your answers to Exercises 57–64.

117. *Energy Expenditure.* On the basis of the information below, what burns more energy: walking $4\frac{1}{2}$ mph for two hours or bicycling 14 mph for one hour?

Approximate Energy Expenditure by a 150-Pound Person in Various Activities

Activity	Calories per Hour
Walking, $2\frac{1}{2}$ mph	210
Bicycling, $5\frac{1}{2}$ mph	210
Walking, $3\frac{3}{4}$ mph	300
Bicycling, 13 mph	660

Source: Based on material prepared by Robert E. Johnson, M.D., Ph.D., and colleagues, University of Illinois.

YOUR TURN ANSWERS: SECTION 2.5

1.

2. $y - (-6) = -5(x - 1)$
3. $y = x + 5$
4. $y - 2 = -\frac{4}{3}(x - 8)$
5. $y = \frac{1}{2}x - 7$
6. $f(x) = \frac{1}{4}x - \frac{5}{2}$
7. $x = 2$ 8. 37%
9. $r(t) = 37t + 610$, where $r(t)$ is the gross revenue, in millions of dollars, t seasons after 1999–2000; $1091 million

QUICK QUIZ: SECTIONS 2.1–2.5

1. Find a linear function whose graph has slope -5 and y-intercept $(0, 25)$. [2.3]

2. Find slope–intercept form for the equation of the line containing $(-1, 6)$ and $(-4, -3)$. [2.5]

3. Determine the slope and the y-intercept of the line given by $3x - y = 6$. [2.3]

4. Determine whether $3x - 7 = 8y$ is linear. [2.4]

5. Determine whether this graph is that of a function. [2.2]

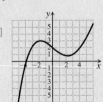

PREPARE TO MOVE ON

Simplify. [1.3]

1. $(2x^2 - x) + (3x - 5)$ 2. $(2t - 1) - (t - 3)$

Find the domain of each function. [2.2]

3. $f(x) = \dfrac{x}{x - 3}$ 4. $g(x) = x^2 - 1$

2.6   The Algebra of Functions

The Sum, Difference, Product, or Quotient of Two Functions   ■   Domains and Graphs

We now examine four ways in which functions can be combined.

The Sum, Difference, Product, or Quotient of Two Functions

Suppose that a is in the domain of two functions, f and g. The input a is paired with $f(a)$ by f and with $g(a)$ by g. The outputs can then be added to get $f(a) + g(a)$.

EXAMPLE 1   Let $f(x) = x + 4$ and $g(x) = x^2 + 1$. Find $f(2) + g(2)$.

SOLUTION   We visualize two function machines, as shown below. Because 2 is in the domain of each function, we can compute $f(2)$ and $g(2)$.

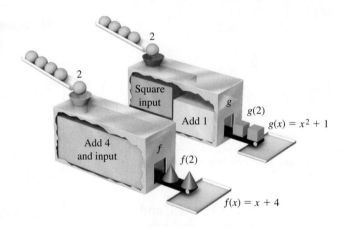

Since

$$f(2) = 2 + 4 = 6 \quad \text{and} \quad g(2) = 2^2 + 1 = 5,$$

we have

$$f(2) + g(2) = 6 + 5 = 11.$$

YOUR TURN

1. Using the functions defined in Example 1, find $f(-5) + g(-5)$.

$(f + g)(2)$

$(f + g)(x) = x^2 + x + 5$

In Example 1, suppose that we were to write $f(x) + g(x)$ as $(x + 4) + (x^2 + 1)$, or $f(x) + g(x) = x^2 + x + 5$. This could then be regarded as a "new" function and visualized as one function machine, shown at left, combining the two machines from Example 1. The notation $(f + g)(x)$ is generally used to indicate the output of a function formed in this manner. Similar notations exist for subtraction, multiplication, and division of functions.

THE ALGEBRA OF FUNCTIONS

If f and g are functions and x is in the domain of both functions, then:
1. $(f + g)(x) = f(x) + g(x)$;
2. $(f - g)(x) = f(x) - g(x)$;
3. $(f \cdot g)(x) = f(x) \cdot g(x)$;
4. $(f/g)(x) = f(x)/g(x)$, provided $g(x) \neq 0$.

EXAMPLE 2 For $f(x) = x^2 - x$ and $g(x) = x + 2$, find the following.

a) $(f + g)(4)$

b) $(f - g)(x)$ and $(f - g)(-1)$

c) $(f/g)(x)$ and $(f/g)(-3)$

d) $(f \cdot g)(4)$

SOLUTION

a) Since $f(4) = 4^2 - 4 = 12$ and $g(4) = 4 + 2 = 6$, we have

$$(f + g)(4) = f(4) + g(4)$$
$$= 12 + 6 \quad \text{Substituting}$$
$$= 18.$$

Alternatively, we could first find $(f + g)(x)$:

$$(f + g)(x) = f(x) + g(x)$$
$$= x^2 - x + x + 2$$
$$= x^2 + 2. \quad \text{Combining like terms}$$

Thus,

$$(f + g)(4) = 4^2 + 2 = 18. \quad \text{Our results match.}$$

b) We have

$$(f - g)(x) = f(x) - g(x)$$
$$= x^2 - x - (x + 2) \quad \text{Substituting}$$
$$= x^2 - 2x - 2. \quad \text{Removing parentheses and combining like terms}$$

Then,

$$(f - g)(-1) = (-1)^2 - 2(-1) - 2 \quad \text{Using } (f - g)(x) \text{ is faster than using } f(x) - g(x).$$
$$= 1. \quad \text{Simplifying}$$

c) We have

$$(f/g)(x) = f(x)/g(x)$$
$$= \frac{x^2 - x}{x + 2}. \quad \text{We assume that } x \neq -2.$$

Then,

$$(f/g)(-3) = \frac{(-3)^2 - (-3)}{-3 + 2} \quad \text{Substituting}$$
$$= \frac{12}{-1} = -12.$$

d) Using our work in part (a), we have

$$(f \cdot g)(4) = f(4) \cdot g(4)$$
$$= 12 \cdot 6$$
$$= 72.$$

2. Using the functions defined in Example 2, find $(g - f)(x)$ and $(g - f)(-1)$.

YOUR TURN

Domains and Graphs

Applications involving sums or differences of functions often appear in print. For example, the following graphs are similar to those published by the California Department of Education to promote breakfast programs in which students eat a balanced meal of fruit or juice, toast or cereal, and 2% or whole milk. The combination of carbohydrate, protein, and fat gives a sustained release of energy, delaying the onset of hunger for several hours.

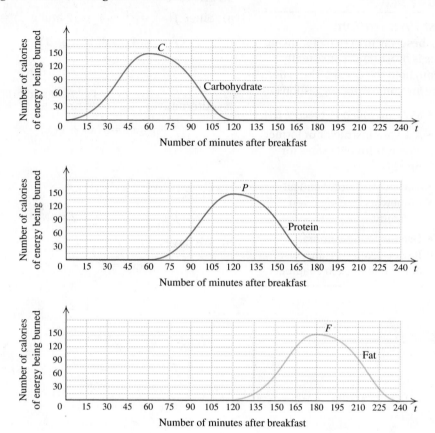

When the three graphs are superimposed, and the calorie expenditures added, it becomes clear that a balanced meal results in a steady, sustained supply of energy.

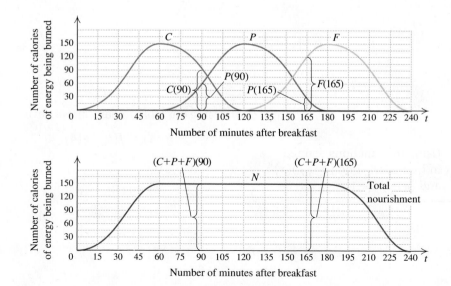

For any point $(t, N(t))$, we have

$$N(t) = (C + P + F)(t) = C(t) + P(t) + F(t).$$

To find $(f + g)(a)$, $(f - g)(a)$, or $(f \cdot g)(a)$, we must know that $f(a)$ and $g(a)$ exist. This means that a must be in the domain of both f and g.

EXAMPLE 3 Let

$$f(x) = \frac{5}{x} \quad \text{and} \quad g(x) = \frac{2x - 6}{x + 1}.$$

Find the domain of $f + g$, the domain of $f - g$, and the domain of $f \cdot g$.

SOLUTION Note that because division by 0 is undefined, we have

$$\text{Domain of } f = \{x \,|\, x \text{ is a real number } and\, x \neq 0\}$$

and

$$\text{Domain of } g = \{x \,|\, x \text{ is a real number } and\, x \neq -1\}.$$

In order to find $f(a) + g(a), f(a) - g(a)$, or $f(a) \cdot g(a)$, we must know that a is in *both* of the above domains. Thus,

$$\text{Domain of } f + g = \text{Domain of } f - g = \text{Domain of } f \cdot g$$
$$= \{x \,|\, x \text{ is a real number } and\, x \neq 0 \, and\, x \neq -1\}.$$

3. Let $f(x) = \dfrac{x}{x - 8}$ and $g(x) = x^2 - 7$. Find the domain of $f + g$, the domain of $f - g$, and the domain of $f \cdot g$.

↳ YOUR TURN

Suppose that for $f(x) = x^2 - x$ and $g(x) = x + 2$, we want to find $(f/g)(-2)$. Finding $f(-2)$ and $g(-2)$ poses no problem:

$$f(-2) = 6 \quad \text{and} \quad g(-2) = 0;$$

but then

$$(f/g)(-2) = f(-2)/g(-2)$$
$$= 6/0. \quad \text{Division by 0 is undefined.}$$

Thus, although -2 is in the domain of both f and g, it is not in the domain of f/g. We can also see this by writing $(f/g)(x)$:

$$(f/g)(x) = \frac{f(x)}{g(x)} = \frac{x^2 - x}{x + 2}.$$

Since $x + 2 = 0$ when $x = -2$, the domain of f/g must exclude -2.

Student Notes

The concern over a denominator being 0 arises throughout this course. Try to develop the habit of checking for any possible input values that would create a denominator of 0 whenever you work with functions.

DETERMINING THE DOMAIN

The domain of $f + g$, $f - g$, or $f \cdot g$ is the set of all values common to the domains of f and g.

The domain of f/g is the set of all values common to the domains of f and g, excluding any values for which $g(x)$ is 0.

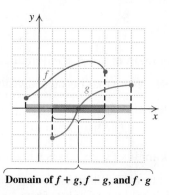

Domain of $f + g$, $f - g$, and $f \cdot g$

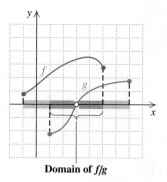

Domain of f/g

EXAMPLE 4   Given

$$f(x) = \frac{1}{x - 3} \quad \text{and} \quad g(x) = 2x - 7,$$

find the domains of $f + g, f - g, f \cdot g,$ and f/g.

Find the domains of f and g.

SOLUTION   We first find the domain of f and the domain of g:

The domain of f is $\{x \mid x \text{ is a real number } and\ x \neq 3\}$.

The domain of g is $\mathbb{R}$.

Find the domains of $f + g, f - g,$ and $f \cdot g$.

The domains of $f + g$, $f - g$, and $f \cdot g$ are the set of all elements common to the domains of f and g. This consists of all real numbers except 3.

The domain of $f + g$ = the domain of $f - g$ = the domain of $f \cdot g$

$= \{x \mid x \text{ is a real number } and\ x \neq 3\}$.

Find any values of x for which $g(x) = 0$.

Because we cannot divide by 0, the domain of f/g must also exclude any values of x for which $g(x)$ is 0. We determine those values by solving $g(x) = 0$:

$$g(x) = 0$$
$$2x - 7 = 0 \quad \text{Replacing } g(x) \text{ with } 2x - 7$$
$$2x = 7$$
$$x = \tfrac{7}{2}.$$

Find the domain of f/g.

The domain of f/g is the domain of the sum, the difference, and the product of f and g, found above, excluding $\tfrac{7}{2}$.

The domain of f/g = $\left\{x \mid x \text{ is a real number } and\ x \neq 3 \text{ } and\ x \neq \tfrac{7}{2}\right\}$.

YOUR TURN

4. Given $f(x) = \dfrac{3}{2x + 1}$ and $g(x) = x - 4$, find the domains of $f + g, f - g, f \cdot g,$ and f/g.

Technology Connection

A partial check of Example 4 can be performed by setting up a table so the TBLSTART is 1 and the increment of change (ΔTbl) is 0.5. (Other choices, like 0.1, will also work.) Next, we let $y_1 = \dfrac{1}{x-3}$ and $y_2 = 2x - 7$. Using Y-VARS to write $y_3 = y_1 + y_2$ and $y_4 = y_1/y_2$, we can create the table of values shown here. Note that when x is 3.5, a value for y_3 can be found, but y_4 is undefined. If we "de-select" y_1 and y_2 as we enter them, the columns for y_3 and y_4 appear without scrolling through the table.

X	Y3	Y4
1	−5.5	.1
1.5	−4.667	.16667
2	−4	.33333
2.5	−4	1
3	ERROR	ERROR
3.5	2	ERROR
4	2	1

X = 3.5

Use a similar approach to partially check Example 3.

Chapter Resources:
Collaborative Activity, p. 140;
Decision Making: Connection, p. 140

Division by 0 is not the only condition that can force restrictions on the domain of a function. When we examine functions similar to that given by $f(x) = \sqrt{x}$, the concern is taking the square root of a negative number.

2.6 EXERCISE SET

FOR EXTRA HELP   MyMathLab® Math XL   PRACTICE   WATCH   READ   REVIEW

Vocabulary and Reading Check

Make each of the following statements true by selecting the correct word for each blank.

1. If f and g are functions, then $(f + g)(x)$ is the _____ of the functions.
   sum/difference

2. One way to compute $(f - g)(2)$ is to _____ $g(2)$ from $f(2)$.
   erase/subtract

3. One way to compute $(f - g)(2)$ is to simplify $f(x) - g(x)$ and then _____ the result for
   evaluate/substitute
   $x = 2$.

4. The domain of $f + g, f - g$, and $f \cdot g$ is the set of all values common to the _____ of f and g.
   domains/ranges

5. The domain of f/g is the set of all values common to the domains of f and g, _____ any
   including/excluding
   values for which $g(x)$ is 0.

6. The height of $(f + g)(a)$ on a graph is the _____ of the heights of $f(a)$ and $g(a)$.
   product/sum

The Sum, Difference, Product, or Quotient of Two Functions

Let $f(x) = -2x + 3$ and $g(x) = x^2 - 5$. Find each of the following.

7. $f(3) + g(3)$ 8. $f(4) + g(4)$

9. $f(1) - g(1)$ 10. $f(2) - g(2)$

11. $f(-2) \cdot g(-2)$ 12. $f(-1) \cdot g(-1)$

13. $f(-4)/g(-4)$ 14. $f(3)/g(3)$

15. $g(1) - f(1)$ 16. $g(-3)/f(-3)$

17. $(f + g)(x)$

18. $(f - g)(x)$

19. $(g - f)(x)$

20. $(g/f)(x)$

Let $F(x) = x^2 - 2$ and $G(x) = 5 - x$. Find each of the following.

21. $(F + G)(x)$ 22. $(F + G)(a)$

23. $(F - G)(3)$ 24. $(F - G)(2)$

25. $(F \cdot G)(-3)$ 26. $(F \cdot G)(-4)$

27. $(F/G)(x)$ 28. $(G - F)(x)$

29. $(G/F)(-2)$

30. $(F/G)(-1)$

31. $(F + F)(1)$

32. $(G \cdot G)(6)$

Domains and Graphs

The following graph shows the number of births in the United States, in millions, from 1970–2010. Here, $C(t)$ represents the number of Caesarean section births, $B(t)$ the number of non-Caesarean section births, and $N(t)$ the total number of births in year t.

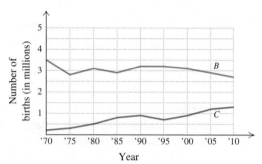

Source: National Center for Health Statistics

33. Use estimates of $C(2005)$ and $B(2005)$ to estimate $N(2005)$.

34. Use estimates of $C(1985)$ and $B(1985)$ to estimate $N(1985)$.

In the following graph, $S(t)$ represents the number of gallons of carbonated soft drinks, $M(t)$ the number of gallons of milk, $J(t)$ the number of gallons of fruit juice, and $W(t)$ the number of gallons of bottled water consumed by the average American in year t.

Beverage consumption

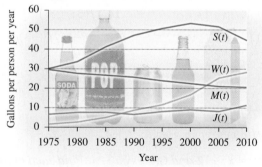

Source: Economic Research Service, U. S. Department of Agriculture

35. Use estimates of $S(2010)$ and $W(2010)$ to estimate $(S - W)(2010)$.

36. Use estimates of $M(2010)$ and $J(2010)$ to estimate $(M - J)(2010)$.

Often function addition is represented by stacking the graphs of individual functions directly on top of each other. The following graph indicates how U.S. municipal solid waste has been managed. The braces indicate the values of the individual functions.

Talking trash

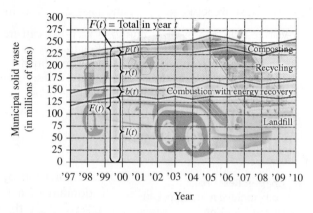

Source: Environmental Protection Agency

37. Estimate $(p + r)('09)$. What does it represent?

38. Estimate $(p + r + b)('09)$. What does it represent?

39. Estimate $F('00)$. What does it represent?

40. Estimate $F('10)$. What does it represent?

41. Estimate $(F - p)('08)$. What does it represent?

42. Estimate $(F - l)('07)$. What does it represent?

For each pair of functions f and g, determine the domain of the sum, the difference, and the product of the two functions.

43. $f(x) = x^2$,
$g(x) = 7x - 4$

44. $f(x) = 5x - 1$,
$g(x) = 2x^2$

45. $f(x) = \dfrac{1}{x + 5}$,
$g(x) = 4x^3$

46. $f(x) = 3x^2$,
$g(x) = \dfrac{1}{x - 9}$

47. $f(x) = \dfrac{2}{x}$,
$g(x) = x^2 - 4$

48. $f(x) = x^3 + 1$,
$g(x) = \dfrac{5}{x}$

49. $f(x) = x + \dfrac{2}{x - 1}$,
$g(x) = 3x^3$

50. $f(x) = 9 - x^2$,
$g(x) = \dfrac{3}{x + 6} + 2x$

51. $f(x) = \dfrac{3}{2x + 9}$,
$g(x) = \dfrac{5}{1 - x}$

52. $f(x) = \dfrac{5}{3 - x}$,
$g(x) = \dfrac{1}{4x - 1}$

For each pair of functions f and g, determine the domain of f/g.

53. $f(x) = x^4,$
$g(x) = x - 3$

54. $f(x) = 2x^3,$
$g(x) = 5 - x$

55. $f(x) = 3x - 2,$
$g(x) = 2x + 8$

56. $f(x) = 5 + x,$
$g(x) = 6 - 2x$

57. $f(x) = \dfrac{3}{x - 4},$
$g(x) = 5 - x$

58. $f(x) = \dfrac{1}{2 - x},$
$g(x) = 7 + x$

59. $f(x) = \dfrac{2x}{x + 1},$
$g(x) = 2x + 5$

60. $f(x) = \dfrac{7x}{x - 2},$
$g(x) = 3x + 7$

For Exercises 61–68, consider the functions F and G as shown.

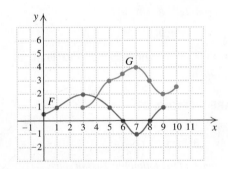

61. Determine $(F + G)(5)$ and $(F + G)(7)$.

62. Determine $(F \cdot G)(6)$ and $(F \cdot G)(9)$.

63. Determine $(G - F)(7)$ and $(G - F)(3)$.

64. Determine $(F/G)(3)$ and $(F/G)(7)$.

65. Find the domains of F, G, $F + G$, and F/G.

66. Find the domains of $F - G$, $F \cdot G$, and G/F.

67. Graph $F + G$.

68. Graph $G - F$.

69. Examine the graph for Exercises 35 and 36. Between what years did the average American drink more soft drinks than juice, bottled water, and milk combined? Explain how you determined this.

70. Examine the graph for Exercises 33 and 34. Did the total number of births increase or decrease from 1970 to 2010? Did the percent of births by Caesarean section increase or decrease from 1970 to 2010? Explain how you determined your answers.

Skill Review

Solve.

71. Ruth's scores on three tests are 85, 72, and 81. What must Ruth score on the fourth test so that her average will be 80?   [1.4]

72. One angle of a triangle measures twice the second angle. The third angle measures three times the second angle. Find the measures of the angles.   [1.4]

73. In one basketball game, Terrence scored 5 fewer points than Isaiah. Together, they scored 27 points. How many points did Terrence score?   [1.4]

74. A *mole* of a substance contains 6.022×10^{23} molecules. If a mole of water weighs 18.015 g, how much does each molecule weigh?   [1.7]

Synthesis

75. Examine the graphs following Example 2 and explain how similar graphs could be drawn to represent the absorption into the bloodstream of 200 mg of Advil® taken four times a day.

76. If $f(x) = c$, where c is some positive constant, describe how the graphs of $y = g(x)$ and $y = (f + g)(x)$ will differ.

77. Find the domain of F/G, if
$$F(x) = \frac{1}{x - 4} \quad \text{and} \quad G(x) = \frac{x^2 - 4}{x - 3}.$$

78. Find the domain of f/g, if
$$f(x) = \frac{3x}{2x + 5} \quad \text{and} \quad g(x) = \frac{x^4 - 1}{3x + 9}.$$

79. Sketch the graph of two functions f and g such that the domain of f/g is
$$\{x \mid -2 \le x \le 3 \text{ and } x \ne 1\}.$$
Answers may vary.

80. Find the domains of $f + g, f - g, f \cdot g$, and f/g, if
$f = \{(-2, 1), (-1, 2), (0, 3), (1, 4), (2, 5)\}$
and
$g = \{(-4, 4), (-3, 3), (-2, 4), (-1, 0), (0, 5), (1, 6)\}.$

81. Find the domain of m/n, if
$$m(x) = 3x \quad \text{for } -1 < x < 5$$
and
$$n(x) = 2x - 3.$$

82. For f and g as defined in Exercise 80, find $(f + g)(-2)$, $(f \cdot g)(0)$, and $(f/g)(1)$.

83. Write equations for two functions f and g such that the domain of $f + g$ is

$$\{x \,|\, x \text{ is a real number } and \; x \neq -2 \; and \; x \neq 5\}.$$

Answers may vary.

84. Let $y_1 = 2.5x + 1.5$, $y_2 = x - 3$, and $y_3 = y_1/y_2$. Depending on whether the CONNECTED or DOT mode is used, the graph of y_3 appears as follows. Use algebra to determine which graph more accurately represents y_3.

CONNECTED MODE
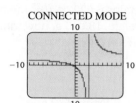

DOT MODE
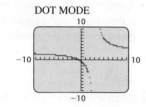

85. Using the window $[-5, 5, -1, 9]$, graph $y_1 = 5$, $y_2 = x + 2$, and $y_3 = \sqrt{x}$. Then predict what shape the graphs of $y_1 + y_2$, $y_1 + y_3$, and $y_2 + y_3$ will take. Use a graphing calculator to check each prediction.

86. Use the TABLE feature on a graphing calculator to check your answers to Exercises 47, 49, 57, and 59.)

➤ YOUR TURN ANSWERS: SECTION 2.6

1. 25   **2.** $(g - f)(x) = -x^2 + 2x + 2$; $(g - f)(-1) = -1$

3. All three domains are $\{x \,|\, x \text{ is a real number } and \; x \neq 8\}$.

4. Domain of $f + g$, $f - g$, and $f \cdot g = \{x \,|\, x \text{ is a real number } and \; x \neq -\frac{1}{2}\}$; domain of $f/g = \{x \,|\, x \text{ is a real number } and \; x \neq -\frac{1}{2} \; and \; x \neq 4\}$

QUICK QUIZ: SECTIONS 2.1–2.6

1. In 1972, the amount spent on Medicaid was about 2% of all federal spending. Medicaid spending is projected to be 11% of all federal spending in 2020. Find the rate of change.

Source: Congressional Budget Office   [2.3]

2. The number of Americans, in millions, ages 65 and older can be estimated by $n(t) = 1.2t + 40$, where t is the number of years since 2010. What do the numbers 1.2 and 40 signify?   [2.3]

Source: Based on data from U.S. Census Bureau

3. Find the intercepts of the line given by $2x - 5y = 20$.   [2.4]

4. Determine whether the graphs of the following equations are parallel, perpendicular, or neither:

$$y = \tfrac{1}{2}x - 8$$
$$x = 2y + 6.   [2.4]$$

5. Let $g(x) = 5x - 7$ and $h(x) = 6 - 4x$. Find $(g - h)(x)$.   [2.6]

PREPARE TO MOVE ON

Solve.   [1.5]

1. $x - 6y = 3$, for y **2.** $3x + 8y = 5$, for y

Translate each of the following. Do not solve.   [1.4]

3. Five more than twice a number is 49.

4. Three less than half of some number is 57.

5. The sum of two consecutive integers is 145.

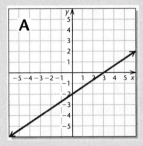

A

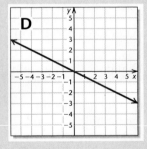

B

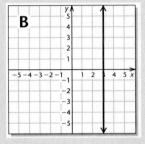

C

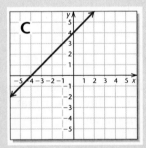

D

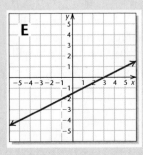

E

Visualizing for Success

Use after Section 2.4.

Match each equation or function with its graph.

1. $y = x + 4$

2. $y = 2x$

3. $y = 3$

4. $x = 3$

5. $f(x) = -\frac{1}{2}x$

6. $2x - 3y = 6$

7. $f(x) = -3x - 2$

8. $3x + 2y = 6$

9. $y - 3 = 2(x - 1)$

10. $y + 2 = \frac{1}{2}(x + 1)$

Answers on page A-11

An alternate, animated version of this activity appears in MyMathLab. *To use* MyMathLab, *you need a course ID and a student access code. Contact your instructor for more information.*

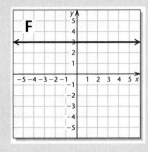

F

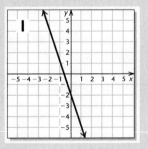

G

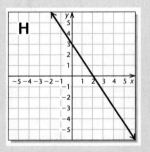

H

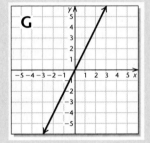

I

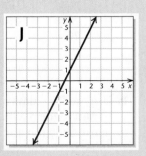

J

Collaborative Activity   *Time on Your Hands*

Focus: The algebra of functions
Use after: Section 2.6
Time: 10–15 minutes
Group size: 2

The graph and the data at right chart the age at which older workers withdraw from the work force, or average effective retirement age, and the life expectancy at retirement age for both men and women.

Activity

1. One group member, focusing on the data for men, should perform the appropriate calculations and then graph $E_M - R_M$. The other member, focusing on the data for women, should perform the appropriate calculations and then graph $E_W - R_W$.

2. What does $(E_M - R_M)(x)$ represent? What does $(E_W - R_W)(x)$ represent? In what fields

of study or business might the functions $E_M - R_M$ and $E_W - R_W$ prove useful?

3. What advice would you give to someone considering early retirement?

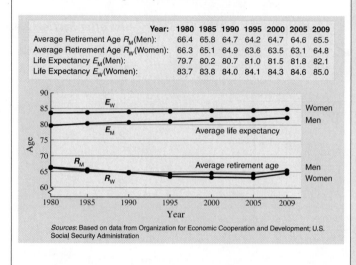

Year:	1980	1985	1990	1995	2000	2005	2009
Average Retirement Age R_M(Men):	66.4	65.8	64.7	64.2	64.7	64.6	65.5
Average Retirement Age R_W(Women):	66.3	65.1	64.9	63.6	63.5	63.1	64.8
Life Expectancy E_M(Men):	79.7	80.2	80.7	81.0	81.5	81.8	82.1
Life Expectancy E_W(Women):	83.7	83.8	84.0	84.1	84.3	84.6	85.0

Sources: Based on data from Organization for Economic Cooperation and Development; U.S. Social Security Administration

Decision Making Connection   *(Use after Section 2.6.)*

Career Choices. When deciding what career to pursue, one concern is often salary. If you plan to first earn a degree, that concern shifts to what the salary will be several years in the future.

Future predictions are made on the basis of past and current trends. If a trend appears linear, a linear function can be used as a model.

Suppose you are interested in becoming a registered nurse or a speech language pathologist. The following table lists the annual median salary for several years for each profession. Assume that both salaries can be modeled as a linear function of time.

	Registered Nurse	Speech Language Pathologist	
2006	$59,710	2006	$57,700
2008	62,450	2008	62,930
2010	67,720	2009	68,350

Source: U.S. Bureau of Labor Statistics

1. Use the 2006 data and the 2010 data for registered nurses to form a linear function r that can be used to estimate the annual median salary t years after 2006.

2. Use the 2006 data and the 2009 data for speech language pathologists to form a linear function p that can be used to estimate the annual median salary t years after 2006.

3. Examine the functions from Exercises 1 and 2.
   a) Which profession had a higher salary in 2006 $(t = 0)$?
   b) Which profession had a higher rate of growth in salary?

4. Form the function $(p - r)$. What does this function represent?

5. According to your models, how much more, on average, will a speech language pathologist make than a registered nurse in 2014?

6. *Research.* Find past and current salaries for one or more professions in which you are interested. If the trend appears linear, form a linear function that could be used to model the salary. Then use the model to predict the salary in the year that you will graduate from college.

Study Summary

KEY TERMS AND CONCEPTS	EXAMPLES	PRACTICE EXERCISES

SECTION 2.1: *Graphs*

We can **graph** an equation in two variables by selecting values for one variable and finding the corresponding values for the other variable. We plot the resulting ordered pairs and draw the graph.

Graph: $y = x^2 - 3$.

x	y	(x, y)
0	-3	$(0, -3)$
-1	-2	$(-1, -2)$
1	-2	$(1, -2)$
-2	1	$(-2, 1)$
2	1	$(2, 1)$

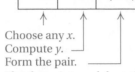

Choose any x.
Compute y.
Form the pair.
Plot the points and draw the graph.

1. Graph: $y = 2x + 1$.

SECTION 2.2: *Functions*

A **function** is a correspondence between a first set, called the **domain**, and a second set, called the **range**, such that each member of the domain corresponds to *exactly one* member of the range.

The correspondence f: $\left\{ \left(-1, \frac{1}{2}\right), (0, 1), (1, 2), (2, 4), (3, 8) \right\}$ is a function.

The domain of $f = \{-1, 0, 1, 2, 3\}$.

The range of $f = \left\{ \frac{1}{2}, 1, 2, 4, 8 \right\}$.

$f(-1) = \frac{1}{2}$

The input -1 corresponds to the output $\frac{1}{2}$.

2. Find $f(-1)$ for $f(x) = 2 - 3x$.

The Vertical-Line Test
If it is possible for a vertical line to cross a graph more than once, then the graph is not the graph of a function.

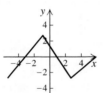

This *is* the graph of a function.

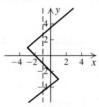

This *is not* the graph of a function.

3. Determine whether the following graph represents a function.

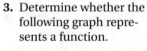

The domain of a function is the set of all first coordinates of the points on the graph.

The range of a function is the set of all second coordinates of the points on the graph.

Consider the function given by $f(x) = |x| - 3$.

The domain of the function is $\mathbb{R}$.

The range of the function is $\{y | y \geq -3\}$.

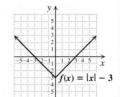

$f(x) = |x| - 3$

4. Determine the domain and the range of the function represented in the following graph.

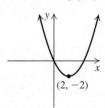

$(2, -2)$

| Unless otherwise stated, the domain of a function is the set of all numbers for which function values can be calculated. | Consider the function given by $f(x) = \dfrac{x + 2}{x - 7}$.

Function values cannot be calculated when the denominator is 0. Since $x - 7 = 0$ when $x = 7$, the domain of f is
$$\{x \mid x \text{ is a real number } and\, x \neq 7\}.$$ | **5.** Determine the domain of the function given by $f(x) = \frac{1}{4}x - 5$. |

SECTION 2.3:   *Linear Functions: Slope, Graphs, and Models*

Slope $$\text{Slope} = m = \dfrac{\text{change in } y}{\text{change in } x}$$ $$= \dfrac{\text{rise}}{\text{run}} = \dfrac{y_2 - y_1}{x_2 - x_1}$$	The slope of the line containing $(-1, -4)$ and $(2, -6)$ is $$m = \dfrac{-6 - (-4)}{2 - (-1)} = \dfrac{-2}{3} = -\dfrac{2}{3}.$$	**6.** Find the slope of the line containing $(1, 4)$ and $(-9, 3)$.
Slope–Intercept Form $$y = mx + b$$ The slope of the line is m. The y-intercept of the line is $(0, b)$.	For the line given by $y = \frac{2}{3}x - 8$: The slope is $\frac{2}{3}$ and the y-intercept is $(0, -8)$.	**7.** Find the slope and the y-intercept of the line given by $y = -4x + \frac{2}{5}$.
To graph a line written in slope–intercept form, plot the y-intercept and count off the slope.	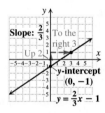	**8.** Graph: $y = \frac{1}{2}x + 2$.

SECTION 2.4:   *Another Look at Linear Graphs*

| The slope of a horizontal line is 0. The graph of $f(x) = b$ or $y = b$ is a horizontal line with y-intercept $(0, b)$.

The slope of a vertical line is undefined. The graph of $x = a$ is a vertical line, with x-intercept $(a, 0)$. | 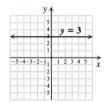 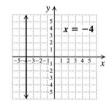 | **9.** Graph: $y = -2$.

10. Graph: $x = 3$. |
| **Intercepts**
To find a y-intercept $(0, b)$, let $x = 0$ and solve for y.
To find an x-intercept $(a, 0)$, let $y = 0$ and solve for x. | 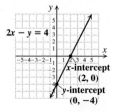 | **11.** Find the x-intercept and the y-intercept of the line given by $10x - y = 10$. |

| **Parallel Lines** Two lines are parallel if they have the same slope or if both are vertical. | Determine whether the graphs of $y = \frac{2}{3}x - 5$ and $3y - 2x = 7$ are parallel. $y = \frac{2}{3}x - 5 \qquad 3y - 2x = 7$ The slope is $\frac{2}{3}$. $\qquad 3y = 2x + 7$ $\qquad\qquad y = \frac{2}{3}x + \frac{7}{3}$ $\qquad\qquad$ The slope is $\frac{2}{3}$. Since the slopes are the same, the graphs are parallel. | **12.** Determine whether the graphs of $y = 4x - 12$ and $4y = x - 9$ are parallel. |
| **Perpendicular Lines** Two lines are perpendicular if the product of their slopes is -1 or if one line is vertical and the other line is horizontal. | Determine whether the graphs of $y = \frac{2}{3}x - 5$ and $2y + 3x = 1$ are perpendicular. $y = \frac{2}{3}x - 5 \qquad 2y + 3x = 1$ The slope is $\frac{2}{3}$. $\qquad 2y = -3x + 1$ $\qquad\qquad y = -\frac{3}{2}x + \frac{1}{2}$ $\qquad\qquad$ The slope is $-\frac{3}{2}$. Since $\frac{2}{3}\left(-\frac{3}{2}\right) = -1$, the graphs are perpendicular. | **13.** Determine whether the graphs of $y = x - 7$ and $x + y = 3$ are perpendicular. |

SECTION 2.5:   *Equations of Lines and Modeling*

| **Point–Slope Form** $y - y_1 = m(x - x_1)$ The slope of the line is m. The line passes through (x_1, y_1). | Write a point–slope equation for the line with slope -2 that contains $(3, -5)$. $y - y_1 = m(x - x_1)$ $y - (-5) = -2(x - 3)$ | **14.** Write a point–slope equation for the line with slope $\frac{1}{4}$ and containing $(-1, 6)$. |

SECTION 2.6:   *The Algebra of Functions*

| $(f + g)(x) = f(x) + g(x)$ $(f - g)(x) = f(x) - g(x)$ $(f \cdot g)(x) = f(x) \cdot g(x)$ $(f/g)(x) = f(x)/g(x)$, provided $g(x) \neq 0$ | For $f(x) = x^2 + 3x$ and $g(x) = x - 5$: $(f + g)(x) = f(x) + g(x)$ $= x^2 + 3x + x - 5$ $= x^2 + 4x - 5;$ $(f - g)(x) = f(x) - g(x)$ $= x^2 + 3x - (x - 5)$ $= x^2 + 2x + 5;$ $(f \cdot g)(1) = f(1) \cdot g(1)$ $= 4 \cdot (-4)$ $= -16$ $(f/g)(x) = f(x)/g(x)$, provided $g(x) \neq 0$ $= \frac{x^2 + 3x}{x - 5}$, provided $x \neq 5$. | For Exercises 15–18, let $f(x) = x - 2$ and $g(x) = x - 7$. **15.** Find $(f + g)(x)$. **16.** Find $(f - g)(x)$. **17.** Find $(f \cdot g)(5)$. **18.** Find $(f/g)(x)$. |

Review Exercises: Chapter 2

⤷ Concept Reinforcement

Classify each of the following statements as either true or false.

1. The slope of a line is a measure of how the line is slanted or tilted.   [2.3]

2. Every line has a *y*-intercept.   [2.4]

3. Every vertical line has an *x*-intercept.   [2.4]

4. No member of a function's range can be used in two different ordered pairs.   [2.2]

5. The horizontal-line test is a quick way to determine whether a graph represents a function.   [2.2]

6. The slope of a vertical line is undefined.   [2.4]

7. The slope of the graph of a constant function is 0.   [2.4]

8. Extrapolation is done to predict future values.   [2.5]

9. If two lines are perpendicular, the slope of one line is the opposite of the slope of the second line.   [2.4]

10. In order for $(f/g)(a)$ to exist, we must have $g(a) \neq 0$.   [2.6]

Determine whether the ordered pair is a solution of the given equation.   [2.1]

11. $(-2, 8)$, $x = 2y + 12$

12. $\left(0, -\frac{1}{2}\right)$, $3a - 4b = 2$

13. In which quadrant or on what axis is $(-3, 5)$ located?   [2.1]

14. Graph: $y = -x^2 + 1$.   [2.1]

15. Find the rate of change for the graph shown. Be sure to use appropriate units.   [2.3]

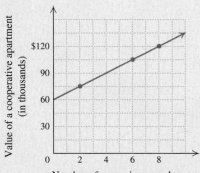

16. In 2011, there were 71,000 new homes sold in the United States by the end of March and 209,000 sold by the end of August. Calculate the rate at which new homes were being sold.   [2.3]
    Source: www.census.gov

Find the slope of each line. If the slope is undefined, state this.

17. Containing the points $(4, 5)$ and $(-3, 1)$ [2.3]

18. Containing the points $(-16.4, 2.8)$ and $(-16.4, 3.5)$ [2.4]

19. Containing the points $(-1, -2)$ and $(-5, -1)$ [2.3]

20. Containing the points $\left(\frac{1}{3}, \frac{1}{2}\right)$ and $\left(\frac{1}{6}, \frac{1}{2}\right)$ [2.3]

Find the slope and the y-intercept of the graph of each equation.   [2.3]

21. $g(x) = -5x - 11$

22. $-6y + 5x = 10$

23. The percentage of Americans $s(t)$ who say they have "a great deal of stress" can be estimated by $s(t) = -2.5t + 32$, where t is the number of years after 2007. What do the numbers -2.5 and 32 signify?   [2.3]
    Sources: American Psychological Association; Harris Interactive

Find the slope of the graph of each equation. If the slope is undefined, state this.   [2.4]

24. $y + 3 = 7$ 25. $-2x = 9$

26. Find the intercepts of the line given by $3x - 2y = 8$.   [2.4]

Graph.

27. $y = -3x + 2$ [2.3]

28. $-2x + 4y = 8$ [2.4]

29. $y = 6$ [2.4]

30. $y + 1 = \frac{3}{4}(x - 5)$ [2.5]

31. $8x + 32 = 0$ [2.4]   32. $g(x) = 15 - x$ [2.3]

33. $f(x) = \frac{1}{2}x - 3$ [2.3]   34. $f(x) = 0$ [2.4]

35. Solve $2 - x = 5 + 2x$ graphically. Then check your answer by solving the equation algebraically.   [2.4]

36. Tee Prints charges $120 to print 5 custom-designed tee shirts. Each additional tee shirt costs $8. Use a graph to estimate the number of tee shirts printed if the total cost of the order was $200.   [2.4]

Determine whether the graphs of each pair of equations are parallel, perpendicular, or neither.   [2.4]

37. $y + 5 = -x$,   38. $3x - 5 = 7y$,
    $x - y = 2$ $7y - 3x = 7$

39. Find a linear function whose graph has slope $\frac{2}{9}$ and *y*-intercept $(0, -4)$.   [2.3]

40. Find an equation in point–slope form of the line with slope -5 and containing $(1, 10)$.   [2.5]

41. Using function notation, write a slope–intercept equation for the line containing $(2, 5)$ and $(-2, 6)$. [2.5]

Find an equation of the line.   [2.5]

42. Containing the point $(2, -5)$ and parallel to the line $3x - 5y = 9$

43. Containing the point $(2, -5)$ and perpendicular to the line $3x - 5y = 9$

The following table shows the number of acres in the United States planted with organic cotton.

Input, Year	Output, Number of Acres Planted with Organic Cotton
2006	6,000
2008	8,500
2010	11,800
2012	16,400

Source: Organic Trade Association

44. Use the data in the table to draw a graph and to estimate the number of acres planted with organic cotton in 2009.   [2.5]

45. Use the graph from Exercise 44 to estimate the number of acres planted with organic cotton in 2014.   [2.5]

46. *Records in the 200-meter Run.*   In 1983, the record for the 200-m run was 19.75 sec.* In 2011, it was 19.19 sec. Let $R(t)$ represent the record in the 200-m run t years after 1980.   [2.5]

Source: International Association of Athletics Federation

a) Find a linear function that fits the data.
b) Use the function of part (a) to predict the record in 2015 and in 2020.

Determine whether each of these is a linear equation.   [2.4]

47. $2x - 7 = 0$

48. $3x - \dfrac{y}{8} = 7$

---

* Records are for elevations less than 1000 m.

49. $2x^3 - 7y = 5$

50. $\dfrac{2}{x} = y$

51. For the following graph of f, determine **(a)** $f(2)$; **(b)** the domain of f; **(c)** any x-values for which $f(x) = 2$; and **(d)** the range of f.   [2.2]

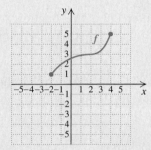

52. Determine the domain and the range of the function g represented below.   [2.2]

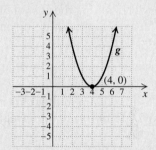

Determine whether each of the following is the graph of a function.   [2.2]

53.

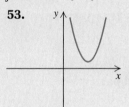

54.

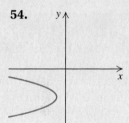

Let $g(x) = 3x - 6$ and $h(x) = x^2 + 1$. Find each of the following.

55. $g(0)$ [2.2]

56. $h(-5)$ [2.2]

57. $g(a + 5)$ [2.2]

58. $(g \cdot h)(4)$ [2.6]

59. $(g/h)(-1)$ [2.6]

60. $(g + h)(x)$ [2.6]

61. The domain of g [2.2]

62. The domain of $g + h$ [2.6]

63. The domain of h/g [2.6]

Synthesis

64. Explain the difference between $f(a) + h$ and $f(a + h)$.   [2.2], [2.6]

65. Explain why the slope of a vertical line is undefined whereas the slope of a horizontal line is 0.   [2.4]

66. Find the *y*-intercept of the function given by
$$f(x) + 3 = 0.17x^2 + (5 - 2x)^x - 7.$$
[1.6], [2.4]

67. Determine the value of *a* such that the graphs of
$$3x - 4y = 12 \quad \text{and} \quad ax + 6y = -9$$
are parallel.   [2.4]

68. Treasure Tea charges \$7.99 for each package of loose tea. Shipping charges are \$2.95 per package plus \$20 per order for overnight delivery. Find a linear function for determining the cost of one order of *x* packages of tea, including shipping and overnight delivery.   [2.5]

69. Match each sentence with the most appropriate of the four graphs below.   [2.3]

a) Joni walks for 10 min to the train station, rides the train for 15 min, and then walks 5 min to the office.

b) During a workout, Carter bikes for 10 min, runs for 15 min, and then walks for 5 min.

c) Andrew pilots his motorboat for 10 min to the middle of the lake, fishes for 15 min, and then motors for another 5 min to another spot.

d) Patti waits 10 min for her train, rides the train for 15 min, and then runs for 5 min to her job.

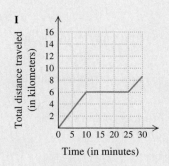

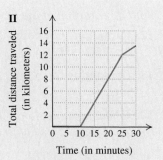

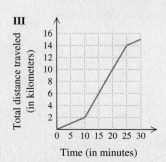

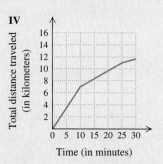

Test: Chapter 2

For step-by-step test solutions, access the Chapter Test Prep Videos in MyMathLab®, on YouTube (search " Bittinger Inter Alg CA" and click on "Channels"), or by scanning the code.

1. Determine whether the ordered pair is a solution of the given equation.
$$(12, -3), \ x + 4y = -20$$

2. Graph: $f(x) = x^2 + 3$.

3. Find the rate of change for the graph below. Use appropriate units.

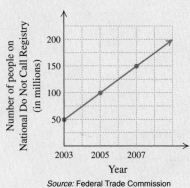

Source: Federal Trade Commission

Find the slope of the line containing the following points. If the slope is undefined, state this.

4. $(-2, -2)$ and $(6, 3)$

5. $(-3.1, 5.2)$ and $(-4.4, 5.2)$

Find the slope and the y-intercept.

6. $f(x) = -\frac{3}{5}x + 12$

7. $-5y - 2x = 7$

Find the slope. If the slope is undefined, state this.

8. $f(x) = -3$

9. $x - 5 = 11$

10. Find the intercepts of the line given by $5x - y = 15$.

Graph.

11. $f(x) = -3x + 4$

12. $y - 1 = -\frac{1}{2}(x + 4)$

13. $-2x + 5y = 20$

14. $3 - x = 9$

15. Solve $x + 3 = 2x$ graphically. Then check your answer by solving the equation algebraically.

16. There were 51 million international visitors to the United States in 2006 and 60 million visitors in 2010. Draw a graph and estimate the number of international visitors in 2009.

   Source: U.S. Department of Commerce

17. Which of these are linear equations?

   a) $8x - 7 = 0$
   b) $4x - 9y^2 = 12$
   c) $2x - 5y = 3$

Determine without graphing whether the graphs of each pair of equations are parallel, perpendicular, or neither.

18. $4y + 2 = 3x$,
   $-3x + 4y = -12$

19. $y = -2x + 5$,
   $2y - x = 6$

20. Find a linear function whose graph has slope -5 and y-intercept $(0, -1)$.

21. Find an equation in point–slope form of the line with slope 4 and containing $(-2, -4)$.

22. Using function notation, write a slope–intercept equation for the line containing $(3, -1)$ and $(4, -2)$.

Find an equation of the line.

23. Containing $(-3, 2)$ and parallel to the line $2x - 5y = 8$

24. Containing $(-3, 2)$ and perpendicular to the line $2x - 5y = 8$

25. If you rent a truck for one day and drive it 250 mi, the cost is \$100. If you rent it for one day and drive it 300 mi, the cost is \$115. Let $C(m)$ represent the cost, in dollars, of driving m miles.

   a) Find a linear function that fits the data.
   b) Use the function to determine how much it will cost to rent the truck for one day and drive it 500 mi.

26. For the following graph of f, determine **(a)** $f(-2)$; **(b)** the domain of f; **(c)** any x-value for which $f(x) = \frac{1}{2}$; and **(d)** the range of f.

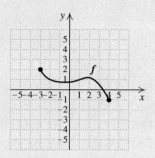

Find the following, given that $g(x) = \dfrac{1}{x}$ *and* $h(x) = 2x + 1$.

27. $h(-5)$

28. $(g + h)(x)$

29. The domain of g

30. The domain of $g + h$

31. The domain of g/h

Synthesis

32. The function $f(t) = 5 + 15t$ can be used to determine a bicycle racer's location, in miles from the starting line, measured t hours after passing the 5-mi mark.

   a) How far from the start will the racer be 100 min (1 hr 40 min) after passing the 5-mi mark?
   b) Assuming a constant rate, how fast is the racer traveling?

33. The graph of $f(x) = mx + b$ contains the points $(r, 3)$ and $(7, s)$. Express s in terms of r if the graph is parallel to the graph of $3x - 2y = 7$.

34. Given that $f(x) = 5x^2 + 1$ and $g(x) = 4x - 3$, find an expression for $h(x)$ so that the domain of $f/g/h$ is

   $$\left\{ x \mid x \text{ is a real number } and \ x \neq \tfrac{3}{4} \ and \ x \neq \tfrac{2}{7} \right\}.$$

   Answers may vary.

Cumulative Review: Chapters 1–2

Perform the indicated operation.   [1.2]

1. $-3 - (-10)$

2. $12 \div (-4)$

3. $-\frac{6}{5} \div \left(-\frac{2}{15}\right)$

4. $-\frac{1}{3} + \frac{5}{6}$

5. Simplify: $3x - 5(x - 7) + 2$.   [1.3]

Solve. If appropriate, classify the equation as an identity or as a contradiction.   [1.3]

6. $2(3t + 1) - t = 5(t + 6)$

7. $-2(4 - x) + 3 = 8 - 6x$

8. Solve for y: $8x - 3y = 12$.   [1.5]

Simplify. Do not use negative exponents in the answer. [1.6]

9. $(3x^2y^{-1})^{-1}$

10. -2^{-3}

11. $(10a^2b)^0$

12. $\left(\dfrac{3x^2y^{-2}}{15x^{-1}y^{-1}}\right)^2$

13. Find the slope of the line containing $(2, 5)$ and $(1, 10)$.   [2.3]

14. Find the slope of the line given by $f(x) = 8x + 3$. [2.3]

15. Find the slope of the line given by $y + 6 = -4$. [2.4]

16. Find a linear function whose graph has slope -1 and y-intercept $\left(0, \frac{1}{5}\right)$.   [2.3]

17. Find a linear function whose graph contains $(-1, 3)$ and $(-3, -5)$.   [2.5]

18. Find an equation of the line containing $(5, -2)$ and perpendicular to the line given by $x - y = 5$.   [2.5]

Find the following, given that $f(x) = x + 5$ and $g(x) = x^2 - 1$.

19. $g(-10)$ [2.2]

20. $(f \cdot g)(-5)$ [2.6]

21. $(g/f)(x)$ [2.6]

22. Find the domain of f if $f(x) = \dfrac{x}{x + 6}$.   [2.2]

Graph.

23. $f(x) = 5$ [2.4]

24. $y - 2 = \frac{1}{3}(x + 1)$ [2.5]

25. $x = -3$ [2.4]

26. $y = |x| + 1$ [2.1]

27. $y = -x - 3$ [2.3]

28. $10x + y = -20$ [2.4]

29. *Enrollment.*   In 2009, there were 7.6 million undergraduates enrolled in U.S. four-year public colleges. The number of full-time students was 2.8 times the number of part-time students. How many full-time students and how many part-time students were enrolled in four-year public colleges in 2009?   [1.4]

Source: National Center for Education Statistics

30. *Tuition.*   The average yearly in-state tuition at four-year public colleges was \$6185 in 2007–2008. In 2011–2012, the average yearly tuition was \$8244. Let $c(t)$ represent the average yearly in-state tuition at a four-year public college t years after 1999–2000.   [2.5]

Source: National Center for Education Statistics

a) Find a linear function that fits the data.

b) Use the function from part (a) to predict the average yearly in-state tuition at a four-year public college in 2014–2015.

31. *Organic Sales.*   Consumer sales of organic nonfood goods, in million of dollars, is given by $s(t) = 245.8t + 745$, where t is the number of years after 2005.

a) Find the consumer sales of organic nonfood goods in 2012.   [2.2]

b) What do the numbers 245.8 and 745 signify? [2.3]

Synthesis

Translate to an algebraic expression.   [1.1]

32. The difference of two squares

33. The product of the sum and the difference of the same two numbers

34. Find an equation for a linear function f if $f(2) = 4$ and $f(0) = 3$.   [2.3]

Systems of Linear Equations and Problem Solving

SERVING SIZE	CALORIES	PROTEIN (in grams)	VITAMIN C (in milligrams)
Roast beef, 3 oz	300	20	0
Baked potato, 1	100	5	20
Broccoli, 156 g	50	5	100
Asparagus, 180 g	50	5	44

What's Good for You Ought to Be Good to You.

Dieticians and nutritionists face the daily challenge of making nutritious meals both attractive and appetizing. Using nutrition information such as that in the table above, they design meal plans that meet individual needs and tastes.

In this chapter, we design two simple meals with equivalent requirements using the foods listed here. You may find that you would prefer eating one of these meals over the other. (*See Exercises 23 and 24 in Section 3.5.*)

Having a solid background in math enables me to understand new research that is critical in my ever-changing field.

Jonah Soolman, a registered dietitian at Soolman Nutrition and Wellness LLC, uses math to analyze the nutrient composition of his patients' diets and to calculate their caloric needs and energy expenditure.

I n fields ranging from business to zoology, problems arise that are most easily solved using a *system of equations*. In this chapter, we solve systems and applications using graphing, substitution, elimination, and matrices.

3.1 Systems of Equations in Two Variables

Translating ◾ Identifying Solutions ◾ Solving Systems Graphically

The whooping crane was one of the first animals on the U.S. endangered list. Conservation efforts have increased the number of whooping cranes from 16 in 1941 to over 500 in 2012.

1. Refer to Example 1. In 2012, there were 25 species of amphibians in the United States that were considered threatened or endangered. The number of species considered endangered was 1.5 times the number considered threatened. Write a system of equations that models the number of amphibian species considered endangered or threatened in 2012.

Source: U.S. Fish and Wildlife Service

Translating

Problems involving two unknown quantities are often translated most easily using two equations in two unknowns. Together, these equations form a **system of equations**. We look for a solution to the problem by attempting to find a pair of numbers for which *both* equations are true.

EXAMPLE 1 *Endangered Species.* In 2012, there were 92 species of birds in the United States that were considered threatened (likely to become endangered) or endangered (in danger of becoming extinct). The number of species considered threatened was 3 less than one-fourth of the number considered endangered. Write a system of equations that models the number of U.S. bird species considered endangered or threatened in 2012.

Source: U.S. Fish and Wildlife Service

SOLUTION

1. **Familiarize.** We let t represent the number of threatened bird species and d the number of endangered bird species in 2012.

2. **Translate.** There are two statements to translate. First, we look at the total number of endangered or threatened species of birds:

Rewording:	The number of threatened species	plus	the number of endangered species	was	92.
Translating:	t	$+$	d	$=$	92

The second statement compares the two amounts, d and t:

Rewording:	The number of threatened species	was	3 less than one-fourth of the number of endangered species.
Translating:	t	$=$	$\frac{1}{4}d - 3$

We have now translated the problem to a pair, or **system, of equations**:

$$t + d = 92,$$
$$t = \tfrac{1}{4}d - 3.$$

YOUR TURN

> ### SYSTEM OF EQUATIONS
>
> A *system of equations* is a set of two or more equations, in two or more variables, for which a common solution is sought.

Problems like Example 1 *can* be solved using one variable; however, as problems become complicated, you will find that using more than one variable (and more than one equation) is often the preferable approach.

EXAMPLE 2   *Jewelry Design.* Star Bright Jewelry Design purchased 80 beads for a total of $39 (excluding tax) to make a necklace. Some of the beads were sterling silver beads costing 40¢ each and the rest were gemstone beads costing 65¢ each. How many of each type were bought? Translate to a system of equations.

SOLUTION

1. **Familiarize.** To familiarize ourselves with this problem, let's guess that the designer bought 20 beads at 40¢ each and 60 beads at 65¢ each. The total cost would then be

   $$20 \cdot 40¢ + 60 \cdot 65¢ = 800¢ + 3900¢, \quad \text{or} \quad 4700¢.$$

   Since 4700¢ = $47 and $47 ≠ $39, our guess is incorrect. Rather than guess again, let's see how algebra can be used to translate the problem.

2. **Translate.** We let s = the number of silver beads and g = the number of gemstone beads. Since the cost of each bead is given in cents and the total cost is in dollars, we must choose one of the units to use throughout the problem. We choose to work in cents, so the total cost is 3900¢. The information can be organized in a table, which will help with the translating.

Type of Bead	Silver	Gemstone	Total
Number Bought	s	g	80
Price	40¢	65¢	
Amount	40s¢	65g¢	3900¢

$s + g = 80$

$40s + 65g = 3900$

The first row of the table and the first sentence of the problem indicate that a total of 80 beads were bought:

$$s + g = 80.$$

Since each silver bead cost 40¢ and s of them were bought, 40s cents was paid for the silver beads. Similarly, 65g cents was paid for the gemstone beads. This leads to a second equation:

$$40s + 65g = 3900.$$

We now have the following system of equations as the translation:

$$s + g = 80,$$
$$40s + 65g = 3900.$$

2. Refer to Example 2. For another necklace, Star Bright Jewelry Design purchased 60 sterling silver beads and gemstone beads for a total of $30. How many of each type did the designer buy? Translate to a system of equations.

 YOUR TURN

Student Notes

We complete the solutions of Examples 1 and 2 in Section 3.3.

> **CAUTION!** Be sure to check the ordered pair in *both* equations.

3. Determine whether $(6, -1)$ is a solution of the system

$$x + y = 5,$$
$$x - 3y = 3.$$

Identifying Solutions

A *solution* of a system of two equations in two variables is an ordered pair of numbers that makes *both* equations true.

EXAMPLE 3   Determine whether $(-4, 7)$ is a solution of the system

$$x + y = 3,$$
$$5x - y = -27.$$

SOLUTION   Unless stated otherwise, we use alphabetical order of the variables. Thus we replace x with -4 and y with 7:

$$\begin{array}{c|c} x + y = 3 \\ \hline -4 + 7 & 3 \\ 3 \overset{?}{=} 3 & \text{TRUE} \end{array} \qquad \begin{array}{c|c} 5x - y = -27 \\ \hline 5(-4) - 7 & -27 \\ -20 - 7 & \\ -27 \overset{?}{=} -27 & \text{TRUE} \end{array}$$

The pair $(-4, 7)$ makes both equations true, so it is a solution of the system. We can also describe the solution by writing $x = -4$ and $y = 7$. Set notation can also be used to list the solution set $\{(-4, 7)\}$.

YOUR TURN

Solving Systems Graphically

Recall that the graph of an equation is a drawing that represents its solution set. If we graph the equations in Example 3, we find that $(-4, 7)$ is the only point common to both lines. Thus one way to solve a system of two equations is to graph both equations and identify any points of intersection. **The coordinates of each point of intersection represent a solution of that system.**

$y = mx + b$

$$x + y = 3,$$
$$5x - y = -27$$

The point of intersection of the graphs is $(-4, 7)$.

The solution of the system is $(-4, 7)$.

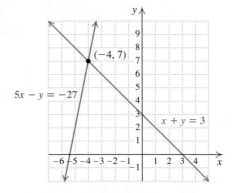

Most pairs of lines have exactly one point in common. We will soon see, however, that this is not always the case.

EXAMPLE 4   Solve each system graphically.

a) $y - x = 1,$
    $y + x = 3$

b) $y = -3x + 5,$
    $y = -3x - 2$

c) $3y - 2x = 6,$
    $-12y + 8x = -24$

SOLUTION

a) We begin by graphing the equations. All ordered pairs from line L_1 are solutions of the first equation. All ordered pairs from line L_2 are solutions of the second equation. The point of intersection has coordinates that make *both* equations true. Apparently, $(1, 2)$ is the solution. Graphs are not always accurate, so solving by graphing may yield approximate answers. Our check below shows that $(1, 2)$ is indeed the solution.

Graph both equations.

Look for any points in common.

$$y - x = 1,$$
$$y + x = 3$$

The solution of the system is $(1, 2)$.

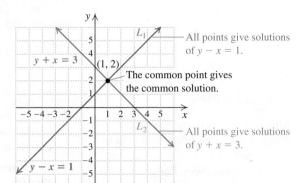

All points give solutions of $y - x = 1$.

The common point gives the common solution.

All points give solutions of $y + x = 3$.

Check.

Check:

$y - x = 1$		$y + x = 3$	
$2 - 1$	1	$2 + 1$	3
$1 \overset{?}{=} 1$	TRUE	$3 \overset{?}{=} 3$	TRUE

b) We graph the equations. The lines have the same slope, -3, and different y-intercepts, so they are parallel. There is no point at which they cross, so the system has no solution.

$$y = -3x + 5,$$
$$y = -3x - 2$$

The system has no solution.

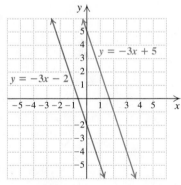

$y = -3x + 5$

$y = -3x - 2$

c) We graph the equations and find that the same line is drawn twice. Thus any solution of one equation is a solution of the other. Each point on the line is a solution of both equations, so the system itself has an infinite number of solutions. We check one solution, $(0, 2)$, which is the y-intercept of each equation.

$$3y - 2x = 6,$$
$$-12y + 8x = -24$$

The solution of the system is $\{(x, y) \mid 3y - 2x = 6\}$.

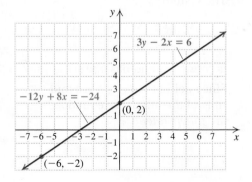

$3y - 2x = 6$

$-12y + 8x = -24$

$(0, 2)$

$(-6, -2)$

Check:

$3y - 2x = 6$		$-12y + 8x = -24$	
$3(2) - 2(0)$	6	$-12(2) + 8(0)$	-24
$6 - 0$		$-24 + 0$	
$6 \overset{?}{=} 6$	TRUE	$-24 \overset{?}{=} -24$	TRUE

Technology Connection

On most graphing calculators, an INTERSECT option allows us to find the coordinates of the intersection directly.

To illustrate, consider the following system:

$$3.45x + 4.21y = 8.39,$$
$$7.12x - 5.43y = 6.18.$$

After solving for y in each equation, we obtain the graph below. Using INTERSECT, we see that, to the nearest hundredth, the coordinates of the intersection are $(1.47, 0.79)$.

$y_1 = (8.39 - 3.45x)/4.21,$
$y_2 = (6.18 - 7.12x)/(-5.43)$

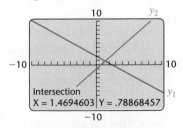

Intersection
X = 1.4694603 Y = .78868457

Use a graphing calculator to solve each of the following systems. Round all x- and y-coordinates to the nearest hundredth.

1. $y = -5.43x + 10.89,$
    $y = 6.29x - 7.04$
2. $-9.25x - 12.94y = -3.88,$
    $21.83x + 16.33y = 13.69$

4. Solve graphically:

$$x = y,$$
$$y = 2x - 4.$$

Student Notes

Although the system in Example 4(c) is true for an infinite number of ordered pairs, those pairs must be of a certain form. Only pairs that are solutions of $3y - 2x = 6$ or $-12y + 8x = -24$ are solutions of the system. It is incorrect to think that *all* ordered pairs are solutions.

You can check that $(-6, -2)$ is another solution of both equations. In fact, any pair that is a solution of one equation is a solution of the other equation as well. Thus the solution set is $\{(x, y) \mid 3y - 2x = 6\}$ or, in words, "the set of all pairs (x, y) for which $3y - 2x = 6$." Since the two equations are equivalent, we could have written instead $\{(x, y) \mid -12y + 8x = -24\}$.

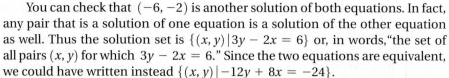 YOUR TURN

When we graph a system of two linear equations in two variables, one of the following three outcomes will occur.

1. The lines have one point in common, and that point is the only solution of the system (see Example 4a). Any system that has *at least one solution* is said to be **consistent**.

2. The lines are parallel, with no point in common, and the system has *no solution* (see Example 4b). This type of system is called **inconsistent**.

3. The lines coincide, sharing the same graph. Because every solution of one equation is a solution of the other, the system has an infinite number of solutions (see Example 4c). Since it has at least one solution, this type of system is also **consistent**.

When one equation in a system can be obtained by multiplying both sides of another equation by a constant, the two equations are said to be **dependent**. Thus the equations in Example 4(c) are dependent, but those in Examples 4(a) and 4(b) are **independent**. For systems of three or more equations, the definitions of dependent and independent will be slightly modified.

Algebraic 🔗 Graphical Connection

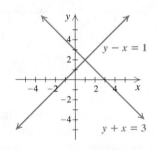

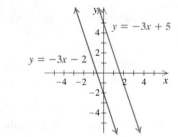

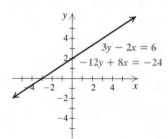

Graphs intersect at one point.

The system

$$y - x = 1,$$
$$y + x = 3$$

is *consistent* and has one solution.

Since neither equation is a multiple of the other, the equations are *independent*.

Graphs are parallel.

The system

$$y = -3x - 2,$$
$$y = -3x + 5$$

is *inconsistent* because there is no solution.

Since neither equation is a multiple of the other, the equations are *independent*.

Equations have the same graph.

The system

$$3y - 2x = 6,$$
$$-12y + 8x = -24$$

is *consistent* and has an infinite number of solutions.

Since one equation is a multiple of the other, the equations are *dependent*.

 Chapter Resource:
Visualizing for Success, p. 213

Graphing is helpful when solving systems because it allows us to "see" the solution. It can also be used on systems of nonlinear equations, and in many applications, it provides a satisfactory answer. However, graphing often lacks precision, especially when fraction solutions or decimal solutions are involved.

3.1 EXERCISE SET

FOR EXTRA HELP

MyMathLab® Math XL
PRACTICE   WATCH   READ   REVIEW

Vocabulary and Reading Check

Classify each of the following statements as either true or false.

1. Every system of equations has at least one solution.

2. It is possible for a system of equations to have an infinite number of solutions.

3. Every point of intersection of the graphs of the equations in a system corresponds to a solution of the system.

4. The graphs of the equations in a system of two equations may coincide.

5. The graphs of the equations in a system of two equations could be parallel lines.

6. Any system of equations that has at most one solution is said to be consistent.

7. Any system of equations that has more than one solution is said to be inconsistent.

8. The equations $x + y = 5$ and $2(x + y) = 2(5)$ are dependent.

Identifying Solutions

Determine whether the ordered pair is a solution of the given system of equations. Remember to use alphabetical order of variables.

9. $(2, 3)$; $2x - y = 1$,
   $5x - 3y = 1$

10. $(4, 0)$; $2x + 7y = 8$,
    $x - 9y = 4$

11. $(-5, 1)$; $x + 5y = 0$,
    $y = 2x + 9$

12. $(-1, -2)$; $x + 3y = -7$,
    $3x - 2y = 12$

13. $(0, -5)$; $x - y = 5$,
    $y = 3x - 5$

14. $(5, 2)$; $a + b = 7$,
    $2a - 8 = b$

Aha! 15. $(3, -1)$; $3x - 4y = 13$,
    $6x - 8y = 26$

16. $(4, -2)$; $-3x - 2y = -8$,
    $8 = 3x + 2y$

Solving Systems Graphically

Solve each system graphically. Be sure to check your solution. If a system has an infinite number of solutions, use set-builder notation to write the solution set. If a system has no solution, state this.

17. $x - y = 1$,
    $x + y = 5$

18. $x + y = 6$,
    $x - y = 4$

19. $3x + y = 5$,
    $x - 2y = 4$

20. $2x - y = 4$,
    $5x - y = 13$

21. $2y = 3x + 5$,
    $x = y - 3$

22. $4x - y = 9$,
    $x - 3y = 16$

23. $x = y - 1$,
    $2x = 3y$

24. $a = 1 + b$,
    $b = 5 - 2a$

25. $y = -1$,
    $x = 3$

26. $y = 2$,
    $x = -4$

27. $t + 2s = -1$,
    $s = t + 10$

28. $b + 2a = 2$,
    $a = -3 - b$

29. $2b + a = 11$,
    $a - b = 5$

30. $y = -\frac{1}{3}x - 1$,
    $4x - 3y = 18$

31. $y = -\frac{1}{4}x + 1$,
    $2y = x - 4$

32. $6x - 2y = 2$,
    $9x - 3y = 1$

33. $y - x = 5$,
    $2x - 2y = 10$

34. $y = x + 2$,
    $3y - 2x = 4$

35. $y = 3 - x$,
    $2x + 2y = 6$

36. $2x - 3y = 6$,
    $3y - 2x = -6$

37. For the systems in the odd-numbered exercises 17–35, which are consistent?

38. For the systems in the even-numbered exercises 18–36, which are consistent?

39. For the systems in the odd-numbered exercises 17–35, which contain dependent equations?

40. For the systems in the even-numbered exercises 18–36, which contain dependent equations?

Translating

Translate each problem situation to a system of equations. Do not attempt to solve, but save for later use.

41. The sum of two numbers is 10. The first number is $\frac{2}{3}$ of the second number. What are the numbers?

42. The sum of two numbers is 30. The first number is twice the second number. What are the numbers?

43. *e-Mail Usage.* In 2012, the average e-mail user sent or received 115 e-mails each day. The number of e-mails received was 1 more than twice the number sent. How many were sent and how many were received each day?

Source: The Radicati Group, Inc.

44. *Nontoxic Furniture Polish.* A nontoxic wood furniture polish can be made by mixing mineral (or olive) oil with vinegar. To make a 16-oz batch for a squirt bottle, Jazmun uses an amount of mineral oil that is 4 oz more than twice the amount of vinegar. How much of each ingredient is required?

Sources: Based on information from Chittenden Solid Waste District and *Clean House, Clean Planet* by Karen Logan

45. *Geometry.* Two angles are supplementary.* One angle is 3° less than twice the other. Find the measures of the angles.

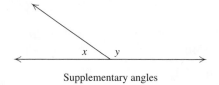

Supplementary angles

46. *Geometry.* Two angles are complementary.† The sum of the measures of the first angle and half the second angle is 64°. Find the measures of the angles.

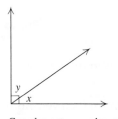

Complementary angles

47. *Basketball Scoring.* Wilt Chamberlain once scored 100 points, setting a record for points scored in an NBA game. Chamberlain took only two-point shots and (one-point) foul shots and made a total of 64 shots. How many shots of each type did he make?

48. *Basketball Scoring.* The Fenton College Cougars made 40 field goals in a recent basketball game, some 2-pointers and the rest 3-pointers. Altogether, the 40 baskets counted for 89 points. How many of each type of field goal was made?

---

*The sum of the measures of two supplementary angles is 180°.
†The sum of the measures of two complementary angles is 90°.

49. *Retail Sales.* Simply Souvenirs sold 45 hats and tee shirts. The hats sold for $14.50 each and the tee shirts for $19.50 each. In all, $697.50 was taken in for the souvenirs. How many of each type of souvenir were sold?

50. *Retail Sales.* Cool Treats sold 60 ice cream cones. Single-dip cones sold for $2.50 each and double-dip cones for $4.15 each. In all, $179.70 was taken in for the cones. How many of each size cone were sold?

51. *Fitness Equipment.* In 2012, the Mio® Classic Heart Rate watch cost $39.99 and the Mio® Drive Plus Heart Rate Watch cost $59.99. A nonprofit community health organization purchased 35 heart-rate watches for use at a wellness center. If the organization spent $1639.65 for the watches, how many of each type did they purchase?

52. *Fundraising.* The Buck Creek Fire Department served 250 dinners. A child's plate cost $5.50 and an adult's plate cost $9.00. A total of $1935 was collected. How many of each type of plate were served?

53. *Lacrosse.* The perimeter of an NCAA men's lacrosse field is 340 yd. The length is 50 yd longer than the width. Find the dimensions.

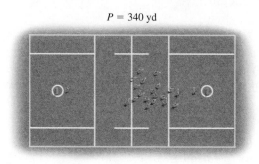

P = 340 yd

54. *Tennis.* The perimeter of a standard tennis court used for doubles is 228 ft. The width is 42 ft less than the length. Find the dimensions.

55. Write a problem for a classmate to solve that requires writing a system of two equations. Devise the problem so that the solution is "The Bucks made 6 three-point baskets and 31 two-point baskets."

56. Write a problem for a classmate to solve that can be translated into a system of two equations. Devise the problem so that the solution is "In 2012, Diana took five 3-credit classes and two 4-credit classes."

Skill Review

Simplify. Do not use negative exponents in the answer.

57. $-\frac{1}{2} - \frac{3}{10}$ [1.2]

58. $(0.05)(-1.2)$ [1.2]

59. $(-3)^2 - 2 - 4 \cdot 6 \div 2 \cdot 3$ [1.2]

60. $(-3x^2y^{-4})(-2x^{-7}y^{12})$ [1.6]

61. -10^{-2} [1.6]

62. $(5a^2b^6)^0$ [1.6]

Synthesis

Advertising Media. For Exercises 63 and 64, consider the following graph showing the U.S. market share for various advertising media.

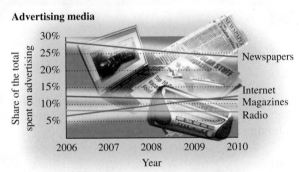

Source: ZenithOptimedia

63. In what year did the Internet and radio have the same advertising market share? Explain.

64. Will Internet advertising ever exceed that of newspapers? If so, when? Explain your answers.

65. For each of the following conditions, write a system of equations. Answers may vary.
   a) $(5, 1)$ is a solution.
   b) There is no solution.
   c) There is an infinite number of solutions.

66. A system of linear equations has $(1, -1)$ and $(-2, 3)$ as solutions. Determine:
   a) a third point that is a solution (answers may vary), and
   b) how many solutions there are.

67. The solution of the following system is $(4, -5)$. Find A and B.
$$Ax - 6y = 13,$$
$$x - By = -8.$$

Translate to a system of equations. Do not solve.

68. *Ages.* Tyler is twice as old as his son. Ten years ago, Tyler was three times as old as his son. How old are they now?

69. *Work Experience.* Dell and Juanita are mathematics professors at a state university. Together, they have 46 years of service. Two years ago, Dell had taught 2.5 times as many years as Juanita. How long has each taught at the university?

70. *Design.* A piece of posterboard has a perimeter of 156 in. If you cut 6 in. off the width, the length becomes four times the width. What are the dimensions of the original piece of posterboard?

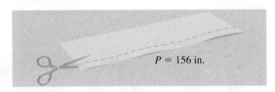

$P = 156$ in.

71. *Nontoxic Scouring Powder.* A nontoxic scouring powder is made up of 4 parts baking soda and 1 part vinegar. How much of each ingredient is needed for a 16-oz mixture?

72. Solve Exercise 41 graphically.

73. Solve Exercise 44 graphically.

Solve graphically.

74.
$$y = |x|,$$
$$3y - x = 8$$

75. $x - y = 0,$
$$y = x^2$$

In Exercises 76–79, use a graphing calculator to solve each system of linear equations for x and y. Round all coordinates to the nearest hundredth.

76. $y = 8.23x + 2.11,$
$\quad y = -9.11x - 4.66$

77. $y = -3.44x - 7.72,$
$\quad y = 4.19x - 8.22$

78. $14.12x + 7.32y = 2.98,$
$21.88x - 6.45y = -7.22$

79. $5.22x - 8.21y = -10.21,$
$-12.67x + 10.34y = 12.84$

80. Solve graphically using the grid below:
$$2x - 3y = 0,$$
$$-4x + 3y = -1.$$

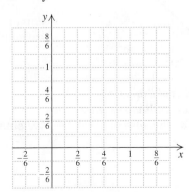

81. *Research.* Refer to Exercise 63. Find advertising market share for a recent year and extend the graph before Exercise 63 to include the new data. Do the 2006–2010 trends continue? If any new trends emerge, try to find reasons for the change.

PREPARE TO MOVE ON

Solve. [1.3]

1. $3x + 2(5x - 1) = 6$

2. $4(3y + 2) - 7y = 3$

3. $2x - (x - 7) = 18$

Solve. [1.5]

4. $3x - y = 4$, for y

5. $5y - 2x = 7$, for x

 YOUR TURN ANSWERS: SECTION 3.1

1. Let t represent the number of threatened amphibian species and d the number of endangered amphibian species; $t + d = 25, d = 1.5t$
2. $s + g = 60, 40s + 65g = 3000$　**3.** No　**4.** $(4, 4)$

3.2 Solving by Substitution or Elimination

The Substitution Method　▪　The Elimination Method

Study Skills

Learn from Your Mistakes

Immediately after each quiz or test, write out a step-by-step solution to any questions you missed. Visit your professor during office hours or consult a tutor for help with problems that still give you trouble. Misconceptions tend to persist if not corrected as soon as possible.

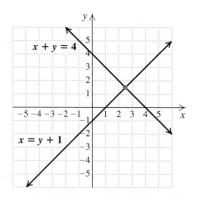

A visualization of Example 1. Note that the coordinates of the intersection are not obvious.

The Substitution Method

Algebraic (nongraphical) methods for solving systems are often superior to graphing, especially when fractions are involved. One algebraic method, the *substitution method*, relies on having a variable isolated.

EXAMPLE 1 Solve the system

$$x + y = 4, \quad (1)$$

For easy reference, we have numbered the equations.

$$x = y + 1. \quad (2)$$

SOLUTION Equation (2) says that x and $y + 1$ name the same number. Thus we can substitute $y + 1$ for x in equation (1):

$$x + y = 4 \qquad \text{Equation (1)}$$
$$(y + 1) + y = 4. \qquad \text{Substituting } y + 1 \text{ for } x$$

We solve this last equation, using methods learned earlier:

$$(y + 1) + y = 4$$
$$2y + 1 = 4 \qquad \text{Removing parentheses and combining like terms}$$
$$2y = 3 \qquad \text{Subtracting 1 from both sides}$$
$$y = \tfrac{3}{2}. \qquad \text{Dividing both sides by 2}$$

We now return to the original pair of equations and substitute $\tfrac{3}{2}$ for y in either equation so that we can solve for x. For this problem, calculations are slightly easier if we use equation (2):

$$x = y + 1 \qquad \text{Equation (2)}$$
$$= \tfrac{3}{2} + 1 \qquad \text{Substituting } \tfrac{3}{2} \text{ for } y$$
$$= \tfrac{3}{2} + \tfrac{2}{2} = \tfrac{5}{2}.$$

We obtain the ordered pair $\left(\tfrac{5}{2}, \tfrac{3}{2}\right)$. A check ensures that it is a solution.

1. Solve the system

$$x - y = 3,$$
$$x = 2 - y.$$

Check:

$$\begin{array}{c|c} x + y = 4 \\ \hline \frac{5}{2} + \frac{3}{2} & 4 \\ \frac{8}{2} & \\ 4 \overset{?}{=} 4 & \text{TRUE} \end{array} \qquad \begin{array}{c|c} x = y + 1 \\ \hline \frac{5}{2} & \frac{3}{2} + 1 \\ & \frac{3}{2} + \frac{2}{2} \\ \frac{5}{2} \overset{?}{=} \frac{5}{2} & \text{TRUE} \end{array}$$

Since $\left(\frac{5}{2}, \frac{3}{2}\right)$ checks, it is the solution.

YOUR TURN

The exact solution to Example 1 is difficult to find graphically because it involves fractions. The graph shown serves as a partial check and provides a visualization of the problem.

If neither equation in a system has a variable alone on one side, we first isolate a variable in one equation and then substitute.

EXAMPLE 2 Solve the system

$$2x + y = 6, \quad (1)$$
$$3x + 4y = 4. \quad (2)$$

SOLUTION First, we select an equation and solve for one variable. We can isolate y by subtracting $2x$ from both sides of equation (1):

$$2x + y = 6 \quad (1)$$
$$y = 6 - 2x. \quad (3) \qquad \text{Subtracting } 2x \text{ from both sides}$$

Next, we proceed as in Example 1, by substituting:

$$3x + 4(6 - 2x) = 4 \qquad \text{Substituting } 6 - 2x \text{ for } y \text{ in equation (2).}$$
$$\text{Use parentheses!}$$
$$3x + 24 - 8x = 4 \qquad \text{Distributing to remove parentheses}$$
$$3x - 8x = 4 - 24 \qquad \text{Subtracting 24 from both sides}$$
$$-5x = -20$$
$$x = 4. \qquad \text{Dividing both sides by } -5$$

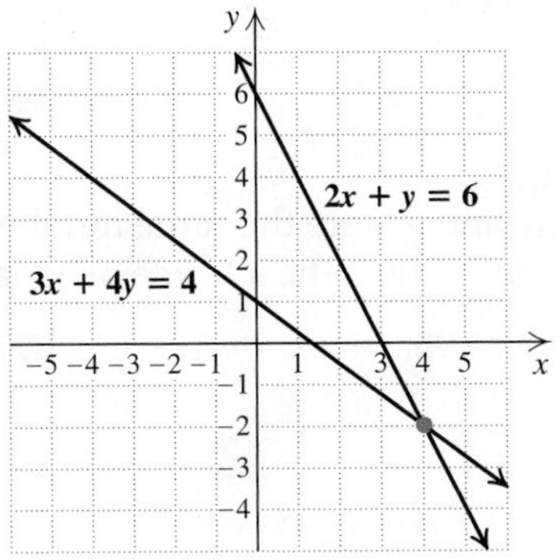

Next, we substitute 4 for x in equation (1), (2), or (3). It is easiest to use equation (3) because it has already been solved for y:

$$y = 6 - 2x$$
$$= 6 - 2(4)$$
$$= 6 - 8 = -2.$$

The pair $(4, -2)$ appears to be the solution. We check in equations (1) and (2).

A visualization of Example 2

Check:

$$\begin{array}{c|c} 2x + y = 6 \\ \hline 2(4) + (-2) & 6 \\ 8 - 2 & \\ 6 \overset{?}{=} 6 & \text{TRUE} \end{array} \qquad \begin{array}{c|c} 3x + 4y = 4 \\ \hline 3(4) + 4(-2) & 4 \\ 12 - 8 & \\ 4 \overset{?}{=} 4 & \text{TRUE} \end{array}$$

Since $(4, -2)$ checks, it is the solution.

2. Solve the system

$$x + 2y = 4,$$
$$2x + 3y = 1.$$

YOUR TURN

For a system with no solution, the graphs of the equations do not intersect. How do we recognize such systems when solving by an algebraic method?

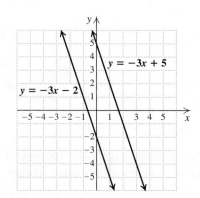

A visualization of Example 3

3. Solve the system

$$x + y = 1,$$
$$x + y = 2.$$

EXAMPLE 3 Solve the system

$$y = -3x + 5 \quad (1)$$
$$y = -3x - 2. \quad (2)$$

SOLUTION If we solve this system graphically, we see that the lines are parallel and the system has no solution. Let's now try to solve the system by substitution. Proceeding as in Example 1, we substitute $-3x - 2$ for y in the first equation:

$$-3x - 2 = -3x + 5 \quad \text{Substituting } -3x - 2 \text{ for } y \text{ in equation (1)}$$
$$-2 = 5. \qquad \text{Adding } 3x \text{ to both sides; } -2 = 5 \text{ is a contradiction. The equation is always false.}$$

Since there is no solution of $-2 = 5$, there is no solution of the system. We state that there is no solution.

 YOUR TURN

When solving a system algebraically yields a contradiction, the system has no solution. As we will see in Example 7, when solving a system of two equations algebraically yields an equation that is always true, the system has an infinite number of solutions.

The Elimination Method

The *elimination method* for solving systems of equations makes use of the *addition principle*: If $a = b$, then $a + c = b + c$.

EXAMPLE 4 Solve the system

$$2x - 3y = 0, \quad (1)$$
$$-4x + 3y = -1. \quad (2)$$

SOLUTION According to equation (2), $-4x + 3y$ and -1 are the same number. Thus we can use the addition principle and add $-4x + 3y$ to the left side of equation (1) and -1 to the right side:

$$2x - 3y = 0 \quad (1)$$
$$\underline{-4x + 3y = -1} \quad (2)$$
$$-2x + 0y = -1. \quad \text{Adding. Note that } y \text{ has been "eliminated."}$$

The resulting equation has just one variable, x, for which we solve:

$$-2x = -1$$
$$x = \tfrac{1}{2}.$$

A visualization of Example 4

4. Solve the system

$$-2x - 7y = 3,$$
$$6x + 7y = -2.$$

Next, we substitute $\tfrac{1}{2}$ for x in equation (1) and solve for y:

$$2 \cdot \tfrac{1}{2} - 3y = 0 \quad \text{Substituting. We also could have used equation (2).}$$
$$1 - 3y = 0$$
$$-3y = -1, \text{ so } y = \tfrac{1}{3}.$$

Check:

$$\begin{array}{c|c} 2x - 3y = 0 \\ \hline 2\left(\tfrac{1}{2}\right) - 3\left(\tfrac{1}{3}\right) & 0 \\ 1 - 1 & \\ & 0 \overset{?}{=} 0 \quad \text{TRUE} \end{array}$$

$$\begin{array}{c|c} -4x + 3y = -1 \\ \hline -4\left(\tfrac{1}{2}\right) + 3\left(\tfrac{1}{3}\right) & -1 \\ -2 + 1 & \\ & -1 \overset{?}{=} -1 \quad \text{TRUE} \end{array}$$

Since $\left(\tfrac{1}{2}, \tfrac{1}{3}\right)$ checks, it is the solution. See also the graph at left.

 YOUR TURN

Adding in Example 4 eliminated the variable y because two terms, $-3y$ in equation (1) and $3y$ in equation (2), are opposites. For some systems, we must multiply before adding.

EXAMPLE 5   Solve the system

$$5x + 4y = 22, \quad (1)$$
$$-3x + 8y = 18. \quad (2)$$

SOLUTION   If we add the left sides of the two equations, we will not eliminate a variable. However, if the $4y$ in equation (1) were changed to $-8y$, we would. To accomplish this change, we multiply both sides of equation (1) by -2:

$$-10x - 8y = -44 \qquad \text{Multiplying both sides of equation (1) by } -2$$
$$\underline{-3x + 8y = \quad 18}$$
$$-13x + 0 = -26 \qquad \text{Adding}$$
$$x = 2. \qquad \text{Solving for } x$$

Then
$$-3 \cdot 2 + 8y = 18 \qquad \text{Substituting 2 for } x \text{ in equation (2)}$$
$$-6 + 8y = 18$$
$$\left.\begin{array}{r} 8y = 24 \\ y = 3. \end{array}\right\} \text{Solving for } y$$

We obtain $(2, 3)$, or $x = 2, y = 3$. We leave it to the student to confirm that this checks and is the solution.

YOUR TURN

Student Notes

It is wise to double-check each step of your work as you go, rather than checking all steps at the end of a problem. One common error is to forget to multiply *both* sides of an equation when using the multiplication principle.

5. Solve the system

$$2x - 3y = 8,$$
$$6x + 5y = 4.$$

Sometimes we must multiply twice in order to make two terms become opposites.

EXAMPLE 6   Solve the system

$$2x + 3y = 17, \quad (1)$$
$$5x + 7y = 29. \quad (2)$$

SOLUTION   We multiply so that the x-terms will be eliminated when we add.

$$2x + 3y = 17, \xrightarrow{\text{Multiplying both sides by 5}} 10x + 15y = 85$$
$$5x + 7y = 29 \xrightarrow{\text{Multiplying both sides by } -2} \underline{-10x - 14y = -58}$$
$$0 + y = 27 \qquad \text{Adding}$$
$$y = 27$$

Multiply to get terms that are opposites.

Solve for one variable.

Next, we substitute to find x:

$$2x + 3 \cdot 27 = 17 \qquad \text{Substituting 27 for } y \text{ in equation (1)}$$
$$2x + 81 = 17$$
$$\left.\begin{array}{r} 2x = -64 \\ x = -32. \end{array}\right\} \text{Solving for } x$$

Substitute.

Solve for the other variable.

Check in both equations.

State the solution as an ordered pair.

Check:
$$\begin{array}{c|c} 2x + 3y = 17 \\ \hline 2(-32) + 3(27) & 17 \\ -64 + 81 & \\ 17 \overset{?}{=} 17 & \text{TRUE} \end{array} \qquad \begin{array}{c|c} 5x + 7y = 29 \\ \hline 5(-32) + 7(27) & 29 \\ -160 + 189 & \\ 29 \overset{?}{=} 29 & \text{TRUE} \end{array}$$

6. Solve the system

$$4x + 3y = 11,$$
$$3x + 2y = 7.$$

We obtain $(-32, 27)$, or $x = -32, y = 27$, as the solution.

YOUR TURN

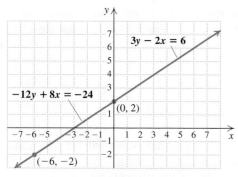

A visualization of Example 7

7. Solve the system

$$x - 3y = 2,$$
$$-5x + 15y = -10.$$

EXAMPLE 7   Solve the system

$$3y - 2x = 6, \qquad (1)$$
$$-12y + 8x = -24. \qquad (2)$$

SOLUTION   If we solve this system graphically, as shown at left, we find that the lines coincide and the system has an infinite number of solutions. Suppose we were to solve this system using the elimination method:

$$12y - 8x = 24 \qquad \text{Multiplying both sides of equation (1) by 4}$$
$$\underline{-12y + 8x = -24}$$
$$0 = 0. \qquad \text{We obtain an identity; } 0 = 0 \text{ is always true.}$$

Note that both variables have been eliminated and what remains is an identity—that is, an equation that is always true. Any pair that is a solution of equation (1) is also a solution of equation (2). The equations are dependent and the solution set is infinite:

$$\{(x, y) \mid 3y - 2x = 6\}, \text{ or equivalently, } \{(x, y) \mid -12y + 8x = -24\}.$$

YOUR TURN

Example 3 and Example 7 illustrate how to tell algebraically whether a system of two equations is inconsistent or whether the equations are dependent.

> **RULES FOR SPECIAL CASES**
>
> When solving a system of two linear equations in two variables:
>
> 1. If we obtain an identity such as $0 = 0$, then the system has an infinite number of solutions. The equations are dependent and, since a solution exists, the system is consistent.*
> 2. If we obtain a contradiction such as $0 = 7$, then the system has no solution. The system is inconsistent.

When decimals or fractions appear, it often helps to *clear* before solving.

EXAMPLE 8   Solve the system

$$0.2x + 0.3y = 1.7,$$
$$\tfrac{1}{7}x + \tfrac{1}{5}y = \tfrac{29}{35}.$$

SOLUTION   We have

$$0.2x + 0.3y = 1.7, \quad \rightarrow \text{ Multiplying both sides by 10} \rightarrow 2x + 3y = 17$$
$$\tfrac{1}{7}x + \tfrac{1}{5}y = \tfrac{29}{35} \quad \rightarrow \text{ Multiplying both sides by 35} \rightarrow 5x + 7y = 29.$$

8. Solve the system

$$\tfrac{1}{2}x - \tfrac{1}{3}y = \tfrac{1}{6},$$
$$0.3x + 0.4y = 1.2.$$

We multiplied both sides of the first equation by 10 to clear the decimals. Multiplication by 35, the least common denominator, clears the fractions in the second equation. The problem now happens to be identical to Example 6. The solution is $(-32, 27)$, or $x = -32, y = 27$.

YOUR TURN

---

*Consistent systems and dependent equations are discussed in greater detail in Section 3.4.

The steps in each algebraic method for solving systems of two equations are given below. Note that in both methods, we find the value of one variable and then substitute to find the corresponding value of the other variable.

TO SOLVE A SYSTEM USING SUBSTITUTION

1. Isolate a variable in one of the equations (unless one is already isolated).
2. Substitute for that variable in the other equation, using parentheses.
3. Solve the equation in which the substitution was made.
4. Substitute the solution from step (3) in any of the equations, and solve for the other variable.
5. Form an ordered pair and check in the original equations.

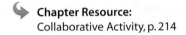

Chapter Resource:
Collaborative Activity, p. 214

TO SOLVE A SYSTEM USING ELIMINATION

1. Write both equations in standard form.
2. Multiply both sides of one or both equations by a constant, if necessary, so that the coefficients of one of the variables are opposites.
3. Add the left sides and the right sides of the resulting equations. One variable should be eliminated in the sum.
4. Solve for the remaining variable.
5. Substitute the value of the second variable in any of the equations, and solve for the other variable.
6. Form an ordered pair and check in the original equations.

CONNECTING 🔗 THE CONCEPTS

We now have three different methods for solving systems of two linear equations. Each method has certain strengths and weaknesses, as outlined below.

Method	Strengths	Weaknesses
Graphical	Solutions are displayed graphically. Can be used with any system that can be graphed.	For some systems, only approximate solutions can be found graphically. The graph drawn may not be large enough to show the solution.
Substitution	Yields exact solutions. Easy to use when a variable has a coefficient of 1.	Introduces extensive computations with fractions when solving more complicated systems. Solutions are not displayed graphically.
Elimination	Yields exact solutions. Easy to use when fractions or decimals appear in the system.	Solutions are not displayed graphically.

(continued)

EXERCISES

Solve using an appropriate method.

1. $x = y,$
$x + y = 2$

2. $x + y = 10,$
$x - y = 8$

3. $y = \frac{1}{2}x + 1,$
$y = 2x - 5$

4. $y = 2x - 3,$
$x + y = 12$

5. $12x - 19y = 13,$
$8x + 19y = 7$

6. $2x - 5y = 1,$
$3x + 2y = 11$

7. $y = \frac{5}{3}x + 7,$
$y = \frac{5}{3}x - 8$

8. $x = 2 - y,$
$3x + 3y = 6$

3.2   EXERCISE SET

FOR EXTRA HELP

MyMathLab® Math
PRACTICE   WATCH   READ   REVIEW

➤ Vocabulary and Reading Check

Choose the word from the following list that best completes each statement. Not all words will be used.

consistent   opposite
elimination   substitution
inconsistent

1. To use the _____ method, a variable must be isolated.

2. The _____ method makes use of the addition principle.

3. To eliminate a variable by adding, two terms must be _____.

4. A(n) _____ system has no solution.

➤ Concept Reinforcement

In each of Exercises 5–10, match the system listed with the choice from the column on the right that would be a subsequent step in solving the system.

5. ____ $3x - 4y = 6,$
$5x + 4y = 1$

6. ____ $2x - y = 8,$
$y = 5x + 3$

7. ____ $x - 2y = 3,$
$5x + 3y = 4$

8. ____ $8x + 6y = -15,$
$5x - 3y = 8$

9. ____ $y = 4x - 7,$
$6x + 3y = 19$

10. ____ $y = 4x - 1,$
$y = -\frac{2}{3}x - 1$

a) $-5x + 10y = -15,$
$5x + 3y = 4$

b) The lines intersect at $(0, -1)$.

c) $6x + 3(4x - 7) = 19$

d) $8x = 7$

e) $2x - (5x + 3) = 8$

f) $8x + 6y = -15,$
$10x - 6y = 16$

For Exercises 11–58, if a system has an infinite number of solutions, use set-builder notation to write the solution set. If a system has no solution, state this.

The Substitution Method

Solve using the substitution method.

11. $y = 3 - 2x,$
$3x + y = 5$

12. $3y + x = 4,$
$x = 2y - 1$

13. $3x + 5y = 3,$
$x = 8 - 4y$

14. $9x - 2y = 3,$
$3x - 6 = y$

15. $3s - 4t = 14,$
$5s + t = 8$

16. $m - 2n = 16,$
$4m + n = 1$

17. $4x - 2y = 6,$
$2x - 3 = y$

18. $t = 4 - 2s,$
$t + 2s = 6$

19. $-5s + t = 11,$
$4s + 12t = 4$

20. $5x + 6y = 14,$
$-3y + x = 7$

21. $2x + 2y = 2,$
$3x - y = 1$

22. $4p - 2q = 16,$
$5p + 7q = 1$

23. $2a + 6b = 4,$
$3a - b = 6$

24. $3x - 4y = 5,$
$2x - y = 1$

25. $2x - 3 = y,$
$y - 2x = 1$

26. $a - 2b = 3,$
$3a = 6b + 9$

The Elimination Method

Solve using the elimination method.

27. $x + 3y = 7,$
$-x + 4y = 7$

28. $2x + y = 6,$
$x - y = 3$

29. $x - 2y = 11,$
$3x + 2y = 17$

30. $5x - 3y = 8,$
$-5x + y = 4$

31. $9x + 3y = -3,$
$2x - 3y = -8$

32. $6x - 3y = 18,$
$6x + 3y = -12$

33. $5x + 3y = 19,$
$\quad x - 6y = 11$

34. $3x + 2y = 3,$
$\quad 9x - 8y = -2$

35. $5r - 3s = 24,$
$\quad 3r + 5s = 28$

36. $5x - 7y = -16,$
$\quad 2x + 8y = 26$

37. $6s + 9t = 12,$
$\quad 4s + 6t = 5$

38. $10a + 6b = 8,$
$\quad 5a + 3b = 2$

39. $\frac{1}{2}x - \frac{1}{6}y = 10,$
$\quad \frac{2}{5}x + \frac{1}{2}y = 8$

40. $\frac{1}{3}x + \frac{1}{5}y = 7,$
$\quad \frac{1}{6}x - \frac{2}{5}y = -4$

41. $\frac{x}{2} + \frac{y}{3} = \frac{7}{6},$
$\quad \frac{2x}{3} + \frac{3y}{4} = \frac{5}{4}$

42. $\frac{2x}{3} + \frac{3y}{4} = \frac{11}{12},$
$\quad \frac{x}{3} + \frac{7y}{18} = \frac{1}{2}$

Aha! **43.** $12x - 6y = -15,$
$\quad -4x + 2y = 5$

44. $8s + 12t = 16,$
$\quad 6s + 9t = 12$

45. $0.3x + 0.2y = 0.3,$
$\quad 0.5x + 0.4y = 0.4$

46. $0.3x + 0.2y = 5,$
$\quad 0.5x + 0.4y = 11$

The Substitution and Elimination Methods

Solve using any algebraic method.

47. $a - 2b = 16,$
$\quad b + 3 = 3a$

48. $5x - 9y = 7,$
$\quad 7y - 3x = -5$

49. $10x + y = 306,$
$\quad 10y + x = 90$

50. $3(a - b) = 15,$
$\quad 4a = b + 1$

51. $6x - 3y = 3,$
$\quad 4x - 2y = 2$

52. $x + 2y = 8,$
$\quad x = 4 - 2y$

53. $3s - 7t = 5,$
$\quad 7t - 3s = 8$

54. $2s - 13t = 120,$
$\quad -14s + 91t = -840$

55. $0.05x + 0.25y = 22,$
$\quad 0.15x + 0.05y = 24$

56. $2.1x - 0.9y = 15,$
$\quad -1.4x + 0.6y = 10$

57. $13a - 7b = 9,$
$\quad 2a - 8b = 6$

58. $3a - 12b = 9,$
$\quad 4a - 5b = 3$

59. Describe a procedure that can be used to write an inconsistent system of equations.

60. Describe a procedure that can be used to write a system that has an infinite number of solutions.

Skill Review

61. Use an associative law to write an equation equivalent to $(4 + m) + n.$ [1.2]

62. Combine like terms: $a^2 - 4a - 3a^2 + 4a + 7.$ [1.3]

63. Simplify: $8x - 3[5x + 2(6 - 9x)].$ [1.3]

64. Evaluate $-p^2$ for $p = -1.$ [1.6]

65. Convert 30,050,000 to scientific notation.   [1.7]

66. Convert 6.1×10^{-4} to decimal notation.   [1.7]

Synthesis

67. Some systems are more easily solved by substitution and some are more easily solved by elimination. What guidelines could be used to help someone determine which method to use?

68. Explain how it is possible to solve Exercise 43 mentally.

69. If $(1, 2)$ and $(-3, 4)$ are two solutions of $f(x) = mx + b,$ find m and $b.$

70. If $(0, -3)$ and $\left(-\frac{3}{2}, 6\right)$ are two solutions of $px - qy = -1,$ find p and $q.$

71. Determine a and b for which $(-4, -3)$ is a solution of the system
$$ax + by = -26,$$
$$bx - ay = 7.$$

72. Solve for x and y in terms of a and $b:$
$$5x + 2y = a,$$
$$x - y = b.$$

Solve.

73. $\dfrac{x + y}{2} - \dfrac{x - y}{5} = 1,$

$\quad \dfrac{x - y}{2} + \dfrac{x + y}{6} = -2$

74. $3.5x - 2.1y = 106.2,$
$\quad 4.1x + 16.7y = -106.28$

Each of the following is a system of nonlinear equations. However, each is reducible to linear, *since an appropriate substitution (say, u for $1/x$ and v for $1/y$) yields a linear system. Make such a substitution, solve for the new variables, and then solve for the original variables.*

75. 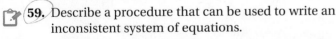 $\dfrac{2}{x} + \dfrac{1}{y} = 0,$

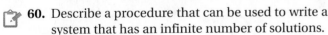 $\quad \dfrac{5}{x} + \dfrac{2}{y} = -5$

76. $\dfrac{1}{x} - \dfrac{3}{y} = 2,$

$\quad \dfrac{6}{x} + \dfrac{5}{y} = -34$

Solve using a system of equations.

77. Energy Consumption. With average use, a toaster oven and a convection oven together consume 15 kilowatt hours (kWh) of electricity each month. A convection oven uses four times as much electricity as a toaster oven. How much does each use per month?

Source: Lee County Electric Cooperative

78. Communication. Terri has two monthly bills: one for her cell phone and one for the data package for her tablet. The total of the two bills is $69.98 per month. Her cell-phone bill is $40 more per month than the bill for her tablet's data package. How much is each bill?

79. To solve the system

$$17x + 19y = 102,$$
$$136x + 152y = 826$$

Preston graphs both equations on a graphing calculator and gets the following screen. He then (incorrectly) concludes that the equations are dependent and the solution set is infinite. How can algebra be used to convince Preston that a mistake has been made?

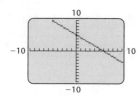

 YOUR TURN ANSWERS: SECTION 3.2

1. $\left(\frac{5}{2}, -\frac{1}{2}\right)$ **2.** $(-10, 7)$ **3.** No solution   **4.** $\left(\frac{1}{4}, -\frac{1}{2}\right)$

5. $\left(\frac{13}{7}, -\frac{10}{7}\right)$ **6.** $(-1, 5)$ **7.** $\{(x, y) \mid x - 3y = 2\}$

8. $\left(\frac{14}{9}, \frac{11}{6}\right)$

QUICK QUIZ: SECTIONS 3.1–3.2

1. Determine whether $(4, -1)$ is a solution of

$$x + y = 3,$$
$$x - y = 5. \quad [3.1]$$

2. Solve graphically:

$$x + y = 4,$$
$$y = 2x - 5. \quad [3.1]$$

3. Solve using substitution:

$$3x - y = 1,$$
$$y = 2x - 4. \quad [3.2]$$

4. Solve using elimination:

$$x - y = 2,$$
$$2x + 3y = 1. \quad [3.2]$$

5. Solve using any appropriate method:

$$2x = 1 - 3y,$$
$$y - 3x = 0. \quad [3.1], [3.2]$$

PREPARE TO MOVE ON

Solve. [1.4]

1. After her condominium had been on the market for 6 months, Gilena reduced the price to $94,500. This was $\frac{9}{10}$ of the original asking price. How much did Gilena originally ask for her condominium?

2. Ellia needs to average 80 on her tests in order to earn a B in her math class. Her average after four tests is 77.5. What score is needed on the fifth test in order to raise the average to 80?

3. North American Truck and Trailer rents vans for $53 per day plus 5¢ per mile. Anazi rented a van for 2 days. The bill was $120.50. How far did Anazi drive the van?

Source: www.northamericantrucktrailer.com

<table>
<tr><td>**3.3**</td><td>**Solving Applications: Systems of Two Equations**</td></tr>
</table>

Total-Value Problems and Mixture Problems ▪ Motion Problems

You are in a much better position to solve problems now that you know how systems of equations can be used. Using systems often makes the translating step easier.

EXAMPLE 1 *Endangered Species.* In 2012, there were 92 species of birds in the United States that were considered threatened or endangered. The number considered threatened was 3 less than one-fourth of the number considered endangered. How many U.S. bird species were considered endangered or threatened in 2012?

Source: U.S. Fish and Wildlife Service

SOLUTION The *Familiarize* and *Translate* steps were completed in Example 1 of Section 3.1. The resulting system of equations is

$$t + d = 92,$$
$$t = \tfrac{1}{4}d - 3,$$

where d is the number of endangered bird species and t is the number of threatened bird species in the United States in 2012.

3. **Carry out.** We solve the system of equations. Since one equation already has a variable isolated, let's use the substitution method:

$$t + d = 92$$
$$\tfrac{1}{4}d - 3 + d = 92 \qquad \text{Substituting } \tfrac{1}{4}d - 3 \text{ for } t$$
$$\tfrac{5}{4}d - 3 = 92 \qquad \text{Combining like terms}$$
$$\tfrac{5}{4}d = 95 \qquad \text{Adding 3 to both sides}$$
$$d = \tfrac{4}{5} \cdot 95 \qquad \text{Multiplying both sides by } \tfrac{4}{5}: \tfrac{4}{5} \cdot \tfrac{5}{4} = 1$$
$$d = 76. \qquad \text{Simplifying}$$

Next, using either of the original equations, we substitute and solve for t:

$$t = \tfrac{1}{4} \cdot 76 - 3 = 19 - 3 = 16.$$

4. **Check.** The sum of 76 and 16 is 92, so the total number of species is correct. Since 3 less than one-fourth of 76 is $19 - 3$, or 16, the numbers check.

5. **State.** In 2012, there were 76 bird species considered endangered and 16 considered threatened.

 YOUR TURN

1. Refer to Example 1. In 2012, there were 25 species of amphibians in the United States that were considered threatened or endangered. The number of species considered endangered was 1.5 times the number considered threatened. How many U.S. amphibian species were considered endangered and how many were considered threatened in 2012?

Source: U.S. Fish and Wildlife Service

Total-Value Problems and Mixture Problems

EXAMPLE 2 *Jewelry Design.* In order to make a necklace, Star Bright Jewelry Design purchased 80 beads for a total of $39 (excluding tax). Some of the beads were sterling silver beads costing 40¢ each and the rest were gemstone beads costing 65¢ each. How many of each type were bought?

Student Notes

It is very important that you clearly label precisely what each variable represents. Not only will this help with writing equations, but it will help you identify and state solutions.

2. Refer to Example 2. For another necklace, the jewelry designer purchased 60 sterling silver beads and gemstone beads for a total of $30. How many of each type did the designer buy?

SOLUTION   The *Familiarize* and *Translate* steps were completed in Example 2 of Section 3.1.

3. Carry out. We are to solve the system of equations

$$s + g = 80, \qquad (1)$$
$$40s + 65g = 3900, \qquad (2) \qquad \text{Working in cents rather than dollars}$$

where *s* is the number of sterling silver beads bought and *g* is the number of gemstone beads bought. Because both equations are in the form $Ax + By = C$, let's use the elimination method to solve the system. We can eliminate *s* by multiplying both sides of equation (1) by -40 and adding them to the corresponding sides of equation (2):

$$\begin{aligned} -40s - 40g &= -3200 \qquad \text{Multiplying both sides of equation (1) by } -40\\ 40s + 65g &= 3900\\ \hline 25g &= 700 \qquad \text{Adding}\\ g &= 28. \qquad \text{Solving for } g \end{aligned}$$

To find *s*, we substitute 28 for *g* in equation (1) and solve for *s*:

$$\begin{aligned} s + g &= 80 \qquad \text{Equation (1)}\\ s + 28 &= 80 \qquad \text{Substituting 28 for } g\\ s &= 52. \qquad \text{Solving for } s \end{aligned}$$

We obtain $(28, 52)$, or $g = 28$ and $s = 52$.

4. Check. We check in the original problem. Recall that *g* is the number of gemstone beads and *s* is the number of silver beads.

Number of beads:	$g + s = 28 + 52 = 80$
Cost of gemstone beads:	$65g = 65 \times 28 = 1820$¢
Cost of silver beads:	$40s = 40 \times 52 = \underline{2080}$¢
	Total $= 3900$¢

The numbers check.

5. State. The designer bought 28 gemstone beads and 52 sterling silver beads.

YOUR TURN

Example 2 involved two types of items (sterling silver beads and gemstone beads), the quantity of each type bought, and the total value of the items. We refer to this type of problem as a *total-value problem*.

EXAMPLE 3   *Blending Teas.*   TeaPots n Treasures sells loose Oolong tea for $2.15 per ounce. Donna mixed Oolong tea with shaved almonds that sell for $0.95 per ounce to create the Market Street Oolong blend that sells for $1.85 per ounce. One week, she made 300 oz of Market Street Oolong. How much tea and how much shaved almonds did Donna use?

SOLUTION

1. Familiarize. Since we know the price per ounce of the blend, we can find the total value of the blend by multiplying 300 ounces times $1.85 per ounce. We let $l =$ the number of ounces of Oolong tea and $a =$ the number of ounces of shaved almonds.

2. Translate. Since a 300-oz batch was made, we must have

$$l + a = 300.$$

To find a second equation, note that the total value of the 300-oz blend must match the combined value of the separate ingredients:

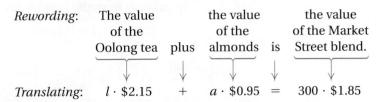

Rewording: The value of the Oolong tea plus the value of the almonds is the value of the Market Street blend.

Translating: $l \cdot \$2.15 + a \cdot \$0.95 = 300 \cdot \$1.85$

These equations can also be obtained from a table.

	Oolong Tea	Almonds	Market Street Blend	
Number of Ounces	l	a	300	→ $l + a = 300$
Price per Ounce	$2.15	$0.95	$1.85	
Value of Tea	2.15l$	0.95a$	$300 \cdot \$1.85$, or $555	→ $2.15l + 0.95a = 555$

3. Refer to Example 3. TeaPots n Treasures also sells loose rooibos tea for $1.50 per ounce. Donna mixed rooibos tea with shaved almonds to create the State Street Rooibos blend that sells for $1.39 per ounce. One week, she made 200 oz of State Street Rooibos. How much tea and how much shaved almonds did Donna use?

Clearing decimals in the second equation, we have $215l + 95a = 55,500$. We have translated to a system of equations:

$$l + a = 300, \quad (1)$$
$$215l + 95a = 55,500. \quad (2)$$

3. **Carry out.** We can solve using substitution. When equation (1) is solved for l, we have $l = 300 - a$. Substituting $300 - a$ for l in equation (2), we find a:

$215(300 - a) + 95a = 55,500$	Substituting
$64,500 - 215a + 95a = 55,500$	Using the distributive law
$-120a = -9000$	Combining like terms; subtracting 64,500 from both sides
$a = 75.$	Dividing both sides by -120

We have $a = 75$ and, from equation (1) above, $l + a = 300$. Thus, $l = 225$.

4. **Check.** Combining 225 oz of Oolong tea and 75 oz of almonds will give a 300-oz blend. The value of 225 oz of Oolong tea is $225(\$2.15)$, or $483.75. The value of 75 oz of almonds is $75(\$0.95)$, or $71.25. Thus the combined value of the blend is $483.75 + \$71.25$, or $555. A 300-oz blend priced at $1.85 per ounce would also be worth $555, so our answer checks.

5. **State.** The Market Street blend was made by combining 225 oz of Oolong tea and 75 oz of almonds.

YOUR TURN

EXAMPLE 4 *Student Loans.* Rani's student loans totaled $9600. Part was a PLUS loan made at 7.9% interest, and the rest was a Stafford loan made at 6.8% interest. After one year, Rani's loans accumulated $702.30 in interest. What was the original amount of each loan?

SOLUTION

1. **Familiarize.** We begin with a guess. If $3000 was borrowed at 7.9% and $6600 was borrowed at 6.8%, the two loans would total $9600. The interest would then be 0.079($3000), or $237, and 0.068($6600), or $448.80, for a total of only $685.80 in interest. Our guess was wrong, but checking the guess familiarized us with the problem. More than $3000 was borrowed at the higher rate.

2. **Translate.** We let p = the amount of the PLUS loan and s = the amount of the Stafford loan. Next, we organize a table in which the entries in each column come from the formula for simple interest:

$$Principal \cdot Rate \cdot Time = Interest.$$

	PLUS Loan	Stafford Loan	Total	
Principal	p	s	$9600	→ $p + s = 9600$
Rate of Interest	7.9%	6.8%		
Time	1 year	1 year		
Interest	$0.079p$	$0.068s$	$702.30	→ $0.079p + 0.068s = 702.30$

The total amount borrowed is found in the first row of the table:

$$p + s = 9600.$$

A second equation, representing accumulated interest, is found in the last row:

$$0.079p + 0.068s = 702.30, \quad \text{or} \quad 79p + 68s = 702,300. \qquad \text{Clearing decimals}$$

3. **Carry out.** The system can be solved by elimination:

$$
\begin{aligned}
p + s &= 9600, \\
79p + 68s &= 702,300
\end{aligned}
\quad \xrightarrow[\text{sides by } -79]{\text{Multiplying both}} \quad
\begin{aligned}
-79p - 79s &= -758,400 \\
79p + 68s &= 702,300 \\
\hline
-11s &= -56,100
\end{aligned}
$$

$$
\begin{aligned}
p + s &= 9600 \quad \xleftarrow{\hspace{2cm}} \quad s = 5100 \\
p + 5100 &= 9600 \\
p &= 4500.
\end{aligned}
$$

We find that $p = 4500$ and $s = 5100$.

4. **Check.** The total amount borrowed is $4500 + $5100, or $9600. The interest on $4500 at 7.9% for 1 year is 0.079($4500), or $355.50. The interest on $5100 at 6.8% for 1 year is 0.068($5100), or $346.80. The total amount of interest is $355.50 + $346.80, or $702.30, so the numbers check.

5. **State.** The PLUS loan was for $4500, and the Stafford loan was for $5100.

4. Refer to Example 4. Chin-Sun's student loans totaled $8400. Part was a PLUS loan at 7.9% interest, and the rest was a Stafford loan at 6.8% interest. After one year, Chin-Sun's loans accumulated $620.70 in interest. What was the original amount of each loan?

↩ YOUR TURN

Before proceeding to Example 5, briefly scan Examples 2–4 for similarities. Note that in each case, one of the equations in the system is a simple sum while the other equation represents a sum of products. Example 5 continues this pattern with what is commonly called a *mixture problem*.

> **PROBLEM-SOLVING TIP**
>
> When solving a problem, see if it is patterned or modeled after a problem that you have already solved.

EXAMPLE 5 *Mixing Fertilizers.* Nature's Green Gardening, Inc., carries two brands of fertilizer containing nitrogen and water. "Gentle Grow" is 3% nitrogen and "Sun Saver" is 8% nitrogen. Nature's Green needs to combine the two types of solution into a 90-L mixture that is 6% nitrogen. How much of each brand should be used?

SOLUTION

1. **Familiarize.** We must consider not only the size of the mixture, but also its strength.

EXPLORING 🔍 THE CONCEPT

In order to have a 90-L mixture of Gentle Grow and Sun Saver that is 6% nitrogen, two things must be true:

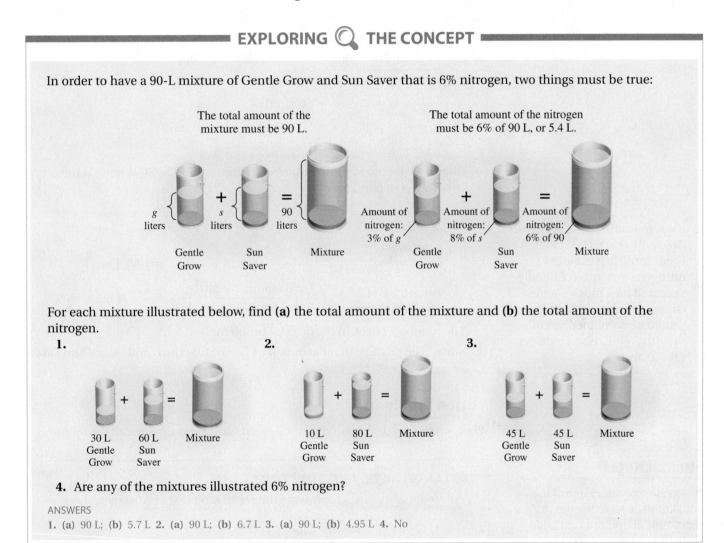

For each mixture illustrated below, find **(a)** the total amount of the mixture and **(b)** the total amount of the nitrogen.

1.   **2.**   **3.**

30 L Gentle Grow + 60 L Sun Saver = Mixture   10 L Gentle Grow + 80 L Sun Saver = Mixture   45 L Gentle Grow + 45 L Sun Saver = Mixture

4. Are any of the mixtures illustrated 6% nitrogen?

ANSWERS
1. **(a)** 90 L; **(b)** 5.7 L 2. **(a)** 90 L; **(b)** 6.7 L 3. **(a)** 90 L; **(b)** 4.95 L 4. No

2. **Translate.** We let g = the number of liters of Gentle Grow and s = the number of liters of Sun Saver. The information can be organized in a table.

	Gentle Grow	Sun Saver	Mixture	
Number of Liters	g	s	90	→ $g + s = 90$
Percent of Nitrogen	3%	8%	6%	
Amount of Nitrogen	$0.03g$	$0.08s$	0.06×90, or 5.4 liters	→ $0.03g + 0.08s = 5.4$

If we add g and s in the first row, we get one equation. It represents the total amount of mixture: $g + s = 90$.

If we add the amounts of nitrogen listed in the third row, we get a second equation. This equation represents the amount of nitrogen in the mixture: $0.03g + 0.08s = 5.4$.

After clearing decimals, we have translated the problem to the system

$$g + s = 90, \quad (1)$$
$$3g + 8s = 540. \quad (2)$$

3. Carry out. We use the elimination method to solve the system:

$$
\begin{array}{ll}
-3g - 3s = -270 & \text{Multiplying both sides of equation (1) by } -3 \\
\underline{3g + 8s = 540} & \\
5s = 270 & \text{Adding} \\
s = 54; & \text{Solving for } s \\
g + 54 = 90 & \text{Substituting into equation (1)} \\
g = 36. & \text{Solving for } g
\end{array}
$$

4. Check. Remember, g is the number of liters of Gentle Grow and s is the number of liters of Sun Saver.

Total amount of mixture:	$g + s = 36 + 54 = 90$
Total amount of nitrogen:	3% of 36 + 8% of 54 = 1.08 + 4.32 = 5.4
Percentage of nitrogen in mixture:	$\dfrac{\text{Total amount of nitrogen}}{\text{Total amount of mixture}} = \dfrac{5.4}{90} = 6\%$

The numbers check in the original problem.

5. State. The mixture should contain 36 L of Gentle Grow and 54 L of Sun Saver.

↩ YOUR TURN

5. Refer to Example 5. Nature's Green also carries "Friendly Feed" fertilizer that is 9% nitrogen. How much Friendly Feed and how much Gentle Grow, containing 3% nitrogen, should be combined in order to form a 60-L mixture that is 4% nitrogen?

Motion Problems

When a problem deals with distance, speed (rate), and time, recall the following.

Student Notes

Be sure to remember one of the equations shown in the box at right. You can multiply or divide on both sides, as needed, to obtain the others.

DISTANCE, RATE, AND TIME EQUATIONS

If r represents rate, t represents time, and d represents distance, then:

$$d = rt, \quad r = \frac{d}{t}, \quad \text{and} \quad t = \frac{d}{r}.$$

EXAMPLE 6 *Train Travel.* A Vermont Railways freight train, loaded with logs, leaves Boston, heading to Washington D.C., at a speed of 60 km/h. Two hours later, an Amtrak® Metroliner leaves Boston, bound for Washington D.C., on a parallel track at 90 km/h. At what point will the Metroliner catch up to the freight train?

SOLUTION

1. Familiarize. Let's make a guess and check to see if it is correct. Suppose the trains meet after traveling 180 km. We can calculate the time for each train.

	Distance	Rate	Time
Freight Train	180 km	60 km/h	$\frac{180}{60} = 3$ hr
Metroliner	180 km	90 km/h	$\frac{180}{90} = 2$ hr

We see that the distance cannot be 180 km, since the difference in travel times for the trains is *not* 2 hr. Although our guess is wrong, we can use a similar chart to organize the information in this problem.

The distance at which the trains meet is unknown, but we do know that the trains will have traveled the same distance when they meet. We let *d* = this distance.

The time that the trains are running is also unknown, but we do know that the freight train has a 2-hr head start. Thus if we let *t* = the number of hours that the freight train is running before they meet, then *t* − 2 is the number of hours that the Metroliner runs before catching up to the freight train.

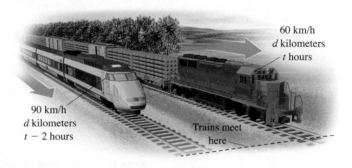

60 km/h
d kilometers
t hours

90 km/h
d kilometers
t − 2 hours

Trains meet here

2. **Translate.** We can organize the information in a chart. The formula *Distance = Rate · Time* guides our choice of rows and columns.

	Distance	Rate	Time
Freight Train	*d*	60	*t*
Metroliner	*d*	90	*t* − 2

$\longrightarrow d = 60t$
$\longrightarrow d = 90(t - 2)$

Using *Distance = Rate · Time* twice, we get two equations:

$$d = 60t, \qquad (1)$$
$$d = 90(t - 2). \qquad (2)$$

3. **Carry out.** We solve the system using substitution:

$$60t = 90(t - 2) \qquad \text{Substituting } 60t \text{ for } d \text{ in equation (2)}$$
$$60t = 90t - 180$$
$$-30t = -180$$
$$t = 6.$$

The time for the freight train is 6 hr, which means that the time for the Metroliner is 6 − 2, or 4 hr. Remember that it is distance, not time, that the problem asked for. Thus for *t* = 6, we have $d = 60 \cdot 6 = 360$ km.

4. **Check.** At 60 km/h, the freight train travels $60 \cdot 6$, or 360 km, in 6 hr. At 90 km/h, the Metroliner travels $90 \cdot (6 - 2) = 360$ km in 4 hr. The numbers check.

5. **State.** The Metroliner catches up to the freight train 360 km from Boston.

Student Notes

Always be careful to answer the question asked in the problem. In Example 6, the problem asks for distance, not time. Answering "6 hr" would be incorrect.

6. An Amtrak® Metroliner traveling 90 mph leaves Washington, D.C., 3 hr after a freight train traveling 60 mph. If they travel on parallel tracks, at what point will the Metroliner catch up to the freight train?

YOUR TURN

EXAMPLE 7 *Jet Travel.* A Boeing 747-400 jet flies 4 hr west with a 60-mph tailwind. Returning *against* the wind takes 5 hr. Find the speed of the jet with no wind.

SOLUTION

1. **Familiarize.** We imagine the situation and make a drawing. Note that the wind *speeds up* the outbound flight but *slows down* the return flight.

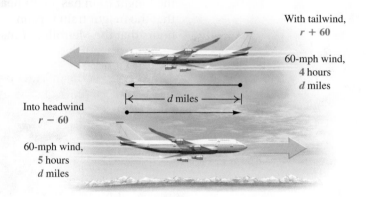

Let's make a guess of 400 mph for the jet's speed if there were no wind. Note that the distances traveled each way must be the same.

Speed with no wind:	400 mph
Speed with the wind:	$400 + 60 = 460$ mph
Speed against the wind:	$400 - 60 = 340$ mph
Distance with the wind:	$460 \cdot 4 = 1840$ mi ⎫
Distance against the wind:	$340 \cdot 5 = 1700$ mi ⎭ These must match.

Since the distances are not the same, our guess of 400 mph is incorrect.

We let $r =$ the speed, in miles per hour, of the jet in still air. Then $r + 60 =$ the jet's speed with the wind and $r - 60 =$ the jet's speed against the wind. We also let $d =$ the distance traveled, in miles.

2. **Translate.** The information can be organized in a chart. The distances traveled are the same, so we use *Distance = Rate* (or *Speed*) *· Time*. Each row of the chart gives an equation.

	Distance	Rate	Time	
With Wind	d	$r + 60$	4	→ $d = (r + 60)4$
Against Wind	d	$r - 60$	5	→ $d = (r - 60)5$

We now have a system of equations:

$$d = (r + 60)4, \quad (1)$$
$$d = (r - 60)5. \quad (2)$$

3. **Carry out.** We solve the system using substitution:

$(r - 60)5 = (r + 60)4$	Substituting $(r - 60)5$ for d in equation (1)
$5r - 300 = 4r + 240$	Using the distributive law
$r = 540.$	Solving for r

7. A motorboat travels 30 min upstream against a 4-mph current. Returning with the current takes 18 min. Find the speed of the motorboat in still water.

4. Check. When $r = 540$, the speed with the wind is $540 + 60 = 600$ mph, and the speed against the wind is $540 - 60 = 480$ mph. The distance with the wind, $600 \cdot 4 = 2400$ mi, matches the distance into the wind, $480 \cdot 5 = 2400$ mi, so we have a check.

5. State. The speed of the jet with no wind is 540 mph.

↪ YOUR TURN

TIPS FOR SOLVING MOTION PROBLEMS

1. Draw a diagram using an arrow or arrows to represent distance and the direction of each object in motion.
2. Organize the information in a chart.
3. Look for times, distances, or rates that are the same. These often can lead to an equation.
4. Translating to a system of equations allows for the use of two variables. Label the variables carefully.
5. Always make sure that you have answered the question asked.

3.3 EXERCISE SET

FOR EXTRA HELP

MyMathLab® Math PRACTICE   WATCH   READ   REVIEW

↪ Vocabulary and Reading Check

Choose the word from the following list that best completes each statement. Not every word will be used.

difference   distance   mixture
principal   sum   total value

1. If 10 coffee mugs are sold for $8 each, the _____ of the mugs is $80.

2. To find simple interest, multiply the _____ by the rate and the time.

3. To solve a motion problem, we often use the fact that _____ divided by rate equals time.

4. When a boat travels downstream, its speed is the _____ of the speed of the current and the speed of the boat in still water.

Applications

5.–18. *For Exercises 5–18, solve Exercises 41–54, respectively, from Section 3.1.*

19. *Renewable Energy.* In 2010, solar and wind energy generation totaled 1033 trillion Btu's. Wind generated 52 trillion Btu's more than eight times that generated by solar energy. How much was generated by each source?

Source: U.S. Energy Information Administration

20. *Snowmen.* The tallest snowman ever recorded—really a snow *woman* named Olympia—was built by residents of Bethel, Maine, and surrounding towns. Her body and head together made up her total record height of 122 ft. The body was 2 ft longer than 14 times the height of the head. What were the separate heights of Olympia's head and body?

Source: Based on information from *Guinness World Records*

Total-Value Problems and Mixture Problems

21. *College Credits.* Each course at Mt. Regis College is worth either 3 or 4 credits. The members of the men's swim team are taking a total of 48 courses that are worth a total of 155 credits. How many 3-credit courses and how many 4-credit courses are being taken?

22. *College Credits.* Each course at Pease County Community College is worth either 3 or 4 credits. The members of the women's golf team are taking a total of 27 courses that are worth a total of 89 credits. How many 3-credit courses and how many 4-credit courses are being taken?

23. *Recycled Paper.*   Staples® recently charged $38.39 per case of regular paper and $44.79 per case of paper made of recycled fibers. Last semester, Valley College Copy Center spent $1106.93 for 27 cases of paper. How many of each type were purchased?

24. *Photocopying.*   Quick Copy recently charged 49¢ per page for color copies and 9¢ per page for black-and-white copies. If Shirlee's bill for 90 copies was $12.90, how many copies of each type were made?

25. *Lighting.*   The Home Depot® recently sold 23-watt CFL bulbs for $2.50 each and 42-watt CFL bulbs for $8.25 each. If River Memorial Hospital purchased 200 such bulbs for a total of $1305, how many of each type did they purchase?

26. *Office Supplies.*   Hancock County Social Services is preparing materials for a seminar. They purchase a combination of 80 large binders and small binders. The large binders cost $8.49 each and the small ones cost $5.99 each. If the total cost of the binders was $544.20, how many of each size were purchased?

27. *Museum Admission.*   The Indianapolis Children's Museum charges $12.50 for a one-day youth admission and $17.50 for a one-day adult admission. One Friday, the museum collected $1535 from a total of 110 youths and adults. How many admissions of each type were sold?

28. *Amusement Park Admission.*   Cedar Point Amusement Park charges $51.99 for an adult admission and $26.99 for a junior admission. One Thursday, the park collected $15,591.40 from a total of 360 adults and juniors. How many admissions of each type were sold?

Aha! 29. *Blending Coffees.*   The Roasted Bean charges $17.00 per pound for Fair Trade Organic Mexican coffee and $15.00 per pound for Fair Trade Organic Peruvian coffee. How much of each type should be used to make a 28-lb blend that sells for $16.00 per pound?

30. *Mixed Nuts.*   Oh Nuts! sells pistachio kernels for $10.00 per pound and almonds for $8.00 per pound. How much of each type should be used to make a 50-lb mixture that sells for $8.80 per pound?

31. *Event Planning.*   As part of the refreshments for Yvette's 25th birthday party, Kim plans to provide a bowl of M&M candies. She wants to mix custom-printed M&Ms costing 60¢ per ounce with bulk

M&Ms costing 25¢ per ounce to create 20 lb of a mixture costing 32¢ per ounce. How much of each type of M&M should she use?

Source: www.mymms.com

32. *Blending Spices.*   Spice of Life sells ground sumac for $1.35 per ounce and ground thyme for $1.85 per ounce. Aman wants to make a 20-oz Zahtar seasoning blend using the two spices that sells for $1.65 per ounce. How much of each spice should Aman use?

33. *Catering.*   Cati's Catering is planning an office reception. The office administrator has requested a candy mixture that is 25% chocolate. Cati has available mixtures that are either 50% chocolate or 10% chocolate. How much of each type should be mixed to get a 20-lb mixture that is 25% chocolate?

34. *Ink Remover.*   Etch Clean Graphics uses one cleanser that is 25% acid and a second that is 50% acid. How many liters of each should be mixed to get 30 L of a solution that is 40% acid?

35. *Blending Granola.*   Deep Thought Granola is 25% nuts and dried fruit. Oat Dream Granola is 10% nuts and dried fruit. How much of Deep Thought and how much of Oat Dream should be mixed to form a 20-lb batch of granola that is 19% nuts and dried fruit?

36. *Livestock Feed.*   Soybean meal is 16% protein and corn meal is 9% protein. How many pounds of each should be mixed to get a 350-lb mixture that is 12% protein?

37. *Student Loans.*   Stacey's two student loans totaled $12,000. One of her loans was at 6.5% simple interest and the other at 7.2%. After one year, Stacey owed $811.50 in interest. What was the amount of each loan?

38. *Investments.* A self-employed contractor nearing retirement made two investments totaling $15,000. In one year, these investments yielded $573 in simple interest. Part of the money was invested at 3% and the rest at 4.5%. How much was invested at each rate?

39. *Automotive Maintenance.* "Steady State" antifreeze is 18% alcohol and "Even Flow" is 10% alcohol. How many liters of each should be mixed to get 20 L of a mixture that is 15% alcohol?

40. *Chemistry.* E-Chem Testing has a solution that is 80% base and another that is 30% base. A technician needs 150 L of a solution that is 62% base. The 150 L will be prepared by mixing the two solutions on hand. How much of each should be used?

41. *Octane Ratings.* The octane rating of a gasoline is a measure of the amount of isooctane in a gallon of gas. Manufacturers recommend using 93-octane gasoline on retuned motors. How much 87-octane gas and how much 95-octane gas should Yousef mix in order to make 10 gal of 93-octane gas for his retuned Ford F-150?

Source: Champlain Electric and Petroleum Equipment

42. *Octane Ratings.* The octane rating of a gasoline is a measure of the amount of isooctane in a gallon of gas. Subaru recommends 91-octane gasoline for the 2008 Legacy 3.0 R. How much 87-octane gas and how much 93-octane gas should Kelsey mix in order to make 12 gal of 91-octane gas for her Legacy?

Sources: Champlain Electric and Petroleum Equipment; Dean Team Ballwin

43. *Food Science.* The following bar graph shows the milk fat percentages in three dairy products.

How many pounds each of whole milk and cream should be mixed to form 200 lb of milk for cream cheese?

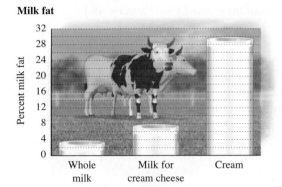

44. *Food Science.* How much lowfat milk (1% fat) and how much whole milk (4% fat) should be mixed to make 5 gal of reduced fat milk (2% fat)?

Motion Problems

45. *Train Travel.* A train leaves Danville Union and travels north at 75 km/h. Two hours later, an express train leaves on a parallel track and travels north at 125 km/h. How far from the station will they meet?

46. *Car Travel.* Two cars leave Salt Lake City, traveling in opposite directions. One car travels at a speed of 80 km/h and the other at 96 km/h. In how many hours will they be 528 km apart?

47. *Canoeing.* Kahla paddled for 4 hr with a 6-km/h current to reach a campsite. The return trip against the same current took 10 hr. Find the speed of Kahla's canoe in still water.

48. *Boating.* Chen's motorboat took 3 hr to make a trip downstream with a 6-mph current. The return trip against the same current took 5 hr. Find the speed of the boat in still water.

49. *Point of No Return.* A plane flying the 3458-mi trip from New York City to London has a 50-mph tailwind. The flight's *point of no return* is the point at which the flight time required to return to New York is the same as the time required to continue to London. If the speed of the plane in still air is 360 mph, how far is New York from the point of no return?

50. *Point of No Return.*   A plane is flying the 2553-mi trip from Los Angeles to Honolulu into a 60-mph headwind. If the speed of the plane in still air is 310 mph, how far from Los Angeles is the plane's point of no return? (See Exercise 49.)

Applications

51. *Architecture.*   The rectangular ground floor of the John Hancock building has a perimeter of 860 ft. The length is 100 ft more than the width. Find the length and the width.

$x + 100$

x

52. *Real Estate.*   The perimeter of a rectangular ocean-front lot is 190 m. The width is one-fourth of the length. Find the dimensions.

53. By February 2012, Nintendo had sold one and one-half as many Wii game machines worldwide as Sony had sold PlayStation 3 consoles. Together, they had sold 154 million game machines. How many of each were sold?

Source: vgchartz.com

54. *Hockey Rankings.*   Hockey teams receive 2 points for a win and 1 point for a tie. The Wildcats once won a championship with 60 points. They won 9 more games than they tied. How many wins and how many ties did the Wildcats have?

55. At one time, Netflix offered members an unlimited number of movies and television episodes streamed to their TVs and computers for $7.99 per month. For $7.99 more per month, Netflix offered an unlimited one DVD-at-a-time by mail rental plan in addition to the unlimited streaming. During one week, 280 new subscribers paid a total of $2956.30 for their plans. How many $7.99 plans and how many $15.98 plans were purchased?

56. *Radio Airplay.*   Akio must play 12 commercials during his 1-hr radio show. Each commercial is either 30 sec or 60 sec long. If the total commercial time during that hour is 10 min, how many commercials of each type does Akio play?

57. *Making Change.*   Monica buys a $9.25 book using a $20 bill. The store has no bills and gives change in quarters and fifty-cent pieces. There are 30 coins in all. How many of each kind are there?

58. *Teller Work.*   Sabina goes to a bank and changes a $50 bill for $5 bills and $1 bills. There are 22 bills in all. How many of each kind are there?

59. In what ways are Examples 3 and 4 similar? In what sense are their systems of equations similar?

60. Write at least three study tips of your own for someone beginning this exercise set.

Skill Review

Let $h(x) = x - 7$ and $f(x) = x^2 + 2$. Find the following.

61. $h(0)$ [2.2]
62. $f(-10)$ [2.2]

63. $(h \cdot f)(7)$ [2.6]
64. $(h + f)(x)$ [2.6]

65. The domain of $h + f$ [2.6]

66. The domain of f/h [2.6]

Synthesis

67. Suppose that in Example 3 you are asked only for the amount of almonds needed for the Market Street blend. Would the method of solving the problem change? Why or why not?

68. Write a problem similar to Example 2 for a classmate to solve. Design the problem so that the solution is "The bakery sold 24 loaves of bread and 18 packages of sandwich rolls."

69. *Recycled Paper.*   Unable to purchase 60 reams of paper that contains 20% post-consumer fiber, the Naylor School bought paper that was either 0% post-consumer fiber or 30% post-consumer fiber. How many reams of each should be purchased in order to use the same amount of post-consumer fiber as if the 20% post-consumer fiber paper were available?

70. *Automotive Maintenance.*   The radiator in Natalie's car contains 6.3 L of antifreeze and water. This mixture is 30% antifreeze. How much of this mixture should she drain and replace with pure antifreeze so that there will be a mixture of 50% antifreeze?

71. *Metal Alloys.* For a metal to be labeled "sterling silver," the silver alloy must contain at least 92.5% pure silver. Nicole has 32 oz of coin silver, which is 90% pure silver. How much pure silver must she add to the coin silver in order to have a sterling-silver alloy?

Source: The Jewelry Repair Manual, R. Allen Hardy, Courier Dover Publications, 1996, p. 271.

72. *Exercise.* Huan jogs and walks to school each day. She averages 4 km/h walking and 8 km/h jogging. From home to school is 6 km and Huan makes the trip in 1 hr. How far does she jog in a trip?

73. *Bakery.* Gigi's Cupcakes offers a gift box with six cupcakes for $15.99. Gigi's also sells cupcakes individually for $3 each. Gigi's sold a total of 256 cupcakes one Saturday for a total of $701.67 in sales (excluding tax). How many six-cupcake gift boxes were included in that day's sales total?

74. The tens digit of a two-digit positive integer is 2 more than three times the units digit. If the digits are interchanged, the new number is 13 less than half the given number. Find the given integer. (*Hint*: Let x = the tens-place digit and y = the units-place digit; then $10x + y$ is the number.)

75. *Wood Stains.* Williams' Custom Flooring has 0.5 gal of stain that is 20% brown and 80% neutral. A customer orders 1.5 gal of a stain that is 60% brown and 40% neutral. How much pure brown stain and how much neutral stain should be added to the original 0.5 gal in order to make up the order?*

76. *Train Travel.* A train leaves Union Station for Central Station, 216 km away, at 9 A.M. One hour later, a train leaves Central Station for Union Station. They meet at noon. If the second train had started at 9 A.M. and the first train at 10:30 A.M., they would still have met at noon. Find the speed of each train.

77. *Fuel Economy.* Grady's station wagon gets 18 miles per gallon (mpg) in city driving and 24 mpg in highway driving. The car is driven 465 mi on 23 gal of gasoline. How many miles were driven in the city and how many were driven on the highway?

78. *Biochemistry.* Industrial biochemists routinely use a machine to mix a buffer of 10% acetone by adding 100% acetone to water. One day, instead of adding 5 L of acetone to a vat of water to create the buffer, a machine added 10 L. How much additional water was needed to bring the concentration down to 10%?

 79. See Exercise 75 above. Let x = the amount of pure brown stain added to the original 0.5 gal. Find a function $P(x)$ that can be used to determine the percentage of brown stain in the 1.5-gal mixture. On a graphing calculator, draw the graph of P and use INTERSECT to confirm the answer to Exercise 75.

80. *Siblings.* Fred and Phyllis are twins. Phyllis has twice as many brothers as she has sisters. Fred has the same number of brothers as sisters. How many girls and how many boys are in the family?

 YOUR TURN ANSWERS: SECTION 3.3

1. Endangered species: 15; threatened species: 10
2. Silver beads: 36; gemstone beads: 24   **3.** Rooibos: 160 oz; almonds: 40 oz   **4.** PLUS loan: $4500; Stafford loan: $3900   **5.** Friendly Feed: 10 L; Gentle Grow: 50 L
6. 540 mi from Washington D.C.   **7.** 16 mph

QUICK QUIZ: SECTIONS 3.1–3.3

Solve. If a system has an infinite number of solutions, use set-builder notation to write the solution set. If a system has no solution, state this. [3.1], [3.2]

1. $x - 2y = 7,$
$\quad x = 2y - 5$

2. $3x - 4y = 11,$
$\quad x + 4y = 12$

3. $y = 2x - 4,$
$\quad y = \frac{1}{2}x + 2$

4. $\quad x + 3y = 3,$
$\quad 2x + 5y = 6$

5. In order to raise funds for a concert tour, Arie's choir sold rolls of trash bags. Large trash bags sold for $17 per roll and small trash bags sold for $12 per roll. If Arie sold 28 rolls and collected $441, how many rolls of each type of trash bag did he sell? [3.3]

PREPARE TO MOVE ON

Evaluate. [1.1], [1.2]

1. $2x - 3y - z,$ for $x = 5, y = 2,$ and $z = 3$

2. $4x + y - 6z,$ for $x = \frac{1}{2}, y = \frac{1}{2},$ and $z = \frac{1}{3}$

3. $3a - b + 2c,$ for $a = 1, b = -6,$ and $c = 4$

4. $a - 2b - 3c,$ for $a = -2, b = 3,$ and $c = -5$

---

*This problem was suggested by Professor Chris Burditt of Yountville, California, and is based on a real-world situation.

3.4 Systems of Equations in Three Variables

Identifying Solutions ■ Solving Systems in Three Variables ■ Dependency, Inconsistency, and Geometric Considerations

Some problems naturally call for a translation to three or more equations. In this section, we learn how to solve systems of three linear equations. Later, we will use such systems in problem-solving situations.

Identifying Solutions

A **linear equation in three variables** is an equation equivalent to one of the form $Ax + By + Cz = D$, where A, B, C, and D are real numbers. We refer to the form $Ax + By + Cz = D$ as *standard form* for a linear equation in three variables.

A solution of a system of three equations in three variables is an ordered triple (x, y, z) that makes *all three* equations true. The numbers in an ordered triple correspond to the variables in alphabetical order unless otherwise indicated.

EXAMPLE 1 Determine whether $\left(\frac{3}{2}, -4, 3\right)$ is a solution of the system

$$4x - 2y - 3z = 5,$$
$$-8x - y + z = -5,$$
$$2x + y + 2z = 5.$$

SOLUTION We substitute $\left(\frac{3}{2}, -4, 3\right)$ into all three equations, using alphabetical order:

$$
\begin{array}{c|c}
4x - 2y - 3z = 5 & \\
\hline
4 \cdot \frac{3}{2} - 2(-4) - 3 \cdot 3 & 5 \\
6 + 8 - 9 & \\
5 \overset{?}{=} 5 & \text{TRUE}
\end{array}
\qquad
\begin{array}{c|c}
-8x - y + z = -5 & \\
\hline
-8 \cdot \frac{3}{2} - (-4) + 3 & -5 \\
-12 + 4 + 3 & \\
-5 \overset{?}{=} -5 & \text{TRUE}
\end{array}
\qquad
\begin{array}{c|c}
2x + y + 2z = 5 & \\
\hline
2 \cdot \frac{3}{2} + (-4) + 2 \cdot 3 & 5 \\
3 - 4 + 6 & \\
5 \overset{?}{=} 5 & \text{TRUE}
\end{array}
$$

The triple makes all three equations true, so it is a solution.

YOUR TURN

1. Determine whether $\left(-2, \frac{1}{2}, 5\right)$ is a solution of the system

$$x - 2y + z = 2,$$
$$3x - 4y + 2z = 3,$$
$$x + 6y - z = -10.$$

Solving Systems in Three Variables

The graph of a linear equation in three variables is a plane. Because a three-dimensional coordinate system is required, solving systems in three variables graphically is difficult. The substitution method *can* be used but is generally cumbersome. Fortunately, the elimination method works well for any system of three equations in three variables.

EXAMPLE 2 Solve the following system of equations:

$$
\begin{aligned}
x + y + z &= 4, & (1) \\
x - 2y - z &= 1, & (2) \\
2x - y - 2z &= -1. & (3)
\end{aligned}
$$

SOLUTION We select *any* two of the three equations and work to get an equation in two variables. Let's add equations (1) and (2):

$$
\begin{array}{ll}
x + y + z = 4 & (1) \\
\underline{x - 2y - z = 1} & (2) \\
2x - y \phantom{{}+z} = 5. & (4) \quad \text{Adding to eliminate } z
\end{array}
$$

Next, we select a different pair of equations and eliminate the *same variable* that we did above. Let's use equations (1) and (3) to again eliminate z. Be careful! A common error is to eliminate a different variable in this step.

$$x + y + z = 4, \quad \xrightarrow{\begin{array}{c}\text{Multiplying both sides}\\\text{of equation (1) by 2}\end{array}} \quad \begin{array}{r} 2x + 2y + 2z = 8 \\ 2x - y - 2z = -1 \\ \hline 4x + y \phantom{{}+2z} = 7 \quad (5) \end{array}$$
$$2x - y - 2z = -1$$

Study Skills

Helping Others Will Help You Too

When you thoroughly understand a topic, don't hesitate to help classmates experiencing trouble. Your understanding and retention of the material will deepen and your classmate will appreciate your help.

Now we solve the resulting system of equations (4) and (5). That solution will give us two of the numbers in the solution of the original system.

$$\begin{array}{rl} 2x - y = 5 & (4) \\ 4x + y = 7 & (5) \\ \hline 6x \phantom{{}+y} = 12 & \text{Adding} \\ x = 2 \end{array}$$

Note that we now have two equations in two variables. Had we not eliminated the *same* variable in both of the above steps, this would not be the case.

We can use either equation (4) or (5) to find y. We choose equation (5):

$$\begin{aligned} 4x + y &= 7 \quad (5) \\ 4 \cdot 2 + y &= 7 \quad \text{Substituting 2 for } x \text{ in equation (5)} \\ 8 + y &= 7 \\ y &= -1. \end{aligned}$$

We now have $x = 2$ and $y = -1$. To find the value for z, we use any of the original three equations and substitute to find the third number, z. Let's use equation (1) and substitute our two numbers in it:

$$\begin{aligned} x + y + z &= 4 \quad (1) \\ 2 + (-1) + z &= 4 \quad \text{Substituting 2 for } x \text{ and } -1 \text{ for } y \\ 1 + z &= 4 \\ z &= 3. \end{aligned}$$

We have obtained the triple $(2, -1, 3)$. It should check in *all three* equations:

$$\begin{array}{c|c} x + y + z = 4 \\ \hline 2 + (-1) + 3 \;\vrule\; 4 \\ 4 \overset{?}{=} 4 \quad \text{TRUE} \end{array} \qquad \begin{array}{c|c} x - 2y - z = 1 \\ \hline 2 - 2(-1) - 3 \;\vrule\; 1 \\ 1 \overset{?}{=} 1 \quad \text{TRUE} \end{array} \qquad \begin{array}{c|c} 2x - y - 2z = -1 \\ \hline 2 \cdot 2 - (-1) - 2 \cdot 3 \;\vrule\; -1 \\ -1 \overset{?}{=} -1 \quad \text{TRUE} \end{array}$$

The solution is $(2, -1, 3)$.

YOUR TURN

2. Solve the following system of equations:

$$\begin{aligned} x + y + z &= 6, \\ 2x - y - z &= 3, \\ x - 2y + 2z &= 13. \end{aligned}$$

SOLVING SYSTEMS OF THREE LINEAR EQUATIONS

To use the elimination method to solve systems of three linear equations:

1. Write all equations in standard form $Ax + By + Cz = D$.
2. Clear any decimals or fractions.
3. Choose a variable to eliminate. Then select two of the three equations and multiply and add, as needed, to produce one equation in which the selected variable is eliminated.
4. Next, use a different pair of equations and eliminate the same variable as in step (3).
5. Solve the system of equations resulting from steps (3) and (4).
6. Substitute the solution from step (5) into one of the original three equations and solve for the third variable. Then check.

EXAMPLE 3 Solve the system

$$4x - 2y - 3z = 5, \quad (1)$$
$$-8x - y + z = -5, \quad (2)$$
$$2x + y + 2z = 5. \quad (3)$$

SOLUTION

Write in standard form.

1., 2. The equations are already in standard form with no fractions or decimals.

3. Select a variable to eliminate. We decide on y because the y-terms are opposites of each other in equations (2) and (3). We add:

Eliminate a variable. (We choose y).

$$\begin{array}{rl} -8x - y + z = -5 & (2) \\ \underline{2x + y + 2z = 5} & (3) \\ -6x + 3z = 0. & (4) \quad \text{Adding} \end{array}$$

4. We use another pair of equations to create a second equation in x and z. That is, we again eliminate y. To do so, we use equations (1) and (3):

Eliminate the same variable using a different pair of equations.

$$4x - 2y - 3z = 5,$$
$$2x + y + 2z = 5 \quad \xrightarrow[\text{of equation (3) by 2}]{\text{Multiplying both sides}}$$

$$\begin{array}{rl} 4x - 2y - 3z = & 5 \\ \underline{4x + 2y + 4z = 10} \\ 8x + z = 15. & (5) \end{array}$$

5. Now we solve the resulting system of equations (4) and (5). That allows us to find two of the three variables.

Solve the system of two equations in two variables.

$$-6x + 3z = 0,$$
$$8x + z = 15 \quad \xrightarrow[\text{of equation (5) by } -3]{\text{Multiplying both sides}}$$

$$\begin{array}{rl} -6x + 3z = & 0 \\ \underline{-24x - 3z = -45} \\ -30x = -45 \\ x = \frac{-45}{-30} = \frac{3}{2} \end{array}$$

We use equation (5) to find z:

$$8x + z = 15$$
$$8 \cdot \tfrac{3}{2} + z = 15 \quad \text{Substituting } \tfrac{3}{2} \text{ for } x$$
$$12 + z = 15$$
$$z = 3.$$

Solve for the remaining variable and check.

6. Finally, we use any of the original equations and substitute to find the third number, y. To do so, we choose equation (3):

$$2x + y + 2z = 5 \quad (3)$$
$$2 \cdot \tfrac{3}{2} + y + 2 \cdot 3 = 5 \quad \text{Substituting } \tfrac{3}{2} \text{ for } x \text{ and } 3 \text{ for } z$$
$$3 + y + 6 = 5$$
$$y + 9 = 5$$
$$y = -4.$$

3. Solve the system

$$x - 3y + z = 13,$$
$$2x + 3y + 2z = 20,$$
$$-3x - 6y + z = 3.$$

The solution is $\left(\tfrac{3}{2}, -4, 3\right)$. The check was performed as Example 1.

YOUR TURN

Sometimes, certain variables are missing at the outset.

EXAMPLE 4 Solve the system

$$x + y + z = 180, \quad (1)$$
$$x - z = -70, \quad (2)$$
$$2y - z = 0. \quad (3)$$

SOLUTION

1., 2. The equations appear in standard form with no fractions or decimals.

3., 4. Note that there is no y in equation (2). Thus, at the outset, we already have y eliminated from one equation. We need another equation with no y-term, so we work with equations (1) and (3):

$$x + y + z = 180, \xrightarrow{\text{Multiplying both sides of equation (1) by } -2}$$

$$\begin{array}{r} -2x - 2y - 2z = -360 \\ 2y - z = 0 \\ \hline -2x \quad - 3z = -360. \end{array} \quad (4)$$

with $2y - z = 0$

5., 6. Now we solve the resulting system of equations (2) and (4):

$$x - z = -70, \xrightarrow{\text{Multiplying both sides of equation (2) by 2}}$$

$$\begin{array}{r} 2x - 2z = -140 \\ -2x - 3z = -360 \\ \hline -5z = -500 \\ z = 100. \end{array}$$

with $-2x - 3z = -360$

4. Solve the system

$$\begin{array}{rl} x - y & = 8, \\ x + y + z & = 17, \\ x \quad + z & = 5. \end{array}$$

Continuing as in Examples 2 and 3, we get the solution $(30, 50, 100)$. The check is left to the student.

 YOUR TURN

Dependency, Inconsistency, and Geometric Considerations

Each equation in Examples 2, 3, and 4 has a graph that is a plane in three dimensions. The solutions are points common to the planes of each system. Since three planes can have an infinite number of points in common or no points at all in common, we need to generalize the concept of *consistency*.

Planes intersect at one point. System is *consistent* and has one solution.

Planes intersect along a common line. System is *consistent* and has an infinite number of solutions.

Three parallel planes. System is *inconsistent;* it has no solution.

Planes intersect two at a time, with no point common to all three. System is *inconsistent;* it has no solution.

CONSISTENCY

A system of equations that has at least one solution is said to be **consistent**.

A system of equations that has no solution is said to be **inconsistent**.

EXAMPLE 5 Solve:

$$\begin{array}{rl} y + 3z & = 4, \quad (1) \\ -x - y + 2z & = 0, \quad (2) \\ x + 2y + z & = 1. \quad (3) \end{array}$$

SOLUTION   There is no x-term in equation (1). By adding equations (2) and (3), we can find a second equation in which x is again absent:

$$-x - \ y + 2z = 0 \quad (2)$$
$$\underline{x + 2y + \ z = 1} \quad (3)$$
$$y + 3z = 1. \quad (4) \qquad \text{Adding}$$

Equations (1) and (4) form a system in y and z. We solve as before:

$$y + 3z = 4, \qquad \xrightarrow{\substack{\text{Multiplying both sides} \\ \text{of equation (1) by } -1}} \qquad -y - 3z = -4$$
$$y + 3z = 1 \qquad \qquad \qquad \underline{y + 3z = \quad 1}$$
$$\text{This is a contradiction.} \xrightarrow{\qquad} 0 = -3. \qquad \text{Adding}$$

Since we end up with a *false* equation, or contradiction, we state that the system has no solution. It is *inconsistent*.

5. Solve:

$$x - 2y + 2z = 6,$$
$$2x + 3y \quad\quad = 1,$$
$$-3x - 8y + 2z = 0.$$

YOUR TURN

The notion of *dependency* can also be extended to systems of three equations.

EXAMPLE 6   Solve:

$$2x + \ y + \ z = 3, \quad (1)$$
$$x - 2y - \ z = 1, \quad (2)$$
$$3x + 4y + 3z = 5. \quad (3)$$

SOLUTION   Our plan is to first use equations (1) and (2) to eliminate z. Then we will select another pair of equations and again eliminate z:

$$2x + \ y + z = 3$$
$$\underline{x - 2y - z = 1}$$
$$3x - \ y \quad\quad = 4. \quad (4)$$

Next, we use equations (2) and (3) to eliminate z again:

$$x - 2y - \ z = 1, \quad \xrightarrow{\substack{\text{Multiplying both sides} \\ \text{of equation (2) by } 3}} \quad 3x - 6y - 3z = 3$$
$$3x + 4y + 3z = 5 \qquad\qquad\qquad \underline{3x + 4y + 3z = 5}$$
$$6x - 2y \quad\quad = 8. \quad (5)$$

We now solve the resulting system of equations (4) and (5):

$$3x - \ y = 4, \quad \xrightarrow{\substack{\text{Multiplying both sides} \\ \text{of equation (4) by } -2}} \quad -6x + 2y = -8$$
$$6x - 2y = 8 \qquad\qquad\qquad \underline{6x - 2y = \quad 8}$$
$$0 = \quad 0. \quad (6)$$

6. Solve:

$$x - y + \ z = -1,$$
$$2x + y + 2z = 5,$$
$$4x - y + 4z = 3.$$

Equation (6), which is an identity, indicates that equations (1), (2), and (3) are *dependent*. This means that the original system of three equations is equivalent to a system of two equations. One way to see this is to note that two times equation (1), minus equation (2), is equation (3). Thus removing equation (3) from the system does not affect the solution of the system.* In writing an answer to this problem, we simply state that "the equations are dependent."

YOUR TURN

---

* A set of equations is dependent if at least one equation can be expressed as a sum of multiples of other equations in that set.

In a system of two equations, when equations are dependent the system is consistent. This is not always the case for systems of three or more equations. The following figures illustrate some possibilities geometrically.

The planes intersect along a common line. The equations are *dependent* and the system is *consistent*. There is an infinite number of solutions.

The planes coincide. The equations are *dependent* and the system is *consistent*. There is an infinite number of solutions.

Two planes coincide. The third plane is parallel. The equations are *dependent* and the system is *inconsistent*. There is no solution.

3.4 EXERCISE SET

FOR EXTRA HELP   MyMathLab® Math XL

PRACTICE   WATCH   READ   REVIEW

Vocabulary and Reading Check

Classify each of the following statements as either true or false.

1. $3x + 5y + 4z = 7$ is a linear equation in three variables.

2. Every system of three equations in three unknowns has at least one solution.

3. It is not difficult to solve a system of three equations in three unknowns by graphing.

4. If, when we are solving a system of three equations, a false equation results from adding a multiple of one equation to another, the system is inconsistent.

5. If, when we are solving a system of three equations, an identity results from adding a multiple of one equation to another, the equations are dependent.

6. Whenever a system of three equations contains dependent equations, there is an infinite number of solutions.

Identifying Solutions

7. Determine whether $(2, -1, -2)$ is a solution of the system
$$x + y - 2z = 5,$$
$$2x - y - z = 7,$$
$$-x - 2y - 3z = 6.$$

8. Determine whether $(-1, -3, 2)$ is a solution of the system
$$x - y + z = 4,$$
$$x - 2y - z = 3,$$
$$3x + 2y - z = 1.$$

Solving Systems in Three Variables

Solve each system. If a system's equations are dependent or if there is no solution, state this.

9. $x - y - z = 0,$
$2x - 3y + 2z = 7,$
$-x + 2y + z = 1$

10. $x + y - z = 0,$
$2x - y + z = 3,$
$-x + 5y - 3z = 2$

11. $x - y - z = 1,$
$2x + y + 2z = 4,$
$x + y + 3z = 5$

12. $x + y - 3z = 4,$
$2x + 3y + z = 6,$
$2x - y + z = -14$

13. $3x + 4y - 3z = 4,$
$5x - y + 2z = 3,$
$x + 2y - z = -2$

14. $2x - 3y + z = 5,$
$x + 3y + 8z = 22,$
$3x - y + 2z = 12$

15. $x + y + z = 0,$
$2x + 3y + 2z = -3,$
$-x - 2y - z = 1$

16. $3a - 2b + 7c = 13,$
$a + 8b - 6c = -47,$
$7a - 9b - 9c = -3$

17. $2x - 3y - z = -9,$
$2x + 5y + z = 1,$
$x - y + z = 3$

18. $4x + y + z = 17,$
$x - 3y + 2z = -8,$
$5x - 2y + 3z = 5$

Aha! **19.** $a + b + c = 5,$
$2a + 3b - c = 2,$
$2a + 3b - 2c = 4$

20. $u - v + 6w = 8,$
$3u - v + 6w = 14,$
$-u - 2v - 3w = 7$

21. $-2x + 8y + 2z = 4,$
$x + 6y + 3z = 4,$
$3x - 2y + z = 0$

22. $x - y + z = 4,$
$5x + 2y - 3z = 2,$
$4x + 3y - 4z = -2$

23. $2u - 4v - w = 8,$
$3u + 2v + w = 6,$
$5u - 2v + 3w = 2$

24. $4p + q + r = 3,$
$2p - q + r = 6,$
$2p + 2q - r = -9$

25. $r + \frac{3}{2}s + 6t = 2,$
$2r - 3s + 3t = 0.5,$
$r + s + t = 1$

26. $5x + 3y + \frac{1}{2}z = \frac{7}{2},$
$0.5x - 0.9y - 0.2z = 0.3,$
$3x - 2.4y + 0.4z = -1$

27. $4a + 9b = 8,$
$8a + 6c = -1,$
$6b + 6c = -1$

28. $3p + 2r = 11,$
$q - 7r = 4,$
$p - 6q = 1$

29. $x + y + z = 57,$
$-2x + y = 3,$
$x - z = 6$

30. $x + y + z = 105,$
$10y - z = 11,$
$2x - 3y = 7$

31. $a - 3c = 6,$
$b + 2c = 2,$
$7a - 3b - 5c = 14$

32. $2a - 3b = 2,$
$7a + 4c = \frac{3}{4},$
$2c - 3b = 1$

Aha! **33.** $x + y + z = 83,$
$y = 2x + 3,$
$z = 40 + x$

34. $l + m = 7,$
$3m + 2n = 9,$
$4l + n = 5$

35. $x + z = 0,$
$x + y + 2z = 3,$
$y + z = 2$

36. $x + y = 0,$
$x + z = 1,$
$2x + y + z = 2$

37. $x + y + z = 1,$
$-x + 2y + z = 2,$
$2x - y = -1$

38. $y + z = 1,$
$x + y + z = 1,$
$x + 2y + 2z = 2$

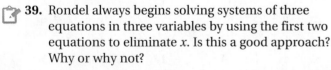 **39.** Rondel always begins solving systems of three equations in three variables by using the first two equations to eliminate x. Is this a good approach? Why or why not?

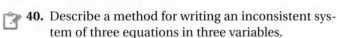 **40.** Describe a method for writing an inconsistent system of three equations in three variables.

Skill Review

41. Find the slope and the y-intercept of the graph of $x - 3y = 7$.   [2.3]

42. Find the slope of the graph of $f(x) = 8$. If the slope is undefined, state this.   [2.4]

43. Find the intercepts of the graph of $2x - 5y = 20$.   [2.4]

44. Find the slope of the line containing $(6, 9)$ and $(-2, 4)$.   [2.3]

Determine whether each pair of lines is parallel, perpendicular, or neither.   [2.4]

45. $3x - y = 12,$
$y = 3x + 7$

46. $2x - 5y = 6,$
$2x + 5y = 1$

Synthesis

47. Is it possible for a system of three linear equations to have exactly two ordered triples in its solution set? Why or why not?

48. Kadi and Ahmed both correctly solve the system

$$x + 2y - z = 1,$$
$$-x - 2y + z = 3,$$
$$2x + 4y - 2z = 2.$$

Kadi states "the equations are dependent" while Ahmed states "there is no solution." How is this possible?

Solve.

49. $\dfrac{x+2}{3} - \dfrac{y+4}{2} + \dfrac{z+1}{6} = 0,$
$\dfrac{x-4}{3} + \dfrac{y+1}{4} - \dfrac{z-2}{2} = -1,$
$\dfrac{x+1}{2} + \dfrac{y}{2} + \dfrac{z-1}{4} = \dfrac{3}{4}$

50. $w + x - y + z = 0,$
$w - 2x - 2y - z = -5,$
$w - 3x - y + z = 4,$
$2w - x - y + 3z = 7$

51. $w + x + y + z = 2,$
$w + 2x + 2y + 4z = 1,$
$w - x + y + z = 6,$
$w - 3x - y + z = 2$

For Exercises 52 and 53, let u represent $1/x$, v represent $1/y$, and w represent $1/z$. Solve for u, v, and w, and then solve for x, y, and z.

52. $\dfrac{2}{x} + \dfrac{2}{y} - \dfrac{3}{z} = 3,$
$\dfrac{1}{x} - \dfrac{2}{y} - \dfrac{3}{z} = 9,$
$\dfrac{7}{x} - \dfrac{2}{y} + \dfrac{9}{z} = -39$

53. $\dfrac{2}{x} - \dfrac{1}{y} - \dfrac{3}{z} = -1,$
$\dfrac{2}{x} - \dfrac{1}{y} + \dfrac{1}{z} = -9,$
$\dfrac{1}{x} + \dfrac{2}{y} - \dfrac{4}{z} = 17$

Determine k so that each system is dependent.

54. $x - 3y + 2z = 1,$
$2x + y - z = 3,$
$9x - 6y + 3z = k$

55. $5x - 6y + kz = -5,$
$x + 3y - 2z = 2,$
$2x - y + 4z = -1$

In each case, three solutions of an equation in x, y, and z are given. Find the equation.

56. $Ax + By + Cz = 12;$
$\left(1, \frac{3}{4}, 3\right), \left(\frac{4}{3}, 1, 2\right),$ and $(2, 1, 1)$

57. $z = b - mx - ny;$
$(1, 1, 2), (3, 2, -6),$ and $\left(\frac{3}{2}, 1, 1\right)$

58. Write an inconsistent system of equations that contains dependent equations.

➤ YOUR TURN ANSWERS: SECTION 3.4

1. No   **2.** $(3, -1, 4)$ **3.** $\left(3, -\frac{2}{3}, 8\right)$ **4.** $(20, 12, -15)$

5. No solution   **6.** The equations are dependent.

QUICK QUIZ: SECTIONS 3.1–3.4

Solve. If a system has an infinite number of solutions, use set-builder notation to write the solution set. If a system has no solution, state this. [3.1], [3.2]

1. $3x + 2y = 9,$ **2.** $2x - y = 4,$
$x - 6y = 1$ $3y = 6x - 12$

3. Solve: $x + y + z = 8,$
$2x + y - z = -5,$
$x - 2y - z = 8.$ [3.4]

4. Jared's motorboat took 2 hr to make a trip downstream with a 4-mph current. The return trip against the same current took 3 hr. Find the speed of the boat in still water. [3.3]

5. Julia has paint with 15% red pigment and paint with 10% red pigment. How much of each should she use to form 3 gal of paint with 12% red pigment? [3.3]

PREPARE TO MOVE ON

Translate each statement to an equation. [1.1]

1. The sum of three consecutive numbers is 100.

2. The sum of three numbers is 100.

3. The product of two numbers is five times a third number.

4. The product of two numbers is twice their sum.

Mid-Chapter Review

Systems of two equations can be solved using graphical or algebraic methods. Since graphing in three dimensions is difficult, algebraic methods are used to solve systems of three equations. Both substitution and elimination work well for systems of two equations, but elimination is usually the preferred method for systems of three equations.

GUIDED SOLUTIONS

Solve. [3.2]

1. $2x - 3y = 5,$
   $y = x - 1$

Solution

$2x - 3\left(\boxed{}\right) = 5$ Substituting $x - 1$ for y

$2x - \boxed{} + \boxed{} = 5$ Using the distributive law

$\boxed{} + 3 = 5$ Combining like terms

$-x = \boxed{}$ Subtracting 3 from both sides

$x = \boxed{}$ Dividing both sides by -1

$y = x - 1$

$y = \boxed{} - 1$ Substituting

$y = \boxed{}$

The solution is $\left(\boxed{}, \boxed{}\right).$

2. $2x - 5y = 1,$
   $x + 5y = 8$

Solution

$2x - 5y = 1$
$x + 5y = 8$
$\overline{\boxed{} = \boxed{}}$

$x = \boxed{}$

$x + 5y = 8$

$\boxed{} + 5y = 8$ Substituting

$5y = \boxed{}$

$y = \boxed{}$

The solution is $\left(\boxed{}, \boxed{}\right).$

MIXED REVIEW

Solve using any appropriate method. [3.1], [3.2], [3.4]

3. $x = y,$
   $x + y = 2$

4. $x + y = 10,$
   $x - y = 8$

5. $y = \frac{1}{2}x + 1,$
   $y = 2x - 5$

6. $y = 2x - 3,$
   $x + y = 12$

7. $x = 5,$
   $y = 10$

8. $3x + 5y = 8,$
   $3x - 5y = 4$

9. $2x - y = 1,$
   $2y - 4x = 3$

10. $x = 2 - y,$
   $3x + 3y = 6$

11. $1.1x - 0.3y = 0.8,$
   $2.3x + 0.3y = 2.6$

12. $\frac{1}{4}x = \frac{1}{3}y,$
   $\frac{1}{2}x - \frac{1}{15}y = 2$

13. $3x + y - z = -1,$
   $2x - y + 4z = 2,$
   $x - y + 3z = 3$

14. $2x + y - 3z = -4,$
   $4x + y + 3z = -1,$
   $2x - y + 6z = 7$

15. $3x + 5y - z = 8,$
   $x + 6y = 4,$
   $x - 7y - z = 3$

16. $x - y = 4,$
   $2x + y - z = 5,$
   $3x - z = 9$

Solve. [3.3]

17. In 2010, 12.3% of the amount spent on Internet advertising went to Yahoo! and Microsoft. Yahoo! received 0.3% more of the amount spent on Internet advertising than twice the amount Microsoft received. What percent of Internet advertising went to each company?
Source: ZenithOptimedia

18. As part of a fundraiser, the Cobblefield Daycare collected 430 returnable bottles and cans, some worth 5 cents each and the rest worth 10 cents each. If the total value of the cans and bottles was $26.20, how many 5-cent bottles or cans and how many 10-cent bottles or cans were collected?

19. Pecan Morning granola is 25% nuts and dried fruit. Oat Dream granola is 10% nuts and dried fruit. How much of Pecan Morning and how much of Oat Dream should be mixed in order to form a 20-lb batch of granola that is 19% nuts and dried fruit?

20. The Grand Royale cruise ship takes 3 hr to make a trip up the Amazon River against a 6-mph current. The return trip with the same current takes 1.5 hr. Find the speed of the ship in still water.

| **3.5** | **Solving Applications: Systems of Three Equations** |

Applications of Three Equations in Three Unknowns

Applications of Three Equations in Three Unknowns

Systems of three or more equations arise in the natural and social sciences, business, and engineering. To begin, let's first look at a purely numerical application.

EXAMPLE 1 The sum of three numbers is 4. The first number minus twice the second, minus the third is 1. Twice the first number minus the second, minus twice the third is −1. Find the numbers.

SOLUTION

1. **Familiarize.** There are three statements involving the same three numbers. Let's label these numbers x, y, and z.

2. **Translate.** We can translate directly as follows.

   The sum of the three numbers ┄┄ is ┄ 4.
   $$x + y + z \qquad\qquad = 4$$

   The first number ┄ minus ┄ twice the second ┄ minus ┄ the third ┄ is ┄ 1.
   $$x \qquad\qquad - \qquad\qquad 2y \qquad\qquad - \qquad z \qquad = 1$$

   Twice the first number ┄ minus ┄ the second ┄ minus ┄ twice the third ┄ is ┄ −1.
   $$2x \qquad\qquad - \qquad y \qquad\qquad - \qquad 2z \qquad = -1$$

   We now have a system of three equations:

   $$x + y + z = 4,$$
   $$x - 2y - z = 1,$$
   $$2x - y - 2z = -1.$$

3. **Carry out.** The solution of this system is $(2, -1, 3)$, as we found in Example 2 of Section 3.4.

4. **Check.** The first statement of the problem says that the sum of the three numbers is 4. That checks, because $2 + (-1) + 3 = 4$. The second statement says that the first number minus twice the second, minus the third is 1: $2 - 2(-1) - 3 = 1$. That checks. The check of the third statement is left to the student.

5. **State.** The three numbers are 2, −1, and 3.

 YOUR TURN

1. The sum of three numbers is 10. Twice the first number plus the second equals the third. Half the first number plus the second plus the third is 6. Find the numbers.

EXAMPLE 2 *Architecture.* In a triangular cross section of a roof, the largest angle is 70° greater than the smallest angle. The largest angle is twice as large as the remaining angle. Find the measure of each angle.

SOLUTION

1. **Familiarize.** The first thing we do is make a drawing, or a sketch.

Student Notes

It is quite likely that you are expected to remember that the sum of the measures of the angles in any triangle is 180°. You may want to ask your instructor which other formulas from geometry and elsewhere you are expected to know.

Since we don't know the size of any angle, we use x, y, and z to represent the three measures, from smallest to largest. Recall that the measures of the angles in any triangle add up to 180°.

2. **Translate.** This geometric fact about triangles gives us one equation:

$$x + y + z = 180.$$

Two of the statements can be translated almost directly.

The largest angle is 70° greater than the smallest angle.
$$z \quad = \quad x + 70$$

The largest angle is twice as large as the remaining angle.
$$z \quad = \quad 2y$$

We now have a system of three equations:

$x + y + z = 180,$	$x + y + z = 180,$
$x + 70 = z,$ or	$x \qquad - z = -70,$ Rewriting in
$2y = z;$	$2y - z = 0.$ standard form

3. **Carry out.** This system was solved in Example 4 of Section 3.4. The solution is $(30, 50, 100)$.

4. **Check.** The sum of the numbers is 180, so that checks. The measure of the largest angle, 100°, is 70° greater than the measure of the smallest angle, 30°, so that checks. The measure of the largest angle is also twice the measure of the remaining angle, 50°. Thus we have a check.

5. **State.** The angles in the triangle measure 30°, 50°, and 100°.

2. In a triangular cross section of a roof, the largest angle is twice the smallest angle. The remaining angle is 20° smaller than the largest angle. Find the measure of each angle.

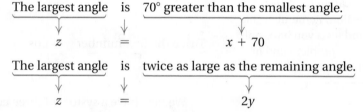 YOUR TURN

EXAMPLE 3 *Downloads.* Kaya frequently downloads music, TV shows, and movies. In January, she downloaded 5 songs, 10 TV shows, and 3 movies for a total of $55. In February, she spent $195 on 25 songs, 25 TV shows, and 12 movies. In March, she spent $81 on 15 songs, 8 TV shows, and 5 movies. Assuming each song is the same price, each TV show is the same price, and each movie is the same price, how much does each type of download cost?

SOLUTION

1. **Familiarize.** We let s = the cost, in dollars, per song, t = the cost, in dollars, per TV show, and m = the cost, in dollars, per movie. The total cost is the sum of the cost per item times the number of items purchased.

2. **Translate.** In January, Kaya spent $5 \cdot s$ for songs, $10 \cdot t$ for TV shows, and $3 \cdot m$ for movies. The total of these amounts was \$55. Each month's downloads will translate to an equation. We can organize the information in a table.

	Cost of Songs	Cost of TV Shows	Cost of Movies	Total Cost	
January	$5s$	$10t$	$3m$	55	$\longrightarrow 5s + 10t + 3m = 55$
February	$25s$	$25t$	$12m$	195	$\longrightarrow 25s + 25t + 12m = 195$
March	$15s$	$8t$	$5m$	81	$\longrightarrow 15s + 8t + 5m = 81$

We now have a system of three equations:

$$5s + 10t + 3m = 55, \quad (1)$$
$$25s + 25t + 12m = 195, \quad (2)$$
$$15s + 8t + 5m = 81. \quad (3)$$

3. **Carry out.** We begin by using equations (1) and (2) to eliminate s.

$$
\begin{aligned}
5s + 10t + 3m &= 55, \\
25s + 25t + 12m &= 195
\end{aligned}
\quad
\xrightarrow[\text{of equation (1) by } -5]{\text{Multiplying both sides}}
\quad
\begin{aligned}
-25s - 50t - 15m &= -275 \\
25s + 25t + 12m &= 195 \\
\hline
-25t - 3m &= -80 \quad (4)
\end{aligned}
$$

We then use equations (1) and (3) to again eliminate s.

$$
\begin{aligned}
5s + 10t + 3m &= 55, \\
15s + 8t + 5m &= 81
\end{aligned}
\quad
\xrightarrow[\text{of equation (1) by } -3]{\text{Multiplying both sides}}
\quad
\begin{aligned}
-15s - 30t - 9m &= -165 \\
15s + 8t + 5m &= 81 \\
\hline
-22t - 4m &= -84 \quad (5)
\end{aligned}
$$

Now we solve the resulting system of equations (4) and (5).

$$
\begin{aligned}
-25t - 3m &= -80, \\
\\
-22t - 4m &= -84
\end{aligned}
\quad
\begin{aligned}
\xrightarrow[\text{of equation (4) by } -4]{\text{Multiplying both sides}} \\
\xrightarrow[\text{of equation (5) by } 3]{\text{Multiplying both sides}}
\end{aligned}
\quad
\begin{aligned}
100t + 12m &= 320 \\
-66t - 12m &= -252 \\
\hline
34t &= 68 \\
t &= 2
\end{aligned}
$$

To find m, we use equation (4):

$$
\begin{aligned}
-25t - 3m &= -80 \\
-25 \cdot 2 - 3m &= -80 \qquad \text{Substituting 2 for } t \\
-50 - 3m &= -80 \\
-3m &= -30 \\
m &= 10.
\end{aligned}
$$

3. Eli frequently downloads music, HDTV shows, and games. In April, he downloaded 3 albums, 10 HDTV shows, and 8 games for a total of $74. In May, he spent $100 for 5 albums, 12 HDTV shows, and 4 games. In June, he spent $79 for 2 albums, 15 HDTV shows, and 10 games. Assuming each album is the same price, each HDTV show is the same price, and each game is the same price, how much does each type of download cost?

Finally, we use equation (1) to find s:

$$5s + 10t + 3m = 55$$
$$5s + 10 \cdot 2 + 3 \cdot 10 = 55 \quad \text{Substituting 2 for } t \text{ and 10 for } m$$
$$5s + 20 + 30 = 55$$
$$5s + 50 = 55$$
$$5s = 5$$
$$s = 1.$$

4. Check. If a song costs $1, a TV show costs $2, and a movie costs $10, then the total cost for each month's downloads is as follows:

January: $5 \cdot \$1 + 10 \cdot \$2 + 3 \cdot \$10 = \$5 + \$20 + \$30 = \$55$;

February: $25 \cdot \$1 + 25 \cdot \$2 + 12 \cdot \$10 = \$25 + \$50 + \$120 = \$195$;

March: $15 \cdot \$1 + 8 \cdot \$2 + 5 \cdot \$10 = \$15 + \$16 + \$50 = \$81$.

This checks with the information given in the problem.

5. State. A song costs $1, a TV show costs $2, and a movie costs $10.

YOUR TURN

3.5 EXERCISE SET

FOR EXTRA HELP

MyMathLab® Math XL

PRACTICE   WATCH   READ   REVIEW

Vocabulary and Reading Check

Match each statement with a translation from the list below.

a) $x + y + z = 50$ c) $x - y + z = 50$
b) $x - y - z = -50$ d) $x - y - z = 50$

1. ____   The sum of three numbers is 50.

2. ____   The first number minus the second plus the third is 50.

3. ____   The first number is 50 more than the sum of the other two numbers.

4. ____   The first number is 50 less than the sum of the other two numbers.

Applications of Three Equations in Three Unknowns

Solve.

5. The sum of three numbers is 85. The second is 7 more than the first. The third is 2 more than four times the second. Find the numbers.

6. The sum of three numbers is 5. The first number minus the second plus the third is 1. The first minus the third is 3 more than the second. Find the numbers.

7. The sum of three numbers is 26. Twice the first minus the second is 2 less than the third. The third is the second minus three times the first. Find the numbers.

8. The sum of three numbers is 105. The third is 11 less than ten times the second. Twice the first is 7 more than three times the second. Find the numbers.

9. *Geometry.* In triangle *ABC*, the measure of angle *B* is three times that of angle *A*. The measure of angle *C* is 20° more than that of angle *A*. Find the angle measures.

10. *Geometry.* In triangle *ABC*, the measure of angle *B* is twice the measure of angle *A*. The measure of angle *C* is 80° more than that of angle *A*. Find the angle measures.

11. *Scholastic Aptitude Test.* Many high-school students take the Scholastic Aptitude Test (SAT). Those taking the SAT receive three scores: a critical reading score, a mathematics score, and a writing score. In 2011, the average total score of high-school seniors was 1500. The average mathematics score exceeded the reading score by 17 points, and the average writing score was 8 points less than the reading score. What was the average score for each category?

Source: College Entrance Examination Board

12. *Advertising.* In 2011, U.S. companies spent a total of $111 billion on newspaper, television, and Internet ads. The amount spent on television ads was $11 billion more than the amount spent on newspaper and Internet ads combined. The amount spent on newspaper ads was $8 billion less than the amount spent on Internet ads. How much was spent on each form of advertising?

Source: eMarketer

13. *Nutrition.* Most nutritionists now agree that a healthy adult diet includes 25–35 g of fiber each day. A breakfast of 2 bran muffins, 1 banana, and a 1-cup serving of Wheaties® contains 9 g of fiber; a breakfast of 1 bran muffin, 2 bananas, and a 1-cup serving of Wheaties® contains 10.5 g of fiber; and a breakfast of 2 bran muffins and a 1-cup serving of Wheaties® contains 6 g of fiber. How much fiber is in each of these foods?

Sources: usda.gov and InteliHealth.com

14. *Nutrition.* Refer to Exercise 13. A breakfast consisting of 2 pancakes and a 1-cup serving of strawberries contains 4.5 g of fiber, whereas a breakfast of 2 pancakes and a 1-cup serving of Cheerios® contains 4 g of fiber. When a meal consists of 1 pancake, a 1-cup serving of Cheerios®, and a 1-cup serving of strawberries, it contains 7 g of fiber. How much fiber is in each of these foods?

Source: InteliHealth.com

Aha! 15. *Automobile Pricing.* The base model of a 2012 Jeep Grand Cherokee Laredo (2WD) with a tow package cost $29,515. When equipped with a tow package and a sunroof, the vehicle's price rose to $30,365. The cost of the base model with a sunroof was $29,770. Find the base price, the cost of a tow package, and the cost of a sunroof.

Source: www.jeep.com

16. *Telemarketing.* Sven, Tina, and Laurie can process 740 telephone orders per day. Sven and Tina together can process 470 orders, while Tina and Laurie together can process 520 orders per day. How many orders can each person process alone?

17. *Coffee Prices.* Reba works at a Starbucks® coffee shop where a 12-oz cup of coffee costs $1.75, a 16-oz cup costs $1.95, and a 20-oz cup costs $2.25. During one busy period, Reba served 55 cups of coffee, emptying six 144-oz "brewers" while collecting a total of $107.75. How many cups of each size did Reba fill?

18. *Restaurant Management.* Chick-fil-A® recently sold 14-oz lemonades for $1.49 each, 20-oz lemonades for $1.69 each, and 32-oz lemonades for $2.05 each. During a lunchtime rush, Chris sold 40 lemonades, using $6\frac{1}{4}$ gal of lemonade while collecting a total of $67.40. How many drinks of each size were sold? (*Hint:* 1 gal = 128 oz.)

19. *Small-Business Loans.* Chelsea took out three loans for a total of $120,000 to start an organic orchard. Her business-equipment loan was at an interest rate of 11%, the small-business loan was at an interest rate of 5%, and her home-equity loan was at an interest rate of 5.5%. The total simple interest due on the loans in one year was $7250. The annual simple interest on the home-equity loan was $2200 more than the interest on the business-equipment loan. How much did she borrow from each source?

20. *Investments.* A business class divided an imaginary investment of $80,000 among three mutual funds. The first fund grew by 4%, the second by 6%, and the third by 8%. Total earnings were $4400. The earnings from the third fund were $200 more than the earnings from the first. How much was invested in each fund?

21. *Gold Alloys.* Gold used to make jewelry is often a blend of gold, silver, and copper. The relative amounts of the metals determine the color of the alloy. Red gold is 75% gold, 5% silver, and 20% copper. Yellow gold is 75% gold, 12.5% silver, and 12.5%

copper. White gold is 37.5% gold and 62.5% silver. If 100 g of red gold costs $4177.15, 100 g of yellow gold costs $4185.25, and 100 g of white gold costs $2153.875, how much do gold, silver, and copper cost per gram?

Source: World Gold Council

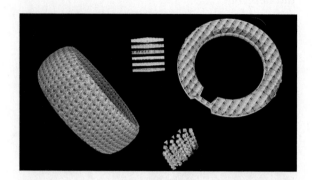

22. *Gardening.*   Dana is designing three large perennial flower beds for her yard and is planning to use combinations of three types of flowers. In her traditional cottage-style garden, Dana will include 7 purple coneflower plants, 6 yellow foxglove plants, and 8 white lupine plants at a total cost of $93.31. The flower bed around her deck will be planted with 12 yellow foxglove plants and 12 white lupine plants at a total cost of $126.00. A third garden area in a corner of Dana's yard will contain 4 purple coneflower plants, 5 yellow foxglove plants, and 7 white lupine plants at a total cost of $72.82. What is the price per plant for the coneflowers, the foxgloves, and the lupines?

23. *Nutrition.*   A dietician in a hospital prepares meals under the guidance of a physician. Suppose that for a particular patient a physician prescribes a meal to have 800 calories, 55 g of protein, and 220 mg of vitamin C. The dietician prepares a meal of roast beef, baked potatoes, and broccoli according to the data in the following table.

Serving Size	Calories	Protein (in grams)	Vitamin C (in milligrams)
Roast Beef, 3 oz	300	20	0
Baked Potato, 1	100	5	20
Broccoli, 156 g	50	5	100

How many servings of each food are needed in order to satisfy the doctor's orders?

24. *Nutrition.*   Repeat Exercise 23 but replace the broccoli with asparagus, for which a 180-g serving contains 50 calories, 5 g of protein, and 44 mg of vitamin C. Which meal would you prefer eating?

25. Students in a Listening Responses class bought 40 tickets for a piano concert. The number of tickets purchased for seats in either the first mezzanine or the main floor was the same as the number purchased for seats in the second mezzanine. First mezzanine seats cost $52 each, main floor seats cost $38 each, and second mezzanine seats cost $28 each. The total cost of the tickets was $1432. How many of each type of ticket were purchased?

26. *Basketball Scoring.*   The New York Knicks recently scored a total of 92 points on a combination of 2-point field goals, 3-point field goals, and 1-point foul shots. Altogether, the Knicks made 50 baskets and 19 more 2-pointers than foul shots. How many shots of each kind were made?

27. *World Population Growth.*   The world population is projected to be 9.4 billion in 2050. At that time, there is expected to be approximately 2.9 billion more people in Asia than in Africa. The population for the rest of the world will be approximately 0.6 billion more than one-fourth of the population of Asia. Find the projected populations of Asia, Africa, and the rest of the world in 2050.

Source: U.S. Census Bureau

28. *History.*   Find the year in which the first U.S. transcontinental railroad was completed. The sum of the digits in the year is 24. The ones digit is 1 more than the hundreds digit. Both the tens and the ones digits are multiples of 3.

29. Jaci knows that one angle in a triangle is twice as large as another. Does she have enough information to find the measures of the angles in the triangle? Why or why not?

30. Write a problem for a classmate to solve. Design the problem so that it translates to a system of three equations in three variables.

Skill Review

Graph.

31. $y = 4$ [2.4]

32. $y = -\frac{2}{5}x + 3$ [2.3]

33. $y - 3x = 3$ [2.4]

34. $2x = -8$ [2.4]

35. $f(x) = 2x - 1$ [2.3]

36. $3x - y = 2$ [2.3]

Synthesis

37. Consider Exercise 26. Suppose there were no foul shots made. Would there still be a solution? Why or why not?

38. Consider Exercise 17. Suppose Reba collected $50. Could the problem still be solved? Why or why not?

39. *Health Insurance.*   In 2012, UnitedHealthOne Plan 100 health insurance for a 35-year-old and his or her spouse cost $304 per month. That rate increased to $420 per month if a child were included and $537 per month if two children were included. The rate dropped to $266 per month for just the applicant and one child. Find the separate costs for insuring the applicant, the spouse, the first child, and the second child.

Source: UNICARE Life and Health Insurance Company® through www.ehealth.com

40. Find a three-digit number such that the sum of the digits is 14, the tens digit is 2 more than the ones digit, and if the digits are reversed, the number is unchanged.

41. *Ages.*   Tammy's age is the sum of the ages of Carmen and Dennis. Carmen's age is 2 more than the sum of the ages of Dennis and Mark. Dennis's age is four times Mark's age. The sum of all four ages is 42. How old is Tammy?

42. *Ticket Revenue.*   A magic show's audience of 100 people consists of adults, students, and children. The ticket prices are $10 each for adults, $3 each for students, and 50¢ each for children. A total of $100 is taken in. How many adults, students, and children are in attendance? Does there seem to be some information missing? Do some more careful reasoning.

43. *Sharing Raffle Tickets.*   Hal gives Tom as many raffle tickets as Tom first had and Gary as many as Gary first had. In like manner, Tom then gives Hal and Gary as many tickets as each then has. Similarly, Gary gives Hal and Tom as many tickets as each then has. If each finally has 40 tickets, with how many tickets does Tom begin?

44. Find the sum of the angle measures at the tips of the star in this figure.

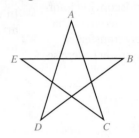

QUICK QUIZ: SECTIONS 3.1–3.5

Solve.   [3.1], [3.2], [3.4]

1. $y = 2x - 5,$
$y = \frac{1}{2}x + 1$

2. $x + 2y = 3,$
$3x = 4 - y$

3. $10x + 20y = 40,$
$x - \quad y = 7$

4. $9x + 8y = 0,$
$11x - 7y = 0$

5. $2x + \quad y + \quad z = 3,$
$x + \quad y - 4z = 13,$
$4x + 3y + 2z = 11$

PREPARE TO MOVE ON

Simplify.   [1.2]

1. $-2(2x - 3y)$

2. $-6(x - 2y) + (6x - 5y)$

3. $-(2a - b - 6c)$

4. $-2(3x - y + z) + 3(-2x + y - 2z)$

5. $(8x - 10y + 7z) + 5(3x + 2y - 4z)$

3.6   Elimination Using Matrices

Matrices and Systems   ■   Row-Equivalent Operations

Study Skills

Double-Check the Numbers

Solving problems is challenging enough, without miscopying information. Always double-check that you have accurately transferred numbers from the correct exercise in the exercise set.

In solving systems of equations, we perform computations with the constants. If we agree to keep all like terms in the same column, we can simplify writing a system by omitting the variables. For example, if we do not write the variables, the operation of addition, and the equal signs, the system

$$\begin{array}{l} 3x + 4y = 5, \\ x - 2y = 1 \end{array} \quad \text{simplifies to} \quad \begin{array}{rrr} 3 & 4 & 5 \\ 1 & -2 & 1 \end{array}.$$

Matrices and Systems

In the example above, we have written a rectangular array of numbers. Such an array is called a **matrix** (plural, **matrices**). We ordinarily write brackets around matrices. The following are matrices:

$$\begin{bmatrix} -3 & 1 \\ 0 & 5 \end{bmatrix}, \quad \begin{bmatrix} 2 & 0 & -1 & 3 \\ -5 & 2 & 7 & -1 \\ 4 & 5 & 3 & 0 \end{bmatrix}, \quad \begin{bmatrix} 2 & 3 \\ 7 & 15 \\ -2 & 23 \\ 4 & 1 \end{bmatrix}.$$

The individual numbers are called *elements*, or *entries*.

The **rows** of a matrix are horizontal, and the **columns** are vertical.

$$\begin{bmatrix} 5 & -2 & 2 \\ 1 & 0 & 1 \\ 0 & 1 & 2 \end{bmatrix} \begin{array}{l} \longrightarrow \text{row 1} \\ \longrightarrow \text{row 2} \\ \longrightarrow \text{row 3} \end{array}$$

$$\begin{array}{ccc} \uparrow & \uparrow & \uparrow \\ \text{column 1} & \text{column 2} & \text{column 3} \end{array}$$

Let's see how matrices can be used to solve a system.

EXAMPLE 1   Solve the system

$$\begin{array}{l} 5x - 4y = -1, \\ -2x + 3y = 2. \end{array}$$

To better explain each step, we list the corresponding system in the margin.

$$\begin{array}{l} 5x - 4y = -1, \\ -2x + 3y = 2 \end{array}$$

SOLUTION   We write a matrix using only coefficients and constants, listing x-coefficients in the first column and y-coefficients in the second. A dashed line separates the coefficients from the constants:

$$\begin{bmatrix} 5 & -4 & \vdots & -1 \\ -2 & 3 & \vdots & 2 \end{bmatrix}.$$

Refer to the notes in the margin for further information.

Our goal is to transform

$$\begin{bmatrix} 5 & -4 & \vdots & -1 \\ -2 & 3 & \vdots & 2 \end{bmatrix} \quad \text{into the form} \quad \begin{bmatrix} a & b & \vdots & c \\ 0 & d & \vdots & e \end{bmatrix}.$$

We can then reinsert the variables x and y, form equations, and complete the solution.

Our calculations are similar to those done if we wrote the entire equations. The first step is to multiply and/or interchange the rows so that each number in the first column below the first number is a multiple of that number. Here that means multiplying Row 2 by 5. This corresponds to multiplying both sides of the second equation by 5.

$$5x - 4y = -1,$$
$$-10x + 15y = 10$$

$$\begin{bmatrix} 5 & -4 & \vdots & -1 \\ -10 & 15 & \vdots & 10 \end{bmatrix}$$

New Row 2 = 5(Row 2 from the step above)
$$= 5(-2 \quad 3 \,\vdots\, 2) = (-10 \quad 15 \,\vdots\, 10)$$

Next, we multiply the first row by 2, add this to Row 2, and write that result as the "new" Row 2. This corresponds to multiplying the first equation by 2 and adding the result to the second equation in order to eliminate a variable. Write out these computations as necessary.

$$5x - 4y = -1,$$
$$7y = 8$$

$$\begin{bmatrix} 5 & -4 & \vdots & -1 \\ 0 & 7 & \vdots & 8 \end{bmatrix}$$

2(Row 1) $= 2(5 \quad -4 \,\vdots\, -1) = (10 \quad -8 \,\vdots\, -2)$
New Row 2 $= (10 \quad -8 \,\vdots\, -2) + (-10 \quad 15 \,\vdots\, 10)$
$= (0 \quad 7 \,\vdots\, 8)$

If we now reinsert the variables, we have

$$5x - 4y = -1, \qquad (1) \qquad \text{From Row 1}$$
$$7y = 8. \qquad (2) \qquad \text{From Row 2}$$

Solving equation (2) for y gives us

$$7y = 8 \qquad (2)$$
$$y = \tfrac{8}{7}.$$

Next, we substitute $\tfrac{8}{7}$ for y in equation (1):

$$5x - 4y = -1 \qquad (1)$$
$$5x - 4 \cdot \tfrac{8}{7} = -1 \qquad \text{Substituting } \tfrac{8}{7} \text{ for } y \text{ in equation (1)}$$
$$x = \tfrac{5}{7}. \qquad \text{Solving for } x$$

The solution is $\left(\tfrac{5}{7}, \tfrac{8}{7}\right)$. The check is left to the student.

1. Solve using matrices:

$$-x + 5y = 4,$$
$$3x - y = 6.$$

YOUR TURN

EXAMPLE 2 Solve the system

$$2x - y + 4z = -3,$$
$$x \qquad - 4z = 5,$$
$$6x - y + 2z = 10.$$

SOLUTION We first write a matrix, using only the constants. Where there are missing terms, we must write 0's:

$$2x - y + 4z = -3,$$
$$x \qquad - 4z = 5,$$
$$6x - y + 2z = 10$$

$$\begin{bmatrix} 2 & -1 & 4 & \vdots & -3 \\ 1 & 0 & -4 & \vdots & 5 \\ 6 & -1 & 2 & \vdots & 10 \end{bmatrix}.$$

Our goal is to transform the matrix to one of the form

$$ax + by + cz = d,$$
$$ey + fz = g,$$
$$hz = i$$

$$\begin{bmatrix} a & b & c & \vdots & d \\ 0 & e & f & \vdots & g \\ 0 & 0 & h & \vdots & i \end{bmatrix}.$$

This matrix is in a form called *row-echelon form.*

A matrix of this form can be rewritten as a system of equations that is equivalent to the original system, and from which a solution can be easily found.

The first step is to multiply and/or interchange the rows so that each number in the first column is a multiple of the first number in the first row. In this case, we begin by interchanging Rows 1 and 2:

$$\begin{array}{c}x \quad\;\; - 4z = 5,\\ 2x - y + 4z = -3,\\ 6x - y + 2z = 10\end{array}$$

$$\left[\begin{array}{ccc|c} 1 & 0 & -4 & 5 \\ 2 & -1 & 4 & -3 \\ 6 & -1 & 2 & 10 \end{array}\right].$$

This corresponds to interchanging the first two equations.

Next, we multiply the first row by -2, add it to the second row, and replace Row 2 with the result:

$$\begin{array}{c}x \quad\;\; - 4z = 5,\\ - y + 12z = -13,\\ 6x - y + 2z = 10\end{array}$$

$$\left[\begin{array}{ccc|c} 1 & 0 & -4 & 5 \\ 0 & -1 & 12 & -13 \\ 6 & -1 & 2 & 10 \end{array}\right].$$

$-2(1\;\;0\;\;-4\;\vdots\;5) = (-2\;\;0\;\;8\;\vdots\;-10)$ and
$(-2\;\;0\;\;8\;\vdots\;-10) + (2\;\;-1\;\;4\;\vdots\;-3) =$
$(0\;\;-1\;\;12\;\vdots\;-13)$

Now we multiply the first row by -6, add it to the third row, and replace Row 3 with the result:

$$\begin{array}{c}x \quad\;\; - 4z = 5,\\ -y + 12z = -13,\\ -y + 26z = -20\end{array}$$

$$\left[\begin{array}{ccc|c} 1 & 0 & -4 & 5 \\ 0 & -1 & 12 & -13 \\ 0 & -1 & 26 & -20 \end{array}\right].$$

$-6(1\;\;0\;\;-4\;\vdots\;5) = (-6\;\;0\;\;24\;\vdots\;-30)$ and
$(-6\;\;0\;\;24\;\vdots\;-30) + (6\;\;-1\;\;2\;\vdots\;10) =$
$(0\;\;-1\;\;26\;\vdots\;-20)$

Next, we multiply Row 2 by -1, add it to the third row, and replace Row 3 with the result:

$$\begin{array}{c}x \quad\;\; - 4z = 5,\\ -y + 12z = -13,\\ 14z = -7\end{array}$$

$$\left[\begin{array}{ccc|c} 1 & 0 & -4 & 5 \\ 0 & -1 & 12 & -13 \\ 0 & 0 & 14 & -7 \end{array}\right].$$

$-1(0\;\;-1\;\;12\;\vdots\;-13) = (0\;\;1\;\;-12\;\vdots\;13)$ and
$(0\;\;1\;\;-12\;\vdots\;13) + (0\;\;-1\;\;26\;\vdots\;-20) =$
$(0\;\;0\;\;14\;\vdots\;-7)$

Reinserting the variables gives us

$$\begin{array}{c}x \quad\;\; - 4z = \;\;\;5,\\ -y + 12z = -13,\\ 14z = \;\;-7.\end{array}$$

Solving this last equation for z, we get $z = -\frac{1}{2}$. Next, we substitute $-\frac{1}{2}$ for z in the preceding equation and solve for y: $-y + 12\left(-\frac{1}{2}\right) = -13$, so $y = 7$. Since there is no y-term in the first equation of this last system, we need only substitute $-\frac{1}{2}$ for z to solve for x: $x - 4\left(-\frac{1}{2}\right) = 5$, so $x = 3$. The solution is $\left(3, 7, -\frac{1}{2}\right)$. The check is left to the student.

2. Solve using matrices:

$$\begin{array}{c}2x + \;\; y + 3z = 1,\\ x + 2y + 4z = 6,\\ -2x \quad\quad\;\; - \;\; z = 7.\end{array}$$

YOUR TURN

The operations used in the preceding example correspond to those used to produce equivalent systems of equations, that is, systems of equations that have the same solution. We call the matrices **row-equivalent** and the operations that produce them **row-equivalent operations.**

Row-Equivalent Operations

Student Notes

Note that row-equivalent matrices are not *equal*. It is the solutions of the corresponding systems that are the same.

ROW-EQUIVALENT OPERATIONS

Each of the following row-equivalent operations produces a row-equivalent matrix:

a) Interchanging any two rows.
b) Multiplying all elements of a row by a nonzero constant.
c) Replacing a row with the sum of that row and a multiple of another row.

Computers solve systems of equations using row-equivalent matrices. Matrices are part of a branch of mathematics known as linear algebra. They are also studied in many courses in finite mathematics.

Technology Connection

Row-equivalent operations can be performed on a graphing calculator. For example, to interchange the first and second rows of the matrix, as in step (1) of Example 2 above, we enter the matrix as matrix **A** and select "rowSwap" from the MATRIX MATH menu.

To store the result of the operation as **B**, we use **STO►**, as shown in the window at right.

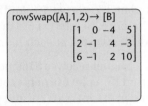
```
rowSwap([A],1,2)→ [B]
        [1   0 -4   5]
        [2  -1  4  -3]
        [6  -1  2  10]
```

1. Use a graphing calculator to proceed through all the steps in Example 2.

3.6 EXERCISE SET

FOR EXTRA HELP

MyMathLab® Math XL
PRACTICE   WATCH   READ   REVIEW

◆ Vocabulary and Reading Check

Complete each of the following statements.

1. A(n) _____ is a rectangular array of numbers.

2. The _____ of a matrix are horizontal, and the columns are _____.

3. Each number in a matrix is called a(n) _____ or element.

4. The plural of the word matrix is _____.

5. As part of solving a system using matrices, we can interchange any two _____.

6. When a matrix is in row-echelon form, the leftmost column in the matrix has zeros in all rows except the _____ one.

Matrices and Systems

Solve using matrices.

7. $x + 2y = 11,$
$3x - y = 5$

8. $x + 3y = 16,$
$6x + y = 11$

9. $3x + y = -1,$
$6x + 5y = 13$

10. $2x - y = 6,$
$8x + 2y = 0$

11. $6x - 2y = 4,$
$7x + y = 13$

12. $3x + 4y = 7,$
$-5x + 2y = 10$

13. $3x + 2y + 2z = 3,$
$x + 2y - z = 5,$
$2x - 4y + z = 0$

14. $4x - y - 3z = 19,$
$8x + y - z = 11,$
$2x + y + 2z = -7$

15. $p - 2q - 3r = 3,$
$2p - q - 2r = 4,$
$4p + 5q + 6r = 4$

16. $x + 2y - 3z = 9,$
$2x - y + 2z = -8,$
$3x - y - 4z = 3$

17. $3p + 2r = 11,$
$q - 7r = 4,$
$p - 6q = 1$

18. $4a + 9b = 8,$
$8a + 6c = -1,$
$6b + 6c = -1$

19. $2x + 2y - 2z - 2w = -10,$
$w + y + z + x = -5,$
$x - y + 4z + 3w = -2,$
$w - 2y + 2z + 3x = -6$

20. $-w - 3y + z + 2x = -8,$
$x + y - z - w = -4,$
$w + y + z + x = 22,$
$x - y - z - w = -14$

Solve using matrices.

21. *Coin Value.* A collection of 42 coins consists of dimes and nickels. The total value is \$3.00. How many dimes and how many nickels are there?

22. *Coin Value.* A collection of 43 coins consists of dimes and quarters. The total value is \$7.60. How many dimes and how many quarters are there?

23. *Snack Mix.* Bree sells a dried-fruit mixture for \$5.80 per pound and Hawaiian macadamia nuts for \$14.75 per pound. She wants to blend the two to get a 15-lb mixture that she will sell for \$9.38 per pound. How much of each should she use?

24. *Mixing Paint.* Higher quality paint typically contains more solids. Alex has available paint that contains 45% solids and paint that contains 25% solids. How much of each should he use to create 20 gal of paint that contains 39% solids?

25. *Investments.* Elena receives $112 per year in simple interest from three investments totaling $2500. Part is invested at 3%, part at 4%, and part at 5%. There is $1100 more invested at 5% than at 4%. Find the amount invested at each rate.

26. *Investments.* Miguel receives $160 per year in simple interest from three investments totaling $3200. Part is invested at 2%, part at 3%, and part at 6%. There is $1900 more invested at 6% than at 3%. Find the amount invested at each rate.

27. Explain how you can recognize dependent equations when solving with matrices.

28. Explain how you can recognize an inconsistent system when solving with matrices.

Skill Review

Simplify. [1.2]

29. $(-7)^2$

30. $-(-7)^2$

31. -7^2

32. $|-7^2|$

Synthesis

33. If the matrices

$$\begin{bmatrix} a_1 & b_1 & \vdots & c_1 \\ d_1 & e_1 & \vdots & f_1 \end{bmatrix} \text{ and } \begin{bmatrix} a_2 & b_2 & \vdots & c_2 \\ d_2 & e_2 & \vdots & f_2 \end{bmatrix}$$

share the same solution, does it follow that the corresponding entries are all equal to each other ($a_1 = a_2, b_1 = b_2$, etc.)? Why or why not?

34. Explain how the row-equivalent operations make use of the addition, multiplication, and distributive properties.

35. The sum of the digits in a four-digit number is 10. Twice the sum of the thousands digit and the tens digit is 1 less than the sum of the other two digits. The tens digit is twice the thousands digit. The ones digit equals the sum of the thousands digit and the hundreds digit. Find the four-digit number.

36. Solve for x and y:

$$ax + by = c,$$
$$dx + ey = f.$$

⬇ YOUR TURN ANSWERS: SECTION 3.6

1. $\left(\frac{17}{7}, \frac{9}{7}\right)$ **2.** $(-10, -18, 13)$

QUICK QUIZ: SECTIONS 3.1–3.6

Solve. If a system has an infinite number of solutions, use set-builder notation to write the solution set. If a system has no solution, state this.

1. $2x + y = 3,$
    $6x + 2y = 4$ [3.2]

2. $y = \dfrac{5}{3}x + 7,$

    $y = \dfrac{5}{3}x - 8$ [3.1], [3.2]

3. Solve: $x + 5y = 6,$
    $x + 2z = 3,$
    $5y + 2z = 8.$ [3.4]

4. Network Community College bought 42 packages of dry-erase markers. Some packages contained 4 markers and some contained 6 markers. If they purchased a total of 200 markers, how many of each size package did they buy?   [3.3]

5. Drink Fresh contains 30% juice, and Summer Light contains 5% juice. How much of each should be mixed to obtain 6 L of a beverage that contains 10% juice?   [3.3]

PREPARE TO MOVE ON

Simplify. [1.2]

1. $3(-1) - (-4)(5)$

2. $7(-5) - 2(-8)$

3. $-2(5 \cdot 3 - 4 \cdot 6) - 3(2 \cdot 7 - 15) + 4(3 \cdot 8 - 5 \cdot 4)$

4. $6(2 \cdot 7 - 3(-4)) - 4(3(-8) - 10) + 5(4 \cdot 3 - (-2)7)$

3.7   Determinants and Cramer's Rule

Determinants of 2 × 2 Matrices   ▪   Cramer's Rule: 2 × 2 Systems   ▪   Cramer's Rule: 3 × 3 Systems

Study Skills

Find the Highlights

If you do not already own one, consider purchasing a highlighter to use as you read this text and work on the exercises. Often the best time to highlight an important sentence or step in an example is after you have read through the section the first time.

Determinants of 2 × 2 Matrices

When a matrix has m rows and n columns, it is called an "m by n" matrix, and its *dimensions* are $m \times n$. If a matrix has the same number of rows and columns, it is called a **square matrix**. Associated with every square matrix is a number called its **determinant**, defined as follows for 2×2 matrices.

---

2 × 2 DETERMINANTS

The determinant of a two-by-two matrix $\begin{bmatrix} a & c \\ b & d \end{bmatrix}$ is denoted $\begin{vmatrix} a & c \\ b & d \end{vmatrix}$

and is defined as follows:

$$\begin{vmatrix} a & c \\ b & d \end{vmatrix} = ad - bc.$$

EXAMPLE 1   Evaluate: $\begin{vmatrix} 2 & -5 \\ 6 & 7 \end{vmatrix}$.

SOLUTION   We multiply and subtract as follows:

$$\begin{vmatrix} 2 & -5 \\ 6 & 7 \end{vmatrix} = 2 \cdot 7 - 6 \cdot (-5) = 14 + 30 = 44.$$

1. Evaluate: $\begin{vmatrix} -3 & -1 \\ 5 & 7 \end{vmatrix}$.

↩ YOUR TURN

Cramer's Rule: 2 × 2 Systems

One of the many uses for determinants is in solving systems of linear equations in which the number of variables is the same as the number of equations and the constants are not all 0. Let's consider a system of two equations:

$$a_1 x + b_1 y = c_1,$$
$$a_2 x + b_2 y = c_2.$$

If we use the elimination method, a series of steps can show that

$$x = \frac{c_1 b_2 - c_2 b_1}{a_1 b_2 - a_2 b_1} \quad \text{and} \quad y = \frac{a_1 c_2 - a_2 c_1}{a_1 b_2 - a_2 b_1}.$$

These fractions can be rewritten using determinants.

CRAMER'S RULE: 2 × 2 SYSTEMS

The solution of the system

$$a_1 x + b_1 y = c_1,$$
$$a_2 x + b_2 y = c_2,$$

if it is unique, is given by

$$x = \dfrac{\begin{vmatrix} c_1 & b_1 \\ c_2 & b_2 \end{vmatrix}}{\begin{vmatrix} a_1 & b_1 \\ a_2 & b_2 \end{vmatrix}}, \qquad y = \dfrac{\begin{vmatrix} a_1 & c_1 \\ a_2 & c_2 \end{vmatrix}}{\begin{vmatrix} a_1 & b_1 \\ a_2 & b_2 \end{vmatrix}}.$$

These formulas apply only if the denominator is not 0.

To use Cramer's rule, we find the determinants and compute x and y as shown above. Note that in the denominators, which are identical, the coefficients of x and y appear in the same position as in the original equations. In the numerator of x, the constants c_1 and c_2 replace a_1 and a_2. In the numerator of y, the constants c_1 and c_2 replace b_1 and b_2.

EXAMPLE 2   Solve using Cramer's rule:

$$2x + 5y = 7,$$
$$5x - 2y = -3.$$

SOLUTION   We have

$$x = \dfrac{\begin{vmatrix} 7 & 5 \\ -3 & -2 \end{vmatrix}}{\begin{vmatrix} 2 & 5 \\ 5 & -2 \end{vmatrix}} \quad \begin{matrix} \leftarrow \text{The constants } \begin{smallmatrix} 7 \\ -3 \end{smallmatrix} \text{ replace } \begin{smallmatrix} a_1 \\ a_2 \end{smallmatrix} \text{ in the first column.} \\[6pt] \leftarrow \text{The columns are the coefficients, } \begin{smallmatrix} a_1 & b_1 \\ a_2 & b_2 \end{smallmatrix}. \end{matrix}$$

$$= \frac{7(-2) - (-3)5}{2(-2) - 5 \cdot 5} = \frac{1}{-29} = -\frac{1}{29}$$

and

$$y = \dfrac{\begin{vmatrix} 2 & 7 \\ 5 & -3 \end{vmatrix}}{\begin{vmatrix} 2 & 5 \\ 5 & -2 \end{vmatrix}} \quad \begin{matrix} \leftarrow \text{The constants } \begin{smallmatrix} 7 \\ -3 \end{smallmatrix} \text{ replace } \begin{smallmatrix} b_1 \\ b_2 \end{smallmatrix} \text{ in the second column.} \\[6pt] \leftarrow \text{The denominator is the same as in the expression for } x. \end{matrix}$$

$$= \frac{2(-3) - 5 \cdot 7}{-29} = \frac{-41}{-29} = \frac{41}{29}.$$

The solution is $\left(-\frac{1}{29}, \frac{41}{29}\right)$. The check is left to the student.

↩ YOUR TURN

2. Solve using Cramer's rule:

$$-2x - y = 7,$$
$$3x + 4y = 1.$$

Cramer's Rule: 3 × 3 Systems

Cramer's rule can be extended for systems of three linear equations. However, before doing so, we must define what a 3 × 3 determinant is.

3 × 3 DETERMINANTS

The determinant of a three-by-three matrix can be defined as follows:

$$
\begin{vmatrix} a_1 & b_1 & c_1 \\ a_2 & b_2 & c_2 \\ a_3 & b_3 & c_3 \end{vmatrix} = a_1 \begin{vmatrix} b_2 & c_2 \\ b_3 & c_3 \end{vmatrix} \overset{\text{Subtract.}}{-} a_2 \begin{vmatrix} b_1 & c_1 \\ b_3 & c_3 \end{vmatrix} \overset{\text{Add.}}{+} a_3 \begin{vmatrix} b_1 & c_1 \\ b_2 & c_2 \end{vmatrix}.
$$

Student Notes

Cramer's rule and the evaluation of determinants rely on patterns. Recognizing and remembering the patterns will help you understand and use the definitions.

Note that the a's come from the first column. Note too that the 2×2 determinants above can be obtained by crossing out the row and the column in which the a occurs.

For a_1:

$$\begin{vmatrix} \cancel{a_1} & \cancel{b_1} & \cancel{c_1} \\ a_2 & b_2 & c_2 \\ a_3 & b_3 & c_3 \end{vmatrix}$$

For a_2:

$$\begin{vmatrix} a_1 & b_1 & c_1 \\ \cancel{a_2} & \cancel{b_2} & \cancel{c_2} \\ a_3 & b_3 & c_3 \end{vmatrix}$$

For a_3:

$$\begin{vmatrix} a_1 & b_1 & c_1 \\ a_2 & b_2 & c_2 \\ \cancel{a_3} & \cancel{b_3} & \cancel{c_3} \end{vmatrix}$$

EXAMPLE 3   Evaluate:

$$\begin{vmatrix} -1 & 0 & 1 \\ -5 & 1 & -1 \\ 4 & 8 & 1 \end{vmatrix}.$$

SOLUTION   We have

$$
\begin{vmatrix} -1 & 0 & 1 \\ -5 & 1 & -1 \\ 4 & 8 & 1 \end{vmatrix} = -1 \begin{vmatrix} 1 & -1 \\ 8 & 1 \end{vmatrix} \overset{\text{Subtract.}}{-} (-5) \begin{vmatrix} 0 & 1 \\ 8 & 1 \end{vmatrix} \overset{\text{Add.}}{+} 4 \begin{vmatrix} 0 & 1 \\ 1 & -1 \end{vmatrix}
$$

$$= -1(1 + 8) + 5(0 - 8) + 4(0 - 1) \qquad \text{Evaluating the three determinants}$$

$$= -9 - 40 - 4 = -53.$$

3. Evaluate: $\begin{vmatrix} 2 & -1 & 0 \\ 1 & 0 & -3 \\ 6 & 2 & 4 \end{vmatrix}.$

 YOUR TURN

Technology Connection

Determinants can be evaluated on most graphing calculators using **2ND** (MATRIX). After entering a matrix, we select the determinant operation from the MATRIX MATH menu and enter the name of the matrix. The graphing calculator will return the value of the determinant of the matrix. For example, if

$$\mathbf{A} = \begin{bmatrix} 1 & 6 & -1 \\ -3 & -5 & 3 \\ 0 & 4 & 2 \end{bmatrix},$$

we have

det([A])

                26

1. Confirm the calculations in Example 4.

CRAMER'S RULE: 3 × 3 SYSTEMS

The solution of the system

$$a_1x + b_1y + c_1z = d_1,$$
$$a_2x + b_2y + c_2z = d_2,$$
$$a_3x + b_3y + c_3z = d_3$$

can be found using the following determinants:

$$D = \begin{vmatrix} a_1 & b_1 & c_1 \\ a_2 & b_2 & c_2 \\ a_3 & b_3 & c_3 \end{vmatrix}, \qquad D_x = \begin{vmatrix} d_1 & b_1 & c_1 \\ d_2 & b_2 & c_2 \\ d_3 & b_3 & c_3 \end{vmatrix},$$

$$D_y = \begin{vmatrix} a_1 & d_1 & c_1 \\ a_2 & d_2 & c_2 \\ a_3 & d_3 & c_3 \end{vmatrix}, \qquad D_z = \begin{vmatrix} a_1 & b_1 & d_1 \\ a_2 & b_2 & d_2 \\ a_3 & b_3 & d_3 \end{vmatrix}.$$

D contains only coefficients. In D_x the d's replace the a's. In D_y, the d's replace the b's. In D_z, the d's replace the c's.

If a unique solution exists, it is given by

$$x = \frac{D_x}{D}, \qquad y = \frac{D_y}{D}, \qquad z = \frac{D_z}{D}.$$

These formulas apply only if $D \neq 0$.

EXAMPLE 4   Solve using Cramer's rule:

$$x - 3y + 7z = 13,$$
$$x + y + z = 1,$$
$$x - 2y + 3z = 4.$$

SOLUTION   We compute D, D_x, D_y, and D_z:

$$D = \begin{vmatrix} 1 & -3 & 7 \\ 1 & 1 & 1 \\ 1 & -2 & 3 \end{vmatrix} = -10; \qquad D_x = \begin{vmatrix} 13 & -3 & 7 \\ 1 & 1 & 1 \\ 4 & -2 & 3 \end{vmatrix} = 20;$$

$$D_y = \begin{vmatrix} 1 & 13 & 7 \\ 1 & 1 & 1 \\ 1 & 4 & 3 \end{vmatrix} = -6; \qquad D_z = \begin{vmatrix} 1 & -3 & 13 \\ 1 & 1 & 1 \\ 1 & -2 & 4 \end{vmatrix} = -24.$$

4. Solve using Cramer's rule:

$$x - y + 2z = 5,$$
$$2x + y - z = 6,$$
$$-x + 2y - 2z = 3.$$

Then

$$x = \frac{D_x}{D} = \frac{20}{-10} = -2; \qquad y = \frac{D_y}{D} = \frac{-6}{-10} = \frac{3}{5}; \qquad z = \frac{D_z}{D} = \frac{-24}{-10} = \frac{12}{5}.$$

The solution is $\left(-2, \frac{3}{5}, \frac{12}{5}\right)$. The check is left to the student.

YOUR TURN

In Example 4, we need not have evaluated D_z. Once x and y were found, we could have substituted them into one of the equations to find z.

When we are using Cramer's rule, if the denominator is 0 and at least one of the other determinants is not 0, the system is inconsistent. If *all* the determinants are 0, then the equations in the system are dependent.

3.7   EXERCISE SET

FOR EXTRA HELP   MyMathLab® Math XL

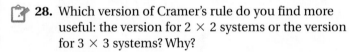
PRACTICE   WATCH   READ   REVIEW

Vocabulary and Reading Check

Classify each of the following statements as either true or false.

1. A square matrix has the same number of rows and columns.

2. A 3×4 matrix has 3 rows and 4 columns.

3. A determinant is a number.

4. Cramer's rule exists only for 2×2 systems.

5. Whenever Cramer's rule yields a denominator that is 0, the system has no solution.

6. Whenever Cramer's rule yields a numerator that is 0, the equations are dependent.

Determinants

Evaluate.

7. $\begin{vmatrix} 3 & 5 \\ 4 & 8 \end{vmatrix}$

8. $\begin{vmatrix} 3 & 2 \\ 2 & -3 \end{vmatrix}$

9. $\begin{vmatrix} 10 & 8 \\ -5 & -9 \end{vmatrix}$

10. $\begin{vmatrix} 3 & 2 \\ -7 & 11 \end{vmatrix}$

11. $\begin{vmatrix} 1 & 4 & 0 \\ 0 & -1 & 2 \\ 3 & -2 & 1 \end{vmatrix}$

12. $\begin{vmatrix} 2 & 4 & -2 \\ 1 & 0 & 2 \\ 0 & 1 & 3 \end{vmatrix}$

13. $\begin{vmatrix} -1 & -2 & -3 \\ 3 & 4 & 2 \\ 0 & 1 & 2 \end{vmatrix}$

14. $\begin{vmatrix} 5 & 2 & 2 \\ 0 & 1 & -1 \\ 3 & 3 & 1 \end{vmatrix}$

15. $\begin{vmatrix} -4 & -2 & 3 \\ -3 & 1 & 2 \\ 3 & 4 & -2 \end{vmatrix}$

16. $\begin{vmatrix} 2 & -1 & 1 \\ 1 & 2 & -1 \\ 3 & 4 & -3 \end{vmatrix}$

Cramer's Rule

Solve using Cramer's rule.

17. $5x + 8y = 1,$
    $3x + 7y = 5$

18. $3x - 4y = 6,$
    $5x + 9y = 10$

19. $5x - 4y = -3,$
    $7x + 2y = 6$

20. $-2x + 4y = 3,$
    $3x - 7y = 1$

21. $3x - y + 2z = 1,$
    $x - y + 2z = 3,$
    $-2x + 3y + z = 1$

22. $3x + 2y - z = 4,$
    $3x - 2y + z = 5,$
    $4x - 5y - z = -1$

23. $2x - 3y + 5z = 27,$
    $x + 2y - z = -4,$
    $5x - y + 4z = 27$

24. $x - y + 2z = -3,$
    $x + 2y + 3z = 4,$
    $2x + y + z = -3$

25. $r - 2s + 3t = 6,$
    $2r - s - t = -3,$
    $r + s + t = 6$

26. $a - 3c = 6,$
    $b + 2c = 2,$
    $7a - 3b - 5c = 14$

27. Describe at least one of the patterns that you see in Cramer's rule.

28. Which version of Cramer's rule do you find more useful: the version for 2×2 systems or the version for 3×3 systems? Why?

Skill Review

For each of Exercises 29–32, find a linear function whose graph has the given characteristics. [2.5]

29. Slope: $\frac{1}{2}$; y-intercept: $(0, -10)$

30. Slope: 3; passes through $(1, -6)$

31. Passes through $(-2, 8)$ and $(3, 0)$

32. Parallel to $y = -x + 5$; y-intercept: $(0, 4)$

Synthesis

33. Cramer's rule states that if $a_1x + b_1y = c_1$ and $a_2x + b_2y = c_2$ are dependent, then

$$\begin{vmatrix} a_1 & b_1 \\ a_2 & b_2 \end{vmatrix} = 0.$$

Explain why this will always happen.

34. Under what conditions can a 3×3 system of linear equations be consistent but unable to be solved using Cramer's rule?

Solve.

35. $\begin{vmatrix} y & -2 \\ 4 & 3 \end{vmatrix} = 44$

36. $\begin{vmatrix} 2 & x & -1 \\ -1 & 3 & 2 \\ -2 & 1 & 1 \end{vmatrix} = -12$

37. $\begin{vmatrix} m+1 & -2 \\ m-2 & 1 \end{vmatrix} = 27$

38. Show that an equation of the line through (x_1, y_1) and (x_2, y_2) can be written

$$\begin{vmatrix} x & y & 1 \\ x_1 & y_1 & 1 \\ x_2 & y_2 & 1 \end{vmatrix} = 0.$$

YOUR TURN ANSWERS: SECTION 3.7

1. -16 **2.** $\left(-\frac{29}{5}, \frac{23}{5}\right)$ **3.** 34 **4.** $\left(\frac{9}{5}, 8, \frac{28}{5}\right)$

QUICK QUIZ: SECTIONS 3.1–3.7

1. Solve graphically:

$$y = 2x - 1,$$
$$y = \tfrac{1}{3}x + 4. \quad [3.1]$$

2. Solve using the substitution method:

$$2x - y = 7,$$
$$y = 3x + 1. \quad [3.2]$$

3. Solve using the elimination method:

$$4x + 3y = 1,$$
$$2x + 3y = 5. \quad [3.2]$$

4. Solve using matrices:

$$5x + 3y = 5,$$
$$x + 2y = 1. \quad [3.6]$$

5. Solve using Cramer's rule:

$$8x + 5y = 4,$$
$$9x - 6y = 1. \quad [3.7]$$

PREPARE TO MOVE ON

For $f(x) = 80x + 2500$ and $g(x) = 150x$, find the following.

1. $(g - f)(x)$ [2.6]

2. $(g - f)(100)$ [2.6]

3. All values of x for which $f(x) = g(x)$ [1.3], [2.2]

4. All values of x for which $(g - f)(x) = 0$ [1.3], [2.6]

3.8 Business and Economics Applications

Break-Even Analysis ▪ Supply and Demand

Study Skills

Try to Look Ahead

If you are able to at least skim through an upcoming section before your instructor covers that lesson, you will be better able to focus on what is being emphasized in class. Similarly, if you can begin studying for a quiz or test a day or two before you really must, you will reap great rewards for doing so.

Break-Even Analysis

The money that a business spends to manufacture a product is its *cost*. The **total cost** of production can be thought of as a function C, where $C(x)$ is the cost of producing x units. When a company sells its product, it takes in money. This is *revenue* and can be thought of as a function R, where $R(x)$ is the **total revenue** from the sale of x units. **Total profit** is the money taken in less the money spent, or total revenue minus total cost. Total profit from the production and sale of x units is a function P given by

$$\textbf{Profit} = \textbf{Revenue} - \textbf{Cost,} \quad \text{or} \quad P(x) = R(x) - C(x).$$

If $R(x)$ is greater than $C(x)$, there is a gain and $P(x)$ is positive. If $C(x)$ is greater than $R(x)$, there is a loss and $P(x)$ is negative. When $R(x) = C(x)$, the company breaks even.

There are two kinds of costs. First, there are costs like rent, insurance, machinery, and so on. These costs, which must be paid regardless of how many items are produced, are called *fixed costs*. Second, costs for labor, materials, marketing, and so on are called *variable costs*, because they vary according to the amount being produced. The sum of the fixed cost and the variable cost gives the **total cost**.

CAUTION! Do not confuse "cost" with "price." When we discuss the *cost* of an item, we are referring to what it costs to produce the item. The *price* of an item is what a consumer pays to purchase the item and is used when calculating revenue.

EXAMPLE 1 *Manufacturing Chairs.* Renewable Designs is planning to make a new chair. Fixed costs will be $90,000, and it will cost $150 to produce each chair. Each chair sells for $400.

a) Find the total cost $C(x)$ of producing x chairs.

b) Find the total revenue $R(x)$ from the sale of x chairs.

c) Find the total profit $P(x)$ from the production and sale of x chairs.

d) What profit will the company realize from the production and sale of 300 chairs? of 800 chairs?

e) Graph the total-cost, total-revenue, and total-profit functions using the same set of axes. Determine the break-even point.

SOLUTION

a) Total cost, in dollars, is given by

$$C(x) = \text{(Fixed costs) plus (Variable costs),}$$

or $\quad C(x) = \quad 90{,}000 \quad + \quad 150x,$

where x is the number of chairs produced.

b) Total revenue, in dollars, is given by

$$R(x) = 400x. \qquad \text{\$400 times the number of chairs sold. We assume}$$
$$\text{that every chair produced is sold.}$$

c) Total profit, in dollars, is given by

$$P(x) = R(x) - C(x) \qquad \text{Profit is revenue minus cost.}$$
$$= 400x - (90{,}000 + 150x) \qquad \text{Parentheses are important.}$$
$$= 250x - 90{,}000.$$

d) Profits are

$$P(300) = 250 \cdot 300 - 90{,}000 = -\$15{,}000$$

when 300 chairs are produced and sold, and

$$P(800) = 250 \cdot 800 - 90{,}000 = \$110{,}000$$

when 800 chairs are produced and sold. Thus the company loses money if only 300 chairs are sold, but makes money if 800 are sold.

e) The graphs of each of the three functions are shown below:

$C(x) = 90{,}000 + 150x,$ This represents the cost function.

$R(x) = 400x,$ This represents the revenue function.

$P(x) = 250x - 90{,}000.$ This represents the profit function.

$C(x)$, $R(x)$, and $P(x)$ are all in dollars.

The revenue function has a graph that goes through the origin and has a slope of 400. The cost function has an intercept on the $-axis of 90,000 and has a slope of 150. The profit function has an intercept on the $-axis of $-90{,}000$ and has a slope of 250. It is shown by the red and black dashed line. The red portion of the dashed line shows a "negative" profit, which is a loss. (That is what is known as "being in the red.") The black portion of the dashed line shows a "positive" profit, or gain. (That is what is known as "being in the black.")

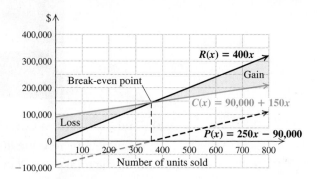

Gains occur when revenue exceeds cost. Losses occur when revenue is less than cost. The **break-even point** occurs where the graphs of R and C cross. Thus to find the break-even point, we solve a system:

$R(x) = 400x,$

$C(x) = 90{,}000 + 150x.$

Since revenue and cost are equal at the break-even point, the system can be rewritten as

$d = 400x,$ (1)

$d = 90{,}000 + 150x,$ (2)

where d is the dollar figure at the break-even point. We solve using substitution:

$400x = 90{,}000 + 150x$ Substituting $400x$ for d in equation (2)

$250x = 90{,}000$

$x = 360.$

Renewable Designs breaks even if it produces and sells 360 chairs and takes in a total of $R(360) = 400(360) = \$144{,}000$ in revenue. Note that the x-coordinate of the break-even point can also be found by solving $P(x) = 0$. The break-even point is (360 chairs, \$144,000).

Student Notes

If you plan to study business or economics, you may want to consult the material in this section when these topics arise in your other courses.

---

1. Refer to Example 1. Renewable Designs is also planning to make a new table. Fixed costs will be $70,000, and it will cost $250 to produce each table. Each table sells for $600.

a) Find the total cost $C(x)$ of producing x tables.

b) Find the total revenue $R(x)$ from the sale of x tables.

c) Find the total profit $P(x)$ from the production and sale of x tables.

d) What profit will the company realize from the production and sale of 500 tables?

e) Determine the break-even point.

 YOUR TURN

Supply and Demand

As the price of coffee varies, so too does the amount sold. The table and the graph below show that *consumers will demand less as the price goes up.*

Demand Function, D

Price, p, per Kilogram	Quantity, $D(p)$ (in millions of kilograms)
$16.00	25
18.00	20
20.00	15
22.00	10
24.00	5

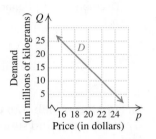

As the price of coffee varies, the amount made available varies as well. The table and the graph below show that *sellers will supply more as the price goes up.*

Supply Function, S

Price, p, per Kilogram	Quantity, $S(p)$ (in millions of kilograms)
$18.00	5
19.00	10
20.00	15
21.00	20
22.00	25

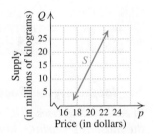

Considering demand and supply together, we see that as price increases, demand decreases. As price increases, supply increases. The point of intersection is called the **equilibrium point**. At that price, the amount that the seller will supply is the same amount that the consumer will buy. The situation is similar to a buyer and a seller negotiating the price of an item. The equilibrium point is the price and quantity that they finally agree on.

Any ordered pair of coordinates from the graph is (price, quantity), because the horizontal axis is the price axis and the vertical axis is the quantity axis. If D is a demand function and S is a supply function, then the equilibrium point is where demand equals supply:

$$D(p) = S(p).$$

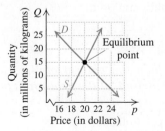

EXAMPLE 2   Find the equilibrium point for the demand and supply functions given:

$$D(p) = 1000 - 60p, \quad (1)$$
$$S(p) = \ 200 + \ 4p. \quad (2)$$

SOLUTION   Since both demand and supply are *quantities* and they are equal at the equilibrium point, we rewrite the system as

$$q = 1000 - 60p, \quad (1)$$
$$q = 200 + 4p. \quad (2)$$

Chapter Resource:
Decision Making: Connection,
p. 214

We substitute $200 + 4p$ for q in equation (1) and solve:

$$200 + 4p = 1000 - 60p \qquad \text{Substituting } 200 + 4p \text{ for } q \text{ in equation (1)}$$
$$200 + 64p = 1000 \qquad \text{Adding } 60p \text{ to both sides}$$
$$64p = 800 \qquad \text{Adding } -200 \text{ to both sides}$$
$$p = \tfrac{800}{64} = 12.5.$$

2. Find the equilibrium point for the demand and supply functions given:

$$D(p) = 850 - 30p,$$
$$S(p) = 550 + 10p.$$

Thus the equilibrium price is $12.50 per unit.
To find the equilibrium quantity, we substitute $12.50 into either $D(p)$ or $S(p)$. We use $S(p)$:

$$S(12.5) = 200 + 4(12.5) = 200 + 50 = 250.$$

The equilibrium quantity is 250 units, and the equilibrium point is ($12.50, 250$).

YOUR TURN

3.8   **EXERCISE SET**   FOR EXTRA HELP   MyMathLab® *Math* XL
PRACTICE   WATCH   READ   REVIEW

Vocabulary and Reading Check

In each of Exercises 1–8, match the word or phrase with the most appropriate choice from the list below.

1. _____   Total cost

2. _____   Fixed costs

3. _____   Variable costs

4. _____   Total revenue

5. _____   Total profit

6. _____   Price

7. _____   Break-even point

8. _____   Equilibrium point

a)   The amount of money that a company takes in
b)   The sum of fixed costs and variable costs
c)   The point at which total revenue equals total cost
d)   What consumers pay per item
e)   The difference between total revenue and total cost
f)   What companies spend whether or not a product is produced
g)   The point at which supply equals demand
h)   The costs that vary according to the number of items produced

Break-Even Analysis

For each of the following pairs of total-cost and total-revenue functions, find **(a)** *the total-profit function and* **(b)** *the break-even point.*

9. $C(x) = 35x + 200,000,$
$R(x) = 55x$

10. $C(x) = 20x + 500,000,$
$R(x) = 70x$

11. $C(x) = 15x + 3100,$
$R(x) = 40x$

12. $C(x) = 30x + 49,500,$
$R(x) = 85x$

13. $C(x) = 40x + 22,500,$
$R(x) = 85x$

14. $C(x) = 20x + 10,000,$
$R(x) = 100x$

15. $C(x) = 24x + 50,000,$
$R(x) = 40x$

16. $C(x) = 40x + 8010,$
$R(x) = 58x$

Aha! **17.** $C(x) = 75x + 100,000,$
$R(x) = 125x$

18. $C(x) = 20x + 120,000,$
$R(x) = 50x$

Supply and Demand

Find the equilibrium point for each of the following pairs of demand and supply functions.

19. $D(p) = 2000 - 15p,$
$S(p) = 740 + 6p$

20. $D(p) = 1000 - 8p,$
$S(p) = 350 + 5p$

21. $D(p) = 760 - 13p,$
$S(p) = 430 + 2p$

22. $D(p) = 800 - 43p,$
$S(p) = 210 + 16p$

23. $D(p) = 7500 - 25p,$
$S(p) = 6000 + 5p$

24. $D(p) = 8800 - 30p,$
$S(p) = 7000 + 15p$

25. $D(p) = 1600 - 53p,$
$S(p) = 320 + 75p$

26. $D(p) = 5500 - 40p,$
$S(p) = 1000 + 85p$

Solve.

27. *Manufacturing.* SoundGen, Inc., plans to manufacture a new type of cell phone. The fixed costs are $45,000, and the variable costs are estimated to be $40 per unit. The revenue from each cell phone is to be $130. Find the following.

   a) The total cost $C(x)$ of producing x cell phones
   b) The total revenue $R(x)$ from the sale of x cell phones
   c) The total profit $P(x)$ from the production and sale of x cell phones
   d) The profit or loss from the production and sale of 3000 cell phones; of 400 cell phones
   e) The break-even point

28. *Computer Manufacturing.* Current Electronics plans to introduce a new laptop computer. The fixed costs are $125,300, and the variable costs are $450 per unit. The revenue from each computer is $800. Find the following.

   a) The total cost $C(x)$ of producing x computers
   b) The total revenue $R(x)$ from the sale of x computers
   c) The total profit $P(x)$ from the production and sale of x computers
   d) The profit or loss from the production and sale of 100 computers; of 400 computers
   e) The break-even point

29. *Pet Safety.* Christine designed and is now producing a pet car seat. The fixed costs for setting up production are $10,000, and the variable costs are $30 per unit. The revenue from each seat is to be $80. Find the following.

   a) The total cost $C(x)$ of producing x seats
   b) The total revenue $R(x)$ from the sale of x seats
   c) The total profit $P(x)$ from the production and sale of x seats
   d) The profit or loss from the production and sale of 2000 seats; of 50 seats
   e) The break-even point

30. *Manufacturing Caps.* Martina's Custom Printing is adding painter's caps to its product line. For the first year, the fixed costs for setting up production are $16,404. The variable costs for producing a dozen caps are $6.00. The revenue on each dozen caps will be $18.00. Find the following.

   a) The total cost $C(x)$ of producing x dozen caps
   b) The total revenue $R(x)$ from the sale of x dozen caps
   c) The total profit $P(x)$ from the production and sale of x dozen caps
   d) The profit or loss from the production and sale of 3000 dozen caps; of 1000 dozen caps
   e) The break-even point

31. In Example 1, the slope of the line representing Revenue is the sum of the slopes of the other two lines. This is not a coincidence. Explain why.

32. Variable costs and fixed costs are often compared to the slope and the y-intercept, respectively, of an equation for a line. Explain why you feel this analogy is or is not valid.

Skill Review

Simplify. [1.7]

33. $(1.25 \times 10^{-15})(8 \times 10^4)$ **34.** $\dfrac{1.2 \times 10^3}{2.4 \times 10^{-17}}$

35. Solve $C = \frac{2}{3}(x - y)$ for y.   [1.5]

36. Determine whether $(-5, -9)$ is a solution of $n - 2m = 1$.   [2.1]

Synthesis

37. Bernadette claims that since her fixed costs are $3000, she need sell only 10 custom birdbaths at $300 each in order to break even. Is her reasoning valid? Why or why not?

38. In this section, we examined supply and demand functions for coffee. Does it seem realistic to you for the graph of D to have a constant slope? Why or why not?

39. *Yo-yo Production.*   Bing Boing Hobbies is willing to produce 100 yo-yo's at $2.00 each and 500 yo-yo's at $8.00 each. Research indicates that the public will buy 500 yo-yo's at $1.00 each and 100 yo-yo's at $9.00 each. Find the equilibrium point.

40. *Loudspeaker Production.*   Sonority Speakers, Inc., has fixed costs of $15,400 and variable costs of $100 for each pair of speakers produced. If the speakers sell for $250 per pair, how many pairs of speakers must be produced (and sold) in order to have enough profit to cover the fixed costs of two additional facilities? Assume that all fixed costs are identical.

Use a graphing calculator to solve.

41. *Dog Food Production.*   Puppy Love, Inc., is producing a new line of puppy food. The marketing department predicts that the demand function will be $D(p) = -14.97p + 987.35$ and the supply function will be $S(p) = 98.55p - 5.13$.

  a) To the nearest cent, what price per unit should be charged in order to have equilibrium between supply and demand?

  b) The production of the puppy food involves $87,985 in fixed costs and $5.15 per unit in variable costs. If the price per unit is the value you found in part (a), how many units must be sold in order to break even?

42. *Computer Production.*   Brushstroke Computers, Inc., is planning a new line of computers, each of which will sell for $970. The fixed costs in setting up production are $1,235,580, and the variable costs for each computer are $697.

  a) What is the break-even point?

  b) The marketing department at Brushstroke is not sure that $970 is the best price. Their demand function for the new computers is given by $D(p) = -304.5p + 374,580$ and their supply function is given by $S(p) = 788.7p - 576,504$. To the nearest dollar, what price p would result in equilibrium between supply and demand?

  c) If the computers are sold for the equilibrium price found in part (b), what is the break-even point?

YOUR TURN ANSWERS: SECTION 3.8

1. (a) $C(x) = 70,000 + 250x$; **(b)** $R(x) = 600x$;
(c) $P(x) = 350x - 70,000$; **(d)** $105,000;
(e) (200 units, $120,000)   **2.** ($7.50, 625)

QUICK QUIZ: SECTIONS 3.1–3.8

1. The perimeter of a rectangular classroom is 140 ft. The width is 10 ft shorter than the length. Find the dimensions.   [3.3]

2. Joanna has in her refrigerator low-fat milk, containing 1% fat, and whole milk, containing 3.5% fat. How much of each should she mix to obtain 16 oz of milk containing 2% fat?   [3.3]

3. The measure of the largest angle in a triangle is equal to the sum of the measures of the other two angles. The smallest angle is one-third the size of the middle angle. Find the measures of the angles.   [3.5]

Evaluate.   [3.7]

4. $\begin{vmatrix} -5 & -2 \\ 3 & -4 \end{vmatrix}$ **5.** $\begin{vmatrix} 2 & 1 & 0 \\ 3 & -1 & 5 \\ 0 & 2 & 1 \end{vmatrix}$

PREPARE TO MOVE ON

Solve.   [1.3]

1. $4x - 3 = 21$ **2.** $5 - x = 7$

3. $x - 4 = 9x - 10$ **4.** $3 - (x + 2) = 7$

5. $1 - 3(2x + 1) = 3 - 5x$

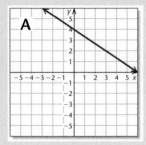

A

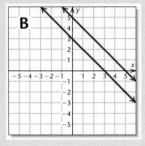

B

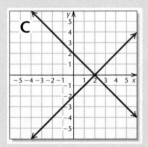

C

D

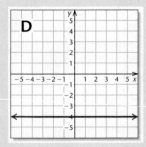

E

Visualizing for Success

Use after Section 3.1.

Match each equation or system of equations with its graph.

1. $x + y = 2,$
$x - y = 2$

2. $y = \frac{1}{3}x - 5$

3. $4x - 2y = -8$

4. $2x + y = 1,$
$x + 2y = 1$

5. $8y + 32 = 0$

6. $f(x) = -x + 4$

7. $\frac{2}{3}x + y = 4$

8. $x = 4,$
$y = 3$

9. $y = \frac{1}{2}x + 3,$
$2y - x = 6$

10. $y = -x + 5,$
$y = 3 - x$

Answers on page A-16

An additional, animated version of this activity appears in MyMathLab. *To use MyMathLab, you need a course ID and a student access code. Contact your instructor for more information.*

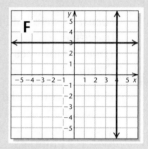

F

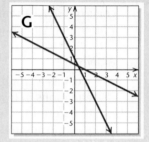

G

H

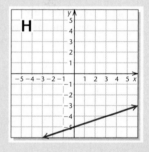

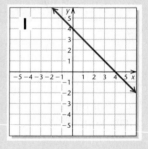

I

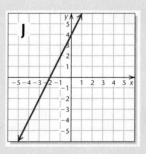

J

Collaborative Activity   *How Many Two's? How Many Three's?*

Focus: Systems of linear equations
Use after: Section 3.2
Time: 20 minutes
Group size: 3

The box score at right, from the 2012 NBA All-Star game, contains information on how many field goals (worth either 2 or 3 points) and free throws (worth 1 point) each player attempted and made. For example, the line "Durant 14–25 5–7 36" means that the West's Kevin Durant made 14 field goals out of 25 attempts and 5 free throws out of 7 attempts, for a total of 36 points.

Activity

1. Work as a group to develop a system of two equations in two unknowns that can be used to determine how many 2-pointers and how many 3-pointers were made by the West.

2. Each group member should solve the system from part (1) in a different way: one person algebraically, one person by making a table and methodically checking all combinations of 2- and 3-pointers, and one person by

guesswork. Compare answers when this has been completed.

3. Determine, as a group, how many 2-pointers and how many 3-pointers the East made.

West (152)
Bryant 9–17 7–8 27, Paul 3–7 0–0 8, Bynum 0–3 0–1 1, Durant 14–25 5–7 36, Griffin 9–12 3–6 22, Westbrook 10–17 0–2 21, Love 7–12 1–3 17, Nowitzki 3–8 0–0 7, Gasol 2–5 0–0 4, Parker 3–5 0–0 6, Aldridge 2–5 0–0 4, Nash 0–0 0–0 0
Totals 62–116 16–26 152

East (149)
Wade 11–15 2–2 24, Rose 6–8 0–0 14, Howard 4–9 1–2 9, James 15–23 0–0 36, Anthony 7–15 5–7 19, Bosh 3–9 0–0 7, Williams 8–11 0–0 20, Rondo 1–3 0–0 2, Iguodala 6–7 0–0 12, Pierce 1–8 0–0 3, Hibbert 1–3 1–1 3, Deng 0–2 0–0 0
Totals 63–113 9–12 149

West 39 49 36 28 — 152
East 28 41 43 37 — 149

Decision Making & Connection   (*Use after Section 3.8.*)

Solar Energy. A photovoltaic (PV) electric system capable of generating 7 kW per hour of electricity cost approximately $45,000 in Vermont in 2011. Because the PV system is tied to the electric grid, local utilities in Vermont pay 20¢ per kilowatt-hour (kWh) for the electricity generated.

Source: Based on data from Bill Heigis, Habie Guion, and David Ellenbogen

1. If the system sends to the grid, on average, 8000 kW per year, how long will it take to break even on the PV investment?

2. In 2011, a Federal Tax Credit of 30% was available to homeowners who installed a PV system. Also, Vermont offered a state rebate of $0.75 per system watt. What was the total tax credit and rebate available for the 7-kW system described? (*Hint*: There are 1000 watts in a kilowatt.)

3. Given the incentives described in Exercise 2, what is the final cost to the owner of the 7-kW system? How long will it take to break even on the investment?

4. Assuming no maintenance costs and a lifespan of 20 years, how much will the system generate in profit?

5. *Research.* Determine the rate that your local utility, or a nearby utility, pays for electricity generated by a home PV system. Use an online calculator to estimate the amount of electricity a 7-kW PV system will generate per year in the area in which you live. Use this information to estimate how long it will take you to break even on a $45,000 PV investment.

Study Summary

KEY TERMS AND CONCEPTS	EXAMPLES	PRACTICE EXERCISES

SECTION 3.1: *Systems of Equations in Two Variables*

A solution of a **system of two equations** is an ordered pair that makes both equations true. The intersection of the graphs of the equations represents the solution of the system.

A system is **consistent** if it has at least one solution. Otherwise it is **inconsistent.**

The equations in a system are **dependent** if one of them can be written as a multiple and/or a sum of the other equation(s). Otherwise, they are **independent.**

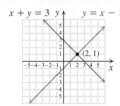

$$x + y = 3,$$
$$y = x - 1$$

The graphs intersect at $(2, 1)$.
The solution is $(2, 1)$.
The system is consistent.
The equations are independent.

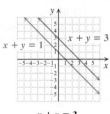

$$x + y = 3,$$
$$x + y = 1$$

The graphs do not intersect.
There is no solution.
The system is inconsistent.
The equations are independent.

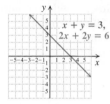

$$x + y = 3,$$
$$2x + 2y = 6$$

The graphs are the same.
The solution set is
$\{(x, y) \mid x + y = 3\}$.
The system is consistent.
The equations are dependent.

1. Solve by graphing:
$$x - y = 3,$$
$$y = 2x - 5.$$

SECTION 3.2: *Solving by Substitution or Elimination*

To use the **substitution method,** we solve one equation for a variable and substitute the expression for that variable in the other equation.

Solve:
$$2x + 3y = 8,$$
$$x = y + 1.$$

Substitute and solve for y:

$$2(y + 1) + 3y = 8$$
$$2y + 2 + 3y = 8$$
$$y = \tfrac{6}{5}.$$

Substitute and solve for x:

$$x = y + 1$$
$$= \tfrac{6}{5} + 1$$
$$= \tfrac{11}{5}.$$

The solution is $\left(\tfrac{11}{5}, \tfrac{6}{5}\right)$.

2. Solve by substitution:
$$x = 3y - 2,$$
$$y - x = 1.$$

To use the **elimination method,** we add to eliminate a variable.

Solve:
$$4x - 2y = 6,$$
$$3x + y = 7.$$

Eliminate y and solve for x:

$$4x - 2y = 6$$
$$\underline{6x + 2y = 14}$$
$$10x \quad\; = 20$$
$$x = 2.$$

Substitute and solve for y:

$$3x + y = 7$$
$$3 \cdot 2 + y = 7$$
$$y = 1.$$

The solution is $(2, 1)$.

3. Solve by elimination:
$$2x - y = 5,$$
$$x + 3y = 1.$$

SECTION 3.3: *Solving Applications: Systems of Two Equations*

Total-value, mixture, and **motion problems** often translate directly to systems of equations.

Motion problems use one of the following relationships:

$$d = rt, \;\; r = \frac{d}{t}, \;\; t = \frac{d}{r}.$$

Simple-interest problems use the formula

Principal · Rate · Time
= Interest.

Total Value
In order to make a necklace, Star Bright Jewelry Design purchased 80 beads for a total of $39 (excluding tax). Some of the beads were sterling silver beads costing 40¢ each and the rest were gemstone beads costing 65¢ each. How many of each type were bought? (See Example 2 in Section 3.3 for a solution.)

Mixture
Nature's Green Gardening, Inc., carries two brands of fertilizer containing nitrogen and water. "Gentle Grow" is 3% nitrogen and "Sun Saver" is 8% nitrogen. Nature's Green needs to combine the two types of solutions into a 90-L mixture that is 6% nitrogen. How much of each brand should be used? (See Example 5 in Section 3.3 for a solution.)

Motion
A Boeing 747-400 jet flies 4 hr west with a 60-mph tailwind. Returning against the wind takes 5 hr. Find the speed of the jet with no wind. (See Example 7 in Section 3.3 for a solution.)

4. Sure Supply charges $17.49 for a box of gel pens and $16.49 for a box of mechanical pencils. If Valley College purchased 120 such boxes for $2010.80, how many boxes of each type did they purchase?

5. A cleaning solution that is 40% nitric acid is being mixed with a solution that is 15% nitric acid in order to create 2 L of a solution that is 25% nitric acid. How much 40%-acid and how much 15%-acid should be used?

6. Ruth paddled for $1\frac{1}{2}$ hr with a 2-mph current. The return trip against the same current took $2\frac{1}{2}$ hr. Find the speed of Ruth's canoe in still water.

SECTION 3.4: *Systems of Equations in Three Variables*

Systems of three equations in three variables are usually most easily solved using elimination.

Solve:

$$
\begin{array}{ll}
x + y - z = 3, & (1) \\
-x + y + 2z = -5, & (2) \\
2x - y - 3z = 9. & (3)
\end{array}
$$

Eliminate x using two equations:

$$
\begin{array}{ll}
x + y - z = 3 & (1) \\
\underline{-x + y + 2z = -5} & (2) \\
2y + z = -2 &
\end{array}
$$

Eliminate x again using two different equations:

$$
\begin{array}{ll}
-2x - 2y + 2z = -6 & (1) \\
\underline{2x - y - 3z = 9} & (3) \\
-3y - z = 3 &
\end{array}
$$

Solve the system of two equations for y and z:

$$
\begin{array}{l}
2y + z = -2 \\
\underline{-3y - z = 3} \\
-y = 1 \\
y = -1
\end{array}
$$

$$2(-1) + z = -2$$
$$z = 0.$$

Substitute and solve for x:

$$
\begin{array}{l}
x + y - z = 3 \\
x + (-1) - 0 = 3 \\
x = 4.
\end{array}
$$

The solution is $(4, -1, 0)$.

7. Solve:

$$
\begin{array}{l}
x - 2y - z = 8, \\
2x + 2y - z = 8, \\
x - 8y + z = 1.
\end{array}
$$

SECTION 3.5: *Solving Applications: Systems of Three Equations*

Many problems with three unknowns can be solved after translating to a system of three equations.

In a triangular cross section of a roof, the largest angle is 70° greater than the smallest angle. The largest angle is twice as large as the remaining angle. Find the measure of each angle.

The angles in the triangle measure 30°, 50°, and 100°.

(See Example 2 in Section 3.5 for a complete solution.)

8. The sum of three numbers is 9. The third number is half the sum of the first and second numbers. The second number is 2 less than the sum of the first and third numbers. Find the numbers.

SECTION 3.6: *Elimination Using Matrices*

A **matrix** (plural, **matrices**) is a rectangular array of numbers. The individual numbers are called **entries**, or **elements.**

By using **row-equivalent** operations, we can solve systems of equations using matrices.

Solve: $x + 4y = 1,$
$\quad\quad 2x - y = 3.$

Write as a matrix in row-echelon form:

$$\begin{bmatrix} 1 & 4 & \vdots & 1 \\ 2 & -1 & \vdots & 3 \end{bmatrix} \longrightarrow \begin{bmatrix} 1 & 4 & \vdots & 1 \\ 0 & -9 & \vdots & 1 \end{bmatrix}.$$

Rewrite as equations and solve:

$-9y = 1 \longrightarrow x + 4\left(-\frac{1}{9}\right) = 1$
$y = -\frac{1}{9} \quad\quad\quad x = \frac{13}{9}.$

The solution is $\left(\frac{13}{9}, -\frac{1}{9}\right).$

9. Solve using matrices:
$3x - 2y = 10,$
$\quad x + y = 5.$

SECTION 3.7: *Determinants and Cramer's Rule*

Determinant of a 2 × 2 Matrix

$$\begin{vmatrix} a & c \\ b & d \end{vmatrix} = ad - bc$$

Determinant of a 3 × 3 Matrix

$$\begin{vmatrix} a_1 & b_1 & c_1 \\ a_2 & b_2 & c_2 \\ a_3 & b_3 & c_3 \end{vmatrix} =$$

$$a_1 \begin{vmatrix} b_2 & c_2 \\ b_3 & c_3 \end{vmatrix} - a_2 \begin{vmatrix} b_1 & c_1 \\ b_3 & c_3 \end{vmatrix} + a_3 \begin{vmatrix} b_1 & c_1 \\ b_2 & c_2 \end{vmatrix}$$

$$\begin{vmatrix} 2 & 3 \\ -1 & 5 \end{vmatrix} = 2 \cdot 5 - (-1)(3) = 13$$

$$\begin{vmatrix} 2 & 3 & 2 \\ 0 & 1 & 0 \\ -1 & 5 & -4 \end{vmatrix}$$

$$= 2\begin{vmatrix} 1 & 0 \\ 5 & -4 \end{vmatrix} - 0\begin{vmatrix} 3 & 2 \\ 5 & -4 \end{vmatrix} + (-1)\begin{vmatrix} 3 & 2 \\ 1 & 0 \end{vmatrix}$$

$$= 2(-4 - 0) - 0 - 1(0 - 2)$$

$$= -8 + 2 = -6$$

Evaluate.

10. $\begin{vmatrix} 3 & -5 \\ 2 & 6 \end{vmatrix}$

11. $\begin{vmatrix} 1 & 2 & -1 \\ 2 & 0 & 3 \\ 0 & 1 & 5 \end{vmatrix}$

We can use determinants and **Cramer's rule** to solve systems of equations.

Cramer's rule for 2 × 2 matrices and for 3 × 3 matrices can be found in Section 3.7.

Solve: $x - 3y = 7,$
$\quad\quad 2x + 5y = 4.$

$$x = \frac{\begin{vmatrix} 7 & -3 \\ 4 & 5 \end{vmatrix}}{\begin{vmatrix} 1 & -3 \\ 2 & 5 \end{vmatrix}}; \quad y = \frac{\begin{vmatrix} 1 & 7 \\ 2 & 4 \end{vmatrix}}{\begin{vmatrix} 1 & -3 \\ 2 & 5 \end{vmatrix}}$$

$x = \frac{47}{11} \quad\quad y = \frac{-10}{11}$

The solution is $\left(\frac{47}{11}, -\frac{10}{11}\right).$

12. Solve using Cramer's rule:
$3x - 5y = 12,$
$2x + 6y = 1.$

SECTION 3.8:   *Business and Economics Applications*

The **break-even point** occurs where the **revenue** equals the **cost,** or where **profit** is 0.

Find **(a)** the total-profit function and **(b)** the break-even point for the total-cost and total-revenue functions

$$C(x) = 38x + 4320 \quad \text{and} \quad R(x) = 62x.$$

a) Profit = Revenue − Cost

$$P(x) = R(x) - C(x)$$
$$P(x) = 62x - (38x + 4320)$$
$$P(x) = 24x - 4320$$

b) $C(x) = R(x)$ At the break-even point, revenue = cost.

$$38x + 4320 = 62x$$
$$180 = x \qquad \text{Solving for } x$$
$$R(180) = 11{,}160 \qquad \text{Finding the revenue (or cost) at the break-even point}$$

The break-even point is (180, $11,160).

13. Find **(a)** the total-profit function and **(b)** the break-even point for the total-cost and total-revenue functions

$$C(x) = 15x + 9000,$$
$$R(x) = 90x.$$

An **equilibrium point** occurs where **supply** equals **demand**.

Find the equilibrium point for the demand and supply functions

$$S(p) = 60 + 7p \quad \text{and} \quad D(p) = 90 - 13p.$$

$$S(p) = D(p) \qquad \text{At the equilibrium point, supply = demand.}$$

$$60 + 7p = 90 - 13p$$
$$20p = 30$$
$$p = 1.5 \qquad \text{Solving for } p$$
$$S(1.5) = 70.5 \qquad \text{Finding the supply (or demand) at the equilibrium point}$$

The equilibrium point is ($1.50, 70.5).

14. Find the equilibrium point for the supply and demand functions

$$S(p) = 60 + 9p,$$
$$D(p) = 195 - 6p.$$

Review Exercises: Chapter 3

 Concept Reinforcement

Choose the word from the list below that best completes each statement.

contradiction	inconsistent
dependent	parallel
determinant	square
elimination	substitution
graphical	zero

1. The system

$$5x + 3y = 7,$$
$$y = 2x + 1$$

is most easily solved using the _____ method.   [3.2]

2. The system

$$-2x + 3y = 8,$$
$$2x + 2y = 7$$

is most easily solved using the _____ method.   [3.2]

3. Of the methods used to solve systems of equations, the _____ method may yield only approximate solutions.   [3.1], [3.2]

4. When one equation in a system is a multiple of another equation in that system, the equations are said to be _____.   [3.1]

5. A system for which there is no solution is said to be _____.   [3.1]

6. When we are using an algebraic method to solve a system of equations, obtaining a(n) _____ tells us that the system is inconsistent.   [3.2]

7. When we are graphing to solve a system of two equations, if there is no solution, the lines will be _____ .   [3.1]

8. When a matrix has the same number of rows and columns, it is said to be _____.   [3.7]

9. Cramer's rule is a formula in which the numerator and the denominator of each fraction is a(n) _____.   [3.7]

10. At the break-even point, the value of the profit function is _____ .   [3.8]

For Exercises 11–20, if a system has an infinite number of solutions, use set-builder notation to write the solution set. If a system has no solution, state this.

Solve graphically.   [3.1]

11. $y = x - 3,$
    $y = \frac{1}{4}x$

12. $2x - 3y = 12,$
    $4x + y = 10$

Solve using the substitution method.   [3.2]

13. $5x - 2y = 4,$
    $x = y - 2$

14. $y = x + 2,$
    $y - x = 8$

Solve using the elimination method.   [3.2]

15. $2x + 5y = 8,$
    $6x - 5y = 10$

16. $3x - 5y = 9,$
    $5x - 3y = -1$

Solve using any appropriate method.   [3.1], [3.2]

17. $x - 3y = -2,$
    $7y - 4x = 6$

18. $4x - 7y = 18,$
    $9x + 14y = 40$

19. $1.5x - 3 = -2y,$
    $3x + 4y = 6$

20. $y = 2x - 5,$
    $y = \frac{1}{2}x + 1$

Solve.   [3.3]

21. Jillian charges $25 for a private guitar lesson and $18 for a group guitar lesson. One day in August, Jillian earned $265 from 12 students. How many students of each type did Jillian teach?

22. A freight train leaves Houston at midnight traveling north at 44 mph. One hour later, a passenger train, going 55 mph, travels north from Houston on a parallel track. How long will it take the passenger train to overtake the freight train?

23. D'Andre wants 14 L of fruit punch that is 10% juice. At the store, he finds only punch that is 15% juice or punch that is 8% juice. How much of each should he purchase?

Solve. If a system's equations are dependent or if there is no solution, state this.   [3.4]

24. $x + 4y + 3z = 2,$
    $2x + y + z = 10,$
    $-x + y + 2z = 8$

25. $4x + 2y - 6z = 34,$
    $2x + y + 3z = 3,$
    $6x + 3y - 3z = 37$

26. $2x - 5y - 2z = -4,$
    $7x + 2y - 5z = -6,$
    $-2x + 3y + 2z = 4$

27. $3x + y = 2,$
    $x + 3y + z = 0,$
    $x + z = 2$

28. $2x - 3y + z = 1,$
    $x - y + 2z = 5,$
    $3x - 4y + 3z = -2$

Solve.   [3.5]

29. In triangle ABC, the measure of angle A is four times the measure of angle C, and the measure of angle B is 45° more than the measure of angle C. What are the measures of the angles of the triangle?

30. The sum of the average number of times a man, a woman, and a one-year-old child cry each month is 56.7. A woman cries 3.9 more times than a man. The average number of times a one-year-old cries per month is 43.3 more than the average number of times combined that a man and a woman cry. What is the average number of times per month that each cries?

Solve using matrices. Show your work.   [3.6]

31. $3x + 4y = -13,$
    $5x + 6y = 8$

32. $3x - y + z = -1,$
    $2x + 3y + z = 4,$
    $5x + 4y + 2z = 5$

Evaluate.   [3.7]

33. $\begin{vmatrix} -2 & -5 \\ 3 & 10 \end{vmatrix}$

34. $\begin{vmatrix} 2 & 3 & 0 \\ 1 & 4 & -2 \\ 2 & -1 & 5 \end{vmatrix}$

Solve using Cramer's rule. Show your work.   [3.7]

35. $2x + 3y = 6,$
    $x - 4y = 14$

36. $2x + y + z = -2,$
    $2x - y + 3z = 6,$
    $3x - 5y + 4z = 7$

37. Find **(a)** the total-profit function and **(b)** the break-even point for the total-cost and total-revenue functions
$$C(x) = 30x + 15,800,$$
$$R(x) = 50x. \quad [3.8]$$

38. Find the equilibrium point for the demand and supply functions

$$S(p) = 60 + 7p$$

and

$$D(p) = 120 - 13p. \quad [3.8]$$

39. Danae is beginning to produce organic honey. For the first year, the fixed costs for setting up production are $54,000. The variable costs for producing each pint of honey are $4.75. The revenue from each pint of honey is $9.25. Find the following. [3.8]

a) The total cost $C(x)$ of producing x pints of honey

b) The total revenue $R(x)$ from the sale of x pints of honey

c) The total profit $P(x)$ from the production and sale of x pints of honey

d) The profit or loss from the production and sale of 5000 pints of honey; of 15,000 pints of honey

e) The break-even point

Synthesis

 40. How would you go about solving a problem that involves four variables? [3.5]

41. Explain how a system of equations can be both dependent and inconsistent. [3.4]

42. Danae is leaving a job that pays $36,000 per year to make honey (see Exercise 39). How many pints of honey must she produce and sell in order to make as much money as she earned at her previous job? [3.8]

43. Solve graphically:

$$y = x + 2,$$
$$y = x^2 + 2. \quad [3.1]$$

Test: Chapter 3

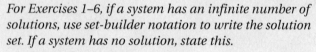

For step-by-step test solutions, access the Chapter Test Prep Videos in MyMathLab®, on YouTube® (search "Bittinger Inter Alg CA" and click on "Channels"), or by scanning the code.

For Exercises 1–6, if a system has an infinite number of solutions, use set-builder notation to write the solution set. If a system has no solution, state this.

1. Solve graphically:

$$2x + y = 8,$$
$$y - x = 2.$$

2. Solve using the substitution method:

$$x + 3y = -8,$$
$$4x - 3y = 23.$$

Solve using the elimination method.

3. $3x - y = 7,$
   $x + y = 1$

4. $4y + 2x = 18,$
   $3x + 6y = 26$

Solve using any appropriate method.

5. $2x - 4y = -6,$
   $x = 2y - 3$

6. $4x - 6y = 3,$
   $6x - 4y = -3$

7. The perimeter of a standard basketball court is 288 ft. The length is 44 ft longer than the width. Find the dimensions.

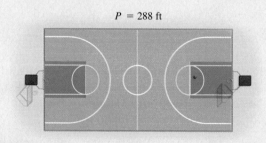

$P = 288$ ft

8. Pepperidge Farm® Goldfish is a snack food for which 40% of its calories come from fat. Rold Gold® Pretzels receive 9% of their calories from fat. How many grams of each would be needed to make 620 g of a snack mix for which 15% of the calories are from fat?

9. Kylie's motorboat took 3 hr to make a trip downstream on a river flowing at 5 mph. The return trip against the same current took 5 hr. Find the speed of the boat in still water.

Solve. If a system's equations are dependent or if there is no solution, state this.

10. $-3x + y - 2z = 8,$
    $-x + 2y - z = 5,$
    $2x + y + z = -3$

11. $6x + 2y - 4z = 15,$
    $-3x - 4y + 2z = -6,$
    $4x - 6y + 3z = 8$

12. $2x + 2y = 0,$
    $4x + 4z = 4,$
    $2x + y + z = 2$

13. $3x + 3z = 0,$
    $2x + 2y = 2,$
    $3y + 3z = 3$

Solve using matrices.

14. $4x + y = 12,$
    $3x + 2y = 2$

15. $x + 3y - 3z = 12,$
    $3x - y + 4z = 0,$
    $-x + 2y - z = 1$

Evaluate.

16. $\begin{vmatrix} 4 & -2 \\ 3 & -5 \end{vmatrix}$

17. $\begin{vmatrix} 3 & 4 & 2 \\ -2 & -5 & 4 \\ 0 & 5 & -3 \end{vmatrix}$

18. Solve using Cramer's rule:
    $3x + 4y = -1,$
    $5x - 2y = 4.$

19. An electrician, a carpenter, and a plumber are hired to work on a house. The electrician earns $30 per hour, the carpenter $28.50 per hour, and the plumber $34 per hour. The first day on the job, they worked a total of 21.5 hr and earned a total of $673.00. If the plumber worked 2 more hours than the carpenter, how many hours did each work?

20. Find the equilibrium point for the demand and supply functions
    $$D(p) = 79 - 8p \quad \text{and} \quad S(p) = 37 + 6p,$$
    where p is the price, in dollars, $D(p)$ is the number of units demanded, and $S(p)$ is the number of units supplied.

21. Kick Back, Inc., is producing a new hammock. For the first year, the fixed costs for setting up production are $44,000. The variable costs for producing each hammock are $25. The revenue from each hammock is $80. Find the following.
    a) The total cost $C(x)$ of producing x hammocks
    b) The total revenue $R(x)$ from the sale of x hammocks
    c) The total profit $P(x)$ from the production and sale of x hammocks
    d) The profit or loss from the production and sale of 300 hammocks; of 900 hammocks
    e) The break-even point

Synthesis

22. The graph of the function $f(x) = mx + b$ contains the points $(-1, 3)$ and $(-2, -4)$. Find m and b.

23. Some of the world's best and most expensive coffee is Hawaii's Kona coffee. In order for coffee to be labeled "Kona Blend," it must contain at least 30% Kona beans. Bean Town Roasters has 40 lb of Mexican coffee. How much Kona coffee must they add if they wish to market it as Kona Blend?

Cumulative Review: Chapters 1–3

Simplify. Do not leave negative exponents in your answers.

1. $x^4 \cdot x^{-6} \cdot x^{13}$ [1.6]

2. $\dfrac{-10a^7 b^{-11}}{25a^{-4}b^{22}}$ [1.6]

3. $\left(\dfrac{3x^4 y^{-2}}{4x^{-5}}\right)^4$ [1.6]

4. $\dfrac{2.42 \times 10^5}{6.05 \times 10^{-2}}$ [1.7]

5. $(1.95 \times 10^{-3})(5.73 \times 10^8)$ [1.7]

6. Solve $A = \frac{1}{2}h(b + t)$ for b. [1.5]

Solve.

7. $\frac{3}{8}x + 7 = -14$ [1.3]

8. $-3 + 5x = 2x + 15$ [1.3]

9. $3n - (4n - 2) = 7$ [1.3]

10. $6y - 5(3y - 4) = 10$ [1.3]

11. $9c - [3 - 4(2 - c)] = 10$ [1.3]

12. $3x + y = 4,$
$y = 6x - 5$ [3.2]

13. $6x - 10y = -22,$
$-11x - 15y = 27$
[3.2]

14. $x + y + z = -5,$
$2x + 3y - 2z = 8,$
$x - y + 4z = -21$
[3.4]

15. $2x + 5y - 3z = -11,$
$-5x + 3y - 2z = -7,$
$3x - 2y + 5z = 12$ [3.4]

Graph.

16. $f(x) = -2x + 8$ [2.3]

17. $y = x^2 - 1$ [2.1]

18. $4x + 16 = 0$ [2.4]

19. $-3x + 2y = 6$ [2.3]

20. Find the slope and the y-intercept of the line with equation $-4y + 9x = 12$. [2.3]

21. Find an equation in slope–intercept form of the line containing the points $(-6, 3)$ and $(4, 2)$. [2.5]

22. Determine whether the lines given by the following equations are parallel, perpendicular, or neither:
$2x = 4y + 7,$
$x - 2y = 5.$ [2.4]

23. Find an equation of the line containing the point $(2, 1)$ and perpendicular to the line $x - 2y = 5$.
[2.5]

24. Determine the domain of the function given by
$$f(x) = \frac{7}{x + 10}.$$ [2.2]

Given $g(x) = 4x - 3$ and $h(x) = -2x^2 + 1$, find the following.

25. $h(4)$ [2.2]

26. $-g(0)$ [2.2]

27. $(g \cdot h)(-1)$ [2.6]

28. $(g - h)(a)$ [2.6]

Solve.

29. *ATM Fees.* The average out-of-network ATM (automated teller machine) convenience fee charged by banks to customers from other banks can be approximated by $f(t) = 0.15t + 1.37$, where t is the number of years after 2004.

Source: Based on data from Bankrate

a) Find the average out-of-network ATM fee in 2011. [2.2]

b) What do the numbers 0.15 and 1.37 signify?
[2.3]

30. *Bank Tellers.* According to the Bureau of Labor Statistics, there were 600,500 bank tellers in 2008. This number was projected to grow to 638,000 by 2018. Let $B(t)$ represent the number of bank tellers, in thousands, t years after 2000.

a) Find a linear function that fits the data. [2.5]

b) Use the function from part (a) to estimate the number of bank tellers in 2016. [2.5]

c) In what year will there be 650,000 bank tellers?
[2.5]

31. *Professional Memberships.* A one-year professional membership in Investigative Reporters and Editors, Inc. (IRE), costs $70. A one-year student membership costs $25. If, in May, 150 members joined IRE for a total of $6810 in membership dues, how many professionals and how many students joined that month? [3.3]

Source: Investigative Reporters and Editors, Inc.

32. *Saline Solutions.* "Sea Spray" is 25% salt and the rest water. "Ocean Mist" is 5% salt and the rest water. How many ounces of each would be needed to obtain 120 oz of a mixture that is 20% salt? [3.3]

33. *Test Scores.* Franco's scores on four tests are 93, 85, 100, and 86. What must the score be on the fifth test for his average to be 90? [1.4]

Synthesis

34. Simplify: $(6x^{a+2}y^{b+2})(-2x^{a-2}y^{y+1})$. [1.6]

35. Given that $f(x) = mx + b$ and that $f(5) = -3$ when $f(-4) = 2$, find m and b. [2.5], [3.3]

Inequalities and Problem Solving

YEAR	MUNICIPAL SOLID WASTE GENERATED (in pounds per person per day)
2000	4.72
2005	4.67
2007	4.64
2008	4.53
2009	4.35
2010	4.43

Reduce, Reuse, Recycle!

All three of these strategies are effective in reducing the amount of waste that ends up in landfills. Although it is difficult to measure how many items are reused, the Environmental Protection Agency (EPA) does measure how much waste is produced and how much is recycled. As the data in the table above indicate, the amount of solid waste *generated* per person per day has slowly been decreasing. Although not shown here, the EPA has evidence that the amount of solid waste *recovered* per person per day has been increasing. In this chapter, we will use functions that model waste recovery and recycling. *(See Example 7 and Exercise 93 in Section 4.1.)*

I use math all the time both to do my work correctly and to provide accurate accounting for reimbursement.

Gonzalo Calderone, a supervisor for a recycling center in California, uses math to calculate percentages of mandatory diversion rates for all recyclable materials. He must have accurate and detailed accounting for all products that can be recycled.

nequalities are mathematical sentences containing symbols such as $<$ (is less than). We solve inequalities using principles similar to those used to solve equations. In this chapter, we solve a variety of inequalities, systems of inequalities, and real-world applications.

4.1   Inequalities and Applications

Solutions of Inequalities ▪ Interval Notation ▪ The Addition Principle for Inequalities ▪
The Multiplication Principle for Inequalities ▪ Using the Principles Together ▪ Problem Solving

Solutions of Inequalities

We now modify our equation-solving skills for the solving of *inequalities*. An **inequality** is any sentence containing $<, >, \leq, \geq,$ or $\neq$. Some examples are

$$-2 < a, \qquad x > 4, \qquad x + 3 \leq 6, \qquad 6 - 7y \geq 10y - 4, \quad \text{and} \quad 5x \neq 10.$$

Any value for the variable that makes an inequality true is called a **solution**. The set of all solutions is called the **solution set**. When all solutions of an inequality are found, we say that we have **solved** the inequality.

EXAMPLE 1   Determine whether the given number is a solution of the inequality.

a) $x + 3 < 6;$ 5 **b)** $-3 > -9 - 2x;$ -1

SOLUTION

a) We substitute to get $5 + 3 < 6$, or $8 < 6$, a false sentence. Thus, 5 *is not* a solution.

b) We substitute to get $-3 > -9 - 2(-1)$, or $-3 > -7$, a true sentence. Thus, -1 *is* a solution.

YOUR TURN

1. Determine whether 4 is a solution of $x - 7 < 0$.

The *graph* of an inequality is a visual representation of the inequality's solution set. Inequalities in one variable can be graphed on the number line. Inequalities in two variables are graphed on a coordinate plane, and appear later in this chapter.

The solution set of an inequality is often an infinite set. For example, the solution set of $x < 3$ is the set containing all numbers less than 3. To graph this set, we shade the number line to the left of 3. To indicate that 3 is not in the solution set, we use a parenthesis. If 3 were included in the solution set, we would use a bracket.

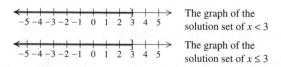

The graph of the solution set of $x < 3$

The graph of the solution set of $x \leq 3$

Interval Notation

To write the solution set of $x < 3$, we can use *set-builder notation*:

$$\{x \mid x < 3\}.$$

This is read "The set of all x such that x is less than 3."

The Multiplication Principle for Inequalities

The multiplication principle for inequalities differs from the multiplication principle for equations. To see this, consider the true inequality $4 < 9$. If we multiply both sides by 2, we get another true inequality:

$$4 \cdot 2 < 9 \cdot 2, \quad \text{or} \quad 8 < 18.$$

If we multiply both sides of $4 < 9$ by -2, we get a false inequality:

FALSE $\longrightarrow 4(-2) < 9(-2), \quad \text{or} \quad -8 < -18. \longleftarrow$ FALSE

Multiplication (or division) by a negative number changes the sign of the number being multiplied (or divided). When the signs of both numbers in an inequality are changed, the position of the numbers with respect to each other is reversed.

$$-8 > -18. \longleftarrow \quad \text{TRUE}$$

The $<$ symbol has been reversed!

THE MULTIPLICATION PRINCIPLE FOR INEQUALITIES

For any real numbers a and b, and for any *positive* number c,

$a < b$ is equivalent to $ac < bc$;
$a > b$ is equivalent to $ac > bc$.

For any real numbers a and b, and for any *negative* number c,

$a < b$ is equivalent to $ac > bc$;
$a > b$ is equivalent to $ac < bc$.

Similar statements hold for $\leq$ and $\geq$.

Since division by c is the same as multiplication by $1/c$, there is no need for a separate division principle. Note that c and $1/c$ have the same sign.

CAUTION!   Remember that whenever we multiply or divide both sides of an inequality by a negative number, we must reverse the inequality symbol.

EXAMPLE 4   Solve and graph: **(a)** $3y < \frac{3}{4}$; **(b)** $-5x \geq -80$.

SOLUTION

a) $3y < \frac{3}{4}$

The symbol stays the same.

$\frac{1}{3} \cdot 3y < \frac{1}{3} \cdot \frac{3}{4}$ Multiplying both sides by $\frac{1}{3}$ or dividing both sides by 3

$y < \frac{1}{4}$

Any number less than $\frac{1}{4}$ is a solution. The solution set is $\left\{y \mid y < \frac{1}{4}\right\}$, or $\left(-\infty, \frac{1}{4}\right)$. The graph is shown below. As a partial check, note that $3y < \frac{3}{4}$ is true for $y = 0$: $3 \cdot 0 < \frac{3}{4}$.

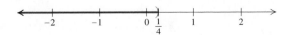

Student Notes

Try to remember to reverse the inequality symbol as soon as both sides are multiplied or divided by a negative number. Don't wait until after the multiplication or division has been completed to reverse the symbol.

4. Solve and graph: $-2x > 10$.

b) $-5x \geq -80$

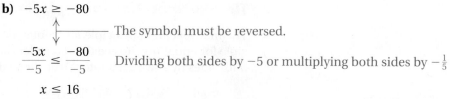

The symbol must be reversed.

$$\frac{-5x}{-5} \leq \frac{-80}{-5}$$ Dividing both sides by -5 or multiplying both sides by $-\frac{1}{5}$

$$x \leq 16$$

The solution set is $\{x \mid x \leq 16\}$, or $(-\infty, 16]$. The graph is shown below. As a partial check, note that $-5x \geq -80$ is true for $x = 10$: $-5 \cdot 10 \geq -80$.

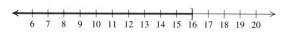

YOUR TURN

Using the Principles Together

We use the addition and multiplication principles together when solving inequalities, much as we did when solving equations.

EXAMPLE 5   Solve: $16 - 7y \geq 10y - 4$.

SOLUTION   We have

$$16 - 7y \geq 10y - 4$$
$$-16 + 16 - 7y \geq -16 + 10y - 4$$ Adding -16 to both sides
$$-7y \geq 10y - 20$$
$$-10y + (-7y) \geq -10y + 10y - 20$$ Adding $-10y$ to both sides
$$-17y \geq -20$$

The symbol must be reversed.

$$-\tfrac{1}{17} \cdot (-17y) \leq -\tfrac{1}{17} \cdot (-20)$$ Multiplying both sides by $-\frac{1}{17}$ or dividing both sides by -17
$$y \leq \tfrac{20}{17}.$$

5. Solve: $3n - 6 < 7n + 4$.

The solution set is $\left\{y \mid y \leq \frac{20}{17}\right\}$, or $\left(-\infty, \frac{20}{17}\right]$. The check is left to the student.

YOUR TURN

EXAMPLE 6   Let $f(x) = -3(x + 8) - 5x$ and $g(x) = 4x - 9$. Find all x for which $f(x) > g(x)$.

SOLUTION   We have

$$f(x) > g(x)$$
$$-3(x + 8) - 5x > 4x - 9$$ Substituting for $f(x)$ and $g(x)$
$$-3x - 24 - 5x > 4x - 9$$ Using the distributive law
$$-24 - 8x > 4x - 9$$
$$-24 - 8x + 8x > 4x - 9 + 8x$$ Adding $8x$ to both sides
$$-24 > 12x - 9$$
$$-24 + 9 > 12x - 9 + 9$$ Adding 9 to both sides
$$-15 > 12x$$

The symbol stays the same.

$$-\tfrac{5}{4} > x.$$ Dividing by 12 and simplifying

6. Let $f(x) = 5 - x$ and $g(x) = 2 - 4(x + 1)$. Find all x for which $f(x) \leq g(x)$.

The solution set is $\left\{x \mid -\frac{5}{4} > x\right\}$, or $\left\{x \mid x < -\frac{5}{4}\right\}$, or $\left(-\infty, -\frac{5}{4}\right)$. The check is left to the student.

YOUR TURN

Problem Solving

Many problem-solving situations translate to inequalities. In addition to "is less than" and "is more than," other phrases are commonly used.

Important Words	Sample Sentence	Definition of Variables	Translation
is at least	Kelby walks at least 1.5 mi a day.	Let k represent the length of Kelby's walk, in miles.	$k \geq 1.5$
is at most	At most 5 students dropped the course.	Let n represent the number of students who dropped the course.	$n \leq 5$
cannot exceed	The cost cannot exceed $12,000.	Let c represent the cost, in dollars.	$c \leq 12{,}000$
must exceed	The speed must exceed 40 mph.	Let s represent the speed, in miles per hour.	$s > 40$
is less than	Hamid's weight is less than 130 lb.	Let w represent Hamid's weight, in pounds.	$w < 130$
is more than	Boston is more than 200 mi away.	Let d represent the distance to Boston, in miles.	$d > 200$
is between	The film is between 90 min and 100 min long.	Let t represent the length of the film, in minutes.	$90 < t < 100$
minimum	Thea drank a minimum of 5 glasses of water a day.	Let w represent the number of glasses of water.	$w \geq 5$
maximum	The maximum penalty is $100.	Let p represent the penalty, in dollars.	$p \leq 100$
no more than	Alan consumes no more than 1500 calories.	Let c represent the number of calories Alan consumes.	$c \leq 1500$
no less than	Patty scored no less than 80.	Let s represent Patty's score.	$s \geq 80$

Year	Municipal Solid Waste Generated (in pounds per person per day)
2000	4.72
2005	4.67
2007	4.64
2008	4.53
2009	4.35
2010	4.43

Source: U.S. Environmental Protection Agency

EXAMPLE 7 *Solid Waste.* The table at left shows the amount of municipal solid waste generated per person per day in the United States for several years. Although the data are not exactly linear, the function given by

$$m(t) = -0.03t + 4.72$$

is a good model. Here, $m(t)$ is the amount of municipal solid waste generated, in number of pounds per person per day, t years after 2000. Using an inequality, determine those years for which less than 4 lb of municipal solid waste will be generated per person per day.

SOLUTION

1. **Familiarize.** By examining the formula, we see that in 2000, there were 4.72 lb of municipal solid waste generated per person per day, and this number was changing at a rate of -0.03 lb per year.

2. **Translate.** We are asked to find the years for which *less than* 4 lb of municipal solid waste will be generated per person per day. Thus we must have

$$m(t) < 4$$
$$-0.03t + 4.72 < 4.$$

3. **Carry out.** We solve the inequality:

$$-0.03t + 4.72 < 4$$
$$-0.03t < -0.72$$
$$t > 24.$$

Note that this corresponds to years after 2024.

7. Refer to Example 7. Determine the years for which 4.3 lb or less of municipal solid waste will be generated per person per day.

4. Check. We can partially check our answer by finding $m(t)$ for a value of t greater than 24. For example,

$$m(25) = -0.03(25) + 4.72 = 3.97, \quad \text{and} \quad 3.97 < 4.$$

5. State. Less than 4 lb of municipal solid waste will be generated per person per day for years after 2024.

YOUR TURN

EXAMPLE 8 *Job Offers.* After graduation, Jessica had two job offers in sales:

Uptown Fashions: A salary of $1500 per month, plus a commission of 4% of sales;

Ergo Designs: A salary of $1700 per month, plus a commission of 6% of sales in excess of $10,000.

If sales always exceed $10,000, for what amount of sales would Uptown Fashions provide higher pay?

SOLUTION

1. Familiarize. Suppose that Jessica sold a certain amount—say, $12,000—in one month. Which plan would be better? Working for Uptown, she would earn $1500 plus 4% of $12,000, or $1500 + 0.04($12,000) = $1980.

Since with Ergo Designs commissions are paid only on sales in excess of $10,000, Jessica would earn $1700 plus 6% of ($12,000 − $10,000), or $1700 + 0.06($2000) = $1820.

Thus for monthly sales of $12,000, Uptown pays more. Similar calculations show that for sales of $30,000 per month, Ergo pays more. To determine *all* values for which Uptown pays more, we solve an inequality based on the above calculations.

Let S = the amount of monthly sales, in dollars, and assume $S > 10,000$ as stated above. We list the given information in a table.

Uptown Fashions Monthly Income	Ergo Designs Monthly Income
$1500 salary 4% of sales = 0.04S *Total*: 1500 + 0.04S	$1700 salary 6% of sales over $10,000 = 0.06(S − 10,000) *Total*: 1700 + 0.06(S − 10,000)

Chapter Resources:
Collaborative Activity, p. 272;
Decision Making: Connection, p. 272

2. Translate. We want to find all values of S for which

Income from Uptown   is greater than   income from Ergo.

$$1500 + 0.04S \quad > \quad 1700 + 0.06(S - 10,000)$$

8. Refer to Example 8. Suppose that after salary negotiations, Uptown Fashions offers a salary of $1800 per month, plus a commission of 3% of sales, and Ergo Designs offers a salary of $1900 per month, plus a commission of 5% of sales in excess of $15,000. If sales always exceed $15,000, for what amount of sales would Ergo Designs provide higher pay?

3. Carry out. We solve the inequality:

$$1500 + 0.04S > 1700 + 0.06(S - 10{,}000)$$

$1500 + 0.04S > 1700 + 0.06S - 600$	Using the distributive law
$1500 + 0.04S > 1100 + 0.06S$	Combining like terms
$400 > 0.02S$	Subtracting 1100 and 0.04S from both sides
$20{,}000 > S$, or $S < 20{,}000$.	Dividing both sides by 0.02

4. Check. The above steps indicate that income from Uptown Fashions is higher than income from Ergo Designs for sales less than $20,000. In the *Familiarize* step, we saw that for sales of $12,000, Uptown pays more. Since $12{,}000 < 20{,}000$, this is a partial check.

5. State. When monthly sales are less than $20,000, Uptown Fashions provides the higher pay (assuming sales are greater than $10,000).

 YOUR TURN

4.1   EXERCISE SET   FOR EXTRA HELP   MyMathLab® Math XL
PRACTICE   WATCH   READ   REVIEW

➔ Vocabulary and Reading Check

Choose the word from the following list that best completes each statement. Not every word will be used.

closed   open
half-open   positive
negative   solution

1. Because $-8 < -1$ is true, -8 is a(n) _____ of $x < -1$.

2. The interval $[4, 9]$ is a(n) _____ interval.

3. The interval $(-7, 1\,]$ is a(n) _____ interval.

4. We reverse the direction of the inequality symbol when we multiply both sides of an inequality by a(n) _____ number.

➔ Concept Reinforcement

Classify each of the following as equivalent inequalities, equivalent equations, equivalent expressions, or not equivalent.

5. $5x + 7 = 6 - 3x$, $8x + 7 = 6$

6. $2(4x + 1)$, $8x + 2$

7. $x - 7 > -2$, $x > 5$

8. $-4t \le 12$, $t \le -3$

9. $\frac{3}{5}a + \frac{1}{5} = 2$, $3a + 1 = 10$

10. $-\frac{1}{3}t \le -5$, $t \ge 15$

Solutions of Inequalities

Determine whether the given numbers are solutions of the inequality.

11. $x - 4 \ge 1$
   a) -4 **b)** 4 **c)** 5 **d)** 8

12. $3x + 1 \le -5$
   a) -5 **b)** -2 **c)** 0 **d)** 3

13. $2y + 3 < 6 - y$
   a) 0 **b)** 1 **c)** -1 **d)** 4

14. $5t - 6 > 1 - 2t$
   a) 6 **b)** 0 **c)** -3 **d)** 1

Interval Notation

Graph each inequality, and write the solution set using both set-builder notation and interval notation.

15. $y < 6$ **16.** $x > 4$

17. $x \ge -4$ **18.** $t \le 6$

19. $t > -3$ **20.** $y < -3$

21. $x \le -7$ **22.** $x \ge -6$

The Addition Principle for Inequalities

Solve. Then graph. Write the solution set using both set-builder notation and interval notation.

23. $x + 2 > 1$ **24.** $x + 9 > 6$

25. $t - 6 \le 4$ **26.** $t - 1 \ge 5$

27. $x - 12 \geq -11$ **28.** $x - 11 \leq -2$

The Multiplication Principle for Inequalities

Solve. Then graph. Write the solution set using both set-builder notation and interval notation.

29. $9t < -81$ **30.** $8x \geq 24$

31. $-0.3x > -15$ **32.** $-0.5x < -30$

33. $-9x \geq 8.1$ **34.** $-8y \leq 3.2$

35. $\frac{3}{4}y \geq -\frac{5}{8}$ **36.** $\frac{5}{6}x \leq -\frac{3}{4}$

Using the Principles Together

Solve. Then graph. Write the solution set using both set-builder notation and interval notation.

37. $3x + 1 < 7$ **38.** $2x - 5 \geq 9$

39. $3 - x \geq 12$ **40.** $8 - x < 15$

41. $\dfrac{2x + 7}{5} < -9$ **42.** $\dfrac{5y + 13}{4} > -2$

43. $\dfrac{3t - 7}{-4} \leq 5$ **44.** $\dfrac{2t - 9}{-3} \geq 7$

45. $\dfrac{9 - x}{-2} \geq -6$ **46.** $\dfrac{3 - x}{-5} < -2$

47. Let $f(x) = 7 - 3x$ and $g(x) = 2x - 3$. Find all values of x for which $f(x) \leq g(x)$.

48. Let $f(x) = 8x - 9$ and $g(x) = 3x - 11$. Find all values of x for which $f(x) \leq g(x)$.

49. Let $f(x) = 2x - 7$ and $g(x) = 5x - 9$. Find all values of x for which $f(x) < g(x)$.

50. Let $f(x) = 0.4x + 5$ and $g(x) = 1.2x - 4$. Find all values of x for which $g(x) \geq f(x)$.

51. Let $y_1 = \frac{3}{8} + 2x$ and $y_2 = 3x - \frac{1}{8}$. Find all values of x for which $y_2 \geq y_1$.

52. Let $y_1 = 2x + 1$ and $y_2 = -\frac{1}{2}x + 6$. Find all values of x for which $y_1 < y_2$.

Solve. Write the solution set using both set-builder notation and interval notation.

53. $3 - 8y \geq 9 - 4y$

54. $4m + 7 \geq 9m - 3$

55. $5(t - 3) + 4t < 2(7 + 2t)$

56. $2(4 + 2x) > 2x + 3(2 - 5x)$

57. $5[3m - (m + 4)] > -2(m - 4)$

58. $8x - 3(3x + 2) - 5 \geq 3(x + 4) - 2x$

59. $19 - (2x + 3) \leq 2(x + 3) + x$

60. $13 - (2c + 2) \geq 2(c + 2) + 3c$

61. $\frac{1}{4}(8y + 4) - 17 < -\frac{1}{2}(4y - 8)$

62. $\frac{1}{3}(6x + 24) - 20 > -\frac{1}{4}(12x - 72)$

63. $2[8 - 4(3 - x)] - 2 \geq 8[2(4x - 3) + 7] - 50$

64. $5[3(7 - t) - 4(8 + 2t)] - 20 \leq -6[2(6 + 3t) - 4]$

Problem Solving

Translate to an inequality.

65. A number is less than 10.

66. A number is greater than or equal to 4.

67. The temperature is at most $-3°C$.

68. The credit-card debt of the average college freshman is at least $2000.

69. The age of the Mayan altar exceeds 1200 years.

70. The time of the test was between 45 min and 55 min.

71. Normandale Community College is no more than 15 mi away.

72. Angenita earns no less than $12 per hour.

73. To rent a car, a driver must have a minimum of 5 years of driving experience.

74. The maximum safe level for chronic inhalation of formaldehyde is 0.003 parts per million.

Source: U.S. Environmental Protection Agency

75. The costs of production of the software cannot exceed $12,500.

76. The number of volunteers was at most 20.

Solve.

77. *Photography.* Eli will photograph a wedding for a flat fee of $900 or for an hourly rate of $120. For what lengths of time would the hourly rate be less expensive?

78. *Truck Rentals.* Jenn can rent a moving truck for either $99 with unlimited mileage or $49 plus 80¢ per mile. For what mileages would the unlimited mileage plan save money?

79. *Graduate School.* An unconditional acceptance into the Master of Business Administration (MBA) program at Arkansas State University is given to students whose GMAT score plus 200 times their undergraduate grade point average is at least 1030. Chloe's GMAT score was 500. What must her grade point average be in order to be unconditionally accepted into the program?

Source: Arkansas State University

80. *Car Payments.* As a rule of thumb, debt payments (other than mortgages) should be less than 8% of a consumer's monthly gross income. Oliver makes $54,000 per year and has a $100 student-loan payment every month. What size car payment can he afford?

Source: money.cnn.com

81. *Exam Scores.* There are 80 questions on a college entrance examination. Two points are awarded for each correct answer, and one-half point is deducted for each incorrect answer. How many questions does Tami need to answer correctly in order to score at least 100 on the test? Assume that Tami answers every question.

82. *Insurance Claims.* After a serious automobile accident, most insurance companies will replace the damaged car with a new one if repair costs exceed 80% of the NADA, or "blue-book," value of the car. Lorenzo's car recently sustained $9200 worth of damage but was not replaced. What was the blue-book value of his car?

83. *Well Drilling.* All Seasons Well Drilling offers two plans. Under the "pay-as-you-go" plan, they charge $500 plus $8 per foot for a well of any depth. Under their "guaranteed-water" plan, they charge a flat fee of $4000 for a well that is guaranteed to provide adequate water for a household. For what depths would it save a customer money to use the pay-as-you-go plan?

84. *Legal Fees.* Bridgewater Legal Offices charges a $250 retainer fee for real estate transactions plus $180 per hour. Dockside Legal charges a $100 retainer fee plus $230 per hour. For what number of hours does Bridgewater charge more?

85. *Wages.* Toni can be paid in one of two ways:

Plan A: A salary of $400 per month, plus a commission of 8% of gross sales;
Plan B: A salary of $610 per month, plus a commission of 5% of gross sales.

For what amount of gross sales should Toni select plan A?

86. *Wages.* Eric can be paid for his masonry work in one of two ways:

Plan A: $300 plus $15.00 per hour;
Plan B: Straight $17.50 per hour.

Suppose that the job takes n hours. For what values of n is plan B better for Eric?

87. *ATM Rates.* The Intercity Bank offers two account plans. Their Local plan charges a $5 monthly service fee plus $3.00 for each ATM (automated teller machine) transaction beyond 4 transactions per month. Their Anywhere plan charges a

$15 monthly service fee plus $1.75 per ATM transaction. For what number of ATM transactions per month will the Anywhere plan cost less?

88. *Checking Accounts.* North Bank charges $10 per month for a student checking account. The first 8 checks are free, and each additional check costs $0.75. South Bank offers a student checking account with no monthly charge. The first 8 checks are free, and each additional check costs $3. For what numbers of checks is the South Bank plan more expensive? (Assume that the student will always write more than 8 checks.)

89. *Digital Music.* The amount of money spent worldwide for streaming music services t years after 2010 can be approximated by

$$m(t) = 0.42t + 0.532,$$

where $m(t)$ is in billions of dollars. Determine (using an inequality) those years for which more than $3 billion will be spent for streaming music services.

Source: Based on data from Gartner

90. *Registered Nurses.* The number of registered nurses $R(t)$ employed in the United States, in millions, t years after 2000 can be approximated by

$$R(t) = 0.05t + 2.2.$$

Determine (using an inequality) those years for which more than 3 million registered nurses will be employed in the United States.

Source: Based on data from the U.S. Department of Health and Human Services and the Bureau of Labor Statistics

91. *Body Fat Percentage.* The function given by
$$F(d) = (4.95/d - 4.50) \times 100$$
can be used to estimate the body fat percentage $F(d)$ of a person with an average body density d, in kilograms per liter.

a) A man is considered obese if his body fat percentage is at least 25%. Find the body densities of an obese man.

b) A woman is considered obese if her body fat percentage is at least 32%. Find the body densities of an obese woman.

92. *Temperature Conversion.* The function
$$C(F) = \tfrac{5}{9}(F - 32)$$
can be used to find the Celsius temperature $C(F)$ that corresponds to $F°$ Fahrenheit.

a) Gold is solid at Celsius temperatures less than 1063°C. Find the Fahrenheit temperatures for which gold is solid.

b) Silver is solid at Celsius temperatures less than 960.8°C. Find the Fahrenheit temperatures for which silver is solid.

93. *Solid Waste.* The number of pounds of municipal solid waste generated per person per day in the United States is approximated by

$$m(t) = -0.03t + 4.72,$$

where *t* is the number of years after 2000. (See Example 7.) The amount of solid waste recovered for recycling or composting is approximated by

$$r(t) = 0.016t + 1.35,$$

where $r(t)$ is in number of pounds per person per day and *t* is the number of years after 2000. Determine (using an equality) those years for which more than half the solid waste generated will be recovered.

Source: Based on data from the U.S. Environmental Protection Agency

94. *Music.* The amount spent for digital music world-wide, in billions of dollars, can be approximated by

$$d(t) = 0.36t + 5.9,$$

where *t* is the number of years after 2010. The amount spent worldwide for music in physical for-mats, in billions of dollars, can be approximated by

$$p(t) = -t + 15,$$

where *t* is the number of years after 2010. Determine (using an inequality) those years for which more will be spent on digital music than on music in physical formats.

Source: Based on data from Gartner

95. *Manufacturing.* Bright Ideas is planning to make a new kind of lamp. Fixed costs are $90,000, and variable costs are $25 per lamp. The total-cost function for *x* lamps is

$$C(x) = 90,000 + 25x.$$

The company makes $48 in revenue for each lamp sold. The total-revenue function for *x* lamps is

$$R(x) = 48x.$$

a) When $R(x) < C(x)$, the company loses money. Find the values of *x* for which the company loses money.

b) When $R(x) > C(x)$, the company makes a profit. Find the values of *x* for which the company makes a profit.

96. *Publishing.* The demand and supply functions for a locally produced poetry book are approximated by

$$D(p) = 2000 - 60p \quad \text{and}$$
$$S(p) = 460 + 94p,$$

where *p* is the price, in dollars.

a) Find those values of *p* for which demand exceeds supply.

b) Find those values of *p* for which demand is less than supply.

97. How is the solution of $x + 3 = 8$ related to the solution sets of

$$x + 3 > 8 \quad \text{and} \quad x + 3 < 8?$$

98. Why isn't roster notation used to write solutions of inequalities?

Skill Review

Solve.

99. $x - (9 - x) = -3(x + 5)$ [1.3]

100. $\frac{2}{3}y - 1 = \frac{1}{4}$ [1.3]

101. $2x - 3y = 5,$
  $x + 3y = -1$ [3.2]

102. $4x - y = 1,$
  $y = 7 - x$ [3.2]

103. Solve $ar = b - cr$ for *r*. [1.5]

104. Solve $y = \dfrac{a + bn}{t}$ for *n*. [1.5]

Synthesis

105. The amount of waste recovered cannot exceed the amount of waste generated. How should the domain of the functions in Exercise 93 be adjusted to reflect this?

106. Explain how the addition principle can be used to avoid ever needing to multiply or divide both sides of an inequality by a negative number.

Solve for x and y. Assume that a, b, c, d, and m are positive constants.

107. $3ax + 2x \geq 5ax - 4$; assume $a > 1$

108. $6by - 4y \leq 7by + 10$

109. $a(by - 2) \geq b(2y + 5)$; assume $a > 2$

110. $c(6x - 4) < d(3 + 2x)$; assume $3c > d$

111. $c(2 - 5x) + dx > m(4 + 2x)$; assume $5c + 2m < d$

112. $a(3 - 4x) + cx < d(5x + 2)$; assume $c > 4a + 5d$

Determine whether each statement is true or false. If false, give an example that shows this.

113. For any real numbers *a*, *b*, *c*, and *d*, if $a < b$ and $c < d$, then $a - c < b - d$.

114. For all real numbers *x* and *y*, if $x < y$, then $x^2 < y^2$.

115. Are the inequalities

$$x < 3 \quad \text{and} \quad x + \frac{1}{x} < 3 + \frac{1}{x}$$

equivalent? Why or why not?

116. Are the inequalities

$$x < 3 \quad \text{and} \quad 0 \cdot x < 0 \cdot 3$$

equivalent? Why or why not?

Solve. Then graph.

117. $x + 5 \le 5 + x$

118. $x + 8 < 3 + x$

119. $x^2 > 0$

120. Abriana rented a compact car for a business trip. At the time of rental, she was given the option of prepaying for an entire tank of gasoline at \$4.099 per gallon, or waiting until her return and paying \$7.34 per gallon for enough gasoline to fill the tank. If the tank holds 14 gal, how many gallons can she use and still save money by choosing the second option? (Assume that Abriana does not put any gasoline in the car.)

121. Refer to Exercise 120. If Abriana's rental car gets 30 mpg, how many miles must she drive in order to make the first option more economical?

122. *Fundraising.* Michelle is planning a fundraising dinner for Happy Hollow Children's Camp. The banquet facility charges a rental fee of \$1500, but will waive the rental fee if more than \$6000 is spent on catering. Michelle knows that 150 people will attend the dinner.

a) How much should each dinner cost in order for the rental fee to be waived?

b) For what costs per person will the total cost (including the rental fee) exceed \$6000?

123. Assume that the graphs of $y_1 = -\frac{1}{2}x + 5$, $y_2 = x - 1$, and $y_3 = 2x - 3$ are as shown below. Solve each of the following inequalities, referring only to the figure.

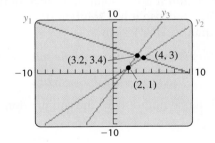

a) $-\frac{1}{2}x + 5 > x - 1$

b) $x - 1 \le 2x - 3$

c) $2x - 3 \ge -\frac{1}{2}x + 5$

124. Use a graphing calculator to check your answers to Exercises 23, 47, and 63.

YOUR TURN ANSWERS: SECTION 4.1

1. Yes   **2.** $\{t \mid t > 1\}$, or $(1, \infty)$

3. $\{n \mid n \le 14\}$, or $(-\infty, 14]$

4. $\{x \mid x < -5\}$, or $(-\infty, -5)$

5. $\{n \mid n > -\frac{5}{2}\}$, or $\left(-\frac{5}{2}, \infty\right)$

6. $\{x \mid x \le -\frac{7}{3}\}$, or $\left(-\infty, -\frac{7}{3}\right]$

7. For years after 2014

8. For monthly sales greater than \$32,500

PREPARE TO MOVE ON

Find the domain of f. [2.2]

1. $f(x) = \dfrac{5}{x}$

2. $f(x) = \dfrac{x + 3}{5x - 7}$

3. $f(x) = \dfrac{x + 10}{8}$

4. $f(x) = \dfrac{3}{x} + 5$

<table>
<tr><td>**4.2**</td><td>**Intersections, Unions, and Compound Inequalities**</td></tr>
</table>

Intersections of Sets and Conjunctions of Sentences   ▪   Unions of Sets and Disjunctions of Sentences   ▪
Interval Notation and Domains

Two inequalities joined by the word "and" or the word "or" are called **compound inequalities**. To solve compound inequalities, we must know how sets can be combined.

Intersections of Sets and Conjunctions of Sentences

The **intersection** of two sets A and B is the set of all elements that are common to both A and B. We denote the intersection of sets A and B as

$$A \cap B.$$

The intersection of two sets is represented by the purple region shown in the figure at left. For example, if $A = \{$all students who are taking a math class$\}$ and $B = \{$all students who are taking a history class$\}$, then $A \cap B = \{$all students who are taking a math class *and* a history class$\}$.

EXAMPLE 1 Find the intersection: $\{1, 2, 3, 4, 5\} \cap \{-2, -1, 0, 1, 2, 3\}$.

SOLUTION The numbers 1, 2, and 3 are common to both sets, so the intersection is $\{1, 2, 3\}$.

YOUR TURN

When two or more sentences are joined by the word *and* to make a compound sentence, the new sentence is called a **conjunction** of the sentences. The following is a conjunction of inequalities:

$$-2 < x \quad and \quad x < 1.$$

A number is a solution of a conjunction if it is a solution of *both* of the separate parts. For example, -1 is a solution because it is a solution of $-2 < x$ as well as $x < 1$; that is, -1 is *both* greater than -2 *and* less than 1.

The solution set of a conjunction is the intersection of the solution sets of the individual sentences.

EXAMPLE 2 Graph and write interval notation for the conjunction

$$-2 < x \quad and \quad x < 1.$$

SOLUTION We first graph $-2 < x$, then $x < 1$, and finally the conjunction $-2 < x$ and $x < 1$.

$\{x \mid -2 < x\}$ ⟨—————————————⟩ $(-2, \infty)$
  −7 −6 −5 −4 −3 −2 −1 0 1 2 3 4 5 6 7

$\{x \mid x < 1\}$ ⟨—————————————⟩ $(-\infty, 1)$
  −7 −6 −5 −4 −3 −2 −1 0 1 2 3 4 5 6 7

$\{x \mid -2 < x\} \cap \{x \mid x < 1\}$
$= \{x \mid -2 < x \text{ and } x < 1\}$ ⟨—————————————⟩ $(-2, 1)$
  −7 −6 −5 −4 −3 −2 −1 0 1 2 3 4 5 6 7

---

1. Find the intersection:

   $\{4, 5, 6, 7\} \cap \{2, 3, 5, 7\}$.

$A \cap B$

2. Graph and write interval notation for the conjunction

$$1 < x \quad and \quad x \le 4.$$

Study Skills

Two Books Are Better Than One

Many students find it helpful to use a second book as a reference when studying. Perhaps you or a friend own a text from a previous math course that can serve as a resource. Often professors have older texts that they will happily give away. Library book sales and thrift shops can also be excellent sources for extra books. Saving your text when you finish a math course can provide you with an excellent aid for your next course.

The solution set of the conjunction $-2 < x$ and $x < 1$ is the interval $(-2, 1)$. In set-builder notation, this is written $\{x \mid -2 < x < 1\}$, the set of all numbers that are *simultaneously* greater than -2 *and* less than 1.

YOUR TURN

For $a < b$,

$$a < x \quad and \quad x < b \quad \text{can be abbreviated} \quad a < x < b;$$

and, equivalently,

$$b > x \quad and \quad x > a \quad \text{can be abbreviated} \quad b > x > a.$$

MATHEMATICAL USE OF THE WORD "AND"

The word "and" corresponds to "intersection" and to the symbol "∩". Any solution of a conjunction must make each part of the conjunction true.

EXAMPLE 3 Solve and graph: $-1 \le 2x + 5 < 13$.

SOLUTION This inequality is an abbreviation for the conjunction

$$-1 \le 2x + 5 \quad and \quad 2x + 5 < 13.$$

The word *and* corresponds to set *intersection*. To solve the conjunction, we solve each inequality separately and then find the intersection of the solution sets:

$$-1 \le 2x + 5 \quad and \quad 2x + 5 < 13$$
$$-6 \le 2x \qquad and \qquad 2x < 8 \qquad \text{Subtracting 5 from both sides of each inequality}$$
$$-3 \le x \qquad and \qquad x < 4. \qquad \text{Dividing both sides of each inequality by 2}$$

The solution of the conjunction is the intersection of the separate solution sets.

$\{x \mid -3 \le x\}$

$[-3, \infty)$

$\{x \mid x < 4\}$

$(-\infty, 4)$

$\{x \mid -3 \le x\} \cap \{x \mid x < 4\}$
$= \{x \mid -3 \le x < 4\}$

$[-3, 4)$

3. Solve and graph:

$$-5 < 3x - 1 < 0.$$

We can now abbreviate the answer as $-3 \le x < 4$. The solution set is $\{x \mid -3 \le x < 4\}$, or, in interval notation, $[-3, 4)$.

YOUR TURN

The steps in Example 3 are often combined as follows:

$$-1 \le 2x + 5 < 13$$
$$-1 - 5 \le 2x + 5 - 5 < 13 - 5 \qquad \text{Subtracting 5 from all three regions}$$
$$-6 \le 2x < 8$$
$$-3 \le x < 4. \qquad \text{Dividing by 2 in all three regions}$$

CAUTION! The abbreviated form of a conjunction, like $-3 \le x < 4$, can be written only if both inequality symbols point in the same direction.

EXAMPLE 4   Solve and graph: $2x - 5 \geq -3$ *and* $5x + 2 \geq 17$.

SOLUTION   We first solve each inequality, retaining the word *and*:

$$2x - 5 \geq -3 \quad and \quad 5x + 2 \geq 17$$
$$2x \geq 2 \quad and \quad 5x \geq 15$$
$$x \geq 1 \quad and \quad x \geq 3.$$

Keep the word "and."

Next, we find the intersection of the two separate solution sets.

$\{x\,|\,x \geq 1\}$ — number line from -7 to 7, shaded from 1 to the right — $[1, \infty)$

$\{x\,|\,x \geq 3\}$ — number line from -7 to 7, shaded from 3 to the right — $[3, \infty)$

$\{x\,|\,x \geq 1\} \cap \{x\,|\,x \geq 3\}$
$= \{x\,|\,x \geq 3\}$ — number line from -7 to 7, shaded from 3 to the right — $[3, \infty)$

The numbers common to both sets are those greater than or equal to 3. Thus the solution set is $\{x\,|\,x \geq 3\}$, or, in interval notation, $[3, \infty)$. You should check that any number in $[3, \infty)$ satisfies the conjunction whereas numbers outside $[3, \infty)$ do not.

4. Solve and graph:

$$5x < 10 \quad and \quad x + 3 \leq 1.$$

YOUR TURN

Sometimes there is no way to solve both parts of a conjunction at once.

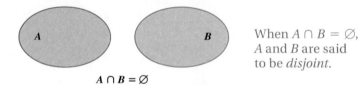

When $A \cap B = \varnothing$, A and B are said to be *disjoint*.

$A \cap B = \varnothing$

EXAMPLE 5   Solve and graph: $2x - 3 > 1$ *and* $3x - 1 < 2$.

SOLUTION   We solve each inequality separately:

$$2x - 3 > 1 \quad and \quad 3x - 1 < 2$$
$$2x > 4 \quad and \quad 3x < 3$$
$$x > 2 \quad and \quad x < 1.$$

The solution set is the intersection of the individual inequalities.

$\{x\,|\,x > 2\}$ — number line from -7 to 7, shaded from 2 to the right — $(2, \infty)$

$\{x\,|\,x < 1\}$ — number line from -7 to 7, shaded from 1 to the left — $(-\infty, 1)$

$\{x\,|\,x > 2\} \cap \{x\,|\,x < 1\}$
$= \{x\,|\,x > 2 \textit{ and } x < 1\} = \varnothing$ — number line from -7 to 7, no shading — $\varnothing$

5. Solve and graph:

$$x + 6 < 5 \quad and \quad 3x + 1 > 7.$$

Since no number is both greater than 2 and less than 1, the solution set is the empty set, $\varnothing$.

YOUR TURN

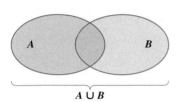

$A \cup B$

Unions of Sets and Disjunctions of Sentences

The **union** of two sets A and B is the collection of elements belonging to A and/or B. We denote the union of A and B by

$$A \cup B.$$

The union of two sets is often pictured as shown at left. For example, if $A = \{$all students who are taking a math class$\}$ and $B = \{$all students who are taking a history class$\}$, then $A \cup B = \{$all students who are taking a math class *or* a history class$\}$. Note that this set includes students who are taking a math class *and* a history class. Mathematically, the word "or" can be regarded as "and/or."

6. Find the union:

$\{4, 6, 8\} \cup \{7, 8, 9\}.$

_____

EXAMPLE 6   Find the union: $\{2, 3, 4\} \cup \{3, 5, 7\}$.

SOLUTION   The numbers in either or both sets are 2, 3, 4, 5, and 7, so the union is $\{2, 3, 4, 5, 7\}$.

YOUR TURN

When two or more sentences are joined by the word *or* to make a compound sentence, the new sentence is called a **disjunction** of the sentences. Here is an example:

$$x < -3 \quad or \quad x > 3.$$

Student Notes

Remember that the union or the intersection of two sets is itself a set and is written using set notation.

_____

A number is a solution of a disjunction if it is a solution of at least one of the separate parts. For example, -5 is a solution of this disjunction since -5 is a solution of $x < -3$.

The solution set of a disjunction is the union of the solution sets of the individual sentences.

EXAMPLE 7   Graph and write interval notation for the disjunction

$$x < -3 \quad or \quad x > 3.$$

SOLUTION   We graph $x < -3$, then $x > 3$, and finally $x < -3$ *or* $x > 3$.

$\{x \mid x < -3\}$ ⟵———————)—————————⟶ $(-\infty, -3)$
                        -6 -5 -4 -3 -2 -1 0 1 2 3 4 5 6

$\{x \mid x > 3\}$ ⟵—————————————(————⟶ $(3, \infty)$
                        -6 -5 -4 -3 -2 -1 0 1 2 3 4 5 6

$\{x \mid x < -3\} \cup \{x \mid x > 3\}$ ⟵———————)————(————⟶ $(-\infty, -3) \cup (3, \infty)$
$= \{x \mid x < -3 \text{ or } x > 3\}$ -6 -5 -4 -3 -2 -1 0 1 2 3 4 5 6

7. Graph and write interval notation for the disjunction

$x < -2 \quad or \quad x > 3.$

_____

The solution set of $x < -3$ *or* $x > 3$ is $\{x \mid x < -3 \text{ or } x > 3\}$, or, in interval notation, $(-\infty, -3) \cup (3, \infty)$. There is no simpler way to write the solution.

YOUR TURN

MATHEMATICAL USE OF THE WORD "OR"

The word "or" corresponds to "union" and to the symbol "∪". For a number to be a solution of a disjunction, it must be in *at least one* of the solution sets of the individual sentences.

EXAMPLE 8   Solve and graph: $7 + 3x < 3$ *or* $13 - 5x \leq 3$.

SOLUTION   We solve each inequality separately, retaining the word *or*:

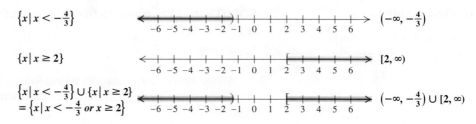

$$7 + 3x < 3 \quad or \quad 13 - 5x \leq 3$$

$$3x < -4 \quad or \quad -5x \leq -10$$

Keep the word "or."

Dividing by a negative number and reversing the symbol

$$x < -\tfrac{4}{3} \quad or \quad x \geq 2.$$

To find the solution set of the disjunction, we consider the individual graphs. We graph $x < -\frac{4}{3}$ and then $x \geq 2$. Then we take the union of these graphs.

$\left\{x \mid x < -\frac{4}{3}\right\}$ $\qquad\qquad$ $\left(-\infty, -\frac{4}{3}\right)$

$\left\{x \mid x \geq 2\right\}$ $\qquad\qquad$ $[2, \infty)$

$\left\{x \mid x < -\frac{4}{3}\right\} \cup \left\{x \mid x \geq 2\right\}$
$= \left\{x \mid x < -\frac{4}{3} \ or \ x \geq 2\right\}$ $\qquad$ $\left(-\infty, -\frac{4}{3}\right) \cup [2, \infty)$

8. Solve and graph:

$$2 - x > 1 \quad or \quad 4x - 9 > 7.$$

The solution set is $\left\{x \mid x < -\frac{4}{3} \ or \ x \geq 2\right\}$, or $\left(-\infty, -\frac{4}{3}\right) \cup [2, \infty)$.

YOUR TURN

CAUTION!   A compound inequality like

$$x < -4 \quad or \quad x \geq 2$$

cannot be expressed as $2 \leq x < -4$. Doing so would be to say that x is *simultaneously* less than -4 and greater than or equal to 2. No number is both less than -4 *and* greater than 2, but many are less than -4 *or* greater than 2.

EXAMPLE 9   Solve: $3x - 11 < 4$ *or* $4x + 9 \geq 1$.

SOLUTION   We solve the individual inequalities separately, retaining the word *or*:

$$3x - 11 < 4 \quad or \quad 4x + 9 \geq 1$$

$$3x < 15 \quad or \quad 4x \geq -8$$

$$x < 5 \quad or \quad x \geq -2.$$

Keep the word "or."

To find the solution set, we first look at the individual graphs.

$\left\{x \mid x < 5\right\}$ $\qquad\qquad$ $(-\infty, 5)$

$\left\{x \mid x \geq -2\right\}$ $\qquad\qquad$ $[-2, \infty)$

$\left\{x \mid x < 5\right\} \cup \left\{x \mid x \geq -2\right\}$
$= \left\{x \mid x < 5 \ or \ x \geq -2\right\}$ $\qquad$ $(-\infty, \infty) = \mathbb{R}$

9. Solve:

$$6x - 2 > 4 \quad or \quad 3x - 5 < 1.$$

Since *all* numbers are less than 5 or greater than or equal to -2, the two sets fill the entire number line. Thus the solution set is $\mathbb{R}$, the set of all real numbers.

YOUR TURN

Interval Notation and Domains

If $g(x) = \dfrac{5x - 2}{x - 3}$, then the number 3 is not in the domain of g. We can represent the domain of g using set-builder notation or interval notation.

EXAMPLE 10   Use interval notation to write the domain of g if $g(x) = \dfrac{5x - 2}{x - 3}$.

SOLUTION   The expression $\dfrac{5x - 2}{x - 3}$ is not defined when the denominator is 0. We set $x - 3$ equal to 0 and solve:

$$x - 3 = 0$$
$$x = 3. \qquad \text{The number 3 is } not \text{ in the domain.}$$

We have the domain of $g = \{x \,|\, x \text{ is a real number } and\ x \neq 3\}$. If we graph this set, we see that the domain can be written as a union of two intervals.

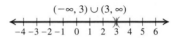

$(-\infty, 3) \cup (3, \infty)$

Thus the domain of $g = (-\infty, 3) \cup (3, \infty)$.

10. Use interval notation to write the domain of f if $f(x) = \dfrac{x}{2x + 1}$.

YOUR TURN

Only nonnegative numbers have square roots that are real numbers. Thus finding the domain of a radical function often involves solving an inequality.

EXAMPLE 11   Find the domain of f if $f(x) = \sqrt{7 - x}$.

SOLUTION   In order for $\sqrt{7 - x}$ to exist as a real number, $7 - x$ must be nonnegative. Thus we solve $7 - x \geq 0$:

$$7 - x \geq 0 \qquad 7 - x \text{ must be nonnegative.}$$
$$-x \geq -7 \qquad \text{Subtracting 7 from both sides}$$
$$\qquad\qquad \text{The symbol must be reversed.}$$
$$x \leq 7. \qquad \text{Multiplying both sides by } -1$$

For $x \leq 7$, we have $7 - x \geq 0$. Thus the domain of f is $\{x \,|\, x \leq 7\}$, or $(-\infty, 7]$.

11. Find the domain of g if $g(x) = \sqrt{x + 3}$.

YOUR TURN

4.2   EXERCISE SET

FOR EXTRA HELP

MyMathLab® MathXL
PRACTICE   WATCH   READ   REVIEW

Vocabulary and Reading Check

Complete each statement using the word intersection *or the word* union.

1. The _____ of two sets is the set of all elements that are in both sets.

2. The symbol $\cup$ indicates _____.

3. The word "and" corresponds to _____.

4. The symbol $\cap$ indicates _____.

5. The solution of a disjunction is the _____ of the solution sets of the individual sentences.

6. The _____ of two sets is the set of all elements that are in either set or in both sets.

Concept Reinforcement

In each of Exercises 7–16, match the set with the most appropriate choice below.

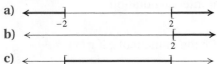

i) $\mathbb{R}$

j) $\varnothing$

7. ____ $\{x \mid x < -2 \text{ or } x > 2\}$

8. ____ $\{x \mid x < -2 \text{ and } x > 2\}$

9. ____ $\{x \mid x > -2\} \cap \{x \mid x < 2\}$

10. ____ $\{x \mid x \le -2\} \cup \{x \mid x \ge 2\}$

11. ____ $\{x \mid x \le -2\} \cup \{x \mid x \le 2\}$

12. ____ $\{x \mid x \le -2\} \cap \{x \mid x \le 2\}$

13. ____ $\{x \mid x \ge -2\} \cap \{x \mid x \ge 2\}$

14. ____ $\{x \mid x \ge -2\} \cup \{x \mid x \ge 2\}$

15. ____ $\{x \mid x \le 2\} \text{ and } \{x \mid x \ge -2\}$

16. ____ $\{x \mid x \le 2\} \text{ or } \{x \mid x \ge -2\}$

Intersections of Sets and Unions of Sets

Find each indicated intersection or union.

17. $\{2, 4, 16\} \cap \{4, 16, 256\}$

18. $\{1, 2, 4\} \cup \{4, 6, 8\}$

19. $\{0, 5, 10, 15\} \cup \{5, 15, 20\}$

20. $\{2, 5, 9, 13\} \cap \{5, 8, 10\}$

21. $\{a, b, c, d, e, f\} \cap \{b, d, f\}$

22. $\{u, v, w\} \cup \{u, w\}$

23. $\{x, y, z\} \cup \{u, v, x, y, z\}$

24. $\{m, n, o, p\} \cap \{m, o, p\}$

25. $\{3, 6, 9, 12\} \cap \{5, 10, 15\}$

26. $\{1, 5, 9\} \cup \{4, 6, 8\}$

27. $\{1, 3, 5\} \cup \varnothing$

28. $\{1, 3, 5\} \cap \varnothing$

Conjunctions of Sentences and Disjunctions of Sentences

Graph and write interval notation for each compound inequality.

29. $1 < x < 3$

30. $0 \le y \le 5$

31. $-6 \le y \le 0$

32. $-8 < x \le -2$

33. $x < -1 \text{ or } x > 4$

34. $x < -5 \text{ or } x > 1$

35. $x \le -2 \text{ or } x > 1$

36. $x \le -5 \text{ or } x > 2$

37. $-4 \le -x < 2$

38. $x > -7 \text{ and } x < -2$

39. $x > -2 \text{ and } x < 4$

40. $3 > -x \ge -1$

41. $5 > a \text{ or } a > 7$

42. $t \ge 2 \text{ or } -3 > t$

43. $x \ge 5 \text{ or } -x \ge 4$

44. $-x < 3 \text{ or } x < -6$

45. $7 > y \text{ and } y \ge -3$

46. $6 > -x \ge 0$

47. $-x < 7 \text{ and } -x \ge 0$

48. $x \ge -3 \text{ and } x < 3$

Aha! 49. $t < 2 \text{ or } t < 5$

50. $t > 4 \text{ or } t > -1$

Solve and graph each solution set.

51. $-3 \le x + 2 < 9$

52. $-1 < x - 3 < 5$

53. $0 < t - 4 \text{ and } t - 1 \le 7$

54. $-6 \le t + 1 \text{ and } t + 8 < 2$

55. $-7 \le 2a - 3 \text{ and } 3a + 1 < 7$

Aha! 56. $-4 \le 3n + 5 \text{ and } 2n - 3 \le 7$

57. $x + 3 \le -1 \text{ or } x + 3 > -2$

58. $x + 5 < -3 \text{ or } x + 5 \ge 4$

59. $-10 \le 3x - 1 \le 5$

60. $-18 \le 4x + 2 \le 30$

61. $5 > \dfrac{x - 3}{4} > 1$

62. $3 \ge \dfrac{x - 1}{2} \ge -4$

63. $-2 \le \dfrac{x + 2}{-5} \le 6$

64. $-10 \le \dfrac{x + 6}{-3} \le -8$

65. $2 \le f(x) \le 8$, where $f(x) = 3x - 1$

66. $7 \ge g(x) \ge -2$, where $g(x) = 3x - 5$

67. $-21 \le f(x) < 0$, where $f(x) = -2x - 7$

68. $4 > g(t) \ge 2$, where $g(t) = -3t - 8$

69. $f(t) < 3 \ or \ f(t) > 8$, where $f(t) = 5t + 3$

70. $g(x) \leq -2 \ or \ g(x) \geq 10$, where $g(x) = 3x - 5$

71. $6 > 2a - 1 \ or \ -4 \leq -3a + 2$

72. $3a - 7 > -10 \ or \ 5a + 2 \leq 22$

73. $a + 3 < -2 \ and \ 3a - 4 < 8$

74. $1 - a < -2 \ and \ 2a + 1 > 9$

75. $3x + 2 < 2 \ and \ 3 - x < 1$

76. $2x - 1 > 5 \ and \ 2 - 3x > 11$

77. $2t - 7 \leq 5 \ or \ 5 - 2t > 3$

78. $5 - 3a \leq 8 \ or \ 2a + 1 > 7$

Interval Notation and Domains

For $f(x)$ as given, use interval notation to write the domain of f.

79. $f(x) = \dfrac{9}{x + 6}$

80. $f(x) = \dfrac{2}{x - 5}$

81. $f(x) = \dfrac{1}{x}$

82. $f(x) = -\dfrac{6}{x}$

83. $f(x) = \dfrac{x + 3}{2x - 8}$

84. $f(x) = \dfrac{x - 1}{3x + 6}$

85. $f(x) = \sqrt{x - 10}$

86. $f(x) = \sqrt{x + 2}$

87. $f(x) = \sqrt{3 - x}$

88. $f(x) = \sqrt{11 - x}$

89. $f(x) = \sqrt{2x + 7}$

90. $f(x) = \sqrt{8 - 5x}$

91. $f(x) = \sqrt{8 - 2x}$

92. $f(x) = \sqrt{2x - 10}$

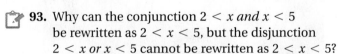 **93.** Why can the conjunction $2 < x \ and \ x < 5$ be rewritten as $2 < x < 5$, but the disjunction $2 < x \ or \ x < 5$ cannot be rewritten as $2 < x < 5$?

94. Can the solution set of a disjunction be empty? Why or why not?

Skill Review

Simplify.

95. $-\frac{2}{15}\left(-\frac{5}{8}\right)$ [1.2]

96. $-3.85 + 4.2$ [1.2]

97. $-2 - 6^2 \div 4(-3) - (8 - 12)$ [1.2]

98. $3(6 - w) - [9 - 2(w - 3)]$ [1.3]

Synthesis

99. What can you conclude about a, b, c, and d, if $[a, b] \cup [c, d] = [a, d]$? Why?

 100. What can you conclude about a, b, c, and d, if $[a, b] \cap [c, d] = [a, b]$? Why?

101. Use the following graph of $f(x) = 2x - 5$ to solve $-7 < 2x - 5 < 7$.

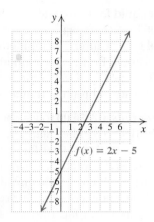

102. Use the following graph of $g(x) = 4 - x$ to solve $4 - x < -2 \ or \ 4 - x > 7$.

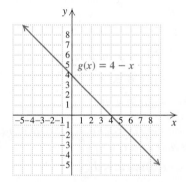

103. *Pressure at Sea Depth.* The function given by

$$P(d) = 1 + \frac{d}{33}$$

gives the pressure, in atmospheres (atm), at a depth of d feet in the sea. For what depths d is the pressure at least 1 atm and at most 7 atm?

104. *Converting Dress Sizes.* The function given by

$$f(x) = 2(x + 10)$$

can be used to convert dress sizes x in the United States to dress sizes $f(x)$ in Italy. For what dress sizes in the United States will dress sizes in Italy be between 32 and 46?

105. *Body Fat Percentage.* The function given by

$$F(d) = (4.95/d - 4.50) \times 100$$

can be used to estimate the body fat percentage $F(d)$ of a person with an average body density d, in kilograms per liter. A woman's body fat percentage is considered healthy if $25 \leq F(d) \leq 31$. What body densities are considered healthy for a woman?

106. *Temperatures of Liquids.* The formula

$$C = \tfrac{5}{9}(F - 32)$$

is used to convert Fahrenheit temperatures F to Celsius temperatures C.

a) Gold is liquid for Celsius temperatures C such that $1063° \le C < 2660°$. Find a comparable inequality for Fahrenheit temperatures.

b) Silver is liquid for Celsius temperatures C such that $960.8° \le C < 2180°$. Find a comparable inequality for Fahrenheit temperatures.

107. *Minimizing Tolls.* A $6.00 toll is charged to cross the bridge to Sanibel Island from mainland Florida. A six-month reduced-fare pass costs $50 and reduces the toll to $2.00. A six-month unlimited-trip pass costs $300 and allows for free crossings. How many crossings in six months does it take for the reduced-fare pass to be the more economical choice?

Source: www.leewayinfo.com

Solve and graph.

108. $4a - 2 \le a + 1 \le 3a + 4$

109. $4m - 8 > 6m + 5$ *or* $5m - 8 < -2$

110. $x - 10 < 5x + 6 \le x + 10$

111. $3x < 4 - 5x < 5 + 3x$

Determine whether each sentence is true or false for all real numbers a, b, and c.

112. If $-b < -a$, then $a < b$.

113. If $a \le c$ and $c \le b$, then $b > a$.

114. If $a < c$ and $b < c$, then $a < b$.

115. If $-a < c$ and $-c > b$, then $a > b$.

For f(x) as given, use interval notation to write the domain of f.

116. $f(x) = \dfrac{\sqrt{5 + 2x}}{x - 1}$

117. $f(x) = \dfrac{\sqrt{3 - 4x}}{x + 7}$

118. For $f(x) = \sqrt{x - 5}$ and $g(x) = \sqrt{9 - x}$, use interval notation to write the domain of $f + g$.

119. Let $y_1 = -1$, $y_2 = 2x + 5$, and $y_3 = 13$. Then use the graphs of y_1, y_2, and y_3 to check the solution to Example 3.

120. Let $y_1 = 3x - 11$, $y_2 = 4$, $y_3 = 4x + 9$, and $y_4 = 1$. Then use the graphs of y_1, y_2, y_3, and y_4 to check the solution to Example 9.

Readability. The reading difficulty of a textbook can be estimated by the Flesch Reading Ease Formula

$$r = 206.835 - 1.015n - 84.6s,$$

where r is the reading ease, n is the average number of words in a sentence, and s is the average number of syllables in a word. Sample reading-level scores are shown in the following table. Use this information for Exercises 121 and 122.

Score	Reading Ease
$90 \le r \le 100$	5th grade
$60 \le r \le 70$	8th and 9th grades
$0 \le r \le 30$	College graduates

Source: readabilityformulas.com

121. Bryan is writing a book for 5th-graders using an average of 1.2 syllables per word. How long should his average sentence length be?

122. The reading score for Alexa's new book for young adults indicates that it should be read with ease by 8th- and 9th-graders. If she averages 8 words per sentence, what is the average number of syllables per word?

123. A machine filling water bottles pours 16 oz of water into each bottle, with a margin of error of 0.1 oz. Write an inequality and interval notation for the amount of water that the machine pours into a bottle.

124. At one point during the 2012 presidential campaign, a Gallup poll indicated that Barack Obama had an approval rating of 46%, with a margin of error of 1%. Write an inequality and interval notation for Obama's approval rating.

Source: www.upi.com

125. Use a graphing calculator to check your answers to Exercises 51–54 and Exercises 69–72.

126. On many graphing calculators, the TEST key provides access to inequality symbols, while the LOGIC option of that same key accesses the conjunction *and* and the disjunction *or*. Thus, if $y_1 = x > -2$ and $y_2 = x < 4$, Exercise 39 can be checked by forming the expression $y_3 = y_1$ *and* y_2. The interval(s) in the solution set appears as a horizontal line 1 unit above the x-axis. (Be careful to "deselect" y_1 and y_2 so that only y_3 is drawn.) Use the TEST key to check Exercises 41, 45, 47, and 49.

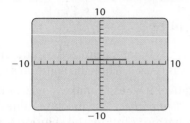

127. Use a graphing calculator to confirm the domains of the functions in Exercises 85, 87, and 91.

128. *Research.* Find a formula for body mass index (BMI), and find the range for which your BMI would be considered healthy. For your height, what weights will result in an acceptable BMI?

129. *Research.* Find what a "95% confidence interval" means, and explain it in writing or to your class.

QUICK QUIZ: SECTIONS 4.1–4.2

Solve. Write the solution set using both set-builder notation and interval notation.

1. $5 - 6x < x + 3$ [4.1]

2. $x - (9 - x) \ge 3(7 - x)$ [4.1]

3. $-\frac{2}{3}m - 5 > 7$ [4.1]

4. $3 > 7 - 2y$ *or* $6y < y$ [4.2]

5. $-1 < 7 - x < 4$ [4.2]

PREPARE TO MOVE ON

Find the absolute value.   [1.2]

1. $\left|\frac{2}{3}\right|$ **2.** $|-16|$ **3.** $|0|$ **4.** $|8 - 15|$

5. Given that $f(x) = 3x - 10$, find all x for which $f(x) = 8$.   [2.2]

4.3   Absolute-Value Equations and Inequalities

Equations with Absolute Value   ▪   Inequalities with Absolute Value

Equations with Absolute Value

The following is a formal definition of absolute value.

Study Skills

Understand Your Mistakes

When a graded quiz, test, or assignment is returned, it is important to review and understand your mistakes. Don't simply file away old papers without first understanding how to correct your mistakes.

ABSOLUTE VALUE

The absolute value of x, denoted $|x|$, is defined as

$$|x| = \begin{cases} x, & \text{if } x \ge 0, \\ -x, & \text{if } x < 0. \end{cases}$$

(When x is nonnegative, the absolute value of x is x. When x is negative, the absolute value of x is the opposite of x.)

To better understand this definition, suppose x is -5. Then $|x| = |-5| = 5$, and 5 is the opposite of -5. This shows that when x represents a negative number, the absolute value of x is the opposite of x (which is positive).

Since distance is always nonnegative, we can think of a number's absolute value as its distance from zero on the number line.

EXAMPLE 1   Find each solution set: **(a)** $|x| = 4$; **(b)** $|x| = 0$; **(c)** $|x| = -7$.

SOLUTION

a) We interpret $|x| = 4$ to mean that the number x is 4 units from zero on the number line. There are two such numbers, 4 and -4. Thus the solution set is $\{-4, 4\}$.

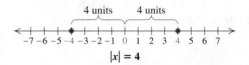

$$|x| = 4$$

b) We interpret $|x| = 0$ to mean that x is 0 units from zero on the number line. The only number that satisfies this is 0 itself. Thus the solution set is $\{0\}$.

c) Since distance is always nonnegative, it doesn't make sense to talk about a number that is -7 units from zero. *Remember*: The absolute value of a number is never negative. Thus, $|x| = -7$ has no solution; the solution set is $\varnothing$.

1. Find the solution set: $|x| = 6$.

 YOUR TURN

Example 1 leads us to the following principle for solving equations.

THE ABSOLUTE-VALUE PRINCIPLE FOR EQUATIONS

For any positive number p and any algebraic expression X:

a) The solutions of $|X| = p$ are those numbers that satisfy

$$X = -p \quad or \quad X = p.$$

b) The equation $|X| = 0$ is equivalent to the equation $X = 0$.
c) The equation $|X| = -p$ has no solution.

EXAMPLE 2   Find each solution set: **(a)** $|2x + 5| = 13$; **(b)** $|4 - 7x| = -8$.

SOLUTION

a) We use the absolute-value principle, knowing that $2x + 5$ is either 13 or -13:

$$|X| = p$$
$$|2x + 5| = 13 \qquad \text{Substituting}$$
$$2x + 5 = -13 \quad or \quad 2x + 5 = 13$$
$$2x = -18 \quad or \qquad 2x = 8$$
$$x = -9 \quad or \qquad x = 4.$$

Check: For -9:

$$\frac{|2x + 5| = 13}{\begin{array}{c|c} |2(-9) + 5| & 13 \\ |-18 + 5| & \\ |-13| & \\ & 13 \overset{?}{=} 13 \quad \text{TRUE} \end{array}}$$

For 4:

$$\frac{|2x + 5| = 13}{\begin{array}{c|c} |2 \cdot 4 + 5| & 13 \\ |8 + 5| & \\ |13| & \\ & 13 \overset{?}{=} 13 \quad \text{TRUE} \end{array}}$$

The number $2x + 5$ is 13 units from zero if x is replaced with -9 or 4. The solution set is $\{-9, 4\}$.

2. Find the solution set:

$$|3x - 5| = 7.$$

b) The absolute-value principle reminds us that absolute value is never negative. The equation $|4 - 7x| = -8$ has no solution. The solution set is $\varnothing$.

 YOUR TURN

To use the absolute-value principle, we must be sure that the absolute-value expression is alone on one side of the equation.

EXAMPLE 3   Given that $f(x) = 2|x + 3| + 1$, find all x for which $f(x) = 15$.

SOLUTION   Since we are looking for $f(x) = 15$, we substitute:

$$f(x) = 15$$
$$2|x + 3| + 1 = 15 \qquad \text{Replacing } f(x) \text{ with } 2|x + 3| + 1$$
$$2|x + 3| = 14 \qquad \text{Subtracting 1 from both sides}$$
$$|x + 3| = 7 \qquad \text{Dividing both sides by 2}$$
$$x + 3 = -7 \quad or \quad x + 3 = 7 \qquad \text{Using the absolute-value principle} \\ \text{for equations}$$
$$x = -10 \quad or \qquad x = 4.$$

We leave it to the student to check that $f(-10) = f(4) = 15$. The solution set is $\{-10, 4\}$.

3. Given that $g(x) = 2|5x| - 4$, find all x for which $g(x) = 10$.

YOUR TURN

EXAMPLE 4   Solve: $|x - 2| = 3$.

SOLUTION   Because this is of the form $|a - b| = c$, it can be solved in two ways.

Method 1.   We interpret $|x - 2| = 3$ as stating that the number $x - 2$ is 3 units from zero. Using the absolute-value principle, we replace X with $x - 2$ and p with 3:

$$|X| = p$$
$$|x - 2| = 3 \qquad \text{We use this approach in Examples 1–3.}$$
$$x - 2 = -3 \quad or \quad x - 2 = 3 \qquad \text{Using the absolute-value principle}$$
$$x = -1 \quad or \qquad x = 5.$$

> **CAUTION!**   There are two solutions of $|x - 2| = 3$. Simply solving $x - 2 = 3$ will yield only one of those solutions.

Method 2.   The expressions $|a - b|$ and $|b - a|$ both represent the *distance between* a and b on the number line. For example, the distance between 7 and 8 is given by $|8 - 7|$ or $|7 - 8|$. From this viewpoint, the equation $|x - 2| = 3$ states that the distance between x and 2 is 3 units. We draw the number line and locate all numbers that are 3 units from 2.

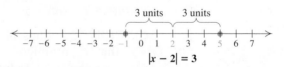

The solutions of $|x - 2| = 3$ are -1 and 5.

Check: The check consists of observing that both methods give the same solutions. The solution set is $\{-1, 5\}$.

4. Solve: $|x - 5| = 1$.

YOUR TURN

Some equations contain two absolute-value expressions. Consider $|a| = |b|$. This means that a and b are the same distance from zero. If a and b are the same distance from zero, they are either the same number or opposites.

For any algebraic expressions X and Y:

If $|X| = |Y|$,   then $X = Y$ or $X = -Y$.

EXAMPLE 5   Solve: $|2x - 3| = |x + 5|$.

SOLUTION   The equation tells us that $2x - 3$ and $x + 5$ are the same distance from zero. This means that they are either the same number or opposites:

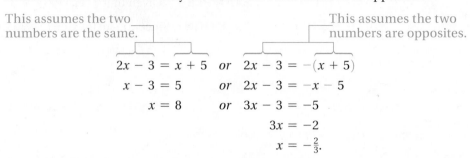

This assumes the two numbers are the same.   This assumes the two numbers are opposites.

$$2x - 3 = x + 5 \quad or \quad 2x - 3 = -(x + 5)$$
$$x - 3 = 5 \quad or \quad 2x - 3 = -x - 5$$
$$x = 8 \quad or \quad 3x - 3 = -5$$
$$3x = -2$$
$$x = -\tfrac{2}{3}.$$

The check is left to the student. The solutions are 8 and $-\tfrac{2}{3}$, and the solution set is $\left\{ -\tfrac{2}{3}, 8 \right\}$.

5. Solve: $|4x - 3| = |3x + 5|$.

YOUR TURN

Inequalities with Absolute Value

Our methods for solving equations with absolute value can be adapted for solving inequalities.

EXAMPLE 6   Solve $|x| < 4$. Then graph.

SOLUTION   The solutions of $|x| < 4$ are all numbers whose *distance from zero is less than* 4. By substituting or by looking at the number line, we can see that $-3, -2, -1, -\tfrac{1}{2}, -\tfrac{1}{4}, 0, \tfrac{1}{4}, \tfrac{1}{2}, 1, 2,$ and 3 are all solutions. In fact, all numbers between -4 and 4 are solutions. The solution set is $\{x | -4 < x < 4\}$, or, in interval notation, $(-4, 4)$. The graph is as follows:

$|x| < 4$

6. Solve $|x| < 2$. Then graph.

YOUR TURN

EXAMPLE 7   Solve $|x| \geq 4$. Then graph.

SOLUTION   Solutions of $|x| \geq 4$ are numbers that are at least 4 units from zero—that is, numbers x for which $x \leq -4$ *or* $4 \leq x$. The solution set is $\{x | x \leq -4 \, or \, x \geq 4\}$. In interval notation, the solution set is $(-\infty, -4] \cup [4, \infty)$. We can check mentally with numbers like $-4.1, -5, 4.1,$ and 5. The graph is as follows:

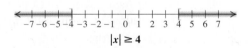

$|x| \geq 4$

7. Solve $|x| \geq 2$. Then graph.

YOUR TURN

    Examples 1, 6, and 7 illustrate three situations in which absolute-value symbols appear. The principles for finding solutions follow.

Student Notes

You may be familiar with the following form of the principles for solving absolute-value inequalities.

If $|X| < p$, then

$$X < p \quad and \quad X > -p.$$

If $|X| > p$, then

$$X > p \quad or \quad X < -p.$$

These statements are equivalent to those stated in the text.

PRINCIPLES FOR SOLVING ABSOLUTE-VALUE PROBLEMS

For any positive number p and any expression X:

a) The solutions of $|X| = p$ are those numbers that satisfy

$$X = -p \quad or \quad X = p.$$

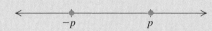

b) The solutions of $|X| < p$ are those numbers that satisfy

$$-p < X < p.$$

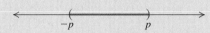

c) The solutions of $|X| > p$ are those numbers that satisfy

$$X < -p \quad or \quad p < X.$$

If p is negative, any value of X will satisfy the inequality $|X| > p$ because absolute value is never negative. Thus, $|2x - 7| > -3$ is true for any real number x, and the solution set is $\mathbb{R}$.

If p is not positive, the inequality $|X| < p$ has no solution. Thus, $|2x - 7| < -3$ has no solution, and the solution set is $\varnothing$.

EXPLORING 🔍 THE CONCEPT

We can solve several equations and inequalities by examining the graphs of $f(x) = |x|$ and $g(x) = 3$.

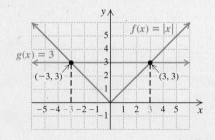

- Where the graph of $f(x)$ intersects the graph of $g(x)$, $|x| = 3$.
- Where the graph of $f(x)$ lies below the graph of $g(x)$, $|x| < 3$.
- Where the graph of $f(x)$ lies above the graph of $g(x)$, $|x| > 3$.

Use the graphs above to match each equation or inequality to its solution set.

1. $|x| = 3$
2. $|x| < 3$
3. $|x| > 3$
4. $|x| \leq 3$
5. $|x| \geq 3$

a) $(-3, 3)$
b) $[-3, 3]$
c) $(-\infty, -3) \cup (3, \infty)$
d) $(-\infty, -3] \cup [3, \infty)$
e) $\{-3, 3\}$

ANSWERS
1. (e)
2. (a)
3. (c)
4. (b)
5. (d)

EXAMPLE 8   Solve $|3x - 2| < 4$. Then graph.

SOLUTION   The number $3x - 2$ must be less than 4 units from zero. This is of the form $|X| < p$, so part (b) of the principles listed above applies:

$$|X| < p \qquad \text{This corresponds to } -p < X < p.$$
$$|3x - 2| < 4 \qquad \text{Replacing } X \text{ with } 3x - 2 \text{ and } p \text{ with } 4$$
$$-4 < 3x - 2 < 4 \qquad \text{The number } 3x - 2 \text{ must be within 4 units of zero.}$$
$$-2 < \quad 3x \quad < 6 \qquad \text{Adding 2}$$
$$-\tfrac{2}{3} < \quad x \quad < 2. \qquad \text{Multiplying by } \tfrac{1}{3}$$

The solution set is $\left\{x \mid -\tfrac{2}{3} < x < 2\right\}$, or, in interval notation, $\left(-\tfrac{2}{3}, 2\right)$. The graph is as follows:

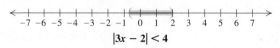

$|3x - 2| < 4$

8. Solve $|8x + 5| < 13$. Then graph.

_____

YOUR TURN

EXAMPLE 9   Given that $f(x) = |4x + 2|$, find all x for which $f(x) \geq 6$.

SOLUTION   We have

$$f(x) \geq 6,$$
or $\qquad |4x + 2| \geq 6. \qquad$ Substituting

To solve, we use part (c) of the principles listed above. Here, X is $4x + 2$ and p is 6:

$$|X| \geq p \qquad \text{This corresponds to } X < -p \text{ or } p < X.$$
$$|4x + 2| \geq 6 \qquad \text{Replacing } X \text{ with } 4x + 2 \text{ and } p \text{ with } 6$$
$$4x + 2 \leq -6 \quad or \quad 6 \leq 4x + 2 \qquad \text{The number } 4x + 2 \text{ must be at least 6 units from zero.}$$
$$4x \leq -8 \quad or \quad 4 \leq 4x \qquad \text{Adding } -2$$
$$x \leq -2 \quad or \quad 1 \leq x. \qquad \text{Multiplying by } \tfrac{1}{4}$$

The solution set is $\{x \mid x \leq -2 \text{ or } x \geq 1\}$, or $(-\infty, -2] \cup [1, \infty)$. The graph is as follows:

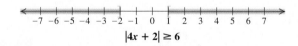

$|4x + 2| \geq 6$

9. Given that $g(x) = |2x - 5|$, find all x for which $g(x) > 7$.

_____

YOUR TURN

Technology Connection

To enter an absolute-value function on a graphing calculator, we press **MATH** and use the ABS(option in the NUM menu. To solve $|4x + 2| = 6$, we graph $y_1 = |4x + 2|$ and $y_2 = 6$.

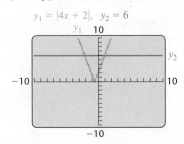

Using the INTERSECT option of the CALC menu, we find that the graphs intersect at $(-2, 6)$ and $(1, 6)$. The x-coordinates -2 and 1 are the solutions. To solve $|4x + 2| \geq 6$, note where the graph of y_1 is *on or above* the line $y = 6$. The corresponding x-values are the solutions of the inequality.

1. How can the same graph be used to solve $|4x + 2| < 6$?
2. Solve Example 8.
3. Use a graphing calculator to show that $|4x + 2| = -6$ has no solution.

4.3 EXERCISE SET

FOR EXTRA HELP

MyMathLab® Math XL
PRACTICE   WATCH   READ   REVIEW

Vocabulary and Reading Check

Classify each of the following statements as either true or false.

1. $|x|$ is never negative.
2. $|x|$ is always positive.
3. If x is negative, then $|x| = -x$.
4. The number a is $|a|$ units from 0.
5. The distance between a and b is $|a - b|$.
6. There are two solutions of $|3x - 8| = 17$.
7. There is no solution of $|4x + 9| > -5$.
8. All real numbers are solutions of $|2x - 7| < -3$.

Concept Reinforcement

Match each equation or inequality with an equivalent statement from the column on the right. Letters may be used more than once or not at all.

9. $|x - 3| = 5$ a) The solution set is $\varnothing$.
10. $|x - 3| < 5$ b) The solution set is $\mathbb{R}$.
11. $|x - 3| > 5$ c) $x - 3 > 5$
12. $|x - 3| < -5$ d) $x - 3 < -5$ or $x - 3 > 5$
13. $|x - 3| = -5$ e) $x - 3 = 5$
14. $|x - 3| > -5$ f) $x - 3 < 5$
                        g) $x - 3 = -5$ or $x - 3 = 5$
                        h) $-5 < x - 3 < 5$

Equations with Absolute Value

Solve.

15. $|x| = 10$ 16. $|x| = 5$
Aha! 17. $|x| = -1$ 18. $|x| = -8$
19. $|p| = 0$ 20. $|y| = 7.3$
21. $|2x - 3| = 4$ 22. $|5x + 2| = 7$
23. $|3x + 5| = -8$ 24. $|7x - 2| = -9$
25. $|x - 2| = 6$ 26. $|x - 3| = 11$
27. $|x - 7| = 1$ 28. $|x - 4| = 5$
29. $|t| + 1.1 = 6.6$ 30. $|m| + 3 = 3$
31. $|5x| - 3 = 37$ 32. $|2y| - 5 = 13$
33. $7|q| + 2 = 9$ 34. $5|z| + 2 = 17$

35. $\left|\dfrac{2x - 1}{3}\right| = 4$ 36. $\left|\dfrac{4 - 5x}{6}\right| = 3$
37. $|5 - m| + 9 = 16$
38. $|t - 7| + 1 = 4$
39. $5 - 2|3x - 4| = -5$
40. $3|2x - 5| - 7 = -1$
41. Let $f(x) = |2x + 6|$. Find all x for which $f(x) = 8$.
42. Let $f(x) = |2x - 4|$. Find all x for which $f(x) = 10$.
43. Let $f(x) = |x| - 3$. Find all x for which $f(x) = 5.7$.
44. Let $f(x) = |x| + 7$. Find all x for which $f(x) = 18$.
45. Let $f(x) = \left|\dfrac{1 - 2x}{5}\right|$. Find all x for which $f(x) = 2$.
46. Let $f(x) = \left|\dfrac{3x + 4}{3}\right|$. Find all x for which $f(x) = 1$.

Solve.

47. $|x - 7| = |2x + 1|$
48. $|3x + 2| = |x - 6|$
49. $|x + 4| = |x - 3|$
50. $|x - 9| = |x + 6|$
51. $|3a - 1| = |2a + 4|$
52. $|5t + 7| = |4t + 3|$
Aha! 53. $|n - 3| = |3 - n|$
54. $|y - 2| = |2 - y|$
55. $|7 - 4a| = |4a + 5|$
56. $|6 - 5t| = |5t + 8|$

Inequalities with Absolute Value

Solve and graph.

57. $|a| \leq 3$ 58. $|x| < 5$
59. $|t| > 0$ 60. $|t| \geq 1$
61. $|x - 1| < 4$ 62. $|x - 1| < 3$
63. $|n + 2| \leq 6$ 64. $|a + 4| \leq 0$
65. $|x - 3| + 2 > 7$ 66. $|x - 4| + 5 > 10$
Aha! 67. $|2y - 9| > -5$ 68. $|3y - 4| > -8$
69. $|3a + 4| + 2 \geq 8$ 70. $|2a + 5| + 1 \geq 9$
71. $|y - 3| < 12$ 72. $|p - 2| < 3$

73. $9 - |x + 4| \leq 5$

74. $12 - |x - 5| \leq 9$

75. $6 + |3 - 2x| > 10$

76. $|7 - 2y| < -8$

Aha! **77.** $|5 - 4x| < -6$

78. $7 + |4a - 5| \leq 26$

79. $\left| \dfrac{1 + 3x}{5} \right| > \dfrac{7}{8}$

80. $\left| \dfrac{2 - 5x}{4} \right| \geq \dfrac{2}{3}$

81. $|m + 3| + 8 \leq 14$

82. $|t - 7| + 3 \geq 4$

83. $25 - 2|a + 3| > 19$

84. $30 - 4|a + 2| > 12$

85. Let $f(x) = |2x - 3|$. Find all x for which $f(x) \leq 4$.

86. Let $f(x) = |5x + 2|$. Find all x for which $f(x) \leq 3$.

87. Let $f(x) = 5 + |3x - 4|$. Find all x for which $f(x) \geq 16$.

88. Let $f(x) = |2 - 9x|$. Find all x for which $f(x) \geq 25$.

89. Let $f(x) = 7 + |2x - 1|$. Find all x for which $f(x) < 16$.

90. Let $f(x) = 5 + |3x + 2|$. Find all x for which $f(x) < 19$.

91. Explain in your own words why $[6, \infty)$ is only part of the solution of $|x| \geq 6$.

92. Explain in your own words why -7 is not a solution of $|x| < 5$.

Skill Review

93. Find a linear function whose graph has slope $\frac{1}{3}$ and y-intercept $(0, -2)$. [2.3]

94. Find an equation in point–slope form of the line with slope -8 that contains $(3, 7)$. [2.5]

95. Find a linear function whose graph contains $(-4, -3)$ and $(-1, 3)$. [2.5]

96. Find the slope–intercept form of the equation of the line perpendicular to $x - y = 6$ that contains $(8, -9)$. [2.5]

Synthesis

97. Describe a procedure that could be used to solve any equation of the form $g(x) < c$ graphically.

98. Explain why the inequality $|x + 5| \geq 2$ can be interpreted as "the number x is at least 2 units from -5."

Solve.

99. $|3x - 5| = x$

100. $|x + 2| > x$

101. $2 \leq |x - 1| \leq 5$

102. $|5t - 3| = 2t + 4$

103. $t - 2 \leq |t - 3|$

104. From the definition of absolute value, $|x| = x$ only for $x \geq 0$. Solve $|3t - 5| = 3t - 5$ using this same reasoning.

Write an equivalent inequality using absolute value.

105. $-3 < x < 3$

106. $-5 \leq y \leq 5$

107. $x \leq -6 \text{ or } 6 \leq x$

108. $x < -4 \text{ or } 4 < x$

109. $x < -8 \text{ or } 2 < x$

110. $-5 < x < 1$

111. x is less than 2 units from 7.

112. x is less than 1 unit from 5.

Write an absolute-value inequality for which the interval shown is the solution.

113. ‹‒+‒+‒+‒+‒+‒+‒+═══════+‒›
      −7 −6 −5 −4 −3 −2 −1 0 1 2 3 4 5 6 7

114. ‹+‒(═══════)‒+‒+‒+‒+‒+‒+‒›
      −5 −4 −3 −2 −1 0 1 2 3 4 5 6 7 8 9

115. ‹(═══════)‒+‒+‒+‒+‒+‒+‒+‒›
      −7 −6 −5 −4 −3 −2 −1 0 1 2 3 4 5 6 7

116. ‹+‒+‒(═══════════)‒+‒+›
      0 1 2 3 4 5 6 7 8 9 10 11 12 13 14

117. *Bungee Jumping.* A bungee jumper is bouncing up and down so that her distance d above a river satisfies the inequality $|d - 60\text{ ft}| \leq 10\text{ ft}$ (see the figure below). If the bridge from which she jumped is 150 ft above the river, how far is the bungee jumper from the bridge at any given time?

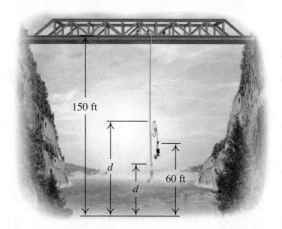

150 ft

d

60 ft

d

118. Use the following graph of $f(x) = |2x - 6|$ to solve $|2x - 6| \leq 4$.

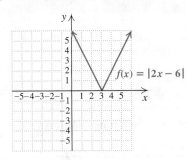

119. Is it possible for an equation in x of the form $|ax + b| = c$ to have exactly one solution? Why or why not?

120. Isabel is using the following graph to solve $|x - 3| < 4$. How can you tell that a mistake has been made in entering $y = |x - 3|$?

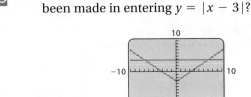

YOUR TURN ANSWERS: SECTION 4.3

1. $\{-6, 6\}$ **2.** $\{-\frac{2}{3}, 4\}$ **3.** $\{-\frac{7}{5}, \frac{7}{5}\}$ **4.** $\{4, 6\}$
5. $\{-\frac{2}{7}, 8\}$ **6.** $\{x \mid -2 < x < 2\}$, or $(-2, 2)$

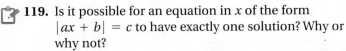

7. $\{x \mid x \leq -2 \text{ or } x \geq 2\}$, or $(-\infty, -2] \cup [2, \infty)$

8. $\{x \mid -\frac{9}{4} < x < 1\}$, or $(-\frac{9}{4}, 1)$

9. $\{x \mid x < -1 \text{ or } x > 6\}$, or $(-\infty, -1) \cup (6, \infty)$

Mid-Chapter Review

Thus far in this chapter, we have encountered several types of equations and inequalities. The following table summarizes the approaches used to solve each type.

Type of Equation or Inequality	Approach
Conjunction of inequalities *Example:* $-3 < x - 5 < 6$	Find the intersection of the separate solution sets.
Disjunction of inequalities *Example:* $x + 8 < 2 \text{ or } x - 4 > 9$	Find the union of the separate solution sets.
Absolute-value equation *Example:* $|x - 4| = 10$	Translate to two equations: If $|X| = p$, then $X = -p \text{ or } X = p$.
Absolute-value inequality including < *Example:* $|x + 2| < 5$	Translate to a conjunction: If $|X| < p$, then $-p < X < p$.
Absolute-value inequality including > *Example:* $|x - 1| > 9$	Translate to a disjunction: If $|X| > p$, then $X < -p \text{ or } p < X$.

GUIDED SOLUTIONS

1. Solve: $-3 \leq x - 5 \leq 6$.   [4.2]

   Solution

   $\boxed{} \leq x \leq \boxed{}$ Adding 5

   The solution is $\left[\boxed{}, \boxed{}\right]$.

2. Solve: $|x - 1| > 9$.   [4.3]

   Solution

   $x - 1 < \boxed{}$ *or* $\boxed{} < x - 1$

   $x < \boxed{}$ *or* $\boxed{} < x$ Adding 1

   The solution is $\left(-\infty, \boxed{}\right) \cup \left(\boxed{}, \infty\right)$.

MIXED REVIEW

Solve.

3. $|x| = 15$ [4.3]

4. $|t| < 10$ [4.3]

5. $|p| > 15$ [4.3]

6. $|2x + 1| = 7$ [4.3]

7. $-1 < 10 - x < 8$ [4.2]

8. $5|t| < 20$ [4.3]

9. $x + 8 < 2 \text{ or } x - 4 > 9$ [4.2]

10. $|x + 2| \leq 5$ [4.3]

11. $2 + |3x| = 10$ [4.3]

12. $2(x - 7) - 5x > 4 - (x + 5)$ [4.1]

13. $-12 < 2n + 6 \text{ and } 3n - 1 \leq 7$ [4.2]

14. $|2x + 5| + 1 \geq 13$ [4.3]

15. $\frac{1}{2}(2x - 6) \leq \frac{1}{3}(9x + 3)$ [4.1]

16. $\left|\dfrac{x + 2}{5}\right| = 8$ [4.3]

17. $|8x - 11| + 6 < 2$ [4.3]

18. $8 - 5|a + 6| > 3$ [4.3]

19. $|5x + 7| + 9 \geq 4$ [4.3]

20. $3x - 7 < 5 \text{ or } 2x + 1 > 0$ [4.2]

4.4   Inequalities in Two Variables

Graphs of Linear Inequalities   ▪   Systems of Linear Inequalities

We have graphed inequalities in one variable on the number line. Now we graph inequalities in two variables on a plane.

Graphs of Linear Inequalities

When the equal sign in a linear equation is replaced with an inequality sign, a **linear inequality** is formed. Solutions of linear inequalities are ordered pairs.

Student Notes

Pay careful attention to the inequality symbol when determining whether an ordered pair is a solution of an inequality. Writing the symbol at the end of the check, as in Example 1, will help you compare the numbers correctly.

1. Determine whether $(4, -5)$ is a solution of $3x + 2y < 4$.

EXAMPLE 1   Determine whether $(-3, 2)$ and $(6, -7)$ are solutions of $5x - 4y > 13$.

SOLUTION   Below, on the left, we replace x with -3 and y with 2. On the right, we replace x with 6 and y with -7.

$$\begin{array}{c|c} \multicolumn{2}{c}{5x - 4y > 13} \\ \hline 5(-3) - 4 \cdot 2 & 13 \\ -15 - 8 & \\ -23 \overset{?}{>} 13 & \text{FALSE} \end{array}$$

Since $-23 > 13$ is false, $(-3, 2)$ *is not* a solution.

$$\begin{array}{c|c} \multicolumn{2}{c}{5x - 4y > 13} \\ \hline 5(6) - 4(-7) & 13 \\ 30 + 28 & \\ 58 \overset{?}{>} 13 & \text{TRUE} \end{array}$$

Since $58 > 13$ is true, $(6, -7)$ *is* a solution.

YOUR TURN

The graph of a linear equation is a straight line. The graph of a linear inequality is a **half-plane**, with a **boundary** that is a straight line. To find the equation of the boundary, we replace the inequality sign with an equal sign.

EXAMPLE 2   Graph: $y \le x$.

SOLUTION   We first graph the equation of the boundary, $y = x$. Every solution of $y = x$ is an ordered pair, like $(3, 3)$, in which both coordinates are the same. The graph of $y = x$ is shown on the left below. Since the inequality symbol is $\le$, the line is drawn solid and is part of the graph of $y \le x$.

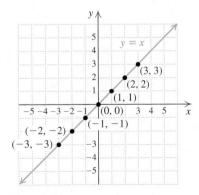   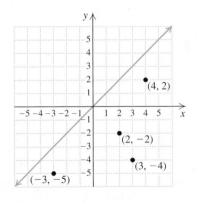

Note that in the graph on the right above each ordered pair on the half-plane below $y = x$ contains a y-coordinate that is less than the x-coordinate. All these pairs represent solutions of $y \le x$. We check one pair, $(4, 2)$, as follows:

$$\begin{array}{c} y \le x \\ \hline 2 \overset{?}{\le} 4 \quad \text{TRUE} \end{array}$$

Study Skills

Add Your Voice to the Author's

If you own your text, consider using it as a notebook. Since many instructors' lesson plans closely parallel the book, you may find it helpful to make notes on the appropriate page as he or she is lecturing.

2. Graph: $y \leq x + 1$.

↩ YOUR TURN

It turns out that *any* point on the same side of $y = x$ as $(4, 2)$ is also a solution. Thus, if one point in a half-plane is a solution, then *all* points in that half-plane are solutions.

We finish drawing the solution set by shading the half-plane below $y = x$. The solution set consists of the shaded half-plane as well as the boundary line itself.

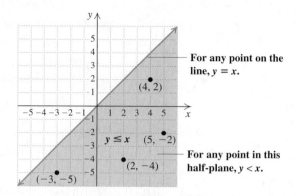

For any point on the line, $y = x$.

$y \leq x$

For any point in this half-plane, $y < x$.

From Example 2, we see that for any inequality of the form $y \leq f(x)$ or $y < f(x)$, we shade *below* the graph of $y = f(x)$.

EXAMPLE 3 Graph: $8x + 3y > 24$.

SOLUTION First, we sketch the graph of $8x + 3y = 24$. A convenient way to graph this equation is to use the x-intercept, $(3, 0)$, and the y-intercept, $(0, 8)$. Since the inequality sign is $>$, points on this line do not represent solutions of the inequality, and the line is drawn dashed. Points representing solutions of $8x + 3y > 24$ are in either the half-plane above the line or the half-plane below the line. To determine which, we select a point that is not on the line and check whether it is a solution of $8x + 3y > 24$. Let's use $(0, 0)$ as this *test point*:

$$\frac{8x + 3y > 24}{8(0) + 3(0) \mid 24}$$
$$0 \overset{?}{>} 24 \quad \text{FALSE}$$

Since $0 > 24$ is *false*, $(0, 0)$ is not a solution. Thus no point in the half-plane containing $(0, 0)$ is a solution. The points in the other half-plane *are* solutions, so we shade that half-plane and obtain the graph shown below.

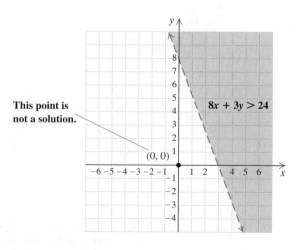

This point is not a solution.

$8x + 3y > 24$

$(0, 0)$

3. Graph: $2x + y < 6$.

↩ YOUR TURN

> **STEPS FOR GRAPHING LINEAR INEQUALITIES**
> 1. Replace the inequality sign with an equal sign and graph this line as the boundary. If the inequality symbol is < or >, draw the line dashed. If the symbol is ≤ or ≥, draw the line solid.
> 2. The graph of the inequality consists of a half-plane on one side of the line and, if the line is solid, the line as well.
>   a) For an inequality of the form $y < mx + b$ or $y \le mx + b$, shade *below* the line.
>   For an inequality of the form $y > mx + b$ or $y \ge mx + b$, shade *above* the line.
>   b) If y is not isolated, use a test point not on the line as in Example 3. If the test point *is* a solution, shade the half-plane containing the point. If it is *not* a solution, shade the other half-plane. Additional test points can also be used as a check.

EXAMPLE 4   Graph: $6x - 2y < 12$.

SOLUTION   We could graph $6x - 2y = 12$ and use a test point, as in Example 3. Instead, let's solve $6x - 2y < 12$ for y:

$6x - 2y < 12$

$\quad -2y < -6x + 12$ Adding $-6x$ to both sides

$\quad\quad y > 3x - 6.$ Dividing both sides by -2 and reversing the < symbol

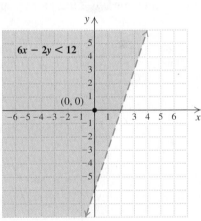

The graph consists of the half-plane above the dashed boundary line $y = 3x - 6$ (see the graph at right). As a check, note that the test point $(0, 0)$ is a solution of the inequality and is in the half-plane that we shaded.

4. Graph: $x > 6y - 6$.

YOUR TURN

EXAMPLE 5   Graph $x > -3$ on a plane.

SOLUTION   There is only one variable in this inequality. If we graph the inequality on a line, its graph is as follows:

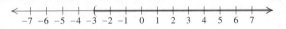

However, we can also write this inequality as $x + 0y > -3$ and graph it on a plane. Using the same technique as in the examples above, we graph the boundary $x = -3$ in the plane, using a dashed line. Then we test some point, say, $(2, 5)$:

$$\frac{x + 0y > -3}{2 + 0 \cdot 5 \mid -3}$$
$$2 \overset{?}{>} -3 \quad \text{TRUE}$$

Since $(2, 5)$ is a solution, all points in the half-plane containing $(2, 5)$ are solutions. We shade that half-plane. We can also simply note that solutions of $x > -3$ are pairs with first coordinates greater than -3.

5. Graph $x \le 2$ on a plane.

YOUR TURN

EXAMPLE 6   Graph $y \leq 4$ on a plane.

SOLUTION   The inequality is of the form $y \leq mx + b$ (with $m = 0$), so we shade below the solid horizontal line representing $y = 4$.

This inequality can also be graphed by drawing $y = 4$ and testing a point above or below the line. We can also simply note that solutions of $y \leq 4$ are pairs with second coordinates less than or equal to 4.

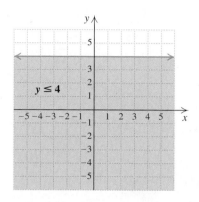

6. Graph $y > -4$ on a plane.

YOUR TURN

Technology Connection

On most graphing calculators, an inequality like $y < \frac{6}{5}x + 3.49$ can be drawn by entering $\frac{6}{5}x + 3.49$ as y_1, moving the cursor to the GraphStyle icon just to the left of y_1, pressing **ENTER** until ◣ appears, and then pressing **GRAPH**.

Many calculators have an INEQUALZ program that is accessed using the **APPS** key. Running this program allows us to write inequalities at the **Y=** screen by pressing **ALPHA** and then one of the five keys just below the screen.

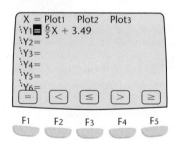

When we are using INEQUALZ, the boundary line appears dashed when < or > is selected.

$$y_1 < \tfrac{6}{5}x + 3.49, \text{ or}$$
$$◣\, y_1 = \tfrac{6}{5}x + 3.49$$

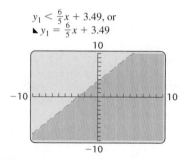

Graph each of the following. Solve for y first if necessary.

1. $y > x + 3.5$ **2.** $7y \leq 2x + 5$
3. $8x - 2y < 11$ **4.** $11x + 13y + 4 \geq 0$

Systems of Linear Inequalities

To graph a system of equations, we graph the individual equations and then find the intersection of the graphs. We work similarly with a system of inequalities: We graph each inequality and find the intersection of the graphs.

EXAMPLE 7   Graph the system

$$x + y \leq 4,$$
$$x - y < 4.$$

SOLUTION   To graph $x + y \leq 4$, we graph $x + y = 4$ using a solid line. Since the test point $(0, 0)$ *is* a solution and $(0, 0)$ is below the line, we shade the half-plane below the graph red. The arrows near the ends of the line are a helpful way of indicating the half-plane containing solutions.

Next, we graph $x - y < 4$. We graph $x - y = 4$ using a dashed line and consider $(0, 0)$ as a test point. Again, $(0, 0)$ is a solution, so we shade that side of the line blue. The solution set of the system is the region that is shaded purple (both red and blue) and part of the line $x + y = 4$.

| Graph the first inequality. | Graph the second inequality. | Shade the intersection. |

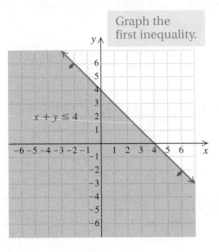
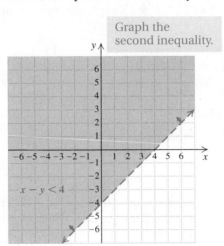
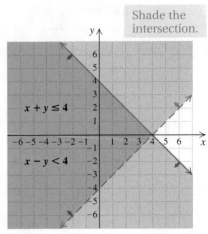

7. Graph the system

$$2x - y < 1,$$
$$x + y \leq 3.$$

——————————— ➥ YOUR TURN

EXAMPLE 8 Graph: $-2 < x \leq 3$.

SOLUTION This is a system of inequalities:

$$-2 < x,$$
$$x \leq 3.$$

We graph the equation $-2 = x$, and see that the graph of the first inequality is the half-plane to the right of the boundary $-2 = x$. It is shaded red.

We graph the second inequality, starting with the boundary line $x = 3$. The inequality's graph is the line and the half-plane to its left. It is shaded blue.

The solution set of the system is the intersection of the individual graphs. Since it is shaded both blue and red, it appears to be purple. All points in this region have x-coordinates that are greater than -2 but do not exceed 3.

Student Notes

If you don't use differently colored pencils or pens to shade regions, consider using a pencil to make slashes that tilt in different directions in each region. You may also find it useful to draw arrowheads indicating the appropriate half-plane, as in the graphs shown.

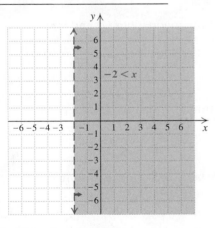
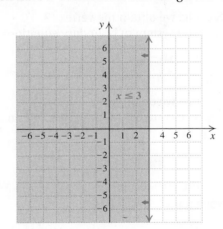
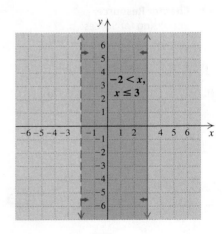

8. Graph: $-1 \leq y \leq 4$.

——————————— ➥ YOUR TURN

A system of inequalities may have a graph consisting of a polygon and its interior. In some applications, we will need the coordinates of the corners, or *vertices* (singular, *vertex*), of such a graph.

EXAMPLE 9 Graph the system of inequalities. Find the coordinates of any vertices formed.

$$6x - 2y \leq 12, \quad (1)$$
$$y - 3 \leq 0, \quad (2)$$
$$x + y \geq 0 \quad (3)$$

SOLUTION We graph the boundaries

$$6x - 2y = 12,$$
$$y - 3 = 0,$$
and $$x + y = 0$$

using solid lines. The regions for each inequality are indicated by arrows near the ends of the lines. We note where the regions overlap and shade the region of solutions purple.

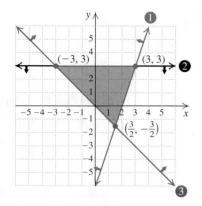

To find the vertices, we solve three different systems of two equations. The system of boundary equations from inequalities (1) and (2) is

$$6x - 2y = 12, \quad \text{The student can use graphing, substitution, or}$$
$$y - 3 = 0. \quad \text{elimination to solve these systems.}$$

Solving, we obtain the vertex $(3, 3)$.

The system of boundary equations from inequalities (1) and (3) is

$$6x - 2y = 12,$$
$$x + y = 0.$$

Solving, we obtain the vertex $\left(\frac{3}{2}, -\frac{3}{2}\right)$.

The system of boundary equations from inequalities (2) and (3) is

$$y - 3 = 0,$$
$$x + y = 0.$$

Solving, we obtain the vertex $(-3, 3)$. We label the graph as shown.

 YOUR TURN

Technology Connection

We can graph systems of inequalities using the INEQUALZ application. We enter the inequalities (solving for y if needed), press $\boxed{\text{GRAPH}}$, and then press $\boxed{\text{ALPHA}}$ and Shades ($\boxed{\text{F1}}$ or $\boxed{\text{F2}}$). At the SHADES menu, we select Ineq Intersection to see the final graph. To find any vertices, or points of intersection, we select PoI-Trace from the graph menu.

$$y_1 \geq 3x - 6, \quad y_2 \leq 3, \quad y_3 \geq -x$$

1. Use a graphing calculator to check the solution of Example 7.

Chapter Resource:
Visualizing for Success, p. 271

9. Graph the system of inequalities. Find the coordinates of any vertices formed.

$$x - y \leq 3,$$
$$y \leq 2,$$
$$y \geq 2 - x.$$

CONNECTING 🔗 THE CONCEPTS

We have now solved a variety of equations, inequalities, systems of equations, and systems of inequalities. Below is a list of the different types of problems we have solved, illustrations of each type, and descriptions of the solutions. Note that a solution set may be empty.

Type	Example	Solution	Graph
Linear equations in one variable	$2x - 8 = 3(x + 5)$	A number	
Linear inequality in one variable	$-3x + 5 > 2$	A set of numbers; an interval	
Linear equation in two variables	$2x + y = 7$	A set of ordered pairs; a line	
Linear inequality in two variables	$x + y \geq 4$	A set of ordered pairs; a half-plane	
System of equations in two variables	$x + y = 3,$ $5x - y = -27$	An ordered pair or a set of ordered pairs	
System of inequalities in two variables	$6x - 2y \leq 12,$ $y - 3 \leq 0,$ $x + y \geq 0$	A set of ordered pairs; a region of a plane	

(continued)

EXERCISES

Graph each solution on the number line.

1. $x + 2 = 7$ **2.** $x + 2 > 7$

3. $x + 2 \leq 7$

Graph on a plane.

4. $x + y = 2$ **5.** $x + y < 2$

6. $x + y \geq 2$ **7.** $x + 2 \leq 7$

8. $y = x - 1,$
    $y = -x + 1$

9. $y \geq 1 - x,$
    $y \leq x - 3,$
    $y \leq 2$

4.4 EXERCISE SET

FOR EXTRA HELP

MyMathLab® Math XL

PRACTICE   WATCH   READ   REVIEW

Vocabulary and Reading Check

In each of Exercises 1–6, match the phrase with the most appropriate choice from the column on the right.

1. _____ The solution set of a linear inequality

2. _____ The graph of a linear inequality

3. _____ The graph of a system of linear inequalities

4. _____ Often a convenient test point

5. _____ The name for the corners of a graph of a system of linear inequalities

6. _____ A dashed line

a) $(0, 0)$

b) Vertices

c) A half-plane

d) The intersection of two or more half-planes

e) All ordered pairs that satisfy the inequality

f) Indicates the line is not part of the solution

Graphs of Linear Inequalities

Determine whether each ordered pair is a solution of the given inequality.

7. $(-2, 3); \ 2x - y > -4$

8. $(1, -6); \ 3x + y \geq -3$

9. $(5, 8); \ 3y - 5x \leq 0$

10. $(6, 20); \ 5y - 8x < 40$

Graph on a plane.

11. $y \geq \frac{1}{2}x$ **12.** $y \leq 3x$

13. $y > x - 3$ **14.** $y < x + 3$

15. $y \leq x + 2$ **16.** $y \geq x - 5$

17. $x - y \leq 4$ **18.** $x + y < 4$

19. $2x + 3y < 6$ **20.** $3x + 4y \leq 12$

21. $2y - x \leq 4$ **22.** $2y - 3x > 6$

23. $2x - 2y \geq 8 + 2y$ **24.** $3x - 2 \leq 5x + y$

25. $x > -2$ **26.** $x \geq 3$

27. $y \leq 6$ **28.** $y < -1$

Systems of Linear Inequalities

Graph.

29. $-2 < y < 7$ **30.** $-4 < y < -1$

31. $-5 \leq x < 4$ **32.** $-2 < y \leq 1$

33. $0 \leq y \leq 3$ **34.** $0 \leq x \leq 6$

35. $y > x,$ **36.** $y < x,$
    $y < -x + 3$ $y > -x + 1$

37. $y \leq x,$ **38.** $y \geq x,$
    $y \leq 2x - 5$ $y \leq -x + 4$

39. $y \leq -3,$ **40.** $y \geq -3,$
    $x \geq -1$ $x \geq 1$

41. $x > -4,$ **42.** $x < 3,$
    $y < -2x + 3$ $y > -3x + 2$

43. $y \leq 5,$ **44.** $y \geq -2,$
    $y \geq -x + 4$ $y \geq x + 3$

45. $x + y \leq 6,$ **46.** $x + y < 1,$
    $x - y \leq 4$ $x - y < 2$

47. $y + 3x > 0,$ **48.** $y - 2x \geq 1,$
    $y + 3x < 2$ $y - 2x \leq 3$

Graph each system of inequalities. Find the coordinates of any vertices formed.

49. $y \leq 2x - 3,$ **50.** $2y - x \leq 2,$
    $y \geq -2x + 1,$ $y - 3x \geq -4,$
    $x \leq 5$ $y \geq -1$

51. $x + 2y \leq 12,$ **52.** $x - y \leq 2,$
    $2x + y \leq 12,$ $x + 2y \geq 8,$
    $x \geq 0,$ $y \leq 4$
    $y \geq 0$

53. $8x + 5y \leq 40,$
$x + 2y \leq 8,$
$x \geq 0,$
$y \geq 0$

54. $4y - 3x \geq -12,$
$4y + 3x \geq -36,$
$y \leq 0,$
$x \leq 0$

55. $y - x \geq 2,$
$y - x \leq 4,$
$2 \leq x \leq 5$

56. $3x + 4y \geq 12,$
$5x + 6y \leq 30,$
$1 \leq x \leq 3$

57. Explain in your own words why the boundary line is drawn dashed for the symbols $<$ and $>$ and why it is drawn solid for the symbols $\leq$ and $\geq$.

58. When graphing linear inequalities, Ron makes a habit of always shading above the line when the symbol $\geq$ is used. Is this wise? Why or why not?

Skill Review

59. In which quadrant or on which axis is $(-152, 0)$ located? [2.1]

60. Find the slope of the line containing $(5, 7)$ and $(-5, 3)$. [2.3]

61. Find the slope and the y-intercept of the graph of $f(x) = \frac{4}{3}x + 15$. [2.3]

62. Find the slope of the graph of $y = -7$. If the slope is undefined, state this. [2.4]

63. Determine whether the equations represent lines that are parallel, perpendicular, or neither. [2.4]
$$2x = 4 - 3y,$$
$$3x - 2y = 10$$

64. Determine whether $y = x^2 + 1$ is a linear equation. [2.4]

Synthesis

65. Explain how a system of linear inequalities could have a solution set consisting of one ordered pair.

66. In Example 7 of this section, is the point $(4, 0)$ part of the solution set? Why or why not?

Graph.

67. $x + y > 8,$
$x + y \leq -2$

68. $x + y \geq 1,$
$-x + y \geq 2,$
$x \geq -2,$
$y \geq 2,$
$y \leq 4,$
$x \leq 2$

69. $x - 2y \leq 0,$
$-2x + y \leq 2,$
$x \leq 2,$
$y \leq 2,$
$x + y \leq 4$

70. Write four systems of four inequalities that describe a 2-unit by 2-unit square that has $(0, 0)$ as one of the vertices.

71. *Luggage Size.* Unless an additional fee is paid, most major airlines will not check any luggage for which the sum of the item's length, width, and height exceeds 62 in. The U.S. Postal Service will ship a package only if the sum of the package's length and girth (distance around its midsection) does not exceed 130 in. Video Promotions is ordering several 30-in. long cases that will be both mailed and checked as luggage. Using w and h for width and height (in inches), respectively, write and graph an inequality that represents all acceptable combinations of width and height.

Sources: U.S. Postal Service; www.case2go.com

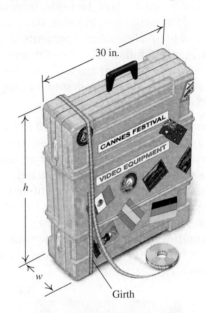

72. *Hockey Wins and Losses.* The Skating Stars believe they need at least 60 points for the season in order to make the playoffs. A win is worth 2 points, a tie is worth 1 point, and a loss is worth 0 points. The team plays no more than 50 games. Graph a system of inequalities that describes the situation. Let w represent the number of wins and t the number of ties.

73. *Graduate-School Admissions.* Students entering a master's degree program at the University of Louisiana at Lafayette must meet minimum score requirements on the Graduate Records Examination (GRE). The GRE Verbal score must be at least 145 and the sum of the GRE Quantitative and Verbal scores must be at least 287. Each score has a maximum of 170. Using q for the quantitative score and v for the verbal score, write and graph a system of inequalities that represents all combinations that meet the requirements for entrance into the program.

Source: University of Louisiana at Lafayette

74. *Widths of a Basketball Floor.* Sizes of basketball floors vary due to building sizes and other constraints such as cost. The length L is to be at most 94 ft and the width W is to be at most 50 ft. Graph a system of inequalities that describes the possible dimensions of a basketball floor.

75. *Elevators.* Many elevators have a capacity of 1 metric ton (1000 kg). Suppose that c children, each weighing 35 kg, and a adults, each 75 kg, are on an elevator. Graph a system of inequalities that indicates when the elevator is overloaded.

76. *Age of Marriage.* The following rule of thumb for determining an appropriate difference in age between a bride and a groom appears in many Internet blogs: *The younger spouse's age should be at least seven more than half the age of the older spouse.* Let b = the age of the bride, in years, and g = the age of the groom, in years. Write and graph a system of inequalities that represents all combinations of ages that follow this rule of thumb. Should a minimum or maximum age for marriage exist? How would the graph of the system of inequalities change with such a requirement?

77. *Waterfalls.* In order for a waterfall to be classified as a classical waterfall, its height must be less than twice its crest width, and its crest width cannot exceed one-and-a-half times its height. The tallest waterfall in the world is about 3200 ft high. Let h represent a waterfall's height, in feet, and w the crest width, in feet. Write and graph a system of inequalities that represents all possible combinations of heights and crest widths of classical waterfalls.

78. Use a graphing calculator to check your answers to Exercises 35–48. Then use INTERSECT to determine any point(s) of intersection.

79. Use a graphing calculator to graph each inequality.

a) $3x + 6y > 2$
b) $x - 5y \leq 10$
c) $13x - 25y + 10 \leq 0$
d) $2x + 5y > 0$

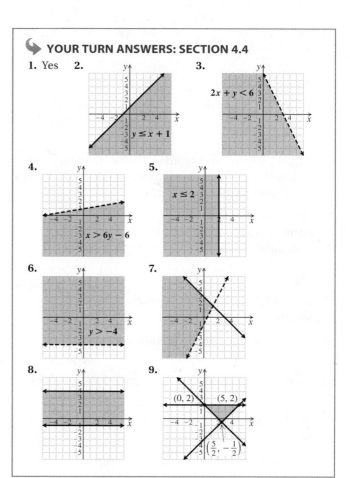
QUICK QUIZ: SECTIONS 4.1–4.4

Solve. Then graph each solution set. Write solution sets using both set-builder notation and interval notation.

1. $0.1a - 4 \leq 2.6a + 5$ [4.1]

2. $2x < 9 \ or -3x < -3$ [4.2]

3. $|4x + 5| - 8 > 9$ [4.3]

4. $|9x + 4| \leq -3$ [4.3]

5. Graph $2x - y > 4$ on a plane. [4.4]

PREPARE TO MOVE ON

1. Gina invested $10,000 in two accounts, one paying 3% simple interest and one paying 5% simple interest. After one year, she had earned $428 from both accounts. How much did she invest in each? [3.3]

2. There were 170 tickets sold for the Ridgefield vs Maplewood basketball game. Tickets were $3 each for students and $5 each for adults. The total amount of money collected was $726. How many of each type of ticket were sold? [3.3]

3. Josh planted 400 acres in corn and soybeans. He planted 80 more acres in corn than he did in soybeans. How many acres of each did he plant? [3.3]

4.5   Applications Using Linear Programming

Linear Programming

Many real-world situations require finding a greatest value (a maximum) or a least value (a minimum). For example, businesses want to make the *most* profit with the *least* expense possible. Some such problems can be solved using systems of inequalities.

Linear Programming

Often a quantity we want to maximize depends on two or more other quantities. For example, a gardener's profits P might depend on the number of shrubs s and the number of trees t that are planted. If the gardener makes a $10 profit from each shrub and an $18 profit from each tree, the total profit, in dollars, is given by the **objective function**

$$P = 10s + 18t.$$

Thus the gardener might be tempted to simply plant lots of trees since they yield the greater profit. This would be a good idea were it not for the fact that the number of trees and shrubs planted—and thus the total profit—is subject to the demands, or **constraints**, of the situation. For example, to improve drainage, the gardener might be required to plant at least 3 shrubs. Thus the objective function would be subject to the *constraint*

$$s \geq 3.$$

Because of limited space, the gardener might also be required to plant no more than 10 plants. This would subject the objective function to a *second* constraint:

$$s + t \leq 10.$$

Finally, the gardener might be told to spend no more than $700 on the plants. If the shrubs cost $40 each and the trees cost $100 each, the objective function is subject to a *third* constraint:

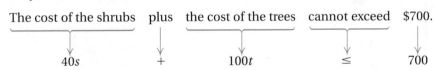

The cost of the shrubs | plus | the cost of the trees | cannot exceed | $700.

$$40s \qquad + \qquad 100t \qquad \leq \qquad 700$$

In short, the gardener wishes to maximize the objective function

$$P = 10s + 18t,$$

subject to the constraints

$$s \geq 3,$$
$$s + t \leq 10,$$
$$40s + 100t \leq 700,$$
$$\left.\begin{array}{l} s \geq 0, \\ t \geq 0. \end{array}\right\} \text{ Because the number of trees and shrubs cannot be negative}$$

These constraints form a system of linear inequalities that can be graphed.

The gardener's problem is "How many shrubs and trees should be planted, subject to the constraints listed, in order to maximize profit?" To solve such a problem, we use a result from a branch of mathematics known as **linear programming**.

THE CORNER PRINCIPLE

Suppose an objective function $F = ax + by + c$ depends on x and y (with a, b, and c constant). Suppose also that F is subject to constraints on x and y, which form a system of linear inequalities. If F has a minimum or a maximum value, then it can be found as follows:

1. Graph the system of inequalities and find the vertices.
2. Find the value of the objective function at each vertex. The greatest and the least of those values are the maximum and the minimum of the function, respectively.
3. The ordered pair (x, y) at which the maximum or the minimum occurs indicates the values at which the maximum or the minimum occurs.

This result was proven during World War II, when linear programming was developed to help allocate troops and supplies bound for Europe.

EXAMPLE 1   Solve the gardener's problem discussed above.

SOLUTION   We are asked to maximize $P = 10s + 18t$, subject to the constraints

$$s \geq 3,$$
$$s + t \leq 10,$$
$$40s + 100t \leq 700,$$
$$s \geq 0,$$
$$t \geq 0.$$

We graph the system. The portion of the graph that is shaded represents all pairs that satisfy the constraints and is called the *feasible region*.

According to the corner principle, P is maximized at one of the vertices of the shaded region. To determine the coordinates of the vertices, we solve the following systems:

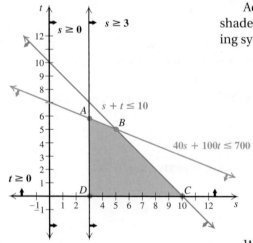

A: $\left.\begin{array}{r} 40s + 100t = 700, \\ s = 3; \end{array}\right\}$ The student can verify that the solution of this system is $(3, 5.8)$. The coordinates of point A are $(3, 5.8)$.

B: $\left.\begin{array}{r} s + t = 10, \\ 40s + 100t = 700; \end{array}\right\}$ The student can verify that the solution of this system is $(5, 5)$. The coordinates of point B are $(5, 5)$.

C: $\left.\begin{array}{r} s + t = 10, \\ t = 0; \end{array}\right\}$ The solution of this system is $(10, 0)$. The coordinates of point C are $(10, 0)$.

D: $\left.\begin{array}{r} t = 0, \\ s = 3. \end{array}\right\}$ The solution of this system is $(3, 0)$. The coordinates of point D are $(3, 0)$.

We now find the value of P at each vertex.

Vertex (s, t)	Profit $P = 10s + 18t$	
A $(3, 5.8)$	$10(3) + 18(5.8) = 134.4$	
B $(5, 5)$	$10(5) + 18(5) = 140$	⟵ Maximum
C $(10, 0)$	$10(10) + 18(0) = 100$	
D $(3, 0)$	$10(3) + 18(0) = 30$	⟵ Minimum

1. Refer to Example 1. Suppose that the gardener is allowed to spend only $580. How many shrubs and how many trees should be planted in order to maximize profit?

The greatest value of *P* occurs at $(5, 5)$. Thus profit is maximized at $140 when the gardener plants 5 shrubs and 5 trees. Incidentally, we have also shown that profit is minimized at $30 when 3 shrubs and 0 trees are planted.

YOUR TURN

EXAMPLE 2 *Grading.* For his history course, Cy can submit book summaries for 70 points each or research projects for 80 points each. He estimates that each book summary will take 9 hr and each research project will take 15 hr and that he will have at most 120 hr to spend. He may submit no more than 12 assignments. How many of each should he complete in order to receive the greatest number of points?

SOLUTION

1. **Familiarize.** We let $b =$ the number of book summaries and $r =$ the number of research projects. Cy is limited by the number of hours he can spend and by the number of summaries and projects he can submit. These two limits are the constraints.

2. **Translate.** We organize the information in a table.

Type	Number of Points for Each	Time Required for Each	Number Completed	Total Time for Each Type	Total Points for Each Type
Book summary	70	9 hr	b	$9b$	$70b$
Research project	80	15 hr	r	$15r$	$80r$
Total			$b + r \leq 12$	$9b + 15r \leq 120$	$70b + 80r$

↑ Because no more than 12 may be submitted ↑ Because the time cannot exceed 120 hr ↑ Cy wants to maximize the total number of points.

Student Notes

It is very important that you clearly label what each variable represents. It is also important to clearly define the objective function.

We let *T* represent the total number of points. We see from the table that

$T = 70b + 80r.$

We wish to maximize *T* subject to the number and time constraints:

$b + r \leq 12,$
$9b + 15r \leq 120,$
$b \geq 0,$
$r \geq 0.$ We include this because the number of summaries and projects cannot be negative.

3. **Carry out.** We graph the system and evaluate *T* at each vertex. The graph is as shown at right.

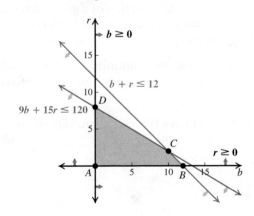

We find the coordinates of each vertex by solving a system of two linear equations. The coordinates of point A are obviously $(0, 0)$. To find the coordinates of point C, we solve the system

$$b + r = 12, \quad (1)$$
$$9b + 15r = 120. \quad (2)$$

We multiply both sides of equation (1) by -9 and add:

$$-9b - 9r = -108$$
$$\underline{9b + 15r = 120}$$
$$6r = 12$$
$$r = 2.$$

Vertex (b, r)	Total Number of Points $T = 70b + 80r$
A $(0, 0)$	0
B $(12, 0)$	840
C $(10, 2)$	860
D $(0, 8)$	640

Substituting, we find that $b = 10$. Thus the coordinates of C are $(10, 2)$. Point B is the intersection of $b + r = 12$ and $r = 0$, so B is $(12, 0)$. Point D is the intersection of $9b + 15r = 120$ and $b = 0$, so D is $(0, 8)$. Computing the score for each ordered pair, we obtain the table at left. The greatest value in the table is 860, obtained when $b = 10$ and $r = 2$.

2. Refer to Example 2. Suppose Cy may turn in no more than 10 summaries and/or projects. How many of each should he submit in order to receive the greatest number of points?

4. Check. We can check that $T \leq 860$ for several other points in the shaded region. This is left to the student.

5. State. In order to maximize his points, Cy should submit 10 book summaries and 2 research projects.

YOUR TURN

4.5 EXERCISE SET

FOR EXTRA HELP

MyMathLab® Math XL

PRACTICE   WATCH   READ   REVIEW

Vocabulary and Reading Check

Complete each of the following statements.

1. In linear programming, the quantity we wish to maximize or minimize is represented by the _____ function.

2. In linear programming, the demands arising from the given situation are known as _____.

3. To solve a linear programming problem, we make use of the _____ principle.

4. The shaded portion of a graph that represents all points that satisfy a problem's constraints is known as the _____ region.

5. In linear programming, the corners of the shaded portion of the graph are referred to as _____.

6. If it exists, the maximum value of an objective function occurs at a(n) _____ of the feasible region.

Linear Programming

Find the maximum and the minimum values of each objective function and the values of x and y at which they occur.

7. $F = 2x + 14y$,
subject to
$5x + 3y \leq 34$,
$3x + 5y \leq 30$,
$x \geq 0$,
$y \geq 0$

8. $G = 7x + 8y$,
subject to
$3x + 2y \leq 12$,
$2y - x \leq 4$,
$x \geq 0$,
$y \geq 0$

9. $P = 8x - y + 20$,
subject to
$6x + 8y \leq 48$,
$0 \leq y \leq 4$,
$0 \leq x \leq 7$

10. $Q = 24x - 3y + 52$,
subject to
$5x + 4y \leq 20$,
$0 \leq y \leq 4$,
$0 \leq x \leq 3$

11. $F = 2y - 3x$,
subject to
$y \leq 2x + 1$,
$y \geq -2x + 3$,
$x \leq 3$

12. $G = 5x + 2y + 4$,
subject to
$y \leq 2x + 1$,
$y \geq -x + 3$,
$x \leq 5$

13. *Transportation Cost.* It takes Caroline 1 hr to ride the train to work and 1.5 hr to ride the bus. Every week, she must make at least 5 trips to work, and she plans to spend no more than 6 hr in travel time. If a train trip costs $5 and a bus trip costs $4, how many times per week should she ride each in order to minimize her cost?

14. *Food Service.* Chad sells shrimp gumbo and shrimp sandwiches. He uses 3 oz of shrimp in each bowl of gumbo and 5 oz of shrimp in each sandwich. One Saturday morning, he realizes that he has only 120 oz of shrimp and that he must make a total of at least 30 shrimp meals. If his profit is $2 per gumbo order and $3 per sandwich, how many of each item should Chad make in order to maximize profit? (Assume that he sells everything that he makes.)

15. *Photo Albums.* Photo Perfect prints pages of photographs for albums. A page containing 4 photos costs $3 and a page containing 6 photos costs $5. Ann can spend no more than $90 for photo pages of her recent vacation, and she can use no more than 20 pages in her album. What combination of 4-photo pages and 6-photo pages will maximize the number of photos she can display? What is the maximum number of photos that she can display?

16. *Recycling.* Mack collects bottles and cans from trash cans to turn in at the recycling center. It takes him 1.5 min to prepare a large container for return and 0.5 min to prepare a small container. He has at most 30 min per day to spend cleaning containers, and he is allowed to return no more than 30 containers per day. If he receives 10¢ for every large container and 5¢ for every small container, how many of each should he return in order to maximize his daily income? What is the maximum amount that he can make each day?

17. *Investing.* Rosa is planning to invest up to $40,000 in corporate or municipal bonds, or both. She must invest from $6000 to $22,000 in corporate bonds, and she won't invest more than $30,000 in municipal bonds. The interest on corporate bonds is 4% and on municipal bonds is $3\frac{1}{2}$ %. This is simple interest for one year. How much should Rosa invest in each type of bond in order to earn the most interest? What is the maximum interest?

18. *Investing.* Jamaal is planning to invest up to $22,000 in City Bank or the Southwick Credit Union, or both. He wants to invest at least $2000 but no more than $14,000 in City Bank. He will invest no more than $15,000 in the Southwick Credit Union. Interest is 6% at City Bank and is $6\frac{1}{2}$% at the Credit Union. This is simple interest for one year. How much should Jamaal invest in each bank in order to earn the most interest? What is the maximum interest?

19. *Test Scores.* Corinna is taking a test in which short-answer questions are worth 10 points each and essay questions are worth 15 points each. She estimates that it takes 3 min to answer each short-answer question and 6 min to answer each essay question. The total time allowed is 60 min, and no more than 16 questions can be answered. Assuming that all her answers are correct, how many questions of each type should Corinna answer in order to get the best score?

20. *Test Scores.* Edy is about to take a test that contains short-answer questions worth 4 points each and word problems worth 7 points each. Edy must do at least 5 short-answer questions, but time restricts doing more than 10. She must do at least 3 word problems, but time restricts doing more than 10. Edy can do no more than 18 questions in total. How many of each type of question should Edy do in order to maximize her score? What is this maximum score?

21. *Grape Growing.* Auggie's vineyard consists of 240 acres upon which he wishes to plant Merlot grapes and Cabernet grapes. Profit per acre of Merlot is $400, and profit per acre of Cabernet is $300. The number of hours of labor available is 3200. Each acre of Merlot requires 20 hr of labor, and each acre of Cabernet requires 10 hr of labor. Determine how the land should be divided between Merlot and Cabernet in order to maximize profit.

Aha!

22. *Coffee Blending.*   The Coffee Peddler has 1440 lb of Sumatran coffee and 700 lb of Kona coffee. A batch of Hawaiian Blend requires 8 lb of Kona and 12 lb of Sumatran, and yields a profit of $90. A batch of Classic Blend requires 4 lb of Kona and 16 lb of Sumatran, and yields a $55 profit. How many batches of each kind should be made in order to maximize profit? What is the maximum profit?

23. *Nutrition.*   Becca must have at least 15 mg but no more than 45 mg of iron each day. She should also have at least 1500 mg but no more than 2500 mg of calcium per day. One serving of goat cheese contains 1 mg of iron, 500 mg of calcium, and 264 calories. One serving of hazelnuts contains 5 mg of iron, 100 mg of calcium, and 628 calories. How many servings of goat cheese and how many servings of hazelnuts should Becca eat in order to meet the daily requirements of iron and calcium but minimize the total number of calories?

24. *Textile Production.*   It takes Cosmic Stitching 2 hr of cutting and 4 hr of sewing to make a knit suit. To make a worsted suit, it takes 4 hr of cutting and 2 hr of sewing. At most 20 hr per day are available for cutting, and at most 16 hr per day are available for sewing. The profit on a knit suit is $68 and on a worsted suit is $62. How many of each kind of suit should be made in order to maximize profit?

25. Before a student begins work in this section, what three sections of the text would you suggest he or she study? Why?

26. What does the use of the word "constraint" in this section have in common with the use of the word in everyday speech?

Skill Review

Simplify. Do not leave negative exponents in your answer. [1.6]

27. 10^{-2}

28. $y^{18}y^{-2}$

29. $\dfrac{-6x^2}{3x^{-10}}$

30. $(-2a^{-3}b^{-4})^3$

31. $\left(\dfrac{4c^2d}{6cd^4}\right)^{-1}$

32. $(-2x^6x^{18})^0$

Synthesis

33. Explain how Exercises 17 and 18 can be answered by logical reasoning without linear programming.

34. Write a linear programming problem for a classmate to solve. Devise the problem so that profit must be maximized subject to at least two (nontrivial) constraints.

35. *Airplane Production.*   Alpha Tours has two types of airplanes, the T3 and the S5, and contracts requiring accommodations for a minimum of 2000 first-class, 1500 tourist-class, and 2400 economy-class passengers. The T3 costs $60 per mile to operate and can accommodate 40 first-class, 40 tourist-class, and 120 economy-class passengers, whereas the S5 costs $50 per mile to operate and can accommodate 80 first-class, 30 tourist-class, and 40 economy-class passengers. How many of each type of airplane should be used in order to minimize the operating cost?

36. *Furniture Production.*   P. J. Edward Furniture Design produces chairs and sofas. The chairs require 20 ft of wood, 5 lb of foam rubber, and 4 sq yd of fabric. The sofas require 100 ft of wood, 50 lb of foam rubber, and 20 sq yd of fabric. The company has 1500 ft of wood, 500 lb of foam rubber, and 240 sq yd of fabric. The chairs can be sold for $400 each and the sofas for $1500 each. How many of each should be produced in order to maximize income?

YOUR TURN ANSWERS: SECTION 4.5
1. 7 shrubs and 3 trees for a maximum profit of $124
2. 5 book summaries and 5 research projects for a maximum of 750 points

QUICK QUIZ: SECTIONS 4.1–4.5
Solve.

1. $4 \le 3 - 5x < 7$ [4.2]   **2.** $|6x - 8| = 12$ [4.3]

3. $3|2x + 1| + 5 < 8$ [4.3]

4. Graph the following system of inequalities. Find the coordinates of any vertices formed. [4.4]

$$y \le 2x + 3,$$
$$x \le 2,$$
$$y \ge 0$$

5. Find the maximum and minimum values of $F = 4x - y$ subject to

$$y \le 2x + 3,$$
$$x \le 2,$$
$$y \ge 0.$$ [4.5]

PREPARE TO MOVE ON
Evaluate.

1. $3x^3 - 5x^2 - 8x + 7$, for $x = -1$ [1.1], [1.2]

2. $t^3 + 6t^2 - 10$, for $t = 2$ [1.1]

Simplify. [1.3]

3. $3(2t - 7) + 5(3t + 1)$ **4.** $(8t + 6) - (7t + 6)$

A
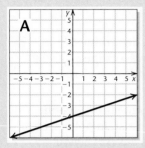

B
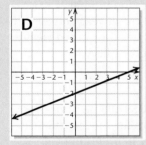

C
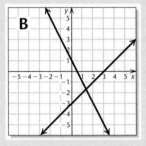

D
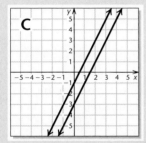

E
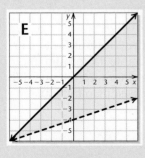

Visualizing for Success

Use after Section 4.4.

Match each equation, inequality, or system of equations or inequalities with its graph.

1. $x - y = 3,$
   $2x + y = 1$

2. $3x - y \leq 5$

3. $x > -3$

4. $y = \dfrac{1}{3}x - 4$

5. $y > \dfrac{1}{3}x - 4,$
   $y \leq x$

6. $x = y$

7. $y = 2x - 1,$
   $y = 2x - 3$

8. $2x - 5y = 10$

9. $x + y \leq 3,$
   $2y \leq x + 1$

10. $y = \dfrac{3}{2}$

Answers on page A-22

An additional, animated version of this activity appears in MyMathLab. To use MyMathLab, you need a course ID and a student access code. Contact your instructor for more information.

F
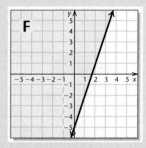

G
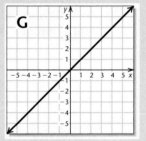

H
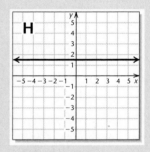

I
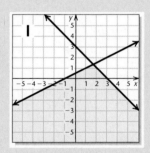

J
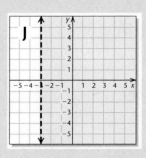

Collaborative Activity   *Saving on Shipping Costs*

Focus: Inequalities and problem solving
Use after: Section 4.1
Time: 20–30 minutes
Group size: 2–3

For overnight delivery packages weighing 10 lb or more sent by Express Mail, the United States Postal Service charges $44.55 (as of March 2012) for a 10-lb package mailed to Zone 3 plus $3.62 for each pound or part of a pound over 10 lb. UPS Next Day Air charges $49.25 for a 10-lb package mailed to Zone 3 plus $1.93 for each pound or part of a pound over 10 lb.*

---

* This activity is based on an article by Michael Contino in *Mathematics Teacher*, May 1995.

Activity

1. One group member should determine the function p, where $p(x)$ represents the cost, in dollars, of mailing x pounds using Express Mail.

2. One member should determine the function r, where $r(x)$ represents the cost, in dollars, of shipping x pounds using UPS Next Day Air.

3. A third member should graph p and r on the same set of axes.

4. Finally, working together, use the graph to determine those weights for which Express Mail is less expensive than UPS Next Day Air shipping. Express your answer in both set-builder notation and interval notation.

Decision Making ⟂ Connection   *(Use after Section 4.1.)*

Choosing an Insurance Plan.   There are many factors to consider when choosing an insurance plan, some of which are the amount of the monthly premium, the yearly deductible, and the percentage of the bill that the insured is expected to pay.

1. Elisabeth, 21, is single, is a nonsmoker, and has no children. Under a United Health One medical insurance plan, she would pay the first $2500 of her medical bills each year and 20% of all remaining bills. Under SmartSense Plus, she would pay the first $1500 of her medical bills each year and 30% of the remaining bills. For what amount of medical bills will the United Health One plan save Elisabeth money? (Assume that Elisabeth's bills will exceed $2500.)

   Source: Quotes from ehealthinsurance.com

2. The monthly premiums for the plans in Exercise 1 are $97.98 per month for the United Health One plan and $197.17 per month for the SmartSense Plus plan. For each plan, how much will Elisabeth pay in premiums per year?

3. Considering the information in Exercises 1 and 2, for what amount of medical bills will the United Health One plan save Elisabeth money?

4. *Research.*   Find the deductibles, copays, and monthly premiums for two or more insurance plans for which you or a friend are eligible. Determine the amount of medical bills for which each plan will save you money. If possible, estimate your annual medical bills and decide on the best plan for you.

Study Summary

SECTION 4.1: *Inequalities and Applications*

An **inequality** is any sentence containing $<$, $>$, $\leq$, $\geq$, or $\neq$. Solution sets of inequalities can be **graphed** and written in **set-builder notation** or **interval notation**.

Interval Notation	Set-builder Notation	Graph
(a, b)	$\{x \mid a < x < b\}$	
$[a, b]$	$\{x \mid a \leq x \leq b\}$	
$[a, b)$	$\{x \mid a \leq x < b\}$	
$(a, b]$	$\{x \mid a < x \leq b\}$	
(a, ∞)	$\{x \mid a < x\}$	
$(-\infty, a)$	$\{x \mid x < a\}$	

1. Write using interval notation:
$$\{x \mid x \leq 0\}.$$

The Addition Principle for Inequalities

For any real numbers a, b, and c,

$a < b$ is equivalent to $a + c < b + c$;

$a > b$ is equivalent to $a + c > b + c$.

Similar statements hold for $\leq$ and $\geq$.

Solve: $x + 3 \leq 5$.
$$x + 3 \leq 5$$
$$x + 3 - 3 \leq 5 - 3 \qquad \text{Subtracting 3 from both sides}$$
$$x \leq 2$$
The solution set is $\{x \mid x \leq 2\}$, or $(-\infty, 2]$.

2. Solve: $x - 11 > -4$.

The Multiplication Principle for Inequalities

For any real numbers a and b, and for any *positive* number c,

$a < b$ is equivalent to $ac < bc$;

$a > b$ is equivalent to $ac > bc$.

For any real numbers a and b, and for any *negative* number c,

$a < b$ is equivalent to $ac > bc$;

$a > b$ is equivalent to $ac < bc$.

Similar statements hold for $\leq$ and $\geq$.

Solve: $3x > 9$.
$$3x > 9$$
$$\tfrac{1}{3} \cdot 3x > \tfrac{1}{3} \cdot 9 \qquad \text{The inequality symbol does not change because } \tfrac{1}{3} \text{ is positive.}$$
$$x > 3$$
The solution set is $\{x \mid x > 3\}$, or $(3, \infty)$.

Solve: $-4x \geq 20$.
$$-4x \geq 20$$
$$-\tfrac{1}{4} \cdot (-4x) \leq -\tfrac{1}{4} \cdot 20 \qquad \text{The inequality symbol is reversed because } -\tfrac{1}{4} \text{ is negative.}$$
$$x \leq -5$$
The solution set is $\{x \mid x \leq -5\}$, or $(-\infty, -5]$.

3. Solve: $-8x \leq 2$.

Many real-world problems can be solved by translating the problem to an inequality and applying the five-step problem-solving strategy.	Translate to an inequality.
	
	    The test score must exceed 85. $s > 85$
	    At most 15 volunteers greeted visitors. $v \leq 15$
	    Ona makes no more than \$100 per week. $w \leq 100$
		4. Translate to an inequality:
		
		Luke runs no less than 3 mi per day.

SECTION 4.2: *Intersections, Unions, and Compound Inequalities*

A **conjunction** consists of two or more sentences joined by the word *and*. The solution set of the conjunction is the **intersection** of the solution sets of the individual sentences.	$-4 \leq x - 1 \leq 5$ $-4 \leq x - 1$ *and* $x - 1 \leq 5$ $-3 \leq x$ *and* $x \leq 6$ The solution set is $\{x \mid -3 \leq x \leq 6\}$, or $[-3, 6]$. 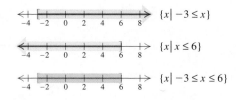	**5.** Solve: $-5 < 4x + 3 \leq 0$.

A **disjunction** consists of two or more sentences joined by the word *or*. The solution set of the disjunction is the **union** of the solution sets of the individual sentences.	$2x + 9 < 1$ *or* $5x - 2 \geq 3$
	$2x < -8$ *or* $5x \geq 5$
	$x < -4$ *or* $x \geq 1$
	The solution set is $\{x \mid x < -4 \text{ or } x \geq 1\}$, or $(-\infty, -4) \cup [1, \infty)$.
	
	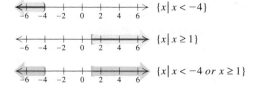
		6. Solve:
		    $x - 3 \leq 10$ *or*
		    $25 - x < 3$.

SECTION 4.3: *Absolute-Value Equations and Inequalities*

For any positive number p and any algebraic expression X: **a)** The solutions of $\lvert X \rvert = p$ are those numbers that satisfy $X = -p$ *or* $X = p$. **b)** The solutions of $\lvert X \rvert < p$ are those numbers that satisfy $-p < X < p$. **c)** The solutions of $\lvert X \rvert > p$ are those numbers that satisfy $X < -p$ *or* $p < X$.	$\lvert x + 3 \rvert = 4$ $x + 3 = 4$ *or* $x + 3 = -4$ Using part (a) $x = 1$ *or* $x = -7$ The solution set is $\{-7, 1\}$. $\lvert x + 3 \rvert < 4$ $-4 < x + 3 < 4$ Using part (b) $-7 < x < 1$ The solution set is $\{x \mid -7 < x < 1\}$, or $(-7, 1)$. $\lvert x + 3 \rvert \geq 4$ $x + 3 \leq -4$ *or* $4 \leq x + 3$ Using part (c) $x \leq -7$ *or* $1 \leq x$ The solution set is $\{x \mid x \leq -7 \text{ or } x \geq 1\}$, or $(-\infty, -7] \cup [1, \infty)$.	Solve. **7.** $\lvert 4x - 7 \rvert = 11$ **8.** $\lvert x - 12 \rvert \leq 1$ **9.** $\lvert 2x + 3 \rvert > 7$

SECTION 4.4:	*Inequalities in Two Variables*

<table>
<tr>
<td>

To graph a linear inequality:

1. Graph the **boundary line.** Draw a dashed line if the inequality symbol is $<$ or $>$, and draw a solid line if the inequality symbol is $\leq$ or $\geq$.

2. Determine which side of the boundary line contains the solution set, and shade that **half-plane.**

</td>
<td>

Graph: $x + y < -1$.

1. Graph $x + y = -1$ using a dashed line.

2. Choose a test point not on the line: $(0, 0)$.

$$\frac{x + y < -1}{0 + 0 \mid -1}$$
$$0 \overset{?}{<} -1 \quad \text{FALSE}$$

Since $0 < -1$ is false, shade the half-plane that does *not* contain $(0, 0)$.

</td>
<td>

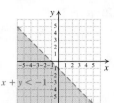

</td>
<td>

10. Graph: $2x - y < 5$.

</td>
</tr>
</table>

SECTION 4.5:	*Applications Using Linear Programming*

<table>
<tr>
<td>

The Corner Principle
The maximum or minimum value of an **objective function** over a *feasible region* is the maximum or minimum value of the function at a **vertex** of that region.

</td>
<td>

Maximize $F = x + 2y$ subject to

$$x + y \leq 5,$$
$$x \geq 0,$$
$$y \geq 1.$$

1. Graph the feasible region.

2. Find the value of F at the vertices.

Vertex	$F = x + 2y$
$(0, 1)$	2
$(0, 5)$	10
$(4, 1)$	6

← The maximum value of F is 10.

</td>
<td>

11. Maximize $F = 2x - y$ subject to

$$x + y \leq 4,$$
$$x \geq 1,$$
$$y \geq 0.$$

</td>
</tr>
</table>

Review Exercises: Chapter 4

⬥ Concept Reinforcement

Classify each of the following statements as either true or false.

1. If x cannot exceed 10, then $x \leq 10$.   [4.1]

2. It is always true that if $a > b$, then $ac > bc$.   [4.1]

3. The solution of $|3x - 5| \leq 8$ is a closed interval. [4.3]

4. The inequality $2 < 5x + 1 < 9$ is equivalent to $2 < 5x + 1$ *or* $5x + 1 < 9$.   [4.2]

5. The solution set of a disjunction is the union of two solution sets.   [4.2]

6. The equation $|x| = r$ has no solution when r is negative.   [4.3]

7. $|f(x)| > 3$ is equivalent to $f(x) < -3$ *or* $f(x) > 3$. [4.3]

8. The inequality symbol is used to determine whether the line in a linear inequality is drawn solid or dashed. [4.4]

9. The graph of a system of linear inequalities is always a half-plane.   [4.4]

10. The corner principle states that every objective function has a maximum or a minimum value.   [4.5]

Graph each inequality and write the solution set using both set-builder notation and interval notation.   [4.1]

11. $x \le 1$

12. $a + 3 \le 7$

13. $4y > -15$

14. $-0.2y < 6$

15. $-6x - 5 < 4$

16. $-\frac{1}{2}x - \frac{1}{4} > \frac{1}{2} - \frac{1}{4}x$

17. $0.3y - 7 < 2.6y + 15$

18. $-2(x - 5) \ge 6(x + 7) - 12$

19. Let $f(x) = 3x + 2$ and $g(x) = 10 - x$. Find all values of x for which $f(x) \le g(x)$.   [4.1]

Solve.   [4.1]

20. Mariah has two offers for a summer job. She can work in a sandwich shop for $8.40 per hour, or she can do carpentry work for $16 per hour. In order to do the carpentry work, she must spend $950 for tools. For how many hours must Mariah work in order for carpentry to be more profitable than the sandwich shop?

21. Clay is going to invest $9000, part at 3% and the rest at 3.5%. What is the most he can invest at 3% and still be guaranteed $300 in interest each year?

22. Find the intersection:
$$\{a, b, c, d\} \cap \{a, c, e, f, g\}.   [4.2]$$

23. Find the union:
$$\{a, b, c, d\} \cup \{a, c, e, f, g\}.   [4.2]$$

Graph and write interval notation.   [4.2]

24. $x \le 2$ *and* $x > -3$

25. $x \le 3$ *or* $x > -5$

Solve and graph each solution set.   [4.2]

26. $-3 < x + 5 \le 5$

27. $-15 < -4x - 5 < 0$

28. $3x < -9$ *or* $-5x < -5$

29. $2x + 5 < -17$ *or* $-4x + 10 \le 34$

30. $2x + 7 \le -5$ *or* $x + 7 \ge 15$

31. $f(x) < -5$ *or* $f(x) > 5$, where $f(x) = 3 - 5x$

For $f(x)$ as given, use interval notation to write the domain of f.

32. $f(x) = \dfrac{2x}{x + 3}$ [4.2]

33. $f(x) = \sqrt{5x - 10}$ [4.1]

34. $f(x) = \sqrt{1 - 4x}$ [4.1]

Solve.   [4.3]

35. $|x| = 11$

36. $|t| \ge 21$

37. $|x - 8| = 3$

38. $|4a + 3| < 11$

39. $|3x - 4| \ge 15$

40. $|2x + 5| = |x - 9|$

41. $|5n + 6| = -11$

42. $\left| \dfrac{x + 4}{6} \right| \le 2$

43. $2|x - 5| - 7 > 3$

44. $19 - 3|x + 1| \ge 4$

45. Let $f(x) = |8x - 3|$. Find all x for which $f(x) < 0$.   [4.3]

46. Graph $x - 2y \ge 6$ on a plane.   [4.4]

Graph each system of inequalities. Find the coordinates of any vertices formed.   [4.4]

47. $x + 3y > -1$,
$x + 3y < 4$

48. $x - 3y \le 3$,
$x + 3y \ge 9$,
$y \le 6$

49. Find the maximum and the minimum values of
$$F = 3x + y + 4$$
subject to
$$y \le 2x + 1,$$
$$x \le 7,$$
$$y \ge 3.   [4.5]$$

50. Better Books orders at least 100 copies per week of the current best-selling fiction book. It costs $2 to ship each book from their East-coast supplier and $4 to ship each book from their West-coast supplier, and they can spend no more than $320 per week for shipping. Because of the shipping methods used, it takes 5 days for shipments from the East coast to arrive but only 2 days for shipments from the West coast to arrive. How many books should they order from each supplier in order to minimize shipping time?

Synthesis

51. Explain in your own words why $|X| = p$ has two solutions when p is positive and no solution when p is negative.   [4.3]

52. Explain why the graph of the solution of a system of linear inequalities is the intersection, not the union, of the individual graphs.   [4.4]

53. Solve: $|2x + 5| \le |x + 3|$.   [4.3]

54. Classify as true or false: If $x < 3$, then $x^2 < 9$. If false, give an example showing why.   [4.1]

55. Super Lock manufactures brass doorknobs with a 2.5-in. diameter and a ± 0.003-in. manufacturing tolerance, or allowable variation in diameter. Write the tolerance as an inequality with absolute value.   [4.3]

Test: Chapter 4

For step-by-step test solutions, access the Chapter Test Prep Videos in My MathLab®, on YouTube® (search "Bittinger Inter Alg CA" and click on "Channels"), or by scanning the code.

Graph each inequality and write the solution set using both set-builder notation and interval notation.

1. $x - 3 < 8$

2. $-\frac{1}{2}t < 12$

3. $-4y - 3 \geq 5$

4. $3a - 5 \leq -2a + 6$

5. $3(7 - x) < 2x + 5$

6. $-2(3x - 1) - 5 \geq 6x - 4(3 - x)$

7. Let $f(x) = -5x - 1$ and $g(x) = -9x + 3$. Find all values of x for which $f(x) > g(x)$.

8. Dani can rent a van for either $80 with unlimited mileage or $45 plus 40¢ per mile. For what numbers of miles traveled would the unlimited mileage plan save Dani money?

9. A refrigeration repair company charges $80 for the first half-hour of work and $60 for each additional hour. Blue Mountain Camp has budgeted $200 to repair its walk-in cooler. For what lengths of a service call will the budget not be exceeded?

10. Find the intersection:

$$\{a, e, i, o, u\} \cap \{a, b, c, d, e\}.$$

11. Find the union:

$$\{a, e, i, o, u\} \cup \{a, b, c, d, e\}.$$

For $f(x)$ as given, use interval notation to write the domain of f.

12. $f(x) = \sqrt{6 - 3x}$

13. $f(x) = \dfrac{x}{x - 7}$

Solve and graph each solution set.

14. $-5 < 4x + 1 \leq 3$

15. $3x - 2 < 7 \ or \ x - 2 > 4$

16. $-3x > 12 \ or \ 4x \geq -10$

17. $1 \leq 3 - 2x \leq 9$

18. $|n| = 15$

19. $|a| > 5$

20. $|3x - 1| < 7$

21. $|-5t - 3| \geq 10$

22. $|2 - 5x| = -12$

23. Let $g(x) = 4 - 2x$. Find all values of x for which $g(x) < -3 \ or \ g(x) > 3$.

24. Let $f(x) = |2x - 1|$ and $g(x) = |2x + 7|$. Find all values of x for which $f(x) = g(x)$.

25. Graph $y \leq 2x + 1$ on a plane.

Graph each system of inequalities. Find the coordinates of any vertices formed.

26. $x + y \geq 3,$
$x - y \geq 5$

27. $2y - x \geq -7,$
$2y + 3x \leq 15,$
$y \leq 0,$
$x \leq 0$

28. Find the maximum and the minimum values of

$$F = 5x + 3y$$

subject to

$$x + y \leq 15,$$
$$1 \leq x \leq 6,$$
$$0 \leq y \leq 12.$$

29. Swift Cuts makes $12 on each manicure and $18 on each haircut. A manicure takes 30 min and a haircut takes 50 min, and there are 5 stylists who each work 6 hr per day. If the salon can schedule 50 appointments per day, how many should be manicures and how many haircuts in order to maximize profit? What is the maximum profit?

Synthesis

Solve. Write each solution set using interval notation.

30. $|2x - 5| \leq 7 \ and \ |x - 2| \geq 2$

31. $7x < 8 - 3x < 6 + 7x$

32. Write an absolute-value inequality for which the interval shown is the solution.

Cumulative Review: Chapters 1–4

Simplify. Do not use negative exponents in your answers.

1. $3 + 24 \div 2^2 \cdot 3 - (6 - 7)$ [1.2]

2. $3c - [8 - 2(1 - c)]$ [1.3]

3. $(3xy^{-4})(-2x^3y)$ [1.6]

4. $\left(\dfrac{18a^2b^{-1}}{12a^{-1}b}\right)^2$ [1.6]

Solve.

5. $3(x - 2) = 14 - x$ [1.3]

6. $|2x - 1| = 8$ [4.3]

7. $|4t| > 12$ [4.3]

8. $|3x - 2| \le 8$ [4.3]

9. $x - 2 < 6 \text{ or } 2x + 1 > 5$ [4.2]

10. $2x + 5y = 2,$
$3x - y = 4$ [3.2]

11. $y = \frac{1}{2}x - 7,$
$2x - 4y = 3$ [3.2]

12. $9(x - 3) - 4x < 2 - (3 - x)$ [4.1]

Graph on a plane.

13. $y = \frac{2}{3}x - 4$ [2.3]

14. $x = -3$ [2.4]

15. $3x - y = 3$ [2.4]

16. $x + y \ge -2$ [4.4]

17. $f(x) = -x + 1$ [2.3]

18. $x - 2y > 4,$
$x + 2y \ge -2$ [4.4]

19. Find the slope and the y-intercept of the line given by $4x - 9y = 18$. [2.3]

20. Using function notation, write a slope–intercept equation for the line with slope -7 that contains $(-3, -4)$. [2.5]

21. Find an equation of the line with y-intercept $(0, 4)$ and perpendicular to the line given by $3x + 2y = 1$. [2.5]

22. For f as shown, determine the domain and the range. [2.2], [4.2]

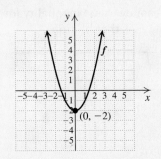

23. Find $g(-2)$ if $g(x) = 3x^2 - 5x$. [2.2]

24. Find $(f - g)(x)$ if $f(x) = x^2 + 3x$ and $g(x) = 9 - 3x$. [2.6]

25. Graph the solution set of $-3 \le f(x) \le 2$, where $f(x) = 1 - x$. [4.2]

26. Find the domain of h/g if $h(x) = \dfrac{1}{x}$ and $g(x) = 3x - 1$. [2.6]

27. Solve for t: $at - dt = c$. [1.5]

28. *Water Usage.* On average, it takes about 750,000 gal of water to create an acre of machine-made snow. Resorts in the Alps make about 60,000 acres of machine-made snow each year. Using scientific notation, find the amount of water used each year to make machine-made snow in the Alps. [1.7]

Sources: Swiss Federal Institute for Snow and Avalanche Research; www.telegraph.co.uk

29. *Water Usage.* In dry climates, it takes about 11,600 gal of water to produce one pound of beef and one pound of wheat. The pound of beef requires 7000 more gallons of water than the pound of wheat. How much water does it take to produce each? [3.3]

30. *Advertising.* Facebook's U.S. advertising revenue was $0.56 billion in 2009 and was estimated to be $3.36 billion in 2013. Let $f(t)$ represent U.S. advertising revenue for Facebook, in billions of dollars, t years after 2009.

Source: eMarketer.com

a) Find a linear function that fits the data. [2.5]

b) Use the function from part (a) to predict U.S. advertising revenue for Facebook in 2015. [2.5]

c) In what year will Facebook's U.S. advertising revenue be $7 billion? [2.5]

Synthesis

31. If $(2, 6)$ and $(-1, 5)$ are two solutions of $f(x) = mx + b$, find m and b. [3.2]

32. Use interval notation to write the domain of the function given by

$$f(x) = \frac{\sqrt{x + 4}}{x}.$$ [4.2]

Polynomials and Polynomial Functions

SPEED (in miles per hour)	GAS MILEAGE (in miles per gallon)
10	15
35	29
50	30
65	27

Speed Limits Save Lives— and Money.

Not only is it safer to observe the speed limit when driving, it can also save on fuel cost. A vehicle's gas mileage generally decreases significantly for speeds over 60 mph. The table shown lists gas mileage for a particular vehicle for several speeds. These data can be modeled using a *polynomial function*, which in turn can be used to estimate mileages not given in the table. (*See Example 7 in Section 5.1.*)

Source: Based on information from www.fueleconomy.gov

Without mathematics, the world of transportation would be literally at a standstill.

Bryan Swank, an engine designer from Columbus, Indiana, says that the application of math is critical in creating successful engine designs. He uses math to calculate basic hardware capabilities, to develop complex equations used by the engine control module to manage the engine's performance, and to understand product reliability and performance.

In many ways, polynomials, like $2x + 5$ or $x^2 - 3$, are central to the study of algebra. We have already used polynomials in this text, without labeling them as such. In this chapter, we will clearly define what polynomials are and discuss how to manipulate them. We will then use polynomials and polynomial functions in problem solving.

5.1   Introduction to Polynomials and Polynomial Functions

Terms and Polynomials ▪ Degree and Coefficients ▪ Polynomial Functions ▪ Adding Polynomials ▪
Opposites and Subtraction

In this section, we define a type of algebraic expression known as a *polynomial*. After developing some vocabulary, we study addition and subtraction of polynomials, and evaluate *polynomial functions*.

Terms and Polynomials

We have seen a variety of algebraic expressions like

$$3a^2b^4, \qquad 2l + 2w, \quad \text{and} \quad 5x^2 + x - 2.$$

Within these expressions, $3a^2b^4$, $2l$, $2w$, $5x^2$, x, and -2 are examples of *terms*. A **term** can be a number (like -2), a variable (like x), a product of numbers and/or variables (like $3a^2b^4$, $2l$, or $5x^2$), or a quotient of numbers and/or variables (like $7/t$).

If a term is a product of constants and/or variables, it is called a **monomial**. A term, but not a monomial, can include division by a variable. A **polynomial** is a monomial or a sum of monomials.

Examples of monomials: $3, \quad n, \quad 2w, \quad 5x^2y^3z, \quad \frac{1}{3}t^{10}$

Examples of polynomials: $3a + 2, \quad \frac{1}{2}x^2, \quad -3t^2 + t - 5, \quad x, \quad 0$

The following algebraic expressions are *not* polynomials:

$$\textbf{(1) } \frac{x + 3}{x - 4}, \qquad \textbf{(2) } 5x^3 - 2x^2 + \frac{1}{x}, \qquad \textbf{(3) } \frac{1}{x^3 - 2}.$$

Expressions (1) and (3) are not polynomials because they represent quotients, not sums. Expression (2) is not a polynomial because $1/x$ is not a monomial.

When a polynomial is written as a sum of monomials, each monomial is called a *term of the polynomial*.

EXAMPLE 1   Identify the terms of the polynomial $3t^4 - 5t^6 - 4t + 2$.

SOLUTION   The terms are $3t^4$, $-5t^6$, $-4t$, and 2. We can see this by rewriting all subtractions as additions of opposites:

$$3t^4 - 5t^6 - 4t + 2 = 3t^4 + (-5t^6) + (-4t) + 2.$$

These are the terms of the polynomial.

1. Identify the terms of $-y^4 + 7y^2 - 2y - 1$.

YOUR TURN

A polynomial with two terms is called a **binomial**, and those with three terms are called **trinomials**. Polynomials with four or more terms have no special name.

Monomials	Binomials	Trinomials	No Special Name
$4x^2$	$2x + 4$	$3t^3 + 4t + 7$	$4x^3 - 5x^2 + xy - 8$
9	$3a^5 + 6bc$	$6x^7 - 8z^2 + 4$	$z^5 + 2z^4 - z^3 + 7z + 3$
$-7a^{19}b^5$	$-9x^7 - 6$	$4x^2 - 6x - \frac{1}{2}$	$4x^6 - 3x^5 + x^4 - x^3 + 2x - 1$

Degree and Coefficients

The **degree of a term** of a polynomial is the number of variable factors in that term. Thus the degree of $7t^2$ is 2 because $7t^2$ has two variable factors: $7t^2 = 7 \cdot t \cdot t$. If a term contains more than one variable, we can find the degree by adding the exponents of the variables.

EXAMPLE 2 Determine the degree of each term: **(a)** $8x^4$; **(b)** $3x$; **(c)** 7; **(d)** $9x^2yz^4$.

SOLUTION

a) The degree of $8x^4$ is 4. x^4 represents 4 variable factors: $x \cdot x \cdot x \cdot x$.

b) The degree of $3x$ is 1.   There is 1 variable factor.

c) The degree of 7 is 0.   There is no variable factor.

d) The degree of $9x^2yz^4$ is 7. $9x^2yz^4 = 9x^2y^1z^4$, and $2 + 1 + 4 = 7$. Note that $9x^2yz^4$ has 7 variable factors: $9 \cdot x \cdot x \cdot y \cdot z \cdot z \cdot z \cdot z$.

2. Determine the degree of $5a^6b^8$.

YOUR TURN

The degree of a constant polynomial, such as 7, is 0, since there are no variable factors. Also, note that $7 = 7 \cdot x^0$. The polynomial 0 is an exception, since $0 = 0x = 0x^2 = 0x^3$, and so on. We say that the polynomial 0 has *no* degree.

The part of a term that is a constant factor is the **coefficient** of that term. Thus the coefficient of $3x$ is 3, and the coefficient of the term 7 is simply 7.

EXAMPLE 3 Identify the coefficient of each term in the polynomial

$$4x^3 - 7x^2y + x - 8.$$

SOLUTION

The coefficient of $4x^3$ is 4.

The coefficient of $-7x^2y$ is -7.

The coefficient of x is 1, since $x = 1x$.

The coefficient of -8 is simply -8.

3. Identify the coefficient of $6a^2b^5$.

YOUR TURN

The **leading term** of a polynomial is the term of highest degree. Its coefficient is called the **leading coefficient**, and its degree is referred to as the **degree of the polynomial**. To see how this terminology is used, consider the polynomial

$$3x^2 - 8x^3 + 5x^4 + 7x - 6.$$

The *terms* are $3x^2$, $-8x^3$, $5x^4$, $7x$, and -6.

The *coefficients* are   3, -8, 5, 7, and -6.

The *degree of each term* is   2,   3,   4,   1,   and 0.

The *leading term* is $5x^4$ and the *leading coefficient* is 5.

The *degree of the polynomial* is 4.

We usually arrange polynomials in one variable so that exponents *decrease* from left to right. This is called **descending order**. Some polynomials may be written in **ascending order**, with exponents *increasing* from left to right. Generally, if an exercise is written in one kind of order, the answer is written in that same order.

EXAMPLE 4   Arrange in ascending order: $12 + 2x^3 - 7x + x^2$.

SOLUTION

$$12 + 2x^3 - 7x + x^2 = 12 - 7x + x^2 + 2x^3$$

4. Arrange the polynomial in Example 4 in descending order.

YOUR TURN

Polynomials in several variables can be arranged with respect to the powers of one of the variables.

EXAMPLE 5   Arrange in descending powers of x:

$$y^4 + 2 - 5x^2 + 3x^3y + 7xy^2.$$

SOLUTION   Using a commutative law, we have

$$y^4 + 2 - 5x^2 + 3x^3y + 7xy^2 = \underbrace{3x^3y - 5x^2 + 7xy^2 + y^4 + 2}.$$

The powers of x decrease
from left to right.

5. Arrange the polynomial in Example 5 in ascending powers of y.

YOUR TURN

Polynomial Functions

In a *polynomial function*, such as $P(x) = 5x^4 - 6x^2 + x - 7$, outputs are determined by evaluating a polynomial. Polynomial functions are classified by the degree of the polynomial used to define the function, as shown below.

Type of Function	Degree	Example
Linear	1	$f(x) = 2x + 5$
Quadratic	2	$g(x) = x^2$
Cubic	3	$p(x) = 5x^3 - \frac{1}{3}x + 2$
Quartic	4	$h(x) = 9x^4 - 6x^3$

To evaluate a polynomial, we substitute a number for the variable.

EXAMPLE 6   For the polynomial function $P(x) = -x^2 + 4x - 1$, find $P(5)$ and $P(-5)$.

SOLUTION

$$P(5) = -5^2 + 4(5) - 1 \qquad \text{We square each input before taking its opposite.}$$

$$= -25 + 20 - 1 = -6$$

$$P(-5) = -(-5)^2 + 4(-5) - 1 \qquad \text{Use parentheses when an input is negative.}$$

$$= -25 - 20 - 1 = -46$$

CAUTION! Note that $-(-5)^2 = -25$. We square the input first and then take its opposite.

6. For the polynomial function $P(x) = x - 2x^2$, find $P(-3)$.

YOUR TURN

Speed (in miles per hour)	Gas mileage (in miles per gallon)
10	15
35	29
50	30
65	27

EXAMPLE 7 *Fuel Efficiency.* One fuel-saving tip is to observe the speed limit. A vehicle's gas mileage generally decreases significantly for speeds over 50 mph, as suggested by the data in the table at left. The polynomial function

$$F(x) = 0.000095x^3 - 0.0225x^2 + 1.421x + 3.26$$

can be used to estimate the fuel economy, in miles per gallon (mpg), for a particular vehicle traveling x miles per hour (mph).

Source: Based on information from www.fueleconomy.gov

a) What is the gas mileage for this vehicle at 60 mph?

b) Use the graph below to estimate $F(75)$.

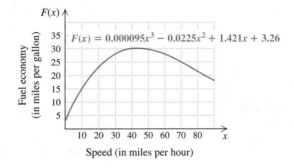

Technology Connection

One way to evaluate a function is to enter and graph it as y_1 and then select TRACE. We can then enter any x-value that appears in that window and the corresponding y-value will appear. We use this approach to check Example 7(a).

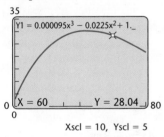

Xscl = 10, Yscl = 5

Use this approach to check Example 7(b).

SOLUTION

a) We evaluate the function for $x = 60$:

$$F(60) = 0.000095(60)^3 - 0.0225(60)^2 + 1.421(60) + 3.26$$
$$= 20.52 - 81 + 85.26 + 3.26$$
$$= 28.04.$$

The gas mileage for this vehicle at 60 mph is approximately 28 mpg.

b) To estimate $F(75)$, the gas mileage at 75 mph, we locate 75 on the horizontal axis on the following graph. From there, we move vertically to the graph of the function and then horizontally to the $F(x)$-axis, as shown. This locates a value for $F(75)$ of about 23.

The gas mileage for this vehicle at 75 mph is approximately 23 mpg.

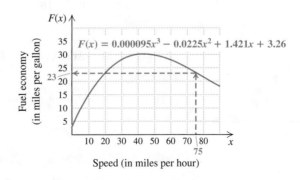

7. Use the function given in Example 7 to estimate the gas mileage at 45 mph.

 YOUR TURN

Adding Polynomials

When two terms have the same variable(s) raised to the same power(s), they are **similar**, or **like**, **terms** and can be "combined" or "collected."

EXAMPLE 8   Combine like terms.

a) $3x^2 - 4y + 2x^2$ **b)** $4t^3 - 6t - 8t^2 + t^3 + 9t^2$

c) $3x^2y + 5xy^2 - 3x^2y - xy^2$

SOLUTION

a) $3x^2 - 4y + 2x^2 = 3x^2 + 2x^2 - 4y$ Rearranging terms using the commutative law for addition

$= (3 + 2)x^2 - 4y$ Using the distributive law

$= 5x^2 - 4y$

b) $4t^3 - 6t - 8t^2 + t^3 + 9t^2 = 5t^3 + t^2 - 6t$ We usually perform the middle steps mentally and write just the answer.

8. Combine like terms:

$3n - n^3 + 2n + 5 - n^3 + 6.$

c) $3x^2y + 5xy^2 - 3x^2y - xy^2 = 4xy^2$

YOUR TURN

We add polynomials by combining like terms.

EXAMPLE 9   Add: $(-3x^3 + 2x - 4) + (4x^3 + 3x^2 + 2).$

9. Add:

$(y^2 - 2y + 3) + (y^2 + 2y - 7).$

SOLUTION

$(-3x^3 + 2x - 4) + (4x^3 + 3x^2 + 2) = x^3 + 3x^2 + 2x - 2$

YOUR TURN

To add using columns, we write the polynomials one under the other, listing like terms under one another and leaving spaces for any missing terms.

EXAMPLE 10   Add $4n^3 + 4n - 5$ and $-n^3 + 7n^2 - 2.$

SOLUTION

$$
\begin{array}{l}
4n^3 \qquad\;\; + 4n - 5 \\
\underline{-n^3 + 7n^2 \qquad - 2} \\
3n^3 + 7n^2 + 4n - 7
\end{array}
$$
   Combining like terms

10. Add $2x^4 + x^2 + x$
and $-3x^3 + 2x^2 - 7.$

YOUR TURN

EXAMPLE 11   Add $13x^3y + 3x^2y - 5y$ and $x^3y + 4x^2y - 3xy.$

SOLUTION

11. Add $6c^2d - 8cd + d^2$
and $3cd + 5cd^2 - d^2.$

$(13x^3y + 3x^2y - 5y) + (x^3y + 4x^2y - 3xy) = 14x^3y + 7x^2y - 3xy - 5y$

YOUR TURN

Opposites and Subtraction

If the sum of two polynomials is 0, the polynomials are *opposites*, or *additive inverses*, of each other. For example,

$$(3x^2 - 5x + 2) + (-3x^2 + 5x - 2) = 0,$$

so the opposite of $(3x^2 - 5x + 2)$ must be $(-3x^2 + 5x - 2)$. We can say the same thing using algebraic symbolism, as follows:

$$\underbrace{\text{The opposite of}}_{-} \quad \underbrace{(3x^2 - 5x + 2)}_{(3x^2 - 5x + 2)} \quad \underbrace{\text{is}}_{=} \quad \underbrace{-3x^2 + 5x - 2.}_{-3x^2 + 5x - 2}$$

To form the opposite of a polynomial, we can think of distributing the "−" sign, or multiplying each term of the polynomial by −1, and removing the parentheses. The effect is to change the sign of each term in the polynomial.

> **THE OPPOSITE OF A POLYNOMIAL**
>
> The *opposite* of a polynomial P can be written as $-P$ or, equivalently, by replacing each term in P with its opposite.

EXAMPLE 12 Write two equivalent expressions for the opposite of $7xy^2 - 6xy - 4y + 3$.

SOLUTION

a) The opposite of $7xy^2 - 6xy - 4y + 3$ can be written with parentheses as

$$-(7xy^2 - 6xy - 4y + 3).\qquad \text{Writing the opposite of } P \text{ as } -P$$

12. Write two equivalent expressions for the opposite of $-3y^4 - y^2 + y + 1$.

b) The opposite of $7xy^2 - 6xy - 4y + 3$ can be written without parentheses as

$$-7xy^2 + 6xy + 4y - 3.\qquad \text{Multiplying each term by } -1$$

YOUR TURN

To subtract a polynomial, we add its opposite.

EXAMPLE 13 Subtract: $(-3x^2 + 4xy) - (2x^2 - 5xy + 7y^2)$.

SOLUTION

$$\begin{aligned}
(-3x^2 &+ 4xy) - (2x^2 - 5xy + 7y^2) \\
&= (-3x^2 + 4xy) + (-2x^2 + 5xy - 7y^2) \qquad \text{Adding the opposite} \\
&= -3x^2 + 4xy - 2x^2 + 5xy - 7y^2 \qquad \text{Try to go directly to this step.} \\
&= -5x^2 + 9xy - 7y^2 \qquad \text{Combining like terms}
\end{aligned}$$

13. Subtract:
$(x^2 - x + 1) - (3x^2 - 2x - 7)$.

YOUR TURN

With practice, you may find that you can skip some steps, by mentally taking the opposite of each term being subtracted and then combining like terms.

To use columns for subtraction, we mentally change the signs of the terms being subtracted.

EXAMPLE 14 Subtract: $(3x^4 - 2x^3 + 6x - 1) - (3x^4 - 9x^3 - x^2 + 7)$.

SOLUTION

Write: (Subtract)

$$\begin{array}{r}
3x^4 - 2x^3 \phantom{{}+ x^2} + 6x - 1 \\
-(3x^4 - 9x^3 - x^2 \phantom{{}+ 6x} \mp 7) \\
\hline
\end{array}$$

Think: (Add)

$$\begin{array}{r}
3x^4 - 2x^3 \phantom{{}+ x^2} + 6x - 1 \\
-3x^4 + 9x^3 + x^2 \phantom{{}+ 6x} - 7 \\
\hline
7x^3 + x^2 + 6x - 8 \end{array}$$

14. Subtract:
$(-2n^3 - n^2 - 6n) - (3n^3 - n^2 + 5)$.

Take the opposite of each term mentally and add.

YOUR TURN

Technology Connection

By pressing **2ND** (TBLSET) and selecting AUTO, we can use a table to check addition and subtraction of polynomials. To check Example 9, we enter $y_1 = (-3x^3 + 2x - 4) + (4x^3 + 3x^2 + 2)$ and $y_2 = x^3 + 3x^2 + 2x - 2$. If the addition is correct, the values of y_1 and y_2 will match, regardless of the x-values used.

X	Y₁	Y₂
-2	-2	-2
-1	-2	-2
0	-2	-2
1	4	4
2	22	22
3	58	58
4	118	118

X = -2

Use a table to determine whether each sum or difference is correct.

1. $(x^3 - 2x^2 + 3x - 7) + (3x^2 - 4x + 5) \stackrel{?}{=}$
   $x^3 + x^2 - x - 2$
2. $(2x^2 + 3x - 6) + (5x^2 - 7x + 4) \stackrel{?}{=}$
   $7x^2 + 4x - 2$
3. $(x^4 + 2x^2 + x) - (3x^4 - 5x + 1) \stackrel{?}{=}$
   $-2x^4 + 2x^2 + 6x - 1$
4. $(3x^4 - 2x^2 - 1) - (2x^4 - 3x^2 - 4) \stackrel{?}{=}$
   $x^4 + x^2 - 5$

5.1 EXERCISE SET

FOR EXTRA HELP

MyMathLab® Math XL

 PRACTICE   WATCH   READ   REVIEW

Vocabulary and Reading Check

In each of Exercises 1–10, match the description with the appropriate expression from the column on the right.

1. ____ A binomial
2. ____ A trinomial
3. ____ A monomial
4. ____ A sixth-degree polynomial
5. ____ A polynomial written in ascending powers of t
6. ____ A term that is not a monomial
7. ____ A polynomial with a leading term of degree 5
8. ____ A polynomial with a leading coefficient of 5
9. ____ A cubic polynomial in one variable
10. ____ A polynomial containing similar terms

a) $9a^7$

b) $6s^2 - 2t + 4st^2 - st^3$

c) $4t^{-2}$

d) $t^4 - st + s^3$

e) $7t^3 - 13 + 5t^4 - 2t$

f) $4t^3 + 12t^2 + 9t - 7$

g) $5 + a$

h) $4t^6 + 7t - 8t^2 + 5$

i) $8st^3 - 6s^2t + 4st^3 - 2s$

j) $7s^3t^2 - 4s^2t + 3st^2 + 1$

Terms and Polynomials

Identify the terms of each polynomial.

11. $7x^4 + x^3 - 5x + 8$ 12. $5a^3 + 4a^2 - a - 7$

13. $-t^6 + 7t^3 - 3t^2 + 6$ 14. $n^5 - 4n^3 + 2n - 8$

Classify each polynomial as a monomial, a binomial, a trinomial, or a polynomial with no special name.

15. $x^2 - 23x + 17$ 16. $-9x^2$

17. $x^3 - 7x^2 + 2x - 4$ 18. $t^3 + 4t$

19. $y + 5$ 20. $4x^2 + 12x + 9$

21. 17

22. $2x^4 - 7x^3 + x^2 + x$

Degree and Coefficients

Determine the degree of each term in each polynomial.

23. $3x^2 - 5x$ 24. $9a^3 + 4a^2$

25. $2t^5 - t^2 + 1$ 26. $x^5 - x^4 + x + 6$

27. $8x^2y - 3x^4y^3 + y^4$ 28. $5a^2b^5 - ab + a^2b$

Determine the coefficient of each term in each polynomial.

29. $4x^5 + 7x - 3$ 30. $8x^3 - x^2 + 7$

31. $x^4 - x^3 + 4x$ 32. $3a^5 - a^3 + a$

33. $a^2b^3 - 5ab + 7b^2 + 1$

34. $10xy - x^2y + x^3 - 11$

In Exercises 35–38, for each polynomial given, answer the following questions.

a) *How many terms are there?*
b) *What is the degree of each term?*
c) *What is the degree of the polynomial?*
d) *What is the leading term?*
e) *What is the leading coefficient?*

35. $-5x^6 + x^4 + 7x^3 - 2x - 10$

36. $t^3 + 5t^2 - t + 9$

37. $7a^4 + a^3b^2 - 5a^2b + 3$

38. $-uv + 8v^4 + 9u^2v^5 - 6u^2 - 1$

Determine the degree of each polynomial.

39. $8y^2 + y^5 - 9 - 2y + 3y^4$

40. $3x^2 - 5x + 8x^4 + 12$

41. $3p^4 - 5pq + 2p^3q^3 + 8pq^2 - 7$

42. $2xy^3 + 9y^2 - 8x^3 + 7x^2y^2 + y^7$

Arrange in descending order. Then find the leading term and the leading coefficient.

43. $4 - 8t + 5t^2 + 2t^3 - 15t^4$

44. $4 - 7y^2 + 6y^4 - 2y - y^5$

45. $3x + 6x^5 - 5 - x^6 + 7x^2$

46. $a - a^2 + 12a^7 + 3a^4 - 15$

Arrange in ascending powers of x.

47. $4x + 5x^3 - x^6 - 9$

48. $7 - x + 3x^6 + 2x^4$

49. $2x^2y + 5xy^3 - x^3 + 8y$

50. $2ax - 9ab + 4x^5 - 7bx^2$

Polynomial Functions

Find g(3) for each polynomial function.

51. $g(x) = x - 5x^2 + 4$

52. $g(x) = 2 - x + 4x^2$

Find f(−1) for each polynomial function.

53. $f(x) = -3x^4 + 5x^3 + 6x - 2$

54. $f(x) = -5x^3 + 4x^2 - 7x + 9$

55. Find $F(2)$ and $F(5)$: $F(x) = 2x^2 - 6x - 9$.

56. Find $P(4)$ and $P(0)$: $P(x) = 3x^2 - 2x + 7$.

57. Find $Q(-3)$ and $Q(0)$:
$$Q(y) = -8y^3 + 7y^2 - 4y - 9.$$

58. Find $G(3)$ and $G(-1)$:
$$G(x) = -6x^2 - 5x + x^3 - 1.$$

Electing Officers.   For a club consisting of p people, the number of ways N(p) in which a president, a vice president, and a treasurer can be elected can be found using
$$N(p) = p^3 - 3p^2 + 2p.$$

59. The Southside Rugby Club has 20 members. In how many ways can they elect a president, a vice president, and a treasurer?

60. The Stage Right Drama Club has 12 members. In how many ways can a president, a vice president, and a treasurer be elected?

Horsepower.   The amount of horsepower needed to overcome air resistance by a car traveling v miles per hour can be approximated by the polynomial function given by
$$h(v) = \frac{0.354}{8250}v^3.$$

Source: "The Physics of Racing," Brian Beckman, www.miata.net

61. How much horsepower does a race car traveling 180 mph need to overcome air resistance?

62. How much horsepower does a car traveling 65 mph need to overcome air resistance?

Wind Energy.   The number of watts of power P(x) generated by a particular home-sized turbine at a wind speed of x miles per hour can be approximated by
$$P(x) = 0.0157x^3 + 0.1163x^2 - 1.3396x + 3.7063.$$
Use the following graph for Exercises 63–66.

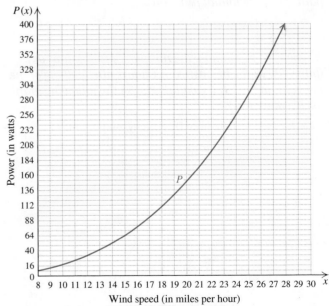

Source: Based on data from *QST*, November 2006

63. Estimate the power, in watts, generated by a 10-mph wind.

64. Estimate the power, in watts, generated by a 25-mph wind.

65. Approximate $P(20)$.

66. Approximate $P(15)$.

67. *Stacking Spheres.* In 2004, the journal *Annals of Mathematics* accepted a proof of the so-called Kepler Conjecture: that the most efficient way to pack spheres is in the shape of a square pyramid.

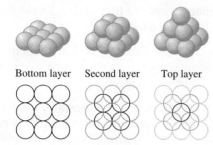

Bottom layer   Second layer   Top layer

The number of balls in the stack $N(x)$ is given by

$$N(x) = \tfrac{1}{3}x^3 + \tfrac{1}{2}x^2 + \tfrac{1}{6}x,$$

where x is the number of layers. Use both the function and the figure to find $N(3)$. Then calculate the number of oranges in a pyramid with 5 layers.

Source: The New York Times 4/6/04

68. *Stacking Cannonballs.* The function in Exercise 67 was discovered by Thomas Harriot, assistant to Sir Walter Raleigh, when preparing for an expedition at sea. How many cannonballs did they pack if there were 10 layers to their pyramid?

Source: The New York Times 4/7/04

Medicine. Ibuprofen is a medication used to relieve pain. The polynomial function

$$M(t) = 0.5t^4 + 3.45t^3 - 96.65t^2 + 347.7t,$$
$$0 \le t \le 6,$$

can be used to estimate the number of milligrams of ibuprofen in the bloodstream t hours after 400 mg of the medication has been swallowed. Use the following graph for Exercises 69–72.

Source: Based on data from Dr. P. Carey, Burlington, VT

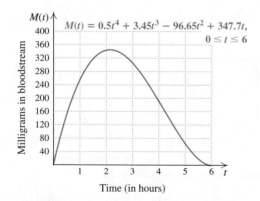

69. How many milligrams of ibuprofen are in the bloodstream 1 hr after 400 mg has been swallowed?

70. How many milligrams of ibuprofen are in the bloodstream 3 hr after 400 mg has been swallowed?

71. Estimate $M(2)$.　　　**72.** Estimate $M(4)$.

Surface Area of a Right Circular Cylinder. *The surface area of a right circular cylinder is given by the polynomial*

$$2\pi r h + 2\pi r^2,$$

where h is the height and r is the radius of the base.

73. A 16-oz beverage can has height 6.3 in. and radius 1.2 in. Find the surface area of the can. (Use a calculator with a $\boxed{\pi}$ key or use 3.141592654 for π.)

74. A 12-oz beverage can has height 4.7 in. and radius 1.2 in. Find the surface area of the can. (Use a calculator with a $\boxed{\pi}$ key or use 3.141592654 for π.)

Total Revenue. *Phinstar Electronics is marketing a tablet computer. The firm determines that when it sells x tablet computers, its total revenue is*

$$R(x) = 280x - 0.4x^2 \text{ dollars.}$$

75. What is the total revenue from the sale of 75 tablet computers?

76. What is the total revenue from the sale of 100 tablet computers?

Total Cost. *Phinstar Electronics determines that the total cost, in dollars, of producing x tablet computers is given by*

$$C(x) = 5000 + 0.6x^2.$$

77. What is the total cost of producing 75 tablet computers?

78. What is the total cost of producing 100 tablet computers?

Adding Polynomials

Combine like terms to write an equivalent expression.

79. $8x + 2 - 5x + 3x^3 - 4x - 1$

80. $2a + 11 - 8a + 5a + 7a^2 + 9$

81. $3a^2b + 4b^2 - 9a^2b - 7b^2$

82. $5x^2y^2 + 4x^3 - 8x^2y^2 - 12x^3$

83. $9x^2 - 3xy + 12y^2 + x^2 - y^2 + 5xy + 4y^2$

84. $a^2 - 2ab + b^2 + 9a^2 + 5ab - 4b^2 + a^2$

Add.

85. $(5t^4 - 2t^3 + t) + (-t^4 - t^3 + 6t^2)$

86. $(3x^3 - 2x - 4) + (-5x^3 + x^2 - 10)$

87. $(x^2 + 2x - 3xy - 7) + (-3x^2 - x + 2y^2 + 6)$

88. $(3a^2 - 2b + a + 6) + (-a^2 + 5b - 5ab - 5)$

89. $(8x^2y - 3xy^2 + 4xy) + (-2x^2y - xy^2 + xy)$

90. $(9ab - 3ac + 5bc) + (13ab - 15ac - 8bc)$

91. $(2r^2 + 12r - 11) + (6r^2 - 2r + 4) + (r^2 - r - 2)$

92. $(5x^2 + 19x - 23) + (7x^2 - 2x + 1) + (-x^2 - 9x + 8)$

93. $\left(\frac{1}{8}xy - \frac{3}{5}x^3y^2 + 4.3y^3\right) + \left(-\frac{1}{3}xy - \frac{3}{4}x^3y^2 - 2.9y^3\right)$

94. $\left(\frac{2}{3}xy + \frac{5}{6}xy^2 + 5.1x^2y\right) + \left(-\frac{4}{5}xy + \frac{3}{4}xy^2 - 3.4x^2y\right)$

Opposites and Subtraction

Write two expressions, one with parentheses and one without, for the opposite of each polynomial.

95. $3t^4 + 8t^2 - 7t - 1$

96. $-4x^5 - 3x^2 - x + 11$

97. $-12y^5 + 4ay^4 - 7by^2$

98. $7ax^3y^2 - 8by^4 - 7abx - 12ay$

Subtract.

99. $(-3x^2 + 2x + 9) - (x^2 + 5x - 4)$

100. $(-7y^2 + 5y + 6) - (4y^2 + 3y - 2)$

101. $(8a - 3b + c) - (2a + 3b - 4c)$

102. $(9r - 5s - t) - (7r - 5s + 3t)$

103. $(6a^2 + 5ab - 4b^2) - (8a^2 - 7ab + 3b^2)$

104. $(4y^2 - 13yz - 9z^2) - (9y^2 - 6yz + 3z^2)$

105. $(6ab - 4a^2b + 6ab^2) - (3ab^2 - 10ab - 12a^2b)$

106. $(10xy - 4x^2y^2 - 3y^3) - (-9x^2y^2 + 4y^3 - 7xy)$

107. $\left(\frac{5}{8}x^4 - \frac{1}{4}x^2 - \frac{1}{2}\right) - \left(-\frac{3}{8}x^4 + \frac{3}{4}x^2 + \frac{1}{2}\right)$

108. $\left(\frac{5}{6}y^4 - \frac{1}{2}y^2 - 7.8y\right) - \left(-\frac{3}{8}y^4 + \frac{3}{4}y^2 + 3.4y\right)$

Perform the indicated operations.

109. $(6t^2 + 7) - (2t^2 + 3) + (t^2 + t)$

110. $(9x^2 + 1) - (x^2 + 7) + (4x^2 - 3x)$

111. $(8r^2 - 6r) - (2r - 6) + (5r^2 - 7)$

112. $(7s^2 - 5s) - (4s - 1) + (3s^2 - 5)$

Aha! **113.** $(x^2 - 4x + 7) + (3x^2 - 9) - (x^2 - 4x + 7)$

114. $(t^2 - 5t + 6) + (5t - 8) - (t^2 + 3t - 4)$

Total Profit. *Total profit is defined as total revenue minus total cost. In Exercises 115 and 116, let $R(x)$ and $C(x)$ represent the revenue and the cost in dollars, respectively, from the sale of x cell phones.*

115. If $R(x) = 280x - 0.4x^2$ and $C(x) = 5000 + 0.6x^2$, find the profit from the sale of 70 cell phones.

116. If $R(x) = 280x - 0.7x^2$ and $C(x) = 8000 + 0.5x^2$, find the profit from the sale of 100 cell phones.

117. Is the sum of two binomials always a binomial? Why or why not?

118. Ani claims that she can easily add polynomials but finds subtraction difficult. What advice would you offer her?

Skill Review

Simplify. [1.2]

119. $-\dfrac{3}{20} - \dfrac{1}{8}$

120. $|1.3 - (-2.48)|$

121. $(-120)(-2)$

122. $-\dfrac{2}{3} \div \dfrac{4}{9}$

123. $3 - (4 - 10)^2 \div 3(2 - 4)$

124. $\dfrac{2(8 - 3) - 6 + 2}{3^2 - 2^3}$

Synthesis

125. For $P(x)$ as given in Exercises 63–66, calculate
$$\frac{P(20) - P(10)}{20 - 10}.$$
Explain what this number represents graphically and what meaning it has in the application.

126. Give a reasonable domain for the function in Example 7. Explain why you chose this domain.

For $P(x)$ and $Q(x)$ as given, find the following.
$P(x) = 13x^5 - 22x^4 - 36x^3 + 40x^2 - 16x + 75,$
$Q(x) = 42x^5 - 37x^4 + 50x^3 - 28x^2 + 34x + 100$

127. $2[P(x)] + Q(x)$

128. $3[P(x)] - Q(x)$

129. $2[Q(x)] - 3[P(x)]$

130. $4[P(x)] + 3[Q(x)]$

131. *Volume of a Display.* The number of spheres in a triangular pyramid with x layers is given by
$$N(x) = \tfrac{1}{6}x^3 + \tfrac{1}{2}x^2 + \tfrac{1}{3}x.$$
The volume of a sphere of radius r is given by
$$V(r) = \tfrac{4}{3}\pi r^3,$$
where π can be approximated as 3.14.

Greta's Chocolate has a window display of truffles piled in a triangular pyramid formation 5 layers deep. If the diameter of each truffle is 3 cm, find the volume of chocolate in the display.

132. If one large truffle were to have the same volume as the display of truffles in Exercise 131, what would be its diameter?

133. Find a polynomial function that gives the outside surface area of the box shown, with an open top and dimensions as shown.

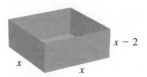

Perform the indicated operation.

134. $(2x^{2a} + 4x^a + 3) + (6x^{2a} + 3x^a + 4)$

135. $(2x^{5b} + 4x^{4b} + 3x^{3b} + 8) -$
$(x^{5b} + 2x^{3b} + 6x^{2b} + 9x^b + 8)$

136. Use a graphing calculator to check your answers to Exercises 53, 59, and 79.

137. A student who is trying to graph
$$p(x) = 0.05x^4 - x^2 + 5$$
gets the following screen. How can the student tell at a glance that a mistake has been made?

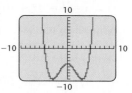

138. *Research.* Using a vehicle with a display showing instantaneous gas mileage, collect data giving gas mileage, in miles per gallon, at various speeds. Graph the data to see if they can be modeled using a polynomial equation. If possible, use regression to find an equation that fits the data.

YOUR TURN ANSWERS: SECTION 5.1

1. $-y^4, 7y^2, -2y, -1$ **2.** 14 **3.** 6
4. $2x^3 + x^2 - 7x + 12$ **5.** $2 - 5x^2 + 3x^3y + 7xy^2 + y^4$
6. -21 **7.** 30 mpg **8.** $-2n^3 + 5n + 11$
9. $2y^2 - 4$ **10.** $2x^4 - 3x^3 + 3x^2 + x - 7$
11. $6c^2d + 5cd^2 - 5cd$ **12.** $-(-3y^4 - y^2 + y + 1)$;
$3y^4 + y^2 - y - 1$ **13.** $-2x^2 + x + 8$
14. $-5n^3 - 6n - 5$

PREPARE TO MOVE ON

Simplify. [1.6]

1. $x^5 \cdot x^3$

2. $(a^2b^3)(a^4b)$

3. $(t^4)^2$

4. $(5y^3)^2$

5. $(2x^5y)^2$

5.2   Multiplication of Polynomials

Multiplying Monomials ▪ Multiplying Monomials and Binomials ▪ Multiplying Any Two Polynomials ▪
The Product of Two Binomials: FOIL ▪ Squares of Binomials ▪ Products of Sums and Differences ▪ Function Notation

Just like numbers, polynomials can be multiplied. We begin by finding products of monomials.

Multiplying Monomials

To multiply two monomials, we multiply coefficients and we multiply variables using the rules for exponents and the commutative and associative laws. With practice, we can work mentally, writing only the answer.

Student Notes

If the meaning of a word is unclear to you, take the time to look it up before continuing your reading. In Example 1, the word "coefficient" appears. The *coefficient* of $-8x^4y^7$ is -8.

1. Multiply and simplify:

$$(6nm^8)(-n^2m^3).$$

EXAMPLE 1 Multiply and simplify.

a) $(-8x^4y^7)(5x^3y^2)$

b) $(-3a^5bc^6)(-4a^2b^5c^8)$

SOLUTION

a) $(-8x^4y^7)(5x^3y^2) = -8 \cdot 5 \cdot x^4 \cdot x^3 \cdot y^7 \cdot y^2$ Using the associative and commutative laws

$$= -40x^{4+3}y^{7+2}$$ Multiplying coefficients; adding exponents

$$= -40x^7y^9$$

b) $(-3a^5bc^6)(-4a^2b^5c^8) = (-3)(-4) \cdot a^5 \cdot a^2 \cdot b \cdot b^5 \cdot c^6 \cdot c^8$

$$= 12a^7b^6c^{14}$$ Multiplying coefficients; adding exponents

YOUR TURN

Multiplying Monomials and Binomials

The distributive law is the basis for multiplying polynomials other than monomials.

EXAMPLE 2 Multiply: **(a)** $2t(3t - 5)$; **(b)** $3a^2b(a^2 - b^2)$.

SOLUTION

a) $2t(3t - 5) = 2t \cdot 3t - 2t \cdot 5$ Using the distributive law

$$= 6t^2 - 10t$$ Multiplying monomials

2. Multiply: $5x^2y^3(3x - 4y^2)$.

b) $3a^2b(a^2 - b^2) = 3a^2b \cdot a^2 - 3a^2b \cdot b^2$ Using the distributive law

$$= 3a^4b - 3a^2b^3$$

YOUR TURN

The distributive law is also used for multiplying two binomials. In this case, however, we begin by distributing a *binomial* rather than a monomial.

EXAMPLE 3 Multiply: $(y^3 - 5)(2y^3 + 4)$.

SOLUTION

$(y^3 - 5)(2y^3 + 4) = (y^3 - 5)\,2y^3 + (y^3 - 5)\,4$ Distributing the $y^3 - 5$

$$= 2y^3(y^3 - 5) + 4(y^3 - 5)$$ Using the commutative law for multiplication. Try to do this step mentally.

$$= 2y^3 \cdot y^3 - 2y^3 \cdot 5 + 4 \cdot y^3 - 4 \cdot 5$$ Using the distributive law (twice)

$$= 2y^6 - 10y^3 + 4y^3 - 20$$ Multiplying the monomials

$$= 2y^6 - 6y^3 - 20$$ Combining like terms

3. Multiply: $(a^2 - 2)(3a^2 + 5)$.

 YOUR TURN

Multiplying Any Two Polynomials

Repeated use of the distributive law enables us to multiply *any* two polynomials, regardless of how many terms are in each.

EXAMPLE 4 Multiply: $(p + 2)(p^4 - 2p^3 + 3)$.

SOLUTION We can use the distributive law from right to left if we wish:

$$(p + 2)(p^4 - 2p^3 + 3) = p(p^4 - 2p^3 + 3) + 2(p^4 - 2p^3 + 3)$$
$$= p \cdot p^4 - p \cdot 2p^3 + p \cdot 3 + 2 \cdot p^4 - 2 \cdot 2p^3 + 2 \cdot 3$$
$$= p^5 - 2p^4 + 3p + 2p^4 - 4p^3 + 6$$
$$= p^5 - 4p^3 + 3p + 6. \qquad \text{Combining like terms}$$

4. Multiply:

$(x + 3)(x^3 - 5x - 1)$.

YOUR TURN

> **THE PRODUCT OF TWO POLYNOMIALS**
>
> To find the product of two polynomials P and Q, multiply each term of P by every term of Q and then combine like terms.

It is also possible to stack the polynomials, multiplying each term at the top by every term below, keeping like terms in columns, and leaving spaces for missing terms. Then we add just as we do in long multiplication with numbers.

EXAMPLE 5 Multiply: $(5x^3 + x - 4)(-2x^2 + 3x + 6)$.

SOLUTION

$$
\begin{array}{r}
5x^3 + x - 4 \\
-2x^2 + 3x + 6 \\
\hline
30x^3 + 6x - 24 \qquad \text{Multiplying by 6}\\
15x^4 + 3x^2 - 12x \qquad \text{Multiplying by } 3x\\
-10x^5 - 2x^3 + 8x^2 \qquad \text{Multiplying by } -2x^2\\
\hline
-10x^5 + 15x^4 + 28x^3 + 11x^2 - 6x - 24 \qquad \text{Adding}
\end{array}
$$

5. Multiply:

$(2x^2 + 8x - 7)(x^2 + x - 4)$.

YOUR TURN

The Product of Two Binomials: FOIL

We now consider what are called *special products*. These products of polynomials occur often and can be simplified using shortcuts that we now develop.

To find a special-product rule for the product of any two binomials, consider $(x + 7)(x + 4)$. We multiply each term of $(x + 7)$ by each term of $(x + 4)$:

$$(x + 7)(x + 4) = x \cdot x + x \cdot 4 + 7 \cdot x + 7 \cdot 4.$$

This multiplication illustrates a pattern that occurs anytime two binomials are multiplied:

A visualization of $(x + 7)(x + 4)$ using areas

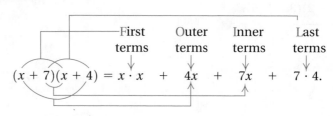

We use the mnemonic device FOIL to remember this method for multiplying.

THE FOIL METHOD

To multiply two binomials $A + B$ and $C + D$, multiply the First terms AC, the Outer terms AD, the Inner terms BC, and then the Last terms BD. Then combine like terms, if possible.

$$(A + B)(C + D) = AC + AD + BC + BD$$

1. Multiply First terms: AC.
2. Multiply Outer terms: AD.
3. Multiply Inner terms: BC.
4. Multiply Last terms: BD.

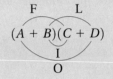

$\downarrow$

FOIL

EXAMPLE 6 Multiply.

a) $(x + 5)(x - 8)$

b) $(2x + 3y)(x - 4y)$

c) $(t + 2)(t - 4)(t + 5)$

SOLUTION

$$\qquad\qquad\qquad\quad \text{F}\quad \text{O}\quad \text{I}\quad \text{L}$$

a) $(x + 5)(x - 8) = x^2 - 8x + 5x - 40$

$$\qquad\qquad\qquad\quad = x^2 - 3x - 40 \qquad \text{Combining like terms}$$

b) $(2x + 3y)(x - 4y) = 2x^2 - 8xy + 3xy - 12y^2 \qquad \text{Using FOIL}$

$$\qquad\qquad\qquad\qquad = 2x^2 - 5xy - 12y^2 \qquad \text{Combining like terms}$$

c) $(t + 2)(t - 4)(t + 5) = (t^2 - 4t + 2t - 8)(t + 5) \qquad \text{Using FOIL}$

$$\qquad\qquad\qquad\qquad = (t^2 - 2t - 8)(t + 5)$$

$$\qquad\qquad\qquad\qquad = (t^2 - 2t - 8) \cdot t + (t^2 - 2t - 8) \cdot 5 \qquad \begin{array}{l}\text{Using the}\\\text{distributive}\\\text{law}\end{array}$$

$$\qquad\qquad\qquad\qquad = t^3 - 2t^2 - 8t + 5t^2 - 10t - 40$$

$$\qquad\qquad\qquad\qquad = t^3 + 3t^2 - 18t - 40 \qquad \text{Combining like terms}$$

6. Multiply:

$$(2m - p)(5m + 6p).$$

 YOUR TURN

Squares of Binomials

A fast method for squaring any binomial can be developed using FOIL:

$$(A + B)^2 = (A + B)(A + B)$$

$$\qquad\qquad = A^2 + AB + AB + B^2 \qquad \text{Note that } AB \text{ occurs twice.}$$

$$\qquad\qquad = A^2 + 2AB + B^2;$$

$$(A - B)^2 = (A - B)(A - B)$$

$$\qquad\qquad = A^2 - AB - AB + B^2 \qquad \text{Note that } -AB \text{ occurs twice.}$$

$$\qquad\qquad = A^2 - 2AB + B^2.$$

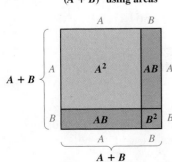

A visualization of $(A + B)^2$ using areas

Technology Connection

To verify that $(x + 3)^2 \neq x^2 + 9$, let $y_1 = (x + 3)^2$ and $y_2 = x^2 + 9$. Then compare y_1 and y_2 using a table of values or a graph. Here we show a table of values created using a graphing calculator app. Note that, in general, $y_1 \neq y_2$.

x	$y_1 = (x+3)^2$	$y_2 = x^2 + 9$
-10.00	49.00	109.00
-9.00	36.00	90.00
-8.00	25.00	73.00
-7.00	16.00	58.00
-6.00	9.00	45.00
-5.00	4.00	34.00
-4.00	1.00	25.00
-3.00	0.00	18.00
-2.00	1.00	13.00
-1.00	4.00	10.00
0.00	9.00	9.00
1.00	16.00	10.00
2.00	25.00	13.00
3.00	36.00	18.00
4.00	49.00	25.00
5.00	64.00	34.00
6.00	81.00	45.00
7.00	100.00	58.00
8.00	121.00	73.00
9.00	144.00	90.00

7. Multiply: $(7x - 3)^2$.

SQUARING A BINOMIAL

$$(A + B)^2 = A^2 + 2AB + B^2;$$
$$(A - B)^2 = A^2 - 2AB + B^2$$

The square of a binomial is the square of the first term, plus twice the product of the two terms, plus the square of the last term.

Trinomials that can be written in the form $A^2 + 2AB + B^2$ or $A^2 - 2AB + B^2$ are called *perfect-square trinomials*.

It can help to say the words of the rules while multiplying.

EXAMPLE 7   Multiply: **(a)** $(y - 5)^2$; **(b)** $(2x + 3y)^2$; **(c)** $\left(\frac{1}{2}x - 3y^4\right)^2$.

SOLUTION

$$(A - B)^2 = A^2 - 2 \cdot A \cdot B + B^2$$

a) $(y - 5)^2 = y^2 - 2 \cdot y \cdot 5 + 5^2$ Note that $-2 \cdot y \cdot 5$ is twice the product of y and -5.

$\qquad\qquad = y^2 - 10y + 25$ The square of a binomial is always a trinomial.

b) $(2x + 3y)^2 = (2x)^2 + 2 \cdot 2x \cdot 3y + (3y)^2$

$\qquad\qquad = 4x^2 + 12xy + 9y^2$ Raising a product to a power

c) $\left(\frac{1}{2}x - 3y^4\right)^2 = \left(\frac{1}{2}x\right)^2 - 2 \cdot \frac{1}{2}x \cdot 3y^4 + (3y^4)^2$ $2 \cdot \frac{1}{2}x \cdot (-3y^4) = -2 \cdot \frac{1}{2}x \cdot 3y^4$

$\qquad\qquad = \frac{1}{4}x^2 - 3xy^4 + 9y^8$ Raising a product to a power; multiplying exponents

YOUR TURN

> **CAUTION!**   Note that $(y - 5)^2 \neq y^2 - 5^2$. (For example, if y is 6, then $(y - 5)^2 = 1$, whereas $y^2 - 5^2 = 11$.) More generally,
>
> $$(A + B)^2 \neq A^2 + B^2 \quad \text{and} \quad (A - B)^2 \neq A^2 - B^2.$$

Products of Sums and Differences

Another pattern emerges when we multiply a sum and a difference of the same two terms. Note the following:

$$\begin{array}{cccc} \text{F} & \text{O} & \text{I} & \text{L} \\ \downarrow & \downarrow & \downarrow & \downarrow \end{array}$$
$$(A + B)(A - B) = A^2 - AB + AB - B^2$$
$$= A^2 - B^2. \qquad -AB + AB = 0$$

THE PRODUCT OF A SUM AND A DIFFERENCE

$$(A + B)(A - B) = A^2 - B^2 \qquad \text{This is called a } \textit{difference of two squares.}$$

The product of the sum and the difference of the same two terms is the square of the first term minus the square of the second term.

EXAMPLE 8 Multiply.

a) $(t + 5)(t - 5)$ **b)** $(2xy^2 + 3x)(2xy^2 - 3x)$

c) $(0.2t - 1.4m)(0.2t + 1.4m)$ **d)** $\left(\frac{2}{3}n - m^3\right)\left(\frac{2}{3}n + m^3\right)$

SOLUTION

$$(A + B)(A - B) = A^2 - B^2$$

a) $(t + 5)(t - 5) = t^2 - 5^2$ Replacing A with t and B with 5
$$= t^2 - 25$$

b) $(2xy^2 + 3x)(2xy^2 - 3x) = (2xy^2)^2 - (3x)^2$
$$= 4x^2y^4 - 9x^2 \quad \text{Raising a product to a power (twice)}$$

c) $(0.2t - 1.4m)(0.2t + 1.4m) = (0.2t)^2 - (1.4m)^2$
$$= 0.04t^2 - 1.96m^2$$

d) $\left(\frac{2}{3}n - m^3\right)\left(\frac{2}{3}n + m^3\right) = \left(\frac{2}{3}n\right)^2 - (m^3)^2$
$$= \frac{4}{9}n^2 - m^6$$

8. Multiply: $(5x + 6y)(5x - 6y)$.

YOUR TURN

Technology Connection

One way to check problems like Example 8(a) is to note that if the multiplication is correct, then $(t + 5)(t - 5) = t^2 - 25$ is an identity and $t^2 - 25 - (t + 5)(t - 5)$ must be 0. In the window below, we set the MODE to G-T so that we can view both a graph and a table. We use a heavy line to distinguish the graph from the x-axis.

$y_1 = x^2 - 25 - (x + 5)(x - 5)$

X	Y1
-2	0
-1	0
0	0
1	0
2	0
3	0
4	0

X = -2

Had we found $y_1 \neq 0$, we would have known that a mistake had been made.

1. Use this procedure to show that $(x - 3)(x + 3) = x^2 - 9$.
2. Use this procedure to show that $(t - 4)^2 = t^2 - 8t + 16$.
3. Show that the graphs of $y_1 = x^2 - 4$ and $y_2 = (x + 2)(x - 2)$ coincide, using the Sequential MODE with a heavier-weight line for y_2. Then, use the Y-VARS option of the **VARS** key to enter $y_3 = y_2 - y_1$. What do you expect the graph of y_3 to look like?

Student Notes

To remember the special products, look for differences between the rules. When we square a binomial, after combining like terms, the result is a trinomial.

In the product of a sum and a difference, the binomials are not the same; one is a sum and the other a difference. After like terms have been combined, the result is a binomial.

$(A + B)(A + B) = A^2 + 2AB + B^2;$
$(A - B)(A - B) = A^2 - 2AB + B^2;$
$(A + B)(A - B) = A^2 - B^2$

EXAMPLE 9 Multiply and simplify.

a) $(5y + 4 + 3x)(5y + 4 - 3x)$ **b)** $(3xy^2 + 4y)(-3xy^2 + 4y)$

c) $(2t + 3)^2 - (t - 1)(t + 1)$

SOLUTION

a) The easiest way to multiply $(5y + 4 + 3x)(5y + 4 - 3x)$ is to note that it is in the form $(A + B)(A - B)$:

$$(5y + 4 + 3x)(5y + 4 - 3x) = (5y + 4)^2 - (3x)^2$$
$$= 25y^2 + 40y + 16 - 9x^2.$$

We can also multiply $(5y + 4 + 3x)(5y + 4 - 3x)$ using columns, but not as quickly.

b) $(3xy^2 + 4y)(-3xy^2 + 4y) = (4y + 3xy^2)(4y - 3xy^2)$ Using a
commutative law

$$= (4y)^2 - (3xy^2)^2$$
$$= 16y^2 - 9x^2y^4$$

c) $(2t + 3)^2 - (t - 1)(t + 1) = 4t^2 + 12t + 9 - (t^2 - 1)$ Squaring a
binomial;
simplifying
$(t - 1)(t + 1)$

$$= 4t^2 + 12t + 9 - t^2 + 1$$ Subtracting

$$= 3t^2 + 12t + 10$$ Combining like
terms

9. Multiply and simplify:

$(p + 6 + 2w)(p + 6 - 2w)$.

 YOUR TURN

Function Notation

We have shown in this section that

$$(x - 2)(x + 2) = x^2 - 4,$$

that is, $x^2 - 4$ and $(x - 2)(x + 2)$ are equivalent expressions.
From the viewpoint of functions, if we have

$$f(x) = x^2 - 4 \quad \text{and} \quad g(x) = (x - 2)(x + 2),$$

then for any given input x, the outputs $f(x)$ and $g(x)$ are identical. Thus the graphs of these functions are identical and we say that f and g represent the same function.

Student Notes

Sometimes it is helpful to consider equivalent expressions representing the same function. For example, the form $f(x) = x^2 - 4$ indicates that $(0, -4)$ is the y-intercept, and the form $g(x) = (x - 2)(x + 2)$ indicates that $(2, 0)$ and $(-2, 0)$ are the x-intercepts.

x	$f(x)$	$g(x)$
3	5	5
2	0	0
1	-3	-3
0	-4	-4
-1	-3	-3
-2	0	0
-3	5	5

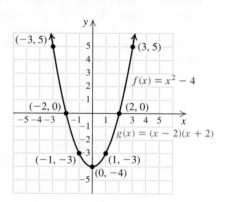

Our work with multiplying can be used when evaluating functions.

EXAMPLE 10 Given $f(x) = x^2 - 4x + 5$, find and simplify each of the following.

a) $f(a) + 3$ **b)** $f(a + 3)$ **c)** $f(a + h) - f(a)$

SOLUTION

a) To find $f(a) + 3$, we replace x with a to find $f(a)$. Then we add 3 to the result:

$$f(a) + 3 = (a^2 - 4 \cdot a + 5) + 3$$ Evaluating $f(a)$
$$= a^2 - 4a + 8.$$ Simplifying

CAUTION! Note from parts (a) and (b) that, in general,

$$f(a + 3) \neq f(a) + 3.$$

b) To find $f(a + 3)$, we replace x with $a + 3$. Then we simplify:

$$f(a + 3) = (a + 3)^2 - 4(a + 3) + 5$$ Replacing each occurrence of x
with $(a + 3)$

$$= a^2 + 6a + 9 - 4a - 12 + 5$$
$$= a^2 + 2a + 2.$$

Chapter Resource:
Collaborative Activity, p. 345

c) To find $f(a + h)$ and $f(a)$, we replace x with $a + h$ and a, respectively:

$$f(a + h) - f(a) = [(a + h)^2 - 4(a + h) + 5] - [a^2 - 4a + 5]$$
$$= [a^2 + 2ah + h^2 - 4a - 4h + 5] - [a^2 - 4a + 5]$$
$$= a^2 + 2ah + h^2 - 4a - 4h + 5 - a^2 + 4a - 5$$
$$= 2ah + h^2 - 4h.$$

10. Given $g(x) = x^2 - x - 2$, find and simplify $g(a - 6)$.

YOUR TURN

5.2 EXERCISE SET

FOR EXTRA HELP

MyMathLab® Math
PRACTICE   WATCH   READ   REVIEW

Vocabulary and Reading Check

Classify each of the following statements as either true or false.

1. The coefficient of $3x^5$ is 5.

2. The product of two monomials is a monomial.

3. The product of a monomial and a binomial is found using the distributive law.

4. To simplify the product of two binomials, we often need to combine like terms.

5. FOIL can be used whenever two monomials are multiplied.

6. The square of a binomial is a difference of two squares.

7. The product of the sum and the difference of the same two terms is a binomial.

8. In general, $f(a + 5) \neq f(a) + 5$.

Multiplying Monomials

Multiply.

9. $3x^4 \cdot 5x$

10. $-2x^3 \cdot 4x$

11. $6a^2(-8ab^2)$

12. $-3uv^2(5u^2v^2)$

13. $(-4x^3y^2)(-9x^2y^4)$

14. $(-7a^2bc^4)(-8ab^3c^2)$

Multiplying Monomials and Binomials

Multiply.

15. $7x(3 - x)$

16. $3a(a^2 - 4a)$

17. $5cd(4c^2d - 5cd^2)$

18. $a^2(2a^2 - 5a^3)$

Multiplying Any Two Polynomials

Multiply.

19. $(x + 3)(x + 5)$

20. $(t - 1)(t - 4)$

21. $(2a + 3)(4a - 1)$

22. $(3r - 4)(2r + 1)$

23. $(x + 2)(x^2 - 3x + 1)$

24. $(a + 3)(a^2 - 4a + 2)$

25. $(t - 5)(t^2 + 2t - 3)$

26. $(x - 4)(x^2 + x - 7)$

27. $(a^2 + a - 1)(a^2 + 4a - 5)$

28. $(x^2 - 2x + 1)(x^2 + x + 2)$

29. $(x + 3)(x^2 - 3x + 9)$

30. $(y + 4)(y^2 - 4y + 16)$

31. $(a - b)(a^2 + ab + b^2)$

32. $(x - y)(x^2 + xy + y^2)$

The Product of Two Binomials: FOIL

Multiply.

33. $(t - 3)(t + 2)$

34. $(x + 6)(x - 1)$

35. $(5x + 2y)(4x + y)$

36. $(3t + 2)(2t + 7)$

37. $\left(t - \frac{1}{3}\right)\left(t - \frac{1}{4}\right)$

38. $\left(x - \frac{1}{2}\right)\left(x - \frac{1}{5}\right)$

39. $(1.2t + 3s)(2.5t - 5s)$

40. $(30a - 0.5b)(0.2a + 10b)$

41. $(r + 3)(r + 2)(r - 1)$

42. $(t + 4)(t + 1)(t - 2)$

Squares of Binomials

Multiply.

43. $(x + 5)^2$ **44.** $(t + 6)^2$

45. $(2y - 7)^2$ **46.** $(3x - 4)^2$

47. $(5c - 2d)^2$ **48.** $(8x - 3y)^2$

49. $(3a^3 - 10b^2)^2$ **50.** $(3s^2 + 4t^3)^2$

51. $(x^3y^4 + 5)^2$ **52.** $(a^4b^2 - 3)^2$

Products of Sums and Differences

Multiply.

53. $(c + 7)(c - 7)$

54. $(x - 3)(x + 3)$

55. $(1 - 4x)(1 + 4x)$

56. $(5 + 2y)(5 - 2y)$

57. $\left(3m - \frac{1}{2}n\right)\left(3m + \frac{1}{2}n\right)$

58. $(0.4c - 0.5d)(0.4c + 0.5d)$

59. $(x^3 + yz)(x^3 - yz)$

60. $(2a^4 + ab)(2a^4 - ab)$

61. $(-mn + 3m^2)(mn + 3m^2)$

62. $(-6u + v^2)(6u + v^2)$

Multiplying Polynomials

Multiply.

63. $(x + 7)^2 - (x + 3)(x - 3)$

64. $(t + 5)^2 - (t - 4)(t + 4)$

65. $(2m - n)(2m + n) - (m - 2n)^2$

66. $(3x + y)(3x - y) - (2x + y)^2$

Aha! **67.** $(a + b + 1)(a + b - 1)$

68. $(m + n + 2)(m + n - 2)$

69. $(2x + 3y + 4)(2x + 3y - 4)$

70. $(3a - 2b + c)(3a - 2b - c)$

71. *Compounding Interest.* Suppose that P dollars is invested in a savings account at interest rate r, compounded annually, for 2 years. The amount A in the account after 2 years is given by

$$A = P(1 + r)^2,$$

where r is in decimal form. Find an equivalent expression for A.

72. *Compounding Interest.* Suppose that P dollars is invested in a savings account at interest rate r, compounded semiannually, for 1 year. The amount A in the account after 1 year is given by

$$A = P\left(1 + \frac{r}{2}\right)^2,$$

where r is in decimal form. Find an equivalent expression for A.

Function Notation

73. Let $P(x) = 3x^2 - 5$ and $Q(x) = 4x^2 - 7x + 1$. Find $P(x) \cdot Q(x)$.

74. Let $P(x) = x^2 - x + 1$ and $Q(x) = x^3 + x^2 + 5$. Find $P(x) \cdot Q(x)$.

75. Let $P(x) = 5x - 2$. Find $P(x) \cdot P(x)$.

76. Let $Q(x) = 3x^2 + 1$. Find $Q(x) \cdot Q(x)$.

77. Let $F(x) = 2x - \frac{1}{3}$. Find $[F(x)]^2$.

78. Let $G(x) = 5x - \frac{1}{2}$. Find $[G(x)]^2$.

79. Given $f(x) = x^2 + 5$, find and simplify the following.

    a) $f(t - 1)$
    b) $f(a + h) - f(a)$
    c) $f(a) - f(a - h)$

80. Given $f(x) = x^2 + 7$, find and simplify the following.

    a) $f(p + 1)$
    b) $f(a + h) - f(a)$
    c) $f(a) - f(a - h)$

81. Given $f(x) = x^2 + x$, find and simplify the following.

    a) $f(a) + f(-a)$
    b) $f(a + h)$
    c) $f(a + h) - f(a)$

82. Given $f(x) = x^2 - x$, find and simplify the following.

    a) $f(a) - f(-a)$
    b) $f(a + h)$
    c) $f(a + h) - f(a)$

83. Find two binomials whose product is $x^2 - 25$ and explain how you decided on those two binomials.

84. Find two binomials whose product is $x^2 - 6x + 9$ and explain how you decided on those two binomials.

Skill Review

Solve.

85. $\dfrac{x}{3} - 7 = \dfrac{1}{4}$ [1.3]

86. $7 - 2x < 3(x - 1) + 2$ [4.1]

87. $|3x - 6| > 8$ [4.3]

88. $4 \le 1 - x \le 6$ [4.2]

89. $2x - 3y = 4,$
$x + 2y = 5$ [3.2]

90. $x - y - z = 5,$
$2x + y - z = 4,$
$x + 2y + 2z = 8$ [3.4]

Synthesis

91. We have seen that $(a - b)(a + b) = a^2 - b^2$. Explain how this result can be used to develop a fast way of calculating $95 \cdot 105$.

92. A student *incorrectly* claims that since $2x^2 \cdot 2x^2 = 4x^4$, it follows that $5x^5 \cdot 5x^5 = 25x^{25}$. How could you convince the student that a mistake has been made?

Multiply. Assume that variables in exponents represent natural numbers.

93. $(x^2 + y^n)(x^2 - y^n)$

94. $(a^n + b^n)^2$

95. $x^2y^3(5x^n + 4y^n)$

96. $(x^n - 4)(x^{2n} + 3x^n - 2)$

Aha! **97.** $(a - b + c - d)(a + b + c + d)$

98. $[(a + b)(a - b)][5 - (a + b)][5 + (a + b)]$

99. $(x^2 - 3x + 5)(x^2 + 3x + 5)$

100. $\left(\frac{2}{3}x + \frac{1}{3}y + 1\right)\left(\frac{2}{3}x - \frac{1}{3}y - 1\right)$

101. $(x - 1)(x^2 + x + 1)(x^3 + 1)$

102. $(x^a + y^b)(x^a - y^b)(x^{2a} + y^{2b})$

103. $(x^{a-b})^{a+b}$

104. $(M^{x+y})^{x+y}$

Aha! **105.** $(x - a)(x - b)(x - c) \cdots (x - z)$

Multiply.

106. $(3x^{-4} + 1)(2x^{-3} - 5)$

107. $(2x^{-2} + 3x^{-1})(5x^{-3} - x^2)$

108. Given $f(x) = x^2 + 7$, find and simplify
$$\dfrac{f(a + h) - f(a)}{h}.$$

109. Given $g(x) = x^2 - 9$, find and simplify
$$\dfrac{g(a + h) - g(a)}{h}.$$

110. Draw rectangles similar to those before Example 6 to show that $(x + 2)(x + 5) = x^2 + 7x + 10$.

111. Draw rectangles similar to those before Example 7 to show that $(A - B)^2 = A^2 - 2AB + B^2$.

112. Use a graphing calculator to check your answers to Exercises 15, 33, and 77.

113. Use a graphing calculator to determine which of the following are identities.

a) $(x - 1)^2 = x^2 - 1$
b) $(x - 2)(x + 3) = x^2 + x - 6$
c) $(x - 1)^3 = x^3 - 3x^2 + 3x - 1$
d) $(x + 1)^4 = x^4 + 1$
e) $(x + 1)^4 = x^4 + 4x^3 + 8x^2 + 4x + 1$

YOUR TURN ANSWERS: SECTION 5.2

1. $-6n^3m^{11}$ 2. $15x^3y^3 - 20x^2y^5$
3. $3a^4 - a^2 - 10$ 4. $x^4 + 3x^3 - 5x^2 - 16x - 3$
5. $2x^4 + 10x^3 - 7x^2 - 39x + 28$
6. $10m^2 + 7mp - 6p^2$ 7. $49x^2 - 42x + 9$
8. $25x^2 - 36y^2$ 9. $p^2 + 12p + 36 - 4w^2$
10. $a^2 - 13a + 40$

QUICK QUIZ: SECTIONS 5.1–5.2

1. Determine the degree of $-7x^3 + 5x + 4 + 2x^6$.   [5.1]
2. Find $f(-2)$ for $f(x) = x^3 - x^2 - 7x$.   [5.1]
3. Combine like terms: $-y^3 + 5y - y + 2y^3$.   [5.1]
4. Multiply: $(5c^3 - d)^2$.   [5.2]
5. Given $f(x) = x^2 - 6$, find and simplify $f(a + 3)$.   [5.2]

PREPARE TO MOVE ON

Find an equivalent expression by factoring.   [1.2]

1. $5x + 15y - 5$
2. $14a + 35b + 42c$
3. $ax + bx - cx$ 4. $bx + by + b$

5.3 | Common Factors and Factoring by Grouping

Terms with Common Factors ■ Factoring by Grouping

Factoring is the reverse of multiplying.

> **FACTORING**
>
> To *factor* a polynomial is to find an equivalent expression that is a product of polynomials. An equivalent expression of this type is called a *factorization* of the polynomial.

Terms with Common Factors

When factoring a polynomial, we always look for a factor common to every term. If one exists, we then use the distributive law to write an equivalent product.

EXAMPLE 1 Factor out a common factor: $6y^2 - 18$.

SOLUTION We have

$$6y^2 - 18 = 6 \cdot y^2 - 6 \cdot 3 \qquad \text{Noting that 6 is a common factor}$$
$$= 6(y^2 - 3). \qquad \text{Using the distributive law}$$

Check: $6(y^2 - 3) = 6y^2 - 18$.

 YOUR TURN

1. Factor out a common factor:

$$5x^2 - 30.$$

It is standard practice to factor out the *largest*, or *greatest, common factor*, so that the resulting polynomial factor cannot be factored any further. In Example 1, the numbers 2 and 3 are also common factors, but 6 is the greatest common factor.

The greatest common factor of a polynomial is the greatest common factor of the coefficients times the greatest common factor of the variable(s) in the terms. Thus, to find the greatest common factor of $30x^4 + 20x^5$, we multiply the greatest common factor of 30 and 20, which is 10, by the greatest common factor of x^4 and x^5, which is x^4:

$$30x^4 + 20x^5 = 10 \cdot 3 \cdot x^4 + 10 \cdot 2 \cdot x^4 \cdot x$$
$$= 10x^4(3 + 2x). \qquad \text{The greatest common factor is } 10x^4.$$

EXAMPLE 2 Write an expression equivalent to $8p^6q^2 - 4p^5q^3 + 10p^4q^4$ by factoring out the greatest common factor.

SOLUTION First, we look for the greatest positive common factor of the coefficients of $8p^6q^2 - 4p^5q^3 + 10p^4q^4$:

$$8, -4, 10 \longrightarrow \text{Greatest common factor} = 2.$$

Second, we look for the greatest common factor of the powers of p:

$$p^6, p^5, p^4 \longrightarrow \text{Greatest common factor} = p^4.$$

Third, we look for the greatest common factor of the powers of q:

$$q^2, q^3, q^4 \longrightarrow \text{Greatest common factor} = q^2.$$

Thus, $2p^4q^2$ is the greatest common factor of the given polynomial. Then

$$8p^6q^2 - 4p^5q^3 + 10p^4q^4 = 2p^4q^2 \cdot 4p^2 - 2p^4q^2 \cdot 2pq + 2p^4q^2 \cdot 5q^2$$
$$= 2p^4q^2(4p^2 - 2pq + 5q^2).$$

We can always check a factorization by multiplying:

$$2p^4q^2(4p^2 - 2pq + 5q^2) = 2p^4q^2 \cdot 4p^2 - 2p^4q^2 \cdot 2pq + 2p^4q^2 \cdot 5q^2$$
$$= 8p^6q^2 - 4p^5q^3 + 10p^4q^4.$$

The factorization is $2p^4q^2(4p^2 - 2pq + 5q^2)$.

2. Write an expression equivalent to $6a^2x^3 + 20a^3x^8 - 4a^5x^3$ by factoring out the greatest common factor.

YOUR TURN

The polynomials in Examples 1 and 2 have been **factored completely**. They cannot be factored further. The factors in the resulting factorizations are said to be **prime polynomials**.

When the leading coefficient is a negative number, we generally factor out a common factor with a negative coefficient.

EXAMPLE 3   Write an equivalent expression by factoring out a common factor with a negative coefficient.

a) $-4x - 24$ **b)** $-2x^3 + 6x^2 - 2x$

SOLUTION

3. Write an expression equivalent to $-3a^4 - 6a^2 + 3a$ by factoring out a common factor with a negative coefficient.

a) $-4x - 24 = -4(x + 6)$ *Check:* $-4(x + 6) = -4 \cdot x + (-4) \cdot 6$
$$= -4x + (-24)$$
$$= -4x - 24$$

b) $-2x^3 + 6x^2 - 2x = -2x(x^2 - 3x + 1)$ The 1 is essential.

YOUR TURN

EXAMPLE 4   *Height of a Thrown Object.*   Suppose that a baseball is thrown upward with an initial velocity of 64 ft/sec. Its height in feet, $h(t)$, after t seconds is given by

$$h(t) = -16t^2 + 64t.$$

Find an equivalent expression for $h(t)$ by factoring out a common factor.

4. Find an equivalent expression for $h(t) = -16t^2 + 100t$ by factoring out a common factor.

SOLUTION   We factor out $-16t$ as follows:

$$h(t) = -16t^2 + 64t = -16t(t - 4).$$ *Check:*
$$-16t(t - 4) = -16t \cdot t - (-16t) \cdot 4$$
$$= -16t^2 + 64t$$

YOUR TURN

Note in Example 4 that we can obtain function values using either expression for $h(t)$, since factoring forms equivalent expressions. For example,

$$h(1) = -16 \cdot 1^2 + 64 \cdot 1 = 48$$

and $\quad h(1) = -16 \cdot 1(1 - 4) = 48.$ Using the factorization

When we evaluate $-16t^2 + 64t$ and $-16t(t - 4)$, the results should always match. Thus a quick partial check of any factorization is to evaluate the factorization and the original polynomial for one or two convenient replacements. The check for Example 4 becomes foolproof if three replacements are used. In general, an nth-degree factorization is correct if it checks for $n + 1$ different replacements. The proof of this useful result is beyond the scope of this text.

Factoring by Grouping

The largest common factor is sometimes a binomial.

EXAMPLE 5   Factor: $(a - b)(x + 5) + (a - b)(x - y^2)$.

5. Write an equivalent expression by factoring:

$(x + y)(2m + n) + (x + y)(3m - 8n).$

SOLUTION   Here the largest common factor is the binomial $a - b$:

$$(a - b)(x + 5) + (a - b)(x - y^2) = (a - b)[(x + 5) + (x - y^2)]$$
$$= (a - b)[2x + 5 - y^2].$$

🔁 YOUR TURN

Often, in order to identify a common binomial factor in a polynomial with four terms, we must regroup into two groups of two terms each.

EXAMPLE 6   Write an equivalent expression by factoring.

a) $y^3 + 3y^2 + 4y + 12$ **b)** $4x^3 - 15 + 20x^2 - 3x$

SOLUTION

a) $y^3 + 3y^2 + 4y + 12 = (y^3 + 3y^2) + (4y + 12)$ Each grouping has a common factor.

$\qquad = y^2(y + 3) + 4(y + 3)$ Factoring out a common factor from each binomial

$\qquad = (y + 3)(y^2 + 4)$ Factoring out $y + 3$

b) When we try grouping $4x^3 - 15 + 20x^2 - 3x$ as $(4x^3 - 15) + (20x^2 - 3x)$, we are unable to factor $4x^3 - 15$. When this happens, we can rearrange the polynomial and try a different grouping:

$4x^3 - 15 + 20x^2 - 3x = 4x^3 + 20x^2 - 3x - 15$ Using a commutative law

$\qquad = 4x^2(x + 5) - 3(x + 5)$ By factoring out -3 instead of 3, we see that $x + 5$ is a common factor.

6. Write an equivalent expression by factoring:

$a^3 + 5a^2 + 6a + 30.$

$\qquad = (x + 5)(4x^2 - 3).$

🔁 YOUR TURN

Factoring out -1 allows us to "reverse the order" of subtraction.

FACTORING OUT -1

$b - a = -1(a - b) = -(a - b)$

Student Notes

In Example 7, make certain that you understand why -1 or $-y$ is factored from $by - ay$.

7. Write an equivalent expression by factoring:

$$xc - xd - 2c + 2d.$$

EXAMPLE 7 Write an equivalent expression by factoring: $ax - bx + by - ay$.

SOLUTION We have

$$
\begin{aligned}
ax - bx + by - ay &= (ax - bx) + (by - ay) && \text{Grouping} \\
&= x(a - b) + y(b - a) && \text{Factoring each binomial} \\
&= x(a - b) + y(-1)(a - b) && \text{Factoring out } -1 \text{ to reverse } b - a \\
&= x(a - b) - y(a - b) && \text{Simplifying} \\
&= (a - b)(x - y). && \text{Factoring out } a - b
\end{aligned}
$$

Check: To check, note that $a - b$ and $x - y$ are both prime and that

$$(a - b)(x - y) = ax - ay - bx + by = ax - bx + by - ay.$$

YOUR TURN

Many polynomials with four terms, like $x^3 + x^2 + 3x - 3$, are prime. Not only is there no common monomial factor, but no matter how we group terms, there is no common binomial factor:

$$
\begin{aligned}
x^3 + x^2 + 3x - 3 &= x^2(x + 1) + 3(x - 1); && \text{No common factor} \\
x^3 + 3x + x^2 - 3 &= x(x^2 + 3) + (x^2 - 3); && \text{No common factor} \\
x^3 - 3 + x^2 + 3x &= (x^3 - 3) + x(x + 3). && \text{No common factor}
\end{aligned}
$$

5.3 EXERCISE SET

FOR EXTRA HELP

MyMathLab® Math XL
PRACTICE WATCH READ REVIEW

Vocabulary and Reading Check

Classify each of the following statements as either true or false.

1. It is possible for a polynomial to contain several different common factors.

2. The largest common factor of $10x^4 + 15x^2$ is $5x$.

3. When the leading coefficient of a polynomial is negative, we generally factor out a common factor with a negative coefficient.

4. The polynomial $3x + 40$ is prime.

5. A binomial can be a common factor.

6. Every polynomial with four terms can be factored by grouping.

7. The expressions $b - a$, $-(a - b)$, and $-1(a - b)$ are all equivalent.

8. The complete factorization of $12x^3 - 20x^2$ is $4x(3x^2 - 5x)$.

Terms with Common Factors

Write an equivalent expression by factoring out the greatest common factor.

9. $10x^2 + 35$

10. $8y^2 + 20$

11. $2y^2 - 18y$

12. $6t^2 - 12t$

13. $5t^3 - 15t + 5$

14. $9x^2 - 3x + 3$

15. $a^6 + 2a^4 - a^3$

16. $3y^7 - y^6 - y^2$

17. $12x^4 - 30x^3 + 42x$

18. $16t^8 + 40t^6 - 24t$

19. $6a^2b - 2ab - 9b$

20. $4x^2y + 10xy + 5y$

21. $15m^4n + 30m^5n^2 + 25m^3n^3$

22. $24s^2t^4 - 18st^3 - 42s^4t^5$

23. $9x^3y^6z^2 - 12x^4y^4z^4 + 15x^2y^5z^3$

24. $14a^4b^3c^5 + 21a^3b^5c^4 - 35a^4b^4c^3$

Write an equivalent expression by factoring out a factor with a negative coefficient.

25. $-5x - 40$

26. $-5x - 35$

27. $-16t^2 + 96$

28. $-16t^2 + 128$

29. $-2x^2 + 12x + 40$

30. $-2x^2 + 4x - 12$

31. $5 - 10y$

32. $7 - 35t$

33. $8d^2 - 12cd$

34. $12q^2 - 21pq$

35. $-m^3 + 8$

36. $-x^2 + 100$

37. $-p^3 - 2p^2 - 5p + 2$

38. $-a^5 - 5a^4 - 11a + 10$

Factoring by Grouping

Write an equivalent expression by factoring.

39. $a(b - 5) + c(b - 5)$ **40.** $r(t - 3) - s(t - 3)$

41. $(x + 7)(x - 1) + (x + 7)(x - 2)$

42. $(a + 5)(a - 2) + (a + 5)(a + 1)$

43. $a^2(x - y) + 5(y - x)$

44. $5x^2(x - 6) + 2(6 - x)$

45. $xy + xz + wy + wz$ **46.** $ac + ad + bc + bd$

47. $y^3 - y^2 + 3y - 3$ **48.** $b^3 - b^2 + 2b - 2$

49. $t^3 + 6t^2 - 2t - 12$ **50.** $a^3 - 3a^2 + 6 - 2a$

51. $12a^4 - 21a^3 - 9a^2$ **52.** $72x^3 - 36x^2 + 24x$

53. $y^8 - 1 - y^7 + y$ **54.** $t^6 - 1 - t^5 + t$

55. $2xy + 3x - x^2y - 6$ **56.** $2y^5 + 15 - 6y^4 - 5y$

57. *Height of a Baseball.* A baseball is popped up with an upward velocity of 72 ft/sec. Its height in feet, $h(t)$, after t seconds is given by
$$h(t) = -16t^2 + 72t.$$

  a) Find an equivalent expression for $h(t)$ by factoring out a common factor with a negative coefficient.

  b) Perform a partial check of part (a) by evaluating both expressions for $h(t)$ at $t = 1$.

58. *Height of a Rocket.* A water rocket is launched upward with an initial velocity of 96 ft/sec. Its height in feet, $h(t)$, after t seconds is given by
$$h(t) = -16t^2 + 96t.$$

  a) Find an equivalent expression for $h(t)$ by factoring out a common factor with a negative coefficient.

  b) Check your factoring by evaluating both expressions for $h(t)$ at $t = 1$.

59. *Surface Area of a Silo.* A silo is a right circular cylinder with a half sphere on top. The surface area of a silo of height h and radius r (including the area of the base) is given by the polynomial $2\pi rh + \pi r^2$. Find an equivalent expression by factoring out a common factor.

60. *Airline Routes.* When an airline links n cities so that from any one city it is possible to fly directly to each of the other cities, the total number of direct routes is given by
$$R(n) = n^2 - n.$$
Find an equivalent expression for $R(n)$ by factoring out a common factor.

61. *Total Profit.* After t weeks of production, Pedal Up, Inc., is making a profit of $P(t) = t^2 - 5t$ from sales of their bicycle decals. Find an equivalent expression by factoring out a common factor.

62. *Total Profit.* When x hundred cameras are sold, Digital Electronics collects a profit of $P(x)$, where
$$P(x) = x^2 - 3x,$$
and $P(x)$ is in thousands of dollars. Find an equivalent expression by factoring out a common factor.

63. *Total Revenue.* Urban Connections is marketing a new cell phone. The firm determines that when it sells x units, the total revenue $R(x)$, in dollars, is given by the polynomial function
$$R(x) = 280x - 0.4x^2.$$
Find an equivalent expression for $R(x)$ by factoring out $0.4x$.

64. *Total Cost.* Urban Connections determines that the total cost $C(x)$, in dollars, of producing x cell phones is given by the polynomial function
$$C(x) = 0.18x + 0.6x^2.$$
Find an equivalent expression for $C(x)$ by factoring out $0.6x$.

65. *Number of Diagonals.* The number of diagonals of a polygon having n sides is given by the polynomial function
$$P(n) = \tfrac{1}{2}n^2 - \tfrac{3}{2}n.$$
Find an equivalent expression for $P(n)$ by factoring out $\tfrac{1}{2}$.

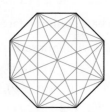

66. *Number of Games in a League.* If there are n teams in a league and each team plays every other team once, we can find the total number of games played by using the polynomial function $f(n) = \tfrac{1}{2}n^2 - \tfrac{1}{2}n$. Find an equivalent expression by factoring out $\tfrac{1}{2}$.

67. *Counting Spheres in a Pile.* The number N of spheres in a triangular pile like the one shown here is given by the polynomial function

$$N(x) = \frac{1}{6}x^3 + \frac{1}{2}x^2 + \frac{1}{3}x,$$

where x is the number of layers and $N(x)$ is the number of spheres. Find an equivalent expression for $N(x)$ by factoring out $\frac{1}{6}$.

68. *High-Fives.* When a team of n players all give each other high-fives, a total of $H(n)$ hand slaps occurs, where

$$H(n) = \frac{1}{2}n^2 - \frac{1}{2}n.$$

Find an equivalent expression by factoring out $\frac{1}{2}n$.

69. What is the *prime factorization* of a polynomial? How does it correspond to the prime factorization of a number?

70. Explain in your own words why $-(a - b) = b - a$.

Skill Review

Graph.

71. $f(x) = -\frac{1}{2}x + 3$ [2.3]   **72.** $3x - y = 9$ [2.4]

73. $y - 1 = 2(x + 3)$ [2.5]   **74.** $y = -3$ [2.4]

75. $6x = 3$ [2.4]   **76.** $3x = 2y - 4$ [2.3]

Synthesis

77. Under what conditions would it be easier to evaluate a polynomial *after* it has been factored?

78. Following Example 4, we stated that checking the factorization of a second-degree polynomial by making a single replacement is only a *partial* check. Write an *incorrect* factorization and explain how evaluating both the polynomial and the factorization might not catch the mistake.

Complete each of the following.

79. $x^5y^4 + \underline{\quad} = x^3y(\underline{\quad} + xy^5)$

80. $a^3b^7 - \underline{\quad} = \underline{\quad}(ab^4 - c^2)$

Write an equivalent expression by factoring.

81. $rx^2 - rx + 5r + sx^2 - sx + 5s$

82. $3a^2 + 6a + 30 + 7a^2b + 14ab + 70b$

83. $a^4x^4 + a^4x^2 + 5a^4 + a^2x^4 + a^2x^2 + 5a^2 + 5x^4 + 5x^2 + 25$

(*Hint*: Use three groups of three.)

Write an equivalent expression by factoring out the smallest power of x in each of the following.

84. $x^{-8} + x^{-4} + x^{-6}$ **85.** $x^{-6} + x^{-9} + x^{-3}$

86. $x^{3/4} + x^{1/2} - x^{1/4}$ **87.** $x^{1/3} - 5x^{1/2} + 3x^{3/4}$

88. $x^{-3/2} + x^{-1/2}$ **89.** $x^{-5/2} + x^{-3/2}$

90. $x^{-3/4} - x^{-5/4} + x^{-1/2}$ **91.** $x^{-4/5} - x^{-7/5} + x^{-1/3}$

Write an equivalent expression by factoring.

92. $2x^{3a} + 8x^a + 4x^{2a}$

93. $3a^{n+1} + 6a^n - 15a^{n+2}$

94. $4x^{a+b} + 7x^{a-b}$

95. $7y^{2a+b} - 5y^{a+b} + 3y^{a+2b}$

96. Use the TABLE feature of a graphing calculator to check your answers to Exercises 25, 35, and 41.

97. Use a graphing calculator to show that

$$(x^2 - 3x + 2)^4 = x^8 + 81x^4 + 16$$

is *not* an identity.

YOUR TURN ANSWERS: SECTION 5.3

1. $5(x^2 - 6)$ **2.** $2a^2x^3(3 + 10ax^5 - 2a^3)$
3. $-3a(a^3 + 2a - 1)$ **4.** $h(t) = -4t(4t - 25)$
5. $(x + y)(5m - 7n)$ **6.** $(a + 5)(a^2 + 6)$
7. $(c - d)(x - 2)$

QUICK QUIZ: SECTIONS 5.1–5.3

1. Add: $(3xy - y^2 + x) + (2y - 4x - xy)$. [5.1]

2. Subtract: $(5a^3 - a - 2) - (3a^3 - a + 7)$. [5.1]

3. Multiply: $\left(t + \frac{1}{3}\right)\left(t - \frac{1}{3}\right)$. [5.2]

Factor. [5.3]

4. $7x^3 - 6xy$ **5.** $2x^3 - 6x^2 + x - 3$

PREPARE TO MOVE ON

Multiply. [5.2]

1. $(x + 5)(x + 3)$ **2.** $(x - 5)(x - 3)$

3. $(x + 5)(x - 3)$ **4.** $(x - 5)(x + 3)$

5. $(2x + 5)(x + 3)$ **6.** $(x + 5)(2x + 3)$

5.4 Factoring Trinomials

Factoring Trinomials of the Type $x^2 + bx + c$ ▪ Factoring Trinomials of the Type $ax^2 + bx + c, a \neq 1$

Our study of how to factor trinomials begins with trinomials of the type $x^2 + bx + c$. We then move on to the form $ax^2 + bx + c$, where $a \neq 1$.

Factoring Trinomials of the Type $x^2 + bx + c$

When trying to factor trinomials of the type $x^2 + bx + c$, we can use a trial-and-error procedure.

CONSTANT TERM POSITIVE

Recall the FOIL method of multiplying two binomials:

$$(x + 3)(x + 5) = x^2 + \underset{F\quad O\quad I\quad L}{5x + 3x} + 15$$
$$= x^2 + 8x + 15.$$

To factor $x^2 + 8x + 15$, we think of FOIL: The term x^2 is the product of the First terms in each of two binomial factors, so the first term in each binomial is x. The challenge is to then find two numbers p and q such that

$$x^2 + 8x + 15 = (x + p)(x + q)$$
$$= x^2 + qx + px + pq.$$

Note that the Outer and Inner products, qx and px, are like terms and can be combined as $(p + q)(x)$. The Last product, pq, is a constant. Thus we need two numbers, p and q, whose product is 15 and whose sum is 8. These numbers are 3 and 5. The factorization is

$$(x + 3)(x + 5), \quad \text{or} \quad (x + 5)(x + 3). \qquad \text{Using a commutative law}$$

When the constant term of a trinomial is positive, the product pq must be positive. Thus the constant terms in the binomial factors must be either both positive or both negative. The sign used is that of the trinomial's middle term.

EXAMPLE 1 Write an equivalent expression by factoring: $x^2 + 9x + 8$.

SOLUTION We think of FOIL in reverse. The first term of each factor is x. We look for numbers p and q such that

$$x^2 + 9x + 8 = (x + p)(x + q) = x^2 + (p + q)x + pq.$$

We list pairs of factors of 8 and choose the pair whose sum is 9.

Pair of Factors	Sum of Factors
2, 4	6
1, 8	9

Both factors are positive.

The numbers we need are 1 and 8, forming the factorization $(x + 1)(x + 8)$.

Check: $(x + 1)(x + 8) = x^2 + 8x + x + 8 = x^2 + 9x + 8.$

The factorization is $(x + 1)(x + 8)$.

Note that we found the factorization by listing all pairs of factors of 8 along with their sums. If, instead, you form binomial factors without calculating sums, you must carefully check that possible factorization. For example, if we attempt the factorization

$$x^2 + 9x + 8 \overset{?}{=} (x + 2)(x + 4),$$

a check reveals that

$$(x + 2)(x + 4) = x^2 + 6x + 8 \neq x^2 + 9x + 8.$$

1. Write an equivalent expression by factoring: $y^2 + 5y + 6$.

YOUR TURN

EXAMPLE 2 Factor: $y^2 - 9y + 20$.

SOLUTION Since the constant term is positive and the coefficient of the middle term is negative, we look for a factorization of 20 in which both factors are negative. Their sum must be -9.

Pair of Factors	Sum of Factors
$-1, -20$	-21
$-2, -10$	-12
$-4, \ -5$	-9 ←

Both factors are negative.

The numbers we need are -4 and -5.

Check: $(y - 4)(y - 5) = y^2 - 5y - 4y + 20 = y^2 - 9y + 20$.

The factorization of $y^2 - 9y + 20$ is $(y - 4)(y - 5)$.

2. Factor: $x^2 - 7x + 12$.

YOUR TURN

TO FACTOR $x^2 + bx + c$ WHEN c IS POSITIVE

When the constant term c of a trinomial is positive, look for two numbers with the same sign. Select pairs of numbers with the sign of b, the coefficient of the middle term.

$$x^2 - 7x + 10 = (x - 2)(x - 5);$$
$$x^2 + 7x + 10 = (x + 2)(x + 5)$$

CONSTANT TERM NEGATIVE

When the constant term of a trinomial is negative, one factor will be negative and one will be positive.

EXAMPLE 3 Factor: $x^3 - x^2 - 30x$.

SOLUTION *Always* look first for a common factor! This time there is one, x. We factor it out:

$$x^3 - x^2 - 30x = x(x^2 - x - 30).$$

Student Notes

Factoring a polynomial often requires more than one step. *Always* look first for a common factor. If one exists, factor out the *greatest* common factor. Then focus on factoring the polynomial in the parentheses.

Now we consider $x^2 - x - 30$. We need a factorization of -30 for which the sum of the factors is -1. Since both the product and the sum are to be negative, we need one positive factor and one negative factor, and the negative factor must have the greater absolute value. This assures a negative sum.

Pair of Factors	Sum of Factors
1, −30	−29
2, −15	−13
3, −10	−7
5, −6	−1

Each pair of factors gives a negative product and a negative sum.

The numbers we need are 5 and −6.

The factorization of $x^2 - x - 30$ is $(x + 5)(x - 6)$. *Don't forget the factor that was factored out in our first step!* We check $x(x + 5)(x - 6)$.

$$\text{Check:} \quad x(x + 5)(x - 6) = x[x^2 - 6x + 5x - 30]$$
$$= x[x^2 - x - 30]$$
$$= x^3 - x^2 - 30x.$$

The factorization of $x^3 - x^2 - 30x$ is $x(x + 5)(x - 6)$.

YOUR TURN

3. Factor: $c^3 - 2c^2 - 35c$.

Technology Connection

To check Example 4, we let

$y_1 = 2x^2 + 34x - 220$,

$y_2 = 2(x - 5)(x + 22)$, and

$y_3 = y_2 - y_1$.

1. How should the graphs of y_1 and y_2 compare?
2. What should the graph of y_3 look like?
3. Use graphs to show that $(2x + 5)(x - 3)$ is *not* a factorization of $2x^2 + x - 15$.

4. Factor: $5y^2 + 35y - 150$.

EXAMPLE 4 Factor: $2x^2 + 34x - 220$.

SOLUTION *Always* look first for a common factor! This time we can factor out 2:

$$2x^2 + 34x - 220 = 2(x^2 + 17x - 110).$$

We next look for a factorization of -110 for which the sum of the factors is 17. Since the product is to be negative and the sum positive, we need one positive factor and one negative factor, and the positive factor must have the larger absolute value.

Pair of Factors	Sum of Factors
−1, 110	109
−2, 55	53
−5, 22	17

Each pair of factors gives a negative product and a positive sum.

The numbers we need are −5 and 22. We stop listing pairs of factors when we have found the correct sum.

Thus, $x^2 + 17x - 110 = (x - 5)(x + 22)$. The factorization of the original trinomial, $2x^2 + 34x - 220$, is $2(x - 5)(x + 22)$. The check is left to the student.

YOUR TURN

TO FACTOR $x^2 + bx + c$ WHEN c IS NEGATIVE

When the constant term c of a trinomial is negative, look for a positive number and a negative number that multiply to c. Select pairs of numbers for which the number with the larger absolute value has the same sign as b, the coefficient of the middle term.

$$x^2 - 4x - 21 = (x + 3)(x - 7);$$
$$x^2 + 4x - 21 = (x - 3)(x + 7)$$

Some polynomials are not factorable using integers.

EXAMPLE 5   Factor: $x^2 - x + 7$.

SOLUTION   Since 7 has very few factors, we can easily check all possibilities.

Pair of Factors	Sum of Factors
7, 1	8
−7, −1	−8

No pair gives a sum of −1.

5. Factor: $t^2 + 2t + 5$.

The polynomial is not factorable using integer coefficients; it is **prime**.

YOUR TURN

TO FACTOR $x^2 + bx + c$

1. If necessary, rewrite the trinomial in descending order.
2. Find a pair of factors that have c as their product and b as their sum.

   • If c is positive, both factors will have the same sign as b.

   • If c is negative, one factor will be positive and the other will be negative. The factor with the larger absolute value will be the factor with the same sign as b.

   • If the sum of the two factors is the opposite of b, changing the signs of both factors will give the desired factors whose sum is b.

3. Check by multiplying.

Trinomials in two variables can be factored using a similar approach.

EXAMPLE 6   Factor: $x^2 - 2xy - 48y^2$.

SOLUTION   We look for numbers p and q such that

$$x^2 - 2xy - 48y^2 = (x + py)(x + qy).$$

The x's and y's can be written in the binomials first: $(x + \blacksquare y)(x + \blacksquare y)$.

Our thinking is much the same as if we were factoring $x^2 - 2x - 48$. We look for factors of -48 whose sum is -2. Those factors are 6 and -8. Thus,

$$x^2 - 2xy - 48y^2 = (x + 6y)(x - 8y).$$

The check is left to the student.

6. Factor: $x^2 - 6xy - 40y^2$.

YOUR TURN

Factoring Trinomials of the Type $ax^2 + bx + c, a \neq 1$

To factor trinomials in which the leading coefficient is not 1, we consider two methods. Use what works best for you or what your instructor chooses.

METHOD 1: FACTORING WITH FOIL

We first consider the **FOIL method** for factoring trinomials of the type $ax^2 + bx + c$, where $a \neq 1$.

Consider the following multiplication.

$$\begin{array}{cccc} \text{F} & \text{O} & \text{I} & \text{L} \end{array}$$
$$(3x + 2)(4x + 5) = 12x^2 + 15x + 8x + 10$$
$$= 12x^2 + 23x + 10$$

To factor $12x^2 + 23x + 10$, we could "reverse" the multiplication and look for two binomials whose product is this trinomial. The product of the First terms must be $12x^2$. The product of the Outer terms plus the product of the Inner terms must be $23x$. The product of the Last terms must be 10. Our first approach to finding such a factorization relies on FOIL.

TO FACTOR $ax^2 + bx + c$ USING FOIL

1. Factor out the largest common factor, if one exists. Here we assume none does.

2. List possible **First** terms whose product is ax^2:

   $$(\blacksquare x + \blacksquare)(\blacksquare x + \blacksquare) = ax^2 + bx + c.$$
   $$\underline{} \text{FOIL}$$

3. List possible **Last** terms whose product is c:

   $$(\blacksquare x + \blacksquare)(\blacksquare x + \blacksquare) = ax^2 + bx + c.$$
   $$\underline{} \text{FOIL}$$

4. Using the possibilities from steps (2) and (3), find a combination for which the sum of the **Outer** and **Inner** products is bx:

   $$(\blacksquare x + \blacksquare)(\blacksquare x + \blacksquare) = ax^2 + bx + c.$$
   $$\underline{\text{I}}$$
   $$\underline{\text{O}} \text{FOIL}$$

If no correct combination exists, then the polynomial is prime.

EXAMPLE 7   Factor: $3x^2 - 10x - 8$.

SOLUTION

1. First, note that there is no common factor (other than 1 or -1).

2. Next, factor the first term, $3x^2$. The only possibility for factors is $3x \cdot x$. Thus, if a factorization exists, it must be of the form

   $$(3x + \blacksquare)(x + \blacksquare).$$

3. The constant term, -8, can be factored as

$(-1)(8)$,	$(8)(-1)$,	When $a \neq 1$, the order
$(1)(-8)$,	$(-8)(1)$,	of the factors can affect
$(-2)(4)$, as well as	$(4)(-2)$,	the middle term.
$(2)(-4)$,	$(-4)(2)$.	

4. Find binomial factors for which the sum of the Outer and Inner products is the middle term, $-10x$. Check each possibility by multiplying:

Pair of Factors	Corresponding Trial	Product	
$-1,\ 8$	$(3x-1)(x+8)$	$3x^2+24x-x-8$	
		$=3x^2+23x-8$	Wrong middle term
$1,-8$	$(3x+1)(x-8)$	$3x^2-24x+x-8$	
		$=3x^2-23x-8$	Wrong middle term
$-2,\ 4$	$(3x-2)(x+4)$	$3x^2+12x-2x-8$	
		$=3x^2+10x-8$	Wrong middle term
$2,-4$	$(3x+2)(x-4)$	$3x^2-12x+2x-8$	
		$=3x^2-10x-8$	Correct middle term!
$8,-1$	$(3x+8)(x-1)$	$3x^2-3x+8x-8$	
		$=3x^2+5x-8$	Wrong middle term
$-8,\ 1$	$(3x-8)(x+1)$	$3x^2+3x-8x-8$	
		$=3x^2-5x-8$	Wrong middle term
$4,-2$	$(3x+4)(x-2)$	$3x^2-6x+4x-8$	
		$=3x^2-2x-8$	Wrong middle term
$-4,\ 2$	$(3x-4)(x+2)$	$3x^2+6x-4x-8$	
		$=3x^2+2x-8$	Wrong middle term

The correct factorization is $(3x+2)(x-4)$.

7. Factor: $2y^2-y-6$.

YOUR TURN

Two observations can be made from Example 7. First, we listed all possible trials even though we generally stop after finding the correct factorization. We did this to show that **each trial differs only in the middle term of the product.** Second, note that **only the sign of the middle term's coefficient changes when the signs in the binomials are reversed.**

EXAMPLE 8 Factor: $6x^6-19x^5+10x^4$.

SOLUTION

1. First, factor out the greatest common factor x^4:

$$x^4(6x^2-19x+10).$$

2. Note that $6x^2=6x\cdot x$ and $6x^2=3x\cdot 2x$. Thus, $6x^2-19x+10$ may factor into

$$(3x+\blacksquare)(2x+\blacksquare)\quad\text{or}\quad(6x+\blacksquare)(x+\blacksquare).$$

3. We factor the last term, 10. Since the middle term's coefficient is negative, we need consider only the factorizations with negative factors:

$$(-10)(-1),\qquad\qquad(-1)(-10),$$

as well as

$$(-5)(-2),\qquad\qquad(-2)(-5).$$

4. There are 4 possibilities for *each* factorization in step (2). The sum of the Outer and Inner products must be the middle term, $-19x$. We first try these factors with $(3x + \quad)(2x + \quad)$. If none gives the correct factorization, then we will consider $(6x + \quad)(x + \quad)$.

Trial	Product
$(3x - 1)(2x - 10)$	$6x^2 - 30x - 2x + 10$
	$= 6x^2 - 32x + 10$ Wrong middle term
$(3x - 10)(2x - 1)$	$6x^2 - 3x - 20x + 10$
	$= 6x^2 - 23x + 10$ Wrong middle term
$(3x - 2)(2x - 5)$	$6x^2 - 15x - 4x + 10$
	$= 6x^2 - 19x + 10$ Correct middle term!

Since we have a correct factorization, we need not consider any additional trials. The factorization of $6x^2 - 19x + 10$ is $(3x - 2)(2x - 5)$. *But do not forget the common factor!* We must include it in the complete factorization of the original trinomial:

$$6x^6 - 19x^5 + 10x^4 = x^4(3x - 2)(2x - 5).$$

YOUR TURN

8. Factor: $15t^4 - 22t^3 + 8t^2$.

Student Notes

Keep your work organized so that you can see what you have already considered. For example, when factoring $6x^2 - 19x + 10$, we can list all possibilities and cross out those in which a common factor appears:

$(3x - 1)(2x - 10),$

$(3x - 10)(2x - 1),$

$(3x - 2)(2x - 5),$

$(3x - 5)(2x - 2),$

$(6x - 1)(x - 10),$

$(6x - 10)(x - 1),$

$(6x - 2)(x - 5),$

$(6x - 5)(x - 2).$

By being organized and not erasing, we can see that there are only four possible factorizations.

In Example 8, look again at the trial $(3x - 1)(2x - 10)$. Without multiplying, we can dismiss this. To see why, note that

$$(3x - 1)(2x - 10) = (3x - 1)2(x - 5).$$

The expression $2x - 10$ has a common factor, 2. But we removed the *largest* common factor in step (1). If $2x - 10$ were one of the factors, then 2 would be *another* common factor in addition to the original, x^4. Thus, $(2x - 10)$ cannot be part of the factorization of $6x^2 - 19x + 10$. Similar reasoning can be used to reject $(3x - 5)(2x - 2)$ as a possible factorization.

Once the largest common factor is factored out, no remaining factor can have a common factor.

TIPS FOR FACTORING $ax^2 + bx + c$ WITH FOIL

1. Once the largest common factor is factored out of the original trinomial, no binomial factor can contain a common factor (other than 1 or −1).
2. If necessary, factor out a −1 so that a is positive. Then if c is also positive, the signs in the factors must match the sign of b.
3. Reversing the two signs in the binomials reverses the sign of the middle term of their product.
4. Organize your work so that you can keep track of those possibilities that you have checked.

METHOD 2: THE GROUPING METHOD

The second method for factoring trinomials of the type $ax^2 + bx + c, a \neq 1$, is known as the *grouping method*, or the *ac-method*. This method relies on rewriting $ax^2 + bx + c$ as $ax^2 + px + qx + c$ and then factoring by grouping.

We find p and q by looking for two numbers whose sum is b and whose product is ac.*

Consider $6x^2 + 23x + 20$.

$$6x^2 + 23x + 20$$

(1) Multiply 6 and 20: $6 \cdot 20 = 120$.

(2) Factor 120: $120 = 8 \cdot 15$, and $8 + 15 = 23$.

(3) Split the middle term: $23x = 8x + 15x$.

(4) Factor by grouping.

We factor by grouping as follows:

$$6x^2 + 23x + 20 = 6x^2 + 8x + 15x + 20 \qquad \text{Writing } 23x \text{ as } 8x + 15x$$
$$= 2x(3x + 4) + 5(3x + 4)$$
$$= (3x + 4)(2x + 5). \qquad \left.\right\} \text{ Factoring by grouping}$$

TO FACTOR $ax^2 + bx + c$ USING GROUPING

1. Make sure that any common factors have been factored out.
2. Multiply the leading coefficient a and the constant c.
3. Find a pair of factors of ac whose sum is b.
4. Rewrite the trinomial's middle term, bx, as $px + qx$.
5. Factor by grouping.

EXAMPLE 9 Factor: $3x^2 + 10x - 8$.

SOLUTION

1. First, look for a common factor. There is none (other than 1 or -1).
2. Multiply the leading coefficient and the constant, 3 and -8:

   $$3(-8) = -24.$$

3. Factor -24 so that the sum of the factors is 10:

   $$-24 = 12(-2) \quad \text{and} \quad 12 + (-2) = 10.$$

4. Split $10x$ using the results of step (3):

   $$10x = 12x - 2x.$$

5. Finally, factor by grouping:

   $$3x^2 + 10x - 8 = 3x^2 + 12x - 2x - 8 \qquad \text{Substituting } 12x - 2x \text{ for } 10x$$
   $$= 3x(x + 4) - 2(x + 4)$$
   $$= (x + 4)(3x - 2). \qquad \left.\right\} \text{ Factoring by grouping}$$

Pair of Factors	Sum of Factors
1, -24	-23
-1, 24	23
2, -12	-10
-2, 12	10
3, -8	-5
-3, 8	5
4, -6	-2
-4, 6	2

The check is left to the student. The factorization is $(x + 4)(3x - 2)$.

9. Factor: $8x^2 + 6x - 5$.

YOUR TURN

EXAMPLE 10 Factor: $6x^4 - 116x^3 - 80x^2$.

SOLUTION

1. First, factor out the greatest common factor $2x^2$:

   $$6x^4 - 116x^3 - 80x^2 = 2x^2(3x^2 - 58x - 40).$$

---

*The rationale behind these steps is outlined in Exercise 111.

Pair of Factors	Sum of Factors
1, −120	−119
2, −60	−58
3, −40	−37
4, −30	−26
5, −24	−19
6, −20	−14
8, −15	−7
10, −12	−2

2. To factor $3x^2 - 58x - 40$, multiply the leading coefficient, 3, and the constant, −40: $3(-40) = -120$.

3. Next, look for factors of −120 that add to −58. Since −58 is negative, the negative factor of −120 must have the larger absolute value. We see from the table at left that the factors we need are 2 and −60.

4. Split the middle term, $-58x$, using the results of step (3): $-58x = 2x - 60x$.

5. Factor by grouping:

$$3x^2 - 58x - 40 = 3x^2 + 2x - 60x - 40 \qquad \text{Substituting } 2x - 60x \text{ for } -58x$$
$$= x(3x + 2) - 20(3x + 2) \Big\}$$
$$= (3x + 2)(x - 20). \qquad \text{Factoring by grouping}$$

The factorization of $3x^2 - 58x - 40$ is $(3x + 2)(x - 20)$. *But don't forget the common factor!*

$$6x^4 - 116x^3 - 80x^2 = 2x^2(3x + 2)(x - 20)$$

The check is left to the student. The factorization is $2x^2(3x + 2)(x - 20)$.

10. Factor: $18t^3 - 21t^2 - 60t$.

YOUR TURN

5.4 EXERCISE SET

FOR EXTRA HELP

MyMathLab® MathXL
PRACTICE   WATCH   READ   REVIEW

 ### Vocabulary and Reading Check

Classify each of the following statements as either true or false.

1. The first step in factoring any polynomial is to look for a common factor.

2. A common factor may have a negative coefficient.

3. If c is prime, $x^2 + bx + c$ cannot be factored.

4. If a trinomial contains a common factor, it cannot be factored using binomials.

5. Whenever the product of two numbers is negative, those numbers have the same sign.

6. If $p + q = -17$, then $-p + (-q) = 17$.

7. If a trinomial has no common factor, then neither of its binomial factors can have a common factor.

8. Trinomials in more than one variable cannot be factored.

Factoring Trinomials of the Type $x^2 + bx + c$

Factor. If a polynomial is prime, state this.

9. $x^2 + 5x + 4$
10. $x^2 + 7x + 12$
11. $y^2 - 12y + 27$
12. $t^2 - 8t + 15$
13. $t^2 - 2t - 8$
14. $y^2 - 3y - 10$
15. $a^2 + a - 2$
16. $n^2 + n - 20$
17. $2x^2 + 6x - 108$
18. $3p^2 - 9p - 120$
19. $14a + a^2 + 45$
20. $11y + y^2 + 24$
21. $p^3 - p^2 - 72p$
22. $x^3 + 2x^2 - 63x$
23. $a^2 - 11a + 28$
24. $t^2 - 14t + 45$
25. $x + x^2 - 6$
26. $3x + x^2 - 10$
27. $5y^2 + 40y + 35$
28. $3x^2 + 15x + 18$
29. $32 + 4y - y^2$
30. $56 + x - x^2$
31. $56x + x^2 - x^3$
32. $32y + 4y^2 - y^3$
33. $y^4 + 5y^3 - 84y^2$
34. $x^4 + 11x^3 - 80x^2$
35. $x^2 - 3x + 5$
36. $x^2 + 12x + 13$
37. $x^2 + 12xy + 27y^2$
38. $p^2 - 5pq - 24q^2$
39. $x^2 - 14xy + 49y^2$
40. $y^2 + 8yz + 16z^2$
41. $n^5 - 80n^4 + 79n^3$
42. $t^5 - 50t^4 + 49t^3$
43. $x^6 + 2x^5 - 63x^4$
44. $x^6 + 7x^5 - 18x^4$

Factoring Trinomials of the Type $ax^2 + bx + c, a \neq 1$

Factor.

45. $3x^2 - 4x - 4$

46. $2x^2 - x - 10$

47. $6t^2 + t - 15$

48. $10y^2 + 7y - 12$

49. $6p^2 - 20p + 16$

50. $24a^2 - 14a + 2$

51. $9a^2 + 18a + 8$

52. $35y^2 + 34y + 8$

53. $8y^2 + 30y^3 - 6y$

54. $4t^2 + 10t^3 - 6t$

55. $18x^2 - 24 - 6x$

56. $8x^2 - 16 - 28x$

57. $t^8 + 5t^7 - 14t^6$

58. $a^6 + a^5 - 6a^4$

59. $70x^4 - 68x^3 + 16x^2$

60. $14x^4 - 19x^3 - 3x^2$

61. $18y^2 - 9y - 20$

62. $20x^2 + x - 30$

63. $16x^2 + 24x + 5$

64. $2y^2 + 9y + 9$

Aha! **65.** $5x^2 + 24x + 16$
(*Hint:* See Exercise 63.)

66. $9y^2 + 9y + 2$

67. $-8t^2 - 8t + 30$

68. $-36a^2 + 21a - 3$

69. $18xy^3 + 3xy^2 - 10xy$

70. $3x^3y^2 - 5x^2y^2 - 2xy^2$

71. $24x^2 - 2 - 47x$

72. $15y^2 - 10 - 47y$

73. $63x^3 + 111x^2 + 36x$

74. $50y^3 + 115y^2 + 60y$

75. $48x^4 + 4x^3 - 30x^2$

76. $40y^4 + 4y^3 - 12y^2$

77. $12a^2 - 17ab + 6b^2$

78. $20p^2 - 23pq + 6q^2$

79. $2x^2 + xy - 6y^2$

80. $8m^2 - 6mn - 9n^2$

81. $6x^2 - 29xy + 28y^2$

82. $10p^2 + 7pq - 12q^2$

83. $9x^2 - 30xy + 25y^2$

84. $4p^2 + 12pq + 9q^2$

85. $9x^2y^2 + 5xy - 4$

86. $7a^2b^2 + 13ab + 6$

87. How can one conclude that $x^2 + 5x + 200$ is prime without performing any trials?

88. Asked to factor $4x^2 + 28x + 48$, Aziz *incorrectly* answers

$$4x^2 + 28x + 48 = (2x + 6)(2x + 8)$$
$$= 2(x + 3)(x + 4).$$

If this were a 10-point quiz question, how many points would you take off? Why?

Skill Review

Simplify. Do not use negative exponents in the answer.
[1.6]

89. $(2a^{-6}b)^{-3}$

90. $(3x^{-6}y)(-4x^{-1}y^{-2})$

91. $\dfrac{12t^{-11}}{8t^6}$

92. $\left(\dfrac{2x^{-5}y}{3x^2y^{-4}}\right)^{-2}$

93. Convert 0.000607 to scientific notation.   [1.7]

94. Convert 3.1875×10^8 to decimal notation.   [1.7]

Synthesis

95. Describe in your own words an approach that can be used to factor any trinomial of the form $ax^2 + bx + c$ that is not prime.

96. Suppose $(rx - p)(sx - q) = ax^2 - bx + c$ is true. Explain how this can be used to factor $ax^2 + bx + c$.

Factor.

97. $60x^8y^6 + 35x^4y^3 + 5$

98. $x^2 + \frac{3}{5}x - \frac{4}{25}$

99. $y^2 - \frac{8}{49} + \frac{2}{7}y$

100. $y^2 + 0.4y - 0.05$

101. $20a^3b^6 - 3a^2b^4 - 2ab^2$

102. $4x^{2a} - 4x^a - 3$

103. $x^{2a} + 5x^a - 24$

104. $bdx^2 + adx + bcx + ac$

105. $2ar^2 + 4asr + as^2 - asr$

106. $a^2p^{2a} + a^2p^a - 2a^2$

Aha! **107.** $(x + 3)^2 - 2(x + 3) - 35$

108. $6(x - 7)^2 + 13(x - 7) - 5$

109. Find all integers m for which $x^2 + mx + 75$ can be factored.

110. Find all integers q for which $x^2 + qx - 32$ can be factored.

111. To better understand factoring $ax^2 + bx + c$ by grouping, suppose that

$$ax^2 + bx + c = (mx + r)(nx + s).$$

Show that if $p = ms$ and $q = rn$, then $p + q = b$ and $pq = ac$.

112. One factor of $x^2 - 345x - 7300$ is $x + 20$. Find the other factor.

113. Use the TABLE feature to check your answers to Exercises 15, 57, and 99.

114. Let $y_1 = 3x^2 + 10x - 8$, $y_2 = (x + 4)(3x - 2)$, and $y_3 = y_2 - y_1$ to check Example 9 graphically.

115. Explain how the following graph of

$$y = x^2 + 3x - 2 - (x - 2)(x + 1)$$

can be used to show that

$$x^2 + 3x - 2 \neq (x - 2)(x + 1).$$

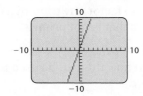

116. Draw three different rectangles that have an area of 12 square units. Use one of those rectangles to complete a rectangle similar to those drawn in Section 5.2 that illustrates the factorization of $x^2 + 7x + 12$.

 YOUR TURN ANSWERS: SECTION 5.4

1. $(y + 2)(y + 3)$ **2.** $(x - 3)(x - 4)$ **3.** $c(c + 5)(c - 7)$
4. $5(y - 3)(y + 10)$ **5.** Prime   **6.** $(x + 4y)(x - 10y)$
7. $(2y + 3)(y - 2)$ **8.** $t^2(3t - 2)(5t - 4)$
9. $(4x + 5)(2x - 1)$ **10.** $3t(2t - 5)(3t + 4)$

QUICK QUIZ: SECTIONS 5.1–5.4

1. Identify the leading term and the leading coefficient of $8 - 5a^3 + 6a + a^2$.   [5.1]

2. Multiply: $(xy - 4)(xy - 5)$.   [5.2]

Factor.

3. $9x^3y^5 + 3x^2y^6 - 15x^4y^5$ [5.3]

4. $t^2 - 6t - 40$ [5.4]

5. $6n^3 - 11n^2 - 10n$ [5.4]

PREPARE TO MOVE ON

Simplify.   [1.6]

1. $(5a)^2$ **2.** $(3x^4)^2$

Multiply.   [5.2]

3. $(x + 3)^2$ **4.** $(2t - 5)^2$

5. $(y + 1)(y - 1)$

6. $(4x^2 + 3y)(4x^2 - 3y)$

Mid-Chapter Review

GUIDED SOLUTIONS

1. Multiply: $(2x - 3)(x + 4)$.   [5.2]

$$\qquad\qquad\quad F \quad\ O \quad\ I \quad\ L$$
$$(2x - 3)(x + 4) = \boxed{} + \boxed{} - \boxed{} - \boxed{}$$
$$= 2x^2 + \boxed{} - 12$$

2. Factor: $3x^3 + 7x^2 + 2x$.   [5.4]

$$3x^3 + 7x^2 + 2x = \boxed{}(3x^2 + 7x + 2)$$
$$= \boxed{}\left(3x + \boxed{}\right)\left(x + \boxed{}\right)$$

MIXED REVIEW

Perform the indicated operation.

3. $(4t^3 - 2t + 6) + (8t^2 - 11t - 7)$ [5.1]

4. $4x^2y(3xy - 2x^3 + 6y^2)$ [5.2]

5. $(8n^2 + 5n - 2) - (-n^2 + 6n - 2)$ [5.1]

6. $(x + 1)(x + 7)$ [5.2]

7. $(2x - 3)(5x - 1)$ [5.2]

8. $\left(\frac{1}{2}x^2 + \frac{1}{3}x - \frac{3}{2}\right) + \left(\frac{2}{3}x^2 - \frac{1}{2}x - \frac{1}{3}\right)$ [5.1]

9. $(3m - 10)^2$ [5.2]

10. $(1.2x^2 - 3.7x) - (2.8x^2 - x + 1.4)$ [5.1]

11. $(a + 2)(a^2 - a - 6)$ [5.2]

12. $(c + 9)(c - 9)$ [5.2]

Factor.

13. $8x^2y^3z + 12x^3y^2 - 16x^2yz^3$ [5.3]

14. $3t^3 - 3t^2 - 1 + t$ [5.3]

15. $x^2 - x - 90$ [5.4]

16. $6x^3 + 60x^2 + 126x$ [5.4]

17. $5x^2 + 7x - 6$ [5.4]

18. $2x + 2y + ax + ay$ [5.3]

5.5   Factoring Perfect-Square Trinomials and Differences of Squares

Perfect-Square Trinomials ▪ Differences of Squares ▪ More Factoring by Grouping

Student Notes

If you're not already quick to recognize squares, this is a good time to memorize these numbers:

$$1 = 1^2, \qquad 49 = 7^2,$$
$$4 = 2^2, \qquad 64 = 8^2,$$
$$9 = 3^2, \qquad 81 = 9^2,$$
$$16 = 4^2, \qquad 100 = 10^2,$$
$$25 = 5^2, \qquad 121 = 11^2,$$
$$36 = 6^2, \qquad 144 = 12^2.$$

Study Skills

Fill In Your Blanks

Don't hesitate to write out any missing steps that you'd like to see included. For instance, in Example 1(c), we state that $100y^2$ is a square. To solidify your understanding, you may want to write in $10y \cdot 10y = 100y^2$.

---

1. Determine whether $25x^2 - 60x + 36$ is a perfect-square trinomial.

We now introduce a faster way to factor trinomials that are squares of binomials. A method for factoring differences of squares is also developed.

Perfect-Square Trinomials

Consider the trinomial $x^2 + 6x + 9$. To factor it, we can look for factors of 9 that add to 6. These factors are 3 and 3:

$$x^2 + 6x + 9 = (x + 3)(x + 3) = (x + 3)^2.$$

Note that the result is the square of a binomial. Because of this, we call $x^2 + 6x + 9$ a **perfect-square trinomial**. Once recognized, a perfect-square trinomial can be quickly factored.

> **TO RECOGNIZE A PERFECT-SQUARE TRINOMIAL**
> - Two terms must be squares, such as A^2 and B^2.
> - The remaining term must be $2AB$ or its opposite, $-2AB$.

EXAMPLE 1   Determine whether each polynomial is a perfect-square trinomial.

a) $x^2 + 10x + 25$ **b)** $4x + 16 + 3x^2$ **c)** $100y^2 + 81 - 180y$

SOLUTION

a)   • Two of the terms in $x^2 + 10x + 25$ are squares: x^2 and 25.
   • Twice the product of the square roots is $2 \cdot x \cdot 5$, or $10x$. This is the remaining term in the trinomial.

Thus, $x^2 + 10x + 25$ *is* a perfect-square trinomial.

b) In $4x + 16 + 3x^2$, only one term, 16, is a square ($3x^2$ is not a square because 3 is not a perfect square; $4x$ is not a square because x is not a square).

Thus, $4x + 16 + 3x^2$ *is not* a perfect-square trinomial.

c) It can help to first write the polynomial in descending order: $100y^2 - 180y + 81$.
   • Two of the terms, $100y^2$ and 81, are squares.
   • Twice the product of the square roots is $2(10y)(9)$, or $180y$. The remaining term in the trinomial is the opposite of $180y$.

Thus, $100y^2 + 81 - 180y$ *is* a perfect-square trinomial.

 YOUR TURN

To factor a perfect-square trinomial, we use the following patterns.

> **FACTORING A PERFECT-SQUARE TRINOMIAL**
> $$A^2 + 2AB + B^2 = (A + B)^2;$$
> $$A^2 - 2AB + B^2 = (A - B)^2$$

EXAMPLE 2 Factor.

a) $x^2 - 10x + 25$

b) $16y^2 + 49 + 56y$

c) $-20xy + 4y^2 + 25x^2$

SOLUTION

a) $x^2 - 10x + 25 = (x - 5)^2$.   We have a perfect square of the form $A^2 - 2AB + B^2$, with $A = x$ and $B = 5$. We write the square roots with a minus sign between them.

Note the sign!

Check: $(x - 5)^2 = (x - 5)(x - 5)$
$$= x^2 - 5x - 5x + 25$$
$$= x^2 - 10x + 25.$$

The factorization is $(x - 5)^2$.

b) $16y^2 + 49 + 56y = 16y^2 + 56y + 49$ Using a commutative law

$$= (4y + 7)^2$$ We have a perfect square of the form $A^2 + 2AB + B^2$, with $A = 4y$ and $B = 7$. We write the square roots with a plus sign between them.

The check is left to the student.

c) $-20xy + 4y^2 + 25x^2 = 4y^2 - 20xy + 25x^2$ Writing descending order with respect to y

$$= (2y - 5x)^2$$

This square can also be expressed as

$$25x^2 - 20xy + 4y^2 = (5x - 2y)^2.$$

The student should confirm that both factorizations check.

2. Factor: $25x^2 - 60x + 36$.

YOUR TURN

When factoring, always look first for a factor common to all the terms.

EXAMPLE 3 Factor: $-4y^2 - 144y^8 + 48y^5$.

SOLUTION We check for a common factor. By factoring out $-4y^2$, we see that the leading coefficient of the polynomial inside the parentheses is positive:

Factor out the common factor.

Factor the perfect-square trinomial.

$-4y^2 - 144y^8 + 48y^5 = -4y^2(1 + 36y^6 - 12y^3)$

$$= -4y^2(36y^6 - 12y^3 + 1)$$ Using a commutative law

$$= -4y^2(6y^3 - 1)^2$$ $36y^6 = (6y^3)^2$

Check: $-4y^2(6y^3 - 1)^2 = -4y^2(6y^3 - 1)(6y^3 - 1)$
$$= -4y^2(36y^6 - 12y^3 + 1)$$
$$= -144y^8 + 48y^5 - 4y^2$$
$$= -4y^2 - 144y^8 + 48y^5$$

The factorization is $-4y^2(6y^3 - 1)^2$.

3. Factor: $-50t^2 + 20pt - 2p^2$.

YOUR TURN

Differences of Squares

An expression of the form $A^2 - B^2$ is a **difference of squares**. Note that, unlike a perfect-square trinomial, $A^2 - B^2$ has a square that is subtracted.

A difference of squares can be factored as the product of two binomials.

FACTORING A DIFFERENCE OF TWO SQUARES

$$A^2 - B^2 = (A + B)(A - B)$$

We often refer to $(A + B)(A - B)$ as the product of the sum and the difference of A and B.

EXAMPLE 4 Factor: **(a)** $x^2 - 9$; **(b)** $25y^6 - 49x^2$.

SOLUTION

a) $x^2 - 9 = x^2 - 3^2 = (x + 3)(x - 3)$ *Check:* $(x + 3)(x - 3)$
$$= x^2 - 3x + 3x - 9$$
$$= x^2 - 9$$

$$
\begin{array}{ccccccc}
A^2 & - & B^2 & = & (A & + B)(A & - B) \\
\downarrow & & \downarrow & & \downarrow & \downarrow \quad \downarrow & \downarrow
\end{array}
$$

4. Factor: $n^{10} - 4$.

b) $25y^6 - 49x^2 = (5y^3)^2 - (7x)^2 = (5y^3 + 7x)(5y^3 - 7x)$

YOUR TURN

As always, the first step in factoring is to look for common factors. Remember too that factoring is complete only when no factor with more than one term can be factored further.

EXAMPLE 5 Factor: **(a)** $5 - 5p^2q^6$; **(b)** $16x^4y - 81y$.

SOLUTION

a) $5 - 5p^2q^6 = 5(1 - p^2q^6)$ Factoring out the common factor
$$= 5[1^2 - (pq^3)^2]$$ Rewriting p^2q^6 as a quantity squared
$$= 5(1 + pq^3)(1 - pq^3)$$ Factoring the difference of squares

Check: $5(1 + pq^3)(1 - pq^3) = 5(1 - pq^3 + pq^3 - p^2q^6)$
$$= 5(1 - p^2q^6) = 5 - 5p^2q^6$$

The factorization is $5(1 + pq^3)(1 - pq^3)$.

Factor out a common factor.

b) $16x^4y - 81y = y(16x^4 - 81)$ Factoring out the common factor
$$= y[(4x^2)^2 - 9^2]$$

Factor a difference of squares.

$$= y(4x^2 + 9)(4x^2 - 9)$$ Factoring the difference of squares

Factor another difference of squares.

$$= y(4x^2 + 9)(2x + 3)(2x - 3)$$ Factoring $4x^2 - 9$, which is itself a difference of squares

5. Factor: $16ap^4 - a$.

The check is left to the student.

YOUR TURN

Note in Example 5(b) that $4x^2 + 9$ is a *sum* of squares that cannot be factored.

More Factoring by Grouping

Sometimes, when factoring a polynomial with four terms, we may be able to factor further.

EXAMPLE 6   Factor: $x^3 + 3x^2 - 4x - 12$.

SOLUTION

$$x^3 + 3x^2 - 4x - 12 = x^2(x + 3) - 4(x + 3) \qquad \text{Factoring by grouping}$$

$$= (x + 3)(x^2 - 4) \qquad \text{Factoring out } x + 3$$

$$= (x + 3)(x + 2)(x - 2) \qquad \text{Factoring } x^2 - 4$$

6. Factor: $y^3 + y^2 - 25y - 25$.

 YOUR TURN

A difference of squares can have four or more terms. For example, one of the squares may be a trinomial. In this case, a new type of grouping can be used.

EXAMPLE 7   Factor: **(a)** $x^2 + 6x + 9 - y^2$; **(b)** $a^2 - b^2 + 8b - 16$.

SOLUTION

a) $x^2 + 6x + 9 - y^2 = (x^2 + 6x + 9) - y^2 \qquad$ Grouping as a perfect-square trinomial minus y^2 to show a difference of squares

$$= (x + 3)^2 - y^2$$

$$= (x + 3 + y)(x + 3 - y)$$

b) Grouping $a^2 - b^2 + 8b - 16$ into two groups of two terms does not yield a common binomial factor, so we look for a perfect-square trinomial. In this case, the perfect-square trinomial is being subtracted from a^2:

$$a^2 - b^2 + 8b - 16 = a^2 - (b^2 - 8b + 16) \qquad \text{Factoring out } -1 \text{ and rewriting as subtraction}$$

$$= a^2 - (b - 4)^2 \qquad \text{Factoring the perfect-square trinomial}$$

$$= (a + (b - 4))(a - (b - 4)) \qquad \text{Factoring a difference of squares}$$

7. Factor:

$4m^2 + 20m + 25 - a^2$.

$$= (a + b - 4)(a - b + 4). \qquad \text{Removing parentheses}$$

YOUR TURN

5.5 EXERCISE SET

FOR EXTRA HELP

MyMathLab® Math XL
PRACTICE   WATCH   READ   REVIEW

Vocabulary and Reading Check

Classify each of the following as a perfect-square trinomial, a difference of two squares, a polynomial having a common factor, or none of these.

1. $x^2 - 100$

2. $t^2 - 18t + 81$

3. $36x^2 - 12x + 1$

4. $36a^2 - 25$

5. $4r^2 + 8r + 9$

6. $9x^2 - 12$

7. $4x^2 + 8x + 10$

8. $t^2 - 6t + 8$

9. $4t^2 + 9s^2 + 12st$

10. $9rt^2 - 5rt + 6r$

Perfect-Square Trinomials

Factor.

11. $x^2 + 20x + 100$

12. $x^2 + 14x + 49$

13. $t^2 - 2t + 1$

14. $t^2 - 4t + 4$

15. $4a^2 - 24a + 36$

16. $9a^2 + 18a + 9$

17. $y^2 + 36 + 12y$

18. $y^2 + 36 - 12y$

19. $-18y^2 + y^3 + 81y$

20. $24a^2 + a^3 + 144a$

21. $2x^2 - 40x + 200$

22. $32x^2 + 48x + 18$

23. $1 - 8d + 16d^2$

24. $64 + 25y^2 - 80y$

25. $-y^3 - 8y^2 - 16y$

26. $-a^3 + 10a^2 - 25a$

27. $0.25x^2 + 0.30x + 0.09$

28. $0.04x^2 - 0.28x + 0.49$

29. $p^2 - 2pq + q^2$

30. $m^2 + 2mn + n^2$

31. $25a^2 + 30ab + 9b^2$

32. $49p^2 - 84pq + 36q^2$

33. $5a^2 + 10ab + 5b^2$

34. $4t^2 - 8tr + 4r^2$

Differences of Squares

Factor completely.

35. $x^2 - 25$

36. $x^2 - 16$

37. $m^2 - 64$

38. $p^2 - 49$

39. $4a^2 - 81$

40. $100c^2 - 1$

41. $12c^2 - 12d^2$

42. $6x^2 - 6y^2$

43. $7xy^4 - 7xz^4$

44. $25ab^4 - 25az^4$

45. $4a^3 - 49a$

46. $9x^4 - 25x^2$

47. $3x^8 - 3y^8$

48. $9a^4 - a^2b^2$

49. $p^2q^2 - 100$

50. $a^2b^2 - 121$

51. $9a^4 - 25a^2b^4$

52. $16x^6 - 81x^2y^4$

53. $y^2 - \frac{1}{4}$

54. $x^2 - \frac{1}{9}$

55. $\frac{1}{100} - x^2$

56. $\frac{1}{16} - y^2$

More Factoring by Grouping

Factor completely.

57. $(a + b)^2 - 36$

58. $(p + q)^2 - 64$

59. $x^2 - 6x + 9 - y^2$

60. $a^2 - 8a + 16 - b^2$

61. $t^3 + 8t^2 - t - 8$

62. $x^3 - 7x^2 - 4x + 28$

63. $r^3 - 3r^2 - 9r + 27$

64. $t^3 + 2t^2 - 4t - 8$

65. $m^2 - 2mn + n^2 - 25$

66. $x^2 + 2xy + y^2 - 9$

67. $81 - (x + y)^2$

68. $49 - (a + b)^2$

69. $r^2 - 2r + 1 - 4s^2$

70. $c^2 + 4cd + 4d^2 - 9p^2$

Aha! **71.** $16 - a^2 - 2ab - b^2$

72. $100 - x^2 - 2xy - y^2$

73. $x^3 + 5x^2 - 4x - 20$

74. $t^3 + 6t^2 - 9t - 54$

75. $a^3 - ab^2 - 2a^2 + 2b^2$

76. $p^2q - 25q + 3p^2 - 75$

77. Describe a procedure that could be used to determine whether a polynomial is a difference of squares.

78. Why are the product and power rules for exponents important when factoring differences of squares?

Skill Review

79. Find $-x$ if $x = -16$. [1.2]

80. Use a commutative law to write an expression equivalent to $x + w$. [1.2]

81. Solve for x: $3x = y - ax$. [1.5]

82. Simplify: $\left(\frac{1}{3}\right)^{-2}$. [1.6]

83. Find the intersection: $\{1, 2, 3\} \cap \{1, 3, 5, 7\}$. [4.2]

84. Find the union: $\{1, 2, 3\} \cup \{1, 3, 5, 7\}$. [4.2]

Synthesis

85. Gretchen plans to use FOIL to factor polynomials rather than looking for perfect-square trinomials or differences of squares. How might you convince her that it is worthwhile to learn the factoring techniques of this section?

86. Without finding the entire factorization, determine the number of factors of $x^{256} - 1$. Explain how you arrived at your answer.

Factor completely.

87. $-\frac{8}{27}r^2 - \frac{10}{9}rs - \frac{1}{6}s^2 + \frac{2}{3}rs$

88. $\frac{1}{36}x^8 + \frac{2}{9}x^4 + \frac{4}{9}$

89. $0.09x^8 + 0.48x^4 + 0.64$

90. $a^2 + 2ab + b^2 - c^2 + 6c - 9$

91. $r^2 - 8r - 25 - s^2 - 10s + 16$

92. $x^{2a} - y^2$

93. $25y^{2a} - (x^{2b} - 2x^b + 1)$

94. $(a - 3)^2 - 8(a - 3) + 16$

95. $3(x + 1)^2 + 12(x + 1) + 12$

96. $m^2 + 4mn + 4n^2 + 5m + 10n$

97. $s^2 - 4st + 4t^2 + 4s - 8t + 4$

98. $5c^{100} - 80d^{100}$

99. $9x^{2n} - 6x^n + 1$

100. $x^{-4} + 2x^{-5} + x^{-6}$

101. If $P(x) = x^2$, use factoring to simplify
$$P(a + h) - P(a).$$

102. If $P(x) = x^4$, use factoring to simplify
$$P(a + h) - P(a).$$

103. *Volume of Carpeting.* The volume of a carpet that is rolled up can be estimated by the polynomial $\pi R^2 h - \pi r^2 h$.

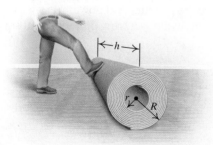

a) Factor the polynomial.
b) Use both the original and the factored forms to find the volume of a roll for which $R = 50$ cm, $r = 10$ cm, and $h = 4$ m. Use 3.14 for π.

104. Use a graphing calculator to check your answers to Exercises 11, 35, and 45 graphically by examining $y_1 =$ the original polynomial, $y_2 =$ the factored polynomial, and $y_3 = y_2 - y_1$.

105. Check your answers to Exercises 11, 35, and 45 by using tables of values. (See Exercise 104.)

YOUR TURN ANSWERS: SECTION 5.5

1. Yes   **2.** $(5x - 6)^2$ **3.** $-2(5t - p)^2$
4. $(n^5 + 2)(n^5 - 2)$ **5.** $a(4p^2 + 1)(2p + 1)(2p - 1)$
6. $(y + 1)(y + 5)(y - 5)$ **7.** $(2m + 5 + a)(2m + 5 - a)$

QUICK QUIZ: SECTIONS 5.1–5.5

1. Arrange in descending powers of x:
$$3x^2y^5 - 9x^4y + x + 2x^3y. \quad [5.1]$$

2. Multiply: $(p + 2)(p^2 - 7p - 3)$.   [5.2]

Factor.

3. $ac + 2a - bc - 2b$ [5.3]

4. $3d^2 - 21d + 30$ [5.4]

5. $x^2y^2 - z^4$ [5.5]

PREPARE TO MOVE ON

Simplify.   [1.6]

1. $(2x^2y^4)^3$ **2.** $(-10x^{10})^3$

Multiply.   [5.2]

3. $(x + 1)(x + 1)(x + 1)$

4. $(x - 1)^3$

5. $(x + 1)(x^2 - x + 1)$

6. $(x - 1)(x^2 + x + 1)$

5.6 Factoring Sums or Differences of Cubes

Factoring Sums or Differences of Cubes

Factoring Sums or Differences of Cubes

Study Skills

A Good Night's Sleep

Plan to study only when you are alert. A good night's sleep or a 10-minute "power nap" can often make a problem suddenly seem much easier to solve.

Factoring Sums or Differences of Cubes

We have seen that a difference of two squares can be factored but a *sum* of two squares is usually prime. The situation is different with cubes: The difference *or* *sum* of two cubes can be factored. To see this, consider the following products:

$$(A + B)(A^2 - AB + B^2) = A(A^2 - AB + B^2) + B(A^2 - AB + B^2)$$
$$= A^3 - A^2B + AB^2 + A^2B - AB^2 + B^3$$
$$= A^3 + B^3 \quad \text{Combining like terms}$$

and

$$(A - B)(A^2 + AB + B^2) = A(A^2 + AB + B^2) - B(A^2 + AB + B^2)$$
$$= A^3 + A^2B + AB^2 - A^2B - AB^2 - B^3$$
$$= A^3 - B^3. \quad \text{Combining like terms}$$

These products allow us to factor a sum or a difference of two cubes. Note how the location of the $+$ and $-$ signs changes.

N	N³
0.1	0.001
0.2	0.008
1	1
2	8
3	27
4	64
5	125
6	216

FACTORING A SUM OR A DIFFERENCE OF TWO CUBES

$$A^3 + B^3 = (A + B)(A^2 - AB + B^2);$$
$$A^3 - B^3 = (A - B)(A^2 + AB + B^2)$$

Remembering the list of cubes shown at left may prove helpful when factoring. Since 2 cubed is 8 and 3 cubed is 27, we say that 2 is the *cube root* of 8, that 3 is the cube root of 27, and so on.

EXAMPLE 1 Write an equivalent expression by factoring: $x^3 + 27$.

SOLUTION We first note that

$$x^3 + 27 = x^3 + 3^3. \quad \text{This is a sum of cubes.}$$

Next, in one set of parentheses, we write the first cube root, x, plus the second cube root, 3:

$$(x + 3)(\quad).$$

To get the other factor, we think of $x + 3$ and do the following:

Square the first term: x^2.
Multiply the terms and then change the sign: $-3x$.
Square the second term: 3^2, or 9.

$$(x + 3)(x^2 - 3x + 9). \quad \text{This is factored completely.}$$

Check: $(x + 3)(x^2 - 3x + 9) = x^3 - 3x^2 + 9x + 3x^2 - 9x + 27$
$$= x^3 + 27. \quad \text{Combining like terms}$$

Thus, $x^3 + 27 = (x + 3)(x^2 - 3x + 9)$.

1. Write an equivalent expression by factoring: $t^3 + 125$.

YOUR TURN

EXAMPLE 2   Factor.

a) $125x^3 - y^3$

b) $m^6 + 64$

c) $128y^7 - 250x^6y$

d) $r^6 - s^6$

SOLUTION

a) We have

$$125x^3 - y^3 = (5x)^3 - y^3.$$ Recognizing this as a difference of cubes

In one set of parentheses, we write the cube root of the first term, $5x$, minus the cube root of the second term, y:

$$(5x - y)(\quad).$$ This can be regarded as $5x$ *plus* the cube root of $(-y)^3$, since $-y^3 = (-y)^3$.

To get the other factor, we think of $5x + y$ and do the following:

Square the first term: $(5x)^2$, or $25x^2$.
Multiply the terms and then change the sign: $5xy$.
Square the second term: $(-y)^2 = y^2$.

$$(5x - y)(25x^2 + 5xy + y^2).$$

Check:

$$(5x - y)(25x^2 + 5xy + y^2) = 125x^3 + 25x^2y + 5xy^2 - 25x^2y - 5xy^2 - y^3$$
$$= 125x^3 - y^3.$$ Combining like terms

Thus, $125x^3 - y^3 = (5x - y)(25x^2 + 5xy + y^2)$.

b) We have

$$m^6 + 64 = (m^2)^3 + 4^3.$$ Rewriting as a sum of quantities cubed

Next, we use the pattern for a sum of cubes:

$$A^3 + B^3 = (A + B)(A^2 - A \cdot B + B^2)$$
$$(m^2)^3 + 4^3 = (m^2 + 4)((m^2)^2 - m^2 \cdot 4 + 4^2)$$ Regarding m^2 as A and 4 as B
$$= (m^2 + 4)(m^4 - 4m^2 + 16).$$

The check is left to the student. We have

$$m^6 + 64 = (m^2 + 4)(m^4 - 4m^2 + 16).$$

c) We have

$$128y^7 - 250x^6y = 2y(64y^6 - 125x^6)$$ Remember: *Always* look for a common factor.
$$= 2y[(4y^2)^3 - (5x^2)^3].$$ Rewriting as a difference of quantities cubed

To factor $(4y^2)^3 - (5x^2)^3$, we use the pattern for a difference of cubes:

$$A^3 - B^3 = (A - B)(A^2 + A \cdot B + B^2)$$
$$(4y^2)^3 - (5x^2)^3 = (4y^2 - 5x^2)((4y^2)^2 + 4y^2 \cdot 5x^2 + (5x^2)^2)$$
$$= (4y^2 - 5x^2)(16y^4 + 20x^2y^2 + 25x^4).$$

The check is left to the student. We have

$$128y^7 - 250x^6y = 2y(4y^2 - 5x^2)(16y^4 + 20x^2y^2 + 25x^4).$$

Student Notes

If you think of $A^3 - B^3$ as $A^3 + (-B)^3$, you need remember only the pattern for factoring a sum of two cubes. Be sure to simplify your result if you do this.

d) We have

$$r^6 - s^6 = (r^3)^2 - (s^3)^2$$
$$= (r^3 + s^3)(r^3 - s^3) \quad \text{Factoring a difference of two } squares$$
$$= (r + s)(r^2 - rs + s^2)(r - s)(r^2 + rs + s^2). \quad \text{Factoring the sum and the difference of two cubes}$$

2. Factor: $y^7 - y$.

To check, read the steps in reverse order and inspect the multiplication.

↩ YOUR TURN

In Example 2(d), suppose we first factored $r^6 - s^6$ as a difference of two cubes:

$$(r^2)^3 - (s^2)^3 = (r^2 - s^2)(r^4 + r^2s^2 + s^4)$$
$$= (r + s)(r - s)(r^4 + r^2s^2 + s^4).$$

In this case, we might have missed some factors; $r^4 + r^2s^2 + s^4$ can be factored as $(r^2 - rs + s^2)(r^2 + rs + s^2)$, but we probably would never have suspected that such a factorization exists. **Given a choice, it is generally better to factor as a difference of squares before factoring as a sum or a difference of cubes.**

USEFUL FACTORING FACTS

Sum of cubes:	$A^3 + B^3 = (A + B)(A^2 - AB + B^2)$
Difference of cubes:	$A^3 - B^3 = (A - B)(A^2 + AB + B^2)$
Difference of squares:	$A^2 - B^2 = (A + B)(A - B)$

In general, a sum of two squares cannot be factored.

5.6 EXERCISE SET

FOR EXTRA HELP   MyMathLab® Math XL

PRACTICE   WATCH   READ   REVIEW

↩ Vocabulary and Reading Check

Classify each binomial as either a sum of cubes, a difference of cubes, a difference of squares, or none of these.

1. $x^3 - 1$

2. $8 + t^3$

3. $9x^4 - 25$

4. $9x^2 + 25$

5. $1000t^3 + 1$

6. $x^3y^3 - 27z^3$

7. $25x^2 + 8x$

8. $100y^8 - 25x^4$

9. $s^{21} - t^{15}$

10. $14x^3 - 2x$

Factoring Sums or Differences of Cubes

Factor completely.

11. $x^3 - 64$

12. $t^3 - 27$

13. $z^3 + 1$

14. $x^3 + 8$

15. $t^3 - 1000$

16. $m^3 + 125$

17. $27x^3 + 1$

18. $8a^3 + 1$

19. $64 - 125x^3$

20. $27 - 8t^3$

21. $x^3 - y^3$

22. $y^3 - z^3$

23. $a^3 + \frac{1}{8}$

24. $x^3 + \frac{1}{27}$

25. $8t^3 - 8$

26. $2y^3 - 128$

27. $54x^3 + 2$

28. $8a^3 + 1000$

29. $rs^4 + 64rs$

30. $ab^5 + 1000ab^2$

31. $5x^3 - 40z^3$

32. $2y^3 - 54z^3$

33. $y^3 - \frac{1}{1000}$

34. $x^3 - \frac{1}{8}$

35. $x^3 + 0.001$

36. $y^3 + 0.125$

37. $64x^6 - 8t^6$

38. $125c^6 - 8d^6$

39. $54y^4 - 128y$

40. $3z^5 - 3z^2$

41. $z^6 - 1$

42. $t^6 + 1$

43. $t^6 + 64y^6$

44. $p^6 - q^6$

45. $x^{12} - y^3z^{12}$

46. $a^9 + b^{12}c^{15}$

47. How could you use factoring to convince someone that $x^3 + y^3 \neq (x + y)^3$?

48. Explain how to use the pattern for factoring $A^3 + B^3$ to factor $A^3 - B^3$.

Skill Review

Solve.

49. *Geometry.* A regular pentagon (all five sides have same length) has the same perimeter as a regular octagon (all eight sides have the same length). One side of the regular pentagon is 1 cm less than twice the length of one side of the regular octagon. Find the perimeter of each shape. [3.3]

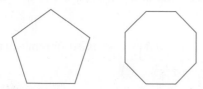

50. *Value of Coins.* There are 50 dimes in a roll of dimes, 40 nickels in a roll of nickels, and 40 quarters in a roll of quarters. Jenna has 10 rolls of coins, which have a total value of $77. There are twice as many rolls of quarters as there are rolls of dimes. How many of each type of roll does she have? [3.5]

51. *Lab Time.* Kyle needs to average at least 45 min each weekday in the math lab. One week, he spent 30 min in the lab on Monday, no time on Tuesday, 50 min on Wednesday, and 80 min on Thursday. How much time must he spend in the lab on Friday in order to average at least 45 min per day for the week? [4.1]

52. *Conservation.* Using helicopters, Ken and Kathy counted alligator nests in a 285-mi^2 area. Ken found 8 more nests than Kathy did, and together they counted 100 nests. How many nests did each count? [1.4]

Synthesis

53. Explain how the geometric model below can be used to verify the formula for factoring $a^3 - b^3$.

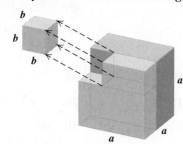

54. Explain how someone could construct a binomial that is both a difference of two cubes and a difference of two squares.

Factor.

55. $x^{6a} - y^{3b}$

56. $2x^{3a} + 16y^{3b}$

Aha! **57.** $(x + 5)^3 + (x - 5)^3$

58. $\frac{1}{16}x^{3a} + \frac{1}{2}y^{6a}z^{9b}$

59. $5x^3y^6 - \frac{5}{8}$

60. $x^3 - (x + y)^3$

61. $x^{6a} - (x^{2a} + 1)^3$

62. $(x^{2a} - 1)^3 - x^{6a}$

63. $t^4 - 8t^3 - t + 8$

64. If $P(x) = x^3$, use factoring to simplify
$$P(a + h) - P(a).$$

65. If $Q(x) = x^6$, use factoring to simplify
$$Q(a + h) - Q(a).$$

66. Using one set of axes, graph the following.
   a) $f(x) = x^3$
   b) $g(x) = x^3 - 8$
   c) $h(x) = (x - 2)^3$

67. Use a graphing calculator to check Example 1: Let $y_1 = x^3 + 27$, $y_2 = (x + 3)(x^2 - 3x + 9)$, and $y_3 = y_1 - y_2$.

YOUR TURN ANSWERS: SECTION 5.6
1. $(t + 5)(t^2 - 5t + 25)$
2. $y(y + 1)(y^2 - y + 1)(y - 1)(y^2 + y + 1)$

QUICK QUICK: SECTIONS 5.1–5.6
1. Determine the degree of $-4ab^5 + a^4b + 8b + 5a^3b^2$. [5.1]

2. Subtract: $(7x - 2y + z) - (3x - 6y - 9z)$. [5.2]

3. Multiply: $(9ab + 7x)(4ab - x)$. [5.3]

Factor.
4. $p^2 - w^2$ [5.5]

5. $p^3 - w^3$ [5.6]

PREPARE TO MOVE ON
Complete each statement.

1. When factoring, always check first for a(n) _____ factor. [5.3]

2. To factor a trinomial of the form $ax^2 + bx + c$, we can use FOIL or the _____ method. [5.4]

3. The formula for factoring a difference of squares is $A^2 - B^2 = $ _____. [5.5]

4. A formula for factoring a perfect-square trinomial is $A^2 + 2AB + B^2 = $ _____. [5.5]

5. The formula for factoring a sum of cubes is $A^3 + B^3 = $ _____. [5.6]

5.7   Factoring: A General Strategy

Mixed Factoring

Student Notes

Try to remember that whenever a new set of parentheses is created while factoring, the expression within it must be checked to see if it can be factored further.

1. Write an equivalent expression by factoring:

   $54x^2y - 24y$.

Factoring is an important algebraic skill. The following strategy for factoring emphasizes recognizing the type of expression being factored.

A STRATEGY FOR FACTORING

A. Always factor out the greatest common factor, if possible.

B. Once the greatest common factor has been factored out, *count the number of terms* in the other factor:

   Two terms: Try factoring as a difference of squares first. Next, try factoring as a sum or a difference of cubes.

   Three terms: If it is a perfect-square trinomial, factor it as such. If not, try factoring using FOIL or the grouping method.

   Four or more terms: Try factoring by grouping and factoring out a common binomial factor. Alternatively, try grouping into a difference of squares, one of which is a perfect-square trinomial.

C. Always *factor completely*. If a factor with more than one term can itself be factored further, do so.

D. Write the complete factorization and check by multiplying. If the original polynomial is prime, state this.

EXAMPLE 1   Write an equivalent expression by factoring: $10a^2x - 40b^2x$.

SOLUTION

A. Factor out the greatest common factor:

   $$10a^2x - 40b^2x = 10x(a^2 - 4b^2).$$

B. The factor $a^2 - 4b^2$ has two terms and is a difference of squares. We factor it, rewriting the common factor from the previous step:

   $$10a^2x - 40b^2x = 10x(a + 2b)(a - 2b).$$

C. No factor with more than one term can be factored further.

D. *Check:* $10x(a + 2b)(a - 2b) = 10x(a^2 - 4b^2) = 10a^2x - 40b^2x$.

YOUR TURN

EXAMPLE 2   Factor: $x^6 - 64$.

SOLUTION

A. Look for a common factor. There is none (other than 1 or −1).

B. There are two terms, a difference of squares: $(x^3)^2 - (8)^2$. We factor it:

   $$x^6 - 64 = (x^3 + 8)(x^3 - 8).   \text{Note that } x^6 = (x^3)^2.$$

C. One factor is a sum of two cubes, and the other factor is a difference of two cubes. We factor both:

$$x^6 - 64 = (x + 2)(x^2 - 2x + 4)(x - 2)(x^2 + 2x + 4).$$

The factorization is complete because no factor can be factored further.

D. The factorization is $(x + 2)(x^2 - 2x + 4)(x - 2)(x^2 + 2x + 4)$. The check is left to the student.

2. Factor: $64a^6 - 1$.

YOUR TURN

EXAMPLE 3 Factor: $7x^6 + 35y^2$.

SOLUTION

A. Factor out the greatest common factor:

$$7x^6 + 35y^2 = 7(x^6 + 5y^2).$$

B. The binomial $x^6 + 5y^2$ is not a difference of squares, a difference of cubes, or a sum of cubes. It cannot be factored.

3. Factor:

$6ax^3 - 4a^2x^2 + 2a^3x^2$.

C. We cannot factor further.

D. *Check:* $7(x^6 + 5y^2) = 7x^6 + 35y^2$.

YOUR TURN

EXAMPLE 4 Factor: $2x^2 + 50a^2 - 20ax$.

SOLUTION

A. Factor out the greatest common factor: $2(x^2 + 25a^2 - 10ax)$.

B. Next, we rearrange the trinomial in descending powers of x:

$$2(x^2 - 10ax + 25a^2).$$

The trinomial is a perfect-square trinomial:

$$2x^2 + 50a^2 - 20ax = 2(x^2 - 10ax + 25a^2) = 2(x - 5a)^2.$$

C. We cannot factor further. Had we used descending powers of a, we would have discovered an equivalent factorization, $2(5a - x)^2$.

4. Factor: $4y^2 + 28y + 49$.

D. The factorization is $2(x - 5a)^2$. The check is left to the student.

YOUR TURN

EXAMPLE 5 Factor: $12x^2 - 40x - 32$.

SOLUTION

A. Factor out the largest common factor: $4(3x^2 - 10x - 8)$.

B. The trinomial factor is not a square. We factor into two binomials:

$$12x^2 - 40x - 32 = 4(x - 4)(3x + 2).$$

C. We cannot factor further.

D. *Check:* $4(x - 4)(3x + 2) = 4(3x^2 + 2x - 12x - 8)$
$$= 4(3x^2 - 10x - 8)$$
$$= 12x^2 - 40x - 32.$$

5. Factor: $8n^2 + 2n - 15$.

YOUR TURN

EXAMPLE 6   Factor: $3x + 12 + ax^2 + 4ax$.

SOLUTION

A. There is no common factor (other than 1 or −1).

B. There are four terms. We try grouping to find a common binomial factor:

$$3x + 12 + ax^2 + 4ax = 3(x + 4) + ax(x + 4) \qquad \text{Factoring two grouped binomials}$$

$$= (x + 4)(3 + ax). \qquad \text{Removing the common binomial factor}$$

C. We cannot factor further.

D. *Check:* $(x + 4)(3 + ax) = 3x + ax^2 + 12 + 4ax = 3x + 12 + ax^2 + 4ax$.

6. Factor: $ct + dt + 4c + 4d$.

YOUR TURN

EXAMPLE 7   Factor: $y^2 - 9a^2 + 12y + 36$.

SOLUTION

A. There is no common factor (other than 1 or −1).

B. There are four terms. We try grouping to remove a common binomial factor, but find none. Next, we try grouping as a difference of squares:

$$(y^2 + 12y + 36) - 9a^2 \qquad \text{Grouping}$$

$$= (y + 6)^2 - (3a)^2 \qquad \text{Factoring the perfect-square trinomial}$$

$$= (y + 6 + 3a)(y + 6 - 3a). \qquad \text{Factoring the difference of squares}$$

C. No factor with more than one term can be factored further.

D. The factorization is $(y + 6 + 3a)(y + 6 - 3a)$. The check is left to the student.

7. Factor: $x^2 - 4x - 16y^2 + 4$.

YOUR TURN

EXAMPLE 8   Factor: $x^3 - xy^2 + x^2y - y^3$.

SOLUTION

A. There is no common factor (other than 1 or −1).

B. There are four terms. We try grouping to remove a common binomial factor:

$$x^3 - xy^2 + x^2y - y^3$$

$$= x(x^2 - y^2) + y(x^2 - y^2) \qquad \text{Factoring two grouped binomials}$$

$$= (x^2 - y^2)(x + y). \qquad \text{Removing the common binomial factor}$$

C. The factor $x^2 - y^2$ can be factored further:

$$x^3 - xy^2 + x^2y - y^3 = (x + y)(x - y)(x + y), \text{ or } (x + y)^2(x - y).$$

No factor can be factored further, so we have factored completely.

8. Factor: $d^3 - d^2 - 9d + 9$.

D. The factorization is $(x + y)^2(x - y)$. The check is left to the student.

YOUR TURN

5.7 EXERCISE SET

FOR EXTRA HELP

MyMathLab®
PRACTICE   WATCH   READ   REVIEW

Vocabulary and Reading Check

Choose the item from the column on the right that corresponds to the type of polynomial. Choices may be used more than once.

1. ____ $25y^2 - 49$

2. ____ $36x^2y - 9xy$

3. ____ $9y^6 + 16x^8$

4. ____ $8a^3 - b^6c^9$

5. ____ $c^{12} + 1$

6. ____ $4t^2 - 12t + 9$

7. ____ $4a^2 + 8a + 16$

8. ____ $9x^2 + 24x - 16$

a) Polynomial with a common factor

b) Difference of two squares

c) Sum of two cubes

d) Difference of two cubes

e) Perfect-square trinomial

f) None of these

Concept Reinforcement

For each polynomial, tell what type of factoring is needed. Then give the factorization of the polynomial. For example, for $3x^2 - 6x$, write "Factor out a common factor; $3x(x - 2)$."

9. $x^2 - 3x - 4$

10. $x^3 - 1$

11. $4x^3 - 10x^2 - 2x + 5$

12. $t^2 + 100 - 20t$

13. $24a^3 - 16a - 8$

14. $a^2b^2 - c^2$

Mixed Factoring

Factor completely.

15. $x^2 - 81$

16. $a^2 - 4$

17. $9m^4 - 900$

18. $t^5 - 49t$

19. $2x^3 + 12x^2 + 16x$

20. $10x^2 - 40x + 40$

21. $a^2 + 25 + 10a$

22. $8a^3 - 18a^2 - 5a$

23. $2y^2 - 11y + 12$

24. $6y^2 - 13y - 5$

25. $3x^2 + 15x - 252$

26. $2y^2 + 10y - 132$

27. $25x^2 - 9y^2$

28. $16a^2 - 81b^2$

29. $t^6 + 1$

30. $64t^6 - 1$

31. $x^2 + 6x - y^2 + 9$

32. $t^2 + 10t - p^2 + 25$

33. $128a^3 + 250b^3$

34. $343x^3 + 27y^3$

35. $7x^3 - 14x^2 - 105x$

36. $2t^3 + 20t^2 - 48t$

37. $-9t^2 + 16t^4$

38. $-24x^6 + 6x^4$

39. $8m^3 + m^6 - 20$

40. $-37x^2 + x^4 + 36$

41. $ac + cd - ab - bd$

42. $xw - yw + xz - yz$

43. $4c^2 - 4cd + d^2$

44. $70b^2 - 3ab - a^2$

45. $40x^2 + 3xy - y^2$

46. $p^2 - 10pq + 25q^2$

47. $4a - 5a^2 - 10 + 2a^3$

48. $24 + 3t^3 - 9t^2 - 8t$

49. $2x^3 + 6x^2 - 8x - 24$

50. $3x^3 + 6x^2 - 27x - 54$

51. $54a^3 - 16b^3$

52. $54x^3 - 250y^3$

53. $36y^2 - 35 + 12y$

54. $2b - 28a^2b + 10ab$

55. $4m^4 - 64n^4$

56. $2x^4 - 32$

57. $a^5b - 16ab^5$

58. $x^3y - 25xy^3$

59. $34t^3 - 6t$

60. $13t^3 - 26t$

Aha! 61. $(a - 3)(a + 7) + (a - 3)(a - 1)$

62. $x^2(x + 3) - 4(x + 3)$

63. $7a^4 - 14a^3 + 21a^2 - 7a$

64. $a^3 - ab^2 + a^2b - b^3$

65. $42ab + 27a^2b^2 + 8$

66. $-23xy + 20x^2y^2 + 6$

67. $-10t^3 + 15t$

68. $-9x^3 + 12x$

69. $-6x^4 + 8x^3 - 12x$

70. $-15t^4 + 10t$

71. $p - 64p^4$

72. $125a - 8a^4$

Aha! 73. $a^2 - b^2 - 6b - 9$

74. $m^2 - n^2 - 8n - 16$

75. Emily has factored a polynomial as $(a - b)(x - y)$, while Jorge has factored the same polynomial as $(b - a)(y - x)$. Can they both be correct? Why or why not?

76. In your own words, outline a procedure that can be used to factor any polynomial.

Skill Review

Let $f(x) = 3x + 1$ and $g(x) = x^2 - 2$. Find the following.

77. $g(-10)$ [2.2]

78. $f(a + h)$ [2.2]

79. $(f + g)(5)$ [2.6]

80. $(g - f)(x)$ [2.6]

81. The domain of f [2.1]

82. The domain of g/f [2.6], [4.2]

Synthesis

83. Explain how one could construct a polynomial that is a difference of squares that contains a sum of two cubes and a difference of two cubes as factors.

84. Explain how one could construct a polynomial with four terms that can be factored by grouping three terms together.

Factor completely.

85. $28a^3 - 25a^2bc + 3ab^2c^2$

86. $-16 + 17(5 - y^2) - (5 - y^2)^2$

Aha! 87. $(x - p)^2 - p^2$

88. $a^4 - 50a^2b^2 + 49b^4$

89. $(y - 1)^4 - (y - 1)^2$

90. $x^6 - 2x^5 + x^4 - x^2 + 2x - 1$

91. $4x^2 + 4xy + y^2 - r^2 + 6rs - 9s^2$

92. $(1 - x)^3 - (x - 1)^6$

93. $\dfrac{x^{27}}{1000} - 1$

94. $a - by^8 + b - ay^8$

95. $3(x + 1)^2 - 9(x + 1) - 12$

96. $3a^2 + 3b^2 - 3c^2 - 3d^2 + 6ab - 6cd$

97. $3(a + 2)^2 + 30(a + 2) + 75$

98. $(m - 1)^3 - (m + 1)^3$

99. $2x^{-1} - 2x^{-3} - 12x^{-5}$

100. $24t^{2a} - 6$

101. $a^{2w+1} + 2a^{w+1} + a$

102. If $\left(x + \dfrac{2}{x}\right)^2 = 6$, find $x^3 + \dfrac{8}{x^3}$.

YOUR TURN ANSWERS: SECTION 5.7

1. $6y(3x + 2)(3x - 2)$
2. $(2a + 1)(4a^2 - 2a + 1)(2a - 1)(4a^2 + 2a + 1)$
3. $2ax^2(3x - 2a + a^2)$ **4.** $(2y + 7)^2$
5. $(2n + 3)(4n - 5)$ **6.** $(c + d)(t + 4)$
7. $(x - 2 + 4y)(x - 2 - 4y)$ **8.** $(d + 3)(d - 3)(d - 1)$

QUICK QUIZ: SECTIONS 5.1–5.7

1. Arrange $9n + 26 - 3n^5 + 7n^3$ in ascending order. [5.1]

2. Given $f(x) = x^2 + 6x$, find and simplify $f(a + h) - f(a)$. [5.2]

3. Multiply: $5x^2y(2xy + 3x^4y^2 - 6x^2)$. [5.2]

Factor.

4. $6a^2 + 17ac + 5c^2$ [5.4]   **5.** $50c^2 + 40c + 8$ [5.5]

PREPARE TO MOVE ON

Solve. [1.3]

1. $x + 2 = 0$

2. $2x - 5 = 0$

3. $4x = 0$

Find the domain of each function. [2.2], [4.2]

4. $f(x) = \dfrac{2x}{3x - 2}$

5. $f(x) = \dfrac{x + 5}{2x + 1}$

5.8   Applications of Polynomial Equations

The Principle of Zero Products   ▪   Problem Solving

We now turn our focus to solving a new type of equation in which factoring plays an important role.

Whenever two polynomials are set equal to each other, we have a **polynomial equation**. Some examples of polynomial equations are

$$4x^3 + x^2 + 5x = 6x - 3, \quad x^2 - x = 6, \quad \text{and} \quad 3y^4 + 2y^2 + 2 = 0.$$

The *degree of a polynomial equation* is the same as the highest degree of any term in the equation. Thus, from left to right, the degree of each equation listed above is 3, 2, and 4. A second-degree polynomial equation in one variable is called a **quadratic equation**. Of the equations listed above, only $x^2 - x = 6$ is a quadratic equation.

Polynomial equations occur frequently in applications, so the ability to solve them is an important skill. One way of solving certain polynomial equations involves factoring.

The Principle of Zero Products

When we multiply two or more numbers, the product is 0 if any one of those numbers (factors) is 0. Conversely, if a product is 0, then at least one of the factors must be 0. This property of 0 gives us a new principle for solving equations.

THE PRINCIPLE OF ZERO PRODUCTS

For any real numbers a and b:

If $ab = 0$, then $a = 0$ or $b = 0$. If $a = 0$ or $b = 0$, then $ab = 0$.

Thus, if $(t - 7)(2t + 5) = 0$, then $t - 7 = 0$ or $2t + 5 = 0$. To solve a quadratic equation using the principle of zero products, we first write it in *standard form*: with 0 on one side of the equation and the leading coefficient positive. We then factor and determine when each factor is 0.

EXAMPLE 1   Solve: $x^2 - x = 6$.

SOLUTION   To apply the principle of zero products, we need 0 on one side of the equation. Thus we subtract 6 from both sides:

Get 0 on one side.

$$x^2 - x - 6 = 0. \quad \text{Getting 0 on one side}$$

To express the polynomial as a product, we factor:

Factor.

$$(x - 3)(x + 2) = 0. \quad \text{Factoring}$$

The principle of zero products says that since $(x - 3)(x + 2)$ is 0, then

Set each factor equal to 0.

$$x - 3 = 0 \quad or \quad x + 2 = 0. \quad \text{Using the principle of zero products}$$

Each of these linear equations is then solved separately:

Solve.

$$x = 3 \quad or \quad x = -2.$$

We check as follows:

Check.

Check:

$x^2 - x = 6$	
$3^2 - 3$	6
$9 - 3$	
$6 \overset{?}{=} 6$	TRUE

$x^2 - x = 6$	
$(-2)^2 - (-2)$	6
$4 + 2$	
$6 \overset{?}{=} 6$	TRUE

Both 3 and -2 are solutions.

1. Solve: $x^2 = 2x + 8$.

YOUR TURN

TO USE THE PRINCIPLE OF ZERO PRODUCTS

1. Write an equivalent equation with 0 on one side, using the addition principle.
2. Factor the nonzero side of the equation.
3. Set each factor that is not a constant equal to 0.
4. Solve the resulting equations.

CAUTION! To use the principle of zero products, we must have 0 on one side of the equation. If neither side of the equation is 0, the procedure will not work.

To see this, consider $x^2 - x = 6$ in Example 1 as $x(x - 1) = 6$. Knowing that the product of two numbers is 6 tells us little about either number. The factors could be $2 \cdot 3$ or $6 \cdot 1$ or $-12 \cdot \left(-\frac{1}{2}\right)$, and so on.

EXAMPLE 2 Solve: **(a)** $5b^2 = 10b$; **(b)** $x^2 - 6x + 9 = 0$.

SOLUTION

a) We have

$$5b^2 = 10b$$
$$5b^2 - 10b = 0 \qquad \text{Getting 0 on one side}$$
$$5b(b - 2) = 0 \qquad \text{Factoring}$$
$$5b = 0 \quad or \quad b - 2 = 0 \qquad \text{Using the principle of zero products}$$
$$b = 0 \quad or \qquad b = 2. \qquad \text{The checks are left to the student.}$$

The solutions are 0 and 2.

b) We have

$$x^2 - 6x + 9 = 0$$
$$(x - 3)(x - 3) = 0 \qquad \text{Factoring}$$
$$x - 3 = 0 \quad or \quad x - 3 = 0 \qquad \text{Using the principle of zero products}$$
$$x = 3 \quad or \qquad x = 3. \qquad \textit{Check:}$$
$$3^2 - 6 \cdot 3 + 9 = 9 - 18 + 9 = 0.$$

There is only one solution, 3.

2. Solve: $y^2 + 10y = 0$.

 YOUR TURN

EXAMPLE 3 Given that $f(x) = 3x^2 - 4x$, find all values of a for which $f(a) = 4$.

SOLUTION We want all numbers a for which $f(a) = 4$. Since $f(a) = 3a^2 - 4a$, we must have

$$3a^2 - 4a = 4 \quad \text{Setting } f(a) \text{ equal to 4}$$
$$3a^2 - 4a - 4 = 0 \quad \text{Getting 0 on one side}$$
$$(3a + 2)(a - 2) = 0 \quad \text{Factoring}$$
$$3a + 2 = 0 \quad or \quad a - 2 = 0$$
$$a = -\tfrac{2}{3} \quad or \quad a = 2.$$

Check: $f\left(-\tfrac{2}{3}\right) = 3\left(-\tfrac{2}{3}\right)^2 - 4\left(-\tfrac{2}{3}\right) = 3 \cdot \tfrac{4}{9} + \tfrac{8}{3} = \tfrac{4}{3} + \tfrac{8}{3} = \tfrac{12}{3} = 4.$

$f(2) = 3(2)^2 - 4(2) = 3 \cdot 4 - 8 = 12 - 8 = 4.$

To have $f(a) = 4$, we must have $a = -\tfrac{2}{3}$ or $a = 2$.

YOUR TURN

3. Given that $g(x) = 6x^2 + x$, find all values of a for which $g(a) = 1$.

EXAMPLE 4 Let $f(x) = 3x^3 - 30x$ and $g(x) = 9x^2$. Find all x-values for which $f(x) = g(x)$.

SOLUTION We substitute the polynomial expressions for $f(x)$ and $g(x)$ and solve the resulting equation:

$$f(x) = g(x)$$
$$3x^3 - 30x = 9x^2 \quad \text{Substituting}$$
$$3x^3 - 9x^2 - 30x = 0 \quad \text{Getting 0 on one side and writing in descending order}$$
$$3x(x^2 - 3x - 10) = 0 \quad \text{Factoring out a common factor}$$
$$3x(x + 2)(x - 5) = 0 \quad \text{Factoring the trinomial}$$
$$3x = 0 \quad or \quad x + 2 = 0 \quad or \quad x - 5 = 0 \quad \text{Using the principle of zero products}$$
$$x = 0 \quad or \quad x = -2 \quad or \quad x = 5.$$

To check, the student can confirm that

$f(0) = g(0) = 0;$
$f(-2) = g(-2) = 36;$ and
$f(5) = g(5) = 225.$

For $x = 0, -2,$ and 5, we have $f(x) = g(x)$.

YOUR TURN

EXAMPLE 5 Find the domain of F if $F(x) = \dfrac{x - 2}{x^2 + 2x - 15}$.

SOLUTION The domain of F is the set of all values for which $F(x)$ is a real number. Since division by 0 is undefined, $F(x)$ cannot be calculated at x-values for which the denominator, $x^2 + 2x - 15$, is 0. To make sure these values are *excluded*, we solve:

$$x^2 + 2x - 15 = 0 \quad \text{Setting the denominator equal to 0}$$
$$(x - 3)(x + 5) = 0 \quad \text{Factoring}$$
$$x - 3 = 0 \quad or \quad x + 5 = 0$$
$$x = 3 \quad or \quad x = -5. \quad \text{These are the values to } exclude.$$

The domain of F is $\{x \,|\, x \text{ is a real number } and \ x \neq -5 \ and \ x \neq 3\}$. In interval notation, the domain is $(-\infty, -5) \cup (-5, 3) \cup (3, \infty)$.

YOUR TURN

Study Skills

Finishing a Chapter

Try to be aware of when you near the end of a chapter. Sometimes the end of a chapter signals the arrival of a quiz or a test. Almost always, the end of a chapter indicates the end of a particular area of study. Make use of the chapter summary, review, and test to solidify your understanding before moving forward.

4. Let $f(x) = 5x^3 + 20x$ and $g(x) = 20x^2$. Find all x-values for which $f(x) = g(x)$.

5. Find the domain of G if $G(x) = \dfrac{x^2}{x^2 - 5x - 14}$.

EXPLORING 🔍 THE CONCEPT

We can use the graph of $f(x) = x^2 - x$ to solve the equation $x^2 - x = 6$. To do so, we look for any x-value that is paired with 6, as shown on the left below. It appears that $f(x) = 6$ when $x = -2$ or $x = 3$.

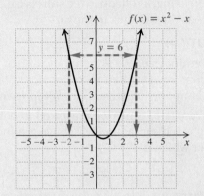
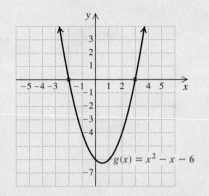

Equivalently, we could use the graph of $g(x) = x^2 - x - 6$ and look for values of x for which $g(x) = 0$. Using this method, we can visualize what we call the *roots*, or *zeros*, of a polynomial function. It appears from the graph on the right above that $g(x) = 0$ when $x = -2$ or $x = 3$.

1. Use the graph of $f(x) = x^2 + 2x$ to solve the equation $x^2 + 2x = 3$. Check your answers.

2. Use the graph of $g(x) = x^2 - 3x - 4$ to solve the equation $x^2 - 3x - 4 = 0$. Check your answers.

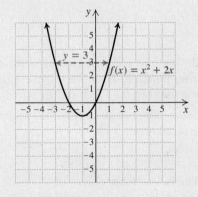
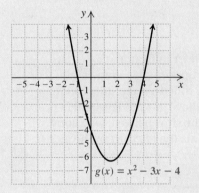

ANSWERS

1. $-3, 1$ **2.** $-1, 4$

Problem Solving

Some problems translate to quadratic equations, which we can now solve. The problem-solving process is the same as for other kinds of problems.

EXAMPLE 6 *Prize Tee Shirts.* During intermission at sporting events, it has become common for team mascots to use a powerful slingshot to launch tightly rolled tee shirts into the stands. The height $h(t)$, in feet, of an airborne tee shirt t seconds after being launched can be approximated by

$$h(t) = -15t^2 + 75t + 10.$$

After peaking, a rolled-up tee shirt is caught by a fan 70 ft above ground level. How long was the tee shirt in the air?

SOLUTION

1. **Familiarize.** We sketch the graph of the function and label the given information. If we wanted to, we could evaluate $h(t)$ for a few values of t. Note that t cannot be negative, since it represents time from launch.

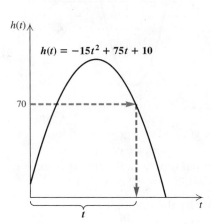

2. **Translate.** Since we are asked to determine how long it will take for the shirt to reach someone 70 ft above ground level, we are interested in the value of t for which $h(t) = 70$:

$$-15t^2 + 75t + 10 = 70. \quad \text{Setting } h(t) \text{ equal to 70}$$

3. **Carry out.** We solve the quadratic equation:

$$-15t^2 + 75t + 10 = 70$$
$$-15t^2 + 75t - 60 = 0 \quad \text{Subtracting 70 from both sides}$$
$$\left.\begin{array}{l} -15(t^2 - 5t + 4) = 0 \\ -15(t - 4)(t - 1) = 0 \end{array}\right\} \text{Factoring}$$
$$t - 4 = 0 \quad or \quad t - 1 = 0$$
$$t = 4 \quad or \quad t = 1.$$

The solutions appear to be 4 and 1.

4. **Check.** We have

$$h(4) = -15 \cdot 4^2 + 75 \cdot 4 + 10 = -240 + 300 + 10 = 70 \text{ ft};$$
$$h(1) = -15 \cdot 1^2 + 75 \cdot 1 + 10 = -15 + 75 + 10 = 70 \text{ ft}.$$

Both 4 and 1 check. However, the problem states that the tee shirt is caught after peaking. Thus we reject 1 since that would indicate when the height of the tee shirt was 70 ft on the way *up*.

5. **State.** The tee shirt was in the air for 4 sec before being caught 70 ft above ground level.

6. Refer to Example 6. Suppose that a fan caught a rolled-up tee shirt 100 ft above ground level after the shirt peaked. How long was the tee shirt in the air?

 YOUR TURN

The following problem involves the **Pythagorean theorem**, which relates the lengths of the sides of a right triangle. A **right triangle** has a 90°, or right, angle, which is indicated in the triangle by the symbol ⌐ or ⌐. The longest side, called the **hypotenuse**, is opposite the 90° angle. The other sides, called **legs**, form the two sides of the right angle.

THE PYTHAGOREAN THEOREM

In any right triangle, if a and b are the lengths of the legs and c is the length of the hypotenuse, then

$$a^2 + b^2 = c^2.$$

The symbol ⌐ denotes a 90° angle.

Technology Connection

To use the INTERSECT option to check Example 1, we let $y_1 = x^2 - x$ and $y_2 = 6$. One intersection occurs at $(-2, 6)$. You should confirm that the other occurs at $(3, 6)$.

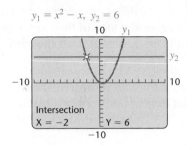

$y_1 = x^2 - x, \ y_2 = 6$

Intersection
X = -2   Y = 6

Another approach is to find where the graph of $y_3 = x^2 - x - 6$ crosses the x-axis, using the ZERO option of the CALC menu. One zero is -2. You should confirm that the other zero is 3.

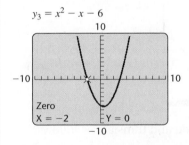

$y_3 = x^2 - x - 6$

Zero
X = -2   Y = 0

To visualize Example 4, we let $y_1 = 3x^3 - 30x$ and $y_2 = 9x^2$ and use a viewing window of $[-4, 6, -110, 360]$.

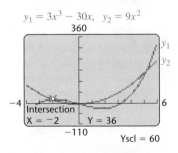

$y_1 = 3x^3 - 30x, \ y_2 = 9x^2$

Intersection
X = -2   Y = 36

Yscl = 60

1. Use the INTERSECT option of the CALC menu to confirm all three solutions of Example 4.

2. As a second check, show that $y_3 = 3x^3 - 9x^2 - 30x$ has zeros at 0, -2, and 5.

The converse of the Pythagorean theorem is also true. For positive numbers a, b, and c, if $a^2 + b^2 = c^2$, then a triangle with sides of lengths a, b, and c is a right triangle.

EXAMPLE 7 *Carpentry.* In order to build a deck at a right angle to their house, Lucinda and Felipe decide to hammer a stake in the ground a precise distance from the back wall of their house. This stake will combine with two marks on the house to form a right triangle. From a course in geometry, Lucinda remembers that there are three consecutive integers that can work as sides of a right triangle. Find the measurements of that triangle.

SOLUTION

1. **Familiarize.** Recall that $x, x + 1$, and $x + 2$ can be used to represent three unknown consecutive integers. Since $x + 2$ is the largest number, it must represent the hypotenuse. The legs serve as the sides of the right angle, so one leg must be formed by the marks on the house. We make a drawing in which

   $x =$ the distance between the marks on the house,

   $x + 1 =$ the length of the other leg, and

   $x + 2 =$ the length of the hypotenuse.

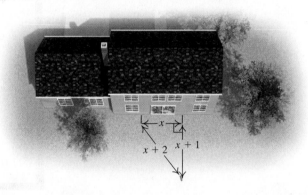

2. **Translate.** Applying the Pythagorean theorem, we translate as follows:

   $$a^2 + b^2 = c^2$$
   $$x^2 + (x + 1)^2 = (x + 2)^2.$$

3. **Carry out.** We solve the equation as follows:

$x^2 + (x^2 + 2x + 1) = x^2 + 4x + 4$	Squaring the binomials
$2x^2 + 2x + 1 = x^2 + 4x + 4$	Combining like terms
$x^2 - 2x - 3 = 0$	Subtracting $x^2 + 4x + 4$ from both sides
$(x - 3)(x + 1) = 0$	Factoring
$x - 3 = 0 \ \ or \ \ x + 1 = 0$	Using the principle of zero products
$x = 3 \ \ or \ \ \ \ \ \ \ \ x = -1.$	

7. Refer to Example 7. One leg of a right triangle is 5 ft long. The lengths, in feet, of the other two sides are consecutive integers. Find the lengths of the other two sides of the triangle.

4. Check. The integer -1 cannot be a length of a side because it is negative. For $x = 3$, we have $x + 1 = 4$, and $x + 2 = 5$. Since $3^2 + 4^2 = 5^2$, the lengths 3, 4, and 5 determine a right triangle. Thus, 3, 4, and 5 check.

5. State. Lucinda and Felipe should use a triangle with sides having a ratio of 3:4:5. Thus, if the marks on the house are 3 yd apart, they should locate the stake at the point in the yard that is precisely 4 yd from one mark and 5 yd from the other mark.

 YOUR TURN

EXAMPLE 8 *Display of a Sports Card.* A sports card is 4 cm wide and 5 cm long. The card is to be encased by Lucite that is $5\frac{1}{2}$ times the area of the card. Find the dimensions of the Lucite that will ensure a uniform border around the card.

SOLUTION

1. Familiarize. We make a drawing and label it, using x to represent the width of the border, in centimeters. Since the border extends uniformly around the entire card, the length of the Lucite must be $5 + 2x$ and the width must be $4 + 2x$.

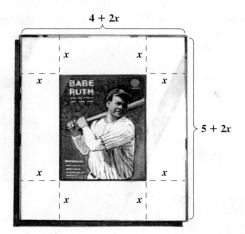

2. Translate. We rephrase the information given and translate as follows:

Area of Lucite is $5\frac{1}{2}$ times area of card.

$$(5 + 2x)(4 + 2x) = 5\frac{1}{2} \cdot 5 \cdot 4$$

3. Carry out. We solve the equation:

$$(5 + 2x)(4 + 2x) = 5\frac{1}{2} \cdot 5 \cdot 4$$
$$20 + 10x + 8x + 4x^2 = 110 \quad \text{Multiplying}$$
$$4x^2 + 18x - 90 = 0 \quad \text{Finding standard form}$$
$$2(2x^2 + 9x - 45) = 0$$
$$2(2x + 15)(x - 3) = 0 \quad \text{Factoring}$$
$$2x + 15 = 0 \quad or \quad x - 3 = 0 \quad \text{Principle of zero products}$$
$$x = -7\frac{1}{2} \quad or \quad x = 3.$$

4. Check. We check 3 in the original problem. (Note that $-7\frac{1}{2}$ is not a solution because measurements cannot be negative.) If the border is 3 cm wide, the Lucite will have a length of $5 + 2 \cdot 3$, or 11 cm, and a width of $4 + 2 \cdot 3$, or 10 cm. The area of the Lucite is thus $11 \cdot 10$, or 110 cm². Since the area of the card is 20 cm² and 110 cm² is $5\frac{1}{2}$ times 20 cm², the number 3 checks.

5. State. The Lucite should be 11 cm long and 10 cm wide.

 YOUR TURN

Chapter Resources:
Visualizing for Success, p. 344;
Decision Making: Connection, p. 345

8. Refer to Example 8. Suppose that the Lucite is $2\frac{1}{10}$ times the area of the card. Find the dimensions of the Lucite that will ensure a uniform border around the card.

CONNECTING ✏ THE CONCEPTS

Be careful not to confuse a polynomial expression with a polynomial equation. We can form equivalent polynomial expressions by combining like terms, adding, subtracting, multiplying, and factoring. An expression cannot be "solved." Compare the following.

Factor: $x^2 - x - 12$.

$$x^2 - x - 12 = (x - 4)(x + 3)$$

The factorization is $(x - 4)(x + 3)$.
The expressions $x^2 - x - 12$ and $(x - 4)(x + 3)$ are equivalent.

Solve: $x^2 - x - 12 = 0$.

$$x^2 - x - 12 = 0$$
$$(x - 4)(x + 3) = 0 \qquad \text{Factoring}$$
$$x - 4 = 0 \quad or \quad x + 3 = 0 \qquad \text{Using the principle of zero products}$$
$$x = 4 \quad or \quad x = -3$$

The solutions are -3 and 4.

EXERCISES

1. Factor: $x^2 + 5x + 6$.

2. Solve: $x^2 + 5x + 6 = 0$.

3. Solve: $x^2 + 6 = 5x$.

4. Combine like terms: $3x^2 - x + x^2 - 5$.

5. Subtract: $(3x^2 - x) - (x^2 - 5)$.

6. Factor: $a^2 - 1$.

7. Multiply: $(a + 1)(a - 1)$.

8. Solve: $(a + 1)(a - 1) = 24$.

5.8 EXERCISE SET

FOR EXTRA HELP

MyMathLab® Math XL
PRACTICE   WATCH   READ   REVIEW

➦ Vocabulary and Reading Check

Classify each of the following statements as either true or false.

1. Not every polynomial equation is quadratic.

2. The equations $3x^2 - 5x + 2 = 0$, $4t^2 = t + 2$, and $9n^2 = 4$ are all examples of quadratic equations.

3. Every quadratic equation has two solutions.

4. To use the principle of zero products, we must have an equation with 0 on one side.

5. Every triangle has a hypotenuse.

6. When we are solving an applied problem, a solution of the translated equation may not be a solution of the problem.

The Principle of Zero Products

Solve.

7. $(x - 2)(x - 5) = 0$

8. $(x + 3)(x + 7) = 0$

9. $x^2 + 8x + 7 = 0$

10. $x^2 - 12x + 20 = 0$

11. $9t(2t + 1) = 0$

12. $3t(2t - 5) = 0$

13. $15t^2 - 12t = 0$

14. $24t^2 + 8t = 0$

15. $(2t + 5)(t - 7) = 0$

16. $(2t - 7)(t + 2) = 0$

17. $x^2 - 3x - 18 = 0$

18. $x^2 + 2x - 35 = 0$

19. $t^2 - 10t = 0$

20. $t^2 - 8t = 0$

21. $(3x - 1)(4x - 5) = 0$

22. $(5x + 3)(2x - 7) = 0$

23. $4a^2 = 10a$

24. $6a^2 = 8a$

25. $t^2 - 6t - 16 = 0$

26. $t^2 - 3t - 18 = 0$

27. $t^2 - 3t = 28$

28. $x^2 - 4x = 45$

29. $r^2 + 16 = 8r$

30. $a^2 + 1 = 2a$

31. $a^2 + 20a + 100 = 0$

32. $z^2 + 6z + 9 = 0$

33. $8y + y^2 + 15 = 0$

34. $9x + x^2 + 20 = 0$

Aha! **35.** $n^2 - 81 = 0$

36. $b^2 - 144 = 0$

37. $x^3 - 2x^2 = 63x$

38. $a^3 - 3a^2 = 40a$

39. $t^2 = 25$

40. $r^2 = 4$

41. $(a - 4)(a + 4) = 20$

42. $(t - 6)(t + 6) = 45$

43. $-9x^2 + 15x - 4 = 0$

44. $3x^2 - 8x + 4 = 0$

45. $-8y^3 - 10y^2 - 3y = 0$

46. $-4t^3 - 11t^2 - 6t = 0$

47. $(z + 4)(z - 2) = -5$

48. $(y - 3)(y + 2) = 14$

49. $x(5 + 12x) = 28$

50. $a(1 + 21a) = 10$

51. $a^2 - \frac{1}{100} = 0$

52. $x^2 - \frac{1}{64} = 0$

53. $t^4 - 26t^2 + 25 = 0$

54. $t^4 - 13t^2 + 36 = 0$

55. Let $f(x) = x^2 + 12x + 40$. Find a such that $f(a) = 8$.

56. Let $f(x) = x^2 + 14x + 50$. Find a such that $f(a) = 5$.

57. Let $g(x) = 2x^2 + 5x$. Find a such that $g(a) = 12$.

58. Let $g(x) = 2x^2 - 15x$. Find a such that $g(a) = -7$.

59. Let $h(x) = 12x + x^2$. Find a such that $h(a) = -27$.

60. Let $h(x) = 4x - x^2$. Find a such that $h(a) = -32$.

61. If $f(x) = 12x^2 - 15x$ and $g(x) = 8x - 5$, find all x-values for which $f(x) = g(x)$.

62. If $f(x) = 10x^2 + 20$ and $g(x) = 43x - 8$, find all x-values for which $f(x) = g(x)$.

63. If $f(x) = 2x^3 - 5x$ and $g(x) = 10x - 7x^2$, find all x-values for which $f(x) = g(x)$.

64. If $f(x) = 3x^3 - 4x$ and $g(x) = 8x^2 + 12x$, find all x-values for which $f(x) = g(x)$.

Find the domain of the function f given by each of the following.

65. $f(x) = \dfrac{3}{x^2 - 3x - 4}$

66. $f(x) = \dfrac{2}{x^2 - 7x + 10}$

67. $f(x) = \dfrac{x}{6x^2 - 54}$

68. $f(x) = \dfrac{2x}{5x^2 - 20}$

69. $f(x) = \dfrac{x - 5}{9x - 18x^2}$

70. $f(x) = \dfrac{1 + x}{3x - 15x^2}$

71. $f(x) = \dfrac{7}{5x^3 - 35x^2 + 50x}$

72. $f(x) = \dfrac{3}{2x^3 - 2x^2 - 12x}$

Problem Solving

Solve.

73. *Postage Rates.* The maximum size envelope that can be mailed at the Large Envelope rate is 3 in. longer than it is wide. The area is 180 in^2. Find the length and the width.

Source: USPS

74. *Photo Size.* A photo is 3 cm longer than it is wide. Find the length and the width if the area is 108 cm^2.

75. *Geometry.* If each of the sides of a square is lengthened by 4 m, the area becomes 49 m^2. Find the length of a side of the original square.

76. *Geometry.* If each side of a square is lengthened by 6 cm, the area becomes 144 cm^2. Find the length of a side of the original square.

77. *Framing a Picture.* A picture frame measures 14 cm by 20 cm, and 160 cm^2 of picture shows. Find the width of the frame.

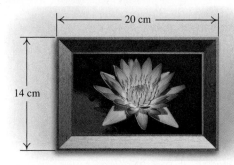

78. *Framing a Picture.* A picture frame measures 12 cm by 20 cm, and 84 cm^2 of picture shows. Find the width of the frame.

79. *Catering.* A rectangular table is 60 in. long and 40 in. wide. A tablecloth that is twice the area of the table will be centered on the table. How far will the tablecloth hang down on each side?

80. *Landscaping.* A rectangular garden is 30 ft by 40 ft. Part of the garden is removed in order to install a walkway of uniform width around it. The area of the new garden is one-half the area of the old garden. How wide is the walkway?

81. Three consecutive even integers are such that the square of the third is 76 more than the square of the second. Find the three integers.

82. Three consecutive even integers are such that the square of the first plus the square of the third is 136. Find the three integers.

83. *Furniture.* The base of a triangular tabletop is 20 in. longer than the height. The area is 750 in². Find the height and the base.

84. *Tent Design.* The triangular entrance to a tent is 2 ft taller than it is wide. The area of the entrance is 12 ft². Find the height and the base.

85. *Building Lots.* A lot for sale in New York City is in the shape of a right triangle, as shown below. The side not bordering a street is 25 ft long. One of the other sides is 5 ft longer than the remaining side. Find the lengths of the sides.

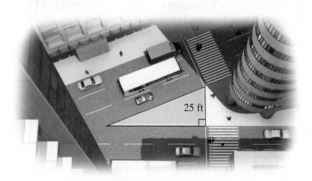

86. *Antenna Wires.* A wire is stretched from the ground to the top of an antenna tower, as shown. The wire is 20 m long. The height of the tower is 4 m greater than the distance d from the tower's base to the bottom of the wire. Find the distance d and the height of the tower.

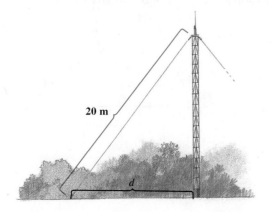

87. *Ladder Location.* The foot of an extension ladder is 9 ft from a wall. The height that the ladder reaches on the wall and the length of the ladder are consecutive integers. How long is the ladder?

88. *Ladder Location.* The foot of an extension ladder is 10 ft from a wall. The ladder is 2 ft longer than the height that it reaches on the wall. How far up the wall does the ladder reach?

89. *Garden Design.* Ignacio is planning a rectangular garden that is 25 m longer than it is wide. The garden will have an area of 7500 m². What will its dimensions be?

90. *Garden Design.* A rectangular flower bed is to be 3 m longer than it is wide. The flower bed will have an area of 108 m². What will its dimensions be?

91. *Home Audio Systems.* Custom Sounds determines that the revenue R, in thousands of dollars, from the sale of x home audio systems is given by $R(x) = 2x^2 + x$. If the cost C, in thousands of dollars, of producing x home audio systems is given by $C(x) = x^2 - 2x + 10$, how many systems must be produced and sold in order for the company to break even?

92. *Violin Production.* Suppose that the cost of making x violins is $C(x) = \frac{1}{9}x^2 + 2x + 1$, where $C(x)$ is in thousands of dollars. If the revenue from the sale of x violins is given by $R(x) = \frac{5}{36}x^2 + 2x$, where $R(x)$ is in thousands of dollars, how many violins must be sold in order for the instrument maker to break even?

93. *Prize Tee Shirts.* Using the model in Example 6, determine how long a tee shirt has been airborne if it is caught on the way *up* by a fan 100 ft above ground level.

94. *Prize Tee Shirts.* Using the model in Example 6, determine how long a tee shirt has been airborne if it is caught on the way *down* by a fan 10 ft above ground level.

95. *Fireworks Displays.* Fireworks are typically launched from a mortar with an upward velocity (initial speed) of about 64 ft/sec. The height $h(t)$, in feet, of a "weeping willow" display, t seconds after having been launched from an 80-ft high rooftop, is given by

$$h(t) = -16t^2 + 64t + 80.$$

How long will it take the cardboard shell from the fireworks to reach the ground?

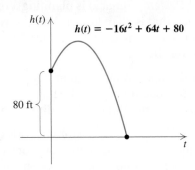

96. *Safety Flares.* Suppose that a flare is launched upward with an initial velocity of 80 ft/sec from a height of 224 ft. Its height in feet, $h(t)$, after t seconds is given by

$$h(t) = -16t^2 + 80t + 224.$$

How long will it take the flare to reach the ground?

97. *Fertility.* The fertility rate in the United States can be approximated by

$$f(t) = -0.5t^2 + 1.9t + 67.3,$$

where $f(t)$ is in births per 1000 women ages 15–44 and t is the number of years after 2005. In what year after 2005 were there 64.3 births per 1000 women ages 15–44?

Source: Based on data from National Center for Health Statistics

98. *Paper Consumption.* The amount of paper used daily for U.S. newspapers can be approximated by

$$p(t) = -\frac{1}{35}t^2 + \frac{3}{4}t + \frac{11}{2},$$

where $p(t)$ is in millions of tons and t is the number of years after 1980. In what year after 1980

were $4\frac{1}{10}$ million tons of paper used daily for U.S. newspapers?

Sources: Based on information from American Forest and Paper Association; Bureau of Labor Statistics; Vertical Research Partners

99. Suppose you are given a detailed graph of $y = p(x)$, where $p(x)$ is some polynomial in x. How could the graph be used to help solve the equation $p(x) = 0$?

100. Can the number of solutions of a quadratic equation exceed two? Why or why not?

Skill Review

101. Find the slope of the line containing $(1, -6)$ and $(3, 10)$. [2.3]

102. Find the slope and the y-intercept of the graph of $2x - 3y = 6$. [2.3]

103. Find the intercepts of the line given by $x - 5y = 20$. [2.4]

104. Find a linear function whose graph has slope $-\frac{1}{2}$ and contains $(3, 7)$. [2.5]

105. Find a linear function whose graph contains $(4, -5)$ and $(6, -10)$. [2.5]

106. Find an equation in slope–intercept form of the line through $(0, 7)$ that is parallel to the line $y = 5x - 6$. [2.5]

Synthesis

107. Explain how one could write a quadratic equation that has -3 and 5 as solutions.

108. If the graph of $f(x) = ax^2 + bx + c$ has no x-intercepts, what can you conclude about the equation $ax^2 + bx + c = 0$?

Solve.

109. $(8x + 11)(12x^2 - 5x - 2) = 0$

110. $(x + 1)^3 = (x - 1)^3 + 26$

111. Use the following graph of $g(x) = -x^2 - 2x + 3$ to solve $-x^2 - 2x + 3 = 0$ and to solve $-x^2 - 2x + 3 \geq -5$.

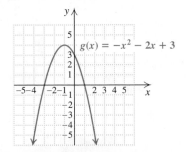

112. Use the following graph of $f(x) = x^2 - 2x - 3$ to solve $x^2 - 2x - 3 = 0$ and to solve $x^2 - 2x - 3 < 5$.

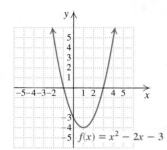

113. Find a polynomial function f for which $f(2) = 0, f(-1) = 0, f(3) = 0$, and $f(0) = 30$.

114. Find a polynomial function g for which $g(-3) = 0, g(1) = 0, g(5) = 0$, and $g(0) = 45$.

115. *Box Construction.* A rectangular piece of tin is twice as long as it is wide. Squares 2 cm on a side are cut out of each corner, and the ends are turned up to make a box whose volume is 480 cm³. What are the dimensions of the original piece of tin?

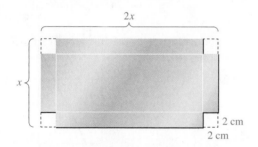

116. *Navigation.* A tugboat and a freighter leave the same port at the same time at right angles. The freighter travels 7 km/h slower than the tugboat. After 4 hr, they are 68 km apart. Find the speed of each boat.

117. *Skydiving.* During the first 13 sec of a jump, a skydiver falls approximately $11.12t^2$ feet in t seconds. A small heavy object (with less wind resistance) falls about $15.4t^2$ feet in t seconds. Suppose that a skydiver jumps from 30,000 ft, and 1 sec later a camera falls out of the airplane. How long will it take the camera to catch up to the skydiver?

118. Use the TABLE feature of a graphing calculator to check that -5 and 3 are not in the domain of F as shown in Example 5.

119. Use the TABLE feature of a graphing calculator to check your answers to Exercises 67, 69, and 71.

 In Exercises 120–123, use a graphing calculator to find any real-number solutions that exist accurate to two decimal places.

120. $x^2 - 2x - 8 = 0$ (Check by factoring.)

121. $-x^2 + 13.80x = 47.61$

122. $-x^2 + 3.63x + 34.34 = x^2$

123. $x^3 - 3.48x^2 + x = 3.48$

 124. Mary Louise is attempting to solve $x^3 + 20x^2 + 4x + 80 = 0$ with a graphing calculator. Unfortunately, when she graphs $y_1 = x^3 + 20x^2 + 4x + 80$ in a standard $[-10, 10, -10, 10]$ window, she sees no graph at all, let alone any x-intercept. Can this problem be solved graphically? If so, how? If not, why not?

125. A Pythagorean triple is a set of three numbers that satisfy the Pythagorean equation. They can be generated by choosing natural numbers n and m, $n > m$, and forming the following three numbers: $n^2 + m^2$, $n^2 - m^2$, and $2mn$. Show that these three expressions satisfy the Pythagorean equation.

⤵ YOUR TURN ANSWERS: SECTION 5.8

1. $-2, 4$ **2.** $-10, 0$ **3.** $-\frac{1}{2}, \frac{1}{3}$ **4.** $0, 2$
5. $\{x \mid x$ is a real number *and* $x \neq -2$ *and* $x \neq 7\}$, or $(-\infty, -2) \cup (-2, 7) \cup (7, \infty)$ **6.** 3 sec **7.** 12 ft, 13 ft
8. 7 cm long and 6 cm wide

QUICK QUIZ: SECTIONS 5.1–5.8

1. Find $P(0)$ if $P(t) = t^4 - 8t^2 + 2t + 7$. [5.1]

2. Simplify: $(4x - 1)^2 - (x + 3)(x - 3)$. [5.2]

3. Factor: $8p^4 - 27p$. [5.6]

4. Solve: $9y^2 = 18y$. [5.8]

5. Find the domain of the function given by

$$f(x) = \frac{2x + 3}{x^2 + x - 20}.$$ [5.8]

PREPARE TO MOVE ON

Find the reciprocal. [1.2]

1. $\dfrac{2}{3}$ 2. -2

Simplify. [1.2]

3. $\dfrac{5}{12} \cdot \left(-\dfrac{45}{8}\right)$ 4. $\dfrac{5}{12} \div \left(-\dfrac{45}{8}\right)$

5. $\dfrac{240}{280}$ 6. $\dfrac{2 \cdot 3 - 5 \cdot 6}{7 - 5^2}$

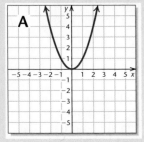

A

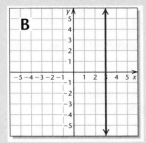

B

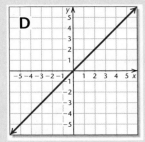

C

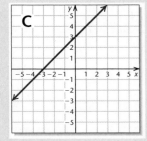

D

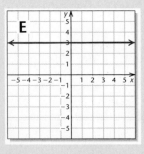

E

Visualizing for Success

Use after Section 5.8.

Match each equation or function with its graph.

1. $f(x) = x$

2. $f(x) = |x|$

3. $f(x) = x^2$

4. $f(x) = 3$

5. $x = 3$

6. $y = x + 3$

7. $y = x - 3$

8. $y = 2x$

9. $y = -2x$

10. $y = 2x + 3$

Answers on page A-27

An alternate, animated version of this activity appears in MyMathlab. To use MyMathlab, you need a course ID and a student access code. Contact your instructor for more information.

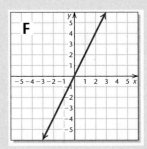

F

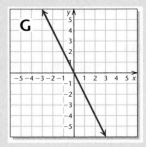

G

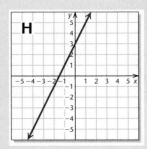

H

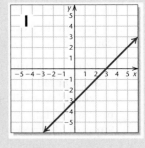

I

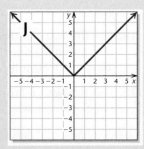

J

Collaborative Activity   *How Many Handshakes?*

Focus: Polynomial functions
Use after: Section 5.2
Time: 20 minutes
Group size: 5

Activity

1. All group members should shake hands with each other. Without "double counting," determine how many handshakes occurred.

2. Complete the table in the next column.

3. Join another group to determine the number of handshakes for a group of size 10.

4. Try to find a function of the form $H(n) = an^2 + bn$, for which $H(n)$ is the number of different handshakes that are possible in a group of n people. Make sure that

$H(n)$ produces all of the values in the table below. (*Hint*: Use the table to twice select n and $H(n)$. Then solve the resulting system of equations for a and b.)

Group Size	Number of Handshakes
1	
2	
3	
4	
5	

Decision Making ⊘ Connection   (*Use after Section 5.8.*)

Home Improvement.   Area and volume formulas are used in a variety of applications, and many such formulas are polynomials. The formulas needed for the following activities can be found in the back of the text or on the Internet.

1. A propane tank is in the shape of a cylinder with one-half of a sphere on each end.

   a) Find the volume of the tank shown, in cubic feet.

   b) How many gallons of propane will it take to fill the tank? (There are approximately 7.5 gal in one cubic foot.)

2. One gallon of paint will cover about 350 ft². How many coats of paint could you give the propane tank with one gallon of paint?

3. How many gallons of paint would you need to paint the house shown below with one coat of paint? What assumptions did you make when calculating this amount?

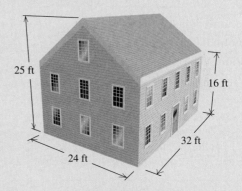

4. A *square* of shingles covers 100 ft² of surface area. How many squares will be needed to reshingle the house shown?

5. *Research.*   Find an online calculator that calculates roof area given the width, length, and height of a roof.

   a) The roof shown in the house above is a gable roof. Enter the dimensions of this roof in the online calculator and compare the result with the roof area calculated in Exercise 4.

   b) Develop a formula that could be used by the online calculator to determine roof area.

Study Summary

KEY TERMS AND CONCEPTS   EXAMPLES   PRACTICE EXERCISES

SECTION 5.1: *Introduction to Polynomials and Polynomial Functions*

A **polynomial** is a monomial or a sum of monomials.

When a polynomial is written as a sum of monomials, each monomial is a **term** of the polynomial.

The **degree of a term** of a polynomial is the number of variable factors in that term.

The **coefficient** of a term is the part of the term that is a constant factor.

The **leading term** of a polynomial is the term of highest degree.

The **leading coefficient** is the coefficient of the leading term.

The **degree of the polynomial** is the degree of the leading term.

Polynomial: $10x - x^3 + 4x^5 + 7$

Term	$10x$	$-x^3$	$4x^5$	7
Degree of Term	1	3	5	0
Coefficient of Term	10	-1	4	7
Leading Term		$4x^5$		
Leading Coefficient		4		
Degree of Polynomial		5		

For Exercises 1–6, consider the polynomial $x^2 - 10 + 5x - 8x^6$.

1. List the terms of the polynomial.

2. What is the degree of the term $5x$?

3. What is the coefficient of the term x^2?

4. What is the leading term of the polynomial?

5. What is the leading coefficient of the polynomial?

6. What is the degree of the polynomial?

A **monomial** has one term.

A **binomial** has two terms.

A **trinomial** has three terms.

Monomial: $4x^3$

Binomial: $x^2 - 5$

Trinomial: $3t^3 + 2t - 10$

7. Classify the polynomial
$$8x - 3 - x^4$$
as a monomial, a binomial, a trinomial, or a polynomial with no special name.

Add polynomials by combining like terms.

$(2x^2 - 3x + 7) + (5x^3 + 3x - 9)$
$\quad = 5x^3 + 2x^2 - 2$

8. Add: $(9x^2 - 3x) + (4x - x^2)$.

Subtract polynomials by adding the opposite of the polynomial being subtracted.

$(2x^2 - 3x + 7) - (5x^3 + 3x - 9)$
$\quad = 2x^2 - 3x + 7 - 5x^3 - 3x + 9$
$\quad = -5x^3 + 2x^2 - 6x + 16$

9. Subtract: $(9x^2 - 3x) - (4x - x^2)$.

SECTION 5.2: *Multiplication of Polynomials*

Multiply polynomials by multiplying each term of one polynomial by each term of the other. Then, if possible, combine like terms.

$(x + 2)(x^2 - x - 1)$
$\quad = x \cdot x^2 - x \cdot x - x \cdot 1 + 2 \cdot x^2 - 2 \cdot x - 2 \cdot 1$
$\quad = x^3 - x^2 - x + 2x^2 - 2x - 2$
$\quad = x^3 + x^2 - 3x - 2$

10. Multiply:
$$(x - 1)(x^2 - x - 2).$$

Special Products $(A+B)(C+D) =$ $\quad AC+AD+BC+BD$ $(A+B)^2 = A^2 + 2AB + B^2$ $(A-B)^2 = A^2 - 2AB + B^2$ $(A+B)(A-B) = A^2 - B^2$	$(y^3-2)(y+4) = y^4 + 4y^3 - 2y - 8$ $(t+6)^2 = t^2 + 2\cdot t\cdot 6 + 36 = t^2 + 12t + 36$ $(c-5d)^2 = c^2 - 2\cdot c\cdot 5d + (5d)^2 = c^2 - 10cd + 25d^2$ $(x^2+3)(x^2-3) = (x^2)^2 - (3)^2 = x^4 - 9$	**11.** Multiply: $(x-2y)(x+2y)$.

SECTION 5.3: *Common Factors and Factoring by Grouping*

To **factor** a polynomial means to write it as a product of polynomials. Whenever possible, begin by factoring out the **largest common factor**.	$12x^4 - 30x^3 = 6x^3(2x-5)$	**12.** Factor: $12x^4 - 18x^3 + 30x$.
Some polynomials with four terms can be **factored by grouping**.	$3x^3 - x^2 - 6x + 2 = x^2(3x-1) - 2(3x-1)$ $\qquad\qquad = (3x-1)(x^2-2)$	**13.** Factor: $2x^3 - 6x^2 - x + 3$.

SECTION 5.4: *Factoring Trinomials*

Some trinomials can be written as the product of two binomials.	$x^2 - 11x + 18 = (x-2)(x-9);$ $6x^2 - 5x - 6 = (3x+2)(2x-3)$ Using FOIL or the grouping method	Factor. **14.** $x^2 - 7x - 18$ **15.** $6x^2 + x - 2$

SECTION 5.5: *Factoring Perfect-Square Trinomials and Differences of Squares*

Factoring a Perfect-Square Trinomial $A^2 + 2AB + B^2 = (A+B)^2$ $A^2 - 2AB + B^2 = (A-B)^2$	$y^2 + 20y + 100 = (y+10)^2$ $m^2 - 14mn + 49n^2 = (m-7n)^2$	**16.** Factor: $100n^2 + 81 + 180n$.
Factoring a Difference of Squares $A^2 - B^2 = (A+B)(A-B)$	$9t^2 - 1 = (3t+1)(3t-1)$	**17.** Factor: $144t^2 - 25$.

SECTION 5.6: *Factoring Sums or Differences of Cubes*

Factoring a Sum or a Difference of Cubes $A^3 + B^3$ $\quad = (A+B)(A^2 - AB + B^2)$ $A^3 - B^3$ $\quad = (A-B)(A^2 + AB + B^2)$	$x^3 + 1000 = (x+10)(x^2 - 10x + 100)$ $z^6 - 8w^3 = (z^2 - 2w)(z^4 + 2wz^2 + 4w^2)$	**18.** Factor: $a^3 - 1$.

SECTION 5.7: *Factoring: A General Strategy*

To factor a polynomial:

A. Factor out the largest common factor.

B. Look at the number of terms.

Two terms: Try to factor as a difference of squares, a sum of cubes, or a difference of cubes.

Three terms: Try to factor as a trinomial square. Then try FOIL or grouping.

Four terms: Try factoring by grouping.

C. Factor completely.

D. Check by multiplying.

$5x^5 - 80x = 5x(x^4 - 16)$ Factoring out a common factor

$\qquad\qquad = 5x(x^2 + 4)(x^2 - 4)$ Factoring a difference of squares

$\qquad\qquad = 5x(x^2 + 4)(x + 2)(x - 2)$ Factoring a difference of squares

Check:
$$5x(x^2 + 4)(x + 2)(x - 2) = 5x(x^2 + 4)(x^2 - 4)$$
$$= 5x(x^4 - 16)$$
$$= 5x^5 - 80x$$

19. Factor:
$$-x^2 y^3 - 3xy^2 + 10y.$$

SECTION 5.8: *Applications of Polynomial Equations*

The Principle of Zero Products

For any real numbers a and b:

If $ab = 0$, then $a = 0$ or $b = 0$.

If $a = 0$ or $b = 0$, then $ab = 0$.

Solve: $x^2 + 7x = 30$.

$x^2 + 7x - 30 = 0$ Getting 0 on one side

$(x + 10)(x - 3) = 0$ Factoring

$x + 10 = 0 \quad or \quad x - 3 = 0$ Using the principle of zero products

$\qquad x = -10 \quad or \qquad x = 3$

The solutions are -10 and 3.

20. Solve: $8x = 6x^2$.

Pythagorean Theorem

In any right triangle, if a and b are the lengths of the legs and c is the length of the hypotenuse, then $a^2 + b^2 = c^2$.

Find the lengths of the legs in this triangle.

$x^2 + (x + 1)^2 = 5^2$

$x^2 + x^2 + 2x + 1 = 25$

$2x^2 + 2x - 24 = 0$

$x^2 + x - 12 = 0$

$(x + 4)(x - 3) = 0$

$x + 4 = 0 \quad or \quad x - 3 = 0$

$\quad x = -4 \quad or \qquad x = 3$

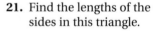

Since lengths are not negative, -4 is not a solution. The lengths of the legs are 3 and 4.

21. Find the lengths of the sides in this triangle.

Review Exercises: Chapter 5

➲ Concept Reinforcement

In each of Exercises 1–10, match the item with the most appropriate choice from the column on the right.

1. _____ A polynomial with four terms [5.1]

2. _____ A term that is not a monomial [5.1]

3. _____ A polynomial written in ascending order [5.1]

4. _____ A polynomial that cannot be factored [5.7]

5. _____ A difference of two squares [5.5]

6. _____ A perfect-square trinomial [5.5]

7. _____ A difference of two cubes [5.6]

8. _____ The principle of zero products [5.8]

9. _____ A quadratic equation [5.8]

10. _____ The longest side in any right triangle [5.8]

a) $5x + 2x^2 - 4x^3$

b) $3x^{-1}$

c) If $a \cdot b = 0$, then $a = 0$ or $b = 0$.

d) Prime

e) $t^2 - 9$

f) Hypotenuse

g) $8x^3 - 4x^2 + 12x + 14$

h) $t^3 - 27$

i) $2x^2 - 4x = 7$

j) $4a^2 - 12a + 9$

11. Determine the degree of $2xy^6 - 7x^8y^3 + 2x^3 + 9$. [5.1]

12. Arrange $3x - 5x^3 + 2x^2 + 9$ in descending order and determine the leading term and the leading coefficient. [5.1]

13. Arrange in ascending powers of x:
$$8x^6y - 7x^8y^3 + 2x^3 - 3x^2. \quad [5.1]$$

14. Find $P(0)$ and $P(-1)$:
$$P(x) = x^3 - x^2 + 4x. \quad [5.1]$$

15. Given $P(x) = x^2 + 10x$, find and simplify
$$P(a + h) - P(a). \quad [5.2]$$

Combine like terms. [5.1]

16. $6 - 4a + a^2 - 2a^3 - 10 + a$

17. $4x^2y - 3xy^2 - 5x^2y + xy^2$

Add. [5.1]

18. $(-7x^3 - 4x^2 + 3x + 2) + (5x^3 + 2x + 6x^2 + 1)$

19. $(4n^3 + 2n^2 - 12n + 7) + (-6n^3 + 9n + 4 + n)$

20. $(-9xy^2 - xy - 6x^2y) + (-5x^2y - xy + 4xy^2)$

Subtract. [5.1]

21. $(8x - 5) - (-6x + 2)$

22. $(4a - b - 3c) - (6a - 7b - 3c)$

23. $(8x^2 - 4xy + y^2) - (2x^2 - 3y^2 - 9y)$

Simplify as indicated. [5.2]

24. $(3x^2y)(-6xy^3)$

25. $(x^4 - 2x^2 + 3)(x^4 + x^2 - 1)$

26. $(4ab + 3c)(2ab - c)$

27. $(7t + 1)(7t - 1)$

28. $(3x - 4y)^2$

29. $(x + 3)(2x - 1)$

30. $(x^2 + 4y^3)^2$

31. $(3t - 5)^2 - (2t + 3)^2$

32. $\left(x - \frac{1}{3}\right)\left(x - \frac{1}{6}\right)$

Factor completely. If a polynomial is prime, state this.

33. $-3y^4 - 9y^2 + 12y$ [5.3]

34. $a^2 - 12a + 27$ [5.4]

35. $3m^2 - 10m - 8$ [5.4]

36. $25x^2 + 20x + 4$ [5.5]

37. $4y^2 - 16$ [5.5]

38. $5x^2 + x^3 - 14x$ [5.4]

39. $ax + 2bx - ay - 2by$ [5.3]

40. $3y^3 + 6y^2 - 5y - 10$ [5.3]

41. $a^4 - 81$ [5.5]

42. $4x^4 + 4x^2 + 20$ [5.3]

43. $27x^3 + 8$ [5.6]

44. $\frac{1}{125}b^3 - \frac{1}{8}c^6$ [5.6]

45. $a^2b^4 - 64$ [5.5]

46. $3x + x^2 + 5$ [5.4]

47. $0.01x^4 - 1.44y^6$ [5.5]

48. $4x^2y + 100y - 40xy$ [5.5]

49. $6t^2 + 17pt + 5p^2$ [5.4]

50. $x^3 + 2x^2 - 9x - 18$ [5.5]

51. $a^2 - 2ab + b^2 - 4t^2$ [5.5]

Solve.   [5.8]

52. $x^2 - 12x + 36 = 0$

53. $6b^2 + 6 = 13b$

54. $8y^2 = 14y$

55. $3r^2 = 12$

56. $a^3 = 4a^2 + 21a$

57. $(y - 1)(y - 4) = 10$

58. Let $f(x) = x^2 - 7x - 40$. Find a such that $f(a) = 4$.   [5.8]

59. Find the domain of the function f given by

$$f(x) = \frac{x - 5}{x^2 - x - 56}.$$ [5.8]

60. The gable of St. Bridget's Convent Ruins in Estonia is $\frac{3}{4}$ as tall as it is wide. Its area is 216 m². Find the height and the base.   [5.8]

61. A photograph is 3 in. longer than it is wide. When a 2-in. border is placed around the photograph, the total area of the photograph and the border is 108 in². Find the dimensions of the photograph.   [5.8]

62. Hassan is designing a garden in the shape of a right triangle. One leg of the triangle is 8 ft long. The other leg and the hypotenuse are consecutive odd integers. How long are the other two sides of the garden?   [5.8]

63. The labor-force participation rate of males ages 65 and older in the United States can be approximated by $p(t) = \frac{1}{50}t^2 - \frac{9}{10}t + 27$, where $p(t)$ is the percentage of males ages 65 and older in the labor force t years after 1970. In what years were 17% of the males ages 65 and older in the labor force?   [5.8]

Source: Based on information from U.S. Bureau of Labor Statistics

Synthesis

64. Explain the difference between multiplying two polynomials and factoring a polynomial. [5.2], [5.7]

65. Explain in your own words why there must be a 0 on one side of an equation before you can use the principle of zero products.   [5.8]

Factor.   [5.6]

66. $128x^6 - 2y^6$

67. $(x - 1)^3 - (x + 1)^3$

68. $3x^{-6} - 12x^{-4} + 15x^{-3}$ [5.3]

Solve.   [5.8]

69. $(x + 1)^3 = x^2(x + 1)$

Aha! **70.** $x^2 + 100 = 0$

Test: Chapter 5

For step-by-step test solutions, access the Chapter Test Prep Videos in MyMathLab®, on YouTube® (search "Bittinger Inter Alg CA" and click on "Channels"), or by scanning the code.

Given the polynomial $8xy^3 - 14x^2y + 5x^5y^4 - 9x^4y$.

1. Determine the degree of the polynomial.

2. Arrange in descending powers of x.

3. Determine the leading term of the polynomial
$7a - 12 + a^2 - 5a^3$.

4. Given $P(x) = 2x^3 + 3x^2 - x + 4$, find $P(0)$ and $P(-2)$.

5. Given $P(x) = x^2 - 3x$, find and simplify
$$P(a + h) - P(a).$$

6. Combine like terms:
$$6xy - 2xy^2 - 2xy + 5xy^2.$$

Add.

7. $(-4y^3 + 6y^2 - y) + (3y^3 - 9y - 7)$

8. $(2m^3 - 4m^2n - 5n^2) + (8m^3 - 3mn^2 + 6n^2)$

Subtract.

9. $(8a - 4b) - (3a + 4b)$

10. $(9y^2 - 2y - 5y^3) - (4y^2 - 2y - 6y^3)$

Multiply.

11. $(-4x^2y^3)(-16xy^5)$

12. $(6a - 5b)(2a + b)$

13. $(x - y)(x^2 - xy - y^2)$

14. $(4t - 3)^2$

15. $(5a^3 + 9)^2$

16. $(x - 2y)(x + 2y)$

Factor completely. If a polynomial is prime, state this.

17. $x^2 - 10x + 25$

18. $y^3 + 5y^2 - 4y - 20$

19. $p^2 - 12p - 28$

20. $t^7 - 3t^5$

21. $12m^2 + 20m + 3$

22. $9y^2 - 25$

23. $3r^3 - 3$

24. $45x^2 + 20 + 60x$

25. $3x^4 - 48y^4$

26. $y^2 + 8y + 16 - 100t^2$

27. $x^2 + 3x + 6$

28. $20a^2 - 5b^2$

29. $24x^2 - 46x + 10$

30. $16a^7b + 54ab^7$

Solve.

31. $x^2 - 3x - 18 = 0$

32. $5t^2 = 125$

33. $2x^2 + 21 = -17x$

34. $9x^2 + 3x = 0$

35. $x^2 + 81 = 18x$

36. Let $f(x) = 3x^2 - 15x + 11$. Find a such that $f(a) = 11$.

37. Find the domain of the function f given by
$$f(x) = \frac{8 - x}{x^2 + 2x + 1}.$$

38. A photograph is 3 cm longer than it is wide. Its area is 40 cm². Find its length and its width.

39. To celebrate Ripton's bicentennial, fireworks are launched off a dam 36 ft above Lake Marley. The height of a display, t seconds after it has been launched, is given by
$$h(t) = -16t^2 + 64t + 36.$$
After how long will the shell from the fireworks reach the water?

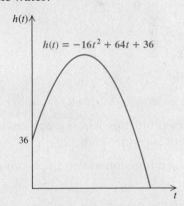

40. The foot of an extension ladder is 10 ft from a wall. The ladder is 2 ft longer than the height that it reaches on the wall. How far up the wall does the ladder reach?

Synthesis

41. Factor: $(a + 3)^2 - 2(a + 3) - 35$.

42. Solve: $20x(x + 2)(x - 1) = 5x^3 - 24x - 14x^2$.

Cumulative Review: Chapters 1–5

Perform the indicated operations and simplify.

1. $-120 \div 2 \cdot 5 - 3 \cdot (-1)^3$ [1.2]

2. $(2a^2b - b^2 - 3a^2) - (5a^2 - 4a^2b - 4b^2)$ [5.1]

3. $(2x + 9)(2x - 9)$ [5.2]

4. $(2x + 9y)^2$ [5.2]

5. $(5m^3 + n)(2m^2 - n)$ [5.2]

Factor.

6. $3t^2 - 48$ [5.5]

7. $a^2 - 14a + 49$ [5.5]

8. $36x^3y^2 - 27x^4y + 45x^2y^3$ [5.3]

9. $125a^3 + 64b^3$ [5.6]

10. $12y^2 + 7y - 10$ [5.4]

11. $d^2 - a^2 + 2ab - b^2$ [5.5]

Solve.

12. $2x - 3(x + 2) = 6 - (x - 1)$ [1.3]

13. $x - 2y = 7,$
$y = \frac{1}{4}x$ [3.2]

14. $x - 2y + 3z = 16,$
$3x + y + z = -5,$
$x - y - z = -3$ [3.4]

15. $3x - 7(x + 1) \geq 4x - 5$ [4.1]

16. $-2 < x + 6 < 0$ [4.2]

17. $|2x - 1| \leq 5$ [4.3]

18. $x^2 = 10x + 24$ [5.8]

19. $5n^2 = 30n$ [5.8]

20. Solve $3a - 5b = 6 + b$ for b.   [1.5]

21. Find a linear function whose graph has slope $\frac{1}{3}$ and
y-intercept $\left(0, -\frac{1}{4}\right)$.   [2.3]

22. Find a linear function whose graph contains $(-2, 5)$
and $(-1, -4)$.   [2.5]

Graph on a plane.

23. $y = \frac{2}{3}x - 1$ [2.3]

24. $4y = -2$ [2.4]

25. $x = y + 3$ [2.3]

26. $4x - 3y \leq 12$ [4.4]

27. Find the domain of f if

$$f(x) = \frac{x + 1}{x^2 - 3x + 2}.$$ [5.8]

28. Write the domain of f using interval notation if
$f(x) = \sqrt{x + 3}$.   [4.1]

29. *Utility Bills.*   In the summer, Georgia Power
charges residential customers approximately
5.1¢ per kilowatt-hour (kWh) for the first 650 kWh of
electricity and 8.4¢ per kWh for usage over 650 kWh
but under 1000 kWh. Eileen never uses more than
1000 kWh per month, and her bill is always between
\$49.95 and \$58.35 each month. How much electricity does Eileen use each month?   [4.1], [4.2]

Source: Georgia Power

30. *Scrapbooking.*   A photo is cut so that its length is
2 cm longer than its width. It is then centered on
a background to form a 3-cm border around the
photo. The area of the background paper is 168 cm^2.
What are the dimensions of the photo?   [5.8]

31. *Scrapbooking.*   A photo is cropped so that its
length is 2 cm longer than its width. It is then placed
on a page and a narrow ribbon is glued around the
perimeter of the photo. If the length of the ribbon
is 50 cm, what are the dimensions of the cropped
photo?   [1.4]

32. *Ordering Pizza.*   The Westville Marching Band
ordered 48 pizzas for their annual party. Single-
topping pizzas were \$16.95 each, and two-topping
pizzas were \$19.95 each. If the total cost of the
pizzas was \$870.60, how many of each type did the
band order?   [3.3]

Synthesis

33. Solve $\dfrac{c + 2d}{c - d} = t$ for d.   [1.5]

34. Write an absolute-value inequality for which the
interval shown is the solution.   [4.3]

Rational Expressions, Equations, and Functions

DISTANCE FROM LIGHT SOURCE (in feet)	ILLUMINATION (in foot-candles)
1	400
2	100
4	25

Source: Based on information from Winchip, Susan M., *Fundamentals of Lighting,* Chapter 4. New York: Fairfield Publications, 2008.

Shine a Little Light on Me!

Interior designers consider many details when choosing lighting fixtures, including the amount of illumination that the fixtures provide. Factors such as bulb wattage, type of bulb, the cleanliness of the environment, and the distance from the light source all influence illumination.

As we can see from the table above, illumination decreases as the distance from the light source increases. However, it does not appear to decrease linearly. In this chapter, we will write a formula that relates the distance from a light source and the illumination that source provides. (*See Exercise 79 in Section 6.8.*)

Math becomes an essential tool in all phases of a successful project.

As Project Interior Designer at Gensler in Washington, D.C., Chen-Hui Li Spicer uses math in order to understand scale and proportion in an interior architectural space.

L ike a fraction in arithmetic, a rational expression is a ratio of two expressions. In this chapter, we add, subtract, multiply, and divide rational expressions, and use them in equations, functions, and applications.

6.1 | Rational Expressions and Functions: Multiplying and Dividing

Rational Functions   ▪   Simplifying Rational Expressions and Functions   ▪   Multiplying and Simplifying   ▪
Dividing and Simplifying

A **rational expression** consists of a polynomial divided by a nonzero polynomial. The following are examples of rational expressions:

$$\frac{3}{4}, \quad \frac{x}{y}, \quad \frac{9}{a+b}, \quad \frac{x^2 + 7xy - 4}{x^3 - y^3}.$$

Rational Functions

Functions described by rational expressions are called **rational functions**.

EXAMPLE 1   The function given by

$$H(t) = \frac{t^2 + 5t}{2t + 5}$$

gives the time, in hours, for two machines, working together, to complete a job that the first machine could do alone in t hours and the second machine could do in $t + 5$ hours. How long will the two machines, working together, require for the job if the first machine alone would take **(a)** 1 hr? **(b)** 6 hr?

SOLUTION

a) $H(1) = \dfrac{1^2 + 5 \cdot 1}{2 \cdot 1 + 5} = \dfrac{1 + 5}{2 + 5} = \dfrac{6}{7}\,\text{hr}$

b) $H(6) = \dfrac{6^2 + 5 \cdot 6}{2 \cdot 6 + 5} = \dfrac{36 + 30}{12 + 5} = \dfrac{66}{17}\,\text{or } 3\dfrac{15}{17}\,\text{hr}$

🔁 YOUR TURN

1. Use the function given in Example 1 to determine how long it would take the two machines, working together, to complete the job if the first machine alone would take 3 hr.

Since division by 0 is undefined, the domain of a rational function must exclude any numbers for which the denominator is 0. For the function H above, the denominator is 0 when t is $-\frac{5}{2}$, so

the domain of H is $\left(-\infty, -\frac{5}{2}\right) \cup \left(-\frac{5}{2}, \infty\right).$ *H is undefined when $2t + 5 = 0$.*

A graph of the above function is shown at left. Note that the graph consists of two unconnected "branches." Since $-\frac{5}{2}$ is not in the domain of H, the graph of H does not touch a vertical line passing through $-\frac{5}{2}$.

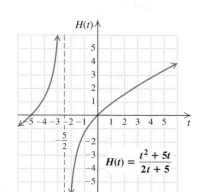

$$H(t) = \frac{t^2 + 5t}{2t + 5}$$

Simplifying Rational Expressions and Functions

The calculations that are performed with rational expressions resemble those performed with fractions in arithmetic.

Recall from arithmetic that multiplication by 1 can be used to find equivalent expressions:

$$\frac{3}{5} = \frac{3}{5} \cdot \frac{2}{2} \qquad \text{Multiplying by } \tfrac{2}{2}, \text{ which is 1}$$

$$= \frac{3 \cdot 2}{5 \cdot 2} \qquad \text{Multiplying numerators and multiplying denominators}$$

$$= \frac{6}{10}. \qquad \tfrac{3}{5} \text{ and } \tfrac{6}{10} \text{ represent the same number.}$$

Recall too that in arithmetic, fractions are *simplified* by "removing" a factor equal to 1. This reverses the process shown above:

$$\frac{6}{10} = \frac{3 \cdot 2}{5 \cdot 2} = \frac{3}{5} \cdot \frac{2}{2} = \frac{3}{5}. \qquad \text{We "removed" the factor that equals 1: } \tfrac{2}{2} = 1.$$

To multiply rational expressions, we multiply numerators and multiply denominators.

PRODUCTS OF RATIONAL EXPRESSIONS

To multiply two rational expressions, multiply numerators and multiply denominators:

$$\frac{A}{B} \cdot \frac{C}{D} = \frac{AC}{BD}, \qquad \text{where } B \neq 0, D \neq 0.$$

EXAMPLE 2 Multiply.

a) $\dfrac{x}{10y} \cdot \dfrac{3}{3}$

b) $\dfrac{2x + 15}{7y} \cdot \dfrac{4}{4}$

SOLUTION

a) $\dfrac{x}{10y} \cdot \dfrac{3}{3} = \dfrac{x \cdot 3}{10y \cdot 3}$ Multiplying numerators and multiplying denominators

$$= \frac{3x}{30y}$$

b) $\dfrac{2x + 15}{7y} \cdot \dfrac{4}{4} = \dfrac{(2x + 15)4}{(7y)4}$ Use parentheses when multiplying a polynomial with more than one term.

$$= \frac{8x + 60}{28y} \qquad \text{Using the commutative, associative, and distributive laws}$$

2. Multiply: $\dfrac{2x + 3}{y} \cdot \dfrac{5}{5}$.

YOUR TURN

To simplify a rational expression, we remove a factor that equals 1.

EXAMPLE 3 Simplify by removing a factor equal to 1.

a) $\dfrac{24}{30y}$

b) $\dfrac{8x + 60}{28y}$

SOLUTION We factor the numerator, factor the denominator, and look for the largest factor common to both.

a) $\dfrac{24}{30y} = \dfrac{6 \cdot 4}{6 \cdot 5 \cdot y}$ Factoring. The greatest common factor is 6.

$= \dfrac{6}{6} \cdot \dfrac{4}{5y}$ Rewriting as a product of two rational expressions

$= 1 \cdot \dfrac{4}{5y}$ $\dfrac{6}{6} = 1$

$= \dfrac{4}{5y}$ Removing a factor that equals 1

b) $\dfrac{8x + 60}{28y} = \dfrac{4(2x + 15)}{4(7y)}$ Factoring. The greatest common factor is 4.

$= \dfrac{4}{4} \cdot \dfrac{2x + 15}{7y}$ $\dfrac{4}{4} = 1$

$= \dfrac{2x + 15}{7y}$ Removing a factor equal to 1

3. Simplify by removing a factor equal to 1:

$\dfrac{18x}{36y}.$

YOUR TURN

It is important that the domain of a rational function not be changed as a result of simplifying. For example, since

$$\frac{(x - 5)(x + 3)}{(x + 2)(x + 3)} = \frac{x - 5}{x + 2} \cdot \frac{x + 3}{x + 3} = \frac{x - 5}{x + 2},$$

We "removed" the factor that equals 1: $\dfrac{x + 3}{x + 3} = 1.$

it is tempting to state that

$$F(x) = \frac{(x - 5)(x + 3)}{(x + 2)(x + 3)} \quad \text{and} \quad G(x) = \frac{x - 5}{x + 2}$$

represent the same function. However, the domain of each function is assumed to be all real numbers for which the denominator is nonzero. Thus,

Domain of $F = \{x \,|\, x \neq -2, x \neq -3\}$, and

Domain of $G = \{x \,|\, x \neq -2\}$. *

Thus, as presently written, the domain of G includes -3, but the domain of F does not. This difficulty is easily addressed by specifying

$$F(x) = \frac{(x - 5)(x + 3)}{(x + 2)(x + 3)} = \frac{x - 5}{x + 2} \quad \text{with } x \neq -3.$$

---

* This use of set-builder notation assumes that, apart from the restrictions listed, all other real numbers are in the domain.

EXAMPLE 4 Write the function given by $f(t) = \dfrac{7t^2 + 21t}{14t}$ in simplified form.

SOLUTION We begin by noting that the domain of $f = \{t \mid t \neq 0\}$.

$$f(t) = \frac{7t^2 + 21t}{14t}$$

$$= \frac{7t(t + 3)}{7 \cdot 2 \cdot t} \qquad \text{Factoring. The greatest common factor is } 7t.$$

$$= \frac{7t}{7t} \cdot \frac{t + 3}{2} \qquad \text{Rewriting as a product of two rational expressions. For } t \neq 0, \text{ we have } (7t)/(7t) = 1.$$

$$= \frac{t + 3}{2}, \; t \neq 0 \qquad \text{Removing a factor equal to 1. To preserve the domain, we specify that } t \neq 0.$$

4. Write the function given by

$$f(x) = \frac{6x^2 - 9x}{15x}$$

in simplified form.

Thus simplified form is $f(t) = \dfrac{t + 3}{2}$, *with* $t \neq 0$.

YOUR TURN

A rational expression is said to be **simplified** when no factors equal to 1 can be removed. Often, this requires two or more steps. For example, suppose we remove $7/7$ instead of $(7t)/(7t)$ in Example 4. We would then have

$$\frac{7t^2 + 21t}{14t} = \frac{7(t^2 + 3t)}{7 \cdot 2t} \left. \begin{array}{l} \\ \\ \end{array} \right\} \quad \text{Removing a factor equal to 1: } \tfrac{7}{7} = 1.$$
$$= \frac{t^2 + 3t}{2t}. \qquad \text{Note that } t \neq 0.$$

Here, since another common factor remains, we need to simplify further:

$$\frac{t^2 + 3t}{2t} = \frac{t(t + 3)}{t \cdot 2} \left. \begin{array}{l} \\ \\ \end{array} \right\} \quad \text{Removing another factor equal to 1: } t/t = 1. \text{ The rational expression is now}$$
$$= \frac{t + 3}{2}, \; t \neq 0. \qquad \text{simplified. We state that } t \neq 0.$$

Student Notes

Note that factoring the numerator and the denominator is the first step in simplifying a rational expression. Operations with rational expressions involve polynomials; if you are not comfortable with factoring polynomials, it will be worth your time to review that topic.

5. Write the function given by

$$g(x) = \frac{x^2 - 25}{x^2 + x - 30}$$

in simplified form, and list all restrictions on the domain.

EXAMPLE 5 Write the function given by

$$g(x) = \frac{x^2 + 3x - 10}{2x^2 - 3x - 2}$$

in simplified form, and list all restrictions on the domain.

SOLUTION We have

$$g(x) = \frac{x^2 + 3x - 10}{2x^2 - 3x - 2}$$

$$= \frac{(x - 2)(x + 5)}{(2x + 1)(x - 2)} \qquad \text{Factoring the numerator and the denominator. Note that } x \neq -\tfrac{1}{2} \text{ and } x \neq 2.$$

$$= \frac{x - 2}{x - 2} \cdot \frac{x + 5}{2x + 1} \qquad \text{Rewriting as a product of two rational expressions}$$

$$= \frac{x + 5}{2x + 1}, x \neq -\frac{1}{2}, 2. \qquad \text{Removing a factor equal to 1: } \tfrac{x-2}{x-2} = 1. \text{ We list both restrictions.}$$

Thus, $g(x) = \dfrac{x + 5}{2x + 1}, x \neq -\dfrac{1}{2}, 2.$

YOUR TURN

We generally list restrictions only when using function notation.

EXAMPLE 6   Simplify: **(a)** $\dfrac{9x^2 + 6xy - 3y^2}{12x^2 - 12y^2}$; **(b)** $\dfrac{4 - t}{3t - 12}$.

SOLUTION

a) $\dfrac{9x^2 + 6xy - 3y^2}{12x^2 - 12y^2} = \dfrac{3(x + y)(3x - y)}{12(x + y)(x - y)}$ Factoring the numerator and the denominator

$= \dfrac{3(x + y)}{3(x + y)} \cdot \dfrac{3x - y}{4(x - y)}$ Rewriting as a product of two rational expressions

$= \dfrac{3x - y}{4(x - y)}$ Removing a factor equal to 1: $\dfrac{3(x + y)}{3(x + y)} = 1$

For purposes of later work, we usually do not multiply out the numerator and the denominator of the simplified expression.

b) $\dfrac{4 - t}{3t - 12} = \dfrac{4 - t}{3(t - 4)}$ Factoring

Since $4 - t$ is the opposite of $t - 4$, we factor out -1 to reverse the subtraction:

$\dfrac{4 - t}{3t - 12} = \dfrac{-1(t - 4)}{3(t - 4)}$ $4 - t = -1(-4 + t) = -1(t - 4)$

$= \dfrac{-1}{3} \cdot \dfrac{t - 4}{t - 4}$

$= -\dfrac{1}{3}$.   Removing a factor equal to 1: $(t - 4)/(t - 4) = 1$

6. Simplify:

$\dfrac{8x^3 - 2x^2 - 15x}{2x^3 + 7x^2 - 15x}$.

YOUR TURN

CANCELING

"Canceling" is a shortcut often used for removing a factor equal to 1 when working with fractions. With caution, we mention it as a possible way to speed up your work. Canceling removes factors equal to 1 in products. It *cannot* be done inside a sum or when adding expressions together. If your instructor permits canceling (not all do), it must be done with care and understanding. Example 6(a) might have been done with less writing as follows:

$\dfrac{9x^2 + 6xy - 3y^2}{12x^2 - 12y^2} = \dfrac{3(x + y)(3x - y)}{3 \cdot 4(x + y)(x - y)}$ When a factor that equals 1 is found, it is "canceled" as shown.

$= \dfrac{3x - y}{4(x - y)}$.   Removing a factor equal to 1: $\dfrac{3(x + y)}{3(x + y)} = 1$

CAUTION!   Canceling is often performed incorrectly:

$\dfrac{x + 3}{x} = 3$, $\dfrac{4x + 3}{2} = 2x + 3$, $\dfrac{5}{5 + x} = \dfrac{1}{x}$ To check that these are not equivalent, substitute a number for x.

Incorrect!   Incorrect!   Incorrect!

In each incorrect situation, one of the expressions canceled is *not* a factor. Factors are parts of products. If it's not a factor, it can't be canceled!

Multiplying and Simplifying

After multiplying rational expressions, we simplify, if possible.

EXAMPLE 7 Find each product and, if possible, simplify.

a) $\dfrac{10}{7} \cdot \dfrac{14x}{15}$

b) $\dfrac{2x}{3} \cdot \dfrac{x+7}{10}$

SOLUTION

a) $\dfrac{10}{7} \cdot \dfrac{14x}{15} = \dfrac{10 \cdot 14x}{7 \cdot 15}$ Forming the product of the numerators and the product of the denominators

$= \dfrac{2 \cdot 5 \cdot 2 \cdot 7 \cdot x}{7 \cdot 3 \cdot 5}$ Factoring the numerator and the denominator

$= \dfrac{2 \cdot \cancel{5} \cdot 2 \cdot \cancel{7} \cdot x}{\cancel{7} \cdot 3 \cdot \cancel{5}}$ Removing a factor equal to 1: $\dfrac{5 \cdot 7}{5 \cdot 7} = 1$

$= \dfrac{4x}{3}$ Simplifying

b) $\dfrac{2x}{3} \cdot \dfrac{x+7}{10} = \dfrac{2x(x+7)}{3 \cdot 10}$ Multiplying numerators and multiplying denominators

$= \dfrac{2 \cdot x \cdot (x+7)}{3 \cdot 2 \cdot 5}$ Factoring the numerator and the denominator

$= \dfrac{\cancel{2} \cdot x \cdot (x+7)}{3 \cdot \cancel{2} \cdot 5}$ Removing a factor equal to 1: $\dfrac{2}{2} = 1$

$= \dfrac{x(x+7)}{15}$ We leave the numerator in factored form.

7. Find the product and, if possible, simplify:

$\dfrac{5x}{4y} \cdot \dfrac{8y}{9x}.$

YOUR TURN

EXAMPLE 8 Multiply. (Write the product as a single rational expression.) Then simplify, if possible, by removing a factor equal to 1:

$$x + 2 \cdot \dfrac{x^2 - 4}{x^2 + x - 2}.$$

SOLUTION We have

$x + 2 \cdot \dfrac{x^2 - 4}{x^2 + x - 2} = \dfrac{x+2}{1} \cdot \dfrac{x^2 - 4}{x^2 + x - 2}$ Writing $x + 2$ as a rational expression

Multiply. $= \dfrac{(x+2)(x^2-4)}{1(x^2+x-2)}$ Forming the product of the numerators and the product of the denominators

Factor. $= \dfrac{(x+2)(x-2)(x+2)}{(x+2)(x-1)}$ Factoring the numerator and the denominator

Remove a factor equal to 1. $= \dfrac{(x+2)(x-2)\cancel{(x+2)}}{\cancel{(x+2)}(x-1)}$ Removing a factor equal to 1: $\dfrac{x+2}{x+2} = 1$

8. Multiply. Then simplify by removing a factor equal to 1.

$\dfrac{t^2 - 1}{3t^2 + 12t + 12} \cdot \dfrac{t+2}{2t^2 + 3t + 1}$

$= \dfrac{(x+2)(x-2)}{x-1}.$ Simplifying

There is no need for us to multiply out the numerator of the final result.

YOUR TURN

Dividing and Simplifying

Two expressions are reciprocals of each other if their product is 1. To find the reciprocal of a rational expression, we interchange numerator and denominator.

The reciprocal of $\dfrac{x}{x^2 + 3}$ is $\dfrac{x^2 + 3}{x}$.

The reciprocal of $y - 8$ is $\dfrac{1}{y - 8}$.

Student Notes

The procedures covered in this chapter are by their nature rather long. Use plenty of paper and, if you are using lined paper, consider using two spaces at a time, writing the fraction bar on a line of the paper. Write any equal signs at the same height as the fraction bars.

> **QUOTIENTS OF RATIONAL EXPRESSIONS**
>
> For any rational expressions A/B and C/D, with $B, C, D \neq 0$,
>
> $$\frac{A}{B} \div \frac{C}{D} = \frac{A}{B} \cdot \frac{D}{C}.$$
>
> (To divide two rational expressions, multiply by the reciprocal of the divisor. We often say that we "*invert* and multiply.")

EXAMPLE 9 Divide. Simplify, if possible, by removing a factor equal to 1.

$$\frac{x - 2}{x + 1} \div \frac{x + 5}{x - 3}$$

SOLUTION

$$\frac{x - 2}{x + 1} \div \frac{x + 5}{x - 3} = \frac{x - 2}{x + 1} \cdot \frac{x - 3}{x + 5}$$ Multiplying by the reciprocal of the divisor

$$= \frac{(x - 2)(x - 3)}{(x + 1)(x + 5)}$$ Multiplying the numerators and the denominators

9. Divide. Simplify, if possible, by removing a factor equal to 1.

$$\frac{x^2 - x}{2x + 1} \div \frac{3x - 3}{x - 2}$$

↻ YOUR TURN

EXAMPLE 10 Write the function in simplified form, and list all restrictions on the domain.

$$g(a) = \frac{a^2 - 2a + 1}{a + 5} \div \frac{a^3 - a}{a - 2}$$

SOLUTION A number is not in the domain of a rational function if it makes a divisor zero. There are three divisors in this rational function:

$$a + 5, \quad a - 2, \quad \text{and} \quad \frac{a^3 - a}{a - 2}.$$

Student Notes

When listing restrictions on the domain of a rational function, we set denominators equal to zero. Note in Example 10 that $a + 5$ and $a - 2$ are denominators. Then, when we rewrite the division as multiplication, $a^3 - a$ becomes a denominator.

None of these can be zero:

$$a + 5 = 0 \text{ when } a = -5,$$
$$a - 2 = 0 \text{ when } a = 2, \text{ and}$$
$$\frac{a^3 - a}{a - 2} = 0 \text{ when } a^3 - a = 0.$$

We solve $a^3 - a = 0$:

$$a^3 - a = 0$$
$$a(a^2 - 1) = 0$$
$$a(a + 1)(a - 1) = 0$$
$$a = 0 \quad \text{or} \quad a + 1 = 0 \quad \text{or} \quad a - 1 = 0$$
$$a = 0 \quad \text{or} \qquad a = -1 \quad \text{or} \qquad a = 1.$$

Thus the domain of g does not contain $-5, -1, 0, 1,$ or 2.

$$g(a) = \frac{a^2 - 2a + 1}{a + 5} \div \frac{a^3 - a}{a - 2} = \frac{a^2 - 2a + 1}{a + 5} \cdot \frac{a - 2}{a^3 - a}$$ Multiplying by the reciprocal of the divisor

$$= \frac{(a^2 - 2a + 1)(a - 2)}{(a + 5)(a^3 - a)}$$ Multiplying the numerators and the denominators

$$= \frac{(a - 1)(a - 1)(a - 2)}{(a + 5)a(a + 1)(a - 1)}$$ Factoring the numerator and the denominator

$$= \frac{(a - 1)\cancel{(a - 1)}(a - 2)}{(a + 5)a(a + 1)\cancel{(a - 1)}}$$ Removing a factor equal to 1: $\dfrac{a - 1}{a - 1} = 1$

$$= \frac{(a - 1)(a - 2)}{a(a + 5)(a + 1)}, \quad a \neq -5, -1, 0, 1, 2 \quad \text{Simplifying}$$

◀ **Chapter Resource:**
Visualizing for Success, p. 423

10. Write the function given by

$$f(a) = \frac{a^2}{a^2 - 9} \div \frac{a^3 + 3a^2}{4a^2 - 11a - 3}$$

in simplified form, and list all restrictions on the domain.

↪ **YOUR TURN**

6.1 EXERCISE SET

FOR EXTRA HELP

MyMathLab® Math XL

PRACTICE   WATCH   READ   REVIEW

◀ **Vocabulary and Reading Check**

Choose the word from the following list that best completes each statement. Words may be used more than once or not at all.

domain   range
factor   rational
invert   reciprocal

1. The expression $\dfrac{x - 4}{5x}$ is an example of a(n)

_____ expression.

2. Any value that makes the denominator of a rational function 0 is not in the _____ of that function.

3. When simplifying rational expressions, remove a(n) _____ equal to 1.

4. To divide rational expressions, multiply by the _____ of the divisor.

◀ **Concept Reinforcement**

In each of Exercises 5–10, match the function described with the appropriate domain listed below.

a) $\{x \mid x \neq 3\}$ **b)** $\{x \mid x \neq -3\}$

c) $\{x \mid x \neq 5\}$ **d)** $\{x \mid x \neq -2, x \neq 5\}$

e) $\{x \mid x \neq 2\}$ **f)** $\{x \mid x \neq 2, x \neq 5\}$

5. ___ $f(x) = \dfrac{2 - x}{x - 5}$

6. ___ $h(x) = \dfrac{x - 5}{x - 2}$

7. ___ $g(x) = \dfrac{x - 3}{(x - 2)(x - 5)}$

8. ___ $g(x) = \dfrac{x + 3}{(x + 2)(x - 5)}$

9. ___ $h(x) = \dfrac{(x - 2)(x - 3)}{x + 3}$

10. ___ $f(x) = \dfrac{(x + 2)(x + 3)}{x - 3}$

Rational Functions

For each rational function, find the function values indicated, provided the value exists.

11. $f(x) = \dfrac{2x^2 - x - 5}{x - 1}$; **(a)** $f(0)$; **(b)** $f(-1)$; **(c)** $f(3)$

12. $v(t) = \dfrac{t^2 + 5t - 9}{t + 4}$; **(a)** $v(0)$; **(b)** $v(-3)$; **(c)** $v(6)$

13. $r(t) = \dfrac{t^2 - 8t - 9}{t^2 - 4}$; **(a)** $r(0)$; **(b)** $r(2)$; **(c)** $r(-1)$

14. $g(x) = \dfrac{2x^3 - x}{x^2 - 6x + 9}$; **(a)** $g(0)$; **(b)** $g(-2)$; **(c)** $g(3)$

Restaurant Management.   *Gregory usually takes 1 hr more than Alayna does to prepare the day's soups for Roux Palace. If Alayna takes t hr to prepare the day's soups, the function given by*

$$H(t) = \frac{t^2 + t}{2t + 1}$$

can be used to determine how long it would take if they worked together.

15. How long will it take them, working together, to prepare the soups if Alayna can prepare them alone in 5 hr?

16. How long will it take them, working together, to prepare the soups if Alayna can prepare them alone in 2 hr?

Simplifying Rational Expressions and Functions

Simplify by removing a factor equal to 1.

17. $\dfrac{8t^4}{40t}$

18. $\dfrac{35n}{5n^2}$

19. $\dfrac{24x^3y}{30x^5y^8}$

20. $\dfrac{10yz^4}{40y^2z^9}$

21. $\dfrac{2a - 10}{2}$

22. $\dfrac{3a + 12}{3}$

23. $\dfrac{5}{25y - 30}$

24. $\dfrac{21}{6x - 9}$

25. $\dfrac{3x - 12}{3x + 15}$

26. $\dfrac{4y - 20}{4y + 12}$

Write simplified form for each of the following. Be sure to list all restrictions on the domain, as in Example 5.

27. $f(x) = \dfrac{5x + 30}{x^2 + 6x}$

28. $f(x) = \dfrac{3x + 30}{x^2 + 10x}$

29. $g(x) = \dfrac{x^2 - 9}{5x + 15}$

30. $g(x) = \dfrac{8x - 16}{x^2 - 4}$

31. $h(x) = \dfrac{2 - x}{7x - 14}$

32. $h(x) = \dfrac{4 - x}{12x - 48}$

33. $f(t) = \dfrac{t^2 - 16}{t^2 - 8t + 16}$

34. $f(t) = \dfrac{t^2 - 25}{t^2 + 10t + 25}$

35. $g(t) = \dfrac{21 - 7t}{3t - 9}$

36. $g(t) = \dfrac{12 - 6t}{5t - 10}$

37. $h(t) = \dfrac{t^2 + 5t + 4}{t^2 - 8t - 9}$

38. $h(t) = \dfrac{t^2 - 3t - 4}{t^2 + 9t + 8}$

39. $f(x) = \dfrac{9x^2 - 4}{3x - 2}$

40. $f(x) = \dfrac{4x^2 - 1}{2x - 1}$

41. $g(t) = \dfrac{16 - t^2}{t^2 - 8t + 16}$

42. $g(p) = \dfrac{25 - p^2}{p^2 + 10p + 25}$

Multiplying and Simplifying

Multiply and, if possible, simplify.

43. $\dfrac{3y^3}{5z} \cdot \dfrac{10z^4}{7y^6}$

44. $\dfrac{20y}{9z^7} \cdot \dfrac{6z^4}{5y^2}$

45. $\dfrac{8x - 16}{5x} \cdot \dfrac{x^3}{5x - 10}$

46. $\dfrac{5t^3}{4t - 8} \cdot \dfrac{6t - 12}{10t}$

47. $\dfrac{y^2 - 9}{y^2} \cdot \dfrac{y^2 - 3y}{y^2 - y - 6}$

48. $\dfrac{y^2 + 10y + 25}{y^2 - 9} \cdot \dfrac{y^2 + 3y}{y + 5}$

49. $\dfrac{7a - 14}{4 - a^2} \cdot \dfrac{5a^2 + 6a + 1}{35a + 7}$

50. $\dfrac{a^2 - 1}{2 - 5a} \cdot \dfrac{15a - 6}{a^2 + 5a - 6}$

Aha! **51.** $\dfrac{t^3 - 4t}{t - t^4} \cdot \dfrac{t^4 - t}{4t - t^3}$

52. $\dfrac{x^2 - 6x + 9}{12 - 4x} \cdot \dfrac{x^6 - 9x^4}{x^3 - 3x^2}$

53. $\dfrac{c^3 + 8}{c^5 - 4c^3} \cdot \dfrac{c^6 - 4c^5 + 4c^4}{c^2 - 2c + 4}$

54. $\dfrac{t^3 - 27}{t^4 - 9t^2} \cdot \dfrac{t^5 - 6t^4 + 9t^3}{t^2 + 3t + 9}$

55. $\dfrac{a^3 - b^3}{3a^2 + 9ab + 6b^2} \cdot \dfrac{a^2 + 2ab + b^2}{a^2 - b^2}$

56. $\dfrac{x^3 + y^3}{x^2 + 2xy - 3y^2} \cdot \dfrac{x^2 - y^2}{3x^2 + 6xy + 3y^2}$

Dividing and Simplifying

Divide and, if possible, simplify.

57. $\dfrac{12a^3}{5b^2} \div \dfrac{4a^2}{15b}$

58. $\dfrac{9x^7}{8y} \div \dfrac{15x^2}{4y}$

59. $\dfrac{5x + 20}{x^6} \div \dfrac{x + 4}{x^2}$

60. $\dfrac{3a + 15}{a^9} \div \dfrac{a + 5}{a^8}$

61. $\dfrac{25x^2 - 4}{x^2 - 9} \div \dfrac{2 - 5x}{x + 3}$

62. $\dfrac{4a^2 - 1}{a^2 - 4} \div \dfrac{2a - 1}{2 - a}$

63. $\dfrac{5y - 5x}{15y^3} \div \dfrac{x^2 - y^2}{3x + 3y}$

64. $\dfrac{x^2 - y^2}{4x + 4y} \div \dfrac{3y - 3x}{12x^2}$

65. $\dfrac{y^2 - 36}{y^2 - 8y + 16} \div \dfrac{3y - 18}{y^2 - y - 12}$

66. $\dfrac{x^2-16}{x^2-10x+25} \div \dfrac{3x-12}{x^2-3x-10}$

67. $\dfrac{x^3-64}{x^3+64} \div \dfrac{x^2-16}{x^2-4x+16}$

68. $\dfrac{8y^3-27}{64y^3-1} \div \dfrac{4y^2-9}{16y^2+4y+1}$

Multiplying and Dividing

Write simplified form for each of the following. Be sure to list all restrictions on the domain.

69. $f(t) = \dfrac{t^2-100}{5t+20} \cdot \dfrac{t+4}{t-10}$

70. $g(n) = \dfrac{n+5}{n-5} \cdot \dfrac{n^2-25}{2n+2}$

71. $g(x) = \dfrac{x^2-2x-35}{2x^3-3x^2} \cdot \dfrac{4x^3-9x}{7x-49}$

72. $h(t) = \dfrac{t^2-10t+9}{t^2-1} \cdot \dfrac{1-t^2}{t^2-5t-36}$

73. $f(x) = \dfrac{x^2-4}{x^3} \div \dfrac{x^5-2x^4}{x+4}$

74. $g(x) = \dfrac{x^2-9}{x^2} \div \dfrac{x^5+3x^4}{x+2}$

75. $h(n) = \dfrac{n^3+3n}{n^2-9} \div \dfrac{n^2+5n-14}{n^2+4n-21}$

76. $f(x) = \dfrac{x^3+4x}{x^2-16} \div \dfrac{x^2+8x+15}{x^2+x-20}$

Perform the indicated operations and, if possible, simplify.

77. $\dfrac{4x^2-9y^2}{8x^3-27y^3} \div \dfrac{4x+6y}{3x-9y} \cdot \dfrac{4x^2+6xy+9y^2}{4x^2-8xy+3y^2}$

78. $\dfrac{5x^2-5y^2}{27x^3+8y^3} \div \dfrac{x^2-2xy+y^2}{9x^2-6xy+4y^2} \cdot \dfrac{6x+4y}{10x-15y}$

79. $\dfrac{a^3-ab^2}{2a^2+3ab+b^2} \cdot \dfrac{4a^2-b^2}{a^2-2ab+b^2} \div \dfrac{a^2+a}{a-1}$

80. $\dfrac{2x+4y}{2x^2+5xy+2y^2} \cdot \dfrac{4x^2-y^2}{8x^2-8} \div \dfrac{x^2+4xy+4y^2}{x^2-6xy+9y^2}$

81. Evelyn *incorrectly* simplifies $\dfrac{x+2}{x}$ as

$$\dfrac{x+2}{x} = \dfrac{\cancel{x}+2}{\cancel{x}} = 1+2 = 3.$$

She insists that this is correct because it checks when x is replaced with 1. Explain her misconception.

82. Give a step-by-step procedure that a classmate could use to divide rational expressions.

Skill Review

Factor completely.

83. $2n^2-11n+12$ [5.4]

84. $7y^2-28$ [5.5]

85. $8x^3+125$ [5.6]

86. $100a^2+60ab+9b^2$ [5.5]

87. t^3+8t^2-33t [5.4]

88. $z^{12}-1$ [5.6]

Synthesis

89. Explain why the graphs of $f(x) = 5x$ and $g(x) = \dfrac{5x^2}{x}$ differ.

90. Todd *incorrectly* argues that since

$$\dfrac{a^2-4}{a-2} = \dfrac{a^2}{a} + \dfrac{-4}{-2} = a+2$$

is correct, it follows that

$$\dfrac{x^2+9}{x+1} = \dfrac{x^2}{x} + \dfrac{9}{1} = x+9.$$

Explain his misconception.

91. Calculate the slope of the line passing through $(a, f(a))$ and $(a+h, f(a+h))$ for the function f given by $f(x) = x^2+5$. Be sure your answer is simplified.

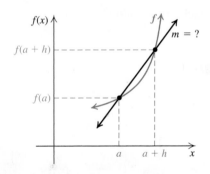

92. Calculate the slope of the line passing through the points $(a, f(a))$ and $(a+h, f(a+h))$ for the function f given by $f(x) = 3x^2$. Be sure your answer is simplified.

93. Graph the function given by

$$f(x) = \dfrac{x^2-9}{x-3}.$$

(*Hint*: Determine the domain of f and simplify.)

Perform the indicated operations and simplify.

94. $\dfrac{r^2-4s^2}{r+2s} \div (r+2s)^2 \left(\dfrac{2s}{r-2s}\right)^2$

95. $\dfrac{d^2-d}{d^2-6d+8} \cdot \dfrac{d-2}{d^2+5d} \div \left(\dfrac{5d^2}{d^2-9d+20}\right)^2$

Aha! 96. $\dfrac{6t^2-26t+30}{8t^2-15t-21} \cdot \dfrac{5t+15}{t^2-4} \div \dfrac{5t+15}{t^2-4}$

Simplify.

97. $\dfrac{m^2 - t^2}{m^2 + t^2 + m + t + 2mt}$

98. $\dfrac{a^3 - 2a^2 + 2a - 4}{a^3 - 2a^2 - 3a + 6}$

99. $\dfrac{x^3 + x^2 - y^3 - y^2}{x^2 - 2xy + y^2}$

100. $\dfrac{u^6 + v^6 + 2u^3v^3}{u^3 - v^3 + u^2v - uv^2}$

101. $\dfrac{x^5 - x^3 + x^2 - 1 - (x^3 - 1)(x + 1)^2}{(x^2 - 1)^2}$

102. Let

$$g(x) = \dfrac{2x + 3}{4x - 1}.$$

Determine each of the following.

a) $g(x + h)$
b) $g(2x - 2) \cdot g(x)$
c) $g\!\left(\tfrac{1}{2}x + 1\right) \cdot g(x)$

103. Let

$$f(x) = \dfrac{4}{x^2 - 1} \quad \text{and} \quad g(x) = \dfrac{4x^2 + 8x + 4}{x^3 - 1}.$$

Find each of the following.

a) $(f \cdot g)(x)$
b) $(f/g)(x)$
c) $(g/f)(x)$

104. Use a graphing calculator to show that

$$\dfrac{x^2 - 16}{x + 2} \neq x - 8.$$

 105. To check Example 4, Kara graphs

$$y_1 = \dfrac{7x^2 + 21x}{14x} \quad \text{and} \quad y_2 = \dfrac{x + 3}{2}.$$

Since the graphs of y_1 and y_2 appear to be identical, Kara believes that the domains of the functions described by y_1 and y_2 are the same, $\mathbb{R}$. How could you convince Kara otherwise?

↱ YOUR TURN ANSWERS: SECTION 6.1

1. $2\frac{2}{11}$ hr   **2.** $\dfrac{10x + 15}{5y}$ **3.** $\dfrac{x}{2y}$ **4.** $f(x) = \dfrac{2x - 3}{5}, x \neq 0$

5. $g(x) = \dfrac{x + 5}{x + 6}, x \neq -6, 5$ **6.** $\dfrac{4x + 5}{x + 5}$ **7.** $\dfrac{10}{9}$

8. $\dfrac{t - 1}{3(t + 2)(2t + 1)}$ **9.** $\dfrac{x(x - 2)}{3(2x + 1)}$

10. $f(a) = \dfrac{4a + 1}{(a + 3)^2}, a \neq -3, -\frac{1}{4}, 0, 3$

PREPARE TO MOVE ON

Simplify.

1. $\dfrac{7}{12} - \dfrac{2}{15}$ [1.2]   **2.** $\dfrac{1}{5} \cdot \dfrac{3}{4} - \dfrac{7}{10} \cdot \dfrac{3}{5}$ [1.2]

3. $(5x^2 - 6x + 1) - (x^2 - 6x + 3)$ [5.1]

4. $(2x + 1)(x + 3) - (x - 7)(3x - 1)$ [5.1]

6.2   Rational Expressions and Functions: Adding and Subtracting

When Denominators Are the Same   ▪   When Denominators Are Different

Rational expressions are added and subtracted in much the same way as the fractions of arithmetic.

When Denominators Are the Same

Recall that $\dfrac{2}{7} + \dfrac{3}{7} = \dfrac{2 + 3}{7} = \dfrac{5}{7}$.

ADDITION AND SUBTRACTION WITH LIKE DENOMINATORS

To add or subtract when denominators are the same, add or subtract the numerators and keep the same denominator.

$$\dfrac{A}{C} + \dfrac{B}{C} = \dfrac{A + B}{C} \quad \text{and} \quad \dfrac{A}{C} - \dfrac{B}{C} = \dfrac{A - B}{C}, \quad \text{where} \quad C \neq 0.$$

Study Skills

Anticipate the Future

It is never too soon to begin reviewing for an exam. Take a few minutes each week to review important problems, formulas, and properties. Working a few problems from material covered earlier in the course will help you keep your skills fresh.

EXAMPLE 1 Add: $\dfrac{3+x}{x} + \dfrac{4}{x}$.

SOLUTION

$$\dfrac{3+x}{x} + \dfrac{4}{x} = \dfrac{3+x+4}{x} = \dfrac{x+7}{x}$$

YOUR TURN

> **CAUTION!** Because x is a *term* in the numerator and not a *factor*, $\dfrac{x+7}{x}$ cannot be simplified.

1. Add: $\dfrac{a}{a+1} + \dfrac{5}{a+1}$.

As in arithmetic, we can sometimes simplify after we add or subtract.

EXAMPLE 2 Add: $\dfrac{4x^2 - 5xy}{x^2 - y^2} + \dfrac{2xy - y^2}{x^2 - y^2}$.

SOLUTION

Add the numerators. Keep the denominators.

$$\dfrac{4x^2 - 5xy}{x^2 - y^2} + \dfrac{2xy - y^2}{x^2 - y^2} = \dfrac{4x^2 - 3xy - y^2}{x^2 - y^2}$$

Adding the numerators and combining like terms. The denominator is unchanged.

Factor.

$$= \dfrac{(x-y)(4x+y)}{(x-y)(x+y)}$$

Factoring the numerator and the denominator and looking for common factors

Remove a factor equal to 1.

$$= \dfrac{\cancel{(x-y)}(4x+y)}{\cancel{(x-y)}(x+y)}$$

Removing a factor equal to 1: $\dfrac{x-y}{x-y} = 1$

2. Add: $\dfrac{2n^2 - 5}{n^2 - n} + \dfrac{n+2}{n^2 - n}$.

$$= \dfrac{4x+y}{x+y}$$

Simplifying

YOUR TURN

Technology Connection

Example 3 can be checked by comparing the graphs of

$$y_1 = \dfrac{4x+5}{x+3} - \dfrac{x-2}{x+3}$$

and

$$y_2 = \dfrac{3x+7}{x+3}$$

on the same set of axes. Since the equations are equivalent, one graph (containing two branches) should appear. Equivalently, you can show that $y_3 = y_2 - y_1$ is 0 for all x not equal to -3.

3. If

$$g(x) = \dfrac{x+4}{x-1} - \dfrac{2-3x}{x-1},$$

find a simplified form for $g(x)$ and list all restrictions on the domain.

When a numerator is subtracted, care must be taken to subtract, or change the sign of, *each* term in that polynomial.

EXAMPLE 3 If

$$f(x) = \dfrac{4x+5}{x+3} - \dfrac{x-2}{x+3},$$

find a simplified form for $f(x)$ and list all restrictions on the domain.

SOLUTION

$$f(x) = \dfrac{4x+5}{x+3} - \dfrac{x-2}{x+3}$$

Note that $x \neq -3$.

$$= \dfrac{4x+5 - (x-2)}{x+3}$$

The parentheses remind us to subtract *both* terms.

$$= \dfrac{4x+5 - x + 2}{x+3}$$

$$= \dfrac{3x+7}{x+3}, \; x \neq -3$$

YOUR TURN

When Denominators Are Different

Recall that when adding fractions with different denominators, we first find common denominators:

$$\frac{1}{6} + \frac{4}{15} = \frac{1}{6} \cdot \frac{5}{5} + \frac{4}{15} \cdot \frac{2}{2} = \frac{5}{30} + \frac{8}{30} = \frac{13}{30}.$$

Our work is easier when we use the *least common multiple* (LCM) of the denominators.

LEAST COMMON MULTIPLE

To find the least common multiple (LCM) of two or more expressions:

1. Find the prime factorization of each expression.
2. Form a product that contains each factor the greatest number of times that it occurs in any one prime factorization.

EXAMPLE 4   Find the least common multiple of each pair of polynomials.

a) $21x$ and $3x^2$

b) $x^2 + x - 12$ and $x^2 - 16$

SOLUTION

a) We write the prime factorizations of $21x$ and $3x^2$:

$$21x = 3 \cdot 7 \cdot x \quad \text{and} \quad 3x^2 = 3 \cdot x \cdot x.$$

The factors $3, 7,$ and x must appear in the LCM if $21x$ is to be a factor of the LCM. The factors $3, x,$ and x must appear in the LCM if $3x^2$ is to be a factor of the LCM. These do not all appear in $3 \cdot 7 \cdot x$. However, if $3 \cdot 7 \cdot x$ is multiplied by another factor of x, a product is formed that contains both $21x$ and $3x^2$ as factors:

$$\text{LCM} = 3 \cdot 7 \cdot x \cdot x = 21x^2.$$

$21x$ is a factor.
$3x^2$ is a factor.

Note that each factor $(3, 7,$ and $x)$ is used the greatest number of times that it occurs as a factor of either $21x$ or $3x^2$. The LCM is $3 \cdot 7 \cdot x \cdot x$, or $21x^2$.

b) We factor both expressions:

$$x^2 + x - 12 = (x - 3)(x + 4),$$
$$x^2 - 16 = (x + 4)(x - 4).$$

The LCM must contain each polynomial as a factor. By multiplying the factors of $x^2 + x - 12$ by $x - 4$, we form a product that contains both $x^2 + x - 12$ and $x^2 - 16$ as factors:

$x^2 + x - 12$ is a factor.
$$\text{LCM} = (x - 3)(x + 4)(x - 4).$$ There is no need to multiply this out.
$x^2 - 16$ is a factor.

4. Find the least common multiple of $t^2 + 10t + 25$ and $2t^2 + 7t - 15$.

YOUR TURN

TO ADD OR SUBTRACT RATIONAL EXPRESSIONS

1. Determine the *least common denominator* (LCD) by finding the least common multiple of the denominators.
2. Rewrite each of the original rational expressions, as needed, in an equivalent form that has the LCD.
3. Add or subtract the resulting rational expressions, as indicated.
4. Simplify the result, if possible, and list any restrictions on the domain of functions.

EXAMPLE 5 Add: $\dfrac{2}{21x} + \dfrac{5}{3x^2}$.

SOLUTION In Example 4(a), we found that the LCM of the denominators is $3 \cdot 7 \cdot x \cdot x$, or $21x^2$. We now multiply each rational expression by 1, using expressions for 1 that give us the LCD in each expression. To determine what to use, ask "$21x$ times what is $21x^2$?" and "$3x^2$ times what is $21x^2$?" The answers are x and 7, respectively, so we multiply by x/x and $7/7$:

$$\frac{2}{21x} \cdot \frac{x}{x} + \frac{5}{3x^2} \cdot \frac{7}{7} = \frac{2x}{21x^2} + \frac{35}{21x^2} \qquad \text{We now have a common denominator.}$$

$$= \frac{2x + 35}{21x^2}. \qquad \text{This expression cannot be simplified.}$$

5. Add: $\dfrac{4}{15n} + \dfrac{1}{6n^3}$.

YOUR TURN

Student Notes

When working with rational expressions, it is usually helpful to begin by factoring all numerators and denominators. For addition and subtraction, this will allow you to identify any expressions that can be simplified as well as to identify the LCD.

EXAMPLE 6 Add: $\dfrac{x^2}{x^2 + 2xy + y^2} + \dfrac{2x - 2y}{x^2 - y^2}$.

SOLUTION To find the LCD, we first factor the denominators. Although the numerators need not always be factored, doing so may enable us to simplify. In this case, the rightmost rational expression can be simplified:

$$\frac{x^2}{x^2 + 2xy + y^2} + \frac{2x - 2y}{x^2 - y^2} = \frac{x^2}{(x + y)(x + y)} + \frac{2(x - y)}{(x + y)(x - y)} \qquad \text{Factoring}$$

$$= \frac{x^2}{(x + y)(x + y)} + \frac{2}{x + y}. \qquad \begin{array}{l} \text{Removing a factor} \\ \text{equal to 1: } \dfrac{x - y}{x - y} = 1 \end{array}$$

Note that the LCM of $(x + y)(x + y)$ and $(x + y)$ is $(x + y)(x + y)$. To get the LCD in the second expression, we multiply by 1, using $(x + y)/(x + y)$. Then we add and, if possible, simplify.

$$\frac{x^2}{(x + y)(x + y)} + \frac{2}{x + y} = \frac{x^2}{(x + y)(x + y)} + \frac{2}{x + y} \cdot \frac{x + y}{x + y}$$

$$= \frac{x^2}{(x + y)(x + y)} + \frac{2x + 2y}{(x + y)(x + y)} \qquad \begin{array}{l} \text{We have} \\ \text{the LCD.} \end{array}$$

$$= \frac{x^2 + 2x + 2y}{(x + y)(x + y)} \qquad \begin{array}{l} \text{Since the numerator} \\ \text{cannot be factored, we} \\ \text{cannot simplify further.} \end{array}$$

6. Add:

$$\frac{a^2 + 3a}{a^2 - 5a} + \frac{a + 2}{a^2 - 4a - 5}.$$

YOUR TURN

EXAMPLE 7   Subtract: $\dfrac{2y + 1}{y^2 - 7y + 6} - \dfrac{y + 3}{y^2 - 5y - 6}$.

SOLUTION

$$\dfrac{2y + 1}{y^2 - 7y + 6} - \dfrac{y + 3}{y^2 - 5y - 6}$$

Determine the LCD.

$$= \dfrac{2y + 1}{(y - 6)(y - 1)} - \dfrac{y + 3}{(y - 6)(y + 1)} \qquad \text{The LCD is } (y - 6)(y - 1)(y + 1).$$

Multiply by 1 so that both expressions have the LCD.

$$= \dfrac{2y + 1}{(y - 6)(y - 1)} \cdot \dfrac{y + 1}{y + 1} - \dfrac{y + 3}{(y - 6)(y + 1)} \cdot \dfrac{y - 1}{y - 1}$$

⎿ Multiplying by 1 to get ⏌
the LCD in each expression

Subtract the second numerator from the first.

$$= \dfrac{(2y + 1)(y + 1) - (y + 3)(y - 1)}{(y - 6)(y - 1)(y + 1)}$$

$$= \dfrac{2y^2 + 3y + 1 - (y^2 + 2y - 3)}{(y - 6)(y - 1)(y + 1)} \qquad \begin{array}{l}\text{Performing the multiplica-}\\ \text{tions in the numerator. The}\\ \text{parentheses are important.}\end{array}$$

Factor and, if possible, simplify.

$$= \dfrac{2y^2 + 3y + 1 - y^2 - 2y + 3}{(y - 6)(y - 1)(y + 1)}$$

7. Subtract:

$$\dfrac{2x}{x^2 - 2x - 3} - \dfrac{x - 1}{2x^2 - 5x - 3}.$$

$$= \dfrac{y^2 + y + 4}{(y - 6)(y - 1)(y + 1)} \qquad \begin{array}{l}\text{This numerator cannot be factored.}\\ \text{We leave the denominator in factored}\\ \text{form.}\end{array}$$

YOUR TURN

EXAMPLE 8   Add: $\dfrac{3}{8a} + \dfrac{1}{-8a}$.

SOLUTION

$$\dfrac{3}{8a} + \dfrac{1}{-8a} = \dfrac{3}{8a} + \dfrac{-1}{-1} \cdot \dfrac{1}{-8a}$$

> When denominators are opposites, we multiply one rational expression by $-1/-1$ to get the LCD.

$$= \dfrac{3}{8a} + \dfrac{-1}{8a} = \dfrac{2}{8a}$$

8. Add: $\dfrac{5}{-3t} + \dfrac{2}{3t}$.

$$= \dfrac{2 \cdot 1}{2 \cdot 4a} = \dfrac{1}{4a} \qquad \begin{array}{l}\text{Simplifying by removing a factor}\\ \text{equal to 1: } \frac{2}{2} = 1\end{array}$$

YOUR TURN

EXAMPLE 9   Subtract: $\dfrac{5x}{x - 2y} - \dfrac{3y - 7}{2y - x}$.

SOLUTION

$$\dfrac{5x}{x - 2y} - \dfrac{3y - 7}{2y - x} = \dfrac{5x}{x - 2y} - \dfrac{-1}{-1} \cdot \dfrac{3y - 7}{2y - x} \qquad \begin{array}{l}\text{Note that } x - 2y \text{ and}\\ 2y - x \text{ are opposites.}\end{array}$$

$$= \dfrac{5x}{x - 2y} - \dfrac{7 - 3y}{x - 2y} \qquad \begin{array}{l}\text{Performing the multiplication.}\\ \textit{Note: } -1(2y - x) = -2y + x\\ \qquad\qquad\qquad\quad = x - 2y.\end{array}$$

$$= \dfrac{5x - (7 - 3y)}{x - 2y}$$

9. Subtract:

$$\dfrac{x + 2}{2x - 1} - \dfrac{x - 2}{1 - 2x}.$$

$$= \dfrac{5x - 7 + 3y}{x - 2y} \qquad \begin{array}{l}\text{Subtracting. The parentheses}\\ \text{are important.}\end{array}$$

YOUR TURN

In Example 9, you may have noticed that when $3y - 7$ is multiplied by -1 and then subtracted, the result is $-7 + 3y$, which is equivalent to the original $3y - 7$. Sometimes it is convenient to subtract by adding the opposite.

$$\frac{5x}{x - 2y} - \frac{3y - 7}{2y - x} = \frac{5x}{x - 2y} + \left(-\frac{3y - 7}{2y - x}\right) \qquad \text{Rewriting subtraction as addition}$$

$$= \frac{5x}{x - 2y} + \frac{3y - 7}{-(2y - x)} \qquad -\frac{a}{b} = \frac{a}{-b}$$

$$= \frac{5x}{x - 2y} + \frac{3y - 7}{x - 2y} \qquad \begin{array}{l}\text{The opposite of } 2y - x \text{ is}\\ x - 2y.\end{array}$$

$$= \frac{5x + 3y - 7}{x - 2y} \qquad \begin{array}{l}\text{This checks with the}\\ \text{answer to Example 9.}\end{array}$$

EXAMPLE 10   Find simplified form for the function given by

$$f(x) = \frac{2x}{x^2 - 4} + \frac{5}{2 - x} - \frac{1}{2 + x}$$

and list all restrictions on the domain.

SOLUTION   We have

$$\frac{2x}{x^2 - 4} + \frac{5}{2 - x} - \frac{1}{2 + x} = \frac{2x}{(x - 2)(x + 2)} + \frac{5}{2 - x} - \frac{1}{2 + x} \qquad \begin{array}{l}\text{Factoring.}\\ \text{Note that}\\ x \neq -2, 2.\end{array}$$

$$= \frac{2x}{(x - 2)(x + 2)} + \frac{-1}{-1} \cdot \frac{5}{(2 - x)} - \frac{1}{x + 2} \qquad \begin{array}{l}\text{Multiplying by } -1/-1 \text{ since}\\ 2 - x \text{ is the opposite of}\\ x - 2\end{array}$$

$$= \frac{2x}{(x - 2)(x + 2)} + \frac{-5}{x - 2} - \frac{1}{x + 2} \qquad \text{The LCD is } (x - 2)(x + 2).$$

$$= \frac{2x}{(x - 2)(x + 2)} + \frac{-5}{x - 2} \cdot \frac{x + 2}{x + 2} - \frac{1}{x + 2} \cdot \frac{x - 2}{x - 2} \qquad \begin{array}{l}\text{Multiplying by 1 to}\\ \text{get the LCD}\end{array}$$

$$= \frac{2x - 5(x + 2) - (x - 2)}{(x - 2)(x + 2)}$$

$$= \frac{2x - 5x - 10 - x + 2}{(x - 2)(x + 2)}$$

$$= \frac{-4x - 8}{(x - 2)(x + 2)}$$

$$= \frac{-4(x + 2)}{(x - 2)(x + 2)}$$

$$= \frac{-4\cancel{(x + 2)}}{(x - 2)\cancel{(x + 2)}} \qquad \begin{array}{l}\text{Removing a factor equal to 1:}\\ \dfrac{x + 2}{x + 2} = 1, x \neq -2\end{array}$$

$$= \frac{-4}{x - 2}, \text{ or } -\frac{4}{x - 2}, x \neq -2, 2.$$

YOUR TURN

Student Notes

Your answer may differ slightly from the answer found at the back of the book and still be correct. For example, an equivalent answer to Example 10 is $\dfrac{4}{2 - x}$:

$$-\frac{4}{x - 2} = \frac{4}{-(x - 2)} = \frac{4}{2 - x}.$$

Before reworking an exercise, be sure that your answer is indeed different from the correct answer.

10. Find simplified form for the function given by

$$g(x) = \frac{x}{x + 3} - \frac{1}{3 - x} - \frac{6}{x^2 - 9}$$

and list all restrictions on the domain.

Our work in Example 10 indicates that for

$$f(x) = \frac{2x}{x^2 - 4} + \frac{5}{2 - x} - \frac{1}{2 + x} \quad \text{and} \quad g(x) = \frac{-4}{x - 2},$$

with $x \neq -2$ and $x \neq 2$, we have $f = g$. Note that whereas the domain of f includes all real numbers except -2 or 2, the domain of g excludes only 2. This is illustrated in the following graphs. Methods for drawing such graphs by hand are discussed in more advanced courses. The graphs are for visualization only.

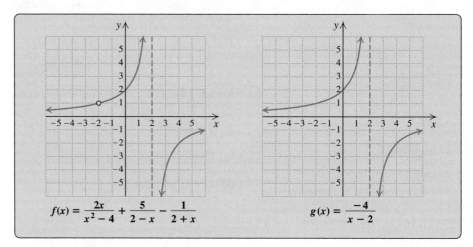

$$f(x) = \frac{2x}{x^2 - 4} + \frac{5}{2 - x} - \frac{1}{2 + x} \qquad g(x) = \frac{-4}{x - 2}$$

A computer-generated visualization of Example 10

A quick, partial check of any simplification is to evaluate both the original expression and the simplified expression for a convenient choice of x. For instance, to check Example 10, if $x = 1$, we have

$$f(1) = \frac{2 \cdot 1}{1^2 - 4} + \frac{5}{2 - 1} - \frac{1}{2 + 1} = \frac{2}{-3} + \frac{5}{1} - \frac{1}{3} = 5 - \frac{3}{3} = 4$$

and

$$g(1) = \frac{-4}{1 - 2} = \frac{-4}{-1} = 4.$$

Since both functions include the pair $(1, 4)$, our algebra was *probably* correct. Although this is only a partial check (on occasion, expressions that are not equivalent may "check" for a particular choice of the variable), because it is so easy to perform, it is quite useful. Further evaluation provides a more definitive check.

6.2 EXERCISE SET   FOR EXTRA HELP

MyMathLab® MathXL
PRACTICE   WATCH   READ   REVIEW

Vocabulary and Reading Check

Classify each of the following statements as either true or false.

1. A common denominator is required in order to add or subtract rational expressions.

2. To find the least common denominator, we find the least common multiple of the denominators.

3. It is never necessary to factor in order to find a common denominator.

4. The sum of two rational expressions is the sum of the numerators over the sum of the denominators.

5. The least common multiple of two expressions is always the product of those two expressions.

6. To add two rational expressions, it is often necessary to multiply at least one of those expressions by a form of 1.

7. After two rational expressions are added, it is unnecessary to simplify the result.

8. Parentheses are particularly important when we are subtracting rational expressions.

When Denominators Are the Same

Perform the indicated operations. Simplify when possible.

9. $\dfrac{4}{3a} + \dfrac{11}{3a}$

10. $\dfrac{2}{5n} + \dfrac{8}{5n}$

11. $\dfrac{5}{3m^2n^2} - \dfrac{4}{3m^2n^2}$

12. $\dfrac{1}{4a^2b} - \dfrac{5}{4a^2b}$

13. $\dfrac{x - 3y}{x + y} + \dfrac{x + 5y}{x + y}$

14. $\dfrac{a - 5b}{a + b} + \dfrac{a + 7b}{a + b}$

15. $\dfrac{3t + 2}{t - 4} - \dfrac{t - 2}{t - 4}$

16. $\dfrac{4y + 2}{y - 2} - \dfrac{y - 3}{y - 2}$

17. $\dfrac{5 - 7x}{x^2 - 3x - 10} + \dfrac{8x - 3}{x^2 - 3x - 10}$

18. $\dfrac{4 - 2x}{x^2 - 9} + \dfrac{3x - 1}{x^2 - 9}$

19. $\dfrac{a - 2}{a^2 - 25} - \dfrac{2a - 7}{a^2 - 25}$

20. $\dfrac{5a - 4}{a^2 - 6a - 7} - \dfrac{6a - 11}{a^2 - 6a - 7}$

Find simplified form for $f(x)$ and list all restrictions on the domain.

21. $f(x) = \dfrac{2x + 1}{x^2 + 6x + 5} + \dfrac{x - 2}{x^2 + 6x + 5}$

22. $f(x) = \dfrac{x - 6}{x^2 - 4x + 3} + \dfrac{5x - 1}{x^2 - 4x + 3}$

23. $f(x) = \dfrac{x - 4}{x^2 - 1} - \dfrac{2x + 1}{x^2 - 1}$

24. $f(x) = \dfrac{3x + 11}{x^2 - 4} - \dfrac{2x - 8}{x^2 - 4}$

When Denominators Are Different

Find the least common multiple of each pair of polynomials.

25. $8x^2,\ 12x^5$

26. $15y,\ 18y^3$

27. $x^2 - 9,\ x^2 - 6x + 9$

28. $x^2 - x - 12,\ x^2 - 16$

Perform the indicated operations. Simplify when possible.

29. $\dfrac{2}{15x^2} + \dfrac{3}{5x}$

30. $\dfrac{8}{9y} - \dfrac{5}{18y^2}$

31. $\dfrac{y + 1}{y - 2} - \dfrac{y - 1}{2y - 4}$

32. $\dfrac{x - 3}{2x + 6} + \dfrac{x + 2}{x + 3}$

33. $\dfrac{4xy}{x^2 - y^2} + \dfrac{x - y}{x + y}$

34. $\dfrac{5ab}{a^2 - b^2} + \dfrac{a + b}{a - b}$

35. $\dfrac{8}{2x^2 - 7x + 5} + \dfrac{3x + 2}{2x^2 - x - 10}$

36. $\dfrac{3y + 2}{y^2 + 5y - 24} + \dfrac{7}{y^2 + 4y - 32}$

37. $\dfrac{5ab}{a^2 - b^2} - \dfrac{a - b}{a + b}$

38. $\dfrac{6xy}{x^2 - y^2} - \dfrac{x + y}{x - y}$

39. $\dfrac{x}{x^2 + 9x + 20} - \dfrac{4}{x^2 + 7x + 12}$

40. $\dfrac{x}{x^2 + 11x + 30} - \dfrac{5}{x^2 + 9x + 20}$

41. $\dfrac{3}{t} - \dfrac{6}{-t}$

42. $\dfrac{8}{p} - \dfrac{7}{-p}$

43. $\dfrac{s^2}{r - s} + \dfrac{r^2}{s - r}$

44. $\dfrac{a^2}{a - b} + \dfrac{b^2}{b - a}$

45. $\dfrac{a + 2}{a - 4} + \dfrac{a - 2}{a + 3}$

46. $\dfrac{a + 3}{a - 5} + \dfrac{a - 2}{a + 4}$

47. $4 + \dfrac{x - 3}{x + 1}$

48. $3 + \dfrac{y + 2}{y - 5}$

49. $\dfrac{x + 6}{5x + 10} - \dfrac{x - 2}{4x + 8}$

50. $\dfrac{a + 3}{5a + 25} - \dfrac{a - 1}{3a + 15}$

51. $\dfrac{4}{x + 1} + \dfrac{x + 2}{x^2 - 1} + \dfrac{3}{x - 1}$

52. $\dfrac{-2}{y+2} + \dfrac{5}{y-2} + \dfrac{y+3}{y^2-4}$

53. $\dfrac{y-4}{y^2-25} - \dfrac{9-2y}{25-y^2}$

54. $\dfrac{x-7}{x^2-16} - \dfrac{x-1}{16-x^2}$

55. $\dfrac{y^2-5}{y^4-81} + \dfrac{4}{81-y^4}$

56. $\dfrac{t^2+3}{t^4-16} + \dfrac{7}{16-t^4}$

57. $\dfrac{r-6s}{r^3-s^3} - \dfrac{5s}{s^3-r^3}$

58. $\dfrac{m-3n}{m^3-n^3} - \dfrac{2n}{n^3-m^3}$

59. $\dfrac{3y}{y^2-7y+10} - \dfrac{2y}{y^2-8y+15}$

60. $\dfrac{n}{3n^2+7n-6} - \dfrac{1}{3n^2-5n+2}$

61. $\dfrac{2x+1}{x-y} + \dfrac{5x^2-5xy}{x^2-2xy+y^2}$

62. $\dfrac{2-3a}{a-b} + \dfrac{3a^2+3ab}{a^2-b^2}$

63. $\dfrac{2y-6}{y^2-9} - \dfrac{y}{y-1} + \dfrac{y^2+2}{y^2+2y-3}$

64. $\dfrac{x-1}{x^2-1} - \dfrac{x}{x-2} + \dfrac{x^2+2}{x^2-x-2}$

Aha! 65. $\dfrac{5y}{1-4y^2} - \dfrac{2y}{2y+1} + \dfrac{5y}{4y^2-1}$

66. $\dfrac{4x}{x^2-1} + \dfrac{3x}{1-x} - \dfrac{4}{x-1}$

Find simplified form for $f(x)$ and list all restrictions on the domain.

67. $f(x) = 2 + \dfrac{x}{x-3} - \dfrac{18}{x^2-9}$

68. $f(x) = 5 + \dfrac{x}{x+2} - \dfrac{8}{x^2-4}$

69. $f(x) = \dfrac{3x-1}{x^2+2x-3} - \dfrac{x+4}{x^2-16}$

70. $f(x) = \dfrac{3x-2}{x^2+2x-24} - \dfrac{x-3}{x^2-9}$

Aha! 71. $f(x) = \dfrac{1}{x^2+5x+6} - \dfrac{2}{x^2+3x+2} - \dfrac{1}{x^2+5x+6}$

72. $f(x) = \dfrac{2}{x^2-5x+6} - \dfrac{4}{x^2-2x-3} + \dfrac{2}{x^2+4x+3}$

73. Badar found that the sum of two rational expressions was $(3-x)/(x-5)$. The answer given at the back of the book is $(x-3)/(5-x)$. Is Badar's answer incorrect? Why or why not?

74. When two rational expressions are added or subtracted, should the numerator of the result be factored? Why or why not?

Skill Review

Solve.

75. $x^2 = 6 + x$ [5.8]

76. $\frac{1}{3} - x \le \frac{2}{5}$ [4.1]

77. $-5 < 2x + 1 < 0$ [4.2]

78. $|x-2| \ge 4$ [4.3]

79. $y = x + 2,$
     $2x = y - 4$ [3.2]

80. $2x - 3y = 5,$
     $x + 2y = 6$ [3.2]

Synthesis

81. Some students always multiply denominators when looking for a common denominator. Use Example 7 to explain why this approach can yield results that are more difficult to simplify.

82. Is the sum of two rational expressions always a rational expression? Why or why not?

83. *Prescription Drugs.*   After visiting her doctor, Jinney went to the pharmacy for a two-week supply of Clarinex, a 20-day supply of Albuterol, and a 30-day supply of Pepcid®. Jinney refills each prescription as soon as her supply runs out. How long will it be until she can refill all three prescriptions on the same day?

84. *Astronomy.*   Earth, Jupiter, Saturn, and Uranus all revolve around the sun. Earth takes 1 year, Jupiter 12 years, Saturn 30 years, and Uranus 84 years. How frequently do these four planets line up with each other?

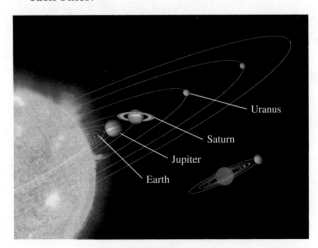

85. *Music.* To duplicate a common African poly-rhythm, a drummer needs to play sextuplets (6 beats per measure) on a tom-tom while simultaneously playing quarter notes (4 beats per measure) on a bass drum. Into how many equally sized parts must a measure be divided, in order to precisely execute this rhythm?

 86. *Research.* Find the average life of three different home appliances, such as electric ranges, room air conditioners, and refrigerators. If you replaced all three appliances this year, in how many years would you once again need to replace all three appliances, assuming that each has an average life?

Find the LCM.

87. $x^8 - x^4$, $x^5 - x^2$, $x^5 - x^3$, $x^5 + x^2$

88. $2a^3 + 2a^2b + 2ab^2$, $a^6 - b^6$, $2b^2 + ab - 3a^2$, $2a^2b + 4ab^2 + 2b^3$

89. The LCM of two expressions is $8a^4b^7$. One of the expressions is $2a^3b^7$. List all the possibilities for the other expression.

90. Determine the domain and the range of the function graphed below.

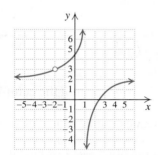

If

$$f(x) = \frac{x^3}{x^2 - 4} \quad and \quad g(x) = \frac{x^2}{x^2 + 3x - 10},$$

find each of the following.

91. $(f + g)(x)$ **92.** $(f - g)(x)$

93. $(f \cdot g)(x)$ **94.** $(f/g)(x)$

Perform the indicated operations and simplify.

95. $x^{-2} + 2x^{-1}$

96. $a^{-3}b - ab^{-3}$

97. $5(x - 3)^{-1} + 4(x + 3)^{-1} - 2(x + 3)^{-2}$

98. $4(y - 1)(2y - 5)^{-1} + 5(2y + 3)(5 - 2y)^{-1} + (y - 4)(2y - 5)^{-1}$

99. $\dfrac{x + 4}{6x^2 - 20x} \cdot \left(\dfrac{x}{x^2 - x - 20} + \dfrac{2}{x + 4} \right)$

100. $\dfrac{x^2 - 7x + 12}{x^2 - x - 29/3} \cdot \left(\dfrac{3x + 2}{x^2 + 5x - 24} + \dfrac{7}{x^2 + 4x - 32} \right)$

101. $\dfrac{8t^5}{2t^2 - 10t + 12} \div \left(\dfrac{2t}{t^2 - 8t + 15} - \dfrac{3t}{t^2 - 7t + 10} \right)$

102. $\dfrac{9t^3}{3t^3 - 12t^2 + 9t} \div \left(\dfrac{t + 4}{t^2 - 9} - \dfrac{3t - 1}{t^2 + 2t - 3} \right)$

103. Use a graphing calculator to check your answers to Exercises 23, 53, and 63.

104. Let

$$f(x) = 2 + \frac{x - 3}{x + 1}.$$

Use algebra, together with a graphing calculator, to determine the domain and the range of f.

YOUR TURN ANSWERS: SECTION 6.2

1. $\dfrac{a + 5}{a + 1}$ **2.** $\dfrac{2n + 3}{n}$ **3.** $g(x) = \dfrac{2(2x + 1)}{x - 1}, x \neq 1$

4. $(t + 5)^2(2t - 3)$ **5.** $\dfrac{8n^2 + 5}{30n^3}$ **6.** $\dfrac{a^2 + 5a + 5}{(a - 5)(a + 1)}$

7. $\dfrac{3x^2 + 2x + 1}{(x - 3)(x + 1)(2x + 1)}$ **8.** $-\dfrac{1}{t}$ **9.** $\dfrac{2x}{2x - 1}$

10. $g(x) = \dfrac{x + 1}{x + 3}, x \neq -3, 3$

QUICK QUIZ: SECTIONS 6.1–6.2

1. Find $f(-2)$ for $f(x) = \dfrac{x^2 - 5}{x + 3}$.   [6.1]

Perform the indicated operation and, if possible, simplify.

2. $\dfrac{a^2 - 3a - 4}{a^2 - 4} \cdot \dfrac{a^2 - 1}{3a - 12}$ [6.1]

3. $\dfrac{12x^2}{5y^4} \div \dfrac{2x^5}{15y}$ [6.1]

4. $\dfrac{1}{2x^2 + 3x + 1} + \dfrac{3}{2x^2 - x - 1}$ [6.2]

5. $\dfrac{a}{a - 4} - \dfrac{3}{4 - a}$ [6.2]

PREPARE TO MOVE ON

Simplify. Use only positive exponents in your answer.
[1.6]

1. $2x^{-1}$ **2.** $ab(a + b)^{-2}$

Multiply and simplify.   [1.2], [1.6]

3. $9x^3 \left(\dfrac{1}{x^2} - \dfrac{2}{3x^3} \right)$

4. $8a^2b^5 \left(\dfrac{3}{8ab^2} + \dfrac{a}{4b^5} \right)$

6.3   Complex Rational Expressions

Multiplying by 1   ▪   Dividing Two Rational Expressions

A **complex rational expression** is a rational expression that contains rational expressions within its numerator and/or its denominator. Here are some examples:

$$\dfrac{x + \dfrac{5}{x}}{4x}, \quad \dfrac{\dfrac{x - y}{x + y}}{\dfrac{2x - y}{3x + y}}, \quad \dfrac{\dfrac{7x}{3} - \dfrac{4}{x}}{\dfrac{5x}{6} + \dfrac{8}{3}}.$$

The rational expressions within each complex rational expression are red.

When we simplify a complex rational expression, we rewrite it so that it is no longer complex. We will consider two methods for simplifying complex rational expressions. Determining restrictions on variables may now require the solution of equations not yet studied. *Thus for this section we will not state restrictions on variables.*

Method 1: Multiplying by 1

Study Skills

Multiple Methods

When more than one method is presented, as is the case for simplifying complex rational expressions, it can be helpful to learn both methods. If two different approaches yield the same result, you can be confident that your answer is correct.

One method of simplifying a complex rational expression is to multiply the entire expression by 1. To write 1, we use the LCD of the rational expressions within the complex rational expression.

USING MULTIPLICATION BY 1 TO SIMPLIFY A COMPLEX RATIONAL EXPRESSION

1. Find the LCD of all rational expressions *within* the complex rational expression.
2. Multiply the complex rational expression by an expression equal to 1. Write 1 as the LCD divided by itself (LCD/LCD).
3. Distribute and simplify so that the numerator and the denominator of the complex rational expression are polynomials.
4. Factor and, if possible, simplify.

EXAMPLE 1   Simplify: $\dfrac{\dfrac{1}{2} + \dfrac{3}{4}}{\dfrac{5}{6} - \dfrac{3}{8}}.$

SOLUTION

1. The LCD of $\frac{1}{2}, \frac{3}{4}, \frac{5}{6}$, and $\frac{3}{8}$ is 24.

2. We multiply by an expression equal to 1:

$$\dfrac{\dfrac{1}{2} + \dfrac{3}{4}}{\dfrac{5}{6} - \dfrac{3}{8}} = \dfrac{\dfrac{1}{2} + \dfrac{3}{4}}{\dfrac{5}{6} - \dfrac{3}{8}} \cdot \dfrac{24}{24}.$$

Multiplying by an expression equal to 1, using the LCD: $\dfrac{24}{24} = 1$

3. Using the distributive law, we perform the multiplication:

$$\frac{\frac{1}{2}+\frac{3}{4}}{\frac{5}{6}-\frac{3}{8}}\cdot\frac{24}{24}=\frac{\left(\frac{1}{2}+\frac{3}{4}\right)24}{\left(\frac{5}{6}-\frac{3}{8}\right)24}$$ ← Multiplying the numerator by 24
Don't forget the parentheses!
← Multiplying the denominator by 24

$$=\frac{\frac{1}{2}(24)+\frac{3}{4}(24)}{\frac{5}{6}(24)-\frac{3}{8}(24)}$$ Using the distributive law

1. Simplify: $\dfrac{\frac{1}{3}+\frac{1}{6}}{\frac{1}{2}-\frac{2}{3}}$.

$$=\frac{12+18}{20-9},\text{ or }\frac{30}{11}.$$ Simplifying

4. The result, $\frac{30}{11}$, cannot be factored or simplified, so we are done.

YOUR TURN

Note that after multiplying by the form of 1 and simplifying, the expression is no longer complex.

EXAMPLE 2 Simplify:

$$\frac{\frac{1}{a^3b}+\frac{1}{b}}{\frac{1}{a^2b^2}-\frac{1}{b^2}}.$$

Find the LCD.

SOLUTION The denominators within the complex rational expression are a^3b, b, a^2b^2, and b^2. Thus the LCD is a^3b^2. We multiply by 1, using $(a^3b^2)/(a^3b^2)$:

Using the LCD, multiply by 1.

$$\frac{\frac{1}{a^3b}+\frac{1}{b}}{\frac{1}{a^2b^2}-\frac{1}{b^2}}=\frac{\frac{1}{a^3b}+\frac{1}{b}}{\frac{1}{a^2b^2}-\frac{1}{b^2}}\cdot\frac{a^3b^2}{a^3b^2}$$ Multiplying by 1, using the LCD

$$=\frac{\left(\frac{1}{a^3b}+\frac{1}{b}\right)a^3b^2}{\left(\frac{1}{a^2b^2}-\frac{1}{b^2}\right)a^3b^2}$$ Multiplying in the numerator and in the denominator. Remember to use parentheses.

Distribute and simplify.

$$=\frac{\frac{1}{a^3b}\cdot a^3b^2+\frac{1}{b}\cdot a^3b^2}{\frac{1}{a^2b^2}\cdot a^3b^2-\frac{1}{b^2}\cdot a^3b^2}$$ Using the distributive law to carry out the multiplications

$$=\frac{\frac{a^3b}{a^3b}\cdot b+\frac{b}{b}\cdot a^3b}{\frac{a^2b^2}{a^2b^2}\cdot a-\frac{b^2}{b^2}\cdot a^3}$$ Removing factors that equal 1. Study this carefully.

$$=\frac{b+a^3b}{a-a^3}$$ Simplifying

$$= \frac{b(1 + a^3)}{a(1 - a^2)} \qquad \text{Factoring}$$

Factor and simplify.

$$= \cdot\frac{b(1 + a)(1 - a + a^2)}{a(1 + a)(1 - a)} \qquad \text{Factoring further and identifying a factor that equals 1}$$

2. Simplify: $\dfrac{\dfrac{x^2}{y} - \dfrac{5}{x}}{xz}$.

$$= \frac{b(1 - a + a^2)}{a(1 - a)}. \qquad \text{Simplifying}$$

↪ YOUR TURN

EXAMPLE 3   Simplify:

$$\frac{\dfrac{3}{2x - 2} - \dfrac{1}{x + 1}}{\dfrac{1}{x - 1} + \dfrac{x}{x^2 - 1}}.$$

SOLUTION   In this case, to find the LCD, we must first factor:

$$\frac{\dfrac{3}{2x - 2} - \dfrac{1}{x + 1}}{\dfrac{1}{x - 1} + \dfrac{x}{x^2 - 1}} = \frac{\dfrac{3}{2(x - 1)} - \dfrac{1}{x + 1}}{\dfrac{1}{x - 1} + \dfrac{x}{(x - 1)(x + 1)}} \qquad \text{The LCD is } 2(x - 1)(x + 1).$$

Student Notes

Writing $2(x - 1)(x + 1)$ as

$$\frac{2(x - 1)(x + 1)}{1}$$

may help with multiplying.

$$= \frac{\dfrac{3}{2(x - 1)} - \dfrac{1}{x + 1}}{\dfrac{1}{x - 1} + \dfrac{x}{(x - 1)(x + 1)}} \cdot \frac{2(x - 1)(x + 1)}{2(x - 1)(x + 1)} \qquad \text{Multiplying by 1, using the LCD}$$

$$= \frac{\dfrac{3}{2(x - 1)} \cdot 2(x - 1)(x + 1) - \dfrac{1}{x + 1} \cdot 2(x - 1)(x + 1)}{\dfrac{1}{x - 1} \cdot 2(x - 1)(x + 1) + \dfrac{x}{(x - 1)(x + 1)} \cdot 2(x - 1)(x + 1)} \qquad \text{Using the distributive law}$$

$$= \frac{\dfrac{2(x - 1)}{2(x - 1)} \cdot 3(x + 1) - \dfrac{x + 1}{x + 1} \cdot 2(x - 1)}{\dfrac{x - 1}{x - 1} \cdot 2(x + 1) + \dfrac{(x - 1)(x + 1)}{(x - 1)(x + 1)} \cdot 2x} \qquad \text{Removing factors that equal 1}$$

$$= \frac{3(x + 1) - 2(x - 1)}{2(x + 1) + 2x} \qquad \text{Simplifying}$$

3. Simplify:

$$\frac{\dfrac{1}{x - 2} + \dfrac{2}{x + 1}}{\dfrac{2}{3x + 3} - \dfrac{1}{x - 2}}.$$

$$= \frac{3x + 3 - 2x + 2}{2x + 2 + 2x} \qquad \text{Using the distributive law}$$

$$= \frac{x + 5}{4x + 2} \qquad \text{Combining like terms}$$

$$= \frac{x + 5}{2(2x + 1)}. \qquad \text{We factor, but it is not possible to simplify further.}$$

↪ YOUR TURN

Method 2: Dividing Two Rational Expressions

Another method for simplifying complex rational expressions involves rewriting the expression as a quotient of two rational expressions.

EXAMPLE 4   Simplify: $\dfrac{\dfrac{x}{x-3}}{\dfrac{4}{5x-15}}$.

SOLUTION   The numerator and the denominator are single rational expressions. We divide the numerator by the denominator:

$$\frac{\dfrac{x}{x-3}}{\dfrac{4}{5x-15}} = \frac{x}{x-3} \div \frac{4}{5x-15} \qquad \text{Rewriting with a division symbol}$$

$$= \frac{x}{x-3} \cdot \frac{5x-15}{4} \qquad \text{Multiplying by the reciprocal of the divisor (inverting and multiplying)}$$

$$= \frac{x}{x-3} \cdot \frac{5(x-3)}{4} \qquad \text{Factoring and removing a factor equal to 1: } \frac{x-3}{x-3} = 1$$

$$= \frac{5x}{4}.$$

4. Simplify:

$$\frac{\dfrac{3}{m-n}}{\dfrac{5}{m+n}}.$$

↩ YOUR TURN

We can use this method even when the numerator and the denominator are not (yet) written as single rational expressions. The key is to express a complex rational expression as one rational expression divided by another.

> **USING DIVISION TO SIMPLIFY A COMPLEX RATIONAL EXPRESSION**
>
> 1. Add or subtract, as necessary, to get one rational expression in the numerator.
> 2. Add or subtract, as necessary, to get one rational expression in the denominator.
> 3. Divide the numerator by the denominator (invert the divisor and multiply).
> 4. Simplify, if possible, by removing any factors that equal 1.

EXAMPLE 5   Simplify:

$$\frac{1 + \dfrac{2}{x}}{1 - \dfrac{4}{x^2}}.$$

Technology Connection

To check Example 5, we can show that the graphs of

$$y_1 = \dfrac{1 + \dfrac{2}{x}}{1 - \dfrac{4}{x^2}}$$

and

$$y_2 = \dfrac{x}{x - 2}$$

coincide, or we can show that (except for $x = -2$ or 0) their tables of values are identical. We can also check by showing that (except for $x = -2$ or 0 or 2) $y_2 - y_1 = 0$.

1. Use a graphing calculator to check Example 3. What values, if any, can x not equal?

5. Simplify:

$$\dfrac{y + \dfrac{1}{y}}{2 - \dfrac{1}{y^2}}.$$

SOLUTION We have

$$\dfrac{1 + \dfrac{2}{x}}{1 - \dfrac{4}{x^2}} \quad \left.\dfrac{\dfrac{x}{x} + \dfrac{2}{x}}{\dfrac{x^2}{x^2} - \dfrac{4}{x^2}}\right\} \quad \begin{array}{l}\text{Finding a common denominator}\\[1em]\text{Finding a common denominator}\end{array}$$

$$= \dfrac{\dfrac{x + 2}{x}}{\dfrac{x^2 - 4}{x^2}} \quad \left.\begin{array}{l}\\\end{array}\right\} \quad \begin{array}{l}\text{Adding in the numerator}\\[1em]\text{Subtracting in the denominator}\end{array}$$

$$= \dfrac{x + 2}{x} \cdot \dfrac{x^2}{x^2 - 4} \quad \begin{array}{l}\text{Multiplying by the reciprocal of}\\\text{the divisor}\end{array}$$

$$= \dfrac{(x + 2) \cdot x^2}{x(x + 2)(x - 2)} \quad \begin{array}{l}\text{Factoring. Remember to simplify}\\\text{when possible.}\end{array}$$

$$= \dfrac{\cancel{(x + 2)}x \cdot x}{\cancel{x(x + 2)}(x - 2)} \quad \begin{array}{l}\text{Removing a factor equal to 1:}\\\dfrac{(x + 2)x}{(x + 2)x} = 1\end{array}$$

$$= \dfrac{x}{x - 2}. \quad \text{Simplifying}$$

As a quick, partial check, we select a convenient value for x—say, 1—and evaluate both the original expression and the simplified expression.

Evaluating the Original Expression for $x = 1$

$$\dfrac{1 + \dfrac{2}{1}}{1 - \dfrac{4}{1^2}} = \dfrac{1 + 2}{1 - 4} = \dfrac{3}{-3} = -1$$

Evaluating the Simplified Expression for $x = 1$

$$\dfrac{1}{1 - 2} = \dfrac{1}{-1} = -1$$

The value of both expressions is -1, so the simplification is probably correct. Evaluating the expression for more values of x would make the check more certain.

▶ YOUR TURN

If negative exponents occur, we first find an equivalent expression using positive exponents.

EXAMPLE 6 Simplify:

$$\dfrac{a^{-1} + b^{-1}}{a^{-3} + b^{-3}}.$$

SOLUTION

$$\dfrac{a^{-1} + b^{-1}}{a^{-3} + b^{-3}} = \dfrac{\dfrac{1}{a} + \dfrac{1}{b}}{\dfrac{1}{a^3} + \dfrac{1}{b^3}} \quad \begin{array}{l}\text{Rewriting with positive}\\\text{exponents}\end{array}$$

$$= \dfrac{\dfrac{1}{a} \cdot \dfrac{b}{b} + \dfrac{1}{b} \cdot \dfrac{a}{a}}{\dfrac{1}{a^3} \cdot \dfrac{b^3}{b^3} + \dfrac{1}{b^3} \cdot \dfrac{a^3}{a^3}} \quad \left.\begin{array}{l}\\[1em]\\\end{array}\right\} \quad \begin{array}{l}\text{Finding a common}\\\text{denominator}\\[1em]\text{Finding a common}\\\text{denominator}\end{array}$$

$$= \dfrac{\dfrac{b}{ab} + \dfrac{a}{ab}}{\dfrac{b^3}{a^3b^3} + \dfrac{a^3}{a^3b^3}}$$

Add or subtract in the numerator.

$$= \left.\frac{\dfrac{b + a}{ab}}{\dfrac{b^3 + a^3}{a^3 b^3}}\right\} \begin{array}{l}\text{Adding in the}\\\text{numerator}\end{array}$$

Add or subtract in the denominator.

Adding in the denominator

Invert the divisor and multiply.

$$= \frac{b + a}{ab} \cdot \frac{a^3 b^3}{b^3 + a^3}$$

Multiplying by the reciprocal of the divisor

$$= \frac{(b + a) \cdot ab \cdot a^2 b^2}{ab(b + a)(b^2 - ab + a^2)}$$

Factoring and looking for common factors

Factor and simplify.

$$= \frac{\cancel{(b + a)} \cdot \cancel{ab} \cdot a^2 b^2}{\cancel{ab}\,\cancel{(b + a)}(b^2 - ab + a^2)}$$

Removing a factor equal to 1: $\dfrac{(b + a)ab}{ab(b + a)} = 1$

6. Simplify:

$$\frac{3x^{-1} + xy^{-1}}{y^{-1} - 2x^{-1}}.$$

$$= \frac{a^2 b^2}{b^2 - ab + a^2}$$

 YOUR TURN

There is no one method that is always best to use. When it is little or no work to write an expression as a quotient of two rational expressions, the second method is probably easier. On the other hand, some expressions require fewer steps if we use the first method. Either method can be used with any complex rational expression.

6.3 EXERCISE SET

FOR EXTRA HELP

MyMathLab® Math XL
PRACTICE   WATCH   READ   REVIEW

Vocabulary and Reading Check

In each of Exercises 1–6, choose the expression(s) below that best matches the description with reference to the complex rational expression

$$\frac{\dfrac{x - 6}{x^2} + \dfrac{2}{5x}}{\dfrac{x}{x + 1} - \dfrac{x}{x - 1}}.$$

a) $\dfrac{x}{x + 1} - \dfrac{x}{x - 1}$ **b)** $\dfrac{x - 6}{x^2}, \dfrac{2}{5x}, \dfrac{x}{x + 1}, \dfrac{x}{x - 1}$

c) $5x^2$ **d)** $(x + 1)(x - 1)$

e) $5x^2(x + 1)(x - 1)$ **f)** $x^2, 5x, x + 1, x - 1$

1. ____ The rational expressions within the complex rational expression

2. ____ The denominator of the complex rational expression

3. ____ The denominators within the complex rational expression

4. ____ The LCD of the rational expressions in the numerator

5. ____ The LCD of the rational expressions in the denominator

6. ____ The LCD of all the rational expressions within the complex rational expression

Simplifying Complex Rational Expressions

Simplify. If possible, use a second method or evaluation as a check.

7. $\dfrac{\dfrac{1}{2} + \dfrac{1}{3}}{\dfrac{1}{4} - \dfrac{1}{6}}$

8. $\dfrac{\dfrac{2}{5} - \dfrac{1}{10}}{\dfrac{7}{20} - \dfrac{4}{15}}$

9. $\dfrac{1 + \dfrac{1}{4}}{2 + \dfrac{3}{4}}$

10. $\dfrac{3 + \dfrac{1}{4}}{1 + \dfrac{1}{2}}$

11. $\dfrac{\dfrac{x}{4} + x}{\dfrac{4}{x} + x}$

12. $\dfrac{\dfrac{1}{c} + 2}{\dfrac{1}{c} - 5}$

13. $\dfrac{\dfrac{x + 5}{x - 3}}{\dfrac{x - 2}{x + 1}}$

14. $\dfrac{\dfrac{x - 3}{x + 4}}{\dfrac{x + 6}{x - 1}}$

15. $\dfrac{\dfrac{3}{x} + \dfrac{2}{x^3}}{\dfrac{5}{x} - \dfrac{3}{x^2}}$

16. $\dfrac{\dfrac{5}{y^2} + \dfrac{3}{y^4}}{\dfrac{3}{y} - \dfrac{2}{y^3}}$

37. $\dfrac{\dfrac{3}{ab^4} + \dfrac{4}{a^3 b}}{ab}$

38. $\dfrac{\dfrac{2}{x^2 y} + \dfrac{3}{xy^2}}{xy}$

17. $\dfrac{\dfrac{6}{r} - \dfrac{1}{s}}{\dfrac{2}{r} + \dfrac{3}{s}}$

18. $\dfrac{\dfrac{9}{a} - \dfrac{5}{b}}{\dfrac{4}{a} + \dfrac{1}{b}}$

39. $\dfrac{x - y}{x^{-3} - y^{-3}}$

40. $\dfrac{a^{-1} + b^{-1}}{a^{-3} + b^{-3}}$

19. $\dfrac{\dfrac{3}{z^2} + \dfrac{2}{yz}}{\dfrac{4}{zy^2} - \dfrac{1}{y}}$

20. $\dfrac{\dfrac{6}{x^3} + \dfrac{7}{y}}{\dfrac{7}{xy^2} - \dfrac{6}{x^2 y^2}}$

41. $\dfrac{\dfrac{1}{x - 2} + \dfrac{3}{x - 1}}{\dfrac{2}{x - 1} + \dfrac{5}{x - 2}}$

42. $\dfrac{\dfrac{2}{y - 3} + \dfrac{1}{y + 1}}{\dfrac{3}{y + 1} + \dfrac{4}{y - 3}}$

21. $\dfrac{\dfrac{a^2 - b^2}{ab}}{\dfrac{a - b}{b}}$

22. $\dfrac{\dfrac{x^2 - y^2}{xy}}{\dfrac{x + y}{y}}$

43. $\dfrac{a(a + 3)^{-1} - 2(a - 1)^{-1}}{a(a + 3)^{-1} - (a - 1)^{-1}}$

44. $\dfrac{a(a + 2)^{-1} - 3(a - 3)^{-1}}{a(a + 2)^{-1} - (a - 3)^{-1}}$

23. $\dfrac{1 - \dfrac{2}{3x}}{x - \dfrac{4}{9x}}$

24. $\dfrac{\dfrac{3x}{y} - x}{2y - \dfrac{y}{x}}$

45. $\dfrac{\dfrac{2}{a^2 - 1} + \dfrac{1}{a + 1}}{\dfrac{3}{a^2 - 1} + \dfrac{2}{a - 1}}$

46. $\dfrac{\dfrac{3}{a^2 - 9} + \dfrac{2}{a + 3}}{\dfrac{4}{a^2 - 9} + \dfrac{1}{a + 3}}$

25. $\dfrac{y^{-1} - x^{-1}}{\dfrac{x^2 - y^2}{xy}}$

26. $\dfrac{a^{-1} + b^{-1}}{\dfrac{a^2 - b^2}{ab}}$

47. $\dfrac{\dfrac{5}{x^2 - 4} - \dfrac{3}{x - 2}}{\dfrac{4}{x^2 - 4} - \dfrac{2}{x + 2}}$

48. $\dfrac{\dfrac{4}{x^2 - 1} - \dfrac{3}{x + 1}}{\dfrac{5}{x^2 - 1} - \dfrac{2}{x - 1}}$

27. $\dfrac{\dfrac{1}{x + h} - \dfrac{1}{x}}{h}$

28. $\dfrac{\dfrac{1}{a - h} - \dfrac{1}{a}}{h}$

49. $\dfrac{\dfrac{y^3}{y^2 - 4} + \dfrac{125}{4 - y^2}}{\dfrac{y}{y^2 - 4} + \dfrac{5}{4 - y^2}}$

50. $\dfrac{\dfrac{y}{y^2 - 1} - \dfrac{3}{1 - y^2}}{\dfrac{y^3}{y^2 - 1} - \dfrac{27}{1 - y^2}}$

29. $\dfrac{\dfrac{a^2 - 4}{a^2 + 3a + 2}}{\dfrac{a^2 - 5a - 6}{a^2 - 6a - 7}}$

30. $\dfrac{\dfrac{x^2 - x - 12}{x^2 - 2x - 15}}{\dfrac{x^2 + 8x + 12}{x^2 - 5x - 14}}$

51. $\dfrac{\dfrac{y^2}{y^2 - 25} - \dfrac{y}{y - 5}}{\dfrac{y}{y^2 - 25} - \dfrac{1}{y + 5}}$

52. $\dfrac{\dfrac{y^2}{y^2 - 9} - \dfrac{y}{y + 3}}{\dfrac{y}{y^2 - 9} - \dfrac{1}{y - 3}}$

31. $\dfrac{\dfrac{x}{x^2 + 3x - 4} - \dfrac{1}{x^2 + 3x - 4}}{\dfrac{x}{x^2 + 6x + 8} + \dfrac{3}{x^2 + 6x + 8}}$

32. $\dfrac{\dfrac{x}{x^2 + 5x - 6} + \dfrac{6}{x^2 + 5x - 6}}{\dfrac{x}{x^2 - 5x + 4} - \dfrac{2}{x^2 - 5x + 4}}$

53. $\dfrac{\dfrac{a}{a + 2} + \dfrac{5}{a}}{\dfrac{a}{2a + 4} + \dfrac{1}{3a}}$

54. $\dfrac{\dfrac{a}{a + 3} + \dfrac{4}{5a}}{\dfrac{a}{2a + 6} + \dfrac{3}{a}}$

33. $\dfrac{\dfrac{1}{y} + 2}{\dfrac{1}{y} - 3}$

34. $\dfrac{7 + \dfrac{1}{a}}{\dfrac{1}{a} - 3}$

55. $\dfrac{\dfrac{1}{x^2 - 3x + 2} + \dfrac{1}{x^2 - 4}}{\dfrac{1}{x^2 + 4x + 4} + \dfrac{1}{x^2 - 4}}$

35. $\dfrac{y + y^{-2}}{y - y^{-2}}$

36. $\dfrac{x - x^{-2}}{x + x^{-2}}$

56. $\dfrac{\dfrac{1}{x^2 + 3x + 2} + \dfrac{1}{x^2 - 1}}{\dfrac{1}{x^2 - 1} + \dfrac{1}{x^2 - 4x + 3}}$

57. $\dfrac{\dfrac{3}{a^2 - 4a + 3} + \dfrac{3}{a^2 - 5a + 6}}{\dfrac{3}{a^2 - 3a + 2} + \dfrac{3}{a^2 + 3a - 10}}$

58. $\dfrac{\dfrac{1}{a^2 + 7a + 12} + \dfrac{1}{a^2 + a - 6}}{\dfrac{1}{a^2 + 2a - 8} + \dfrac{1}{a^2 + 5a + 4}}$

Aha! **59.** $\dfrac{\dfrac{y}{y^2 - 4} - \dfrac{2y}{y^2 + y - 6}}{\dfrac{2y}{y^2 + y - 6} - \dfrac{y}{y^2 - 4}}$

60. $\dfrac{\dfrac{y}{y^2 - 1} - \dfrac{3y}{y^2 + 5y + 4}}{\dfrac{3y}{y^2 - 1} - \dfrac{y}{y^2 - 4y + 3}}$

61. $\dfrac{t + 5 + \dfrac{3}{t}}{t + 2 + \dfrac{1}{t}}$

62. $\dfrac{a + 3 + \dfrac{2}{a}}{a + 2 + \dfrac{5}{a}}$

63. Parker *incorrectly* simplifies
$$\frac{a + b^{-1}}{a + c^{-1}} \quad as \quad \frac{a + c}{a + b}.$$
What mistake is he probably making and how could you convince him that this is incorrect?

64. Write your own complex rational expression that is most easily simplified using method 1 and a second expression that is most easily simplified using method 2. Explain how you created each.

Skill Review

Perform the indicated operations and, if possible, simplify.

65. $(2x^2 - x - 7) - (x^3 - x + 16)$ [5.1]

66. $(3x^2 + y)(3x^2 - y)$ [5.2]

67. $(m - 6)^2$ [5.2]

68. Determine the degree of $4 - y^3 + 5y + 8y$.   [5.1]

Synthesis

69. In arithmetic, we are taught that
$$\frac{a}{b} \div \frac{c}{d} = \frac{a}{b} \cdot \frac{d}{c}.$$
(To divide by a fraction, we invert the divisor and multiply.) Use method 1 to explain *why* this is the correct approach.

70. An LCD is used in both method 1 and method 2. Explain how the use of the LCD differs in these methods.

Simplify.

71. $\dfrac{5x^{-2} + 10x^{-1}y^{-1} + 5y^{-2}}{3x^{-2} - 3y^{-2}}$

72. $1 + \dfrac{1}{1 + \dfrac{1}{1 + \dfrac{1}{x}}}$

73. Are the following expressions equivalent? Justify your answer.
$$\dfrac{\dfrac{a}{b}}{c} \quad and \quad \dfrac{a}{\dfrac{b}{c}}$$

74. The formula
$$\dfrac{P\left(1 + \dfrac{i}{12}\right)^2}{\dfrac{\left(1 + \dfrac{i}{12}\right)^2 - 1}{\dfrac{i}{12}}},$$
where P is a loan amount and i is an interest rate, arises in certain business situations. Simplify this expression. (*Hint*: Expand the binomials.)

75. *Financial Planning.* Michael wishes to invest a portion of each month's pay in an account that pays 7.5% interest. If he wants to have $30,000 in the account after 10 years, the amount invested each month is given by
$$\dfrac{30,000 \cdot \dfrac{0.075}{12}}{\left(1 + \dfrac{0.075}{12}\right)^{120} - 1}.$$
Find the amount of Michael's monthly investment.

76. *Astronomy.* When two galaxies are moving in opposite directions at velocities v_1 and v_2, an observer in one of the galaxies would see the other galaxy receding at speed

$$\frac{v_1 + v_2}{1 + \dfrac{v_1 v_2}{c^2}},$$

where c is the speed of light. Determine the observed speed if v_1 and v_2 are both one-fourth the speed of light.

77. Find simplified form for the reciprocal of

$$x^2 + x + 1 + \frac{1}{x} + \frac{1}{x^2}.$$

Find and simplify

$$\frac{f(x + h) - f(x)}{h}$$

for each rational function f in Exercises 78 and 79.

78. $f(x) = \dfrac{x}{1 - x}$　　　**79.** $f(x) = \dfrac{3}{x}$

80. If

$$F(x) = \frac{3 + \dfrac{1}{x}}{2 - \dfrac{8}{x^2}},$$

find the domain of F.

81. For $f(x) = \dfrac{2}{2 + x}$, find $f(f(a))$.

82. For $g(x) = \dfrac{x + 3}{x - 1}$, find $g(g(a))$.

83. Let

$$f(x) = \left[\frac{\dfrac{x + 3}{x - 3} + 1}{\dfrac{x + 3}{x - 3} - 1} \right]^4 .$$

Find a simplified form of $f(x)$ and specify the domain of f.

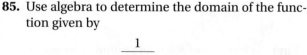
84. Use a graphing calculator to check your answers to Exercises 33, 41, and 53.

85. Use algebra to determine the domain of the function given by

$$f(x) = \frac{\dfrac{1}{x - 2}}{\dfrac{x}{x - 2} - \dfrac{5}{x - 2}}.$$

Then explain how a graphing calculator could be used to check your answer.

YOUR TURN ANSWERS: SECTION 6.3

1. -3　**2.** $\dfrac{x^3 - 5y}{x^2 yz}$　**3.** $\dfrac{-9(x - 1)}{x + 7}$　**4.** $\dfrac{3(m + n)}{5(m - n)}$

5. $\dfrac{y(y^2 + 1)}{2y^2 - 1}$　**6.** $\dfrac{x^2 + 3y}{x - 2y}$

QUICK QUIZ: SECTIONS 6.1–6.3

1. Write simplified form for $f(x) = \dfrac{x^2 - 1}{x^2 - x - 2}$ and list all restrictions on the domain. [6.1]

Perform the indicated operation and, if possible, simplify.

2. $\dfrac{6a^3 - 6}{5a^4} \cdot \dfrac{10a^2}{9a - 9}$　[6.1]

3. $\dfrac{t^2 - t - 12}{t^2 + 5t + 4} \div \dfrac{t^2 + 6t + 9}{t^2 - 1}$　[6.1]

4. $\dfrac{n - 10}{n^2 - 2n - 8} + \dfrac{1}{n - 4}$　[6.2]

5. Simplify: $\dfrac{\dfrac{5}{a + b}}{\dfrac{3}{a - b}}$. [6.3]

PREPARE TO MOVE ON

Solve.

1. $2(y + 3) - 5(y - 1) = 10y$　[1.3]

2. $x^2 = 25$　[5.8]

3. $a^2 + 8 = 6a$　[5.8]

4. $\dfrac{1}{3}x - \dfrac{1}{4} = \dfrac{1}{6} - \dfrac{1}{2}x$　[1.3]

6.4 Rational Equations

Solving Rational Equations

In this section, we learn to solve a new type of equation. A **rational equation** is an equation that contains one or more rational expressions. Here are some examples:

$$\frac{2}{3} - \frac{5}{6} = \frac{1}{t}, \qquad \frac{a-1}{a-5} = \frac{4}{a^2-25}, \qquad x^3 + \frac{6}{x} = 5.$$

Solving Rational Equations

Recall that we can *clear fractions* from an equation by multiplying both sides of the equation by the LCM of the denominators.

Most of the rational equations that we will encounter contain a variable in at least one denominator. Since division by 0 is undefined, any replacement for the variable that makes a denominator 0 cannot be a solution of the equation. We can rule out these numbers before we even attempt to find a solution. After we have solved the equation, we must check that no possible solution makes a denominator 0.

TO SOLVE A RATIONAL EQUATION

1. List any numbers that will make a denominator 0. State that the variable *cannot* equal these numbers.
2. Clear fractions by multiplying both sides of the equation by the LCM of the denominators.
3. Solve the equation.
4. Check possible solutions against the list of numbers that cannot be solutions and in the original, rational equation.

EXAMPLE 1 Solve: $\dfrac{2}{3x} + \dfrac{1}{x} = 10$.

SOLUTION

1. Because the left side of this equation is undefined when x is 0, we state at the outset that $x \neq 0$.

2. We multiply both sides of the equation by the LCM, $3x$:

$$3x\left(\frac{2}{3x} + \frac{1}{x}\right) = 3x(10) \qquad \text{Multiplying by the LCM to clear fractions}$$

$$3x \cdot \frac{2}{3x} + 3x \cdot \frac{1}{x} = 3x(10) \qquad \text{Using the distributive law}$$

$$\frac{3 \cdot x \cdot 2}{3 \cdot x} + \frac{3 \cdot x}{x} = 30x \qquad \text{Multiplying and factoring}$$

$$2 + 3 = 30x. \qquad \text{Removing factors equal to 1: } \frac{3x}{3x} = 1; \frac{x}{x} = 1$$

We have solved equations like this before.

Student Notes

Not all checking is for finding errors in computation. For these equations, the solution process itself can introduce numbers that do not check.

3. We solve the equation:

$$2 + 3 = 30x$$
$$5 = 30x$$
$$\tfrac{1}{6} = x.$$

4. We stated that $x \neq 0$, so $\tfrac{1}{6}$ should check.

Check:

$$\frac{2}{3x} + \frac{1}{x} = 10$$

$$\begin{array}{c|c} \dfrac{2}{3 \cdot \frac{1}{6}} + \dfrac{1}{\frac{1}{6}} & 10 \\[2ex] \dfrac{2}{\frac{1}{2}} + \dfrac{1}{\frac{1}{6}} & \\[2ex] 2 \cdot \frac{2}{1} + 1 \cdot \frac{6}{1} & \\[1ex] 4 + 6 & \\[1ex] 10 \stackrel{?}{=} 10 & \text{TRUE} \end{array}$$

The solution is $\tfrac{1}{6}$.

1. Solve: $\dfrac{5}{t} = \dfrac{4}{3}$.

🢒 YOUR TURN

EXAMPLE 2   Solve: $1 + \dfrac{3x}{x+2} = \dfrac{-6}{x+2}$.

SOLUTION   If $x = -2$, the rational expressions are undefined, so $x \neq -2$. We clear fractions and solve:

1. List restrictions.

2. Clear fractions.

$$(x+2)\left(1 + \frac{3x}{x+2}\right) = (x+2)\frac{-6}{x+2}$$
Multiplying both sides by the LCM. *Don't forget the parentheses!*

$$(x+2)\cdot 1 + (x+2)\cdot \frac{3x}{x+2} = (x+2)\frac{-6}{x+2}$$
Using the distributive law

$$x + 2 + \frac{(x+2)(3x)}{x+2} = \frac{(x+2)(-6)}{x+2}$$
Multiplying

3. Solve.

$$x + 2 + 3x = -6$$
Removing factors equal to 1: $(x+2)/(x+2)=1$

$$4x + 2 = -6$$
$$4x = -8$$
$$x = -2.$$
Above, we stated that $x \neq -2$.

Because of the above restriction, -2 must be rejected as a solution. The check below simply confirms this.

4. Check.

Check:

$$1 + \frac{3x}{x+2} = \frac{-6}{x+2}$$

$$\begin{array}{c|c} 1 + \dfrac{3(-2)}{-2+2} & \dfrac{-6}{-2+2} \\[2ex] 1 + \dfrac{-6}{0} \stackrel{?}{=} \dfrac{-6}{0} & \text{FALSE} \end{array}$$

2. Solve: $\dfrac{x-1}{x-5} = \dfrac{4}{x-5}$.

The equation has no solution.

🢒 YOUR TURN

CONNECTING 🔗 THE CONCEPTS

Simplifying an expression is different from solving an equation. An equation contains an equal sign; an expression does not. When expressions are simplified, the result is an equivalent expression. When equations are solved, the result is a solution. Compare the following.

Simplify: $\dfrac{x}{x+1} + \dfrac{2}{3}.$ There is no equal sign.

SOLUTION

$$\dfrac{x}{x+1} + \dfrac{2}{3} = \dfrac{x}{x+1} \cdot \dfrac{3}{3} + \dfrac{2}{3} \cdot \dfrac{x+1}{x+1}$$

$$= \dfrac{3x}{3(x+1)} + \dfrac{2x+2}{3(x+1)}$$ Writing with the LCD, $3(x+1)$

$$= \dfrac{5x+2}{3(x+1)}$$

> The equal signs indicate that all the expressions are equivalent.

The result is an expression equivalent to $\dfrac{x}{x+1} + \dfrac{2}{3}.$

Solve: $\dfrac{x}{x+1} = \dfrac{2}{3}.$ There is an equal sign.

SOLUTION   Note that $x \neq -1.$

$$\dfrac{x}{x+1} = \dfrac{2}{3}$$

$$3(x+1) \cdot \dfrac{x}{x+1} = 3(x+1) \cdot \dfrac{2}{3}$$ Multiplying by the LCM, $3(x+1)$

$$3x = 2(x+1)$$

$$3x = 2x + 2$$

$$x = 2$$

> Each line is an equivalent equation.

The result is a solution; 2 is the solution of
$$\dfrac{x}{x+1} = \dfrac{2}{3}.$$

EXERCISES

1. Simplify: $\dfrac{5x^2 - 10x}{5x^2 + 5x}.$

2. Add and, if possible, simplify: $\dfrac{5}{3t} + \dfrac{1}{2t-1}.$

3. Solve: $\dfrac{t}{2} + \dfrac{t}{3} = 5.$

4. Solve: $\dfrac{1}{y} - \dfrac{1}{2} = \dfrac{5}{6y}.$

5. Simplify: $\dfrac{\dfrac{1}{z} + 1}{\dfrac{1}{z^2} - 1}.$

6. Solve: $\dfrac{5}{x+3} = \dfrac{3}{x+2}.$

6.4 EXERCISE SET

MyMathLab® Math XL
PRACTICE WATCH READ REVIEW

🍫 Vocabulary and Reading Check

Classify each of the following statements as either true or false.

1. Every rational equation has at least one solution.

2. If one expression in a rational equation has a denominator of x, then 0 cannot be a solution of the equation.

3. It is possible to make no mistakes when solving a rational equation and still obtain a possible solution that does not check.

4. It may be necessary to use the principle of zero products when solving a rational equation.

🍫 Concept Reinforcement

Classify each of the following as either an expression or an equation.

5. $\dfrac{2}{3} = \dfrac{1}{x}$

6. $\dfrac{2}{x+1} - \dfrac{x-5}{x^2 - 2x}$

7. $\dfrac{4}{t^2 - 1} + \dfrac{3}{t+1}$

8. $\dfrac{2}{t^2 - 1} = \dfrac{3}{t+1}$

9. $\dfrac{5x}{x^2 - 4} \cdot \dfrac{7}{x^2 - 5x + 4}$

10. $\dfrac{7x}{2-x} = \dfrac{3}{4-x}$

Solving Rational Equations

Solve. If no solution exists, state this.

11. $\dfrac{t}{10} + \dfrac{t}{15} = 1$

12. $\dfrac{t}{45} + \dfrac{t}{30} = 1$

13. $\dfrac{1}{8} + \dfrac{1}{10} = \dfrac{1}{t}$

14. $\dfrac{1}{6} + \dfrac{1}{8} = \dfrac{1}{t}$

15. $\dfrac{d}{7} - \dfrac{7}{d} = 0$

16. $\dfrac{c}{5} - \dfrac{5}{c} = 0$

17. $\dfrac{3}{4} - \dfrac{1}{x} = \dfrac{7}{8}$

18. $\dfrac{2}{3} - \dfrac{1}{y} = \dfrac{5}{6}$

19. $\dfrac{n+3}{n-5} = \dfrac{1}{2}$

20. $\dfrac{n-1}{n+6} = \dfrac{3}{10}$

21. $\dfrac{9}{x} = \dfrac{x}{4}$

22. $\dfrac{x}{5} = \dfrac{20}{x}$

23. $\dfrac{1}{3t} + \dfrac{1}{t} = \dfrac{1}{2}$

24. $\dfrac{1}{t} + \dfrac{1}{2t} = \dfrac{1}{5}$

25. $\dfrac{3}{x-1} + \dfrac{3}{10} = \dfrac{5}{2x-2}$

26. $\dfrac{3}{2n+10} + \dfrac{5}{4} = \dfrac{7}{n+5}$

Aha! 27. $\dfrac{2}{6} + \dfrac{1}{2x} = \dfrac{1}{3}$

28. $\dfrac{12}{15} - \dfrac{1}{3x} = \dfrac{4}{5}$

29. $y + \dfrac{4}{y} = -5$

30. $t + \dfrac{6}{t} = -5$

31. $x - \dfrac{12}{x} = 4$

32. $y - \dfrac{14}{y} = 5$

33. $\dfrac{9}{10} = \dfrac{1}{y}$

34. $-\dfrac{5}{6} = \dfrac{1}{x}$

35. $\dfrac{t-1}{t-3} = \dfrac{2}{t-3}$

36. $\dfrac{x-2}{x-4} = \dfrac{2}{x-4}$

37. $\dfrac{x}{x-5} = \dfrac{25}{x^2-5x}$

38. $\dfrac{t}{t-6} = \dfrac{36}{t^2-6t}$

39. $\dfrac{n+1}{n+2} = \dfrac{n-3}{n+1}$

40. $\dfrac{n+2}{n-3} = \dfrac{n+1}{n-2}$

Aha! 41. $\dfrac{x^2+4}{x-1} = \dfrac{5}{x-1}$

42. $\dfrac{x^2-1}{x+2} = \dfrac{3}{x+2}$

43. $\dfrac{6}{a+1} = \dfrac{a}{a-1}$

44. $\dfrac{4}{a-7} = \dfrac{-2a}{a+3}$

45. $\dfrac{60}{t-5} - \dfrac{18}{t} = \dfrac{40}{t}$

46. $\dfrac{50}{t-2} - \dfrac{16}{t} = \dfrac{30}{t}$

47. $\dfrac{4}{y^2+y-12} = \dfrac{1}{y+4} - \dfrac{2}{y-3}$

48. $\dfrac{3}{a^2-7a+10} = \dfrac{2}{a-2} + \dfrac{1}{a-5}$

49. $\dfrac{3}{x-3} + \dfrac{5}{x+2} = \dfrac{5x}{x^2-x-6}$

50. $\dfrac{2}{x-2} + \dfrac{1}{x+4} = \dfrac{x}{x^2+2x-8}$

51. $\dfrac{3}{x} + \dfrac{x}{x+2} = \dfrac{4}{x^2+2x}$

52. $\dfrac{x}{x+1} + \dfrac{5}{x} = \dfrac{1}{x^2+x}$

53. $\dfrac{2}{t-4} + \dfrac{1}{t} = \dfrac{t}{4-t}$

54. $\dfrac{2t}{3-t} - \dfrac{4}{t} = \dfrac{1}{t-3}$

55. $\dfrac{5}{x+2} - \dfrac{3}{x-2} = \dfrac{2x}{4-x^2}$

56. $\dfrac{y+3}{y+2} - \dfrac{y}{y^2-4} = \dfrac{y}{y-2}$

57. $\dfrac{1}{x^2+2x+1} = \dfrac{x-1}{3x+3} + \dfrac{x+2}{5x+5}$

58. $\dfrac{3}{x^2-6x+9} + \dfrac{x-2}{3x-9} = \dfrac{x}{2x-6}$

59. $\dfrac{3-2y}{y+1} - \dfrac{10}{y^2-1} = \dfrac{2y+3}{1-y}$

60. $\dfrac{1-2x}{x+2} + \dfrac{20}{x^2-4} = \dfrac{2x+1}{2-x}$

In Exercises 61–66, a rational function f is given. Find all values of a for which f(a) is the indicated value.

61. $f(x) = 2x - \dfrac{15}{x}$; $f(a) = 7$

62. $f(x) = 2x - \dfrac{6}{x}$; $f(a) = 1$

63. $f(x) = \dfrac{x-5}{x+1}$; $f(a) = \dfrac{3}{5}$

64. $f(x) = \dfrac{x-3}{x+2}$; $f(a) = \dfrac{1}{5}$

65. $f(x) = \dfrac{12}{x} - \dfrac{12}{2x}$; $f(a) = 8$

66. $f(x) = \dfrac{6}{x} - \dfrac{6}{2x}$; $f(a) = 5$

For each pair of functions f and g, find all values of a for which f(a) = g(a).

67. $f(x) = \dfrac{3x - 1}{x^2 - 7x + 10}$,

$g(x) = \dfrac{x - 1}{x^2 - 4} + \dfrac{2x + 1}{x^2 - 3x - 10}$

68. $f(x) = \dfrac{2x + 5}{x^2 + 4x + 3}$,

$g(x) = \dfrac{x + 2}{x^2 - 9} + \dfrac{x - 1}{x^2 - 2x - 3}$

69. $f(x) = \dfrac{2}{x^2 - 8x + 7}$,

$g(x) = \dfrac{3}{x^2 - 2x - 3} - \dfrac{1}{x^2 - 1}$

70. $f(x) = \dfrac{4}{x^2 + 3x - 10}$,

$g(x) = \dfrac{3}{x^2 - x - 12} + \dfrac{1}{x^2 + x - 6}$

71. Explain why it is essential to check any possible solutions of rational equations.

72. Explain the difference between adding rational expressions and solving rational equations.

Skill Review

73. Evaluate $2x - y^2 \div 3x$ for $x = 3$ and $y = -6$. [1.1], [1.2]

74. Convert to scientific notation: 391,000,000. [1.7]

Simplify. Do not use negative exponents in the answer. [1.6]

75. -3^{-2}

76. $(-4x^3)^0$

77. $\dfrac{24a^{-4}c^{-8}}{16a^5c^{-7}}$

78. $(-5x^2y^{-6})^{-3}$

Synthesis

79. Is the following statement true or false: "For any real numbers a, b, and c, if $ac = bc$, then $a = b$"? Explain why you answered as you did.

80. When checking a possible solution of a rational equation, is it sufficient to check that the "solution" does not make any denominator equal to 0? Why or why not?

For each pair of functions f and g, find all values of a for which f(a) = g(a).

81. $f(x) = \dfrac{2 - \dfrac{x}{4}}{2}$, $g(x) = \dfrac{\dfrac{x}{4} - 2}{\dfrac{x}{2} + 2}$

82. $f(x) = \dfrac{x + 3}{x + 2} - \dfrac{x + 4}{x + 3}$, $g(x) = \dfrac{x + 5}{x + 4} - \dfrac{x + 6}{x + 5}$

83. $f(x) = \dfrac{1}{1 + x} + \dfrac{x}{1 - x}$, $g(x) = \dfrac{1}{1 - x} - \dfrac{x}{1 + x}$

84. $f(x) = \dfrac{0.793}{x} + 18.15$, $g(x) = \dfrac{6.034}{x} - 43.17$

Solve.

85. $\dfrac{\dfrac{1}{x} + 1}{x} = \dfrac{\dfrac{1}{x}}{2}$

86. $\dfrac{\dfrac{1}{3}}{x} = \dfrac{1 - \dfrac{1}{x}}{x}$

87. Use a graphing calculator to check your answers to Exercises 11, 53, and 63.

88. The reciprocal of a number is the number itself. What is the number?

YOUR TURN ANSWERS: SECTION 6.4

1. $\dfrac{15}{4}$ **2.** No solution   **3.** $-6, 6$ **4.** 0   **5.** $-\dfrac{1}{3}, \dfrac{1}{2}$

QUICK QUIZ: SECTIONS 6.1–6.4

Perform the indicated operation and, if possible, simplify.

1. $\dfrac{a + 1}{6a} \cdot \dfrac{8a^2}{a^2 - 1}$ [6.1]

2. $\dfrac{2a}{a + 1} - \dfrac{4a}{1 - a^2}$ [6.2]

3. $\dfrac{27a^2}{8} \div \dfrac{12}{5a}$ [6.1]

4. Simplify: $\dfrac{a^{-1} + b^{-1}}{ab^{-1} - ba^{-1}}$.   [6.3]

5. Solve: $\dfrac{15}{x} - \dfrac{15}{x + 2} = 2$.   [6.4]

PREPARE TO MOVE ON

Solve.

1. Gail's Cessna travels 135 mph in still air. With a tailwind of 15 mph, how long will it take Gail to travel 200 mi?   [1.4]

2. Brenton paddles 55 m per minute in still water. If he paddles 135 m upstream in 3 min, what is the speed of the current?   [1.4]

3. The area of a wooden shutter is 900 in². The shutter is four times as long as it is wide. Find the length and the width of the shutter.   [5.8]

4. Find two consecutive even integers whose product is 120.   [5.8]

Mid-Chapter Review

To add, subtract, multiply, divide, and simplify rational expressions, we use the same steps that we use for fractions in arithmetic.

To solve rational equations, we clear fractions and solve the resulting equation. Any possible solutions *must* be checked in the original equation.

GUIDED SOLUTIONS

1. Add: $\dfrac{2}{x} + \dfrac{1}{x^2 + x}$. Simplify, if possible. [6.2]

Solution

$$\dfrac{2}{x} + \dfrac{1}{x^2 + x}$$

$$= \dfrac{2}{x} + \dfrac{1}{x\,\square} \qquad \text{Factoring denominators. The LCD is } x(x+1).$$

$$= \dfrac{2}{x} \cdot \dfrac{\square}{\square} + \dfrac{1}{x(x+1)} \qquad \text{Multiplying by 1 to get the LCD in the first denominator}$$

$$= \dfrac{\square}{x(x+1)} + \dfrac{1}{x(x+1)} \qquad \text{Multiplying}$$

$$= \dfrac{\square}{x(x+1)} \qquad \text{Adding numerators. We cannot simplify.}$$

2. Solve: $\dfrac{10}{t^2 - 1} = \dfrac{t+4}{t^2 - t}$. [6.4]

Solution

$$\dfrac{10}{(t+1)(t-1)} = \dfrac{t+4}{t(\square)} \qquad t \neq -1, t \neq 1, \; t \neq 0.$$

$$t(t+1)(t-1)\left(\dfrac{10}{(t+1)(t-1)}\right) \qquad \text{The LCM is } t(t+1)(t-1).$$

$$= t(t+1)(t-1)\left(\dfrac{t+4}{t(t-1)}\right)$$

$$\square = (t+1)(t+4) \qquad \text{Removing factors equal to 1}$$

$$\square = t^2 + 5t + \square \qquad \text{Multiplying}$$

$$0 = t^2 - \square + 4 \qquad \text{Getting 0 on one side}$$

$$0 = (t-1)(\square) \qquad \text{Factoring}$$

$$t - 1 = 0 \quad or \quad \square = 0 \qquad \text{Using the principle of zero products}$$

$$t = 1 \quad or \quad t = \square$$

One restriction stated above is $t \neq 1$; thus, 1 is not a solution. Since 4 checks in the original equation, the solution is $\square$.

MIXED REVIEW

Perform the indicated operation and, if possible, simplify.

3. $\dfrac{2x - 6}{5x + 10} \cdot \dfrac{x + 2}{6x - 12}$ [6.1]

4. $\dfrac{2}{x - 5} \div \dfrac{6}{x - 5}$ [6.1]

5. $\dfrac{x}{x + 2} - \dfrac{1}{x - 1}$ [6.2]

6. $\dfrac{2}{x + 3} + \dfrac{3}{x + 4}$ [6.2]

7. $\dfrac{3}{x - 4} - \dfrac{2}{4 - x}$ [6.2]

8. $\dfrac{x^2 - 16}{x^2 - x} \cdot \dfrac{x^2}{x^2 - 5x + 4}$ [6.1]

9. $\dfrac{x + 1}{x^2 - 7x + 10} + \dfrac{3}{x^2 - x - 2}$ [6.2]

10. $(t^2 + t - 20) \cdot \dfrac{t + 5}{t - 4}$ [6.1]

11. $\dfrac{a^2 - 2a + 1}{a^2 - 4} \div (a^2 - 3a + 2)$ [6.1]

(continued)

12. $\dfrac{\dfrac{3}{z}+\dfrac{2}{y}}{\dfrac{4}{z}-\dfrac{1}{y}}$ [6.3]

13. $\dfrac{xy^{-1}+x^{-1}}{2x^{-1}+4y^{-1}}$ [6.3]

Solve. [6.4]

16. $\dfrac{a+1}{3}+\dfrac{a-4}{5}=\dfrac{2a}{9}$

17. $\dfrac{5}{4t}=\dfrac{7}{5t-2}$

18. $\dfrac{2}{1-x}=\dfrac{-4}{x^2-1}$

19. $\dfrac{3}{x}+\dfrac{2}{x-2}=1$

14. $\dfrac{\dfrac{y}{y^2-4}+\dfrac{5}{4-y^2}}{\dfrac{y^2}{y^2-4}+\dfrac{25}{4-y^2}}$ [6.3]

15. $\dfrac{\dfrac{1}{a}-\dfrac{1}{b}}{\dfrac{1}{a^3}-\dfrac{1}{b^3}}$ [6.3]

20. $\dfrac{t-1}{t^2-3t}-\dfrac{4}{t^2-9}=0$

6.5 Solving Applications Using Rational Equations

Problems Involving Work ▪ Problems Involving Motion

Now that we are able to solve rational equations, we can also solve new types of applications. The five problem-solving steps remain the same.

Problems Involving Work

EXAMPLE 1 The roof of Finn and Paige's townhouse needs to be reshingled. Finn can do the job alone in 8 hr and Paige can do the job alone in 10 hr. How long will it take the two of them, working together, to reshingle the roof?

SOLUTION

1. **Familiarize.** This *work problem* is a type of problem we have not yet encountered. Work problems are often *incorrectly* translated to mathematical language in several ways.

   a) Add the times together: 8 hr + 10 hr = 18 hr.   ← Incorrect

   This cannot be the correct approach because Finn and Paige working together should take *less* time than either of them working alone.

   b) Average the times: (8 hr + 10 hr)/2 = 9 hr.   ← Incorrect

   Again, this is longer than it would take Finn to do the job alone.

   c) Assume that each person does half the job.   ← Incorrect

   Finn would reshingle $\frac{1}{2}$ the roof in $\frac{1}{2}$(8 hr), or 4 hr, and Paige would reshingle $\frac{1}{2}$ the roof in $\frac{1}{2}$(10 hr), or 5 hr, so Finn would finish an hour before Paige. The problem assumes that the two are working together, so Finn will help Paige after completing his half. This tells us that the job will take between 4 hr and 5 hr.

   Each incorrect approach started with the time that it took each worker to do the job. The correct approach instead focuses on the *rate* of work, or the amount of the job that each person completes in 1 hr.

   Since it takes Finn 8 hr to reshingle the entire roof, in 1 hr he reshingles $\frac{1}{8}$ of the roof. Since it takes Paige 10 hr to reshingle the entire roof, in 1 hr she reshingles $\frac{1}{10}$ of the roof. Together, they reshingle $\frac{1}{8}+\frac{1}{10}=\frac{5}{40}+\frac{4}{40}=\frac{9}{40}$ of the roof per hour. The rates are thus

   Finn: $\frac{1}{8}$ roof per hour,

   Paige: $\frac{1}{10}$ roof per hour,

   Together: $\frac{9}{40}$ roof per hour.

We are looking for the time required to reshingle 1 entire roof.

Time	Fraction of the Roof Reshingled		
	By Finn	By Paige	Together
1 hr	$\frac{1}{8}$	$\frac{1}{10}$	$\frac{1}{8} + \frac{1}{10}$, or $\frac{9}{40}$
2 hr	$\frac{1}{8} \cdot 2$	$\frac{1}{10} \cdot 2$	$\frac{1}{8} \cdot 2 + \frac{1}{10} \cdot 2$
3 hr	$\frac{1}{8} \cdot 3$	$\frac{1}{10} \cdot 3$	$\frac{1}{8} \cdot 3 + \frac{1}{10} \cdot 3$
t hr	$\frac{1}{8} \cdot t$	$\frac{1}{10} \cdot t$	$\frac{1}{8} \cdot t + \frac{1}{10} \cdot t$

Study Skills

Take Advantage of Free Checking

It is always wise to check an answer, if it is possible to do so. When an applied problem is being solved, do check an answer in the equation from which it came. However, it is even more important to check the answer with the words of the original problem.

2. Translate. From the table, we see that t must be some number for which

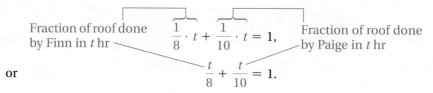

or

$$\frac{t}{8} + \frac{t}{10} = 1.$$

3. Carry out. We solve the equation:

$$\frac{t}{8} + \frac{t}{10} = 1$$

$$40\left(\frac{t}{8} + \frac{t}{10}\right) = 40 \cdot 1 \qquad \text{Multiplying by the LCM to clear fractions}$$

$$\frac{40t}{8} + \frac{40t}{10} = 40$$

$$5t + 4t = 40 \qquad \text{Simplifying}$$

$$9t = 40$$

$$t = \frac{40}{9}, \text{ or } 4\frac{4}{9}.$$

1. Refer to Example 1. Instead of working with Finn, Paige works with Clen, who can reshingle the roof in 12 hr. How long will it take the two of them, working together, to reshingle the roof?

4. Check. In $\frac{40}{9}$ hr, Finn reshingles $\frac{1}{8} \cdot \frac{40}{9}$, or $\frac{5}{9}$, of the roof and Paige reshingles $\frac{1}{10} \cdot \frac{40}{9}$, or $\frac{4}{9}$, of the roof. Together, they reshingle $\frac{5}{9} + \frac{4}{9}$, or 1 roof. The fact that our solution is between 4 hr and 5 hr (see step 1 above) is also a check.

5. State. It will take $4\frac{4}{9}$ hr for Finn and Paige, working together, to reshingle the roof.

YOUR TURN

EXAMPLE 2   It takes Manuel 9 hr longer than it does Zoe to rebuild an engine. Working together, they can do the job in 20 hr. How long would it take each, working alone, to rebuild an engine?

SOLUTION

1. Familiarize. Unlike Example 1, this problem does not provide us with the times required by the individuals to do the job alone. We let $z =$ the number of hours it would take Zoe working alone and $z + 9 =$ the number of hours it would take Manuel working alone.

2. Translate. Using the same reasoning as in Example 1, we see that Zoe completes $\frac{1}{z}$ of the job in 1 hr and Manuel completes $\frac{1}{z + 9}$ of the job in 1 hr. Thus, in 20 hr, Zoe completes $\frac{1}{z} \cdot 20$ of the job and Manuel completes $\frac{1}{z + 9} \cdot 20$ of the job. Together, Zoe and Manuel can complete the entire job in 20 hr. This gives the following:

Fraction of job done by Zoe in 20 hr $\dfrac{1}{z} \cdot 20 + \dfrac{1}{z + 9} \cdot 20 = 1,$ Fraction of job done by Manuel in 20 hr

or $\dfrac{20}{z} + \dfrac{20}{z + 9} = 1.$

2. Oliver can paint the trim on the Polinskis' Queen Anne house in 12 fewer days than it takes Tammy to do the same job. Working together, they can do the job in 8 days. How long would it take each, working alone, to paint the house?

3. Carry out. We solve the equation:

$$\frac{20}{z} + \frac{20}{z + 9} = 1 \qquad \text{The LCM is } z(z + 9).$$

$$z(z + 9)\left(\frac{20}{z} + \frac{20}{z + 9}\right) = z(z + 9)1 \qquad \text{Multiplying to clear fractions}$$

$$(z + 9)20 + z \cdot 20 = z(z + 9) \qquad \text{Distributing and simplifying}$$

$$40z + 180 = z^2 + 9z$$

$$0 = z^2 - 31z - 180 \qquad \text{Getting 0 on one side}$$

$$0 = (z - 36)(z + 5) \qquad \text{Factoring}$$

$$z - 36 = 0 \quad \text{or} \quad z + 5 = 0 \qquad \text{Principle of zero products}$$

$$z = 36 \quad \text{or} \qquad z = -5.$$

4. Check. Since negative time has no meaning in the problem, −5 is not a solution to the original problem. The number 36 checks since, if Zoe takes 36 hr alone and Manuel takes 36 + 9 = 45 hr alone, in 20 hr they would have finished

$$\tfrac{20}{36} + \tfrac{20}{45} = \tfrac{5}{9} + \tfrac{4}{9} = 1 \text{ complete rebuild.}$$

5. State. It would take Zoe 36 hr to rebuild an engine alone, and Manuel 45 hr.

YOUR TURN

The equations used in Examples 1 and 2 can be generalized as follows.

MODELING WORK PROBLEMS

If

a = the time needed for A to complete the work alone,

b = the time needed for B to complete the work alone, and

t = the time needed for A and B to complete the work together,

then

$$\frac{t}{a} + \frac{t}{b} = 1.$$

The following are equivalent equations that can also be used:

$$\frac{1}{a} \cdot t + \frac{1}{b} \cdot t = 1, \qquad \left(\frac{1}{a} + \frac{1}{b}\right)t = 1, \quad \text{and} \quad \frac{1}{a} + \frac{1}{b} = \frac{1}{t}.$$

Problems Involving Motion

Problems dealing with distance, rate (or speed), and time are called **motion problems**. To translate them, we use either the basic motion formula, $d = rt$, or the formulas $r = d/t$ or $t = d/r$, which can be derived from $d = rt$.

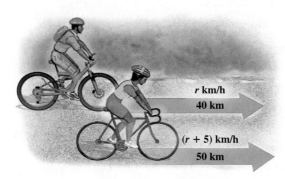

r km/h
40 km

(r + 5) km/h
50 km

EXAMPLE 3   On her road bike, Olivia bikes 5 km/h faster than Jason does on his mountain bike. In the time it takes Olivia to travel 50 km, Jason travels 40 km. Find the speed of each bicyclist.

SOLUTION

1. **Familiarize.** Let's make a guess and check it.

   Guess: Jason's speed: 10 km/h

   Olivia's speed: $10 + 5$, or 15 km/h

   Jason's time: $40/10 = 4$ hr ⎫   The times are
   Olivia's time: $50/15 = 3\frac{1}{3}$ hr ⎭   not the same.

   Our guess is wrong, but we can make some observations. If Jason's speed $= r$, in kilometers/hour, then Olivia's speed $= r + 5$. Jason's travel time is the same as Olivia's travel time.
   We can also make a sketch and label it to help us visualize the situation.

2. **Translate.** We organize the information in a table. By looking at how we checked our guess, we see that we can fill in the **Time** column of the table using the formula *Time = Distance/Rate.*

	Distance	Speed	Time
Jason's Mountain Bike	40	r	$40/r$
Olivia's Road Bike	50	$r + 5$	$50/(r + 5)$

Since we know that the times are the same, we can write an equation:

$$\frac{40}{r} = \frac{50}{r + 5}.$$

3. **Carry out.** We solve the equation:

   $$\frac{40}{r} = \frac{50}{r + 5}$$ The LCM is $r(r + 5)$.

   $$r(r + 5)\frac{40}{r} = r(r + 5)\frac{50}{r + 5}$$ Multiplying to clear fractions

   $$40r + 200 = 50r$$ Simplifying

   $$200 = 10r$$

   $$20 = r.$$

4. **Check.** If our answer checks, Jason's mountain bike is going 20 km/h and Olivia's road bike is going $20 + 5 = 25$ km/h.
   Traveling 50 km at 25 km/h, Olivia is riding for $\frac{50}{25} = 2$ hr. Traveling 40 km at 20 km/h, Jason is riding for $\frac{40}{20} = 2$ hr. Our answer checks since the two times are the same.

5. **State.** Olivia's speed is 25 km/h, and Jason's speed is 20 km/h.

    YOUR TURN

Student Notes

You need remember only the motion formula $d = rt$. Then you can divide both sides by t to get $r = d/t$, or you can divide both sides by r to get $t = d/r$.

3. Peter can drive 25 mph faster on the highway than he can on county roads. In the time it would take Peter to drive 70 mi on county roads, he could drive 120 mi on the highway. How fast can he drive on each type of road?

EXAMPLE 4   A Hudson River tugboat goes 10 mph in still water. It travels 24 mi upstream and 24 mi back in a total time of 5 hr. What is the speed of the current?

Sources: Based on information from the Department of the Interior, U.S. Geological Survey, and *The Tugboat Captain*, Montgomery County Community College

SOLUTION

1. **Familiarize.** Let's make a guess and check it.

   Guess: Speed of current: 4 mph

                 Tugboat's speed upstream: $10 - 4 = 6$ mph

                 Tugboat's speed downstream: $10 + 4 = 14$ mph

                 Travel time upstream: $24/6 = 4$ hr   ⎫ The total time

                 Travel time downstream: $24/14 = 1\frac{5}{7}$ hr ⎭ is not 5 hr.

   Our guess is wrong, but we can make some observations. If $c =$ the current's rate, in miles per hour, we have the following.

       The tugboat's speed upstream is $(10 - c)$ mph.

       The tugboat's speed downstream is $(10 + c)$ mph.

       The total travel time is 5 hr.

   We make a sketch and label it, using the information we know.

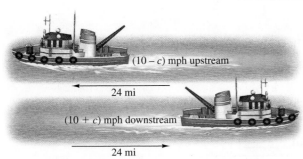

(10 − c) mph upstream

24 mi

(10 + c) mph downstream

24 mi

2. **Translate.** We organize the information in a table. From examining our guess, we see that the time traveled can be represented using the formula *Time = Distance/Rate.*

	Distance	Speed	Time
Upstream	24	$10 - c$	$24/(10 - c)$
Downstream	24	$10 + c$	$24/(10 + c)$

Since the total time upstream and back is 5 hr, we use the last column of the table to form an equation:

$$\frac{24}{10 - c} + \frac{24}{10 + c} = 5.$$

3. **Carry out.** We solve the equation:

$$\frac{24}{10 - c} + \frac{24}{10 + c} = 5 \qquad \text{The LCM is} \atop (10 - c)(10 + c).$$

$$(10 - c)(10 + c)\left[\frac{24}{10 - c} + \frac{24}{10 + c}\right] = (10 - c)(10 + c)5 \qquad \text{Multiplying to} \atop \text{clear fractions}$$

$$24(10 + c) + 24(10 - c) = (100 - c^2)5$$

$$480 = 500 - 5c^2 \qquad \text{Simplifying}$$

$$5c^2 - 20 = 0$$

$$5(c^2 - 4) = 0$$

$$5(c - 2)(c + 2) = 0$$

$$c = 2 \quad or \quad c = -2.$$

4. **Check.** Since speed cannot be negative in this problem, −2 cannot be a solution. You should confirm that 2 checks in the original problem.

5. **State.** The speed of the current is 2 mph.

 YOUR TURN

Chapter Resource:
Collaborative Activity, p. 424

4. At a length of 4 km, the Comal River in Texas is the shortest navigable river in the United States. Tristan paddled his canoe up and back down the river in $1\frac{1}{3}$ hr. If he paddles 8 km/h in still water, what was the speed of the current?

EXPLORING 🔍 THE CONCEPT

Motion problems are often much simpler to solve if the information is organized in a table. For each motion problem, fill in the missing entries in the table using the list of options given below.

1. Tara runs 1 km/h faster than Cassie. Tara can run 20 km in the same time that it takes Cassie to run 18 km. Find the speed of each runner.

	Distance	Speed	Time
Tara	20	(a)_____	(b)_____
Cassie	(c)_____	r	(d)_____

Options: $\quad 18 \quad r+1 \quad \dfrac{20}{r+1} \quad \dfrac{18}{r}$

2. Damon rode 50 mi to a state park at a certain speed. Had he been able to ride 3 mph faster, the trip would have been $\frac{1}{4}$ hr shorter. How fast did he ride?

	Distance	Speed	Time
Actual Trip	(a)_____	r	(b)_____
Faster Trip	50	(c)_____	(d)_____

Options: $\quad 50 \quad r+3 \quad \dfrac{50}{r+3} \quad \dfrac{50}{r}$

3. The speed of the Green River current is 4 mph. In the same time that it takes Blair to motor 48 mi downstream, he can travel only 32 mi upstream. What is the speed of the boat in still water?

	Distance	Speed	Time
Downstream	48	(a)_____	$\dfrac{48}{x+4}$
Upstream	(b)_____	(c)_____	(d)_____

Options: $\quad 32 \quad \dfrac{32}{x-4} \quad x+4 \quad x-4$

ANSWERS

1. (a) $r+1$; (b) $\dfrac{20}{r+1}$; (c) 18; (d) $\dfrac{18}{r}$

2. (a) 50; (b) $\dfrac{50}{r}$; (c) $r+3$; (d) $\dfrac{50}{r+3}$

3. (a) $x+4$; (b) 32; (c) $x-4$; (d) $\dfrac{32}{x-4}$

6.5 EXERCISE SET

FOR EXTRA HELP

MyMathLab® Math XL
PRACTICE   WATCH   READ   REVIEW

Vocabulary and Reading Check

Classify each of the following statements as either true or false.

1. In order to find the time that it would take two people to complete a job working together, we average the times that it takes each of them to complete the job working separately.

2. To find the rate at which two people work together, we add the rates at which they work separately.

3. Distance equals rate times time.

4. Rate equals distance divided by time.

5. Time equals distance divided by rate.

6. To find a boat's speed downstream, we add the speed of the boat in still water to the speed of the current.

Concept Reinforcement

Find each rate.

7. If Sandy can decorate a cake in 2 hr, what is her rate?

8. If Eric can decorate a cake in 3 hr, what is his rate?

9. If Sandy can decorate a cake in 2 hr and Eric can decorate the same cake in 3 hr, what is their rate, working together?

10. If Lisa and Mark can mow their lawn together in 1 hr, what is their rate?

11. If Lisa can mow their lawn by herself in 3 hr, what is her rate?

12. If Lisa and Mark can mow their lawn together in 1 hr, and Lisa can mow the lawn by herself in 3 hr, what is Mark's rate, working alone?

Solve.

13. The reciprocal of 3, plus the reciprocal of 6, is the reciprocal of what number?

14. The reciprocal of 10, plus the reciprocal of 15, is the reciprocal of what number?

15. The sum of a number and six times its reciprocal is -5. Find the number.

16. The sum of a number and twenty-one times its reciprocal is -10. Find the number.

17. The reciprocal of the product of two consecutive integers is $\frac{1}{90}$. Find the two integers.

18. The reciprocal of the product of two consecutive integers is $\frac{1}{30}$. Find the two integers.

Problems Involving Work

19. *Custom Embroidery.* Chandra can embroider logos on a team's sweatshirts in 6 hr. Traci, a new employee, needs 9 hr to complete the same job. Working together, how long will it take them to do the job?

20. *Filling a Pool.* The San Paulo community swimming pool can be filled in 12 hr if water enters through a pipe alone or in 30 hr if water enters through a hose alone. If water is entering through both the pipe and the hose, how long will it take to fill the pool?

21. *Pumping Water.* A $\frac{1}{2}$-hp Gempler's sump pump can remove water from Martha's flooded basement in 48 min. A 250 series Liberty sump pump can complete the same job in 30 min. How long would it take the two pumps together to pump out the basement?

    Source: Based on data from manufacturers

22. *Hotel Management.* The Winix Plasmawave 9000 air purifier can clean the air in a conference room in 8 min. The Blueair 203 Air Purifier can clean the air in the same room in 12 min. How long would it take the two machines together to clean the air in the room?

    Source: Based on data from manufacturers

23. *Scanners.* The Epson WorkForce G7-1500 takes twice the time required by the Epson WorkForce Pro GT-S50 to scan the manuscript for a book. If, working together, the two machines can complete the job in 5 min, how long would it take each machine, working alone, to scan the manuscript?

    Source: epson.com

24. *Cutting Firewood.* Kent can cut and split a cord of wood twice as fast as Brent can. When they work together, it takes them 4 hr. How long would it take each of them to do the job alone?

25. *Mulching.* Anita can mulch the college gardens in 3 fewer days than it takes Tori to mulch the same areas. When they work together, it takes them 2 days. How long would it take each of them to do the job alone?

26. *Photo Printing.* It takes the Canon PIXMA MP495 15 min longer to print a set of photo proofs than it takes the Canon PIXMA MG8220. Together it would take them 10 min to print the photos. How long would it take each machine, working alone, to print the photos?

Source: www.usa.canon.com

27. *Software Development.* Tristan, an experienced programmer, can write video-game software three times as fast as Sara, who is just learning to program. Working together on one project, it took them 1 month to complete the job. How long would it take each of them to complete the project alone?

28. *Forest Fires.* The Erickson Air-Crane helicopter can scoop water and douse a certain forest fire four times as fast as an S-58T helicopter. Working together, the two helicopters can douse the fire in 8 hr. How long would it take each helicopter, working alone, to douse the fire?

Sources: Based on information from www.emergency.com and www.arishelicopters.com

29. *Baking.* Zeno takes 20 min longer to decorate a dozen cupcakes than it takes Lia. When they work together, it takes them 10.5 min. How long would each take to do the job alone?

30. *Waxing a Car.* It takes Valerie 48 min longer to wax the family car than it takes Gretchen. When they work together, they can wax the car in 45 min. How long would it take Gretchen, working by herself, to wax the car?

31. *Sorting Recyclables.* Together, it takes Kim and Chris 2 hr 55 min to sort recyclables. Alone, Chris would require 2 fewer hours than Kim. How long would it take Chris to do the job alone? (*Hint*: Convert minutes to hours or hours to minutes.)

32. *Paving.* Together, Steve and Bill require 4 hr 48 min to pave a driveway. Alone, Steve would require 4 hr more than Bill. How long would it take Bill to do the job alone? (*Hint*: Convert minutes to hours.)

Problems Involving Motion

33. *Kayaking.* The speed of the current in Catamount Creek is 3 mph. Sean can kayak 4 mi upstream in the same time that it takes him to kayak 10 mi downstream. What is the speed of Sean's kayak in still water?

34. *Boating.* The current in the Lazy River moves at a rate of 4 mph. Mickie's dinghy motors 6 mi upstream in the same time that it takes to motor 12 mi downstream. What is the speed of the dinghy in still water?

35. *Moving Sidewalks.* The moving sidewalk at O'Hare Airport in Chicago moves 1.8 ft/sec. Walking on the moving sidewalk, Roslyn travels 105 ft forward in the time that it takes to travel 51 ft in the opposite direction. How fast does Roslyn walk on a nonmoving sidewalk?

36. *Moving Sidewalks.* Newark Airport's moving sidewalk moves at a speed of 1.7 ft/sec. Walking on the moving sidewalk, Drew can travel 120 ft forward in the same time that it takes to travel 52 ft in the opposite direction. What is Drew's walking speed on a nonmoving sidewalk?

37. *Train Speed.* The speed of the A&M freight train is 14 mph less than the speed of the A&M passenger train. The passenger train travels 400 mi in the same time that the freight train travels 330 mi. Find the speed of each train.

38. *Walking.* Courtney walks 1 mph slower than Brandi. In the time that it takes Brandi to walk 6.5 mi, Courtney walks 5 mi. Find the speed of each person.

Aha! **39.** *Bus Travel.* A local bus travels 7 mph slower than the express. The express travels 45 mi in the time that it takes the local to travel 38 mi. Find the speed of each bus.

40. *Moped Speed.* Cameron's moped travels 8 km/h faster than Ellia's. Cameron travels 69 km in the same time that Ellia travels 45 km. Find the speed of each person's moped.

41. *Boating.* Annette's paddleboat travels 2 km/h in still water. The boat is paddled 4 km downstream in the same time that it takes to go 1 km upstream. What is the speed of the river?

42. *Shipping.* A barge moves 7 km/h in still water. It travels 45 km upriver and 45 km downriver in a total time of 14 hr. What is the speed of the current?

43. *Aviation.* A Citation CV jet travels 460 mph in still air and flies 525 mi into the wind and 525 mi with the wind in a total of 2.3 hr. Find the wind speed.

Source: Blue Star Jets, Inc.

44. *Canoeing.* Chad paddles 55 m/min in still water. He paddles 150 m upstream and 150 m downstream in a total time of 5.5 min. What is the speed of the current?

45. *Train Travel.* A freight train covers 120 mi at its typical speed. If the train travels 10 mph faster, the trip is 2 hr shorter. How fast does the train typically travel?

46. *Boating.* Fiona's Boston Whaler cruised 45 mi upstream and 45 mi back in a total of 8 hr. The speed of the river is 3 mph. Find the speed of the boat in still water.

47. Two steamrollers are paving a parking lot. Working together, will the two steamrollers take less than half as long as the slower steamroller would working alone? Why or why not?

48. Two fuel lines are filling a freighter with oil. Will the faster fuel line take more or less than twice as long to fill the freighter by itself? Why?

Skill Review

49. Omar invested $5000 in two accounts. He put $2200 in an account paying 4% simple interest and the rest in an account paying 5% simple interest. How much interest did he earn in one year from both accounts? [1.4]

50. In January 2006, *The Phantom of the Opera* became the longest-running Broadway show, with 7486 performances. By January 2012, the musical had been performed 9988 times. Calculate the rate at which the number of performances was rising [2.3]

51. A nontoxic floor wax can be made from lemon juice and food-grade linseed oil. The amount of oil should be twice the amount of lemon juice. How much of each ingredient is needed to make 32 oz of floor wax? (The mix should be spread with a rag and buffed when dry.) [3.3]

52. Together, Magic Kingdom, Disneyland, and California Adventure have 107 rides and attractions. Magic Kingdom has 7 fewer rides and attractions than half the total at California Adventure and Disneyland. Disneyland has 1 more ride and attraction than the total number at Magic Kingdom and California Adventure. How many rides and attractions does each amusement park have? [3.5]
Source: The Walt Disney Company

Synthesis

53. Write a work problem for a classmate to solve. Devise the problem so that the solution is "Beth and Leanne will take 4 hr to complete the job, working together."

54. Write a work problem for a classmate to solve. Devise the problem so that the solution is "Rosa takes 5 hr and Egberto takes 6 hr to complete the job alone."

55. *Filling a Bog.* The Norwich cranberry bog can be filled in 9 hr and drained in 11 hr. How long will it take to fill the bog if the drainage gate is left open?

56. *Filling a Tub.* Anju's hot tub can be filled in 10 min and drained in 8 min. How long will it take to empty a full tub if the water is left on?

57. Refer to Exercise 33. How long will it take Sean to kayak 5 mi downstream?

58. Refer to Exercise 34. How long will it take Mickie to motor 3 mi downstream?

59. *Escalators.* Together, a 100-cm wide escalator and a 60-cm wide escalator can empty a 1575-person auditorium in 14 min. The wider escalator moves twice as many people as the narrower one. How many people per hour does the 60-cm wide escalator move?
Source: McGraw-Hill Encyclopedia of Science and Technology

60. *Aviation.* A Coast Guard plane has enough fuel to fly for 6 hr, and its speed in still air is 240 mph. The plane departs with a 40-mph tailwind and returns to the same airport flying into the same wind. How far can the plane travel under these conditions?

61. *Boating.* Shoreline Travel operates a 3-hr paddleboat cruise on the Missouri River. If the speed of the boat in still water is 12 mph, how far upriver can the pilot travel against a 5-mph current before it is time to turn around?

62. *Travel by Car.* Angenita drives to work at 50 mph and arrives 1 min late. She drives to work at 60 mph and arrives 5 min early. How far does Angenita live from work?

63. *Photocopying.* The printer in an admissions office can print a 500-page document in 5 min, while the printer in the business office can print the same document in 4 min. If the two printers work together to print the document, with the faster machine starting on page 1 and the slower machine working backwards from page 500, at what page will the two machines meet to complete the job?

64. At what time after 4:00 will the minute hand and the hour hand of a clock first be in the same position?

65. At what time after 10:30 will the hands of a clock first be perpendicular?

Average speed is defined as total distance divided by total time.

66. Ferdaws drove 200 km. For the first 100 km of the trip, she drove at a speed of 40 km/h. For the second half of the trip, she traveled at a speed of 60 km/h. What was the average speed of the entire trip? (It was *not* 50 km/h.)

67. For the first 50 mi of a 100-mi trip, Garry drove 40 mph. What speed would he have to travel for the last half of the trip so that the average speed for the entire trip would be 45 mph?

 YOUR TURN ANSWERS: SECTION 6.5

1. $5\frac{5}{11}$ hr   **2.** Oliver: 12 days; Tammy: 24 days
3. County roads: 35 mph; highway: 60 mph   **4.** 4 km/h

QUICK QUIZ: SECTIONS 6.1–6.5

1. Add and, if possible, simplify:

$$\frac{4}{x^2 - 6x - 16} + \frac{x}{x^2 - x - 6}. \quad [6.2]$$

2. Divide and, if possible, simplify:

$$\frac{n^3 + 1}{15n} \div \frac{n^2 + n}{25}. \quad [6.1]$$

Solve. [6.4]

3. $y - 6 = \dfrac{16}{y}$

4. $\dfrac{5}{x + 2} - \dfrac{1}{x + 3} = \dfrac{2}{x^2 + 5x + 6}$

5. Kendra can refinish the floor of an apartment in 8 hr. Dominique can refinish the floor in 6 hr. How long will it take them, working together, to refinish the floor? [6.5]

PREPARE TO MOVE ON

Simplify.

1. $\dfrac{42x^8y^9}{7x^2y}$ [1.6]   **2.** $\dfrac{-20a^4b^3}{4a^3b^3}$ [1.6]

3. $\begin{array}{r} 4x^3 - 3x^2 \phantom{{}+8x^2} - 7 \\ -(4x^3 - 8x^2 + 4x) \\ \hline \end{array}$ **4.** $\begin{array}{r} -3x^2 - 2x + 1 \\ -(-3x^2 - x + 6) \\ \hline \end{array}$
[5.1]   [5.1]

6.6   Division of Polynomials

Dividing by a Monomial   ▪   Dividing by a Polynomial

Study Skills

Get a Terrific Seat

Your classtime is very precious. To make the most of that time, try to seat yourself at or near the front of the classroom. Studies have shown that students sitting near the front of the classroom generally outperform those students sitting further back.

A rational expression indicates division. Division of polynomials, like division of real numbers, relies on our multiplication and subtraction skills.

Dividing by a Monomial

To divide a monomial by a monomial, we divide coefficients and, if the bases are the same, subtract exponents.

Dividend $\longrightarrow \dfrac{45x^{10}}{3x^4} \longleftarrow$ Divisor $= 15x^{10-4} = 15x^6, \qquad \dfrac{8a^2b^5}{-2ab^2} = -4a^{2-1}b^{5-2} = -4ab^3.$

$\uparrow$ Quotient

To divide a polynomial by a monomial, we regard the division as a sum of quotients of monomials. This uses the fact that since

$$\frac{A}{C} + \frac{B}{C} = \frac{A + B}{C}, \quad \text{we know that} \quad \frac{A + B}{C} = \frac{A}{C} + \frac{B}{C}.$$

EXAMPLE 1   Divide $12x^3 + 8x^2 + x + 4$ by $4x$.

SOLUTION

$$(12x^3 + 8x^2 + x + 4) \div (4x) = \frac{12x^3 + 8x^2 + x + 4}{4x}$$ Writing a rational expression

$$= \frac{12x^3}{4x} + \frac{8x^2}{4x} + \frac{x}{4x} + \frac{4}{4x}$$ Writing as a sum of quotients

$$= 3x^2 + 2x + \frac{1}{4} + \frac{1}{x}$$ Performing the four indicated divisions

> **CAUTION!** The coefficients are divided but the exponents are subtracted.

1. Divide $6t^3 - 12t^2 + 2t - 9$ by $3t$.

YOUR TURN

EXAMPLE 2   Divide: $(8x^4y^5 - 3x^3y^4 + 5x^2y^3) \div (-x^2y^3)$.

SOLUTION

$$\frac{8x^4y^5 - 3x^3y^4 + 5x^2y^3}{-x^2y^3} = \frac{8x^4y^5}{-x^2y^3} - \frac{3x^3y^4}{-x^2y^3} + \frac{5x^2y^3}{-x^2y^3}$$ Try to perform this step mentally.

$$= -8x^2y^2 + 3xy - 5$$

2. Divide:

$$(6c^2d^3 - 16c^4d^5 - 2c^2d^2) \div (-2c^2d^2).$$

YOUR TURN

> **DIVISION BY A MONOMIAL**
>
> To divide a polynomial by a monomial, divide each term of the polynomial by the monomial.

Dividing by a Polynomial

When the divisor has more than one term, we use a procedure very similar to long division in arithmetic.

EXAMPLE 3   Divide $2x^2 - 7x - 15$ by $x - 5$.

SOLUTION   We have

$$
\require{enclose}
\begin{array}{r}
2x \\
x - 5 \enclose{longdiv}{2x^2 - 7x - 15} \\
-(2x^2 - 10x) \\
\hline
3x - 15
\end{array}
$$

Divide $2x^2$ by x: $2x^2/x = 2x$.

Multiply $x - 5$ by $2x$.

Subtract by mentally changing signs and adding:
$-7x - (-10x) = -7x + 10x = 3x.$

We next divide the leading term of this remainder, $3x$, by the leading term of the divisor, x.

$$
\require{enclose}
\begin{array}{r}
2x + 3 \\
x - 5 \enclose{longdiv}{2x^2 - 7x - 15} \\
2x^2 - 10x \\
\hline
3x - 15 \\
-(3x - 15) \\
\hline
0
\end{array}
$$

Divide $3x$ by x: $3x/x = 3$.

Multiply $x - 5$ by 3.

Subtract.   Our remainder is now 0.

To check, we multiply the quotient and divisor and add any remainder. The result should be the dividend.

Check: $(x - 5)(2x + 3) = 2x^2 - 7x - 15$. The answer checks.

The quotient is $2x + 3$.

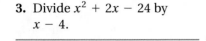 YOUR TURN

3. Divide $x^2 + 2x - 24$ by $x - 4$.

To understand why we perform long division as we do, note that Example 3 amounts to "filling in" an unknown polynomial:

$$(x - 5)(\quad ? \quad) = 2x^2 - 7x - 15.$$

We see that $2x$ must be in the unknown polynomial if we are to get the first term, $2x^2$, from the multiplication:

$$(x - 5)(2x \quad) = 2x^2 - 10x \neq 2x^2 - 7x - 15.$$

The $2x$ can be regarded as a (poor) approximation of the quotient that we are seeking. To see how far off the approximation is, we subtract:

$$
\begin{array}{r}
2x^2 - 7x - 15 \\
-(2x^2 - 10x) \\
\hline
3x - 15
\end{array}
\left.\begin{array}{l} \\ \\ \\ \end{array}\right\}
$$
Note where this appeared in the long division above.
←—This is the first remainder.

To get the needed terms, $3x - 15$, we need another term in the unknown polynomial. We use 3 because $(x - 5) \cdot 3$ is $3x - 15$:

$$
\begin{aligned}
(x - 5)(2x + 3) &= 2x^2 - 10x + 3x - 15 \\
&= 2x^2 - 7x - 15.
\end{aligned}
$$

Now when we subtract the product $(x - 5)(2x + 3)$ from $2x^2 - 7x - 15$, the remainder is 0.

If the remainder is not 0, we continue dividing until the degree of the remainder is less than the degree of the divisor.

EXAMPLE 4 Divide $x^2 + 5x + 8$ by $x + 3$.

SOLUTION We have

$$
\begin{array}{r}
x \\
x + 3 \overline{)x^2 + 5x + 8} \\
\underline{x^2 + 3x} \\
2x + 8.
\end{array}
$$

Divide the leading term of the dividend by the leading term of the divisor: $x^2/x = x$.
←—— Multiply x above by $x + 3$.
Subtract: $(x^2 + 5x) - (x^2 + 3x) = 2x$.
—— This is the first remainder.

We now focus on the current remainder, $2x + 8$, and repeat the process:

$$
\begin{array}{r}
x + 2 \\
x + 3 \overline{)x^2 + 5x + 8} \\
\underline{x^2 + 3x} \\
2x + 8 \\
\underline{2x + 6} \\
2
\end{array}
$$

←—— Divide the leading term of the first remainder by the leading term of the divisor: $2x/x = 2$.
→ $2x + 8$ ←— $2x + 8$ is the first remainder.
←— Multiply 2 by $x + 3$.
←— Subtract: $(2x + 8) - (2x + 6)$.
—— This is the final remainder.

The quotient is $x + 2$, with remainder 2. Note that the degree of the remainder, 2, is 0 and the degree of the divisor, $x + 3$, is 1. Since $0 < 1$, the process stops.

Student Notes

After studying Examples 3 and 4, try to explain why long division of numbers is performed as it is.

Check: $(x + 3)(x + 2) + 2 = x^2 + 5x + 6 + 2$ Add the remainder to the product.

$$= x^2 + 5x + 8$$ This is the original dividend.

We write our answer as $x + 2$, R 2, or as

$$\text{Quotient} + \frac{\text{Remainder}}{\text{Divisor}}$$

$$x + 2 \quad + \quad \frac{2}{x + 3}.$$

This is how answers are listed at the back of the book.

This last form of the answer can also be checked by multiplying:

$$(x + 3)\left[(x + 2) + \frac{2}{x + 3}\right] = (x + 3)(x + 2) + (x + 3)\frac{2}{x + 3}$$

         Using the distributive law

$$= x^2 + 5x + 6 + 2$$

$$= x^2 + 5x + 8.$$ This is the original dividend.

4. Divide $3x^2 + x - 5$ by $x + 1$.

_____ **↩ YOUR TURN**

TIPS FOR DIVIDING POLYNOMIALS

1. Arrange polynomials in descending order.
2. If there are missing terms in the dividend, either write them with 0 coefficients or leave space for them.
3. Perform the long-division process until the degree of the remainder is less than the degree of the divisor.

EXAMPLE 5   Divide: $(9a^2 + a^3 - 5) \div (a^2 - 1)$.

SOLUTION   We rewrite the polynomials in descending order:

$$(a^3 + 9a^2 - 5) \div (a^2 - 1).$$ There is no a-term in the dividend.

Thus we have

$$
\begin{array}{r}
a + 9 \\
a^2 - 1 \overline{)a^3 + 9a^2 + 0a - 5} \\
\underline{a^3 - a} \\
9a^2 + a - 5 \\
\underline{9a^2 - 9} \\
a + 4.
\end{array}
$$

When there is a missing term in the dividend, we can write it in, as shown here, or leave space, as in Example 6 below.

Subtracting: $0a - (-a) = a$.

The degree of the remainder is less than the degree of the divisor, so we are finished.

5. Divide:

$$(y^3 + 3 - 2y^2) \div (y^2 + 2).$$

_____

The answer is $a + 9 + \dfrac{a + 4}{a^2 - 1}$.

↩ YOUR TURN

EXAMPLE 6 Let $f(x) = 125x^3 - 8$ and $g(x) = 5x - 2$. If $F(x) = (f/g)(x)$, find a simplified expression for $F(x)$ and list all restrictions on the domain.

SOLUTION Recall that $(f/g)(x) = f(x)/g(x)$. Thus,

$$F(x) = \frac{125x^3 - 8}{5x - 2}.$$

We have

$$
\begin{array}{r}
25x^2 + 10x + 4 \\
5x - 2 \overline{)125x^3 \qquad\qquad - 8} \\
\underline{125x^3 - 50x^2} \\
50x^2 \qquad - 8 \\
\underline{50x^2 - 20x} \\
20x - 8 \\
\underline{20x - 8} \\
0.
\end{array}
$$

Leaving space for the missing terms

Subtracting:
$125x^3 - (125x^3 - 50x^2) = 50x^2$

Subtracting

6. Let $f(x) = 6x^2 + 7x - 3$ and $g(x) = 3x - 1$. If $F(x) = (f/g)(x)$, find a simplified expression for $F(x)$ and list all restrictions on the domain.

Note that, because $F(x) = f(x)/g(x)$, it follows that $g(x)$ cannot be 0. Since $g(x)$ is 0 for $x = \frac{2}{5}$ (check this), we have

$$F(x) = 25x^2 + 10x + 4, \quad \text{provided } x \neq \tfrac{2}{5}.$$

YOUR TURN

6.6 EXERCISE SET

FOR EXTRA HELP

MyMathLab® Math XL
PRACTICE   WATCH   READ   REVIEW

Vocabulary and Reading Check

Fill in each blank by referencing the following division.

$$
\begin{array}{r}
x + 2 \\
x - 3 \overline{)x^2 - x - 1} \\
\underline{x^2 - 3x} \\
2x - 1 \\
\underline{2x - 6} \\
5
\end{array}
$$

1. The divisor is _____ .

2. The dividend is _____ .

3. The quotient is _____ .

4. The remainder is _____ .

5. The degree of the divisor is _____ .

6. The degree of the remainder is _____ .

Dividing by a Monomial

Divide and check.

7. $\dfrac{36x^6 + 18x^5 - 27x^2}{9x^2}$

8. $\dfrac{30y^8 - 15y^6 + 40y^4}{5y^4}$

9. $\dfrac{21a^3 + 7a^2 - 3a - 14}{-7a}$

10. $\dfrac{-25x^3 + 20x^2 - 3x + 7}{-5x}$

11. $\dfrac{16y^4z^2 - 8y^6z^4 + 12y^8z^3}{-4y^4z}$

12. $\dfrac{6p^2w^2 - 9p^2w + 12pw^2}{-3pw}$

13. $(16y^3 - 9y^2 - 8y) \div (2y^2)$

14. $(6a^4 + 9a^2 - 8) \div (2a)$

15. $(15x^7 - 21x^4 - 3x^2) \div (-3x^2)$

16. $(36y^6 - 18y^4 - 12y^2) \div (-6y)$

17. $(a^2b - a^3b^3 - a^5b^5) \div (a^2b)$

18. $(x^3y^2 - x^3y^3 - x^4y^2) \div (x^2y^2)$

Dividing by a Polynomial

Divide and check.

19. $(x^2 + 10x + 21) \div (x + 7)$

20. $(y^2 - 8y + 16) \div (y - 4)$

21. $(y^2 - 10y - 25) \div (y - 5)$

22. $(a^2 - 8a - 16) \div (a + 4)$

23. $(x^2 - 9x + 21) \div (x - 4)$

24. $(y^2 - 11y + 33) \div (y - 6)$

Aha! **25.** $(y^2 - 25) \div (y + 5)$

26. $(a^2 - 81) \div (a - 9)$

27. $\dfrac{a^3 + 8}{a + 2}$ **28.** $\dfrac{t^3 + 27}{t + 3}$

29. $\dfrac{5x^2 - 14x}{5x + 1}$

30. $\dfrac{3x^2 - 7x}{3x - 1}$

31. $(y^3 - 4y^2 + 3y - 6) \div (y - 2)$

32. $(x^3 - 2x^2 + 4x - 5) \div (x + 3)$

33. $(2x^3 + 3x^2 - x - 3) \div (x + 2)$

34. $(3x^3 - 5x^2 - 3x - 2) \div (x - 2)$

35. $(a^3 - 10a + 24) \div (a + 4)$

36. $(x^3 - x + 6) \div (x + 2)$

37. $(6y^2 - 9y + 10y^3 + 10) \div (5y - 2)$

38. $(6x^3 + 11x - 11x^2 - 2) \div (2x - 3)$

39. $(3x^4 + x^3 - 8x^2 - 3x - 3) \div (x^2 - 3)$

40. $(2x^4 - 2x^3 + 3x^2 - 2x + 1) \div (x^2 + 1)$

41. $(2x^4 - x^3 - 5x^2 + x - 6) \div (x^2 + 2)$

42. $(3x^4 + 2x^3 - 11x^2 - 2x + 5) \div (x^2 - 2)$

For Exercises 43–50, $f(x)$ and $g(x)$ are as given. Find a simplified expression for $F(x)$ if $F(x) = (f/g)(x)$. (See Example 6.) Be sure to list all restrictions on the domain of $F(x)$.

43. $f(x) = 6x^2 - 11x - 10,\ g(x) = 3x + 2$

44. $f(x) = 8x^2 - 22x - 21,\ g(x) = 2x - 7$

45. $f(x) = 8x^3 - 27,\ g(x) = 2x - 3$

46. $f(x) = 64x^3 + 8,\ g(x) = 4x + 2$

47. $f(x) = x^4 - 24x^2 - 25,\ g(x) = x^2 - 25$

48. $f(x) = x^4 - 3x^2 - 54,\ g(x) = x^2 - 9$

49. $f(x) = 8x^2 - 3x^4 - 2x^3 + 2x^5 - 5,\ g(x) = x^2 - 1$

50. $f(x) = 4x - x^3 - 10x^2 + 3x^4 - 8,\ g(x) = x^2 - 4$

51. Explain how factoring could be used to solve Example 6.

52. Explain how to construct a polynomial of degree 4 that has a remainder of 3 when divided by $x + 1$.

Skill Review

Graph on a plane.

53. $3x - y = 9$ [2.4]

54. $5y = -15$ [2.4]

55. $y < \frac{5}{2}x$ [4.4]

56. $x + y \geq 3$ [4.4]

57. $y = -\frac{3}{4}x + 1$ [2.3]

58. $x \geq -1$ [4.4]

Synthesis

59. Explain how to construct a polynomial of degree 4 that has a remainder of 2 when divided by $x + c$.

60. Do addition, subtraction, and multiplication of polynomials always result in a polynomial? Does division? Why or why not?

Divide.

61. $(4a^3b + 5a^2b^2 + a^4 + 2ab^3) \div (a^2 + 2b^2 + 3ab)$

62. $(x^4 - x^3y + x^2y^2 + 2x^2y - 2xy^2 + 2y^3) \div (x^2 - xy + y^2)$

63. $(a^7 + b^7) \div (a + b)$

64. Find k such that when $x^3 - kx^2 + 3x + 7k$ is divided by $x + 2$, the remainder is 0.

65. When $x^2 - 3x + 2k$ is divided by $x + 2$, the remainder is 7. Find k.

66. Let
$$f(x) = \frac{3x + 7}{x + 2}.$$
a) Use division to find an expression equivalent to $f(x)$. Then graph f.
b) On the same set of axes, sketch the graphs of both $g(x) = 1/(x + 2)$ and $h(x) = 1/x$.
c) How do the graphs of f, g, and h compare?

67. Frank incorrectly states that
$$(x^3 + 9x^2 - 6) \div (x^2 - 1) = x + 9 + \frac{x + 4}{x^2 - 1}.$$
Without performing any long division, how could you show Frank that his division cannot possibly be correct?

68. Use a graphing calculator to check Example 3 by setting $y_1 = (2x^2 - 7x - 15)/(x - 5)$ and $y_2 = 2x + 3$. Then use either TRACE (after selecting the ZOOM ZINTEGER option) or TABLE (with TblMin = 0 and ΔTbl = 1) to show that $y_1 \neq y_2$ for $x = 5$.

69. Use a graphing calculator to check Example 5. Perform the check using
$$y_1 = (9x^2 + x^3 - 5)/(x^2 - 1),$$
$$y_2 = x + 9 + (x + 4)/(x^2 - 1), \text{ and } y_3 = y_2 - y_1.$$

70. *Business.* A company's **revenue** *R* from an item is defined as the price paid per item times the quantity of items sold.

a) Easy on the Eyes sells high-quality reproductions of original watercolors. Find an expression for the price paid per reproduction if the revenue from the sale of *q* reproductions is $(80q - q^2)$ dollars.

b) Find an expression for the price paid per reproduction if one more reproduction is sold but the revenue remains $(80q - q^2)$ dollars.

↳ YOUR TURN ANSWERS: SECTION 6.6

1. $2t^2 - 4t + \dfrac{2}{3} - \dfrac{3}{t}$ **2.** $-3d + 8c^2d^3 + 1$

3. $x + 6$ **4.** $3x - 2 + \dfrac{-3}{x + 1}$

5. $y - 2 + \dfrac{-2y + 7}{y^2 + 2}$ **6.** $2x + 3, x \neq \frac{1}{3}$

QUICK QUIZ: SECTIONS 6.1–6.6

1. Multiply and, if possible, simplify:
$$\frac{8t + 8}{2t^2 + t - 1} \cdot \frac{t^2 - 1}{t^2 - 2t + 1}. \quad [6.1]$$

2. Subtract and, if possible, simplify:
$$\frac{2n - 1}{n - 2} - \frac{n - 3}{n + 1}. \quad [6.2]$$

3. Divide: $(x^3 - x - 3) \div (x^2 + 1)$. [6.6]

Solve. [6.4]

4. $\dfrac{t + 4}{t - 1} = \dfrac{5}{t - 1}$ **5.** $\dfrac{10}{x + 1} = \dfrac{x}{x - 2}$

PREPARE TO MOVE ON

Given $f(x) = x^3 - 3x^2 - 6x + 10$, find each of the following. [2.2]

1. $f(1)$ **2.** $f(-1)$

3. $f(2)$ **4.** $f(-2)$

5. $f(-5)$

6.7 Synthetic Division and the Remainder Theorem

Synthetic Division ▪ The Remainder Theorem

Study Skills

If a Question Stumps You

Don't be surprised if a quiz or a test includes a question that you feel unsure about. Should this happen, simply skip the question and continue with the quiz or test, returning to the trouble spot after you have answered the other questions.

Synthetic Division

To divide a polynomial by a binomial of the type $x - a$, we can streamline the usual procedure to develop a process called *synthetic division*.

Compare the following. In each stage, we attempt to write less than in the previous stage, while retaining enough essentials to solve the problem.

When a polynomial is written in descending order, the coefficients provide the essential information.

Long Division

$$
\require{enclose}
\begin{array}{r}
4x^2 + 5x + 11 \\
x - 2 \enclose{longdiv}{4x^3 - 3x^2 + x + 7} \\
\underline{4x^3 - 8x^2} \\
5x^2 + x \\
\underline{5x^2 - 10x} \\
11x + 7 \\
\underline{11x - 22} \\
29
\end{array}
$$

Stage 1

$$
\begin{array}{r}
4 + 5 + 11 \\
1 - 2 \enclose{longdiv}{4 - 3 + 1 + 7} \\
\underline{4 - 8} \\
5 + 1 \\
\underline{5 - 10} \\
11 + 7 \\
\underline{11 - 22} \\
29
\end{array}
$$

Because the leading coefficient in $x - 2$ is 1, each time we multiply it by a term in the answer, the leading coefficient of that product is the same as the coefficient in the answer. In stage 2, rather than duplicate these numbers, we focus on where -2 is used. We also drop the 1 from the divisor. To simplify further, in stage 3, we reverse the sign of the -2 in the divisor and, in exchange, *add* at each step in the long division.

Stage 2

$$4 + 5 + 11$$
$$-2\,\overline{)4 - 3 + 1 + 7}$$

-8 ← Multiply: $-2 \cdot 4 = -8$.
$5 + 1$ Subtract: $-3 - (-8) = 5$.
-10 ← Multiply: $-2 \cdot 5 = -10$.
$11 + 7$ Subtract: $1 - (-10) = 11$.
-22 ← Multiply: $-2 \cdot 11 = -22$.
29 ← Subtract: $7 - (-22) = 29$.

Stage 3

$$4 + 5 + 11$$
$$2\,\overline{)4 - 3 + 1 + 7}$$ Replace the -2 with 2.

8 ← Multiply: $2 \cdot 4 = 8$.
$5 + 1$ Add: $-3 + 8 = 5$.
10 ← Multiply: $2 \cdot 5 = 10$.
$11 + 7$ Add: $1 + 10 = 11$.
22 ← Multiply: $2 \cdot 11 = 22$.
29 ← Add: $7 + 22 = 29$.

Student Notes

You will not need to write out all five stages when performing synthetic division on your own. We show the steps to help you understand the reasoning behind the method.

The blue numbers can be eliminated if we look at the red numbers instead, as shown in stage 4 below. Note that the 5 and the 11 preceding the remainder 29 coincide with the 5 and the 11 following the 4 on the top line. By writing a 4 to the left of 5 on the bottom line, we can eliminate the top line in stage 4 and read our answer from the bottom line. This final stage is commonly called **synthetic division**.

Stage 4

$$\begin{array}{r}4 \quad 5 \quad 11 \\ 2\,\overline{)4 \; -3 \; 1 \; 7} \\ 8 \; 10 \; 22 \\ \hline 5 \; 11 \; 29\end{array}$$

Stage 5

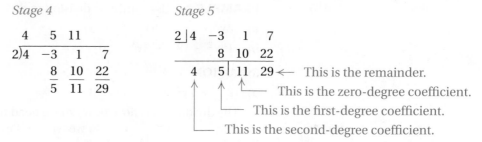

The quotient is $4x^2 + 5x + 11$ with a remainder of 29.

Remember that in order for this method to work, the divisor must be of the form $x - a$, that is, a variable minus a constant. Note that the coefficient of the variable is 1.

Before using synthetic division in Example 1, let's review how stage 5, above, is formed.

This is the constant being subtracted in the divisor. → $2\,|\,4 \; -3 \; 1 \; 7$ ← These are the coefficients in the dividend.

This line is formed by adding in each column. → $8 \; 10 \; 22$ ← These are found by multiplying 2 by the number below and to the left.
$4 \; 5 \; 11 \,|\, 29$

This is found first by bringing down the 4.

EXAMPLE 1 Use synthetic division to divide: $(x^3 + 6x^2 - x - 30) \div (x - 2)$.

SOLUTION

$$\underline{2 \,|}\; 1 \quad 6 \quad -1 \quad -30$$

Write the 2 of $x - 2$ and the coefficients of the dividend.

$$1$$

Bring down the first coefficient.

$$\underline{2 \,|}\; 1 \quad 6 \quad -1 \quad -30$$
$$\quad 2$$
$$1 \quad 8$$

Multiply 1 by 2 to get 2.
Add 6 and 2.

$$\underline{2 \,|}\; 1 \quad 6 \quad -1 \quad -30$$
$$\quad 2 \quad 16$$
$$1 \quad 8 \quad 15$$

Multiply 8 by 2.
Add −1 and 16.

$$\underline{2 \,|}\; 1 \quad 6 \quad -1 \quad -30$$
$$\quad 2 \quad 16 \quad 30$$
$$1 \quad 8 \quad 15 \quad 0$$

Multiply 15 by 2 and add.

The answer is $x^2 + 8x + 15$ with R 0, or just $x^2 + 8x + 15$.

YOUR TURN

1. Use synthetic division to divide:

$(x^3 + 2x^2 - 17x + 6) \div (x - 3)$.

EXAMPLE 2 Use synthetic division to divide.
a) $(2x^3 + 7x^2 - 5) \div (x + 3)$
b) $(10x^2 - 13x + 3x^3 - 20) \div (4 + x)$

SOLUTION

a) $(2x^3 + 7x^2 - 5) \div (x + 3)$

The dividend has no x-term, so we need to write 0 as the coefficient of x. Note that $x + 3 = x - (-3)$, so we write −3 inside the ⌋.

$$\underline{-3 \,|}\; 2 \quad 7 \quad 0 \quad -5$$
$$\quad -6 \quad -3 \quad 9$$
$$2 \quad 1 \quad -3 \;|\; 4$$

The answer is $2x^2 + x - 3$, with R 4, or $2x^2 + x - 3 + \dfrac{4}{x + 3}$.

b) We first rewrite $(10x^2 - 13x + 3x^3 - 20) \div (4 + x)$ in descending order:

$$(3x^3 + 10x^2 - 13x - 20) \div (x + 4).$$

Next, we use synthetic division. Note that $x + 4 = x - (-4)$.

$$\underline{-4 \,|}\; 3 \quad 10 \quad -13 \quad -20$$
$$\quad -12 \quad 8 \quad 20$$
$$3 \quad -2 \quad -5 \;|\; 0$$

The answer is $3x^2 - 2x - 5$.

YOUR TURN

2. Use synthetic division to divide:

$(3x - 9 + 2x^3) \div (x + 2)$.

The Remainder Theorem

When a polynomial function $f(x)$ is divided by $x - a$, the remainder is related to the function value $f(a)$. Compare the following from Examples 1 and 2.

Polynomial Function	Divisor	Remainder	Function Value
Example 1: $f(x) = x^3 + 6x^2 - x - 30$	$x - 2$	0	$f(2) = 0$
Example 2(a): $g(x) = 2x^3 + 7x^2 - 5$	$x + 3$, or $x - (-3)$	4	$g(-3) = 4$
Example 2(b): $p(x) = 10x^2 - 13x + 3x^3 - 20$	$4 + x$, or $x - (-4)$	0	$p(-4) = 0$

When the remainder is 0, the divisor $x - a$ is a factor of the polynomial being divided. We can write $f(x) = x^3 + 6x^2 - x - 30$ in Example 1 as

$$f(x) = (x - 2)(x^2 + 8x + 15).$$

Then

$$\left. \begin{array}{l} f(2) = (2 - 2)(2^2 + 8 \cdot 2 + 15) \\ = 0(2^2 + 8 \cdot 2 + 15) = 0. \end{array} \right\} \quad \text{Since } x - 2 \text{ is a factor, } f(2) = 0.$$

Thus, when the remainder is 0 after division by $x - a$, the function value $f(a)$ is also 0. Remarkably, this pattern extends to nonzero remainders as well. The fact that the remainder and the function value coincide is predicted by the remainder theorem.

THE REMAINDER THEOREM
The remainder obtained by dividing $P(x)$ by $x - r$ is $P(r)$.

A proof of this result is outlined in Exercise 37.

EXAMPLE 3 Let $f(x) = 8x^5 - 6x^3 + x - 8$. Use synthetic division to find $f(2)$.

SOLUTION The remainder theorem tells us that $f(2)$ is the remainder when $f(x)$ is divided by $x - 2$. We use synthetic division to find that remainder:

$$\begin{array}{r|rrrrrr} 2 & 8 & 0 & -6 & 0 & 1 & -8 \\ & & 16 & 32 & 52 & 104 & 210 \\ \hline & 8 & 16 & 26 & 52 & 105 & 202 \end{array}$$

Although the bottom line can be used to find the quotient for the division $(8x^5 - 6x^3 + x - 8) \div (x - 2)$, what we are really interested in is the remainder. It tells us that $f(2) = 202$. The calculations are easier than the more typical calculation of $f(2)$.

YOUR TURN

The remainder theorem is often used to check division. Thus Example 2(a) can be checked by computing $g(-3) = 2(-3)^3 + 7(-3)^2 - 5$. Since $g(-3) = 4$ and the remainder in Example 2(a) is also 4, our division was probably correct.

3. Let
$$f(x) = 5x^4 + 3x^2 + 2x - 7.$$
Use synthetic division to find $f(-2)$.

6.7 EXERCISE SET

FOR EXTRA HELP

MyMathLab® MathXL
PRACTICE   WATCH   READ   REVIEW

Vocabulary and Reading Check

Classify each of the following statements as either true or false.

1. If $P(-5) = 39$ and $P(x) = x^3 + 7x^2 + 3x + 4$, then

$$\begin{array}{r|rrrr} -5 & 1 & 7 & 3 & 4 \\ & & -5 & -10 & 35 \\ \hline & 1 & 2 & -7 & 39 \end{array}$$

2. In order to use synthetic division, we must be sure that the divisor is of the form $x - a$.

3. Synthetic division can be used in problems in which long division could not be used.

4. In order for $f(x)/g(x)$ to exist, $g(x)$ must be 0.

5. If $x - 2$ is a factor of some polynomial $P(x)$, then $P(2) = 0$.

6. If $p(3) = 0$ for some polynomial $p(x)$, then $x - 3$ is a factor of $p(x)$.

Synthetic Division

Use synthetic division to divide.

7. $(x^3 - 4x^2 - 2x + 5) \div (x - 1)$

8. $(x^3 - 4x^2 + 5x - 6) \div (x - 3)$

9. $(a^2 + 8a + 11) \div (a + 3)$

10. $(a^2 + 8a + 11) \div (a + 5)$

11. $(2x^3 - x^2 - 7x + 14) \div (x + 2)$

12. $(3x^3 - 10x^2 - 9x + 15) \div (x - 4)$

13. $(a^3 - 10a + 12) \div (a - 2)$

14. $(a^3 - 14a + 15) \div (a - 3)$

15. $(3y^3 - 7y^2 - 20) \div (y - 3)$

16. $(2x^3 - 3x^2 + 8) \div (x + 2)$

17. $(x^5 - 32) \div (x - 2)$

18. $(y^5 - 1) \div (y - 1)$

19. $(3x^3 + 1 - x + 7x^2) \div \left(x + \frac{1}{3}\right)$

20. $(8x^3 - 1 + 7x - 6x^2) \div \left(x - \frac{1}{2}\right)$

The Remainder Theorem

Use synthetic division to find the indicated function value.

21. $f(x) = 5x^4 + 12x^3 + 28x + 9;\ f(-3)$

22. $g(x) = 3x^4 - 25x^2 - 18;\ g(3)$

23. $P(x) = 2x^4 - x^3 - 7x^2 + x + 2;\ P(-3)$

24. $F(x) = 3x^4 + 8x^3 + 2x^2 - 7x - 4;\ F(-2)$

25. $f(x) = x^4 - 6x^3 + 11x^2 - 17x + 20;\ f(4)$

26. $p(x) = x^4 + 7x^3 + 11x^2 - 7x - 12;\ p(2)$

27. Why is it that we *add* when performing synthetic division, but *subtract* when performing long division?

28. Explain how synthetic division could be useful when attempting to factor a polynomial.

Skill Review

Find a linear function whose graph has the given characteristics.

29. Slope: 3; y-intercept: $(0, -4)$ [2.3]

30. Slope: $\frac{1}{2}$; contains $(-6, 3)$ [2.5]

31. Contains $(1, 7)$ and $(4, 2)$ [2.5]

32. Parallel to the graph of $2x - y = 7$; contains $(8, 0)$ [2.4]

33. Perpendicular to the graph of $2x - y = 7$; contains $(0, 6)$ [2.4]

34. Horizontal; contains $(-9, 5)$ [2.4]

Synthesis

35. Let $Q(x)$ be a polynomial function with $p(x)$ a factor of $Q(x)$. If $p(3) = 0$, does it follow that $Q(3) = 0$? Why or why not? If $Q(3) = 0$, does it follow that $p(3) = 0$? Why or why not?

36. What adjustments must be made if synthetic division is to be used to divide a polynomial by a binomial of the form $ax + b$, with $a > 1$?

37. To prove the remainder theorem, note that any polynomial $P(x)$ can be rewritten as $(x - r) \cdot Q(x) + R$, where $Q(x)$ is the quotient polynomial that arises when $P(x)$ is divided by $x - r$, and R is some constant (the remainder).

    a) How do we know that R must be a constant?
    b) Show that $P(r) = R$ (this says that $P(r)$ is the remainder when $P(x)$ is divided by $x - r$).

38. Let $f(x) = 6x^3 - 13x^2 - 79x + 140$. Find $f(4)$ and then solve the equation $f(x) = 0$.

39. Let $f(x) = 4x^3 + 16x^2 - 3x - 45$. Find $f(-3)$ and then solve the equation $f(x) = 0$.

40. Use the TRACE feature on a graphing calculator to check your answer to Exercise 38.

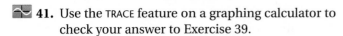 **41.** Use the TRACE feature on a graphing calculator to check your answer to Exercise 39.

Nested Evaluation. *One way to evaluate a polynomial function like* $P(x) = 3x^4 - 5x^3 + 4x^2 - 1$ *is to successively factor out x as shown:*

$$P(x) = x(x(x(3x - 5) + 4) + 0) - 1.$$

Computations are then performed using this "nested" form of P(x).

42. Use nested evaluation to find $f(4)$ in Exercise 38. Note the similarities to the calculations performed with synthetic division.

43. Use nested evaluation to find $f(-3)$ in Exercise 39. Note the similarities to the calculations performed with synthetic division.

↘ **YOUR TURN ANSWERS: SECTION 6.7**

1. $x^2 + 5x - 2$ **2.** $2x^2 - 4x + 11 + \dfrac{-31}{x + 2}$ **3.** 81

QUICK QUIZ: SECTIONS 6.1–6.7

1. Simplify: $\dfrac{x^{-1} - y^{-1}}{2xy^{-2}}$.   [6.3]

2. Divide: $(2x^3 - x^2 - 11x - 5) \div (2x + 1)$.   [6.6]

3. Use synthetic division to divide: [6.7]
$$(x^4 - 5x^2 + 3) \div (x - 3).$$

Solve.

4. $\dfrac{x + 2}{x + 3} = \dfrac{2x + 1}{x - 1}$ [6.4]

5. LeBron's Mercruiser travels 15 km/h in still water. He motors 140 km downstream in the same time that it takes to travel 35 km upstream. What is the speed of the river?   [6.5]

PREPARE TO MOVE ON

Solve.   [1.5]

1. $ac = b$, for c **2.** $x - wz = y$, for w

3. $pq - rq = st$, for q **4.** $ab = d - cb$, for b

5. $ab - cd = 3b + d$, for b

6.8 Formulas, Applications, and Variation

Formulas ▪ Direct Variation ▪ Inverse Variation ▪ Joint Variation and Combined Variation

Formulas

Formulas occur frequently as mathematical models. Many formulas contain rational expressions, and to solve such formulas for a specified letter, we proceed as we do when solving rational equations.

EXAMPLE 1 *Electronics.* The formula

$$\frac{1}{R} = \frac{1}{r_1} + \frac{1}{r_2}$$

is used by electricians to determine the resistance R of two resistors r_1 and r_2 connected in parallel. Solve for r_1.

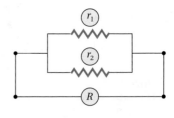

SOLUTION We use the same approach as in Section 6.4:

$$Rr_1r_2 \cdot \frac{1}{R} = Rr_1r_2 \cdot \left(\frac{1}{r_1} + \frac{1}{r_2} \right) \qquad \text{Multiplying both sides by the LCM of the denominators}$$

$$Rr_1r_2 \cdot \frac{1}{R} = Rr_1r_2 \cdot \frac{1}{r_1} + Rr_1r_2 \cdot \frac{1}{r_2} \qquad \text{Multiplying to remove parentheses}$$

$$r_1r_2 = Rr_2 + Rr_1. \qquad \text{Simplifying by removing factors equal to 1: } \frac{R}{R} = 1; \frac{r_1}{r_1} = 1; \frac{r_2}{r_2} = 1$$

Student Notes

The subscripts 1 and 2 in Example 1 indicate that r_1 and r_2 are different variables representing similar quantities.

At this point it is tempting to multiply by $1/r_2$ to get r_1 alone on the left, *but* note that there is an r_1 on the right. We must get all the terms involving r_1 on the *same side* of the equation.

$$r_1r_2 - Rr_1 = Rr_2 \qquad \text{Subtracting } Rr_1 \text{ from both sides}$$

$$r_1(r_2 - R) = Rr_2 \qquad \text{Factoring out } r_1 \text{ in order to combine like terms}$$

$$r_1 = \frac{Rr_2}{r_2 - R} \qquad \text{Dividing both sides by } r_2 - R \text{ to get } r_1 \text{ alone}$$

1. Solve $\dfrac{1}{x} = \dfrac{1}{y} + \dfrac{1}{z}$ for x.

This formula can be used to calculate r_1 whenever R and r_2 are known.

 YOUR TURN

EXAMPLE 2   *Astronomy.*   The formula

$$\frac{V^2}{R^2} = \frac{2g}{R + h}$$

is used to find a satellite's *escape velocity V*, where R is a planet's radius, h is the satellite's height above the planet, and g is the planet's gravitational constant. Solve for h.

SOLUTION   We first multiply by the LCM, $R^2(R + h)$, to clear fractions:

$$\frac{V^2}{R^2} = \frac{2g}{R + h}$$

$$R^2(R + h)\frac{V^2}{R^2} = R^2(R + h)\frac{2g}{R + h} \qquad \text{Multiplying to clear fractions}$$

$$\frac{R^2(R + h)V^2}{R^2} = \frac{R^2(R + h)2g}{R + h}$$

$$(R + h)V^2 = R^2 \cdot 2g. \qquad \begin{array}{l}\text{Removing factors equal to 1:} \\ \dfrac{R^2}{R^2} = 1 \text{ and } \dfrac{R + h}{R + h} = 1\end{array}$$

Remember: We are solving for h. Although we could distribute V^2, since h appears only within the factor $R + h$, it is easier to divide both sides by V^2:

$$\frac{(R + h)V^2}{V^2} = \frac{2R^2g}{V^2} \qquad \text{Dividing both sides by } V^2$$

$$R + h = \frac{2R^2g}{V^2} \qquad \text{Removing a factor equal to 1: } \frac{V^2}{V^2} = 1$$

2. Solve

$$\frac{a}{x} = \frac{b}{x + c}$$

for c.

$$h = \frac{2R^2g}{V^2} - R. \qquad \text{Subtracting } R \text{ from both sides}$$

The last equation can be used to determine the height of a satellite above a planet when the planet's radius and gravitational constant, along with the satellite's escape velocity, are known.

 YOUR TURN

EXAMPLE 3   *Acoustics (the Doppler Effect).*   The formula

$$f = \frac{sg}{s + v}$$

is used to determine the frequency f of a sound that is moving at velocity v toward a listener who hears the sound as frequency g. Here s is the speed of sound in a particular medium. Solve for s.

Student Notes

The steps used to solve equations are precisely the same steps used to solve formulas. If you feel "rusty" in this regard, study the earlier section in which this type of equation first appears. When you can consistently solve those equations, you are ready to work with formulas.

3. Solve

$$w = \frac{tv}{v + x}$$

for v.

SOLUTION   We first clear fractions by multiplying by the LCM, $s + v$:

$$f \cdot (s + v) = \frac{sg}{s + v}(s + v)$$

$$fs + fv = sg. \qquad \text{The variable for which we are solving, } s, \text{ appears on both sides, forcing us to distribute on the left side.}$$

Next, we must get all terms containing s on one side:

$$fv = sg - fs \qquad \text{Subtracting } fs \text{ from both sides}$$

$$fv = s(g - f) \qquad \text{Factoring out } s. \text{ This is like combining like terms.}$$

$$\frac{fv}{g - f} = s. \qquad \text{Dividing both sides by } g - f$$

Since s is isolated on one side, we have solved for s. This last equation can be used to determine the speed of sound whenever f, v, and g are known.

YOUR TURN

TO SOLVE A RATIONAL EQUATION FOR A SPECIFIED VARIABLE

1. Multiply both sides by the LCM of the denominators to clear fractions, if necessary.
2. Multiply to remove parentheses, if necessary.
3. Get all terms with the specified variable alone on one side.
4. Factor out the specified variable if it is in more than one term.
5. Multiply or divide on both sides to isolate the specified variable.

Variation

We now examine three real-world situations: direct variation, inverse variation, and combined variation.

DIRECT VARIATION

A fitness trainer earns $22 per hour. In 1 hr, $22 is earned. In 2 hr, $44 is earned. In 3 hr, $66 is earned, and so on. This gives rise to a set of ordered pairs:

$$(1, 22), (2, 44), (3, 66), (4, 88), \quad \text{and so on.}$$

Note that the ratio of earnings E to time t is $\frac{22}{1}$ in every case.

If a situation is modeled by pairs for which the ratio is constant, we say there is **direct variation**. Here earnings *vary directly* as the time:

We have $\dfrac{E}{t} = 22$, so $E = 22t$ or, using function notation, $E(t) = 22t$.

DIRECT VARIATION

When a situation is modeled by a linear function of the form $f(x) = kx$, or $y = kx$, where k is a nonzero constant, we say that there is *direct variation*, that *y varies directly* as *x*, or that *y is proportional to x*. The number k is called the *variation constant*, or the *constant of proportionality*.

Note that for $k > 0$, any equation of the form $y = kx$ indicates that as x increases, y increases as well.

EXAMPLE 4   Find the variation constant and an equation of variation if y varies directly as x, and $y = 32$ when $x = 2$.

SOLUTION   We know that $(2, 32)$ is a solution of $y = kx$. Therefore,

$$32 = k \cdot 2 \qquad \text{Substituting}$$

$$\frac{32}{2} = k, \quad \text{or} \quad k = 16. \qquad \text{Solving for } k$$

4. Find the variation constant and an equation of variation if y varies directly as x, and $y = 5$ when $x = 10$.

The variation constant is 16. The equation of variation is $y = 16x$. The notation $y(x) = 16x$ or $f(x) = 16x$ is also used.

YOUR TURN

EXAMPLE 5   *Ocean Waves.*   The speed v of a train of ocean waves varies directly as the swell period t, the time between successive waves. Waves with a swell period of 12 sec are traveling 21 mph. How fast are waves traveling that have a swell period of 20 sec?

Source: www.rodntube.com

SOLUTION

1. **Familiarize.** Because of the phrase "v ... varies directly as ... t," we express the speed of the wave v, in miles per hour, as a function of the swell period t, in seconds. Thus, $v(t) = kt$, where k is the variation constant. Because we are using ratios, we can use the units "seconds" and "miles per hour" without converting sec to hr or hr to sec. Knowing that waves with a swell period of 12 sec are traveling 21 mph, we have $v(12) = 21$.

2. **Translate.** We find the variation constant using the data and then use it to write the equation of variation:

$$v(t) = kt$$

$$v(12) = k \cdot 12 \qquad \text{Replacing } t \text{ with 12}$$

$$21 = k \cdot 12 \qquad \text{Replacing } v(12) \text{ with 21}$$

$$\frac{21}{12} = k \qquad \text{Solving for } k$$

$$1.75 = k. \qquad \text{This is the variation constant.}$$

5. The amount of vegetables produced varies directly as the amount spent on seeds and fertilizer. According to a recent survey, an investment of \$25 for seeds and fertilizer can produce vegetables worth \$625 at a grocery store. What is the market value of vegetables produced with a \$40 investment in seeds and fertilizer?

Source: W. Atlee Burpee & Co.

The equation of variation is $v(t) = 1.75t$. This is the translation.

3. **Carry out.** To find the speed of waves with a swell period of 20 sec, we compute $v(20)$:

$$v(t) = 1.75t$$

$$v(20) = 1.75(20) \qquad \text{Substituting 20 for } t$$

$$= 35.$$

4. **Check.** To check, we could reexamine all our calculations. Note that our answer seems reasonable since the ratios $21/12$ and $35/20$ are both 1.75.

5. **State.** Waves with a swell period of 20 sec are traveling 35 mph.

YOUR TURN

Study Skills

Visualize the Steps

If you have completed all assignments and are studying for a quiz or a test, a productive use of your time is to reread the assigned problems, making certain that you can visualize the steps that lead to a solution. When you are unsure of how to solve a problem, work out that problem in its entirety, seeking outside help as needed.

INVERSE VARIATION

Suppose a bus travels 20 mi. At 20 mph, the trip takes 1 hr. At 40 mph, it takes $\frac{1}{2}$ hr. At 60 mph, it takes $\frac{1}{3}$ hr, and so on. This gives pairs of numbers, all having the same product:

$$(20, 1), \left(40, \tfrac{1}{2}\right), \left(60, \tfrac{1}{3}\right), \left(80, \tfrac{1}{4}\right), \quad \text{and so on.}$$

Note that the product of each pair is 20. When a situation is modeled by pairs for which the product is constant, we say that there is **inverse variation**. Since $r \cdot t = 20$, we have

$$t = \frac{20}{r} \quad \text{or, using function notation,} \quad t(r) = \frac{20}{r}.$$

INVERSE VARIATION

When a situation is modeled by a rational function of the form $f(x) = k/x$, or $y = k/x$, where k is a nonzero constant, we say that there is *inverse variation,* that *y varies inversely as x,* or that *y is inversely proportional to x.* The number k is called the *variation constant,* or the *constant of proportionality.*

Note that for $k > 0$, any equation of the form $y = k/x$ indicates that as x increases, y decreases.

EXAMPLE 6 Find the variation constant and an equation of variation if y varies inversely as x, and $y = 32$ when $x = 0.2$.

SOLUTION   We know that $(0.2, 32)$ is a solution of

$$y = \frac{k}{x}.$$

Therefore,

$$32 = \frac{k}{0.2} \qquad \text{Substituting}$$

$$(0.2)32 = k \qquad \text{Multiplying both sides by 0.2}$$

$$6.4 = k. \qquad \text{Solving for } k$$

The variation constant is 6.4. The equation of variation is

$$y = \frac{6.4}{x}.$$

6. Find the variation constant and an equation of variation if y varies inversely as x, and $y = \frac{1}{2}$ when $x = 10$.

 YOUR TURN

There are many real-life quantities that vary inversely.

EXAMPLE 7 *Fuel Efficiency.* The number of gallons of fuel that a vehicle uses varies inversely as its fuel efficiency. Maria's 2011 Ford Escape gets 25 mpg (miles per gallon) in combined city and highway driving. Last year, her vehicle used 480 gal of gasoline. How many gallons of gasoline would she have used if she had driven a 2012 Ford Focus, which gets 30 mpg, instead?

Source: www.fueleconomy.gov

SOLUTION

1. **Familiarize.** Because of the phrase ". . . varies inversely as its fuel efficiency," we express the number of gallons used g as a function of the fuel efficiency m, in miles per gallon. Thus we have $g(m) = k/m$.

2. **Translate.** We use the given information to solve for k. We will then use that result to write the equation of variation.

$$g(m) = \frac{k}{m}$$

$$g(25) = \frac{k}{25} \quad \text{Replacing } m \text{ with 25}$$

$$480 = \frac{k}{25} \quad \text{Replacing } g(25) \text{ with 480}$$

$$12,000 = k.$$

The equation of variation is $g(m) = 12,000/m$. This is the translation.

3. **Carry out.** To determine the number of gallons of gasoline used by a vehicle getting 30 mpg, we calculate $g(30)$:

$$g(30) = \frac{12,000}{30} = 400.$$

4. **Check.** Note that, as expected, as the number of miles per gallon goes *up*, the gasoline usage goes *down*. Also, note that the product of the number of gallons of gasoline used and the number of miles per gallon gives us the number of miles driven. Both (480 gal) (25 mpg) and (400 gal) (30 mpg) are 12,000 miles.

5. **State.** If Maria had driven the 2012 Focus, she would have used 400 gal of gasoline.

YOUR TURN

7. The time t that it takes to download a movie file varies inversely as the transfer speed x of the Internet connection. A typical full-length movie file will transfer in 48 min at a transfer speed of 256 KB/s (kilobytes per second). How long will it take to transfer the same movie file at a transfer speed of 32 KB/s?

Source: www.xsvidmovies.com

JOINT VARIATION AND COMBINED VARIATION

When a variable varies directly with more than one other variable, we say that there is *joint variation*. For example, in the formula for the volume of a right circular cylinder, $V = \pi r^2 h$, we say that V varies *jointly* as h and the square of r.

> **JOINT VARIATION**
> y varies *jointly* as x and z if, for some nonzero constant k, $y = kxz$.

EXAMPLE 8 Find an equation of variation if y varies jointly as x and z, and $y = 30$ when $x = 2$ and $z = 3$.

SOLUTION We have

$$y = kxz$$
$$30 = k \cdot 2 \cdot 3 \quad \text{Substituting}$$
$$k = 5. \quad \text{The variation constant is 5.}$$

The equation of variation is $y = 5xz$.

8. Find an equation of variation if y varies jointly as x and z, and $y = 16$ when $x = 8$ and $z = 6$.

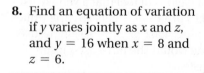 YOUR TURN

Joint variation is one form of *combined variation*. In general, when a variable varies directly and/or inversely, at the same time, with more than one other variable, there is **combined variation**. Examples 8 and 9 are both examples of combined variation.

EXAMPLE 9 Find an equation of variation if y varies jointly as x and z and inversely as the square of w, and $y = 105$ when $x = 3$, $z = 20$, and $w = 2$.

SOLUTION The equation of variation is of the form

$$y = k \cdot \frac{xz}{w^2}.$$ "Inversely as w^2" indicates that w^2 is in the denominator.

Substituting, we have

$$105 = k \cdot \frac{3 \cdot 20}{2^2}$$

$$105 = k \cdot 15 \qquad \text{Simplifying}$$

$$k = 7. \qquad \text{Solving for } k$$

The equation of variation is

$$y = 7 \cdot \frac{xz}{w^2}.$$

Chapter Resource:
Decision Making: Connection, p. 424

9. Find an equation of variation if y varies jointly as x and the square of z and inversely as w, and $y = 12$ when $x = 5$, $z = 3$, and $w = 15$.

YOUR TURN

6.8 EXERCISE SET

FOR EXTRA HELP

MyMathLab® Math XP
PRACTICE   WATCH   READ   REVIEW

Vocabulary and Reading Check

Complete each statement with the correct term from the following list.

a) Directly
b) Inversely
c) Jointly
d) LCM
e) Product
f) Ratio

1. To clear fractions, we can multiply both sides of an equation by the ____.

2. With direct variation, pairs of numbers have a constant ____.

3. With inverse variation, pairs of numbers have a constant ____.

4. If $y = k/x$, then y varies ____ as x.

5. If $y = kx$, then y varies ____ as x.

6. If $y = kxz$, then y varies ____ as x and z.

Concept Reinforcement

Determine whether each situation represents direct variation or inverse variation.

7. Two painters can scrape a house in 9 hr, whereas three painters can scrape the house in 6 hr.

8. Andres planted 5 bulbs in 20 min and 7 bulbs in 28 min.

9. Salma swam 2 laps in 7 min and 6 laps in 21 min.

10. It took 2 band members 80 min to set up for a show; with 4 members working, it took 40 min.

11. It took 3 hr for 4 volunteers to wrap the campus' collection of Toys for Tots, but only 1.5 hr with 8 volunteers working.

12. Ayana's air conditioner cooled off 1000 ft³ in 10 min and 3000 ft³ in 30 min.

Formulas

Solve each formula for the specified variable.

13. $f = \dfrac{L}{d}$; d

14. $\dfrac{W_1}{W_2} = \dfrac{d_1}{d_2}$; W_1

15. $s = \dfrac{(v_1 + v_2)t}{2}$; v_1

16. $s = \dfrac{(v_1 + v_2)t}{2}$; t

17. $\dfrac{t}{a} + \dfrac{t}{b} = 1$; b

18. $\dfrac{1}{R} = \dfrac{1}{r_1} + \dfrac{1}{r_2}$; R

19. $R = \dfrac{gs}{g + s}$; g

20. $K = \dfrac{rt}{r - t}$; t

21. $I = \dfrac{nE}{R + nr}$; n

22. $I = \dfrac{nE}{R + nr}$; r

23. $\dfrac{1}{p} + \dfrac{1}{q} = \dfrac{1}{f}$; q

24. $\dfrac{1}{p} + \dfrac{1}{q} = \dfrac{1}{f}$; p

25. $S = \dfrac{H}{m(t_1 - t_2)}$; t_1

26. $S = \dfrac{H}{m(t_1 - t_2)}$; H

27. $\dfrac{E}{e} = \dfrac{R + r}{r}$; r

28. $\dfrac{E}{e} = \dfrac{R + r}{R}$; R

29. $S = \dfrac{a}{1 - r}$; r

30. $S = \dfrac{a - ar^n}{1 - r}$; a

Aha! **31.** $c = \dfrac{f}{(a + b)c}$; $a + b$

32. $d = \dfrac{g}{d(c + f)}$; $c + f$

33. *Interest.* The formula

$$P = \dfrac{A}{1 + r}$$

is used to determine what principal P should be invested for one year at $(100 \cdot r)\%$ simple interest in order to have A dollars after one year. Solve for r.

34. *Taxable Interest.* The formula

$$I_t = \dfrac{I_f}{1 - T}$$

gives the *taxable interest rate* I_t equivalent to the *tax-free interest rate* I_f for a person in the $(100 \cdot T)\%$ tax bracket. Solve for T.

35. *Average Speed.* The formula

$$v = \dfrac{d_2 - d_1}{t_2 - t_1}$$

gives an object's average speed v when that object has traveled d_1 miles in t_1 hours and d_2 miles in t_2 hours. Solve for t_1.

36. *Average Acceleration.* The formula

$$a = \dfrac{v_2 - v_1}{t_2 - t_1}$$

gives a vehicle's *average acceleration* when its velocity changes from v_1 at time t_1 to v_2 at time t_2. Solve for t_2.

37. *Work Rate.* The formula

$$\dfrac{1}{t} = \dfrac{1}{a} + \dfrac{1}{b}$$

gives the total time t required for two workers to complete a job, if the workers' individual times are a and b. Solve for t.

38. *Planetary Orbits.* The formula

$$\dfrac{x^2}{a^2} + \dfrac{y^2}{b^2} = 1$$

can be used to plot a planet's elliptical orbit of width $2a$ and length $2b$. Solve for b^2.

39. *Semester Average.* The formula

$$A = \dfrac{2Tt + Qq}{2T + Q}$$

gives a student's average A after T tests and Q quizzes, where each test counts as 2 quizzes, t is the test average, and q is the quiz average. Solve for Q.

40. *Astronomy.* The formula

$$L = \dfrac{dR}{D - d},$$

where D is the diameter of the sun, d is the diameter of the earth, R is the earth's distance from the sun, and L is some fixed distance, is used in calculating when lunar eclipses occur. Solve for D.

41. *Body-Fat Percentage.* The YMCA calculates men's body-fat percentage p using the formula

$$p = \dfrac{-98.42 + 4.15c - 0.082w}{w},$$

where c is the waist measurement, in inches, and w is the weight, in pounds. Solve for w.

Source: YMCA guide to Physical Fitness Assessment

42. *Preferred Viewing Distance.* Researchers model the distance D from which an observer prefers to watch television in "picture heights"—that is, multiples of the height of the viewing screen. The preferred viewing distance is given by

$$D = \dfrac{3.55H + 0.9}{H},$$

where D is in picture heights and H is in meters. Solve for H.

Source: www.tid.es, Telefonica Investigación y Desarrollo, S.A. Unipersonal

Direct Variation

Find the variation constant and an equation of variation if y varies directly as x and the following conditions apply.

43. $y = 30$ when $x = 5$

44. $y = 80$ when $x = 16$

45. $y = 3.4$ when $x = 2$

46. $y = 2$ when $x = 5$

47. $y = 2$ when $x = \frac{1}{5}$

48. $y = 0.9$ when $x = 0.5$

Inverse Variation

Find the variation constant and an equation of variation in which y varies inversely as x, and the following conditions exist.

49. $y = 5$ when $x = 20$

50. $y = 40$ when $x = 8$

51. $y = 11$ when $x = 4$

52. $y = 9$ when $x = 10$

53. $y = 27$ when $x = \frac{1}{3}$

54. $y = 81$ when $x = \frac{1}{9}$

Direct Variation and Inverse Variation

55. *Hooke's Law.* Hooke's law states that the distance d that a spring is stretched by a hanging object varies directly as the mass m of the object. If the distance is 20 cm when the mass is 3 kg, what is the distance when the mass is 5 kg?

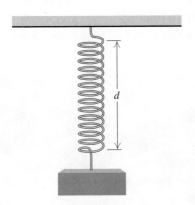

56. *Ohm's Law.* The electric current I, in amperes, in a circuit varies directly as the voltage V. When 15 volts are applied, the current is 5 amperes. What is the current when 18 volts are applied?

57. *Work Rate.* The time T required to do a job varies inversely as the number of people P working. It takes 5 hr for 7 volunteers to pick up rubbish from 1 mi of roadway. How long would it take 10 volunteers to complete the job?

58. *Pumping Rate.* The time t required to empty a tank varies inversely as the rate r of pumping. If a Briggs and Stratton pump can empty a tank in 45 min at the rate of 600 kL/min, how long will it take the pump to empty the tank at 1000 kL/min?

59. *Charitable Giving.* The cost of a dinner varies directly as the number of people fed. In 2011, a request for donations for a Thanksgiving dinner read, "You can feed seven people with a gift of $15.75." How many people could be fed with a gift of $27.00?

Source: Wheeler Mission Ministries, Indianapolis, Indiana

60. *Dancing.* The number of calories burned while dancing is directly proportional to the time spent. It takes 25 min to burn 110 calories. How long would it take to burn 176 calories when dancing?

Source: www.nutristrategy.com

Aha! **61.** *Road Maintenance.* The amount of salt needed per season to control ice on roadways varies directly as the number of storms. During a winter with 8 storms, Green County used 1200 tons of salt on roadways. How many tons would they need during a winter with 4 storms?

Source: www.saltinstitute.org

62. *Weight on Mars.* The weight M of an object on Mars varies directly as its weight E on Earth. A person who weighs 95 lb on Earth weighs 38 lb on Mars. How much would a 100-lb person weigh on Mars?

63. *String Length and Frequency.* The frequency of a string is inversely proportional to its length. A violin string that is 33 cm long vibrates with a frequency of 260 Hz. What is the frequency when the string is shortened to 30 cm?

64. *Wavelength and Frequency.* The wavelength W of a radio wave varies inversely as its frequency F. A wave with a frequency of 1200 kilohertz has a length of 300 meters. What is the length of a wave with a frequency of 800 kilohertz?

65. *Ultraviolet Index.* At an ultraviolet, or UV, rating of 4, those people who are less sensitive to the sun will burn in 75 min. Given that the number of minutes it takes to burn, t, varies inversely with the UV rating, u, how long will it take less sensitive people to burn when the UV rating is 14?

Source: The Electronic Textbook of Dermatology at www.telemedicine.org

66. *Current and Resistance.* The current I in an electrical conductor varies inversely as the resistance R of the conductor. If the current is $\frac{1}{2}$ ampere when the resistance is 240 ohms, what is the current when the resistance is 540 ohms?

67. *Fabric Manufacturing.* Knitted fabric is described in terms of wales per inch (for the fabric width) and courses per inch (CPI) (for the fabric length). The CPI c is inversely proportional to the stitch length l. For a specific fabric with a stitch length of 0.166 in., the CPI is 34.85. What would the CPI be if the stitch length were increased to 0.175 in.?

Source: Based on information from "Engineered Knitting Program for 100% Cotton Knit Fabrics" found on cottoninc.com

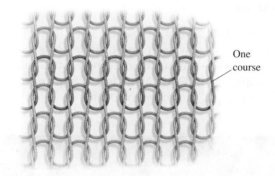

One course

68. *Relative Aperture.* The relative aperture, or f-stop, of a 23.5-mm lens is directly proportional to the focal length F of the lens. If a lens with a 150-mm focal length has an f-stop of 6.3, find the f-stop of a 23.5-mm lens with a focal length of 80 mm.

Joint Variation and Combined Variation

Find an equation of variation in which:

69. y varies directly as the square of x, and $y = 50$ when $x = 10$.

70. y varies directly as the square of x, and $y = 0.15$ when $x = 0.1$.

71. y varies inversely as the square of x, and $y = 50$ when $x = 10$.

72. y varies inversely as the square of x, and $y = 0.15$ when $x = 0.1$.

73. y varies jointly as x and z, and $y = 105$ when $x = 14$ and $z = 5$.

74. y varies jointly as x and z, and $y = \frac{3}{2}$ when $x = 2$ and $z = 10$.

75. y varies jointly as w and the square of x and inversely as z, and $y = 49$ when $w = 3$, $x = 7$, and $z = 12$.

76. y varies directly as x and inversely as w and the square of z, and $y = 4.5$ when $x = 15$, $w = 5$, and $z = 2$.

77. *Stopping Distance of a Car.* The stopping distance d of a car after the brakes have been applied varies directly as the square of the speed r. Once the brakes are applied, a car traveling 60 mph can stop in 138 ft. What stopping distance corresponds to a speed of 40 mph?

Source: Based on data from Edmunds.com

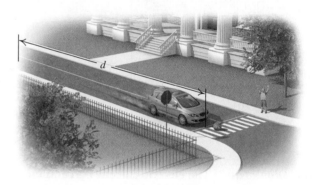

78. *Reverberation Time.* A sound's reverberation time T is the time that it takes for the sound level to decrease by 60 dB (decibels) after the sound has been turned off. Reverberation time varies directly as the volume V of a room and inversely as the sound absorption A of the room. A given sound has a reverberation time of 1.5 sec in a room with a volume of 90 m³ and a sound absorption of 9.6. What is the reverberation time of the same sound in a room with a volume of 84 m³ and a sound absorption of 10.5?

Source: www.isover.co.uk

79. *Intensity of Light.* The intensity I of light varies inversely as the square of the distance d from the light source. The following table shows the illumination from a light source at several distances from the source. What is the illumination 2.5 ft from the source?

Distance from Light Source (in feet)	Illumination (in foot-candles)
1	400
2	100
4	25

Source: Based on information from Winchip, Susan M., *Fundamentals of Lighting*, Chapter 4. New York: Fairfield Publications, 2008.

80. *Volume of a Gas.* The volume V of a given mass of a gas varies directly as the temperature T and inversely as the pressure P. If $V = 231$ cm^3 when $T = 300°$K (Kelvin) and $P = 20$ lb/cm^2, what is the volume when $T = 320°$K and $P = 16$ lb/cm^2?

81. *Atmospheric Drag.* Wind resistance, or atmospheric drag, tends to slow down moving objects. Atmospheric drag W varies jointly as an object's surface area A and velocity v. If a car traveling at a speed of 40 mph with a surface area of 37.8 ft^2 experiences a drag of 222 N (Newtons), how fast must a car with 51 ft^2 of surface area travel in order to experience a drag force of 430 N?

82. *Drag Force.* The drag force F on a boat varies jointly as the wetted surface area A and the square of the velocity of the boat. If a boat traveling 6.5 mph experiences a drag force of 86 N when the wetted surface area is 41.2 ft^2, find the wetted surface area of a boat traveling 8.2 mph with a drag force of 94 N.

83. If y varies directly as x, does doubling x cause y to be doubled as well? Why or why not?

84. Which exercise did you find easier to work: Exercise 15 or Exercise 19? Why?

Skill Review

85. If $f(x) = 4x - 7$, find $f(a) + h$. [2.2]

86. If $f(x) = 4x - 7$, find $f(a + h)$. [2.2]

Find the domain of f.

87. $f(x) = \dfrac{x - 5}{2x + 1}$ [2.2], [4.2]

88. $f(x) = \dfrac{3x}{x^2 + 1}$ [2.2]

89. $f(x) = \sqrt{2x + 8}$ [4.1]

90. $f(x) = \dfrac{3x}{x^2 - 1}$ [5.8]

Synthesis

91. In Example 7, we saw that the amount of fuel that a vehicle uses is inversely proportional to its mpg rating. Will a person decrease fuel use more by trading in a car that gets 20 mpg for one that gets 25 mpg or by trading in a car that gets 35 mpg for one that gets 40 mpg? Explain your reasoning.

92. Why do you think subscripts are used in Exercises 15 and 25 but not in Exercises 27 and 28?

93. *Escape Velocity.* A satellite's escape velocity is 6.5 mi/sec, the radius of the earth is 3960 mi, and the earth's gravitational constant is 32.2 ft/sec^2. How far is the satellite from the surface of the earth? (See Example 2.)

94. The *harmonic mean* of two numbers a and b is a number M such that the reciprocal of M is the average of the reciprocals of a and b. Find a formula for the harmonic mean.

95. *Health Care.* Young's rule for determining the size of a particular child's medicine dosage c is

$$c = \frac{a}{a + 12} \cdot d,$$

where a is the child's age and d is the typical adult dosage. If a child's age is doubled, the dosage increases. Find the ratio of the larger dosage to the smaller dosage. By what percent does the dosage increase?

Source: Olsen, June Looby, Leon J. Ablon, and Anthony Patrick Giangrasso, *Medical Dosage Calculations*, 6th ed.

96. Solve for x:

$$x^2\left(1 - \frac{2pq}{x}\right) = \frac{2p^2q^3 - pq^2x}{-q}.$$

97. *Average Acceleration.* The formula

$$a = \frac{\dfrac{d_4 - d_3}{t_4 - t_3} - \dfrac{d_2 - d_1}{t_2 - t_1}}{t_4 - t_2}$$

can be used to approximate average acceleration, where the d's are distances and the t's are the corresponding times. Solve for t_1.

98. If y varies inversely as the cube of x and x is multiplied by 0.5, what is the effect on y?

99. *Intensity of Light.* The intensity I of light from a bulb varies directly as the wattage of the bulb and inversely as the square of the distance d from the bulb. If the wattage of a light source and its distance from reading matter are both doubled, how does the intensity change?

100. Describe in words the variation represented by $W = \dfrac{km_1M_1}{d^2}$. Assume k is a constant.

101. *Tension of a Musical String.* The tension T on a string in a musical instrument varies jointly as the string's mass per unit length m, the square of its length l, and the square of its fundamental frequency f. A 2-m long string of mass 5 gm/m with a fundamental frequency of 80 has a tension of 100 N (Newtons). How long should the same string be if its tension is going to be changed to 72 N?

102. *Volume and Cost.* A peanut butter jar in the shape of a right circular cylinder is 4 in. high and 3 in. in diameter and sells for $2.40. If we assume that cost is proportional to volume, how much should a jar 6 in. high and 6 in. in diameter cost?

103. *Golf Distance Finder.* A device used in golf to estimate the distance d to a hole measures the size s that the 7-ft pin *appears* to be in a viewfinder. The viewfinder uses the principle, diagrammed here, that s gets bigger when d gets smaller. If $s = 0.56$ in. when $d = 50$ yd, find an equation of variation that expresses d as a function of s. What is d when $s = 0.40$ in.?

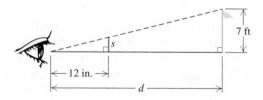

HOW IT WORKS:
Just sight the flagstick through the viewfinder…
fit flag between top dashed line and the solid line below…
…read the distance, 50 – 220 yards.

50 70 90 110 130 150 170 190 210
RANGE YARDS

QUICK QUIZ: SECTIONS 6.1–6.8

1. Write simplified form for
$$f(x) = \frac{3x^4 - 3x^2}{6x^3 - 9x^2 + 3x}$$
and list all restrictions on the domain. [6.1]

2. Add and, if possible, simplify:
$$\frac{2}{x^2+x-6} + \frac{1}{x^2+7x+12}.\quad [6.2]$$

3. Solve: $\dfrac{4}{x^2+x-6} = \dfrac{1}{x^2+7x+12}.$ [6.4]

4. Divide: $\dfrac{6x^2y^3 - 8xy^2 + 4x^3y^2}{4xy^2}.$ [6.6]

5. The number of kilograms W of water in a human body varies directly as the mass of the body. A 96-kg person contains 64 kg of water. How many kilograms of water are in a 60-kg person? [6.8]

PREPARE TO MOVE ON
Simplify.
1. $(a^6)^2$ [1.6] **2.** $(3t^5)^2$ [1.6]
3. $(x+2)^2$ [5.2] **4.** $(3a-1)^2$ [5.2]

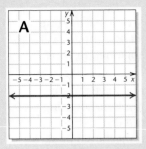

A

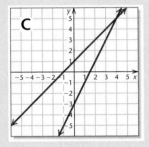

B

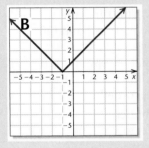

C

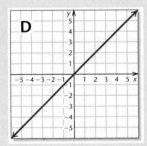

D

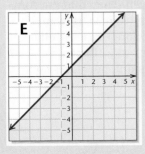

E

Visualizing for Success

Use after Section 6.1.

Match each equation, inequality, set of equations, or set of inequalities with its graph.

1. $y = -2$

2. $y = x$

3. $y = x^2$

4. $y = \dfrac{1}{x}$

5. $y = x + 1$

6. $y = |x + 1|$

7. $y \leq x + 1$

8. $y > 2x - 3$

9. $y = x + 1,$
$y = 2x - 3$

10. $y \leq x + 1,$
$y > 2x - 3$

Answers on page A-32

An alternate, animated version of this activity appears in MyMathLab. To use MyMathLab, you need a course ID and a student access code. Contact your instructor for more information.

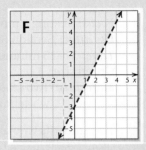

F

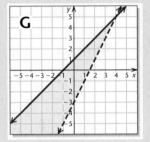

G

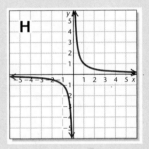

H

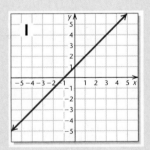

I

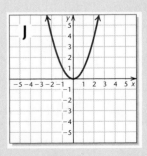

J

Collaborative Activity   *Can We Count on the Model?*

Focus: Testing a mathematical model
Use after: Section 6.5
Time: 10–15 minutes
Group size: 4
Materials: Two stopwatches, a blackboard or a
whiteboard, chalk or markers

Two students who are willing to write on the board
should each take a piece of chalk or a marker and
stand at the chalkboard or whiteboard. Two other
students, each with a stopwatch, should be ready
to time each of the students at the board.

Activity

1. The "job" is to write the numbers from 1 to
   100 on the board. One student at the board
   should work carefully and neatly and at a
   steady pace. The other should work as quickly
   as possible. They should begin at the same
   time, and the time it takes each to complete
   the job should be recorded.

2. Using the times found in part (1), each group
   should use algebra to predict how long it will
   take the students, working together, to write
   the numbers from 1 to 100.

3. In order to do the job together, have the student working neatly write numbers counting
   up from 1 and the other count down from
   100 until they meet. Each group should predict
   the number at which the students will meet.

4. The students at the board should again write
   the numbers from 1 to 100. This time, the
   one working quickly should start at 1 and the
   one working more carefully should start at
   100 and count down. Again, they should start
   at the same time, and the time at which the
   numbers meet should be recorded.

5. How accurate were your predictions? What
   may have caused the results to differ from
   your predictions?

Decision Making & Connection   (*Use after Section 6.8.*)

Vehicle Costs.

1. Fuel usage varies directly as the number of miles driven. One month, Vicki
   drove 1500 mi and used 64 gal of gasoline. How many gallons would she use
   during a month in which she drove 1200 mi?

2. Fuel costs vary directly as the price of fuel per gallon and inversely as the
   mpg rating of the vehicle driven. One year, Vicki spent an average of $4.00
   per gallon for gasoline, drove a car that got 25 mpg, and spent a total of
   $2400 for gasoline. How much would she have spent if the gasoline cost
   $4.50 per gallon and her car got 30 mpg?

3. Vicki wants to trade in her car for one that gets better gas mileage. Her current car gets 25 mpg, and last year she spent $1920 for gasoline for the car.
   How much will she save in gasoline costs if her new car gets 30 mpg?
   40 mpg? 50 mpg?

4. *Research.*   Estimate how much you (or a friend) spent on gasoline in the
   past year, as well as the mpg rating of the vehicle driven. Then find the combined city and highway gas mileage of a more efficient vehicle. How much
   less would you have paid last year for gasoline if the more efficient vehicle
   were driven? How much would it cost to trade for the more efficient vehicle?
   Assuming your gas usage and the gas prices stay constant, how long would it
   take to pay for the more efficient vehicle?

Study Summary

KEY TERMS AND CONCEPTS	EXAMPLES	PRACTICE EXERCISES

SECTION 6.1: *Rational Expressions and Functions: Multiplying and Dividing*

KEY TERMS AND CONCEPTS	EXAMPLES	PRACTICE EXERCISES
A **rational expression** can be written as a quotient of two polynomials, and is undefined when the denominator is zero. To simplify rational expressions, we remove a factor equal to 1. We list any restrictions when simplifying functions.	Simplify: $f(x) = \dfrac{x^2 - 3x - 4}{x^2 - 1}$. $f(x) = \dfrac{x^2 - 3x - 4}{x^2 - 1} = \dfrac{(x+1)(x-4)}{(x+1)(x-1)}$ Factoring. We list the restrictions $x \neq -1$, $x \neq 1$. $f(x) = \dfrac{x-4}{x-1},\ x \neq -1, x \neq 1$ $\dfrac{(x+1)}{(x+1)} = 1$	**1.** Simplify. List all restrictions on the domain. $f(x) = \dfrac{x^2 - 5x}{x^2 - 25}$
The Product of Two Rational Expressions $\dfrac{A}{B} \cdot \dfrac{C}{D} = \dfrac{AC}{BD}$	$\dfrac{5v + 5}{v - 2} \cdot \dfrac{2v^2 - 8v + 8}{v^2 - 1}$ $= \dfrac{5(v+1) \cdot 2(v-2)(v-2)}{(v-2)(v+1)(v-1)}$ Multiplying and factoring $= \dfrac{10(v-2)}{v-1}$ $\dfrac{(v+1)(v-2)}{(v+1)(v-2)} = 1$	**2.** Multiply and, if possible, simplify: $\dfrac{6x - 12}{2x^2 + 3x - 2} \cdot \dfrac{x^2 - 4}{8x - 8}$.
The Quotient of Two Rational Expressions $\dfrac{A}{B} \div \dfrac{C}{D} = \dfrac{A}{B} \cdot \dfrac{D}{C} = \dfrac{AD}{BC}$	$(x^2 - 5x - 6) \div \dfrac{x^2 - 1}{x + 6}$ $= \dfrac{x^2 - 5x - 6}{1} \cdot \dfrac{x + 6}{x^2 - 1}$ Multiplying by the reciprocal of the divisor $= \dfrac{(x - 6)(x + 1)(x + 6)}{(x + 1)(x - 1)}$ Multiplying and factoring $= \dfrac{(x - 6)(x + 6)}{x - 1}$ $\dfrac{x + 1}{x + 1} = 1$	**3.** Divide and, if possible, simplify: $\dfrac{t - 3}{6} \div \dfrac{t + 1}{15}$.

SECTION 6.2: *Rational Expressions and Functions: Adding and Subtracting*

KEY TERMS AND CONCEPTS	EXAMPLES	PRACTICE EXERCISES
Addition and Subtraction with Like Denominators $\dfrac{A}{B} + \dfrac{C}{B} = \dfrac{A + C}{B}$ $\dfrac{A}{B} - \dfrac{C}{B} = \dfrac{A - C}{B}$	$\dfrac{7x - 6}{2x + 1} - \dfrac{x - 9}{2x + 1} = \dfrac{7x - 6 - (x - 9)}{2x + 1}$ Subtracting numerators; keeping the denominator $= \dfrac{7x - 6 - x + 9}{2x + 1}$ $= \dfrac{6x + 3}{2x + 1}$ $= \dfrac{3(2x + 1)}{1(2x + 1)}$ Factoring $= 3$ $\dfrac{2x + 1}{2x + 1} = 1$	**4.** Add and, if possible, simplify: $\dfrac{5x + 4}{x + 3} + \dfrac{4x + 1}{x + 3}$.

425

Addition and Subtraction with Unlike Denominators 1. Determine the least common denominator (LCD). 2. Rewrite each expression, using the LCD. 3. Add or subtract, as indicated. 4. Simplify, if possible.	$$\frac{2x}{x^2-16} + \frac{x}{x-4}$$ $$= \frac{2x}{(x+4)(x-4)} + \frac{x}{x-4} \qquad \text{The LCD is} \\ (x+4)(x-4).$$ $$= \frac{2x}{(x+4)(x-4)} + \frac{x}{x-4} \cdot \frac{x+4}{x+4}$$ $$= \frac{2x}{(x+4)(x-4)} + \frac{x^2+4x}{(x+4)(x-4)}$$ $$= \frac{x^2+6x}{(x+4)(x-4)}$$	**5.** Subtract and, if possible, simplify: $$\frac{t}{t-1} - \frac{t-2}{t+1}.$$

SECTION 6.3: *Complex Rational Expressions*

Complex rational expressions contain one or more rational expressions within the numerator and/or the denominator. They can be simplified either by multiplying by a form of 1 to clear the fractions or by division.	**Multiplying by 1:** $$\frac{\frac{4}{x}}{\frac{3}{x}+\frac{2}{x^2}} = \frac{\frac{4}{x}}{\frac{3}{x}+\frac{2}{x^2}} \cdot \frac{x^2}{x^2} \qquad \text{Multiplying by } \frac{x^2}{x^2}$$ $$= \frac{\frac{4 \cdot x \cdot x}{x}}{\frac{3 \cdot x \cdot x}{x}+\frac{2 \cdot x^2}{x^2}} = \frac{4x}{3x+2}$$ **Using division to simplify:** $$\frac{\frac{1}{6}-\frac{1}{x}}{\frac{6-x}{6}} = \frac{\frac{1}{6}\cdot\frac{x}{x}-\frac{1}{x}\cdot\frac{6}{6}}{\frac{6-x}{6}} = \frac{\frac{x-6}{6x}}{\frac{6-x}{6}} \quad \begin{array}{l}\text{Subtracting to get} \\ \text{a single rational} \\ \text{expression in the} \\ \text{numerator}\end{array}$$ $$= \frac{x-6}{6x} \div \frac{6-x}{6} = \frac{x-6}{6x} \cdot \frac{6}{6-x} \qquad \text{Dividing}$$ $$= \frac{6(x-6)}{6x(-1)(x-6)} = \frac{1}{-x} = -\frac{1}{x} \qquad \frac{6(x-6)}{6(x-6)}=1$$	**6.** Simplify: $$\frac{\frac{4}{x}-4}{\frac{7}{x}-7}.$$

SECTION 6.4: *Rational Equations*

To Solve a Rational Equation 1. List any numbers that will make a denominator zero. 2. Clear the equation of fractions. 3. Solve the resulting equation. 4. Check the possible solution(s) in the original equation.	Solve: $\dfrac{2}{x+1} = \dfrac{1}{x-2}$. $$\frac{2}{x+1} = \frac{1}{x-2} \qquad \begin{array}{l}\text{The restrictions are} \\ x \neq -1, x \neq 2.\end{array}$$ $$(x+1)(x-2) \cdot \frac{2}{x+1} = (x+1)(x-2) \cdot \frac{1}{x-2}$$ $$2(x-2) = x+1$$ $$2x-4 = x+1$$ $$x = 5$$ *Check:* Since $\dfrac{2}{5+1} = \dfrac{1}{5-2}$, the solution is 5.	**7.** Solve: $$\frac{3}{x+4} = \frac{1}{x-1}.$$

SECTION 6.5: *Solving Applications Using Rational Equations*

Modeling Work Problems

If a = the time needed for A to complete the work alone,

 b = the time needed for B to complete the work alone, and

 t = the time needed for A and B to complete the work together, then:

$$\frac{t}{a} + \frac{t}{b} = 1;$$

$$\left(\frac{1}{a} + \frac{1}{b}\right)t = 1;$$

$$\frac{1}{a} \cdot t + \frac{1}{b} \cdot t = 1;$$

$$\frac{1}{a} + \frac{1}{b} = \frac{1}{t}.$$

It takes Manuel 9 hr longer than Zoe to rebuild an engine. Working together, they can do the job in 20 hr. How long would it take each, working alone, to rebuild an engine?

We model the situation using the work principle, with $a = z$, $b = z + 9$, and $t = 20$:

$$\frac{20}{z} + \frac{20}{z + 9} = 1.$$

See Example 2 in Section 6.5 for the complete solution.

8. Jackson can sand the oak floors and stairs in a home in 12 hr. Charis can do the same job in 9 hr. How long would it take if they worked together? (Assume that two sanders are available.)

The Motion Formula

$$d = r \cdot t, \quad r = \frac{d}{t}, \quad \text{or}$$

$$t = \frac{d}{r}$$

On her road bike, Olivia bikes 5 km/h faster than Jason does on his mountain bike. In the time it takes Olivia to travel 50 km, Jason travels 40 km. Find the speed of each bicyclist.

    Jason's speed: r km/h

    Olivia's speed: $(r + 5)$ km/h

    Jason's time: $40/r$ hr

    Olivia's time: $50/(r + 5)$ hr

The times are equal:

$$\frac{40}{r} = \frac{50}{r + 5}.$$

See Example 3 in Section 6.5 for the complete solution.

9. The current in the South River is 4 mph. Oscar's boat travels 35 mi downstream in the same time that it takes to travel 15 mi upstream. What is the speed of Oscar's boat in still water?

SECTION 6.6: *Division of Polynomials*

To divide a polynomial by a monomial, divide each term by the monomial. Divide coefficients and subtract exponents.

$$\frac{18x^2y^3 - 9xy^2 - 3x^2y}{9xy} = 2xy^2 - y - \frac{1}{3}x$$

10. Divide:

$(32x^6 + 18x^5 - 27x^2) \div (6x^2).$

To divide a polynomial by a binomial, use long division.

$$
\begin{array}{r}
x + 11 \\
x - 5 \overline{)x^2 + 6x - 8} \\
\underline{x^2 - 5x} \\
11x - 8 \\
\underline{11x - 55} \\
47
\end{array}
$$

$$(x^2 + 6x - 8) \div (x - 5) = x + 11 + \frac{47}{x - 5}$$

11. Divide:

$(x^2 - 9x + 21) \div (x - 4).$

SECTION 6.7: *Synthetic Division and the Remainder Theorem*		
The Remainder Theorem The remainder obtained by dividing a polynomial $P(x)$ by $x - r$ is $P(r)$. The remainder can be found using **synthetic division**.	$\underline{2\rfloor}\ 3\quad -4\quad 0\quad\ \ 6 \quad (3x^3 - 4x^2 + 6) \div (x - 2)$ $\ \ \underline{\ \ 6\quad 4\quad 8\ }\qquad = 3x^2 + 2x + 4 + \dfrac{14}{x-2}$ $\ 3\quad\ 2\quad 4\ \rfloor\ 14$ For $P(x) = 3x^3 - 4x^2 + 6$, we have $P(2) = 14$.	**12.** Use synthetic division to find $f(4)$ if $f(x) = x^4 - x^3 - 19x^2 + 49x - 30$.

SECTION 6.8: *Formulas, Applications, and Variation*		
Direct Variation $y = kx$	If y varies directly as x, and $y = 45$ when $x = 0.15$, find the equation of variation. $$y = kx$$ $$45 = k(0.15)$$ $$300 = k$$ The equation of variation is $y = 300x$.	**13.** If y varies directly as x, and $y = 10$ when $x = 0.2$, find the equation of variation.
Inverse Variation $y = \dfrac{k}{x}$	If y varies inversely as x, and $y = 45$ when $x = 0.15$, find the equation of variation. $$y = \dfrac{k}{x}$$ $$45 = \dfrac{k}{0.15}$$ $$6.75 = k$$ The equation of variation is $y = \dfrac{6.75}{x}$.	**14.** If y varies inversely as x, and $y = 5$ when $x = 8$, find the equation of variation.
Joint Variation $y = kxz$	If y varies jointly as x and z, and $y = 40$ when $x = 5$ and $z = 4$, find the equation of variation. $$y = kxz$$ $$40 = k \cdot 5 \cdot 4$$ $$2 = k$$ The equation of variation is $y = 2xz$.	**15.** If y varies jointly as x and z, and $y = 2$ when $x = 5$ and $z = 4$, find the equation of variation.

Review Exercises: Chapter 6

◆ Concept Reinforcement

Classify each of the following statements as either true or false.

1. If $f(x) = \dfrac{x-3}{x^2-4}$, the domain of f is assumed to be $\{x \mid x \neq -2, x \neq 2\}$. [6.1]

2. We write numerators and denominators in factored form before we simplify rational expressions. [6.1]

3. The LCM of $x - 3$ and $3 - x$ is $(x - 3)(3 - x)$. [6.2]

4. Checking the solution of a rational equation is no more important than checking the solution of a linear equation. [6.4]

5. If Camden can do a job alone in t_1 hr and Jacob can do the same job in t_2 hr, then working together it will take them $(t_1 + t_2)/2$ hr. [6.5]

6. If Skye swims 5 km/h in still water and heads into a current of 2 km/h, her speed changes to 3 km/h. [6.5]

7. The formulas $d = rt$, $r = d/t$, and $t = d/r$ are equivalent. [6.5]

8. A remainder of 0 indicates that the divisor is a factor of the quotient. [6.7]

9. To divide two polynomials using synthetic division, we must make sure that the divisor is of the form $x - a$. [6.7]

10. If x varies inversely as y, then there exists some constant k for which $x = k/y$. [6.8]

11. If
$$f(t) = \frac{t^2 - 3t + 2}{t^2 - 9},$$
find the following function values. [6.1]
   a) $f(0)$
   b) $f(-1)$
   c) $f(1)$

Find the LCM of the polynomials. [6.2]

12. $20x^3$, $24x^2$

13. $x^2 + 8x - 20$, $x^2 + 7x - 30$

Perform the indicated operations and, if possible, simplify.

14. $\dfrac{x^2}{x - 8} - \dfrac{64}{x - 8}$ [6.2]

15. $\dfrac{12a^2b^3}{5c^3d^2} \cdot \dfrac{25c^9d^4}{9a^7b}$ [6.1]

16. $\dfrac{5}{6m^2n^3p} + \dfrac{7}{9mn^4p^2}$ [6.2]

17. $\dfrac{x^3 - 8}{x^2 - 25} \cdot \dfrac{x^2 + 10x + 25}{x^2 + 2x + 4}$ [6.1]

18. $\dfrac{x^2 - 4x - 12}{x^2 - 6x + 8} \div \dfrac{x^2 - 4}{x^3 - 64}$ [6.1]

19. $\dfrac{x}{x^2 + 5x + 6} - \dfrac{2}{x^2 + 3x + 2}$ [6.2]

20. $\dfrac{-4xy}{x^2 - y^2} + \dfrac{x + y}{x - y}$ [6.2]

21. $\dfrac{5a^2}{a - b} + \dfrac{5b^2}{b - a}$ [6.2]

22. $\dfrac{3}{y + 4} - \dfrac{y}{y - 1} + \dfrac{y^2 + 3}{y^2 + 3y - 4}$ [6.2]

Find simplified form for $f(x)$ and list all restrictions on the domain.

23. $f(x) = \dfrac{4x - 2}{x^2 - 5x + 4} - \dfrac{3x + 2}{x^2 - 5x + 4}$ [6.2]

24. $f(x) = \dfrac{x + 8}{x + 5} \cdot \dfrac{2x + 10}{x^2 - 64}$ [6.1]

25. $f(x) = \dfrac{9x^2 - 1}{x^2 - 9} \div \dfrac{3x + 1}{x + 3}$ [6.1]

Simplify. [6.3]

26. $\dfrac{\frac{4}{x} - 4}{\frac{9}{x} - 9}$

27. $\dfrac{\frac{3}{a} + \frac{3}{b}}{\frac{6}{a^3} + \frac{6}{b^3}}$

28. $\dfrac{\frac{y^2 + 4y - 77}{y^2 - 10y + 25}}{\frac{y^2 - 5y - 14}{y^2 - 25}}$

29. $\dfrac{\frac{5}{x^2 - 9} - \frac{3}{x + 3}}{\frac{4}{x^2 + 6x + 9} + \frac{2}{x - 3}}$

Solve. [6.4]

30. $\dfrac{3}{x} + \dfrac{7}{x} = 5$

31. $\dfrac{5}{3x + 2} = \dfrac{3}{2x}$

32. $\dfrac{4x}{x + 1} + \dfrac{4}{x} + 9 = \dfrac{4}{x^2 + x}$

33. $\dfrac{x + 6}{x^2 + x - 6} + \dfrac{x}{x^2 + 4x + 3} = \dfrac{x + 2}{x^2 - x - 2}$

34. $\dfrac{x}{x - 3} - \dfrac{3x}{x + 2} = \dfrac{5}{x^2 - x - 6}$

35. If
$$f(x) = \frac{2}{x - 1} + \frac{2}{x + 2},$$
find all a for which $f(a) = 1$. [6.4]

Solve. [6.5]

36. Meg can arrange the books for a book sale in 9 hr. Kelly can set up for the same book sale in 12 hr. How long would it take them, working together, to set up for the book sale?

37. Ben and Jon are working for the summer building trails in a state park. Ben can build one section of the trail in 15 hr less time than Jon. Working together, they can build the section of the trail in 18 hr. How long does it take each to build the section?

38. The Black River's current is 6 mph. A boat travels 50 mi downstream in the same time that it takes to travel 30 mi upstream. What is the speed of the boat in still water?

39. Jennifer's home is 105 mi from her college dorm, and Elizabeth's home is 93 mi away. One Friday afternoon, they left school at the same time and arrived at their homes at the same time. If Jennifer drove 8 mph faster than Elizabeth, how fast did each drive?

Divide. [6.6]

40. $(30r^2s^3 + 25r^2s^2 - 20r^3s^3) \div (10r^2s)$

41. $(y^3 + 8) \div (y + 2)$

42. $(4x^3 + 3x^2 - 5x - 2) \div (x^2 + 1)$

43. Divide using synthetic division:
$(x^3 + 3x^2 + 2x - 6) \div (x - 3)$. [6.7]

44. If $f(x) = 4x^3 - 6x^2 - 9$, use synthetic division to find $f(5)$. [6.7]

Solve. [6.8]

45. $I = \dfrac{2V}{R + 2r}$, for r

46. $S = \dfrac{H}{m(t_1 - t_2)}$, for m

47. $\dfrac{1}{ac} = \dfrac{2}{ab} - \dfrac{3}{bc}$, for c

48. $T = \dfrac{A}{v(t_2 - t_1)}$, for t_1

49. For a triangle with a fixed area, the base of the triangle varies inversely as the height. If the base of a triangle with area A is 8 cm when the height is 10 cm, what is the base when the height is 4 cm?

50. The number of centimeters W of water produced from melting snow varies directly as the number of centimeters S of snow. Meterologists know that under certain conditions, 150 cm of snow will melt to 16.8 cm of water. The average annual snowfall in Alta, Utah, is about 500 in. Assuming the above conditions, how much water will replace the 500 in. of snow?

51. *Electrical Safety.* The amount of time t needed for an electrical shock to stop a 150-lb person's heart varies inversely as the square of the current flowing through the body. A 0.089-amp current is deadly to a 150-lb person after 3.4 sec. How long would it take a 0.096-amp current to be deadly?

Source: Safety Consulting Services

Synthesis

52. Discuss at least three different uses of the LCD studied in this chapter. [6.2], [6.3], [6.4]

53. Explain the difference between a rational expression and a rational equation. [6.1], [6.4]

Solve.

54. $\dfrac{5}{x - 13} - \dfrac{5}{x} = \dfrac{65}{x^2 - 13x}$ [6.4]

55. $\dfrac{\dfrac{x}{x^2 - 25} + \dfrac{2}{x - 5}}{\dfrac{3}{x - 5} - \dfrac{4}{x^2 - 10x + 25}} = 1$ [6.3], [6.4]

56. Andrew can build the section of the trail in Exercise 37 in 20 hr. How long would it take Andrew working together with Jon and Ben to build the section of the trail? [6.5]

Test: Chapter 6

For step-by-step test solutions, access the Chapter Test Prep Videos in MyMathLab®, on YouTube® (search "Bittinger inter Alg CA" and click on "Channels"), or by scanning the code.

Simplify.

1. $\dfrac{t+1}{t+3} \cdot \dfrac{5t+15}{4t^2-4}$

2. $\dfrac{x^3+27}{x^2-16} \div \dfrac{x^2+8x+15}{x^2+x-20}$

Perform the indicated operation and simplify when possible.

3. $\dfrac{25x}{x+5} + \dfrac{x^3}{x+5}$

4. $\dfrac{3a^2}{a-b} - \dfrac{3b^2-6ab}{b-a}$

5. $\dfrac{4ab}{a^2-b^2} + \dfrac{a^2+b^2}{a+b}$

6. $\dfrac{6}{x^3-64} - \dfrac{4}{x^2-16}$

Find simplified form for $f(x)$ and list all restrictions on the domain.

7. $f(x) = \dfrac{4}{x+3} - \dfrac{x}{x-2} + \dfrac{x^2+4}{x^2+x-6}$

8. $f(x) = \dfrac{x^2-1}{x+2} \div \dfrac{x^2-2x}{x^2+x-2}$

Simplify.

9. $\dfrac{\dfrac{2}{a}+\dfrac{3}{b}}{\dfrac{5}{ab}+\dfrac{1}{a^2}}$

10. $\dfrac{\dfrac{x^2-5x-36}{x^2-36}}{\dfrac{x^2+x-12}{x^2-12x+36}}$

11. $\dfrac{\dfrac{x}{8}-\dfrac{8}{x}}{\dfrac{1}{8}+\dfrac{1}{x}}$

Solve.

12. $\dfrac{1}{t} + \dfrac{1}{3t} = \dfrac{1}{2}$

13. $\dfrac{t+11}{t^2-t-12} + \dfrac{1}{t-4} = \dfrac{4}{t+3}$

14. $\dfrac{15}{x} - \dfrac{15}{x-2} = -2$

For Exercises 15 and 16, let $f(x) = \dfrac{x+5}{x-1}$.

15. Find $f(0)$ and $f(-3)$.

16. Find all a for which $f(a) = 10$.

Divide.

17. $(16a^4b^3c - 10a^5b^2c^2 + 12a^2b^2c) \div (4a^2b)$

18. $(y^2 - 20y + 64) \div (y - 6)$

19. $(6x^4 + 3x^2 + 5x + 4) \div (x^2 + 2)$

20. Divide using synthetic division:
$$(x^3 + 5x^2 + 4x - 7) \div (x - 2).$$

21. If $f(x) = 3x^4 - 5x^3 + 2x - 7$, use synthetic division to find $f(4)$.

22. Solve $R = \dfrac{gs}{g+s}$ for s.

23. Ella can install a countertop in 5 hr. Sari can perform the same job in 4 hr. How long will it take them, working together, to install the countertop?

24. Terrel bicycles 12 mph with no wind. Against the wind, he bikes 8 mi in the same time that it takes to bike 14 mi with the wind. What is the speed of the wind?

25. Katie and Tyler work together to prepare a meal at a soup kitchen in $2\frac{6}{7}$ hr. Working alone, it would take Katie 6 hr more than it would take Tyler. How long would it take each of them to complete the meal, working alone?

26. The number of workers n needed to clean a stadium after a game varies inversely as the amount of time t allowed for the cleanup. If it takes 25 workers to clean the stadium when there are 6 hr allowed for the job, how many workers are needed if the stadium must be cleaned in 5 hr?

27. The surface area of a balloon varies directly as the square of its radius. The area is 325 in² when the radius is 5 in. What is the area when the radius is 7 in.?

Synthesis

28. Let
$$f(x) = \dfrac{1}{x+3} + \dfrac{5}{x-2}.$$
Find all a for which $f(a) = f(a+5)$.

29. Solve: $\dfrac{6}{x-15} - \dfrac{6}{x} = \dfrac{90}{x^2-15x}$.

30. Simplify: $1 - \dfrac{1}{1 - \dfrac{1}{1 - \dfrac{1}{a}}}$.

31. One summer, Alex mowed 4 lawns for every 3 lawns mowed by his brother Ryan. Together, they mowed 98 lawns. How many lawns did each mow?

Cumulative Review: Chapters 1–6

1. Determine the slope and the y-intercept for the line given by $7x - 4y = 12$.   [2.3]

2. Find an equation for the line that passes through the points $(-1, 7)$ and $(4, -3)$.   [2.5]

3. If
$$f(x) = \frac{x - 3}{x^2 - 11x + 30},$$
find **(a)** $f(3)$ and **(b)** the domain of f.   [2.2], [5.8]

4. Write the domain of f using interval notation if $f(x) = \sqrt{x - 9}$.   [4.1]

Graph on a plane.

5. $5x = y$ [2.3]

6. $8y + 2x = 16$ [2.4]

7. $4x \geq 5y + 12$ [4.4]

8. $y = \frac{1}{3}x - 2$ [2.3]

Perform the indicated operations and simplify.

9. $(3x^2 + y)^2$ [5.2]

10. $(2x^2 - 9)(2x^2 + 9)$ [5.2]

11. $\dfrac{y^2 - 36}{2y + 8} \cdot \dfrac{y + 4}{y + 6}$ [6.1]

12. $\dfrac{x^4 - 1}{x^2 - x - 2} \div \dfrac{x^2 + 1}{x - 2}$ [6.1]

13. $\dfrac{5ab}{a^2 - b^2} + \dfrac{a + b}{a - b}$ [6.2]

14. Simplify: $\dfrac{\dfrac{1}{x} - \dfrac{1}{y}}{x + y}$.   [6.3]

15. Divide: $(9x^3 + 5x^2 + 2) \div (x + 2)$.   [6.6]

Factor.

16. $x^2 + 8x - 84$ [5.4]

17. $16y^2 - 25$ [5.5]

18. $64x^3 + 8$ [5.6]

19. $t^2 - 16t + 64$ [5.5]

20. $\frac{1}{8}b^3 - c^3$ [5.6]

21. $3t^2 + 17t - 28$ [5.4]

Solve.

22. $8x = 1 + 16x^2$ [5.8]

23. $288 = 2y^2$ [5.8]

24. $\frac{1}{3}x - \frac{1}{5} \geq \frac{1}{5}x - \frac{1}{3}$ [4.1]

25. $-13 < 3x + 2 < -1$ [4.2]

26. $|x| > 6.4$ [4.3]

27. $\dfrac{6}{x - 5} = \dfrac{2}{2x}$ [6.4]

28. $5x - 2y = -23,$
$3x + 4y = 7$ [3.2]

29. $-3x + 4y + z = -5,$
$x - 3y - z = 6,$
$2x + 3y + 5z = -8$ [3.4]

30. $P = \dfrac{4a}{a + b}$, for a [6.8]

31. *Trail Mix.*   Kenny mixes Himalayan Diamonds trail mix, which contains 40% nuts, with Alpine Gold trail mix, which contains 25% nuts, to create 20 lb of a mixture that is 30% nuts. How much of each type of trail mix does he use?   [3.3]

32. *Quilting.*   A rectangular quilted wall hanging is 4 in. longer than it is wide. The area of the quilt is 320 in². Find the perimeter of the quilt.   [5.8]

33. *Hotel Management.*   The IQAir HealthPro Plus air purifier can clean the air in a 20-ft by 25-ft meeting room in 5 fewer minutes than it takes the Austin Healthmate HM400 to do the same job. Together the two machines can purify the air in the room in 6 min. How long would it take each machine, working alone, to purify the air in the room?   [6.5]

Source: Manufacturers' and retailers' websites

34. *Driving Time.*   The time t that it takes for Johann to drive to work varies inversely as his speed. On a day when Johann averages 45 mph, it takes him 20 min to drive to work. How long will it take him to drive to work when he averages only 40 mph?   [6.8]

Synthesis

Solve.

35. $4 \leq |3 - x| \leq 6$ [4.2], [4.3]

36. $\dfrac{18}{x - 9} + \dfrac{10}{x + 5} = \dfrac{28x}{x^2 - 4x - 45}$ [6.4]

37. $16x^3 = x$ [5.8]

Exponents and Radicals

HEIGHT (in centimeters)	MASS (in kilograms)			
	55	60	65	70
100	1.11	1.15	1.19	1.23
110	1.19	1.24	1.28	1.32
120	1.27	1.32	1.36	1.41

Your Largest Organ System Renews Itself Every Month.

Your skin is your largest organ system, but just how big is it? The total surface area of the human body is called the body surface area (BSA) and is used by professionals such as doctors and pharmacologists to determine drug dosage, how much fluid to administer intravenously, and thermal energy loss. BSA cannot be measured directly but is calculated from a person's mass and height using one of several formulas. In this chapter, we will calculate the BSA of a child using the DuBois formula. (*See Exercise 123 in Exercise Set 7.2.*)

I use math every day to calculate important factors related to patient care.

Beth Bennett Lopez, MPAS, PA-C, a dermatology physician assistant in Phoenix, Arizona, uses math when calculating pediatric drug doses, measuring patient's moles, and calculating body surface area for a psoriasis patient.

In this chapter, we learn about square roots, cube roots, fourth roots, and so on. These roots can be expressed in radical notation or in exponential notation using exponents that are fractions. The chapter closes with an introduction to the complex-number system.

7.1   Radical Expressions and Functions

Square Roots and Square-Root Functions ▪ Expressions of the Form $\sqrt{a^2}$ ▪ Cube Roots ▪ Odd and Even nth Roots

In this section, we consider roots, such as square roots and cube roots, and the *radical* expressions involving such roots.

Square Roots and Square-Root Functions

When a number is multiplied by itself, we say that the number is squared. If we can find a number that was squared in order to produce some value a, we call that first number a *square root* of a.

> **SQUARE ROOT**
>
> The number c is a *square root* of a if $c^2 = a$.

For example,

9 has −3 and 3 as square roots because $(-3)^2 = 9$ and $3^2 = 9$.

25 has −5 and 5 as square roots because $(-5)^2 = 25$ and $5^2 = 25$.

−4 does not have a real-number square root because there is no real number c for which $c^2 = -4$.

Every positive number has two square roots, and 0 has only itself as a square root. Negative numbers do not have real-number square roots; however, they do have nonreal square roots. Later in this chapter we introduce the *complex-number* system in which such square roots exist.

EXAMPLE 1   Find the two square roots of 36.

SOLUTION   The square roots are 6 and −6, because $6^2 = 36$ and $(-6)^2 = 36$.

YOUR TURN

Whenever we refer to *the* square root of a number, we mean the nonnegative square root of that number. This is also called the *principal square root* of the number.

> **PRINCIPAL SQUARE ROOT**
>
> The *principal square root* of a nonnegative number is its nonnegative square root. The symbol $\sqrt{\ }$ is called a *radical sign* and is used to indicate the principal square root of the number over which it appears.

Study Skills

Advance Planning Pays Off

The best way to prepare for a final exam is to do so over a period of at least two weeks. First, review each chapter, studying the terminology, formulas, problems, properties, and procedures in the Study Summaries. Then retake your quizzes and tests. If you miss any questions, spend extra time reviewing the corresponding topics. Also consider participating in a study group or attending a tutoring or review session.

1. Find the two square roots of 49.

Student Notes

It is important to remember the difference between *the* square root of 9 and *a* square root of 9. *A* square root of 9 means either 3 or −3, but *the* square root of 9, denoted $\sqrt{9}$, means the principal square root of 9, or 3.

EXAMPLE 2   Simplify each of the following.

a) $\sqrt{25}$ b) $\sqrt{\dfrac{25}{64}}$ c) $-\sqrt{64}$ d) $\sqrt{0.0049}$

SOLUTION

a) $\sqrt{25} = 5$ $\sqrt{}$ indicates the principal square root. Note that $\sqrt{25} \neq -5$.

b) $\sqrt{\dfrac{25}{64}} = \dfrac{5}{8}$ Since $\left(\dfrac{5}{8}\right)^2 = \dfrac{25}{64}$

c) $-\sqrt{64} = -8$ Since $\sqrt{64} = 8, -\sqrt{64} = -8$.

d) $\sqrt{0.0049} = 0.07$ $(0.07)(0.07) = 0.0049$. Note too that

$$\sqrt{0.0049} = \sqrt{\dfrac{49}{10{,}000}} = \dfrac{7}{100}.$$

2. Simplify: $-\sqrt{\dfrac{1}{9}}$.

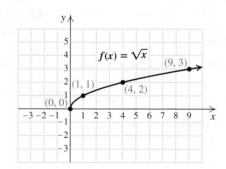

YOUR TURN

In addition to being read as "the principal square root of a," $\sqrt{a}$ is also read as "the square root of a," "root a," or "radical a." Any expression in which a radical sign appears is called a *radical expression*. The following are radical expressions:

$$\sqrt{5}, \quad \sqrt{a}, \quad -\sqrt{3x}, \quad \sqrt{\dfrac{y^2 + 7}{y}}, \quad \sqrt{x} + 8.$$

The expression under the radical sign is called the **radicand**. In the expressions above, the radicands are 5, a, $3x$, $(y^2 + 7)/y$, and x, respectively.

Values for square roots found on calculators are, for the most part, approximations. For example, a calculator will show a number like

2.23606798

for $\sqrt{5}$. The exact value of $\sqrt{5}$ is not given by any repeating or terminating decimal. In general, for any number a that is not a perfect square, $\sqrt{a}$ is a nonterminating, nonrepeating decimal or an *irrational number*.

The square-root function, given by

$$f(x) = \sqrt{x},$$

has $[0, \infty)$ as its domain and $[0, \infty)$ as its range. We can draw its graph by selecting convenient values for x and calculating the corresponding outputs. Once these ordered pairs have been graphed, a smooth curve can be drawn.

$$f(x) = \sqrt{x}$$

x	$\sqrt{x}$	$(x, f(x))$
0	0	$(0, 0)$
1	1	$(1, 1)$
4	2	$(4, 2)$
9	3	$(9, 3)$

EXAMPLE 3   For each function, find the indicated function value.

a) $f(x) = \sqrt{3x - 2}$; $f(1)$ b) $g(z) = -\sqrt{6z + 4}$; $g(3)$

SOLUTION

a) $f(1) = \sqrt{3 \cdot 1 - 2}$ Substituting

   $= \sqrt{1} = 1$ Simplifying

3. If $f(x) = \sqrt{1-x}$, find $f(-3)$.

b) $g(3) = -\sqrt{6 \cdot 3 + 4}$ Substituting

$= -\sqrt{22}$ Simplifying. This answer is exact.

≈ -4.69041576 Using a calculator to write an approximation

YOUR TURN

Expressions of the Form $\sqrt{a^2}$

As the next example shows, $\sqrt{a^2}$ does not always simplify to a.

EXAMPLE 4   Evaluate $\sqrt{a^2}$ for each of the following values: **(a)** 5; **(b)** 0; **(c)** -5.

SOLUTION

a) $\sqrt{5^2} = \sqrt{25} = 5$
    └────────────── Same

b) $\sqrt{0^2} = \sqrt{0} = 0$
    └────────────── Same

4. Evaluate $\sqrt{t^2}$ for $t = -3$ and for $t = 3$.

c) $\sqrt{(-5)^2} = \sqrt{25} = 5$
    └────────────── Opposites   Note that $\sqrt{(-5)^2} \neq -5$.

YOUR TURN

You may have noticed that evaluating $\sqrt{a^2}$ is just like evaluating $|a|$.

<table>
<tr><td>

Technology Connection

To see the necessity of absolute-value signs, let y_1 represent the left side and y_2 the right side of each of the following equations. Then use a graph or a table to determine whether these equations are true.

1. $\sqrt{x^2} \overset{?}{=} x$
2. $\sqrt{x^2} \overset{?}{=} |x|$
3. $x \overset{?}{=} |x|$

</td></tr>
</table>

<table>
<tr><td>

SIMPLIFYING $\sqrt{a^2}$

For any real number a,
$$\sqrt{a^2} = |a|.$$
(The principal square root of a^2 is the absolute value of a.)

</td></tr>
</table>

When a radicand is the square of a variable expression, like $(x+1)^2$ or $36t^2$, absolute-value signs are needed when simplifying. **We use absolute-value signs unless we know that the expression being squared is nonnegative.** This ensures that our result is never negative.

Student Notes

Some absolute-value notation can be simplified.

- $|ab| = |a| \cdot |b|$, so an expression like $|3x|$ can be written

$$|3x| = |3| \cdot |x| = 3|x|.$$

- Even powers of real numbers are never negative, so

$$|x^2| = x^2,$$
$$|x^4| = x^4,$$
$$|x^6| = x^6, \quad \text{and so on.}$$

EXAMPLE 5   Simplify each expression. Assume that the variable can represent any real number.

a) $\sqrt{36t^2}$ **b)** $\sqrt{(x+1)^2}$ **c)** $\sqrt{x^2 - 8x + 16}$
d) $\sqrt{t^6}$ **e)** $\sqrt{a^8}$

SOLUTION

a) $\sqrt{36t^2} = \sqrt{(6t)^2} = |6t|$, or $6|t|$ Since t can be negative, absolute-value notation is necessary.

b) $\sqrt{(x+1)^2} = |x+1|$ Since $x+1$ can be negative (for example, if $x = -3$), absolute-value notation is necessary.

c) $\sqrt{x^2 - 8x + 16} = \sqrt{(x-4)^2} = |x-4|$ Since $x-4$ can be negative, absolute-value notation is necessary.

5. Simplify $\sqrt{(3x - 7)^2}$. Assume that x can represent any real number.

d) Recall that $(a^m)^n = a^{mn}$, and note that $(t^3)^2 = t^6$. Thus,

$$\sqrt{t^6} = |t^3|.$$ Since t^3 can be negative, absolute-value notation is necessary.

e) $\sqrt{a^8} = \sqrt{(a^4)^2} = |a^4| = a^4$ Since a^4 is never negative, $|a^4| = a^4$.

YOUR TURN

If we assume that the expression being squared is nonnegative, then absolute-value notation is not necessary.

EXAMPLE 6   Simplify each expression. Assume that the expressions being squared are nonnegative. Thus absolute-value notation is not necessary.

a) $\sqrt{y^2}$ **b)** $\sqrt{a^{10}}$ **c)** $\sqrt{9x^2 - 6x + 1}$

SOLUTION

a) $\sqrt{y^2} = y$ We assume that y is nonnegative, so no absolute-value notation is necessary. When y *is* negative, $\sqrt{y^2} \neq y$.

b) $\sqrt{a^{10}} = a^5$ Assuming that a^5 is nonnegative. Note that $(a^5)^2 = a^{10}$.

6. Simplify $\sqrt{n^{14}}$. Assume that n is nonnegative.

c) $\sqrt{9x^2 - 6x + 1} = \sqrt{(3x - 1)^2} = 3x - 1$ Assuming that $3x - 1$ is nonnegative

YOUR TURN

Cube Roots

We often need to know what number cubed produces a certain value. When such a number is found, we say that we have found a *cube root*. For example,

2 is the cube root of 8 because $2^3 = 2 \cdot 2 \cdot 2 = 8$;

-4 is the cube root of -64 because $(-4)^3 = (-4)(-4)(-4) = -64$.

CUBE ROOT

The number c is the *cube root* of a if $c^3 = a$. In symbols, we write $\sqrt[3]{a}$ to denote the cube root of a.

Each real number has only one real-number cube root. The cube-root function, given by

$$f(x) = \sqrt[3]{x},$$

has $\mathbb{R}$ as its domain and $\mathbb{R}$ as its range. To draw its graph, we select convenient values for x and calculate the corresponding outputs. Once these ordered pairs have been graphed, a smooth curve is drawn. Note that the cube root of a positive number is positive, and the cube root of a negative number is negative.

$$f(x) = \sqrt[3]{x}$$

x	$\sqrt[3]{x}$	$(x, f(x))$
0	0	$(0, 0)$
1	1	$(1, 1)$
8	2	$(8, 2)$
-1	-1	$(-1, -1)$
-8	-2	$(-8, -2)$

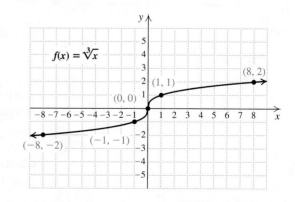

EXAMPLE 7 For each function, find the indicated function value.

a) $f(y) = \sqrt[3]{y}$; $f(125)$ **b)** $g(x) = \sqrt[3]{x-1}$; $g(-26)$

SOLUTION

a) $f(125) = \sqrt[3]{125} = 5$ Since $5 \cdot 5 \cdot 5 = 125$

b) $g(-26) = \sqrt[3]{-26-1}$

$= \sqrt[3]{-27}$

$= -3$ Since $(-3)(-3)(-3) = -27$

7. If $g(x) = \sqrt[3]{2x}$, find $g(-4)$.

YOUR TURN

EXAMPLE 8 Simplify: $\sqrt[3]{-8y^3}$.

SOLUTION

$$\sqrt[3]{-8y^3} = -2y \quad \text{Since } (-2y)^3 = (-2y)(-2y)(-2y) = -8y^3$$

8. Simplify: $\sqrt[3]{1000t^3}$.

YOUR TURN

Odd and Even *n*th Roots

The 4th root of a number a is the number c for which $c^4 = a$. There are also 5th roots, 6th roots, and so on. We write $\sqrt[n]{a}$ for the principal *n*th root. The number n is called the *index* (plural, *indices*). When the index is 2, we do not write it.

When the index n is odd, we are taking an *odd root*.

> Every number has exactly one real root when n is odd. Odd roots of positive numbers are positive and odd roots of negative numbers are negative.
>
> Absolute-value signs are not appropriate when finding odd roots.

EXAMPLE 9 Simplify each expression.

a) $\sqrt[5]{32}$ **b)** $\sqrt[5]{-32}$ **c)** $-\sqrt[5]{32}$

d) $-\sqrt[5]{-32}$ **e)** $\sqrt[7]{x^7}$ **f)** $\sqrt[9]{(t-1)^9}$

SOLUTION

a) $\sqrt[5]{32} = 2$ Since $2^5 = 32$

b) $\sqrt[5]{-32} = -2$ Since $(-2)^5 = -32$

c) $-\sqrt[5]{32} = -2$ Taking the opposite of $\sqrt[5]{32}$

d) $-\sqrt[5]{-32} = -(-2) = 2$ Taking the opposite of $\sqrt[5]{-32}$

e) $\sqrt[7]{x^7} = x$ Absolute-value signs are not correct here.

f) $\sqrt[9]{(t-1)^9} = t-1$

9. Simplify: $\sqrt[5]{(3n)^5}$.

YOUR TURN

When the index n is even, we are taking an *even root*.

> Every positive real number has two real *n*th roots when n is even—one positive and one negative. Negative numbers do not have real *n*th roots when n is even.
>
> When n is even, the notation $\sqrt[n]{a}$ indicates the nonnegative *n*th root. Thus, when we simplify even *n*th roots, absolute-value signs are often required.

Technology Connection

To enter cube roots or higher roots on a graphing calculator, enter the index and then choose the $\sqrt[x]{}$ option in the MATH MATH menu. The characters $6\sqrt[x]{}$ indicate the sixth root.

1. Use a $\boxed{\text{TABLE}}$ or $\boxed{\text{GRAPH}}$ and $\boxed{\text{TRACE}}$ to check the solution of Example 11.

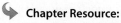

Chapter Resource:
Visualizing for Success, p. 494

10. Simplify $\sqrt[6]{64x^6}$, if possible. Assume that x can represent any real number.

Compare the following.

Odd Root	Even Root
$\sqrt[3]{8} = 2$	$\sqrt[4]{16} = 2$
$\sqrt[3]{-8} = -2$	$\sqrt[4]{-16}$ is not a real number.
$\sqrt[3]{x^3} = x$	$\sqrt[4]{x^4} = |x|$

EXAMPLE 10 Simplify each expression, if possible. Assume that variables can represent any real number.

a) $\sqrt[4]{81}$ b) $-\sqrt[4]{81}$ c) $\sqrt[4]{-81}$

d) $\sqrt[4]{81x^4}$ e) $\sqrt[6]{(y+7)^6}$

SOLUTION

a) $\sqrt[4]{81} = 3$ Since $3^4 = 81$

b) $-\sqrt[4]{81} = -3$ Taking the opposite of $\sqrt[4]{81}$

c) $\sqrt[4]{-81}$ cannot be simplified. $\sqrt[4]{-81}$ is not a real number.

d) $\sqrt[4]{81x^4} = |3x|$, or $3|x|$ Use absolute-value notation since x could represent a negative number.

e) $\sqrt[6]{(y+7)^6} = |y+7|$ Use absolute-value notation since $y + 7$ is negative for $y < -7$.

YOUR TURN

EXAMPLE 11 Determine the domain of g if $g(x) = \sqrt[6]{7 - 3x}$.

SOLUTION Since the index is even, the radicand, $7 - 3x$, must be nonnegative. We solve the inequality:

$7 - 3x \geq 0$ We cannot find the 6th root of a negative number.

$-3x \geq -7$

$x \leq \frac{7}{3}.$ Multiplying both sides by $-\frac{1}{3}$ and reversing the inequality

Thus,

$$\text{Domain of } g = \left\{x \mid x \leq \tfrac{7}{3}\right\}$$ Using set-builder notation

$$= \left(-\infty, \tfrac{7}{3}\right].$$ Using interval notation

11. Determine the domain of f if $f(x) = \sqrt{2x + 3}$.

YOUR TURN

7.1 **EXERCISE SET**   FOR EXTRA HELP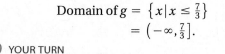

PRACTICE   WATCH   READ   REVIEW

Vocabulary and Reading Check

Select the appropriate word to complete each of the following.

1. Every positive number has _____ square
   one/two
   root(s).

2. The principal square root is never _____.
   negative/positive

3. For any _____ number a, we have
   negative/positive
   $\sqrt{a^2} = a$.

4. For any _____ number a, we have
   negative/positive
   $\sqrt{a^2} = -a$.

5. If a is a whole number that is not a perfect square, then $\sqrt{a}$ is a(n) _____ number.

irrational/rational

6. The domain of the function f given by $f(x) = \sqrt[3]{x}$ is the set of all _____ numbers.

whole/real/positive

7. If $\sqrt[4]{x}$ is a real number, then x must be _____ .

negative/positive/nonnegative.

8. If $\sqrt[3]{x}$ is negative, then x must be _____ .

negative/positive

Square Roots and Square-Root Functions

For each number, find all of its square roots.

9. 64

10. 81

11. 100

12. 121

13. 400

14. 2500

15. 625

16. 225

Simplify.

17. $\sqrt{49}$

18. $\sqrt{144}$

19. $-\sqrt{16}$

20. $-\sqrt{100}$

21. $\sqrt{\dfrac{36}{49}}$

22. $\sqrt{\dfrac{4}{9}}$

23. $-\sqrt{\dfrac{16}{81}}$

24. $-\sqrt{\dfrac{81}{144}}$

25. $\sqrt{0.04}$

26. $\sqrt{0.36}$

27. $\sqrt{0.0081}$

28. $\sqrt{0.0016}$

For each function, find the specified function value, if it exists.

29. $f(t) = \sqrt{5t - 10}$; $f(3), f(2), f(1), f(-1)$

30. $g(x) = \sqrt{x^2 - 25}$; $g(-6), g(3), g(6), g(13)$

31. $t(x) = -\sqrt{2x^2 - 1}$; $t(5), t(0), t(-1), t\left(-\frac{1}{2}\right)$

32. $p(z) = \sqrt{2z - 20}$; $p(4), p(10), p(12), p(0)$

33. $f(t) = \sqrt{t^2 + 1}$; $f(0), f(-1), f(-10)$

34. $g(x) = -\sqrt{(x + 1)^2}$; $g(-3), g(4), g(-5)$

Expressions of the Form $\sqrt{a^2}$

Simplify. Variables may represent any real number, so remember to use absolute-value notation when necessary. If a root cannot be simplified, state this.

35. $\sqrt{100x^2}$

36. $\sqrt{16t^2}$

37. $\sqrt{(-4b)^2}$

38. $\sqrt{(-7c)^2}$

39. $\sqrt{(8 - t)^2}$

40. $\sqrt{(a + 3)^2}$

41. $\sqrt{y^2 + 16y + 64}$

42. $\sqrt{x^2 - 4x + 4}$

43. $\sqrt{4x^2 + 28x + 49}$

44. $\sqrt{9x^2 - 30x + 25}$

45. $\sqrt{a^{22}}$

46. $\sqrt{x^{10}}$

47. $\sqrt{-25}$

48. $\sqrt{-16}$

Cube Roots

Simplify.

49. $\sqrt[3]{-1}$

50. $-\sqrt[3]{-1000}$

51. $-\sqrt[3]{64}$

52. $\sqrt[3]{27}$

53. $-\sqrt[3]{-125y^3}$

54. $\sqrt[3]{-64x^3}$

Odd and Even nth Roots

Identify the radicand and the index for each expression.

55. $5\sqrt{p^2 + 4}$

56. $-7\sqrt{y^2 - 8}$

57. $x^2 y^3 \sqrt[5]{\dfrac{x}{y + 4}}$

58. $\dfrac{a^2}{b} \sqrt[6]{a(a + b)}$

Simplify. Use absolute-value notation when necessary.

59. $-\sqrt[4]{256}$

60. $-\sqrt[4]{625}$

61. $-\sqrt[5]{-\dfrac{32}{243}}$

62. $\sqrt[5]{-\dfrac{1}{32}}$

63. $\sqrt[6]{x^6}$

64. $\sqrt[8]{y^8}$

65. $\sqrt[9]{t^9}$

66. $\sqrt[5]{a^5}$

67. $\sqrt[4]{(6a)^4}$

68. $\sqrt[4]{(7b)^4}$

69. $\sqrt[10]{(-6)^{10}}$

70. $\sqrt[12]{(-10)^{12}}$

71. $\sqrt[414]{(a + b)^{414}}$

72. $\sqrt[1976]{(2a + b)^{1976}}$

Simplify. Assume that no radicands were formed by raising negative quantities to even powers. Thus absolute-value notation is not necessary.

73. $\sqrt{16x^2}$

74. $\sqrt{25t^2}$

75. $-\sqrt{(3t)^2}$

76. $-\sqrt{(7c)^2}$

77. $\sqrt{(-5b)^2}$

78. $\sqrt{(-10a)^2}$

79. $\sqrt{a^2 + 2a + 1}$

80. $\sqrt{9 - 6y + y^2}$

81. $\sqrt[4]{16x^4}$

82. $\sqrt[4]{81x^4}$

83. $\sqrt[3]{(x - 1)^3}$

84. $-\sqrt[3]{(7y)^3}$

85. $\sqrt{t^{18}}$

86. $\sqrt{a^{14}}$

87. $\sqrt{(x - 2)^8}$

88. $\sqrt{(x + 3)^{10}}$

For each function, find the specified function value, if it exists.

89. $f(x) = \sqrt[3]{x + 1}$; $f(7), f(26), f(-9), f(-65)$

90. $g(x) = -\sqrt[3]{2x - 1}$; $g(0), g(-62), g(-13), g(63)$

91. $g(t) = \sqrt[4]{t - 3}$; $g(19), g(-13), g(1), g(84)$

92. $f(t) = \sqrt[4]{t + 1}$; $f(0), f(15), f(-82), f(80)$

Determine the domain of each function described.

93. $f(x) = \sqrt{x - 6}$

94. $g(x) = \sqrt{x + 8}$

95. $g(t) = \sqrt[4]{t + 8}$

96. $f(x) = \sqrt[4]{x - 9}$

97. $g(x) = \sqrt[4]{10 - 2x}$

98. $g(t) = \sqrt[3]{2t - 6}$

99. $f(t) = \sqrt[5]{2t + 7}$

100. $f(t) = \sqrt[6]{4 + 3t}$

101. $h(z) = -\sqrt[6]{5z + 2}$

102. $d(x) = -\sqrt[4]{5 - 7x}$

Aha! **103.** $f(t) = 7 + \sqrt[8]{t^8}$

104. $g(t) = 9 + \sqrt[6]{t^6}$

105. Explain how to write the negative square root of a number using radical notation.

106. Does the square root of a number's absolute value always exist? Why or why not?

Skill Review

Let $f(x) = 3x - 1$ and $g(x) = \dfrac{1}{x}$.

107. Find $f\left(\frac{1}{3}\right)$.   [2.2]

108. Find the domain of f.   [2.2]

109. Find the domain of g.   [2.2]

110. Find $(f + g)(x)$.   [2.6]

111. Find $(fg)(x)$.   [2.6], [6.1]

112. Which function, f or g, is a linear function?   [2.3]

Synthesis

113. Under what conditions does the nth root of x^3 exist as a real number? Explain your reasoning.

114. Under what conditions does the nth root of x^2 exist? Explain your reasoning.

Firefighting.   The number of gallons per minute discharged from a fire hose depends on the diameter of the hose and the nozzle pressure. For a 2-in. diameter solid bore nozzle, the discharge can be modeled by

$$f(p) = 118.8\sqrt{p},$$

where p is the nozzle pressure, in pounds per square inch (psi), and $f(p)$ is the water flow, in gallons per minute (GPM). Use this function for Exercises 115 and 116.

115. Estimate the water flow when the nozzle pressure is 50 psi.

116. Estimate the water flow when the nozzle pressure is 175 psi.

117. *Biology.*   The number of species S of plants in Guyana in an area of A hectares can be estimated using the formula

$$S = 88.63\sqrt[4]{A}.$$

The Kaieteur National Park in Guyana has an area of 63,000 hectares. How many species of plants are in the park?

Source: Hans ter Steege, "A Perspective on Guyana and its Plant Richness," as found on www.bio.uu.nl

118. *Spaces in a Parking Lot.*   A parking lot has attendants to park the cars. The number of spaces N needed for waiting cars before attendants can get to them is given by the formula $N = 2.5\sqrt{A}$, where A is the number of arrivals in peak hours. Find the number of spaces needed for the given number of arrivals in peak hours: **(a)** 25; **(b)** 36; **(c)** 49; **(d)** 64.

Determine the domain of each function described. Then draw the graph of each function.

119. $f(x) = \sqrt{x} + 5$

120. $g(x) = \sqrt{x + 5}$

121. $g(x) = \sqrt{x} - 2$

122. $f(x) = \sqrt{x - 2}$

123. Find the domain of f if
$$f(x) = \frac{\sqrt{x + 3}}{\sqrt[4]{2 - x}}.$$

124. Find the domain of g if
$$g(x) = \frac{\sqrt[4]{5 - x}}{\sqrt[6]{x + 4}}.$$

125. Find the domain of F if
$$F(x) = \frac{x}{\sqrt{x^2 - 5x - 6}}.$$

126. Use a graphing calculator to check your answers to Exercises 39, 45, and 67. On some graphing calculators, a MATH key is needed to enter higher roots.

127. *Length of Marriage.*   John Tierney and Garth Sundem have developed an equation that can be used to predict the probability that a celebrity marriage will last. The probability P that a couple will still be married in T years is given by

$$P = 50\sqrt[15]{\frac{\text{NYT}(\text{Ah} + \text{Aw})}{\text{ENQ}(\text{Sc} + 5)} \cdot \text{Md} \cdot \left[\frac{\text{Md}}{(\text{Md} + 2)}\right]^{T^2}}$$

where NYT = the number of times since 1990 that the wife's name appeared in the *New York Times*, ENQ = the number of times since 1990 that the wife's name appeared in the *National Enquirer*, Ah = the husband's age, in years, Aw = the wife's age, in years, Md = the number of months that the couple dated before marriage, and Sc = the number of the top five photos returned by a Google images search for the wife's name in which she was "scantily clad."

Kate Middleton and Prince William were both 29 when they were married and had dated for 120 months. Kate appeared 258 times in the *New York Times* and 44 times in the *National Enquirer*, and her Sc value was 0. What is the probability that the marriage will last 5 years?

Source: "Refining the Formula That Predicts Celebrity Marriages' Doom," www.nytimes.com, March 12, 2012

PREPARE TO MOVE ON

Simplify. Do not use negative exponents in your answer.
[1.6]

1. $(3xy^8)(5x^2y)$

2. $(2a^{-1}b^2c)^{-3}$

3. $\left(\dfrac{10x^{-1}y^5}{5x^2y^{-1}}\right)^{-1}$

4. $\left(\dfrac{8x^3y^{-2}}{2xz^4}\right)^{-2}$

YOUR TURN ANSWERS: SECTION 7.1

1. $-7, 7$ 2. $-\frac{1}{3}$ 3. 2 4. $3; 3$ 5. $|3x - 7|$
6. n^7 7. -2 8. $10t$ 9. $3n$ 10. $|2x|$, or $2|x|$
11. $\left\{x \mid x \geq -\frac{3}{2}\right\}$, or $\left[-\frac{3}{2}, \infty\right)$

7.2   Rational Numbers as Exponents

Rational Exponents ▪ Negative Rational Exponents ▪ Laws of Exponents ▪ Simplifying Radical Expressions

We have already considered natural-number exponents and integer exponents. We now expand the study of exponents further to include all rational numbers. This will give meaning to expressions like $7^{1/3}$ and $(2x)^{-4/5}$.

Study Skills

This Looks Familiar

Sometimes a topic seems familiar and students are tempted to assume that they already know the material. Try to avoid this tendency. Often new extensions or applications are included when a topic reappears.

Rational Exponents

When defining rational exponents, we want the rules for exponents to hold for them just as they do for integer exponents. For example, we want to continue to add exponents when multiplying expressions with the same base. Then $a^{1/2} \cdot a^{1/2} = a^{1/2+1/2} = a^1$ suggests that $a^{1/2}$ means $\sqrt{a}$, and $a^{1/3} \cdot a^{1/3} \cdot a^{1/3} = a^{1/3+1/3+1/3} = a^1$ suggests that $a^{1/3}$ means $\sqrt[3]{a}$.

$$a^{1/n} = \sqrt[n]{a}$$

$a^{1/n}$ means $\sqrt[n]{a}$. When a is nonnegative, n can be any natural number greater than 1. When a is negative, n can be any odd natural number greater than 1.

Thus, $a^{1/5} = \sqrt[5]{a}$ and $a^{1/10} = \sqrt[10]{a}$.

The denominator of the exponent is the index of the radical expression.

EXAMPLE 1   Write an equivalent expression using radical notation and, if possible, simplify.

a) $16^{1/2}$ b) $(-8)^{1/3}$ c) $(abc)^{1/5}$ d) $(25x^{16})^{1/2}$

1. Write an expression equivalent to $49^{1/2}$ using radical notation and, if possible, simplify.

> *CAUTION!* When we are converting from radical notation to exponential notation, parentheses are often needed to indicate the base.
>
> $$\sqrt{5x} = (5x)^{1/2}$$

2. Write an expression equivalent to $\sqrt[4]{2ac}$ using exponential notation.

SOLUTION

a) $16^{1/2} = \sqrt{16} = 4$

b) $(-8)^{1/3} = \sqrt[3]{-8} = -2$

c) $(abc)^{1/5} = \sqrt[5]{abc}$

> The denominator of the exponent becomes the index. The base becomes the radicand. Recall that for square roots, the index 2 is understood without being written.

d) $(25x^{16})^{1/2} = 25^{1/2}x^8 = \sqrt{25} \cdot x^8 = 5x^8$

YOUR TURN

EXAMPLE 2 Write an equivalent expression using exponential notation.

a) $\sqrt[5]{7ab}$　　　　b) $\sqrt[7]{\dfrac{x^3y}{4}}$　　　　c) $\sqrt{5x}$

SOLUTION　Parentheses are required to indicate the base.

a) $\sqrt[5]{7ab} = (7ab)^{1/5}$

b) $\sqrt[7]{\dfrac{x^3y}{4}} = \left(\dfrac{x^3y}{4}\right)^{1/7}$

> The index becomes the denominator of the exponent. The radicand becomes the base.

c) $\sqrt{5x} = (5x)^{1/2}$ 　The index 2 is understood without being written. We assume $x \ge 0$.

YOUR TURN

How shall we define $a^{2/3}$? If the property for multiplying exponents is to hold, we must have $a^{2/3} = (a^{1/3})^2$, as well as $a^{2/3} = (a^2)^{1/3}$. This would suggest that $a^{2/3} = (\sqrt[3]{a})^2$ and $a^{2/3} = \sqrt[3]{a^2}$. We make our definition accordingly.

Student Notes

It is important to remember both meanings of $a^{m/n}$. When the root of the base a is known, $(\sqrt[n]{a})^m$ is generally easier to work with. When $\sqrt[n]{a}$ is not known, $\sqrt[n]{a^m}$ is often more convenient.

> **POSITIVE RATIONAL EXPONENTS**
>
> For any natural numbers m and n ($n \ne 1$) and any real number a for which $\sqrt[n]{a}$ exists,
>
> $$a^{m/n} \quad \text{means} \quad (\sqrt[n]{a})^m, \quad \text{or} \quad \sqrt[n]{a^m}.$$

EXAMPLE 3 Write an equivalent expression using radical notation and simplify.

a) $27^{2/3}$　　　　　　　　　　　b) $25^{3/2}$

SOLUTION

a) $27^{2/3}$ means $(\sqrt[3]{27})^2$ or, equivalently, $\sqrt[3]{27^2}$. Let's see which is easier to simplify:

$$(\sqrt[3]{27})^2 = 3^2 \qquad \sqrt[3]{27^2} = \sqrt[3]{729}$$
$$\qquad\quad = 9; \qquad\qquad\qquad = 9.$$

The simplification on the left is probably easier for most people.

3. Write an expression equivalent to $1000^{2/3}$ using radical notation and simplify.

b) $25^{3/2}$ means $(\sqrt{25})^3$ or, equivalently, $\sqrt[2]{25^3}$ (the index 2 is normally omitted). Since $\sqrt{25}$ is more commonly known than $\sqrt{25^3}$, we use that form:

$$25^{3/2} = (\sqrt{25})^3 = 5^3 = 125.$$

YOUR TURN

EXAMPLE 4 Write an equivalent expression using exponential notation.

a) $\sqrt[3]{9^4}$

b) $\left(\sqrt[4]{7xy}\right)^5$

SOLUTION

4. Write an expression equivalent to $\left(\sqrt[5]{4x}\right)^3$ using exponential notation.

a) $\sqrt[3]{9^4} = 9^{4/3}$

b) $\left(\sqrt[4]{7xy}\right)^5 = (7xy)^{5/4}$

The index becomes the denominator of the fraction that is the exponent.

YOUR TURN

Negative Rational Exponents

Recall that $x^{-2} = 1/x^2$. Negative rational exponents behave similarly.

> **NEGATIVE RATIONAL EXPONENTS**
>
> For any rational number m/n and any nonzero real number a for which $a^{m/n}$ exists,
>
> $$a^{-m/n} \quad \text{means} \quad \frac{1}{a^{m/n}}.$$

> **CAUTION!** A negative exponent does not indicate that the expression in which it appears is negative: $a^{-1} \neq -a$.

Technology Connection

To approximate $7^{2/3}$ on many calculators, we enter 7 ⌃ (2/3).

1. Why are the parentheses needed in the expression above?
2. Compare the graphs of $y_1 = x^{1/2}$, $y_2 = x$, and $y_3 = x^{3/2}$ and determine those x-values for which $y_1 > y_3$.

EXAMPLE 5 Write an equivalent expression with positive exponents and, if possible, simplify.

a) $9^{-1/2}$

b) $(5xy)^{-4/5}$

c) $64^{-2/3}$

d) $4x^{-2/3}y^{1/5}$

e) $\left(\dfrac{3r}{7s}\right)^{-5/2}$

SOLUTION

a) $9^{-1/2} = \dfrac{1}{9^{1/2}}$ $9^{-1/2}$ is the reciprocal of $9^{1/2}$.

Since $9^{1/2} = \sqrt{9} = 3$, the answer simplifies to $\dfrac{1}{3}$.

b) $(5xy)^{-4/5} = \dfrac{1}{(5xy)^{4/5}}$ $(5xy)^{-4/5}$ is the reciprocal of $(5xy)^{4/5}$.

c) $64^{-2/3} = \dfrac{1}{64^{2/3}}$ $64^{-2/3}$ is the reciprocal of $64^{2/3}$.

Since $64^{2/3} = \left(\sqrt[3]{64}\right)^2 = 4^2 = 16$, the answer simplifies to $\dfrac{1}{16}$.

d) $4x^{-2/3}y^{1/5} = 4 \cdot \dfrac{1}{x^{2/3}} \cdot y^{1/5} = \dfrac{4y^{1/5}}{x^{2/3}}$

e) Recall that $(a/b)^{-n} = (b/a)^n$. This property holds for *any* negative exponent:

$$\left(\frac{3r}{7s}\right)^{-5/2} = \left(\frac{7s}{3r}\right)^{5/2}.$$

Writing the reciprocal of the base and changing the sign of the exponent

5. Write an expression equivalent to $16^{-1/2}p^{2/3}w^{-3/5}$ with positive exponents and, if possible, simplify.

YOUR TURN

Laws of Exponents

The same laws hold for rational exponents as for integer exponents.

LAWS OF EXPONENTS

For any real numbers a and b and any rational exponents m and n for which a^m, a^n, and b^m are defined:

1. $a^m \cdot a^n = a^{m+n}$ When multiplying, add exponents if the bases are the same.

2. $\dfrac{a^m}{a^n} = a^{m-n}$ When dividing, subtract exponents if the bases are the same. (Assume $a \neq 0$.)

3. $(a^m)^n = a^{m \cdot n}$ To raise a power to a power, multiply the exponents.

4. $(ab)^m = a^m b^m$ To raise a product to a power, raise each factor to the power and multiply.

EXAMPLE 6   Use the laws of exponents to simplify.

a) $3^{1/5} \cdot 3^{3/5}$

b) $\dfrac{a^{1/4}}{a^{1/2}}$

c) $(7.2^{2/3})^{3/4}$

d) $(a^{-1/3}b^{2/5})^{1/2}$

SOLUTION

a) $3^{1/5} \cdot 3^{3/5} = 3^{1/5+3/5} = 3^{4/5}$ Adding exponents

b) $\dfrac{a^{1/4}}{a^{1/2}} = a^{1/4-1/2} = a^{1/4-2/4}$ Subtracting exponents after finding a common denominator

$\qquad\qquad = a^{-1/4}, \text{ or } \dfrac{1}{a^{1/4}}$ $a^{-1/4}$ is the reciprocal of $a^{1/4}$.

c) $(7.2^{2/3})^{3/4} = 7.2^{(2/3)(3/4)} = 7.2^{6/12}$ Multiplying exponents

$\qquad\qquad\qquad = 7.2^{1/2}$ Using arithmetic to simplify the exponent

d) $(a^{-1/3}b^{2/5})^{1/2} = a^{(-1/3)(1/2)} \cdot b^{(2/5)(1/2)}$ Raising a product to a power and multiplying exponents

$\qquad\qquad\qquad = a^{-1/6}b^{1/5}, \text{ or } \dfrac{b^{1/5}}{a^{1/6}}$

6. Use the laws of exponents to simplify $\dfrac{8^{3/4}}{8^{-1/6}}$.

YOUR TURN

Simplifying Radical Expressions

Many radical expressions can be simplified by using rational exponents.

TO SIMPLIFY RADICAL EXPRESSIONS

1. Convert radical expressions to exponential expressions.
2. Use arithmetic and the laws of exponents to simplify.
3. Convert back to radical notation as needed.

One way to check Example 7(a) is to let $y_1 = (5x)^{3/6}$ and $y_2 = \sqrt{5x}$. Then we use $\boxed{\text{GRAPH}}$ or $\boxed{\text{TABLE}}$ to see if $y_1 = y_2$. An alternative check is to let $y_3 = y_2 - y_1$ and see if $y_3 = 0$. Check Example 7(a) using one of these two methods.

1. Why are rational exponents especially useful when working on a graphing calculator?

7. Use rational exponents to simplify $\left(\sqrt[3]{ab}\right)^{15}$. Do not use fraction exponents in the final answer.

EXAMPLE 7 Use rational exponents to simplify. Do not use exponents that are fractions in the final answer.

a) $\sqrt[6]{(5x)^3}$

b) $\sqrt[5]{t^{20}}$

c) $\left(\sqrt[3]{ab^2c}\right)^{12}$

d) $\sqrt{\sqrt[3]{x}}$

SOLUTION

a) $\sqrt[6]{(5x)^3} = (5x)^{3/6}$ Converting to exponential notation

$= (5x)^{1/2}$ Simplifying the exponent

$= \sqrt{5x}$ Returning to radical notation

b) $\sqrt[5]{t^{20}} = t^{20/5}$ Converting to exponential notation

$= t^4$ Simplifying the exponent

c) $\left(\sqrt[3]{ab^2c}\right)^{12} = (ab^2c)^{12/3}$ Converting to exponential notation

$= (ab^2c)^4$ Simplifying the exponent

$= a^4b^8c^4$ Using the laws of exponents

d) $\sqrt{\sqrt[3]{x}} = \sqrt{x^{1/3}}$ Converting the radicand to exponential notation

$= (x^{1/3})^{1/2}$ Try to go directly to this step.

$= x^{1/6}$ Using the laws of exponents

$= \sqrt[6]{x}$ Returning to radical notation

YOUR TURN

7.2 EXERCISE SET

FOR EXTRA HELP   MyMathLab®   Math XL
   PRACTICE   WATCH   READ   REVIEW

Vocabulary and Reading Check

Choose the word from the list below that best completes each statement. Not every word will be used.

add   radical
equivalent   rational
multiply   subtract

1. The expression $\sqrt{3x}$ is an example of a(n) _____ expression.

2. When dividing one exponential expression by another with the same base, we _____ exponents.

3. The expressions $\sqrt[3]{5mn}$ and $(5mn)^{1/3}$ are _____.

4. The exponent in the expression $7^{2/3}$ is a(n) _____ exponent.

Concept Reinforcement

In Exercises 5–12, match each expression with the equivalent expression from the column on the right.

5. ____ $x^{2/5}$

6. ____ $x^{5/2}$

7. ____ $x^{-5/2}$

8. ____ $x^{-2/5}$

9. ____ $x^{1/5} \cdot x^{2/5}$

10. ____ $(x^{1/5})^{5/2}$

11. ____ $\sqrt[5]{x^4}$

12. ____ $\left(\sqrt[4]{x}\right)^5$

a) $x^{3/5}$

b) $\left(\sqrt[5]{x}\right)^4$

c) $\sqrt{x^5}$

d) $x^{1/2}$

e) $\dfrac{1}{(\sqrt{x})^5}$

f) $\sqrt[4]{x^5}$

g) $\sqrt[5]{x^2}$

h) $\dfrac{1}{(\sqrt[5]{x})^2}$

Note: Assume for all exercises that all variables are non-negative and that all denominators are nonzero.

Rational Exponents

Write an equivalent expression using radical notation and, if possible, simplify.

13. $y^{1/3}$ 14. $t^{1/4}$

15. $36^{1/2}$ 16. $125^{1/3}$

17. $32^{1/5}$

18. $81^{1/4}$

19. $64^{1/2}$

20. $100^{1/2}$

21. $(xyz)^{1/2}$

22. $(ab)^{1/4}$

23. $(a^2b^2)^{1/5}$

24. $(x^3y^3)^{1/4}$

25. $t^{5/6}$

26. $a^{3/2}$

27. $16^{3/4}$

28. $4^{7/2}$

29. $125^{4/3}$

30. $9^{5/2}$

31. $(81x)^{3/4}$

32. $(125a)^{2/3}$

33. $(25x^4)^{3/2}$

34. $(9y^6)^{3/2}$

Write an equivalent expression using exponential notation.

35. $\sqrt[3]{18}$

36. $\sqrt[4]{10}$

37. $\sqrt{30}$

38. $\sqrt{22}$

39. $\sqrt{x^7}$

40. $\sqrt{a^3}$

41. $\sqrt[5]{m^2}$

42. $\sqrt[5]{n^4}$

43. $\sqrt[4]{xy}$

44. $\sqrt[3]{cd}$

45. $\sqrt[5]{xy^2z}$

46. $\sqrt[7]{x^3y^2z^2}$

47. $(\sqrt{3mn})^3$

48. $(\sqrt[3]{7xy})^4$

49. $(\sqrt[7]{8x^2y})^5$

50. $(\sqrt[6]{2a^5b})^7$

51. $\dfrac{2x}{\sqrt[3]{z^2}}$

52. $\dfrac{3a}{\sqrt[5]{c^2}}$

Negative Rational Exponents

Write an equivalent expression with positive exponents and, if possible, simplify.

53. $8^{-1/3}$

54. $10{,}000^{-1/4}$

55. $(2rs)^{-3/4}$

56. $(5xy)^{-5/6}$

57. $\left(\dfrac{1}{16}\right)^{-3/4}$

58. $\left(\dfrac{1}{8}\right)^{-2/3}$

59. $\dfrac{8c}{a^{-3/5}}$

60. $\dfrac{3b}{a^{-5/7}}$

61. $2a^{3/4}b^{-1/2}c^{2/3}$

62. $5x^{-2/3}y^{4/5}z$

63. $3^{-5/2}a^3b^{-7/3}$

64. $2^{-1/3}x^4y^{-2/7}$

65. $\left(\dfrac{2ab}{3c}\right)^{-5/6}$

66. $\left(\dfrac{7x}{8yz}\right)^{-3/5}$

67. $xy^{-1/4}$

68. $aw^{-1/3}$

Laws of Exponents

Use the laws of exponents to simplify. Do not use negative exponents in any answers.

69. $11^{1/2} \cdot 11^{1/3}$

70. $5^{1/4} \cdot 5^{1/8}$

71. $\dfrac{3^{5/8}}{3^{-1/8}}$

72. $\dfrac{8^{7/11}}{8^{-2/11}}$

73. $\dfrac{4.3^{-1/5}}{4.3^{-7/10}}$

74. $\dfrac{2.7^{-11/12}}{2.7^{-1/6}}$

75. $(10^{3/5})^{2/5}$

76. $(5^{5/4})^{3/7}$

77. $a^{2/3} \cdot a^{5/4}$

78. $x^{3/4} \cdot x^{1/3}$

Aha! **79.** $(64^{3/4})^{4/3}$

80. $(27^{-2/3})^{3/2}$

81. $(m^{2/3}n^{-1/4})^{1/2}$

82. $(x^{-1/3}y^{2/5})^{1/4}$

Simplifying Radical Expressions

Use rational exponents to simplify. Do not use fraction exponents in the final answer. Write answers using radical notation.

83. $\sqrt[9]{x^3}$

84. $\sqrt[12]{a^3}$

85. $\sqrt[3]{y^{15}}$

86. $\sqrt[4]{y^{40}}$

87. $\sqrt[12]{a^6}$

88. $\sqrt[30]{x^5}$

89. $(\sqrt[7]{xy})^{14}$

90. $(\sqrt[3]{ab})^{15}$

91. $\sqrt[4]{(7a)^2}$

92. $\sqrt[8]{(3x)^2}$

93. $(\sqrt[8]{2x})^6$

94. $(\sqrt[10]{3a})^5$

95. $\sqrt{\sqrt[5]{m}}$

96. $\sqrt[6]{\sqrt{n}}$

97. $\sqrt[4]{(xy)^{12}}$

98. $\sqrt{(ab)^6}$

99. $(\sqrt[5]{a^2b^4})^{15}$

100. $(\sqrt[3]{x^2y^5})^{12}$

101. $\sqrt[3]{\sqrt[4]{xy}}$

102. $\sqrt[5]{\sqrt[3]{2a}}$

103. If $f(x) = (x + 5)^{1/2}(x + 7)^{-1/2}$, find the domain of f. Explain how you found your answer.

104. Let $f(x) = 5x^{-1/3}$. Under what condition will we have $f(x) > 0$? Why?

Skill Review

Solve.

105. $2(t + 3) - 5 = 1 - (6 - t)$ [1.3]

106. $10 - 5y > 4$ [4.1]

107. $-3 \le 5x + 7 \le 10$ [4.2]

108. $x^2 = x + 6$ [5.8]

109. $\dfrac{15}{x} - \dfrac{15}{x + 2} = 2$ [6.4]

110. $2x - y = 3,$
$x = 1 - y$ [3.2]

Synthesis

111. Explain why $\sqrt[3]{x^6} = x^2$ for any value of x, whereas $\sqrt[2]{x^6} = x^3$ only when $x \ge 0$.

112. If $g(x) = x^{3/n}$, why does the domain of g depend on whether n is odd or even?

Use rational exponents to simplify.

113. $\sqrt{x\sqrt[3]{x^2}}$

114. $\sqrt[4]{\sqrt[3]{8x^3y^6}}$

115. $\sqrt[14]{c^2 - 2cd + d^2}$

*Music.   The function given by $f(x) = k2^{x/12}$ can be used to determine the frequency, in cycles per second, of a musical note that is x half-steps above a note with frequency k.**

116. The frequency of concert A for a trumpet is 440 cycles per second. Find the frequency of the A that is two octaves (24 half-steps) above concert A. (Skilled trumpeters can reach this note.)

117. Show that the G that is 7 half-steps (a "perfect fifth") above middle C (262 cycles per second) has a frequency that is about 1.5 times that of middle C.

118. Show that the C sharp that is 4 half-steps (a "major third") above concert A (see Exercise 116) has a frequency that is about 25% greater than that of concert A.

119. *Road Pavement Messages.*   In a psychological study, it was determined that the proper length L of the letters of a word printed on pavement is given by

$$L = \frac{0.000169d^{2.27}}{h},$$

where d is the distance of a car from the lettering and h is the height of the eye above the surface of the road. All units are in meters. This formula says that from a vantage point h meters above the surface of the road, if a driver is to be able to recognize a message d meters away, that message will be the most recognizable if the length of the letters is L. Find L to the nearest tenth of a meter, given d and h.

---

*This application was inspired by information provided by Dr. Homer B. Tilton of Pima Community College East.

a) $h = 1$ m, $d = 60$ m
b) $h = 0.9906$ m, $d = 75$ m
c) $h = 2.4$ m, $d = 80$ m
d) $h = 1.1$ m, $d = 100$ m

120. *Baseball.*   The statistician Bill James has found that a baseball team's winning percentage P can be approximated by

$$P = \frac{r^{1.83}}{r^{1.83} + \sigma^{1.83}},$$

where r is the total number of runs scored by that team and σ (sigma) is the total number of runs scored by their opponents. During a recent season, the San Francisco Giants scored 799 runs and their opponents scored 749 runs. Use James's formula to predict the Giants' winning percentage. (The team actually won 55.6% of their games.)

Source: Bittinger, M., *One Man's Journey Through Mathematics.* Boston: Addison-Wesley, 2004

121. *Forestry.*   The total wood volume T, in cubic feet, in a California black oak can be estimated using the formula

$$T = 0.936\, d^{1.97}h^{0.85},$$

where d is the diameter of the tree at breast height and h is the total height of the tree. How much wood is in a California black oak that is 3 ft in diameter at breast height and 80 ft high?

Source: Norman H. Pillsbury and Michael L. Kirkley, 1984. Equations for total, wood, and saw-log volume for thirteen California hardwoods, USDA Forest Service PNW Research Note No. 414: 52 p.

122. *Physics.*   The equation $m = m_0(1 - v^2c^{-2})^{-1/2}$, developed by Albert Einstein, is used to determine the mass m of an object that is moving v meters per second and has mass m_0 before the motion begins. The constant c is the speed of light, approximately 3×10^8 m/sec. Suppose that a particle with mass 8 mg is accelerated to a speed of $\frac{9}{5} \times 10^8$ m/sec. Without using a calculator, find the new mass of the particle.

123. The total surface area of the human body is called the body surface area (BSA). Values of the BSA can be read from tables such as the following or approximated using a formula. A person's body surface area (BSA) can be approximated by the DuBois formula

$$\text{BSA} = 0.007184w^{0.425}h^{0.725},$$

where w is mass, in kilograms, h is height, in centimeters, and BSA is in square meters. What is the

BSA of a child who is 122 cm tall and has a mass of 29.5 kg?

Source: www.halls.md

Height (in centimeters)	Mass (in kilograms)			
	55	**60**	**65**	**70**
100	1.11	1.15	1.19	1.23
110	1.19	1.24	1.28	1.32
120	1.27	1.32	1.36	1.41

124. Using a graphing calculator, select **MODE** SIMUL and **FORMAT** EXPROFF. Then graph

$$y_1 = x^{1/2}, \quad y_2 = 3x^{2/5},$$
$$y_3 = x^{4/7}, \quad \text{and} \quad y_4 = \tfrac{1}{5}x^{3/4}.$$

Looking only at coordinates, match each graph with its equation.

 125. a) Graph $f(x) = x^{1/3}$, $g(x) = (x^{1/6})^2$, and $h(x) = (x^2)^{1/6}$. How do the graphs and the domains differ?

b) Study the definition of $a^{m/n}$ carefully and then predict which of the graphs in part (a), if any, would best represent the graph of $k(x) = x^{2/6}$. Explain why.

c) Check using a graphing calculator.

YOUR TURN ANSWERS: SECTION 7.2
1. $\sqrt{49}$, or 7   **2.** $(2ac)^{1/4}$ **3.** $\left(\sqrt[3]{1000}\right)^2$, or 100
4. $(4x)^{3/5}$ **5.** $\dfrac{p^{2/3}}{16^{1/2}w^{3/5}}$, or $\dfrac{p^{2/3}}{4w^{3/5}}$ **6.** $8^{11/12}$ **7.** a^5b^5

QUICK QUIZ: SECTIONS 7.1–7.2
Simplify. Assume that a variable can represent any real number.
1. $-\sqrt{81}$ [7.1]   **2.** $\sqrt[6]{x^6}$ [7.1]
3. $16^{1/4}$ [7.2]   **4.** $\left(\sqrt[4]{2}\right)^8$ [7.2]
5. Find $f(-5)$ if $f(t) = \sqrt{10 - 3t}$. [7.1]

PREPARE TO MOVE ON
Multiply. [5.2]
1. $(x + 5)(x - 5)$ **2.** $(x - 2)(x^2 + 2x + 4)$
Factor. [5.5]
3. $9a^2 - 24a + 16$ **4.** $3n^2 + 12n + 12$

7.3 Multiplying Radical Expressions

Multiplying Radical Expressions ▪ Simplifying by Factoring ▪ Multiplying and Simplifying

Multiplying Radical Expressions

Note that $\sqrt{4}\sqrt{25} = 2 \cdot 5 = 10$ and $\sqrt{4 \cdot 25} = \sqrt{100} = 10$. Likewise,
$$\sqrt[3]{27}\,\sqrt[3]{8} = 3 \cdot 2 = 6 \quad \text{and} \quad \sqrt[3]{27 \cdot 8} = \sqrt[3]{216} = 6.$$
These examples suggest the following.

THE PRODUCT RULE FOR RADICALS
For any real numbers $\sqrt[n]{a}$ and $\sqrt[n]{b}$,
$$\sqrt[n]{a} \cdot \sqrt[n]{b} = \sqrt[n]{a \cdot b}.$$
(The product of two *n*th roots is the *n*th root of the product of the two radicands.)

> **CAUTION!** The product rule for radicals applies only when radicals have the same index.

Rational exponents can be used to derive this rule:

$$\sqrt[n]{a} \cdot \sqrt[n]{b} = a^{1/n} \cdot b^{1/n} = (a \cdot b)^{1/n} = \sqrt[n]{a \cdot b}.$$

EXAMPLE 1   Multiply.

a) $\sqrt{2} \cdot \sqrt{7}$ **b)** $\sqrt{x + 3}\,\sqrt{x - 3}$ **c)** $\sqrt[4]{\dfrac{y}{5}} \cdot \sqrt[4]{\dfrac{7}{x}}$

SOLUTION

a) When no index is written, roots are understood to be square roots with an unwritten index of two. We apply the product rule:

$$\sqrt{2} \cdot \sqrt{7} = \sqrt{2 \cdot 7} = \sqrt{14}.$$

> **CAUTION!**
> $\sqrt{x^2 - 9} \neq \sqrt{x^2} - \sqrt{9}.$

b) $\sqrt{x + 3}\sqrt{x - 3} = \sqrt{(x + 3)(x - 3)}$ The product of two square roots is the square root of the product.

$$= \sqrt{x^2 - 9}$$

c) The index in each radical expression is 4, so in order to multiply we can use the product rule:

$$\sqrt[4]{\frac{y}{5}} \cdot \sqrt[4]{\frac{7}{x}} = \sqrt[4]{\frac{y}{5} \cdot \frac{7}{x}} = \sqrt[4]{\frac{7y}{5x}}.$$

1. Multiply: $\sqrt[3]{4} \cdot \sqrt[3]{5}$.

YOUR TURN

Simplifying by Factoring

The number p is a *perfect square* if there exists a rational number q for which $q^2 = p$. We say that p is a *perfect cube* if $q^3 = p$ for some rational number q. In general, p is a *perfect nth power* if $q^n = p$ for some rational number q. The product rule allows us to simplify $\sqrt[n]{ab}$ when a or b is a perfect nth power.

> **USING THE PRODUCT RULE TO SIMPLIFY**
> $$\sqrt[n]{ab} = \sqrt[n]{a} \cdot \sqrt[n]{b}.$$
> ($\sqrt[n]{a}$ and $\sqrt[n]{b}$ must both be real numbers.)

To illustrate, suppose we wish to simplify $\sqrt{20}$. Since this is a *square* root, we check to see if there is a factor of 20 that is a perfect *square*. There is one, 4, so we express 20 as $4 \cdot 5$ and use the product rule:

$$\sqrt{20} = \sqrt{4 \cdot 5} \qquad \text{Factoring the radicand (4 is a perfect square)}$$
$$= \sqrt{4} \cdot \sqrt{5} \qquad \text{Factoring into two radicals}$$
$$= 2\sqrt{5}. \qquad \text{Finding the square root of 4}$$

> **TO SIMPLIFY A RADICAL EXPRESSION WITH INDEX n BY FACTORING**
>
> 1. If possible, express the radicand as a product in which one or more factors are perfect nth powers.
> 2. Rewrite the expression as the nth root of each factor.
> 3. Simplify any expressions containing perfect nth powers.
> 4. Simplification is complete when no radicand has a factor that is a perfect nth power. (All factors in the radicand can be written with exponents less than the index.)

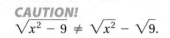

Technology Connection

To check Example 1(b), let $y_1 = \sqrt{x + 3}\sqrt{x - 3}$ and $y_2 = \sqrt{x^2 - 9}$ and compare:

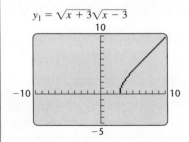

$y_1 = \sqrt{x + 3}\sqrt{x - 3}$

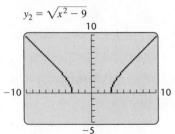

$y_2 = \sqrt{x^2 - 9}$

Because $y_1 = y_2$ for all x-values that can be used in *both* y_1 and y_2, Example 1(b) *is* correct.

1. Why do the graphs above differ in appearance?

It is often safe to assume that a radicand does not represent a negative number raised to an even power. We will henceforth make this assumption—unless functions are involved—and discontinue use of absolute-value notation when taking even roots.

EXAMPLE 2 Simplify by factoring: **(a)** $\sqrt{200}$; **(b)** $\sqrt{50x^2y}$; **(c)** $\sqrt[3]{-72}$; **(d)** $\sqrt[4]{162x^6}$.

SOLUTION

a) $\sqrt{200} = \sqrt{100 \cdot 2}$ 100 is the largest perfect-square factor of 200.

$\qquad\qquad = \sqrt{100} \cdot \sqrt{2} = 10\sqrt{2}$ $\sqrt{100} = 10$

Express the radicand as a product.

b) $\sqrt{50x^2y} = \sqrt{25 \cdot 2 \cdot x^2 \cdot y}$ $25x^2$ is the largest perfect-square factor of $50x^2y$.

Rewrite as the *n*th root of each factor.

$\qquad\qquad = \sqrt{25x^2} \cdot \sqrt{2y}$ Factoring into two radicals

Simplify.

$\qquad\qquad = 5x\sqrt{2y}$ $\sqrt{25x^2} = 5x$. We have assumed that $x \geq 0$.

c) $\sqrt[3]{-72} = \sqrt[3]{-8 \cdot 9}$ -8 is a perfect-cube (third-power) factor of -72.

$\qquad\qquad = \sqrt[3]{-8} \cdot \sqrt[3]{9} = -2\sqrt[3]{9}$ Taking the cube root of -8

d) $\sqrt[4]{162x^6} = \sqrt[4]{81 \cdot 2 \cdot x^4 \cdot x^2}$ $81 \cdot x^4$ is the largest perfect fourth-power factor of $162x^6$.

$\qquad\qquad = \sqrt[4]{81x^4} \cdot \sqrt[4]{2x^2}$ Factoring into two radicals

$\qquad\qquad = 3x\sqrt[4]{2x^2}$ $\sqrt[4]{81x^4} = 3x$. We have assumed that $x \geq 0$.

Let's look at this example another way. We write a complete factorization and look for quadruples of factors. Each quadruple makes a perfect fourth power:

$$\sqrt[4]{162x^6} = \sqrt[4]{\boxed{3 \cdot 3 \cdot 3 \cdot 3} \cdot 2 \cdot \boxed{x \cdot x \cdot x \cdot x} \cdot x \cdot x} \quad \begin{array}{l} 3 \cdot 3 \cdot 3 \cdot 3 = 3^4 \text{ and} \\ x \cdot x \cdot x \cdot x = x^4 \end{array}$$

$$= 3 \cdot x \cdot \sqrt[4]{2 \cdot x \cdot x} \quad \sqrt[4]{3^4} = 3 \text{ and } \sqrt[4]{3^4} = x$$

$$= 3x\sqrt[4]{2x^2}.$$

2. Simplify by factoring:

$\sqrt{18ab^2}$.

YOUR TURN

EXAMPLE 3 If $f(x) = \sqrt{3x^2 - 6x + 3}$, find a simplified form for $f(x)$. Because we are working with a function, assume that x can be any real number.

SOLUTION

$$f(x) = \sqrt{3x^2 - 6x + 3}$$

$$\left. \begin{array}{l} = \sqrt{3(x^2 - 2x + 1)} \\ = \sqrt{(x-1)^2 \cdot 3} \end{array} \right\} \quad \begin{array}{l} \text{Factoring the radicand; } x^2 - 2x + 1 \text{ is a} \\ \text{perfect square.} \end{array}$$

$$= \sqrt{(x-1)^2} \cdot \sqrt{3} \quad \text{Factoring into two radicals}$$

$$= |x - 1|\sqrt{3} \quad \text{Finding the square root of } (x-1)^2$$

3. If $f(x) = \sqrt{10x^2 + 60x + 90}$, find a simplified form for $f(x)$. Assume that x can be any real number.

YOUR TURN

Technology Connection

To check Example 3, let
$y_1 = \sqrt{3x^2 - 6x + 3}$,
$y_2 = |x - 1|\sqrt{3}$, and
$y_3 = (x - 1)\sqrt{3}$. Do the graphs all coincide? Why or why not?

EXAMPLE 4 Simplify: **(a)** $\sqrt{x^7y^{11}z^9}$; **(b)** $\sqrt[3]{16a^7b^{14}}$.

SOLUTION

a) There are many ways to factor $x^7y^{11}z^9$. Because of the square root (index of 2), we identify the largest exponents that are multiples of 2:

$$\sqrt{x^7y^{11}z^9} = \sqrt{x^6 \cdot x \cdot y^{10} \cdot y \cdot z^8 \cdot z} \quad \begin{array}{l} \text{The largest perfect-square} \\ \text{factor is } x^6y^{10}z^8. \end{array}$$

$$= \sqrt{x^6}\sqrt{y^{10}}\sqrt{z^8}\sqrt{xyz} \quad \text{Factoring into several radicals}$$

$$= x^{6/2}y^{10/2}z^{8/2}\sqrt{xyz} \quad \begin{array}{l} \text{Converting to rational exponents.} \\ \text{Try to do this mentally.} \end{array}$$

$$= x^3y^5z^4\sqrt{xyz}. \quad \text{Simplifying}$$

Check: $\left(x^3y^5z^4\sqrt{xyz}\right)^2 = (x^3)^2(y^5)^2(z^4)^2\left(\sqrt{xyz}\right)^2$

$$= x^6 \cdot y^{10} \cdot z^8 \cdot xyz = x^7y^{11}z^9.$$

Our check shows that $x^3y^5z^4\sqrt{xyz}$ is the square root of $x^7y^{11}z^9$.

b) There are many ways to factor $16a^7b^{14}$. Because of the cube root (index of 3), we identify factors with the largest exponents that are multiples of 3:

$$\sqrt[3]{16a^7b^{14}} = \sqrt[3]{8 \cdot 2 \cdot a^6 \cdot a \cdot b^{12} \cdot b^2} \qquad \text{The largest perfect-cube factor is } 8a^6b^{12}.$$

$$= \sqrt[3]{8}\sqrt[3]{a^6}\sqrt[3]{b^{12}}\sqrt[3]{2ab^2} \qquad \text{Rewriting as a product of cube roots}$$

$$= 2a^2b^4\sqrt[3]{2ab^2}. \qquad \text{Simplifying the expressions containing perfect cubes}$$

As a check, let's redo the problem using a complete factorization of the radicand:

$$\sqrt[3]{16a^7b^{14}} = \sqrt[3]{2 \cdot 2 \cdot 2 \cdot 2 \cdot a \cdot a \cdot a \cdot a \cdot a \cdot a \cdot a \cdot b \cdot b \cdot b \cdot b \cdot b \cdot b \cdot b \cdot b \cdot b \cdot b \cdot b \cdot b \cdot b \cdot b}$$

$$= 2 \cdot a \cdot a \cdot b \cdot b \cdot b \cdot b \cdot \sqrt[3]{2 \cdot a \cdot b \cdot b} = 2a^2b^4\sqrt[3]{2ab^2}. \qquad \text{Each triple of factors makes a cube.} \qquad \text{Our answer checks.}$$

4. Simplify: $\sqrt[3]{5000x^{12}y^{13}z^2}$.   YOUR TURN

> To simplify an *n*th root, identify factors in the radicand with exponents that are multiples of *n*.

Multiplying and Simplifying

We have used the product rule for radicals to find products and also to simplify radical expressions. For some radical expressions, it is possible to do both: First find a product and then simplify.

EXAMPLE 5   Multiply and simplify.

a) $\sqrt{15}\sqrt{6}$ **b)** $3\sqrt[3]{25} \cdot 2\sqrt[3]{5}$ **c)** $\sqrt[4]{8x^3y^5}\sqrt[4]{4x^2y^3}$

SOLUTION

a) $\sqrt{15}\sqrt{6} = \sqrt{15 \cdot 6}$ Multiplying radicands

$\qquad\qquad = \sqrt{90} = \sqrt{9}\sqrt{10}$ 9 is a perfect square.

$\qquad\qquad = 3\sqrt{10}$

b) $3\sqrt[3]{25} \cdot 2\sqrt[3]{5} = 3 \cdot 2 \cdot \sqrt[3]{25 \cdot 5}$ Using a commutative law; multiplying radicands

$\qquad\qquad\qquad = 6 \cdot \sqrt[3]{125}$ 125 is a perfect cube.

$\qquad\qquad\qquad = 6 \cdot 5, \text{ or } 30$

c) $\sqrt[4]{8x^3y^5}\sqrt[4]{4x^2y^3} = \sqrt[4]{32x^5y^8}$ Multiplying radicands

$\qquad\qquad\qquad = \sqrt[4]{16x^4y^8 \cdot 2x}$ Identifying the largest perfect fourth-power factor

$\qquad\qquad\qquad = \sqrt[4]{16x^4y^8}\sqrt[4]{2x}$ Factoring into radicals

$\qquad\qquad\qquad = 2xy^2\sqrt[4]{2x}$ Finding the fourth root. We assume that $x \geq 0$.

Student Notes

To multiply $\sqrt{x} \cdot \sqrt{x}$, remember what $\sqrt{x}$ represents and go directly to the product, x. For $x \geq 0$,

$$\sqrt{x} \cdot \sqrt{x} = x,$$

$$\left(\sqrt{x}\right)^2 = x, \text{ and}$$

$$\sqrt{x^2} = x.$$

To check, we can use complete factorizations of the radicands. Checking part (c), we have

$$\sqrt[4]{8x^3y^5}\sqrt[4]{4x^2y^3} = \sqrt[4]{2 \cdot 2 \cdot 2 \cdot x \cdot x \cdot x \cdot y \cdot y \cdot y \cdot y \cdot y \cdot 2 \cdot 2 \cdot x \cdot x \cdot y \cdot y \cdot y}$$

$$= 2 \cdot x \cdot y \cdot y \sqrt[4]{2x} = 2xy^2\sqrt[4]{2x}.$$

5. Multiply and simplify:

$$\sqrt{10x}\sqrt{15x}.$$

 YOUR TURN

7.3 EXERCISE SET

FOR EXTRA HELP

MyMathLab® Math XL
PRACTICE   WATCH   READ   REVIEW

Vocabulary and Reading Check

Classify each of the following statements as either true or false.

1. For any real numbers $\sqrt[n]{a}$ and $\sqrt[n]{b}$,
   $\sqrt[n]{a} \cdot \sqrt[n]{b} = \sqrt[n]{ab}$.

2. For any real numbers $\sqrt[n]{a}$ and $\sqrt[n]{b}$,
   $\sqrt[n]{a} + \sqrt[n]{b} = \sqrt[n]{a + b}$.

3. For $x > 0$, $\sqrt{x^2 - 9} = x - 3$.

4. Since $(-10)^3 = -1000$, the number -1000 is a perfect cube.

5. The expression $\sqrt[3]{X}$ is not simplified if X contains a factor that is a perfect cube.

6. When no index is written, as in $\sqrt{5}$, the root is understood to be a square root.

Multiplying Radical Expressions

Multiply.

7. $\sqrt{3} \sqrt{10}$

8. $\sqrt{6} \sqrt{5}$

9. $\sqrt[3]{7} \sqrt[3]{5}$

10. $\sqrt[3]{2} \sqrt[3]{3}$

11. $\sqrt[4]{6} \sqrt[4]{9}$

12. $\sqrt[4]{4} \sqrt[4]{10}$

13. $\sqrt{2x} \sqrt{13y}$

14. $\sqrt{5a} \sqrt{6b}$

15. $\sqrt[5]{8y^3} \sqrt[5]{10y}$

16. $\sqrt[5]{9t^2} \sqrt[5]{2t}$

17. $\sqrt{y - b} \sqrt{y + b}$

18. $\sqrt{x - a} \sqrt{x + a}$

19. $\sqrt[3]{0.7y} \sqrt[3]{0.3y}$

20. $\sqrt[3]{0.5x} \sqrt[3]{0.2x}$

21. $\sqrt[5]{x - 2} \sqrt[5]{(x - 2)^2}$

22. $\sqrt[4]{x - 1} \sqrt[4]{x^2 + x + 1}$

23. $\sqrt{\dfrac{2}{t}} \sqrt{\dfrac{3s}{11}}$

24. $\sqrt{\dfrac{7p}{6}} \sqrt{\dfrac{5}{q}}$

25. $\sqrt[7]{\dfrac{x - 3}{4}} \sqrt[7]{\dfrac{5}{x + 2}}$

26. $\sqrt[6]{\dfrac{a}{b - 2}} \sqrt[6]{\dfrac{3}{b + 2}}$

Simplifying by Factoring

Simplify. Assume that no radicands were formed by raising negative numbers to even powers.

27. $\sqrt{12}$

28. $\sqrt{300}$

29. $\sqrt{45}$

30. $\sqrt{27}$

31. $\sqrt{8x^9}$

32. $\sqrt{75y^5}$

33. $\sqrt{120}$

34. $\sqrt{350}$

35. $\sqrt{36a^4b}$

36. $\sqrt{175y^8}$

37. $\sqrt[3]{8x^3y^2}$

38. $\sqrt[3]{27ab^6}$

39. $\sqrt[3]{-16x^6}$

40. $\sqrt[3]{-32a^6}$

Find a simplified form of $f(x)$. Assume that x can be any real number.

41. $f(x) = \sqrt[3]{40x^6}$

42. $f(x) = \sqrt[3]{27x^5}$

43. $f(x) = \sqrt{49(x - 3)^2}$

44. $f(x) = \sqrt{81(x - 1)^2}$

45. $f(x) = \sqrt{5x^2 - 10x + 5}$

46. $f(x) = \sqrt{2x^2 + 8x + 8}$

Simplify. Assume that no radicands were formed by raising negative numbers to even powers.

47. $\sqrt{a^{10}b^{11}}$

48. $\sqrt{x^8y^7}$

49. $\sqrt[3]{x^5y^6z^{10}}$

50. $\sqrt[3]{a^6b^7c^{13}}$

51. $\sqrt[4]{16x^5y^{11}}$

52. $\sqrt[5]{-32a^7b^{11}}$

53. $\sqrt[5]{x^{13}y^8z^{17}}$

54. $\sqrt[5]{a^6b^8c^9}$

55. $\sqrt[3]{-80a^{14}}$

56. $\sqrt[4]{810x^9}$

Multiplying and Simplifying

Multiply and simplify. Assume that no radicands were formed by raising negative numbers to even powers.

57. $\sqrt{5} \sqrt{10}$

58. $\sqrt{2} \sqrt{6}$

59. $\sqrt{6} \sqrt{33}$

60. $\sqrt{10} \sqrt{35}$

61. $\sqrt[3]{9} \sqrt[3]{3}$

62. $\sqrt[3]{2} \sqrt[3]{4}$

Aha! 63. $\sqrt{24y^5} \sqrt{24y^5}$

64. $\sqrt{120t^9} \sqrt{120t^9}$

65. $\sqrt[3]{5a^2} \sqrt[3]{2a}$

66. $\sqrt[3]{7x} \sqrt[3]{3x^2}$

67. $3\sqrt{2x^5} \cdot 4\sqrt{10x^2}$

68. $3\sqrt{5a^7} \cdot 2\sqrt{15a^3}$

69. $\sqrt[3]{s^2t^4} \sqrt[3]{s^4t^6}$

70. $\sqrt[3]{x^2y^4} \sqrt[3]{x^2y^6}$

71. $\sqrt[3]{(x - y)^2} \sqrt[3]{(x - y)^{10}}$

72. $\sqrt[3]{(t + 4)^5} \sqrt[3]{(t + 4)}$

73. $\sqrt[4]{20a^3b^7} \sqrt[4]{4a^2b^5}$

74. $\sqrt[4]{9x^7y^2} \sqrt[4]{9x^2y^9}$

75. $\sqrt[5]{x^3(y + z)^6} \sqrt[5]{x^3(y + z)^4}$

76. $\sqrt[5]{a^3(b - c)^4} \sqrt[5]{a^7(b - c)^4}$

77. Explain how you could convince a friend that $\sqrt{x^2 - 16} \neq \sqrt{x^2} - \sqrt{16}$.

78. Why is it incorrect to say that, in general, $\sqrt{x^2} = x$?

Skill Review

Perform the indicated operation and, if possible, simplify.

79. $\dfrac{15a^2x}{8b} \cdot \dfrac{24b^2x}{5a}$ [6.1]

80. $\dfrac{x^2 - 1}{x^2 - 4} \div \dfrac{x^2 - x - 2}{x^2 + x - 2}$ [6.1]

81. $\dfrac{x - 3}{2x - 10} - \dfrac{3x - 5}{x^2 - 25}$ [6.2]

82. $\dfrac{6x}{25y^2} + \dfrac{3y}{10x}$ [6.2]

83. $\dfrac{a^{-1} + b^{-1}}{ab}$ [6.3]

84. $\dfrac{\dfrac{1}{x + 1} - \dfrac{2}{x}}{\dfrac{3}{x} + \dfrac{1}{x + 1}}$ [6.3]

Synthesis

85. Explain why it is true that $\sqrt[n]{ab} = \sqrt[n]{a} \cdot \sqrt[n]{b}$ for any real numbers $\sqrt[n]{a}$ and $\sqrt[n]{b}$.

86. Is $\sqrt{(2x + 3)^8} = (2x + 3)^4$ always, sometimes, or never true? Why?

87. *Radar Range.* The function given by

$$R(x) = \frac{1}{2}\sqrt[4]{\frac{x \cdot 3.0 \times 10^6}{\pi^2}}$$

can be used to determine the maximum range $R(x)$, in miles, of an ARSR-3 surveillance radar with a peak power of x watts. Determine the maximum radar range when the peak power is 5×10^4 watts.

Source: Introduction to RADAR Techniques, Federal Aviation Administration, 1988

88. *Speed of a Skidding Car.* Police can estimate the speed at which a car was traveling by measuring its skid marks. The function given by

$$r(L) = 2\sqrt{5L}$$

can be used, where L is the length of a skid mark, in feet, and $r(L)$ is the speed, in miles per hour. Find the exact speed and an estimate (to the nearest tenth mile per hour) for the speed of a car that left skid marks **(a)** 20 ft long; **(b)** 70 ft long; **(c)** 90 ft long. See also Exercise 102.

89. *Wind Chill Temperature.* When the temperature is T degrees Celsius and the wind speed is v meters per second, the *wind chill temperature*, T_w, is the temperature (with no wind) that it feels like. Here is a formula for finding wind chill temperature:

$$T_w = 33 - \frac{(10.45 + 10\sqrt{v} - v)(33 - T)}{22}.$$

Estimate the wind chill temperature (to the nearest tenth of a degree) for the given actual temperatures and wind speeds.

a) $T = 7°C, v = 8\,\text{m/sec}$
b) $T = 0°C, v = 12\,\text{m/sec}$
c) $T = -5°C, v = 14\,\text{m/sec}$
d) $T = -23°C, v = 15\,\text{m/sec}$

Simplify. Assume that all variables are nonnegative.

90. $\left(\sqrt{r^3 t}\right)^7$ **91.** $\left(\sqrt[3]{25x^4}\right)^4$

92. $\left(\sqrt[3]{a^2b^4}\right)^5$ **93.** $\left(\sqrt{a^3b^5}\right)^7$

Draw and compare the graphs of each group of equations.

94. $f(x) = \sqrt{x^2 - 2x + 1},$
$g(x) = x - 1,$
$h(x) = |x - 1|$

95. $f(x) = \sqrt{x^2 + 2x + 1},$
$g(x) = x + 1,$
$h(x) = |x + 1|$

96. If $f(t) = \sqrt{t^2 - 3t - 4}$, what is the domain of f?

97. What is the domain of g, if $g(x) = \sqrt{x^2 - 6x + 8}$?

Solve.

98. $\sqrt[3]{5x^{k+1}} \sqrt[3]{25x^k} = 5x^7$, for k

99. $\sqrt[5]{4a^{3k+2}} \sqrt[5]{8a^{6-k}} = 2a^4$, for k

100. Use a graphing calculator to check your answers to Exercises 21 and 41.

101. Antonio is puzzled. When he uses a graphing calculator to graph $y = \sqrt{x} \cdot \sqrt{x}$, he gets the following screen. Explain why Antonio did not get the complete line $y = x$.

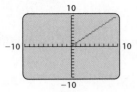

102. Does a car traveling twice as fast as another car leave a skid mark that is twice as long? (See Exercise 88.) Why or why not?

YOUR TURN ANSWERS: SECTION 7.3

1. $\sqrt[3]{20}$ **2.** $3b\sqrt{2a}$ **3.** $f(x) = |x + 3|\sqrt{10}$
4. $10x^4y^4\sqrt[3]{5yz^2}$ **5.** $5x\sqrt{6}$

QUICK QUIZ: SECTIONS 7.1–7.3

Simplify. Assume that no radicands were formed by raising negative quantities to even powers.

1. $\sqrt{(5 - y)^{10}}$ [7.1]

2. $\sqrt[3]{40x^5y^6}$ [7.3]

3. Use rational exponents to simplify: $\sqrt[20]{c^4}$. [7.2]

4. Use the laws of exponents to simplify: $3^{2/5} \cdot 3^{1/2}$. [7.2]

5. Multiply and simplify: $\sqrt[3]{4x^2y}\sqrt[3]{2xy^2}$. [7.3]

PREPARE TO MOVE ON

Simplify. [1.6]

1. $\dfrac{82ab}{2}$ 2. $\dfrac{120m^2}{64n^3}$

3. $\dfrac{34xy^5}{2y}$ 4. $\dfrac{45x^5y^2}{3xy^2}$

7.4 Dividing Radical Expressions

Dividing and Simplifying   ▪   Rationalizing Denominators or Numerators with One Term

Dividing and Simplifying

Just as the root of a product can be expressed as the product of two roots, the root of a quotient can be expressed as the quotient of two roots. For example,

$$\sqrt[3]{\frac{27}{8}} = \frac{3}{2} \quad \text{and} \quad \frac{\sqrt[3]{27}}{\sqrt[3]{8}} = \frac{3}{2}.$$

This example suggests the following.

THE QUOTIENT RULE FOR RADICALS

For any real numbers $\sqrt[n]{a}$ and $\sqrt[n]{b}$, $b \neq 0$,

$$\sqrt[n]{\frac{a}{b}} = \frac{\sqrt[n]{a}}{\sqrt[n]{b}}.$$

Remember that an nth root is simplified when its radicand has no factors that are perfect nth powers. **Unless functions are involved, we assume that no radicands represent negative quantities raised to an even power.**

EXAMPLE 1 Simplify by taking the roots of the numerator and the denominator.

a) $\sqrt[3]{\dfrac{27}{125}}$

b) $\sqrt{\dfrac{25}{y^2}}$

SOLUTION

1. Simplify by taking the roots of the numerator and the denominator: $\sqrt{\frac{25}{49}}$.

a) $\sqrt[3]{\frac{27}{125}} = \frac{\sqrt[3]{27}}{\sqrt[3]{125}} = \frac{3}{5}$ Taking the cube roots of the numerator and the denominator

b) $\sqrt{\frac{25}{y^2}} = \frac{\sqrt{25}}{\sqrt{y^2}} = \frac{5}{y}$ Taking the square roots of the numerator and the denominator. Assume $y > 0$.

YOUR TURN

Any radical expressions appearing in the answers should be simplified as much as possible.

EXAMPLE 2   Simplify: **(a)** $\sqrt{\frac{16x^3}{y^8}}$; **(b)** $\sqrt[3]{\frac{27y^{14}}{8x^3}}$.

SOLUTION

a) $\sqrt{\frac{16x^3}{y^8}} = \frac{\sqrt{16x^3}}{\sqrt{y^8}}$

$= \frac{\sqrt{16x^2 \cdot x}}{\sqrt{y^8}} = \frac{4x\sqrt{x}}{y^4}$ Simplifying the numerator and the denominator

2. Simplify: $\sqrt[3]{\frac{16x^5}{27y^6}}$.

b) $\sqrt[3]{\frac{27y^{14}}{8x^3}} = \frac{\sqrt[3]{27y^{14}}}{\sqrt[3]{8x^3}}$

$= \frac{\sqrt[3]{27y^{12}y^2}}{\sqrt[3]{8x^3}}$ y^{12} is the largest perfect-cube factor of y^{14}.

$= \frac{\sqrt[3]{27y^{12}} \sqrt[3]{y^2}}{\sqrt[3]{8x^3}}$

$= \frac{3y^4\sqrt[3]{y^2}}{2x}$ Simplifying the numerator and the denominator

YOUR TURN

If we read from right to left, the quotient rule tells us that to divide two radical expressions that have the same index, we can divide the radicands.

EXAMPLE 3   Divide and, if possible, simplify.

a) $\frac{\sqrt{80}}{\sqrt{5}}$

b) $\frac{5\sqrt[3]{32}}{\sqrt[3]{2}}$

c) $\frac{\sqrt{72xy}}{2\sqrt{2}}$

d) $\frac{\sqrt[4]{18a^9b^5}}{\sqrt[4]{3b}}$

Student Notes

When writing radical signs, be careful what you include in the radicand. The following represent *different* numbers:

$$\sqrt{\frac{5 \cdot 2}{3}}, \quad \frac{\sqrt{5 \cdot 2}}{3}, \quad \frac{\sqrt{5} \cdot 2}{3}.$$

SOLUTION

a) $\frac{\sqrt{80}}{\sqrt{5}} = \sqrt{\frac{80}{5}} = \sqrt{16} = 4$ Because the indices match, we can divide the radicands.

b) $\frac{5\sqrt[3]{32}}{\sqrt[3]{2}} = 5\sqrt[3]{\frac{32}{2}} = 5\sqrt[3]{16}$

$= 5\sqrt[3]{8 \cdot 2}$ 8 is the largest perfect-cube factor of 16.

$= 5\sqrt[3]{8}\sqrt[3]{2} = 5 \cdot 2\sqrt[3]{2}$

$= 10\sqrt[3]{2}$

c) $\dfrac{\sqrt{72xy}}{2\sqrt{2}} = \dfrac{1}{2}\sqrt{\dfrac{72xy}{2}}$

> Because the indices match, we can divide the radicands.

$= \dfrac{1}{2}\sqrt{36xy} = \dfrac{1}{2}\cdot 6\sqrt{xy}$

$= 3\sqrt{xy}$

3. Divide and, if possible, simplify:

$\dfrac{3\sqrt{8x}}{\sqrt{2}}.$

d) $\dfrac{\sqrt[4]{18a^9b^5}}{\sqrt[4]{3b}} = \sqrt[4]{\dfrac{18a^9b^5}{3b}}$

$= \sqrt[4]{6a^9b^4} = \sqrt[4]{a^8b^4}\,\sqrt[4]{6a}$

$= a^2b\sqrt[4]{6a}$

Note that 8 is the largest power less than 9 that is a multiple of the index 4.

YOUR TURN

Rationalizing Denominators or Numerators with One Term*

The expressions

$\dfrac{1}{\sqrt{2}}$ and $\dfrac{\sqrt{2}}{2}$

are equivalent, but the second expression does not have a radical expression in the denominator.[†] We can **rationalize the denominator** of a radical expression if we multiply by 1 in either of two ways.

One way is to multiply by 1 *under* the radical to make the denominator of the radicand a perfect power.

EXAMPLE 4 Rationalize each denominator.

a) $\sqrt{\dfrac{7}{3}}$ **b)** $\sqrt[3]{\dfrac{5}{16}}$

SOLUTION

a) We multiply by 1 under the radical, using $\tfrac{3}{3}$. We do this so that the denominator of the radicand will be a perfect square:

$\sqrt{\dfrac{7}{3}} = \sqrt{\dfrac{7}{3}\cdot\dfrac{3}{3}}$ Multiplying by 1 under the radical

$= \sqrt{\dfrac{21}{9}}$ The denominator, 9, is now a perfect square.

$= \dfrac{\sqrt{21}}{\sqrt{9}} = \dfrac{\sqrt{21}}{3}.$

Student Notes

When we rationalize a denominator, the resulting expression is equivalent to the original expression. You can check this in Example 4(a) by approximating $\sqrt{7/3}$ and $\sqrt{21}/3$ using a calculator.

b) Note that the index is 3, so we want the denominator to be a perfect cube. Since $16 = 4^2$, we multiply under the radical by $\tfrac{4}{4}$:

$\sqrt[3]{\dfrac{5}{16}} = \sqrt[3]{\dfrac{5}{4\cdot 4}\cdot\dfrac{4}{4}}$ Since the index is 3, we need 3 identical factors in the denominator.

$= \sqrt[3]{\dfrac{20}{4^3}}$ The denominator is now a perfect cube.

$= \dfrac{\sqrt[3]{20}}{\sqrt[3]{4^3}} = \dfrac{\sqrt[3]{20}}{4}.$

4. Rationalize the denominator:

$\sqrt[3]{\dfrac{5x^2}{y}}.$

YOUR TURN

*Denominators and numerators with two terms are rationalized in Section 7.5.
[†]See Exercise 75 in Exercise Set 7.4.

Another way to rationalize a denominator is to multiply by 1 *outside* the radical.

EXAMPLE 5 Rationalize each denominator.

a) $\dfrac{\sqrt{4}}{5\sqrt{x}}$
                                        b) $\dfrac{3b\sqrt[3]{a}}{\sqrt[3]{25bc^5}}$

SOLUTION

a) $\dfrac{\sqrt{4}}{5\sqrt{x}} = \dfrac{2}{5\sqrt{x}}$ Simplifying. We assume $x > 0$.

$\phantom{\dfrac{\sqrt{4}}{5\sqrt{x}}} = \dfrac{2}{5\sqrt{x}} \cdot \dfrac{\sqrt{x}}{\sqrt{x}}$ Multiplying by 1. Since the factor 5 in the denominator is rational, we do not need to include it in the form of 1.

$\phantom{\dfrac{\sqrt{4}}{5\sqrt{x}}} = \dfrac{2\sqrt{x}}{5(\sqrt{x})^2}$ Try to do this step mentally.

$\phantom{\dfrac{\sqrt{4}}{5\sqrt{x}}} = \dfrac{2\sqrt{x}}{5x}$

b) Note that the radicand $25bc^5$ is $5 \cdot 5 \cdot b \cdot c \cdot c \cdot c \cdot c \cdot c$. In order for this to be a cube, we need another factor of 5, two more factors of b, and one more factor of c. Thus we multiply by 1, using $\sqrt[3]{5b^2c}/\sqrt[3]{5b^2c}$:

$\dfrac{3b\sqrt[3]{a}}{\sqrt[3]{25bc^5}} = \dfrac{3b\sqrt[3]{a}}{\sqrt[3]{25bc^5}} \cdot \dfrac{\sqrt[3]{5b^2c}}{\sqrt[3]{5b^2c}}$ Multiplying by 1

$\phantom{\dfrac{3b\sqrt[3]{a}}{\sqrt[3]{25bc^5}}} = \dfrac{3b\sqrt[3]{5ab^2c}}{\sqrt[3]{125b^3c^6}}$ ←——— This radicand is now a perfect cube.

$\phantom{\dfrac{3b\sqrt[3]{a}}{\sqrt[3]{25bc^5}}} = \dfrac{3b\sqrt[3]{5ab^2c}}{5bc^2}$

$\phantom{\dfrac{3b\sqrt[3]{a}}{\sqrt[3]{25bc^5}}} = \dfrac{3\sqrt[3]{5ab^2c}}{5c^2}.$ Always simplify if possible.

5. Rationalize the denominator:

$\dfrac{\sqrt{6a}}{2\sqrt{10ab}}.$

 YOUR TURN

Sometimes it is necessary to rationalize a numerator. To do so, we multiply by 1 to make the radicand in the *numerator* a perfect power.

EXAMPLE 6 Rationalize the numerator: $\dfrac{\sqrt[3]{4a^2}}{\sqrt[3]{5b}}.$

SOLUTION

$\dfrac{\sqrt[3]{4a^2}}{\sqrt[3]{5b}} = \dfrac{\sqrt[3]{4a^2}}{\sqrt[3]{5b}} \cdot \dfrac{\sqrt[3]{2a}}{\sqrt[3]{2a}}$ Multiplying by 1

$\phantom{\dfrac{\sqrt[3]{4a^2}}{\sqrt[3]{5b}}} = \dfrac{\sqrt[3]{8a^3}}{\sqrt[3]{10ba}}$ ←——— This radicand is now a perfect cube.

6. Rationalize the numerator:

$\sqrt{\dfrac{7}{20}}.$

$\phantom{\dfrac{\sqrt[3]{4a^2}}{\sqrt[3]{5b}}} = \dfrac{2a}{\sqrt[3]{10ab}}$

 YOUR TURN

7.4 EXERCISE SET

Vocabulary and Reading Check

Give a justification for each equation by indicating either the quotient rule for radicals or multiplying by 1.

1. $\sqrt{\dfrac{x}{100}} = \dfrac{\sqrt{x}}{\sqrt{100}}$ _____

2. $\sqrt{\dfrac{5}{2}} = \sqrt{\dfrac{5}{2} \cdot \dfrac{2}{2}}$ _____

3. $\dfrac{\sqrt[3]{y^4}}{\sqrt[3]{x}} = \dfrac{\sqrt[3]{y^4}}{\sqrt[3]{x}} \cdot \dfrac{\sqrt[3]{x^2}}{\sqrt[3]{x^2}}$ _____

4. $\dfrac{\sqrt{50a^3}}{\sqrt{2a}} = \sqrt{\dfrac{50a^3}{2a}}$ _____

Concept Reinforcement

In Exercises 5–10, match each expression with an equivalent expression from the column on the right. Assume $a, b > 0$.

5. ___ $\sqrt[4]{\dfrac{16a^6}{a^2}}$

6. ___ $\dfrac{\sqrt[3]{a^6}}{\sqrt[3]{b^9}}$

7. ___ $\sqrt[5]{\dfrac{a^6}{b^4}}$

8. ___ $\sqrt{\dfrac{a}{b^3}}$

9. ___ $\dfrac{\sqrt{5a^4}}{\sqrt{5a^3}}$

10. ___ $\sqrt[3]{\dfrac{a^2}{b^6}}$

a) $\dfrac{a^2}{b^3}$

b) $\sqrt{\dfrac{a \cdot b}{b^3 \cdot b}}$

c) $\sqrt{a}$

d) $\dfrac{\sqrt[3]{a^2}}{b^2}$

e) $\sqrt[5]{\dfrac{a^6 b}{b^4 \cdot b}}$

f) $2a$

Dividing and Simplifying

Simplify by taking the roots of the numerator and the denominator. Assume that all variables represent positive numbers.

11. $\sqrt{\dfrac{49}{100}}$

12. $\sqrt{\dfrac{81}{25}}$

13. $\sqrt[3]{\dfrac{125}{8}}$

14. $\sqrt[3]{\dfrac{1000}{27}}$

15. $\sqrt{\dfrac{121}{t^2}}$

16. $\sqrt{\dfrac{144}{p^2}}$

17. $\sqrt{\dfrac{36y^3}{x^4}}$

18. $\sqrt[4]{\dfrac{25a^5}{b^6}}$

19. $\sqrt[3]{\dfrac{27a^4}{8b^3}}$

20. $\sqrt[3]{\dfrac{64x^7}{216y^6}}$

21. $\sqrt[4]{\dfrac{32a^4}{2b^4c^8}}$

22. $\sqrt[4]{\dfrac{81x^4}{y^8z^4}}$

23. $\sqrt[4]{\dfrac{a^5b^8}{c^{10}}}$

24. $\sqrt[4]{\dfrac{x^9y^{12}}{z^6}}$

25. $\sqrt[5]{\dfrac{32x^6}{y^{11}}}$

26. $\sqrt[5]{\dfrac{243a^9}{b^{13}}}$

27. $\sqrt[6]{\dfrac{x^6y^8}{z^{15}}}$

28. $\sqrt[6]{\dfrac{a^9b^{12}}{c^{13}}}$

Divide and, if possible, simplify. Assume that all variables represent positive numbers.

29. $\dfrac{\sqrt{18y}}{\sqrt{2y}}$

30. $\dfrac{\sqrt{700x}}{\sqrt{7x}}$

31. $\dfrac{\sqrt[3]{26}}{\sqrt[3]{13}}$

32. $\dfrac{\sqrt[3]{35}}{\sqrt[3]{5}}$

33. $\dfrac{\sqrt{40xy^3}}{\sqrt{8x}}$

34. $\dfrac{\sqrt{56ab^3}}{\sqrt{7a}}$

35. $\dfrac{\sqrt[3]{96a^4b^2}}{\sqrt[3]{12a^2b}}$

36. $\dfrac{\sqrt[3]{189x^5y^7}}{\sqrt[3]{7x^2y^2}}$

37. $\dfrac{\sqrt{100ab}}{5\sqrt{2}}$

38. $\dfrac{\sqrt{75ab}}{3\sqrt{3}}$

39. $\dfrac{\sqrt[4]{48x^9y^{13}}}{\sqrt[4]{3xy^{-2}}}$

40. $\dfrac{\sqrt[5]{64a^{11}b^{28}}}{\sqrt[5]{2ab^{-2}}}$

41. $\dfrac{\sqrt[3]{x^3 - y^3}}{\sqrt[3]{x - y}}$ ⟵

42. ⟶ $\dfrac{\sqrt[3]{r^3 + s^3}}{\sqrt[3]{r + s}}$

Hint: Factor and then simplify.

Rationalizing Denominators or Numerators with One Term

Rationalize each denominator. Assume that all variables represent positive numbers.

43. $\sqrt{\dfrac{2}{5}}$

44. $\sqrt{\dfrac{7}{2}}$

45. $\dfrac{2\sqrt{5}}{7\sqrt{3}}$

46. $\dfrac{3\sqrt{5}}{2\sqrt{7}}$

47. $\sqrt[3]{\dfrac{5}{4}}$

48. $\sqrt[3]{\dfrac{2}{9}}$

49. $\dfrac{\sqrt[3]{3a}}{\sqrt[3]{5c}}$

50. $\dfrac{\sqrt[3]{7x}}{\sqrt[3]{3y}}$

51. $\dfrac{\sqrt[4]{5y^6}}{\sqrt[4]{9x}}$

52. $\dfrac{\sqrt[5]{3a^4}}{\sqrt[5]{2b^7}}$

53. $\sqrt[3]{\dfrac{2}{x^2y}}$

54. $\sqrt[3]{\dfrac{5}{ab^2}}$

55. $\sqrt{\dfrac{7a}{18}}$

56. $\sqrt{\dfrac{3x}{20}}$

57. $\sqrt[5]{\dfrac{9}{32x^5y}}$

58. $\sqrt[4]{\dfrac{7}{64a^2b^4}}$ **Aha!** 59. $\sqrt{\dfrac{10ab^2}{72a^3b}}$

60. $\sqrt{\dfrac{21x^2y}{75xy^5}}$

Rationalize each numerator. Assume that all variables represent positive numbers.

61. $\sqrt{\dfrac{5}{11}}$ **62.** $\sqrt{\dfrac{2}{3}}$ **63.** $\dfrac{2\sqrt{6}}{5\sqrt{7}}$

64. $\dfrac{3\sqrt{10}}{2\sqrt{3}}$ **65.** $\dfrac{\sqrt{8}}{2\sqrt{3x}}$ **66.** $\dfrac{\sqrt{12}}{\sqrt{5y}}$

67. $\dfrac{\sqrt[3]{7}}{\sqrt[3]{2}}$ **68.** $\dfrac{\sqrt[3]{5}}{\sqrt[3]{4}}$ **69.** $\sqrt{\dfrac{7x}{3y}}$

70. $\sqrt{\dfrac{7a}{6b}}$ **71.** $\sqrt[3]{\dfrac{2a^5}{5b}}$ **72.** $\sqrt[3]{\dfrac{2a^4}{7b}}$

73. $\sqrt{\dfrac{x^3y}{2}}$ **74.** $\sqrt{\dfrac{ab^5}{3}}$

75. Explain why it is easier to approximate

$$\dfrac{\sqrt{2}}{2} \quad \text{than} \quad \dfrac{1}{\sqrt{2}}$$

if no calculator is available and $\sqrt{2} \approx 1.414213562$.

76. A student *incorrectly* claims that

$$\dfrac{5 + \sqrt{2}}{\sqrt{18}} = \dfrac{5 + \sqrt{1}}{\sqrt{9}} = \dfrac{5 + 1}{3}.$$

How could you convince the student that a mistake has been made? How would you explain the correct way of rationalizing the denominator?

Skill Review

Perform the indicated operations and, if possible, simplify.

77. $-\frac{2}{9} \div \frac{4}{6}$ [1.2] **78.** $-\frac{2}{9}\left(-\frac{4}{6}\right)$ [1.2]

79. $12 - 100 \div 5 \cdot (-2)^2 - 3(6 - 7)$ [1.2]

80. $\left(9x^3 - 3x - \frac{1}{2}\right) - \left(x^2 - 12x - \frac{1}{2}\right)$ [5.1]

81. $(12x^3 - 6x - 8) \div (x + 1)$ [6.6], [6.7]

82. $(7m - 2n)^2$ [5.5]

Synthesis

83. Is the quotient of two irrational numbers always an irrational number? Why or why not?

84. Is it possible to understand how to rationalize a denominator without knowing how to multiply rational expressions? Why or why not?

85. *Pendulums.* The *period* of a pendulum is the time it takes the pendulum to complete one cycle, swinging to and fro. For a pendulum that is L centimeters long, the period T is given by the formula

$$T = 2\pi\sqrt{\dfrac{L}{980}},$$

where T is in seconds. Find, to the nearest hundredth of a second, the period of a pendulum of length **(a)** 65 cm; **(b)** 98 cm; **(c)** 120 cm. Use a calculator's π key if possible.

Perform the indicated operations.

86. $\dfrac{7\sqrt{a^2b}\,\sqrt{25xy}}{5\sqrt{a^{-4}b^{-1}}\sqrt{49x^{-1}y^{-3}}}$ **87.** $\dfrac{(\sqrt[3]{81mn^2})^2}{(\sqrt[3]{mn})^2}$

88. $\dfrac{\sqrt{44x^2y^9z}\,\sqrt{22y^9z^6}}{(\sqrt{11xy^8z^2})^2}$

89. $\sqrt{a^2 - 3} - \dfrac{a^2}{\sqrt{a^2 - 3}}$

90. $5\sqrt{\dfrac{x}{y}} + 4\sqrt{\dfrac{y}{x}} - \dfrac{3}{\sqrt{xy}}$

91. Provide a reason for each step in the following derivation of the quotient rule:

$$\sqrt[n]{\dfrac{a}{b}} = \left(\dfrac{a}{b}\right)^{1/n} \quad \underline{\hspace{2em}}$$

$$= \dfrac{a^{1/n}}{b^{1/n}} \quad \underline{\hspace{2em}}$$

$$= \dfrac{\sqrt[n]{a}}{\sqrt[n]{b}} \quad \underline{\hspace{2em}}$$

92. Show that $\dfrac{\sqrt[n]{a}}{\sqrt[n]{b}}$ is the nth root of $\dfrac{a}{b}$ by raising it to the nth power and simplifying.

93. Let $f(x) = \sqrt{18x^3}$ and $g(x) = \sqrt{2x}$. Find $(f/g)(x)$ and specify the domain of f/g.

94. Let $f(t) = \sqrt{2t}$ and $g(t) = \sqrt{50t^3}$. Find $(f/g)(t)$ and specify the domain of f/g.

95. Let $f(x) = \sqrt{x^2 - 9}$ and $g(x) = \sqrt{x - 3}$. Find $(f/g)(x)$ and specify the domain of f/g.

96. *Research.* A *Foucault pendulum* is designed to demonstrate the earth's rotation.

a) Find the lengths of several Foucault pendulums, typically found in museums or universities. Calculate the period of each pendulum.

b) Explain how a Foucault pendulum demonstrates the earth's rotation using words, pictures, or a model.

QUICK QUIZ: SECTIONS 7.1–7.4

Simplify. For Exercises 1–4, assume that all variables represent positive numbers.

1. $\sqrt{\dfrac{9}{100}}$ [7.1], [7.4]   2. $(x^{1/5})^{5/3}$ [7.2]

3. $\sqrt{360m^3n^4p}$ [7.3]   4. $\dfrac{\sqrt{48x^3y}}{\sqrt{3y}}$ [7.4]

5. Find the domain of f, if $f(x) = \sqrt{x+8}$.   [7.1]

PREPARE TO MOVE ON

Perform the indicated operations.

1. $\dfrac{1}{2} - \dfrac{3}{4}$ [1.2]   2. $-\dfrac{2}{9} + \dfrac{5}{6}$ [1.2]

3. $(a+b)(a-b)$ [5.2]   4. $(a^2 - 2y)(a^2 + 2y)$ [5.2]

5. $(3+2x)(5-2x)$ [5.2]   6. $6x^4(x^2+x)$ [5.2]

7.5   Expressions Containing Several Radical Terms

Adding and Subtracting Radical Expressions   ▪   Products of Two or More Radical Terms   ▪
Rationalizing Denominators or Numerators With Two Terms   ▪   Terms with Differing Indices

Radical expressions like $6\sqrt{7} + 4\sqrt{7}$ or $(\sqrt{a} + \sqrt{b})(\sqrt{a} - \sqrt{b})$ contain more than one *radical term* and can sometimes be simplified.

Student Notes

Combining like radicals is similar to combining like terms. Recall the following:

$$3x + 8x = (3+8)x = 11x$$

and

$$6x^2 - 7x^2 = (6-7)x^2 = -x^2.$$

Adding and Subtracting Radical Expressions

When two radical expressions have the same indices and radicands, they are said to be **like radicals**. Like radicals can be combined (added or subtracted) in much the same way that we combine like terms.

EXAMPLE 1   Simplify by combining like radical terms.

a) $6\sqrt{7} + 4\sqrt{7}$ **b)** $6\sqrt[5]{4x} + 3\sqrt[5]{4x} - \sqrt[3]{4x}$

SOLUTION

a) $6\sqrt{7} + 4\sqrt{7} = (6+4)\sqrt{7}$ Using the distributive law (factoring out $\sqrt{7}$)

$\qquad = 10\sqrt{7}$ *Think*: 6 square roots of 7 plus 4 square roots of 7 is 10 square roots of 7.

b) $6\sqrt[5]{4x} + 3\sqrt[5]{4x} - \sqrt[3]{4x} = (6+3)\sqrt[5]{4x} - \sqrt[3]{4x}$ Try to do this step mentally.

$\qquad = 9\sqrt[5]{4x} - \sqrt[3]{4x}$ The indices are different. We cannot combine these terms.

1. Simplify by combining like radical terms:

$$2\sqrt{5} - 7\sqrt{5} + 4\sqrt{5}.$$

 YOUR TURN

Our ability to simplify radical expressions can help us to find like radicals even when, at first, it may appear that there are none.

EXAMPLE 2 Simplify by combining like radical terms, if possible.

a) $3\sqrt{8} - 5\sqrt{2}$ **b)** $9\sqrt{5} - 4\sqrt{3}$

SOLUTION

a)
$$
\begin{aligned}
3\sqrt{8} - 5\sqrt{2} &= 3\sqrt{4 \cdot 2} - 5\sqrt{2} \\
&= 3\sqrt{4} \cdot \sqrt{2} - 5\sqrt{2} \\
&= 3 \cdot 2 \cdot \sqrt{2} - 5\sqrt{2} \\
&= 6\sqrt{2} - 5\sqrt{2} \\
&= \sqrt{2}
\end{aligned}
$$
Simplifying $\sqrt{8}$

Combining like radicals

b) $9\sqrt{5} - 4\sqrt{3}$ cannot be simplified.   The radicands are different.

2. Simplify by combining like radical terms, if possible:

$5\sqrt{2} + 3\sqrt{8} + \sqrt{18}$.

YOUR TURN

If terms contain the same radical factor, we can factor out that radical expression in order to combine like terms, if possible.

EXAMPLE 3 Simplify by combining like terms, if possible.

a) $\sqrt[3]{2} - 7x\sqrt[3]{2} + 5\sqrt[3]{2}$ **b)** $\sqrt[3]{2x^6y^4} + 7\sqrt[3]{2y}$

SOLUTION

a)
$$
\begin{aligned}
\sqrt[3]{2} - 7x\sqrt[3]{2} + 5\sqrt[3]{2} &= (1 - 7x + 5)\sqrt[3]{2} \\
&= (6 - 7x)\sqrt[3]{2}
\end{aligned}
$$
Factoring out $\sqrt[3]{2}$

These parentheses are important!

b)
$$
\begin{aligned}
\sqrt[3]{2x^6y^4} + 7\sqrt[3]{2y} &= \sqrt[3]{x^6y^3 \cdot 2y} + 7\sqrt[3]{2y} \\
&= \sqrt[3]{x^6y^3} \cdot \sqrt[3]{2y} + 7\sqrt[3]{2y} \\
&= x^2y \cdot \sqrt[3]{2y} + 7\sqrt[3]{2y} \\
&= (x^2y + 7)\sqrt[3]{2y}
\end{aligned}
$$
Simplifying $\sqrt[3]{2x^6y^4}$. $\sqrt[3]{2y}$ is a factor of both terms.

Factoring to combine like radical terms

3. Simplify by combining like radical terms, if possible:

$3\sqrt{x} + \sqrt{25x^3}$.

YOUR TURN

Products of Two or More Radical Terms

Radical expressions often contain products with factors that have more than one term. Multiplying such products is similar to multiplying polynomials. Some products will yield like radical terms, which we can now combine.

EXAMPLE 4 Multiply.

a) $\sqrt{3}(x - \sqrt{5})$ **b)** $\sqrt[3]{y}(\sqrt[3]{y^2} + \sqrt[3]{2})$ **c)** $(4 - \sqrt{7})^2$
d) $(4\sqrt{3} + \sqrt{2})(\sqrt{3} - 5\sqrt{2})$ **e)** $(\sqrt{a} + \sqrt{b})(\sqrt{a} - \sqrt{b})$

SOLUTION

a)
$$
\begin{aligned}
\sqrt{3}(x - \sqrt{5}) &= \sqrt{3} \cdot x - \sqrt{3} \cdot \sqrt{5} \\
&= x\sqrt{3} - \sqrt{15}
\end{aligned}
$$
Using the distributive law

Multiplying radicals

b)
$$
\begin{aligned}
\sqrt[3]{y}(\sqrt[3]{y^2} + \sqrt[3]{2}) &= \sqrt[3]{y} \cdot \sqrt[3]{y^2} + \sqrt[3]{y} \cdot \sqrt[3]{2} \\
&= \sqrt[3]{y^3} + \sqrt[3]{2y} \\
&= y + \sqrt[3]{2y}
\end{aligned}
$$
Using the distributive law

Multiplying radicals

Simplifying $\sqrt[3]{y^3}$

c)
$$
(4 - \sqrt{7})^2 = (4 - \sqrt{7})(4 - \sqrt{7})
$$
We could also use the pattern $(A - B)^2 = A^2 - 2AB + B^2$.

F   O   I   L
$$
\begin{aligned}
&= 4^2 - 4\sqrt{7} - 4\sqrt{7} + (\sqrt{7})^2 \\
&= 16 - 8\sqrt{7} + 7 \\
&= 23 - 8\sqrt{7}
\end{aligned}
$$
Squaring and combining like terms

Adding 16 and 7

$$\overset{\text{F} \qquad\quad \text{O} \qquad\quad \text{I} \qquad\quad \text{L}}{}$$

d) $(4\sqrt{3} + \sqrt{2})(\sqrt{3} - 5\sqrt{2}) = 4(\sqrt{3})^2 - 20\sqrt{3} \cdot \sqrt{2} + \sqrt{2} \cdot \sqrt{3} - 5(\sqrt{2})^2$

$\qquad\qquad\qquad\qquad\qquad = 4 \cdot 3 - 20\sqrt{6} + \sqrt{6} - 5 \cdot 2$ Multiplying radicals

$\qquad\qquad\qquad\qquad\qquad = 12 - 20\sqrt{6} + \sqrt{6} - 10$

$\qquad\qquad\qquad\qquad\qquad = 2 - 19\sqrt{6}$ Combining like terms

e) $(\sqrt{a} + \sqrt{b})(\sqrt{a} - \sqrt{b}) = (\sqrt{a})^2 - \sqrt{a}\sqrt{b} + \sqrt{a}\sqrt{b} - (\sqrt{b})^2$ Using FOIL

$\qquad\qquad\qquad\qquad\qquad = a - b$ Combining like terms

4. Multiply: $(7 + \sqrt{x})^2$.

↪ YOUR TURN

In Example 4(e) above, note that the outer and inner products in FOIL are opposites, so that $a - b$ is not itself a radical expression. Pairs of radical expressions like $\sqrt{a} + \sqrt{b}$ and $\sqrt{a} - \sqrt{b}$ are called **conjugates**.

Rationalizing Denominators or Numerators with Two Terms

The use of conjugates allows us to rationalize denominators or numerators that contain two terms.

EXAMPLE 5   Write an equivalent expression with a rationalized denominator.

a) $\dfrac{4}{7 + \sqrt{3}}$ **b)** $\dfrac{4 + \sqrt{2}}{\sqrt{5} - \sqrt{2}}$

SOLUTION

a) $\dfrac{4}{7 + \sqrt{3}} = \dfrac{4}{7 + \sqrt{3}} \cdot \dfrac{7 - \sqrt{3}}{7 - \sqrt{3}}$ Multiplying by 1, using the conjugate of $7 + \sqrt{3}$, which is $7 - \sqrt{3}$

$\qquad\quad = \dfrac{4(7 - \sqrt{3})}{(7 + \sqrt{3})(7 - \sqrt{3})}$ Multiplying numerators and denominators

$\qquad\quad = \dfrac{4(7 - \sqrt{3})}{7^2 - (\sqrt{3})^2}$ Using $(A + B)(A - B) = A^2 - B^2$

$\qquad\quad = \dfrac{28 - 4\sqrt{3}}{49 - 3}$ No radicals remain in the denominator.

$\qquad\quad = \dfrac{28 - 4\sqrt{3}}{46}$

$\qquad\quad = \dfrac{2(14 - 2\sqrt{3})}{2 \cdot 23}$ Simplifying

$\qquad\quad = \dfrac{14 - 2\sqrt{3}}{23}$

b) $\dfrac{4 + \sqrt{2}}{\sqrt{5} - \sqrt{2}} = \dfrac{4 + \sqrt{2}}{\sqrt{5} - \sqrt{2}} \cdot \dfrac{\sqrt{5} + \sqrt{2}}{\sqrt{5} + \sqrt{2}}$ Multiplying by 1, using the conjugate of $\sqrt{5} - \sqrt{2}$, which is $\sqrt{5} + \sqrt{2}$

$\qquad\quad = \dfrac{(4 + \sqrt{2})(\sqrt{5} + \sqrt{2})}{(\sqrt{5} - \sqrt{2})(\sqrt{5} + \sqrt{2})}$ Multiplying numerators and denominators

$\qquad\quad = \dfrac{4\sqrt{5} + 4\sqrt{2} + \sqrt{2}\sqrt{5} + (\sqrt{2})^2}{(\sqrt{5})^2 - (\sqrt{2})^2}$ Multiplying

$\qquad\quad = \dfrac{4\sqrt{5} + 4\sqrt{2} + \sqrt{10} + 2}{5 - 2}$ Squaring in the denominator and the numerator

$\qquad\quad = \dfrac{4\sqrt{5} + 4\sqrt{2} + \sqrt{10} + 2}{3}$ No radicals remain in the denominator.

↪ YOUR TURN

Study Skills

Review Material on Your Own

Never hesitate to review earlier material in which you feel a lack of confidence. For example, if you feel unsure about how to multiply with fraction notation, be sure to review that material before studying any new material that requires that skill. Doing (and checking) some practice problems from that earlier section also helps to sharpen any skills that you may not have used for a while.

5. Rationalize the denominator:

$$\dfrac{\sqrt{2}}{\sqrt{3} - \sqrt{2}}.$$

To rationalize a numerator with two terms, we use the conjugate of the numerator.

EXAMPLE 6 Rationalize the numerator: $\dfrac{4 + \sqrt{2}}{\sqrt{5} - \sqrt{2}}$.

SOLUTION

$$\dfrac{4 + \sqrt{2}}{\sqrt{5} - \sqrt{2}} = \dfrac{4 + \sqrt{2}}{\sqrt{5} - \sqrt{2}} \cdot \dfrac{4 - \sqrt{2}}{4 - \sqrt{2}} \qquad \text{Multiplying by 1, using the conjugate of } 4 + \sqrt{2}, \text{ which is } 4 - \sqrt{2}$$

$$= \dfrac{16 - (\sqrt{2})^2}{4\sqrt{5} - \sqrt{5}\sqrt{2} - 4\sqrt{2} + (\sqrt{2})^2}$$

$$= \dfrac{14}{4\sqrt{5} - \sqrt{10} - 4\sqrt{2} + 2}$$

6. Rationalize the numerator:

$$\dfrac{5 - \sqrt{y}}{\sqrt{7}}.$$

YOUR TURN

CONNECTING 🔗 THE CONCEPTS

To rationalize denominators with one term or those with two terms, we multiply by 1.

One Term

Multiply by 1, using the factor(s) needed to make the denominator a perfect nth power.

$$\dfrac{3}{\sqrt{5}} = \dfrac{3}{\sqrt{5}} \cdot \dfrac{\sqrt{5}}{\sqrt{5}} = \dfrac{3\sqrt{5}}{5}$$

Two Terms

Multiply by 1, using the conjugate of the denominator.

$$\dfrac{3}{7 + \sqrt{5}} = \dfrac{3}{7 + \sqrt{5}} \cdot \dfrac{7 - \sqrt{5}}{7 - \sqrt{5}} = \dfrac{21 - 3\sqrt{5}}{44}$$

EXERCISES

Write an equivalent expression with a rationalized denominator.

1. $\dfrac{6}{\sqrt{7}}$

2. $\dfrac{1}{3 - \sqrt{2}}$

3. $\dfrac{2}{\sqrt{xy}}$

4. $\dfrac{5}{\sqrt{8}}$

5. $\dfrac{\sqrt{2}}{\sqrt{5} + \sqrt{3}}$

6. $\dfrac{2}{1 - \sqrt{5}}$

7. $\dfrac{1}{\sqrt[3]{x^2 y}}$

8. $\dfrac{a}{\sqrt[4]{a^3 b^2}}$

Terms with Differing Indices

If radical terms have different indices, we cannot multiply or divide the terms by simply multiplying or dividing radicands. However, we can often convert to exponential notation, use the rules for exponents, and then convert back to radical notation.

EXAMPLE 7   Divide and, if possible, simplify: $\dfrac{\sqrt[4]{(x+y)^3}}{\sqrt{x+y}}$.

SOLUTION

$$\frac{\sqrt[4]{(x+y)^3}}{\sqrt{x+y}} = \frac{(x+y)^{3/4}}{(x+y)^{1/2}}$$ Converting to exponential notation

$$= (x+y)^{3/4-1/2}$$ Since the bases are identical, we can subtract exponents:
$$\tfrac{3}{4} - \tfrac{1}{2} = \tfrac{3}{4} - \tfrac{2}{4} = \tfrac{1}{4}.$$

$$\left.\begin{array}{l} = (x+y)^{1/4} \\ = \sqrt[4]{x+y} \end{array}\right\}$$ Converting back to radical notation

7. Divide and, if possible, simplify:

$$\frac{\sqrt{3x}}{\sqrt[5]{(3x)^2}}.$$

 YOUR TURN

TO SIMPLIFY PRODUCTS OR QUOTIENTS WITH DIFFERING INDICES

1. Convert all radical expressions to exponential notation.
2. When the bases are identical, subtract exponents to divide and add exponents to multiply. This may require finding a common denominator for the exponents.
3. Convert back to radical notation and, if possible, simplify.

EXAMPLE 8   Multiply and simplify: $\sqrt{x^3}\sqrt[3]{x}$.

SOLUTION

$$\sqrt{x^3}\sqrt[3]{x} = x^{3/2} \cdot x^{1/3}$$ Converting to exponential notation

$$= x^{11/6}$$ Adding exponents: $\tfrac{3}{2} + \tfrac{1}{3} = \tfrac{9}{6} + \tfrac{2}{6}$

$$= \sqrt[6]{x^{11}}$$ Converting back to radical notation

$$\left.\begin{array}{l} = \sqrt[6]{x^6}\sqrt[6]{x^5} \\ = x\sqrt[6]{x^5} \end{array}\right\}$$ Simplifying

8. Multiply and simplify:

$$\sqrt{x}\,\sqrt[3]{x^4}.$$

 YOUR TURN

EXAMPLE 9   If $f(x) = \sqrt[3]{x^2}$ and $g(x) = \sqrt{x} + \sqrt[4]{x}$, find $(f \cdot g)(x)$.

SOLUTION   Recall that $(f \cdot g)(x) = f(x) \cdot g(x)$. Thus,

$$(f \cdot g)(x) = \sqrt[3]{x^2}\left(\sqrt{x} + \sqrt[4]{x}\right)$$ x is assumed to be nonnegative.

$$= x^{2/3}(x^{1/2} + x^{1/4})$$ Converting to exponential notation

$$= x^{2/3} \cdot x^{1/2} + x^{2/3} \cdot x^{1/4}$$ Using the distributive law

$$= x^{2/3+1/2} + x^{2/3+1/4}$$ Adding exponents:

$$= x^{7/6} + x^{11/12}$$ $\tfrac{2}{3} + \tfrac{1}{2} = \tfrac{4}{6} + \tfrac{3}{6}; \tfrac{2}{3} + \tfrac{1}{4} = \tfrac{8}{12} + \tfrac{3}{12}$

$$= \sqrt[6]{x^7} + \sqrt[12]{x^{11}}$$ Converting back to radical notation

$$\left.\begin{array}{l} = \sqrt[6]{x^6}\sqrt[6]{x} + \sqrt[12]{x^{11}} \\ = x\sqrt[6]{x} + \sqrt[12]{x^{11}}. \end{array}\right\}$$ Simplifying

9. If $f(x) = 2\sqrt{x}$ and $g(x) = \sqrt[3]{x} - \sqrt[5]{x^2}$, find $(f \cdot g)(x)$.

 YOUR TURN

We often can write the final result as a single radical expression by finding a common denominator in the exponents.

EXAMPLE 10   Divide and, if possible, simplify: $\dfrac{\sqrt[3]{a^2b^4}}{\sqrt{ab}}$.

SOLUTION

$$\dfrac{\sqrt[3]{a^2b^4}}{\sqrt{ab}} = \dfrac{(a^2b^4)^{1/3}}{(ab)^{1/2}} \qquad \text{Converting to exponential notation}$$

$$= \dfrac{a^{2/3}b^{4/3}}{a^{1/2}b^{1/2}} \qquad \text{Using the product and power rules}$$

10. Divide and, if possible, simplify:

$\dfrac{\sqrt[5]{x^{15}y^7}}{\sqrt{x^2y}}$.

$$= a^{2/3-1/2}b^{4/3-1/2} \qquad \text{Subtracting exponents}$$

$$= a^{1/6}b^{5/6}$$

$$= \sqrt[6]{a}\,\sqrt[6]{b^5} \qquad \text{Converting to radical notation}$$

$$= \sqrt[6]{ab^5} \qquad \text{Using the product rule for radicals}$$

 YOUR TURN

7.5 EXERCISE SET

FOR EXTRA HELP

MyMathLab® Math XL
PRACTICE   WATCH   READ   REVIEW

Vocabulary and Reading Check

For Exercises 1–6, fill in each blank by selecting from the following words (which may be used more than once).

base(s)   indice(s)
conjugate(s)   numerator(s)
denominator(s)   radicand(s)

1. To add radical expressions, both the _____ and the _____ must be the same.

2. To multiply radicands, the _____ must be the same.

3. To find a product by adding exponents, the _____ must be the same.

4. To add the numerators of rational expressions, the _____ must be the same.

5. To rationalize the _____ of $\dfrac{\sqrt{c}-\sqrt{a}}{5}$, we multiply by a form of 1, using the _____ of $\sqrt{c}-\sqrt{a}$, which is $\sqrt{c}+\sqrt{a}$, to write 1.

6. To find a quotient by subtracting exponents, the _____ must be the same.

Adding and Subtracting Radical Expressions

Add or subtract. Simplify by combining like radical terms, if possible. Assume that all variables and radicands represent positive real numbers.

7. $4\sqrt{3} + 7\sqrt{3}$

8. $6\sqrt{5} + 2\sqrt{5}$

9. $7\sqrt[3]{4} - 5\sqrt[3]{4}$

10. $14\sqrt[5]{2} - 8\sqrt[5]{2}$

11. $\sqrt[3]{y} + 9\sqrt[3]{y}$

12. $4\sqrt[4]{t} - \sqrt[4]{t}$

13. $8\sqrt{2} - \sqrt{2} + 5\sqrt{2}$

14. $\sqrt{6} + 3\sqrt{6} - 8\sqrt{6}$

15. $9\sqrt[3]{7} - \sqrt{3} + 4\sqrt[3]{7} + 2\sqrt{3}$

16. $5\sqrt{7} - 8\sqrt[4]{11} + \sqrt{7} + 9\sqrt[4]{11}$

17. $4\sqrt{27} - 3\sqrt{3}$

18. $9\sqrt{50} - 4\sqrt{2}$

19. $3\sqrt{45} - 8\sqrt{20}$

20. $5\sqrt{12} + 16\sqrt{27}$

21. $3\sqrt[3]{16} + \sqrt[3]{54}$

22. $\sqrt[3]{27} - 5\sqrt[3]{8}$

23. $\sqrt{a} + 3\sqrt{16a^3}$

24. $2\sqrt{9x^3} - \sqrt{x}$

25. $\sqrt[3]{6x^4} - \sqrt[3]{48x}$

26. $\sqrt[3]{54x} - \sqrt[3]{2x^4}$

27. $\sqrt{4a-4} + \sqrt{a-1}$

28. $\sqrt{9y+27} + \sqrt{y+3}$

29. $\sqrt{x^3-x^2} + \sqrt{9x-9}$

30. $\sqrt{4x-4} - \sqrt{x^3-x^2}$

Products of Two or More Radical Terms

Multiply. Assume that all variables represent nonnegative real numbers.

31. $\sqrt{2}(5+\sqrt{2})$

32. $\sqrt{3}(6-\sqrt{3})$

33. $3\sqrt{5}(\sqrt{6}-\sqrt{7})$

34. $4\sqrt{2}(\sqrt{3}+\sqrt{5})$

35. $\sqrt{2}(3\sqrt{10}-\sqrt{8})$

36. $\sqrt{3}(2\sqrt{15}-3\sqrt{4})$

37. $\sqrt[3]{3}(\sqrt[3]{9}-4\sqrt[3]{21})$

38. $\sqrt[3]{2}(\sqrt[3]{4}-2\sqrt[3]{32})$

39. $\sqrt[3]{a}\left(\sqrt[3]{a^2}+\sqrt[3]{24a^2}\right)$

40. $\sqrt[3]{x}\left(\sqrt[3]{3x^2}-\sqrt[3]{81x^2}\right)$

41. $(2+\sqrt{6})(5-\sqrt{6})$

42. $(4-\sqrt{5})(2+\sqrt{5})$

43. $(\sqrt{2}+\sqrt{7})(\sqrt{3}-\sqrt{7})$

44. $(\sqrt{7}-\sqrt{2})(\sqrt{5}+\sqrt{2})$

45. $(2-\sqrt{3})(2+\sqrt{3})$

46. $(3+\sqrt{11})(3-\sqrt{11})$

47. $(\sqrt{10}-\sqrt{15})(\sqrt{10}+\sqrt{15})$

48. $(\sqrt{12}+\sqrt{5})(\sqrt{12}-\sqrt{5})$

49. $(3\sqrt{7}+2\sqrt{5})(2\sqrt{7}-4\sqrt{5})$

50. $(4\sqrt{5}-3\sqrt{2})(2\sqrt{5}+4\sqrt{2})$

51. $(4+\sqrt{7})^2$ **52.** $(3+\sqrt{10})^2$

53. $(\sqrt{3}-\sqrt{2})^2$ **54.** $(\sqrt{5}-\sqrt{3})^2$

55. $(\sqrt{2t}+\sqrt{5})^2$ **56.** $(\sqrt{3x}-\sqrt{2})^2$

57. $(3-\sqrt{x+5})^2$ **58.** $(4+\sqrt{x-3})^2$

59. $\left(2\sqrt[4]{7}-\sqrt[4]{6}\right)\left(3\sqrt[4]{9}+2\sqrt[4]{5}\right)$

60. $\left(4\sqrt[3]{3}+\sqrt[3]{10}\right)\left(2\sqrt[3]{7}+5\sqrt[3]{6}\right)$

Rationalizing Denominators or Numerators with Two Terms

Rationalize each denominator. If possible, simplify your result.

61. $\dfrac{6}{3-\sqrt{2}}$ **62.** $\dfrac{5}{4-\sqrt{5}}$

63. $\dfrac{2+\sqrt{5}}{6+\sqrt{3}}$ **64.** $\dfrac{1+\sqrt{2}}{3+\sqrt{5}}$

65. $\dfrac{\sqrt{a}}{\sqrt{a}+\sqrt{b}}$ **66.** $\dfrac{\sqrt{z}}{\sqrt{x}-\sqrt{z}}$

Aha! **67.** $\dfrac{\sqrt{7}-\sqrt{3}}{\sqrt{3}-\sqrt{7}}$ **68.** $\dfrac{\sqrt{7}+\sqrt{5}}{\sqrt{5}+\sqrt{2}}$

69. $\dfrac{3\sqrt{2}-\sqrt{7}}{4\sqrt{2}+2\sqrt{5}}$ **70.** $\dfrac{5\sqrt{3}-\sqrt{11}}{2\sqrt{3}-5\sqrt{2}}$

Rationalize each numerator. If possible, simplify your result.

71. $\dfrac{\sqrt{5}+1}{4}$ **72.** $\dfrac{\sqrt{15}-3}{6}$

73. $\dfrac{\sqrt{6}-2}{\sqrt{3}+7}$ **74.** $\dfrac{\sqrt{10}+4}{\sqrt{2}-3}$

75. $\dfrac{\sqrt{x}-\sqrt{y}}{\sqrt{x}+\sqrt{y}}$ **76.** $\dfrac{\sqrt{a}+\sqrt{b}}{\sqrt{a}-\sqrt{b}}$

77. $\dfrac{\sqrt{a+h}-\sqrt{a}}{h}$ **78.** $\dfrac{\sqrt{x-h}-\sqrt{x}}{h}$

Terms with Differing Indices

Perform the indicated operation and simplify. Assume that all variables represent positive real numbers.

79. $\sqrt[3]{a}\sqrt[6]{a}$ **80.** $\sqrt[10]{a}\sqrt[5]{a^2}$

81. $\sqrt{b^3}\sqrt[5]{b^4}$ **82.** $\sqrt[3]{b^4}\sqrt[4]{b^3}$

83. $\sqrt{xy^3}\sqrt[3]{x^2y}$ **84.** $\sqrt[5]{a^3b}\sqrt{ab}$

85. $\sqrt[4]{9ab^3}\sqrt{3a^4b}$ **86.** $\sqrt{2x^3y^3}\sqrt[3]{4xy^2}$

87. $\sqrt{a^4b^3c^4}\sqrt[3]{ab^2c}$ **88.** $\sqrt[3]{xy^2z}\sqrt{x^3yz^2}$

89. $\dfrac{\sqrt[3]{a^2}}{\sqrt[4]{a}}$ **90.** $\dfrac{\sqrt[3]{x^2}}{\sqrt[5]{x}}$

91. $\dfrac{\sqrt[4]{x^2y^3}}{\sqrt[3]{xy}}$ **92.** $\dfrac{\sqrt[5]{a^4b}}{\sqrt[3]{ab}}$

93. $\dfrac{\sqrt{ab^3}}{\sqrt[5]{a^2b^3}}$ **94.** $\dfrac{\sqrt[5]{x^3y^4}}{\sqrt{xy}}$

95. $\dfrac{\sqrt{(7-y)^3}}{\sqrt[3]{(7-y)^2}}$ **96.** $\dfrac{\sqrt[5]{(y-9)^3}}{\sqrt{y-9}}$

97. $\dfrac{\sqrt[4]{(5+3x)^3}}{\sqrt[3]{(5+3x)^2}}$ **98.** $\dfrac{\sqrt[3]{(2x+1)^2}}{\sqrt[5]{(2x+1)^2}}$

99. $\sqrt[3]{x^2y}\left(\sqrt{xy}-\sqrt[5]{xy^3}\right)$

100. $\sqrt[4]{a^2b}\left(\sqrt[3]{a^2b}-\sqrt[5]{a^2b^2}\right)$

101. $\left(m+\sqrt[3]{n^2}\right)\left(2m+\sqrt[4]{n}\right)$

102. $\left(r-\sqrt[4]{s^3}\right)\left(3r-\sqrt[5]{s}\right)$

Products of Two or More Radical Terms

In Exercises 103–106, $f(x)$ and $g(x)$ are as given. Find $(f\cdot g)(x)$. Assume that all variables represent nonnegative real numbers.

103. $f(x)=\sqrt[4]{x},\ g(x)=2\sqrt{x}-\sqrt[3]{x^2}$

104. $f(x)=\sqrt[4]{2x}+5\sqrt{2x},\ g(x)=\sqrt[3]{2x}$

105. $f(x)=x+\sqrt{7},\ g(x)=x-\sqrt{7}$

106. $f(x)=x-\sqrt{2},\ g(x)=x+\sqrt{6}$

Let $f(x)=x^2$. Find each of the following.

107. $f(3-\sqrt{2})$

108. $f(5-\sqrt{3})$

109. $f(\sqrt{6}+\sqrt{21})$

110. $f(\sqrt{2}+\sqrt{10})$

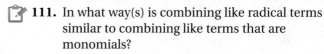 **111.** In what way(s) is combining like radical terms similar to combining like terms that are monomials?

112. Why do we need to know how to multiply radical expressions before learning how to add them?

Skill Review

113. In which quadrant is the point $\left(6, -\frac{1}{2}\right)$ located? [2.1]

114. Find the slope of the line containing the points $(9, 10)$ and $(6, 7)$. [2.3]

115. Find the x-intercept and the y-intercept of the line given by $x - y = 10$. [2.4]

116. Find the slope and the y-intercept of the line given by $3y + 5x = 1$. [2.3]

117. Write the slope–intercept equation of the line that is perpendicular to the line $y = \frac{1}{2}x - 7$ and has a y-intercept of $(0, 12)$. [2.5]

118. Write the slope–intercept equation of the line that contains the points $(-1, -6)$ and $(-4, 0)$. [2.5]

Synthesis

 119. Nadif *incorrectly* writes
$$\sqrt[5]{x^2} \cdot \sqrt{x^3} = x^{2/5} \cdot x^{3/2} = \sqrt[5]{x^3}.$$
What mistake do you suspect he is making?

 120. After examining the expression $\sqrt[4]{25xy^3}\,\sqrt{5x^4y}$, Sofia (correctly) concludes that x and y are both nonnegative. Explain how she could reach this conclusion.

Find a simplified form for $f(x)$. Assume $x \geq 0$.

121. $f(x) = \sqrt{x^3 - x^2} + \sqrt{9x^3 - 9x^2} - \sqrt{4x^3 - 4x^2}$

122. $f(x) = \sqrt{20x^2 + 4x^3} - 3x\sqrt{45 + 9x}$
$+ \sqrt{5x^2 + x^3}$

123. $f(x) = \sqrt[4]{x^5 - x^4} + 3\sqrt[4]{x^9 - x^8}$

124. $f(x) = \sqrt[4]{16x^4 + 16x^5} - 2\sqrt[4]{x^8 + x^9}$

Simplify.

125. $7x\sqrt{(x + y)^3} - 5xy\sqrt{x + y} - 2y\sqrt{(x + y)^3}$

126. $\sqrt{27a^5(b + 1)}\,\sqrt[3]{81a(b + 1)^4}$

127. $\sqrt{8x(y + z)^5}\,\sqrt[3]{4x^2(y + z)^2}$

128. $\frac{1}{2}\sqrt{36a^5bc^4} - \frac{1}{2}\sqrt[3]{64a^4bc^6} + \frac{1}{6}\sqrt{144a^3bc^6}$

129. $\dfrac{\dfrac{1}{\sqrt{w}} - \sqrt{w}}{\dfrac{\sqrt{w} + 1}{\sqrt{w}}}$

130. $\dfrac{1}{4 + \sqrt{3}} + \dfrac{1}{\sqrt{3}} + \dfrac{1}{\sqrt{3} - 4}$

Express each of the following as the product of two radical expressions.

131. $x - 5$ **132.** $y - 7$

133. $x - a$

Multiply.

134. $\sqrt{9 + 3\sqrt{5}}\,\sqrt{9 - 3\sqrt{5}}$

135. $\left(\sqrt{x + 2} - \sqrt{x - 2}\right)^2$

 136. Use a graphing calculator to check your answers to Exercises 25, 39, and 81.

137. A formula for factoring a sum of squares is
$$A^2 + B^2 = (A + B + \sqrt{2AB})(A + B - \sqrt{2AB}).$$

a) Show that this is an identity.
b) What must be true of A and B in order for $A^2 + B^2$ to be written as the product of two binomials with rational coefficients?

➤ YOUR TURN ANSWERS: SECTION 7.5

1. $-\sqrt{5}$ **2.** $14\sqrt{2}$ **3.** $(3 + 5x)\sqrt{x}$ **4.** $49 + 14\sqrt{x} + x$

5. $\sqrt{6} + 2$ **6.** $\dfrac{25 - y}{5\sqrt{7} + \sqrt{7y}}$ **7.** $\sqrt[10]{3x}$ **8.** $x\sqrt[6]{x^5}$

9. $2\sqrt[6]{x^5} - 2\sqrt[10]{x^9}$ **10.** $x^2\sqrt[10]{y^9}$

QUICK QUIZ: SECTIONS 7.1–7.5

1. Simplify $\sqrt{100t^2 + 20t + 1}$. Assume that t can represent any real number. [7.1]

2. Write an equivalent expression using radical notation: $(3xy)^{7/8}$. [7.2]

3. Write an equivalent expression using exponential notation: $\sqrt{17ab}$. [7.2]

4. Rationalize the denominator: $\sqrt[3]{\dfrac{5}{4x^5y}}$. [7.4]

5. Rationalize the numerator: $\dfrac{2 + \sqrt{3}}{\sqrt{10}}$. [7.5]

PREPARE TO MOVE ON

Solve.

1. $3x - 1 = 125$ [1.3]

2. $x^2 + 2x + 1 = 2(11 - x)$ [5.8]

3. $-6 = 5 - x$ [1.3]

4. $x^2 - 10x + 25 = 4(x - 2)$ [5.8]

Mid-Chapter Review

Many radical expressions can be simplified. It is important to know under which conditions radical expressions can be multiplied and divided and radical terms can be combined.

Multiplication and division: The indices must be the same.

$$\frac{\sqrt{50t^5}}{\sqrt{2t^{11}}} = \sqrt{\frac{50t^5}{2t^{11}}} = \sqrt{\frac{25}{t^6}} = \frac{5}{t^3}; \qquad \sqrt[4]{8x^3} \cdot \sqrt[4]{2x} = \sqrt[4]{16x^4} = 2x$$

Combining like radical terms: The indices and the radicands must both be the same.

$$\sqrt{75x} + \sqrt{12x} - \sqrt{3x} = 5\sqrt{3x} + 2\sqrt{3x} - \sqrt{3x} = 6\sqrt{3x}$$

Radical expressions with differing indices can sometimes be simplified using rational exponents.

$$\sqrt[3]{x^2}\sqrt{x} = x^{2/3}x^{1/2} = x^{4/6}x^{3/6} = x^{7/6} = \sqrt[6]{x^7} = x\sqrt[6]{x}$$

GUIDED SOLUTIONS

1. Multiply and simplify: $\sqrt{6x^9} \cdot \sqrt{2xy}$. [7.3]

Solution

$$\sqrt{6x^9} \cdot \sqrt{2xy} = \sqrt{6x^9 \cdot \Box}$$
$$= \sqrt{12x^{10}y}$$
$$= \sqrt{\Box \cdot 3y}$$
$$= \sqrt{\Box} \cdot \sqrt{3y}$$
$$= \Box\sqrt{3y} \quad \text{Taking the square root}$$

2. Combine like radical terms:

$$\sqrt{12} - 3\sqrt{75} + \sqrt{8}. \quad [7.5]$$

Solution

$$\sqrt{12} - 3\sqrt{75} + \sqrt{8}$$
$$= \Box - 3 \cdot 5\sqrt{3} + \Box \quad \text{Simplifying each term}$$
$$= 2\sqrt{3} - \Box + 2\sqrt{2} \quad \text{Multiplying}$$
$$= \Box + 2\sqrt{2} \quad \text{Combining like radical terms}$$

MIXED REVIEW

Simplify. Assume that variables can represent any real number.

3. $\sqrt{81}$ [7.1]

4. $-\sqrt{\frac{9}{100}}$ [7.1]

5. $\sqrt{64t^2}$ [7.1]

6. $\sqrt[5]{x^5}$ [7.1]

7. Find $f(-5)$ if $f(x) = \sqrt[3]{12x - 4}$. [7.1]

8. Determine the domain of g if $g(x) = \sqrt[4]{10 - x}$. [7.1]

9. Write an equivalent expression using radical notation and simplify: $8^{2/3}$. [7.2]

Simplify. Assume for Exercises 10–25 that no radicands were formed by raising negative numbers to even powers.

10. $\sqrt[6]{\sqrt{a}}$ [7.2]

11. $\sqrt[3]{y^{24}}$ [7.2]

12. $\sqrt{(t+5)^2}$ [7.1]

13. $\sqrt[3]{-27a^{12}}$ [7.1]

14. $\sqrt{6x}\sqrt{15x}$ [7.3]

15. $\frac{\sqrt{20y}}{\sqrt{45y}}$ [7.4]

16. $\sqrt{6}(\sqrt{10} - \sqrt{33})$ [7.5]

17. $\frac{\sqrt{t}}{\sqrt[8]{t^3}}$ [7.5]

18. $\sqrt[5]{\frac{3a^{12}}{96a^2}}$ [7.4]

19. $2\sqrt{3} - 5\sqrt{12}$ [7.5]

20. $(\sqrt{5}+3)(\sqrt{5}-3)$ [7.5]

21. $(\sqrt{15}+\sqrt{10})^2$ [7.5]

22. $\sqrt{25x-25} - \sqrt{9x-9}$ [7.5]

23. $\sqrt{x^3y}\sqrt[5]{xy^4}$ [7.5]

24. $\sqrt[3]{5000} + \sqrt[3]{625}$ [7.5]

25. $\sqrt[3]{12x^2y^5}\sqrt[3]{18x^7y}$ [7.3]

7.6   Solving Radical Equations

The Principle of Powers   ▪   Equations with Two Radical Terms

Now that we know how to work with radicals and rational exponents, we can learn how to solve a new type of equation.

The Principle of Powers

A **radical equation** is an equation in which the variable appears in a radicand. Examples are

$$\sqrt[3]{2x} + 1 = 5, \qquad \sqrt{a - 2} = 7, \quad \text{and} \quad 4 - \sqrt{3x + 1} = \sqrt{6 - x}.$$

To solve such equations, we need a new principle. Suppose $a = b$ is true. If we square both sides, we get another true equation: $a^2 = b^2$. This can be generalized.

THE PRINCIPLE OF POWERS

If $a = b$, then $a^n = b^n$ for any exponent n.

Note that the principle of powers is an "if–then" statement. If we interchange the two parts of the sentence, then we have the statement "If $a^n = b^n$ for some exponent n, then $a = b$." **This statement is not always true.** For example, "if $x = 3$, then $x^2 = 9$" is true, but the statement "if $x^2 = 9$, then $x = 3$" is *not* true when x is replaced with -3.

When n is even, every solution of $x = a$ is a solution of $x^n = a^n$, but not every solution of $x^n = a^n$ is a solution of $x = a$.

When we raise both sides of an equation to an even exponent, it is essential that we check the answer in the *original* equation.

EXAMPLE 1   Solve: $\sqrt{x} - 3 = 4$.

SOLUTION   Before using the principle of powers, we must isolate the radical term:

$$\sqrt{x} - 3 = 4$$
$$\sqrt{x} = 7 \qquad \text{Isolating the radical by adding 3 to both sides}$$
$$(\sqrt{x})^2 = 7^2 \qquad \text{Using the principle of powers}$$
$$x = 49.$$

Check:

$$\begin{array}{c|c} \sqrt{x} - 3 = 4 \\ \hline \sqrt{49} - 3 & 4 \\ 7 - 3 & \\ 4 \overset{?}{=} 4 & \text{TRUE} \end{array}$$

The solution is 49.

1. Solve: $\sqrt{x + 1} - 3 = 2$.

YOUR TURN

It is important that we isolate a radical term before using the principle of powers. Suppose in Example 1 that both sides of the equation were squared *before* we isolated the radical. We would have had the expression $(\sqrt{x} - 3)^2$ or $x - 6\sqrt{x} + 9$ on the left side, and the radical would have remained in the problem.

EXAMPLE 2   Solve: $\sqrt{x} + 5 = 3$.

SOLUTION

$$\sqrt{x} + 5 = 3$$
$$\sqrt{x} = -2 \qquad \text{Isolating the radical by adding } -5 \text{ to both sides}$$

> The equation $\sqrt{x} = -2$ has no solution because the principal square root of a number is never negative. We continue as in Example 1 for comparison.

$$(\sqrt{x})^2 = (-2)^2 \qquad \text{Using the principle of powers}$$
$$x = 4$$

Check:
$$\begin{array}{c|c} \sqrt{x} + 5 = 3 \\ \hline \sqrt{4} + 5 & 3 \\ 2 + 5 & \\ 7 \overset{?}{=} 3 & \text{FALSE} \end{array}$$

The number 4 does not check. Thus, $\sqrt{x} + 5 = 3$ has no solution.

2. Solve: $\sqrt{x} + 6 = 1$.

YOUR TURN

Note in Example 2 that $x = 4$ has the solution 4, but $\sqrt{x} + 5 = 3$ has *no* solution. Thus, $x = 4$ and $\sqrt{x} + 5 = 3$ are *not* equivalent equations.

TO SOLVE AN EQUATION WITH A RADICAL TERM
1. Isolate the radical term on one side of the equation.
2. Use the principle of powers and solve the resulting equation.
3. Check any possible solution in the original equation.

EXAMPLE 3   Solve: $x = \sqrt{x + 7} + 5$.

SOLUTION

$$x = \sqrt{x + 7} + 5$$
$$x - 5 = \sqrt{x + 7} \qquad \text{Isolating the radical by subtracting 5 from both sides}$$
$$\left. \begin{array}{l} (x - 5)^2 = (\sqrt{x + 7})^2 \\ x^2 - 10x + 25 = x + 7 \end{array} \right\} \quad \text{Using the principle of powers; squaring both sides}$$
$$x^2 - 11x + 18 = 0 \qquad \text{Adding } -x - 7 \text{ to both sides to form a quadratic equation in standard form}$$
$$(x - 9)(x - 2) = 0 \qquad \text{Factoring}$$
$$x = 9 \quad or \quad x = 2 \qquad \text{Using the principle of zero products}$$

The possible solutions are 9 and 2. Let's check.

Check:

For 9:
$$\frac{x = \sqrt{x + 7} + 5}{9 \mid \sqrt{9 + 7} + 5}$$
$$9 \stackrel{?}{=} 9 \qquad \text{TRUE}$$

For 2:
$$\frac{x = \sqrt{x + 7} + 5}{2 \mid \sqrt{2 + 7} + 5}$$
$$2 \stackrel{?}{=} 8 \qquad \text{FALSE}$$

3. Solve: $x = \sqrt{x + 10} + 10$.

Since 9 checks but 2 does not, the solution is 9.

YOUR TURN

EXAMPLE 4   Solve: $(2x + 1)^{1/3} + 5 = 0$.

SOLUTION   We can use exponential notation to solve:

$$(2x + 1)^{1/3} + 5 = 0$$
$$(2x + 1)^{1/3} = -5 \qquad \text{Subtracting 5 from both sides}$$
$$[(2x + 1)^{1/3}]^3 = (-5)^3 \qquad \text{Cubing both sides}$$
$$(2x + 1)^1 = (-5)^3 \qquad \text{Multiplying exponents}$$
$$2x + 1 = -125$$
$$2x = -126 \qquad \text{Subtracting 1 from both sides}$$
$$x = -63.$$

Student Notes

In Example 4, $(2x + 1)^{1/3}$ can also be written $\sqrt[3]{2x + 1}$. Then cubing both sides would show

$$\left(\sqrt[3]{2x + 1}\right)^3 = (-5)^3$$
$$2x + 1 = -125.$$

Because both sides were raised to an *odd* power, a check is not *essential*. It is wise, however, for the student to confirm that -63 checks and is the solution.

YOUR TURN

4. Solve: $(3x - 5)^{1/5} = 2$.

Equations with Two Radical Terms

A strategy for solving equations with two or more radical terms is as follows.

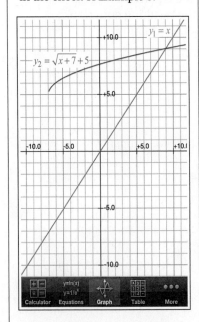

Technology Connection

To solve Example 3, we can graph $y_1 = x$ and $y_2 = \sqrt{x + 7} + 5$ and then find any point(s) of intersection. The intersection occurs at $x = 9$. Note that there is no intersection when $x = 2$, as predicted in the check of Example 3.

> **TO SOLVE AN EQUATION WITH TWO OR MORE RADICAL TERMS**
> **1.** Isolate one of the radical terms.
> **2.** Use the principle of powers.
> **3.** If a radical remains, perform steps (1) and (2) again.
> **4.** Solve the resulting equation.
> **5.** Check possible solutions in the original equation.

EXAMPLE 5   Solve: $\sqrt{2x - 5} = 1 + \sqrt{x - 3}$.

SOLUTION

$$\sqrt{2x - 5} = 1 + \sqrt{x - 3}$$
$$(\sqrt{2x - 5})^2 = (1 + \sqrt{x - 3})^2 \qquad \text{One radical is already isolated. We square both sides.}$$

> This is similar to squaring a binomial. We square 1, then find twice the product of 1 and $\sqrt{x - 3}$, and finally square $\sqrt{x - 3}$. Study this carefully.

$$2x - 5 = 1 + 2\sqrt{x - 3} + (\sqrt{x - 3})^2$$
$$2x - 5 = 1 + 2\sqrt{x - 3} + (x - 3)$$
$$x - 3 = 2\sqrt{x - 3} \qquad \text{Isolating the remaining radical term}$$
$$(x - 3)^2 = (2\sqrt{x - 3})^2 \qquad \text{Squaring both sides}$$
$$x^2 - 6x + 9 = 4(x - 3) \qquad \text{Remember to square both the 2 and the } \sqrt{x - 3} \text{ on the right side.}$$

1. Use a graphing calculator to solve Examples 1, 2, 4, 5, and 6.

$$x^2 - 6x + 9 = 4x - 12$$
$$x^2 - 10x + 21 = 0$$
$$(x - 7)(x - 3) = 0 \qquad \text{Factoring}$$
$$x = 7 \quad or \quad x = 3 \qquad \text{Using the principle of zero products}$$

5. Solve:

$$\sqrt{x - 2} + 2 = \sqrt{2x + 3}.$$

We leave it to the student to show that 7 and 3 both check and are the solutions.

 YOUR TURN

 Chapter Resources:
Collaborative Activity, p. 495;
Decision Making: Connection,
p. 495

CAUTION!   A common error in solving equations like

$$\sqrt{2x - 5} = 1 + \sqrt{x - 3}$$

is to obtain $1 + (x - 3)$ as the square of the right side. This is wrong because $(A + B)^2 \neq A^2 + B^2$. Placing parentheses around each side when squaring serves as a reminder to square the entire expression.

EXAMPLE 6   Let

$$f(x) = \sqrt{x + 5} - \sqrt{x - 7}.$$

Find all x-values for which $f(x) = 2$.

SOLUTION   We must have $f(x) = 2$, or

$$\sqrt{x + 5} - \sqrt{x - 7} = 2. \qquad \text{Substituting for } f(x)$$

To solve, we isolate one radical term and square both sides:

Isolate a radical term.	$\sqrt{x + 5} = 2 + \sqrt{x - 7}$	Adding $\sqrt{x - 7}$ to both sides to isolate a radical term
Raise both sides to the same power.	$(\sqrt{x + 5})^2 = (2 + \sqrt{x - 7})^2$	Using the principle of powers (squaring both sides)
	$x + 5 = 4 + 4\sqrt{x - 7} + (x - 7)$	Using $(A + B)^2 = A^2 + 2AB + B^2$
	$5 = 4\sqrt{x - 7} - 3$	Adding $-x$ to both sides and combining like terms
Isolate a radical term.	$8 = 4\sqrt{x - 7}$	Isolating the remaining radical term
	$2 = \sqrt{x - 7}$	
Raise both sides to the same power.	$2^2 = (\sqrt{x - 7})^2$	Squaring both sides
	$4 = x - 7$	
Solve.	$11 = x.$	

Check: $f(11) = \sqrt{11 + 5} - \sqrt{11 - 7}$
$= \sqrt{16} - \sqrt{4}$
$= 4 - 2 = 2.$

6. Let $f(x) = \sqrt{x} + \sqrt{x + 1}$.
Find all x-values for which
$f(x) = 2$.

We have $f(x) = 2$ when $x = 11$.

 YOUR TURN

7.6 EXERCISE SET

FOR EXTRA HELP

Vocabulary and Reading Check

Choose the word from the following list that best completes each statement. Not every word will be used.

even	radical
isolate	raise
odd	rational
powers	square roots

1. When we "square both sides" of an equation, we are using the principle of _____.

2. The equation $\sqrt{2x - 5} = 7$ is a(n) _____ equation.

3. To solve an equation with a radical term, we first _____ the radical term on one side of the equation.

4. A check is essential when we raise both sides of an equation to a(n) _____ power.

Concept Reinforcement

Classify each of the following statements as either true or false.

5. If $t = 7$, then $t^2 = 49$.

6. If $\sqrt{x} = 3$, then $(\sqrt{x})^2 = 3^2$.

7. If $x^2 = 36$, then $x = 6$.

8. $\sqrt{x} - 8 = 7$ is equivalent to $\sqrt{x} = 15$.

The Principle of Powers

Solve.

9. $\sqrt{5x + 1} = 4$

10. $\sqrt{7x - 3} = 5$

11. $\sqrt{3x} + 1 = 5$

12. $\sqrt{2x} - 1 = 2$

13. $\sqrt{y + 5} - 4 = 1$

14. $\sqrt{x - 2} - 7 = -4$

15. $\sqrt{8 - x} + 7 = 10$

16. $\sqrt{y + 4} + 6 = 7$

17. $\sqrt[3]{y + 3} = 2$

18. $\sqrt[3]{x - 2} = 3$

19. $\sqrt[4]{t - 10} = 3$

20. $\sqrt[4]{t + 5} = 2$

21. $6\sqrt{x} = x$

22. $7\sqrt{y} = y$

23. $2y^{1/2} - 13 = 7$

24. $3x^{1/2} + 12 = 9$

25. $\sqrt[3]{x} = -5$

26. $\sqrt[3]{y} = -4$

27. $z^{1/4} + 8 = 10$

28. $x^{1/4} - 2 = 1$

 29. $\sqrt{n} = -2$

30. $\sqrt{a} = -1$

31. $\sqrt[4]{3x + 1} - 4 = -1$

32. $\sqrt[4]{2x + 3} - 5 = -2$

33. $(21x + 55)^{1/3} = 10$

34. $(5y + 31)^{1/4} = 2$

35. $\sqrt[3]{3y + 6} + 7 = 8$

36. $\sqrt[3]{6x + 9} + 5 = 2$

37. $3 + \sqrt{5 - x} = x$

38. $x = \sqrt{x - 1} + 3$

Equations with Two Radical Terms

Solve.

39. $\sqrt{3t + 4} = \sqrt{4t + 3}$

40. $\sqrt{2t - 7} = \sqrt{3t - 12}$

41. $3(4 - t)^{1/4} = 6^{1/4}$

42. $2(1 - x)^{1/3} = 4^{1/3}$

43. $\sqrt{4x - 3} = 2 + \sqrt{2x - 5}$

44. $3 + \sqrt{z - 6} = \sqrt{z + 9}$

45. $\sqrt{20 - x} + 8 = \sqrt{9 - x} + 11$

46. $4 + \sqrt{10 - x} = 6 + \sqrt{4 - x}$

47. $\sqrt{x + 2} + \sqrt{3x + 4} = 2$

48. $\sqrt{6x + 7} - \sqrt{3x + 3} = 1$

49. If $f(x) = \sqrt{x} + \sqrt{x - 9}$, find any x for which $f(x) = 1$.

50. If $g(x) = \sqrt{x} + \sqrt{x - 5}$, find any x for which $g(x) = 5$.

51. If $f(t) = \sqrt{t - 2} - \sqrt{4t + 1}$, find any t for which $f(t) = -3$.

52. If $g(t) = \sqrt{2t + 7} - \sqrt{t + 15}$, find any t for which $g(t) = -1$.

53. If $f(x) = \sqrt{2x - 3}$ and $g(x) = \sqrt{x + 7} - 2$, find any x for which $f(x) = g(x)$.

54. If $f(x) = 2\sqrt{3x + 6}$ and $g(x) = 5 + \sqrt{4x + 9}$, find any x for which $f(x) = g(x)$.

55. If $f(t) = 4 - \sqrt{t - 3}$ and $g(t) = (t + 5)^{1/2}$, find any t for which $f(t) = g(t)$.

56. If $f(t) = 7 + \sqrt{2t - 5}$ and $g(t) = 3(t + 1)^{1/2}$, find any t for which $f(t) = g(t)$.

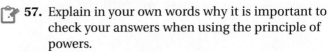 57. Explain in your own words why it is important to check your answers when using the principle of powers.

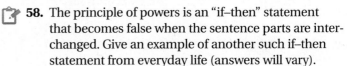 58. The principle of powers is an "if–then" statement that becomes false when the sentence parts are interchanged. Give an example of another such if–then statement from everyday life (answers will vary).

Skill Review

Solve.

59. In order to earn at least a B in her Economics class, Taylor must average at least 80% on the five tests. Her grades on the first four tests are 74%, 88%, 76%, and 78%. What must she score on the last test in order to earn at least a B in the course?   [4.1]

60. The number of building permits issued for single-family homes in the United States decreased from 1,378,000 in 2006 to 419,000 in 2011. What was the rate of change?   [2.3]

Source: U.S. Census Bureau

61. A flood rescue team uses a boat that travels 10 mph in still water. To reach a stranded family, they travel 7 mi against the current and return 7 mi with the current in a total time of $1\frac{2}{3}$ hr. What is the speed of the current?   [6.5]

62. Melted Goodness mixes Swiss chocolate and whipping cream to make a dessert fondue. Swiss chocolate costs $1.20 per ounce and whipping cream costs $0.30 per ounce. How much of each does Melted Goodness use in order to make 65 oz of fondue at a cost of $60.00?   [3.3]

Synthesis

63. Describe a procedure for creating radical equations that have no solution.

64. Is checking essential when the principle of powers is used with an odd power n? Why or why not?

65. *Firefighting.*   The velocity of water flow, in feet per second, from a nozzle is given by
$$v(p) = 12.1\sqrt{p},$$
where p is the nozzle pressure, in pounds per square inch (psi). Find the nozzle pressure if the water flow is 100 feet per second.

Source: Houston Fire Department Continuing Education

66. *Firefighting.*   The velocity of water flow, in feet per second, from a water tank that is h feet high is given by
$$v(h) = 8\sqrt{h}.$$

Find the height of a water tank that provides a water flow of 60 feet per second.

Source: Houston Fire Department Continuing Education

67. *Music.*   The frequency of a violin string varies directly with the square root of the tension on the string. A violin string vibrates with a frequency of 260 Hz when the tension on the string is 28 N. What is the frequency when the tension is 32 N?

68. *Music.*   The frequency of a violin string varies inversely with the square root of the density of the string. A nylon violin string with a density of 1200 kg/m^3 vibrates with a frequency of 250 Hz. What is the frequency of a silk and steel-core violin string with a density of 1300 kg/m^3?

Source: www.speech.kth.se

Steel Manufacturing.   In the production of steel and other metals, the temperature of the molten metal is so great that conventional thermometers melt. Instead, sound is transmitted across the surface of the metal to a receiver on the far side and the speed of the sound is measured. The formula
$$S(t) = 1087.7\sqrt{\frac{9t + 2617}{2457}}$$
gives the speed of sound $S(t)$, in feet per second, at a temperature of t degrees Celsius.

69. Find the temperature of a blast furnace where sound travels 1880 ft/sec.

70. Find the temperature of a blast furnace where sound travels 1502.3 ft/sec.

71. Solve the above equation for t.

Escape Velocity.   A formula for the escape velocity v of a satellite is
$$v = \sqrt{2gr}\sqrt{\frac{h}{r + h}},$$
where g is the force of gravity, r is the planet or star's radius, and h is the height of the satellite above the planet or star's surface.

72. Solve for h.

73. Solve for r.

Solve.

74. $\left(\dfrac{z}{4} - 5\right)^{2/3} = \dfrac{1}{25}$

75. $\dfrac{x + \sqrt{x + 1}}{x - \sqrt{x + 1}} = \dfrac{5}{11}$

76. $\sqrt{\sqrt{\sqrt{y} + 49}} = 7$

77. $(z^2 + 17)^{3/4} = 27$

78. $x^2 - 5x - \sqrt{x^2 - 5x - 2} = 4$
(*Hint*: Let $u = x^2 - 5x - 2$.)

79. $\sqrt{8 - b} = b\sqrt{8 - b}$

Without graphing, determine the x-intercepts of the graphs given by each of the following.

80. $f(x) = \sqrt{x - 2} - \sqrt{x + 2} + 2$

81. $g(x) = 6x^{1/2} + 6x^{-1/2} - 37$

82. $f(x) = (x^2 + 30x)^{1/2} - x - (5x)^{1/2}$

83. Use a graphing calculator to check your answers to Exercises 9, 17, and 33.

84. Use a graphing calculator to check your answers to Exercises 29, 39, and 43.

👉 **YOUR TURN ANSWERS: SECTION 7.6**

1. 24   **2.** No solution   **3.** 15   **4.** $\frac{37}{3}$ **5.** 3, 11   **6.** $\frac{9}{16}$

QUICK QUIZ: SECTIONS 7.1–7.6
Simplify. Write all answers using radical notation. Assume that all variables represent positive numbers.

1. $\sqrt{121n^2}$ [7.1]   **2.** $\sqrt[3]{\sqrt{a}}$ [7.2]

3. $2\sqrt{12} - \sqrt{75}$ [7.5]   **4.** $\sqrt{x}\sqrt[5]{x^3}$ [7.3]

5. Solve: $7 - \sqrt{3x + 1} = 5$.   [7.6]

PREPARE TO MOVE ON
Solve.

1. The largest sign in the United States is a rectangle with a perimeter of 430 ft. The length of the rectangle is 5 ft longer than thirteen times the width. Find the dimensions of the sign.   [1.4]

   Source: Florida Center for Instructional Technology

2. The base of a triangular sign is 4 in. longer than twice the height. The area of the sign is 255 in². Find the dimensions of the sign.   [5.8]

3. The length of a rectangular lawn between classroom buildings is 2 yd less than twice the width of the lawn. A path that is 34 yd long stretches diagonally across the area. What are the dimensions of the lawn?   [5.8]

4. One leg of a right triangle is 5 cm long. The hypotenuse is 1 cm longer than the other leg. Find the length of the hypotenuse.   [5.8]

7.7   The Distance Formula, the Midpoint Formula, and Other Applications

Using the Pythagorean Theorem   ▪   Two Special Triangles   ▪   The Distance Formula and the Midpoint Formula

Study Skills

Making Sketches

One need not be an artist to make highly useful mathematical sketches. That said, it is important to make sure that your sketches are drawn accurately enough to represent the relative sizes within each shape. For example, if one side of a triangle is clearly the longest, make sure your drawing reflects this.

Using the Pythagorean Theorem

There are many kinds of problems that involve powers and roots. Many also involve right triangles and the Pythagorean theorem.

THE PYTHAGOREAN THEOREM*

In any right triangle, if a and b are the lengths of the legs and c is the length of the hypotenuse, then

$$a^2 + b^2 = c^2.$$

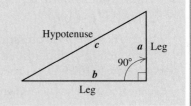

---

*The converse of the Pythagorean theorem also holds. That is, if a, b, and c are the lengths of the sides of a triangle and $a^2 + b^2 = c^2$, then the triangle is a right triangle.

In applying the Pythagorean theorem, we often make use of the following principle.

THE PRINCIPLE OF SQUARE ROOTS

For any nonnegative real number n,

If $x^2 = n$, then $x = \sqrt{n}$ or $x = -\sqrt{n}$.

For most real-world applications involving length or distance, $-\sqrt{n}$ is not needed.

EXAMPLE 1 *Baseball.* A baseball diamond is a square with sides of length 90 ft. Suppose a catcher fields a ball while standing on the third-base line 10 ft from home plate, as shown in the figure. How far is the catcher's throw to first base? Give an exact answer and an approximation to three decimal places.

SOLUTION We make a drawing and let $d =$ the distance, in feet, to first base. Note that a right triangle is formed in which the leg from home plate to first base measures 90 ft and the leg from home plate to where the catcher fields the ball measures 10 ft.

We substitute these values into the Pythagorean theorem to find d:

$$d^2 = 90^2 + 10^2$$
$$= 8100 + 100$$
$$= 8200.$$

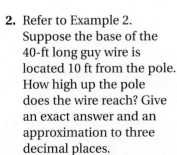

1. Refer to Example 1. How far would the catcher's throw to first base be from a point on the third-base line 20 ft from home plate? Give an exact answer and an approximation to three decimal places.

We now use the principle of square roots: If $d^2 = 8200$, then $d = \sqrt{8200}$ or $d = -\sqrt{8200}$. Since d represents length, it must be the *positive* square root of 8200:

$$d = \sqrt{8200} \text{ ft} = 10\sqrt{82} \text{ ft} \quad \text{This is an exact answer.}$$
$$\approx 90.554 \text{ ft.} \quad \text{Using a calculator for an approximation}$$

 YOUR TURN

EXAMPLE 2 *Guy Wires.* The base of a 40-ft long guy wire is 15 ft from the telephone pole that it anchors. How high up the pole does the guy wire reach? Give an exact answer and an approximation to three decimal places.

SOLUTION We make a drawing and let $h =$ the height, in feet, to which the guy wire reaches. A right triangle is formed in which one leg measures 15 ft and the hypotenuse measures 40 ft. Using the Pythagorean theorem, we have

$$h^2 + 15^2 = 40^2$$
$$h^2 + 225 = 1600$$
$$h^2 = 1375$$
$$h = \sqrt{1375}.$$

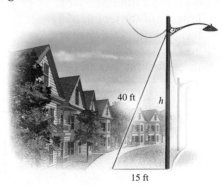

2. Refer to Example 2. Suppose the base of the 40-ft long guy wire is located 10 ft from the pole. How high up the pole does the wire reach? Give an exact answer and an approximation to three decimal places.

Exact answer:

$$h = \sqrt{1375} \text{ ft} \quad \text{Using the positive square root}$$
$$= 5\sqrt{55} \text{ ft}$$

Approximation:

$$h \approx 37.081 \text{ ft} \quad \text{Using a calculator}$$

YOUR TURN

Two Special Triangles

When both legs of a right triangle are the same size, as shown at left, we call the triangle an *isosceles right triangle*, or a 45°–45°–90° triangle. If one leg of an isosceles right triangle has length a, we can find a formula for the length of the hypotenuse as follows:

$$c^2 = a^2 + b^2$$
$$= a^2 + a^2 \qquad \text{Because the triangle is isosceles, both legs are the same size: } a = b.$$
$$= 2a^2. \qquad \text{Combining like terms}$$

Next, we use the principle of square roots. Because a, b, and c are lengths, there is no need to consider negative square roots or absolute values. Thus,

$$c = \sqrt{2a^2} \qquad \text{Using the principle of square roots}$$
$$= \sqrt{a^2 \cdot 2} = a\sqrt{2}. \qquad \text{The equation } c = a\sqrt{2} \text{ is worth remembering.}$$

EXAMPLE 3 One leg of an isosceles right triangle measures 7 cm. Find the length of the hypotenuse. Give an exact answer and an approximation to three decimal places.

3. One leg of an isosceles right triangle measures 5 m. Find the length of the hypotenuse. Give an exact answer and an approximation to three decimal places.

SOLUTION We substitute:

$$c = a\sqrt{2}$$
$$= 7\sqrt{2}.$$

Exact answer: $c = 7\sqrt{2}$ cm

Approximation: $c \approx 9.899$ cm

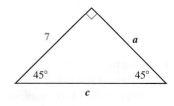

YOUR TURN

When the hypotenuse of an isosceles right triangle is known, the length of the legs can be found.

EXAMPLE 4 The hypotenuse of an isosceles right triangle is 5 ft long. Find the length of a leg. Give an exact answer and an approximation to three decimal places.

SOLUTION We replace c with 5 and solve for a:

$$5 = a\sqrt{2} \qquad \text{Substituting 5 for } c \text{ in } c = a\sqrt{2}$$
$$\frac{5}{\sqrt{2}} = a \qquad \text{Dividing both sides by } \sqrt{2}$$
$$\frac{5\sqrt{2}}{2} = a. \qquad \text{Rationalize the denominator if desired.}$$

4. The hypotenuse of an isosceles right triangle is 12 ft long. Find the length of a leg. Give an exact answer and an approximation to three decimal places.

Exact answer: $a = \dfrac{5}{\sqrt{2}}$ ft, or $\dfrac{5\sqrt{2}}{2}$ ft

Approximation: $a \approx 3.536$ ft   Using a calculator

YOUR TURN

LENGTHS WITHIN ISOSCELES RIGHT TRIANGLES

The length of the hypotenuse in an isosceles right triangle, or a 45°–45°–90° triangle, is the length of a leg times $\sqrt{2}$.

$$c = a\sqrt{2}$$

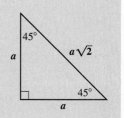

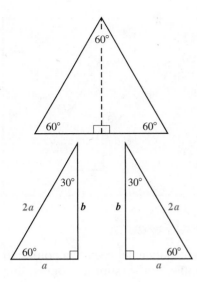

A second special triangle is known as a 30°–60°–90° triangle, so named because of the measures of its angles. Note that in an equilateral triangle, all sides have the same length and all angles are 60°. An altitude, drawn dashed in the figure, bisects, or splits in half, one angle and one side. Two 30°–60°–90° right triangles are thus formed.

If we let a represent the length of the shorter leg in a 30°–60°–90° triangle, then $2a$ represents the length of the hypotenuse. We have

$$a^2 + b^2 = (2a)^2 \qquad \text{Using the Pythagorean theorem}$$
$$a^2 + b^2 = 4a^2$$
$$b^2 = 3a^2 \qquad \text{Subtracting } a^2 \text{ from both sides}$$
$$b = \sqrt{3a^2} \qquad \text{Considering only the positive square root}$$
$$= \sqrt{a^2 \cdot 3}$$
$$= a\sqrt{3}. \qquad \text{This relationship is worth remembering.}$$

EXAMPLE 5 The shorter leg of a 30°–60°–90° triangle measures 8 in. Find the lengths of the other sides. Give exact answers and, where appropriate, an approximation to three decimal places.

SOLUTION The hypotenuse is twice as long as the shorter leg, so we have

$$c = 2a \qquad\qquad \text{This relationship is worth remembering.}$$
$$= 2 \cdot 8 = 16 \text{ in.} \qquad \text{This is the length of the hypotenuse.}$$

The length of the longer leg is the length of the shorter leg times $\sqrt{3}$. This gives us

$$b = a\sqrt{3} \qquad \text{This holds for all 30°–60°–90° triangles.}$$
$$= 8\sqrt{3} \text{ in.} \qquad \text{This is the length of the longer leg.}$$

Exact answer: $c = 16$ in., $b = 8\sqrt{3}$ in.

Approximation: $b \approx 13.856$ in.

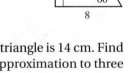

5. The shorter leg of a 30°–60°–90° triangle measures 7 in. Find the lengths of the other sides. Give exact answers and, where appropriate, an approximation to three decimal places.

YOUR TURN

EXAMPLE 6 The length of the longer leg of a 30°–60°–90° triangle is 14 cm. Find the length of the hypotenuse. Give an exact answer and an approximation to three decimal places.

SOLUTION The length of the hypotenuse is twice the length of the shorter leg. We first find a, the length of the shorter leg, by using the length of the longer leg:

$$14 = a\sqrt{3} \qquad \text{Substituting 14 for } b \text{ in } b = a\sqrt{3}$$
$$\frac{14}{\sqrt{3}} = a. \qquad \text{Dividing by } \sqrt{3}$$

Since the hypotenuse is twice as long as the shorter leg, we have

$$c = 2a$$
$$= 2 \cdot \frac{14}{\sqrt{3}} \qquad \text{Substituting}$$
$$= \frac{28}{\sqrt{3}} \text{ cm.}$$

Exact answer: $c = \dfrac{28}{\sqrt{3}}$ cm, or $\dfrac{28\sqrt{3}}{3}$ cm

Approximation: $c \approx 16.166$ cm

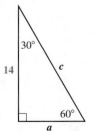

6. The length of the longer leg of a 30°–60°–90° triangle is 6 cm. Find the length of the hypotenuse. Give an exact answer and an approximation to three decimal places.

YOUR TURN

Student Notes

Perhaps the easiest way to remember the important results listed for these special triangles is to write out, on your own, the derivations shown in this section.

LENGTHS WITHIN 30°–60°–90° TRIANGLES

The length of the longer leg in a 30°–60°–90° triangle is the length of the shorter leg times $\sqrt{3}$. The hypotenuse is twice as long as the shorter leg.

$$b = a\sqrt{3},$$
$$c = 2a$$

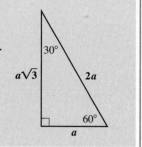

The Distance Formula and the Midpoint Formula

We can use the Pythagorean theorem to find the distance between two points. To find the distance between two points on the number line, we subtract. Depending on the order in which we subtract, the difference may be positive or negative. However, if we take the absolute value of the difference, we obtain the same positive value for the distance regardless of the order in which we subtract:

$$|4 - (-3)| = |7| = 7;$$
$$|-3 - 4| = |-7| = 7.$$

In a plane, if two points are on a horizontal line, they have the same second coordinate. We can find the distance between them by subtracting their first coordinates and taking the absolute value of that difference.

EXPLORING 🔍 THE CONCEPT

1. Find the coordinates of point A.
2. Find the distance from $(4, -4)$ to A.
3. Find the distance from $(1, 2)$ to A.
4. Use the Pythagorean theorem to find the distance from $(1, 2)$ to $(4, -4)$.

ANSWERS

1. $(1, -4)$ **2.** 3 units   **3.** 6 units
4. $\sqrt{45}$ units $= 3\sqrt{5}$ units

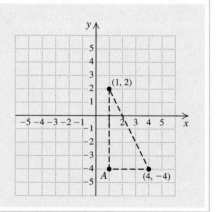

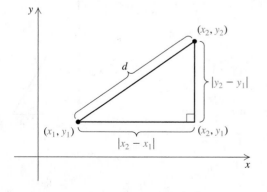

Generalizing, we have that the distance between the points (x_1, y_1) and (x_2, y_1) on a horizontal line is $|x_2 - x_1|$. Similarly, the distance between (x_2, y_1) and (x_2, y_2) on a vertical line is $|y_2 - y_1|$.

So long as $x_1 \neq x_2$ and $y_1 \neq y_2$, the points (x_1, y_1) and (x_2, y_2), along with the point (x_2, y_1), describe a right triangle. The lengths of the legs are $|x_2 - x_1|$ and $|y_2 - y_1|$. We find d, the length of the hypotenuse, by using the Pythagorean theorem:

$$d^2 = |x_2 - x_1|^2 + |y_2 - y_1|^2.$$

Since the square of a number is the same as the square of its opposite, we can replace the absolute-value signs with parentheses:

$$d^2 = (x_2 - x_1)^2 + (y_2 - y_1)^2.$$

Taking the principal square root, we have a formula for distance.

> **THE DISTANCE FORMULA**
> The distance d between any two points (x_1, y_1) and (x_2, y_2) is given by
> $$d = \sqrt{(x_2 - x_1)^2 + (y_2 - y_1)^2}.$$

EXAMPLE 7   Find the distance between $(5, -1)$ and $(-4, 6)$. Find an exact answer and an approximation to three decimal places.

SOLUTION   We substitute into the distance formula:

$$\begin{aligned} d &= \sqrt{(-4 - 5)^2 + [6 - (-1)]^2} \quad &&\text{Substituting. A drawing is optional.}\\ &= \sqrt{(-9)^2 + 7^2} \\ &= \sqrt{130} &&\text{This is exact.} \\ &\approx 11.402. &&\text{Using a calculator for an approximation} \end{aligned}$$

YOUR TURN

7. Find the distance between $(4, 7)$ and $(-8, 2)$.

The distance formula can be used to verify a formula for the coordinates of the *midpoint* of a segment connecting two points. We state the midpoint formula and leave its proof to the exercises.

> **THE MIDPOINT FORMULA**
> If the endpoints of a segment are (x_1, y_1) and (x_2, y_2), then the coordinates of the midpoint are
> $$\left(\frac{x_1 + x_2}{2}, \frac{y_1 + y_2}{2}\right).$$
>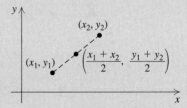
> (To locate the midpoint, average the x-coordinates and average the y-coordinates.)

Student Notes

To help remember the formulas correctly, note that the distance formula is a variation of the Pythagorean theorem and the result is a number. The midpoint formula involves averages and the result is an ordered pair.

EXAMPLE 8   Find the midpoint of the segment with endpoints $(-2, 3)$ and $(4, -6)$.

SOLUTION   Using the midpoint formula, we obtain

$$\left(\frac{-2 + 4}{2}, \frac{3 + (-6)}{2}\right), \quad \text{or} \quad \left(\frac{2}{2}, \frac{-3}{2}\right), \quad \text{or} \quad \left(1, -\frac{3}{2}\right).$$

8. Find the midpoint of the segment with endpoints $(-3, -4)$ and $(0, -10)$.

 YOUR TURN

7.7 EXERCISE SET

Vocabulary and Reading Check

Complete each statement with the best choice from the following list.

1. In any ____ triangle, the square of the length of the hypotenuse is the sum of the squares of the lengths of the legs.

2. The shortest side of a right triangle is always one of the two ____.

3. The principle of ____ states that if $x^2 = n$, then $x = \sqrt{n}$ or $x = -\sqrt{n}$.

4. In a(n) ____ right triangle, both legs have the same length.

5. In a(n) ____ right triangle, the hypotenuse is twice as long as the shorter leg.

6. If both legs in a right triangle have measure a, then the ____ measures $a\sqrt{2}$.

a) Hypotenuse

b) Isosceles

c) Legs

d) Right

e) Square roots

f) 30°–60°–90°

Using the Pythagorean Theorem

In a right triangle, find the length of the side not given. Give an exact answer and, where appropriate, an approximation to three decimal places.

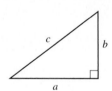

7. $a = 5$, $b = 3$

8. $a = 8$, $b = 10$

Aha! 9. $a = 9$, $b = 9$

10. $a = 10$, $b = 10$

11. $b = 15$, $c = 17$

12. $a = 7$, $c = 25$

In Exercises 13–18, give an exact answer and, where appropriate, an approximation to three decimal places.

13. A right triangle's hypotenuse is 8 m, and one leg is $4\sqrt{3}$ m. Find the length of the other leg.

14. A right triangle's hypotenuse is 6 cm, and one leg is $\sqrt{5}$ cm. Find the length of the other leg.

15. The hypotenuse of a right triangle is $\sqrt{20}$ in., and one leg measures 1 in. Find the length of the other leg.

16. The hypotenuse of a right triangle is $\sqrt{15}$ ft, and one leg measures 2 ft. Find the length of the other leg.

Aha! 17. One leg in a right triangle is 1 m, and the hypotenuse measures $\sqrt{2}$ m. Find the length of the other leg.

18. One leg of a right triangle is 1 yd, and the hypotenuse measures 2 yd. Find the length of the other leg.

In Exercises 19–28, give an exact answer and, where appropriate, an approximation to three decimal places.

19. *Bicycling.* Clare routinely bicycles across a rectangular parking lot on her way to work. If the lot is 200 ft long and 150 ft wide, how far does Clare travel when she rides across the lot diagonally?

20. *Guy Wire.* How long is a guy wire that reaches from the top of a 15-ft pole to a point on the ground 10 ft from the pole?

21. *Zipline.* For Super Bowl XLVI in Indianapolis, a temporary zipline was constructed on Capitol Street. The ride extended 800 ft along the street, and riders dropped 60 ft. How long was the zipline?

22. *Baseball.* Suppose the catcher in Example 1 makes a throw to second base from the same location. How far is that throw?

23. *Television Sets.* What does it mean to refer to a 51-in. TV set? Such units refer to the diagonal of the screen. A 51-in. TV set has a width of 45 in. What is its height?

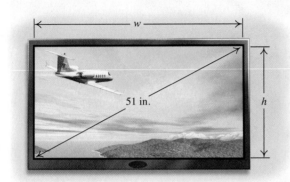

24. *Television Sets.* A 53-in. TV set has a screen with a height of 28 in. What is its width? (See Exercise 23.)

25. *Speaker Placement.* A stereo receiver is in a corner of a 12-ft by 14-ft room. Wire will run under a rug, diagonally, to a subwoofer in the far corner. If 4 ft of slack is required on each end, how long a piece of wire should be purchased?

26. *Distance over Water.* To determine the width of a pond, a surveyor locates two stakes at either end of the pond and uses instrumentation to place a third stake so that the distance across the pond is the length of a hypotenuse. If the third stake is 90 m from one stake and 70 m from the other, what is the distance across the pond?

27. *Walking.* Students at Pohlman Community College have worn a path that cuts diagonally across the campus "quad." If the quad is a rectangle that Marissa measures as 70 paces long and 40 paces wide, how many paces will Marissa save by using the diagonal path?

28. *Crosswalks.* The diagonal crosswalk at the intersection of State St. and Main St. is the hypotenuse of a triangle in which the crosswalks across State St. and Main St. are the legs. If State St. is 28 ft wide and Main St. is 40 ft wide, how much distance is saved by using the diagonal crosswalk rather than crossing both streets?

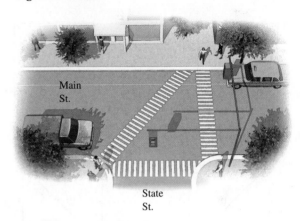
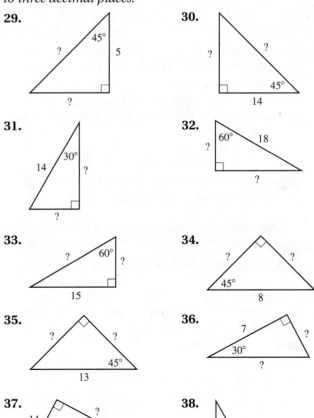

Two Special Triangles

For each triangle, find the missing length(s). Give an exact answer and, where appropriate, an approximation to three decimal places.

29. 45°, 5, ?, ?

30. ?, 45°, 14, ?

31. 14, 30°, ?, ?

32. 60°, 18, ?, ?

33. ?, 60°, ?, 15

34. ?, 45°, ?, 8

35. ?, ?, 45°, 13

36. 7, 30°, ?, ?

37. 14, ?, 30°, ?

38. ?, ?, 60°, 9

39.

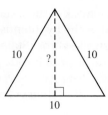

40.

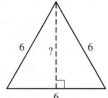

41.

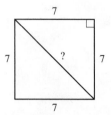

42.

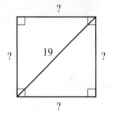

43. The figure is a square.

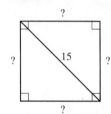

44. The figure is a square.

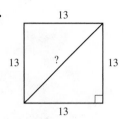

Solving Right Triangles

In Exercises 45–48, give an exact answer and, where appropriate, an approximation to three decimal places.

45. *Bridge Expansion.* During the summer heat, a 2-mi bridge expands 2 ft in length. If we assume that the bulge occurs straight up the middle, how high is the bulge? (The answer may surprise you. Most bridges have expansion spaces to avoid such buckling.)

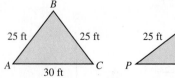

46. Triangle *ABC* has sides of lengths 25 ft, 25 ft, and 30 ft. Triangle *PQR* has sides of lengths 25 ft, 25 ft, and 40 ft. Which triangle, if either, has the greater area and by how much?

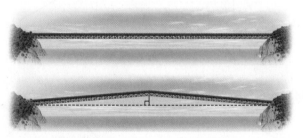

47. *Architecture.* The Rushton Triangular Lodge in Northamptonshire, England, was designed and

constructed by Sir Thomas Tresham between 1593 and 1597. The base of the building is in the shape of an equilateral triangle with walls of length 33 ft. How many square feet of land is covered by the lodge?

Source: The Internet Encyclopedia of Science

48. *Antenna Length.* As part of an emergency radio communication station, Rik sets up an "Inverted-V" antenna. He stretches a copper wire from one point on the ground to a point on a tree and then back down to the ground, forming two 30°–60°–90° triangles. If the wire is fastened to the tree 34 ft above the ground, how long is the copper wire?

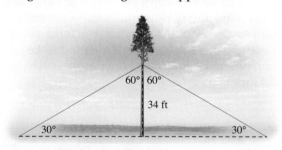

49. Find all points on the *y*-axis of a Cartesian coordinate system that are 5 units from the point (3, 0).

50. Find all points on the *x*-axis of a Cartesian coordinate system that are 5 units from the point (0, 4).

The Distance Formula and the Midpoint Formula

Find the distance between each pair of points. Where appropriate, find an approximation to three decimal places.

51. (4, 5) and (7, 1)

52. (0, 8) and (6, 0)

53. (0, −5) and (1, −2)

54. (−1, −4) and (−3, −5)

55. $(-4, 4)$ and $(6, -6)$

56. $(5, 21)$ and $(-3, 1)$

Aha! **57.** $(8.6, -3.4)$ and $(-9.2, -3.4)$

58. $(5.9, 2)$ and $(3.7, -7.7)$

59. $\left(\frac{1}{2}, \frac{1}{3}\right)$ and $\left(\frac{5}{6}, -\frac{1}{6}\right)$

60. $\left(\frac{5}{7}, \frac{1}{14}\right)$ and $\left(\frac{1}{7}, \frac{11}{14}\right)$

61. $(-\sqrt{6}, \sqrt{6})$ and $(0, 0)$

62. $(\sqrt{5}, -\sqrt{3})$ and $(0, 0)$

63. $(-1, -30)$ and $(-2, -40)$

64. $(0.5, 100)$ and $(1.5, -100)$

In Exercises 65–76, find the midpoint of the segment with the given endpoints.

65. $(-2, 5)$ and $(8, 3)$

66. $(1, 4)$ and $(9, -6)$

67. $(2, -1)$ and $(5, 8)$

68. $(-1, 2)$ and $(1, -3)$

69. $(-8, -5)$ and $(6, -1)$

70. $(8, -2)$ and $(-3, 4)$

71. $(-3.4, 8.1)$ and $(4.8, -8.1)$

72. $(4.1, 6.9)$ and $(5.2, -8.9)$

73. $\left(\frac{1}{6}, -\frac{3}{4}\right)$ and $\left(-\frac{1}{3}, \frac{5}{6}\right)$

74. $\left(-\frac{4}{5}, -\frac{2}{3}\right)$ and $\left(\frac{1}{8}, \frac{3}{4}\right)$

75. $(\sqrt{2}, -1)$ and $(\sqrt{3}, 4)$

76. $(9, 2\sqrt{3})$ and $(-4, 5\sqrt{3})$

77. Are there any right triangles, other than those with sides measuring 3, 4, and 5, that have consecutive numbers for the lengths of the sides? Why or why not?

78. If a 30°–60°–90° triangle and an isosceles right triangle have the same perimeter, which will have the greater area? Why?

Skill Review

Graph on a plane.

79. $y = 2x - 3$ [2.3]

80. $y < x$ [4.4]

81. $8x - 4y = 8$ [2.4]

82. $2y - 1 = 7$ [2.4]

83. $x \geq 1$ [4.4]

84. $x - 5 = 6 - 2y$ [2.3]

Synthesis

85. Describe a procedure that uses the distance formula to determine whether three points, (x_1, y_1), (x_2, y_2), and (x_3, y_3), are vertices of a right triangle.

86. Outline a procedure that uses the distance formula to determine whether three points, (x_1, y_1), (x_2, y_2), and (x_3, y_3), are collinear (lie on the same line).

87. The perimeter of a regular hexagon is 72 cm. Determine the area of the shaded region shown.

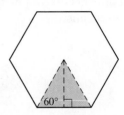

88. If the perimeter of a regular hexagon is 120 ft, what is its area? (*Hint*: See Exercise 87.)

89. Each side of a regular octagon has length *s*.

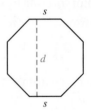

a) Find a formula for the distance *d* between the parallel sides of the octagon.

b) Find a formula for the area of the octagon.

90. *Contracting.*   Oxford Builders has an extension cord on their generator that permits them to work, with electricity, anywhere in a circular area of 3850 ft^2. Find the dimensions of the largest square room on which they could work without having to relocate the generator to reach each corner of the floor.

91. *Contracting.*   Cleary Construction has a hose attached to their insulation blower that permits them to reach anywhere in a circular area of 6160 ft^2. Find the dimensions of the largest square room with 12-ft ceilings in which they could reach all corners with the hose while leaving the blower centrally located. Assume that the blower sits on the floor.

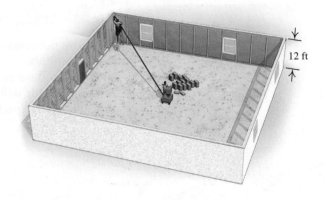

92. The length and the width of a rectangle are given by consecutive integers. The area of the rectangle is 90 cm². Find the length of a diagonal of the rectangle.

93. A cube measures 5 cm on each side. How long is the diagonal that connects two opposite corners of the cube? Give an exact answer.

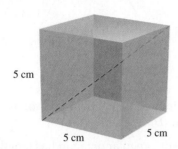

5 cm

5 cm   5 cm

94. Prove the midpoint formula by showing that

i) the distance from (x_1, y_1) to

$$\left(\frac{x_1 + x_2}{2}, \frac{y_1 + y_2}{2}\right)$$

equals the distance from (x_2, y_2) to

$$\left(\frac{x_1 + x_2}{2}, \frac{y_1 + y_2}{2}\right);$$

and

ii) the points

$$(x_1, y_1), \left(\frac{x_1 + x_2}{2}, \frac{y_1 + y_2}{2}\right),$$

and

$$(x_2, y_2)$$

lie on the same line (see Exercise 86).

 YOUR TURN ANSWERS: SECTION 7.7

1. $\sqrt{8500}$ ft $= 10\sqrt{85}$ ft; 92.195 ft
2. $\sqrt{1500}$ ft $= 10\sqrt{15}$ ft; 38.730 ft   **3.** $5\sqrt{2}$ m; 7.071 m
4. $\frac{12}{\sqrt{2}}$ ft $= 6\sqrt{2}$ ft; 8.485 ft   **5.** Leg: $7\sqrt{3}$ in.; 12.124 in.;

hypotenuse: 14 in.   **6.** $\frac{12}{\sqrt{3}}$ cm $= 4\sqrt{3}$ cm; 6.928 cm

7. 13   **8.** $\left(-\frac{3}{2}, -7\right)$

QUICK QUIZ: SECTIONS 7.1–7.7

Simplify. Assume that all variables represent positive numbers.

1. $\sqrt[3]{\dfrac{54x^4}{125t^6}}$ [7.4]   **2.** $\sqrt[4]{32x^5} - \sqrt[4]{2x^{13}}$ [7.5]

3. $\sqrt{40x^3}\sqrt{45x}$ [7.3]

4. Solve: $5 - \sqrt{x+1} = \sqrt{x}$.   [7.6]

5. Find the distance between $(4, -1)$ and $(-3, -7)$. Give an exact answer and an approximation to three decimal places.   [7.7]

PREPARE TO MOVE ON

Find the conjugate of each number.   [7.5]

1. $2 - \sqrt{3}$ **2.** $\sqrt{7} + 6$

Multiply.   [5.2]

3. $(3x - 2y)(3x + 2y)$ **4.** $(5w - 2x)^2$

5. $-4c(2a - 7c)$ **6.** $(4a + p)(6a - 5p)$

7.8 The Complex Numbers

Imaginary Numbers and Complex Numbers   ▪   Addition and Subtraction   ▪   Multiplication   ▪
Conjugates and Division   ▪   Powers of *i*

Imaginary Numbers and Complex Numbers

Negative numbers do not have square roots in the real-number system. However, a larger number system that contains the real-number system is designed so that negative numbers *do* have square roots. That system is called the **complex-number system**, and it will allow us to solve equations like $x^2 + 1 = 0$. The complex-number system makes use of *i*, a number that is, by definition, a square root of -1.

1. Express in terms of i: $\sqrt{-36}$.

THE NUMBER i

i is the unique number for which $i = \sqrt{-1}$ and $i^2 = -1$.

We can now define the square root of a negative number as follows:

$$\sqrt{-p} = \sqrt{-1}\sqrt{p} = i\sqrt{p} \text{ or } \sqrt{p}\,i, \text{ for any positive number } p.$$

EXAMPLE 1   Express in terms of i: **(a)** $\sqrt{-7}$; **(b)** $\sqrt{-16}$; **(c)** $-\sqrt{-50}$.

SOLUTION

a) $\sqrt{-7} = \sqrt{-1 \cdot 7} = \sqrt{-1} \cdot \sqrt{7} = i\sqrt{7}$, or $\sqrt{7}\,i$ | i is *not* under the radical. |

b) $\sqrt{-16} = \sqrt{-1 \cdot 16} = \sqrt{-1} \cdot \sqrt{16} = i \cdot 4 = 4i$

c) $-\sqrt{-50} = -\sqrt{-1} \cdot \sqrt{25} \cdot \sqrt{2} = -i \cdot 5 \cdot \sqrt{2} = -5i\sqrt{2}$, or $-5\sqrt{2}\,i$

YOUR TURN

IMAGINARY NUMBERS

An *imaginary number* is any number that can be written in the form $a + bi$, where a and b are real numbers and $b \neq 0$.

Don't let the name "imaginary" fool you. Imaginary numbers appear in fields such as engineering and the physical sciences. The following are examples of imaginary numbers:

$5 + 4i$,   Here $a = 5$, $b = 4$.

$\sqrt{3} - \pi i$,   Here $a = \sqrt{3}$, $b = -\pi$.

$\sqrt{7}\,i$.   Here $a = 0$, $b = \sqrt{7}$.

The union of the set of all imaginary numbers and the set of all real numbers is the set of all **complex numbers**.

COMPLEX NUMBERS

A *complex number* is any number that can be written in the form $a + bi$, where a and b are real numbers. (Note that a and b both can be 0.)

The following are examples of complex numbers:

$7 + 3i$ (here $a \neq 0$, $b \neq 0$); $4i$ (here $a = 0$, $b \neq 0$);

8 (here $a \neq 0$, $b = 0$); 0 (here $a = 0$, $b = 0$).

Complex numbers like $17i$ or $4i$, in which $a = 0$ and $b \neq 0$, are called *pure imaginary numbers*.

For $b = 0$, we have $a + 0i = a$, so every real number is a complex number. The relationships among various complex numbers are shown below.

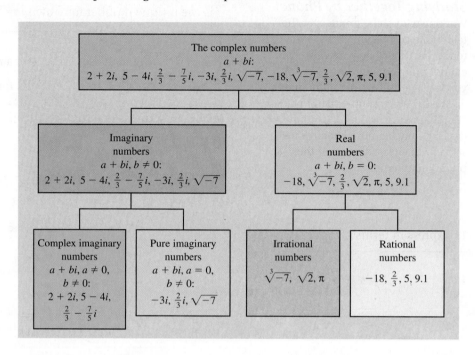

Note that although $\sqrt{-7}$ and $\sqrt[3]{-7}$ are both complex numbers, $\sqrt{-7}$ is imaginary whereas $\sqrt[3]{-7}$ is real.

Addition and Subtraction

We can add and subtract complex numbers just as we do binomials. It is customary to write the result in the form $a + bi$.

EXAMPLE 2   Add or subtract and simplify.

a) $(8 + 6i) + (3 + 2i)$ **b)** $(4 + 5i) - (6 - 3i)$

SOLUTION

a) $(8 + 6i) + (3 + 2i) = (8 + 3) + (6i + 2i)$ Combining the real parts and the imaginary parts

$$= 11 + (6 + 2)i = 11 + 8i$$

b) $(4 + 5i) - (6 - 3i) = (4 - 6) + [5i - (-3i)]$ Note that the 6 and the $-3i$ are *both* being subtracted.

$$= -2 + 8i$$

2. Subtract and simplify:

$(3 - 4i) - (7 - i)$.

YOUR TURN

Student Notes

The rule developed in Section 7.3, $\sqrt[n]{a} \cdot \sqrt[n]{b} = \sqrt[n]{a \cdot b}$, does *not* apply when n is 2 and either a or b is negative. Indeed this condition is stated within that rule when it is specified that $\sqrt[n]{a}$ and $\sqrt[n]{b}$ are both *real* numbers.

Multiplication

To multiply square roots of negative real numbers, we first express them in terms of i. For example,

$$\sqrt{-2} \cdot \sqrt{-5} = \sqrt{-1} \cdot \sqrt{2} \cdot \sqrt{-1} \cdot \sqrt{5}$$
$$= i \cdot \sqrt{2} \cdot i \cdot \sqrt{5}$$
$$= i^2 \cdot \sqrt{10}   \text{Since } i \text{ is a square root of } -1, i^2 = -1.$$
$$= -1\sqrt{10} = -\sqrt{10} \text{ is correct!}$$

> **CAUTION!** With complex numbers, multiplying radicands is *incorrect* when both radicands are negative: $\sqrt{-2} \cdot \sqrt{-5} \neq \sqrt{10}$.

With this in mind, we can now multiply complex numbers.

EXAMPLE 3 Multiply and simplify. Write any imaginary answers in the form $a + bi$.

a) $\sqrt{-4}\sqrt{-25}$ **b)** $\sqrt{-5} \cdot \sqrt{-7}$ **c)** $-3i \cdot 8i$

d) $-4i(3 - 5i)$ **e)** $(1 + 2i)(4 + 3i)$

SOLUTION

a) $\sqrt{-4}\sqrt{-25} = \sqrt{-1} \cdot \sqrt{4} \cdot \sqrt{-1} \cdot \sqrt{25}$
$= i \cdot 2 \cdot i \cdot 5$
$= i^2 \cdot 10$
$= -1 \cdot 10$ $i^2 = -1$
$= -10$

b) $\sqrt{-5} \cdot \sqrt{-7} = \sqrt{-1} \cdot \sqrt{5} \cdot \sqrt{-1} \cdot \sqrt{7}$ Try to do this step mentally.
$= i \cdot \sqrt{5} \cdot i \cdot \sqrt{7}$
$= i^2 \cdot \sqrt{35}$
$= -1 \cdot \sqrt{35}$ $i^2 = -1$
$= -\sqrt{35}$

c) $-3i \cdot 8i = -24 \cdot i^2$
$= -24 \cdot (-1)$ $i^2 = -1$
$= 24$

d) $-4i(3 - 5i) = -4i \cdot 3 + (-4i)(-5i)$ Using the distributive law
$= -12i + 20i^2$
$= -12i - 20$ $i^2 = -1$
$= -20 - 12i$ Writing in the form $a + bi$

e) $(1 + 2i)(4 + 3i) = 4 + 3i + 8i + 6i^2$ Using FOIL
$= 4 + 3i + 8i - 6$ $i^2 = -1$
$= -2 + 11i$ Combining like terms

3. Multiply and simplify:

$(2 + 5i)(1 - 4i)$.

Write the answer in the form $a + bi$.

YOUR TURN

Conjugates and Division

Recall that the conjugate of $4 + \sqrt{2}$ is $4 - \sqrt{2}$. Conjugates of complex numbers are defined in a similar manner.

> **CONJUGATE OF A COMPLEX NUMBER**
> The *conjugate* of a complex number $a + bi$ is $a - bi$, and the *conjugate* of $a - bi$ is $a + bi$.

EXAMPLE 4 Find the conjugate of each number.

a) $-3 - 7i$ **b)** $4i$

4. Find the conjugate of $2 + 9i$.

SOLUTION

a) $-3 - 7i$ The conjugate is $-3 + 7i$.

b) $4i$ The conjugate is $-4i$. Note that $4i = 0 + 4i$.

↩ YOUR TURN

The product of a complex number and its conjugate is a real number.

EXAMPLE 5 Multiply: $(5 + 7i)(5 - 7i)$.

SOLUTION

5. Multiply:

   $(-3 + 2i)(-3 - 2i)$.

$$(5 + 7i)(5 - 7i) = 5^2 - (7i)^2 \qquad \text{Using } (A + B)(A - B) = A^2 - B^2$$
$$= 25 - 49i^2$$
$$= 25 - 49(-1) \qquad i^2 = -1$$
$$= 25 + 49 = 74$$

↩ YOUR TURN

Conjugates are used when dividing by an imaginary number. The procedure is much like that used to rationalize denominators with two terms.

EXAMPLE 6 Divide and simplify to the form $a + bi$.

a) $\dfrac{-2 + 9i}{1 - 3i}$

b) $\dfrac{7 + 4i}{5i}$

SOLUTION

a) To divide and simplify $(-2 + 9i)/(1 - 3i)$, we multiply by 1, using the conjugate of the denominator to form 1:

$$\frac{-2 + 9i}{1 - 3i} = \frac{-2 + 9i}{1 - 3i} \cdot \frac{1 + 3i}{1 + 3i} \qquad \begin{array}{l}\text{Multiplying by 1 using the con-}\\\text{jugate of the denominator in the}\\\text{symbol for 1}\end{array}$$

$$= \frac{(-2 + 9i)(1 + 3i)}{(1 - 3i)(1 + 3i)} \qquad \begin{array}{l}\text{Multiplying numerators;}\\\text{multiplying denominators}\end{array}$$

$$= \frac{-2 - 6i + 9i + 27i^2}{1^2 - 9i^2} \qquad \text{Using FOIL}$$

$$= \frac{-2 + 3i + (-27)}{1 - (-9)} \qquad i^2 = -1$$

$$= \left.\begin{array}{l}\dfrac{-29 + 3i}{10}\\[2ex]= -\dfrac{29}{10} + \dfrac{3}{10}i.\end{array}\right\} \qquad \begin{array}{l}\text{Writing in the form } a + bi\\[1ex]\text{Recall that } \dfrac{M + N}{D} = \dfrac{M}{D} + \dfrac{N}{D}.\end{array}$$

b) The conjugate of $5i$ is $-5i$, so we *could* multiply by $-5i/(-5i)$. However, when the denominator is a pure imaginary number, it is easiest if we multiply by i/i:

$$\frac{7 + 4i}{5i} = \frac{7 + 4i}{5i} \cdot \frac{i}{i} \qquad \begin{array}{l}\text{Multiplying by 1 using } i/i. \text{ We can also use the}\\\text{conjugate of } 5i \text{ to write 1 as } -5i/(-5i).\end{array}$$

$$= \frac{7i + 4i^2}{5i^2} \qquad \text{Multiplying}$$

6. Divide and simplify to the form $a + bi$:

   $\dfrac{2 + 4i}{1 - 2i}$.

$$= \frac{7i + 4(-1)}{5(-1)} \qquad i^2 = -1$$

$$= \frac{7i - 4}{-5} = \frac{-4}{-5} + \frac{7}{-5}i, \text{ or } \frac{4}{5} - \frac{7}{5}i. \qquad \text{Writing in the form } a + bi$$

↩ YOUR TURN

Powers of i

In the following discussion, we show why there is no need to use powers of i (other than 1) when writing answers.

We use the following to simplify powers of i.

- $i^2 = -1$
- $i^n = i \cdot i^{n-1}$
- $(-1)^n = 1$ when n is even
- $(-1)^n = -1$ when n is odd

To simplify i^n when n is even, we rewrite i^n as a power of -1. Even powers of i are 1 or -1.

EXAMPLE 7   Simplify: i^{30}.

SOLUTION

$$
\begin{aligned}
i^{30} &= (i^2)^{15} && (a^m)^n = a^{mn} \\
&= (-1)^{15} && i^2 = -1 \\
&= -1 && (-1)^n = -1 \text{ when } n \text{ is odd.}
\end{aligned}
$$

7. Simplify: i^{26}.

_____

⤴ YOUR TURN

Student Notes

You may notice that the powers of i cycle through $i, -1, -i, 1$:

$$i^1 = i,$$
$$i^2 = -1,$$
$$i^3 = -i,$$
$$i^4 = 1,$$
$$i^5 = i, \text{ and so on.}$$

_____

8. Simplify: i^{33}.

_____

To simplify i^n when n is odd, we rewrite i^n as $i \cdot i^{n-1}$ and simplify i^{n-1}. Odd powers of i are i or $-i$.

EXAMPLE 8   Simplify: i^{49}.

SOLUTION

$$
\begin{aligned}
i^{49} &= i \cdot i^{48} && i^n = i \cdot i^{n-1} \\
&= i(i^2)^{24} && (a^m)^n = a^{mn} \\
&= i(-1)^{24} && i^2 = -1 \\
&= i(1) && (-1)^n = 1 \text{ when } n \text{ is even.} \\
&= i
\end{aligned}
$$

⤴ YOUR TURN

EXAMPLE 9   Simplify: **(a)** i^{24}; **(b)** i^{75}.

SOLUTION

a)
$$
\begin{aligned}
i^{24} &= (i^2)^{12} \\
&= (-1)^{12} = 1
\end{aligned}
$$

b)
$$
\begin{aligned}
i^{75} &= i \cdot i^{74} && \text{Writing } i^n \text{ as } i \cdot i^{n-1} \\
&= i(i^2)^{37} \\
&= i(-1)^{37} \\
&= i(-1) \\
&= -i
\end{aligned}
$$

9. Simplify: i^{60}.

_____

⤴ YOUR TURN

7.8 EXERCISE SET

FOR EXTRA HELP

MyMathLab® Math‾XL

PRACTICE   WATCH   READ   REVIEW

Vocabulary and Reading Check

Classify each of the following statements as either true or false.

1. Imaginary numbers are so named because they have no real-world applications.

2. Every real number is imaginary, but not every imaginary number is real.

3. Every imaginary number is a complex number, but not every complex number is imaginary.

4. Every real number is a complex number, but not every complex number is real.

5. We add complex numbers by combining real parts and combining imaginary parts.

6. The product of a complex number and its conjugate is always a real number.

7. The square of a complex number is always a real number.

8. The quotient of two complex numbers is always a complex number.

Imaginary Numbers and Complex Numbers

Express in terms of i.

9. $\sqrt{-100}$ 10. $\sqrt{-9}$

11. $\sqrt{-5}$ 12. $\sqrt{-7}$

13. $\sqrt{-8}$ 14. $\sqrt{-12}$

15. $-\sqrt{-11}$ 16. $-\sqrt{-17}$

17. $-\sqrt{-49}$ 18. $-\sqrt{-81}$

19. $-\sqrt{-300}$ 20. $-\sqrt{-75}$

21. $6 - \sqrt{-84}$ 22. $4 - \sqrt{-60}$

23. $-\sqrt{-76} + \sqrt{-125}$ 24. $\sqrt{-4} + \sqrt{-12}$

25. $\sqrt{-18} - \sqrt{-64}$ 26. $\sqrt{-72} - \sqrt{-25}$

Addition and Subtraction

Add or subtract and simplify. Write each answer in the form $a + bi$.

27. $(3 + 4i) + (2 - 7i)$ 28. $(5 - 6i) + (8 + 9i)$

29. $(9 + 5i) - (2 + 3i)$ 30. $(8 + 7i) - (2 + 4i)$

31. $(7 - 4i) - (5 - 3i)$

32. $(5 - 3i) - (9 + 2i)$

33. $(-5 - i) - (7 + 4i)$

34. $(-2 + 6i) - (-7 + i)$

Multiplication

Multiply and simplify. Write each answer in the form $a + bi$.

35. $5i \cdot 8i$ 36. $3i \cdot 9i$

37. $(-4i)(-6i)$ 38. $7i \cdot (-8i)$

39. $\sqrt{-36}\sqrt{-9}$ 40. $\sqrt{-49}\sqrt{-16}$

41. $\sqrt{-3}\sqrt{-10}$ 42. $\sqrt{-6}\sqrt{-7}$

43. $\sqrt{-6}\sqrt{-21}$ 44. $\sqrt{-15}\sqrt{-10}$

45. $5i(2 + 6i)$ 46. $2i(7 + 3i)$

47. $-7i(3 + 4i)$ 48. $-4i(6 - 5i)$

49. $(1 + i)(3 + 2i)$ 50. $(4 + i)(2 + 3i)$

51. $(6 - 5i)(3 + 4i)$ 52. $(5 - 6i)(2 + 5i)$

53. $(7 - 2i)(2 - 6i)$ 54. $(-4 + 5i)(3 - 4i)$

55. $(3 + 8i)(3 - 8i)$ 56. $(1 + 2i)(1 - 2i)$

57. $(-7 + i)(-7 - i)$ 58. $(-4 + 5i)(-4 - 5i)$

59. $(4 - 2i)^2$ 60. $(1 - 2i)^2$

61. $(2 + 3i)^2$ 62. $(3 + 2i)^2$

63. $(-2 + 3i)^2$ 64. $(-5 - 2i)^2$

Conjugates and Division

Divide and simplify. Write each answer in the form $a + bi$.

65. $\dfrac{10}{3 + i}$ 66. $\dfrac{26}{5 + i}$

67. $\dfrac{2}{3 - 2i}$ 68. $\dfrac{4}{2 - 3i}$

69. $\dfrac{2i}{5 + 3i}$ 70. $\dfrac{3i}{4 + 2i}$

71. $\dfrac{5}{6i}$ 72. $\dfrac{4}{7i}$

73. $\dfrac{5 - 3i}{4i}$ 74. $\dfrac{2 + 7i}{5i}$

Aha! 75. $\dfrac{7i + 14}{7i}$ 76. $\dfrac{6i + 3}{3i}$

77. $\dfrac{4 + 5i}{3 - 7i}$ 78. $\dfrac{5 + 3i}{7 - 4i}$

79. $\dfrac{2 + 3i}{2 + 5i}$ 80. $\dfrac{3 + 2i}{4 + 3i}$

81. $\dfrac{3 - 2i}{4 + 3i}$ 82. $\dfrac{5 - 2i}{3 + 6i}$

Powers of i

Simplify.

83. i^{32} **84.** i^{19}

85. i^{15} **86.** i^{38}

87. i^{42} **88.** i^{64}

89. i^9 **90.** i^{17}

91. $(-i)^6$ **92.** $(-i)^4$

93. $(5i)^3$ **94.** $(-3i)^5$

95. $i^2 + i^4$ **96.** $5i^5 + 4i^3$

 97. Is the product of two imaginary numbers always an imaginary number? Why or why not?

 98. In what way(s) are conjugates of complex numbers similar to the conjugates used in Section 7.5?

Skill Review

Factor completely.

99. $x^2 - 100$ [5.5]   **100.** $t^3 + 1000$ [5.6]

101. $2x - 63 + x^2$ [5.4]   **102.** $12a^3 - 5a^2 - 3a$ [5.4]

103. $w^3 - 4w + 3w^2 - 12$ [5.3], [5.4]

104. $24x^3y^2 - 60x^2y^4 - 12x^2y^2$ [5.1]

Synthesis

 105. Is the set of real numbers a subset of the set of complex numbers? Why or why not?

 106. Explain why there is no need to use powers of i (other than 1) when writing complex numbers.

Complex numbers are often graphed on a plane. The horizontal axis is the real axis and the vertical axis is the imaginary axis. A complex number such as $5 - 2i$ then corresponds to 5 on the real axis and -2 on the imaginary axis.

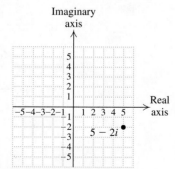

107. Graph each of the following.

  a) $3 + 2i$ **b)** $-1 + 4i$

  c) $3 - i$ **d)** $-5i$

108. Graph each of the following.

  a) $1 - 4i$ **b)** $-2 - 3i$

  c) i **d)** 4

The absolute value of a complex number $a + bi$ is its distance from the origin. (See the graph above.) Using the distance formula, we have $|a + bi| = \sqrt{a^2 + b^2}$. Find the absolute value of each complex number.

109. $|3 + 4i|$ **110.** $|8 - 6i|$

111. $|-1 + i|$ **112.** $|-3 - i|$

Consider the function g given by

$$g(z) = \frac{z^4 - z^2}{z - 1}.$$

113. Find $g(3i)$.   **114.** Find $g(1 + i)$.

115. Find $g(5i - 1)$.   **116.** Find $g(2 - 3i)$.

117. Evaluate

$$\frac{1}{w - w^2} \quad \text{for} \quad w = \frac{1 - i}{10}.$$

Simplify.

118. $\dfrac{i^5 + i^6 + i^7 + i^8}{(1 - i)^4}$ **119.** $(1 - i)^3(1 + i)^3$

120. $\dfrac{5 - \sqrt{5}i}{\sqrt{5}i}$ **121.** $\dfrac{6}{1 + \dfrac{3}{i}}$

122. $\left(\dfrac{1}{2} - \dfrac{1}{3}i\right)^2 - \left(\dfrac{1}{2} + \dfrac{1}{3}i\right)^2$

123. $\dfrac{i - i^{38}}{1 + i}$

YOUR TURN ANSWERS: SECTION 7.8

1. $6i$ **2.** $-4 - 3i$ **3.** $22 - 3i$ **4.** $2 - 9i$ **5.** 13
6. $-\frac{6}{5} + \frac{8}{5}i$ **7.** -1 **8.** i **9.** 1

QUICK QUIZ: SECTIONS 7.1–7.8

Let $f(x) = \sqrt{2x - 1}$.

1. Find $f(5)$. [7.1]

2. Find the domain of f. [7.1]

3. Find all a such that $f(a) = 7$. [7.6]

4. Simplify: $(5 - \sqrt{2})(1 - 3\sqrt{6})$. [7.5]

5. Simplify: $(3 - i)(5 + 2i)$. [7.8]

PREPARE TO MOVE ON

Solve. [5.8]

1. $x^2 - x - 6 = 0$ **2.** $(x - 5)^2 = 0$

3. $2t^2 - 50 = 0$ **4.** $15x^2 = 14x + 8$

A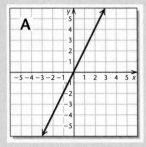

B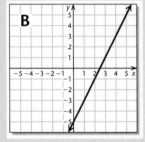

C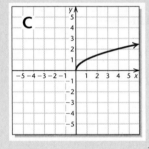

D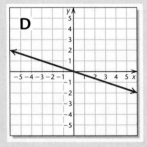

E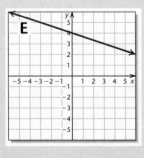

Visualizing for Success

Use after Section 7.1.

Match each function with its graph.

1. $f(x) = 2x - 5$

2. $f(x) = x^2 - 1$

3. $f(x) = \sqrt{x}$

4. $f(x) = x - 2$

5. $f(x) = -\frac{1}{3}x$

6. $f(x) = 2x$

7. $f(x) = 4 - x$

8. $f(x) = |2x - 5|$

9. $f(x) = -2$

10. $f(x) = -\frac{1}{3}x + 4$

Answers on page A-37

An additional, animated version of this activity appears in MyMathLab. *To use MyMathLab, you need a course ID and a student access code. Consult your instructor for more information.*

F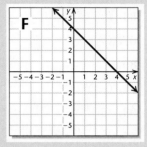

G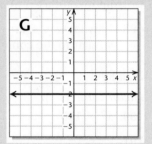

H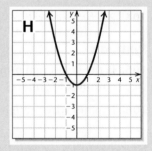

I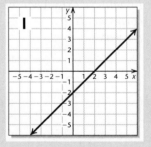

J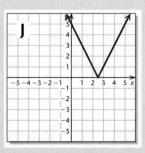

Collaborative Activity   *Tailgater Alert*

Focus: Radical equations and problem solving
Use after: Section 7.6
Time: 15–25 minutes
Group size: 2–3
Materials: Calculators

The faster a car travels, the more distance it needs to stop. Police recommend that for each 10 mph of speed, a driver allow 1 car length between vehicles. Thus a driver traveling at 30 mph should have at least 3 car lengths between his or her vehicle and the one in front.

The function $r(L) = 2\sqrt{5L}$ can be used to find the speed, in miles per hour, that a car was traveling when it left skid marks L feet long.

Activity

1. Each group member should estimate the length of a car in which he or she frequently travels. (Each should use a different length, if possible.)

2. Using a calculator as needed, each group member should complete the table below.

Column 1 gives a car's speed s, and column 2 lists the minimum amount of space between cars traveling s miles per hour, as recommended by police. Column 3 is the speed that a vehicle *could* travel were it forced to stop in the distance listed in column 2, using the above function.

Column 1 s (in miles per hour)	Column 2 $L(s)$ (in feet)	Column 3 $r(L)$ (in miles per hour)
20		
30		
40		
50		
60		

3. Compare tables to determine whether there are any speeds at which the "1 car length per 10 mph" guideline might not suffice. What recommendations would your group make to a new driver?

Decision Making 🔗 Connection   *(Use after Section 7.6.)*

Distance.   The distance D, in miles, that one can see to the horizon from a height of h feet can be approximated by $D = \sqrt{1.5h}$.

1. On a clear day, Max stands at the top of an observation tower on a mountain overlooking the ocean. According to his GPS, he is at an elevation of 1500 ft. How far can he see to the horizon?

2. On Judy's first flight in a hot air balloon, she noticed that she could detect, on the horizon, a landmark that she knew was 30 mi away from the ground under her balloon. How high was the balloon?

3. Cole plans to erect two radio antennas, each of which is 100 ft tall. Assume for this situation that all terrain is at sea level.
   a) How far could one see to the horizon from the top of one antenna?
   b) How far apart should he place the antennas if he wants the top of one antenna to be just visible from the top of the other over the horizon? (This distance is called the visual line-of-sight (VLOS).)
   c) Radio waves can "bend" slightly around the horizon. The radio line-of-sight (RLOS) is

the maximum distance between two antennas in radio communication. Under standard atmospheric conditions, the RLOS is approximately $\frac{4}{3}$ of the VLOS. How far apart can Cole erect the antennas and still maintain radio communication?

4. *Research.*   Because the surface of the earth is spherical, distance between points described by latitude and longitude coordinates is found using trigonometry.
   a) Find the latitude and longitude coordinates of two locations that are between 100 and 1000 mi apart. Use a GPS or an online tool to determine the distance between those points.
   b) If points on the earth's surface are mapped on a plane, they can be described using UTM (Universal Transverse Mercator) coordinates, and the distance between the points can be found using the Pythagorean theorem. Use a GPS or an online tool to convert the coordinates in part (a) to UTM coordinates and to determine the distance between the points.
   c) Explain why the distances found in parts (a) and (b) are not exactly the same.

Study Summary

KEY TERMS AND CONCEPTS	EXAMPLES	PRACTICE EXERCISES
SECTION 7.1: *Radical Expressions and Functions*		

KEY TERMS AND CONCEPTS	EXAMPLES	PRACTICE EXERCISES
c is a **square root** of a if $c^2 = a$. c is a **cube root** of a if $c^3 = a$. $\sqrt{a}$ indicates the **principal** square root of a. $\sqrt[n]{a}$ indicates the **nth root** of a. **index** $\sqrt[n]{a}$—**radicand**	The square roots of 25 are -5 and 5. The cube root of -8 is -2. $\sqrt{25} = 5$ $\sqrt[3]{-8} = -2$ The index of $\sqrt[3]{-8}$ is 3. The radicand of $\sqrt[3]{-8}$ is -8.	*Simplify.* **1.** $-\sqrt{81}$ **2.** $\sqrt[3]{-1}$
For all a, $\sqrt[n]{a^n} = |a|$ when n is even; $\sqrt[n]{a^n} = a$ when n is odd. If a represents a nonnegative number, $\sqrt[n]{a^n} = a$.	Assume that x can represent any real number. $\sqrt{(3 + x)^2} = |3 + x|$ Assume that x represents a nonnegative number. $\sqrt{(7x)^2} = 7x$	**3.** Simplify $\sqrt{36x^2}$. Assume that x can represent any real number. **4.** Simplify $\sqrt[4]{x^4}$. Assume that x represents a nonnegative number.

SECTION 7.2: *Rational Numbers as Exponents*		
$a^{1/n}$ means $\sqrt[n]{a}$. $a^{m/n}$ means $(\sqrt[n]{a})^m$ or $\sqrt[n]{a^m}$. $a^{-m/n}$ means $\dfrac{1}{a^{m/n}}$.	$64^{1/2} = \sqrt{64} = 8$ $125^{2/3} = (\sqrt[3]{125})^2 = 5^2 = 25$ $8^{-1/3} = \dfrac{1}{8^{1/3}} = \dfrac{1}{2}$	**5.** Simplify: $100^{-1/2}$.

SECTION 7.3: *Multiplying Radical Expressions*		
The Product Rule for Radicals For any real numbers $\sqrt[n]{a}$ and $\sqrt[n]{b}$, $\sqrt[n]{a} \cdot \sqrt[n]{b} = \sqrt[n]{a \cdot b}$.	$\sqrt[3]{4x} \cdot \sqrt[3]{5y} = \sqrt[3]{20xy}$ $\sqrt{75x^8y^{11}} = \sqrt{25 \cdot x^8 \cdot y^{10} \cdot 3 \cdot y}$ $= \sqrt{25} \cdot \sqrt{x^8} \cdot \sqrt{y^{10}} \cdot \sqrt{3y}$ $= 5x^4y^5\sqrt{3y}$	**6.** Multiply: $\sqrt{7x} \cdot \sqrt{3y}$. **7.** Simplify: $\sqrt{200x^5y^{18}}$.

SECTION 7.4: *Dividing Radical Expressions*		
The Quotient Rule for Radicals For any real numbers $\sqrt[n]{a}$ and $\sqrt[n]{b}$, $b \neq 0$, $\sqrt[n]{\dfrac{a}{b}} = \dfrac{\sqrt[n]{a}}{\sqrt[n]{b}}$.	$\sqrt[3]{\dfrac{8y^4}{125}} = \dfrac{\sqrt[3]{8y^4}}{\sqrt[3]{125}} = \dfrac{2y\sqrt[3]{y}}{5}$ $\dfrac{\sqrt{18a^9}}{\sqrt{2a^3}} = \sqrt{\dfrac{18a^9}{2a^3}} = \sqrt{9a^6} = 3a^3$ Assuming a is positive	**8.** Simplify: $\sqrt{\dfrac{12x^3}{25}}$.

| We can **rationalize a denominator** by multiplying by 1. | $$\frac{1}{\sqrt{2}} = \frac{1}{\sqrt{2}} \cdot \frac{\sqrt{2}}{\sqrt{2}} = \frac{\sqrt{2}}{2}$$ | **9.** Rationalize the denominator: $$\sqrt{\frac{2x}{3y}}.$$ |

SECTION 7.5: *Expressions Containing Several Radical Terms*

Like radicals have both the same indices and radicands.	$$\sqrt{12} + 5\sqrt{3} = \sqrt{4 \cdot 3} + 5\sqrt{3} = 2\sqrt{3} + 5\sqrt{3} = 7\sqrt{3}$$	**10.** Simplify: $5\sqrt{8} - 3\sqrt{50}.$
Radical expressions are multiplied in much the same way that polynomials are multiplied.	$(1 + 5\sqrt{6})(4 - \sqrt{6})$ $\quad = 1 \cdot 4 - 1\sqrt{6} + 4 \cdot 5\sqrt{6} - 5\sqrt{6} \cdot \sqrt{6}$ $\quad = 4 - \sqrt{6} + 20\sqrt{6} - 5 \cdot 6$ $\quad = -26 + 19\sqrt{6}$	**11.** Simplify: $(2 - \sqrt{3})(5 - 7\sqrt{3}).$
To rationalize a denominator containing two terms, we use the **conjugate** of the denominator to write a form of 1.	$\dfrac{2}{1 - \sqrt{3}} = \dfrac{2}{1 - \sqrt{3}} \cdot \dfrac{1 + \sqrt{3}}{1 + \sqrt{3}}$ $1 + \sqrt{3}$ is the conjugate of $1 - \sqrt{3}$. $\quad = \dfrac{2(1 + \sqrt{3})}{-2} = -1 - \sqrt{3}$	**12.** Rationalize the denominator: $$\frac{\sqrt{15}}{3 + \sqrt{5}}.$$
When terms have different indices, we can often use rational exponents to simplify.	$\sqrt[3]{p} \cdot \sqrt[4]{q^3} = p^{1/3} \cdot q^{3/4}$ $\quad = p^{4/12} \cdot q^{9/12}$ Finding a common denominator $\quad = \sqrt[12]{p^4 q^9}$	**13.** Simplify: $\dfrac{\sqrt{x^5}}{\sqrt[3]{x}}.$

SECTION 7.6: *Solving Radical Equations*

| **The Principle of Powers**
If $a = b$, then $a^n = b^n$.

Solutions found using the principle of powers must be checked in the original equation. | $x - 7 = \sqrt{x - 5}$
$(x - 7)^2 = (\sqrt{x - 5})^2$
$x^2 - 14x + 49 = x - 5$
$x^2 - 15x + 54 = 0$
$(x - 6)(x - 9) = 0$
$\quad x = 6 \quad or \quad x = 9$ Only 9 checks and is the solution. | **14.** Solve: $\sqrt{2x + 3} = x.$ |

SECTION 7.7: *The Distance Formula, the Midpoint Formula, and Other Applications*

| **The Pythagorean Theorem**
In any right triangle, if a and b are the lengths of the legs and c is the length of the hypotenuse, then
$$a^2 + b^2 = c^2.$$
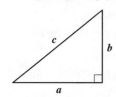 | Find the length of the hypotenuse of a right triangle with legs of lengths 4 and 7. Give an exact answer in radical notation, as well as a decimal approximation to three decimal places.
$a^2 + b^2 = c^2$
$4^2 + 7^2 = c^2$ Substituting
$16 + 49 = c^2$
$65 = c^2$
$\sqrt{65} = c$ This is exact.
$8.062 \approx c$ This is approximate. | **15.** The hypotenuse of a right triangle is 10 m long, and one leg is 7 m long. Find the length of the other leg. Give an exact answer in radical notation, as well as a decimal approximation to three decimal places. |

Special Triangles
The length of the hypotenuse in an isosceles right triangle (45°–45°–90° triangle) is the length of a leg times $\sqrt{2}$.

The length of the longer leg in a 30°–60°–90° triangle is the length of the shorter leg times $\sqrt{3}$. The hypotenuse is twice as long as the shorter leg.

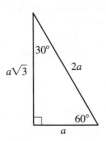

Find the missing lengths. Give an exact answer and, where appropriate, an approximation to three decimal places.

$$a = 10; \quad c = a\sqrt{2}$$
$$c = 10\sqrt{2}$$
$$c \approx 14.142$$

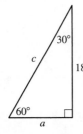

$$b = a\sqrt{3} \qquad c = 2a$$
$$18 = a\sqrt{3} \qquad c = 2\left(\frac{18}{\sqrt{3}}\right)$$
$$\frac{18}{\sqrt{3}} = a \qquad c = \frac{36}{\sqrt{3}} = 12\sqrt{3}$$
$$10.392 \approx a; \qquad c \approx 20.785$$

Find the missing lengths. Give an exact answer and, where appropriate, an approximation to three decimal places.

16.

17.

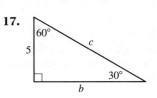

The Distance Formula
The distance d between any two points (x_1, y_1) and (x_2, y_2) is given by $d = \sqrt{(x_2 - x_1)^2 + (y_2 - y_1)^2}$.

Find the distance between $(3, -5)$ and $(-1, -2)$.
$$d = \sqrt{(-1 - 3)^2 + (-2 - (-5))^2}$$
$$= \sqrt{(-4)^2 + (3)^2}$$
$$= \sqrt{16 + 9} = \sqrt{25} = 5$$

18. Find the distance between $(-2, 1)$ and $(6, -10)$. Give an exact answer and an approximation to three decimal places.

The Midpoint Formula
If the endpoints of a segment are (x_1, y_1) and (x_2, y_2), then the coordinates of the midpoint are
$$\left(\frac{x_1 + x_2}{2}, \frac{y_1 + y_2}{2}\right).$$

Find the midpoint of the segment with endpoints $(3, -5)$ and $(-1, -2)$.
$$\left(\frac{3 + (-1)}{2}, \frac{-5 + (-2)}{2}\right), \quad \text{or} \quad \left(1, -\frac{7}{2}\right)$$

19. Find the midpoint of the segment with endpoints $(-2, 1)$ and $(6, -10)$.

SECTION 7.8:   *The Complex Numbers*

A **complex number** is any number that can be written in the form $a + bi$, where a and b are real numbers, $i = \sqrt{-1}$, and $i^2 = -1$.

$$(3 + 2i) + (4 - 7i) = 7 - 5i;$$
$$(8 + 6i) - (5 + 2i) = 3 + 4i;$$
$$(2 + 3i)(4 - i) = 8 - 2i + 12i - 3i^2$$
$$= 8 + 10i - 3(-1) = 11 + 10i;$$

$$\frac{1 - 4i}{3 - 2i} = \frac{1 - 4i}{3 - 2i} \cdot \frac{3 + 2i}{3 + 2i} = \frac{3 + 2i - 12i - 8i^2}{9 + 6i - 6i - 4i^2}$$
$$= \frac{3 - 10i - 8(-1)}{9 - 4(-1)} = \frac{11 - 10i}{13} = \frac{11}{13} - \frac{10}{13}i$$
$$i^{38} = (i^2)^{19} = (-1)^{19} = -1$$

20. Add:
$$(5 - 3i) + (-8 - 9i).$$

21. Subtract:
$$(2 - i) - (-1 + i).$$

22. Multiply:
$$(1 - 7i)(3 - 5i).$$

23. Divide: $\dfrac{1 + i}{1 - i}$.

24. Simplify: i^{39}.

Review Exercises: Chapter 7

⟣ Concept Reinforcement

Classify each of the following statements as either true or false.

1. $\sqrt{ab} = \sqrt{a} \cdot \sqrt{b}$ for any real numbers $\sqrt{a}$ and $\sqrt{b}$.   [7.3]

2. $\sqrt{a + b} = \sqrt{a} + \sqrt{b}$ for any real numbers $\sqrt{a}$ and $\sqrt{b}$.   [7.5]

3. $\sqrt{a^2} = a$, for any real number a.   [7.1]

4. $\sqrt[3]{a^3} = a$, for any real number a.   [7.1]

5. $x^{2/5}$ means $\sqrt[5]{x^2}$ and $\left(\sqrt[5]{x}\right)^2$.   [7.2]

6. The hypotenuse of a right triangle is never shorter than either leg.   [7.7]

7. Some radical equations have no solution.   [7.6]

8. If $f(x) = \sqrt{x - 5}$, then the domain of f is the set of all nonnegative real numbers.   [7.1]

Simplify.   [7.1]

9. $\sqrt{\dfrac{100}{121}}$

10. $-\sqrt{0.36}$

Let $f(x) = \sqrt{x + 10}$. Find the following.   [7.1]

11. $f(15)$

12. The domain of f

Simplify. Assume that each variable can represent any real number.   [7.1]

13. $\sqrt{64t^2}$

14. $\sqrt{(c + 7)^2}$

15. $\sqrt{4x^2 + 4x + 1}$

16. $\sqrt[5]{-32}$

17. Write an equivalent expression using exponential notation: $\left(\sqrt[3]{5ab}\right)^4$.   [7.2]

18. Write an equivalent expression using radical notation: $(3a^4)^{1/5}$.   [7.2]

Use rational exponents to simplify. Assume $x, y \geq 0$.   [7.2]

19. $\sqrt{x^6 y^{10}}$

20. $\left(\sqrt[6]{x^2 y}\right)^2$

Simplify. Do not use negative exponents in the answers.   [7.2]

21. $\left(x^{-2/3}\right)^{3/5}$

22. $\dfrac{7^{-1/3}}{7^{-1/2}}$

23. If $f(x) = \sqrt{25(x - 6)^2}$, find a simplified form for $f(x)$.   [7.3]

Simplify. Write all answers using radical notation. Assume that all variables represent positive numbers.

24. $\sqrt[4]{16x^{20}y^8}$ [7.3]

25. $\sqrt{250x^3y^2}$ [7.3]

26. $\sqrt{5a}\sqrt{7b}$ [7.3]

27. $\sqrt[3]{3x^4b}\sqrt[3]{9xb^2}$ [7.3]

28. $\sqrt[3]{-24x^{10}y^8}\,\sqrt[3]{18x^7y^4}$ [7.3]

29. $\dfrac{\sqrt[3]{60xy^3}}{\sqrt[3]{10x}}$ [7.4]

30. $\dfrac{\sqrt{75x}}{2\sqrt{3}}$ [7.4]

31. $\sqrt[4]{\dfrac{48a^{11}}{c^8}}$ [7.4]

32. $5\sqrt[3]{4y} + 2\sqrt[3]{4y}$ [7.5]

33. $2\sqrt{75} - 9\sqrt{3}$ [7.5]

34. $\sqrt{50} + 2\sqrt{18} + \sqrt{32}$ [7.5]

35. $(3 + \sqrt{10})(3 - \sqrt{10})$ [7.5]

36. $(\sqrt{3} - 3\sqrt{8})(\sqrt{5} + 2\sqrt{8})$ [7.5]

37. $\sqrt[4]{x}\,\sqrt{x}$ [7.5]

38. $\dfrac{\sqrt[3]{x^2}}{\sqrt[4]{x}}$ [7.5]

39. If $f(x) = x^2$, find $f(2 - \sqrt{a})$.   [7.5]

40. Rationalize the denominator:

$$\sqrt{\dfrac{x}{8y}}.\quad [7.4]$$

41. Rationalize the denominator:

$$\dfrac{4\sqrt{5}}{\sqrt{2} + \sqrt{3}}.\quad [7.5]$$

42. Rationalize the numerator of the expression in Exercise 41.   [7.5]

Solve.   [7.6]

43. $\sqrt{y + 6} - 2 = 3$

44. $(x + 1)^{1/3} = -5$

45. $1 + \sqrt{x} = \sqrt{3x - 3}$

46. If $f(x) = \sqrt{x + 2} + x$, find a such that $f(a) = 4$.   [7.6]

Solve. Give an exact answer and, where appropriate, an approximation to three decimal places. [7.7]

47. The diagonal of a square has length 10 cm. Find the length of a side of the square.

48. A skate-park jump has a ramp that is 6 ft long and is 2 ft high. How long is its base?

49. Find the missing lengths. Give exact answers and, where appropriate, an approximation to three decimal places.

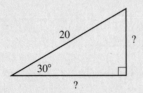

50. Find the distance between $(-6, 4)$ and $(-1, 5)$. Give an exact answer and an approximation to three decimal places. [7.7]

51. Find the midpoint of the segment with endpoints $(-7, -2)$ and $(3, -1)$. [7.7]

52. Express in terms of i and simplify: $\sqrt{-45}$. [7.8]

53. Add: $(-4 + 3i) + (2 - 12i)$. [7.8]

54. Subtract: $(9 - 7i) - (3 - 8i)$. [7.8]

Simplify. [7.8]

55. $(2 + 5i)(2 - 5i)$

56. i^{34}

57. $(6 - 3i)(2 - i)$

58. Divide. Write the answer in the form $a + bi$.

$$\frac{7 - 2i}{3 + 4i} \quad [7.8]$$

Synthesis

59. What makes some complex numbers real and others imaginary? [7.8]

60. Explain why $\sqrt[n]{x^n} = |x|$ when n is even, but $\sqrt[n]{x^n} = x$ when n is odd. [7.1]

61. Write a quotient of two imaginary numbers that is a real number (answers may vary). [7.8]

62. Solve: $\sqrt{11x} + \sqrt{6 + x} = 6$. [7.6]

63. Simplify:

$$\frac{2}{1 - 3i} - \frac{3}{4 + 2i}. \quad [7.8]$$

64. Don's Discount Shoes has two locations. The sign at the original location is shaped like an isosceles right triangle. The sign at the newer location is shaped like a 30°–60°–90° triangle. The hypotenuse of each sign measures 6 ft. Which sign has the greater area and by how much? (Round to three decimal places.) [7.7]

Test: Chapter 7

For step-by-step test solutions, access the Chapter Test Prep Videos in MyMathLab®, on YouTube® (search "Bittinger Inter Alg CA" and click on "Channels"), or by scanning the code.

Simplify. Assume that variables can represent any real number.

1. $\sqrt{50}$

2. $\sqrt[3]{-\dfrac{8}{x^6}}$

3. $\sqrt{81a^2}$

4. $\sqrt{x^2 - 8x + 16}$

5. Write an equivalent expression using exponential notation: $\sqrt{7xy}$.

6. Write an equivalent expression using radical notation: $(4a^3b)^{5/6}$.

7. If $f(x) = \sqrt{2x - 10}$, determine the domain of f.

8. If $f(x) = x^2$, find $f(5 + \sqrt{2})$.

Simplify. Write all answers using radical notation. Assume that all variables represent positive numbers.

9. $\sqrt[5]{32x^{16}y^{10}}$

10. $\sqrt[3]{4w}\,\sqrt[3]{4v^2}$

11. $\sqrt{\dfrac{100a^4}{9b^6}}$

12. $\dfrac{\sqrt[5]{48x^6y^{10}}}{\sqrt[5]{16x^2y^9}}$

13. $\sqrt[4]{x^3}\sqrt{x}$

14. $\dfrac{\sqrt{y}}{\sqrt[10]{y}}$

15. $8\sqrt{2} - 2\sqrt{2}$

16. $\sqrt{50xy} + \sqrt{72xy} - \sqrt{8xy}$

17. $(7 + \sqrt{x})(2 - 3\sqrt{x})$

18. Rationalize the denominator:
$$\dfrac{\sqrt[3]{x}}{\sqrt[3]{4y}}.$$

Solve.

19. $6 = \sqrt{x - 3} + 5$

20. $x = \sqrt{3x + 3} - 1$

21. $\sqrt{2x} = \sqrt{x + 1} + 1$

Solve. For Exercises 22–24, give exact answers and approximations to three decimal places.

22. A referee jogs diagonally from one corner of a 50-ft by 90-ft basketball court to the far corner. How far does she jog?

23. The hypotenuse of a 30°–60°–90° triangle is 10 cm long. Find the lengths of the legs.

24. Find the distance between the points $(3, 7)$ and $(-1, 8)$.

25. Find the midpoint of the segment with endpoints $(2, -5)$ and $(1, -7)$.

26. Express in terms of i and simplify: $\sqrt{-50}$.

27. Subtract: $(9 + 8i) - (-3 + 6i)$.

28. Multiply. Write the answer in the form $a + bi$.
$$(4 - i)^2$$

29. Divide. Write the answer in the form $a + bi$.
$$\dfrac{-2 + i}{3 - 5i}$$

30. Simplify: i^{37}.

Synthesis

31. Solve: $\sqrt{2x - 2} + \sqrt{7x + 4} = \sqrt{13x + 10}$.

32. Simplify:
$$\dfrac{1 - 4i}{4i(1 + 4i)^{-1}}.$$

33. The function $D(h) = 1.2\sqrt{h}$ can be used to approximate the distance D, in miles, that a person can see to the horizon from a height h, in feet. How far above sea level must a pilot fly in order to see a horizon that is 180 mi away?

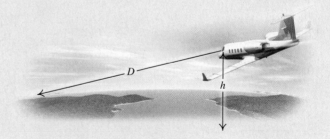

Cumulative Review: Chapters 1–7

Solve.

1. $x(x + 2) = 35$ [5.8]

2. $\dfrac{1}{x} = \dfrac{2}{5}$ [6.4]

3. $\sqrt[3]{t} = -1$ [7.6]

4. $25x^2 - 10x + 1 = 0$ [5.8]

5. $|x - 2| \leq 5$ [4.3]

6. $2x + 5 > 6$ *or* $x - 3 \leq 9$ [4.2]

7. $\dfrac{2x}{x - 1} + \dfrac{x}{x - 3} = 2$ [6.4]

8. $x = \sqrt{2x - 5} + 4$ [7.6]

9. $2x - y + z = 1,$
$\quad x + 2y + z = -3,$
$\quad 5x - y + 3z = 0$ [3.4]

Graph on a plane.

10. $3y = -6$ [2.4]

11. $y = -x + 5$ [2.3]

12. $x + y \leq 2$ [4.4]

13. $2x = y$ [2.3]

14. Find an equation for the line parallel to the line given by $y = 7x$ and passing through the point $(0, -11)$. [2.5]

Perform the indicated operations and, if possible, simplify.

15. $18 \div 3 \cdot 2 - 6^2 \div (2 + 4)$ [1.2]

16. $(2a - 5b)^2$ [5.2]

17. $(c^2 - 3d)(c^2 + 3d)$ [5.2]

18. $\dfrac{x + 3}{x - 2} - \dfrac{x + 5}{x + 1}$ [6.2]

19. $\dfrac{a^2 - a - 6}{a^2 - 1} \div \dfrac{a^2 - 6a + 9}{2a^2 + 3a + 1}$ [6.1]

20. $\dfrac{\dfrac{1}{x} + \dfrac{1}{x + 1}}{\dfrac{x}{x + 1}}$ [6.3]

21. $\sqrt{200} - 5\sqrt{8}$ [7.5]

22. $(1 + \sqrt{5})(4 - \sqrt{5})$ [7.5]

23. $\sqrt[3]{y}\sqrt[5]{y}$ [7.5]

Factor.

24. $x^2 - 5x - 14$ [5.4]

25. $4y^8 - 4y^5$ [5.6]

26. $3t^2 - 5t - 8$ [5.4]

27. $yt - xt - yz^2 + xz^2$ [5.3]

Find the domain of each function.

28. $f(x) = \dfrac{2x - 3}{x^2 - 6x + 9}$ [5.8]

29. $f(x) = \sqrt{2x - 11}$ [7.1]

Find each of the following, if $f(x) = \sqrt{2x - 3}$ and $g(x) = x^2$.

30. $g(1 - \sqrt{5})$ [7.5]

31. $(f + g)(x)$ [2.6], [7.5]

32. *Emergency Shelter.* The entrance to a tent used by a rescue team is the shape of an equilateral triangle. If the base of the tent is 4 ft wide, how tall is the tent? Give an exact answer and an approximation to three decimal places. [7.7]

33. *Age at Marriage.* The median age at first marriage for U.S. men has grown from 25.1 in 2001 to 25.5 in 2006. Let $m(t)$ represent the median age of men at first marriage t years after 2000. [2.5]

Source: U.S. Census Bureau

a) Find a linear function that fits the data.

b) Use the function from part (a) to predict the median age of men at first marriage in 2020.

c) In what year will the median age of men at first marriage reach 28 for the first time?

34. *Salary.* Nell's annual salary is $38,849. This includes a 6% superior performance raise. What would Nell's salary have been without the performance raise? [1.4]

35. *Landscaping.* A rectangular parking lot is 80 ft by 100 ft. Part of the asphalt is removed in order to install a landscaped border of uniform width around it. The area of the new parking lot is 6300 ft². How wide is the landscaped border? [5.8]

Synthesis

36. Give an equation in standard form for the line whose x-intercept is $(-3, 0)$ and whose y-intercept is $(0, 5)$. [2.4]

Solve.

37. $\dfrac{\dfrac{1}{x} + \dfrac{1}{x + 1}}{\dfrac{1}{x} - 1} = 1$ [6.3], [6.4]

38. $2\sqrt{3x - 2} = 2 + \sqrt{7x + 1}$ [7.6]

Quadratic Functions and Equations

YEAR	NUMBER OF SELF-PUBLISHED BOOKS
2006	51,000
2007	63,000
2008	74,000
2009	95,000
2010	133,000

Source: Based on data from Bowker's Books in Print database, in "Secret of Self-Publishing: Success," *The Wall Street Journal*, October 31, 2011

You Too Can Write a Book!

Although publication carries no guarantee of sales, many new and established authors are turning to self-publishing to see their works appear, either in print or as e-books. As the table above indicates, the number of self-published books is increasing dramatically and can be modeled using a *quadratic function*. (*See Example 3 in Section 8.8.*)

A self-published author uses math in some way in almost every stage of production.

Alex Van Wyhe, a writer with Tea Mug Collective from Kenny Lake, Alaska, uses math to determine the format and the layout of a book. To determine price, he uses a sustainable approach, considering cost and enough profit to continue his projects.

The mathematical translation of a problem is often a function or an equation containing a second-degree polynomial in one variable. Such functions or equations are said to be *quadratic*. In this chapter, we examine a variety of ways to solve quadratic equations and look at graphs and applications of quadratic functions.

8.1 Quadratic Equations

The Principle of Square Roots ■ Completing the Square ■ Problem Solving

The general form of a quadratic function is

$$f(x) = ax^2 + bx + c, \quad \text{with } a \neq 0,$$

and its graph is a *parabola*. Such graphs open up or down and can have 0, 1, or 2 *x*-intercepts. We learn to graph quadratic functions later in this chapter.

Algebraic 𝒢 Graphical Connection

The graphs of the quadratic function $f(x) = x^2 + 6x + 8$ and the linear function $g(x) = 0$ are shown below.

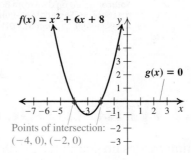

Note that $(-4, 0)$ and $(-2, 0)$ are the points of intersection of the graphs of $f(x) = x^2 + 6x + 8$ and $g(x) = 0$ (the *x*-axis).

As we saw in Section 5.8, we can solve equations like $x^2 + 6x + 8 = 0$ by factoring:

$$x^2 + 6x + 8 = 0$$

$$(x + 4)(x + 2) = 0 \qquad\qquad \text{Factoring}$$

$$x + 4 = 0 \quad or \quad x + 2 = 0 \qquad \text{Using the principle of zero products}$$

$$x = -4 \quad or \qquad\quad x = -2.$$

Note that -4 and -2 are the first coordinates of the points of intersection (or the *x*-intercepts) of the graph of $f(x)$ above.

To solve a quadratic equation by factoring, we write the equation in the *standard form* $ax^2 + bx + c = 0$, factor, and use the principle of zero products.

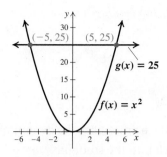

A visualization of Example 1

1. Solve: $64 = y^2$.

EXAMPLE 1   Solve: $x^2 = 25$.

SOLUTION   We have

$$x^2 = 25$$
$$x^2 - 25 = 0 \qquad \text{Writing in standard form}$$
$$(x - 5)(x + 5) = 0 \qquad \text{Factoring}$$
$$x - 5 = 0 \quad or \quad x + 5 = 0 \qquad \text{Using the principle of zero products}$$
$$x = 5 \quad or \qquad x = -5.$$

The solutions are 5 and −5. A graph in which $f(x) = x^2$ represents the left side of the equation and $g(x) = 25$ represents the right side provides a check (see the figure at left). We can also check by substituting 5 and −5 into the original equation.

YOUR TURN

In this section and the next, we develop algebraic methods for solving *any* quadratic equation, including those that cannot be solved by factoring.

The Principle of Square Roots

Let's reconsider $x^2 = 25$. The number 25 has two square roots, 5 and −5, the solutions of the equation. Square roots provide quick solutions for equations of the type $x^2 = k$.

> **THE PRINCIPLE OF SQUARE ROOTS**
> For any real number k, if $x^2 = k$, then
> $$x = \sqrt{k} \quad or \quad x = -\sqrt{k}.$$

EXAMPLE 2   Solve: $3x^2 = 6$. Give exact solutions and approximations to three decimal places.

SOLUTION   We have

$$3x^2 = 6$$
$$x^2 = 2 \qquad \text{Isolating } x^2$$
$$x = \sqrt{2} \quad or \quad x = -\sqrt{2}. \qquad \text{Using the principle of square roots}$$

We can use the symbol $\pm\sqrt{2}$ to represent both of the solutions.

> *CAUTION!*   There are *two* solutions: $\sqrt{2}$ and $-\sqrt{2}$. Don't forget the second solution.

Check: For $\sqrt{2}$:   For $-\sqrt{2}$:

$$\begin{array}{c|c} 3x^2 = 6 & 3x^2 = 6 \\ \hline 3(\sqrt{2})^2 \mid 6 & 3(-\sqrt{2})^2 \mid 6 \\ 3 \cdot 2 \mid & 3 \cdot 2 \mid \\ 6 \stackrel{?}{=} 6 \text{ TRUE} & 6 \stackrel{?}{=} 6 \text{ TRUE} \end{array}$$

The solutions are $\sqrt{2}$ and $-\sqrt{2}$, or $\pm\sqrt{2}$, which round to 1.414 and −1.414.

A visualization of Example 2

2. Solve: $t^2 = 10$. Give exact solutions and approximations to three decimal places.

YOUR TURN

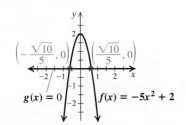

$\left(-\dfrac{\sqrt{10}}{5},0\right)$ $\left(\dfrac{\sqrt{10}}{5},0\right)$

$g(x)=0$ $f(x)=-5x^2+2$

A visualization of Example 3

3. Solve: $3x^2 = 1$.

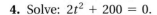

EXAMPLE 3 Solve: $-5x^2 + 2 = 0$.

SOLUTION We have

$$-5x^2 + 2 = 0$$
$$x^2 = \tfrac{2}{5} \qquad \text{Isolating } x^2$$
$$x = \sqrt{\tfrac{2}{5}} \quad or \quad x = -\sqrt{\tfrac{2}{5}}. \qquad \text{Using the principle of square roots}$$

The solutions are $\sqrt{\tfrac{2}{5}}$ and $-\sqrt{\tfrac{2}{5}}$, or simply $\pm\sqrt{\tfrac{2}{5}}$. If we rationalize the denominator, the solutions are written $\pm\dfrac{\sqrt{10}}{5}$. The checks are left to the student.

YOUR TURN

We can now find imaginary-number solutions. Unless the context dictates otherwise, you should find all complex-number solutions of equations.

EXAMPLE 4 Solve: $4x^2 + 9 = 0$. (Find all solutions in the complex-number system.)

SOLUTION We have

$$4x^2 + 9 = 0$$
$$x^2 = -\tfrac{9}{4} \qquad \text{Isolating } x^2$$
$$x = \sqrt{-\tfrac{9}{4}} \quad or \quad x = -\sqrt{-\tfrac{9}{4}} \qquad \text{Using the principle of square roots}$$
$$x = \sqrt{\tfrac{9}{4}}\sqrt{-1} \quad or \quad x = -\sqrt{\tfrac{9}{4}}\sqrt{-1}$$
$$x = \tfrac{3}{2}i \quad or \quad x = -\tfrac{3}{2}i. \qquad \text{Recall that } \sqrt{-1} = i.$$

Check: For $\tfrac{3}{2}i$:

$$\begin{array}{c|c} 4x^2 + 9 = 0 \\ \hline 4\left(\tfrac{3}{2}i\right)^2 + 9 & 0 \\ 4\cdot\tfrac{9}{4}\cdot i^2 + 9 \\ 9(-1) + 9 \\ & 0 \overset{?}{=} 0 \quad \text{TRUE} \end{array}$$

For $-\tfrac{3}{2}i$:

$$\begin{array}{c|c} 4x^2 + 9 = 0 \\ \hline 4\left(-\tfrac{3}{2}i\right)^2 + 9 & 0 \\ 4\cdot\tfrac{9}{4}\cdot i^2 + 9 \\ 9(-1) + 9 \\ & 0 \overset{?}{=} 0 \quad \text{TRUE} \end{array}$$

The solutions are $\tfrac{3}{2}i$ and $-\tfrac{3}{2}i$, or $\pm\tfrac{3}{2}i$. The graph at left confirms that there are no real-number solutions.

YOUR TURN

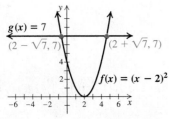

$f(x) = 4x^2 + 9$

$g(x) = 0$ No intersection

A visualization of Example 4

4. Solve: $2t^2 + 200 = 0$.

The principle of square roots can be restated in a more general form.

> **THE PRINCIPLE OF SQUARE ROOTS (GENERALIZED FORM)**
>
> For any real number k and any algebraic expression X:
>
> If $X^2 = k$, then $X = \sqrt{k}$ or $X = -\sqrt{k}$.

EXAMPLE 5 Let $f(x) = (x-2)^2$. Find all x-values for which $f(x) = 7$.

SOLUTION We are asked to find all x-values for which

$$f(x) = 7,$$
$$or \qquad (x-2)^2 = 7. \qquad \text{Substituting } (x-2)^2 \text{ for } f(x)$$

The generalized principle of square roots gives us

$$x - 2 = \sqrt{7} \qquad or \quad x - 2 = -\sqrt{7}$$
$$x = 2 + \sqrt{7} \quad or \qquad x = 2 - \sqrt{7}.$$

$g(x) = 7$

$(2 - \sqrt{7}, 7)$ $(2 + \sqrt{7}, 7)$

$f(x) = (x-2)^2$

A visualization of Example 5

Check: $f(2 + \sqrt{7}) = (2 + \sqrt{7} - 2)^2 = (\sqrt{7})^2 = 7.$

Similarly,

$$f(2 - \sqrt{7}) = (2 - \sqrt{7} - 2)^2 = (-\sqrt{7})^2 = 7.$$

The solutions are $2 + \sqrt{7}$ and $2 - \sqrt{7}$, or simply $2 \pm \sqrt{7}$.

YOUR TURN

5. Let $f(x) = (x + 5)^2$. Find all x-values for which $f(x) = 3$.

Example 5 is of the form $(x - a)^2 = c$, where a and c are constants. Sometimes we must factor in order to obtain this form.

EXAMPLE 6 Solve: $x^2 + 6x + 9 = 2$.

SOLUTION We have

$$x^2 + 6x + 9 = 2 \qquad \text{The left side is the square of a binomial.}$$
$$(x + 3)^2 = 2 \qquad \text{Factoring}$$
$$x + 3 = \sqrt{2} \quad \text{or} \quad x + 3 = -\sqrt{2} \qquad \begin{array}{l}\text{Using the principle} \\ \text{of square roots}\end{array}$$
$$x = -3 + \sqrt{2} \quad \text{or} \quad x = -3 - \sqrt{2}. \qquad \begin{array}{l}\text{Adding } -3 \text{ to both} \\ \text{sides}\end{array}$$

The solutions are $-3 + \sqrt{2}$ and $-3 - \sqrt{2}$, or $-3 \pm \sqrt{2}$. The checks are left to the student.

YOUR TURN

$f(x) = x^2 + 6x + 9$

$(-3 - \sqrt{2}, 2)$ $g(x) = 2$
$(-3 + \sqrt{2}, 2)$

A visualization of Example 6

6. Solve: $t^2 - 10t + 25 = 3$.

Completing the Square

By using a method called *completing the square*, we can use the principle of square roots to solve *any* quadratic equation. To see how this is done, consider

$$x^2 + 6x + 4 = 0.$$

The trinomial $x^2 + 6x + 4$ is not a perfect square. We can, however, create an equivalent equation with a perfect-square trinomial on one side:

$$x^2 + 6x + 4 = 0$$
$$x^2 + 6x = -4 \qquad \text{Only variable terms are on the left side.}$$
$$x^2 + 6x + 9 = -4 + 9 \qquad \begin{array}{l}\text{Adding 9 to both sides. We explain this} \\ \text{decision shortly.}\end{array}$$
$$(x + 3)^2 = 5. \qquad \begin{array}{l}\text{Factoring the perfect-square trinomial. We} \\ \text{could continue to solve as in Example 6.}\end{array}$$

We chose to add 9 to both sides because it creates a perfect-square trinomial on the left side. The 9 was found by taking half of the coefficient of x and squaring it.

To understand why this procedure works, examine the following drawings.

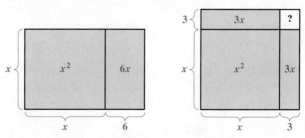

Note that the shaded areas in both figures represent the same area, $x^2 + 6x$. However, only the figure on the right, in which the $6x$ is halved, can be converted into a square with the addition of a constant term. The constant 9 is the "missing" piece that *completes* the square.

To complete the square for $x^2 + bx$, we add $(b/2)^2$.

Example 7, which follows, provides practice in finding numbers that complete the square. We will then use this skill to solve equations.

EXAMPLE 7   Replace the blanks in each equation with constants to form a true equation.

a) $x^2 + 14x + \underline{\quad} = (x + \underline{\quad})^2$

b) $x^2 - 5x + \underline{\quad} = (x - \underline{\quad})^2$

c) $x^2 + \frac{3}{4}x + \underline{\quad} = (x + \underline{\quad})^2$

Student Notes

In problems like Examples 7(b) and (c), it is best to avoid decimal notation. Most students have an easier time recognizing $\frac{9}{64}$ as $\left(\frac{3}{8}\right)^2$ than seeing 0.140625 as 0.375^2.

7. Replace the blanks with constants to form a true equation: $x^2 + 7x + \underline{\quad} = (x + \underline{\quad})^2$.

SOLUTION

a) Take half of the coefficient of x: Half of 14 is 7.

Square this number: $7^2 = 49$.

Add 49 to complete the square: $x^2 + 14x + 49 = (x + 7)^2$.

b) Take half of the coefficient of x: Half of -5 is $-\frac{5}{2}$.

Square this number: $\left(-\frac{5}{2}\right)^2 = \frac{25}{4}$.

Add $\frac{25}{4}$ to complete the square: $x^2 - 5x + \frac{25}{4} = \left(x - \frac{5}{2}\right)^2$.

c) Take half of the coefficient of x: Half of $\frac{3}{4}$ is $\frac{3}{8}$.

Square this number: $\left(\frac{3}{8}\right)^2 = \frac{9}{64}$.

Add $\frac{9}{64}$ to complete the square: $x^2 + \frac{3}{4}x + \frac{9}{64} = \left(x + \frac{3}{8}\right)^2$.

YOUR TURN

We can now use the method of completing the square to solve equations.

EXAMPLE 8   Solve: $x^2 - 8x - 7 = 0$.

SOLUTION   We begin by adding 7 to both sides:

$$x^2 - 8x - 7 = 0$$

$$x^2 - 8x \qquad = 7 \qquad \text{Adding 7 to both sides. We can now complete the square on the left side.}$$

$$x^2 - 8x + 16 = 7 + 16 \qquad \text{Adding 16 to both sides to complete the square: } \frac{1}{2}(-8) = -4, \text{ and } (-4)^2 = 16$$

$$(x - 4)^2 = 23 \qquad \text{Factoring and simplifying}$$

$$x - 4 = \pm\sqrt{23} \qquad \text{Using the principle of square roots}$$

$$x = 4 \pm \sqrt{23}. \qquad \text{Adding 4 to both sides}$$

Technology Connection

One way to check Example 8 is to store $4 + \sqrt{23}$ as x using the STO► key. We can then evaluate $x^2 - 8x - 7$ by entering $x^2 - 8x - 7$ and pressing ENTER.

1. Check Example 8 using the method described above.

Check: For $4 + \sqrt{23}$:

$$
\begin{array}{c|c}
\multicolumn{1}{c}{x^2 - 8x - 7 = 0} & \\
\hline
(4 + \sqrt{23})^2 - 8(4 + \sqrt{23}) - 7 & 0 \\
16 + 8\sqrt{23} + 23 - 32 - 8\sqrt{23} - 7 & \\
16 + 23 - 32 - 7 + 8\sqrt{23} - 8\sqrt{23} & \\
& 0 \overset{?}{=} 0 \quad \text{TRUE}
\end{array}
$$

For $4 - \sqrt{23}$:

$$x^2 - 8x - 7 = 0$$

$$\frac{(4 - \sqrt{23})^2 - 8(4 - \sqrt{23}) - 7 \mid 0}{}$$

$$16 - 8\sqrt{23} + 23 - 32 + 8\sqrt{23} - 7$$

$$16 + 23 - 32 - 7 - 8\sqrt{23} + 8\sqrt{23}$$

$$0 \overset{?}{=} 0 \quad \text{TRUE}$$

8. Solve: $x^2 + 6x - 2 = 0$.

The solutions are $4 + \sqrt{23}$ and $4 - \sqrt{23}$, or $4 \pm \sqrt{23}$.

YOUR TURN

Recall that the value of $f(x)$ must be 0 at any x-intercept of the graph of f. If $f(a) = 0$, then $(a, 0)$ is an x-intercept of the graph.

EXAMPLE 9 Find the x-intercepts of the graph of $f(x) = x^2 + 5x - 3$.

SOLUTION We set $f(x)$ equal to 0 and solve:

$$f(x) = 0$$

$$x^2 + 5x - 3 = 0 \qquad \text{Substituting}$$

$$x^2 + 5x \quad = 3 \qquad \text{Adding 3 to both sides}$$

$$x^2 + 5x + \frac{25}{4} = 3 + \frac{25}{4} \qquad \begin{array}{l}\text{Completing the square:} \\ \frac{1}{2} \cdot 5 = \frac{5}{2}, \text{ and } \left(\frac{5}{2}\right)^2 = \frac{25}{4}\end{array}$$

$$\left(x + \frac{5}{2}\right)^2 = \frac{37}{4} \qquad \text{Factoring and simplifying}$$

$$x + \frac{5}{2} = \pm\frac{\sqrt{37}}{2} \qquad \begin{array}{l}\text{Using the principle of square roots} \\ \text{and the quotient rule for radicals}\end{array}$$

$$x = -\frac{5}{2} \pm \frac{\sqrt{37}}{2}, \quad \text{or} \quad \frac{-5 \pm \sqrt{37}}{2}. \qquad \text{Adding } -\frac{5}{2} \text{ to both sides}$$

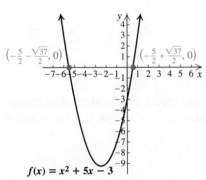

$\left(-\frac{5}{2} - \frac{\sqrt{37}}{2}, 0\right)$ $\left(-\frac{5}{2} + \frac{\sqrt{37}}{2}, 0\right)$

$f(x) = x^2 + 5x - 3$

A visualization of Example 9

The x-intercepts are

$$\left(-\frac{5}{2} - \frac{\sqrt{37}}{2}, 0\right) \quad \text{and} \quad \left(-\frac{5}{2} + \frac{\sqrt{37}}{2}, 0\right).$$

9. Find the x-intercepts of the graph of $f(x) = x^2 + 4x + 1$.

The checks are left to the student.

YOUR TURN

Before we complete the square in a quadratic equation, the leading coefficient must be 1. When it is not 1, we divide both sides of the equation by whatever that coefficient may be.

TO SOLVE A QUADRATIC EQUATION IN x BY COMPLETING THE SQUARE

1. Isolate the terms with variables on one side of the equation, and arrange them in descending order.
2. Divide both sides by the coefficient of x^2 if that coefficient is not 1.
3. Complete the square by taking half of the coefficient of x and adding its square to both sides.
4. Express the trinomial as the square of a binomial (factor the trinomial) and simplify the other side.
5. Use the principle of square roots (find the square roots of both sides).
6. Solve for x by adding or subtracting on both sides.

EXAMPLE 10   Solve: $3x^2 + 7x - 2 = 0$.

SOLUTION   We follow the steps listed above:

$$3x^2 + 7x - 2 = 0$$

Isolate the variable terms.	$3x^2 + 7x = 2$ $\qquad$ Adding 2 to both sides
Divide both sides by the coefficient of x^2.	$x^2 + \frac{7}{3}x = \frac{2}{3}$ $\qquad$ Dividing both sides by 3
Complete the square.	$x^2 + \frac{7}{3}x + \frac{49}{36} = \frac{2}{3} + \frac{49}{36}$ $\qquad$ Completing the square: $\left(\frac{1}{2}\cdot\frac{7}{3}\right)^2 = \frac{49}{36}$
Factor the trinomial.	$\left(x + \frac{7}{6}\right)^2 = \frac{73}{36}$ $\qquad$ Factoring and simplifying
Use the principle of square roots.	$x + \frac{7}{6} = \pm\frac{\sqrt{73}}{6}$ $\qquad$ Using the principle of square roots and the quotient rule for radicals
Solve for x.	$x = -\frac{7}{6} \pm \frac{\sqrt{73}}{6}$, or $\frac{-7 \pm \sqrt{73}}{6}$. $\qquad$ Adding $-\frac{7}{6}$ to both sides

The checks are left to the student. The solutions are $-\frac{7}{6} \pm \frac{\sqrt{73}}{6}$, or $\frac{-7 \pm \sqrt{73}}{6}$.

This can be written as

$$-\frac{7}{6} + \frac{\sqrt{73}}{6} \quad \text{and} \quad -\frac{7}{6} - \frac{\sqrt{73}}{6}, \quad \text{or} \quad \frac{-7 + \sqrt{73}}{6} \quad \text{and} \quad \frac{-7 - \sqrt{73}}{6}.$$

10. Solve: $2x^2 - 8x - 3 = 0$.

YOUR TURN

Any quadratic equation can be solved by completing the square. The procedure is also useful when graphing quadratic equations and will be used to develop a formula for solving quadratic equations.

Problem Solving

After one year, an amount of money P, invested at 4% per year, is worth 104% of P, or $P(1.04)$. If that amount continues to earn 4% interest per year, after the second year the investment will be worth 104% of $P(1.04)$, or $P(1.04)^2$. This is called **compounding interest** since after the first time period, interest is earned on both the initial investment *and* the interest from the first time period. Continuing the above pattern, we see that after the third year, the investment will be worth 104% of $P(1.04)^2$, or $P(1.04)^3$. Generalizing, we have the following.

> **THE COMPOUND-INTEREST FORMULA**
>
> If an amount of money P is invested at interest rate r, compounded annually, then in t years, it will grow to the amount A given by
>
> $$A = P(1 + r)^t. \quad (r \text{ is written in decimal notation.})$$

EXAMPLE 11   *Investment Growth.*   Katia invested $4000 at interest rate r, compounded annually. In 2 years, it grew to $4410. What was the interest rate?

SOLUTION

1. **Familiarize.** The compound-interest formula is given above.
2. **Translate.** The translation consists of substituting into the formula:

$$A = P(1 + r)^t$$
$$4410 = 4000(1 + r)^2. \quad \text{Substituting}$$

3. **Carry out.** We solve for r:

$$4410 = 4000(1 + r)^2$$
$$\frac{4410}{4000} = (1 + r)^2 \qquad \text{Dividing both sides by 4000}$$
$$\frac{441}{400} = (1 + r)^2 \qquad \text{Simplifying}$$
$$\pm\sqrt{\frac{441}{400}} = 1 + r \qquad \text{Using the principle of square roots}$$
$$\pm\frac{21}{20} = 1 + r \qquad \text{Simplifying}$$
$$-\frac{20}{20} \pm \frac{21}{20} = r \qquad \text{Adding } -1, \text{ or } -\frac{20}{20}, \text{ to both sides}$$
$$\frac{1}{20} = r \quad or \quad -\frac{41}{20} = r.$$

11. Max invested $1600 at interest rate r, compounded annually. In 2 years, it grew to $1936. What was the interest rate?

4. **Check.** Since the interest rate cannot be negative, we need check only $\frac{1}{20}$, or 5%. If $4000 were invested at 5% interest, compounded annually, then in 2 years it would grow to $4000(1.05)^2$, or $4410. The rate 5% checks.

5. **State.** The interest rate was 5%.

YOUR TURN

EXAMPLE 12 *Free-Falling Objects.* The formula $s = 16t^2$ is used to approximate the distance s, in feet, that an object falls freely from rest in t seconds. The Grand Canyon Skywalk is 4000 ft above the Colorado River. How long will it take a stone to fall from the Skywalk to the river? Round to the nearest tenth of a second.

Source: www.grandcanyonskywalk.com

SOLUTION

1. **Familiarize.** We agree to disregard air resistance and use the given formula.

2. **Translate.** We substitute into the formula:

$$s = 16t^2$$
$$4000 = 16t^2.$$

3. **Carry out.** We solve for t:

$$4000 = 16t^2$$
$$250 = t^2$$
$$\sqrt{250} = t \qquad \text{Using the principle of square roots; rejecting the negative square root since } t \text{ cannot be negative in this problem}$$
$$15.8 \approx t. \qquad \text{Using a calculator and rounding to the nearest tenth}$$

12. The Willis Tower in Chicago is 1454 ft tall. How long would it take an object to fall freely from the top? Round to the nearest tenth of a second.

4. **Check.** Since $16(15.8)^2 = 3994.24 \approx 4000$, our answer checks.

5. **State.** It takes about 15.8 sec for a stone to fall freely from the Grand Canyon Skywalk to the river.

YOUR TURN

Technology Connection

To check Example 10, we graph $y = 3x^2 + 7x - 2$ and use the ZERO or ROOT option of the CALC menu. We enter a Left Bound, a Right Bound, and a Guess, and a value for the root then appears. Since $-7/6 - \sqrt{73}/6 \approx -2.590667$, the answer checks.

1. Use a graphing calculator to confirm the solutions in Example 9.
2. Use a graphing calculator to confirm that there are no real-number solutions of $x^2 - 6x + 11 = 0$.

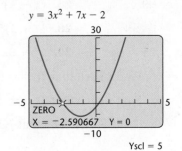

$y = 3x^2 + 7x - 2$

ZERO
X = −2.590667   Y = 0

Yscl = 5

8.1 EXERCISE SET

FOR EXTRA HELP

MyMathLab® Math XL
PRACTICE WATCH READ REVIEW

➲ Vocabulary and Reading Check

Choose the word or phrase from the following list that best completes each statement.

complete the square   square roots
parabola   standard form
quadratic function   zero products

1. The general form of a(n) _____ is $f(x) = ax^2 + bx + c$.

2. The graph of a quadratic function is a(n) _____.

3. The quadratic equation $ax^2 + bx + c = 0$ is written in _____.

4. If $(x - 2)(x + 3) = 0$, we know that $x - 2 = 0$ or $x + 3 = 0$ because of the principle of _____.

5. If $x^2 = 7$, we know that $x = \sqrt{7}$ or $x = -\sqrt{7}$ because of the principle of _____.

6. We add 25 to $x^2 + 10x$ in order to _____.

The Principle of Square Roots

Solve. (Find all complex-number solutions.)

7. $x^2 = 100$

8. $t^2 = 144$

9. $p^2 - 50 = 0$

10. $c^2 - 8 = 0$

11. $5y^2 = 30$

12. $4y^2 = 12$

13. $9x^2 - 49 = 0$

14. $36a^2 - 25 = 0$

15. $6t^2 - 5 = 0$

16. $7x^2 - 5 = 0$

17. $a^2 + 1 = 0$

18. $t^2 + 4 = 0$

19. $4d^2 + 81 = 0$

20. $25y^2 + 16 = 0$

21. $(x - 3)^2 = 16$

22. $(x + 1)^2 = 100$

23. $(t + 5)^2 = 12$

24. $(y - 4)^2 = 18$

25. $(x + 1)^2 = -9$

26. $(x - 1)^2 = -49$

27. $\left(y + \frac{3}{4}\right)^2 = \frac{17}{16}$

28. $\left(t + \frac{3}{2}\right)^2 = \frac{7}{2}$

29. $x^2 - 10x + 25 = 64$

30. $x^2 - 6x + 9 = 100$

31. Let $f(x) = x^2$. Find x such that $f(x) = 19$.

32. Let $f(x) = x^2$. Find x such that $f(x) = 11$.

33. Let $f(x) = (x - 5)^2$. Find x such that $f(x) = 16$.

34. Let $g(x) = (x - 2)^2$. Find x such that $g(x) = 25$.

35. Let $F(t) = (t + 4)^2$. Find t such that $F(t) = 13$.

36. Let $f(t) = (t + 6)^2$. Find t such that $f(t) = 15$.

Aha! 37. Let $g(x) = x^2 + 14x + 49$. Find x such that $g(x) = 49$.

38. Let $F(x) = x^2 + 8x + 16$. Find x such that $F(x) = 9$.

Completing the Square

Replace the blanks in each equation with constants to complete the square and form a true equation.

39. $x^2 + 16x + \underline{\quad} = (x + \underline{\quad})^2$

40. $x^2 + 12x + \underline{\quad} = (x + \underline{\quad})^2$

41. $t^2 - 10t + \underline{\quad} = (t - \underline{\quad})^2$

42. $t^2 - 6t + \underline{\quad} = (t - \underline{\quad})^2$

43. $t^2 - 2t + \underline{\quad} = (t - \underline{\quad})^2$

44. $x^2 + 2x + \underline{\quad} = (x + \underline{\quad})^2$

45. $x^2 + 3x + \underline{\quad} = \left(x + \underline{\quad}\right)^2$

46. $t^2 - 9t + \underline{\quad} = \left(t - \underline{\quad}\right)^2$

47. $x^2 + \frac{2}{5}x + \underline{\quad} = \left(x + \underline{\quad}\right)^2$

48. $x^2 + \frac{2}{3}x + \underline{\quad} = \left(x + \underline{\quad}\right)^2$

49. $t^2 - \frac{5}{6}t + \underline{\quad} = \left(t - \underline{\quad}\right)^2$

50. $t^2 - \frac{5}{3}t - \underline{\quad} = \left(t - \underline{\quad}\right)^2$

Solve by completing the square. Show your work.

51. $x^2 + 6x = 7$

52. $x^2 + 8x = 9$

53. $t^2 - 10t = -23$

54. $t^2 - 4t = -1$

55. $x^2 + 12x + 32 = 0$

56. $x^2 + 16x + 15 = 0$

57. $t^2 + 8t - 3 = 0$

58. $t^2 + 6t - 5 = 0$

Complete the square to find the x-intercepts of each function given by the equation listed.

59. $f(x) = x^2 + 6x + 7$

60. $f(x) = x^2 + 10x - 2$

61. $g(x) = x^2 + 9x - 25$

62. $g(x) = x^2 + 5x + 2$

63. $f(x) = x^2 - 10x - 22$

64. $f(x) = x^2 - 8x - 10$

Solve by completing the square. Remember to first divide, as in Example 10, to make sure that the coefficient of x^2 is 1.

65. $9x^2 + 18x = -8$

66. $4x^2 + 8x = -3$

67. $3x^2 - 5x - 2 = 0$

68. $2x^2 - 5x - 3 = 0$

69. $5x^2 + 4x - 3 = 0$

70. $4x^2 + 3x - 5 = 0$

71. Find the x-intercepts of the graph of
$f(x) = 4x^2 + 2x - 3$.

72. Find the x-intercepts of the graph of
$f(x) = 3x^2 + x - 5$.

73. Find the x-intercepts of the graph of
$g(x) = 2x^2 - 3x - 1$.

74. Find the x-intercepts of the graph of
$g(x) = 3x^2 - 5x - 1$.

Problem Solving

Interest.   Use $A = P(1 + r)^t$ to find the interest rate in Exercises 75–78. Refer to Example 11.

75. $2000 grows to $2205 in 2 years

76. $1000 grows to $1060.90 in 2 years

77. $6250 grows to $6760 in 2 years

78. $6250 grows to $7290 in 2 years

Free-Falling Objects.   Use $s = 16t^2$ for Exercises 79–82. Refer to Example 12 and neglect air resistance. Round answers to the nearest tenth of a second.

79. At 121 ft, Excalibur Tower in Groningen, the Netherlands, is the world's tallest man-made climbing wall. How long would it take an object to fall freely from the top?

Source: http://enpundit.com/2012/excalibur-worlds-highest-climbing-wall

80. At a height of approximately 1200 ft, Tushuk Tash in Xinjiang Uyghur Autonomous Region, China, is the world's highest natural arch. How long would it take an object to fall freely from the top of the arch?

Source: www.naturalarches.org

81. El Capitan in Yosemite National Park is 3593 ft high. How long would it take a carabiner to fall freely from the top?

Source: Guinness World Records 2008

82. At 2063 ft, the KVLY-TV tower in North Dakota is the world's tallest supported tower. How long would it take an object to fall freely from the top?

Source: North Dakota Tourism Division

83. Explain in your own words a sequence of steps that can be used to solve any quadratic equation in the quickest way.

84. Write an interest-rate problem for a classmate to solve. Devise the problem so that the solution is "The loan was made at 7% interest."

Skill Review

Factor completely.

85. $3y^3 - 300y$ [5.5]

86. $12t + 36 + t^2$ [5.5]

87. $6x^2 + 6x + 6$ [5.3]

88. $10a^5 - 10a^4 - 60a^3$
[5.4]

89. $20x^2 + 7x - 6$ [5.4]

90. $n^6 - 1$ [5.6]

Synthesis

91. What would be better: to receive 3% interest every 6 months, or to receive 6% interest every 12 months? Why?

92. Write a problem involving a free-falling object for a classmate to solve (see Example 12). Devise the problem so that the solution is "The object takes about 4.5 sec to fall freely from the top of the structure."

Find b such that each trinomial is a square.

93. $x^2 + bx + 81$

94. $x^2 + bx + 49$

95. If $f(x) = 2x^5 - 9x^4 - 66x^3 + 45x^2 + 280x$ and $x^2 - 5$ is a factor of $f(x)$, find all five values of a for which $f(a) = 0$.

96. If $f(x) = \left(x - \frac{1}{3}\right)(x^2 + 6)$ and $g(x) = \left(x - \frac{1}{3}\right)\left(x^2 - \frac{2}{3}\right)$, find all a for which $(f + g)(a) = 0$.

97. *Boating.*   A barge and a fishing boat leave a dock at the same time, traveling at a right angle to each other. The barge travels 7 km/h slower than the fishing boat. After 4 hr, the boats are 68 km apart. Find the speed of each boat.

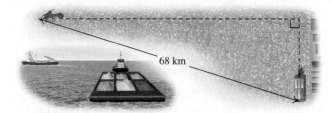

98. Find three consecutive integers such that the square of the first plus the product of the other two is 67.

99. Exercises 29, 33, and 53 can be solved on a graphing calculator without first rewriting in standard form. Simply let y_1 represent the left side of the equation and y_2 the right side. Then use a graphing calculator to determine the x-coordinate of any point of intersection. Use a graphing calculator to solve Exercises 29, 33, and 53 in this manner.

100. Use a graphing calculator to check your answers to Exercises 7, 13, 71, and 73.

101. Example 11 can be solved with a graphing calculator by graphing each side of

$$4410 = 4000(1 + r)^2.$$

How could you determine, from a reading of the problem, a suitable viewing window? What might that window be?

↩ YOUR TURN ANSWERS: SECTION 8.1

1. $8, -8$ **2.** $\sqrt{10}, -\sqrt{10}$, or $3.162, -3.162$

3. $\sqrt{\dfrac{1}{3}}, -\sqrt{\dfrac{1}{3}}$, or $\dfrac{\sqrt{3}}{3}, -\dfrac{\sqrt{3}}{3}$ **4.** $10i, -10i$

5. $-5 + \sqrt{3}, -5 - \sqrt{3}$ **6.** $5 + \sqrt{3}, 5 - \sqrt{3}$

7. $x^2 + 7x + \dfrac{49}{4} = \left(x + \dfrac{7}{2}\right)^2$ **8.** $-3 + \sqrt{11}, -3 - \sqrt{11}$

9. $(-2 - \sqrt{3}, 0), (-2 + \sqrt{3}, 0)$ **10.** $2 + \dfrac{\sqrt{22}}{2}, 2 - \dfrac{\sqrt{22}}{2}$,

or $\dfrac{4 + \sqrt{22}}{2}, \dfrac{4 - \sqrt{22}}{2}$ **11.** 10% **12.** About 9.5 sec

---

PREPARE TO MOVE ON

Evaluate. [1.2]

1. $b^2 - 4ac$, for $a = 3$, $b = 2$, and $c = -5$

2. $b^2 - 4ac$, for $a = 1$, $b = -1$, and $c = 4$

Simplify. [7.3], [7.8]

3. $\sqrt{200}$ **4.** $\sqrt{-4}$

5. $\sqrt{-8}$

8.2   The Quadratic Formula

Solving Using the Quadratic Formula ▪ Approximating Solutions

We can use the process of completing the square to develop a general formula for solving quadratic equations.

Solving Using the Quadratic Formula

Each time we solve by completing the square, the procedure is the same. Here we develop a formula that condenses this work.

We begin with a quadratic equation in standard form,

$$ax^2 + bx + c = 0,$$

with $a > 0$. For $a < 0$, a slightly different derivation is needed (see Exercise 60), but the result is the same. Let's solve by completing the square. As the steps are performed, compare them with Example 10 in Section 8.1.

$$ax^2 + bx = -c \qquad \text{Adding } -c \text{ to both sides}$$

$$x^2 + \frac{b}{a}x = -\frac{c}{a} \qquad \text{Dividing both sides by } a$$

Half of $\dfrac{b}{a}$ is $\dfrac{b}{2a}$ and $\left(\dfrac{b}{2a}\right)^2$ is $\dfrac{b^2}{4a^2}$. We add $\dfrac{b^2}{4a^2}$ to both sides.

$$x^2 + \frac{b}{a}x + \frac{b^2}{4a^2} = -\frac{c}{a} + \frac{b^2}{4a^2}$$

Adding $\dfrac{b^2}{4a^2}$ to complete the square

$$\left(x + \frac{b}{2a}\right)^2 = -\frac{4ac}{4a^2} + \frac{b^2}{4a^2}$$

Factoring on the left side; finding a common denominator on the right side

$$\left(x + \frac{b}{2a}\right)^2 = \frac{b^2 - 4ac}{4a^2}$$

$$x + \frac{b}{2a} = \pm\frac{\sqrt{b^2 - 4ac}}{2a}$$

Using the principle of square roots and the quotient rule for radicals. Since $a > 0$, $\sqrt{4a^2} = 2a$.

$$x = \frac{-b \pm \sqrt{b^2 - 4ac}}{2a}$$

Adding $-\dfrac{b}{2a}$ to both sides

It is important to remember the quadratic formula and know how to use it.

THE QUADRATIC FORMULA

The solutions of $ax^2 + bx + c = 0$, $a \neq 0$, are given by

$$x = \frac{-b \pm \sqrt{b^2 - 4ac}}{2a}.$$

EXAMPLE 1 Solve $5x^2 + 8x = -3$ using the quadratic formula.

SOLUTION We first find standard form and determine a, b, and c:

$$5x^2 + 8x + 3 = 0;$$ Adding 3 to both sides to get 0 on one side
$$a = 5, \quad b = 8, \quad c = 3.$$

Next, we use the quadratic formula:

$$x = \frac{-b \pm \sqrt{b^2 - 4ac}}{2a}$$ It is important to remember this formula.

$$x = \frac{-8 \pm \sqrt{8^2 - 4 \cdot 5 \cdot 3}}{2 \cdot 5}$$ Substituting

$$x = \frac{-8 \pm \sqrt{64 - 60}}{10}$$

Be sure to write the fraction bar all the way across.

$$x = \frac{-8 \pm \sqrt{4}}{10}$$

$$= \frac{-8 \pm 2}{10}$$

$$x = \frac{-8 + 2}{10} \quad or \quad x = \frac{-8 - 2}{10}$$ The symbol $\pm$ indicates two solutions.

$$x = \frac{-6}{10} \quad or \quad x = \frac{-10}{10}$$

$$x = -\frac{3}{5} \quad or \quad x = -1.$$

1. Solve $12x^2 - 8x - 15$ using the quadratic formula.

The solutions are $-\frac{3}{5}$ and -1. The checks are left to the student.

YOUR TURN

Because $5x^2 + 8x + 3$ can be factored, the quadratic formula may not have been the fastest way of solving Example 1. However, because the quadratic formula works for *any* quadratic equation, we need not spend too much time struggling to solve a quadratic equation by factoring.

TO SOLVE A QUADRATIC EQUATION

1. If the equation can be easily written in the form $ax^2 = p$ or $(x + k)^2 = d$, use the principle of square roots.
2. If step (1) does not apply, write the equation in the form $ax^2 + bx + c = 0$.
3. Try factoring and using the principle of zero products.
4. If factoring seems difficult or impossible, use the quadratic formula. Completing the square can also be used.

The solutions of a quadratic equation can always be found using the quadratic formula. They cannot always be found by factoring.

A second-degree polynomial in one variable is said to be quadratic. Similarly, a second-degree polynomial function in one variable is said to be a **quadratic function**.

EXAMPLE 2   For the quadratic function given by $f(x) = 3x^2 - 6x - 4$, find all x for which $f(x) = 0$.

SOLUTION   We substitute and solve for x:

$$f(x) = 0$$
$$3x^2 - 6x - 4 = 0. \qquad \text{Substituting}$$

Since $3x^2 - 6x - 4$ does not factor, we use the quadratic formula with $a = 3$, $b = -6$, and $c = -4$:

$$x = \frac{-(-6) \pm \sqrt{(-6)^2 - 4 \cdot 3 \cdot (-4)}}{2 \cdot 3}$$

$$= \frac{6 \pm \sqrt{36 + 48}}{6} \qquad (-6)^2 - 4 \cdot 3 \cdot (-4) = 36 - (-48) = 36 + 48$$

$$= \frac{6 \pm \sqrt{84}}{6} \qquad \text{Note that 4 is a perfect-square factor of 84.}$$

$$= \frac{6}{6} \pm \frac{\sqrt{84}}{6} \qquad \text{Writing as two fractions to simplify each separately}$$

$$= 1 \pm \frac{\sqrt{4}\sqrt{21}}{6} \qquad 84 = 4 \cdot 21$$

$$\left. \begin{array}{l} = 1 \pm \dfrac{2\sqrt{21}}{2 \cdot 3} \\[2mm] = 1 \pm \dfrac{\sqrt{21}}{3}. \end{array} \right\} \qquad \text{Removing a factor of 1: } \dfrac{2}{2} = 1$$

The solutions are $1 - \dfrac{\sqrt{21}}{3}$ and $1 + \dfrac{\sqrt{21}}{3}$. The checks are left to the student.

 YOUR TURN

When we use the quadratic formula to solve equations, we will be able to find imaginary-number solutions.

---

2. For the quadratic function given by

$$f(x) = 2x^2 - 2x - 3,$$

find all x for which $f(x) = 0$.

EXAMPLE 3   Solve: $x(x + 5) = 2(2x - 1)$. (Find all complex-number solutions.)

SOLUTION   We first find standard form:

$$x^2 + 5x = 4x - 2 \quad \text{Multiplying}$$
$$x^2 + x + 2 = 0. \quad \text{Subtracting } 4x \text{ and adding 2 to both sides}$$

Since we cannot factor $x^2 + x + 2$, we use the quadratic formula with $a = 1$, $b = 1$, and $c = 2$:

$$x = \frac{-1 \pm \sqrt{1^2 - 4 \cdot 1 \cdot 2}}{2 \cdot 1} \quad \text{Substituting}$$

$$= \frac{-1 \pm \sqrt{1 - 8}}{2}$$

$$= \frac{-1 \pm \sqrt{-7}}{2}$$

$$= \frac{-1 \pm i\sqrt{7}}{2}, \text{ or } -\frac{1}{2} \pm \frac{\sqrt{7}}{2}i.$$

The solutions are $-\frac{1}{2} - \frac{\sqrt{7}}{2}i$ and $-\frac{1}{2} + \frac{\sqrt{7}}{2}i$. The checks are left to the student.

A visualization of Example 3

3. Solve: $x(x + 2) = -5$.

YOUR TURN

EXAMPLE 4   If $f(t) = 2 + \dfrac{7}{t}$ and $g(t) = \dfrac{4}{t^2}$, find all t for which $f(t) = g(t)$.

SOLUTION   We set $f(t)$ equal to $g(t)$ and solve:

$$f(t) = g(t)$$
$$2 + \frac{7}{t} = \frac{4}{t^2}. \quad \text{Substituting. Note that } t \neq 0.$$

This is a rational equation. To solve, we multiply both sides by the LCM of the denominators, t^2:

$$t^2\left(2 + \frac{7}{t}\right) = t^2 \cdot \frac{4}{t^2}$$

$$2t^2 + 7t = 4 \quad \text{Simplifying}$$
$$2t^2 + 7t - 4 = 0. \quad \text{Subtracting 4 from both sides}$$

We use the quadratic formula with $a = 2$, $b = 7$, and $c = -4$:

$$t = \frac{-7 \pm \sqrt{7^2 - 4 \cdot 2 \cdot (-4)}}{2 \cdot 2}$$

$$= \frac{-7 \pm \sqrt{49 + 32}}{4} \quad \begin{array}{l} 7^2 - 4 \cdot 2 \cdot (-4) = 49 - (-32) \\ = 49 + 32 \end{array}$$

$$= \frac{-7 \pm \sqrt{81}}{4}$$

$$= \frac{-7 \pm 9}{4}$$

$$t = \frac{-7 + 9}{4} \quad or \quad t = \frac{-7 - 9}{4}$$

$$t = \frac{2}{4} = \frac{1}{2} \quad or \quad t = \frac{-16}{4} = -4. \quad \begin{array}{l} \text{Both answers should check since} \\ t \neq 0. \end{array}$$

You can confirm that $f\left(\frac{1}{2}\right) = g\left(\frac{1}{2}\right)$ and $f(-4) = g(-4)$. The solutions are $\frac{1}{2}$ and -4.

YOUR TURN

Technology Connection

To determine whether quadratic equations are solved more quickly on a graphing calculator or by using the quadratic formula, solve Examples 2 and 4 both ways. Which method is faster? Which method is more precise? Why?

4. If $f(x) = 2 + \dfrac{1}{x}$ and $g(x) = \dfrac{3}{x^2}$, find all x for which $f(x) = g(x)$.

Approximating Solutions

When the solution of an equation is irrational, a rational-number approximation is often useful in real-world applications.

EXAMPLE 5 Use a calculator to approximate, to three decimal places, the solutions of Example 2.

SOLUTION On most calculators, one of the following sequences of keystrokes can be used to approximate $1 + \sqrt{21}/3$:

Similar keystrokes can be used to approximate $1 - \sqrt{21}/3$.

The solutions are approximately 2.527525232 and −0.5275252317. Rounded to three decimal places, the solutions are approximately 2.528 and −0.528.

YOUR TURN

Student Notes

It is important that you understand both the rules for order of operations *and* the manner in which your calculator applies those rules.

5. Use a calculator to approximate, to three decimal places, the solutions of Your Turn Exercise 2.

CONNECTING 🔗 THE CONCEPTS

We have studied four different ways of solving quadratic equations. Each method has advantages and disadvantages, as outlined below. Note that although the quadratic formula can be used to solve *any* quadratic equation, sometimes other methods are faster and easier to use.

Method	Advantages	Disadvantages
Factoring	Can be very fast.	Can be used only on certain equations. Many equations are difficult or impossible to solve by factoring.
The principle of square roots	Fastest way to solve equations of the form $X^2 = k$. Can be used to solve *any* quadratic equation.	Can be slow when original equation is not written in the form $X^2 = k$.
Completing the square	Works well on equations of the form $x^2 + bx = -c$, when b is even. Can be used to solve *any* quadratic equation.	Can be complicated when $a \neq 1$ or when b in $x^2 + bx = -c$ is not even.
The quadratic formula	Can be used to solve *any* quadratic equation.	Can be slower than factoring or using the principle of square roots for certain equations.

EXERCISES

Solve. Examine each exercise carefully, and solve using the easiest method.

1. $x^2 - 3x - 10 = 0$

2. $x^2 = 121$

3. $x^2 + 6x = 10$

4. $x^2 + x - 3 = 0$

5. $(x + 1)^2 = 2$

6. $x^2 - 10x + 25 = 0$

7. $x^2 - 2x = 6$

8. $4t^2 = 11$

8.2 EXERCISE SET

FOR EXTRA HELP

MyMathLab® Math XL
PRACTICE   WATCH READ REVIEW

Vocabulary and Reading Check

Classify each of the following statements as either true or false.

1. The quadratic formula can be used to solve *any* quadratic equation.

2. The steps used to derive the quadratic formula are the same as those used when solving by completing the square.

3. The quadratic formula does not work if solutions are imaginary numbers.

4. Solving by factoring is always slower than using the quadratic formula.

5. A quadratic equation can have as many as four solutions.

6. It is possible for a quadratic equation to have no real-number solutions.

Solving Using the Quadratic Formula

Solve. (Find all complex-number solutions.)

7. $2x^2 + 3x - 5 = 0$

8. $3x^2 - 7x + 2 = 0$

9. $u^2 + 2u - 4 = 0$

10. $u^2 - 2u - 2 = 0$

11. $t^2 + 3 = 6t$

12. $t^2 + 4t = 1$

13. $x^2 = 3x + 5$

14. $x^2 + 5x + 3 = 0$

15. $3t(t + 2) = 1$

16. $2t(t + 2) = 1$

17. $\dfrac{1}{x^2} - 3 = \dfrac{8}{x}$

18. $\dfrac{9}{x} - 2 = \dfrac{5}{x^2}$

19. $t^2 + 10 = 6t$

20. $t^2 + 10t + 26 = 0$

21. $p^2 - p + 1 = 0$

22. $p^2 + p + 4 = 0$

23. $x^2 + 4x + 6 = 0$

24. $x^2 + 11 = 6x$

25. $12t^2 + 17t = 40$

26. $15t^2 + 7t = 2$

27. $25x^2 - 20x + 4 = 0$

28. $36x^2 + 84x + 49 = 0$

29. $7x(x + 2) + 5 = 3x(x + 1)$

30. $5x(x - 1) - 7 = 4x(x - 2)$

31. $14(x - 4) - (x + 2) = (x + 2)(x - 4)$

32. $11(x - 2) + (x - 5) = (x + 2)(x - 6)$

33. $51p = 2p^2 + 72$

34. $72 = 3p^2 + 50p$

35. $x(x - 3) = x - 9$

36. $x(x - 1) = 2x - 7$

37. $x^3 - 8 = 0$ (*Hint:* Factor the difference of cubes. Then use the quadratic formula.)

38. $x^3 + 1 = 0$

39. Let $f(x) = 6x^2 - 7x - 20$. Find x such that $f(x) = 0$.

40. Let $g(x) = 4x^2 - 2x - 3$. Find x such that $g(x) = 0$.

41. Let
$$f(x) = \frac{7}{x} + \frac{7}{x + 4}.$$
Find all x for which $f(x) = 1$.

42. Let
$$g(x) = \frac{2}{x} + \frac{2}{x + 3}.$$
Find all x for which $g(x) = 1$.

43. Let
$$F(x) = \frac{3 - x}{4} \quad \text{and} \quad G(x) = \frac{1}{4x}.$$
Find all x for which $F(x) = G(x)$.

44. Let
$$f(x) = x + 5 \quad \text{and} \quad g(x) = \frac{3}{x - 5}.$$
Find all x for which $f(x) = g(x)$.

Approximating Solutions

Solve using the quadratic formula. Then use a calculator to approximate, to three decimal places, the solutions as rational numbers.

45. $x^2 + 6x + 4 = 0$

46. $x^2 + 4x - 7 = 0$

Aha! 47. $x^2 - 6x + 4 = 0$

48. $x^2 - 4x + 1 = 0$

49. $2x^2 - 3x - 7 = 0$

50. $3x^2 - 3x - 2 = 0$

51. Are there any equations that can be solved by the quadratic formula but not by completing the square? Why or why not?

52. Suppose you are solving a quadratic equation with no constant term ($c = 0$). Would you use factoring or the quadratic formula to solve? Why?

Skill Review

Simplify.

53. $(-3x^2 y^6)^0$ [1.6]

54. $100^{3/2}$ [7.2]

55. $x^{1/4} \cdot x^{2/3}$ [7.2]

56. $(27^{-2})^{1/3}$ [7.2]

57. $\dfrac{18a^5 bc^{10}}{24a^{-5}bc^3}$ [1.6]

58. $\left(\dfrac{2xw^{-3}}{3x^{-4}w}\right)^{-2}$ [1.6]

Synthesis

 59. Explain how you could use the quadratic formula to help factor a quadratic polynomial.

 60. If $a < 0$ and $ax^2 + bx + c = 0$, then $-a$ is positive and the equivalent equation, $-ax^2 - bx - c = 0$, can be solved using the quadratic formula.

   a) Find this solution, replacing a, b, and c in the formula with $-a$, $-b$, and $-c$ from the equation.

   b) Why does the result of part (a) indicate that the quadratic formula "works" regardless of the sign of a?

For Exercises 61–63, let

$$f(x) = \frac{x^2}{x-2} + 1 \quad and \quad g(x) = \frac{4x-2}{x-2} + \frac{x+4}{2}.$$

61. Find the x-intercepts of the graph of f.

62. Find the x-intercepts of the graph of g.

63. Find all x for which $f(x) = g(x)$.

Solve. Approximate the solutions to three decimal places.

64. $x^2 - 0.75x - 0.5 = 0$

65. $z^2 + 0.84z - 0.4 = 0$

Solve.

66. $(1 + \sqrt{3})x^2 - (3 + 2\sqrt{3})x + 3 = 0$

67. $\sqrt{2}x^2 + 5x + \sqrt{2} = 0$

68. $ix^2 - 2x + 1 = 0$

69. One solution of $kx^2 + 3x - k = 0$ is -2. Find the other.

 70. Use a graphing calculator to solve Exercises 9, 27, and 43.

 71. Use a graphing calculator to solve Exercises 11, 33, and 41. Use the method of graphing each side of the equation.

 72. Can a graphing calculator be used to solve *any* quadratic equation? Why or why not?

↪ YOUR TURN ANSWERS: SECTION 8.2

1. $-\dfrac{5}{6}, \dfrac{3}{2}$ **2.** $\dfrac{1}{2} + \dfrac{\sqrt{7}}{2}, \dfrac{1}{2} - \dfrac{\sqrt{7}}{2}$, or $\dfrac{1}{2} \pm \dfrac{\sqrt{7}}{2}$

3. $-1 + 2i, -1 - 2i$, or $-1 \pm 2i$ **4.** $-\dfrac{3}{2}, 1$

5. $1.823, -0.823$

QUICK QUIZ: SECTIONS 8.1–8.2

1. Solve using the principle of zero products:

$$(z + 3)(2z - 5) = 0. \quad [8.1]$$

2. Solve using the principle of square roots:

$$(x - 2)^2 = 3. \quad [8.1]$$

3. Solve by completing the square:

$$t^2 - 6t + 4 = 0. \quad [8.1]$$

4. Solve using the quadratic formula:

$$x^2 - 3x - 1 = 0. \quad [8.2]$$

5. Solve using any appropriate method:

$$2x^2 + x + 5 = 0. \quad [8.2]$$

PREPARE TO MOVE ON

Multiply and simplify.

1. $(x - 2i)(x + 2i)$ [7.8]

2. $(x - 6\sqrt{5})(x + 6\sqrt{5})$ [7.5]

3. $(x - (2 - \sqrt{7}))(x - (2 + \sqrt{7}))$ [7.5]

4. $(x - (-3 + 5i))(x - (-3 - 5i))$ [7.8]

The Discriminant   •   Writing Equations from Solutions

The Discriminant

It is sometimes enough to know what *type* of number a solution will be, without actually solving the equation. Suppose we want to know if $2x^2 + 7x - 15 = 0$ has rational solutions (and thus can be solved by factoring). Using the quadratic formula, we would have

$$x = \frac{-b \pm \sqrt{b^2 - 4ac}}{2a}$$

$$= \frac{-(7) \pm \sqrt{(7)^2 - 4 \cdot 2(-15)}}{2 \cdot 2} = \frac{-7 \pm \sqrt{169}}{4}.$$

Since 169 is a perfect square ($\sqrt{169} = 13$), we know that the solutions of the equation are rational numbers. This means that $2x^2 + 7x - 15 = 0$ *can* be solved by factoring. Note that the radicand, 169, determines what type of number the solutions will be.

The radicand $b^2 - 4ac$ is known as the **discriminant**. If a, b, and c are rational, then we can make the following observations.

Study Skills

Sharpen Your Skills

Every so often, you may encounter a lesson that you remember from a previous math course. When this occurs, make sure that you understand *all* of that lesson. Take time to review any new concepts and to sharpen any old skills.

Discriminant	Observation	Example
$b^2 - 4ac = 0$	We get the same solution twice. There is one *repeated* solution and it is rational.	$9x^2 + 6x + 1 = 0$ $b^2 - 4ac = 6^2 - 4 \cdot 9 \cdot 1 = 0$ Solving, we have $x = \dfrac{-6 \pm \sqrt{0}}{2 \cdot 9}$. The (repeated) solution is $-\frac{1}{3}$.
$b^2 - 4ac$ is positive. 1. $b^2 - 4ac$ *is* a perfect square. 2. $b^2 - 4ac$ is *not* a perfect square.	There are two different real-number solutions. 1. The solutions are rational numbers. 2. The solutions are irrational conjugates.	1. $6x^2 + 5x + 1 = 0$ $b^2 - 4ac = 5^2 - 4 \cdot 6 \cdot 1 = 1$ Solving, we have $x = \dfrac{-5 \pm \sqrt{1}}{2 \cdot 6}$. The solutions are $-\frac{1}{3}$ and $-\frac{1}{2}$. 2. $x^2 + 4x + 2 = 0$ $b^2 - 4ac = 4^2 - 4 \cdot 1 \cdot 2 = 8$ Solving, we have $x = \dfrac{-4 \pm \sqrt{8}}{2 \cdot 1}$. The solutions are $-2 + \sqrt{2}$ and $-2 - \sqrt{2}$.
$b^2 - 4ac$ is negative.	There are two different imaginary-number solutions. They are complex conjugates.	$x^2 + 4x + 5 = 0$ $b^2 - 4ac = 4^2 - 4 \cdot 1 \cdot 5 = -4$ Solving, we have $x = \dfrac{-4 \pm \sqrt{-4}}{2 \cdot 1}$. The solutions are $-2 + i$ and $-2 - i$.

Note that all quadratic equations have either one or two solutions. These solutions can always be found algebraically; only real-number solutions can be found graphically. Also, note that any equation for which $b^2 - 4ac$ is a perfect square can be solved by factoring.

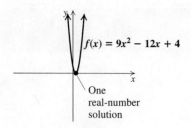

A visualization of part (a)

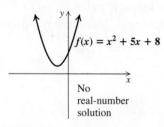

A visualization of part (b)

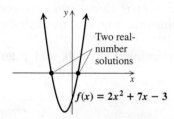

A visualization of part (c)

1. Determine what type of number the solutions are and how many solutions exist:

$$4x^2 - 9x + 2 = 0.$$

EXAMPLE 1   For each equation, determine what type of number the solutions are and how many solutions exist.

a) $9x^2 - 12x + 4 = 0$ **b)** $x^2 + 5x + 8 = 0$ **c)** $2x^2 + 7x - 3 = 0$

SOLUTION

a)   For $9x^2 - 12x + 4 = 0$, we have

$$a = 9, \quad b = -12, \quad c = 4.$$

We substitute and compute the discriminant:

$$b^2 - 4ac = (-12)^2 - 4 \cdot 9 \cdot 4$$
$$= 144 - 144 = 0.$$

There is exactly one solution (it is repeated), and it is rational. Thus,

$$9x^2 - 12x + 4 = 0$$

can be solved by factoring.

b)   For $x^2 + 5x + 8 = 0$, we have

$$a = 1, \quad b = 5, \quad c = 8.$$

We substitute and compute the discriminant:

$$b^2 - 4ac = 5^2 - 4 \cdot 1 \cdot 8$$
$$= 25 - 32 = -7.$$

Since the discriminant is negative, there are two different imaginary-number solutions that are complex conjugates of each other.

c)   For $2x^2 + 7x - 3 = 0$, we have

$$a = 2, \quad b = 7, \quad c = -3.$$

We substitute and compute the discriminant:

$$b^2 - 4ac = 7^2 - 4 \cdot 2(-3)$$
$$= 49 - (-24) = 73.$$

The discriminant is a positive number that is not a perfect square. Thus there are two different irrational solutions that are conjugates of each other.

YOUR TURN

Discriminants can also be used to determine the number of x-intercepts of the graph of a quadratic function.

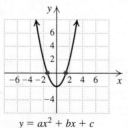

$y = ax^2 + bx + c$
$b^2 - 4ac > 0$
Two real solutions
of $ax^2 + bx + c = 0$
Two x-intercepts

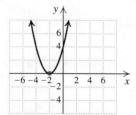

$y = ax^2 + bx + c$
$b^2 - 4ac = 0$
One real solution
of $ax^2 + bx + c = 0$
One x-intercept

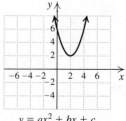

$y = ax^2 + bx + c$
$b^2 - 4ac < 0$
No real solutions
of $ax^2 + bx + c = 0$
No x-intercept

Writing Equations from Solutions

We know by the principle of zero products that $(x - 2)(x + 3) = 0$ has solutions 2 and -3. If we wish for two given numbers to be solutions of an equation, we can create such an equation, using the principle in reverse.

EXAMPLE 2   Find an equation for which the given numbers are solutions.

a) 3 and $-\frac{2}{5}$ **b)** $2i$ and $-2i$

c) $5\sqrt{7}$ and $-5\sqrt{7}$ **d)** $-4, 0,$ and 1

SOLUTION

a)
$$x = 3 \quad or \quad x = -\tfrac{2}{5}$$
$$x - 3 = 0 \quad or \quad x + \tfrac{2}{5} = 0 \qquad \text{Getting 0's on one side}$$
$$(x - 3)\left(x + \tfrac{2}{5}\right) = 0 \qquad \text{Using the principle of zero}$$
$$\text{products (multiplying)}$$
$$x^2 + \tfrac{2}{5}x - 3x - 3 \cdot \tfrac{2}{5} = 0 \qquad \text{Multiplying}$$
$$x^2 - \tfrac{13}{5}x - \tfrac{6}{5} = 0 \qquad \text{Combining like terms}$$
$$5x^2 - 13x - 6 = 0 \qquad \text{Multiplying both sides by 5 to clear fractions}$$

Note that multiplying both sides by 5 clears the equation of fractions. Had we preferred, we could have multiplied $x + \frac{2}{5} = 0$ by 5, thus clearing fractions *before* using the principle of zero products.

b)
$$x = 2i \quad or \quad x = -2i$$
$$x - 2i = 0 \quad or \quad x + 2i = 0 \qquad \text{Getting 0's on one side}$$
$$(x - 2i)(x + 2i) = 0 \qquad \text{Using the principle of zero}$$
$$\text{products (multiplying)}$$
$$x^2 - (2i)^2 = 0 \qquad \text{Finding the product of a sum}$$
$$\text{and a difference}$$
$$x^2 - 4i^2 = 0$$
$$x^2 + 4 = 0 \qquad i^2 = -1$$

c)
$$x = 5\sqrt{7} \quad or \quad x = -5\sqrt{7}$$
$$x - 5\sqrt{7} = 0 \quad or \quad x + 5\sqrt{7} = 0 \qquad \text{Getting 0's on one side}$$
$$(x - 5\sqrt{7})(x + 5\sqrt{7}) = 0 \qquad \text{Using the principle of zero products}$$
$$x^2 - (5\sqrt{7})^2 = 0 \qquad \text{Finding the product of a sum and a difference}$$
$$x^2 - 25 \cdot 7 = 0$$
$$x^2 - 175 = 0$$

d)
$$x = -4 \quad or \quad x = 0 \quad or \quad x = 1$$
$$x + 4 = 0 \quad or \quad x = 0 \quad or \quad x - 1 = 0 \qquad \text{Getting 0's on one side}$$
$$(x + 4)x(x - 1) = 0 \qquad \text{Using the principle of zero}$$
$$\text{products}$$
$$x(x^2 + 3x - 4) = 0 \qquad \text{Multiplying}$$
$$x^3 + 3x^2 - 4x = 0$$

2. Find an equation for which 0 and $\frac{1}{5}$ are solutions.
_____

YOUR TURN

To check any of these equations, we can simply substitute one or more of the given solutions. For example, in Example 2(d) above,
$$(-4)^3 + 3(-4)^2 - 4(-4) = -64 + 3 \cdot 16 + 16$$
$$= -64 + 48 + 16 = 0.$$

8.3 EXERCISE SET

FOR EXTRA HELP   MyMathLab® Math XL
PRACTICE   WATCH   READ   REVIEW

Vocabulary and Reading Check

Match the description of the solution(s) with each discriminant. Answers may be used more than once.

1. ____ $b^2 - 4ac = 9$ **a)** One rational solution

2. ____ $b^2 - 4ac = 0$ **b)** Two different rational solutions

3. ____ $b^2 - 4ac = -1$

4. ____ $b^2 - 4ac = 1$ **c)** Two different irrational solutions

5. ____ $b^2 - 4ac = 8$ **d)** Two different imaginary-number solutions

6. ____ $b^2 - 4ac = 12$

The Discriminant

For each equation, determine what type of number the solutions are and how many solutions exist.

7. $x^2 - 7x + 5 = 0$ 8. $x^2 - 5x + 3 = 0$

9. $x^2 + 11 = 0$ 10. $x^2 + 7 = 0$

11. $x^2 - 11 = 0$ 12. $x^2 - 7 = 0$

13. $4x^2 + 8x - 5 = 0$ 14. $4x^2 - 12x + 9 = 0$

15. $x^2 + 4x + 6 = 0$ 16. $x^2 - 2x + 4 = 0$

17. $9t^2 - 48t + 64 = 0$ 18. $10t^2 - t - 2 = 0$

Aha! 19. $9t^2 + 3t = 0$ 20. $4m^2 + 7m = 0$

21. $x^2 + 4x = 8$ 22. $x^2 + 5x = 9$

23. $2a^2 - 3a = -5$ 24. $3a^2 + 5 = -7a$

25. $7x^2 = 19x$ 26. $5x^2 = 48x$

27. $y^2 + \frac{9}{4} = 4y$ 28. $x^2 = \frac{1}{2}x - \frac{3}{5}$

Writing Equations from Solutions

Write a quadratic equation having the given numbers as solutions.

29. $-5, 4$ 30. $-2, 8$

31. 3, only solution (*Hint:* It must be a repeated solution.)

32. -5, only solution

33. $-1, -3$ 34. $-2, -5$

35. $5, \frac{3}{4}$ 36. $4, \frac{2}{3}$

37. $-\frac{1}{4}, -\frac{1}{2}$ 38. $\frac{1}{2}, \frac{1}{3}$

39. $2.4, -0.4$ 40. $-0.6, 1.4$

41. $-\sqrt{3}, \sqrt{3}$ 42. $-\sqrt{7}, \sqrt{7}$

43. $2\sqrt{5}, -2\sqrt{5}$ 44. $3\sqrt{2}, -3\sqrt{2}$

45. $4i, -4i$ 46. $3i, -3i$

47. $2 - 7i, 2 + 7i$ 48. $5 - 2i, 5 + 2i$

49. $3 - \sqrt{14}, 3 + \sqrt{14}$ 50. $2 - \sqrt{10}, 2 + \sqrt{10}$

51. $1 - \dfrac{\sqrt{21}}{3}, 1 + \dfrac{\sqrt{21}}{3}$ 52. $\dfrac{5}{4} - \dfrac{\sqrt{33}}{4}, \dfrac{5}{4} + \dfrac{\sqrt{33}}{4}$

Write a third-degree equation having the given numbers as solutions.

53. $-2, 1, 5$ 54. $-5, 0, 2$

55. $-1, 0, 3$ 56. $-2, 2, 3$

57. Explain why there are not two different solutions when the discriminant is 0.

58. Describe a procedure that could be used to write an equation having the first 7 natural numbers as solutions.

Skill Review

Simplify.

59. $\sqrt{270a^7b^{12}}$ [7.3]   60. $\sqrt[4]{8w^3}\,\sqrt[4]{4w^7}$ [7.3]

61. $\sqrt[3]{x}\,\sqrt{x}$ [7.5]   62. $\sqrt{-3}\,\sqrt{-2}$ [7.8]

63. $(2 - i)(3 + i)$ [7.8]   64. i^{18} [7.8]

Synthesis

65. If we assume that a quadratic equation has integers for coefficients, will the product of the solutions always be a real number? Why or why not?

66. Can a fourth-degree equation with rational coefficients have exactly three irrational solutions? Why or why not?

67. The graph of an equation of the form
$$y = ax^2 + bx + c$$
is shown below. Determine a, b, and c from the identified points.

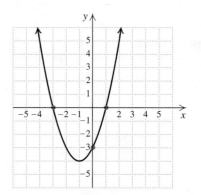

68. Show that the product of the solutions of $ax^2 + bx + c = 0$ is c/a.

For each equation under the given condition, **(a)** *find k and* **(b)** *find the other solution.*

69. $kx^2 - 2x + k = 0$; one solution is -3

70. $x^2 - kx + 2 = 0$; one solution is $1 + i$

71. $x^2 - (6 + 3i)x + k = 0$; one solution is 3

72. Show that the sum of the solutions of $ax^2 + bx + c = 0$ is $-b/a$.

73. Show that whenever there is just one solution of $ax^2 + bx + c = 0$, that solution is of the form $-b/(2a)$.

74. Find h and k such that $3x^2 - hx + 4k = 0$, the sum of the solutions is -12, and the product of the solutions is 20. (*Hint*: See Exercises 68 and 72.)

75. Suppose that $f(x) = ax^2 + bx + c$, with $f(-3) = 0$, $f\left(\frac{1}{2}\right) = 0$, and $f(0) = -12$. Find a, b, and c.

76. Find an equation for which $2 - \sqrt{3}$, $2 + \sqrt{3}$, $5 - 2i$, and $5 + 2i$ are solutions.

Aha! **77.** Write a quadratic equation with integer coefficients for which $-\sqrt{2}$ is one solution.

78. Write a quadratic equation with integer coefficients for which $10i$ is one solution.

79. Find an equation with integer coefficients for which $1 - \sqrt{5}$ and $3 + 2i$ are two of the solutions.

80. A discriminant that is a perfect square indicates that factoring can be used to solve the quadratic equation. Why?

81. While solving an equation of the form $ax^2 + bx + c = 0$ with a graphing calculator, Keisha gets the following screen. How could the sign of the discriminant help her check the graph?

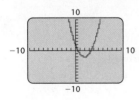

YOUR TURN ANSWERS: SECTION 8.3 **YOUR TURN ANSWERS: SECTION 8.3**

1. Two rational **2.** $5x^2 - x = 0$; answers may vary

QUICK QUIZ: SECTIONS 8.1–8.3

Solve.

1. $x^2 + 16x + 64 = 3$ [8.1]

2. $3x^2 - 1 = x$ [8.2]

3. Find the x-intercepts of the graph of $f(x) = x^2 - 8$.   [8.1]

4. Solve $x^2 + 2x - 1 = 0$. Use a calculator to approximate the solutions with rational numbers. Round to three decimal places.   [8.2]

5. Determine how many different solutions there are of $3x^2 - 5x - 7 = 0$ and whether they are real or imaginary. If they are real, specify whether they are rational or irrational.   [8.3]

PREPARE TO MOVE ON

Solve each formula for the specified variable.   [6.8]

1. $\dfrac{c}{d} = c + d$, for c

2. $x = \dfrac{3}{1 - y}$, for y

Solve.

3. Kiara's motorboat took 4 hr to make a trip downstream with a 2-mph current. The return trip against the same current took 6 hr. Find the speed of the boat in still water.   [3.3]

4. Homer walks 1.5 mph faster than Gladys. In the time it takes Homer to walk 7 mi, Gladys walks 4 mi. Find the speed of each person.   [6.5]

8.4   Applications Involving Quadratic Equations

Solving Formulas ▪ Solving Problems

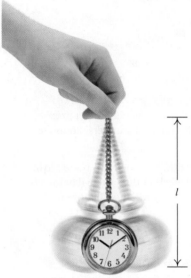

1. Water flow F from a hose, in number of gallons per minute, is given by $F = 118.8\sqrt{x}$, where x is the nozzle pressure, in pounds per square inch. Solve for x.

Source: Based on data from www.firetactics.com

Solving Formulas

To solve a formula for a certain letter, we use the principles for solving equations to get that letter alone on one side.

EXAMPLE 1 *Period of a Pendulum.* The time T required for a pendulum of length l to swing back and forth (complete one period) is given by $T = 2\pi\sqrt{l/g}$, where g is the earth's gravitational constant. Solve for l.

SOLUTION We have

$$T = 2\pi\sqrt{\frac{l}{g}}$$ This is a radical equation.

$$T^2 = \left(2\pi\sqrt{\frac{l}{g}}\right)^2$$ Squaring both sides

$$T^2 = 2^2\pi^2\frac{l}{g}$$

$$gT^2 = 4\pi^2 l$$ Multiplying both sides by g to clear fractions

$$\frac{gT^2}{4\pi^2} = l.$$ Dividing both sides by $4\pi^2$

We now have l alone on one side and l does not appear on the other side, so the formula is solved for l.

↩ YOUR TURN

In formulas for which variables represent only nonnegative numbers, there is no need for absolute-value signs when taking square roots.

EXAMPLE 2 *Hang Time.** An athlete's *hang time* is the amount of time that the athlete can remain airborne when jumping. A formula relating an athlete's vertical leap V, in inches, to hang time T, in seconds, is $V = 48T^2$. Solve for T.

SOLUTION We have

$$48T^2 = V$$

$$T^2 = \frac{V}{48}$$ Dividing by 48 to isolate T^2

$$T = \frac{\sqrt{V}}{\sqrt{48}}$$ Using the principle of square roots and the quotient rule for radicals. We assume $V, T \geq 0$.

$$= \frac{\sqrt{V}}{\sqrt{16}\sqrt{3}}$$

---

*This formula is taken from an article by Peter Brancazio, "The Mechanics of a Slam Dunk," *Popular Mechanics,* November 1991. Courtesy of Professor Peter Brancazio, Brooklyn College.

$$= \frac{\sqrt{V}}{4\sqrt{3}}$$

$$= \frac{\sqrt{V}}{4\sqrt{3}} \cdot \frac{\sqrt{3}}{\sqrt{3}}$$

$$= \frac{\sqrt{3V}}{12} \quad \text{Rationalizing the denominator}$$

2. Solve $y = (x - 2)^2$ for x.

↪ YOUR TURN

3. Solve $d = -16h^2 + 64h$ for h.

EXAMPLE 3 *Falling Distance.*   An object tossed downward with an initial speed (velocity) of v_0 will travel a distance of s meters, where $s = 4.9t^2 + v_0t$ and t is measured in seconds. Solve for t.

SOLUTION   Since t is squared in one term and raised to the first power in the other term, the equation is quadratic in t.

$$4.9t^2 + v_0t = s$$

$$4.9t^2 + v_0t - s = 0 \quad \text{Writing standard form}$$

$$a = 4.9, \quad b = v_0, \quad c = -s$$

$$t = \frac{-v_0 \pm \sqrt{(v_0)^2 - 4(4.9)(-s)}}{2(4.9)} \quad \text{Using the quadratic formula}$$

Since the negative square root would yield a negative value for t, we use only the positive root:

$$t = \frac{-v_0 + \sqrt{(v_0)^2 + 19.6s}}{9.8}.$$

↪ YOUR TURN

The following list of steps should help you when solving formulas for a given letter. Remember that solving a formula requires the same approach as solving an equation.

TO SOLVE A FORMULA FOR A LETTER—SAY, h

1. Clear fractions and use the principle of powers, as needed. Perform these steps until any radicals containing h are gone and h is not in any denominator.
2. Combine all like terms.
3. If the only power of h is h^1, the equation can be solved as a linear equation or a rational equation. (See Example 1.)
4. If h^2 appears but h does not, solve for h^2 and use the principle of square roots to then solve for h. (See Example 2.)
5. If there are terms containing both h and h^2, put the equation in standard form and use the quadratic formula. (See Example 3.)

Solving Problems

Some problems translate to rational equations. The solution of such rational equations can involve quadratic equations.

EXAMPLE 4 *Motorcycle Travel.*   Fiona rode her motorcycle 300 mi at a certain average speed. Had she traveled 10 mph faster, the trip would have taken 1 hr less. Find Fiona's average speed.

SOLUTION

1. **Familiarize.** We make a drawing, labeling it with the information provided, and create a table. To do so, we let r represent the rate, in miles per hour, and t the time, in hours, for Fiona's trip.

Recall that the definition of speed, $r = d/t$, relates rate, time, and distance.

Distance	Speed	Time
300	r	t	$\longrightarrow r = \dfrac{300}{t}$
300	$r + 10$	$t - 1$	$\longrightarrow r + 10 = \dfrac{300}{t-1}$

2. **Translate.** From the table, we obtain

$$r = \frac{300}{t} \quad \text{and} \quad r + 10 = \frac{300}{t-1}.$$

3. **Carry out.** A system of equations has been formed. We solve using substitution:

$$\frac{300}{t} + 10 = \frac{300}{t-1} \qquad \text{Substituting } 300/t \text{ for } r \text{ in the second equation}$$

$$t(t-1) \cdot \left[\frac{300}{t} + 10\right] = t(t-1) \cdot \frac{300}{t-1} \qquad \text{Multiplying by the LCM to clear fractions}$$

$$t(t-1) \cdot \frac{300}{t} + t(t-1) \cdot 10 = t(t-1) \cdot \frac{300}{t-1} \qquad \text{Using the distributive law}$$

$$\frac{\cancel{t}(t-1)}{1} \cdot \frac{300}{\cancel{t}} + t(t-1) \cdot 10 = \frac{t\cancel{(t-1)}}{1} \cdot \frac{300}{\cancel{t-1}} \qquad \begin{array}{l}\text{Removing factors that equal 1:} \\ t/t = 1 \text{ and} \\ (t-1)/(t-1) = 1\end{array}$$

$$\left.\begin{array}{r}300(t-1) + 10(t^2 - t) = 300t \\ 300t - 300 + 10t^2 - 10t = 300t \\ 10t^2 - 10t - 300 = 0\end{array}\right\} \quad \begin{array}{l}\text{Rewriting in} \\ \text{standard form}\end{array}$$

$$t^2 - t - 30 = 0 \qquad \text{Dividing by 10}$$

$$(t-6)(t+5) = 0 \qquad \text{Factoring}$$

$$t = 6 \quad \text{or} \quad t = -5. \qquad \begin{array}{l}\text{Using the principle of} \\ \text{zero products}\end{array}$$

4. Check. Note that we have solved for t, not r as required. Since negative time has no meaning here, we disregard the -5 and use 6 hr to find r:

$$r = \frac{300 \text{ mi}}{6 \text{ hr}} = 50 \text{ mph}.$$

> **CAUTION!** Always make sure that you find the quantity asked for in the problem.

To see if 50 mph checks, we increase the speed 10 mph to 60 mph and see how long the trip would have taken at that speed:

$$t = \frac{d}{r} = \frac{300 \text{ mi}}{60 \text{ mph}} = 5 \text{ hr}.$$

Note that $\text{mi}/\text{mph} = \text{mi} \div \frac{\text{mi}}{\text{hr}} = \cancel{\text{mi}} \cdot \frac{\text{hr}}{\cancel{\text{mi}}} = \text{hr}.$

4. Taryn's Cessna travels 120 mph in still air. She flies 140 mi into the wind and 140 mi with the wind in a total of 2.4 hr. Find the wind speed.

This is 1 hr less than the trip actually took, so the answer checks.

5. State. Fiona traveled at an average speed of 50 mph.

YOUR TURN

8.4 EXERCISE SET

FOR EXTRA HELP

MyMathLab® Math XL

PRACTICE   WATCH   READ   REVIEW

Vocabulary and Reading Check

Match each formula with its description from the column on the right.

1. _____ $T = 2\pi\sqrt{\dfrac{l}{g}}$

2. _____ $V = 48T^2$

3. _____ $s = 4.9t^2 + v_0 t$

4. _____ $t = \dfrac{d}{r}$

a) Falling distance

b) Motion formula

c) Period of a pendulum

d) Vertical leap

Solving Formulas

Solve each formula for the indicated letter. Assume that all variables represent positive numbers.

5. $A = 4\pi r^2$, for r
(Surface area of a sphere)

6. $A = 6s^2$, for s
(Surface area of a cube)

7. $A = 2\pi r^2 + 2\pi rh$, for r
(Surface area of a right cylindrical solid)

8. $N = \dfrac{k^2 - 3k}{2}$, for k
(Number of diagonals of a polygon)

9. $F = \dfrac{Gm_1 m_2}{r^2}$, for r
(Law of gravity)

10. $N = \dfrac{kQ_1 Q_2}{s^2}$, for s
(Number of phone calls between two cities)

11. $c = \sqrt{gH}$, for H
(Velocity of ocean wave)

12. $r = 2\sqrt{5L}$, for L
(Speed of car based on length of skid marks)

13. $a^2 + b^2 = c^2$, for b
(Pythagorean formula in two dimensions)

14. $a^2 + b^2 + c^2 = d^2$, for c
(Pythagorean formula in three dimensions)

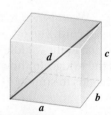

15. $s = v_0 t + \dfrac{gt^2}{2}$, for t
(A motion formula)

16. $A = \pi r^2 + \pi rs$, for r
(Surface area of a cone)

17. $N = \frac{1}{2}(n^2 - n)$, for n
(Number of games if n teams play each other once)

18. $A = A_0(1 - r)^2$, for r
(A business formula)

19. $T = I\sqrt{\dfrac{s}{d}}$, for d
(True airspeed)

20. $W = \sqrt{\dfrac{1}{LC}}$, for L
(An electricity formula)

Aha! 21. $at^2 + bt + c = 0$, for t
(An algebraic formula)

22. $A = P_1(1 + r)^2 + P_2(1 + r)$, for r
(Amount in an account)

Solve.

23. *Falling Distance.* (Use $4.9t^2 + v_0t = s$.)
   a) A bolt falls off an airplane at an altitude of 500 m. Approximately how long does it take the bolt to reach the ground?
   b) A ball is thrown downward at a speed of 30 m/sec from an altitude of 500 m. Approximately how long does it take the ball to reach the ground?
   c) Approximately how far will an object fall in 5 sec, when thrown downward at an initial velocity of 30 m/sec from a plane?

24. *Falling Distance.* (Use $4.9t^2 + v_0t = s$.)
   a) A life preserver is dropped from a helicopter at an altitude of 75 m. Approximately how long does it take the life preserver to reach the water?
   b) A coin is tossed downward with an initial velocity of 30 m/sec from an altitude of 75 m. Approximately how long does it take the coin to reach the ground?
   c) Approximately how far will an object fall in 2 sec, if thrown downward at an initial velocity of 20 m/sec from a helicopter?

25. *Bungee Jumping.* Wyatt is tied to one end of a 40-m elasticized (bungee) cord. The other end of the cord is secured to a winch at the middle of a bridge. If Wyatt jumps off the bridge, for how long will he fall before the cord begins to stretch? (Use $4.9t^2 = s$.)

40 m

26. *Bungee Jumping.* Chika is tied to a bungee cord (see Exercise 25) and falls for 2.5 sec before her cord begins to stretch. How long is the bungee cord?

27. *Hang Time.* The NBA's Russell Westbrook has a vertical leap of 36.5 in. What is his hang time? (Use $V = 48T^2$.)
Source: www.nbadraft.net

28. *League Schedules.* In a bowling league, each team plays each of the other teams once. If a total of 66 games is played, how many teams are in the league? (See Exercise 17.)

For Exercises 29 and 30, use $4.9t^2 + v_0t = s$.

29. *Downward Speed.* A stone thrown downward from a 100-m cliff travels 51.6 m in 3 sec. What was the initial velocity of the object?

30. *Downward Speed.* A pebble thrown downward from a 200-m cliff travels 91.2 m in 4 sec. What was the initial velocity of the object?

For Exercises 31 and 32, use $A = P_1(1 + r)^2 + P_2(1 + r)$. (See Exercise 22.)

31. *Compound Interest.* A firm invests $3200 in a savings account for 2 years. At the beginning of the second year, an additional $1800 is invested. If a total of $5375.48 is in the account at the end of the second year, what is the annual interest rate?

32. *Compound Interest.* A business invests $10,000 in a savings account for 2 years. At the beginning of the second year, an additional $3500 is invested. If a total of $14,822.75 is in the account at the end of the second year, what is the annual interest rate?

Solving Problems

Solve.

33. *Car Trips.* During the first part of a trip, Tara drove 120 mi at a certain speed. Tara then drove another 100 mi at a speed that was 10 mph slower. If the total time of Tara's trip was 4 hr, what was her speed on each part of the trip?

34. *Canoeing.* During the first part of a canoe trip, Ken covered 60 km at a certain speed. He then traveled 24 km at a speed that was 4 km/h slower. If the total time for the trip was 8 hr, what was the speed on each part of the trip?

35. *Car Trips.* Diane's Dodge travels 200 mi averaging a certain speed. If the car had gone 10 mph faster, the trip would have taken 1 hr less. Find Diane's average speed.

36. *Car Trips.* Stan's Subaru travels 280 mi averaging a certain speed. If the car had gone 5 mph faster, the trip would have taken 1 hr less. Find Stan's average speed.

37. *Air Travel.* A Cessna flies 600 mi at a certain speed. A Beechcraft flies 1000 mi at a speed that is 50 mph faster, but takes 1 hr longer. Find the speed of each plane.

38. *Air Travel.* A turbo-jet flies 50 mph faster than a super-prop plane. If a turbo-jet goes 2000 mi in 3 hr less time than it takes the super-prop to go 2800 mi, find the speed of each plane.

39. *Bicycling.* Naoki bikes the 36 mi to Hillsboro averaging a certain speed. The return trip is made at a speed that is 3 mph slower. Total time for the round trip is 7 hr. Find Naoki's average speed on each part of the trip.

40. *Car Speed.* On a sales trip, Samir drives the 600 mi to Richmond averaging a certain speed. The return trip is made at an average speed that is 10 mph slower. Total time for the round trip is 22 hr. Find Samir's average speed on each part of the trip.

41. *Navigation.* The Hudson River flows at a rate of 3 mph. A patrol boat travels 60 mi upriver and returns in a total time of 9 hr. What is the speed of the boat in still water?

42. *Navigation.* The current in a typical Mississippi River shipping route flows at a rate of 4 mph. In order for a barge to travel 24 mi upriver and then return in a total of 5 hr, approximately how fast must the barge be able to travel in still water?

43. *Filling a Pool.* A well and a spring are filling a swimming pool. Together, they can fill the pool in 3 hr. The well, working alone, can fill the pool in 8 hr less time than it would take the spring. How long would the spring take, working alone, to fill the pool?

44. *Filling a Tank.* Two pipes are connected to the same tank. Working together, they can fill the tank in 4 hr. The larger pipe, working alone, can fill the tank in 6 hr less time than it would take the smaller one. How long would the smaller one take, working alone, to fill the tank?

45. *Paddleboats.* Kofi paddles 1 mi upstream and 1 mi back in a total time of 1 hr. The speed of the river is 2 mph. Find the speed of Kofi's paddleboat in still water.

46. *Rowing.* Abby rows 10 km upstream and 10 km back in a total time of 3 hr. The speed of the river is 5 km/h. Find Abby's speed in still water.

47. *Reforestation.* Working together, Katherine and Julianna can plant new trees on their recently reforested land in 6 days. Working alone, it would take Julianna 2 days longer than it would take Katherine to plant the trees. How long would it take Katherine, working alone, to plant the trees?

48. *Team Teaching.* Working together, Tanner and Joel can grade their students' projects in 2 hr. Working alone, it would take Tanner 2 hr longer than it would take Joel to grade the projects. How long would it take Joel, working alone, to grade the projects?

49. Marti is tied to a bungee cord that is twice as long as the cord tied to Rafe. Will Marti's fall take twice as long as Rafe's before their cords begin to stretch? Why or why not? (See Exercises 25 and 26.)

50. Under what circumstances would a negative value for *t*, time, have meaning?

Skill Review

Solve.

51. $\frac{1}{2}(x - 7) = \frac{1}{3}x + 4$ [1.3] **52.** $\frac{1}{x} = -8$ [6.4]

53. $6|3x + 2| = 12$ [4.3] **54.** $2x - y = 4,$
$x + 2y = 3$ [3.2]

55. $\sqrt{2x - 8} = 15$ [7.6] **56.** $2 + \sqrt{t} = \sqrt{t + 5}$ [7.6]

Synthesis

57. Write a problem for a classmate to solve. Devise the problem so that **(a)** the solution is found after solving a rational equation and **(b)** the solution is "The express train travels 90 mph."

58. Sophia has no difficulty creating a table for motion problems but cannot write equations from the table. What suggestion can you offer Sophia?

59. *Biochemistry.* The equation

$$A = 6.5 - \frac{20.4t}{t^2 + 36}$$

is used to calculate the acid level A in a person's blood t minutes after sugar is consumed. Solve for t.

60. *Special Relativity.*   Einstein found that an object with initial mass m_0 and traveling velocity v has mass

$$m = \frac{m_0}{\sqrt{1 - \dfrac{v^2}{c^2}}},$$

where c is the speed of light. Solve the formula for c.

61. Find a number for which the reciprocal of 1 less than the number is the same as 1 more than the number.

62. *Purchasing.*   A discount store bought a quantity of potted plants for $250 and sold all but 15 at a profit of $3.50 per plant. With the total amount received, the manager could buy 4 more than twice as many as were bought before. Find the cost per plant.

63. *Art and Aesthetics.*   For over 2000 years, artists, sculptors, and architects have regarded the proportions of a "golden" rectangle as visually appealing. A rectangle of width w and length l is considered "golden" if

$$\frac{w}{l} = \frac{l}{w + l}.$$

Solve for l.

64. *Diagonal of a Cube.*   Find a formula that expresses the length of the three-dimensional diagonal of a cube as a function of the cube's surface area.

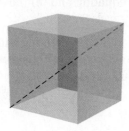

65. Solve for n:

$$mn^4 - r^2pm^3 - r^2n^2 + p = 0.$$

66. *Surface Area.*   Find a formula that expresses the diameter of a right cylindrical solid as a function of its surface area and its height. (See Exercise 7.)

67. A sphere is inscribed in a cube as shown in the figure below. Express the surface area of the sphere as a function of the surface area S of the cube. (See Exercise 5.)

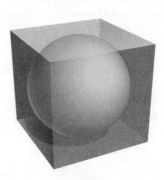

↳ **YOUR TURN ANSWERS: SECTION 8.4**

1. $x = \dfrac{F^2}{14,113.44}$ **2.** $x = 2 \pm \sqrt{y}$

3. $h = 2 \pm \dfrac{\sqrt{64 - d}}{4}$ **4.** 20 mph

QUICK QUIZ: SECTIONS 8.1–8.4

Solve. [8.1], [8.2]

1. $x^2 - 20x = 15$

2. $x(x - 2) = 25$

3. $2500 grows to $2601 in 2 years. Use the formula $A = P(1 + r)^t$ to determine the interest rate.   [8.1]

4. Write a quadratic equation having the solutions $2\sqrt{3}$ and $-2\sqrt{3}$.   [8.3]

5. Solve $n = d^2 + 2d$ for d.   [8.4]

PREPARE TO MOVE ON

Simplify.

1. $(m^{-1})^2$ [1.6]   **2.** $(y^{1/6})^2$ [7.2]

Solve.

3. $t^{-1} = \dfrac{1}{2}$ [6.4]   **4.** $x^{1/4} = 3$ [7.6]

8.5 Equations Reducible to Quadratic

Equations in Quadratic Form ■ Radical Equations and Rational Equations

Student Notes

To identify an equation in quadratic form, look for two variable expressions in the equation. In order for an equation to be in quadratic form, the exponent in one expression must be twice the exponent in the other expression.

CAUTION! A common error when working on problems like Example 1 is to solve for u but forget to solve for x. Remember to solve for the *original* variable!

---

1. Solve: $x^4 - 8x^2 - 9 = 0$.

Equations in Quadratic Form

Certain equations that are not really quadratic can be regarded in a manner that allows us to use the methods developed for quadratic equations. For example, because the square of x^2 is x^4, the equation $x^4 - 9x^2 + 8 = 0$ is said to be "quadratic in x^2":

$$x^4 - 9x^2 + 8 = 0$$
$$(x^2)^2 - 9(x^2) + 8 = 0 \quad \text{Thinking of } x^4 \text{ as } (x^2)^2$$
$$u^2 - 9u + 8 = 0. \quad \text{To make this clearer, write } u \text{ instead of } x^2.$$

The equation $u^2 - 9u + 8 = 0$ can be solved for u by factoring or by the quadratic formula. Then, remembering that $u = x^2$, we can solve for x. Equations that can be solved like this are *reducible to quadratic* and are said to be *in quadratic form*.

EXAMPLE 1 Solve: $x^4 - 9x^2 + 8 = 0$.

SOLUTION We begin by letting $u = x^2$ and finding u^2.

If we let $u = x^2$,
then $u^2 = (x^2)^2 = x^4$.

Next, we substitute u^2 for x^4 and u for x^2:

$$u^2 - 9u + 8 = 0$$
$$(u - 8)(u - 1) = 0 \quad \text{Factoring}$$
$$u - 8 = 0 \quad or \quad u - 1 = 0 \quad \text{Principle of zero products}$$
$$u = 8 \quad or \quad u = 1. \quad \text{We have solved for } u.$$

We now replace u with x^2 and solve these equations:

$$x^2 = 8 \quad or \quad x^2 = 1$$
$$x = \pm\sqrt{8} \quad or \quad x = \pm1$$
$$x = \pm2\sqrt{2} \quad or \quad x = \pm1. \quad \text{We have solved for } x.$$

To check, note that for both $x = 2\sqrt{2}$ and $-2\sqrt{2}$, we have $x^2 = 8$ and $x^4 = 64$. Similarly, for both $x = 1$ and -1, we have $x^2 = 1$ and $x^4 = 1$. Thus instead of making four checks, we need make only two.

Check: For $\pm2\sqrt{2}$:

$$\frac{x^4 - 9x^2 + 8 = 0}{(\pm2\sqrt{2})^4 - 9(\pm2\sqrt{2})^2 + 8 \mid 0}$$
$$64 - 9 \cdot 8 + 8$$
$$0 \stackrel{?}{=} 0 \quad \text{TRUE}$$

For ±1:

$$\frac{x^4 - 9x^2 + 8 = 0}{(\pm1)^4 - 9(\pm1)^2 + 8 \mid 0}$$
$$1 - 9 + 8$$
$$0 \stackrel{?}{=} 0 \quad \text{TRUE}$$

The solutions are $1, -1, 2\sqrt{2}$, and $-2\sqrt{2}$.

YOUR TURN

Technology Connection

We can check Example 2 by graphing $y_1 = (x^2 - 1)^2 - (x^2 - 1) - 2$. We can see that the graph crosses or touches the x-axis at approximately $(-\sqrt{3}, 0)$, $(0, 0)$, and $(\sqrt{3}, 0)$.

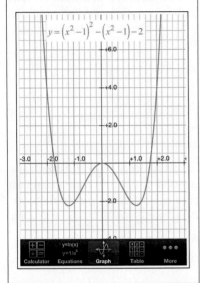

2. Find the x-intercepts of the graph of $f(x) = (x^2 - 2)^2 - 3(x^2 - 2) - 4 = 0$.

Equations like those in Example 1 can be solved by factoring:

$$x^4 - 9x^2 + 8 = 0$$
$$(x^2 - 1)(x^2 - 8) = 0$$
$$x^2 - 1 = 0 \quad or \quad x^2 - 8 = 0$$
$$x^2 = 1 \quad or \quad x^2 = 8$$
$$x = \pm 1 \quad or \quad x = \pm 2\sqrt{2}.$$

However, it can become difficult to solve an equation without first making a substitution.

EXAMPLE 2   Find the x-intercepts of the graph of
$$f(x) = (x^2 - 1)^2 - (x^2 - 1) - 2.$$

SOLUTION   The x-intercepts occur where $f(x) = 0$ so we must have
$$(x^2 - 1)^2 - (x^2 - 1) - 2 = 0. \qquad \text{Setting } f(x) \text{ equal to 0}$$
If we let $\quad u = x^2 - 1,$
then $\qquad u^2 = (x^2 - 1)^2.$

Substituting, we have
$$u^2 - u - 2 = 0 \qquad \begin{array}{l}\text{Substituting in}\\ (x^2 - 1)^2 - (x^2 - 1) - 2 = 0\end{array}$$
$$(u - 2)(u + 1) = 0$$
$$u = 2 \quad or \quad u = -1. \qquad \text{Using the principle of zero products}$$

Next, we replace u with $x^2 - 1$ and solve these equations:
$$x^2 - 1 = 2 \quad or \quad x^2 - 1 = -1$$
$$x^2 = 3 \quad or \quad x^2 = 0 \qquad \text{Adding 1 to both sides}$$
$$x = \pm\sqrt{3} \quad or \quad x = 0. \qquad \begin{array}{l}\text{Using the principle of square}\\ \text{roots}\end{array}$$

The x-intercepts occur at $(-\sqrt{3}, 0)$, $(0, 0)$, and $(\sqrt{3}, 0)$.

YOUR TURN

The following tips may prove useful.

TO SOLVE AN EQUATION THAT IS REDUCIBLE TO QUADRATIC

1. Look for two variable expressions in the equation. One expression should be the square of the other. (The exponent in one expression will be twice the exponent in the other expression.)
2. Write down any substitutions that you are making.
3. Remember to solve for the variable that is used in the original equation.
4. Check possible answers in the original equation.

Radical Equations and Rational Equations

Sometimes rational equations, radical equations, or equations containing exponents that are fractions are reducible to quadratic. It is especially important that answers to these equations be checked in the original equation.

EXAMPLE 3   Solve: $x - 3\sqrt{x} - 4 = 0$.

SOLUTION   This radical equation could be solved using the principle of powers. However, if we note that the square of $\sqrt{x}$ is x, we can regard the equation as "quadratic in $\sqrt{x}$."

$$\text{If we let} \quad u = \sqrt{x},$$
$$\text{then} \quad u^2 = x.$$

Substituting, we have

$$x - 3\sqrt{x} - 4 = 0$$
$$u^2 - 3u - 4 = 0$$
$$(u - 4)(u + 1) = 0$$
$$u = 4 \quad or \quad u = -1. \qquad \text{Using the principle of zero products}$$

Next, we replace u with $\sqrt{x}$ and solve these equations:

$$\sqrt{x} = 4 \quad or \quad \sqrt{x} = -1.$$

Squaring gives us $x = 16$ or $x = 1$ and also makes checking essential.

Check:

For 16:		For 1:	
$x - 3\sqrt{x} - 4 = 0$		$x - 3\sqrt{x} - 4 = 0$	
$16 - 3\sqrt{16} - 4$	0	$1 - 3\sqrt{1} - 4$	0
$16 - 3 \cdot 4 - 4$		$1 - 3 \cdot 1 - 4$	
	$0 \overset{?}{=} 0$ TRUE		$-6 \overset{?}{=} 0$ FALSE

The number 16 checks, but 1 does not. Had we noticed that $\sqrt{x} = -1$ has no solution (since principal square roots are never negative), we could have focused only on $\sqrt{x} = 4$. The solution is 16.

YOUR TURN

EXAMPLE 4   Solve: $2m^{-2} + m^{-1} - 15 = 0$.

SOLUTION   Note that the square of m^{-1} is $(m^{-1})^2$, or m^{-2}.

$$\text{If we let} \quad u = m^{-1},$$
$$\text{then} \quad u^2 = m^{-2}.$$

Substituting, we have

$$2u^2 + u - 15 = 0 \qquad \text{Substituting in } 2m^{-2} + m^{-1} - 15 = 0$$
$$(2u - 5)(u + 3) = 0$$
$$2u - 5 = 0 \quad or \quad u + 3 = 0 \qquad \text{Using the principle of zero products}$$
$$2u = 5 \quad or \quad \qquad u = -3$$
$$u = \tfrac{5}{2} \quad or \quad \qquad u = -3.$$

Now, we replace u with m^{-1} and solve:

$$m^{-1} = \frac{5}{2} \quad or \quad m^{-1} = -3$$

$$\frac{1}{m} = \frac{5}{2} \quad or \quad \frac{1}{m} = -3 \qquad \text{Recall that } m^{-1} = \frac{1}{m}.$$

$$1 = \frac{5}{2}m \quad or \quad 1 = -3m \qquad \text{Multiplying both sides by } m$$

$$\frac{2}{5} = m \quad or \quad -\frac{1}{3} = m. \qquad \text{Solving for } m$$

Student Notes

Note that $x^2 = -1$ has complex solutions but $\sqrt{x} = -1$ does not.

Technology Connection

Check Example 3 with a graphing calculator. Use the ZERO, ROOT, or INTERSECT option, if possible.

3. Solve: $x - 5\sqrt{x} - 14 = 0$.

Determine u and u^2.

Substitute.

Solve for u.

Solve for the original variable.

Check.

Check:

For $\frac{2}{5}$:

$$2m^{-2} + m^{-1} - 15 = 0$$

$$\begin{array}{c|c} 2\left(\frac{2}{5}\right)^{-2} + \left(\frac{2}{5}\right)^{-1} - 15 & 0 \\ 2\left(\frac{5}{2}\right)^2 + \left(\frac{5}{2}\right) - 15 & \\ 2\left(\frac{25}{4}\right) + \frac{5}{2} - 15 & \\ \frac{25}{2} + \frac{5}{2} - 15 & \\ \frac{30}{2} - 15 & \\ & 0 \overset{?}{=} 0 \quad \text{TRUE} \end{array}$$

For $-\frac{1}{3}$:

$$2m^{-2} + m^{-1} - 15 = 0$$

$$\begin{array}{c|c} 2\left(-\frac{1}{3}\right)^{-2} + \left(-\frac{1}{3}\right)^{-1} - 15 & 0 \\ 2\left(-\frac{3}{1}\right)^2 + \left(-\frac{3}{1}\right) - 15 & \\ 2(9) + (-3) - 15 & \\ 18 - 3 - 15 & \\ & 0 \overset{?}{=} 0 \quad \text{TRUE} \end{array}$$

4. Solve: $m^{-4} - 5m^{-2} + 6 = 0$.

Both numbers check. The solutions are $-\frac{1}{3}$ and $\frac{2}{5}$.

YOUR TURN

Note that Example 4 can also be written $2/m^2 + 1/m - 15 = 0$. It can then be solved by letting $u = 1/m$ and $u^2 = 1/m^2$ or by clearing fractions.

EXAMPLE 5 Solve: $t^{2/5} - t^{1/5} - 2 = 0$.

SOLUTION Note that the square of $t^{1/5}$ is $(t^{1/5})^2$, or $t^{2/5}$. The equation is therefore quadratic in $t^{1/5}$.

If we let $\quad u = t^{1/5}$,

then $\quad u^2 = t^{2/5}$.

Substituting, we have

$$u^2 - u - 2 = 0 \qquad \text{Substituting in } t^{2/5} - t^{1/5} - 2 = 0$$
$$(u - 2)(u + 1) = 0$$
$$u = 2 \quad or \quad u = -1. \qquad \text{Using the principle of zero products}$$

Now, we replace u with $t^{1/5}$ and solve:

$$t^{1/5} = 2 \quad or \quad t^{1/5} = -1$$
$$t = 32 \quad or \quad t = -1. \qquad \text{Principle of powers; raising both sides to the 5th power}$$

Check:

For 32:

$$t^{2/5} - t^{1/5} - 2 = 0$$

$$\begin{array}{c|c} 32^{2/5} - 32^{1/5} - 2 & 0 \\ (32^{1/5})^2 - 32^{1/5} - 2 & \\ 2^2 - 2 - 2 & \\ & 0 \overset{?}{=} 0 \quad \text{TRUE} \end{array}$$

For -1:

$$t^{2/5} - t^{1/5} - 2 = 0$$

$$\begin{array}{c|c} (-1)^{2/5} - (-1)^{1/5} - 2 & 0 \\ [(-1)^{1/5}]^2 - (-1)^{1/5} - 2 & \\ (-1)^2 - (-1) - 2 & \\ & 0 \overset{?}{=} 0 \quad \text{TRUE} \end{array}$$

5. Solve: $2t^{2/3} - t^{1/3} - 3 = 0$.

Both numbers check. The solutions are 32 and -1.

YOUR TURN

EXAMPLE 6 Solve: $(5 + \sqrt{r})^2 + 6(5 + \sqrt{r}) + 2 = 0$.

SOLUTION

If we let $\quad u = 5 + \sqrt{r}$,

then $\quad u^2 = (5 + \sqrt{r})^2$.

Substituting, we have

$$u^2 + 6u + 2 = 0$$

$$u = \frac{-6 \pm \sqrt{6^2 - 4 \cdot 1 \cdot 2}}{2 \cdot 1} \quad \text{Using the quadratic formula}$$

$$= \frac{-6 \pm \sqrt{28}}{2}$$

$$= \frac{-6}{2} \pm \frac{2\sqrt{7}}{2} \quad \Big\} \quad \text{Simplifying; } \sqrt{28} = \sqrt{4}\sqrt{7}$$

$$= -3 \pm \sqrt{7}.$$

Now, we replace u with $5 + \sqrt{r}$ and solve for r:

$$u = -3 + \sqrt{7} \quad or \qquad u = -3 - \sqrt{7}$$
$$5 + \sqrt{r} = -3 + \sqrt{7} \quad or \quad 5 + \sqrt{r} = -3 - \sqrt{7}$$
$$\sqrt{r} = -8 + \sqrt{7} \quad or \qquad \sqrt{r} = -8 - \sqrt{7}.$$

6. Solve:

$(3 - \sqrt{x})^2 - 2(3 - \sqrt{x}) - 8 = 0.$

Both $-8 + \sqrt{7}$ and $-8 - \sqrt{7}$ are negative. Since $\sqrt{r}$ is never negative, both values of $\sqrt{r}$ must be rejected. The equation has no solution.

YOUR TURN

8.5 EXERCISE SET

FOR EXTRA HELP

MyMathLab® MathXL
PRACTICE WATCH READ REVIEW

Vocabulary and Reading Check

Classify each of the following statements as either true or false.

1. Some equations that are not really quadratic are quadratic in form.

2. Some radical equations and rational equations are reducible to quadratic.

3. We have not completed solving an equation that is quadratic in form until we have solved for the original variable.

4. When solving an equation that is quadratic in form, we should check any possible solutions in the original equation.

Concept Reinforcement

Write the substitution that could be used to make each equation quadratic in u.

5. For $3p - 4\sqrt{p} + 6 = 0$, use $u =$ _____.

6. For $x^{1/2} - x^{1/4} - 2 = 0$, use $u =$ _____.

7. For $(x^2 + 3)^2 + (x^2 + 3) - 7 = 0$, use $u =$ _____.

8. For $t^{-6} + 5t^{-3} - 6 = 0$, use $u =$ _____.

9. For $(1 + t)^4 + (1 + t)^2 + 4 = 0$, use $u =$ _____.

10. For $w^{1/3} - 3w^{1/6} + 8 = 0$, use $u =$ _____.

Equations in Quadratic Form

Solve.

11. $x^4 - 13x^2 + 36 = 0$

12. $x^4 - 17x^2 + 16 = 0$

13. $t^4 - 7t^2 + 12 = 0$

14. $t^4 - 11t^2 + 18 = 0$

15. $4x^4 - 9x^2 + 5 = 0$

16. $9x^4 - 38x^2 + 8 = 0$

17. $w + 4\sqrt{w} - 12 = 0$

18. $s + 3\sqrt{s} - 40 = 0$

19. $(x^2 - 7)^2 - 3(x^2 - 7) + 2 = 0$

20. $(x^2 - 2)^2 - 12(x^2 - 2) + 20 = 0$

21. $x^4 + 5x^2 - 36 = 0$

22. $x^4 + 5x^2 + 4 = 0$

23. $(n^2 + 6)^2 - 7(n^2 + 6) + 10 = 0$

24. $(m^2 + 7)^2 - 6(m^2 + 7) - 16 = 0$

Radical Equations and Rational Equations

Solve.

25. $r - 2\sqrt{r} - 6 = 0$

26. $s - 4\sqrt{s} - 1 = 0$

27. $(1 + \sqrt{x})^2 + 5(1 + \sqrt{x}) + 6 = 0$

28. $(3 + \sqrt{x})^2 + 3(3 + \sqrt{x}) - 10 = 0$

29. $x^{-2} - x^{-1} - 6 = 0$

30. $2x^{-2} - x^{-1} - 1 = 0$

31. $4t^{-2} - 3t^{-1} - 1 = 0$

32. $2m^{-2} + 7m^{-1} - 15 = 0$

33. $t^{2/3} + t^{1/3} - 6 = 0$

34. $w^{2/3} - 2w^{1/3} - 8 = 0$

35. $y^{1/3} - y^{1/6} - 6 = 0$

36. $t^{1/2} + 3t^{1/4} + 2 = 0$

37. $t^{1/3} + 2t^{1/6} = 3$

38. $m^{1/2} + 6 = 5m^{1/4}$

39. $(3 - \sqrt{x})^2 - 10(3 - \sqrt{x}) + 23 = 0$

40. $(5 + \sqrt{x})^2 - 12(5 + \sqrt{x}) + 33 = 0$

41. $16\left(\dfrac{x - 1}{x - 8}\right)^2 + 8\left(\dfrac{x - 1}{x - 8}\right) + 1 = 0$

42. $9\left(\dfrac{x + 2}{x + 3}\right)^2 - 6\left(\dfrac{x + 2}{x + 3}\right) + 1 = 0$

Equations in Quadratic Form

Find all x-intercepts of the graph of f. If none exists, state this. Do not graph.

43. $f(x) = 5x + 13\sqrt{x} - 6$

44. $f(x) = 3x + 10\sqrt{x} - 8$

45. $f(x) = (x^2 - 3x)^2 - 10(x^2 - 3x) + 24$

46. $f(x) = (x^2 - 6x)^2 - 2(x^2 - 6x) - 35$

47. $f(x) = x^{2/5} + x^{1/5} - 6$

48. $f(x) = x^{1/2} - x^{1/4} - 6$

Aha! **49.** $f(x) = \left(\dfrac{x^2 + 2}{x}\right)^4 + 7\left(\dfrac{x^2 + 2}{x}\right)^2 + 5$

50. $f(x) = \left(\dfrac{x^2 + 1}{x}\right)^4 + 4\left(\dfrac{x^2 + 1}{x}\right)^2 + 12$

51. To solve $25x^6 - 10x^3 + 1 = 0$, Jose lets $u = 5x^3$ and Robin lets $u = x^3$. Are they both correct? Why or why not?

52. Jenn writes that the solutions of $x^4 - 5x^2 + 6 = 0$ are 2 and 3. What mistake is she probably making?

Skill Review

Graph.

53. $2x = -5y$ [2.3]

54. $y = \frac{1}{3}x - 2$ [2.3]

55. $2x - 5y = 10$ [2.4]

56. $3x = -9$ [2.4]

57. $y - 2 = 3(x - 4)$ [2.5]

58. $2(y - 7) = y - 10$ [2.4]

Synthesis

59. Describe a procedure that could be used to solve any equation of the form $ax^4 + bx^2 + c = 0$.

60. Describe a procedure that could be used to write an equation that is quadratic in $3x^2 - 1$. Then explain how the procedure could be adjusted to write equations that are quadratic in $3x^2 - 1$ and have no real-number solution.

Solve.

61. $3x^4 + 5x^2 - 1 = 0$

62. $(x^2 - 5x - 1)^2 - 18(x^2 - 5x - 1) + 65 = 0$

63. $\dfrac{x}{x - 1} - 6\sqrt{\dfrac{x}{x - 1}} - 40 = 0$

64. $a^5(a^2 - 25) + 13a^3(25 - a^2) + 36a(a^2 - 25) = 0$

65. $a^3 - 26a^{3/2} - 27 = 0$

66. $x^6 - 28x^3 + 27 = 0$

67. $x^6 + 7x^3 - 8 = 0$

68. Use a graphing calculator to check your answers to Exercises 11, 13, 37, and 49.

69. Use a graphing calculator to solve
$$x^4 - x^3 - 13x^2 + x + 12 = 0.$$

YOUR TURN ANSWERS: SECTION 8.5

1. $\pm 3, \pm i$ **2.** $(-\sqrt{6}, 0), (-1, 0), (1, 0), (\sqrt{6}, 0)$

3. 49   **4.** $\pm\dfrac{\sqrt{3}}{3}, \pm\dfrac{\sqrt{2}}{2}$ **5.** $-1, \dfrac{27}{8}$ **6.** 25

QUICK QUIZ: SECTIONS 8.1–8.5

Solve.

1. $2t^2 + 1 = 3t$ [8.1]   **2.** $c^2 + c + 1 = 0$ [8.2]

3. $x^4 - 10x^2 + 9 = 0$ [8.5]   **4.** $\dfrac{1}{x} + \dfrac{x}{x - 2} = 3$ [8.2]

5. Solve $3c = 4\sqrt{5xy}$ for x.   [8.4]

PREPARE TO MOVE ON

Graph.

1. $f(x) = x$ [2.3]   **2.** $g(x) = x + 2$ [2.3]

3. $h(x) = x - 2$ [2.3]   **4.** $f(x) = x^2$ [2.1]

5. $g(x) = x^2 + 2$ [2.1]   **6.** $h(x) = x^2 - 2$ [2.1]

Mid-Chapter Review

We have discussed four methods of solving quadratic equations:

- factoring and the principle of zero products;
- the principle of square roots;
- completing the square;
- the quadratic formula.

Any of these may also be appropriate when solving an applied problem or an equation that is reducible to quadratic form.

GUIDED SOLUTIONS

1. Solve: $(x - 7)^2 = 5$.   [8.1]

Solution

$\quad x - 7 = \boxed{}$ Using the principle of square roots

$\quad\quad\quad x = \boxed{}$ Adding 7 to both sides

The solutions are $7 + \boxed{}$ and $7 - \boxed{}$.

2. Solve: $x^2 - 2x - 1 = 0$.   [8.2]

Solution

$\quad a = \boxed{}, \quad b = \boxed{}, \quad c = \boxed{}$

$$x = \frac{-(\boxed{}) \pm \sqrt{(\boxed{})^2 - 4 \cdot 1 \cdot (\boxed{})}}{2 \cdot \boxed{}}$$

$$x = \frac{\boxed{} \pm \sqrt{\boxed{}}}{\boxed{}}$$

$$x = \frac{2}{2} \pm \frac{\boxed{} \sqrt{2}}{2}$$

The solutions are $1 + \boxed{}$ and $1 - \boxed{}$.

MIXED REVIEW

Solve. Examine each exercise carefully, and solve using the easiest method.

3. $x^2 + 4x = 21$ [8.1]

4. $t^2 - 196 = 0$ [8.1]

5. $x^2 = 2x + 5$ [8.2]

6. $x^2 = 2x - 5$ [8.2]

7. $x^4 = 16$ [8.5]

8. $(t + 3)^2 = 7$ [8.1]

9. $n(n - 3) = 2n(n + 1)$ [8.1]

10. $6y^2 - 7y - 10 = 0$ [8.1]

11. $16c^2 = 7c$ [8.1]

12. $3x^2 + 5x = 1$ [8.2]

13. $(t + 4)(t - 3) = 18$ [8.1]

14. $(m^2 + 3)^2 - 4(m^2 + 3) - 5 = 0$ [8.5]

For each equation, determine what type of number the solutions are and how many solutions exist.   [8.3]

15. $x^2 - 8x + 1 = 0$

16. $3x^2 = 4x + 7$

17. $5x^2 - x + 6 = 0$

Solve each formula for the indicated letter. Assume that all variables represent positive numbers.   [8.4]

18. $F = \dfrac{Av^2}{400}$, for v

(Force of wind on a sail)

19. $D^2 - 2Dd - 2hd = 0$, for D

(Dynamic load)

20. Sophie drove 225 mi south through the Smoky Mountains. Snow during her return trip made her average speed on the return trip 30 mph slower. The total driving time was 8 hr. Find Sophie's average speed on each part of the trip.   [8.4]

8.6   Quadratic Functions and Their Graphs

The Graph of $f(x) = ax^2$ ▪ The Graph of $f(x) = a(x - h)^2$ ▪ The Graph of $f(x) = a(x - h)^2 + k$

The graph of any *linear* function $f(x) = mx + b$ is a straight line. In this section and the next, we will see that the graph of any *quadratic* function $f(x) = ax^2 + bx + c$ is a *parabola*. We examine the shape of such graphs by first looking at quadratic functions with $b = 0$ and $c = 0$.

The Graph of $f(x) = ax^2$

The most basic quadratic function is $f(x) = x^2$.

EXAMPLE 1   Graph: $f(x) = x^2$.

SOLUTION   We choose some values for x and compute $f(x)$ for each. Then we plot the ordered pairs and connect them with a smooth curve.

x	$f(x) = x^2$	$(x, f(x))$
-3	9	$(-3, 9)$
-2	4	$(-2, 4)$
-1	1	$(-1, 1)$
0	0	$(0, 0)$
1	1	$(1, 1)$
2	4	$(2, 4)$
3	9	$(3, 9)$

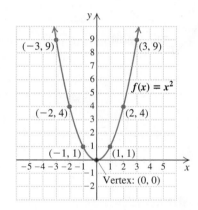

1. Graph: $f(x) = 2x^2$.

 YOUR TURN

All quadratic functions have graphs similar to the one in Example 1. Such curves are called **parabolas**. They are U-shaped and can open upward, as in Example 1, or downward. The "turning point" of the graph is called the **vertex** of the parabola. The vertex of the graph in Example 1 is $(0, 0)$.

A parabola is symmetric with respect to a line that goes through the center of the parabola and the vertex. This line is known as the parabola's **axis of symmetry**. In Example 1, the y-axis (the vertical line $x = 0$) is the axis of symmetry. Were the paper folded on this line, the two halves of the curve would match.

Student Notes

By paying attention to the symmetry of each parabola and the location of the vertex, you save yourself considerable work. Note that the x-values 1 unit to the right or to the left of the vertex are paired with the y-value a units above the vertex. Thus the graph of $y = \frac{3}{2}x^2$ includes the points $\left(-1, \frac{3}{2}\right)$ and $\left(1, \frac{3}{2}\right)$.

The graph of any function of the form $y = ax^2$ has the vertex $(0,0)$ and the axis of symmetry $x = 0$. By plotting points, we can compare the graphs of $g(x) = \frac{1}{2}x^2$ and $h(x) = 2x^2$ with the graph of $f(x) = x^2$.

x	$g(x) = \frac{1}{2}x^2$
-3	$\frac{9}{2}$
-2	2
-1	$\frac{1}{2}$
0	0
1	$\frac{1}{2}$
2	2
3	$\frac{9}{2}$

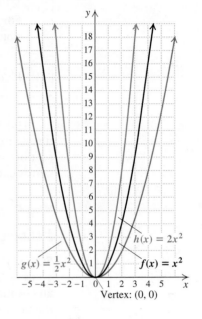

x	$h(x) = 2x^2$
-3	18
-2	8
-1	2
0	0
1	2
2	8
3	18

Note that the graph of $g(x) = \frac{1}{2}x^2$ is "wider" than the graph of $\boldsymbol{f(x) = x^2}$, and the graph of $h(x) = 2x^2$ is "narrower." The vertex and the axis of symmetry, however, remain $(0, 0)$ and the line $x = 0$, respectively.

When we consider the graph of $k(x) = -\frac{1}{2}x^2$, we see that the parabola is the same shape as the graph of $g(x) = \frac{1}{2}x^2$, but opens downward. We say that the graphs of k and g are *reflections* of each other across the x-axis.

Technology Connection

To explore the effect of a on the graph of $y = ax^2$, let $y_1 = x^2$, $y_2 = 3x^2$, and $y_3 = \frac{1}{3}x^2$. Graph the equations and use ⟨TRACE⟩ to see how the y-values compare, using ⌃ or ⌄ to hop the cursor from one curve to the next.

Many graphing calculators include a Transfrm application. If you run that application and let $y_1 = Ax^2$, the graph becomes interactive. The value of A can be entered while viewing the graph, or the values can be stepped up or down by pressing ⟨ or ⟩.

1. Compare the graphs of
   $y_1 = \frac{1}{5}x^2$, $y_2 = x^2$,
   $y_3 = \frac{5}{2}x^2$, $y_4 = -\frac{1}{5}x^2$,
   $y_5 = -x^2$, and
   $y_6 = -\frac{5}{2}x^2$.
2. Describe the effect that A has on each graph.

x	$k(x) = -\frac{1}{2}x^2$
-3	$-\frac{9}{2}$
-2	-2
-1	$-\frac{1}{2}$
0	0
1	$-\frac{1}{2}$
2	-2
3	$-\frac{9}{2}$

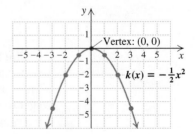

> ### GRAPHING $f(x) = ax^2$
>
> The graph of $f(x) = ax^2$ is a parabola with $x = 0$ as its axis of symmetry. Its vertex is the origin.
>
> For $a > 0$, the parabola opens upward. For $a < 0$, the parabola opens downward.
>
> If $|a|$ is greater than 1, the parabola is narrower than $y = x^2$.
>
> If $|a|$ is between 0 and 1, the parabola is wider than $y = x^2$.

The width of a parabola and whether it opens upward or downward are determined by the coefficient a in $f(x) = ax^2 + bx + c$. In the remainder of this section, we graph quadratic functions that are written in a form from which the vertex can be read directly.

The Graph of $f(x) = a(x - h)^2$

EXAMPLE 2   Graph: $f(x) = (x - 3)^2$.

SOLUTION   We choose some values for x and compute $f(x)$. Since $(x - 3)^2 = 1 \cdot (x - 3)^2$, $a = 1$, and the graph opens upward. It is important to note that for any input that is 3 more than an input for Example 1, the outputs match. We plot the points and draw the curve.

x	$f(x) = (x - 3)^2$	
-1	16	
0	9	
1	4	
2	1	
3	0	← Vertex
4	1	
5	4	
6	9	

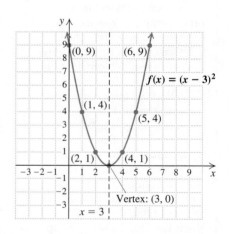

The line $x = 3$ is the axis of symmetry and the point $(3, 0)$ is the vertex. Had we recognized earlier that $x = 3$ is the axis of symmetry, we could have computed some values on one side, such as $(4, 1)$, $(5, 4)$, and $(6, 9)$, and then used symmetry to get their mirror images $(2, 1)$, $(1, 4)$, and $(0, 9)$ without further computation.

2. Graph: $f(x) = (x + 2)^2$.
_____   YOUR TURN

The result of Example 2 can be generalized:

The vertex of the graph of $f(x) = a(x - h)^2$ is $(h, 0)$.

EXAMPLE 3   Graph: $g(x) = -2(x + 4)^2$.

SOLUTION   We choose some values for x and compute $g(x)$. Since $a = -2$, the graph will open downward. If we rewrite the equation as $g(x) = -2(x - (-4))^2$, we see that $(-4, 0)$ is the vertex. The axis of symmetry is then $x = -4$. We plot some points and draw the curve.

x	$g(x) = -2(x + 4)^2$
-6	-8
-5	-2
-4	0
-3	-2
-2	-8

←Vertex

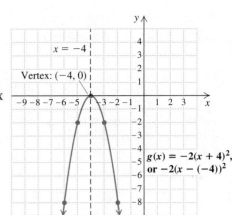

$$g(x) = -2(x + 4)^2,$$
$$\text{or } -2(x - (-4))^2$$

3. Graph: $g(x) = -(x - 3)^2$.

YOUR TURN

In Example 2, the graph of $f(x) = (x - 3)^2$ looks just like the graph of $y = x^2$, except that it is moved, or *translated*, 3 units to the right. In Example 3, the graph of $g(x) = -2(x + 4)^2$ looks like the graph of $y = -2x^2$, except that it is shifted 4 units to the left. These results are generalized as follows.

GRAPHING $f(x) = a(x - h)^2$

The graph of $f(x) = a(x - h)^2$ has the same shape as the graph of $y = ax^2$.

- If h is positive, the graph of $y = ax^2$ is shifted h units to the right.
- If h is negative, the graph of $y = ax^2$ is shifted $|h|$ units to the left.
- The vertex is $(h, 0)$, and the axis of symmetry is $x = h$.

The Graph of $f(x) = a(x - h)^2 + k$

If we add a positive constant k to the right side of $f(x) = a(x - h)^2$, the graph of $f(x)$ is moved up. If we add a negative constant k, the curve is moved down. The axis of symmetry for the parabola remains $x = h$, but the vertex will be at (h, k), or, equivalently, $(h, f(h))$.

Because of the shape of their graphs, quadratic functions have either a *minimum* value or a *maximum* value. Many real-world applications involve finding that value. For example, a business owner is concerned with minimizing cost and maximizing profit. If a parabola opens upward ($a > 0$), the function value, or y-value, at the vertex is a least, or minimum, value. That is, it is less than the y-value at any other point on the graph. If the parabola opens downward ($a < 0$), the function value at the vertex is a greatest, or maximum, value.

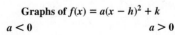

Graphs of $f(x) = a(x - h)^2 + k$

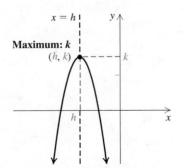

$a < 0$

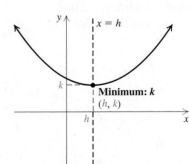

$a > 0$

Technology Connection

To study the effect of k on the graph of $f(x) = a(x - h)^2 + k$, let $y_1 = 7(x - 1)^2$ and $y_2 = 7(x - 1)^2 + 2$. Graph both y_1 and y_2 in the window $[-5, 5, -5, 5]$ and use TRACE or a TABLE to compare the y-values for any given x-value.

1. Let $y_3 = 7(x - 1)^2 - 4$ and compare its graph and y-values with those of y_1 and y_2.

2. Try other values of k, including decimals and fractions. Describe the effect of k on the graph of $f(x) = a(x - h)^2$.

3. If the Transfrm application is available, let $y_1 = A(x - B)^2 + C$ and describe the effect that A, B, and C have on each graph.

GRAPHING $f(x) = a(x - h)^2 + k$

The graph of $f(x) = a(x - h)^2 + k$ has the same shape as the graph of $y = a(x - h)^2$.

- If k is positive, the graph of $y = a(x - h)^2$ is shifted k units up.
- If k is negative, the graph of $y = a(x - h)^2$ is shifted $|k|$ units down.
- The vertex is (h, k), and the axis of symmetry is $x = h$.
- For $a > 0$, the minimum function value is k. For $a < 0$, the maximum function value is k.

EXAMPLE 4   Graph $g(x) = (x - 3)^2 - 5$, and find the vertex and the minimum function value.

SOLUTION   The graph will look like that of $f(x) = (x - 3)^2$ (see Example 2) but shifted 5 units down. You can confirm this by plotting some points.
   The vertex is now $(3, -5)$, and the minimum function value is -5.

x	$g(x) = (x - 3)^2 - 5$
0	4
1	-1
2	-4
3	-5
4	-4
5	-1
6	4

← Vertex

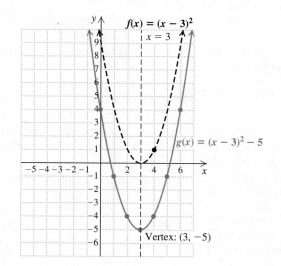

4. Graph $g(x) = (x + 2)^2 - 1$, and find the vertex and the minimum function value.

 YOUR TURN

EXAMPLE 5 Graph $h(x) = \frac{1}{2}(x - 3)^2 + 6$, and find the vertex and the minimum function value.

SOLUTION The graph looks just like that of $f(x) = \frac{1}{2}x^2$ but is shifted 3 units to the right and 6 units up. The vertex is $(3, 6)$, and the axis of symmetry is $x = 3$. We draw $f(x) = \frac{1}{2}x^2$ and then shift the curve over and up. The minimum function value is 6. By plotting some points, we have a check.

x	$h(x) = \frac{1}{2}(x - 3)^2 + 6$
0	$10\frac{1}{2}$
1	8
3	6
5	8
6	$10\frac{1}{2}$

← Vertex

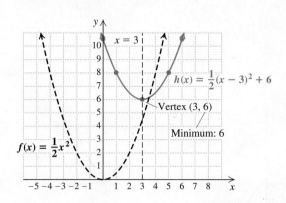

5. Graph
$f(x) = 2(x - 1)^2 + 4$, and find the vertex and the minimum function value.

YOUR TURN

EXAMPLE 6 Graph $y = -2(x + 3)^2 + 5$. Find the vertex, the axis of symmetry, and the maximum or minimum value.

SOLUTION We first express the equation in the equivalent form

$$y = -2[x - (-3)]^2 + 5.$$ This is in the form $y = a(x - h)^2 + k$.

The graph looks like that of $y = -2x^2$ translated 3 units to the left and 5 units up. The vertex is $(-3, 5)$, and the axis of symmetry is $x = -3$. Since -2 is negative, the graph opens downward, and we know that 5, the second coordinate of the vertex, is the maximum y-value.

We compute a few points as needed, selecting convenient x-values on either side of the vertex. The graph is shown here.

Chapter Resource:
Collaborative Activity, p. 575

x	$y = -2(x + 3)^2 + 5$
-4	3
-3	5
-2	3

← Vertex

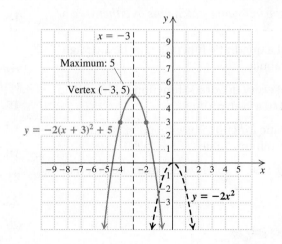

6. Graph $y = -\frac{1}{2}(x - 2)^2 - 1$. Find the vertex, the axis of symmetry, and the maximum or minimum value.

YOUR TURN

EXPLORING THE CONCEPT

The graph shown at right is that of a quadratic function $f(x) = ax^2$.
Match each of the following functions with the appropriate transformation
of the graph of f.

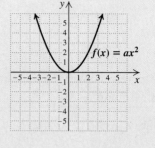

1. $g(x) = ax^2 + 1$
2. $p(x) = a(x + 1)^2$
3. $h(x) = -ax^2$
4. $q(x) = 2ax^2$

a)

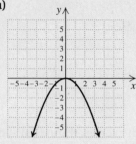

b)

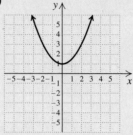

c)

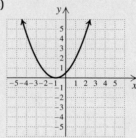

d)

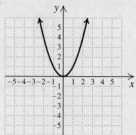

ANSWERS

1. (b)   2. (c)   3. (a)   4. (d)

8.6   EXERCISE SET

FOR EXTRA HELP

MyMathLab® Math XL
PRACTICE   WATCH   READ   REVIEW

Vocabulary and Reading Check

Classify each of the following statements as either true or false.

1. The graph of a quadratic function may be a straight line or a parabola.

2. The graph of every quadratic function is symmetric with respect to a vertical line.

3. Every quadratic function has either a maximum value or a minimum value.

4. The graph of $f(x) = 5x^2$ is wider than the graph of $f(x) = 3x^2$.

The Graph of $f(x) = ax^2$

Graph.

5. $f(x) = x^2$

6. $f(x) = -x^2$

7. $f(x) = -2x^2$

8. $f(x) = -3x^2$

9. $g(x) = \frac{1}{3}x^2$

10. $g(x) = \frac{1}{4}x^2$

Aha! 11. $h(x) = -\frac{1}{3}x^2$

12. $h(x) = -\frac{1}{4}x^2$

13. $f(x) = \frac{5}{2}x^2$

14. $f(x) = \frac{3}{2}x^2$

The Graph of $f(x) = a(x - h)^2$

For each of the following, graph the function, label the vertex, and draw the axis of symmetry.

15. $g(x) = (x + 1)^2$

16. $g(x) = (x + 4)^2$

17. $f(x) = (x - 2)^2$

18. $f(x) = (x - 1)^2$

19. $f(x) = -(x + 1)^2$

20. $f(x) = -(x - 1)^2$

21. $g(x) = -(x - 2)^2$

22. $g(x) = -(x + 4)^2$

23. $f(x) = 2(x + 1)^2$

24. $f(x) = 2(x + 4)^2$

25. $g(x) = 3(x - 4)^2$

26. $g(x) = 3(x - 5)^2$

27. $h(x) = -\frac{1}{2}(x - 4)^2$

28. $h(x) = -\frac{3}{2}(x - 2)^2$

29. $f(x) = \frac{1}{2}(x - 1)^2$

30. $f(x) = \frac{1}{3}(x + 2)^2$

31. $f(x) = -2(x + 5)^2$

32. $f(x) = -3(x + 7)^2$

33. $h(x) = -3\left(x - \frac{1}{2}\right)^2$

34. $h(x) = -2\left(x + \frac{1}{2}\right)^2$

The Graph of $f(x) = a(x - h)^2 + k$

For each of the following, graph the function and find the vertex, the axis of symmetry, and the maximum value or the minimum value.

35. $f(x) = (x - 5)^2 + 2$

36. $f(x) = (x + 3)^2 - 2$

37. $f(x) = (x + 1)^2 - 3$

38. $f(x) = (x - 1)^2 + 2$

39. $g(x) = \frac{1}{2}(x + 4)^2 + 1$

40. $g(x) = -(x - 2)^2 - 4$

41. $h(x) = -2(x - 1)^2 - 3$

42. $h(x) = -2(x + 1)^2 + 4$

43. $f(x) = 2(x + 3)^2 + 1$

44. $f(x) = 2(x - 5)^2 - 3$

45. $g(x) = -\frac{3}{2}(x - 2)^2 + 4$

46. $g(x) = \frac{3}{2}(x + 2)^2 - 1$

Without graphing, find the vertex, the axis of symmetry, and the maximum value or the minimum value.

47. $f(x) = 5(x - 3)^2 + 9$

48. $f(x) = 2(x - 1)^2 - 10$

49. $f(x) = -\frac{3}{7}(x + 8)^2 + 2$

50. $f(x) = -\frac{1}{4}(x + 4)^2 - 12$

51. $f(x) = \left(x - \frac{7}{2}\right)^2 - \frac{29}{4}$

52. $f(x) = -\left(x + \frac{3}{4}\right)^2 + \frac{17}{16}$

53. $f(x) = -\sqrt{2}(x + 2.25)^2 - \pi$

54. $f(x) = 2\pi(x - 0.01)^2 + \sqrt{15}$

55. Explain, without plotting points, why the graph of $y = x^2 - 4$ looks like the graph of $y = x^2$ translated 4 units down.

56. Explain, without plotting points, why the graph of $y = (x + 2)^2$ looks like the graph of $y = x^2$ translated 2 units to the left.

Skill Review

Add or subtract, as indicated. Simplify if possible.

57. $\frac{3}{x} + \frac{x}{x + 2}$ [6.2]

58. $3\sqrt{2x} + \sqrt{50x}$ [7.5]

59. $\sqrt[3]{8t} - \sqrt[3]{27t} + \sqrt{25t}$ [7.5]

60. $(2a^2 - 3a - 7) - (9a^2 - 6a + 1)$ [5.1]

61. $\frac{1}{x - 1} - \frac{x - 2}{x + 3}$ [6.2]

62. $\frac{1}{4 - x} + \frac{8}{x^2 - 16} - \frac{2}{x + 4}$ [6.2]

Synthesis

63. Before graphing a quadratic function, Martha always plots five points. First, she calculates and plots the coordinates of the vertex. Then she plots *four* more points after calculating *two* more ordered pairs. How is this possible?

64. If the graphs of $f(x) = a_1(x - h_1)^2 + k_1$ and $g(x) = a_2(x - h_2)^2 + k_2$ have the same shape, what, if anything, can you conclude about the a's, the h's, and the k's? Why?

Write an equation for a function having a graph with the same shape as the graph of $f(x) = \frac{3}{5}x^2$, but with the given point as the vertex.

65. $(1, 3)$ **66.** $(2, 8)$

67. $(4, -7)$ **68.** $(9, -6)$

69. $(-2, -5)$ **70.** $(-4, -2)$

For each of the following, write the equation of the parabola that has the shape of $f(x) = 2x^2$ or $g(x) = -2x^2$ and has a maximum value or a minimum value at the specified point.

71. Minimum: $(2, 0)$ **72.** Minimum: $(-4, 0)$

73. Maximum: $(0, -5)$ **74.** Maximum: $(3, 8)$

Use the following graph of $f(x) = a(x - h)^2 + k$ for Exercises 75–78.

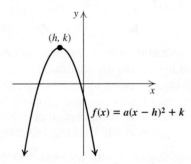

75. Describe what will happen to the graph if h is increased.

76. Describe what will happen to the graph if k is decreased.

77. Describe what will happen to the graph if a is replaced with $-a$.

78. Describe what will happen to the graph if $(x - h)$ is replaced with $(x + h)$.

Find an equation for the quadratic function F that satisfies the following conditions.

79. The graph of F is the same shape as the graph of f, where $f(x) = 3(x + 2)^2 + 7$, and $F(x)$ is a minimum at the same point that $g(x) = -2(x - 5)^2 + 1$ is a maximum.

80. The graph of F is the same shape as the graph of f, where $f(x) = -\frac{1}{3}(x - 2)^2 + 7$, and $F(x)$ is a maximum at the same point that $g(x) = 2(x + 4)^2 - 6$ is a minimum.

Functions other than parabolas can be translated. When calculating $f(x)$, if we replace x with $x - h$, where h is a constant, the graph will be moved horizontally. If we replace $f(x)$ with $f(x) + k$, the graph will be moved vertically. Use the graph below for Exercises 81–86.

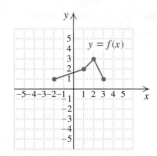

Draw a graph of each of the following.

81. $y = f(x - 1)$

82. $y = f(x + 2)$

83. $y = f(x) + 2$

84. $y = f(x) - 3$

85. $y = f(x + 3) - 2$

86. $y = f(x - 3) + 1$

87. Use the TRACE and/or TABLE features of a graphing calculator to confirm the maximum and minimum values given as answers to Exercises 47, 49, and 51. Be sure to adjust the window appropriately. On many graphing calculators, a maximum or minimum option may be available by using a CALC key.

88. Use a graphing calculator to check your graphs for Exercises 14, 24, and 44.

89. While trying to graph $y = -\frac{1}{2}x^2 + 3x + 1$, Yusef gets the following screen. How can Yusef tell at a glance that a mistake has been made?

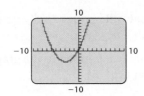

YOUR TURN ANSWERS: SECTION 8.6

1.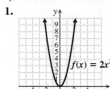

2.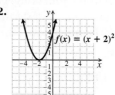

3.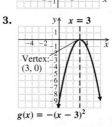

4. Minimum: -1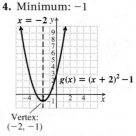

5. Minimum: 4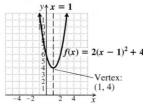

6. Vertex $(2, -1)$; axis of symmetry: $x = 2$; maximum: -1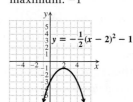

QUICK QUIZ: SECTIONS 8.1–8.6

Solve.

1. $(t + 7)^2 = 13$ [8.1]

2. $x^2 - 3x + 3 = 0$ [8.2]

3. $2m^{-2} - m^{-1} = 15$ [8.5]

4. Solve $t = \dfrac{xy}{3z^2}$ for z.   [8.4]

5. Graph $f(x) = (x - 3)^2 - 2$. Find the vertex, the axis of symmetry, and the maximum or minimum value.   [8.6]

PREPARE TO MOVE ON

Find the x-intercept and the y-intercept.   [2.4]

1. $8x - 6y = 24$

2. $3x + 4y = 8$

Find the x-intercepts of the graph of each equation.   [5.8]

3. $f(x) = x^2 + 8x + 15$

4. $g(x) = 2x^2 - x - 3$

Replace the blanks with constants to form a true equation.   [8.1]

5. $x^2 - 14x + \underline{\quad} = (x - \underline{\quad})^2$

6. $x^2 + 7x + \underline{\quad} = (x + \underline{\quad})^2$

8.7   More About Graphing Quadratic Functions

Graphing $f(x) = ax^2 + bx + c$ • Finding Intercepts

By *completing the square*, we can rewrite any polynomial $ax^2 + bx + c$ in the form $a(x - h)^2 + k$. This will allow us to graph any quadratic function.

Graphing $f(x) = ax^2 + bx + c$

EXAMPLE 1   Graph: $g(x) = x^2 - 6x + 4$. Label the vertex and the axis of symmetry.

SOLUTION   We have

$$g(x) = x^2 - 6x + 4$$
$$= (x^2 - 6x) + 4.$$

To complete the square inside the parentheses, we take half the x-coefficient, $\frac{1}{2} \cdot (-6) = -3$, and square it to get $(-3)^2 = 9$. Then we add $9 - 9$ inside the parentheses:

$$g(x) = (x^2 - 6x + 9 - 9) + 4 \qquad \text{The effect is of adding 0.}$$
$$= (x^2 - 6x + 9) + (-9 + 4) \qquad \text{Using an associative law}$$
$$= (x - 3)^2 - 5. \qquad \text{Factoring and simplifying}$$

The graph is that of $f(x) = x^2$ translated 3 units right and 5 units down. The vertex is $(3, -5)$, and the axis of symmetry is $x = 3$. As a simple check, note that $g(0) = 4$ and $(0, 4)$ is on the graph.

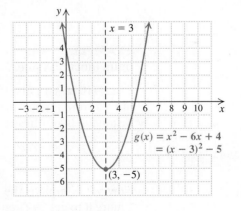

 YOUR TURN

When the leading coefficient is not 1, we factor out that number from the first two terms. Then we complete the square and use the distributive law.

EXAMPLE 2   Graph: $f(x) = 3x^2 + 12x + 13$. Label the vertex and the axis of symmetry.

SOLUTION   Since the coefficient of x^2 is not 1, we need to factor out that number—in this case, 3—from the first two terms. Remember that we want the form $f(x) = a(x - h)^2 + k$:

$$f(x) = 3x^2 + 12x + 13 = 3(x^2 + 4x) + 13.$$

1. Graph: $f(x) = x^2 - 2x + 3$. Label the vertex and the axis of symmetry.

Student Notes

In this section, we add and subtract the same number when completing the square instead of adding the same number to both sides of an equation. The effect is the same with both approaches: An equivalent equation is formed.

Now we complete the square as before. We take half the x-coefficient, $\frac{1}{2} \cdot 4 = 2$, and square it: $2^2 = 4$. Then we add $4 - 4$ inside the parentheses:

$$f(x) = 3(x^2 + 4x + 4 - 4) + 13. \qquad \text{Adding } 4 - 4, \text{ or } 0, \text{ inside the parentheses}$$

The distributive law allows us to separate the -4 from the perfect-square trinomial so long as it is multiplied by 3. *This step is critical*:

$$f(x) = 3(x^2 + 4x + 4 - 4) + 13$$
$$= 3(x^2 + 4x + 4) + 3(-4) + 13 \qquad \text{This leaves a perfect-square trinomial inside the parentheses.}$$
$$= 3(x + 2)^2 + 1. \qquad \text{Factoring and simplifying}$$

The vertex is $(-2, 1)$, and the axis of symmetry is $x = -2$. The coefficient of x^2 is 3, so the graph is narrow and opens upward. We choose x-values on either side of the vertex, compute y-values, and then graph the parabola.

x	$f(x) = 3(x + 2)^2 + 1$
-2	1
-3	4
-1	4

$\longleftarrow$ Vertex

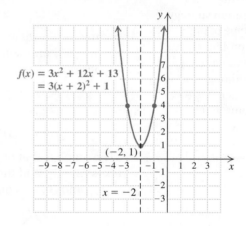

$f(x) = 3x^2 + 12x + 13$
$= 3(x + 2)^2 + 1$

$(-2, 1)$

$x = -2$

2. Graph:

$$f(x) = 2x^2 + 12x + 16.$$

Label the vertex and the axis of symmetry.

YOUR TURN

EXAMPLE 3   Graph $f(x) = -2x^2 + 10x - 7$, and find the maximum or minimum function value.

SOLUTION   We first find the vertex by completing the square. To do so, we factor out -2 from the first two terms of the expression. This makes the coefficient of x^2 inside the parentheses 1:

$$f(x) = -2x^2 + 10x - 7$$
$$= -2(x^2 - 5x) - 7.$$

Factor out a from both variable terms.

Now we complete the square as before. We take half of the x-coefficient and square it to get $\frac{25}{4}$. Then we add $\frac{25}{4} - \frac{25}{4}$ inside the parentheses:

$$f(x) = -2\left(x^2 - 5x + \frac{25}{4} - \frac{25}{4}\right) - 7$$
$$= -2\left(x^2 - 5x + \frac{25}{4}\right) + (-2)\left(-\frac{25}{4}\right) - 7 \qquad \text{Multiplying by } -2, \text{ using the distributive law, and regrouping}$$
$$= -2\left(x - \frac{5}{2}\right)^2 + \frac{11}{2}. \qquad \text{Factoring and simplifying}$$

Complete the square inside the parentheses.

Regroup.

Factor and simplify.

The vertex is $\left(\frac{5}{2}, \frac{11}{2}\right)$, and the axis of symmetry is $x = \frac{5}{2}$. The coefficient of x^2, -2, is negative, so the graph opens downward and the second coordinate of the vertex, $\frac{11}{2}$, is the maximum function value.

We plot a few points on either side of the vertex, including the y-intercept, $f(0)$, and graph the parabola.

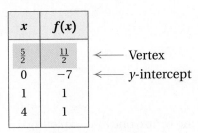

x	$f(x)$	
$\frac{5}{2}$	$\frac{11}{2}$	← Vertex
0	-7	← y-intercept
1	1	
4	1	

3. Graph $f(x) = -2x^2 - 2x - 2$, and find the maximum or minimum function value.

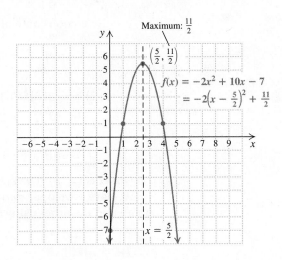

YOUR TURN

The method used in Examples 1–3 can be generalized to find a formula for locating the vertex. We complete the square as follows:

$$f(x) = ax^2 + bx + c$$
$$= a\left(x^2 + \frac{b}{a}x\right) + c.$$

Factoring a out of the first two terms. Check by multiplying.

Half of the x-coefficient, $\frac{b}{a}$, is $\frac{b}{2a}$. We square it to get $\frac{b^2}{4a^2}$ and add $\frac{b^2}{4a^2} - \frac{b^2}{4a^2}$ inside the parentheses. Then we distribute the a and regroup terms:

$$f(x) = a\left(x^2 + \frac{b}{a}x + \frac{b^2}{4a^2} - \frac{b^2}{4a^2}\right) + c$$
$$= a\left(x^2 + \frac{b}{a}x + \frac{b^2}{4a^2}\right) + a\left(-\frac{b^2}{4a^2}\right) + c$$

Using the distributive law

$$= a\left(x + \frac{b}{2a}\right)^2 + \frac{-b^2}{4a} + \frac{4ac}{4a}$$

Factoring and finding a common denominator

$$= a\left[x - \left(-\frac{b}{2a}\right)\right]^2 + \frac{4ac - b^2}{4a}.$$

Thus we have the following.

Student Notes

It is easier to remember a formula when you understand its derivation. Check with your instructor to determine what formulas you will be expected to remember.

THE VERTEX OF A PARABOLA

The vertex of the parabola given by $f(x) = ax^2 + bx + c$ is

$$\left(-\frac{b}{2a}, f\left(-\frac{b}{2a}\right)\right), \quad \text{or} \quad \left(-\frac{b}{2a}, \frac{4ac - b^2}{4a}\right).$$

- The x-coordinate of the vertex is $-b/(2a)$.
- The axis of symmetry is $x = -b/(2a)$.
- The second coordinate of the vertex is most commonly found by computing $f\left(-\frac{b}{2a}\right)$.

Let's reexamine Example 3 to see how we could have found the vertex directly. From the formula above,

$$\text{the } x\text{-coordinate of the vertex is } -\frac{b}{2a} = -\frac{10}{2(-2)} = \frac{5}{2}.$$

Substituting $\frac{5}{2}$ into $f(x) = -2x^2 + 10x - 7$, we find the second coordinate of the vertex:

$$
\begin{aligned}
f\left(\tfrac{5}{2}\right) &= -2\left(\tfrac{5}{2}\right)^2 + 10\left(\tfrac{5}{2}\right) - 7 \\
&= -2\left(\tfrac{25}{4}\right) + 25 - 7 \\
&= -\tfrac{25}{2} + 18 \\
&= -\tfrac{25}{2} + \tfrac{36}{2} = \tfrac{11}{2}.
\end{aligned}
$$

The vertex is $\left(\tfrac{5}{2}, \tfrac{11}{2}\right)$. The axis of symmetry is $x = \tfrac{5}{2}$.

We have developed two methods for finding the vertex, one by completing the square and the other using a formula.

Finding Intercepts

All quadratic functions have a y-intercept and 0, 1, or 2 x-intercepts. For $f(x) = ax^2 + bx + c$, the y-intercept is $(0, f(0))$, or $(0, c)$. To find x-intercepts, if any exist, we look for points where $y = 0$ or $f(x) = 0$. Thus, for $f(x) = ax^2 + bx + c$, the x-intercepts occur at those x-values for which $ax^2 + bx + c = 0$.

Chapter Resource:
Visualizing for Success, p. 574

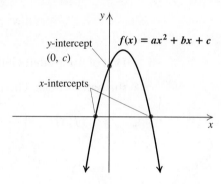

EXAMPLE 4 Find any x-intercepts and the y-intercept of the graph of $f(x) = x^2 - 2x - 2$.

SOLUTION The y-intercept is simply $(0, f(0))$, or $(0, -2)$. To find any x-intercepts, we solve

$$0 = x^2 - 2x - 2.$$

4. Find any x-intercepts and the y-intercept of the graph of $f(x) = 3x^2 + 7x - 20$.

We are unable to factor $x^2 - 2x - 2$, so we use the quadratic formula and get $x = 1 \pm \sqrt{3}$. Thus the x-intercepts are $(1 - \sqrt{3}, 0)$ and $(1 + \sqrt{3}, 0)$.

If graphing, we would approximate, to get $(-0.7, 0)$ and $(2.7, 0)$.

YOUR TURN

If the solutions of $f(x) = 0$ are imaginary, the graph of f has no x-intercepts.

Algebraic & Graphical Connection

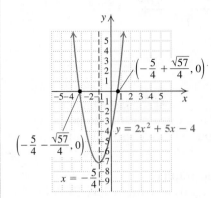

Because the graph of a quadratic equation is symmetric, the x-intercepts of the graph, if they exist, will be symmetric with respect to the axis of symmetry. This symmetry can be seen directly if the x-intercepts are found using the quadratic formula.

For example, the x-intercepts of the graph of $y = 2x^2 + 5x - 4$ are

$$\left(-\frac{5}{4} + \frac{\sqrt{57}}{4}, 0\right) \quad \text{and} \quad \left(-\frac{5}{4} - \frac{\sqrt{57}}{4}, 0\right).$$

For this equation, the axis of symmetry is $x = -\frac{5}{4}$ and the x-intercepts are

$\frac{\sqrt{57}}{4}$ units to the left and right of $-\frac{5}{4}$ on the x-axis.

8.7 EXERCISE SET

FOR EXTRA HELP

MyMathLab® Math XL
PRACTICE   WATCH   READ   REVIEW

Vocabulary and Reading Check

Classify each of the following statements as either true or false.

1. The graph of $f(x) = 3x^2 - x + 6$ opens upward.

2. The function given by $g(x) = -x^2 + 3x + 1$ has a minimum value.

3. The graph of $f(x) = -2(x - 3)^2 + 7$ has its vertex at $(3, 7)$.

4. The graph of $g(x) = 4(x + 6)^2 - 2$ has its vertex at $(-6, -2)$.

5. The graph of $g(x) = \frac{1}{2}(x - \frac{3}{2})^2 + \frac{1}{4}$ has $x = \frac{1}{4}$ as its axis of symmetry.

6. The function given by $f(x) = (x - 2)^2 - 5$ has a minimum value of -5.

7. The y-intercept of the graph of $f(x) = 2x^2 - 6x + 7$ is $(7, 0)$.

8. If the graph of a quadratic function f opens upward and has a vertex of $(1, 5)$, then the graph has no x-intercepts.

Graphing $f(x) = ax^2 + bx + c$

Complete the square to write each function in the form $f(x) = a(x - h)^2 + k$.

9. $f(x) = x^2 - 8x + 2$

10. $f(x) = x^2 - 6x - 1$

11. $f(x) = x^2 + 3x - 5$

12. $f(x) = x^2 + 5x + 3$

13. $f(x) = 3x^2 + 6x - 2$

14. $f(x) = 2x^2 - 20x - 3$

15. $f(x) = -x^2 - 4x - 7$

16. $f(x) = -2x^2 - 8x + 4$

17. $f(x) = 2x^2 - 5x + 10$

18. $f(x) = 3x^2 + 7x - 3$

For each quadratic function, (a) find the vertex and the axis of symmetry and (b) graph the function.

19. $f(x) = x^2 + 4x + 5$

20. $f(x) = x^2 + 2x - 5$

21. $f(x) = x^2 + 8x + 20$

22. $f(x) = x^2 - 10x + 21$

23. $h(x) = 2x^2 - 16x + 25$

24. $h(x) = 2x^2 + 16x + 23$

25. $f(x) = -x^2 + 2x + 5$

26. $f(x) = -x^2 - 2x + 7$

27. $g(x) = x^2 + 3x - 10$

28. $g(x) = x^2 + 5x + 4$

29. $h(x) = x^2 + 7x$

30. $h(x) = x^2 - 5x$

31. $f(x) = -2x^2 - 4x - 6$

32. $f(x) = -3x^2 + 6x + 2$

*For each quadratic function, **(a)** find the vertex, the axis of symmetry, and the maximum or minimum function value and **(b)** graph the function.*

33. $g(x) = x^2 - 6x + 13$

34. $g(x) = x^2 - 4x + 5$

35. $g(x) = 2x^2 - 8x + 3$

36. $g(x) = 2x^2 + 5x - 1$

37. $f(x) = 3x^2 - 24x + 50$

38. $f(x) = 4x^2 + 16x + 13$

39. $f(x) = -3x^2 + 5x - 2$

40. $f(x) = -3x^2 - 7x + 2$

41. $h(x) = \frac{1}{2}x^2 + 4x + \frac{19}{3}$

42. $h(x) = \frac{1}{2}x^2 - 3x + 2$

Finding Intercepts

Find any x-intercepts and the y-intercept. If no x-intercepts exist, state this.

43. $f(x) = x^2 - 6x + 3$

44. $f(x) = x^2 + 5x + 4$

45. $g(x) = -x^2 + 2x + 3$

46. $g(x) = x^2 - 6x + 9$

Aha! 47. $f(x) = x^2 - 9x$

48. $f(x) = x^2 - 7x$

49. $h(x) = -x^2 + 4x - 4$

50. $h(x) = -2x^2 - 20x - 50$

51. $g(x) = x^2 + x - 5$

52. $g(x) = 2x^2 + 3x - 1$

53. $f(x) = 2x^2 - 4x + 6$

54. $f(x) = x^2 - x + 2$

55. The graph of a quadratic function f opens downward and has no x-intercepts. In what quadrant(s) must the vertex lie? Explain your reasoning.

56. Is it possible for the graph of a quadratic function to have only one x-intercept if the vertex is off the x-axis? Why or why not?

Skill Review

Multiply or divide, as indicated. Simplify if possible.

57. $(x^2 - 7)(x^2 + 3)$ [5.2]

58. $\dfrac{x^2 - x - 2}{x^2 - 9} \cdot \dfrac{x^2 + 7x + 12}{x^2 + x}$ [6.1]

59. $\sqrt[3]{18x^4y} \cdot \sqrt[3]{6x^2y}$ [7.3]

60. $(2x^3 - x - 3) \div (x - 1)$ [6.6], [6.7]

61. $\dfrac{4a^2 - b^2}{2ab} \div \dfrac{2a^2 - ab - b^2}{6a^2}$ [6.1]

62. $\dfrac{\sqrt[4]{x^3}}{\sqrt[3]{x^4}}$ [7.5]

Synthesis

63. If the graphs of two quadratic functions have the same x-intercepts, will they also have the same vertex? Why or why not?

64. Suppose that the graph of $f(x) = ax^2 + bx + c$ has $(x_1, 0)$ and $(x_2, 0)$ as x-intercepts. Explain why the graph of $g(x) = -ax^2 - bx - c$ will also have $(x_1, 0)$ and $(x_2, 0)$ as x-intercepts.

*For each quadratic function, find **(a)** the maximum or minimum value and **(b)** any x-intercepts and the y-intercept.*

65. $f(x) = 2.31x^2 - 3.135x - 5.89$

66. $f(x) = -18.8x^2 + 7.92x + 6.18$

67. Graph the function
$$f(x) = x^2 - x - 6.$$
Then use the graph to approximate solutions of the following equations.
   a) $x^2 - x - 6 = 2$
   b) $x^2 - x - 6 = -3$

68. Graph
$$f(x) = \frac{x^2}{2} + x - \frac{3}{2}.$$
Then use the graph to approximate solutions of the following equations.
   a) $\dfrac{x^2}{2} + x - \dfrac{3}{2} = 0$
   b) $\dfrac{x^2}{2} + x - \dfrac{3}{2} = 1$
   c) $\dfrac{x^2}{2} + x - \dfrac{3}{2} = 2$

Find an equivalent equation of the type
$$f(x) = a(x - h)^2 + k.$$

69. $f(x) = mx^2 - nx + p$

70. $f(x) = 3x^2 + mx + m^2$

71. The graph of a quadratic function has $(-1, 0)$ as one intercept and $(3, -5)$ as its vertex. Find an equation for the function.

72. The graph of a quadratic function has $(4, 0)$ as one intercept and $(-1, 7)$ as its vertex. Find an equation for the function.

Graph.

73. $f(x) = |x^2 - 1|$

74. $f(x) = |x^2 - 3x - 4|$

75. $f(x) = |2(x - 3)^2 - 5|$

76. Use a graphing calculator to check your answers to Exercises 25, 41, 53, 65, and 67.

QUICK QUIZ: SECTIONS 8.1–8.7

Solve.

1. $3t^2 = 5$ [8.1]

2. $2x^2 + 3x = 6$ [8.2]

3. Write a quadratic equation having the solutions $\frac{2}{5}$ and -1.   [8.3]

Graph.

4. $f(x) = -3(x + 1)^2$ [8.6]

5. $f(x) = x^2 - 2x + 3$ [8.7]

YOUR TURN ANSWERS: SECTION 8.7

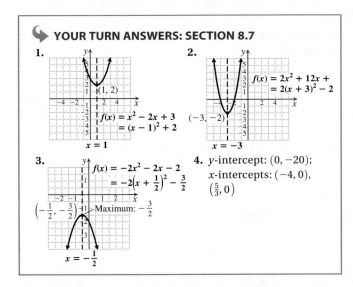

1. $f(x) = x^2 - 2x + 3 = (x - 1)^2 + 2$; $(1, 2)$; $x = 1$

2. $f(x) = 2x^2 + 12x + = 2(x + 3)^2 - 2$; $(-3, -2)$; $x = -3$

3. $f(x) = -2x^2 - 2x - 2 = -2\left(x + \frac{1}{2}\right)^2 - \frac{3}{2}$; $\left(-\frac{1}{2}, -\frac{3}{2}\right)$; Maximum: $-\frac{3}{2}$; $x = -\frac{1}{2}$

4. *y*-intercept: $(0, -20)$; *x*-intercepts: $(-4, 0)$, $\left(\frac{5}{3}, 0\right)$

PREPARE TO MOVE ON

Solve.   [3.4]

1. $x - y + z = -6$,
$2x + y + z = 2$,
$3x + y + z = 0$

2. $z = -5$,
$2x - y + 3z = -27$,
$x + 2y + 7z = -26$

3. $\frac{1}{2} = c$,
$5 = 9a + 6b + 2c$,
$29 = 81a + 9b + c$

8.8   Problem Solving and Quadratic Functions

Maximum and Minimum Problems   •   Fitting Quadratic Functions to Data

Let's look now at some of the many situations in which quadratic functions are used for problem solving.

Maximum and Minimum Problems

We have seen that for any quadratic function f, the value of $f(x)$ at the vertex is either a maximum or a minimum. Thus problems in which a quantity must be maximized or minimized can be solved by finding the coordinates of a vertex, assuming the problem can be modeled with a quadratic function.

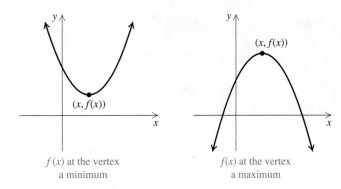

$f(x)$ at the vertex
a minimum

$f(x)$ at the vertex
a maximum

Year	PET Plastic Bottle Recycling Rate
1995	39.7%
2000	22.3
2005	23.1
2010	29.1

Source: National Association for PET Container Resources

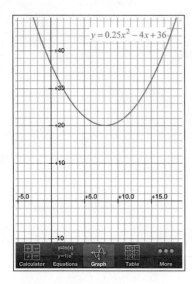

1. The amount of precipitation, in inches, that falls in Seattle, Washington, during the nth month of the year can be approximated by $p(n) = 0.1n^2 - 1.6n + 7$. Here, $n = 1$ represents January, $n = 2$ represents February, and so on. Find the minimum monthly precipitation and the month in which it occurs.

Source: Based on data from city-data.com

EXAMPLE 1 *Recycling.* After dropping for several years, the recyling rate of PET plastic bottles has begun to rise, as indicated by the data in the table at left. The recycling rate, in percent, can be approximated by $r(t) = 0.25t^2 - 4t + 36$, where t is the number of years after 1995. In what year was the recycling rate the lowest, and what percent of PET plastic bottles were recycled that year?

SOLUTION

1., 2. Familiarize and **Translate.** The function given is quadratic. Since the coefficient of the t^2-term is positive, the graph opens upward so a minimum value exists. The calculator-generated graph at left confirms this.

3. Carry out. We can either complete the square or use the formula for the vertex. Completing the square, we have

$$r(t) = 0.25t^2 - 4t + 36$$
$$= 0.25(t^2 - 16t) + 36$$
$$= 0.25(t^2 - 16t + 64 - 64) + 36 \qquad \text{Completing the square}$$
$$= 0.25(t^2 - 16t + 64) + (0.25)(-64) + 36 \qquad \text{Using the distributive law}$$
$$= 0.25(t - 8)^2 + 20. \qquad \text{Factoring and simplifying}$$

There is a minimum value of 20 when $t = 8$.

4. Check. Using the formula, we have $-b/(2a) = -(-4)/(2 \cdot 0.25) = 8$. Then

$$r(8) = 0.25(8)^2 - 4(8) + 36 = 20.$$

Both approaches give the same minimum, and that minimum is also confirmed by the graph. The answer checks.

5. State. The minimum recycling rate was 20%. It occurred 8 years after 1995, or in 2003.

YOUR TURN

EXAMPLE 2 *Swimming Area.* A lifeguard has 100 m of linked flotation devices with which to cordon off a rectangular swimming area at North Beach. If the shoreline forms one side of the rectangle, what dimensions will maximize the size of the area for swimming?

SOLUTION

1. Familiarize. We make a drawing and label it, letting $w =$ the width of the rectangle, in meters, and $l =$ the length of the rectangle, in meters.

Recall that Area $= l \cdot w$ and Perimeter $= 2w + 2l$. Since the beach forms one length of the rectangle, the flotation devices comprise three sides. Thus

$$2w + l = 100.$$

The table below shows some possible dimensions for a rectangular area that can be enclosed with 100 m of flotation devices. All possibilities are chosen so that $2w + l = 100$.

l	w	Rope Length	Area, A
40 m	30 m	100 m	1200 m²
30 m	35 m	100 m	1050 m²
20 m	40 m	100 m	800 m²
⋮	⋮	⋮	⋮

What choice of l and w will maximize A?

Technology Connection

To generate a table of values for Example 2, let x represent the width of the swimming area, in meters. If l represents the length, in meters, we must have $100 = 2x + l$. Next, solve for l and use that expression for y_1. Then let $y_2 = x \cdot y_1$ (to enter y_1, press **VARS** and select Y-VARS and then FUNCTION and then 1) so that y_2 represents the area. Scroll through the resulting table, adjusting the settings as needed, to determine the point at which area is maximized.

2. Refer to Example 2. If the lifeguard has 160 m of flotation devices, what dimensions will maximize the size of the swimming area?

Year	Number of Self-Published Books
2006	51,000
2007	63,000
2008	74,000
2009	95,000
2010	133,000

2. Translate. We have two equations: One guarantees that 100 m of flotation devices are used; the other expresses area in terms of length and width.

$$2w + l = 100,$$
$$A = l \cdot w$$

3. Carry out. We need to express A as a function of either l or w but not both. To do so, we solve for l in the first equation to obtain $l = 100 - 2w$. Substituting for l in the second equation, we get a quadratic function:

$$A = (100 - 2w)w \qquad \text{Substituting for } l$$
$$= 100w - 2w^2 \qquad \text{This represents a parabola opening downward, so a maximum exists.}$$
$$= -2w^2 + 100w.$$

Factoring and completing the square, we get

$$A = -2(w^2 - 50w + 625 - 625) \qquad \text{We could also use the vertex formula.}$$
$$= -2(w - 25)^2 + 1250.$$

There is a maximum value of 1250 when $w = 25$.

4. Check. If $w = 25$ m, then $l = 100 - 2 \cdot 25 = 50$ m. These dimensions give an area of 1250 m². Note that 1250 m² is greater than any of the values for A found in the *Familiarize* step. To be more certain, we could check values other than those used in that step. For example, if $w = 26$ m, then $l = 48$ m, and $A = 26 \cdot 48 = 1248$ m². Since 1250 m² is greater than 1248 m², it appears that we have a maximum.

5. State. The largest rectangular area for swimming that can be enclosed is 25 m by 50 m. Note that the beach is a long side of this rectangle.

YOUR TURN

Fitting Quadratic Functions to Data

Whenever a certain quadratic function fits a situation, that function can be determined if three inputs and their outputs are known.

EXAMPLE 3 *Publishing.* The number of self-published books has increased from 2006 to 2010. As the table and graph suggest, the number of self-published books can be modeled by a quadratic function if we consider the right half of the graph of the function.

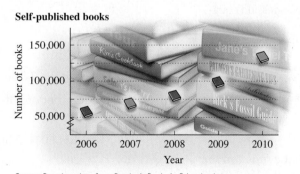

Self-published books

Source: Based on data from Bowker's Books in Print database, in "Secret of Self-Publishing: Success," *The Wall Street Journal*, October 31, 2011.

Chapter Resource:
Decision Making: Connection,
p. 575

a) Let t represent the number of years since 2006 and $p(t)$ the number of self-published books, in thousands. Use the data points $(0, 51), (2, 74),$ and $(4, 133)$ to find a quadratic function that fits the data.

b) Use the function from part (a) to estimate the number of self-published books in 2013.

SOLUTION

a) We are looking for a function of the form $p(t) = at^2 + bt + c$ given that $p(0) = 51, p(2) = 74,$ and $p(4) = 133$. Thus,

$$51 = a \cdot 0^2 + b \cdot 0 + c, \quad \text{Using the data point } (0, 51)$$
$$74 = a \cdot 2^2 + b \cdot 2 + c, \quad \text{Using the data point } (2, 74)$$
$$133 = a \cdot 4^2 + b \cdot 4 + c. \quad \text{Using the data point } (4, 133)$$

After simplifying, we see that we need to solve the system

$$51 = c, \qquad\qquad\qquad (1)$$
$$74 = 4a + 2b + c, \qquad (2)$$
$$133 = 16a + 4b + c. \qquad (3)$$

We know from equation (1) that $c = 51$. Substituting that value into equations (2) and (3), we have

$$74 = 4a + 2b + 51,$$
$$133 = 16a + 4b + 51.$$

Subtracting 51 from both sides of each equation, we have

$$23 = 4a + 2b, \qquad (4)$$
$$82 = 16a + 4b. \qquad (5)$$

To solve, we multiply equation (4) by -2, and then add to eliminate b:

$$-46 = -8a - 4b$$
$$\underline{82 = 16a + 4b}$$
$$36 = 8a$$
$$4.5 = a. \qquad \text{Solving for } a$$

Next, we solve for b, using equation (4) above:

$$23 = 4(4.5) + 2b \qquad \text{Substituting}$$
$$23 = 18 + 2b$$
$$5 = 2b$$
$$2.5 = b. \qquad\qquad \text{Solving for } b$$

We can now write $p(t) = at^2 + bt + c$ as

$$p(t) = 4.5t^2 + 2.5t + 51.$$

b) To find the number of self-published books in 2013, we evaluate the function for $t = 7$, because 2013 is 7 years after 2006:

$$p(7) = 4.5(7)^2 + 2.5(7) + 51$$
$$= 289.$$

We estimate that 289,000 books will be self-published in 2013.

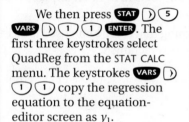

Technology Connection

To use a graphing calculator to fit a quadratic function to the data in Example 3, we first select EDIT in the **STAT** menu and enter the given data.

L1	L2	L3	2
0	51	------	
2	74		
4	133		

L2(4) =

We then press **STAT** ▷ **5**
VARS ▷ **1** **1** **ENTER**. The first three keystrokes select QuadReg from the STAT CALC menu. The keystrokes **VARS** ▷ **1** **1** copy the regression equation to the equation-editor screen as y_1.

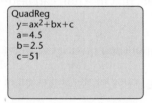

QuadReg
y=ax²+bx+c
a=4.5
b=2.5
c=51

We see that the regression equation is $y = 4.5x^2 + 2.5x + 51$.
　　To check Example 3(b), we set Indpnt to Ask in the Table Setup and enter $X = 7$ in the table. A Y1-value of 289 confirms our answer.

3. Find a quadratic function that fits the points $(0, 6), (1, 8),$ and $(3, 4)$.

 YOUR TURN

8.8 EXERCISE SET

⤷ Vocabulary and Reading Check

In each of Exercises 1–6, match the description with the graph that displays that characteristic.

1. ____ A minimum value of $f(x)$ exists.

2. ____ A maximum value of $f(x)$ exists.

3. ____ No maximum or minimum value of $f(x)$ exists.

4. ____ The data points appear to suggest a linear model for g.

5. ____ The data points appear to suggest that g is a quadratic function with a maximum.

6. ____ The data points appear to suggest that g is a quadratic function with a minimum.

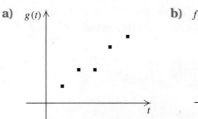

a)

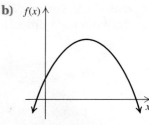

b)

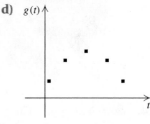

c)

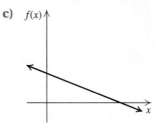

d)

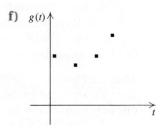

e)

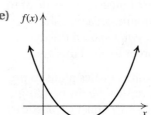

f)

Maximum and Minimum Problems

Solve.

7. *Newborn Calves.* The number of pounds of milk per day recommended for a calf that is x weeks old can be approximated by $p(x)$, where

$$p(x) = -0.2x^2 + 1.3x + 6.2.$$

When is a calf's milk consumption greatest and how much milk does it consume at that time?

Source: C. Chaloux, University of Vermont

8. *Stock Prices.* The value of a share of I. J. Solar can be represented by $V(x) = x^2 - 6x + 13$, where x is the number of months after January 2009. What is the lowest value $V(x)$ will reach, and when did that occur?

9. *Minimizing Cost.* Sweet Harmony Crafts has determined that when x hundred dulcimers are built, the average cost per dulcimer can be estimated by

$$C(x) = 0.1x^2 - 0.7x + 2.425,$$

where $C(x)$ is in hundreds of dollars. What is the minimum average cost per dulcimer and how many dulcimers should be built in order to achieve that minimum?

10. *Maximizing Profit.* Recall that total profit P is the difference between total revenue R and total cost C. Given $R(x) = 1000x - x^2$ and $C(x) = 3000 + 20x$, find the total profit, the maximum value of the total profit, and the value of x at which it occurs.

11. *Architecture.* An architect is designing an atrium for a hotel. The atrium is to be rectangular with a perimeter of 720 ft of brass piping. What dimensions will maximize the area of the atrium?

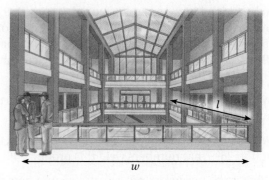

12. *Furniture Design.* A furniture builder is designing a rectangular end table with a perimeter of 128 in. What dimensions will yield the maximum area?

13. *Patio Design.*   A stone mason has enough stones to enclose a rectangular patio with 60 ft of perimeter, assuming that the attached house forms one side of the rectangle. What is the maximum area that the mason can enclose? What should the dimensions of the patio be in order to yield this area?

14. *Garden Design.*   Ginger is fencing in a rectangular garden, using the side of her house as one side of the rectangle. What is the maximum area that she can enclose with 40 ft of fence? What should the dimensions of the garden be in order to yield this area?

15. *Molding Plastics.*   Economite Plastics plans to produce a one-compartment vertical file by bending the long side of an 8-in. by 14-in. sheet of plastic along two lines to form a U shape. How tall should the file be in order to maximize the volume that the file can hold?

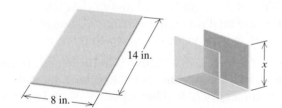

16. *Composting.*   A rectangular compost container is to be formed in a corner of a fenced yard, with 8 ft of chicken wire completing the other two sides of the rectangle. If the chicken wire is 3 ft high, what dimensions of the base will maximize the container's volume?

17. What is the maximum product of two numbers that add to 18? What numbers yield this product?

18. What is the maximum product of two numbers that add to 26? What numbers yield this product?

19. What is the minimum product of two numbers that differ by 8? What are the numbers?

20. What is the minimum product of two numbers that differ by 7? What are the numbers?

Aha! **21.** What is the maximum product of two numbers that add to -10? What numbers yield this product?

22. What is the maximum product of two numbers that add to -12? What numbers yield this product?

Fitting Quadratic Functions to Data

Choosing Models.   For the scatterplots and graphs in Exercises 23–34, determine which, if any, of the following functions might be used as a model for the data: Linear, with $f(x) = mx + b$; quadratic, with $f(x) = ax^2 + bx + c, a > 0$; quadratic, with $f(x) = ax^2 + bx + c, a < 0$; neither quadratic nor linear.

23.   **Sonoma Sunshine**

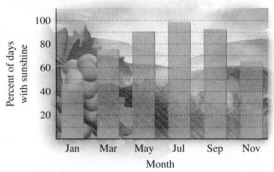

Source: www.city-data.com

24.   **Sonoma Precipitation**

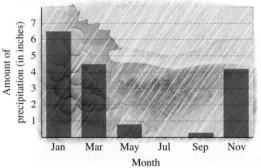

Source: www.city-data.com

25.   **Safe sight distance to the left**

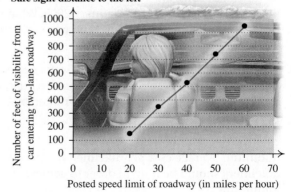

Source: Institute of Traffic Engineers

26. Safe sight distance to the right

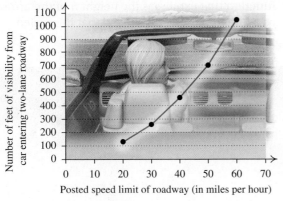

Source: Institute of Traffic Engineers

27. Dow Jones Industrial Average

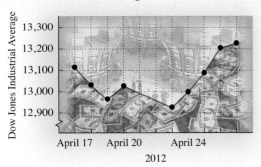

28. U.S. senior population

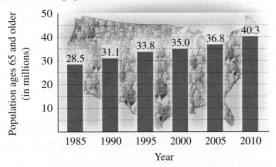

Source: U.S. Bureau of Labor Statistics

29. Working longer

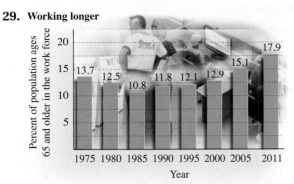

Source: U.S. Department of Labor, Bureau of Labor Statistics

30. Airline bumping rate

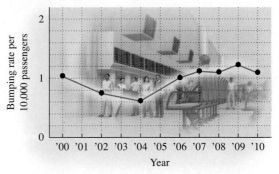

Source: U.S. Department of Transportation

31. Carbon footprint

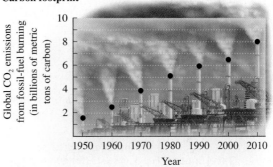

Source: Carbon Dioxide Information Analysis Center, U.S. Department of Energy

32. Angie's List

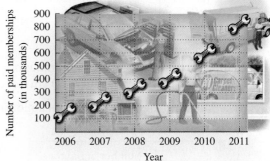

Source: *The Wall Street Journal*, August 30, 2011

33. The worth of a dollar

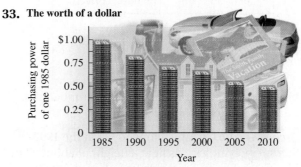

Source: Based on data from U.S. Bureau of Labor Statistics

34. Average number of live births per 1000 women, 2009

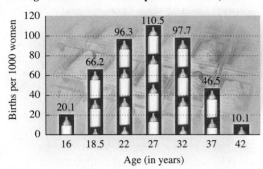

Source: U.S. Centers for Disease Control

Find a quadratic function that fits the set of data points.

35. $(1, 4), (-1, -2), (2, 13)$

36. $(1, 4), (-1, 6), (-2, 16)$

37. $(2, 0), (4, 3), (12, -5)$

38. $(-3, -30), (3, 0), (6, 6)$

39. a) Find a quadratic function that fits the following data.

Travel Speed (in kilometers per hour)	Number of Nighttime Accidents (for every 200 million kilometers driven)
60	400
80	250
100	250

b) Use the function to estimate the number of nighttime accidents that occur at 50 km/h.

40. a) Find a quadratic function that fits the following data.

Travel Speed (in kilometers per hour)	Number of Daytime Accidents (for every 200 million kilometers driven)
60	100
80	130
100	200

b) Use the function to estimate the number of daytime accidents that occur at 50 km/h.

41. *Archery.* The Olympic flame tower at the 1992 Summer Olympics was lit at a height of about 27 m by a flaming arrow that was launched about 63 m from the base of the tower. If the arrow landed about 63 m beyond the tower, find a quadratic function that expresses the height h of the arrow as a function of the distance d that it traveled horizontally.

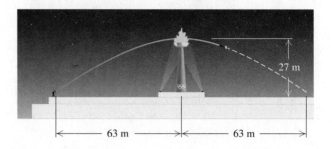

42. *Sump Pump.* The lift distances for a Liberty 250 sump pump moving fluid at various flow rates are shown in the following table.

Gallons per Minute	Lift Distance (in feet)
10	21
20	18
40	8

Source: Liberty Pumps

a) Let x represent the flow rate, in gallons per minute, and $d(x)$ the lift distance, in feet. Find a quadratic function that fits the data.

b) Use the function to find the lift distance for a flow rate of 50 gal per min.

43. Does every nonlinear function have a minimum or a maximum value? Why or why not?

44. Explain how the leading coefficient of a quadratic function can be used to determine whether a maximum or a minimum function value exists.

Skill Review

Find an equation in slope–intercept form of a line with the given characteristics.

45. Slope: $-\frac{1}{3}$; y-intercept: $(0, 16)$ [2.3]

46. Slope: 2; contains $(-3, 7)$ [2.5]

47. Contains $(4, 8)$ and $(10, 0)$ [2.5]

48. Parallel to $y = \frac{2}{3}x + 5$; contains $(-2, -9)$ [2.5]

49. Perpendicular to $2x + y = 3$; y-intercept: $(0, -6)$ [2.5]

50. Horizontal line through $(7, -4)$ [2.5]

Synthesis

The following graphs can be used to compare the base-ball statistics of pitcher Roger Clemens with the 31 other pitchers since 1968 who started at least 10 games in at least 15 seasons and pitched at least 3000 innings. Use the graphs to answer questions 51 and 52.

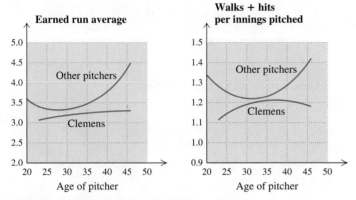

Earned run average

5.0
4.5 Other pitchers
4.0
3.5
3.0 Clemens
2.5
2.0
  20 25 30 35 40 45 50
Age of pitcher

Walks + hits per innings pitched

1.5
1.4 Other pitchers
1.3
1.2 Clemens
1.1
1.0
0.9
  20 25 30 35 40 45 50
Age of pitcher

*Source: The New York Times, February 10, 2008;
Eric Bradlow, Shane Jensen, Justin Wolfers and Adi Wyner*

51. The earned run average describes how many runs a pitcher has allowed per game. The lower the earned run average, the better a pitcher. Compare, in terms of maximums or minimums, the earned run average of Roger Clemens with that of other pitchers. Is there any reason to suspect that the aging process was unusual for Clemens? Explain.

52. The statistic "Walks + hits per innings pitched" is related to how often a pitcher allows a batter to reach a base. The lower this statistic, the better. Compare, in terms of maximums or minimums, the "walks + hits" statistic of Roger Clemens with that of other pitchers.

53. *Bridge Design.* The cables supporting a straight-line suspension bridge are nearly parabolic in shape. Suppose that a suspension bridge is being designed with concrete supports 160 ft apart and with vertical cables 30 ft above road level at the midpoint of the bridge and 80 ft above road level at a point 50 ft from the midpoint of the bridge. How long are the longest vertical cables?

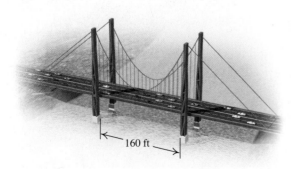

160 ft

54. *Trajectory of a Launched Object.* The height above the ground of a launched object is a quadratic function of the time that it is in the air. Suppose that a flare is launched from a cliff 64 ft above sea level. If 3 sec after being launched the flare is again level with the cliff, and if 2 sec after that it lands in the sea, what is the maximum height that the flare reaches?

55. *Cover Charges.* When the owner of Sweet Sounds charges a $10 cover charge, an average of 80 people will attend a show. For each 25¢ increase in admission price, the average number attending decreases by 1. What should the owner charge in order to make the most money?

56. *Crop Yield.* An orange grower finds that she gets an average yield of 40 bushels (bu) per tree when she plants 20 trees on an acre of ground. Each time she adds one tree per acre, the yield per tree decreases by 1 bu, due to congestion. How many trees per acre should she plant for maximum yield?

57. *Norman Window.* A *Norman window* is a rectangle with a semicircle on top. Big Sky Windows is designing a Norman window that will require 24 ft of trim. What dimensions will allow the maximum amount of light to enter a house?

58. *Minimizing Area.* A 36-in. piece of string is cut into two pieces. One piece is used to form a circle while the other is used to form a square. How should the string be cut so that the sum of the areas is a minimum?

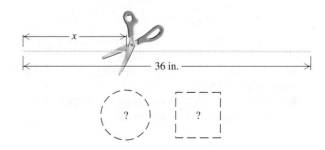

Regression can be used to find the "best"-fitting quadratic function when more than three data points are provided.

59. Business. The numbers of paid subscriptions per year to Angie's List, a company that provides business reviews, are shown in the following table.

Year	Number of Paid Subscriptions (in thousands)
2006	152.1
2007	234.9
2008	333.5
2009	411.5
2010	602.9
2011	821.8*

Source: The Wall Street Journal, August 30, 2011

*As of June 30, 2011

a) Use regression to find a quadratic function that can be used to estimate the number of paid subscriptions $a(x)$ to Angie's List x years after 2006.

b) Use the function found in part (a) to predict the number of paid subscriptions to Angie's List in 2014.

60. Hydrology. The drawing below shows the cross section of a river. Typically rivers are deepest in the middle, with the depth decreasing to 0 at the edges. A hydrologist measures the depths D, in feet, of a river at distances x, in feet, from one bank. The results are listed in the table below.

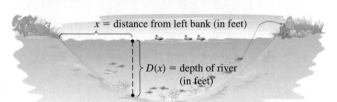

Distance x, from the Left Bank (in feet)	Depth D of the River (in feet)
0	0
15	10.2
25	17
50	20
90	7.2
100	0

a) Use regression to find a quadratic function that fits the data.

b) Use the function to estimate the depth of the river 70 ft from the left bank.

61. Research. Find the number of self-published books in 2013 and compare it with the estimate in Example 3(b).

YOUR TURN ANSWERS: SECTION 8.8

1. Minimum precipitation: 0.6 in. in month 8, or August

2. 40 m by 80 m   **3.** $f(x) = -\frac{4}{3}x^2 + \frac{10}{3}x + 6$

QUICK QUIZ: SECTIONS 8.1–8.8

Solve.   [8.1], [8.2], [8.5]

1. $12x^2 + 7x = 10$

2. $(x - 4)^2 - (x - 4) = 6$

3. Write a quadratic equation having the solutions $5i$ and $-5i$.   [8.3]

4. Solve $V = 3.5\sqrt{h}$ for h.   [8.4]

5. Graph: $f(x) = 2(x - 4)^2 + 3$.   [8.6]

PREPARE TO MOVE ON

Solve.

1. $4 - x \le 7$ [4.1]   **2.** $|4x + 1| < 11$ [4.3]

Subtract to find an equivalent expression for $f(x)$ and list all restrictions on the domain.   [6.2]

3. $f(x) = \dfrac{x - 3}{x + 4} - 5$ **4.** $f(x) = \dfrac{x}{x - 1} - 1$

Solve.   [6.4]

5. $\dfrac{x}{x - 1} = 1$ **6.** $\dfrac{(x + 6)(x - 9)}{x + 5} = 0$

8.9 Polynomial Inequalities and Rational Inequalities

Quadratic and Other Polynomial Inequalities ▪ Rational Inequalities

Quadratic and Other Polynomial Inequalities

Inequalities like the following are called *polynomial inequalities*:

$$x^3 - 5x > x^2 + 7, \quad 4x - 3 < 9, \quad 5x^2 - 3x + 2 \geq 0.$$

Second-degree polynomial inequalities in one variable are called *quadratic inequalities*. To solve polynomial inequalities, we can focus attention on where the outputs of a polynomial function are positive and where they are negative.

EXAMPLE 1   Solve: $x^2 + 3x - 10 > 0$.

SOLUTION   Consider the "related" function $f(x) = x^2 + 3x - 10$. We are looking for those x-values for which $f(x) > 0$. Graphically, function values are positive when the graph is above the x-axis.

The graph of f opens upward since the leading coefficient is positive. Thus function values are positive outside the interval formed by the x-intercepts. To find the intercepts, we set the polynomial equal to 0 and solve:

$$x^2 + 3x - 10 = 0$$
$$(x + 5)(x - 2) = 0$$
$$x + 5 = 0 \quad or \quad x - 2 = 0$$
$$x = -5 \quad or \quad x = 2. \quad \text{The } x\text{-intercepts are } (-5, 0) \text{ and } (2, 0).$$

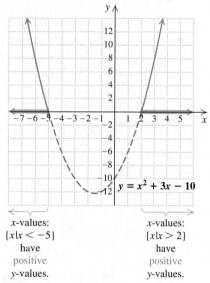

$y = x^2 + 3x - 10$

x-values: $\{x | x < -5\}$ have positive y-values.

x-values: $\{x | x > 2\}$ have positive y-values.

Thus the solution set of the inequality is

$$(-\infty, -5) \cup (2, \infty), \quad \text{or} \quad \{x \,|\, x < -5 \ or \ x > 2\}.$$

1. Solve: $x^2 - 2x - 8 > 0$.

↩ YOUR TURN

Any inequality with 0 on one side can be solved by considering a graph of the related function and finding intercepts as in Example 1.

EXAMPLE 2   Solve: $x^2 - 2x \leq 2$.

SOLUTION   We first write the quadratic inequality in standard form:

$$x^2 - 2x - 2 \leq 0. \qquad \text{This is equivalent to the original inequality.}$$

The graph of $f(x) = x^2 - 2x - 2$ is a parabola opening upward. Values of $f(x)$ are negative for x-values between the x-intercepts. We find the x-intercepts by solving $f(x) = 0$:

$$x = \frac{-b \pm \sqrt{b^2 - 4ac}}{2a}$$

$$= \frac{-(-2) \pm \sqrt{(-2)^2 - 4 \cdot 1(-2)}}{2 \cdot 1}$$

$$= \frac{2 \pm \sqrt{12}}{2} = \frac{2}{2} \pm \frac{2\sqrt{3}}{2} = 1 \pm \sqrt{3}.$$

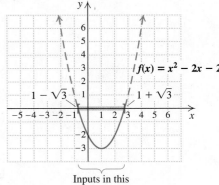

Inputs in this interval have negative or 0 outputs.

At the x-intercepts, $1 - \sqrt{3}$ and $1 + \sqrt{3}$, the value of $f(x)$ is 0. Since the inequality symbol is $\leq$, the solution set will include all values of x for which $f(x)$ is negative *or* $f(x)$ is 0. Thus the solution set of the inequality is

$$[1 - \sqrt{3}, 1 + \sqrt{3}], \quad \text{or} \quad \{x \mid 1 - \sqrt{3} \leq x \leq 1 + \sqrt{3}\}.$$

2. Solve: $x^2 + 4x \leq 2$.

_____   YOUR TURN

In Example 2, it was not essential to draw the graph. The important information came from finding the x-intercepts and the sign of $f(x)$ on each side of those intercepts. We now solve a polynomial inequality, without graphing the related function f, but instead by locating the x-intercepts, or **zeros**, of f and then using *test points* to determine the sign of $f(x)$ over each interval of the x-axis.

EXAMPLE 3   For $f(x) = 5x^3 + 10x^2 - 15x$, find all x-values for which $f(x) > 0$.

SOLUTION   We first solve the related equation:

$$f(x) = 0$$

$$5x^3 + 10x^2 - 15x = 0 \qquad \text{Substituting}$$

$$\left.\begin{array}{l} 5x(x^2 + 2x - 3) = 0 \\ 5x(x + 3)(x - 1) = 0 \end{array}\right\} \quad \begin{array}{l} \text{Since } f(x) \text{ is third-degree, we expect up to} \\ \text{three zeros.} \end{array}$$

$$5x = 0 \quad or \quad x + 3 = 0 \quad or \quad x - 1 = 0$$

$$x = 0 \quad or \quad x = -3 \quad or \quad x = 1.$$

The zeros of f are $-3, 0$, and 1. These zeros divide the number line, or x-axis, into four intervals: A, B, C, and D.

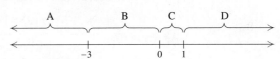

Student Notes

When we are evaluating test values, there is often no need to do lengthy computations since all we need to determine is the sign of the result.

Next, selecting one convenient test value from each interval, we determine the sign of $f(x)$ over that interval. We know that, within each interval, the sign of $f(x)$ cannot change. If it did, there would need to be another zero in that interval. Using the factored form of $f(x)$ eases the computations:

$$f(x) = (5x) \cdot (x + 3) \cdot (x - 1).$$

For interval A,

$$f(-4) = (5(-4)) \cdot (-4 + 3) \cdot (-4 - 1) \qquad \text{-4 is a convenient value in interval A.}$$
$$= (-20) \cdot (-1) \cdot (-5)$$
$$= -100. \qquad \text{$f(-4)$ is negative.}$$

For interval B,

$$f(-1) = (5(-1)) \cdot (-1 + 3) \cdot (-1 - 1) \qquad \text{-1 is a convenient value in interval B.}$$
$$= (-5) \cdot (2) \cdot (-2)$$
$$= 20. \qquad \text{$f(-1)$ is positive.}$$

For interval C,

$$f\left(\tfrac{1}{2}\right) = \left(5 \cdot \tfrac{1}{2}\right) \cdot \left(\tfrac{1}{2} + 3\right) \cdot \left(\tfrac{1}{2} - 1\right) \qquad \text{$\tfrac{1}{2}$ is a convenient value in interval C.}$$
$$= \left(\tfrac{5}{2}\right) \cdot \left(\tfrac{7}{2}\right) \cdot \left(-\tfrac{1}{2}\right)$$
$$= -\tfrac{35}{8}. \qquad \text{$f\left(\tfrac{1}{2}\right)$ is negative.}$$

For interval D,

$$f(2) = (5 \cdot 2) \cdot (2 + 3) \cdot (2 - 1) \qquad \text{2 is a convenient value in interval D.}$$
$$= (10) \cdot (5) \cdot (1)$$
$$= 50. \qquad \text{$f(2)$ is positive.}$$

We indicate on the number line the sign of $f(x)$ in each interval.

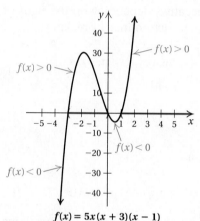

$f(x) > 0$

$f(x) > 0$

$f(x) < 0$

$f(x) < 0$

$f(x) = 5x(x + 3)(x - 1)$

A computer-generated visualization of Example 3

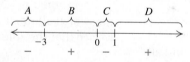

Recall that we are looking for all x for which $5x^3 + 10x^2 - 15x > 0$. The calculations above indicate that $f(x)$ is positive for any number in intervals B and D. The solution set of the original inequality is

$$(-3, 0) \cup (1, \infty), \quad \text{or} \quad \{x \mid -3 < x < 0 \ or \ x > 1\}.$$

3. For $f(x) = 3x^3 + 9x^2 + 6x$, find all x-values for which $f(x) < 0$.

 YOUR TURN

TO SOLVE A POLYNOMIAL INEQUALITY USING FACTORS

1. Add or subtract to get 0 on one side and solve the related polynomial equation.
2. Use the numbers found in step (1) to divide the number line into intervals.
3. Using a test value from each interval, determine the sign of the function over each interval.
4. Select the interval(s) for which the inequality is satisfied and write interval notation or set-builder notation for the solution set. Include endpoints of intervals when $\leq$ or $\geq$ appear.

We need focus only on the *sign* of $f(x)$. By looking at the number of positive and negative factors, we can determine the sign of the polynomial function.

EXAMPLE 4   For $f(x) = 4x^4 - 4x^2$, find all x-values for which $f(x) \le 0$.

SOLUTION   We first solve the related equation:

Solve $f(x) = 0$.

$$f(x) = 0$$
$$4x^4 - 4x^2 = 0 \qquad \text{Substituting}$$
$$\left. \begin{array}{l} 4x^2(x^2 - 1) = 0 \\ 4x^2(x + 1)(x - 1) = 0 \end{array} \right\} \quad \begin{array}{l}\text{We expect up to four zeros of a fourth-}\\ \text{degree polynomial function.}\end{array}$$
$$4x^2 = 0 \quad or \quad x + 1 = 0 \quad or \quad x - 1 = 0$$
$$x = 0 \quad or \qquad\quad x = -1 \quad or \qquad x = 1.$$

Divide the number line into intervals.

Since f has zeros at $-1, 0$, and 1, we divide the number line into four intervals:

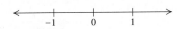

The product $4x^2(x + 1)(x - 1)$ is positive or negative, depending on the signs of $4x^2$, $x + 1$, and $x - 1$. The sign of the product can be determined by making a chart.

Determine the sign of the function over each interval.

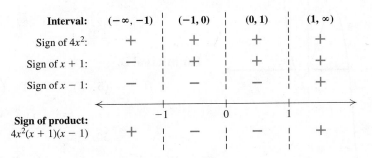

Select the interval(s) for which the inequality is satisfied.

A product is negative when it has an odd number of negative factors. Since the $\le$ sign allows for equality, the zeros $-1, 0$, and 1 are solutions. From the chart, we see that the solution set is

$$[-1, 0] \cup [0, 1], \quad \text{or simply} \quad [-1, 1], \quad \text{or} \quad \{x \mid -1 \le x \le 1\}.$$

4. For $f(x) = 5x^3 - 5x$, find all x-values for which $f(x) \ge 0$.

YOUR TURN

Technology Connection

To solve $2.3x^2 \le 9.11 - 2.94x$, we write the inequality in the form $2.3x^2 + 2.94x - 9.11 \le 0$ and graph the function $f(x) = 2.3x^2 + 2.94x - 9.11$.

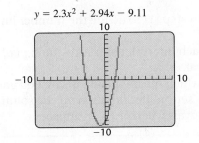

$y = 2.3x^2 + 2.94x - 9.11$

The x-values for which the graph lies *on or below* the x-axis begin somewhere between -3 and -2, and continue to somewhere between 1 and 2. Using the ZERO option of CALC and rounding, we find that these endpoints are -2.73 and 1.45. The solution set is approximately $\{x \mid -2.73 \le x \le 1.45\}$.

Use a graphing calculator to solve each inequality. Round the values of the endpoints to the nearest hundredth.

1. $4.32x^2 - 3.54x - 5.34 \le 0$
2. $7.34x^2 - 16.55x - 3.89 \ge 0$
3. $5.79x^3 - 5.68x^2 + 10.68x$
    $> 2.11x^3 + 16.90x - 11.69$

Rational Inequalities

Inequalities involving rational expressions are called **rational inequalities**. Like polynomial inequalities, rational inequalities can be solved using test values. Unlike polynomials, however, rational expressions often have values for which the expression is undefined. These values must be used when dividing the number line into intervals.

EXAMPLE 5 Solve: $\dfrac{x - 3}{x + 4} \geq 2$.

SOLUTION We write the related equation by changing the $\geq$ symbol to $=$:

$$\frac{x - 3}{x + 4} = 2. \qquad \text{Note that } x \neq -4.$$

Next, we solve this related equation:

$$(x + 4) \cdot \frac{x - 3}{x + 4} = (x + 4) \cdot 2 \qquad \begin{array}{l}\text{Clearing fractions by multiplying both}\\ \text{sides by } x + 4\end{array}$$

$$x - 3 = 2x + 8$$

$$-11 = x. \qquad \text{Solving for } x. \text{ Note that } x \neq -4.$$

Since -11 is a solution of the related equation, we use -11 when dividing the number line into intervals. Since the rational expression is undefined for $x = -4$, we must also use -4:

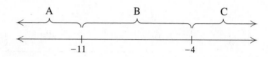

We test a number from each interval to see where the original inequality is satisfied:

$$\frac{x - 3}{x + 4} \geq 2.$$

For interval A,

$$\text{Test } -15, \quad \frac{-15 - 3}{-15 + 4} = \frac{-18}{-11} \qquad \text{Any } x < -11 \text{ could be the test value.}$$

$$= \frac{18}{11} \not\geq 2. \qquad \begin{array}{l}-15 \text{ } is \text{ } not \text{ a solution, so interval A is}\\ \text{not part of the solution set.}\end{array}$$

For interval B,

$$\text{Test } -8, \quad \frac{-8 - 3}{-8 + 4} = \frac{-11}{-4} \qquad \begin{array}{l}\text{Any } x \text{ between } -11 \text{ and } -4 \text{ could be}\\ \text{used.}\end{array}$$

$$= \frac{11}{4} \geq 2. \qquad \begin{array}{l}-8 \text{ } is \text{ a solution, so interval B is part of}\\ \text{the solution set.}\end{array}$$

For interval C,

$$\text{Test } 1, \quad \frac{1 - 3}{1 + 4} = \frac{-2}{5} \qquad \text{Any } x > -4 \text{ could be used.}$$

$$= -\frac{2}{5} \not\geq 2. \qquad \begin{array}{l}1 \text{ } is \text{ } not \text{ a solution, so interval C is not}\\ \text{part of the solution set.}\end{array}$$

The solution set includes interval B. The endpoint -11 is included because the inequality symbol is $\geq$ and -11 is a solution of the related equation. The number -4 is *not* included because $(x - 3)/(x + 4)$ is undefined for $x = -4$. Thus the solution set of the original inequality is

5. Solve: $\dfrac{x + 2}{x - 1} \leq 5$.

$$[-11, -4), \quad \text{or} \quad \{x \mid -11 \leq x < -4\}.$$

↩ YOUR TURN

Algebraic 🔗 Graphical Connection

To compare the algebraic solution of Example 5 with a graphical solution, we graph $f(x) = (x - 3)/(x + 4)$ and the line $y = 2$. The solutions of $(x - 3)/(x + 4) \geq 2$ are found by locating all x-values for which $f(x) \geq 2$.

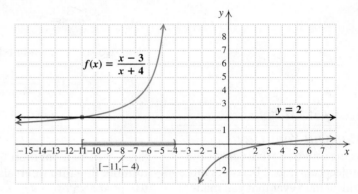

TO SOLVE A RATIONAL INEQUALITY

1. Find any replacements for which the rational expression is undefined.
2. Change the inequality symbol to an equal sign and solve the related equation.
3. Use the numbers found in steps (1) and (2) to divide the number line into intervals.
4. Substitute a test value from each interval into the inequality. If the number is a solution, then the interval to which it belongs is part of the solution set.
5. Select all interval(s) and endpoints for which the inequality is satisfied and use interval notation or set-builder notation to write the solution set. If the inequality symbol is $\leq$ or $\geq$, then solutions from step (2) are included in the solution set. (All numbers found in step (1) are excluded from the solution set.)

8.9 EXERCISE SET

FOR EXTRA HELP   MyMathLab® Math XL
PRACTICE   WATCH   READ   REVIEW

➤ Vocabulary and Reading Check

Classify each of the following statements as either true or false.

1. The solution of $(x - 3)(x + 2) \leq 0$ is $[-2, 3]$.

2. The solution of $(x + 5)(x - 4) \geq 0$ is $[-5, 4]$.

3. The solution of $(x - 1)(x - 6) > 0$ is $\{x \,|\, x < 1 \text{ or } x > 6\}$.

4. The solution of $(x + 4)(x + 2) < 0$ is $(-4, -2)$.

5. To solve $\dfrac{x + 2}{x - 3} < 0$ using intervals, we divide the number line into the intervals $(-\infty, -2)$ and $(-2, \infty)$.

6. To solve $\dfrac{x - 5}{x + 4} \geq 0$ using intervals, we divide the number line into the intervals $(-\infty, -4)$, $(-4, 5)$, and $(5, \infty)$.

Polynomial Inequalities and Rational Inequalities

Solve each inequality using the graph provided.

7. $p(x) \leq 0$

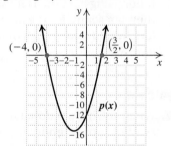

8. $p(x) < 0$

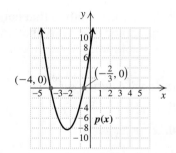

9. $x^4 + 12x > 3x^3 + 4x^2$

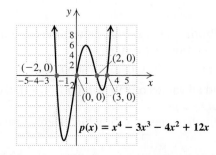

10. $x^4 + x^3 \geq 6x^2$

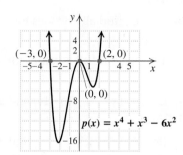

11. $\dfrac{x - 1}{x + 2} < 3$

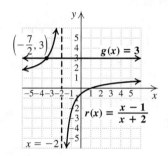

12. $\dfrac{2x - 1}{x - 5} \geq 1$

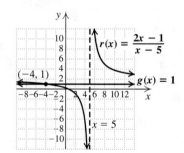

Quadratic and Other Polynomial Inequalities

Solve.

13. $(x - 6)(x - 5) < 0$

14. $(x + 8)(x + 10) > 0$

15. $(x + 7)(x - 2) \geq 0$

16. $(x - 1)(x + 4) \leq 0$

17. $x^2 - x - 2 > 0$

18. $x^2 + x - 2 < 0$

Aha! 19. $x^2 + 4x + 4 < 0$

20. $x^2 + 6x + 9 < 0$

21. $x^2 - 4x \leq 3$

22. $x^2 + 6x \geq 2$

23. $3x(x + 2)(x - 2) < 0$

24. $5x(x + 1)(x - 1) > 0$

25. $(x - 1)(x + 2)(x - 4) \geq 0$

26. $(x + 3)(x + 2)(x - 1) < 0$

27. For $f(x) = 7 - x^2$, find all x-values for which $f(x) \geq 3$.

28. For $f(x) = 14 - x^2$, find all x-values for which $f(x) > 5$.

29. For $g(x) = (x - 2)(x - 3)(x + 1)$, find all x-values for which $g(x) > 0$.

30. For $g(x) = (x + 3)(x - 2)(x + 1)$, find all x-values for which $g(x) < 0$.

31. For $F(x) = x^3 - 7x^2 + 10x$, find all x-values for which $F(x) \leq 0$.

32. For $G(x) = x^3 - 8x^2 + 12x$, find all x-values for which $G(x) \geq 0$.

Rational Inequalities

Solve.

33. $\dfrac{1}{x - 5} < 0$

34. $\dfrac{1}{x + 4} > 0$

35. $\dfrac{x + 1}{x - 3} \geq 0$

36. $\dfrac{x - 2}{x + 4} \leq 0$

37. $\dfrac{x + 1}{x + 6} \geq 1$

38. $\dfrac{x - 1}{x - 2} \leq 1$

39. $\dfrac{(x - 2)(x + 1)}{x - 5} \leq 0$

40. $\dfrac{(x + 4)(x - 1)}{x + 3} \geq 0$

41. $\dfrac{x}{x + 3} \geq 0$

42. $\dfrac{x - 2}{x} \leq 0$

43. $\dfrac{x - 5}{x} < 1$

44. $\dfrac{x}{x - 1} > 2$

45. $\dfrac{x - 1}{(x - 3)(x + 4)} \leq 0$

46. $\dfrac{x + 2}{(x - 2)(x + 7)} \geq 0$

47. For $f(x) = \dfrac{5 - 2x}{4x + 3}$, find all x-values for which $f(x) \geq 0$.

48. For $g(x) = \dfrac{2 + 3x}{2x - 4}$, find all x-values for which $g(x) \geq 0$.

49. For $G(x) = \dfrac{1}{x - 2}$, find all x-values for which $G(x) \leq 1$.

50. For $F(x) = \dfrac{1}{x - 3}$, find all x-values for which $F(x) \leq 2$.

51. Explain how any quadratic inequality can be solved by examining a parabola.

52. Describe a method for creating a quadratic inequality for which there is no solution.

Skill Review

53. On a typical weekday, the average full-time college student spends a total of 7.1 hr in educational or leisure activities. The student spends 0.7 hr more in leisure activities than in educational activities. On an average weekday, how many hours does the student spend on educational activities?   [1.4]
Source: U.S. Bureau of Labor Statistics

54. Kent paddled for 2 hr with a 5-km/h current to reach a campsite. The return trip against the same current took 7 hr. Find the speed of Kent's canoe in still water.   [3.3].

55. Josh and Lindsay plan to rent a moving truck. The truck costs $70 plus 40¢ per mile. They have budgeted $90 for the truck rental. For what mileages will they not exceed their budget? [4.1]

56. It takes Deanna twice as long to set up a fundraising auction as it takes Donna. Together they can set up for the auction in 4 hr. How long would it take each of them to do the job alone?   [6.5]

Synthesis

57. When solving a rational inequality containing the symbol $\leq$ or $\geq$, endpoints of some intervals may not be part of the solution set. Why?

58. Describe a method that could be used to create a quadratic inequality that has $(-\infty, a] \cup [b, \infty)$ as the solution set. Assume $a < b$.

Find each solution set.

59. $x^2 + 2x < 5$

60. $x^4 + 2x^2 \geq 0$

61. $x^4 + 3x^2 \leq 0$

62. $\left| \dfrac{x + 2}{x - 1} \right| \leq 3$

63. *Total Profit.*   Derex, Inc., determines that its total-profit function is given by
$$P(x) = -3x^2 + 630x - 6000.$$
a) Find all values of x for which Derex makes a profit.
b) Find all values of x for which Derex loses money.

64. *Height of a Thrown Object.* The function
$$S(t) = -16t^2 + 32t + 1920$$
gives the height S, in feet, of an object thrown from a cliff that is 1920 ft high. Here t is the time, in seconds, that the object is in the air.

a) For what times does the height exceed 1920 ft?
b) For what times is the height less than 640 ft?

65. *Number of Handshakes.* There are n people in a room. The number N of possible handshakes by the people is given by the function
$$N(n) = \frac{n(n-1)}{2}.$$

For what number of people n is $66 \le N \le 300$?

66. *Number of Diagonals.* A polygon with n sides has D diagonals, where D is given by the function
$$D(n) = \frac{n(n-3)}{2}.$$

Find the number of sides n if
$$27 \le D \le 230.$$

Use a graphing calculator to graph each function and find solutions of $f(x) = 0$. Then solve the inequalities $f(x) < 0$ and $f(x) > 0$.

67. $f(x) = x^3 - 2x^2 - 5x + 6$

68. $f(x) = \frac{1}{3}x^3 - x + \frac{2}{3}$

69. $f(x) = x + \frac{1}{x}$

70. $f(x) = x - \sqrt{x}, x \ge 0$

71. $f(x) = \dfrac{x^3 - x^2 - 2x}{x^2 + x - 6}$

72. $f(x) = x^4 - 4x^3 - x^2 + 16x - 12$

Find the domain of each function.

73. $f(x) = \sqrt{x^2 - 4x - 45}$

74. $f(x) = \sqrt{9 - x^2}$

75. $f(x) = \sqrt{x^2 + 8x}$

76. $f(x) = \sqrt{x^2 + 2x + 1}$

77. Describe a method that could be used to create a rational inequality that has $(-\infty, a] \cup (b, \infty)$ as the solution set. Assume $a < b$.

78. Use a graphing calculator to solve Exercises 43 and 49 by drawing two curves, one for each side of the inequality.

YOUR TURN ANSWERS: SECTION 8.9

1. $(-\infty, -2) \cup (4, \infty)$, or $\{x \mid x < -2 \text{ or } x > 4\}$
2. $[-2 - \sqrt{6}, -2 + \sqrt{6}]$, or $\{x \mid -2 - \sqrt{6} \le x \le -2 + \sqrt{6}\}$
3. $(-\infty, -2) \cup (-1, 0)$, or $\{x \mid x < -2 \text{ or } -1 < x < 0\}$
4. $[-1, 0] \cup [1, \infty)$, or $\{x \mid -1 \le x \le 0 \text{ or } x \ge 1\}$
5. $(-\infty, 1) \cup \left[\frac{7}{4}, \infty\right)$, or $\left\{x \mid x < 1 \text{ or } x \ge \frac{7}{4}\right\}$

QUICK QUIZ: SECTIONS 8.1–8.9

Solve.

1. $3x^2 + 1 = 0$ [8.1]

2. $x - 3\sqrt{x} - 4 = 0$ [8.5]

3. $5c^2 - c - 1 = 0$ [8.2]

4. $2x^2 - x - 3 \ge 0$ [8.9]

5. Find any x-intercepts and the y-intercept of the graph of $f(x) = 4x^2 - 3x$. [8.7]

PREPARE TO MOVE ON

Graph each function. [2.1]

1. $f(x) = x^3 - 2$

2. $g(x) = \dfrac{2}{x}$

3. If $g(x) = x^2 - 3$, find $g(\sqrt{a - 5})$. [2.1], [7.1]

4. If $g(x) = x^2 + 2$, find $g(2a + 5)$. [2.1], [5.2]

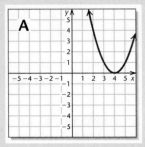

A

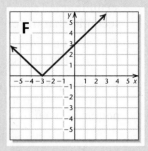

F

Visualizing for Success

Use after Section 8.7.

Match each function with its graph.

1. $f(x) = 3x^2$

2. $f(x) = x^2 - 4$

3. $f(x) = (x - 4)^2$

4. $f(x) = x - 4$

5. $f(x) = -2x^2$

6. $f(x) = x + 3$

7. $f(x) = |x + 3|$

8. $f(x) = (x + 3)^2$

9. $f(x) = \sqrt{x + 3}$

10. $f(x) = (x + 3)^2 - 4$

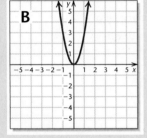

B

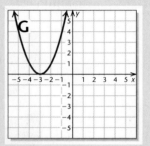

G

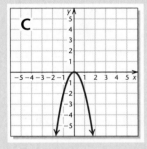

C

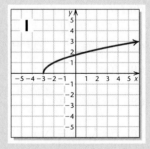

H

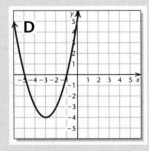

D

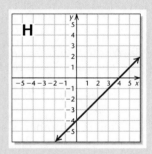

I

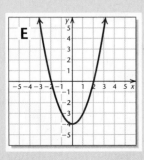

E

Answers on page A-45

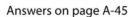

An additional, animated version of this activity appears in MyMathLab. *To use MyMathLab, you need a course ID and a student access code. Contact your instructor for more information.*

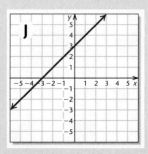

J

Collaborative Activity   *Match the Graph*

Focus: Graphing quadratic functions
Use after: Section 8.6
Time: 15–20 minutes
Group size: 6
Materials: Index cards

Activity

1. On each of six index cards, write one of the following equations:

$y = \frac{1}{2}(x - 3)^2 + 1;$ $y = \frac{1}{2}(x - 1)^2 + 3;$
$y = \frac{1}{2}(x + 1)^2 - 3;$ $y = \frac{1}{2}(x + 3)^2 + 1;$
$y = \frac{1}{2}(x + 3)^2 - 1;$ $y = \frac{1}{2}(x + 1)^2 + 3.$

2. Fold each index card and mix up the six cards in a hat or bag. Then, one by one, each group member should select one of the equations. Do not let anyone see your equation.

3. Each group member should carefully graph the equation selected. Make the graph large enough so that when it is finished, it can be easily viewed by the rest of the group. Be sure to scale the axes and label the vertex, but **do not label the graph with the equation used**.

4. When all group members have drawn a graph, place the graphs in a pile. The group should then match and agree on the correct equation for each graph *with no help from the person who drew the graph*. If a mistake has been made and a graph has no match, determine what its equation *should* be.

5. Compare your group's labeled graphs with those of other groups to reach consensus within the class on the correct label for each graph.

Decision Making & Connection   (*Use after Section 8.8.*)

Pizza Pricing. Papa Romeo's Pizza in Chicago, Illinois, sells a 10-in. diameter cheese pizza for $9, a 14-in. diameter pizza for $13, and an 18-in. diameter pizza for $21. Which better models the price of the pizza: a linear function or a quadratic function of the diameter?

Source: mypaparomeospizza.com

1. Graph ordered pairs from the data above using the form (diameter, price). Do the data appear to be quadratic or linear?

2. Fit each of the following models to the data, where $p(x)$ is the price, in dollars, of an x-inch diameter pizza. Using a different color for each, graph the functions on the same graph as the ordered pairs. Then determine visually which model best fits the data.

   a) Linear function $p(x) = mx + b$, using the points $(10, 9)$ and $(14, 13)$
   b) Linear function $p(x) = mx + b$, using the points $(10, 9)$ and $(18, 21)$
   c) Quadratic function $p(x) = ax^2 + bx + c$, using all three points

3. One way to tell whether a function is a good fit is to see how well it predicts another known value. Papa Romeo's also sells a 22-in. diameter cheese pizza for $28. Which function from part (2) comes closest to predicting the actual value?

4. Because the area of a circle is given by $A = \pi r^2$, would you expect the price of a cheese pizza to be quadratic or linear?

5. *Research.* Find another restaurant that sells at least four sizes of pizza. Listing diameter on the horizontal axis and price on the vertical axis, graph their pizza prices and determine whether a linear model or a quadratic model appears to be the best fit. Use two or three of the prices to find a function that models the data, and test your model by predicting a known price not used to form the model.

Study Summary

KEY TERMS AND CONCEPTS	EXAMPLES	PRACTICE EXERCISES

SECTION 8.1: *Quadratic Equations*

A **quadratic equation in standard form** is written $ax^2 + bx + c = 0$, with a, b, and c constant and $a \neq 0$. Some quadratic equations can be solved by factoring.	$x^2 - 3x - 10 = 0$ $(x + 2)(x - 5) = 0$ $x + 2 = 0 \quad or \quad x - 5 = 0$ $\quad x = -2 \quad or \quad\quad x = 5$	**1.** Solve: $x^2 - 12x + 11 = 0.$
The Principle of Square Roots For any real number k, if $X^2 = k$, then $X = \sqrt{k}$ or $X = -\sqrt{k}$.	$x^2 - 8x + 16 = 25$ $\quad (x - 4)^2 = 25$ $x - 4 = -5 \quad or \quad x - 4 = 5$ $\quad x = -1 \quad or \quad\quad x = 9$	**2.** Solve: $x^2 - 18x + 81 = 5.$
Any quadratic equation can be solved by **completing the square.**	$x^2 + 6x = 1$ $x^2 + 6x + \left(\frac{6}{2}\right)^2 = 1 + \left(\frac{6}{2}\right)^2$ $x^2 + 6x + 9 = 1 + 9$ $(x + 3)^2 = 10$ $x + 3 = \pm\sqrt{10}$ $x = -3 \pm \sqrt{10}$	**3.** Solve by completing the square: $x^2 + 20x = 21.$

SECTION 8.2: *The Quadratic Formula*

The Quadratic Formula The solutions of $ax^2 + bx + c = 0$ are given by $$x = \frac{-b \pm \sqrt{b^2 - 4ac}}{2a}.$$	$3x^2 - 2x - 5 = 0 \quad\quad a = 3, b = -2, c = -5$ $x = \dfrac{-(-2) \pm \sqrt{(-2)^2 - 4 \cdot 3(-5)}}{2 \cdot 3}$ $x = \dfrac{2 \pm \sqrt{4 + 60}}{6}$ $x = \dfrac{2 \pm \sqrt{64}}{6}$ $x = \dfrac{2 \pm 8}{6}$ $x = \dfrac{10}{6} = \dfrac{5}{3} \quad or \quad x = \dfrac{-6}{6} = -1$	**4.** Solve: $2x^2 - 3x - 9 = 0.$

SECTION 8.3: *Studying Solutions of Quadratic Equations*

The **discriminant** of the quadratic formula is $b^2 - 4ac$. $b^2 - 4ac = 0 \rightarrow$ One solution; a rational number	For $4x^2 - 12x + 9 = 0$, $b^2 - 4ac = (-12)^2 - 4(4)(9)$ $= 144 - 144 = 0.$ Thus, $4x^2 - 12x + 9 = 0$ has one rational solution.	**5.** Use the discriminant to determine the number and type of solutions of $2x^2 + 5x + 9 = 0.$

576

| $b^2 - 4ac > 0 \rightarrow$ Two real solutions; both are rational if $b^2 - 4ac$ is a perfect square.

$b^2 - 4ac < 0 \rightarrow$ Two imaginary-number solutions | For $x^2 + 6x - 2 = 0$, $b^2 - 4ac = (6)^2 - 4(1)(-2)$
$\qquad\qquad = 36 + 8 = 44.$
Thus, $x^2 + 6x - 2 = 0$ has two irrational real-number solutions.
For $2x^2 - 3x + 5 = 0$, $b^2 - 4ac = (-3)^2 - 4(2)(5)$
$\qquad\qquad = 9 - 40 = -31.$
Thus, $2x^2 - 3x + 5 = 0$ has two imaginary-number solutions. | |

SECTION 8.4:   *Applications Involving Quadratic Equations*

| To solve a formula for a letter, use the same principles used for solving equations. | Solve $y = pn^2 + dn$ for n.
$\qquad pn^2 + dn - y = 0 \qquad a = p, b = d, c = -y$

$\qquad\qquad n = \dfrac{-d \pm \sqrt{d^2 - 4p(-y)}}{2 \cdot p}$

$\qquad\qquad n = \dfrac{-d \pm \sqrt{d^2 + 4py}}{2p}$ | **6.** Solve $a = n^2 + 1$ for n. |

SECTION 8.5:   *Equations Reducible to Quadratic*

| Equations that are **reducible to quadratic** or in **quadratic form** can be solved by making an appropriate substitution. | $x^4 - 10x^2 + 9 = 0 \qquad$ Let $u = x^2$. Then $u^2 = x^4$.
$u^2 - 10u + 9 = 0 \qquad$ Substituting
$(u - 9)(u - 1) = 0$
$u - 9 = 0 \quad or \quad u - 1 = 0$
$\quad u = 9 \quad or \qquad u = 1 \qquad$ Solving for u
$\quad x^2 = 9 \quad or \qquad x^2 = 1 \qquad$ Replacing u with x^2
$\quad x = \pm 3 \;\; or \qquad x = \pm 1 \qquad$ Solving for x | **7.** Solve:
$\quad x - \sqrt{x} - 30 = 0.$ |

SECTION 8.6:   *Quadratic Functions and Their Graphs*
SECTION 8.7:   *More About Graphing Quadratic Functions*

| The graph of a quadratic function
$f(x) = ax^2 + bx + c$
$\qquad = a(x - h)^2 + k$
is a **parabola.** The graph opens upward for $a > 0$ and downward for $a < 0$.
The **vertex** is (h, k), and the **axis of symmetry** is $x = h$.
If $a > 0$, the function has a **minimum** value of k, and if $a < 0$, the function has a **maximum** value of k.
The vertex and the axis of symmetry occur where
$x = -\dfrac{b}{2a}.$ | 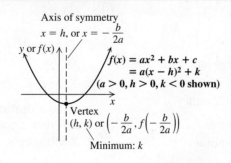 | **8.** Graph
$\quad f(x) = 2x^2 - 12x + 3.$
Label the vertex and the axis of symmetry, and identify the minimum or maximum function value. |

SECTION 8.8: *Problem Solving and Quadratic Functions*

Some problem situations can be **modeled** using quadratic functions. For those problems, a quantity can often be **maximized** or **minimized** by finding the coordinates of a vertex.	A lifeguard has 100 m of linked flotation devices with which to cordon off a rectangular swimming area at North Beach. If the shoreline forms one side of the rectangle, what dimensions will maximize the size of the area for swimming? This problem and its solution appear as Example 2 in Section 8.8.	**9.** Loretta is putting fencing around a rectangular vegetable garden. She can afford to buy 120 ft of fencing. What dimensions should she plan for the garden in order to maximize its area?

SECTION 8.9: *Polynomial Inequalities and Rational Inequalities*

When solving a **polynomial inequality**, use the x-intercepts, or **zeros**, of a function to divide the x-axis into intervals.
 	
When solving a **rational inequality**, use the solutions of a rational equation along with any replacements that make a denominator zero to divide the x-axis into intervals.
	Solve: $x^2 - 2x - 15 > 0$.
	
	$\qquad x^2 - 2x - 15 = 0$ Solving the related equation
	$\qquad (x - 5)(x + 3) = 0$
	$\qquad\qquad x = 5 \quad or \quad x = -3$ -3 and 5 divide the number line into three intervals.
	
	$$\begin{array}{c} + \qquad - \qquad + \\ \leftarrow\!\!\!\!-\!\!\!\!+\!\!\!\!-\!\!\!\!-\!\!\!\!+\!\!\!\!-\!\!\!\!\rightarrow \\ \ \ -3 \quad\ \ 5 \end{array}$$
	For $f(x) = x^2 - 2x - 15 = (x - 5)(x + 3)$:
	  $f(x)$ is positive for $x < -3$;
	  $f(x)$ is negative for $-3 < x < 5$;
	  $f(x)$ is positive for $x > 5$.
	Thus, $x^2 - 2x - 15 > 0$ for $(-\infty, -3) \cup (5, \infty)$, or $\{x \mid x < -3 \ or \ x > 5\}$.
		10. Solve:
		$\qquad x^2 - 11x - 12 < 0$.

Review Exercises: Chapter 8

⤷ Concept Reinforcement

Classify each of the following statements as either true or false.

1. Every quadratic equation has two different solutions. [8.3]

2. Every quadratic equation has at least one solution. [8.3]

3. If an equation cannot be solved by completing the square, it cannot be solved by the quadratic formula. [8.2]

4. A negative discriminant indicates two imaginary-number solutions of a quadratic equation. [8.3]

5. The graph of $f(x) = 2(x + 3)^2 - 4$ has its vertex at $(3, -4)$. [8.6]

6. The graph of $g(x) = 5x^2$ has $x = 0$ as its axis of symmetry. [8.6]

7. The graph of $f(x) = -2x^2 + 1$ has no minimum value. [8.6]

8. The zeros of $g(x) = x^2 - 9$ are -3 and 3. [8.6]

9. If a quadratic function has two different imaginary-number zeros, the graph of the function has two x-intercepts. [8.7]

10. To solve a polynomial inequality, we often must solve a polynomial equation. [8.9]

Solve.

11. $9x^2 - 2 = 0$ [8.1]

12. $8x^2 + 6x = 0$ [8.1]

13. $x^2 - 12x + 36 = 9$ [8.1]

14. $x^2 - 4x + 8 = 0$ [8.2]

15. $x(3x + 4) = 4x(x - 1) + 15$ [8.2]

16. $x^2 + 9x = 1$ [8.2]

17. Solve $x^2 - 5x - 2 = 0$, using a calculator to approximate the solutions to three decimal places.   [8.2]

18. Let $f(x) = 4x^2 - 3x - 1$. Find all x such that $f(x) = 0$.   [8.2]

Replace the blanks with constants to form a true equation.   [8.1]

19. $x^2 - 18x + \underline{\quad} = (x - \underline{\quad})^2$

20. $x^2 + \frac{3}{5}x + \underline{\quad} = (x + \underline{\quad})^2$

21. Solve by completing the square. Show your work.

$x^2 - 6x + 1 = 0$ [8.1]

22. $2500 grows to $2704 in 2 years. Use the formula $A = P(1 + r)^t$ to find the interest rate.   [8.1]

23. The London Eye observation wheel is 443 ft tall. Use $s = 16t^2$ to approximate how long it would take an object to fall from the top.   [8.1]

For each equation, determine whether the solutions are real or imaginary. If they are real, specify whether they are rational or irrational.   [8.3]

24. $x^2 + 3x - 6 = 0$

25. $x^2 + 2x + 5 = 0$

26. Write a quadratic equation having the solutions $3i$ and $-3i$.   [8.3]

27. Write a quadratic equation having -5 as its only solution.   [8.3]

Solve.   [8.4]

28. Horizons has a manufacturing plant 300 mi from company headquarters. Their corporate pilot must fly from headquarters to the plant and back in 4 hr. If there is a 20-mph headwind going and a 20-mph tailwind returning, how fast must the plane be able to travel in still air?

29. Working together, Dani and Cheri can reply to a day's worth of customer-service e-mails in 4 hr. Working alone, Dani takes 6 hr longer than Cheri. How long would it take Cheri alone to reply to the e-mails?

30. Find all x-intercepts of the graph of $f(x) = x^4 - 13x^2 + 36$.   [8.5]

Solve.   [8.5]

31. $15x^{-2} - 2x^{-1} - 1 = 0$

32. $(x^2 - 4)^2 - (x^2 - 4) - 6 = 0$

33. a) Graph: $f(x) = -3(x + 2)^2 + 4$.   [8.6]
  b) Label the vertex.
  c) Draw the axis of symmetry.
  d) Find the maximum or the minimum value.

34. For the function given by $f(x) = 2x^2 - 12x + 23$:   [8.7]

  a) find the vertex and the axis of symmetry;
  b) graph the function.

35. Find any x-intercepts and the y-intercept of the graph of

$$f(x) = x^2 - 9x + 14.$$ [8.7]

36. Solve $N = 3\pi\sqrt{\dfrac{1}{p}}$ for p.   [8.4]

37. Solve $2A + T = 3T^2$ for T.   [8.4]

State whether each graph appears to represent a quadratic function or a linear function.   [8.8]

38.

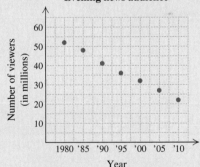

Evening news audience

Source: Nielsen Media Research

39.

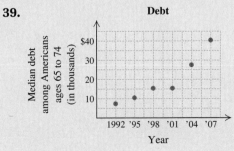

Debt

Source: U.S. Federal Reserve

40. Eastgate Consignments wants to build a rectangular area in a corner for children to play in while their parents shop. They have 30 ft of low fencing. What is the maximum area they can enclose? What dimensions will yield this area? [8.8]

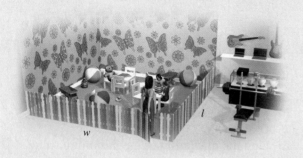

41. The following table lists the median debt among Americans ages 65 to 74 x years after 1992. (See Exercise 39.) [8.8]

Years Since 1992	Median Debt Among Americans Ages 65 to 74 (in thousands)
0	$ 7
6	15
12	40

a) Find the quadratic function that fits the data.
b) Use the function to estimate the median debt among Americans ages 65 to 74 in 2013.

Solve. [8.9]

42. $x^3 - 3x > 2x^2$

43. $\dfrac{x - 5}{x + 3} \le 0$

Synthesis

44. Explain how the x-intercepts of a quadratic function can be used to help find the maximum or minimum value of the function. [8.7], [8.8]

45. Explain how the x-intercepts of a quadratic function can be used to factor a quadratic polynomial. [8.4], [8.7]

46. Discuss two ways in which completing the square was used in this chapter. [8.1], [8.2], [8.7]

47. A quadratic function has x-intercepts at -3 and 5. If the y-intercept is at -7, find an equation for the function. [8.7]

48. Find h and k if, for $3x^2 - hx + 4k = 0$, the sum of the solutions is 20 and the product of the solutions is 80. [8.3]

49. The average of two positive integers is 171. One of the numbers is the square root of the other. Find the integers. [8.5]

Test: Chapter 8

For step-by-step test solutions, access the Chapter Test Prep Videos in MyMathLab®, on YouTube® (search "Bittinger Inter AlgCA"), or by scanning the code.

Solve.

1. $25x^2 - 7 = 0$

2. $4x(x - 2) - 3x(x + 1) = -18$

3. $x^2 + 2x + 3 = 0$

4. $2x + 5 = x^2$

5. $x^{-2} - x^{-1} = \frac{3}{4}$

6. Solve $x^2 + 3x = 5$, using a calculator to approximate the solutions to three decimal places.

7. Let $f(x) = 12x^2 - 19x - 21$. Find x such that $f(x) = 0$.

Replace the blanks with constants to form a true equation.

8. $x^2 - 20x + \underline{\quad} = (x - \underline{\quad})^2$

9. $x^2 + \frac{2}{7}x + \underline{\quad} = (x + \underline{\quad})^2$

10. Solve by completing the square. Show your work.
    $$x^2 + 10x + 15 = 0$$

11. Determine the type of number that the solutions of $x^2 + 2x + 5 = 0$ will be.

12. Write a quadratic equation having solutions $\sqrt{11}$ and $-\sqrt{11}$.

Solve.

13. The Connecticut River flows at a rate of 4 km/h for the length of a popular scenic route. In order for a cruiser to travel 60 km upriver and then return in a total of 8 hr, how fast must the boat be able to travel in still water?

14. Dal and Kim can assemble a swing set in $1\frac{1}{2}$ hr. Working alone, it takes Kim 4 hr longer than Dal to assemble the swing set. How long would it take Dal, working alone, to assemble the swing set?

15. Find all x-intercepts of the graph of
    $$f(x) = x^4 - 15x^2 - 16.$$

16. a) Graph: $f(x) = 4(x - 3)^2 + 5$.
    b) Label the vertex.
    c) Draw the axis of symmetry.
    d) Find the maximum or the minimum function value.

17. For the function $f(x) = 2x^2 + 4x - 6$:
    a) find the vertex and the axis of symmetry;
    b) graph the function.

18. Find the x- and y-intercepts of
    $$f(x) = x^2 - x - 6.$$

19. Solve $V = \frac{1}{3}\pi(R^2 + r^2)$ for r. Assume all variables are positive.

20. State whether the graph appears to represent a linear function, a quadratic function, or neither.

Chicago air quality

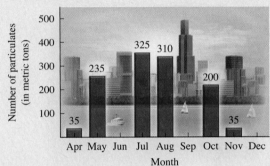

Source: The National Arbor Day Foundation

21. Jay's Metals has determined that when x hundred storage cabinets are built, the average cost per cabinet is given by
    $$C(x) = 0.2x^2 - 1.3x + 3.4025,$$
    where $C(x)$ is in hundreds of dollars. What is the minimum cost per cabinet and how many cabinets should be built to achieve that minimum?

22. Find the quadratic function that fits the data points $(0, 0)$, $(3, 0)$, and $(5, 2)$.

Solve.

23. $x^2 + 5x < 6$

24. $x - \dfrac{1}{x} \geq 0$

Synthesis

25. One solution of $kx^2 + 3x - k = 0$ is -2. Find the other solution.

26. Find a fourth-degree polynomial equation, with integer coefficients, for which $-\sqrt{3}$ and $2i$ are solutions.

27. Solve: $x^4 - 4x^2 - 1 = 0$.

Cumulative Review: Chapters 1–8

Simplify.

1. $-3 \cdot 8 \div (-2)^3 \cdot 4 - 6(5 - 7)$ [1.2]

2. $(5x^2y - 8xy - 6xy^2) - (2xy - 9x^2y + 3xy^2)$ [5.1]

3. $(9p^2q + 8t)(9p^2q - 8t)$ [5.2]

4. $\dfrac{t^2 - 25}{9t^2 + 24t + 16} \div \dfrac{3t^2 - 11t - 20}{t^2 + t}$ [6.1]

5. $(3\sqrt{2} + i)(2\sqrt{2} - i)$ [7.8]

Factor.

6. $12x^4 - 75y^4$ [5.5]

7. $x^3 - 24x^2 + 80x$ [5.4]

8. $100m^6 - 100$ [5.6]

9. $6t^2 + 35t + 36$ [5.4]

Solve.

10. $2(5x - 3) - 8x = 4 - (3 - x)$ [1.3]

11. $2(5x - 3) - 8x < 4 - (3 - x)$ [4.1]

12. $\begin{aligned} 2x - 6y &= 3, \\ -3x + 8y &= -5 \end{aligned}$ [3.2]

13. $x(x - 5) = 66$ [5.8]

14. $\dfrac{2}{t} + \dfrac{1}{t - 1} = 2$ [6.4]

15. $\sqrt{x} = 1 + \sqrt{2x - 7}$ [7.6]

16. $m^2 + 10m + 25 = 2$ [8.1]

17. $3x^2 + 1 = x$ [8.2]

Graph.

18. $9x - 2y = 18$ [2.4]   **19.** $x < \frac{1}{2}y$ [4.4]

20. $y = 2(x - 3)^2 + 1$ [8.6]

21. $f(x) = x^2 + 4x + 3$ [8.7]

22. Find an equation in slope–intercept form whose graph has slope -5 and y-intercept $\left(0, \frac{1}{2}\right)$. [2.3]

23. Find the slope of the line containing $(8, 3)$ and $(-2, 10)$. [2.3]

Find the domain of f.

24. $f(x) = \sqrt{10 - x}$ [4.1]

25. $f(x) = \dfrac{x + 3}{x - 4}$ [2.2], [4.2]

Solve each formula for the specified letter.

26. $b = \dfrac{a + c}{2a}$, for a [6.8]

27. $p = 2\sqrt{\dfrac{r}{3t}}$, for t [8.4]

Solve.

28. *Gold Prices.* Marisa is selling some of her gold jewelry. She has 4 bracelets and 1 necklace that weigh a total of 3 oz. [1.4]

a) Marisa's jewelry is 58% gold. How many ounces of gold does her jewelry contain?

b) A gold dealer offers Marisa $2088 for the jewelry. How much per ounce of gold was she offered?

c) The retail price of gold at the time of Marisa's sale was $1600 per ounce. What percent of the gold price was she offered for her jewelry?

29. *Television Viewing.* The average number of Americans who watch the evening news has decreased from 52 million in 1980 to 22 million in 2010. [2.5]

Source: Nielsen Media Research

a) Let $f(t)$ represent the average number of viewers of the evening news, in millions, t years after 1980. Find a linear function that fits the data.

b) Use the function from part (a) to predict the number of Americans watching the evening news in 2014.

c) If the trend continues, in what year will no one watch the evening news?

30. *Education.* Andres ordered number tiles at $9 per set and alphabet tiles at $15 per set for his classroom. He ordered a total of 36 sets for $384. How many sets of each did he order? [3.3]

31. *Minimizing Cost.* Dormitory Furnishings has determined that when x bunk beds are built, the average cost, in dollars, per bunk bed can be estimated by $c(x) = 0.004375x^2 - 3.5x + 825$. What is the minimum average cost per bunk bed and how many bunk beds should be built to achieve that minimum? [8.8]

Synthesis

Solve.

32. $\dfrac{\dfrac{1}{x}}{2 + \dfrac{1}{x - 1}} = 3$ [6.3], [8.2]

33. $x^4 + 5x^2 \le 0$ [8.9]

34. Find the points of intersection of the graphs of $f(x) = x^2 + 8x + 1$ and $g(x) = 10x + 6$. [8.2]

Exponential Functions and Logarithmic Functions

NOTE	FREQUENCY (in hertz)
A	27.5
A#	29.1352
B	30.8677
C	32.7032

Music Contains Mathematics.

Math and music are closely connected, and much of the beauty and pleasure of music can be described in mathematical terms. From rhythms to harmonies to Bach's fugues, the languages and structure of music and math overlap. One example that we will consider in this chapter deals with frequencies in music. When we hear different musical pitches, we are detecting differences in vibrations of sound, described in oscillations per second, or hertz. The table above lists the frequencies for the lowest four notes on an 88-key piano. The frequencies between notes do not increase at the same rate; instead, they can be modeled using an *exponential function*. (*See Exercise 9 in Exercise Set 9.7.*)

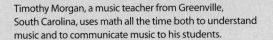

Math is the mechanism that moves music.
Music without math becomes noise.

Timothy Morgan, a music teacher from Greenville, South Carolina, uses math all the time both to understand music and to communicate music to his students.

The *exponential* functions and *logarithmic* functions that we consider in this chapter have rich applications in many fields, such as epidemiology (the study of the spread of disease), population growth, and marketing. Exponential functions have variable exponents, and logarithmic functions are their closely related *inverse* functions.

9.1   Composite Functions and Inverse Functions

Composite Functions   ▪   Inverses and One-to-One Functions   ▪   Finding Formulas for Inverses   ▪
Graphing Functions and Their Inverses   ▪   Inverse Functions and Composition

Later in this chapter, we introduce two closely related types of functions: exponential functions and logarithmic functions. In order to properly understand the link between these functions, we must first understand composite functions and inverse functions.

Composite Functions

Functions frequently occur in which some quantity depends on a variable that, in turn, depends on another variable. For instance, a firm's profits may be a function of the number of items the firm produces, which may in turn be a function of the number of employees hired. In this case, the firm's profits may be considered a **composite function**.

Let's consider an example of a profit function. Tea Mug Collective sells hand-painted tee shirts. The monthly profit p, in dollars, from the sale of m shirts is given by $p = 15m - 1200$. The number of shirts m produced in a month by x employees is given by $m = 40x$.

If Tea Mug Collective employs 10 people, then in one month they can produce $m = 40(10) = 400$ shirts. The profit from selling these 400 shirts would be $p = 15(400) - 1200 = 4800$ dollars. Can we find an equation that would allow us to calculate the monthly profit on the basis of the number of employees? We begin with the profit equation and substitute:

$$p = 15m - 1200$$
$$= 15(40x) - 1200 \qquad \text{Substituting } 40x \text{ for } m$$
$$= 600x - 1200.$$

The equation $p = 600x - 1200$ gives the monthly profit when Tea Mug Collective has x employees.

To find a composition of functions, we follow the same reasoning above using function notation:

$$p(m) = 15m - 1200, \qquad \text{Profit as a function of the number of shirts produced}$$
$$m(x) = 40x; \qquad \text{Number of shirts as a function of the number of employees}$$

$$p(m(x)) = p(40x)$$
$$= 15(40x) - 1200$$
$$= 600x - 1200.$$

If we call this new function P, then $P(x) = 600x - 1200$. This gives profit as a function of the number of employees.

Tea Mug Collective's Shane Kimberlin, Alaskan artist

We call P the *composition* of p and m. In general, the composition of f and g is written $f \circ g$ and is read "the composition of f and g," "f composed with g," or "f circle g."

COMPOSITION OF FUNCTIONS

The *composite function* $f \circ g$, the *composition* of f and g, is defined as

$$(f \circ g)(x) = f(g(x)).$$

It is not uncommon to use the same variable to represent the input in more than one function.

Throughout this chapter, keep in mind that equations such as $m(x) = 40x$ and $m(t) = 40t$ describe the same function. Both equations tell us to find a function value by multiplying the input by 40.

We can visualize the composition of functions as follows.

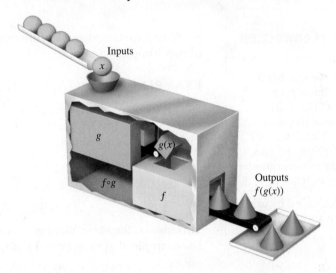

EXAMPLE 1 Given $f(x) = 3x$ and $g(x) = 1 + x^2$:

a) Find $(f \circ g)(5)$ and $(g \circ f)(5)$.   **b)** Find $(f \circ g)(x)$ and $(g \circ f)(x)$.

SOLUTION Consider each function separately:

$$f(x) = 3x \qquad \text{This function multiplies each input by 3.}$$

and

$$g(x) = 1 + x^2. \qquad \text{This function adds 1 to the square of each input.}$$

a) To find $(f \circ g)(5)$, we find $g(5)$ and then use that as an input for f:

$$(f \circ g)(5) = f(g(5)) = f(1 + 5^2) \qquad \text{Using } g(x) = 1 + x^2$$
$$= f(26) = 3 \cdot 26 = 78. \qquad \text{Using } f(x) = 3x$$

To find $(g \circ f)(5)$, we find $f(5)$ and then use that as an input for g:

$$(g \circ f)(5) = g(f(5)) = g(3 \cdot 5) \qquad \text{Note that } f(5) = 3 \cdot 5 = 15.$$
$$= g(15) = 1 + 15^2 = 1 + 225 = 226.$$

b) We find $(f \circ g)(x)$ by substituting $g(x)$ for x in the equation for $f(x)$:

$$(f \circ g)(x) = f(g(x)) = f(1 + x^2) \qquad \text{Using } g(x) = 1 + x^2$$
$$= 3 \cdot (1 + x^2) = 3 + 3x^2. \qquad \text{Using } f(x) = 3x$$

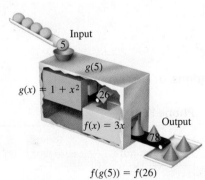

A composition machine for Example 1

To find $(g \circ f)(x)$, we substitute $f(x)$ for x in the equation for $g(x)$:

$$(g \circ f)(x) = g(f(x)) = g(3x) \qquad \text{Substituting } 3x \text{ for } f(x)$$
$$= 1 + (3x)^2 = 1 + 9x^2.$$

We can now find the function values of part (a) using the functions of part (b):

$$(f \circ g)(5) = 3 + 3(5)^2 = 3 + 3 \cdot 25 = 78;$$
$$(g \circ f)(5) = 1 + 9(5)^2 = 1 + 9 \cdot 25 = 226.$$

1. Given $f(x) = x^2 - 2$ and $g(x) = 5x$, find $(f \circ g)(x)$.

↪ YOUR TURN

Example 1 shows that, in general, $(f \circ g)(x) \neq (g \circ f)(x)$.

EXAMPLE 2 Given $f(x) = \sqrt{x}$ and $g(x) = x - 1$, find $(f \circ g)(x)$ and $(g \circ f)(x)$.

SOLUTION

2. Given $f(x) = 2x - 7$ and $g(x) = \sqrt[3]{x}$, find $(g \circ f)(x)$.

$$(f \circ g)(x) = f(g(x)) = f(x - 1) = \sqrt{x - 1}; \qquad \text{Using } g(x) = x - 1$$
$$(g \circ f)(x) = g(f(x)) = g(\sqrt{x}) = \sqrt{x} - 1 \qquad \text{Using } f(x) = \sqrt{x}$$

↪ YOUR TURN

Technology Connection

In Example 3, we see that if $g(x) = 7x + 3$ and $f(x) = x^2$, then $f(g(x)) = (7x + 3)^2$. One way to show this is to let $y_1 = 7x + 3$ and $y_2 = x^2$. If we let $y_3 = (7x + 3)^2$ and $y_4 = y_2(y_1)$, we can use graphs or a table to show that $y_3 = y_4$.

1. Check Example 2 by using the above approach.

When we *decompose* a function, we think of the function as the composition of two "simpler" functions.

EXAMPLE 3 If $h(x) = (7x + 3)^2$, find f and g such that $h(x) = (f \circ g)(x)$.

SOLUTION We can think of $h(x)$ as the result of first evaluating $7x + 3$ and then squaring that. This suggests that we let $g(x) = 7x + 3$ and $f(x) = x^2$. We check by forming the composition:

$$(f \circ g)(x) = f(g(x))$$
$$= f(7x + 3)$$
$$= (7x + 3)^2 = h(x), \text{ as desired.}$$

This may be the most "obvious" solution, but there are other less obvious answers. For example, if $f(x) = (x - 1)^2$ and $g(x) = 7x + 4$, then

$$(f \circ g)(x) = f(g(x))$$
$$= f(7x + 4)$$
$$= (7x + 4 - 1)^2 = (7x + 3)^2 = h(x).$$

3. If $h(x) = \sqrt{3 - x}$, find f and g such that $h(x) = (f \circ g)(x)$. Answers may vary.

↪ YOUR TURN

Inverses and One-to-One Functions

Let's view the following two functions as relations, or correspondences.

Countries and Their Capitals

Domain (Set of Inputs)	Range (Set of Outputs)
Australia	→ Canberra
China	→ Beijing
Germany	→ Berlin
Madagascar	→ Antananariro
Turkey	→ Ankara
United States	→ Washington, D.C.

Phone Keys

Domain (Set of Inputs)	Range (Set of Outputs)
a	
b	2
c	
d	
e	3
f	

Suppose we reverse the arrows. We obtain what is called the **inverse relation**. Are these inverse relations functions?

Countries and Their Capitals

Range (Set of Outputs)		Domain (Set of Inputs)
Australia	←	Canberra
China	←	Beijing
Germany	←	Berlin
Madagascar	←	Antananaviro
Turkey	←	Ankara
United States	←	Washington, D.C.

Phone Keys

Range (Set of Outputs)	Domain (Set of Inputs)
a←	
b←	2
c←	
d←	
e←	3
f←	

Recall that for a function, each input has exactly one output. In some functions, different inputs correspond to the same output. Only when this possibility is *excluded* will the inverse be a function. For the functions listed above, this means the inverse of the "Capitals" correspondence is a function, but the inverse of the "Phone Keys" correspondence is not.

In the Capitals function, each input has its own output, so it is a **one-to-one function**. In the Phone Keys function, a and b are both paired with 2. Thus the Phone Keys function is not a one-to-one function.

> **ONE-TO-ONE FUNCTION**
>
> A function f is *one-to-one* if different inputs have different outputs. That is, f is one-to-one if for any a and b in the domain of f with $a \neq b$, we have $f(a) \neq f(b)$. If a function is one-to-one, then its inverse correspondence is also a function.

How can we tell graphically whether a function is one-to-one?

EXAMPLE 4   Below is the graph of a function. Determine whether the function is one-to-one and thus has an inverse that is a function.

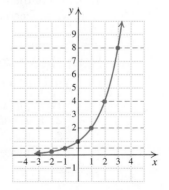

4. Determine whether the function graphed below is one-to-one.

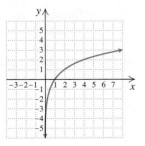

SOLUTION   A function is one-to-one if different inputs have different outputs—that is, if no two x-values share the same y-value. For this function, we cannot find two x-values that have the same y-value. To see this, note that no horizontal line can be drawn so that it crosses the graph more than once. The function is one-to-one so its inverse is a function.

YOUR TURN

The graph of every function must pass the vertical-line test. In order for a function to have an inverse that is a function, it must pass the *horizontal-line test* as well.

> **THE HORIZONTAL-LINE TEST**
>
> If it is impossible to draw a horizontal line that intersects a function's graph more than once, then the function is one-to-one. For every one-to-one function, an inverse function exists.

EXAMPLE 5   Determine whether the function given by $f(x) = x^2$ is one-to-one and thus has an inverse that is a function.

SOLUTION   The graph of $f(x) = x^2$ is shown here. Many horizontal lines cross the graph more than once. For example, the line $y = 4$ crosses where the first coordinates are -2 and 2. Although these are different inputs, they have the same output. That is, $-2 \neq 2$, but

$$f(-2) = (-2)^2 = 4 = 2^2 = f(2).$$

Thus the function is not one-to-one and its inverse is not a function.

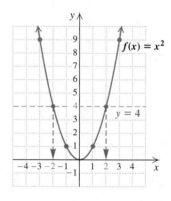

5. Determine whether the function given by $f(x) = (x - 2)^2$ is one-to-one.

YOUR TURN

Student Notes

If you are not certain whether a function is one-to-one, you can try to find a formula for its inverse. If it is possible to solve *uniquely* for *y* after *x* and *y* have been interchanged, then the function is one-to-one.

Finding Formulas for Inverses

When the inverse of f is also a function, it is denoted f^{-1} (read "f-inverse").

> *CAUTION!*   The -1 in f^{-1} is *not* an exponent!

For any equation in two variables, if we interchange the variables, we form an equation of the inverse correspondence. If the inverse correspondence is a function, we proceed as follows to find a formula for f^{-1}.

> **TO FIND A FORMULA FOR f^{-1}**
>
> First make sure that f is one-to-one. Then:
>
> **1.** Replace $f(x)$ with y.
> **2.** Interchange x and y. (This gives the inverse function.)
> **3.** Solve for y.
> **4.** Replace y with $f^{-1}(x)$. (This is inverse function notation.)

EXAMPLE 6   Determine whether each function is one-to-one and if it is, find a formula for $f^{-1}(x)$.

a) $f(x) = 2x - 3$

b) $f(x) = \dfrac{3}{x}$

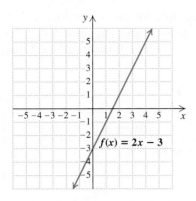

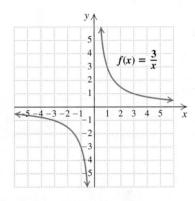

6. Determine whether the function given by $f(x) = x + 2$ is one-to-one. If it is, find a formula for $f^{-1}(x)$.

SOLUTION

a) The graph of $f(x) = 2x - 3$ is shown at left. This function, like any linear function that is not constant, passes the horizontal-line test. Thus, f is one-to-one and we can find a formula for $f^{-1}(x)$.

1. Replace $f(x)$ with y: $y = 2x - 3$.

2. Interchange x and y: $x = 2y - 3$.

3. Solve for y: $x + 3 = 2y$

$$\frac{x + 3}{2} = y.$$

4. Replace y with $f^{-1}(x)$: $f^{-1}(x) = \dfrac{x + 3}{2}$.

In this case, the function f doubles all inputs and then subtracts 3. Thus, to "undo" f, the function f^{-1} adds 3 to each input and then divides by 2.

b) The graph of $f(x) = 3/x$ is shown at left. The function passes the horizontal-line test. Thus it is one-to-one and its inverse is a function.

1. Replace $f(x)$ with y: $y = \dfrac{3}{x}$.

2. Interchange x and y: $x = \dfrac{3}{y}$.

3. Solve for y: $xy = 3$

$$y = \frac{3}{x}.$$

4. Replace y with $f^{-1}(x)$: $f^{-1}(x) = \dfrac{3}{x}$.

Note that this function and its inverse are the same function.

YOUR TURN

Graphing Functions and Their Inverses

How do the graphs of a function and its inverse compare?

EXAMPLE 7   Graph $f(x) = 2x - 3$ and $f^{-1}(x) = (x + 3)/2$ on the same set of axes. Then compare.

SOLUTION   The graph of each function follows. Note that the graph of f^{-1} can be drawn by reflecting the graph of f across the line $y = x$. That is, if we graph $f(x) = 2x - 3$ in wet ink and fold the paper along the line $y = x$, the graph of $f^{-1}(x) = (x + 3)/2$ will appear as the impression made by f.

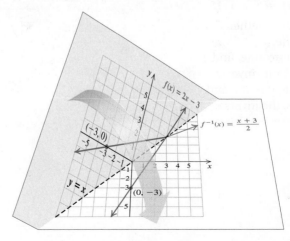

7. Graph $f(x) = \frac{1}{2}x + 1$ and $f^{-1}(x) = 2x - 2$ on the same set of axes.

When x and y are interchanged to find a formula for the inverse, we are, in effect, reflecting or flipping the graph of $f(x) = 2x - 3$ across the line $y = x$. For example, when $(0, -3)$, the coordinates of the y-intercept of the graph of f, are reversed, we get $(-3, 0)$, the x-intercept of the graph of f^{-1}.

 YOUR TURN

VISUALIZING INVERSES

The graph of f^{-1} is a reflection of the graph of f across the line $y = x$.

EXAMPLE 8   Consider $g(x) = x^3 + 2$.

a) Determine whether g is one-to-one.

b) If it is one-to-one, find a formula for its inverse.

c) Graph the inverse, if it exists.

SOLUTION

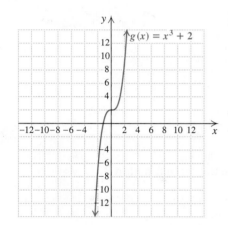

a) The graph of $g(x) = x^3 + 2$ is shown at left. It passes the horizontal-line test and thus is one-to-one and has an inverse that is a function.

b) **1.** Replace $g(x)$ with y: $y = x^3 + 2$.   Using $g(x) = x^3 + 2$

   2. Interchange x and y: $x = y^3 + 2$.   This represents the inverse relation.

   3. Solve for y: $x - 2 = y^3$
                              $\sqrt[3]{x - 2} = y$.   Each real number has only one cube root, so we can solve uniquely for y.

   4. Replace y with $g^{-1}(x)$: $g^{-1}(x) = \sqrt[3]{x - 2}$.

c) To graph g^{-1}, we can reflect the graph of $g(x) = x^3 + 2$ across the line $y = x$, as we did in Example 7. We also could graph $g^{-1}(x) = \sqrt[3]{x - 2}$ by plotting points. As a check, note that $(2, 10)$ is on the graph of g and $(10, 2)$ is on the graph of g^{-1}.

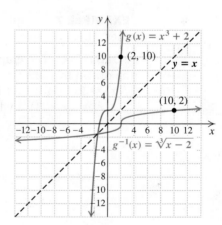

8. Consider $f(x) = \sqrt[3]{x + 1}$.

   a) Determine whether f is one-to-one.

   b) If it is one-to-one, find a formula for its inverse.

   c) Graph the function and, if it exists, the inverse.

 YOUR TURN

EXPLORING 🔍 THE CONCEPT

The graph of an inverse function is the reflection of the graph of the function across the line $y = x$.
Match each function below with the graph of its inverse function.

1.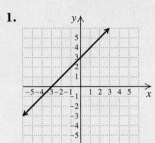

2.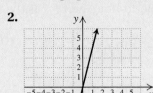

3.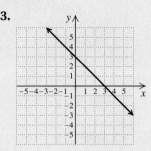

4.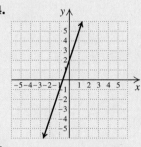

(a)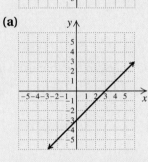

(b)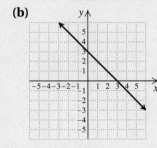

(c)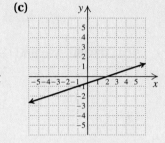

(d)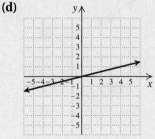

ANSWERS
1. (a)   **2.** (d)   **3.** (b)   **4.** (c)

Inverse Functions and Composition

Let's consider inverses of functions in terms of function machines. Suppose that a one-to-one function f is programmed into a machine. If the machine is run in reverse, it will perform the inverse function f^{-1}. Inputs then enter at the opposite end, and the entire process is reversed.

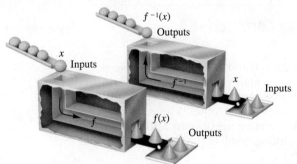

Consider $f(x) = x^3 + 2$ and $f^{-1}(x) = \sqrt[3]{x - 2}$ from Example 8. For the input 3,

$$f(3) = 3^3 + 2 = 27 + 2 = 29.$$

The output, $f(3)$, is 29. Let's now use 29 as an input in the inverse:

$$f^{-1}(29) = \sqrt[3]{29 - 2} = \ = \sqrt[3]{27} = 3.$$

The function f takes 3 to 29. The inverse function f^{-1} takes the number 29 back to 3.

In general, if f is one-to-one, then f^{-1} takes the output $f(x)$ back to x. Similarly, f takes the output $f^{-1}(x)$ back to x.

COMPOSITION AND INVERSES

If a function f is one-to-one, then f^{-1} is the unique function for which

$$(f^{-1} \circ f)(x) = f^{-1}(f(x)) = x$$

and

$$(f \circ f^{-1})(x) = f(f^{-1}(x)) = x.$$

EXAMPLE 9   Let $f(x) = 2x + 1$. Show that

$$f^{-1}(x) = \frac{x - 1}{2}.$$

SOLUTION   We find $(f^{-1} \circ f)(x)$ and $(f \circ f^{-1})(x)$ and check to see that each is x.

$$(f^{-1} \circ f)(x) = f^{-1}(f(x)) = f^{-1}(2x + 1)$$

$$= \frac{(2x + 1) - 1}{2}$$

$$= \frac{2x}{2} = x; \quad \text{Thus, } (f^{-1} \circ f)(x) = x.$$

$$(f \circ f^{-1})(x) = f(f^{-1}(x)) = f\left(\frac{x - 1}{2}\right)$$

$$= 2 \cdot \frac{x - 1}{2} + 1$$

$$= x - 1 + 1 = x \quad \text{Thus, } (f \circ f^{-1})(x) = x.$$

9. Let $f(x) = 3x - 5$. Show that $f^{-1}(x) = \dfrac{x + 5}{3}$.

 YOUR TURN

Technology Connection

To see if $y_1 = 2x + 6$ and $y_2 = \frac{1}{2}x - 3$ are inverses of each other, we can graph both functions, along with the line $y = x$, on a "squared" set of axes. It *appears* that y_1 and y_2 are inverses of each other. A more precise check is achieved by selecting the DRAWINV option of the (DRAW) menu. The resulting graph of the inverse of y_1 should coincide with y_2.

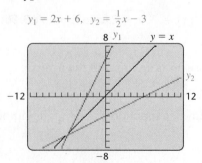

$y_1 = 2x + 6, \quad y_2 = \frac{1}{2}x - 3$

For a more dependable check, examine a TABLE in which $y_1 = 2x + 6$ and $y_2 = \frac{1}{2} \cdot y_1 - 3$. Note that y_2 "undoes" what y_1 does.

TBLSTART = −3   ΔTBL = 1 $y_2 = \frac{1}{2}y_1 - 3$

X	Y₁	Y₂
−3	0	−3
−2	2	−2
−1	4	−1
0	6	0
1	8	1
2	10	2
3	12	3

X = 3

1. Use a graphing calculator to check Examples 7, 8, and 9.
2. Will DRAWINV work for *any* choice of y_1? Why or why not?

9.1 EXERCISE SET

FOR EXTRA HELP

MyMathLab® Math XL
PRACTICE   WATCH   READ   REVIEW

Vocabulary and Reading Check

Classify each of the following statements as either true or false.

1. The composition of two functions f and g is written $f \circ g$.

2. The notation $(f \circ g)(x)$ means $f(g(x))$.

3. If $f(x) = x^2$ and $g(x) = x + 3$, then $(g \circ f)(x) = (x + 3)^2$.

4. For any function h, there is only one way to decompose the function as $h = f \circ g$.

5. The function f is one-to-one if $f(1) = 1$.

6. The -1 in f^{-1} is an exponent.

7. The function f is the inverse of f^{-1}.

8. If g and h are inverses of each other, then $(g \circ h)(x) = x$.

Composite Functions

For each pair of functions, find **(a)** $(f \circ g)(1)$;
(b) $(g \circ f)(1)$; **(c)** $(f \circ g)(x)$; **(d)** $(g \circ f)(x)$.

9. $f(x) = x^2 + 1$; $g(x) = x - 3$

10. $f(x) = x + 4$; $g(x) = x^2 - 5$

11. $f(x) = 5x + 1$; $g(x) = 2x^2 - 7$

12. $f(x) = 3x^2 + 4$; $g(x) = 4x - 1$

13. $f(x) = x + 7$; $g(x) = 1/x^2$

14. $f(x) = 1/x^2$; $g(x) = x + 2$

15. $f(x) = \sqrt{x}$; $g(x) = x + 3$

16. $f(x) = 10 - x$; $g(x) = \sqrt{x}$

17. $f(x) = \sqrt{4x}$; $g(x) = 1/x$

18. $f(x) = \sqrt{x + 3}$; $g(x) = 13/x$

19. $f(x) = x^2 + 4$; $g(x) = \sqrt{x - 1}$

20. $f(x) = x^2 + 8$; $g(x) = \sqrt{x + 17}$

Find $f(x)$ and $g(x)$ such that $h(x) = (f \circ g)(x)$. Answers may vary.

21. $h(x) = (3x - 5)^4$

22. $h(x) = (2x + 7)^3$

23. $h(x) = \sqrt{9x + 1}$

24. $h(x) = \sqrt[3]{4x - 5}$

25. $h(x) = \dfrac{6}{5x - 2}$

26. $h(x) = \dfrac{3}{x} + 4$

Inverses and One-to-One Functions

Determine whether each function is one-to-one.

27. $f(x) = -x$

28. $f(x) = x + 5$

Aha! 29. $f(x) = x^2 + 3$

30. $f(x) = 3 - x^2$

31.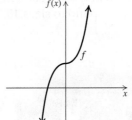

32.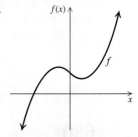

33.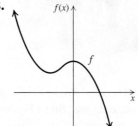

34.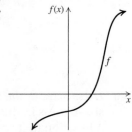

Finding Formulas for Inverses

For each function, **(a)** *determine whether it is one-to-one;*
(b) *if it is one-to-one, find a formula for the inverse.*

35. $f(x) = x + 3$

36. $f(x) = x + 2$

37. $f(x) = 2x$

38. $f(x) = 3x$

39. $g(x) = 3x - 1$

40. $g(x) = 2x - 3$

41. $f(x) = \frac{1}{2}x + 1$

42. $f(x) = \frac{1}{3}x + 2$

43. $g(x) = x^2 + 5$

44. $g(x) = x^2 - 4$

45. $h(x) = -10 - x$

46. $h(x) = 7 - x$

Aha! 47. $f(x) = \dfrac{1}{x}$

48. $f(x) = \dfrac{4}{x}$

49. $g(x) = 1$

50. $h(x) = 8$

51. $f(x) = \dfrac{2x + 1}{3}$

52. $f(x) = \dfrac{3x + 2}{5}$

53. $f(x) = x^3 + 5$

54. $f(x) = x^3 - 4$

55. $g(x) = (x - 2)^3$

56. $g(x) = (x + 7)^3$

57. $f(x) = \sqrt{x}$

58. $f(x) = \sqrt{x - 1}$

59. *Dress Sizes in the United States and Italy.* A size-6 dress in the United States is size 36 in Italy. A function that converts dress sizes in the United States to those in Italy is

$$f(x) = 2(x + 12).$$

a) Find the dress sizes in Italy that correspond to sizes 8, 10, 14, and 18 in the United States.

b) Does f have an inverse that is a function? If so, find a formula for the inverse.

c) Use the inverse function to find dress sizes in the United States that correspond to sizes 40, 44, 52, and 60 in Italy.

60. *Dress Sizes in the United States and France.* A size-6 dress in the United States is size 38 in France. A function that converts dress sizes in the United States to those in France is

$$f(x) = x + 32.$$

a) Find the dress sizes in France that correspond to sizes 8, 10, 14, and 18 in the United States.

b) Does f have an inverse that is a function? If so, find a formula for the inverse.

c) Use the inverse function to find dress sizes in the United States that correspond to sizes 40, 42, 46, and 50 in France.

Graphing Functions and Their Inverses

Graph each function and its inverse using the same set of axes.

61. $f(x) = \frac{2}{3}x + 4$

62. $g(x) = \frac{1}{4}x + 2$

63. $f(x) = x^3 + 1$

64. $f(x) = x^3 - 1$

65. $g(x) = \frac{1}{2}x^3$

66. $g(x) = \frac{1}{3}x^3$

67. $F(x) = -\sqrt{x}$

68. $f(x) = \sqrt{x}$

69. $f(x) = -x^2, x \geq 0$

70. $f(x) = x^2 - 1, x \leq 0$

Inverse Functions and Composition

71. Let $f(x) = \sqrt[3]{x} - 4$. Show that

$$f^{-1}(x) = x^3 + 4.$$

72. Let $f(x) = 3/(x + 2)$. Show that

$$f^{-1}(x) = \frac{3}{x} - 2.$$

73. Let $f(x) = (1 - x)/x$. Show that

$$f^{-1}(x) = \frac{1}{x + 1}.$$

74. Let $f(x) = x^3 - 5$. Show that

$$f^{-1}(x) = \sqrt[3]{x + 5}.$$

75. Is there a one-to-one relationship between items in a store and the price of each of those items? Why or why not?

76. Mathematicians usually try to select "logical" words when forming definitions. Does the term "one-to-one" seem logical? Why or why not?

Skill Review

Simplify.

77. $t^{1/5}t^{2/3}$ [7.2]

78. $\sqrt[3]{40a^5b^{12}}$ [7.3]

79. $(-3x^{-6}y^4)^{-2}$ [1.6]

80. i^{43} [7.8]

81. $3^3 + 2^2 - (32 \div 4 - 16 \div 8)$ [1.1]

82. $(1.5 \times 10^{-3})(4.2 \times 10^{-12})$ [1.7]

Synthesis

83. The function $V(t) = 750(1.2)^t$ is used to predict the value $V(t)$ of a certain rare stamp t years from 2008. Do not calculate $V^{-1}(t)$, but explain how V^{-1} could be used.

84. An organization determines that the cost per person $C(x)$, in dollars, of chartering a bus with x passengers is given by

$$C(x) = \frac{100 + 5x}{x}.$$

Determine $C^{-1}(x)$ and explain how this inverse function could be used.

For Exercises 85 and 86, graph the inverse of f.

85.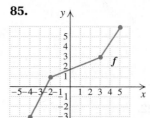

86.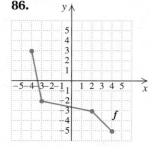

87. *Dress Sizes in France and Italy.* Use the information in Exercises 59 and 60 to find a function for the French dress size that corresponds to a size x dress in Italy.

88. *Dress Sizes in Italy and France.* Use the information in Exercises 59 and 60 to find a function for the Italian dress size that corresponds to a size x dress in France.

 89. What relationship exists between the answers to Exercises 87 and 88? Explain how you determined this.

90. Show that function composition is associative by showing that $((f \circ g) \circ h)(x) = (f \circ (g \circ h))(x)$.

91. Show that if $h(x) = (f \circ g)(x)$, then $h^{-1}(x) = (g^{-1} \circ f^{-1})(x)$. (*Hint*: Use Exercise 90.)

Determine whether or not the given pairs of functions are inverses of each other.

92. $f(x) = 0.75x^2 + 2$; $g(x) = \sqrt{\dfrac{4(x-2)}{3}}$

93. $f(x) = 1.4x^3 + 3.2$; $g(x) = \sqrt[3]{\dfrac{x - 3.2}{1.4}}$

94. $f(x) = \sqrt{2.5x + 9.25}$; $g(x) = 0.4x^2 - 3.7, x \geq 0$

95. $f(x) = 0.8x^{1/2} + 5.23$; $g(x) = 1.25(x^2 - 5.23), x \geq 0$

96. $f(x) = 2.5(x^3 - 7.1)$; $g(x) = \sqrt[3]{0.4x + 7.1}$

97. Match each function in Column A with its inverse from Column B.

Column A   *Column B*

(1) $y = 5x^3 + 10$ **A.** $y = \dfrac{\sqrt[3]{x} - 10}{5}$

(2) $y = (5x + 10)^3$ **B.** $y = \sqrt[3]{\dfrac{x}{5}} - 10$

(3) $y = 5(x + 10)^3$ **C.** $y = \sqrt[3]{\dfrac{x - 10}{5}}$

(4) $y = (5x)^3 + 10$ **D.** $y = \dfrac{\sqrt[3]{x} - 10}{5}$

 98. Examine the following table. Is it possible that f and g are inverses of each other? Why or why not?

x	$f(x)$	$g(x)$
6	6	6
7	6.5	8
8	7	10
9	7.5	12
10	8	14
11	8.5	16
12	9	18

 99. The following window appears on a graphing calculator.

X	Y1	Y2
0	1	−2
1	1.5	0
2	2	2
3	2.5	4
4	3	6
5	3.5	8
6	4	10
X = 0		

a) What evidence is there that the functions Y1 and Y2 are inverses of each other?

b) Find equations for Y1 and Y2, assuming that both are linear functions.

c) On the basis of your answer to part (b), are Y1 and Y2 inverses of each other?

YOUR TURN ANSWERS: SECTION 9.1

1. $(f \circ g)(x) = 25x^2 - 2$ **2.** $(g \circ f)(x) = \sqrt[3]{2x - 7}$
3. $f(x) = \sqrt{x}, g(x) = 3 - x$ **4.** Yes   **5.** No
6. f is one-to-one; $f^{-1}(x) = x - 2$
7.

$f(x) = \frac{1}{2}x + 1$

$f^{-1}(x) = 2x - 2$

$y = x$

8. **(a)** Yes;
(b) $f^{-1}(x) = x^3 - 1$;
(c)

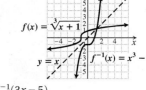

$f(x) = \sqrt[3]{x + 1}$

$f^{-1}(x) = x^3 - 1$

$y = x$

9. $(f^{-1} \circ f)(x) = f^{-1}(f(x)) = f^{-1}(3x - 5)$
$$= \frac{(3x - 5) + 5}{3} = \frac{3x}{3} = x;$$
$$(f \circ f^{-1})(x) = f(f^{-1}(x)) = f\left(\frac{x + 5}{3}\right) = 3\left(\frac{x + 5}{3}\right) - 5$$
$$= (x + 5) - 5 = x$$

PREPARE TO MOVE ON

Simplify.

1. 2^{-3} [1.6]   **2.** $5^{(1-3)}$ [1.6]

3. $4^{5/2}$ [7.2]   **4.** $3^{7/10}$ [7.2]

Graph.   [2.1]

5. $y = x^3$ **6.** $x = y^3$

9.2 Exponential Functions

Graphing Exponential Functions ▪ Equations with *x* and *y* Interchanged ▪ Applications of Exponential Functions

Study Skills

Know Your Machine

Whether you use a scientific calculator or a graphing calculator, it is a wise investment of time to learn how to best use the device. Experimenting by pressing various combinations of keystrokes can be very helpful, as well as reading your user's manual, which is often available online.

In this section, we introduce a new type of function, the *exponential function*. These functions and their inverses, called *logarithmic functions*, have applications in many fields.

Consider the graph below. The rapidly rising curve approximates the graph of an *exponential function*.

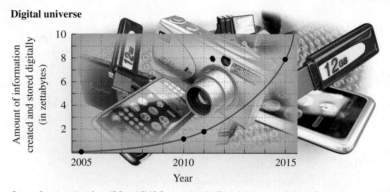

Digital universe

Source: Based on data from IDC and EMC Corporation, in "Digital Universe to Add 1.8 Zettabytes in 2011," Rich Miller, June 28, 2011, datacenterknowledge.com

Graphing Exponential Functions

In Chapter 7, we studied exponential expressions with rational-number exponents. For example, $5^{1.73}$, or $5^{173/100}$, represents the 100th root of 5 raised to the 173rd power. What about expressions with irrational exponents, such as $5^{\sqrt{3}}$ or $7^{-\pi}$? To attach meaning to $5^{\sqrt{3}}$, consider a rational approximation, r, of $\sqrt{3}$. As r gets closer to $\sqrt{3}$, the value of 5^r gets closer to some real number p.

$\underbrace{r \text{ closes in on } \sqrt{3}.}$ $\qquad$ $\underbrace{5^r \text{ closes in on some real number } p.}$

$1.7 < r < 1.8$	$15.426 \approx 5^{1.7} < p < 5^{1.8} \approx 18.119$
$1.73 < r < 1.74$	$16.189 \approx 5^{1.73} < p < 5^{1.74} \approx 16.452$
$1.732 < r < 1.733$	$16.241 \approx 5^{1.732} < p < 5^{1.733} \approx 16.267$

We define $5^{\sqrt{3}}$ to be the number p. To eight decimal places,

$$5^{\sqrt{3}} \approx 16.24245082.$$

Any positive irrational exponent can be interpreted in a similar way. Negative irrational exponents are then defined using reciprocals. Thus, so long as a is positive, a^x has meaning for *any* real number x, and all the laws of exponents still hold. We can now define an *exponential function*.

EXPONENTIAL FUNCTION

The function $f(x) = a^x$, where a is a positive constant, $a \neq 1$, and x is any real number, is called the *exponential function*, base a.

We require the base a to be positive to avoid imaginary numbers that would result from taking even roots of negative numbers. The restriction $a \neq 1$ is made to exclude the constant function $f(x) = 1^x$, or $f(x) = 1$.

The following are examples of exponential functions:

$$f(x) = 2^x, \qquad f(x) = \left(\tfrac{1}{3}\right)^x, \qquad f(x) = 5^{-7x}. \qquad \text{Note that } 5^{-7x} = (5^{-7})^x.$$

Like polynomial functions, the domain of an exponential function is the set of all real numbers. Unlike polynomial functions, exponential functions have a variable exponent. Because of this, graphs of exponential functions either rise or fall dramatically.

EXAMPLE 1 Graph the exponential function given by $y = f(x) = 2^x$.

SOLUTION We compute some function values, thinking of y as $f(x)$, and list the results in a table. It is helpful to start by letting $x = 0$. This gives us the y-intercept.

$$f(0) = 2^0 = 1; \qquad f(-1) = 2^{-1} = \frac{1}{2^1} = \frac{1}{2};$$
$$f(1) = 2^1 = 2;$$
$$f(2) = 2^2 = 4; \qquad f(-2) = 2^{-2} = \frac{1}{2^2} = \frac{1}{4};$$
$$f(3) = 2^3 = 8;$$
$$f(-3) = 2^{-3} = \frac{1}{2^3} = \frac{1}{8}$$

x	y, or $f(x)$
0	1
1	2
2	4
3	8
−1	$\frac{1}{2}$
−2	$\frac{1}{4}$
−3	$\frac{1}{8}$

Next, we plot these points and connect them with a smooth curve.

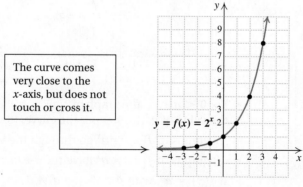

The curve comes very close to the x-axis, but does not touch or cross it.

Be sure to plot enough points to determine how steeply the curve rises.

Note that as x increases, the function values increase without bound. As x decreases, the function values decrease, getting closer to 0. The x-axis, or the line $y = 0$, is a horizontal *asymptote*, meaning that the curve gets closer and closer to this line the further we move to the left.

1. Graph: $y = f(x) = 10^x$.

YOUR TURN

EXAMPLE 2 Graph: $y = f(x) = \left(\tfrac{1}{2}\right)^x$.

SOLUTION We compute some function values, thinking of y as $f(x)$, and list the results in a table. Before we do this, note that

$$y = f(x) = \left(\tfrac{1}{2}\right)^x = (2^{-1})^x = 2^{-x}.$$

Then we have

$$f(0) = 2^{-0} = 1; \qquad f(3) = 2^{-3} = \frac{1}{2^3} = \frac{1}{8};$$
$$f(1) = 2^{-1} = \frac{1}{2^1} = \frac{1}{2}; \qquad f(-1) = 2^{-(-1)} = 2^1 = 2;$$
$$f(-2) = 2^{-(-2)} = 2^2 = 4;$$
$$f(2) = 2^{-2} = \frac{1}{2^2} = \frac{1}{4}; \qquad f(-3) = 2^{-(-3)} = 2^3 = 8.$$

x	y, or $f(x)$
0	1
1	$\frac{1}{2}$
2	$\frac{1}{4}$
3	$\frac{1}{8}$
−1	2
−2	4
−3	8

Next, we plot these points and connect them with a smooth curve. This curve is a mirror image, or *reflection*, of the graph of $y = 2^x$ (see Example 1) across the y-axis. The line $y = 0$ is again the horizontal asymptote.

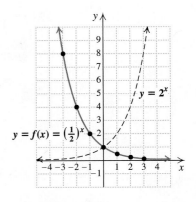

2. Graph: $y = f(x) = \left(\frac{1}{10}\right)^x$.

↩ YOUR TURN

From Examples 1 and 2, we can make the following observations.

- For $a > 1$, the graph of $f(x) = a^x$ increases from left to right. The greater the value of a, the steeper the curve. (See the figure on the left below.)

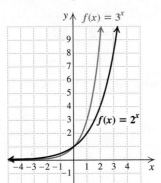
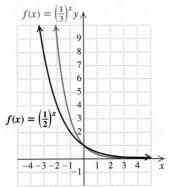

- For $0 < a < 1$, the graph of $f(x) = a^x$ decreases from left to right. The smaller the value of a, the steeper the curve. (See the figure on the right above.)
- All graphs of $f(x) = a^x$ go through the y-intercept $(0, 1)$.
- All graphs of $f(x) = a^x$ have the x-axis as the horizontal asymptote.
- If $f(x) = a^x$, with $a > 0$ and $a \neq 1$, the domain of f is all real numbers, and the range of f is all positive real numbers.
- For $a > 0$ and $a \neq 1$, the function given by $f(x) = a^x$ is one-to-one. Its graph passes the horizontal-line test.

EXAMPLE 3 Graph: $y = f(x) = 2^{x-2}$.

SOLUTION We construct a table of values. Then we plot the points and connect them with a smooth curve. Here $x - 2$ is the *exponent*.

$f(0) = 2^{0-2} = 2^{-2} = \frac{1}{4};$

$f(1) = 2^{1-2} = 2^{-1} = \frac{1}{2};$

$f(2) = 2^{2-2} = 2^0 = 1;$

$f(3) = 2^{3-2} = 2^1 = 2;$

$f(4) = 2^{4-2} = 2^2 = 4;$

$f(-1) = 2^{-1-2} = 2^{-3} = \frac{1}{8};$

$f(-2) = 2^{-2-2} = 2^{-4} = \frac{1}{16}$

x	y, or $f(x)$
0	$\frac{1}{4}$
1	$\frac{1}{2}$
2	1
3	2
4	4
−1	$\frac{1}{8}$
−2	$\frac{1}{16}$

Technology Connection

Graphing calculators are helpful when graphing equations like $y = 5000(1.075)^x$. To set the window, note that y-values are positive and increase rapidly. One suitable window is $[-10, 10, 0, 15000]$, with a y-scale of 1000.

$y = 5000(1.075)^x$

Graph each pair of functions. Select an appropriate window and scale.

1. $y_1 = \left(\frac{5}{2}\right)^x$ and $y_2 = \left(\frac{2}{5}\right)^x$
2. $y_1 = 3.2^x$ and $y_2 = 3.2^{-x}$
3. $y_1 = \left(\frac{3}{7}\right)^x$ and $y_2 = \left(\frac{7}{3}\right)^x$
4. $y_1 = 5000(1.08)^x$ and $y_2 = 5000(1.08)^{x-3}$

Student Notes

When using translations, make sure that you are shifting in the correct direction. When in doubt, substitute a value for x and make some calculations.

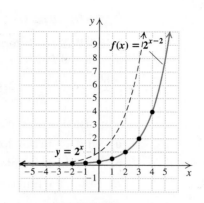

The graph looks just like the graph of $y = 2^x$, but translated 2 units to the right. The y-intercept of $y = 2^x$ is $(0, 1)$. The y-intercept of $y = 2^{x-2}$ is $\left(0, \frac{1}{4}\right)$. The line $y = 0$ is again the horizontal asymptote.

3. Graph: $y = f(x) = 2^{x+2}$.

YOUR TURN

Equations with *x* and *y* Interchanged

It will be helpful in later work to be able to graph an equation in which the x and the y in $y = a^x$ are interchanged.

EXAMPLE 4   Graph: $x = 2^y$.

SOLUTION   Note that x is alone on one side of the equation. To find ordered pairs that are solutions, we choose values for y and then compute values for x.

For $y = 0$, $x = 2^0 = 1$.

For $y = 1$, $x = 2^1 = 2$.

For $y = 2$, $x = 2^2 = 4$.

For $y = 3$, $x = 2^3 = 8$.

For $y = -1$, $x = 2^{-1} = \frac{1}{2}$.

For $y = -2$, $x = 2^{-2} = \frac{1}{4}$.

For $y = -3$, $x = 2^{-3} = \frac{1}{8}$.

x	y
1	0
2	1
4	2
8	3
$\frac{1}{2}$	-1
$\frac{1}{4}$	-2
$\frac{1}{8}$	-3

(1) Choose values for y.
(2) Compute values for x.

We plot the points and connect them with a smooth curve.

This curve does not touch or cross the y-axis, which serves as a vertical asymptote.

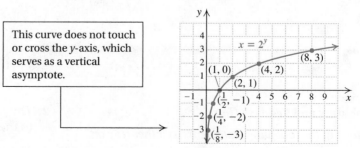

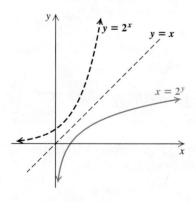

Note too that this curve looks just like the graph of $y = 2^x$, except that it is reflected across the line $y = x$, as shown at left.

4. Graph: $x = 10^y$.

YOUR TURN

Applications of Exponential Functions

EXAMPLE 5   *Interest Compounded Annually.*   The amount of money A that a principal P will be worth after t years at interest rate r, compounded annually, is given by

$$A = P(1 + r)^t.$$

Suppose that \$100,000 is invested at 5% interest, compounded annually.

a) Find a function for the amount in the account after t years.

b) Find the amount of money in the account at $t = 0$, $t = 4$, $t = 8$, and $t = 10$.

c) Graph the function.

SOLUTION

a) If $P = \$100{,}000$ and $r = 5\% = 0.05$, we can substitute these values and form the following function:

$$A(t) = \$100{,}000(1 + 0.05)^t \quad \text{Using } A = P(1 + r)^t$$
$$= \$100{,}000(1.05)^t.$$

b) To find the function values, a calculator with a power key is helpful.

$$A(0) = \$100{,}000(1.05)^0 \qquad\qquad A(8) = \$100{,}000(1.05)^8$$
$$= \$100{,}000(1) \qquad\qquad\qquad \approx \$100{,}000(1.477455444)$$
$$= \$100{,}000; \qquad\qquad\qquad\quad \approx \$147{,}745.54;$$

$$A(4) = \$100{,}000(1.05)^4 \qquad\qquad A(10) = \$100{,}000(1.05)^{10}$$
$$= \$100{,}000(1.21550625) \qquad \approx \$100{,}000(1.628894627)$$
$$\approx \$121{,}550.63; \qquad\qquad\qquad \approx \$162{,}889.46$$

c) We use the function values computed in part (b), and others if we wish, to draw the graph as follows. Note that the axes are scaled differently because of the large numbers.

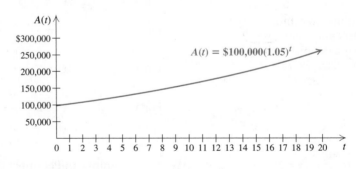

Chapter Resource:
Collaborative Activity, p. 646

5. Refer to Example 5. Suppose that \$20,000 is invested at 4% interest, compounded annually.

a) Find a function for the amount in the account after t years.

b) Find the amount of money in the account at $t = 0$ and $t = 10$.

YOUR TURN

9.2   **EXERCISE SET**   FOR EXTRA HELP   MyMathLab® *Math* XP
PRACTICE   WATCH   READ   REVIEW

Vocabulary and Reading Check

Classify each of the following statements as either true or false.

1. The graph of $f(x) = a^x$ always passes through the point $(0, 1)$.

2. The graph of $g(x) = \left(\frac{1}{2}\right)^x$ gets closer and closer to the x-axis as x gets larger and larger.

3. The graph of $f(x) = 2^{x-3}$ looks just like the graph of $y = 2^x$, but it is translated 3 units to the right.

4. The graph of $g(x) = 2^x - 3$ looks just like the graph of $y = 2^x$, but it is translated 3 units up.

5. The graph of $y = 3^x$ gets close to, but never touches, the y-axis.

6. The graph of $x = 3^y$ gets close to, but never touches, the y-axis.

Graphing Exponential Functions
Graph.

7. $y = f(x) = 3^x$ **8.** $y = f(x) = 4^x$

9. $y = 6^x$ **10.** $y = 5^x$

11. $y = 2^x + 1$ **12.** $y = 2^x + 3$

13. $y = 3^x - 2$ **14.** $y = 3^x - 1$

15. $y = 2^x - 5$ **16.** $y = 2^x - 4$

17. $y = 2^{x-3}$ **18.** $y = 2^{x-1}$

19. $y = 2^{x+1}$ **20.** $y = 2^{x+3}$

21. $y = \left(\frac{1}{4}\right)^x$ **22.** $y = \left(\frac{1}{5}\right)^x$

23. $y = \left(\frac{1}{3}\right)^x$ **24.** $y = \left(\frac{1}{6}\right)^x$

25. $y = 2^{x+1} - 3$ **26.** $y = 2^{x-3} - 1$

Equations with x and y Interchanged
Graph.

27. $x = 6^y$ **28.** $x = 3^y$

29. $x = 3^{-y}$ **30.** $x = 2^{-y}$

31. $x = 4^y$ **32.** $x = 5^y$

33. $x = \left(\frac{4}{3}\right)^y$ **34.** $x = \left(\frac{3}{2}\right)^y$

Graph each pair of equations on the same set of axes.

35. $y = 3^x,\ x = 3^y$ **36.** $y = 2^x,\ x = 2^y$

37. $y = \left(\frac{1}{2}\right)^x,\ x = \left(\frac{1}{2}\right)^y$ **38.** $y = \left(\frac{1}{4}\right)^x,\ x = \left(\frac{1}{4}\right)^y$

Applications of Exponential Functions
Solve.

39. *Digital Music.* Digital music sales, in billions of dollars, t years after 2010 can be approximated by
$$M(t) = 3.2(1.23)^t.$$
Source: Forrester Research

a) Estimate digital music sales in 2010, in 2012, and in 2014.

b) Graph the function.

40. *Growth of Bacteria.* The bacteria *Escherichia coli* are commonly found in the human bladder. Suppose that 3000 bacteria are present at time

$t = 0$. Then t minutes later, the number of bacteria present can be approximated by
$$N(t) = 3000(2)^{t/20}.$$

a) How many bacteria will be present after 10 min? 20 min? 30 min? 40 min? 60 min?

b) Graph the function.

41. *Smoking Cessation.* The percentage of smokers P who receive telephone counseling to quit smoking and are still successful t months later can be approximated by
$$P(t) = 21.4(0.914)^t.$$
Sources: New England Journal of Medicine; data from California's Smokers' Hotline

a) Estimate the percentage of smokers receiving telephone counseling who are successful in quitting for 1 month, 3 months, and 1 year.

b) Graph the function.

42. *Smoking Cessation.* The percentage of smokers P who, without telephone counseling, have successfully quit smoking for t months (see Exercise 41) can be approximated by
$$P(t) = 9.02(0.93)^t.$$
Sources: New England Journal of Medicine; data from California's Smokers' Hotline

a) Estimate the percentage of smokers not receiving telephone counseling who are successful in quitting for 1 month, 3 months, and 1 year.

b) Graph the function.

43. *Marine Biology.* Due to excessive whaling prior to the mid-1970s, the humpback whale is considered an endangered species. The worldwide population of humpbacks, $P(t)$, in thousands, t years after 1900 ($t < 70$) can be approximated by
$$P(t) = 150(0.960)^t.$$
Sources: Based on information from the American Cetacean Society and the ASK Archive

a) How many humpback whales were alive in 1930? in 1960?

b) Graph the function.

44. *Salvage Value.* A laser printer is purchased for $1200. Its value each year is about 80% of the value of the preceding year. Its value, in dollars, after t years is given by the exponential function
$$V(t) = 1200(0.8)^t.$$

a) Find the value of the printer after 0 year, 1 year, 2 years, 5 years, and 10 years.

b) Graph the function.

45. *Marine Biology.* As a result of preservation efforts in most countries in which whaling was common, the humpback whale population has grown since the 1970s. The worldwide population of humpbacks, $P(t)$, in thousands, t years after 1982 can be approximated by

$$P(t) = 5.5(1.047)^t.$$

Sources: Based on information from the American Cetacean Society and the ASK Archive

a) How many humpback whales were alive in 1992? in 2004?

b) Graph the function.

46. *Recycling Aluminum Cans.* About three-fifths of all aluminum cans distributed will be recycled each year. A beverage company distributes 250,000 cans. The number still in use after time t, in years, is given by the exponential function

$$N(t) = 250{,}000\left(\tfrac{3}{5}\right)^t.$$

Source: The Aluminum Association, Inc., 2011

a) The aluminum from how many cans is still in use after 0 year? 1 year? 4 years? 10 years?

b) Graph the function.

47. *Animal Population.* The moose population in New York is growing exponentially. The number of moose in the state t years after 1997 can be approximated by

$$M(t) = 50(1.25)^t.$$

Source: Based on data from the New York State Department of Environmental Conservation

a) Estimate the number of moose in New York in 1997, in 2012, and in 2020.

b) Graph the function.

48. *Vehicle Sales.* The number of cars and light trucks sold in China is increasing exponentially and can be estimated by

$$N(t) = 1.09(1.3)^t,$$

where $N(t)$ is in millions of vehicles and t is the number of years after 2000.

Source: Based on data from Wards Auto

a) Estimate the number of cars and light trucks sold in China in 2000, in 2010, and in 2014.

b) Graph the function.

49. Without using a calculator, explain why 2^π must be greater than 8 but less than 16.

50. Suppose that $1000 is invested for 5 years at 7% interest, compounded annually. In what year will the most interest be earned? Why?

Skill Review

Factor.

51. $3x^2 - 48$ [5.5]

52. $x^2 - 20x + 100$ [5.5]

53. $6x^2 + x - 12$ [5.4]

54. $8x^6 - 64y^6$ [5.6]

55. $t^2 - y^2 + 2y - 1$ [5.5]

56. $5x^4 - 10x^3 - 3x^2 + 6x$ [5.3]

Synthesis

57. Examine Exercise 48. Do you believe that the equation for the number of cars and light trucks sold in China will be accurate 20 years from now? Why or why not?

58. Explain why the graph of $x = 2^y$ is the graph of $y = 2^x$ reflected across the line $y = x$.

Determine which of the two numbers is larger. Do not use a calculator.

59. $\pi^{1.3}$ or $\pi^{2.4}$

60. $\sqrt{8^3}$ or $8^{\sqrt{3}}$

Graph.

61. $f(x) = 2.5^x$

62. $f(x) = 0.5^x$

63. $y = 2^x + 2^{-x}$

64. $y = \left|\left(\tfrac{1}{2}\right)^x - 1\right|$

65. $y = |2^x - 2|$

66. $y = 2^{-(x-1)^2}$

67. $y = \left|2^{x^2} - 1\right|$

68. $y = 3^x + 3^{-x}$

Graph both equations using the same set of axes.

69. $y = 3^{-(x-1)}$, $x = 3^{-(y-1)}$

70. $y = 1^x$, $x = 1^y$

71. *Invasive Species.* Ruffe is a species of freshwater fish that is considered invasive where it is not native. In 1984, there were about 100 ruffe in Loch Lomond, Scotland. By 1988, there were about 3000 ruffe in the lake, and by 1992 there were about 14,000 ruffe. After pressing **STAT** and entering the data, use the EXP REG option in the STAT CALC menu to find an exponential function that models the number of ruffe in Loch Lomond t years after 1984. Then use that function to estimate the number of ruffe in the lake in 1990.

Source: Drake, John M., "Risk Analysis for Species Introductions: Forecasting Population Growth of Eurasian Ruffe (*Gymnocephalus cernus*)," 2005. Retrieved from dragonfly.ecology.uga.edu

72. *Keyboarding Speed.* Trey is studying keyboarding. After he has studied for t hours, Trey's speed, in words per minute, is given by the exponential function

$$S(t) = 200[1 - (0.99)^t].$$

Use a graph and/or table of values to predict Trey's speed after studying for 10 hr, 40 hr, and 80 hr.

73. The following graph shows growth in the height of ocean waves over time, assuming a steady surface wind.

Source: magicseaweed.com

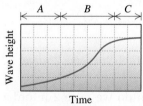

Time
Source: magicseaweed.com

a) Consider the portions of the graph marked A, B, and C. Suppose that each portion can be labeled Exponential Growth, Linear Growth, or Saturation. How would you label each portion?
b) Small vertical movements in wind, surface roughness of water, and gravity are three forces that create waves. How might these forces be related to the shape of the wave-height graph?

74. Consider any exponential function of the form $f(x) = a^x$ with $a > 1$. Will it always follow that $f(3) - f(2) > f(2) - f(1)$, and, in general, $f(n + 2) - f(n + 1) > f(n + 1) - f(n)$? Why or why not? (*Hint:* Think graphically.)

75. On many graphing calculators, it is possible to enter and graph $y_1 = A \wedge (X - B) + C$ after first pressing **APPS** Transfrm. Use this application to graph $f(x) = 2.5^{x-3} + 2, g(x) = 2.5^{x+3} + 2,$ $h(x) = 2.5^{x-3} - 2,$ and $k(x) = 2.5^{x+3} - 2.$

76. *Research.* Ruffe (see Exercise 71) have been introduced into the Great Lakes and several rivers in the United States. What impact does this species now have on fishing and the environment?

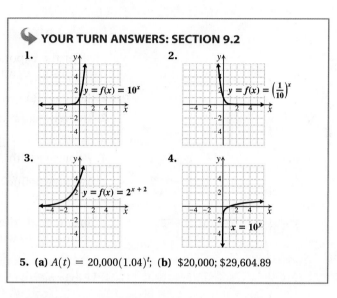

YOUR TURN ANSWERS: SECTION 9.2

1. $y = f(x) = 10^x$
2. $y = f(x) = \left(\frac{1}{10}\right)^x$
3. $y = f(x) = 2^{x+2}$
4. $x = 10^y$

5. **(a)** $A(t) = 20,000(1.04)^t$; **(b)** \$20,000; \$29,604.89

QUICK QUIZ: SECTIONS 9.1–9.2

1. Find $(f \circ g)(x)$ for $f(x) = 5 - x$ and $g(x) = 3x^2 - x - 4$. [9.1]

2. Find $f(x)$ and $g(x)$ such that $(f \circ g)(x) = h(x)$ and $h(x) = (x - 6)^4$. Answers may vary. [9.1]

3. Find a formula for the inverse of $g(x) = 2x + 5$. [9.1]

Graph. [9.2]

4. $y = 2^x - 3$

5. $x = 2^y$

PREPARE TO MOVE ON

Graph.

1. $f(x) = \sqrt{x} - 3$ [7.1]
2. $g(x) = \sqrt[3]{x} + 1$ [7.1]
3. $g(x) = x^3 + 2$ [2.2]
4. $f(x) = 1 - x^2$ [8.7]

9.3   Logarithmic Functions

The Meaning of Logarithms ▪ Graphs of Logarithmic Functions ▪ Equivalent Equations ▪
Solving Certain Logarithmic Equations

We are now ready to study inverses of exponential functions. These functions have many applications and are called *logarithm*, or *logarithmic, functions*.

The Meaning of Logarithms

Consider the exponential function $f(x) = 2^x$. Like all exponential functions, f is one-to-one. Let's attempt to find a formula for $f^{-1}(x)$:

1. Replace $f(x)$ with y: $y = 2^x$.
2. Interchange x and y: $x = 2^y$.
3. Solve for y: $y =$ the exponent to which we raise 2 to get x.
4. Replace y with $f^{-1}(x)$: $f^{-1}(x) =$ the exponent to which we raise 2 to get x.

We now define a new symbol to replace the words "the exponent to which we raise 2 to get x":

> $\log_2 x$, read "the logarithm, base 2, of x," or "log, base 2, of x," means "the exponent to which we raise 2 to get x."

Thus if $f(x) = 2^x$, then $f^{-1}(x) = \log_2 x$. Note that $f^{-1}(8) = \log_2 8 = 3$, because 3 is *the exponent to which we raise* 2 *to get* 8.

EXAMPLE 1   Simplify: **(a)** $\log_2 32$; **(b)** $\log_2 1$; **(c)** $\log_2 \frac{1}{8}$.

SOLUTION

a) Think of $\log_2 32$ as the exponent to which we raise 2 to get 32. That exponent is 5. Therefore, $\log_2 32 = 5$.

b) We ask ourselves: "To what exponent do we raise 2 in order to get 1?" That exponent is 0 (recall that $2^0 = 1$). Thus, $\log_2 1 = 0$.

c) To what exponent do we raise 2 in order to get $\frac{1}{8}$? Since $2^{-3} = \frac{1}{8}$, we have $\log_2 \frac{1}{8} = -3$.

1. Simplify: $\log_2 16$.

YOUR TURN

Although numbers like $\log_2 13$ can be only approximated, we must remember that $\log_2 13$ represents *the exponent to which we raise* 2 *to get* 13. That is, $2^{\log_2 13} = 13$. A calculator indicates that $\log_2 13 \approx 3.7$ and $2^{3.7} \approx 13$.

For any exponential function $f(x) = a^x$, the inverse is called a **logarithmic function, base a**, and is written $f^{-1}(x) = \log_a x$. The graph of the inverse can be drawn by reflecting the graph of $f(x) = a^x$ across the line $y = x$.

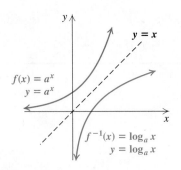

> **THE MEANING OF $\log_a x$**
>
> For $x > 0$ and a a positive constant other than 1, $\log_a x$ is the exponent to which a must be raised in order to get x. Thus,
>
> $$\log_a x = m \quad \text{means} \quad a^m = x$$
>
> or equivalently,
>
> $$\log_a x \text{ is that unique exponent for which } a^{\log_a x} = x.$$

It is important to remember that *a logarithm is an exponent*. It might help to verbalize: "The logarithm, base *a*, of a number *x* is the exponent to which *a* must be raised in order to get *x*."

EXAMPLE 2 Simplify: $7^{\log_7 85}$.

SOLUTION Remember that $\log_7 85$ is the exponent to which 7 is raised to get 85. Raising 7 to that exponent, we have

$$7^{\log_7 85} = 85.$$

2. Simplify: $5^{\log_5 14}$.

 YOUR TURN

Because logarithmic functions and exponential functions are inverses of each other, the result in Example 2 should come as no surprise: If $f(x) = \log_7 x$, then

for $\quad f(x) = \log_7 x$, we have $f^{-1}(x) = 7^x$

and $\quad f^{-1}(f(x)) = f^{-1}(\log_7 x) = 7^{\log_7 x} = x.$

Thus, $f^{-1}(f(85)) = 7^{\log_7 85} = 85.$

The following is a comparison of exponential functions and logarithmic functions.

Exponential Function	Logarithmic Function
$y = a^x$	$x = a^y$
$f(x) = a^x$	$g(x) = \log_a x$
$a > 0, a \neq 1$	$a > 0, a \neq 1$
The domain is $\mathbb{R}$.	The range is $\mathbb{R}$.
$y > 0$ (Outputs are positive.)	$x > 0$ (Inputs are positive.)
$f^{-1}(x) = \log_a x$	$g^{-1}(x) = a^x$

Graphs of Logarithmic Functions

EXAMPLE 3 Graph: $y = f(x) = \log_5 x$.

SOLUTION If $y = \log_5 x$, then $5^y = x$. We can find ordered pairs that are solutions by choosing values for *y* and computing the *x*-values.

For $y = 0, x = 5^0 = 1.$

For $y = 1, x = 5^1 = 5.$

For $y = 2, x = 5^2 = 25.$

For $y = -1, x = 5^{-1} = \frac{1}{5}.$

For $y = -2, x = 5^{-2} = \frac{1}{25}.$

(1) Select *y*.

(2) Compute *x*.

x, or 5^y	y
1	0
5	1
25	2
$\frac{1}{5}$	−1
$\frac{1}{25}$	−2

This table shows the following:

$$\left. \begin{array}{l} \log_5 1 = 0; \\ \log_5 5 = 1; \\ \log_5 25 = 2; \\ \log_5 \frac{1}{5} = -1; \\ \log_5 \frac{1}{25} = -2. \end{array} \right\}$$

These can all be checked using the equations above.

We plot the set of ordered pairs and connect the points with a smooth curve. The graphs of $y = 5^x$ and $y = x$ are shown only for reference.

3. Graph: $y = f(x) = \log_8 x$.

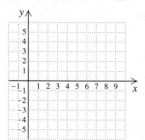

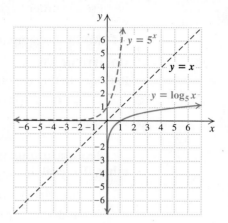

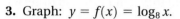 YOUR TURN

Equivalent Equations

We use the definition of logarithm to rewrite a *logarithmic equation* as an equivalent *exponential equation* or the other way around:

$$m = \log_a x \quad \text{is equivalent to} \quad a^m = x.$$

> *CAUTION!*   **Do not forget this relationship!** It may be the most important statement in the chapter. Many times this is used to justify a property we are considering.

EXAMPLE 4   Rewrite each as an equivalent exponential equation.

a) $y = \log_3 5$ **b)** $-2 = \log_a 7$ **c)** $a = \log_b d$

SOLUTION

a) $y = \log_3 5$ is equivalent to $3^y = 5$ The logarithm is the exponent.

The base remains the base.

4. Rewrite $6 = \log_3 x$ as an equivalent exponential equation.

b) $-2 = \log_a 7$ is equivalent to $a^{-2} = 7$

c) $a = \log_b d$ is equivalent to $b^a = d$

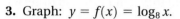 YOUR TURN

EXAMPLE 5   Rewrite each as an equivalent logarithmic equation.

a) $8 = 2^x$ **b)** $y^{-1} = 4$ **c)** $a^b = c$

SOLUTION

a) $8 = 2^x$ is equivalent to $x = \log_2 8$ The exponent is the logarithm.

The base remains the base.

b) $y^{-1} = 4$ is equivalent to $-1 = \log_y 4$

5. Rewrite $t^{-3} = 5$ as an equivalent logarithmic equation.

c) $a^b = c$ is equivalent to $b = \log_a c$

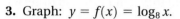 YOUR TURN

Solving Certain Logarithmic Equations

Many logarithmic equations can be solved by rewriting them as equivalent exponential equations.

EXAMPLE 6   Solve: **(a)** $\log_2 x = -3$; **(b)** $\log_x 16 = 2$.

SOLUTION

a) $\log_2 x = -3$

$\quad\quad 2^{-3} = x$ Rewriting as an exponential equation

$\quad\quad \frac{1}{8} = x$ Computing 2^{-3}

Check: $\log_2 \frac{1}{8}$ is the exponent to which 2 is raised to get $\frac{1}{8}$. Since that exponent is -3, we have a check. The solution is $\frac{1}{8}$.

b) $\log_x 16 = 2$

$\quad\quad x^2 = 16$ Rewriting as an exponential equation

$\quad x = 4 \;\; or \;\; x = -4$ Principle of square roots

Check: $\log_4 16 = 2$ because $4^2 = 16$. Thus, 4 is a solution of $\log_x 16 = 2$. Because all logarithmic bases must be positive, -4 cannot be a solution. Logarithmic bases must be positive because logarithms are defined using exponential functions that require positive bases. The solution is 4.

6. Solve: $\log_3 x = 2$.

YOUR TURN

One method for solving certain logarithmic and exponential equations relies on the following property, which results from the fact that exponential functions are one-to-one.

> **THE PRINCIPLE OF EXPONENTIAL EQUALITY**
>
> For any real number b, where $b \neq -1, 0,$ or 1,
>
> $$b^m = b^n \quad \text{is equivalent to} \quad m = n.$$
>
> (Powers of the same base are equal if and only if the exponents are equal.)

EXAMPLE 7   Solve: **(a)** $\log_{10} 1000 = x$; **(b)** $\log_4 1 = t$.

SOLUTION

a) We rewrite $\log_{10} 1000 = x$ in exponential form and solve:

$\quad\quad 10^x = 1000$ Rewriting as an exponential equation

$\quad\quad 10^x = 10^3$ Writing 1000 as a power of 10

$\quad\quad\;\; x = 3.$ Equating exponents

Check: This equation can also be solved directly by determining the exponent to which we raise 10 in order to get 1000. In both cases we find that $\log_{10} 1000 = 3$, so we have a check. The solution is 3.

b) We rewrite $\log_4 1 = t$ in exponential form and solve:

$$4^t = 1 \qquad \text{Rewriting as an exponential equation}$$
$$4^t = 4^0 \qquad \text{Writing 1 as a power of 4}$$
$$t = 0. \qquad \text{Equating exponents}$$

Check: As in part (a), this equation can be solved directly by determining the exponent to which we raise 4 in order to get 1. In both cases we find that $\log_4 1 = 0$, so we have a check. The solution is 0.

7. Solve: $\log_9 9 = x$.

 YOUR TURN

Example 7 illustrates an important property of logarithms.

$\log_a 1$

The logarithm, base a, of 1 is 0: $\log_a 1 = 0$.

This follows from the fact that $a^0 = 1$ is equivalent to the logarithmic equation $\log_a 1 = 0$. Thus, $\log_{10} 1 = 0$, $\log_7 1 = 0$, and so on.

Another property results from the fact that $a^1 = a$. This is equivalent to the logarithmic equation $\log_a a = 1$.

$\log_a a$

The logarithm, base a, of a is 1: $\log_a a = 1$.

Thus, $\log_{10} 10 = 1$, $\log_8 8 = 1$, and so on.

9.3   EXERCISE SET

FOR EXTRA HELP

MyMathLab® *Math* XL

PRACTICE   WATCH   READ   REVIEW

Vocabulary and Reading Check

Choose the word from those listed below the blank that best completes each statement.

1. The inverse of the function given by $f(x) = 3^x$ is a(n) _____ function.
   exponential/logarithmic

2. A logarithm is a(n) _____.
   base/exponent

3. Logarithm bases are _____.
   negative/positive

4. The logarithm, base a, of a is _____.
   0/1

Concept Reinforcement

In each of Exercises 5–12, match the expression or equation with an equivalent expression or equation from the column on the right.

5. ____ $\log_5 25$ a) 1
6. ____ $2^5 = x$ b) x
7. ____ $\log_5 5$ c) $x^5 = 2$
8. ____ $\log_2 1$ d) $\log_2 x = 5$
9. ____ $\log_5 5^x$ e) $\log_2 5 = x$
10. ____ $\log_x 2 = 5$ f) $5^2 = x$
11. ____ $5 = 2^x$ g) 2
12. ____ $\log_5 x = 2$ h) 0

The Meaning of Logarithms

Simplify.

13. $\log_{10} 1000$

14. $\log_{10} 100$

15. $\log_7 49$

16. $\log_3 9$

17. $\log_5 \frac{1}{25}$

18. $\log_5 \frac{1}{5}$

19. $\log_8 \frac{1}{8}$

20. $\log_8 \frac{1}{64}$

21. $\log_5 625$

22. $\log_5 125$

23. $\log_7 7$

24. $\log_9 1$

25. $\log_3 1$

26. $\log_3 3$

Aha! **27.** $\log_6 6^5$

28. $\log_6 6^9$

29. $\log_{10} 0.01$

30. $\log_{10} 0.1$

31. $\log_{16} 4$

32. $\log_{100} 10$

33. $\log_9 27$

34. $\log_4 32$

35. $\log_{1000} 100$

36. $\log_{16} 8$

37. $3^{\log_3 29}$

38. $6^{\log_6 13}$

Graphs of Logarithmic Functions

Graph.

39. $y = \log_{10} x$

40. $y = \log_2 x$

41. $y = \log_3 x$

42. $y = \log_7 x$

43. $f(x) = \log_6 x$

44. $f(x) = \log_4 x$

45. $f(x) = \log_{2.5} x$

46. $f(x) = \log_{1/2} x$

Graph each pair of functions using one set of axes.

47. $f(x) = 3^x, \ f^{-1}(x) = \log_3 x$

48. $f(x) = 4^x, \ f^{-1}(x) = \log_4 x$

Equivalent Equations

Rewrite each of the following as an equivalent exponential equation. Do not solve.

49. $x = \log_{10} 8$

50. $y = \log_8 10$

51. $\log_9 9 = 1$

52. $\log_6 36 = 2$

53. $\log_{10} 0.1 = -1$

54. $\log_{10} 0.01 = -2$

55. $\log_{10} 7 = 0.845$

56. $\log_{10} 3 = 0.4771$

57. $\log_c m = 8$

58. $\log_b n = 23$

59. $\log_r C = t$

60. $\log_m P = a$

61. $\log_e 0.25 = -1.3863$

62. $\log_e 0.989 = -0.0111$

63. $\log_r T = -x$

64. $\log_c M = -w$

Rewrite each of the following as an equivalent logarithmic equation. Do not solve.

65. $10^2 = 100$

66. $10^4 = 10,000$

67. $5^{-3} = \frac{1}{125}$

68. $2^{-5} = \frac{1}{32}$

69. $16^{1/4} = 2$

70. $8^{1/3} = 2$

71. $10^{0.4771} = 3$

72. $10^{0.3010} = 2$

73. $z^m = 6$

74. $m^n = r$

75. $p^t = q$

76. $y^t = x$

77. $e^3 = 20.0855$

78. $e^2 = 7.3891$

Solving Certain Logarithmic Equations

Solve.

79. $\log_6 x = 2$

80. $\log_4 x = 3$

81. $\log_2 32 = x$

82. $\log_5 25 = x$

83. $\log_x 9 = 1$

84. $\log_x 12 = 1$

85. $\log_x 7 = \frac{1}{2}$

86. $\log_x 9 = \frac{1}{2}$

87. $\log_3 x = -2$

88. $\log_2 x = -1$

89. $\log_{32} x = \frac{2}{5}$

90. $\log_8 x = \frac{2}{3}$

91. Explain why a logarithm base must be positive.

92. Is it easier to find x given $x = \log_9 \frac{1}{3}$ or given $9^x = \frac{1}{3}$? Explain your reasoning.

Skill Review

Simplify.

93. $\sqrt{18a^3b}\sqrt{50ab^7}$ [7.3]

94. $(2\sqrt{3} + \sqrt{5})(2\sqrt{3} - \sqrt{10})$ [7.5]

95. $\sqrt{192x} - \sqrt{75x}$ [7.5]

96. $\sqrt[4]{\sqrt[3]{x}}$ [7.2]

97. $\dfrac{\sqrt[3]{24xy^8}}{\sqrt[3]{3xy}}$ [7.4]

98. $\dfrac{\sqrt[5]{a^4y^6}}{\sqrt{ay}}$ [7.5]

Synthesis

99. Would a manufacturer be pleased or unhappy if sales of a product grew logarithmically? Why?

100. Explain why the number $\log_2 13$ must be between 3 and 4.

101. Graph both equations using one set of axes:
$$y = \left(\tfrac{3}{2}\right)^x, \qquad y = \log_{3/2} x.$$

Graph.

102. $y = \log_2 (x - 1)$

103. $y = \log_3 |x + 1|$

Solve.

104. $|\log_3 x| = 2$

105. $\log_4 (3x - 2) = 2$

106. $\log_8 (2x + 1) = -1$

107. $\log_{10} (x^2 + 21x) = 2$

Simplify.

108. $\log_{1/4} \frac{1}{64}$

109. $\log_{1/5} 25$

110. $\log_{81} 3 \cdot \log_3 81$

111. $\log_{10} (\log_4 (\log_3 81))$

112. $\log_2 (\log_2 (\log_4 256))$

113. Show that $b^{x_1} = b^{x_2}$ is *not* equivalent to $x_1 = x_2$ for $b = 0$ or $b = 1$.

114. If $\log_b a = x$, does it follow that $\log_a b = 1/x$? Why or why not?

⮑ YOUR TURN ANSWERS: SECTION 9.3

1. 4   **2.** 14   **3.**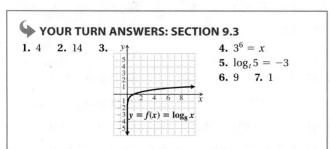

4. $3^6 = x$

5. $\log_t 5 = -3$

6. 9   **7.** 1

$y = f(x) = \log_8 x$

9.4   Properties of Logarithmic Functions

Logarithms of Products ▪ Logarithms of Powers ▪ Logarithms of Quotients ▪ Using the Properties Together

Logarithmic functions are important in many applications and in more advanced mathematics. We now establish some basic properties that are useful in manipulating expressions involving logarithms. As their proofs reveal, the properties of logarithms are related to the properties of exponents.

Logarithms of Products

The first property we discuss is related to the product rule for exponents: $a^m \cdot a^n = a^{m+n}$. Its proof appears immediately after Example 2.

THE PRODUCT RULE FOR LOGARITHMS

For any positive numbers M, N, and a ($a \neq 1$),

$$\log_a (MN) = \log_a M + \log_a N.$$

(The logarithm of a product is the sum of the logarithms of the factors.)

EXAMPLE 1 Express as an equivalent expression that is a sum of logarithms: $\log_2 (4 \cdot 16)$.

SOLUTION We have

$$\log_2 (4 \cdot 16) = \log_2 4 + \log_2 16. \quad \text{Using the product rule for logarithms}$$

As a check, note that

$$\log_2 (4 \cdot 16) = \log_2 64 = 6, \quad \text{since } 2^6 = 64,$$

and that

$$\log_2 4 + \log_2 16 = 2 + 4 = 6, \quad \text{since } 2^2 = 4 \text{ and } 2^4 = 16.$$

YOUR TURN

1. Express as an equivalent expression that is a sum of logarithms: $\log_{10} (10 \cdot 1000)$.

EXAMPLE 2 Express as an equivalent expression that is a single logarithm: $\log_b 7 + \log_b 5$.

SOLUTION

$$\log_b 7 + \log_b 5 = \log_b (7 \cdot 5) \quad \text{Using the product rule for logarithms}$$
$$= \log_b 35.$$

YOUR TURN

2. Express as an equivalent expression that is a single logarithm: $\log_3 x + \log_3 y$.

A Proof of the Product Rule. Let $\log_a M = x$ and $\log_a N = y$. Converting to exponential equations, we have $a^x = M$ and $a^y = N$.

Now we multiply the left side of the first exponential equation by the left side of the second equation and similarly multiply the right sides to obtain

$$MN = a^x \cdot a^y, \quad \text{or} \quad MN = a^{x+y}.$$

Converting back to a logarithmic equation, we get

$$\log_a (MN) = x + y.$$

Recalling what x and y represent, we have

$$\log_a (MN) = \log_a M + \log_a N.$$ ∎

Logarithms of Powers

The second basic property is related to the power rule for exponents: $(a^m)^n = a^{mn}$. Its proof follows Example 3.

THE POWER RULE FOR LOGARITHMS

For any positive numbers M and a ($a \neq 1$), and any real number p,

$$\log_a M^p = p \cdot \log_a M.$$

(The logarithm of a power of M is the exponent times the logarithm of M.)

To better understand the power rule, note that

$$\log_a M^3 = \log_a (M \cdot M \cdot M) = \log_a M + \log_a M + \log_a M = 3 \log_a M.$$

EXAMPLE 3   Use the power rule for logarithms to write an equivalent expression that is a product: **(a)** $\log_a 9^{-5}$; **(b)** $\log_7 \sqrt[3]{x}$.

SOLUTION

3. Use the power rule for logarithms to write an equivalent expression that is a product: $\log_5 25^7$.

a) $\log_a 9^{-5} = -5\log_a 9$ Using the power rule for logarithms

b) $\log_7 \sqrt[3]{x} = \log_7 x^{1/3}$ Writing exponential notation

$\qquad\qquad = \frac{1}{3}\log_7 x$ Using the power rule for logarithms

↩ YOUR TURN

A Proof of the Power Rule.   Let $x = \log_a M$. We then write the equivalent exponential equation, $a^x = M$. Raising both sides to the pth power, we get

$$(a^x)^p = M^p, \quad \text{or} \quad a^{xp} = M^p. \qquad \text{Multiplying exponents}$$

Converting back to a logarithmic equation gives us

$$\log_a M^p = xp.$$

But $x = \log_a M$, so substituting, we have

$$\log_a M^p = (\log_a M)p = p \cdot \log_a M. \qquad ■$$

Student Notes

Without understanding and remembering the rules of this section, it will be extremely difficult to solve exponential and logarithmic equations.

Logarithms of Quotients

The third property that we study is related to the quotient rule for exponents: $a^m/a^n = a^{m-n}$. Its proof follows Example 5.

THE QUOTIENT RULE FOR LOGARITHMS

For any positive numbers M, N, and a ($a \neq 1$),

$$\log_a \frac{M}{N} = \log_a M - \log_a N.$$

(The logarithm of a quotient is the logarithm of the dividend minus the logarithm of the divisor.)

To better understand the quotient rule, note that

$$\log_2 \tfrac{8}{32} = \log_2 \tfrac{1}{4} = -2$$

and $\log_2 8 - \log_2 32 = 3 - 5 = -2.$

EXAMPLE 4   Express as an equivalent expression that is a difference of logarithms: $\log_t (6/U)$.

4. Express as an equivalent expression that is a difference of logarithms: $\log_4 \left(\dfrac{x}{8}\right)$.

SOLUTION

$$\log_t \frac{6}{U} = \log_t 6 - \log_t U \qquad \text{Using the quotient rule for logarithms}$$

↩ YOUR TURN

EXAMPLE 5   Express as an equivalent expression that is a single logarithm:

$$\log_b 17 - \log_b 27.$$

SOLUTION

5. Express as an equivalent expression that is a single logarithm: $\log_3 a - \log_3 16.$

$$\log_b 17 - \log_b 27 = \log_b \frac{17}{27} \qquad \substack{\text{Using the quotient rule for} \\ \text{logarithms "in reverse"}}$$

↩ YOUR TURN

A Proof of the Quotient Rule. Our proof uses both the product rule and the power rule:

$$\log_a \frac{M}{N} = \log_a (MN^{-1}) \qquad \text{Rewriting } \frac{M}{N} \text{ as } MN^{-1}$$

$$= \log_a M + \log_a N^{-1} \qquad \text{Using the product rule for logarithms}$$

$$= \log_a M + (-1)\log_a N \qquad \text{Using the power rule for logarithms}$$

$$= \log_a M - \log_a N. \qquad \blacksquare$$

Using the Properties Together

EXAMPLE 6 Express as an equivalent expression, using the logarithms of x, y, and z.

a) $\log_b \dfrac{x^3}{yz}$

b) $\log_a \sqrt[4]{\dfrac{xy}{z^3}}$

SOLUTION

a) $\log_b \dfrac{x^3}{yz} = \log_b x^3 - \log_b yz$ Using the quotient rule for logarithms

$$= 3\log_b x - \log_b yz \qquad \text{Using the power rule for logarithms}$$

$$= 3\log_b x - (\log_b y + \log_b z) \qquad \text{Using the product rule for logarithms. Because of the subtraction, parentheses are essential.}$$

$$= 3\log_b x - \log_b y - \log_b z \qquad \text{Using the distributive law}$$

b) $\log_a \sqrt[4]{\dfrac{xy}{z^3}} = \log_a \left(\dfrac{xy}{z^3}\right)^{1/4}$ Writing exponential notation

$$= \frac{1}{4} \cdot \log_a \frac{xy}{z^3} \qquad \text{Using the power rule for logarithms}$$

$$= \frac{1}{4}\left(\log_a xy - \log_a z^3\right) \qquad \text{Using the quotient rule for logarithms. Parentheses are important.}$$

$$= \frac{1}{4}\left(\log_a x + \log_a y - 3\log_a z\right) \qquad \text{Using the product rule and the power rule for logarithms}$$

6. Express as an equivalent expression, using the logarithms of x, y, and z:

$$\log_a \frac{x^2 y}{z^4}.$$

YOUR TURN

> **CAUTION!** Because the product rule and the quotient rule replace one term with two, it is often essential to apply the rules within parentheses, as in Example 6.

EXAMPLE 7 Express as an equivalent expression that is a single logarithm.

a) $\dfrac{1}{2}\log_a x - 7\log_a y + \log_a z$

b) $\log_a \dfrac{b}{\sqrt{x}} + \log_a \sqrt{bx}$

SOLUTION

a) $\frac{1}{2}\log_a x - 7\log_a y + \log_a z$

$= \log_a x^{1/2} - \log_a y^7 + \log_a z$ Using the power rule for logarithms

$= (\log_a \sqrt{x} - \log_a y^7) + \log_a z$ Using parentheses to emphasize the order of operations; $x^{1/2} = \sqrt{x}$

$= \log_a \dfrac{\sqrt{x}}{y^7} + \log_a z$ Using the quotient rule for logarithms. Note that all terms have the same base.

$= \log_a \dfrac{z\sqrt{x}}{y^7}$ Using the product rule for logarithms

b) $\log_a \dfrac{b}{\sqrt{x}} + \log_a \sqrt{bx} = \log_a \dfrac{b \cdot \sqrt{bx}}{\sqrt{x}}$ Using the product rule for logarithms

$= \log_a b\sqrt{b}$ Removing a factor equal to 1: $\dfrac{\sqrt{x}}{\sqrt{x}} = 1$

7. Express as an equivalent expression that is a single logarithm:

$2\log_a x - 3\log_a y - \log_a z.$

$= \log_a b^{3/2}$, or $\dfrac{3}{2}\log_a b$ Noting that $b\sqrt{b} = b^1 \cdot b^{1/2}$

YOUR TURN

If we know the logarithms of two different numbers (with the same base), the properties allow us to calculate other logarithms.

EXAMPLE 8 Given $\log_a 2 = 0.431$ and $\log_a 3 = 0.683$, use the properties of logarithms to calculate the value of each of the following. If this is not possible, state so.

a) $\log_a 6$ **b)** $\log_a \frac{2}{3}$ **c)** $\log_a 81$

d) $\log_a \frac{1}{3}$ **e)** $\log_a (2a)$ **f)** $\log_a 5$

SOLUTION

a) $\log_a 6 = \log_a(2 \cdot 3) = \log_a 2 + \log_a 3$ Using the product rule for logarithms

$= 0.431 + 0.683 = 1.114$

Check: $a^{1.114} = a^{0.431} \cdot a^{0.683} = 2 \cdot 3 = 6$

b) $\log_a \frac{2}{3} = \log_a 2 - \log_a 3$ Using the quotient rule for logarithms

$= 0.431 - 0.683 = -0.252$

c) $\log_a 81 = \log_a 3^4 = 4\log_a 3$ Using the power rule for logarithms

$= 4(0.683) = 2.732$

d) $\log_a \frac{1}{3} = \log_a 1 - \log_a 3$ Using the quotient rule for logarithms

$= 0 - 0.683 = -0.683$

8. Given $\log_a 2 = 0.431$ and $\log_a 3 = 0.683$, use the properties of logarithms to calculate the value of $\log_a \frac{9}{4}$.

e) $\log_a (2a) = \log_a 2 + \log_a a$ Using the product rule for logarithms

$= 0.431 + 1 = 1.431$

f) $\log_a 5$ *cannot be found using these properties.* ($\log_a 5 \neq \log_a 2 + \log_a 3$)

YOUR TURN

A final property follows from the product rule: Since $\log_a a^k = k\log_a a$, and $\log_a a = 1$, we have $\log_a a^k = k$.

> ### THE LOGARITHM OF THE BASE TO AN EXPONENT
>
> For any base a,
> $$\log_a a^k = k.$$
> (The logarithm, base a, of a to an exponent is the exponent.)

This property also follows directly from the definition of logarithm: k is the exponent to which you raise a in order to get a^k.

EXAMPLE 9 Simplify: **(a)** $\log_3 3^7$; **(b)** $\log_{10} 10^{-5.2}$.

SOLUTION

a) $\log_3 3^7 = 7$ 7 is the exponent to which you raise 3 in order to get 3^7.

b) $\log_{10} 10^{-5.2} = -5.2$

9. Simplify: $\log_8 8^5$.

YOUR TURN

We summarize the properties of logarithms as follows.

> For any positive numbers M, N, and a $(a \neq 1)$:
> $$\log_a (MN) = \log_a M + \log_a N; \qquad \log_a M^p = p \cdot \log_a M;$$
> $$\log_a \frac{M}{N} = \log_a M - \log_a N; \qquad \log_a a^k = k.$$

> **CAUTION!**   Keep in mind that, in general,
> $$\log_a(M + N) \neq \log_a M + \log_a N, \qquad \log_a (MN) \neq (\log_a M)(\log_a N),$$
> $$\log_a (M - N) \neq \log_a M - \log_a N, \qquad \log_a \frac{M}{N} \neq \frac{\log_a M}{\log_a N}.$$

9.4   EXERCISE SET

FOR EXTRA HELP

MyMathLab® Math XL
PRACTICE   WATCH   READ   REVIEW

Vocabulary and Reading Check

Match each expression with an equivalent expression from the column on the right.

1. ____ $\log_7 20$

2. ____ $\log_7 5^4$

3. ____ $\log_7 \frac{5}{4}$

4. ____ $\log_7 7$

5. ____ $\log_7 7^4$

6. ____ $\log_7 5 + \log_7 6$

a) $\log_7 5 - \log_7 4$

b) 1

c) 4

d) $\log_7 30$

e) $\log_7 5 + \log_7 4$

f) $4 \log_7 5$

Logarithms of Products

Express as an equivalent expression that is a sum of logarithms.

7. $\log_3 (81 \cdot 27)$

8. $\log_2 (16 \cdot 32)$

9. $\log_4 (64 \cdot 16)$

10. $\log_5 (25 \cdot 125)$

11. $\log_c (rst)$

12. $\log_t (3ab)$

Express as an equivalent expression that is a single logarithm.

13. $\log_a 2 + \log_a 10$

14. $\log_b 5 + \log_b 9$

15. $\log_c t + \log_c y$

16. $\log_t H + \log_t M$

Logarithms of Powers

Express as an equivalent expression that is a product.

17. $\log_a r^8$

18. $\log_b t^5$

19. $\log_2 y^{1/3}$

20. $\log_{10} y^{1/2}$

21. $\log_b C^{-3}$

22. $\log_c M^{-5}$

Logarithms of Quotients

Express as an equivalent expression that is a difference of two logarithms.

23. $\log_2 \frac{5}{11}$

24. $\log_3 \frac{29}{13}$

25. $\log_b \dfrac{m}{n}$

26. $\log_a \dfrac{y}{x}$

Express as an equivalent expression that is a single logarithm.

27. $\log_a 19 - \log_a 2$

28. $\log_b 3 - \log_b 32$

29. $\log_b 36 - \log_b 4$

30. $\log_a 26 - \log_a 2$

31. $\log_a x - \log_a y$

32. $\log_b c - \log_b d$

Using the Properties Together

Express as an equivalent expression, using the individual logarithms of w, x, y, and z.

33. $\log_a (xyz)$

34. $\log_a (wxy)$

35. $\log_a (x^3 z^4)$

36. $\log_a (x^2 y^5)$

37. $\log_a (w^2 x^{-2} y)$

38. $\log_a (xy^2 z^{-3})$

39. $\log_a \dfrac{x^5}{y^3 z}$

40. $\log_a \dfrac{x^4}{yz^2}$

41. $\log_b \dfrac{xy^2}{wz^3}$

42. $\log_b \dfrac{w^2 x}{y^3 z}$

43. $\log_a \sqrt{\dfrac{x^7}{y^5 z^8}}$

44. $\log_c \sqrt{\dfrac{x^4}{y^3 z^2}}$

45. $\log_a \sqrt[3]{\dfrac{x^6 y^3}{a^2 z^7}}$

46. $\log_a \sqrt[4]{\dfrac{x^8 y^{12}}{a^3 z^5}}$

Express as an equivalent expression that is a single logarithm and, if possible, simplify.

47. $\log_a x + \log_a 3$

48. $\log_a c - \log_a d$

49. $8 \log_a x + 3 \log_a z$

50. $2 \log_b m + \frac{1}{2} \log_b n$

51. $2 \log_b w - \log_b z - 4 \log_b y$

52. $3 \log_c p + \frac{1}{2} \log_c t - \log_c 7$

53. $\log_a x^2 - 2 \log_a \sqrt{x}$

54. $\log_a \dfrac{a}{\sqrt{x}} - \log_a \sqrt{ax}$

55. $\frac{1}{2} \log_a x + 5 \log_a y - 2 \log_a x$

56. $\log_a 2x + 3(\log_a x - \log_a y)$

57. $\log_a (x^2 - 9) - \log_a (x + 3)$

58. $\log_a (2x + 10) - \log_a (x^2 - 25)$

Given $\log_b 3 = 0.792$ and $\log_b 5 = 1.161$. If possible, use the properties of logarithms to calculate values for each of the following.

59. $\log_b 15$

60. $\log_b \frac{5}{3}$

61. $\log_b \frac{3}{5}$

62. $\log_b \frac{1}{3}$

63. $\log_b \frac{1}{5}$

64. $\log_b \sqrt{b}$

65. $\log_b \sqrt{b^3}$

66. $\log_b 3b$

67. $\log_b 8$

68. $\log_b 45$

Simplify.

Aha! **69.** $\log_t t^{10}$

70. $\log_p p^{-5}$

71. $\log_e e^m$

72. $\log_Q Q^t$

73. Explain the difference between the phrases "the logarithm of a quotient" and "a quotient of logarithms."

74. How could you convince someone that
$$\log_a c \neq \log_c a?$$

Skill Review

Solve.

75. $|3x + 7| \le 4$ [4.3]

76. $24x^2 - 2x = 15$ [5.8]

77. $x^2 + 4x + 5 = 0$ [8.2]

78. $\sqrt[3]{2x} = 1$ [7.6]

79. $x^{1/2} - 6x^{1/4} + 8 = 0$ [8.5]

80. $2y - 7\sqrt{y} + 3 = 0$ [8.5]

Synthesis

81. Bob *incorrectly* reasons that
$$\log_b \frac{1}{x} = \log_b \frac{x}{xx}$$
$$= \log_b x - \log_b x + \log_b x = \log_b x.$$
What mistake has Bob made?

82. Why are properties of logarithms related to properties of exponents?

Express as an equivalent expression that is a single logarithm and, if possible, simplify.

83. $\log_a (x^8 - y^8) - \log_a (x^2 + y^2)$

84. $\log_a (x + y) + \log_a (x^2 - xy + y^2)$

Express as an equivalent expression that is a sum or a difference of logarithms and, if possible, simplify.

85. $\log_a \sqrt{1 - s^2}$

86. $\log_a \dfrac{c - d}{\sqrt{c^2 - d^2}}$

87. If $\log_a x = 2$, $\log_a y = 3$, and $\log_a z = 4$, what is

$$\log_a \frac{\sqrt[3]{x^2 z}}{\sqrt[3]{y^2 z^{-2}}}?$$

88. If $\log_a x = 2$, what is $\log_a (1/x)$?

89. If $\log_a x = 2$, what is $\log_{1/a} x$?

Solve.

90. $\log_{10} 2000 - \log_{10} x = 3$

91. $\log_2 80 + \log_2 x = 5$

Classify each of the following as true or false. Assume a, x, P, and $Q > 0$, $a \neq 1$.

92. $\log_a \left(\dfrac{P}{Q}\right)^x = x \log_a P - \log_a Q$

93. $\log_a (Q + Q^2) = \log_a Q + \log_a (Q + 1)$

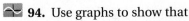 **94.** Use graphs to show that
$$\log x^2 \neq \log x \cdot \log x.$$
(*Note:* log means $\log_{10}$.)

Mid-Chapter Review

We use the following properties to simplify expressions and to rewrite equivalent logarithmic equations and exponential equations.

$$\log_a x = m \text{ means } a^m = x \qquad \log_a a^k = k$$
$$\log_a (MN) = \log_a M + \log_a N \qquad \log_a a = 1$$
$$\log_a \frac{M}{N} = \log_a M - \log_a N \qquad \log_a 1 = 0$$
$$\log_a M^p = p \cdot \log_a M$$

GUIDED SOLUTIONS

1. Find a formula for the inverse of $f(x) = 2x - 5$. [9.1]

Solution

$$y = 2x - 5$$

$$\boxed{} = 2\boxed{} - 5 \qquad \text{Interchanging } x \text{ and } y$$

$$\left.\begin{array}{l} \boxed{} = 2y \\[1em] \dfrac{\boxed{}}{2} = y \end{array}\right\} \qquad \text{Solving for } y$$

$$f^{-1}(x) = \boxed{}$$

2. Solve: $\log_4 x = 1$. [9.3]

Solution

$$\log_4 x = 1$$

$$\boxed{}^{\boxed{}} = x \qquad \text{Rewriting as an exponential equation}$$

$$\boxed{} = x$$

MIXED REVIEW

3. Find $(f \circ g)(x)$ if $f(x) = x^2 + 1$ and $g(x) = x - 5$. [9.1]

4. If $h(x) = \sqrt{5x - 3}$, find $f(x)$ and $g(x)$ such that $h(x) = (f \circ g)(x)$. Answers may vary. [9.1]

5. Find a formula for the inverse of $g(x) = 6 - x$. [9.1]

6. Graph: $f(x) = 2^x + 3$. [9.2]

Simplify.

7. $\log_4 16$ [9.3]

8. $\log_5 \frac{1}{5}$ [9.3]

9. $\log_{100} 10$ [9.3]

10. $\log_b b$ [9.4]

11. $\log_8 8^{19}$ [9.4]

12. $\log_t 1$ [9.4]

Rewrite each of the following as an equivalent exponential equation.

13. $\log_x 3 = m$ [9.3]

14. $\log_2 1024 = 10$ [9.3]

Rewrite each of the following as an equivalent logarithmic equation.

15. $e^t = x$ [9.3]

16. $64^{2/3} = 16$ [9.3]

17. Express as an equivalent expression using $\log x$, $\log y$, and $\log z$:

$$\log \sqrt{\frac{x^2}{yz^3}}. \quad [9.4]$$

18. Express as an equivalent expression that is a single logarithm:

$$\log a - 2 \log b - \log c. \quad [9.4]$$

Solve. [9.3]

19. $\log_x 64 = 3$

20. $\log_3 x = -1$

9.5 Common Logarithms and Natural Logarithms

Common Logarithms on a Calculator ▪ The Base *e* and Natural Logarithms on a Calculator ▪
Changing Logarithmic Bases ▪ Graphs of Exponential Functions and Logarithmic Functions, Base *e*

Study Skills

Is Your Answer Reasonable?

It is always a good idea—especially when using a calculator—to check that your answer is reasonable. It is easy for an incorrect calculation or keystroke to result in an answer that is clearly too big or too small.

Technology Connection

To find log 6500/log 0.007 on a graphing calculator, we must use parentheses with care.

1. What keystrokes are needed to create the following?

log(7)/log(3)
1.771243749

1. Use a calculator to approximate $\dfrac{\log 5}{\log 2}$ to four decimal places.

Any positive number other than 1 can serve as the base of a logarithmic function. However, there are logarithmic bases that fit into certain applications more naturally than others.

Base-10 logarithms, called **common logarithms**, are useful because they have the same base as our "commonly" used decimal system. Before calculators became widely available, common logarithms helped with tedious calculations. In fact, that is why logarithms were devised.

The logarithmic base most widely used today is an irrational number named *e*. We will consider *e* and base *e*, or *natural*, logarithms later in this section. First we examine common logarithms.

Common Logarithms on a Calculator

Before the advent of scientific calculators, printed tables listed common logarithms. Today we find common logarithms using calculators.

The abbreviation **log**, with no base written, is generally understood to mean logarithm base 10, that is, a common logarithm. Thus,

$$\log 17 \quad \text{means} \quad \log_{10} 17.$$

The key for common logarithms is usually marked **LOG**. To find the common logarithm of a number, we key in that number and press **LOG**. With most graphing calculators, we press **LOG**, the number, and then **ENTER**.

EXAMPLE 1 Use a calculator to approximate each number to four decimal places.

a) log 5312

b) $\dfrac{\log 6500}{\log 0.007}$

SOLUTION

a) We enter 5312 and then press **LOG**. On most graphing calculators, we press **LOG**, followed by 5312 and **ENTER**. We find that

$$\log 5312 \approx 3.7253. \quad \text{Rounded to four decimal places}$$

b) We enter 6500 and then press **LOG**. Next, we press **÷**, enter 0.007, and then press **LOG** **=**. On most graphing calculators, we press **LOG**, key in 6500, press **)** **÷** **LOG**, key in 0.007, and then press **)** **ENTER**. Be careful not to round until the end:

$$\frac{\log 6500}{\log 0.007} \approx -1.7694. \quad \text{Rounded to four decimal places}$$

 YOUR TURN

The inverse of a logarithmic function is an exponential function. Because of this, on many calculators the **LOG** key doubles as the **10ˣ** key after a **2ND** or **SHIFT** key is pressed. Calculators lacking a **10ˣ** key may have a key labeled **xʸ**, **aˣ**, or **^**.

EXAMPLE 2   Use a calculator to approximate $10^{3.417}$ to four decimal places.

SOLUTION   We enter 3.417 and then press $\boxed{10^x}$. On most graphing calculators, $\boxed{10^x}$ is pressed first, followed by 3.417 and **ENTER**. Rounding to four decimal places, we have

$$10^{3.417} \approx 2612.1614.$$

 YOUR TURN

2. Use a calculator to approximate $10^{1.5}$ to four decimal places.

The Base *e* and Natural Logarithms on a Calculator

When interest is compounded *n* times a year, the compound interest formula is

$$A = P\left(1 + \frac{r}{n}\right)^{nt},$$

where *A* is the amount that an initial investment *P* is worth after *t* years at interest rate *r*. Suppose that $1 is invested at 100% interest for 1 year (no bank would pay this). The preceding formula becomes a function *A* defined in terms of the number of compounding periods *n*:

$$A(n) = \left(1 + \frac{1}{n}\right)^n.$$

Let's find some function values. We use a calculator and round to six decimal places.

n	$A(n) = \left(1 + \dfrac{1}{n}\right)^n$
1 (compounded annually)	$2.00
2 (compounded semiannually)	2.25
3	2.370370
4 (compounded quarterly)	2.441406
12 (compounded monthly)	2.613035
100	2.704814
365 (compounded daily)	2.714567
8760 (compounded hourly)	2.718127

The numbers in this table approach a very important number in mathematics, called *e*. Because *e* is irrational, its decimal representation does not terminate or repeat.

> **THE NUMBER *e***
>
> $e \approx 2.7182818284\ldots$

Logarithms base *e* are called **natural logarithms**, or **Napierian logarithms**, in honor of John Napier (1550–1617), the "inventor" of logarithms.

The abbreviation "ln" is generally used with natural logarithms. Thus,

$\ln 53$ means $\log_e 53$.

On most calculators, the key for natural logarithms is marked **LN**.

Technology Connection

To visualize the number *e*, let $y_1 = (1 + 1/x)^x$.

$y_1 = (1 + 1/x)^x$

1. Use $\boxed{\text{TRACE}}$ or $\boxed{\text{TABLE}}$ to confirm that as *x* gets larger, the number *e* is more closely approximated.

2. Graph $y_2 = e$ and compare y_1 and y_2 for large values of *x*.

3. Confirm that 0 is not in the domain of this function. Why?

EXAMPLE 3   Use a calculator to approximate ln 4568 to four decimal places.

3. Use a calculator to approximate ln 2.1 to four decimal places.

SOLUTION   We enter 4568 and then press **LN**. On most graphing calculators, we press **LN** first, followed by 4568 and **ENTER**. We find that

$$\ln 4568 \approx 8.4268.$$ Rounded to four decimal places

YOUR TURN

On many calculators, the **LN** key doubles as the **eˣ** key after a **2ND** or SHIFT key has been pressed.

EXAMPLE 4   Use a calculator to approximate $e^{-1.524}$ to four decimal places.

4. Use a calculator to approximate $e^{2.56}$ to four decimal places.

SOLUTION   We enter -1.524 and then press **eˣ**. On most graphing calculators, **eˣ** is pressed first, followed by -1.524 and **ENTER**. We find that

$$e^{-1.524} \approx 0.2178.$$ Rounded to four decimal places

YOUR TURN

Student Notes

When "log" is written without a base, you can always write in a base of 10. Similarly, you can always replace "ln" with "$\log_e$."

Changing Logarithmic Bases

Most calculators can find both common logarithms and natural logarithms. To find a logarithm with some other base, a conversion formula is often used.

> **THE CHANGE-OF-BASE FORMULA**
>
> For any logarithmic bases a and b, and any positive number M,
>
> $$\log_b M = \frac{\log_a M}{\log_a b}.$$
>
> (To find the log, base b, of M, we typically compute $\log M/\log b$ or $\ln M/\ln b$.)

Technology Connection

Some calculators can find logarithms with any base, often through a LOGBASE(option in the MATH MATH submenu. You can fill in the blanks shown below, moving between blanks using the left and right arrow keys, and then press **ENTER** to find the logarithm.

Proof.   Let $x = \log_b M$. Then,

$$b^x = M$$ $\log_b M = x$ is equivalent to $b^x = M$.

$$\log_a b^x = \log_a M$$ Taking the logarithm, base a, on both sides

$$x \log_a b = \log_a M$$ Using the power rule for logarithms

$$x = \frac{\log_a M}{\log_a b}.$$ Dividing both sides by $\log_a b$

At the outset, we stated that $x = \log_b M$. Thus, by substitution, we have

$$\log_b M = \frac{\log_a M}{\log_a b}.$$ This is the change-of-base formula.   ∎

EXAMPLE 5   Find $\log_5 8$ using the change-of-base formula.

SOLUTION   We use the change-of-base formula with $a = 10$, $b = 5$, and $M = 8$:

$$\log_5 8 = \frac{\log_{10} 8}{\log_{10} 5} \qquad \text{Substituting into } \log_b M = \frac{\log_a M}{\log_a b}$$

$$\approx \frac{0.903089987}{0.6989700043} \qquad \text{Using } \boxed{\text{LOG}} \text{ twice}$$

$$\approx 1.2920. \qquad \text{When using a calculator, it is best to wait until the end to round.}$$

To check, note that $\ln 8/\ln 5 \approx 1.2920$. We can also use a calculator to verify that $5^{1.2920} \approx 8$. As a quick partial check, note that since $5^1 = 5$ and $5^2 = 25$, we can expect an answer between 1 and 2.

5. Find $\log_2 6$ using the change-of-base formula.

YOUR TURN

Student Notes

The choice of the logarithm base a in the change-of-base formula should be either 10 or e so that the logarithms can be found using a calculator. Either choice will yield the same end result.

EXAMPLE 6   Find $\log_4 31$.

SOLUTION   As shown in the check of Example 5, base e can also be used.

$$\log_4 31 = \frac{\log_e 31}{\log_e 4} \qquad \text{Substituting into } \log_b M = \frac{\log_a M}{\log_a b}$$

$$= \frac{\ln 31}{\ln 4} \approx \frac{3.433987204}{1.386294361} \qquad \text{Using } \boxed{\text{LN}} \text{ twice}$$

$$\approx 2.4771. \qquad \textit{Check: } 4^{2.4771} \approx 31.$$

6. Find $\log_8 3$.

YOUR TURN

Graphs of Exponential Functions and Logarithmic Functions, Base e

EXAMPLE 7   Graph $f(x) = e^x$ and $g(x) = e^{-x}$ and state the domain and the range of f and g.

SOLUTION   We use a calculator with an $\boxed{e^x}$ key to find approximate values of e^x and e^{-x}. Using these values, we can graph the functions.

x	e^x	e^{-x}
0	1	1
1	2.7	0.4
2	7.4	0.1
−1	0.4	2.7
−2	0.1	7.4

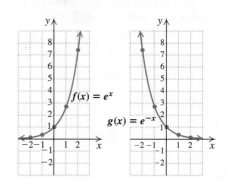

7. Graph $f(x) = 3e^x$ and state the domain and the range of f.

YOUR TURN

The domain of each function is $\mathbb{R}$, and the range of each function is $(0, \infty)$.

EXAMPLE 8   Graph $f(x) = e^{-x} + 2$ and state the domain and the range of f.

SOLUTION   We find some solutions with a calculator, plot them, and then draw the graph. For example, $f(2) = e^{-2} + 2 \approx 0.1 + 2 \approx 2.1$. The graph is exactly like the graph of $g(x) = e^{-x}$, but is translated up 2 units.

x	$e^{-x} + 2$
0	3
1	2.4
2	2.1
−1	4.7
−2	9.4

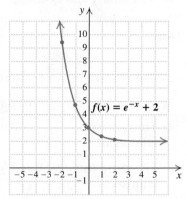

8. Graph $g(x) = e^x + 1$ and state the domain and the range of g.

The domain of f is $\mathbb{R}$, and the range is $(2, \infty)$.

YOUR TURN

Technology Connection

Logarithmic functions with bases other than 10 or e can be drawn using the change-of-base formula. For example, $y = \log_5 x$ can be written $y = \ln x / \ln 5$. If your calculator has a LOGBASE(option, the change-of-base formula is not needed.

  1. Graph $y = \log_7 x$.
  2. Graph $y = \log_5(x + 2)$.
  3. Graph $y = \log_7 x + 2$.

EXAMPLE 9   Graph and state the domain and the range of each function.

a) $g(x) = \ln x$ 　　　　　　　　**b)** $f(x) = \ln(x + 3)$

SOLUTION

a) We find some solutions with a calculator and then draw the graph. As expected, the graph is a reflection across the line $y = x$ of the graph of $y = e^x$.

x	$\ln x$
1	0
4	1.4
7	1.9
0.5	−0.7

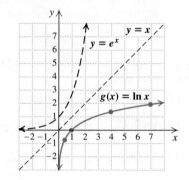

The domain of g is $(0, \infty)$, and the range is $\mathbb{R}$.

b) We find some solutions with a calculator, plot them, and draw the graph.

Chapter Resource:
Visualizing for Success, p. 645

x	$\ln(x + 3)$
0	1.1
1	1.4
2	1.6
3	1.8
4	1.9
−1	0.7
−2	0
−2.5	−0.7

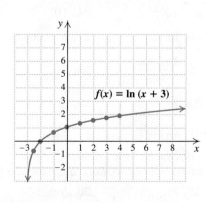

9. Graph $f(x) = \ln x + 2$ and state the domain and the range of the function.

The graph of $y = \ln(x + 3)$ is the graph of $y = \ln x$ translated 3 units to the left. Since $x + 3$ must be positive, the domain is $(-3, \infty)$ and the range is $\mathbb{R}$.

YOUR TURN

9.5 EXERCISE SET

FOR EXTRA HELP   MyMathLab® Math XL

PRACTICE   WATCH   READ   REVIEW

▶ Vocabulary and Reading Check

Classify each of the following statements as either true or false.

1. The expression log 23 means $\log_{10} 23$.

2. The expression ln 7 means $\log_e 7$.

3. The number e is approximately 2.7.

4. The expressions log 9 and log 18/log 2 are equivalent.

5. The expressions log 9 and log 18 − log 2 are equivalent.

6. The expressions $\log_2 9$ and ln 9/ln 2 are equivalent.

7. The expressions ln 81 and 2 ln 9 are equivalent.

8. The domain of the function given by $f(x) = \ln(x + 2)$ is $(-2, \infty)$.

9. The range of the function given by $g(x) = e^x$ is $(0, \infty)$.

10. The range of the function given by $f(x) = \ln x$ is $(-\infty, \infty)$.

Logarithms on a Calculator

Use a calculator to find each of the following to four decimal places.

11. log 7

12. log 2

13. log 13.7

14. log 98.3   Aha!   15. log 1000

16. log 100

17. log 0.75

18. log 0.25

19. $\dfrac{\log 8200}{\log 2}$

20. $\dfrac{\log 5700}{\log 5}$

21. $10^{1.7}$

22. $10^{0.59}$

23. $10^{-2.9523}$

24. $10^{-3.2046}$

25. ln 9

26. ln 13

27. ln 0.0062

28. ln 0.00073

29. $\dfrac{\ln 2300}{0.08}$

30. $\dfrac{\ln 1900}{0.07}$

31. $e^{2.71}$

32. $e^{3.06}$

33. $e^{-3.49}$

34. $e^{-2.64}$

Changing Logarithmic Bases

Find each of the following logarithms using the change-of-base formula. Round answers to four decimal places.

35. $\log_3 28$

36. $\log_6 37$

37. $\log_2 100$

38. $\log_7 100$

39. $\log_4 5$

40. $\log_8 7$

41. $\log_{0.1} 2$

42. $\log_{0.25} 25$

43. $\log_2 0.1$

44. $\log_{25} 0.25$

45. $\log_\pi 10$

46. $\log_\pi 100$

Graphs of Exponential Functions and Logarithmic Functions, Base e

Graph and state the domain and the range of each function.

47. $f(x) = e^x$

48. $f(x) = e^{-x}$

49. $f(x) = e^x + 3$

50. $f(x) = e^x + 2$

51. $f(x) = e^x - 2$

52. $f(x) = e^x - 3$

53. $f(x) = 0.5e^x$

54. $f(x) = 2e^x$

55. $f(x) = 0.5e^{2x}$

56. $f(x) = 2e^{-0.5x}$

57. $f(x) = e^{x-3}$

58. $f(x) = e^{x-2}$

59. $f(x) = e^{x+2}$

60. $f(x) = e^{x+3}$

61. $f(x) = -e^x$

62. $f(x) = -e^{-x}$

63. $g(x) = \ln x + 1$

64. $g(x) = \ln x + 3$

65. $g(x) = \ln x - 2$

66. $g(x) = \ln x - 1$

67. $g(x) = 2 \ln x$

68. $g(x) = 3 \ln x$

69. $g(x) = -2 \ln x$

70. $g(x) = -\ln x$

71. $g(x) = \ln(x + 2)$

72. $g(x) = \ln(x + 1)$

73. $g(x) = \ln(x - 1)$

74. $g(x) = \ln(x - 3)$

75. Using a calculator, Adan gives an *incorrect* approximation for log 79 that is between 4 and 5. How could you convince him, without using a calculator, that he is mistaken?

76. Examine Exercise 75. What mistake do you believe Adan made?

Skill Review

Find each of the following, given that

$$f(x) = \frac{1}{x + 2} \quad and \quad g(x) = 5x - 8.$$

77. $f(-1)$ [2.2]

78. $(f + g)(0)$ [2.6]

79. $(g - f)(x)$ [2.6]

80. The domain of f [2.2], [4.2]

81. The domain of f/g [2.6], [4.2]

82. $gg(x)$ [5.2]

Synthesis

83. Explain how the graph of $f(x) = e^x$ could be used to graph the function given by $g(x) = 1 + \ln x$.

84. How would you explain to a classmate why $\log_2 5 = \log 5/\log 2$ *and* $\log_2 5 = \ln 5/\ln 2$?

Knowing only that $\log 2 \approx 0.301$ *and* $\log 3 \approx 0.477$, *approximate each of the following to three decimal places.*

85. $\log_6 81$ **86.** $\log_9 16$ **87.** $\log_{12} 36$

88. Find a formula for converting common logarithms to natural logarithms.

89. Find a formula for converting natural logarithms to common logarithms.

 Solve for x. Give an approximation to four decimal places.

90. $\log (275x^2) = 38$

91. $\log (492x) = 5.728$

92. $\dfrac{3.01}{\ln x} = \dfrac{28}{4.31}$

93. $\log 692 + \log x = \log 3450$

For each function given below, (a) determine the domain and the range, (b) set an appropriate window, and (c) draw the graph. Graphs may vary, depending on the scale used.

94. $f(x) = 7.4e^x \ln x$

95. $f(x) = 3.4 \ln x - 0.25e^x$

96. $f(x) = x \ln (x - 2.1)$

97. $f(x) = 2x^3 \ln x$

98. Use a graphing calculator to check your answers to Exercises 49, 57, and 71.

99. In an attempt to solve $\ln x = 1.5$, Emma gets the following graph. How can Emma tell at a glance that she has made a mistake?

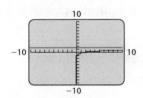

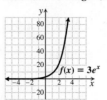
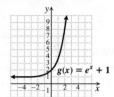
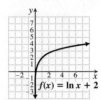
QUICK QUIZ: SECTIONS 9.1–9.5

1. Find $f(x)$ and $g(x)$ such that $h(x) = (f \circ g)(x)$ and $h(x) = \sqrt{3x - 7}$. Answers may vary. [9.1]

Express as an equivalent expression that is a single logarithm and, if possible, simplify. [9.4]

2. $2 \log_a x - 3 \log_a y$

3. $\log_a (x + 1) + \log_a (x - 1)$

Graph.

4. $y = 2^{x-5}$ [9.2]   **5.** $y = \ln x + 2$ [9.5]

PREPARE TO MOVE ON

Solve.

1. $x(x - 3) = 28$ [5.8]   **2.** $5x^2 - 7x = 0$ [5.8]

3. $17x - 15 = 0$ [1.3]   **4.** $\dfrac{5}{3} = 2t$ [1.3]

5. $(x - 5) \cdot 9 = 11$ [1.3]   **6.** $\dfrac{x + 3}{x - 3} = 7$ [6.4]

9.6 Solving Exponential Equations and Logarithmic Equations

Solving Exponential Equations ▪ Solving Logarithmic Equations

Solving Exponential Equations

Equations with variables in exponents, such as $5^x = 12$ and $2^{7x} = 64$, are called **exponential equations**. We can solve certain exponential equations by using the principle of exponential equality, first stated in Section 9.3.

THE PRINCIPLE OF EXPONENTIAL EQUALITY

For any real number b, where $b \neq -1, 0,$ or 1,

$$b^m = b^n \quad \text{is equivalent to} \quad m = n.$$

(Powers of the same base are equal if and only if the exponents are equal.)

EXAMPLE 1   Solve: $4^{3x} = 16$.

SOLUTION   Note that $16 = 4^2$. Thus we can write each side as a power of the same base:

$$4^{3x} = 4^2 \qquad \text{Rewriting 16 as a power of 4}$$
$$3x = 2 \qquad \text{The base on each side is 4, so the exponents must be equal.}$$
$$x = \tfrac{2}{3}. \qquad \text{Solving for } x$$

Since $4^{3x} = 4^{3(2/3)} = 4^2 = 16$, the answer checks. The solution is $\frac{2}{3}$.

YOUR TURN

1. Solve: $3^{4x} = 9$.

In Example 1, we wrote both sides of the equation as powers of 4. When it seems impossible to write both sides of an equation as powers of the same base, we use the following principle and write an equivalent logarithmic equation.

THE PRINCIPLE OF LOGARITHMIC EQUALITY

For any logarithmic base a, and for $x, y > 0$,

$$x = y \quad \text{is equivalent to} \quad \log_a x = \log_a y.$$

(Two expressions are equal if and only if the logarithms of those expressions are equal.)

The principle of logarithmic equality, used together with the power rule for logarithms, allows us to solve equations in which the variable is an exponent.

EXAMPLE 2   Solve: $7^{x-2} = 60$.

SOLUTION   We have

Take the logarithm
of both sides.

$$7^{x-2} = 60$$
$$\log 7^{x-2} = \log 60$$

Using the principle of logarithmic equality to take the common logarithm on both sides. We could also use a logarithm with another base, such as e.

Use the power
rule for logarithms.

$$(x - 2)\log 7 = \log 60$$

Using the power rule for logarithms

$$x - 2 = \frac{\log 60}{\log 7}$$

← **CAUTION!**   This is *not* $\log 60 - \log 7$.

$$x = \frac{\log 60}{\log 7} + 2$$

Adding 2 to both sides

Solve for x.

$$\approx 4.1041.$$

Using a calculator and rounding to four decimal places

Check.

Since $7^{4.1041-2} \approx 60.0027$, we have a check. We can also note that since $7^{4-2} = 49$, we expect a solution greater than 4. The solution is $\frac{\log 60}{\log 7} + 2$, or approximately 4.1041.

2. Solve: $5^{x+1} = 12$.

YOUR TURN

EXAMPLE 3   Solve: $e^{0.06t} = 1500$.

SOLUTION   Since one side is a power of e, it is easiest to take the *natural logarithm* on both sides:

$$\ln e^{0.06t} = \ln 1500$$

Taking the natural logarithm on both sides

$$0.06t = \ln 1500$$

Finding the logarithm of the base to a power: $\log_a a^k = k$. Logarithmic and exponential functions are inverses of each other.

$$t = \frac{\ln 1500}{0.06}$$

Dividing both sides by 0.06

$$\approx 121.8870.$$

Using a calculator and rounding to four decimal places

3. Solve: $e^{-0.03t} = 120$.

YOUR TURN

TO SOLVE AN EQUATION OF THE FORM $a^t = b$ FOR t
1. Take the logarithm (either natural or common) of both sides.
2. Use the power rule for logarithms so that the variable is a factor instead of an exponent.
3. Divide both sides by the coefficient of the variable to isolate the variable.
4. If appropriate, use a calculator to find an approximate solution.

Solving Logarithmic Equations

Certain logarithmic equations can be solved by writing an equivalent exponential equation.

EXAMPLE 4   Solve: **(a)** $\log_4(8x - 6) = 3$; **(b)** $\ln(5x) = 27$.

SOLUTION

a) $\log_4(8x - 6) = 3$

$$4^3 = 8x - 6 \qquad \text{Writing the equivalent exponential equation}$$

$$64 = 8x - 6$$

$$70 = 8x \qquad \text{Adding 6 to both sides}$$

$$x = \tfrac{70}{8}, \text{ or } \tfrac{35}{4}$$

Check:

$$\begin{array}{c|c} \log_4(8x - 6) = 3 \\ \hline \log_4\left(8 \cdot \tfrac{35}{4} - 6\right) & 3 \\ \log_4(2 \cdot 35 - 6) & \\ \log_4 64 & \\ 3 \overset{?}{=} 3 & \text{TRUE} \end{array}$$

The solution is $\tfrac{35}{4}$.

b) $\ln(5x) = 27$ Remember: $\ln(5x)$ means $\log_e(5x)$.

$$e^{27} = 5x \qquad \text{Writing the equivalent exponential equation}$$

$$\frac{e^{27}}{5} = x \qquad \text{This is a very large number.}$$

The solution is $\dfrac{e^{27}}{5}$. The check is left to the student.

4. Solve: $\log(5x - 3) = 2$.

 YOUR TURN

Often the properties for logarithms are needed in order to solve a logarithmic equation. The goal is to first write an equivalent equation in which the variable appears in just one logarithmic expression. We then isolate that expression and solve as in Example 4.

Student Notes

It is essential that you remember the properties of logarithms from Section 9.4. Consider reviewing the properties before attempting to solve equations similar to those in Example 5.

> Find a single logarithm.

> Write an equivalent exponential equation.

> Solve.

EXAMPLE 5   Solve.

a) $\log x + \log(x - 3) = 1$

b) $\log_2(x + 7) - \log_2(x - 7) = 3$

c) $\log_7(x + 1) + \log_7(x - 1) = \log_7 8$

SOLUTION

a) Here, log means $\log_{10}$, so we write in the base, 10, for both logarithmic expressions.

$$\log_{10} x + \log_{10}(x - 3) = 1$$

$$\log_{10}[x(x - 3)] = 1 \qquad \begin{array}{l}\text{Using the product rule for logarithms} \\ \text{to obtain a single logarithm}\end{array}$$

$$x(x - 3) = 10^1 \qquad \begin{array}{l}\text{Writing an equivalent exponential} \\ \text{equation}\end{array}$$

$$x^2 - 3x = 10$$

$$x^2 - 3x - 10 = 0$$

$$(x + 2)(x - 5) = 0 \qquad \text{Factoring}$$

$$x + 2 = 0 \quad or \quad x - 5 = 0 \qquad \begin{array}{l}\text{Using the principle of zero} \\ \text{products}\end{array}$$

$$x = -2 \quad or \qquad\quad x = 5$$

Check.

Check:

For -2:

$$\log x + \log (x - 3) = 1$$
$$\overline{\log (-2) + \log (-2 - 3) \overset{?}{=} 1} \quad \text{FALSE}$$

For 5:

$$\log x + \log (x - 3) = 1$$
$$\begin{array}{c|c} \log 5 + \log (5 - 3) & 1 \\ \log 5 + \log 2 & \\ \log 10 & \\ 1 \overset{?}{=} 1 & \text{TRUE} \end{array}$$

The number -2 *does not check* because the logarithm of a negative number is undefined. The solution is 5.

b) We have

$$\log_2 (x + 7) - \log_2 (x - 7) = 3$$

$$\log_2 \frac{x + 7}{x - 7} = 3 \qquad \begin{array}{l} \text{Using the quotient rule} \\ \text{for logarithms to obtain} \\ \text{a single logarithm} \end{array}$$

$$\frac{x + 7}{x - 7} = 2^3 \qquad \begin{array}{l} \text{Writing an equivalent} \\ \text{exponential equation} \end{array}$$

$$\frac{x + 7}{x - 7} = 8$$

$$x + 7 = 8(x - 7) \qquad \begin{array}{l} \text{Multiplying by } x - 7 \text{ to clear} \\ \text{fractions} \end{array}$$

$$x + 7 = 8x - 56 \qquad \text{Using the distributive law}$$

$$63 = 7x$$

$$9 = x. \qquad \text{Dividing by 7}$$

Check:
$$\begin{array}{c|c} \log_2 (x + 7) - \log_2 (x - 7) = 3 \\ \hline \log_2 (9 + 7) - \log_2 (9 - 7) & 3 \\ \log_2 16 - \log_2 2 & \\ 4 - 1 & \\ 3 \overset{?}{=} 3 & \text{TRUE} \end{array}$$

The solution is 9.

c) We have

$$\log_7 (x + 1) + \log_7 (x - 1) = \log_7 8$$

$$\log_7 [(x + 1)(x - 1)] = \log_7 8 \qquad \begin{array}{l} \text{Using the product rule for} \\ \text{logarithms} \end{array}$$

$$\log_7 (x^2 - 1) = \log_7 8 \qquad \begin{array}{l} \text{Multiplying. Note that } both \\ sides \text{ are base-7 } logarithms. \end{array}$$

$$x^2 - 1 = 8 \qquad \begin{array}{l} \text{Using the principle of} \\ \text{logarithmic equality} \end{array}$$

$$x^2 - 9 = 0$$

$$(x - 3)(x + 3) = 0 \qquad \text{Solving the quadratic equation}$$

$$x = 3 \quad or \quad x = -3.$$

We leave it to the student to show that 3 checks but -3 does not. The solution is 3.

YOUR TURN

Technology Connection

To solve exponential equations and logarithmic equations, we can determine the x-coordinate at any point of intersection. For example, to solve $e^{0.5x} - 7 = 2x + 6$, we graph $y_1 = e^{0.5x} - 7$ and $y_2 = 2x + 6$ as shown.

We find that the x-coordinates at the intersections are approximately -6.48 and 6.52.

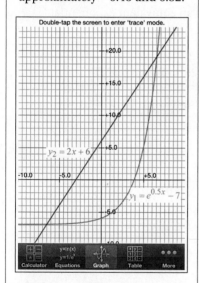

Use a graphing calculator to solve each equation to the nearest hundredth.

1. $e^{7x} = 14$
2. $8e^{0.5x} = 3$
3. $xe^{3x-1} = 5$
4. $4 \ln (x + 3.4) = 2.5$
5. $\ln 3x = 0.5x - 1$
6. $\ln x^2 = -x^2$

5. Solve:

$$\log_3 (x + 4) + \log_3 (x + 2) = 1.$$

CONNECTING 🔗 THE CONCEPTS

We have used several procedures for solving exponential equations and logarithmic equations. Carefully inspecting an equation helps us choose the best method to use. Compare the following.

Equation	Description of Equation	Solution
$8^{2x} = 8^5$	Exponential equation Each side is an exponential expression with the same base.	The expressions $2x$ and 5 must be equal. $2x = 5$ $x = \frac{5}{2}$
$2^t = 7$	Exponential equation Expressions have different bases.	Take the logarithm of both sides. $\log 2^t = \log 7$ $t \cdot \log 2 = \log 7$ $t = \dfrac{\log 7}{\log 2}$
$\log_5 x = 3$	Logarithmic equation One logarithmic expression	Rewrite as an equivalent exponential equation. $5^3 = x$ $125 = x$
$\log_3 (x + 1) = \log_3 (2x)$	Logarithmic equation Each side is a logarithmic expression with the same base.	The expressions $x + 1$ and $2x$ must be equal. $x + 1 = 2x$ $1 = x$

EXERCISES

Solve.

1. $\log (2x) = 3$

2. $2^{x+1} = 2^9$

3. $e^t = 5$

4. $\log_3 (x^2 + 1) = \log_3 26$

5. $\ln 2x = \ln 8$

6. $3^{5x} = 4$

7. $\log_2 (x + 1) = -1$

8. $9^{x+1} = 27^{-x}$

9.6 EXERCISE SET

FOR EXTRA HELP

MyMathLab® Math XP
PRACTICE   WATCH   READ   REVIEW

Vocabulary and Reading Check

Classify each of the following statements as either true or false.

1. The solution of a logarithmic equation is never a negative number.

2. To solve an exponential equation, we can take the common logarithm of both sides of the equation.

3. We cannot calculate the logarithm of a negative number.

4. To solve an exponential equation, we can take the natural logarithm of both sides of the equation.

Concept Reinforcement

Match each equation with an equivalent equation from the list below that could be the next step in the solution process.

a) $5^3 = x$

b) $\log_5 (x^2 - 2x) = 3$

c) $\log_5 \dfrac{x}{x - 2} = 3$

d) $\log 5^x = \log 3$

5. ____ $5^x = 3$

6. ____ $\log_5 x = 3$

7. ____ $\log_5 x + \log_5 (x - 2) = 3$

8. ____ $\log_5 x - \log_5 (x - 2) = 3$

Solving Exponential Equations

Solve. Where appropriate, include approximations to three decimal places.

9. $3^{2x} = 81$

10. $2^{3x} = 64$

11. $4^x = 32$

12. $9^x = 27$

13. $2^x = 10$

14. $2^x = 24$

15. $2^{x+5} = 16$

16. $2^{x-1} = 8$

17. $8^{x-3} = 19$

18. $5^{x+2} = 15$

19. $e^t = 50$

20. $e^t = 20$

21. $e^{-0.02t} = 8$

22. $e^{-0.01t} = 100$

23. $4.9^x - 87 = 0$

24. $7.2^x - 65 = 0$

25. $19 = 2e^{4x}$

26. $29 = 3e^{2x}$

27. $7 + 3e^{-x} = 13$

28. $4 + 5e^{-x} = 9$

Solving Logarithmic Equations

Solve. Where appropriate, include approximations to three decimal places. If no solution exists, state this.

Aha! 29. $\log_3 x = 4$

30. $\log_2 x = 6$

31. $\log_4 x = -2$

32. $\log_5 x = -3$

33. $\ln x = 5$

34. $\ln x = 4$

35. $\ln (4x) = 3$

36. $\ln (3x) = 2$

37. $\log x = 1.2$

38. $\log x = 0.6$

39. $\ln (2x + 1) = 4$

40. $\ln (4x - 2) = 3$

Aha! 41. $\ln x = 1$

42. $\log x = 1$

43. $5 \ln x = -15$

44. $3 \ln x = -3$

45. $\log_2 (8 - 6x) = 5$

46. $\log_5 (7 - 2x) = 3$

47. $\log (x - 9) + \log x = 1$

48. $\log (x + 9) + \log x = 1$

49. $\log x - \log (x + 3) = 1$

50. $\log x - \log (x + 7) = -1$

Aha! 51. $\log (2x + 1) = \log 5$

52. $\log (x + 1) - \log x = 0$

53. $\log_4 (x + 3) = 2 + \log_4 (x - 5)$

54. $\log_2 (x + 3) = 4 + \log_2 (x - 3)$

55. $\log_7 (x + 1) + \log_7 (x + 2) = \log_7 6$

56. $\log_6 (x + 3) + \log_6 (x + 2) = \log_6 20$

57. $\log_5 (x + 4) + \log_5 (x - 4) = \log_5 20$

58. $\log_4 (x + 2) + \log_4 (x - 7) = \log_4 10$

59. $\ln (x + 5) + \ln (x + 1) = \ln 12$

60. $\ln (x - 6) + \ln (x + 3) = \ln 22$

61. $\log_2 (x - 3) + \log_2 (x + 3) = 4$

62. $\log_3 (x - 4) + \log_3 (x + 4) = 2$

63. $\log_{12} (x + 5) - \log_{12} (x - 4) = \log_{12} 3$

64. $\log_6 (x + 7) - \log_6 (x - 2) = \log_6 5$

65. $\log_2 (x - 2) + \log_2 x = 3$

66. $\log_4 (x + 6) - \log_4 x = 2$

67. Madison finds that the solution of $\log_3 (x + 4) = 1$ is -1, but rejects -1 as an answer. What mistake do you suspect she is making?

68. Could Example 3 have been solved by taking the common logarithm on both sides? Why or why not?

Skill Review

Simplify.

69. $\dfrac{a^2 - 4}{a^2 + a} \cdot \dfrac{a^2 - a - 2}{a^3}$ [6.1]

70. $\dfrac{2t^2 - t - 3}{t^2 - 2t} \div \dfrac{2t^2 + 5t + 3}{t^4 - 3t^3 + 2t^2}$ [6.1]

71. $\dfrac{2}{m + 1} + \dfrac{3}{m - 5}$ [6.2]

72. $\dfrac{3}{x - 2} - \dfrac{x + 1}{x}$ [6.2]

73. $\dfrac{\dfrac{3}{x} - \dfrac{2}{xy}}{\dfrac{2}{x^2} + \dfrac{1}{xy}}$ [6.3]

74. $\dfrac{\dfrac{4 + x}{x^2 + 2x + 1}}{\dfrac{3}{x + 1} - \dfrac{2}{x + 2}}$ [6.3]

Synthesis

 75. Can the principle of logarithmic equality be expanded to include all functions? That is, is the statement "$m = n$ is equivalent to $f(m) = f(n)$" true for any function f? Why or why not?

 76. Explain how Exercises 37 and 38 could be solved using the graph of $f(x) = \log x$.

Solve. If no solution exists, state this.

77. $8^x = 16^{3x+9}$

78. $27^x = 81^{2x-3}$

79. $\log_6 (\log_2 x) = 0$

80. $\log_x (\log_3 27) = 3$

81. $\log_5 \sqrt{x^2 - 9} = 1$

82. $x \log \frac{1}{8} = \log 8$

83. $2^{x^2+4x} = \frac{1}{8}$

84. $\log (\log x) = 5$

85. $\log_5 |x| = 4$

86. $\log x^2 = (\log x)^2$

87. $\log \sqrt{2x} = \sqrt{\log 2x}$

88. $1000^{2x+1} = 100^{3x}$

89. $3^{x^2} \cdot 3^{4x} = \frac{1}{27}$

90. $3^{3x} \cdot 3^{x^2} = 81$

91. $\log x^{\log x} = 25$

92. $3^{2x} - 8 \cdot 3^x + 15 = 0$

93. $(81^{x-2})(27^{x+1}) = 9^{2x-3}$

94. $3^{2x} - 3^{2x-1} = 18$

95. Given that $2^y = 16^{x-3}$ and $3^{y+2} = 27^x$, find the value of $x + y$.

96. If $x = (\log_{125} 5)^{\log_5 125}$, what is the value of $\log_3 x$?

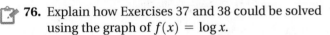 **97.** Find the value of x for which the natural logarithm is the same as the common logarithm.

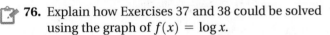 **98.** Use a graphing calculator to check your answers to Exercises 11, 31, 41, and 59.

⤵ YOUR TURN ANSWERS: SECTION 9.6

1. $\dfrac{1}{2}$ **2.** $\dfrac{\log 12}{\log 5} - 1 \approx 0.5440$ **3.** $\dfrac{\ln 120}{-0.03} \approx -159.5831$

4. $\dfrac{103}{5}$ **5.** -1

QUICK QUIZ: SECTIONS 9.1–9.6

1. Find a formula for the inverse of $g(x) = x - 6$.   [9.1]

Simplify.

2. $\log_2 \frac{1}{2}$ [9.3]   **3.** $\log_m m^{11}$ [9.4]

Solve. Where appropriate, include an approximation to three decimal places.

4. $\log_x 8 = 1$ [9.4]   **5.** $\log x = 2.7$ [9.6]

PREPARE TO MOVE ON

Solve.

1. A rectangle is 6 ft longer than it is wide. Its perimeter is 26 ft. Find the length and the width.   [1.4]

2. The average cost of a wedding in the United States was $15,200 in 1990 and $27,000 in 2012.

 a) Find a linear function that fits the data.   [2.5]
 b) Use the function found in part (a) to estimate the average cost of a wedding in 2015.   [2.5]

 Sources: theweddingreport.com; theknot.com

3. Max can key in a musical score in 2 hr. Miles takes 3 hr to key in the same score. How long would it take them, working together, to key in the score?   [6.5]

9.7   Applications of Exponential Functions and Logarithmic Functions

Applications of Logarithmic Functions ▪ Applications of Exponential Functions

Applications of Logarithmic Functions

EXAMPLE 1   *Sound Levels.*   To measure the volume, or "loudness," of a sound, the *decibel* scale is used. The loudness L, in decibels (dB), of a sound is given by

$$L = 10 \cdot \log \frac{I}{I_0},$$

where I is the intensity of the sound, in watts per square meter (W/m^2), and $I_0 = 10^{-12}\,W/m^2$. Here, I_0 is approximately the intensity of the softest sound that can be heard by the human ear.

a) The average maximum intensity of sound in a New York subway car is about $3.2 \times 10^{-3}\,W/m^2$. How loud, in decibels, is the sound level?

   Source: Columbia University Mailman School of Public Health

b) The Occupational Safety and Health Administration (OSHA) considers sustained sound levels of 90 dB and above unsafe. What is the intensity of such sounds?

SOLUTION

a) To find the loudness, in decibels, we use the above formula:

$$L = 10 \cdot \log \frac{I}{I_0}$$

$$= 10 \cdot \log \frac{3.2 \times 10^{-3}}{10^{-12}} \qquad \text{Substituting}$$

$$= 10 \cdot \log (3.2 \times 10^{9}) \qquad \text{Subtracting exponents}$$

$$\approx 95. \qquad \text{Using a calculator and rounding}$$

The volume of the sound in a subway car is about 95 decibels.

b) We substitute and solve for I:

$$L = 10 \cdot \log \frac{I}{I_0}$$

$$90 = 10 \cdot \log \frac{I}{10^{-12}} \qquad \text{Substituting}$$

$$9 = \log \frac{I}{10^{-12}} \qquad \text{Dividing both sides by 10}$$

$$9 = \log I - \log 10^{-12} \qquad \text{Using the quotient rule for logarithms}$$

$$9 = \log I - (-12) \qquad \log 10^{a} = a$$

$$-3 = \log I \qquad \text{Adding } -12 \text{ to both sides}$$

$$10^{-3} = I. \qquad \text{Converting to an exponential equation}$$

Sustained sounds with intensities exceeding $10^{-3}\,W/m^2$ are considered unsafe.

YOUR TURN

Study Skills

Sorting by Type

When a section contains a variety of problems, try to sort them out by type. For instance, interest compounded continuously, population growth, and the spread of a virus can all be regarded as one type of problem: exponential growth. Once you know how to solve this type of problem, you can focus on determining which problems fall into this category. The solution should then follow in a rather straightforward manner.

---

1. Blue whales and fin whales are the loudest animals, capable of producing sound levels up to 180 dB. What is the intensity of such a sound?

   Source: Guinness World Records

EXAMPLE 2   *Chemistry: pH of Liquids.*   In chemistry, the pH of a liquid is a measure of its acidity. We calculate pH as follows:

$$pH = -\log[H^+],$$

where $[H^+]$ is the hydrogen ion concentration in moles per liter.

a)   The hydrogen ion concentration of human blood is normally about 3.98×10^{-8} moles per liter. Find the pH.
Source: www.merck.com

b)   The average pH of seawater is about 8.2. Find the hydrogen ion concentration.
Source: www.seafriends.org.nz

SOLUTION

a)   To find the pH of blood, we use the above formula:

$$
\begin{aligned}
pH &= -\log[H^+] \\
&= -\log[3.98 \times 10^{-8}] \quad &&\text{Substituting} \\
&\approx -(-7.400117) \quad &&\text{Using a calculator} \\
&\approx 7.4.
\end{aligned}
$$

The pH of human blood is normally about 7.4.

b)   We substitute and solve for $[H^+]$:

$$
\begin{aligned}
8.2 &= -\log[H^+] \quad &&\text{Using } pH = -\log[H^+] \\
-8.2 &= \log[H^+] \quad &&\text{Dividing both sides by } -1 \\
10^{-8.2} &= [H^+] \quad &&\text{Converting to an exponential equation} \\
6.31 \times 10^{-9} &\approx [H^+]. \quad &&\text{Using a calculator; writing scientific notation}
\end{aligned}
$$

The hydrogen ion concentration of seawater is about 6.31×10^{-9} moles per liter.

2. The pH of the soil in Jeannette's garden is 6.3. Find the hydrogen ion concentration.

 YOUR TURN

Applications of Exponential Functions

EXAMPLE 3   *Interest Compounded Annually.*   Suppose that $25,000 is invested at 4% interest, compounded annually. In t years, it will grow to the amount A given by

$$A(t) = 25{,}000(1.04)^t.$$

a)   How long will it take to have $80,000 in the account?

b)   Find the amount of time it takes for the $25,000 to double itself.

SOLUTION

a)   We set $A(t) = 80{,}000$ and solve for t:

$$
\begin{aligned}
80{,}000 &= 25{,}000(1.04)^t \\
3.2 &= 1.04^t \quad &&\text{Dividing both sides by 25,000} \\
\log 3.2 &= \log 1.04^t \quad &&\text{Taking the common logarithm on both sides} \\
\log 3.2 &= t \log 1.04 \quad &&\text{Using the power rule for logarithms} \\
\frac{\log 3.2}{\log 1.04} &= t \quad &&\text{Dividing both sides by } \log 1.04 \\
29.7 &\approx t. \quad &&\text{Using a calculator}
\end{aligned}
$$

Student Notes

Study the different steps in the solution of Example 3(b). Note that if 50,000 and 25,000 are replaced with 6000 and 3000, the doubling time is unchanged.

As always, when doing a calculation like this, it is best to wait until the end to round. At an interest rate of 4% per year, it will take about 29.7 years for $25,000 to grow to $80,000.

b) To find the *doubling time*, we replace $A(t)$ with 50,000 and solve for t:

$$50{,}000 = 25{,}000(1.04)^t$$

$$2 = (1.04)^t \qquad \text{Dividing both sides by 25,000}$$

$$\log 2 = \log(1.04)^t \qquad \begin{array}{l}\text{Taking the common logarithm}\\\text{on both sides}\end{array}$$

$$\log 2 = t \log 1.04 \qquad \text{Using the power rule for logarithms}$$

$$t = \frac{\log 2}{\log 1.04} \approx 17.7. \qquad \begin{array}{l}\text{Dividing both sides by } \log 1.04 \text{ and using}\\\text{a calculator}\end{array}$$

At an interest rate of 4% per year, the doubling time is about 17.7 years.

3. If $25,000 is invested at 5% interest, compounded annually, in t years it will grow to the amount A given by $A(t) = 25{,}000(1.05)^t$.

  a) How long will it take to have $80,000 in the account?

  b) Find the amount of time it takes for the $25,000 to double itself.

YOUR TURN

Like investments, populations often grow exponentially.

EXPONENTIAL GROWTH

An **exponential growth model** is a function of the form

$$P(t) = P_0 e^{kt}, \quad k > 0,$$

where P_0 is the population at time 0, $P(t)$ is the population at time t, and k is the **exponential growth rate** for the situation. The **doubling time** is the amount of time necessary for the population to double in size.

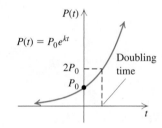

The exponential growth rate is the rate of growth of a population or other quantity at any *instant* in time. Since the change in population is continually growing, the percent of total growth after one year exceeds the exponential growth rate.

EXAMPLE 4 *Invasive Species.* Beginning in 1988, infestations of zebra mussels started spreading through North American waters. These mussels spread with such speed that water treatment facilities, power plants, and entire ecosystems can become threatened. The area of an infestation of zebra mussels can have an annual exponential growth rate of 350%.

Source: Based on information from Dr. Gerald Mackie, Department of Zoology, University of Guelph in Ontario

a) A zoologist discovers an infestation of zebra mussels covering 10 cm². Find the exponential growth function that models the data. Let t represent the number of years since the discovery of the infestation.

b) Use the function found in part (a) to estimate the size of the infestation after 5 years.

SOLUTION

a) At $t = 0$, the size of the infestation is 10 cm². We substitute 10 for P_0 and 350%, or 3.5, for k. This gives the exponential growth function

$$P(t) = 10e^{3.5t}.$$

4. Refer to Example 4. Suppose that the area of infestation increases exponentially at a rate of 400% per year. To what size would a 10-cm² infestation grow after 5 years?

b) To estimate the size of the infestation after 5 years, we compute $P(5)$:

$$P(5) = 10e^{3.5(5)} \qquad \text{Using } P(t) = 10e^{3.5t} \text{ from part (a)}$$
$$= 10e^{17.5} \approx 398{,}247{,}844. \qquad \text{Using a calculator}$$

After 5 years, the zebra mussels will cover about 398,247,844 cm², or about 39,825 m².

 YOUR TURN

EXAMPLE 5 *Social Networking.* The number of unique visitors to the virtual pinboard Pinterest.com increased exponentially in the last six months of 2011. In June 2011, there were 608,000 unique visitors to Pinterest; this number had increased to 11,716,000 by January 2012.

Source: comScore, in "Pinterest Hits 10 Million U.S. Monthly Uniques Faster Than Any Standalone Site Ever," Josh Constine, Feb. 7, 2012, on techcrunch.com

a) Find the exponential growth rate and the exponential growth function.

b) Assuming that the trend continues, estimate the month in which there will be 30 million unique visitors to Pinterest.

SOLUTION

a) We let $P(t) = P_0 e^{kt}$, where t is the number of months since June 2011 and $P(t)$ is the number of unique visitors to Pinterest, in thousands. Next, we substitute 608 for P_0:

$$P(t) = 608e^{kt}.$$

To find the exponential growth rate k, we note that after 7 months, there were 11,716,000, or 11,716 thousand, unique visitors:

$$\left.\begin{array}{l} P(7) = 608e^{k \cdot 7} \\ 11{,}716 = 608e^{7k} \end{array}\right\} \text{ Substituting}$$

$$\frac{11{,}716}{608} = e^{7k} \qquad \text{Dividing both sides by 608}$$

$$\ln\left(11{,}716/608\right) = \ln e^{7k} \qquad \text{Taking the natural logarithm on both sides}$$

$$\ln\left(11{,}716/608\right) = 7k \qquad \ln e^{7k} = \log_e e^{7k} = 7k$$

$$\frac{\ln\left(11{,}716/608\right)}{7} = k \qquad \text{Dividing both sides by 7}$$

$$0.423 \approx k. \qquad \text{Using a calculator and rounding}$$

The exponential growth rate is 42.3%, and the exponential growth function is given by $P(t) = 608e^{0.423t}$.

b) To estimate the month in which Pinterest has 30 million unique visitors, we replace $P(t)$ with 30,000 (since 30 million is 30,000 thousand) and solve for t:

$$30{,}000 = 608e^{0.423t}$$

$$\frac{30{,}000}{608} = e^{0.423t} \qquad \text{Dividing both sides by 608}$$

$$\ln\left(30{,}000/608\right) = \ln e^{0.423t} \qquad \text{Taking the natural logarithm on both sides}$$

$$\ln\left(30{,}000/608\right) = 0.423t \qquad \ln e^a = a$$

$$\frac{\ln\left(30{,}000/608\right)}{0.423} = t \qquad \text{Dividing both sides by 0.423}$$

$$9.2 \approx t. \qquad \text{Using a calculator and rounding}$$

According to this model, Pinterest had 30 million unique visitors 9.2 months after June 2011, or sometime during March 2012.

5. *Malware.* The number of malware variants targeted to Android devices increased exponentially from 10 in the first quarter of 2011 to 37 in the first quarter of 2012. Find the exponential growth rate and the exponential growth function.

Source: crn.com

 YOUR TURN

EXAMPLE 6 *Interest Compounded Continuously.* When an amount of money P_0 is invested at interest rate k, compounded *continuously*, interest is computed every "instant" and added to the original amount. The balance $P(t)$, after t years, is given by the exponential growth model

$$P(t) = P_0 e^{kt}.$$

a) Suppose that \$30,000 is invested and grows to \$44,754.75 in 5 years. Find the exponential growth function.

b) What is the doubling time?

SOLUTION

a) We have $P(0) = 30{,}000$. Thus the exponential growth function is

$$P(t) = 30{,}000 e^{kt}, \quad \text{where } k \text{ must still be determined.}$$

Knowing that for $t = 5$ we have $P(5) = 44{,}754.75$, it is possible to solve for k:

$$44{,}754.75 = 30{,}000 e^{k(5)}$$

$$44{,}754.75 = 30{,}000 e^{5k}$$

$$\frac{44{,}754.75}{30{,}000} = e^{5k} \qquad \text{Dividing both sides by 30,000}$$

$$1.491825 = e^{5k}$$

$$\ln 1.491825 = \ln e^{5k} \qquad \text{Taking the natural logarithm on both sides}$$

$$\ln 1.491825 = 5k \qquad \ln e^a = a$$

$$\frac{\ln 1.491825}{5} = k \qquad \text{Dividing both sides by 5}$$

$$0.08 \approx k. \qquad \text{Using a calculator and rounding}$$

The interest rate is about 0.08, or 8%, compounded continuously. Because interest is being compounded continuously, the yearly interest rate is a bit more than 8%. The exponential growth function is

$$P(t) = 30{,}000 e^{0.08t}.$$

b) To find the doubling time T, we replace $P(T)$ with 60,000 and solve for T:

$$60{,}000 = 30{,}000 e^{0.08T}$$

$$2 = e^{0.08T} \qquad \text{Dividing both sides by 30,000}$$

$$\ln 2 = \ln e^{0.08T} \qquad \text{Taking the natural logarithm on both sides}$$

$$\ln 2 = 0.08T \qquad \ln e^a = a$$

$$\frac{\ln 2}{0.08} = T \qquad \text{Dividing both sides by 0.08}$$

$$8.7 \approx T. \qquad \text{Using a calculator and rounding}$$

Thus the original investment of \$30,000 will double in about 8.7 years.

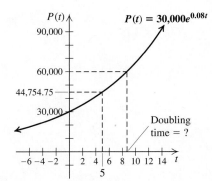

A visualization of Example 6

6. Refer to Example 6. If \$20,000 is invested and grows to \$22,103.42 in 5 years, find the exponential growth function and the doubling time.

YOUR TURN

For any specified interest rate, continuous compounding gives the highest yield and the shortest doubling time.

In some real-life situations, a quantity or population is *decreasing* or *decaying* exponentially.

EXPONENTIAL DECAY

An **exponential decay model** is a function of the form

$$P(t) = P_0 e^{-kt}, \quad k > 0,$$

where P_0 is the quantity present at time 0, $P(t)$ is the amount present at time t, and k is the **decay rate**. The **half-life** is the amount of time necessary for half of the quantity to decay.

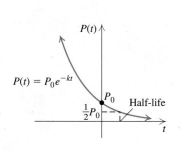

EXAMPLE 7 *Carbon Dating.* The radioactive element carbon-14 has a half-life of 5750 years. The percentage of carbon-14 in the remains of organic matter can be used to determine the age of that material. Recently, near Patuxent River, Maryland, archaeologists discovered charcoal that had lost 8.1% of its carbon-14. The age of this charcoal was evidence that this is the oldest dwelling ever discovered in Maryland. What was the age of the charcoal?

Source: Based on information from *The Baltimore Sun.* "Digging Where Indians Camped Before Columbus," by Frank D. Roylance, July 2, 2009

SOLUTION We first find k. To do so, we use the concept of half-life. When $t = 5750$ (the half-life), $P(t)$ is half of P_0. Then

$0.5P_0 = P_0 e^{-k(5750)}$	Substituting in $P(t) = P_0 e^{-kt}$
$0.5 = e^{-5750k}$	Dividing both sides by P_0
$\ln 0.5 = \ln e^{-5750k}$	Taking the natural logarithm on both sides
$\ln 0.5 = -5750k$	$\ln e^a = a$
$\dfrac{\ln 0.5}{-5750} = k$	Dividing both sides by -5750
$0.00012 \approx k.$	Using a calculator and rounding

Now we have a function for the decay of carbon-14:

$$P(t) = P_0 e^{-0.00012t}. \quad \text{This completes the first part of our solution.}$$

(*Note*: This equation can be used for subsequent carbon-dating problems.) If the charcoal has lost 8.1% of its carbon-14 from an initial amount P_0, then $100\% - 8.1\%$, or 91.9%, of P_0 is still present. To find the age t of the charcoal, we solve this equation for t:

Chapter Resource:
Decision Making: Connection, p. 646

$0.919P_0 = P_0 e^{-0.00012t}$	We want to find t for which $P(t) = 0.919P_0$.
$0.919 = e^{-0.00012t}$	Dividing both sides by P_0
$\ln 0.919 = \ln e^{-0.00012t}$	Taking the natural logarithm on both sides
$\ln 0.919 = -0.00012t$	$\ln e^a = a$
$\dfrac{\ln 0.919}{-0.00012} = t$	Dividing both sides by -0.00012
$700 \approx t.$	Using a calculator

7. In Chaco Canyon, New Mexico, archaelogists found corn pollen that had lost 38.1% of its carbon-14. What was the age of the pollen?

The charcoal is about 700 years old.

 YOUR TURN

9.7 EXERCISE SET

MyMathLab® Math XL

PRACTICE WATCH READ REVIEW

Vocabulary and Reading Check

For the exponential growth model $P(t) = P_0 e^{kt}$, $k > 0$, match each variable with its description.

1. ____ k
2. ____ $P(t)$
3. ____ P_0
4. ____ T,
   where $2P_0 = P_0 e^{kT}$

a) Doubling time
b) Exponential growth rate
c) Population at time 0
d) Population at time t

Applications of Exponential Functions and Logarithmic Functions

5. *Digital Storage.* The amount of information created and stored digitally can be estimated by

   $$A(t) = 0.08(1.6)^t,$$

   where $A(t)$ is in zettabytes (1 zettabyte = 10^{21} bytes) and t is the number of years after 2005.

   Source: Based on data from IDC and EMC Corporation, in "Digital Universe to Add 1.8 Zettabytes in 2011." Rich Miller, June 28, 2011, datacenterknowledge.com

   a) Determine the year in which the amount of information created and stored digitally reached 1 zettabyte.
   b) What is the doubling time for the amount of information created and stored digitally?

6. *Health.* The rate of the number of deaths due to stroke in the United States can be estimated by

   $$S(t) = 180(0.97)^t,$$

   where $S(t)$ is the number of deaths per 100,000 people and t is the number of years after 1960.

   Source: Based on data from Centers for Disease Control and Prevention

   a) In what year was the death rate due to stroke 100 per 100,000 people?
   b) In what year will the death rate due to stroke be 25 per 100,000 people?

7. *Student Loan Repayment.* A college loan of $29,000 is made at 3% interest, compounded annually. After t years, the amount due, A, is given by the function

   $$A(t) = 29,000(1.03)^t.$$

   a) After what amount of time will the amount due reach $35,000?
   b) Find the doubling time.

8. *Spread of a Rumor.* The number of people who have heard a rumor increases exponentially. If each person who hears a rumor repeats it to two people per day, and if 20 people start the rumor, the number of people N who have heard the rumor after t days is given by

   $$N(t) = 20(3)^t.$$

   a) After what amount of time will 1000 people have heard the rumor?
   b) What is the doubling time for the number of people who have heard the rumor?

9. The frequency, in hertz (Hz), of the nth key on an 88-key piano is given by

   $$f(n) = 27.5\left(\sqrt[12]{2}\right)^{n-1},$$

   where $n = 1$ corresponds to the lowest key on the piano keyboard, an A.

   Source: "Piano Key Frequencies," on en.wikipedia.org

   a) What number key on the keyboard has a frequency of 440 Hz?
   b) How many keys does it take for the frequency to double?

10. *Smoking.* The percentage of smokers who receive telephone counseling and successfully quit smoking for t months is given by

$$P(t) = 21.4(0.914)^t.$$

Sources: *New England Journal of Medicine;* data from California's Smoker's Hotline

a) In what month will 15% of those who quit and used telephone counseling still be smoke-free?

b) In what month will 5% of those who quit and used telephone counseling still be smoke-free?

11. *Internet Traffic.* The number of unique visitors $V(t)$ to tumblr.com, a blogging platform, can be approximated by

$$V(t) = 12(1.044)^t,$$

where $V(t)$ is in millions of visitors and t is the number of months after March 2011.

Source: Based on data from siteanalytics.compete.com/tumblr.com/

a) In what month were there 50 million unique visitors to tumblr.com?

b) Find the doubling time.

12. *Marine Biology.* As a result of preservation efforts in countries in which whaling was once common, the humpback whale population has grown since the 1970s. The worldwide population $P(t)$, in thousands, t years after 1982 can be estimated by

$$P(t) = 5.5(1.047)^t.$$

a) In what year will the humpback whale population reach 30,000?

b) Find the doubling time.

Use the pH formula in Example 2 for Exercises 13–16.

13. *Chemistry.* The hydrogen ion concentration of fresh-brewed coffee is about 1.3×10^{-5} moles per liter. Find the pH.

14. *Chemistry.* The hydrogen ion concentration of milk is about 1.6×10^{-7} moles per liter. Find the pH.

15. *Medicine.* When the pH of a patient's blood drops below 7.4, a condition called *acidosis* sets in. Acidosis can be deadly when the pH drops to 7.0. What would the hydrogen ion concentration of the patient's blood be at that point?

16. *Medicine.* When the pH of a patient's blood rises above 7.4, a condition called *alkalosis* sets in. Alkalosis can be deadly when the patient's pH reaches 7.8. What would the hydrogen ion concentration of the patient's blood be at that point?

Use the formula in Example 1 for Exercises 17–20.

17. *Racing.* The intensity of sound from a race car in full throttle is about 10 W/m^2. How loud in decibels is this sound level?

Source: nascar.about.com

18. *Audiology.* The intensity of sound in normal conversation is about $3.2 \times 10^{-6} \text{ W/m}^2$. How loud in decibels is this sound level?

19. *Concerts.* The crowd at a Hearsay concert at Wembley Arena in London cheered at a sound level of 128.8 dB. What is the intensity of such a sound?

Source: www.peterborough.gov.uk

20. *City Ordinances.* In New York City, the maximum allowable sound level of music from a commercial establishment, as measured inside nearby residences, is 42 dB. What is the intensity of such sound?

Source: New York City Department of Environmental Protection, Bureau of Environmental Compliance, Nov. 2011

21. *E-mail Volume.* The SenderBase® Security Network ranks e-mail volume using a logarithmic scale. The magnitude M of a network's daily e-mail volume is given by

$$M = \log \frac{v}{1.34},$$

where v is the number of e-mail messages sent each day. How many e-mail messages are sent each day by a network that has a magnitude of 7.5?

Source: forum.spamcop.net

22. *Richter Scale.* The Richter scale, developed in 1935, has been used for years to measure earthquake magnitude. The Richter magnitude m of an earthquake is given by

$$m = \log \frac{A}{A_0},$$

where A is the maximum amplitude of the earthquake and A_0 is a constant. What is the magnitude on the Richter scale of an earthquake with an amplitude that is a million times A_0?

Use the compound-interest formula in Example 6 for Exercises 23 and 24.

23. *Interest Compounded Continuously.* Suppose that P_0 is invested in a savings account where interest is compounded continuously at 2.5% per year.

a) Express $P(t)$ in terms of P_0 and 0.025.

b) Suppose that $5000 is invested. What is the balance after 1 year? after 2 years?

c) When will an investment of $5000 double itself?

24. *Interest Compounded Continuously.* Suppose that P_0 is invested in a savings account where interest is compounded continuously at 3.1% per year.

  a) Express $P(t)$ in terms of P_0 and 0.031.
  b) Suppose that $1000 is invested. What is the balance after 1 year? after 2 years?
  c) When will an investment of $1000 double itself?

25. *Population Growth.* In 2012, the population of the United States was 314 million and the exponential growth rate was 0.963% per year.

  Source: U.S. Census Bureau

  a) Find the exponential growth function.
  b) Predict the U.S. population in 2020.
  c) When will the U.S. population reach 400 million?

26. *World Population Growth.* In 2012, the world population was 7 billion and the exponential growth rate was 1.12% per year.

  Source: U.S. Census Bureau

  a) Find the exponential growth function.
  b) Predict the world population in 2016.
  c) When will the world population be 10 billion?

27. *Population Growth.* The exponential growth rate of the population of United Arab Emirates is 3.3% per year (one of the highest in the world). What is the doubling time?

  Source: CIA World Factbook

28. *Bacteria Growth.* The number of bacteria in a culture grows at an exponential growth rate of 139% per hour. What is the doubling time for these bacteria?

29. *World Population.* The function

$$Y(x) = 89.29 \ln \frac{x}{7}$$

can be used to estimate the number of years $Y(x)$ after 2012 required for the world population to reach x billion people.

  Sources: Based on data from U.S. Census Bureau; International Data Base

  a) In what year will the world population reach 10 billion?
  b) In what year will the world population reach 12 billion?
  c) Graph the function.

30. *Marine Biology.* The function

$$Y(x) = 21.77 \ln \frac{x}{5.5}$$

can be used to estimate the number of years $Y(x)$ after 1982 required for the world's humpback whale population to reach x thousand whales.

  a) In what year will the whale population reach 15,000?
  b) In what year will the whale population reach 25,000?
  c) Graph the function.

31. *Computer Usage.* The number of tablet PC users in the United States can be estimated by

$$c(t) = 26 + 36 \ln t, \quad t \geq 1,$$

where $c(t)$ is in millions of users and t is the number of years after 2010.

  Source: Based on data from Forrester Research Report, "Tablets Will Grow As Fast As MP3 Players"

  a) How many tablet PC users were there in the United States in 2012?
  b) Graph the function.
  c) In what year will there be 100 million tablet PC users in the United States?

32. *Forgetting.* Students in an English class took a final exam. They took equivalent forms of the exam at monthly intervals thereafter. The average score $S(t)$, in percent, after t months was found to be

$$S(t) = 78 - 20 \log (t + 1), \quad t \geq 0.$$

  a) What was the average score when they initially took the test, $t = 0$?
  b) What was the average score after 4 months? after 24 months?
  c) Graph the function.
  d) After what time t was the average score 60%?

33. *Wind Power.* U.S. wind-power capacity has grown exponentially from 4232 megawatts in 2001 to 46,919 megawatts in 2011.

  Source: U.S. Department of Energy.

  a) Find the exponential growth rate k and write an equation for an exponential function that can be used to predict U.S. wind-power capacity, in megawatts, t years after 2001.
  b) Estimate the year in which wind-power capacity will reach 100,000 megawatts.

34. *Spread of a Computer Virus.* The number of computers infected by a virus t days after it first appears usually increases exponentially. In 2009, the "Conflicker" worm spread from about 2.4 million computers on January 12 to about 3.2 million computers on January 13.

  Source: Based on data from *PC World*

  a) Find the exponential growth rate k and write an equation for an exponential function that can be used to predict the number of computers infected t days after January 12, 2009.
  b) Assuming exponential growth, estimate how long it took the Conflicker worm to infect 10 million computers.

35. *Pharmaceuticals.* The concentration of acetaminophen in the body decreases exponentially after a dosage is given. In one clinical study, adult subjects averaged 11 micrograms/milliliter (mcg/mL) of the drug in their blood plasma 1 hr after a 1000-mg dosage and 2 micrograms/milliliter 6 hr after dosage.

Sources: Based on information from tylenolprofessional.com and Mark Knopp, M.D.

a) Find the value k, and write an equation for an exponential function that can be used to predict the concentration of acetaminophen, in micrograms/milliliter, t hours after a 1000-mg dosage.

b) Estimate the concentration of acetaminophen 3 hr after a 1000-mg dosage.

c) To relieve a fever, the concentration of acetaminophen should go no lower than 4 mcg/mL. After how many hours will a 1000-mg dosage drop to that level?

d) Find the half-life of acetaminophen.

36. *Atmospheric Pressure.* The atmospheric pressure in the lower stratosphere decreases exponentially from 473 lb/ft^2 at 36,152 ft to 51 lb/ft^2 at 82,345 ft.

Source: Based on information from grc.nasa.gov

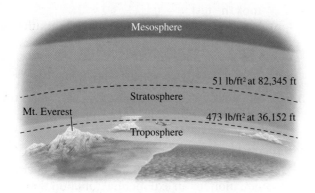

a) Find the exponential decay rate k, and write an equation for an exponential function that can be used to estimate the atmospheric pressure in the stratosphere h feet above 36,152 ft.

b) Estimate the atmospheric pressure at 50,000 ft ($h = 50,000 - 36,152$).

c) At what height is the atmospheric pressure 100 lb/ft^2?

d) What change in altitude will result in atmospheric pressure being halved?

37. *Archaeology.* A date palm seedling is growing in Kibbutz Ketura, Israel, from a seed found in King Herod's palace at Masada. The seed had lost 21% of its carbon-14. How old was the seed? (See Example 7.)

Source: Based on information from www.sfgate.com

38. *Archaeology.* Soil from beneath the Kish Church in Azerbaijan was found to have lost 12% of its carbon-14. How old was the soil? (See Example 7.)

Source: Based on information from www.azer.com

39. *Chemistry.* The exponential decay rate of iodine-131 is 9.6% per day. What is its half-life?

40. *Chemistry.* The decay rate of krypton-85 is 6.3% per year. What is its half-life?

41. *Caffeine.* The half-life of caffeine in the human body for a healthy adult is approximately 5 hr.

a) What is the exponential decay rate?

b) How long will it take 95% of the caffeine consumed to leave the body?

42. *Home Construction.* The chemical urea formaldehyde was used in some insulation in houses built during the mid to late 1960s. Unknown at the time was the fact that urea formaldehyde emitted toxic fumes as it decayed. The half-life of urea formaldehyde is 1 year.

a) What is its decay rate?

b) How long will it take 95% of the urea formaldehyde present to decay?

43. *Art Masterpieces.* As of August 2012, the highest auction price for a sculpture was $104.3 million, paid in 2010 for Alberto Giacometti's bronze sculpture *Walking Man I.* The same sculpture was purchased for about $9 million in 1990.

Source: Based on information from *The New York Times*, 02/02/10

a) Find the exponential growth rate k, and determine the exponential growth function that can be used to estimate the sculpture's value $V(t)$, in millions of dollars, t years after 1990.

b) Estimate the value of the sculpture in 2020.

c) What is the doubling time for the value of the sculpture?

d) How long after 1990 will the value of the sculpture be $1 billion?

44. *Value of a Sports Card.*   Legend has it that because he objected to teenagers smoking, and because his first baseball card was issued in cigarette packs, the great shortstop Honus Wagner halted production of his card before many were produced. One of these cards was purchased in 1991 by hockey great Wayne Gretzky (and a partner) for $451,000. The same card was sold in 2007 for $2.8 million. For the following questions, assume that the card's value increases exponentially, as it has for many years.

a) Find the exponential growth rate k, and determine an exponential function that can be used to estimate the dollar value, $V(t)$, of the card t years after 1991.
b) Predict the value of the card in 2015.
c) What is the doubling time for the value of the card?
d) In what year will the value of the card first exceed $8,000,000?

45. Write a problem for a classmate to solve in which information is provided and the classmate is asked to find an exponential growth function. Make the problem as realistic as possible.

46. Examine the restriction on t in Exercise 32.
a) What upper limit might be placed on t?
b) In practice, would this upper limit ever be enforced? Why or why not?

Skill Review

Find a linear function whose graph has the given characteristics.

47. Slope: 18; y-intercept: $\left(0, \frac{1}{2}\right)$ [2.3]

48. Contains $(6, 11)$ and $(-6, -11)$ [2.5]

49. Parallel to $2x - 3y = 4$; contains $(-3, 7)$ [2.5]

50. Perpendicular to $y = \frac{1}{2}x + 3$; y-intercept: $(0, 8)$ [2.5]

Synthesis

51. Can the model used in Example 5 to predict the number of unique visitors to Pinterest still be valid? Why or why not?

52. *Atmospheric Pressure.*   Atmospheric pressure P at an elevation a feet above sea level can be estimated by
$$P = P_0 e^{-0.00004a},$$
where P_0 is the pressure at sea level, which is approximately 29.9 in. of mercury (Hg). Explain how a barometer, or some other device for measuring atmospheric pressure, can be used to find the height of a skyscraper.

53. *Sports Salaries.*   As of August 2012, Alex Rodriguez of the New York Yankees had the largest contract in sports history. The 10-year $275-million deal, signed in 2007, stipulated that he receive $20 million in 2016. How much money would have been invested in 2007, at 4% interest compounded continuously, in order to have $20 million for Rodriguez in 2016? (This is much like determining what $20 million in 2016 is worth in 2007 dollars.)

Source: The San Francisco Chronicle

54. *Supply and Demand.*   The supply and demand for the sale of stereos by Sound Ideas are given by
$$S(x) = e^x \quad \text{and} \quad D(x) = 162{,}755e^{-x},$$
where $S(x)$ is the price at which the company is willing to supply x stereos and $D(x)$ is the demand price for a quantity of x stereos. Find the equilibrium point. (For reference, see Section 3.8.)

55. *Stellar Magnitude.*   The apparent stellar magnitude m of a star with received intensity I is given by
$$m(I) = -(19 + 2.5 \cdot \log I),$$
where I is in watts per square meter (W/m^2). The smaller the apparent stellar magnitude, the brighter the star appears.

Source: The Columbus Optical SETI Observatory

a) The intensity of light received from the sun is 1390 W/m^2. What is the apparent stellar magnitude of the sun?
b) The 5-m diameter Hale telescope on Mt. Palomar can detect a star with magnitude +23. What is the received intensity of light from such a star?

56. *Growth of Bacteria.*   The bacteria *Escherichia coli* (*E. coli*) are commonly found in the human bladder. Suppose that 3000 of the bacteria are present at time $t = 0$. Then t minutes later, the number of bacteria present is
$$N(t) = 3000(2)^{t/20}.$$
If 100,000,000 bacteria accumulate, a bladder infection can occur. If, at 11:00 A.M., a patient's bladder contains 25,000 *E. coli* bacteria, at what time can infection occur?

57. Show that for exponential growth at rate k, the doubling time T is given by $T = \dfrac{\ln 2}{k}$.

58. Show that for exponential decay at rate k, the half-life T is given by $T = \dfrac{\ln 2}{k}$.

↪ YOUR TURN ANSWERS: SECTION 9.7

1. $10^6\,\text{W/m}^2$ **2.** 5.01×10^{-7} moles per liter
3. **(a)** About 23.8 years; **(b)** about 14.2 years
4. 4,851,651,954 cm^2, or 485,165 m^2
5. $k = 1.308$; $P(t) = 10e^{1.308t}$, where $P(t)$ is the number of malware variants t years after 2011
6. $P(t) = 20{,}000e^{0.02t}$; 34.7 years   **7.** About 4000 years

---

QUICK QUIZ: SECTIONS 9.1–9.7

1. Determine whether $f(x) = 7 - x$ is one-to-one. [9.1]

2. Simplify: $\log_b \sqrt[4]{b^3}$. [9.4]

3. Use a calculator to find $\dfrac{\log 15}{\log 2}$. Round to four decimal places. [9.5]

4. Solve: $\log_2 (x - 3) + \log_2 (x + 3) = 4$. [9.6]

5. Stephanie invests \$10,000 at 3% interest, compounded annually. How long does it take for the investment to double itself? [9.7]

PREPARE TO MOVE ON

1. Find the distance between $(-3, 7)$ and $(-2, 6)$ [7.7]

2. Find the coordinates of the midpoint of the segment connecting $(3, -8)$ and $(5, -6)$. [7.7]

3. Solve by completing the square: $x^2 + 8x = 1$. [8.1]

4. Graph: $y = x^2 - 5x - 6$. [8.7]

A

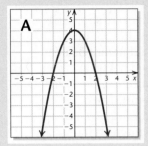

B

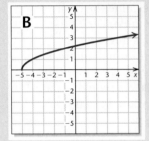

C

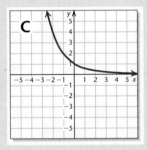

D

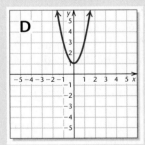

E

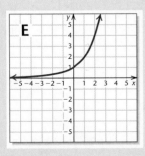

Visualizing for Success

Use after Section 9.5.

Match each function with its graph.

1. $f(x) = 2x - 3$

2. $f(x) = 2x^2 + 1$

3. $f(x) = \sqrt{x + 5}$

4. $f(x) = |x - 4|$

5. $f(x) = \ln x$

6. $f(x) = 2^{-x}$

7. $f(x) = -4$

8. $f(x) = \log x + 3$

9. $f(x) = 2^x$

10. $f(x) = 4 - x^2$

Answers on page A-53

An additional, animated version of this activity appears in MyMathLab. *To use MyMathLab, you need a course ID and a student access code. Contact your instructor for more information.*

F

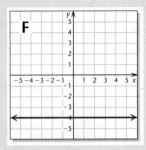

G

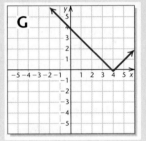

H

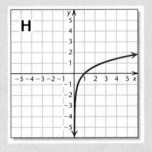

I

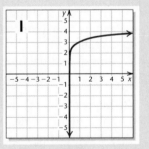

J

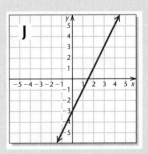

Collaborative Activity   *The True Cost of a New Car*

Focus: Car loans and exponential functions
Use after: Section 9.2
Time: 30 minutes
Group size: 2
Materials: Calculators with exponentiation keys

The formula

$$M = \frac{Pr}{1 - (1 + r)^{-n}}$$

is used to determine the payment size, M, when a loan of P dollars is to be repaid in n equally sized monthly payments. Here, r represents the monthly interest rate. Loans repaid in this fashion are said to be *amortized* (spread out equally) over a period of n months.

Activity

1. Suppose one group member is selling the other a car for $2600, financed at 1% interest per month for 24 months. What should be the size of each monthly payment?

2. Suppose both group members are shopping for the same model new car. To save time, each group member visits a different dealer. One dealer offers the car for $13,000 at 10.5% interest (0.00875 monthly interest) for 60 months (no down payment). The other dealer offers the same car for $12,000, but at 12% interest (0.01 monthly interest) for 48 months (no down payment).

   a) Determine the monthly payment size for each offer. Then determine the total amount paid for the car under each offer. How much of each total is interest?
   b) Work together to find the annual interest rate for which the total cost of 60 monthly payments for the $13,000 car would equal the total amount paid for the $12,000 car (as found in part a above).

Decision Making & Connection   *(Use after Section 9.7.)*

College Costs. It is difficult to plan for future college costs, but you can use historical data of costs from previous years to help predict future costs. Here, we assume that the costs are rising exponentially.

1. The average cost of in-state tuition and fees for a public four-year college, adjusted for inflation, was $4430 in 2000–2001 and $8244 in 2010–2011. Find an exponential function that fits the data.
   Source: collegeboard.com

2. Many students receive financial aid. Although this amount varies widely, the average amount of education tax benefits and grant aid per student, adjusted for inflation, was $2050 in 2000–2001 and $5750 in 2010–2011. Find an exponential function that fits the data.
   Source: collegeboard.com

3. Use the functions found in steps (1) and (2) to estimate the average in-state tuition and fees and the average amount of financial aid awarded for the school year following the one in which you are currently enrolled.

4. Subtract the financial aid awarded from the tuition and fees to estimate the average in-state net tuition cost to a student for the next school year.

5. *Research.* Find the cost of tuition and fees for two years for the college you currently attend. If possible, use the numbers for this year and for a year 5–10 years in the past. Use this information to find an exponential function that fits the data and then estimate the cost of tuition and fees for next year. If you have received financial aid for more than one year, use the data for two years to estimate your financial aid for next year. Then estimate what your net cost for the next school year would be.

Study Summary

KEY TERMS AND CONCEPTS	EXAMPLES	PRACTICE EXERCISES

SECTION 9.1: *Composite Functions and Inverse Functions*

The **composition** of f and g is defined as
$$(f \circ g)(x) = f(g(x)).$$

If $f(x) = \sqrt{x}$ and $g(x) = 2x - 5$, then
$$(f \circ g)(x) = f(g(x)) = f(2x - 5)$$
$$= \sqrt{2x - 5}.$$

1. Find $(f \circ g)(x)$ if $f(x) = 1 - 6x$ and $g(x) = x^2 - 3$.

A function f is **one-to-one** if different inputs always have different outputs. The graph of a one-to-one function passes the **horizontal-line test.**

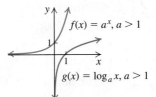

f is *not* one-to-one

f is one-to-one

2. Determine whether $f(x) = 5x - 7$ is one-to-one.

If f is one-to-one, it is possible to find its inverse:

1. Replace $f(x)$ with y.
2. Interchange x and y.
3. Solve for y.

4. Replace y with $f^{-1}(x)$.

If $f(x) = 2x - 3$, find $f^{-1}(x)$.

1. $y = 2x - 3$
2. $x = 2y - 3$
3. $x + 3 = 2y$
   $\dfrac{x + 3}{2} = y$
4. $\dfrac{x + 3}{2} = f^{-1}(x)$

3. If $f(x) = 5x + 1$, find $f^{-1}(x)$.

SECTION 9.2: *Exponential Functions*
SECTION 9.3: *Logarithmic Functions*

For an **exponential function** f:
$f(x) = a^x$, $a > 0$, $a \neq 1$;
Domain of f is $\mathbb{R}$;
$f^{-1}(x) = \log_a x$.

For a **logarithmic function** g:
$g(x) = \log_a x$, $a > 0$, $a \neq 1$;
Domain of g is $(0, \infty)$;
$g^{-1}(x) = a^x$.

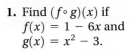

$f(x) = a^x, a > 1$

$g(x) = \log_a x, a > 1$

4. Graph by hand: $f(x) = 2^x$.

5. Graph by hand: $f(x) = \log x$.

$\log_a x = m$ means $a^m = x$.

Solve: $\log_8 x = 2$.
$8^2 = x$ Rewriting as an exponential equation
$64 = x$

6. Rewrite as an equivalent logarithmic equation:
$5^4 = 625$.

647

SECTION 9.4: *Properties of Logarithmic Functions*
SECTION 9.5: *Common Logarithms and Natural Logarithms*

Properties of Logarithms

$\log_a (MN) = \log_a M + \log_a N$

$\log_a \dfrac{M}{N} = \log_a M - \log_a N$

$\log_a M^p = p \cdot \log_a M$

$\log_a 1 = 0$

$\log_a a = 1$

$\log_a a^k = k$

$\log M = \log_{10} M$

$\ln M = \log_e M$

$\log_b M = \dfrac{\log_a M}{\log_a b}$

$\log_7 10 = \log_7 5 + \log_7 2$

$\log_5 \dfrac{14}{3} = \log_5 14 - \log_5 3$

$\log_8 5^{12} = 12 \log_8 5$

$\log_9 1 = 0$

$\log_4 4 = 1$

$\log_3 3^8 = 8$

$\log 43 = \log_{10} 43$

$\ln 37 = \log_e 37$

$\log_6 31 = \dfrac{\log 31}{\log 6} = \dfrac{\ln 31}{\ln 6}$

7. Express as an equivalent expression that is a sum of logarithms: $\log_9 xy$.

8. Express as an equivalent expression that is a difference of logarithms: $\log_6 \frac{7}{10}$.

9. Simplify: $\log_8 1$.

10. Simplify: $\log_6 6^{19}$.

11. Use the change-of-base formula to find $\log_2 5$.

SECTION 9.6: *Solving Exponential Equations and Logarithmic Equations*

The Principle of Exponential Equality
For any real number b, $b \neq -1, 0$, or 1:

$b^x = b^y$ is equivalent to $x = y$.

Solve: $25 = 5^x$.

$$5^2 = 5^x$$
$$2 = x$$

12. Solve: $2^{3x} = 16$.

The Principle of Logarithmic Equality
For any logarithm base a, and for $x, y > 0$:

$x = y$ is equivalent to $\log_a x = \log_a y$.

Solve: $83 = 7^x$.

$$\log 83 = \log 7^x$$
$$\log 83 = x \log 7$$
$$\dfrac{\log 83}{\log 7} = x$$

13. Solve: $e^{0.1x} = 10$.

SECTION 9.7: *Applications of Exponential Functions and Logarithmic Functions*

Exponential Growth Model

$P(t) = P_0 e^{kt}, \quad k > 0$

P_0 is the population at time 0.

$P(t)$ is the population at time t.

k is the **exponential growth rate**.

The **doubling time** is the amount of time necessary for the population to double in size.

If \$1000 is invested at 5%, compounded continuously, then:

• $P_0 = 1000$;
• $k = 0.05$;
• $P(t) = 1000e^{0.05t}$;
• The doubling time is 13.9 years.

14. In 2014, the population of Bridgeford was 15,000, and it was increasing at an exponential growth rate of 2.3% per year.

   a) Write an exponential function that describes the population t years after 2014.
   b) Find the doubling time.

Exponential Decay Model

$P(t) = P_0 e^{-kt}, \quad k > 0$

P_0 is the quantity present at time 0.

$P(t)$ is the quantity present at time t.

k is the **exponential decay rate**.

The **half-life** is the amount of time necessary for half of the quantity to decay.

If the population of Ridgeton is 2000, and the population is decreasing exponentially at a rate of 1.5% per year, then:

• $P_0 = 2000$;
• $k = 0.015$;
• $P(t) = 2000e^{-0.015t}$;
• The half-life is 46.2 years.

15. Argon-37 has an exponential decay rate of 1.98% per day. Find the half-life.

Review Exercises: Chapter 9

Concept Reinforcement

Classify each of the following statements as either true or false.

1. The functions given by $f(x) = 3^x$ and $g(x) = \log_3 x$ are inverses of each other.   [9.3]

2. A function's doubling time is the amount of time t for which $f(t) = 2 \cdot f(0)$.   [9.7]

3. A radioactive isotope's half-life is the amount of time t for which $f(t) = \frac{1}{2} \cdot f(0)$.   [9.7]

4. $\ln(ab) = \ln a - \ln b$ [9.4]

5. $\log x^a = x \ln a$ [9.4]

6. $\log_a \dfrac{m}{n} = \log_a m - \log_a n$ [9.4]

7. For $f(x) = 3^x$, the domain of f is $[0, \infty)$.   [9.2]

8. For $g(x) = \log_2 x$, the domain of g is $[0, \infty)$.   [9.3]

9. The function F is not one-to-one if $F(-2) = F(5)$.   [9.1]

10. The function g is one-to-one if it passes the vertical-line test.   [9.1]

11. Find $(f \circ g)(x)$ and $(g \circ f)(x)$ if $f(x) = x^2 + 1$ and $g(x) = 2x - 3$.   [9.1]

12. If $h(x) = \sqrt{3 - x}$, find $f(x)$ and $g(x)$ such that $h(x) = (f \circ g)(x)$. Answers may vary.   [9.1]

13. Is $f(x) = 4 - x^2$ one-to-one?   [9.1]

Find a formula for the inverse of each function.   [9.1]

14. $f(x) = x - 10$

15. $g(x) = \dfrac{3x + 1}{2}$

16. $f(x) = 27x^3$

Graph.

17. $f(x) = 3^x + 1$ [9.2]

18. $x = \left(\frac{1}{4}\right)^y$ [9.2]

19. $y = \log_5 x$ [9.3]

Simplify.   [9.3]

20. $\log_9 81$

21. $\log_3 \frac{1}{9}$

22. $\log_2 2^{11}$

23. $\log_{16} 4$

Rewrite as an equivalent logarithmic equation.   [9.3]

24. $2^{-3} = \frac{1}{8}$

25. $25^{1/2} = 5$

Rewrite as an equivalent exponential equation.   [9.3]

26. $\log_4 16 = x$

27. $\log_8 1 = 0$

Express as an equivalent expression using the individual logarithms of x, y, and z.   [9.4]

28. $\log_a x^4 y^2 z^3$

29. $\log_a \dfrac{x^5}{yz^2}$

30. $\log \sqrt[4]{\dfrac{z^2}{x^3 y}}$

Express as an equivalent expression that is a single logarithm and, if possible, simplify.   [9.4]

31. $\log_a 5 + \log_a 8$

32. $\log_a 48 - \log_a 12$

33. $\frac{1}{2} \log a - \log b - 2 \log c$

34. $\frac{1}{3}[\log_a x - 2 \log_a y]$

Simplify.   [9.4]

35. $\log_m m$

36. $\log_m 1$

37. $\log_m m^{17}$

Given $\log_a 2 = 1.8301$ and $\log_a 7 = 5.0999$, find each of the following.   [9.4]

38. $\log_a 14$

39. $\log_a \frac{2}{7}$

40. $\log_a 28$

41. $\log_a 3.5$

42. $\log_a \sqrt{7}$

43. $\log_a \frac{1}{4}$

Use a calculator to find each of the following to four decimal places.   [9.5]

44. $\log 75$

45. $10^{1.789}$

46. $\ln 0.3$

47. $e^{-0.98}$

Find each of the following using the change-of-base formula. Round answers to four decimal places.   [9.5]

48. $\log_5 50$

49. $\log_6 5$

Graph and state the domain and the range of each function.   [9.5]

50. $f(x) = e^x - 1$

51. $g(x) = 0.6 \ln x$

Solve. Where appropriate, include approximations to four decimal places.   [9.6]

52. $5^x = 125$

53. $3^{2x} = \frac{1}{9}$

54. $\log_3 x = -4$

55. $\log_x 16 = 4$

56. $\log x = -3$

57. $6 \ln x = 18$

58. $4^{2x-5} = 19$

59. $2^x = 12$

60. $e^{-0.1t} = 0.03$

61. $2 \ln x = -6$

62. $\log (2x - 5) = 1$

63. $\log_4 x - \log_4 (x - 15) = 2$

64. $\log_3 (x - 4) = 2 - \log_3 (x + 4)$

65. In a business class, students were tested at the end of the course with a final exam. They were then tested again 6 months later. The forgetting formula was determined to be

$$S(t) = 82 - 18 \log (t + 1),$$

where $S(t)$ was the average student grade t months after taking the final exam.   [9.7]

   a) Determine the average score when they first took the exam (when $t = 0$).

   b) What was the average score after 6 months?

   c) After what time was the average score 54?

66. A laptop computer is purchased for $1500. Its value each year is about 80% of its value in the preceding year. Its value in dollars after t years is given by the exponential function

$$V(t) = 1500(0.8)^t. \quad [9.7]$$

   a) After what amount of time will the computer's value be $900?

   b) After what amount of time will the computer's value be half the original value?

67. *Mobile Payments.* Paypal's mobile payments grew exponentially from $175 million in 2009 to $4 billion in 2011.

   Source: http://venturebeat.com/2012/01/10/paypals-mobile-payments-4b-2011/

   a) Find the exponential growth rate k, and write a function that describes the amount $A(t)$, in millions of dollars, of Paypal's mobile payments t years after 2009.   [9.7]

   b) Assuming the trend continues, estimate the amount of Paypal's mobile payments in 2013.   [9.7]

   c) In what year will Paypal's mobile payments reach $438 billion?   [9.7]

   d) Find the doubling time.   [9.7]

68. *Mining Safety.* The number of fatalities in coal mining in the United States decreased exponentially from approximately 2500 per year in 1910 to approximately 20 per year in 1990.

   Source: http://www.weitzlux.com/workaccidentshistory_725.html

   a) Find the exponential decay rate k, and write a function that describes the number of fatalities $F(t)$ in coal mining t years after 1910.   [9.7]

   b) Estimate the number of fatalities in coal mining in 2020.   [9.7]

   c) What is the first year in which there will be 5 or fewer coal mining fatalities?   [9.7]

69. The value of Aret's stock market portfolio doubled in 6 years. What was the exponential growth rate? [9.7]

70. How long will it take $7600 to double if it is invested at 4.2%, compounded continuously?   [9.7]

71. How old is a skull that has lost 34% of its carbon-14? (Use $P(t) = P_0 e^{-0.00012t}$.)   [9.7]

72. What is the pH of coffee if its hydrogen ion concentration is 7.9×10^{-6} moles per liter? (Use pH $= -\log [H^+]$.)   [9.7]

73. The roar of a lion can reach a sound intensity of 2.5×10^{-1} W/m². How loud in decibels is this sound level? $\left(\text{Use } L = 10 \cdot \log \dfrac{I}{10^{-12} \, \text{W/m}^2}. \right)$ [9.7]

Source: en.allexperts.com

Synthesis

74. Explain why negative numbers do not have logarithms.   [9.3]

75. Explain why $f(x) = e^x$ and $g(x) = \ln x$ are inverse functions.   [9.5]

Solve.   [9.6]

76. $\ln (\ln x) = 3$

77. $2^{x^2 + 4x} = \frac{1}{8}$

78. Solve the system

$$5^{x+y} = 25,$$
$$2^{2x-y} = 64. \quad [9.6]$$

Test: Chapter 9

For step-by-step test solutions, access the Chapter Test Prep Videos in My MathLab®, on YouTube® (search "Bittinger Inter Alg CA" and click on "Channels"), or by scanning the code.

1. Find $(f \circ g)(x)$ and $(g \circ f)(x)$ if $f(x) = x + x^2$ and $g(x) = 2x + 1$.

2. If
$$h(x) = \frac{1}{2x^2 + 1},$$
find $f(x)$ and $g(x)$ such that $h(x) = (f \circ g)(x)$. Answers may vary.

3. Determine whether $f(x) = x^2 + 3$ is one-to-one.

Find a formula for the inverse of each function.

4. $f(x) = 3x + 4$

5. $g(x) = (x + 1)^3$

Graph.

6. $f(x) = 2^x - 3$

7. $g(x) = \log_7 x$

Simplify.

8. $\log_5 125$

9. $\log_{100} 10$

10. $\log_n n$

11. $\log_c 1$

12. Rewrite as an equivalent logarithmic equation: $5^{-4} = \frac{1}{625}$.

13. Rewrite as an equivalent exponential equation: $m = \log_2 \frac{1}{2}$.

14. Express as an equivalent expression using the individual logarithms of a, b, and c:
$$\log \frac{a^3 b^{1/2}}{c^2}.$$

15. Express as an equivalent expression that is a single logarithm:
$$\tfrac{1}{3} \log_a x + 2 \log_a z.$$

Given $\log_a 2 = 0.301$, $\log_a 6 = 0.778$, and $\log_a 7 = 0.845$, find each of the following.

16. $\log_a 14$

17. $\log_a 3$

18. $\log_a 16$

📱 *Use a calculator to find each of the following to four decimal places.*

19. $\log 25$

20. $10^{-0.8}$

21. $\ln 0.4$

22. $e^{4.8}$

23. Find $\log_3 14$ using the change-of-base formula. Round to four decimal places.

Graph and state the domain and the range of each function.

24. $f(x) = e^x + 3$

25. $g(x) = \ln(x - 4)$

Solve. Where appropriate, include approximations to four decimal places.

26. $2^x = \frac{1}{32}$

27. $\log_4 x = \frac{1}{2}$

28. $\log x = -2$

29. $7^x = 1.2$

30. $\log(x - 3) + \log(x + 1) = \log 5$

📱 31. The average walking speed R of people in a city of population P can be modeled by the equation $R = 0.37 \ln P + 0.05$, where R is in feet per second and P is in thousands.

   a) The population of Tulsa, Oklahoma, is 392,000. Find the average walking speed.

   b) San Diego, California, has an average walking speed of about 2.7 ft/sec. Find the population.

📱 32. The population of Nigeria was about 155 million in 2012, and the exponential growth rate was 1.9% per year.

   a) Write an exponential function describing the population of Nigeria.

   b) What will the population be in 2016? in 2050?

   c) When will the population be 200 million?

   d) What is the doubling time?

📱 33. The average cost of a year at a private four-year college grew exponentially from $30,360 in 2001 to $38,590 in 2011, in constant 2011 dollars.

   Source: "Trends in College Pricing 2011," collegeboard.org

   a) Find the exponential growth rate k, and write a function that approximates the cost $C(t)$ of a year at a private four-year college t years after 2001.

   b) Predict the cost of a year at a private four-year college in 2016.

   c) In what year will the average cost of a year at a private four-year college be $50,000?

📱 34. An investment with interest compounded continuously doubled itself in 16 years. What was the interest rate?

35. The hydrogen ion concentration of water is 1.0×10^{-7} moles per liter. What is the pH? (Use $pH = -\log[H^+]$.)

Synthesis

36. Solve: $\log_5 |2x - 7| = 4$.

37. If $\log_a x = 2$, $\log_a y = 3$, and $\log_a z = 4$, find
$$\log_a \frac{\sqrt[3]{x^2 z}}{\sqrt[3]{y^2 z^{-1}}}.$$

Cumulative Review: Chapters 1–9

Simplify.

1. $(-2x^2y^{-3})^{-4}$ [1.6]

2. $\dfrac{3x^4y^6z^{-2}}{-9x^4y^2z^3}$ [1.6]

3. $(-5x^4y^{-3}z^2)(-4x^2y^2)$ [1.6]

Solve.

4. $3(2x - 3) = 9 - 5(2 - x)$ [1.3]

5. $4x - 3y = 15,$
$3x + 5y = 4$ [3.2]

6. $x + y - 3z = -1,$
$2x - y + z = 4,$
$-x - y + z = 1$ [3.4]

7. $x(x - 3) = 70$ [5.8]

8. $\dfrac{7}{x^2 - 5x} - \dfrac{2}{x - 5} = \dfrac{4}{x}$ [6.4]

9. $\sqrt{4 - 5x} = 2x - 1$ [7.6]

10. $3x^2 + 48 = 0$ [8.1]

11. $x^4 - 13x^2 + 36 = 0$ [8.5]

12. $\log_x 81 = 2$ [9.3]

▦ 13. $3^{5x} = 7$ [9.6]

14. $\ln x - \ln (x - 8) = 1$ [9.6]

15. $x^2 + 4x > 5$ [8.9]

16. If $f(x) = x^2 + 6x$, find a such that $f(a) = 11$. [8.2]

17. If $f(x) = |2x - 3|$, find all x for which $f(x) \geq 7$. [4.3]

Perform the indicated operations and simplify.

18. $\dfrac{a^2 - a - 6}{a^3 - 27} \cdot \dfrac{a^2 + 3a + 9}{6}$ [6.1]

19. $\dfrac{3}{x + 6} - \dfrac{2}{x^2 - 36} + \dfrac{4}{x - 6}$ [6.2]

20. $\sqrt{x + 5}\,\sqrt[5]{x + 5}$ [7.5]

21. $(2 - i\sqrt{3})(6 + i\sqrt{3})$ [7.8]

Factor.

22. $27 + 64n^3$ [5.6]

23. $6x^2 + 8xy - 8y^2$ [5.4]

24. $2m^2 + 12mn + 18n^2$ [5.5]

25. $x^4 - 16y^4$ [5.5]

26. Rationalize the denominator:

$\dfrac{3 - \sqrt{y}}{2 - \sqrt{y}}.$ [7.5]

27. Find the inverse of f if $f(x) = 9 - 2x$. [9.1]

28. Find a linear function with a graph that contains the points $(0, -8)$ and $(-1, 2)$. [2.5]

Graph.

29. $5x = 15 + 3y$ [2.4]

30. $y = \log_3 x$ [9.3]

31. $-2x - 3y \leq 12$ [4.4]

32. Graph: $f(x) = 2x^2 + 12x + 19$. [8.7]

   a) Label the vertex.
   b) Draw the axis of symmetry.
   c) Find the maximum or minimum value.

33. Graph $f(x) = 2e^x$ and determine the domain and the range. [9.5]

Solve.

34. *Colorado River.* The Colorado River delivers 1.5 million acre-feet of water to Mexico each year. This is only 10% of the volume of the river; the remainder is diverted upstream for agricultural use. How much water is diverted each year from the Colorado River? [1.4]

Source: www.sierraclub.org

35. *Desalination.* More cities are supplying some of their fresh water through desalination, the process of removing the salt from ocean water. The world-wide desalination capacity has grown exponentially from 15 million m^3 per day in 1990 to 68.5 million m^3 per day in 2010.

Source: International Desalination Association

   a) Find the exponential growth rate k, and write an equation for an exponential function that can be used to predict the worldwide desalination capacity $D(t)$, in millions of cubic meters per day, t years after 1990. [9.7]
   b) Predict the worldwide desalination capacity in 2016. [9.7]
   c) In what year will worldwide desalination capacity reach 150 million m^3 per day? [9.7]

36. Good's Candies of Indiana makes all their chocolates by hand. It takes Anne 10 min to coat a tray of candies in chocolate. It takes Clay 12 min to coat a tray of candies. How long would it take Anne and Clay, working together, to coat the candies? [6.5]

37. Joe's Thick and Tasty salad dressing gets 45% of its calories from fat. The Light and Lean dressing gets 20% of its calories from fat. How many ounces of each should be mixed in order to get 15 oz of dressing that gets 30% of its calories from fat? [3.3]

Synthesis

38. Solve: $\log \sqrt{3x} = \sqrt{\log 3x}$. [9.6]

39. The Danville Express travels 280 mi at a certain speed. If the speed were increased 5 mph, the trip would take 1 hr less. Find the actual speed. [8.4]

CHAPTER

10

Conic Sections

ROSSIGNOL EXPERIENCE 98 SKI	
Width at tip	139 mm
Width at waist	98 mm
Width at tail	128 mm

Source: rossignol.com

10.1 Conic Sections: Parabolas and Circles

10.2 Conic Sections: Ellipses

10.3 Conic Sections: Hyperbolas

CONNECTING THE CONCEPTS

MID-CHAPTER REVIEW

10.4 Nonlinear Systems of Equations

CHAPTER RESOURCES

  Visualizing for Success
  Collaborative Activity
  Decision Making: Connection

STUDY SUMMARY

REVIEW EXERCISES
CHAPTER TEST
CUMULATIVE REVIEW

There's a Lot of Science in a Ski!

As a boy in Norway in the 1830s, Sondre Norheim skied down mountains (and his parents' roof) on simple pine skis made by his father. Because of his later use of heel bindings and carved sidewalls and his demonstration of the "joy of skiing," he has become known as the "father of modern skiing" (www.sondrenorheim.com). Today, skis are designed and crafted with science, precision, and continual innovation. The table above lists some of the specifications of the Rossignol Experience 98 ski. Although the curves in the side of the ski described above are arcs of two different-size circles, we will model such curves using just one circle. (*See Exercise 82 in Exercise Set 10.1.*)

Having a solid understanding of math is a necessity in ski design.

Howard Wu, Chief Engineer at Wubanger LLC in Salt Lake City, Utah, uses math equations for different shapes to design skis, as well as the concept of slope to change the flex pattern of a ski. All calculations must be done carefully, since being incorrect by even one-hundredth of a centimeter will make a significant dif-ference in the way in which the ski rides.

T here are a variety of applications and equations with graphs that are conic sections. A circle is one example of a *conic section*, meaning that it can be regarded as a cross section of a cone.

| **10.1** | **Conic Sections: Parabolas and Circles** |

Parabolas ▪ Circles

This section and the next two examine curves formed by cross sections of cones. These curves are all graphs of $Ax^2 + By^2 + Cxy + Dx + Ey + F = 0$. The constants $A, B, C, D, E,$ and F determine which of the following shapes serve as the graph.

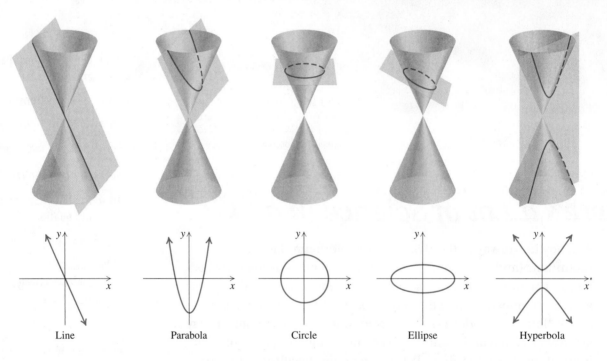

| Line | Parabola | Circle | Ellipse | Hyperbola |

Parabolas

When a cone is sliced as in the second figure above, the conic section formed is a **parabola**. Parabolas have many applications in electricity, mechanics, and optics. A cross section of a contact lens or a satellite dish is a parabola, and arches that support certain bridges are parabolas.

EQUATION OF A PARABOLA

A parabola with a vertical axis of symmetry opens upward or downward and has an equation that can be written in the form

$$y = ax^2 + bx + c.$$

A parabola with a horizontal axis of symmetry opens to the right or to the left and has an equation that can be written in the form

$$x = ay^2 + by + c.$$

Parabolas with equations of the form $f(x) = ax^2 + bx + c$ were graphed in Chapter 8.

EXAMPLE 1   Graph: $y = x^2 - 4x + 9$.

SOLUTION   To locate the vertex, we can use either of two approaches. One way is to complete the square:

$$y = (x^2 - 4x) + 9 \qquad \text{Note that half of } -4 \text{ is } -2, \text{ and } (-2)^2 = 4.$$
$$= (x^2 - 4x + 4 - 4) + 9 \qquad \text{Adding and subtracting 4}$$
$$= (x^2 - 4x + 4) + (-4 + 9) \qquad \text{Regrouping}$$
$$= (x - 2)^2 + 5. \qquad \text{Factoring and simplifying}$$

The vertex is $(2, 5)$.

A second way to find the vertex is to recall that the x-coordinate of the vertex of the parabola given by $y = ax^2 + bx + c$ is $-b/(2a)$:

$$x = -\frac{b}{2a} = -\frac{-4}{2(1)} = 2.$$

To find the y-coordinate of the vertex, we substitute 2 for x:

$$y = x^2 - 4x + 9 = 2^2 - 4(2) + 9 = 5.$$

Either way, the vertex is $(2, 5)$. Next, we calculate and plot some points on each side of the vertex. As expected for a positive coefficient of x^2, the graph opens upward.

x	y
2	5	←Vertex
0	9	←y-intercept
1	6
3	6
4	9

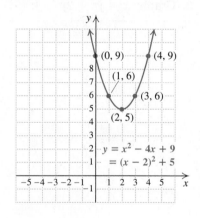

1. Graph: $y = x^2 + 2x - 3$.

YOUR TURN

TO GRAPH AN EQUATION OF THE FORM $y = ax^2 + bx + c$

1. Find the vertex (h, k) either by completing the square to find an equivalent equation

   $$y = a(x - h)^2 + k,$$

   or by using $-b/(2a)$ to find the x-coordinate and substituting to find the y-coordinate.
2. Choose other values for x on each side of the vertex, and compute the corresponding y-values.
3. The graph opens upward for $a > 0$ and downward for $a < 0$.

If we interchange x and y in the equation in Example 1, we obtain an equation for the *inverse* relation, $x = y^2 - 4y + 9$. The graph of this equation will be the reflection of the graph in Example 1 across $y = x$.

Any equation of the form $x = ay^2 + by + c$ represents a horizontal parabola that opens to the right for $a > 0$, opens to the left for $a < 0$, and has an axis of symmetry parallel to the x-axis.

EXAMPLE 2 Graph: $x = y^2 - 4y + 9$.

SOLUTION This equation is like that in Example 1 but with x and y interchanged. The vertex is $(5, 2)$ instead of $(2, 5)$. To find ordered pairs, we choose values for y on each side of the vertex. Then we compute values for x. Note that the x- and y-values of the table in Example 1 are now switched. You should confirm that, by completing the square, we have $x = (y - 2)^2 + 5$.

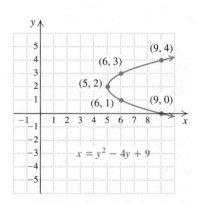

x	y	
5	2	←Vertex
9	0	←x-intercept
6	1	
6	3	
9	4	

(1) Choose values for y.
(2) Compute values for x.

2. Graph: $x = y^2 + 2y - 3$.

 YOUR TURN

TO GRAPH AN EQUATION OF THE FORM $x = ay^2 + by + c$

1. Find the vertex (h, k) either by completing the square to find an equivalent equation

   $$x = a(y - k)^2 + h,$$

   or by using $-b/(2a)$ to find the y-coordinate and substituting to find the x-coordinate.
2. Choose other values for y that are on either side of k and compute the corresponding x-values.
3. The graph opens to the right if $a > 0$ and to the left if $a < 0$.

EXAMPLE 3 Graph: $x = -2y^2 + 10y - 7$.

SOLUTION We find the vertex by completing the square:

$$\begin{aligned}
x &= -2y^2 + 10y - 7 \\
&= -2(y^2 - 5y \quad) - 7 \\
&= -2\left(y^2 - 5y + \tfrac{25}{4}\right) - 7 - (-2)\tfrac{25}{4} \qquad \tfrac{1}{2}(-5) = \tfrac{-5}{2}; \left(\tfrac{-5}{2}\right)^2 = \tfrac{25}{4}; \text{ we} \\
&\qquad\qquad\qquad\qquad\qquad\qquad\qquad\qquad\qquad\quad \text{add and subtract } (-2)\tfrac{25}{4}. \\
&= -2\left(y - \tfrac{5}{2}\right)^2 + \tfrac{11}{2}. \qquad\qquad\quad \text{Factoring and simplifying}
\end{aligned}$$

The vertex is $\left(\tfrac{11}{2}, \tfrac{5}{2}\right)$.

For practice, we also find the vertex by first computing its y-coordinate, $-b/(2a)$, and then substituting to find the x-coordinate:

$$y = -\frac{b}{2a} = -\frac{10}{2(-2)} = \frac{5}{2}$$

$$x = -2y^2 + 10y - 7 = -2\left(\tfrac{5}{2}\right)^2 + 10\left(\tfrac{5}{2}\right) - 7$$

$$= \tfrac{11}{2}.$$

To find ordered pairs, we choose values for y on each side of the vertex and then compute values for x. A table is shown below, together with the graph. The graph opens to the left because the y^2-coefficient, -2, is negative.

x	y	
$\frac{11}{2}$	$\frac{5}{2}$	←Vertex
-7	0	←x-intercept
5	2	
5	3	
1	1	
1	4	
-7	5	

— **(1)** Choose these values for y.

— **(2)** Compute these values for x.

3. Graph: $x = -3y^2 - 6y + 1$.

YOUR TURN

Circles

Another conic section, the **circle**, is the set of points in a plane that are a fixed distance r, called the **radius** (plural, **radii**), from a fixed point (h, k), called the **center**. Note that the word radius can mean either any segment connecting a point on a circle to the center or the length of such a segment. Using the idea of a fixed distance r and the distance formula,

$$d = \sqrt{(x_2 - x_1)^2 + (y_2 - y_1)^2},$$

we can find the equation of a circle.

If (x, y) is on a circle of radius r, centered at (h, k), then by the definition of a circle and the distance formula, it follows that

$$r = \sqrt{(x - h)^2 + (y - k)^2}.$$

Squaring both sides gives the equation of a circle in standard form.

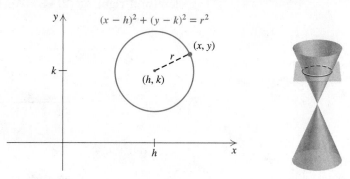

> **EQUATION OF A CIRCLE (STANDARD FORM)**
>
> The equation of a circle, centered at (h, k), with radius r, is given by
> $$(x - h)^2 + (y - k)^2 = r^2.$$

Note that for $h = 0$ and $k = 0$, the circle is centered at the origin. Otherwise, the circle is translated $|h|$ units horizontally and $|k|$ units vertically.

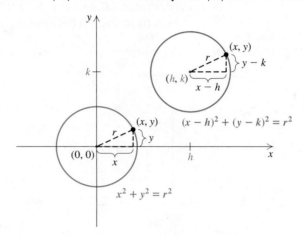

EXAMPLE 4 Find an equation of the circle centered at $(4, -5)$ with radius 6.

SOLUTION Using the standard form, we obtain

$$(x - 4)^2 + (y - (-5))^2 = 6^2, \qquad \text{Using } (x - h)^2 + (y - k)^2 = r^2$$

or $\qquad (x - 4)^2 + (y + 5)^2 = 36.$

4. Find an equation of the circle centered at $(-2, 7)$ with radius 9.

YOUR TURN

EXAMPLE 5 Find the center and the radius and then graph each circle.

a) $(x - 2)^2 + (y + 3)^2 = 4^2$

b) $x^2 + y^2 + 8x - 2y + 15 = 0$

SOLUTION

a) We write standard form:

$$(x - 2)^2 + [y - (-3)]^2 = 4^2.$$

The center is $(2, -3)$ and the radius is 4. To graph, we plot the points $(2, 1)$, $(2, -7)$, $(-2, -3)$, and $(6, -3)$, which are, respectively, 4 units above, below, left, and right of $(2, -3)$. We then either sketch a circle by hand or use a compass.

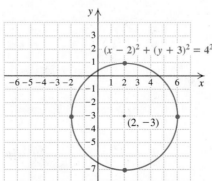

$x^2 + y^2 + 8x - 2y + 15 = 0$
$[x - (-4)]^2 + (y - 1)^2 = (\sqrt{2})^2$

$(-4, 1)$

b) To write the equation $x^2 + y^2 + 8x - 2y + 15 = 0$ in standard form, we complete the square twice, once with $x^2 + 8x$ and once with $y^2 - 2y$:

$$x^2 + y^2 + 8x - 2y + 15 = 0$$

$$x^2 + 8x \qquad + y^2 - 2y \qquad = -15$$

Grouping the x-terms and the y-terms; subtracting 15 from both sides

$$x^2 + 8x + 16 + y^2 - 2y + 1 = -15 + 16 + 1$$

Adding $\left(\frac{8}{2}\right)^2$, or 16, and $\left(-\frac{2}{2}\right)^2$, or 1, to both sides to get standard form

$$(x + 4)^2 + (y - 1)^2 = 2$$

Factoring

$$[x - (-4)]^2 + (y - 1)^2 = (\sqrt{2})^2.$$

Writing standard form

5. Find the center and the radius and then graph the circle:

$$x^2 + (y - 3)^2 = 5.$$

The center is $(-4, 1)$ and the radius is $\sqrt{2}$.

YOUR TURN

Technology Connection

Graphing the equation of a circle using a graphing calculator usually requires two steps:

1. Solve the equation for y. The result will include a $\pm$ sign in front of a radical.
2. Graph two functions, one for the $+$ sign and the other for the $-$ sign, on the same set of axes.

For example, to graph $(x - 3)^2 + (y + 1)^2 = 16$, solve for $y + 1$ and then y:

$$(y + 1)^2 = 16 - (x - 3)^2$$
$$y + 1 = \pm\sqrt{16 - (x - 3)^2}$$
$$y = -1 \pm \sqrt{16 - (x - 3)^2},$$

or $\qquad y_1 = -1 + \sqrt{16 - (x - 3)^2}$

and $\qquad y_2 = -1 - \sqrt{16 - (x - 3)^2}.$

When both functions are graphed in a "squared" window (to eliminate distortion), the result is as follows.

$y_1 = -1 + \sqrt{16 - (x - 3)^2}$,
$y_2 = -1 - \sqrt{16 - (x - 3)^2}$

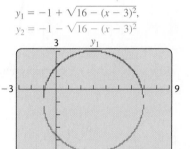

On many calculators, pressing **APPS** and selecting Conics and then Circle accesses a program in which equations in standard form can be graphed directly and then Traced.

Graph each of the following equations.

1. $x^2 + y^2 - 16 = 0$
2. $(x - 1)^2 + (y - 2)^2 = 25$
3. $(x + 3)^2 + (y - 5)^2 = 16$

10.1 EXERCISE SET

FOR EXTRA HELP

MyMathLab® Math XP
PRACTICE WATCH READ REVIEW

Vocabulary and Reading Check

Choose the word from the following list that best completes each statement. Words may be used more than once or not at all.

center horizontal vertex
circle parabola vertical
conic sections radii

1. Parabolas and circles are examples of

_____.

2. A(n) _____ is the set of points in a plane that are a fixed distance from its center.

3. A parabola with a(n) _____ axis of symmetry opens to the right or to the left.

4. In the equation of a parabola, the point (h, k) represents the _____ of the parabola.

5. In the equation of a circle, the point (h, k) represents the _____ of the circle.

6. A radius is the distance from a point on a circle to the _____.

 Concept Reinforcement

In each of Exercises 7–12, match the equation with the graph of that equation from those shown.

7. ____ $(x - 2)^2 + (y + 5)^2 = 9$

8. ____ $(x + 2)^2 + (y - 5)^2 = 9$

9. ____ $y = (x - 2)^2 - 5$

10. ____ $y = (x - 5)^2 - 2$

11. ____ $x = (y - 2)^2 - 5$

12. ____ $x = (y - 5)^2 - 2$

a)

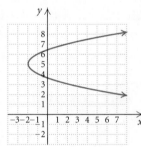

b)

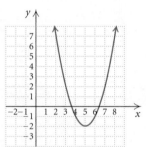

c)

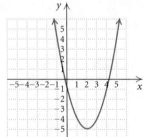

d)

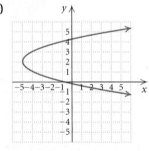

e)

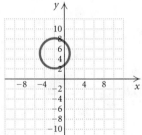

f)

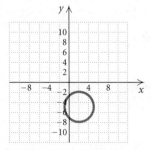

Parabolas

Graph. Be sure to label each vertex.

13. $y = -x^2$

14. $y = 2x^2$

15. $y = -x^2 + 4x - 5$

16. $x = 4 - 3y - y^2$

17. $x = y^2 - 4y + 2$

18. $y = x^2 + 2x + 3$

19. $x = y^2 + 3$

20. $x = -y^2$

21. $x = 2y^2$

22. $x = y^2 - 1$

23. $x = -y^2 - 4y$

24. $x = y^2 + 3y$

25. $y = x^2 - 2x + 1$

26. $y = x^2 + 2x + 1$

27. $x = -\frac{1}{2}y^2$

28. $y = -\frac{1}{2}x^2$

29. $x = -y^2 + 2y - 1$

30. $x = -y^2 - 2y + 3$

31. $x = -2y^2 - 4y + 1$

32. $x = 2y^2 + 4y - 1$

Circles

Find an equation of the circle satisfying the given conditions.

33. Center $(0, 0)$, radius 8

34. Center $(0, 0)$, radius 11

35. Center $(7, 3)$, radius $\sqrt{6}$

36. Center $(5, 6)$, radius $\sqrt{11}$

37. Center $(-4, 3)$, radius $3\sqrt{2}$

38. Center $(-2, 7)$, radius $2\sqrt{5}$

39. Center $(-5, -8)$, radius $10\sqrt{3}$

40. Center $(-7, -2)$, radius $5\sqrt{2}$

Aha! 41. Center $(0, 0)$, passing through $(-3, 4)$

42. Center $(0, 0)$, passing through $(11, -10)$

43. Center $(-4, 1)$, passing through $(-2, 5)$

44. Center $(-1, -3)$, passing through $(-4, 2)$

Find the center and the radius of each circle. Then graph the circle.

45. $x^2 + y^2 = 1$

46. $x^2 + y^2 = 25$

47. $(x + 1)^2 + (y + 3)^2 = 49$

48. $(x - 2)^2 + (y + 3)^2 = 100$

49. $(x - 4)^2 + (y + 3)^2 = 10$

50. $(x + 5)^2 + (y - 1)^2 = 15$

51. $x^2 + y^2 = 8$

52. $x^2 + y^2 = 20$

53. $(x - 5)^2 + y^2 = \frac{1}{4}$

54. $x^2 + (y - 1)^2 = \frac{1}{25}$

55. $x^2 + y^2 + 8x - 6y - 15 = 0$

56. $x^2 + y^2 + 6x - 4y - 15 = 0$

57. $x^2 + y^2 - 8x + 2y + 13 = 0$

58. $x^2 + y^2 + 6x + 4y + 12 = 0$

59. $x^2 + y^2 + 10y - 75 = 0$

60. $x^2 + y^2 - 8x - 84 = 0$

61. $x^2 + y^2 + 7x - 3y - 10 = 0$

62. $x^2 + y^2 - 21x - 33y + 17 = 0$

63. $36x^2 + 36y^2 = 1$

64. $4x^2 + 4y^2 = 1$

65. Does the graph of an equation of a circle include the point that is the center? Why or why not?

66. Is a point a conic section? Why or why not?

Skill Review

Simplify. Assume all variables represent positive numbers.

67. $\sqrt[4]{48x^7y^{12}}$ [7.3]

68. $\sqrt{y}\,\sqrt[3]{y^2}$ [7.5]

69. $\dfrac{\sqrt{200x^4w^2}}{\sqrt{2w}}$ [7.4]

70. $\dfrac{\sqrt[3]{t}}{\sqrt[10]{t}}$ [7.5]

71. $\sqrt{8} - 2\sqrt{2} + \sqrt{12}$ [7.5]

72. $(3 + \sqrt{2})(4\sqrt{3} - \sqrt{2})$ [7.5]

Synthesis

73. On a piece of graph paper, draw a line and a point not on the line. Then plot several points that are equidistant from the point and the line. What shape do the points appear to form? How could you confirm this?

74. If an equation has two variable terms with the same degree, can its graph be a parabola? Why or why not?

Find an equation of a circle satisfying the given conditions.

75. Center $(3, -5)$ and tangent to (touching at one point) the y-axis

76. Center $(-7, -4)$ and tangent to the x-axis

77. The endpoints of a diameter are $(7, 3)$ and $(-1, -3)$.

78. Center $(-3, 5)$ with a circumference of 8π units

79. *Wrestling.* The equation $x^2 + y^2 = \frac{81}{4}$, where x and y represent the number of meters from the center, can be used to draw the outer circle on a wrestling mat used in International, Olympic, and World Championship wrestling. The equation $x^2 + y^2 = 16$ can be used to draw the inner edge of the red zone. Find the area of the red zone.

Source: Based on data from the Government of Western Australia

80. *Snowboarding.* Each side edge of the Burton X8 155 snowboard is an arc of a circle with a "running length" of 1180 mm and a "sidecut depth" of 23 mm (see the figure below).

Source: evogear.com

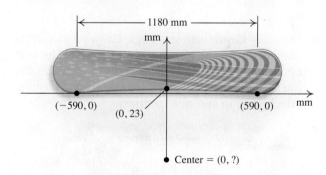

a) Using the coordinates shown, locate the center of the circle. (*Hint:* Equate distances.)

b) What radius is used for the edge of the board?

81. *Snowboarding.* The Never Summer Infinity 149 snowboard has a running length of 1160 mm and a sidecut depth of 23.5 mm (see Exercise 80). What radius is used for the edge of this snowboard?

Source: neversummer.com

82. *Skiing.* The Rossignol Experience 98 ski, when lying flat and viewed from above, has edges that are arcs of a circle. (Actually, each edge is made of two arcs of slightly different radii. The arc for the rear half of the ski edge has a slightly larger radius.)

Source: rossignol.com

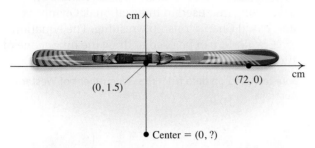

a) Using the coordinates shown, locate the center of the circle. (*Hint*: Equate distances.)
b) What radius is used for the arc passing through $(0, 1.5)$ and $(72, 0)$?

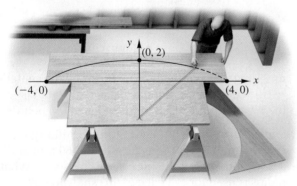

83. *Doorway Construction.* Engle Carpentry needs to cut an arch for the top of an entranceway. The arch needs to be 8 ft wide and 2 ft high. To draw the arch, the carpenters will use a stretched string with chalk attached at an end as a compass.

a) Using a coordinate system, locate the center of the circle.
b) What radius should the carpenters use to draw the arch?

84. *Archaeology.* During an archaeological dig, Estella finds the bowl fragment shown below. What was the original diameter of the bowl?

85. *Ferris Wheel Design.* A ferris wheel has a radius of 24.3 ft. Assuming that the center is 30.6 ft above the base of the ferris wheel and that the origin is below the center, as in the following figure, find an equation of the circle.

86. Use a graph of $x = y^2 - y - 6$ to approximate to the nearest tenth the solutions of each of the following.

a) $y^2 - y - 6 = 2$
b) $y^2 - y - 6 = -3$

87. *Power of a Motor.* The horsepower of a certain kind of engine is given by the formula

$$H = \frac{D^2 N}{2.5},$$

where N is the number of cylinders and D is the diameter, in inches, of each piston. Graph this equation, assuming that $N = 6$ (a six-cylinder engine). Let D run from 2.5 to 8. Then use the graph to estimate the diameter of each piston in a six-cylinder 120-horsepower engine.

88. If the equation $x^2 + y^2 - 6x + 2y - 6 = 0$ is written as $y^2 + 2y + (x^2 - 6x - 6) = 0$, it can be regarded as quadratic in y.

a) Use the quadratic formula to solve for y.
b) Show that the graph of your answer to part (a) coincides with the graph in the Technology Connection on p. 659.

89. Why should a graphing calculator's window be "squared" before graphing a circle?

YOUR TURN ANSWERS: SECTION 10.1

1.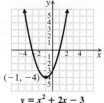
$y = x^2 + 2x - 3$

2.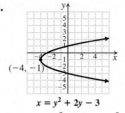
$(-4, -1)$
$x = y^2 + 2y - 3$

3.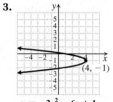
$(4, -1)$
$x = -3y^2 - 6y + 1$

4. $(x - (-2))^2 + (y - 7)^2 = 81,$
or $(x + 2)^2 + (y - 7)^2 = 81$

5. $(0, 3); \sqrt{5}$
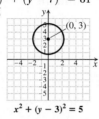
$(0, 3)$
$x^2 + (y - 3)^2 = 5$

10.2 Conic Sections: Ellipses

Ellipses Centered at (0, 0) ▪ Ellipses Centered at (h, k)

Study Skills

Preparing for the Final Exam

It is never too early to begin studying for a final exam. If you have at least three days, consider the following:

- Reviewing the highlighted or boxed information in each chapter;
- Studying the Chapter Tests, Review Exercises, Cumulative Reviews, and Study Summaries;
- Re-taking on your own all quizzes and tests;
- Attending any review sessions being offered;
- Organizing or joining a study group;
- Asking a tutor or a professor about specific trouble spots;
- Asking for previous final exams (and answers) to work for practice.

When a cone is sliced at an angle, as shown below, the conic section formed is an *ellipse*. To draw an ellipse, stick two tacks in a piece of cardboard. Then tie a loose string to the tacks, place a pencil as shown, and draw an oval by moving the pencil while stretching the string tight.

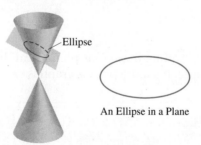
An Ellipse in a Plane

Ellipses Centered at (0, 0)

An **ellipse** is defined as the set of all points in a plane for which the sum of the distances from two fixed points F_1 and F_2 is constant. The points F_1 and F_2 are called **foci** (pronounced fō-sī), the plural of focus. In the figure above, the tacks are at the foci and the length of the string is the constant sum of the distances from the tacks to the pencil. The midpoint of the segment F_1F_2 is the **center**. The equation of an ellipse follows. Its derivation is outlined in Exercise 51.

EQUATION OF AN ELLIPSE CENTERED AT THE ORIGIN

The equation of an ellipse centered at the origin and symmetric with respect to both axes is

$$\frac{x^2}{a^2} + \frac{y^2}{b^2} = 1, \quad a, b > 0. \quad \text{(Standard form)}$$

To graph an ellipse centered at the origin, it helps to first find the intercepts. If we replace x with 0, we can find the y-intercepts:

$$\frac{0^2}{a^2} + \frac{y^2}{b^2} = 1$$

$$\frac{y^2}{b^2} = 1$$

$$y^2 = b^2 \quad \text{or} \quad y = \pm b.$$

Thus the y-intercepts are $(0, b)$ and $(0, -b)$. Similarly, the x-intercepts are $(a, 0)$ and $(-a, 0)$. If $a > b$, the ellipse is said to be horizontal and $(-a, 0)$ and $(a, 0)$ are referred to as the **vertices** (singular, **vertex**). If $b > a$, the ellipse is said to be vertical and $(0, -b)$ and $(0, b)$ are then the vertices.

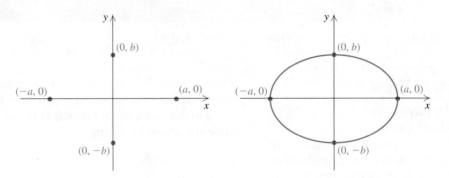

Plotting these four points and drawing an oval-shaped curve, we graph the ellipse. If a more precise graph is desired, we can plot more points.

USING a AND b TO GRAPH AN ELLIPSE

For the ellipse

$$\frac{x^2}{a^2} + \frac{y^2}{b^2} = 1,$$

the x-intercepts are $(-a, 0)$ and $(a, 0)$. The y-intercepts are $(0, -b)$ and $(0, b)$. For $a^2 > b^2$, the ellipse is horizontal. For $b^2 > a^2$, the ellipse is vertical.

EXAMPLE 1 Graph the ellipse

$$\frac{x^2}{4} + \frac{y^2}{9} = 1.$$

SOLUTION Note that

$$\frac{x^2}{4} + \frac{y^2}{9} = \frac{x^2}{2^2} + \frac{y^2}{3^2}. \qquad \text{Identifying } a \text{ and } b. \text{ Since } b^2 > a^2, \text{ the ellipse is vertical.}$$

Since $a = 2$ and $b = 3$, the x-intercepts are $(-2, 0)$ and $(2, 0)$, and the y-intercepts are $(0, -3)$ and $(0, 3)$. We plot these points and connect them with an oval-shaped curve. To plot two other points, we let $x = 1$ and solve for y:

$$\frac{1^2}{4} + \frac{y^2}{9} = 1$$

$$36\left(\frac{1}{4} + \frac{y^2}{9}\right) = 36 \cdot 1$$

$$36 \cdot \frac{1}{4} + 36 \cdot \frac{y^2}{9} = 36$$

$$9 + 4y^2 = 36$$

$$4y^2 = 27$$

$$y^2 = \frac{27}{4}$$

$$y = \pm\sqrt{\frac{27}{4}}$$

$$y \approx \pm 2.6.$$

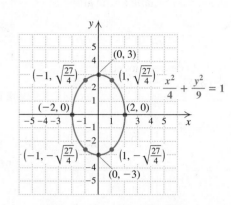

1. Graph the ellipse

$$\frac{x^2}{25} + \frac{y^2}{9} = 1.$$

Thus, $(1, 2.6)$ and $(1, -2.6)$ can also be used to draw the graph. We leave it to the student to confirm that $(-1, 2.6)$ and $(-1, -2.6)$ also appear on the graph.

YOUR TURN

EXAMPLE 2 Graph: $4x^2 + 25y^2 = 100$.

Student Notes

Note that any equation of the form $Ax^2 + By^2 = C$ (with $A \neq B$ and $A, B > 0$) can be rewritten as an equivalent equation in standard form. The graph is an ellipse.

SOLUTION To write the equation in standard form, we divide both sides by 100 to get 1 on the right side:

$$\frac{4x^2 + 25y^2}{100} = \frac{100}{100} \qquad \text{Dividing by 100 to get 1 on the right side}$$

$$\left.\begin{array}{c} \dfrac{4x^2}{100} + \dfrac{25y^2}{100} = 1 \\[2ex] \dfrac{x^2}{25} + \dfrac{y^2}{4} = 1 \end{array}\right\} \qquad \text{Simplifying}$$

$$\frac{x^2}{5^2} + \frac{y^2}{2^2} = 1. \qquad a = 5, b = 2$$

The x-intercepts are $(-5, 0)$ and $(5, 0)$, and the y-intercepts are $(0, -2)$ and $(0, 2)$. We plot the intercepts and connect them with an oval-shaped curve. Other points can also be computed and plotted.

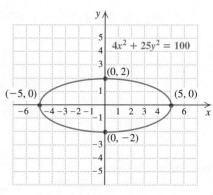

2. Graph: $4x^2 + y^2 = 4$.

YOUR TURN

Ellipses Centered at (h, k)

Horizontal and vertical translations can be used to graph ellipses that are not centered at the origin.

Student Notes

The graph of

$$\frac{(x-h)^2}{a^2} + \frac{(y-k)^2}{b^2} = 1$$

is the same shape as the graph of

$$\frac{x^2}{a^2} + \frac{y^2}{b^2} = 1,$$

with its center moved from $(0,0)$ to (h,k).

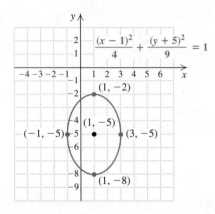

3. Graph the ellipse

$$\frac{(x+2)^2}{9} + \frac{(y-3)^2}{4} = 1.$$

EQUATION OF AN ELLIPSE CENTERED AT (h,k)

The standard form of a horizontal or vertical ellipse centered at (h,k) is

$$\frac{(x-h)^2}{a^2} + \frac{(y-k)^2}{b^2} = 1.$$

The vertices are $(h+a,k)$ and $(h-a,k)$ if horizontal; $(h,k+b)$ and $(h,k-b)$ if vertical.

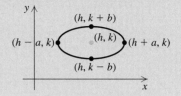

EXAMPLE 3   Graph the ellipse

$$\frac{(x-1)^2}{4} + \frac{(y+5)^2}{9} = 1.$$

SOLUTION   Note that

$$\frac{(x-1)^2}{4} + \frac{(y+5)^2}{9} = \frac{(x-1)^2}{2^2} + \frac{(y+5)^2}{3^2}.$$

Thus, $a=2$ and $b=3$. To determine the center of the ellipse, (h,k), note that

$$\frac{(x-1)^2}{2^2} + \frac{(y+5)^2}{3^2} = \frac{(x-1)^2}{2^2} + \frac{(y-(-5))^2}{3^2}.$$

Thus the center is $(1,-5)$. We plot points 2 units to the left and right of center, as well as 3 units above and below center. These are the points $(3,-5)$, $(-1,-5)$, $(1,-2)$, and $(1,-8)$. The graph of the ellipse is shown at left.

Note that this ellipse is the same as the ellipse in Example 1 but translated 1 unit to the right and 5 units down.

YOUR TURN

Ellipses have many applications. Communications satellites move in elliptical orbits with the earth as a focus while the earth itself follows an elliptical path around the sun. A medical instrument, the lithotripter, uses shock waves originating at one focus to crush a kidney stone located at the other focus.

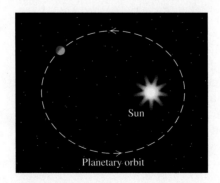
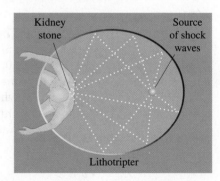

Chapter Resource:
Collaborative Activity, p. 689

In some buildings, an ellipsoidal ceiling creates a "whispering gallery" in which a person at one focus can whisper and still be heard clearly at the other focus. This happens because sound waves coming from one focus are all reflected to the other focus. Similarly, light waves bouncing off an ellipsoidal mirror are used in a dentist's or surgeon's reflector light. The light source is located at one focus while the patient's mouth or surgical field is at the other.

Technology Connection

To graph an ellipse on a graphing calculator, we solve for y and graph two functions.

To illustrate, let's check Example 2:

$$4x^2 + 25y^2 = 100$$
$$25y^2 = 100 - 4x^2$$
$$y^2 = 4 - \tfrac{4}{25}x^2$$
$$y = \pm\sqrt{4 - \tfrac{4}{25}x^2}.$$

Using a squared window, we have our check:

$$y_1 = -\sqrt{4 - \tfrac{4}{25}x^2}, \quad y_2 = \sqrt{4 - \tfrac{4}{25}x^2}$$

On many calculators, pressing **APPS** and selecting Conics and then Ellipse accesses a program in which equations in Standard Form can be graphed directly.

10.2 EXERCISE SET

FOR EXTRA HELP

MyMathLab® MathXL
PRACTICE WATCH READ REVIEW

Vocabulary and Reading Check

For each term, write the letter of the appropriate labeled part of the drawing.

1. ____ Ellipse
2. ____ Focus
3. ____ Center
4. ____ Vertex

Concept Reinforcement

Classify each of the following statements as either true or false.

5. The graph of $\dfrac{x^2}{25} + \dfrac{y^2}{50} = 1$ is a vertical ellipse.

6. The graph of $\dfrac{x^2}{25} - \dfrac{y^2}{9} = 1$ is a horizontal ellipse.

7. The graph of $\dfrac{x^2}{9} + \dfrac{y^2}{25} = 1$ includes the points $(-3, 0)$ and $(3, 0)$.

8. The graph of $\dfrac{(x+3)^2}{25} + \dfrac{(y-2)^2}{36} = 1$ is an ellipse centered at $(-3, 2)$.

Ellipses Centered at $(0, 0)$

Graph.

9. $\dfrac{x^2}{1} + \dfrac{y^2}{4} = 1$

10. $\dfrac{x^2}{4} + \dfrac{y^2}{1} = 1$

11. $\dfrac{x^2}{25} + \dfrac{y^2}{9} = 1$

12. $\dfrac{x^2}{16} + \dfrac{y^2}{25} = 1$

13. $4x^2 + 9y^2 = 36$

14. $9x^2 + 4y^2 = 36$

15. $16x^2 + 9y^2 = 144$

16. $9x^2 + 16y^2 = 144$

17. $2x^2 + 3y^2 = 6$

18. $5x^2 + 7y^2 = 35$

Aha! 19. $5x^2 + 5y^2 = 125$

20. $8x^2 + 5y^2 = 80$

21. $3x^2 + 7y^2 - 63 = 0$

22. $3x^2 + 3y^2 - 48 = 0$

23. $16x^2 = 16 - y^2$

24. $9y^2 = 9 - x^2$

25. $16x^2 + 25y^2 = 1$

26. $9x^2 + 4y^2 = 1$

Ellipses Centered at (h, k)

Graph.

27. $\dfrac{(x - 3)^2}{9} + \dfrac{(y - 2)^2}{25} = 1$

28. $\dfrac{(x - 2)^2}{25} + \dfrac{(y - 4)^2}{9} = 1$

29. $\dfrac{(x + 4)^2}{16} + \dfrac{(y - 3)^2}{49} = 1$

30. $\dfrac{(x + 5)^2}{4} + \dfrac{(y - 2)^2}{36} = 1$

31. $12(x - 1)^2 + 3(y + 4)^2 = 48$
(*Hint*: Divide both sides by 48.)

32. $4(x - 6)^2 + 9(y + 2)^2 = 36$

Aha! **33.** $4(x + 3)^2 + 4(y + 1)^2 - 10 = 90$

34. $9(x + 6)^2 + (y + 2)^2 - 20 = 61$

35. How can you tell from the equation of an ellipse whether its graph is horizontal or vertical?

36. Can an ellipse ever be the graph of a function? Why or why not?

Skill Review

Solve.

37. $x^2 - 5x + 3 = 0$ [8.2]

38. $\log_x 81 = 4$ [9.6]

39. $\dfrac{4}{x + 2} + \dfrac{3}{2x - 1} = 2$ [6.4]

40. $3 - \sqrt{2x - 1} = 1$ [7.6]

41. $x^2 = 11$ [8.1]

42. $x^2 + 4x = 60$ [5.8]

Synthesis

43. Explain how it is possible to recognize that the graph of $9x^2 + 18x + y^2 - 4y + 4 = 0$ is an ellipse.

44. As the foci get closer to the center of an ellipse, what shape does the graph begin to resemble? Explain why this happens.

Find an equation of an ellipse that contains the following points.

45. $(-9, 0), (9, 0), (0, -11)$, and $(0, 11)$

46. $(-7, 0), (7, 0), (0, -10)$, and $(0, 10)$

47. $(-2, -1), (6, -1), (2, -4)$, and $(2, 2)$

48. $(4, 3), (-6, 3), (-1, -1)$, and $(-1, 7)$

49. *Theatrical Lighting.* A spotlight on a violin soloist casts an ellipse of light on the floor below her that is 6 ft wide and 10 ft long. Find an equation of that ellipse if the performer is in its center, x is the distance from the performer to the nearest side of the ellipse, and y is the distance from the performer to the top of the ellipse.

50. *Astronomy.* The maximum distance of Mars from the sun is 2.48×10^8 mi. Its minimum distance is 3.46×10^7 mi. The sun is one focus of the elliptical orbit. Find the distance from the sun to the other focus.

51. Let $(-c, 0)$ and $(c, 0)$ be the foci of an ellipse. Any point $P(x, y)$ is on the ellipse if the sum of the distances from the foci to P is some constant. Use $2a$ to represent this constant.

a) Show that an equation for the ellipse is given by
$$\frac{x^2}{a^2} + \frac{y^2}{a^2 - c^2} = 1.$$

b) Substitute b^2 for $a^2 - c^2$ to get standard form.

52. *President's Office.* The Oval Office of the President of the United States is an ellipse 31 ft wide and 38 ft long. Show in a sketch precisely where the President and an adviser could sit to best hear each other using the room's acoustics. (*Hint*: See Exercise 51(b) and the discussion following Example 3.)

53. *Dentistry.* The light source in some dental lamps shines against a reflector that is shaped like a portion of an ellipse in which the light source is one focus of the ellipse. Reflected light enters a patient's mouth at the other focus of the ellipse. If the ellipse from which the reflector was formed is 2 ft wide and 6 ft long, how far should the patient's mouth be from the light source? (*Hint:* See Exercise 51(b).)

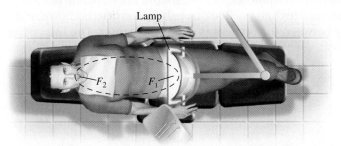

Lamp

54. *Firefighting.* The size and shape of certain forest fires can be approximated as the union of two "half-ellipses." For the blaze modeled below, the equation of the smaller ellipse—the part of the fire moving *into* the wind—is

$$\frac{x^2}{40,000} + \frac{y^2}{10,000} = 1.$$

The equation of the other ellipse—the part moving *with* the wind—is

$$\frac{x^2}{250,000} + \frac{y^2}{10,000} = 1.$$

Determine the width and the length of the fire.

Source for figure: "Predicting Wind-Driven Wild Land Fire Size and Shape," Hal E. Anderson, Research Paper INT-305, U.S. Department of Agriculture, Forest Service, February 1983

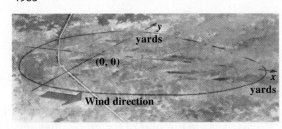

For each of the following equations, complete the square as needed and find an equivalent equation in standard form. Then graph the ellipse.

55. $x^2 - 4x + 4y^2 + 8y - 8 = 0$

56. $4x^2 + 24x + y^2 - 2y - 63 = 0$

57. Use a graphing calculator to check your answers to Exercises 11, 25, 29, and 33.

YOUR TURN ANSWERS: SECTION 10.2

1.

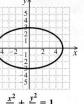

$$\frac{x^2}{25} + \frac{y^2}{9} = 1$$

2.

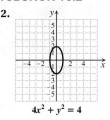

$$4x^2 + y^2 = 4$$

3.

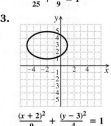

$$\frac{(x + 2)^2}{9} + \frac{(y - 3)^2}{4} = 1$$

QUICK QUIZ: SECTIONS 10.1–10.2

1. Find the center and the radius of the circle:

$$(x + 5)^2 + (y + 1)^2 = 4. \quad [10.1]$$

2. Find an equation of the circle with center $(9, -23)$ and radius $10\sqrt{2}$. [10.1]

Graph.

3. $(x + 3)^2 + (y + 1)^2 = 4$ [10.1]

4. $y = x^2 - 2x - 2$ [10.1]

5. $\dfrac{x^2}{1} + \dfrac{y^2}{25} = 1$ [10.2]

PREPARE TO MOVE ON

1. Solve $xy = 4$ for y. [6.4]

2. Write $6x - 3y = 12$ in the form $y = mx + b$. [2.3]

3. Write $5x = x^2 - 7$ in the form $ax^2 + bx + c = 0, a > 0$. [8.2]

10.3 Conic Sections: Hyperbolas

Hyperbolas   ▪   Hyperbolas (Nonstandard Form)   ▪   Classifying Graphs of Equations

Hyperbolas

A **hyperbola** looks like a pair of parabolas, but the shapes are not quite parabolic. Every hyperbola has two **vertices** and the line through the vertices is known as the **axis**. The point halfway between the vertices is called the **center**. The two curves that comprise a hyperbola are called **branches**.

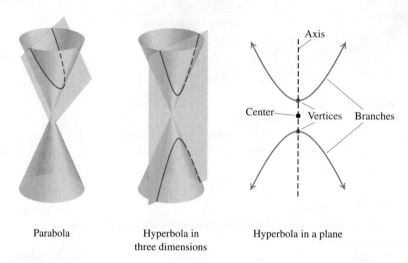

Parabola   Hyperbola in   Hyperbola in a plane
                  three dimensions

EQUATION OF A HYPERBOLA CENTERED AT THE ORIGIN

A hyperbola with its center at the origin* has its equation as follows:

$$\frac{x^2}{a^2} - \frac{y^2}{b^2} = 1 \qquad \text{(Horizontal axis);}$$

$$\frac{y^2}{b^2} - \frac{x^2}{a^2} = 1 \qquad \text{(Vertical axis).}$$

Note that both equations have a positive term and a negative term on the left-hand side. For the discussion that follows, we assume $a, b > 0$.

To graph a hyperbola, it helps to begin by graphing two lines called **asymptotes**. Although the asymptotes themselves are not part of the graph, they serve as guidelines for an accurate drawing.

As a hyperbola gets farther away from the origin, it gets closer and closer to its asymptotes. That is, the larger $|x|$ gets, the closer the graph gets to an asymptote. The asymptotes act to "constrain" the graph of a hyperbola. Parabolas are *not* constrained by any asymptotes.

---

*Hyperbolas with horizontal or vertical axes and centers *not* at the origin are discussed in Exercises 57–62.

ASYMPTOTES OF A HYPERBOLA

For hyperbolas with equations as shown below, the asymptotes are the lines

$$y = \frac{b}{a}x \quad \text{and} \quad y = -\frac{b}{a}x.$$

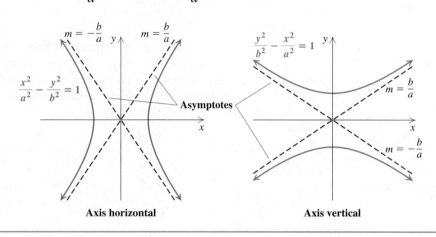

Axis horizontal · Axis vertical

In Section 10.2, we used a and b to determine the width and the length of an ellipse. For hyperbolas, a and b are used to determine the base and the height of a rectangle that can be used as an aid in sketching asymptotes and locating vertices.

EXAMPLE 1 Graph: $\dfrac{x^2}{4} - \dfrac{y^2}{9} = 1.$

SOLUTION Note that

$$\frac{x^2}{4} - \frac{y^2}{9} = \frac{x^2}{2^2} - \frac{y^2}{3^2}, \quad \text{Identifying } a \text{ and } b$$

so $a = 2$ and $b = 3$. The asymptotes are thus

$$y = \frac{3}{2}x \quad \text{and} \quad y = -\frac{3}{2}x.$$

To help us sketch asymptotes and locate vertices, we use a and b—in this case, 2 and 3—to form the pairs $(-2, 3), (2, 3), (2, -3),$ and $(-2, -3)$. We plot these pairs and lightly sketch a rectangle. The asymptotes pass through the corners and, since this is a horizontal hyperbola, the vertices $(-2, 0)$ and $(2, 0)$ are where the rectangle intersects the x-axis. Finally, we draw the hyperbola, as shown below.

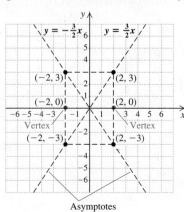

Asymptotes

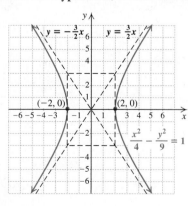

1. Graph: $\dfrac{x^2}{25} - \dfrac{y^2}{4} = 1.$

YOUR TURN

EXAMPLE 2   Graph: $\dfrac{y^2}{36} - \dfrac{x^2}{4} = 1$.

SOLUTION   Note that

$$\frac{y^2}{36} - \frac{x^2}{4} = \frac{y^2}{6^2} - \frac{x^2}{2^2} = 1.$$

> Whether the hyperbola is horizontal or vertical is determined by which term is nonnegative. Here the y^2-term is nonnegative, so the hyperbola is vertical.

Using ± 2 as x-coordinates and ± 6 as y-coordinates, we plot $(2, 6)$, $(2, -6)$, $(-2, 6)$, and $(-2, -6)$, and lightly sketch a rectangle through them. The asymptotes pass through the corners (see the figure on the left below). Since the hyperbola is vertical, its vertices are $(0, 6)$ and $(0, -6)$. Finally, we draw curves through the vertices toward the asymptotes, as shown below.

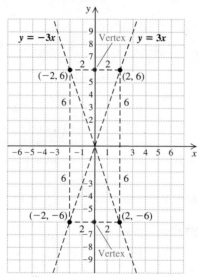 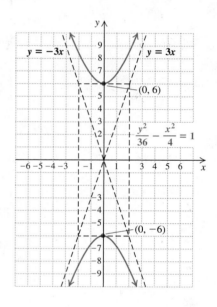

2. Graph: $\dfrac{y^2}{9} - \dfrac{x^2}{16} = 1$.

 YOUR TURN

EXPLORING 🔍 THE CONCEPT

The graphs of $\dfrac{x^2}{4} - \dfrac{y^2}{9} = 1$, $\dfrac{y^2}{9} - \dfrac{x^2}{4} = 1$, and $\dfrac{x^2}{4} + \dfrac{y^2}{9} = 1$ are all related to the rectangle shown at right.

Match each equation with a description of its graph as it relates to the rectangle.

1. $\dfrac{x^2}{4} - \dfrac{y^2}{9} = 1$

2. $\dfrac{y^2}{9} - \dfrac{x^2}{4} = 1$

3. $\dfrac{x^2}{4} + \dfrac{y^2}{9} = 1$

a) The ellipse inscribed in the rectangle

b) The horizontal hyperbola with asymptotes that are diagonals of the rectangle

c) The vertical hyperbola with asymptotes that are diagonals of the rectangle

ANSWERS

1. (b)   **2.** (c)   **3.** (a)

Hyperbolas (Nonstandard Form)

The equations for hyperbolas just examined are the standard ones, but there are other hyperbolas. We consider some of them.

> **EQUATION OF A HYPERBOLA IN NONSTANDARD FORM**
>
> Hyperbolas having the x- and y-axes as asymptotes have equations that can be written in the form
>
> $$xy = c, \quad \text{where } c \text{ is a nonzero constant.}$$

EXAMPLE 3 Graph: $xy = -8$.

SOLUTION We first solve for y:

$$y = -\frac{8}{x}. \qquad \text{Dividing both sides by } x. \text{ Note that } x \neq 0.$$

Next, we find some solutions and form a table. Note that x cannot be 0 and that for large values of $|x|$, the value of y is close to 0. Thus the x- and y-axes serve as asymptotes. We plot the points and draw two curves.

x	y
2	-4
-2	4
4	-2
-4	2
1	-8
-1	8
8	-1
-8	1

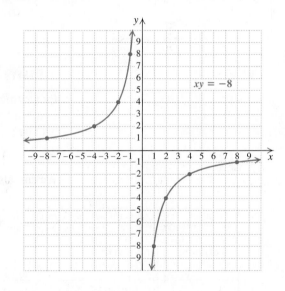

3. Graph: $xy = 6$.

 YOUR TURN

Hyperbolas have many applications. A jet breaking the sound barrier creates a sonic boom with a wave front the shape of a cone. The intersection of the cone with the ground is one branch of a hyperbola. Some comets travel in hyperbolic orbits, and a cross section of many lenses is hyperbolic in shape.

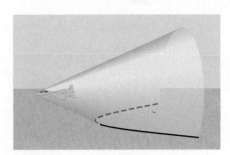

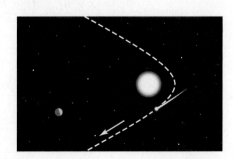

Technology Connection

The procedure used to graph a hyperbola in standard form is similar to that used to draw a circle or an ellipse. Consider the graph of

$$\frac{x^2}{25} - \frac{y^2}{49} = 1.$$

The student should confirm that solving for y yields

$$y_1 = \frac{\sqrt{49x^2 - 1225}}{5} = \frac{7}{5}\sqrt{x^2 - 25}$$

and

$$y_2 = \frac{-\sqrt{49x^2 - 1225}}{5} = -\frac{7}{5}\sqrt{x^2 - 25},$$

or $\quad y_2 = -y_1.$

When the two pieces are drawn on the same squared window, the result is as shown. The gaps occur where the graph is nearly vertical.

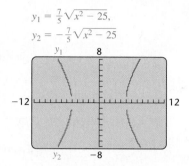

$$y_1 = \frac{7}{5}\sqrt{x^2 - 25},$$
$$y_2 = -\frac{7}{5}\sqrt{x^2 - 25}$$

On many calculators, pressing **APPS** and selecting Conics and then Hyperbola accesses a program in which hyperbolas in standard form can be graphed directly. Graph each of the following.

1. $\dfrac{x^2}{16} - \dfrac{y^2}{60} = 1$ **2.** $16x^2 - 3y^2 = 64$

3. $\dfrac{y^2}{20} - \dfrac{x^2}{64} = 1$ **4.** $45y^2 - 9x^2 = 441$

Classifying Graphs of Equations

By writing an equation of a conic section in a standard form, we can classify its graph as a parabola, a circle, an ellipse, or a hyperbola. Every conic section can also be represented by an equation of the form

$$Ax^2 + By^2 + Cxy + Dx + Ey + F = 0.$$

We can also classify graphs using values of A and B.

Graph	Standard Form	$Ax^2 + By^2 + Cxy + Dx + Ey + F = 0$
Parabola	$y = ax^2 + bx + c;$ Vertical parabola $x = ay^2 + by + c$ Horizontal parabola	Either $A = 0$ or $B = 0$, but not both.
Circle	$x^2 + y^2 = r^2;$ Center at the origin $(x - h)^2 + (y - k)^2 = r^2$ Center at (h, k)	$A = B$, and neither A nor B is 0.
Ellipse	$\dfrac{x^2}{a^2} + \dfrac{y^2}{b^2} = 1;$ Center at the origin $\dfrac{(x - h)^2}{a^2} + \dfrac{(y - k)^2}{b^2} = 1$ Center at (h, k)	$A \neq B$, and A and B have the same sign.
Hyperbola	$\dfrac{x^2}{a^2} - \dfrac{y^2}{b^2} = 1;$ Horizontal hyperbola $\dfrac{y^2}{b^2} - \dfrac{x^2}{a^2} = 1$ Vertical hyperbola	A and B have opposite signs.
	$xy = c$ Asymptotes are axes	Only C and F are nonzero.

Algebraic manipulations are often needed to express an equation in one of the preceding forms.

EXAMPLE 4 Classify the graph of each equation as a circle, an ellipse, a parabola, or a hyperbola. Refer to the above table as needed.

a) $5x^2 = 20 - 5y^2$ **b)** $x + 3 + 8y = y^2$

c) $x^2 = y^2 + 4$ **d)** $x^2 = 16 - 4y^2$

SOLUTION

a) We get the terms with variables on one side by adding $5y^2$ to both sides:

$$5x^2 + 5y^2 = 20.$$

Since *both* x and y are squared, we do not have a parabola. The fact that the squared terms are *added* tells us that we have an ellipse or a circle. Since the coefficients are the same, we factor 5 out of both terms on the left and then divide by 5:

$$5(x^2 + y^2) = 20 \quad \text{Factoring out 5}$$
$$x^2 + y^2 = 4 \quad \text{Dividing both sides by 5}$$
$$x^2 + y^2 = 2^2. \quad \text{This is an equation for a circle.}$$

We see that the graph is a circle centered at the origin with radius 2.

We can also write the equation in the form

$$5x^2 + 5y^2 - 20 = 0. \quad A = 5, B = 5$$

Since $A = B$, the graph is a circle.

b) The equation $x + 3 + 8y = y^2$ has only one variable that is squared, so we solve for the other variable:

$$x = y^2 - 8y - 3. \quad \text{This is an equation for a parabola.}$$

The graph is a horizontal parabola that opens to the right.

We can also write the equation in the form

$$y^2 - x - 8y - 3 = 0. \quad A = 0, B = 1$$

Since $A = 0$ and $B \ne 0$, the graph is a parabola.

c) In $x^2 = y^2 + 4$, both variables are squared, so the graph is not a parabola. We subtract y^2 on both sides and divide by 4 to obtain

$$\frac{x^2}{2^2} - \frac{y^2}{2^2} = 1. \quad \text{This is an equation for a hyperbola.}$$

The subtraction indicates that the graph is a hyperbola. Because it is the x^2-term that is nonnegative, the hyperbola is horizontal.

We can also write the equation in the form

$$x^2 - y^2 - 4 = 0. \quad A = 1, B = -1$$

Since A and B have opposite signs, the graph is a hyperbola.

d) In $x^2 = 16 - 4y^2$, both variables are squared, so the graph cannot be a parabola. We add $4y^2$ to both sides to obtain the following equivalent equation:

$$x^2 + 4y^2 = 16.$$

If the coefficients of the terms were the same, the graph would be a circle, as in part (a). Since they are not, we divide both sides by the constant term, 16:

$$\frac{x^2}{16} + \frac{y^2}{4} = 1. \quad \text{This is an equation for an ellipse.}$$

The graph of this equation is a horizontal ellipse.

Chapter Resource:
Visualizing for Success, p. 688

4. Classify the graph of $7x^2 = 12 + 7y^2$ as a circle, an ellipse, a parabola, or a hyperbola.

We can also write the equation in the form

$$x^2 + 4y^2 - 16 = 0. \qquad A = 1, B = 4$$

Since $A \neq B$ and both A and B are positive, the graph is an ellipse.

↻ YOUR TURN

CONNECTING 🔗 THE CONCEPTS

When graphing equations of conic sections, it is usually helpful to first determine what type of graph the equation represents. We then find the coordinates of key points and equations of lines that determine the shape and the location of the graph.

Graph	Equation	Key Points	Equations of Lines
Parabola	$y = a(x - h)^2 + k$ $x = a(y - k)^2 + h$	Vertex: (h, k) Vertex: (h, k)	Axis of symmetry: $x = h$ Axis of symmetry: $y = k$
Circle	$(x - h)^2 + (y - k)^2 = r^2$	Center: (h, k)	
Ellipse	$\dfrac{x^2}{a^2} + \dfrac{y^2}{b^2} = 1$	x-intercepts: $(-a, 0), (a, 0)$; y-intercepts: $(0, -b), (0, b)$	
Hyperbola	$\dfrac{x^2}{a^2} - \dfrac{y^2}{b^2} = 1$ $\dfrac{y^2}{b^2} - \dfrac{x^2}{a^2} = 1$	Vertices: $(-a, 0), (a, 0)$ Vertices: $(0, -b), (0, b)$	Asymptotes (for both equations): $y = \dfrac{b}{a}x, \quad y = -\dfrac{b}{a}x$
	$xy = c$		Asymptotes: $x = 0, y = 0$

EXERCISES

1. Find the vertex and the axis of symmetry of the graph of
$$y = 3(x - 4)^2 + 1.$$

2. Find the vertex and the axis of symmetry of the graph of
$$x = y^2 + 2y + 3.$$

3. Find the center of the graph of
$$(x - 3)^2 + (y - 2)^2 = 5.$$

4. Find the center of the graph of
$$x^2 + 6x + y^2 + 10y = 12.$$

5. Find the x-intercepts and the y-intercepts of the graph of
$$\frac{x^2}{144} + \frac{y^2}{81} = 1.$$

6. Find the vertices of the graph of
$$\frac{x^2}{9} - \frac{y^2}{121} = 1.$$

7. Find the vertices of the graph of
$$4y^2 - x^2 = 4.$$

8. Find the asymptotes of the graph of
$$\frac{y^2}{9} - \frac{x^2}{4} = 1.$$

10.3 EXERCISE SET

FOR EXTRA HELP

 MyMathLab® Math XL
PRACTICE   WATCH   READ   REVIEW

Vocabulary and Reading Check

For each term, write the letter of the appropriate labeled part of the drawing.

1. ____ Asymptote
2. ____ Axis
3. ____ Branch
4. ____ Center
5. ____ Hyperbola
6. ____ Vertex

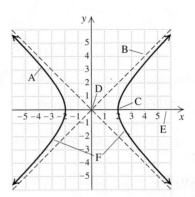

Hyperbolas

Graph each hyperbola. Label all vertices and sketch all asymptotes.

7. $\dfrac{y^2}{16} - \dfrac{x^2}{16} = 1$

8. $\dfrac{x^2}{9} - \dfrac{y^2}{9} = 1$

9. $\dfrac{x^2}{4} - \dfrac{y^2}{25} = 1$

10. $\dfrac{y^2}{16} - \dfrac{x^2}{9} = 1$

11. $\dfrac{y^2}{36} - \dfrac{x^2}{9} = 1$

12. $\dfrac{x^2}{25} - \dfrac{y^2}{36} = 1$

13. $y^2 - x^2 = 25$

14. $x^2 - y^2 = 4$

15. $25x^2 - 16y^2 = 400$

16. $4y^2 - 9x^2 = 36$

Hyperbolas (Nonstandard Form)

Graph.

17. $xy = -6$

18. $xy = 8$

19. $xy = 4$

20. $xy = -9$

21. $xy = -2$

22. $xy = -1$

23. $xy = 1$

24. $xy = 2$

Classifying Graphs of Equations

Classify each of the following as the equation of a circle, an ellipse, a parabola, or a hyperbola.

25. $x^2 + y^2 - 6x + 10y - 40 = 0$

26. $y - 4 = 2x^2$

27. $9x^2 + 4y^2 - 36 = 0$

28. $x + 3y = 2y^2 - 1$

29. $4x^2 - 9y^2 - 72 = 0$

30. $y^2 + x^2 = 8$

31. $y^2 = 20 - x^2$

32. $2y + 13 + x^2 = 8x - y^2$

33. $x - 10 = y^2 - 6y$

34. $y = \dfrac{5}{x}$

35. $x - \dfrac{3}{y} = 0$

36. $9x^2 = 9 - y^2$

37. $y + 6x = x^2 + 5$

38. $x^2 = 49 + y^2$

39. $25y^2 = 100 + 4x^2$

40. $3x^2 + 5y^2 + x^2 = y^2 + 49$

41. $3x^2 + y^2 - x = 2x^2 - 9x + 10y + 40$

42. $4y^2 + 20x^2 + 1 = 8y - 5x^2$

43. $16x^2 + 5y^2 - 12x^2 + 8y^2 - 3x + 4y = 568$

44. $56x^2 - 17y^2 = 234 - 13x^2 - 38y^2$

45. Is it possible for a hyperbola to represent the graph of a function? Why or why not?

46. Explain how the equation of a hyperbola differs from the equation of an ellipse.

Skill Review

Factor completely.

47. $16 - y^4$ [5.5]

48. $9x^2y^2 - 30xy + 25$ [5.5]

49. $10c^3 - 80c^2 + 150c$ [5.4]

50. $x^3 + x^2 + 3x + 3$ [5.3]

51. $8t^4 - 8t$ [5.6]

52. $6a^2 + 11ab - 10b^2$ [5.4]

Synthesis

53. What is it in the equation of a hyperbola that controls how wide open the branches are? Explain your reasoning.

54. If, in

$$\frac{x^2}{a^2} - \frac{y^2}{b^2} = 1,$$

$a = b$, what are the asymptotes of the graph? Why?

Find an equation of a hyperbola satisfying the given conditions.

55. Having intercepts $(0, 6)$ and $(0, -6)$ and asymptotes $y = 3x$ and $y = -3x$

56. Having intercepts $(8, 0)$ and $(-8, 0)$ and asymptotes $y = 4x$ and $y = -4x$

The standard form for equations of horizontal or vertical hyperbolas centered at (h, k) are as follows:

$$\frac{(x - h)^2}{a^2} - \frac{(y - k)^2}{b^2} = 1$$

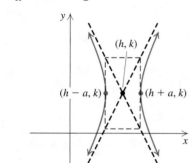

$$\frac{(y - k)^2}{b^2} - \frac{(x - h)^2}{a^2} = 1$$

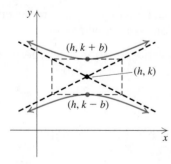

The vertices are as labeled and the asymptotes are

$$y - k = \frac{b}{a}(x - h) \quad \text{and} \quad y - k = -\frac{b}{a}(x - h).$$

For each of the following equations of hyperbolas, complete the square, if necessary, and write in standard form. Find the center, the vertices, and the asymptotes. Then graph the hyperbola.

57. $\dfrac{(x - 5)^2}{36} - \dfrac{(y - 2)^2}{25} = 1$

58. $\dfrac{(x - 2)^2}{9} - \dfrac{(y - 1)^2}{4} = 1$

59. $8(y + 3)^2 - 2(x - 4)^2 = 32$

60. $25(x - 4)^2 - 4(y + 5)^2 = 100$

61. $4x^2 - y^2 + 24x + 4y + 28 = 0$

62. $4y^2 - 25x^2 - 8y - 100x - 196 = 0$

63. Use a graphing calculator to check your answers to Exercises 13, 25, 31, and 57.

64. *Research.* What conic sections have been used to model paths of comets? Find what comets have recently passed close to the earth, and describe the path of each.

➥ **YOUR TURN ANSWERS: SECTION 10.3**

1.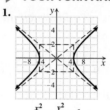
$\dfrac{x^2}{25} - \dfrac{y^2}{4} = 1$

2.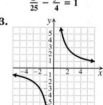
$\dfrac{y^2}{9} - \dfrac{x^2}{16} = 1$

3.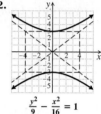
$xy = 6$

4. Hyperbola

QUICK QUIZ: SECTIONS 10.1–10.3

Graph.

1. $(x - 2)^2 + y^2 = 16$ [10.1]   **2.** $xy = 6$ [10.3]

3. $\dfrac{x^2}{36} + \dfrac{y^2}{4} = 1$ [10.2]   **4.** $\dfrac{x^2}{36} - \dfrac{y^2}{4} = 1$ [10.3]

5. $x = -y^2 - 4y + 5$ [10.1]

PREPARE TO MOVE ON

Solve.

1. $5x + 2y = -3,$
 $2x + 3y = 12$ [3.2]

2. $3x - y = 2,$
 $y = 2x + 1$ [3.2]

3. $\frac{3}{4}x^2 + x^2 = 7$ [8.2]

4. $3x^2 + 10x - 8 = 0$ [8.2]

5. $x^2 - 3x - 1 = 0$ [8.2]

6. $x^2 + \dfrac{25}{x^2} = 26$ [8.5]

Mid-Chapter Review

Parabolas, circles, ellipses, and hyperbolas are all conic sections, that is, curves formed by cross sections of cones. Section 10.3 contains a summary of the characteristics of the graphs of these conic sections.

GUIDED SOLUTIONS

1. Find the center and the radius:

$$x^2 + y^2 - 4x + 2y = 6. \quad [10.1]$$

Solution

$$\left(x^2 - 4x\right) + \left(y^2 + \boxed{}\right) = 6$$

$$\left(x^2 - 4x + \boxed{}\right) + \left(y^2 + 2y + \boxed{}\right) = 6 + \boxed{} + \boxed{}$$

$$\left(x - \boxed{}\right)^2 + \left(y + \boxed{}\right)^2 = 11$$

The center of the circle is $(\boxed{}, \boxed{})$.

The radius is $\boxed{}$.

2. Classify the equation as representing a circle, an ellipse, a parabola, or a hyperbola:

$$x^2 - \frac{y^2}{25} = 1. \quad [10.1], [10.2], [10.3]$$

Solution

To answer this, complete the following:

a) Is there both an x^2-term and a y^2-term? $\boxed{}$
b) Do both the x^2-term and the y^2-term have the same sign? $\boxed{}$
c) The graph of the equation is a(n) $\boxed{}$.

MIXED REVIEW

3. Find an equation of the circle with center $(-4, 9)$ and radius $2\sqrt{5}$. [10.1]

4. Find the center and the radius of the graph of $x^2 - 10x + y^2 + 2y = 10$. [10.1]

Classify each of the following as the graph of a parabola, a circle, an ellipse, or a hyperbola. Then graph. [10.1], [10.2], [10.3]

5. $x^2 + y^2 = 36$

6. $y = x^2 - 5$

7. $\dfrac{x^2}{25} + \dfrac{y^2}{49} = 1$

8. $\dfrac{x^2}{25} - \dfrac{y^2}{49} = 1$

9. $x = (y + 3)^2 + 2$

10. $4x^2 + 9y^2 = 36$

11. $xy = -4$

12. $(x + 2)^2 + (y - 3)^2 = 1$

13. $x^2 + y^2 - 8y - 20 = 0$

14. $x = y^2 + 2y$

15. $16y^2 - x^2 = 16$

16. $x = \dfrac{9}{y}$

10.4   Nonlinear Systems of Equations

Systems Involving One Nonlinear Equation   ▪   Systems of Two Nonlinear Equations   ▪   Problem Solving

We now consider systems of two equations in which at least one equation is nonlinear.

Systems Involving One Nonlinear Equation

Suppose that a system consists of an equation of a circle and an equation of a line. The figures below represent three ways in which the circle and the line can intersect. We see that such a system will have 0, 1, or 2 real solutions.

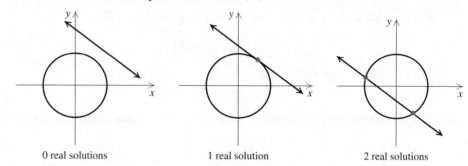

0 real solutions   1 real solution   2 real solutions

Recall that *graphing, elimination,* and *substitution* were all used to solve systems of linear equations. To solve systems in which one equation is of first degree and one is of second degree, it is preferable to use the *substitution* method.

EXAMPLE 1   Solve the system

$$x^2 + y^2 = 25, \quad (1) \qquad \text{(The graph is a circle.)}$$
$$3x - 4y = 0. \quad (2) \qquad \text{(The graph is a line.)}$$

SOLUTION   First, we solve the linear equation, (2), for x:

$$x = \tfrac{4}{3}y. \quad (3) \qquad \text{We could have solved for } y \text{ instead.}$$

Then we substitute $\tfrac{4}{3}y$ for x in equation (1) and solve for y:

$$\left(\tfrac{4}{3}y\right)^2 + y^2 = 25$$
$$\tfrac{16}{9}y^2 + y^2 = 25$$
$$\tfrac{25}{9}y^2 = 25$$
$$y^2 = 9 \qquad \text{Multiplying both sides by } \tfrac{9}{25}$$
$$y = \pm 3. \qquad \text{Using the principle of square roots}$$

Now we substitute these numbers for y in equation (3) and solve for x:

for $y = 3$, $x = \tfrac{4}{3}(3) = 4$;   The ordered pair is $(4, 3)$.

for $y = -3$, $x = \tfrac{4}{3}(-3) = -4$.   The ordered pair is $(-4, -3)$.

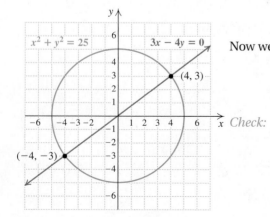

Check: For $(4, 3)$:

$$\begin{array}{c|c} x^2 + y^2 = 25 & \\ \hline 4^2 + 3^2 & 25 \\ 16 + 9 & \\ 25 \overset{?}{=} 25 & \text{TRUE} \end{array} \qquad \begin{array}{c|c} 3x - 4y = 0 & \\ \hline 3(4) - 4(3) & 0 \\ 12 - 12 & \\ 0 \overset{?}{=} 0 & \text{TRUE} \end{array}$$

1. Solve the system

$$x^2 + y^2 = 169,$$
$$x - 2y = 2.$$

We leave it to the student to confirm that $(-4, -3)$ also checks in both equations. The pairs $(4, 3)$ and $(-4, -3)$ check, so they are solutions. The graph on the preceding page serves as a check. Intersections occur at $(4, 3)$ and $(-4, -3)$.

YOUR TURN

EXAMPLE 2 Solve the system

$$y + 3 = 2x, \quad (1) \quad \text{(A first-degree equation)}$$
$$x^2 + 2xy = -1. \quad (2) \quad \text{(A second-degree equation)}$$

SOLUTION First, we solve the linear equation (1) for y:

$$y = 2x - 3. \quad (3)$$

Then we substitute $2x - 3$ for y in equation (2) and solve for x:

$$x^2 + 2x(2x - 3) = -1$$
$$x^2 + 4x^2 - 6x = -1$$
$$5x^2 - 6x + 1 = 0$$
$$(5x - 1)(x - 1) = 0 \quad \text{Factoring}$$
$$5x - 1 = 0 \quad or \quad x - 1 = 0 \quad \text{Using the principle of zero products}$$
$$x = \tfrac{1}{5} \quad or \quad x = 1.$$

Student Notes

Be sure to either list each solution of a system as an ordered pair or separately state the value of each variable.

Now we substitute these numbers for x in equation (3) and solve for y:

for $x = \tfrac{1}{5}$, $y = 2\left(\tfrac{1}{5}\right) - 3 = -\tfrac{13}{5}$; The ordered pair is $\left(\tfrac{1}{5}, -\tfrac{13}{5}\right)$.
for $x = 1$, $y = 2(1) - 3 = -1$. The ordered pair is $(1, -1)$.

You can confirm that $\left(\tfrac{1}{5}, -\tfrac{13}{5}\right)$ and $(1, -1)$ check, so they are both solutions.

YOUR TURN

2. Solve the system

$$y^2 = xy + 3,$$
$$y = 2x - 1.$$

EXAMPLE 3 Solve the system

$$x + y = 5, \quad (1) \quad \text{(The graph is a line.)}$$
$$y = 3 - x^2. \quad (2) \quad \text{(The graph is a parabola.)}$$

SOLUTION We substitute $3 - x^2$ for y in the first equation:

$$x + 3 - x^2 = 5$$
$$-x^2 + x - 2 = 0 \quad \text{Adding } -5 \text{ to both sides and rearranging}$$
$$x^2 - x + 2 = 0. \quad \text{Multiplying both sides by } -1$$

Since $x^2 - x + 2$ does not factor, we need the quadratic formula:

$$x = \frac{-b \pm \sqrt{b^2 - 4ac}}{2a}$$
$$= \frac{-(-1) \pm \sqrt{(-1)^2 - 4 \cdot 1 \cdot 2}}{2(1)} \quad \text{Substituting}$$
$$= \frac{1 \pm \sqrt{1 - 8}}{2} = \frac{1 \pm \sqrt{-7}}{2} = \frac{1}{2} \pm \frac{\sqrt{7}}{2}i.$$

Solving equation (1) for y, we have $y = 5 - x$. Substituting values for x gives

$$y = 5 - \left(\frac{1}{2} + \frac{\sqrt{7}}{2}i\right) = \frac{9}{2} - \frac{\sqrt{7}}{2}i \quad \text{and}$$
$$y = 5 - \left(\frac{1}{2} - \frac{\sqrt{7}}{2}i\right) = \frac{9}{2} + \frac{\sqrt{7}}{2}i.$$

The solutions are

$$\left(\frac{1}{2} + \frac{\sqrt{7}}{2}\,i, \frac{9}{2} - \frac{\sqrt{7}}{2}\,i\right) \quad \text{and} \quad \left(\frac{1}{2} - \frac{\sqrt{7}}{2}\,i, \frac{9}{2} + \frac{\sqrt{7}}{2}\,i\right).$$

There are no real-number solutions. Note in the figure at right that the graphs do not intersect. Getting only nonreal solutions indicates that the graphs do not intersect.

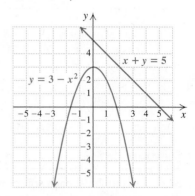

3. Solve the system

$$x = \tfrac{1}{2}y^2,$$
$$x = y - 3.$$

YOUR TURN

Systems of Two Nonlinear Equations

We now consider systems of two second-degree equations. Graphs of such systems can involve any two nonlinear conic sections. The following figure shows some ways in which a circle and a hyperbola can intersect.

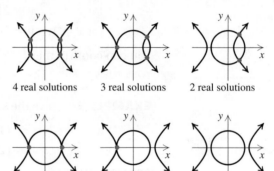

To solve systems of two second-degree equations, we either substitute or eliminate. The elimination method is generally better when both equations are of the form $Ax^2 + By^2 = C$. Then we can eliminate an x^2-term or a y^2-term in a manner similar to the procedure used for systems of linear equations.

EXAMPLE 4   Solve the system

$$2x^2 + 5y^2 = 22, \quad (1) \quad \text{(The graph is an ellipse.)}$$
$$3x^2 - y^2 = -1. \quad (2) \quad \text{(The graph is a hyperbola.)}$$

SOLUTION   Here we multiply equation (2) by 5 and then add:

$$
\begin{array}{ll}
2x^2 + 5y^2 = 22 & \\
\underline{15x^2 - 5y^2 = -5} & \text{Multiplying both sides of equation (2) by 5} \\
17x^2 = 17 & \text{Adding} \\
x^2 = 1 & \\
x = \pm 1. &
\end{array}
$$

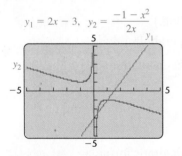

There is no x-term in equation (2), and whether x is -1 or 1, we still have $x^2 = 1$. Thus we can simultaneously substitute 1 and -1 for x in equation (2):

$$\left.\begin{array}{r} 3 \cdot (\pm 1)^2 - y^2 = -1 \\ 3 - y^2 = -1 \\ -y^2 = -4 \end{array}\right\} \quad \text{Since } (-1)^2 = 1^2, \text{ we can evaluate for } x = -1 \text{ and } x = 1 \text{ simultaneously.}$$

$$y^2 = 4, \quad \text{or} \quad y = \pm 2.$$

Thus, if $x = 1$, then $y = 2$ or $y = -2$; and if $x = -1$, then $y = 2$ or $y = -2$. The four possible solutions are $(1, 2)$, $(1, -2)$, $(-1, 2)$, and $(-1, -2)$.

Check: Since $(2)^2 = (-2)^2$ and $(1)^2 = (-1)^2$, we can check all four pairs at once.

$$\begin{array}{c|c} 2x^2 + 5y^2 = 22 \\ \hline 2(\pm 1)^2 + 5(\pm 2)^2 & 22 \\ 2 + 20 & \\ 22 \stackrel{?}{=} 22 & \text{TRUE} \end{array} \qquad \begin{array}{c|c} 3x^2 - y^2 = -1 \\ \hline 3(\pm 1)^2 - (\pm 2)^2 & -1 \\ 3 - 4 & \\ -1 \stackrel{?}{=} -1 & \text{TRUE} \end{array}$$

4. Solve the system

$$x^2 - 6y^2 = 1,$$
$$2x^2 - 5y^2 = 30.$$

The solutions are $(1, 2)$, $(1, -2)$, $(-1, 2)$, and $(-1, -2)$.

YOUR TURN

When a product of variables appears in one equation and the other equation is of the form $Ax^2 + By^2 = C$, we often solve for a variable in the equation with the product and then use substitution.

EXAMPLE 5 Solve the system

$$x^2 + 4y^2 = 20, \quad (1) \quad \text{(The graph is an ellipse.)}$$
$$xy = 4. \quad (2) \quad \text{(The graph is a hyperbola.)}$$

SOLUTION First, we solve equation (2) for y:

$$y = \frac{4}{x}. \quad \text{Dividing both sides by } x. \text{ Note that } x \neq 0.$$

Then we substitute $4/x$ for y in equation (1) and solve for x:

$$x^2 + 4\left(\frac{4}{x}\right)^2 = 20$$

$$x^2 + \frac{64}{x^2} = 20$$

$$x^4 + 64 = 20x^2 \quad \text{Multiplying by } x^2$$

$$x^4 - 20x^2 + 64 = 0 \quad \begin{array}{l}\text{Obtaining standard form.}\\ \text{This equation is reducible}\\ \text{to quadratic.}\end{array}$$

$$(x^2 - 4)(x^2 - 16) = 0 \quad \begin{array}{l}\text{Factoring. If you prefer, let}\\ u = x^2 \text{ and substitute.}\end{array}$$

$$(x - 2)(x + 2)(x - 4)(x + 4) = 0 \quad \text{Factoring again}$$

$$x = 2 \quad or \quad x = -2 \quad or \quad x = 4 \quad or \quad x = -4. \quad \begin{array}{l}\text{Using the principle}\\ \text{of zero products}\end{array}$$

5. Solve the system

$$2x^2 + 3y^2 = 21,$$
$$xy = 3.$$

Since $y = 4/x$, for $x = 2$, we have $y = 4/2$, or 2. Thus, $(2, 2)$ is a solution. Similarly, $(-2, -2)$, $(4, 1)$, and $(-4, -1)$ are solutions. You can show that all four pairs check.

YOUR TURN

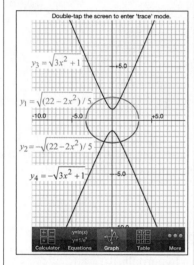
Problem Solving

We now consider applications that can be modeled by a system of equations in which at least one equation is not linear.

EXAMPLE 6   *Architecture.*   For a college fitness center, an architect plans to lay out a rectangular piece of land that has a perimeter of 204 m and an area of 2565 m². Find the dimensions of the piece of land.

SOLUTION

1. Familiarize. We draw and label a sketch, letting l = the length and w = the width, both in meters.

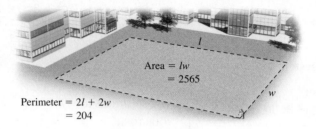

Area = lw
= 2565

Perimeter = $2l + 2w$
= 204

2. Translate. We then have the following translation:

Perimeter: $2w + 2l = 204$;

Area: $lw = 2565$.

3. Carry out. We solve the system

$$2w + 2l = 204,$$

$$lw = 2565.$$

Solving the second equation for l gives us $l = 2565/w$. Then we substitute $2565/w$ for l in the first equation and solve for w:

$$2w + 2\left(\frac{2565}{w}\right) = 204$$

$$2w^2 + 2(2565) = 204w \qquad \text{Multiplying both sides by } w$$

$$2w^2 - 204w + 2(2565) = 0 \qquad \text{Standard form}$$

$$w^2 - 102w + 2565 = 0 \qquad \text{Multiplying by } \tfrac{1}{2}$$

> Factoring could be used instead of the quadratic formula, but the numbers are quite large.

$$w = \frac{-(-102) \pm \sqrt{(-102)^2 - 4 \cdot 1 \cdot 2565}}{2 \cdot 1}$$

$$w = \frac{102 \pm \sqrt{144}}{2} = \frac{102 \pm 12}{2}$$

$$w = 57 \quad or \quad w = 45.$$

If $w = 57$, then $l = 2565/w = 2565/57 = 45$. If $w = 45$, then $l = 2565/w = 2565/45 = 57$. Since length is usually considered to be longer than width, we have the solution $l = 57$ and $w = 45$, or $(57, 45)$.

4. Check. If $l = 57$ and $w = 45$, the perimeter is $2 \cdot 57 + 2 \cdot 45$, or 204. The area is $57 \cdot 45$, or 2565. The numbers check.

5. State. The length is 57 m, and the width is 45 m.

YOUR TURN

6. Gretchen used 94 ft of fencing to enclose her rectangular garden. If the area of the garden is 550 ft², what are the garden's dimensions?

EXAMPLE 7 *HDTV Dimensions.* The Kaplans' new HDTV has a 40-in. diagonal screen with an area of 768 in². Find the width and the length of the screen.

SOLUTION

1. **Familiarize.** We make a drawing and label it. Note the right triangle in the figure. We let l = the length and w = the width, both in inches.

Chapter Resource:
Decision Making: Connection, p. 689

2. **Translate.** We translate to a system of equations:

$l^2 + w^2 = 40^2$,   Using the Pythagorean theorem
$lw = 768$.   Using the formula for the area of a rectangle

3. **Carry out.** We solve the system

$l^2 + w^2 = 1600$,
$lw = 768$ You should complete the solution of this system.

to get $(32, 24)$, $(24, 32)$, $(-32, -24)$, and $(-24, -32)$.

4. **Check.** Measurements must be positive and length is usually greater than width, so we check only $(32, 24)$. In the right triangle, $32^2 + 24^2 = 1024 + 576 = 1600$, or 40^2. The area is $32 \cdot 24 = 768$, so our answer checks.

5. **State.** The length is 32 in., and the width is 24 in.

7. The area of a rectangular stamp is 60 mm², and the length of a diagonal is 13 mm. Find the width and the length of the stamp.

 YOUR TURN

10.4 EXERCISE SET

FOR EXTRA HELP

MyMathLab® Math XL
PRACTICE   WATCH   READ   REVIEW

Vocabulary and Reading Check

Classify each of the following statements as either true or false.

1. A system of equations that represent a line and an ellipse can have 0, 1, or 2 solutions.

2. A system of equations that represent a parabola and a circle can have up to 4 solutions.

3. A system of equations representing a hyperbola and a circle can have no fewer than 2 solutions.

4. A system of equations representing an ellipse and a line has either 0 or 2 solutions.

5. Systems containing one first-degree equation and one second-degree equation are most easily solved using the substitution method.

6. Systems containing two second-degree equations of the form $Ax^2 + By^2 = C$ are most easily solved using the elimination method.

Systems Involving One Nonlinear Equation

Solve. Remember that graphs can be used to confirm all real solutions.

7. $x^2 + y^2 = 41$,
    $y - x = 1$

8. $x^2 + y^2 = 45$,
    $y - x = 3$

9. $4x^2 + 9y^2 = 36$,
    $3y + 2x = 6$

10. $9x^2 + 4y^2 = 36$,
    $3x + 2y = 6$

11. $y^2 = x + 3$,
    $2y = x + 4$

12. $y = x^2$,
    $3x = y + 2$

13. $x^2 - xy + 3y^2 = 27,$
$x - y = 2$

14. $2y^2 + xy + x^2 = 7,$
$x - 2y = 5$

15. $x^2 + 4y^2 = 25,$
$x + 2y = 7$

16. $x^2 - y^2 = 16,$
$x - 2y = 1$

17. $x^2 - xy + 3y^2 = 5,$
$x - y = 2$

18. $m^2 + 3n^2 = 10,$
$m - n = 2$

19. $3x + y = 7,$
$4x^2 + 5y = 24$

20. $2y^2 + xy = 5,$
$4y + x = 7$

21. $a + b = 6,$
$ab = 8$

22. $p + q = -1,$
$pq = -12$

23. $2a + b = 1,$
$b = 4 - a^2$

24. $4x^2 + 9y^2 = 36,$
$x + 3y = 3$

25. $a^2 + b^2 = 89,$
$a - b = 3$

26. $xy = 10,$
$x + y = 7$

Systems of Two Nonlinear Equations

Solve. Remember that graphs can be used to confirm all real solutions.

27. $y = x^2,$
$x = y^2$

28. $x^2 + y^2 = 25,$
$y^2 = x + 5$

Aha! **29.** $x^2 + y^2 = 16,$
$x^2 - y^2 = 16$

30. $y^2 - 4x^2 = 25,$
$4x^2 + y^2 = 25$

31. $x^2 + y^2 = 25,$
$xy = 12$

32. $x^2 - y^2 = 16,$
$x + y^2 = 4$

33. $x^2 + y^2 = 9,$
$25x^2 + 16y^2 = 400$

34. $x^2 + y^2 = 4,$
$9x^2 + 16y^2 = 144$

35. $x^2 + y^2 = 14,$
$x^2 - y^2 = 4$

36. $x^2 + y^2 = 16,$
$y^2 - 2x^2 = 10$

37. $x^2 + y^2 = 10,$
$xy = 3$

38. $x^2 + y^2 = 5,$
$xy = 2$

39. $x^2 + 4y^2 = 20,$
$xy = 4$

40. $x^2 + y^2 = 13,$
$xy = 6$

41. $2xy + 3y^2 = 7,$
$3xy - 2y^2 = 4$

42. $3xy + x^2 = 34,$
$2xy - 3x^2 = 8$

43. $4a^2 - 25b^2 = 0,$
$2a^2 - 10b^2 = 3b + 4$

44. $xy - y^2 = 2,$
$2xy - 3y^2 = 0$

45. $ab - b^2 = -4,$
$ab - 2b^2 = -6$

46. $x^2 - y = 5,$
$x^2 + y^2 = 25$

Problem Solving

Solve.

47. *Art.* Elliot is designing a rectangular stained glass miniature that has a perimeter of 28 cm and a diagonal of length 10 cm. What should the dimensions of the glass be?

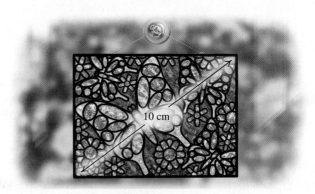

48. *Geometry.* A rectangle has an area of 2 yd^2 and a perimeter of 6 yd. Find its dimensions.

49. *Tile Design.* The Clay Works tile company wants to make a new rectangular tile that has a perimeter of 6 in. and a diagonal of length $\sqrt{5}$ in. What should the dimensions of the tile be?

50. *Geometry.* A rectangle has an area of 20 in^2 and a perimeter of 18 in. Find its dimensions.

51. *Design of a Van.* The cargo area of a delivery van must be 60 ft^2, and the length of a diagonal must accommodate a 13-ft board. Find the dimensions of the cargo area.

52. *Dimensions of a Rug.* The diagonal of a Persian rug is 25 ft, and the area of the rug is 300 ft^2. Find the length and the width of the rug.

53. The product of two numbers is 90. The sum of their squares is 261. Find the numbers.

54. *Investments.* A certain amount of money saved for 1 year at a certain interest rate yielded $125 in simple interest. If $625 more had been invested and the rate had been 1% less, the interest would have been the same. Find the principal and the rate.

55. *Laptop Dimensions.* The screen on Ashley's new laptop has an area of 90 in^2 and a $\sqrt{200.25}$-in. diagonal. Find the width and the length of the screen.

56. *Garden Design.* A garden contains two square flower beds. Find the length of each bed if the sum of their areas is 832 ft² and the difference of their areas is 320 ft².

57. The area of a rectangle is $\sqrt{3}$ m², and the length of a diagonal is 2 m. Find the dimensions.

58. The area of a rectangle is $\sqrt{2}$ m², and the length of a diagonal is $\sqrt{3}$ m. Find the dimensions.

59. How can an understanding of conic sections be helpful when a system of nonlinear equations is being solved algebraically?

60. Write a problem for a classmate to solve. Devise the problem so that a system of two nonlinear equations with exactly one real solution is solved.

Skill Review
Simplify.

61. $(3a^{-4})^2(2a^{-5})^{-1}$ [1.6]

62. $16^{-1/2}$ [7.2]

63. $\log 10,000$ [9.5]

64. i^{71} [7.8]

65. $-10^2 \div 2 \cdot 5 - 3$ [1.2]

66. $\sqrt{500}$ [7.3]

Synthesis

67. Write a problem that translates to a system of two equations. Design the problem so that at least one equation is nonlinear and so that no real solution exists.

68. Suppose a system of equations is comprised of one linear equation and one nonlinear equation. Is it possible for such a system to have three solutions? Why or why not?

Solve.

69. $p^2 + q^2 = 13,$
$\dfrac{1}{pq} = -\dfrac{1}{6}$

70. $a + b = \dfrac{5}{6},$
$\dfrac{a}{b} + \dfrac{b}{a} = \dfrac{13}{6}$

71. *Fence Design.* A roll of chain-link fencing contains 100 ft of fence. The fencing is bent at a 90° angle to enclose a rectangular work area of 2475 ft², as shown. Find the length and the width of the rectangle.

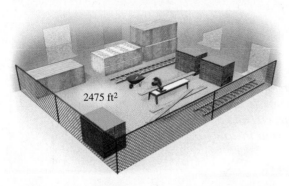

2475 ft²

72. A piece of wire 100 cm long is to be cut into two pieces and those pieces are each to be bent to make a square. The area of one square is to be 144 cm² greater than that of the other. How should the wire be cut?

73. *Box Design.* Four squares with sides 5 in. long are cut from the corners of a rectangular metal sheet that has an area of 340 in². The edges are bent up to form an open box with a volume of 350 in³. Find the dimensions of the box.

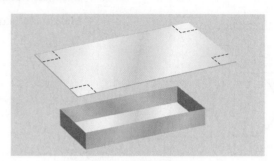

YOUR TURN ANSWERS: SECTION 10.4

1. $(12,5), \left(-\dfrac{56}{5}, -\dfrac{33}{5}\right)$ **2.** $(2,3), \left(-\dfrac{1}{2}, -2\right)$
3. $(-2+\sqrt{5}i, 1+\sqrt{5}i), (-2-\sqrt{5}i, 1-\sqrt{5}i)$
4. $(5,2), (5,-2), (-5,2), (-5,-2)$
5. $(3,1), (-3,-1), \left(\dfrac{\sqrt{6}}{2}, \sqrt{6}\right), \left(-\dfrac{\sqrt{6}}{2}, -\sqrt{6}\right)$
6. Length: 25 ft; width: 22 ft **7.** Length: 12 mm; width: 5 mm

QUICK QUIZ: SECTIONS 10.1–10.4
Classify each of the following as the equation of a circle, an ellipse, a parabola, or a hyperbola. [10.3]
1. $y^2 = 8 - x^2$ **2.** $2x^2 - x - y^2 + 4 = 0$
3. $5x^2 + 10y^2 = 50$ **4.** $4x + y^2 + y = 7$
5. Solve:
$$x - y = 3,$$
$$x^2 + y^2 = 5. \quad [10.4]$$

PREPARE TO MOVE ON
Simplify. [1.2]
1. $(-1)^9(-3)^2$ **2.** $(-1)^{10}(-3)^3$
Evaluate each of the following. [1.2]
3. $\dfrac{(-1)^k}{k-5}$, for $k = 10$ **4.** $\dfrac{n}{2}(3+n)$, for $n = 11$
5. $\dfrac{7(1-r^2)}{1-r}$, for $r = \dfrac{1}{2}$

A

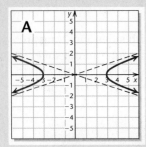

B

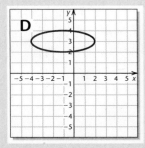

C

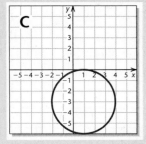

D

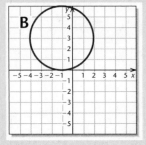

E

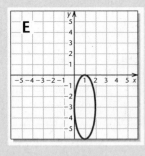

Visualizing for Success

Use after Section 10.3.

Match each equation with its graph.

1. $(x - 1)^2 + (y + 3)^2 = 9$

2. $\dfrac{x^2}{9} - \dfrac{y^2}{1} = 1$

3. $y = (x - 1)^2 - 3$

4. $(x + 1)^2 + (y - 3)^2 = 9$

5. $x = (y - 1)^2 - 3$

6. $\dfrac{(x + 1)^2}{9} + \dfrac{(y - 3)^2}{1} = 1$

7. $xy = 3$

8. $y = -(x + 1)^2 + 3$

9. $\dfrac{y^2}{9} - \dfrac{x^2}{1} = 1$

10. $\dfrac{(x - 1)^2}{1} + \dfrac{(y + 3)^2}{9} = 1$

Answers on page A-60

An additional, animated version of this activity appears in MyMathLab. *To use MyMathLab, you need a course ID and a student access code. Contact your instructor for more information.*

F

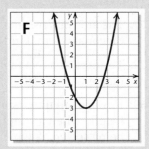

G

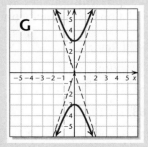

H

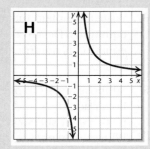

I

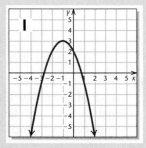

J

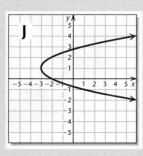

Collaborative Activity   *A Cosmic Path*

Focus: Ellipses
Use after: Section 10.2
Time: 20–30 minutes
Group size: 2
Materials: Scientific calculators

On May 4, 2007, Comet 17P/Holmes was at the point closest to the sun in its orbit. Comet 17P is traveling in an elliptical orbit with the sun as one focus, and one orbit takes about 6.88 years. One astronomical unit (AU) is 93,000,000 mi. One group member should do the following calculations in AU and the other in millions of miles.

Source: Harvard-Smithsonian Center for Astrophysics

Activity

1. At its *perihelion*, a comet with an elliptical orbit is at the point in its orbit closest to the sun. At its *aphelion*, the comet is at the point farthest from the sun. The perihelion distance for Comet 17P is 2.053218 AU, and the aphelion distance is 5.183610 AU. Use these distances to find a. (See the following diagram.)

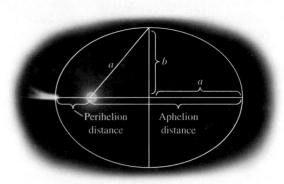

2. Using the figure above, express b^2 as a function of a. Then find b using the value found for a in part (1).

3. One formula for approximating the perimeter of an ellipse is
$$P = \pi\left(3a + 3b - \sqrt{(3a + b)(a + 3b)}\right),$$
developed by the Indian mathematician S. Ramanujan in 1914. How far does Comet 17P travel in one orbit?

4. What is the speed of the comet? Find the answer in AU per year and in miles per hour.

5. Which calculations—AUs or mi—were easier to use? Why?

Decision Making & Connection   (*Use after Section 10.4.*)

Aspect Ratio. Photographs, television screens, and movie screens are often described in terms of an aspect ratio, as follows.

Area: $A = bh$
Diagonal: $d = \sqrt{b^2 + h^2}$
Aspect ratio: $r = \dfrac{b}{h}$

1. Wide-screen televisions have an aspect ratio of $\frac{16}{9}$. If a manufacturer plans a diagonal of length 40 in., what must the dimensions of the screen be?

2. Older televisions had an aspect ratio of $\frac{4}{3}$. For a diagonal of length 40 in., what would the dimensions of this television be?

3. For a fixed diagonal of 40 in., which has more screen area: a wide-screen TV or an older TV?

4. Find a formula for the height of a rectangle given the diagonal and the aspect ratio.

5. *Research.*   The aspect ratio of a rectangle that is considered most pleasing aesthetically is called the *golden ratio*. Find the value of the golden ratio. What aspect ratios used today for photography or video are closest to the golden ratio?

Study Summary

SECTION 10.1: *Conic Sections: Parabolas and Circles*

Parabola

$y = ax^2 + bx + c$

$\quad = a(x - h)^2 + k$

$\quad$ Opens upward $(a > 0)$
or downward $(a < 0)$
Vertex: (h, k)

$x = ay^2 + by + c$

$\quad = a(y - k)^2 + h$

$\quad$ Opens right $(a > 0)$
or left $(a < 0)$
Vertex: (h, k)

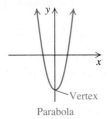

Parabola

$x = -y^2 + 4y - 1$

$\quad = -(y^2 - 4y \quad\quad) - 1$

$\quad = -(y^2 - 4y + 4) - 1 - (-1)(4)$

$\quad = -(y - 2)^2 + 3 \quad\quad a = -1;$ parabola opens left

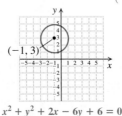

$x = -y^2 + 4y - 1$

---

1. Graph

$\quad x = y^2 + 6y + 7.$

$\quad$ Label the vertex.

Circle

$x^2 + y^2 = r^2$

$\quad$ Radius: r

$\quad$ Center: $(0, 0)$

$(x - h)^2 + (y - k)^2 = r^2$

$\quad$ Radius: r

$\quad$ Center: (h, k)

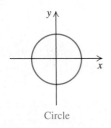

Circle

$\quad\quad x^2 + y^2 + 2x - 6y + 6 = 0$

$x^2 + 2x + \quad\quad y^2 - 6y \quad\quad = -6$

$x^2 + 2x + 1 + y^2 - 6y + 9 = -6 + 1 + 9$

$\quad\quad (x + 1)^2 + (y - 3)^2 = 4$

$\quad [x - (-1)]^2 + (y - 3)^2 = 2^2 \quad\quad$ Radius: 2; center:
$\quad\quad\quad\quad\quad\quad\quad\quad\quad\quad\quad\quad\quad\quad\quad (-1, 3)$

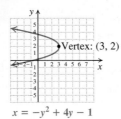

$x^2 + y^2 + 2x - 6y + 6 = 0$

---

2. Find the center and the
radius and then graph

$\quad x^2 + y^2 - 6x + 5 = 0.$

SECTION 10.2: *Conic Sections: Ellipses*

Ellipse

$$\frac{x^2}{a^2} + \frac{y^2}{b^2} = 1 \quad \text{Center: } (0,0)$$

$$\frac{(x-h)^2}{a^2} + \frac{(y-k)^2}{b^2} = 1$$

Center: (h, k)

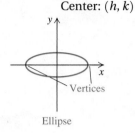

Ellipse

$$\frac{(x-4)^2}{4} + \frac{(y+1)^2}{9} = 1$$

$$\frac{(x-4)^2}{2^2} + \frac{[y-(-1)]^2}{3^2} = 1 \qquad \begin{array}{l} 3 > 2; \text{ ellipse is vertical} \\ \text{with center } (4, -1) \end{array}$$

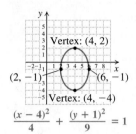

$$\frac{(x-4)^2}{4} + \frac{(y+1)^2}{9} = 1$$

3. Graph: $\dfrac{x^2}{9} + y^2 = 1$.

SECTION 10.3: *Conic Sections: Hyperbolas*

Hyperbola

$$\frac{x^2}{a^2} - \frac{y^2}{b^2} = 1$$

Two branches opening right and left

$$\frac{y^2}{b^2} - \frac{x^2}{a^2} = 1$$

Two branches opening upward and downward

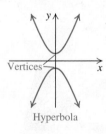

Hyperbola

$$\frac{x^2}{4} - \frac{y^2}{1} = 1$$

$$\frac{x^2}{2^2} - \frac{y^2}{1^2} = 1 \qquad \text{Opens right and left}$$

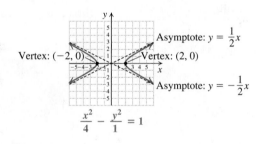

$$\frac{x^2}{4} - \frac{y^2}{1} = 1$$

4. Graph: $\dfrac{y^2}{16} - \dfrac{x^2}{4} = 1$.

SECTION 10.4: *Nonlinear Systems of Equations*

We can solve a system containing at least one nonlinear equation using substitution or elimination.

Solve:

$x^2 - y = -1,$ (1)   (The graph is a parabola.)

$x + 2y = 3.$ (2)   (The graph is a line.)

$x = 3 - 2y$ Solving for x

$(3 - 2y)^2 - y = -1$ Substituting

$9 - 12y + 4y^2 - y = -1$

$4y^2 - 13y + 10 = 0$

$(4y - 5)(y - 2) = 0$

$4y - 5 = 0 \quad or \quad y - 2 = 0$

$y = \frac{5}{4} \quad or \quad y = 2$

If $y = \frac{5}{4}$, then $x = 3 - 2\left(\frac{5}{4}\right) = \frac{1}{2}$. $\left(\frac{1}{2}, \frac{5}{4}\right)$ is a solution.

If $y = 2$, then $x = 3 - 2(2) = -1$. $(-1, 2)$ is a solution.

The solutions are $\left(\frac{1}{2}, \frac{5}{4}\right)$ and $(-1, 2)$.

5. Solve:

$x^2 + y^2 = 41,$

$y - x = 1.$

Review Exercises: Chapter 10

➤ Concept Reinforcement

Classify each of the following statements as either true or false.

1. Any parabola that opens upward or downward represents the graph of a function.   [10.1]

2. The center of a circle is part of the circle itself.   [10.1]

3. The foci of an ellipse are part of the ellipse itself.   [10.2]

4. It is possible for a hyperbola to represent the graph of a function.   [10.3]

5. If an equation of a conic section has only one term of degree 2, its graph cannot be a circle, an ellipse, or a hyperbola.   [10.3]

6. Two nonlinear graphs can intersect in more than one point.   [10.4]

7. Every system of nonlinear equations has at least one real solution.   [10.4]

8. Both substitution and elimination can be used as methods for solving a system of nonlinear equations.   [10.4]

Find the center and the radius of each circle.   [10.1]

9. $(x + 3)^2 + (y - 2)^2 = 16$

10. $(x - 5)^2 + y^2 = 11$

11. $x^2 + y^2 - 6x - 2y + 1 = 0$

12. $x^2 + y^2 + 8x - 6y = 20$

13. Find an equation of the circle with center $(-4, 3)$ and radius 4.   [10.1]

14. Find an equation of the circle with center $(7, -2)$ and radius $2\sqrt{5}$.   [10.1]

Classify each equation as a circle, an ellipse, a parabola, or a hyperbola. Then graph.

15. $5x^2 + 5y^2 = 80$ [10.1], [10.3]

16. $9x^2 + 2y^2 = 18$ [10.2], [10.3]

17. $y = -x^2 + 2x - 3$ [10.1], [10.3]

18. $\dfrac{y^2}{9} - \dfrac{x^2}{4} = 1$ [10.3]

19. $xy = 9$ [10.3]

20. $x = y^2 + 2y - 2$ [10.1], [10.3]

21. $\dfrac{(x + 1)^2}{3} + (y - 3)^2 = 1$ [10.2], [10.3]

22. $x^2 + y^2 + 6x - 8y - 39 = 0$ [10.1], [10.3]

Solve.   [10.4]

23. $x^2 - y^2 = 21,$
    $x + y = 3$

24. $x^2 - 2x + 2y^2 = 8,$
    $2x + y = 6$

25. $x^2 - y = 5,$
    $2x - y = 5$

26. $x^2 + y^2 = 25,$
    $x^2 - y^2 = 7$

27. $x^2 - y^2 = 3,$
    $y = x^2 - 3$

28. $x^2 + y^2 = 18,$
    $2x + y = 3$

29. $x^2 + y^2 = 100,$
    $2x^2 - 3y^2 = -120$

30. $x^2 + 2y^2 = 12,$
    $xy = 4$

31. A rectangular bandstand has a perimeter of 38 m and an area of 84 m². What are the dimensions of the bandstand?   [10.4]

32. One type of carton used by tableproducts.com exactly fits both a rectangular plate of area 108 in² and chopsticks of length 15 in., laid diagonally on top of the plate. Find the length and the width of the carton.   [10.4]

15 in.

33. The perimeter of a square mounting board is 12 cm more than the perimeter of a square mirror. The board's area exceeds the area of the mirror by 39 cm². Find the perimeter of each object.   [10.4]

34. The sum of the areas of two circles is 130π ft². The difference of the circumferences is 16π ft. Find the radius of each circle.   [10.4]

Synthesis

35. How does the graph of a hyperbola differ from the graph of a parabola?   [10.1], [10.3]

36. Explain why function notation rarely appears in this chapter, and list the types of graphs discussed for which function notation could be used.   [10.1], [10.2], [10.3]

37. Solve:
$$4x^2 - x - 3y^2 = 9,$$
$$-x^2 + x + y^2 = 2. \quad [10.4]$$

38. Find an equation of the circle that passes through $(-2, -4)$, $(5, -5)$, and $(6, 2)$.   [10.1], [10.4]

39. Find an equation of the ellipse with the following intercepts: $(-10, 0)$, $(10, 0)$, $(0, -1)$, and $(0, 1)$.   [10.2]

For step-by-step test solutions, access the Chapter Test Prep Videos in MyMathLab, on YouTube (search "Bittinger Inter Alg CA" and click on Channels"), or by scanning the code.

Test: Chapter 10

1. Find an equation of the circle with center $(3, -4)$ and radius $2\sqrt{3}$.

Find the center and the radius of each circle.

2. $(x - 4)^2 + (y + 1)^2 = 5$

3. $x^2 + y^2 + 4x - 6y + 4 = 0$

Classify the equation as a circle, an ellipse, a parabola, or a hyperbola. Then graph.

4. $y = x^2 - 4x - 1$

5. $x^2 + y^2 + 2x + 6y + 6 = 0$

6. $\dfrac{x^2}{16} - \dfrac{y^2}{9} = 1$

7. $16x^2 + 4y^2 = 64$

8. $xy = -5$

9. $x = -y^2 + 4y$

Solve.

10. $x^2 + y^2 = 36,$
$3x + 4y = 24$

11. $x^2 - y = 3,$
$2x + y = 5$

12. $x^2 - 2y^2 = 1,$
$xy = 6$

13. $x^2 + y^2 = 10,$
$x^2 = y^2 + 2$

14. A rectangular bookmark with diagonal of length $5\sqrt{5}$ has an area of 22. Find the dimensions of the bookmark.

15. Two squares are such that the sum of their areas is 8 m^2 and the difference of their areas is 2 m^2. Find the length of a side of each square.

16. A rectangular dance floor has a diagonal of length 40 ft and a perimeter of 112 ft. Find the dimensions of the dance floor.

17. Brett invested a certain amount of money for 1 year and earned $72 in interest. Erin invested $240 more than Brett at an interest rate that was $\frac{5}{6}$ of the rate given to Brett, but she earned the same amount of interest. Find the principal and the interest rate for Brett's investment.

Synthesis

18. Find an equation of the ellipse passing through $(6, 0)$ and $(6, 6)$ with vertices at $(1, 3)$ and $(11, 3)$.

19. The sum of two numbers is 36, and the product is 4. Find the sum of the reciprocals of the numbers.

20. *Theatrical Production.* An E.T.C. spotlight for a college's production of *Hamlet* projects an ellipse of light on a stage that is 8 ft wide and 14 ft long. Find an equation of that ellipse if an actor is in its center and x represents the number of feet, horizontally, from the actor to the edge of the ellipse and y represents the number of feet, vertically, from the actor to the edge of the ellipse.

Cumulative Review: Chapters 1–10

Simplify.

1. $(4t^2 - 5s)^2$ [5.2]

2. $\dfrac{1}{3t} + \dfrac{1}{t-3}$ [6.2]

3. $\sqrt{6t}\,\sqrt{15t^3 w}$ [7.3]

4. $(81a^{2/3}b^{1/4})^{3/4}$ [7.2]

5. $\log_2 \dfrac{1}{16}$ [9.3]

6. $(4 + 3i)(4 - 3i)$ [7.8]

7. -8^{-2} [1.6]

Factor.

8. $100x^2 - 60xy + 9y^2$ [5.5]

9. $3m^6 - 24$ [5.6]

10. $ax + by - ay - bx$ [5.3]

11. $32x^2 - 20x - 3$ [5.4]

Solve. Where appropriate, give an approximation to four decimal places.

12. $3(x - 5) - 4x \geq 2(x + 5)$ [4.1]

13. $16x^2 - 18x = 0$ [5.8]

14. $\dfrac{2}{x} + \dfrac{1}{x-2} = 1$ [6.4]

15. $5x^2 + 5 = 0$ [8.2]

16. $\log_x 64 = 3$ [9.6]

17. $3^x = 1.5$ [9.6]

18. $x = \sqrt{2x - 5} + 4$ [7.6]

19. $x^2 + 2y^2 = 5,$
$2x^2 + \ \ y^2 = 7$ [10.4]

Graph.

20. $3x - y = 9$ [2.4]

21. $y = \log_5 x$ [9.3]

22. $\dfrac{x^2}{25} + \dfrac{y^2}{1} = 1$ [10.2]

23. $f(x) = 2^{x-1}$ [9.2]

24. $x^2 + (y - 3)^2 = 4$ [10.1]

25. $x < 2y + 1$ [4.4]

26. Graph: $f(x) = -(x + 2)^2 + 3$. [8.7]

   a) Label the vertex.
   b) Draw the axis of symmetry.
   c) Find the maximum or minimum value.

27. Find the slope–intercept equation of the line containing the points $(-3, 6)$ and $(1, 2)$. [2.3]

28. Write a quadratic equation having the solutions $\sqrt{3}$ and $-\sqrt{3}$. Answers may vary. [8.3]

Solve.

29. *Aviation.* BlueAir owns two types of airplanes. One type flies 60 mph faster than the other. Laura often rents a plane from BlueAir to visit her parents. The flight takes 4 hr with the faster plane and 4 hr 24 min with the slower plane. What distance does she fly? [3.3]

30. *Retail.* The median size of a grocery store in the United States was 48,750 ft^2 in 2006 and 46,000 ft^2 in 2010. [2.5]

   Source: Food Marketing Industry

   a) Find a linear function that fits the data. Let t represent the number of years since 2006.
   b) Use the function from part (a) to predict the median size of a grocery store in 2014.
   c) Assuming the trend continues, in what year will the median size of a grocery store be 40,000 ft^2?

31. *Population.* The population of Latvia was 2.19 million in 2012 and was decreasing exponentially at a rate of 0.84% per year. [9.7]

   Sources: Encyclopedia of the Nations; The CIA World Factbook

   a) Write an exponential function that could be used to find $P(t)$, the population of Latvia, in millions, t years after 2012.
   b) Predict what the population will be in 2025.
   c) What is the half life of the population?

32. *Art.* Elyse is designing a rectangular tray. She wants to put a row of beads around the tray, and has enough beads to make an edge that is 32 in. long. What dimensions of the tray will give it the greatest area? [8.8]

Synthesis

33. If y varies inversely as the square root of x and x is multiplied by 100, what is the effect on y? [6.8], [7.1]

34. For $f(x) = x - \dfrac{1}{x^2}$, find all x-values for which $f(x) \leq 0$. [8.9]

Sequences, Series, and the Binomial Theorem

YEAR	GLOBAL SOLAR PV CAPACITY (in gigawatts)
2012	70
2013	84
2014	100.8

Sources: Based on data from The European Photovoltaic Industry Association; GlobalData

Get Clean Power from the Sun.

As the cost of generating electricity from sunlight decreases, more solar photovoltaic (PV) panels are being installed around the world. The table above shows one estimate of the total global solar PV capacity for several years. Later in this chapter, we will use a *geometric series* to estimate how much total power will be generated by solar energy from 2012 through 2020. (*See Exercise 67 in Exercise Set 11.3.*)

It would be impossible to track utility usage and costs and to determine savings without using math in my daily work.

Wayne Benbow, an energy management consultant from Lakewood, Colorado, uses math to track energy consumption and the associated costs in over 13 million square feet of floor space in 150 schools.

A *sequence* is, quite simply, an ordered list. When the members of a sequence are numbers, we can discuss their sum. Such a sum is called a *series*.

Section 11.4 presents the *binomial theorem*, which is used to expand expressions of the form $(a + b)^n$. Such an expansion is itself a series.

11.1   Sequences and Series

Sequences ▪ Finding the General Term ▪ Sums and Series ▪ Sigma Notation

Sequences

Suppose that $10,000 is invested at 5%, compounded annually. The value of the account at the start of years 1, 2, 3, 4, and so on, is

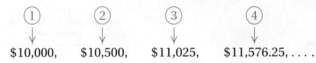

$10,000, $10,500, $11,025, $11,576.25, ...

We can regard this as a function that pairs 1 with $10,000, 2 with $10,500, 3 with $11,025, and so on. This is an example of a **sequence** (or **progression**). The domain of a sequence is a set of counting numbers beginning with 1.

If we stop after a certain number of years, we obtain a **finite sequence**:

$10,000, $10,500, $11,025, $11,576.25.

If we continue listing the amounts in the account, we obtain an **infinite sequence**:

$10,000, $10,500, $11,025, $11,576.25, $12,155.06, ...

The three dots near the end indicate that the sequence goes on without stopping.

SEQUENCES

An *infinite sequence* is a function having for its domain the set of natural numbers: $\{1, 2, 3, 4, 5, \ldots\}$.

A *finite sequence* is a function having for its domain a set of natural numbers: $\{1, 2, 3, 4, 5, \ldots, n\}$, for some natural number n.

As another example, consider the sequence given by

$$a(n) = 2^n, \quad \text{or} \quad a_n = 2^n.$$

The notation a_n means $a(n)$ but is used more commonly with sequences. Some function values (also called *terms* of the sequence) follow:

$$a_1 = 2^1 = 2,$$
$$a_2 = 2^2 = 4,$$
$$a_3 = 2^3 = 8,$$
$$a_6 = 2^6 = 64.$$

Note that n gives the position in the sequence and a_n defines the number that is in that position. The first term of the sequence is a_1, the fifth term is a_5, and the

nth term, or **general term**, is a_n. This sequence can also be denoted in the following ways:

$$2, 4, 8, \ldots;$$

or $\quad 2, 4, 8, \ldots, 2^n, \ldots.$ The 2^n emphasizes that the nth term of this sequence is found by raising 2 to the nth power.

EXAMPLE 1   Find the first four terms and the 57th term of the sequence for which the general term is $a_n = (-1)^n/(n + 1)$.

SOLUTION   We have

$$a_1 = \frac{(-1)^1}{1 + 1} = -\frac{1}{2}, \quad \text{Substituting in } a_n = \frac{(-1)^n}{n + 1}$$

$$a_2 = \frac{(-1)^2}{2 + 1} = \frac{1}{3},$$

$$a_3 = \frac{(-1)^3}{3 + 1} = -\frac{1}{4},$$

$$a_4 = \frac{(-1)^4}{4 + 1} = \frac{1}{5},$$

$$a_{57} = \frac{(-1)^{57}}{57 + 1} = -\frac{1}{58}.$$

Note that the factor $(-1)^n$ causes the signs of the terms to alternate between positive and negative, depending on whether n is even or odd.

YOUR TURN

Finding the General Term

By looking for a pattern, we can often write an expression for the general term of a sequence. When only a few terms are given, more than one pattern may fit.

EXAMPLE 2   For each sequence, predict the general term.

a) $1, 4, 9, 16, 25, \ldots$ **b)** $2, 4, 8, \ldots$ **c)** $-1, 2, -4, 8, -16, \ldots$

SOLUTION

a) $1, 4, 9, 16, 25 \ldots$

These are squares of consecutive positive integers, so the general term could be n^2.

b) $2, 4, 8, \ldots$

We regard the pattern as powers of 2, in which case 16 would be the next term and 2^n the general term.

c) $-1, 2, -4, 8, -16, \ldots$

These are powers of 2 with alternating signs, so the general term may be

$$(-1)^n[2^{n-1}].$$

Making sure the signs of the terms alternate

Raising 2 to a power that is 1 less than the term's position in the sequence

To check, note that -4 is the third term, and $(-1)^3[2^{3-1}] = -1 \cdot 2^2 = -4$.

YOUR TURN

1. Find the first four terms and the 50th term of the sequence for which the general term is $a_n = (-1)^n \cdot n^2$.

2. Predict the general term:

$$5, 10, 15, 20, \ldots.$$

In Example 2(b), if 2^n is the general term, the sequence can be written with more terms as

$$2, \ 4, \ 8, \ 16, \ 32, \ 64, \ 128, \ldots.$$

Suppose instead that the second term is found by adding 2, the third term by adding 4, the next term by adding 6, and so on. In this case, 14 would be the next term and the sequence would be

$$2, 4, 8, 14, 22, 32, 44, 58, \ldots.$$

This illustrates that the fewer terms we are given, the greater the uncertainty about determining the nth term.

Sums and Series

> **SERIES**
>
> Given the infinite sequence
>
> $$a_1, a_2, a_3, a_4, \ldots, a_n, \ldots,$$
>
> the sum of the terms
>
> $$a_1 + a_2 + a_3 + \cdots + a_n + \cdots$$
>
> is called an *infinite series* and is denoted S_∞. The nth *partial sum* is the sum of the first n terms:
>
> $$a_1 + a_2 + a_3 + \cdots + a_n.$$
>
> A partial sum is also called a *finite series* and is denoted S_n.

EXAMPLE 3 For the sequence $-2, 4, -6, 8, -10, 12, -14$, find: **(a)** S_2; **(b)** S_3; **(c)** S_7.

SOLUTION

a) $S_2 = -2 + 4 = 2$ This is the sum of the first 2 terms.

b) $S_3 = -2 + 4 + (-6) = -4$ This is the sum of the first 3 terms.

c) $S_7 = -2 + 4 + (-6) + 8 + (-10) + 12 + (-14) = -8$ This is the sum of the first 7 terms.

3. For the sequence
$$1, -1, 1 -1, \ldots,$$
find S_{12}.

YOUR TURN

Sigma Notation

When the general term of a sequence is known, the Greek letter Σ (uppercase sigma) can be used to write a series. For example, the sum of the first four terms of the sequence $3, 5, 7, 9, 11, \ldots, 2k + 1, \ldots$ can be named as follows, using *sigma notation*, or *summation notation*:

$$\sum_{k=1}^{4} (2k + 1).$$

This represents $(2 \cdot 1 + 1) + (2 \cdot 2 + 1) + (2 \cdot 3 + 1) + (2 \cdot 4 + 1)$, and is read "the sum as k goes from 1 to 4 of $(2k + 1)$." The letter k is called the *index of summation*. The index need not always start at 1.

EXAMPLE 4   Write out and evaluate each sum.

a) $\displaystyle\sum_{k=1}^{5} k^2$ **b)** $\displaystyle\sum_{k=4}^{6} (-1)^k(2k)$

SOLUTION

a) $\displaystyle\sum_{k=1}^{5} k^2 = 1^2 + 2^2 + 3^2 + 4^2 + 5^2 = 1 + 4 + 9 + 16 + 25 = 55$

Evaluate k^2 for all integers from 1 through 5. Then add.

4. Write out and evaluate

$$\sum_{k=0}^{3}(2^k + 5).$$

b) $\displaystyle\sum_{k=4}^{6} (-1)^k(2k) = (-1)^4(2 \cdot 4) + (-1)^5(2 \cdot 5) + (-1)^6(2 \cdot 6)$
$$= 8 - 10 + 12 = 10$$

 YOUR TURN

EXAMPLE 5   Write sigma notation for each sum.

a) $1 + 4 + 9 + 16 + 25$

b) $3 + 9 + 27 + 81 + \cdots$

c) $-1 + 3 - 5 + 7$

SOLUTION

a) $1 + 4 + 9 + 16 + 25$

Note that this is a sum of squares, $1^2 + 2^2 + 3^2 + 4^2 + 5^2$, so the general term is k^2. Sigma notation is

$$\sum_{k=1}^{5} k^2. \qquad \text{The sum starts with } 1^2 \text{ and ends with } 5^2.$$

Answers may vary. For example, another—perhaps less obvious—way of writing $1 + 4 + 9 + 16 + 25$ is

$$\sum_{k=2}^{6} (k - 1)^2.$$

b) $3 + 9 + 27 + 81 + \cdots$

This is a sum of powers of 3, and it is also an infinite series. We use the symbol ∞ for infinity and write the series using sigma notation:

$$\sum_{k=1}^{\infty} 3^k.$$

c) $-1 + 3 - 5 + 7$

Except for the alternating signs, this is the sum of the first four positive odd numbers. It is useful to remember that $2k - 1$ is a formula for the kth positive odd number. It is also important to remember that the factor $(-1)^k$ can be used to create the alternating signs. The general term is thus $(-1)^k(2k - 1)$, beginning with $k = 1$. Sigma notation is

$$\sum_{k=1}^{4} (-1)^k(2k - 1).$$

5. Write sigma notation for

$$10 + 20 + 30 + 40 + \cdots.$$

To check, we can evaluate $(-1)^k(2k - 1)$ using $1, 2, 3,$ and 4, and write the sum of the four terms. We leave this to the student.

 YOUR TURN

11.1 EXERCISE SET

FOR EXTRA HELP

MyMathLab® Math XL

PRACTICE   WATCH   READ   REVIEW

Vocabulary and Reading Check

Indicate whether each description or expression applies to either A or to B.

A. $3, 9, 27, 81, \ldots$ **B.** $3 + 6 + 9 + 12 + 15$

1. ____   A series

2. ____   A sequence

3. ____   Finite

4. ____   Infinite

5. ____   The general term is 3^n.

6. ____ $\displaystyle\sum_{n=1}^{5} 3n$

Concept Reinforcement

In each of Exercises 7–12, match the expression with the most appropriate expression from the column on the right.

7. ____ $\displaystyle\sum_{k=1}^{4} k^2$

8. ____ $\displaystyle\sum_{k=3}^{6} (-1)^k$

9. ____ $5 + 10 + 15 + 20$

10. ____ $a_n = 5^n$

11. ____ $a_n = 3n + 2$

12. ____ $a_1 + a_2 + a_3$

a) $-1 + 1 + (-1) + 1$

b) $a_2 = 25$

c) $a_2 = 8$

d) $\displaystyle\sum_{k=1}^{4} 5k$

e) S_3

f) $1 + 4 + 9 + 16$

Sequences

Find the indicated term of each sequence.

13. $a_n = 5n + 3$; a_8

14. $a_n = 3n - 4$; a_8

15. $a_n = (3n + 1)(2n - 5)$; a_9

16. $a_n = (3n + 2)^2$; a_6

17. $a_n = (-1)^{n-1}(3.4n - 17.3)$; a_{12}

📟 18. $a_n = (-2)^{n-2}(45.68 - 1.2n)$; a_{23}

19. $a_n = 3n^2(9n - 100)$; a_{11}

20. $a_n = 4n^2(2n - 39)$; a_{22}

21. $a_n = \left(1 + \dfrac{1}{n}\right)^2$; a_{20}

22. $a_n = \left(1 - \dfrac{1}{n}\right)^3$; a_{15}

In each of the following, the nth term of a sequence is given. Find the first 4 terms; the 10th term, a_{10}; and the 15th term, a_{15}, of the sequence.

23. $a_n = 3n - 1$ 24. $a_n = 2n + 1$

25. $a_n = n^2 + 2$ 26. $a_n = n^2 - 2n$

27. $a_n = \dfrac{n}{n+1}$ 28. $a_n = \dfrac{n^2 - 1}{n^2 + 1}$

29. $a_n = \left(-\dfrac{1}{2}\right)^{n-1}$ 30. $a_n = (-2)^{n+1}$

31. $a_n = (-1)^n/n$ 32. $a_n = (-1)^n n^2$

33. $a_n = (-1)^n(n^3 - 1)$

34. $a_n = (-1)^{n+1}(3n - 5)$

Finding the General Term

Look for a pattern and then write an expression for the general term, or nth term, a_n, of each sequence. Answers may vary.

35. $2, 4, 6, 8, 10, \ldots$ 36. $1, 3, 5, 7, \ldots$

37. $-1, 1, -1, 1, \ldots$ 38. $1, -1, 1, -1, \ldots$

39. $1, -2, 3, -4, \ldots$

40. $-1, 2, -3, 4, \ldots$

41. $3, 5, 7, 9, \ldots$

42. $4, 6, 8, 10, \ldots$

43. $0, 3, 8, 15, 24, \ldots$

44. $2, 6, 12, 20, 30, \ldots$

45. $\dfrac{1}{2}, \dfrac{2}{3}, \dfrac{3}{4}, \dfrac{4}{5}, \dfrac{5}{6}, \ldots$

46. $1 \cdot 3, 2 \cdot 4, 3 \cdot 5, 4 \cdot 6, \ldots$

47. $0.1, 0.01, 0.001, 0.0001, \ldots$

48. $\dfrac{1}{2}, \dfrac{1}{4}, \dfrac{1}{8}, \dfrac{1}{16}, \ldots$

49. $-1, 4, -9, 16, \ldots$

50. $1, -4, 9, -16, \ldots$

Sums and Series

Find the indicated partial sum for each sequence.

51. $-1, 2, -3, 4, -5, 6, \ldots$; S_{10}

52. $2, -4, 6, -8, 10, -12, \ldots$; S_{10}

53. $1, \dfrac{1}{10}, \dfrac{1}{100}, \dfrac{1}{1000}, \ldots$; S_6

54. $3, 6, 9, 12, 15, \ldots$; S_6

Sigma Notation

Write out and evaluate each sum.

55. $\sum_{k=1}^{5} \dfrac{1}{2k}$

56. $\sum_{k=1}^{6} \dfrac{1}{2k-1}$

57. $\sum_{k=0}^{4} 10^k$

58. $\sum_{k=2}^{6} \sqrt{5k-1}$

59. $\sum_{k=2}^{8} \dfrac{k}{k-1}$

60. $\sum_{k=2}^{5} \dfrac{k-1}{k+1}$

61. $\sum_{k=1}^{8} (-1)^{k+1}2^k$

62. $\sum_{k=1}^{7} (-1)^k 4^{k+1}$

63. $\sum_{k=0}^{5} (k^2-2k+3)$

64. $\sum_{k=0}^{5} (k^2-3k+4)$

65. $\sum_{k=3}^{5} \dfrac{(-1)^k}{k(k+1)}$

66. $\sum_{k=3}^{7} \dfrac{k}{2^k}$

Rewrite each sum using sigma notation. Answers may vary.

67. $\dfrac{2}{3} + \dfrac{3}{4} + \dfrac{4}{5} + \dfrac{5}{6} + \dfrac{6}{7}$

68. $\dfrac{1}{1^2} + \dfrac{1}{2^2} + \dfrac{1}{3^2} + \dfrac{1}{4^2} + \dfrac{1}{5^2}$

69. $1 + 4 + 9 + 16 + 25 + 36$

70. $1 + \sqrt{2} + \sqrt{3} + 2 + \sqrt{5} + \sqrt{6}$

71. $4 - 9 + 16 - 25 + \cdots + (-1)^n n^2$

72. $9 - 16 + 25 + \cdots + (-1)^{n+1}n^2$

73. $6 + 12 + 18 + 24 + \cdots$

74. $11 + 22 + 33 + 44 + \cdots$

75. $\dfrac{1}{1\cdot 2} + \dfrac{1}{2\cdot 3} + \dfrac{1}{3\cdot 4} + \dfrac{1}{4\cdot 5} + \cdots$

76. $\dfrac{1}{1\cdot 2^2} + \dfrac{1}{2\cdot 3^2} + \dfrac{1}{3\cdot 4^2} + \dfrac{1}{4\cdot 5^2} + \cdots$

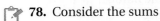 **77.** The sequence $1, 4, 9, 16, \ldots$ can be written as $f(x) = x^2$ with the domain the set of all positive integers. How would the graph of f compare with the graph of $y = x^2$?

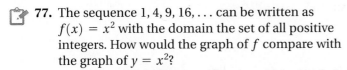 **78.** Consider the sums

$$\sum_{k=1}^{5} 3k^2 \quad \text{and} \quad 3\sum_{k=1}^{5} k^2.$$

Which is easier to evaluate and why?

Skill Review

Simplify.

79. $\dfrac{t^3 + 1}{t + 1}$ [6.1]

80. $\dfrac{x - a^{-1}}{a - x^{-1}}$ [6.3]

Perform the indicated operation and, if possible, simplify.

81. $\dfrac{3}{a^2 + a} + \dfrac{4}{2a^2 - 2}$ [6.2]

82. $\dfrac{t}{t-1} - \dfrac{t}{1-t}$ [6.2]

83. $\dfrac{x^2 - 6x + 8}{4x + 12} \cdot \dfrac{x+3}{x^2 - 4}$ [6.1]

84. $\dfrac{y^3 - y}{3y+1} \div \dfrac{y^2}{9y+3}$ [6.1]

Synthesis

85. Explain why the equation

$$\sum_{k=1}^{n} (a_k + b_k) = \sum_{k=1}^{n} a_k + \sum_{k=1}^{n} b_k$$

is true for any positive integer n. What laws are used to justify this result?

86. Is it true that

$$\sum_{k=1}^{n} ca_k = c\sum_{k=1}^{n} a_k?$$

Why or why not?

Some sequences are given by a recursive *definition. The value of the first term, a_1, is given, and then we are told how to find any subsequent term from the term preceding it. Find the first six terms of each of the following recursively defined sequences.*

87. $a_1 = 1, a_{n+1} = 5a_n - 2$

88. $a_1 = 0, a_{n+1} = (a_n)^2 + 3$

89. *Value of a Projector.* The value of an LCD projector is \$2500. Its resale value each year is 80% of its value the year before. Write a sequence listing the resale value of the machine at the start of each year for a 10-year period.

90. *Cell Biology.* A single cell of bacterium divides into two every 15 min. Suppose that the same rate of division is maintained for 4 hr. Write a sequence listing the number of cells after successive 15-min periods.

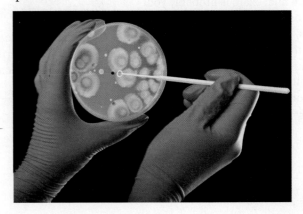

91. Find S_{100} and S_{101} for the sequence in which $a_n = (-1)^n$.

Find the first five terms of each sequence; then find S_5.

92. $a_n = \dfrac{1}{2^n} \log 1000^n$

93. $a_n = i^n, i = \sqrt{-1}$

94. Find all values for x that solve the following:

$$\sum_{k=1}^{x} i^k = -1.$$

95. The nth term of a sequence is given by
$$a_n = n^5 - 14n^4 + 6n^3 + 416n^2 - 655n - 1050.$$

Use a graphing calculator with a TABLE feature to determine which term in the sequence is 6144.

96. To define a sequence recursively on a graphing calculator (see Exercises 87 and 88), we use the SEQ MODE. The general term U_n or V_n can often be expressed in terms of U_{n-1} or V_{n-1} by pressing **2ND** **7** or **2ND** **8**. The starting values of U_n, V_n, and n are set as one of the WINDOW variables.

Use recursion to determine how many different handshakes occur when 50 people shake hands with one another. To develop the recursion formula, begin with a group of 2 and determine how many additional handshakes occur with the arrival of each new person.

⤷ **YOUR TURN ANSWERS: SECTION 11.1**

1. $-1, 4, -9, 16; 2500$ **2.** $a_n = 5n$ **3.** 0
4. $(2^0 + 5) + (2^1 + 5) + (2^2 + 5) + (2^3 + 5) = 35$
5. $\displaystyle\sum_{k=1}^{\infty} 10k$

PREPARE TO MOVE ON

Evaluate. [1.1]

1. $\dfrac{7}{2}(a_1 + a_7)$, for $a_1 = 8$ and $a_7 = 20$

2. $a_1 + (n - 1)d$, for $a_1 = 3$, $n = 10$, and $d = -2$

Simplify. [1.3]

3. $(a_1 + 5d) + (a_n - 5d)$

4. $(a_1 + 8d) - (a_1 + 7d)$

11.2 Arithmetic Sequences and Series

Arithmetic Sequences ▪ Sum of the First n Terms of an Arithmetic Sequence ▪ Problem Solving

Study Skills

Rest Before a Test

The final exam is probably your most important test of the semester. Do yourself a favor and see to it that you get a good night's sleep the night before. Being well rested will help you put forth your best work.

In this section, we concentrate on sequences and series that are said to be arithmetic (pronounced ar-ith-MET-ik).

Arithmetic Sequences

In an **arithmetic sequence** (or **progression**), adding the same number to any term gives the next term in the sequence. For example, the sequence 2, 5, 8, 11, 14, 17, . . . is arithmetic because adding 3 to any term produces the next term.

ARITHMETIC SEQUENCE

A sequence is *arithmetic* if there exists a number d, called the *common difference*, such that $a_{n+1} = a_n + d$ for any integer $n \geq 1$.

EXAMPLE 1 For each arithmetic sequence, identify the first term, a_1, and the common difference, d.

a) $4, 9, 14, 19, 24, \ldots$ b) $27, 20, 13, 6, -1, -8, \ldots$

SOLUTION To find a_1, we simply use the first term listed. To find d, we choose any term other than a_1 and subtract the preceding term from it.

Sequence	First Term, a_1	Common Difference, d
a) $4, 9, 14, 19, 24, \ldots$	4	$5 \leftarrow 9 - 4 = 5$
b) $27, 20, 13, 6, -1, -8, \ldots$	27	$-7 \leftarrow 20 - 27 = -7$

To find the common difference, we subtracted a_1 from a_2. Had we subtracted a_2 from a_3 or a_3 from a_4, we would have found the same values for d.

Check: As a check, note that when d is added to each term, the result is the next term in the sequence.

YOUR TURN

1. Identify the first term, a_1, and the common difference, d, for the arithmetic sequence

$$0, \tfrac{1}{2}, 1, \tfrac{3}{2}, 2, \ldots.$$

To develop a formula for the general, nth, term of any arithmetic sequence, we denote the common difference by d and write out the first few terms:

$a_1,$

$a_2 = a_1 + d,$

$a_3 = a_2 + d = (a_1 + d) + d = a_1 + 2d,$ Substituting $a_1 + d$ for a_2

$a_4 = a_3 + d = (a_1 + 2d) + d = a_1 + 3d.$ Substituting $a_1 + 2d$ for a_3

Note that the coefficient of d in each case is 1 less than the subscript.

Generalizing, we obtain the following formula.

TO FIND a_n FOR AN ARITHMETIC SEQUENCE

The nth term of an arithmetic sequence with common difference d is

$$a_n = a_1 + (n - 1)d, \quad \text{for any integer } n \geq 1.$$

EXAMPLE 2 Find the 14th term of the arithmetic sequence $6, 9, 12, 15, \ldots.$

SOLUTION First we note that $a_1 = 6$, $d = 3$, and $n = 14$. Using the formula for the nth term of an arithmetic sequence, we have

$$a_n = a_1 + (n - 1)d$$
$$a_{14} = 6 + (14 - 1) \cdot 3 = 6 + 13 \cdot 3 = 6 + 39 = 45.$$

The 14th term is 45.

2. Find the 20th term of the arithmetic sequence

$$100, 97, 94, 91, \ldots.$$

YOUR TURN

EXAMPLE 3   For the sequence 6, 9, 12, 15, ..., which term is 300?

SOLUTION   Determining which term is 300 is the same as finding n if $a_n = 300$. In Example 2, we found that for this sequence we have $a_1 = 6$ and $d = 3$. Thus,

$$a_n = a_1 + (n - 1)d \qquad \text{Using the formula for the } n\text{th term}$$
$$\text{of an arithmetic sequence}$$
$$300 = 6 + (n - 1) \cdot 3 \qquad \text{Substituting}$$
$$300 = 6 + 3n - 3$$
$$297 = 3n$$
$$99 = n.$$

3. For the sequence

$100, 97, 94, 91, \ldots,$

which term is -8?

The term 300 is the 99th term of the sequence.

YOUR TURN

Given two terms and their places in an arithmetic sequence, we can construct the sequence.

EXAMPLE 4   The 3rd term of an arithmetic sequence is 14, and the 16th term is 79. Find a_1 and d and construct the sequence.

SOLUTION   We know that $a_3 = 14$ and $a_{16} = 79$. Thus we would have to add d a total of 13 times to get from 14 to 79. That is,

$$14 + 13d = 79. \qquad a_3 \text{ and } a_{16} \text{ are 13 terms apart; } 16 - 3 = 13$$

Solving $14 + 13d = 79$, we obtain

$$13d = 65 \qquad \text{Subtracting 14 from both sides}$$
$$d = 5. \qquad \text{Dividing both sides by 13}$$

We subtract d twice from a_3 to get to a_1. Thus,

4. The 4th term of an arithmetic sequence is 5, and the 21st term is 175. Find a_1 and d and construct the sequence.

$$a_1 = 14 - 2 \cdot 5 = 4. \qquad a_1 \text{ and } a_3 \text{ are 2 terms apart; } 3 - 1 = 2$$

The sequence is 4, 9, 14, 19, ... Note that we could have subtracted d a total of 15 times from a_{16} in order to find a_1.

YOUR TURN

In general, d should be subtracted $(n - 1)$ times from a_n in order to find a_1.

Sum of the First n Terms of an Arithmetic Sequence

When the terms of an arithmetic sequence are added, an **arithmetic series** is formed. To develop a formula for computing S_n when the series is arithmetic, we list the first n terms of the sequence as follows:

This is the next-to-last term. If you add d to this term, the result is a_n.

$$a_1, (a_1 + d), (a_1 + 2d), \ldots, (a_n - 2d), \overline{(a_n - d)}, a_n$$

This term is two terms back from the end. If you add d to this term, you get the next-to-last term, $a_n - d$.

Thus, S_n is given by

$$S_n = a_1 + (a_1 + d) + (a_1 + 2d) + \cdots + (a_n - 2d) + (a_n - d) + a_n.$$

Using a commutative law, we have a second equation:

$$S_n = a_n + (a_n - d) + (a_n - 2d) + \cdots + (a_1 + 2d) + (a_1 + d) + a_1.$$

Adding corresponding terms on each side of the two equations above, we get

$$2S_n = [a_1 + a_n] + [(a_1 + d) + (a_n - d)] + [(a_1 + 2d) + (a_n - 2d)]$$
$$+ \cdots + [(a_n - 2d) + (a_1 + 2d)] + [(a_n - d) + (a_1 + d)] + [a_n + a_1].$$

This simplifies to

$$2S_n = [a_1 + a_n] + [a_1 + a_n] + [a_1 + a_n]$$
$$+ \cdots + [a_n + a_1] + [a_n + a_1] + [a_n + a_1].$$ There are n bracketed sums.

Since $[a_1 + a_n]$ is being added n times, it follows that

$$2S_n = n[a_1 + a_n].$$

Dividing both sides by 2 leads to the following formula.

Student Notes

The formula for the sum of an arithmetic sequence is very useful, but remember that it does not work for sequences that are not arithmetic.

TO FIND S_n FOR AN ARITHMETIC SEQUENCE

The sum of the first n terms of an arithmetic sequence is given by

$$S_n = \frac{n}{2}(a_1 + a_n).$$

EXAMPLE 5   Find the sum of the first 100 positive even numbers.

SOLUTION   The sum is

$$2 + 4 + 6 + \cdots + 198 + 200.$$

This is the sum of the first 100 terms of the arithmetic sequence for which

$$a_1 = 2, \quad n = 100, \quad \text{and} \quad a_n = 200.$$

We use the formula for S_n for an arithmetic sequence:

$$S_n = \frac{n}{2}(a_1 + a_n),$$

5. Find the sum of the first 100 positive odd numbers.

$$S_{100} = \frac{100}{2}(2 + 200) = 50(202) = 10{,}100.$$

 YOUR TURN

The above formula is useful when we know the first and last terms, a_1 and a_n. To find S_n when a_n is unknown, but a_1, n, and d are known, we can use $a_n = a_1 + (n - 1)d$ to calculate a_n and then proceed as in Example 5.

EXAMPLE 6   Find the sum of the first 15 terms of the arithmetic sequence $13, 10, 7, 4, \ldots$.

SOLUTION   Note that

$$a_1 = 13, \quad n = 15, \quad \text{and} \quad d = -3.$$

Before using the formula for S_n, we find a_{15}:

$$a_{15} = 13 + (15 - 1)(-3) \quad \text{Substituting into the formula for } a_n$$
$$= 13 + 14(-3) = -29.$$

6. Find the sum of the first 18 terms of the arithmetic sequence

$$10, 8, 6, 4, \ldots.$$

Knowing that $a_{15} = -29$, we have

$$S_{15} = \frac{15}{2}(13 + (-29)) \quad \text{Using the formula for } S_n$$
$$= \frac{15}{2}(-16) = -120.$$

YOUR TURN

Problem Solving

In problem-solving situations, translation may involve sequences or series. As always, there is often a variety of ways in which a problem can be solved. You should use the approach that is best or easiest for you. In this chapter, however, we will try to emphasize sequences and series and their related formulas.

EXAMPLE 7 *Hourly Wages.* Chris accepts a job managing a music store, starting with an hourly wage of $14.60, and is promised a raise of 25¢ per hour every 2 months for 5 years. After 5 years of work, what will be Chris's hourly wage?

SOLUTION

1. **Familiarize.** It helps to write down the hourly wage for several two-month time periods.

   Beginning: 14.60,

   After two months: 14.85,

   After four months: 15.10,

   and so on.

   What appears is a sequence of numbers: 14.60, 14.85, 15.10, ... Since the same amount is added each time, the sequence is arithmetic.

   Because we want to know a particular term in the sequence, we will use the formula $a_n = a_1 + (n - 1)d$. To do so, we need a_1, n, and d. From our list above, we have

   $$a_1 = 14.60 \quad \text{and} \quad d = 0.25.$$

   What is n? After 1 year, there have been 6 raises, since Chris gets a raise every 2 months. There are 5 years, so the total number of raises will be $5 \cdot 6$, or 30. Altogether, there will be 31 terms: the original wage and 30 increased rates.

2. **Translate.** We want to find a_n for the arithmetic sequence in which $a_1 = 14.60$, $n = 31$, and $d = 0.25$.

3. **Carry out.** Substituting in the formula for a_n gives us

   $$a_{31} = 14.60 + (31 - 1) \cdot 0.25$$
   $$= 22.10.$$

4. **Check.** We can check by redoing the calculations or we can calculate in a slightly different way for another check. For example, at the end of a year, there will be 6 raises, for a total raise of $1.50. At the end of 5 years, the total raise will be $5 \times \$1.50$, or $7.50. If we add that to the original wage of $14.60, we obtain $22.10. The answer checks.

5. **State.** After 5 years, Chris's hourly wage will be $22.10.

7. Refer to Example 7. Chris takes a different job, with a starting hourly wage of $12.80 and a promised raise of 40¢ per hour every 2 months for 5 years. After 5 years of work, what will be Chris's hourly wage?

YOUR TURN

EXAMPLE 8 *Telephone Pole Storage.* A stack of telephone poles has 30 poles in the bottom row. There are 29 poles in the second row, 28 in the next row, and so on. How many poles are in the stack if there are 5 poles in the top row?

SOLUTION

1. **Familiarize.** The following figure shows the ends of the poles.

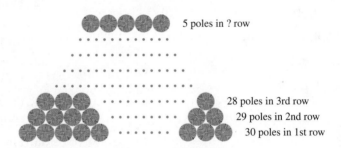

5 poles in ? row

28 poles in 3rd row
29 poles in 2nd row
30 poles in 1st row

Note that there are $30 - 1 = 29$ poles in the 2nd row, $30 - 2 = 28$ poles in the 3rd row, $30 - 3 = 27$ poles in the 4th row, and so on. The pattern leads to $30 - 25 = 5$ poles in the 26th row.

The situation is represented by the equation

$30 + 29 + 28 + \cdots + 5.$ There are 26 terms in this series.

Thus we have an arithmetic series. We recall the formula

$$S_n = \frac{n}{2}(a_1 + a_n).$$

2. **Translate.** We want to find the sum of the first 26 terms of an arithmetic sequence in which $a_1 = 30$ and $a_{26} = 5$.

3. **Carry out.** Substituting into the above formula gives us

$S_{26} = \frac{26}{2}(30 + 5)$

$= 13 \cdot 35 = 455.$

4. **Check.** In this case, we can check the calculations by doing them again. A longer, more difficult way would be to do the entire addition:

$30 + 29 + 28 + \cdots + 5.$

5. **State.** There are 455 poles in the stack.

8. How many poles will be in a pile of telephone poles if there are 50 in the first layer, 49 in the second, and so on, until there are 6 in the top layer?

 YOUR TURN

11.2 EXERCISE SET

FOR EXTRA HELP

MyMathLab® Math XL
PRACTICE   WATCH   READ   REVIEW

➡ Vocabulary and Reading Check

Choose the expression that best completes each statement. Not every expression will be used.

arithmetic sequence   first term
arithmetic series   sum
common difference

1. $5 + 7 + 9 + 11$ is an example of a(n) _____.

2. In an arithmetic sequence, subtracting a_n from a_{n+1} will give the _____.

3. In $5, 7, 9, 11$, the _____ is 5.

4. For $5 + 7 + 9 + 11$, the expression $\frac{4}{2}(5 + 11)$ gives the _____.

Arithmetic Sequences

Find the first term and the common difference.

5. $8, 13, 18, 23, \ldots$

6. $2.5, 3, 3.5, 4, \ldots$

7. $7, 3, -1, -5, \ldots$

8. $-8, -5, -2, 1, \ldots$

9. $\frac{3}{2}, \frac{9}{4}, 3, \frac{15}{4}, \ldots$

10. $\frac{3}{5}, \frac{1}{10}, -\frac{2}{5}, \ldots$

11. $\$8.16, \$8.46, \$8.76, \$9.06, \ldots$

12. $\$825, \$804, \$783, \$762, \ldots$

13. Find the 19th term of the arithmetic sequence $10, 18, 26, \ldots$.

14. Find the 23rd term of the arithmetic sequence $10, 16, 22, \ldots$.

15. Find the 18th term of the arithmetic sequence $8, 2, -4, \ldots$.

16. Find the 14th term of the arithmetic sequence $3, \frac{7}{3}, \frac{5}{3}, \ldots$.

17. Find the 13th term of the arithmetic sequence $\$1200, \$964.32, \$728.64, \ldots$.

18. Find the 10th term of the arithmetic sequence $\$2345.78, \$2967.54, \$3589.30, \ldots$.

19. In the sequence of Exercise 13, what term is 210?

20. In the sequence of Exercise 14, what term is 208?

21. In the sequence of Exercise 15, what term is -328?

22. In the sequence of Exercise 16, what term is -27?

23. Find a_{18} when $a_1 = 8$ and $d = 10$.

24. Find a_{20} when $a_1 = 12$ and $d = -5$.

25. Find a_1 when $d = 4$ and $a_8 = 33$.

26. Find a_1 when $d = 8$ and $a_{11} = 26$.

27. Find n when $a_1 = 5, d = -3$, and $a_n = -76$.

28. Find n when $a_1 = 25, d = -14$, and $a_n = -507$.

29. For an arithmetic sequence in which $a_{17} = -40$ and $a_{28} = -73$, find a_1 and d. Write the first five terms of the sequence.

30. In an arithmetic sequence, $a_{17} = \frac{25}{3}$ and $a_{32} = \frac{95}{6}$. Find a_1 and d. Write the first five terms of the sequence.

Aha! 31. Find a_1 and d if $a_{13} = 13$ and $a_{54} = 54$.

32. Find a_1 and d if $a_{12} = 24$ and $a_{25} = 50$.

Sum of the First *n* Terms of an Arithmetic Sequence

33. Find the sum of the first 20 terms of the arithmetic series $1 + 5 + 9 + 13 + \cdots$.

34. Find the sum of the first 14 terms of the arithmetic series $11 + 7 + 3 + \cdots$.

35. Find the sum of the first 250 natural numbers.

36. Find the sum of the first 400 natural numbers.

37. Find the sum of the even numbers from 2 to 100, inclusive.

38. Find the sum of the odd numbers from 1 to 99, inclusive.

39. Find the sum of all multiples of 6 from 6 to 102, inclusive.

40. Find the sum of all multiples of 4 that are between 15 and 521.

41. An arithmetic series has $a_1 = 4$ and $d = 5$. Find S_{20}.

42. An arithmetic series has $a_1 = 9$ and $d = -3$. Find S_{32}.

Problem Solving

Solve.

43. *Band Formations.* The South Brighton Drum and Bugle Corps has 7 musicians in the front row, 9 in the second row, 11 in the third row, and so on, for 15 rows. How many musicians are in the last row? How many musicians are there altogether?

44. *Gardening.* A gardener is planting tulip bulbs at the entrance to a college. She puts 50 bulbs in the first row, 46 in the second row, 42 in the third row, and so on, for 13 rows. How many bulbs will be in the last row? How many bulbs will she plant altogether?

45. *Archaeology.* Many ancient Mayan pyramids were constructed over a span of several generations. Each layer of the pyramid has a stone perimeter, enclosing a layer of dirt or debris on which a structure once stood. One drawing of such a pyramid indicates that the perimeter of the bottom layer contains 36 stones, the next level up contains 32 stones, and so on, up to the top row, which contains 4 stones. How many stones are in the pyramid?

46. *Auditorium Design.* Theaters are often built with more seats per row as the rows move toward the back. The Community Theater has 20 seats in the first row, 22 in the second, 24 in the third, and so on, for 16 rows. How many seats are in the theater?

47. *Accumulated Savings.* If 10¢ is saved on October 1, another 20¢ on October 2, another 30¢ on October 3, and so on, how much is saved during October? (October has 31 days.)

48. *Accumulated Savings.* Carrie saves money in an arithmetic sequence: $700 for the first year, another $850 the second, and so on, for 20 years. How much does she save in all (disregarding interest)?

49. It is said that as a young child, the mathematician Karl F. Gauss (1777–1855) was able to compute the sum $1 + 2 + 3 + \cdots + 100$ very quickly in his head. Explain how Gauss might have done this and present a formula for the sum of the first n natural numbers. (*Hint*: $1 + 99 = 100$.)

50. Write a problem for a classmate to solve. Devise the problem so that its solution requires computing S_{17} for an arithmetic sequence.

Skill Review

Find an equation of the line satisfying the given conditions.

51. Slope $\frac{1}{3}$, y-intercept $(0, 10)$ [2.3]

52. Containing the points $(2, 3)$ and $(4, -5)$ [2.5]

53. Containing the point $(5, 0)$ and parallel to the line given by $2x + y = 8$ [2.5]

54. Containing the point $(-1, -4)$ and perpendicular to the line given by $3x - 4y = 7$ [2.5]

Find an equation of the circle satisfying the given conditions.   [10.1]

55. Center $(0, 0)$, radius 4

56. Center $(-2, 1)$, radius $2\sqrt{5}$

Synthesis

57. When every term in an arithmetic sequence is an integer, S_n must also be an integer. Given that n, a_1, and a_n may each, at times, be even or odd, explain why $\dfrac{n}{2}(a_1 + a_n)$ is always an integer.

58. The sum of the first n terms of an arithmetic sequence is also given by

$$S_n = \frac{n}{2}\big[2a_1 + (n - 1)d\big].$$

Use the earlier formulas for a_n and S_n to explain how this equation was developed.

59. A frog is at the bottom of a 100-ft well. With each jump, the frog climbs 4 ft, but then slips back 1 ft. How many jumps does it take for the frog to reach the top of the hole?

60. Find a formula for the sum of the first n consecutive odd numbers starting with 1:

$$1 + 3 + 5 + \cdots + (2n - 1).$$

61. Prove that if p, m, and q are consecutive terms in an arithmetic sequence, then

$$m = \frac{p + q}{2}.$$

62. *Straight-Line Depreciation.* A company buys a copier for $5200 on January 1 of a given year. The machine is expected to last for 8 years, at the end of which time its *trade-in*, or *salvage, value* will be $1100. If the company figures the decline in value to be the same each year, then the trade-in values, after t years, $0 \le t \le 8$, form an arithmetic sequence given by

$$a_t = C - t\left(\frac{C - S}{N}\right),$$

where C is the original cost of the item, N the years of expected life, and S the salvage value.

a) Find the formula for a_t for the straight-line depreciation of the copier.

b) Find the trade-in value after 0 year, 1 year, 2 years, 3 years, 4 years, 7 years, and 8 years.

c) Find a formula that expresses a_t recursively. (See Exercises 87 and 88 in Exercise Set 11.1.)

63. Use your answer to Exercise 35 to find the sum of all integers from 501 through 750.

> **YOUR TURN ANSWERS: SECTION 11.2**
> **1.** $a_1 = 0; d = \frac{1}{2}$ **2.** 43 **3.** 37th
> **4.** $a_1 = -25; d = 10; -25, -15, -5, 5, 15, \ldots$ **5.** 10,000
> **6.** -126 **7.** \$24.80 **8.** 1260 poles

QUICK QUIZ: SECTIONS 11.1–11.2

1. Find a_5 for a sequence with the general term $a_n = n^2 - n - 1$. [11.1]

2. Write an expression for the general term, or nth term, a_n, of the sequence $0, 1, 4, 9, 16, \ldots$. Answers may vary. [11.1]

3. Rewrite using sigma notation: $2 + 4 + 6 + 8 + \cdots$. Answers may vary. [11.1]

4. Find the common difference for the arithmetic sequence $10, 9.5, 9, 8.5, \ldots$. [11.2]

5. Find the 19th term of the arithmetic sequence $10, 9.5, 9, 8.5, \ldots$. [11.2]

PREPARE TO MOVE ON

Evaluate.

1. $\dfrac{a_1(1 - r^n)}{1 - r}$, for $a_1 = 5$, $n = 6$, and $r = 2$ [1.6]

2. $\dfrac{a_1}{1 - r}$, for $a_1 = 25$ and $r = \dfrac{1}{2}$ [1.2]

3. $\dfrac{a_1}{1 - r}$, for $a_1 = 0.6$ and $r = 0.1$ [1.2]

11.3 Geometric Sequences and Series

Geometric Sequences ◼ Sum of the First n Terms of a Geometric Sequence ◼ Infinite Geometric Series ◼
Problem Solving

Study Skills

Ask to See Your Final

Once a course is over, many students neglect to find out how they fared on the final exam. Please don't overlook this valuable opportunity to extend your learning. It is important for you to find out what mistakes you may have made and to be certain no grading errors have occurred.

In an arithmetic sequence, a certain number is added to each term to get the next term. When each term in a sequence is *multiplied* by a certain fixed number to get the next term, the sequence is **geometric**. In this section, we examine both geometric sequences (or progressions) and geometric series.

Geometric Sequences

Consider the sequence

$$2, 6, 18, 54, 162, \ldots$$

If we multiply each term by 3, we obtain the next term. The multiplier is called the *common ratio* because it is found by dividing any term by the preceding term.

> **GEOMETRIC SEQUENCE**
>
> A sequence is *geometric* if there exists a number r, called the *common ratio*, for which
>
> $$\frac{a_{n+1}}{a_n} = r, \quad \text{or} \quad a_{n+1} = a_n \cdot r \quad \text{for any integer } n \geq 1.$$

EXAMPLE 1 For each geometric sequence, find the common ratio.

a) $4, 20, 100, 500, 2500, \ldots$

b) $3, -6, 12, -24, 48, -96, \ldots$

c) $\$5200, \$3900, \$2925, \$2193.75, \ldots$

SOLUTION

Sequence		*Common Ratio*	
a) $4, 20, 100, 500, 2500, \ldots$	5	$\frac{20}{4} = 5, \frac{100}{20} = 5$, and so on	
b) $3, -6, 12, -24, 48, -96, \ldots$	-2	$\frac{-6}{3} = -2, \frac{12}{-6} = -2$, and so on	
c) $\$5200, \$3900, \$2925, \$2193.75, \ldots$	0.75	$\frac{\$3900}{\$5200} = 0.75, \frac{\$2925}{\$3900} = 0.75$	

1. Find the common ratio for the geometric sequence

$$20, 10, 5, 2\tfrac{1}{2}, \ldots.$$

 YOUR TURN

Note that when the signs of the terms alternate, the common ratio is negative.

To develop a formula for the general, or *n*th, term of a geometric sequence, let a_1 be the first term and let r be the common ratio. We write out a few terms:

$$a_1,$$
$$a_2 = a_1 r,$$
$$a_3 = a_2 r = (a_1 r)r = a_1 r^2, \qquad \text{Substituting } a_1 r \text{ for } a_2$$
$$a_4 = a_3 r = (a_1 r^2)r = a_1 r^3. \qquad \text{Substituting } a_1 r^2 \text{ for } a_3$$

Note that the exponent is 1 less than the subscript.

Generalizing, we obtain the following.

> **TO FIND a_n FOR A GEOMETRIC SEQUENCE**
>
> The *n*th term of a geometric sequence with common ratio r is given by
>
> $$a_n = a_1 r^{n-1}, \quad \text{for any integer } n \geq 1.$$

EXAMPLE 2 Find the 7th term of the geometric sequence $4, 20, 100, \ldots$.

SOLUTION First, we note that

$$a_1 = 4 \quad \text{and} \quad n = 7.$$

To find the common ratio, we can divide any term (other than the first) by the term preceding it. Since the second term is 20 and the first is 4,

$$r = \frac{20}{4}, \quad \text{or } 5.$$

2. Find the 8th term of the geometric sequence

$$3, 6, 12, \ldots.$$

Substituting in the formula $a_n = a_1 r^{n-1}$, we have

$$a_7 = 4 \cdot 5^{7-1} = 4 \cdot 5^6 = 4 \cdot 15{,}625 = 62{,}500.$$

 YOUR TURN

EXAMPLE 3   Find the 10th term of the geometric sequence

$$64, -32, 16, -8, \ldots.$$

SOLUTION   First, we note that

$$a_1 = 64, \quad n = 10, \quad \text{and} \quad r = \frac{-32}{64} = -\frac{1}{2}.$$

Then, using the formula for the nth term of a geometric sequence, we have

$$a_{10} = 64 \cdot \left(-\frac{1}{2}\right)^{10-1} = 64 \cdot \left(-\frac{1}{2}\right)^9 = 2^6 \cdot \left(-\frac{1}{2^9}\right) = -\frac{1}{2^3} = -\frac{1}{8}.$$

3. Find the 9th term of the geometric sequence

$$-5, 10, -20, 40, \ldots.$$

YOUR TURN

Sum of the First n Terms of a Geometric Sequence

We next develop a formula for S_n when a sequence is geometric:

$$a_1, \ a_1 r, \ a_1 r^2, \ a_1 r^3, \ldots, a_1 r^{n-1}, \ldots.$$

The **geometric series** S_n is given by

$$S_n = a_1 + a_1 r + a_1 r^2 + \cdots + a_1 r^{n-2} + a_1 r^{n-1}. \tag{1}$$

Multiplying both sides by r gives us

$$r S_n = a_1 r + a_1 r^2 + a_1 r^3 + \cdots + a_1 r^{n-1} + a_1 r^n. \tag{2}$$

When we subtract corresponding sides of equation (2) from equation (1), the color terms drop out, leaving

$$S_n - r S_n = a_1 - a_1 r^n$$
$$S_n(1 - r) = a_1(1 - r^n), \quad \text{Factoring}$$

or

$$S_n = \frac{a_1(1 - r^n)}{1 - r}. \quad \text{Dividing both sides by } 1 - r$$

Student Notes

The three determining characteristics of a geometric sequence or series are the first term (a_1), the number of terms (n), and the common ratio (r). Be sure you understand how to use these characteristics to write out a sequence or a series.

TO FIND S_n FOR A GEOMETRIC SEQUENCE

The sum of the first n terms of a geometric sequence with common ratio r is given by

$$S_n = \frac{a_1(1 - r^n)}{1 - r}, \quad \text{for any } r \neq 1.$$

EXAMPLE 4   Find the sum of the first 7 terms of the geometric sequence $3, 15, 75, 375, \ldots.$

SOLUTION   First, we note that

$$a_1 = 3, \quad n = 7, \quad \text{and} \quad r = \frac{15}{3} = 5.$$

Then, substituting in the formula $S_n = \dfrac{a_1(1 - r^n)}{1 - r}$, we have

4. Find the sum of the first 9 terms of the geometric sequence

$$-5, 10, -20, 40, \ldots.$$

$$S_7 = \frac{3(1 - 5^7)}{1 - 5} = \frac{3(1 - 78{,}125)}{-4}$$

$$= \frac{3(-78{,}124)}{-4} = 58{,}593.$$

 YOUR TURN

Infinite Geometric Series

Suppose we consider the sum of the terms of an infinite geometric sequence, such as 3, 6, 12, 24, 48, ... We get what is called an **infinite geometric series**:

$$3 + 6 + 12 + 24 + 48 + \cdots.$$

Here, as n increases, the sum of the first n terms, S_n, increases without bound. There are also infinite series that get closer and closer to some specific number. For example, consider the sequence

$$\frac{1}{2} + \frac{1}{4} + \frac{1}{8} + \frac{1}{16} + \cdots + \frac{1}{2^n} + \cdots,$$

and evaluate S_n for the first four values of n:

$$S_1 = \tfrac{1}{2} \qquad\qquad = \tfrac{1}{2} = 0.5,$$
$$S_2 = \tfrac{1}{2} + \tfrac{1}{4} \qquad\quad = \tfrac{3}{4} = 0.75,$$
$$S_3 = \tfrac{1}{2} + \tfrac{1}{4} + \tfrac{1}{8} \qquad = \tfrac{7}{8} = 0.875,$$
$$S_4 = \tfrac{1}{2} + \tfrac{1}{4} + \tfrac{1}{8} + \tfrac{1}{16} = \tfrac{15}{16} = 0.9375.$$

The denominator of each sum is 2^n, where n is the subscript of S. The numerator is $2^n - 1$.

Thus, for this particular series, we have

$$S_n = \frac{2^n - 1}{2^n} = \frac{2^n}{2^n} - \frac{1}{2^n} = 1 - \frac{1}{2^n}.$$

Note that the value of S_n is less than 1 for any value of n, but as n gets larger and larger, the value of $1/2^n$ gets closer to 0, so the value of S_n gets closer to 1. We can visualize S_n by considering a square with area 1. For S_1, we shade half the square. For S_2, we shade half the square plus half the remaining part, or $\tfrac{1}{4}$. For S_3, we shade the parts shaded in S_2 plus half the remaining part. Again we see that the values of S_n will continue to get close to 1 (shading the complete square). We say that 1 is the **limit** of S_n and that 1 is the sum of this infinite geometric series.

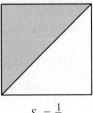 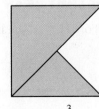 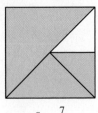 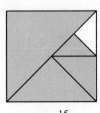

$$S_1 = \tfrac{1}{2} \qquad S_2 = \tfrac{3}{4} \qquad S_3 = \tfrac{7}{8} \qquad S_4 = \tfrac{15}{16}$$

An infinite geometric series is denoted S_∞. It can be shown (but we will not do so here) that the sum of the terms of an infinite geometric sequence exists if and only if $|r| < 1$ (that is, the common ratio's absolute value is less than 1).

To find a formula for the sum of an infinite geometric series, we first consider the sum of the first n terms:

$$S_n = \frac{a_1(1 - r^n)}{1 - r} = \frac{a_1 - a_1 r^n}{1 - r}. \qquad \text{Using the distributive law}$$

For $|r| < 1$, it follows that the value of r^n gets closer to 0 as n gets larger. (Check this by selecting a number between -1 and 1 and finding larger and larger powers on a calculator.) As r^n gets closer to 0, so too does $a_1 r^n$. Thus, S_n gets closer to $a_1/(1 - r)$.

EXPLORING 🔍 THE CONCEPT

Graphically, a geometric series has a limit if the graph of the sequence gets closer to $n = 0$ as n increases.

1. For which of the following sequences does it appear that the series will have a limit?

A.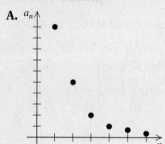

B.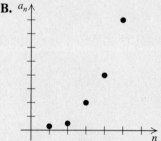

C.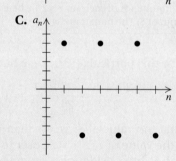

D.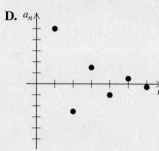

ANSWER
1. A, D

THE LIMIT OF AN INFINITE GEOMETRIC SERIES

For $|r| < 1$, the limit of an infinite geometric series is given by

$$S_\infty = \frac{a_1}{1 - r}. \qquad \text{(For } |r| \geq 1, \text{ no limit exists.)}$$

EXAMPLE 5 Determine whether each series has a limit. If a limit exists, find it.

a) $1 + 3 + 9 + 27 + \cdots$ **b)** $-35 + 7 - \frac{7}{5} + \frac{7}{25} + \cdots$

SOLUTION

a) Here $r = 3$, so $|r| = |3| = 3$. Since $|r| \not< 1$, the series *does not* have a limit.

b) Here $r = -\frac{1}{5}$, so $|r| = |-\frac{1}{5}| = \frac{1}{5}$. Since $|r| < 1$, the series *does* have a limit. We find the limit by substituting into the formula for S_∞:

$$S_\infty = \frac{-35}{1 - \left(-\frac{1}{5}\right)} = \frac{-35}{\frac{6}{5}} = -35 \cdot \frac{5}{6} = \frac{-175}{6} = -29\frac{1}{6}.$$

5. Determine whether $20 + 10 + 5 + \cdots$ has a limit. If a limit exists, find it.

 YOUR TURN

EXAMPLE 6 Find fraction notation for 0.63636363. . . .

SOLUTION We can express this as

$$0.63 + 0.0063 + 0.000063 + \cdots.$$

This is an infinite geometric series, where $a_1 = 0.63$ and $r = 0.01$. Since $|r| < 1$, this series has a limit:

$$S_\infty = \frac{a_1}{1 - r} = \frac{0.63}{1 - 0.01} = \frac{0.63}{0.99} = \frac{63}{99}.$$

6. Find fraction notation for 0.575757. . . .

Thus fraction notation for 0.63636363 . . . is $\frac{63}{99}$, or $\frac{7}{11}$.

YOUR TURN

CONNECTING 🔗 THE CONCEPTS

If a sequence is arithmetic or geometric, the general term and a partial sum can be found using a formula. An infinite sum exists only for geometric series with $|r| < 1$.

Arithmetic Sequences	Geometric Sequences
Common difference: d	Common ratio: r
$a_n = a_1 + (n - 1)d$	$a_n = a_1 r^{n-1}$
$S_n = \frac{n}{2}(a_1 + a_n)$	$S_n = \frac{a_1(1 - r^n)}{1 - r}$; $S_\infty = \frac{a_1}{1 - r}, |r| < 1$

EXERCISES

1. Find the common difference for the arithmetic sequence 115, 112, 109, 106, ...

2. Find the common ratio for the geometric sequence $\frac{1}{3}, -\frac{1}{6}, \frac{1}{12}, -\frac{1}{24}, \ldots$.

3. Find the 21st term of the arithmetic sequence 10, 15, 20, 25, ...

4. Find the 8th term of the geometric sequence 5, 10, 20, 40, ...

5. Find S_{30} for the arithmetic series $2 + 12 + 22 + 32 + \cdots$.

6. Find S_{10} for the geometric series $\$100 + \$100(1.03) + \$100(1.03)^2 + \cdots$.

7. Determine whether the infinite geometric series $0.9 + 0.09 + 0.009 + \cdots$ has a limit. If a limit exists, find it.

8. Determine whether the infinite geometric series $0.9 + 9 + 90 + \cdots$ has a limit. If a limit exists, find it.

Problem Solving

For some problem-solving situations, the translation may involve geometric sequences or series.

EXAMPLE 7 *Loan Repayment.* Francine's student loan is in the amount of $6000. Interest is 9% compounded annually, and the entire amount is to be paid after 10 years. How much is to be paid back?

SOLUTION

1. **Familiarize.** Suppose we let P represent any principal amount. At the end of one year, the amount owed will be $P + 0.09P$, or $1.09P$. That amount will be the principal for the second year. The amount owed at the end of the second year will be $1.09 \times$ New principal $= 1.09(1.09P)$, or 1.09^2P. Thus the amount owed at the beginning of successive years is as follows:

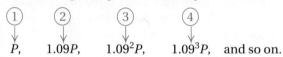

$$P, \quad 1.09P, \quad 1.09^2P, \quad 1.09^3P, \quad \text{and so on.}$$

We have a geometric sequence. The amount owed at the beginning of the 11th year will be the amount owed at the end of the 10th year.

2. **Translate.** This is a geometric sequence with $a_1 = 6000$, $r = 1.09$, and $n = 11$. The appropriate formula for finding the nth term is

$$a_n = a_1 r^{n-1}.$$

3. **Carry out.** We substitute and calculate:

$$a_{11} = \$6000(1.09)^{11-1} = \$6000(1.09)^{10}$$

$$\approx \$14{,}204.18. \qquad \text{Using a calculator and rounding to the nearest hundredth}$$

4. **Check.** A check, by repeating the calculations, is left to the student.

5. **State.** Francine will owe $14,204.18 at the end of 10 years.

7. Refer to Example 7. If Francine's loan amount is $8000, with 5% interest compounded annually, how much is owed after 10 years?

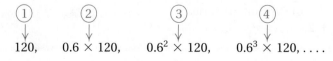
YOUR TURN

EXAMPLE 8 *Bungee Jumping.* A bungee jumper rebounds 60% of the height jumped. Clyde's bungee jump is made using a cord that stretches to 200 ft.

a) After jumping and then rebounding 9 times, how far has Clyde traveled upward (the total rebound distance)?

b) Theoretically, how far will Clyde travel upward (bounce) before coming to rest?

SOLUTION

1. **Familiarize.** Let's do some calculations and look for a pattern.

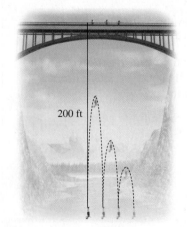

200 ft

First fall:	200 ft
First rebound:	0.6×200, or 120 ft
Second fall:	120 ft, or 0.6×200
Second rebound:	0.6×120, or $0.6(0.6 \times 200)$, which is 72 ft
Third fall:	72 ft, or $0.6(0.6 \times 200)$
Third rebound:	0.6×72, or $0.6(0.6(0.6 \times 200))$, which is 43.2 ft

The rebound distances form a geometric sequence:

$$120, \quad 0.6 \times 120, \quad 0.6^2 \times 120, \quad 0.6^3 \times 120, \ldots.$$

2. **Translate.**

   a) The total rebound distance after 9 bounces is the sum of a geometric sequence. The first term is 120 and the common ratio is 0.6. There will be 9 terms, so we can use the formula

   $$S_n = \frac{a_1(1 - r^n)}{1 - r}.$$

   b) Theoretically, Clyde will never stop bouncing. Realistically, the bouncing will eventually stop. To approximate the actual distance bounced, we consider an infinite number of bounces and use the formula

   $$S_\infty = \frac{a_1}{1 - r}. \qquad \text{Since } r = 0.6 \text{ and } |0.6| < 1, \text{ we know that } S_\infty \text{ exists.}$$

3. **Carry out.**

   a) We substitute into the formula and calculate:

   $$S_9 = \frac{120[1 - (0.6)^9]}{1 - 0.6} \approx 297. \qquad \text{Using a calculator}$$

   b) We substitute and calculate:

   $$S_\infty = \frac{120}{1 - 0.6} = 300.$$

4. **Check.** We can do the calculations again. It makes sense that $S_\infty > S_9$.

5. **State.**

   a) In 9 bounces, Clyde will have traveled upward a total distance of about 297 ft.

   b) Theoretically, Clyde will travel upward a total of 300 ft before coming to rest.

Chapter Resources:
Visualizing for Success, p. 730 ;
Collaborative Activity, p. 731;
Decision Making: Connection,
p. 731

8. Refer to Example 8. If Clyde's cord stretches to 300 ft, how far will he "bounce" or "travel upward" before coming to rest?

YOUR TURN

11.3 EXERCISE SET   FOR EXTRA HELP   MyMathLab® Math XL
PRACTICE   WATCH   READ   REVIEW

Vocabulary and Reading Check

Complete each statement by selecting the appropriate word or expression from those listed below each blank.

1. The list 16, 8, 4, 2, 1, . . . is a(n) _____
   finite/infinite

   _____ _____.
   arithmetic/geometric   sequence/series

2. The sum 16 + 8 + 4 + 2 + 1 is a(n) _____
   finite/infinite

   _____ _____.
   arithmetic/geometric   sequence/series

3. For 16 + 8 + 4 + 2 + 1 + · · ·, the common _____ is _____ than 1, so the
   difference/ratio   less/more
   limit _____ exist.
         does/does not

4. The number $0.\overline{2}$ is equal to _____
   0.2222/0.22 . . .

   and can be written as the _____
   arithmetic/geometric

   _____ 0.2 + 0.02 + 0.002 + · · · .
   sequence/series

Concept Reinforcement

Classify each of the following as an arithmetic sequence, a geometric sequence, an arithmetic series, a geometric series, or none of these.

5. 3, 6, 12, 24, . . .

6. 10, 7, 4, 1, −2, . . .

7. 4 + 20 + 100 + 500 + 2500 + 12,500

8. 10 + 12 + 14 + 16 + 18 + 20

9. $3 - \frac{3}{2} + \frac{3}{4} - \frac{3}{8} + \frac{3}{16} - \cdots$

10. $1 + \frac{1}{2} + \frac{1}{3} + \frac{1}{4} + \frac{1}{5} + \frac{1}{6} + \cdots$

Geometric Sequences

Find the common ratio for each geometric sequence.

11. $10, 20, 40, 80, \ldots$

12. $5, 20, 80, 320, \ldots$

13. $6, -0.6, 0.06, -0.006, \ldots$

14. $-5, -0.5, -0.05, -0.005, \ldots$

15. $\frac{1}{2}, -\frac{1}{4}, \frac{1}{8}, -\frac{1}{16}, \ldots$

16. $\frac{2}{3}, -\frac{4}{3}, \frac{8}{3}, -\frac{16}{3}, \ldots$

17. $75, 15, 3, \frac{3}{5}, \ldots$

18. $12, -4, \frac{4}{3}, -\frac{4}{9}, \ldots$

19. $\dfrac{1}{m}, \dfrac{6}{m^2}, \dfrac{36}{m^3}, \dfrac{216}{m^4}, \ldots$

20. $4, \dfrac{4m}{5}, \dfrac{4m^2}{25}, \dfrac{4m^3}{125}, \ldots$

Find the indicated term for each geometric sequence.

21. $2, 6, 18, \ldots$; the 7th term

22. $2, 8, 32, \ldots$; the 9th term

23. $\sqrt{3}, 3, 3\sqrt{3}, \ldots$; the 10th term

24. $2, 2\sqrt{2}, 4, \ldots$; the 8th term

25. $-\frac{8}{243}, \frac{8}{81}, -\frac{8}{27}, \ldots$; the 14th term

26. $\frac{7}{625}, \frac{-7}{125}, \frac{7}{25}, \ldots$; the 13th term

27. $\$1000, \$1040, \$1081.60, \ldots$; the 10th term

28. $\$1000, \$1050, \$1102.50, \ldots$; the 12th term

Find the nth, or general, term for each geometric sequence.

29. $1, 5, 25, 125, \ldots$

30. $2, 4, 8, \ldots$

31. $1, -1, 1, -1, \ldots$

32. $\frac{1}{4}, \frac{1}{16}, \frac{1}{64}, \ldots$

33. $\dfrac{1}{x}, \dfrac{1}{x^2}, \dfrac{1}{x^3}, \ldots$

34. $5, \dfrac{5m}{2}, \dfrac{5m^2}{4}, \ldots$

Sum of the First *n* Terms of a Geometric Sequence

For Exercises 35–42, use the formula for S_n to find the indicated sum for each geometric series.

35. S_9 for $6 + 12 + 24 + \cdots$

36. S_6 for $16 - 8 + 4 - \cdots$

37. S_7 for $\frac{1}{18} - \frac{1}{6} + \frac{1}{2} - \cdots$

Aha! **38.** S_5 for $7 + 0.7 + 0.07 + \cdots$

39. S_8 for $1 + x + x^2 + x^3 + \cdots$

40. S_{10} for $1 + x^2 + x^4 + x^6 + \cdots$

41. S_{16} for $\$200 + \$200(1.06) + \$200(1.06)^2 + \cdots$

42. S_{23} for $\$1000 + \$1000(1.08) + \$1000(1.08)^2 + \cdots$

Infinite Geometric Series

Determine whether each infinite geometric series has a limit. If a limit exists, find it.

43. $18 + 6 + 2 + \cdots$

44. $80 + 20 + 5 + \cdots$

45. $7 + 3 + \frac{9}{7} + \cdots$

46. $12 + 9 + \frac{27}{4} + \cdots$

47. $3 + 15 + 75 + \cdots$

48. $2 + 3 + \frac{9}{2} + \cdots$

49. $4 - 6 + 9 - \frac{27}{2} + \cdots$

50. $-6 + 3 - \frac{3}{2} + \frac{3}{4} - \cdots$

51. $0.43 + 0.0043 + 0.000043 + \cdots$

52. $0.37 + 0.0037 + 0.000037 + \cdots$

53. $\$500(1.02)^{-1} + \$500(1.02)^{-2} + \$500(1.02)^{-3} + \cdots$

54. $\$1000(1.08)^{-1} + \$1000(1.08)^{-2} + \$1000(1.08)^{-3} + \cdots$

Find fraction notation for each repeating decimal.

55. $0.5555\ldots$ **56.** $0.8888\ldots$

57. $3.4646\ldots$ **58.** $1.2323\ldots$

59. $0.15151515\ldots$ **60.** $0.12121212\ldots$

Problem Solving

Solve. Use a calculator as needed for evaluating formulas.

61. *Rebound Distance.*   A ping-pong ball is dropped from a height of 20 ft and always rebounds one-fourth of the distance fallen. How high does it rebound the 6th time?

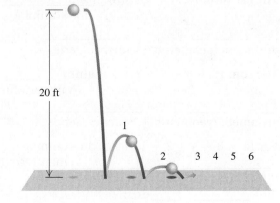

20 ft

62. *Rebound Distance.* Approximate the total of the rebound heights of the ball in Exercise 61.

63. *Population Growth.* Yorktown has a current population of 100,000 that is increasing by 3% each year. What will the population be in 15 years?

64. *Amount Owed.* Gilberto borrows $15,000. The loan is to be repaid in 13 years at 5.5% interest, compounded annually. How much will he owe at the end of 13 years?

65. *Shrinking Population.* A population of 5000 fruit flies is dying off at a rate of 4% per minute. How many flies will be alive after 15 min?

66. *Shrinking Population.* For the population of fruit flies in Exercise 65, how long will it take for only 1800 fruit flies to remain alive? (*Hint*: Use logarithms.) Round to the nearest minute.

67. *Solar Energy.* The global solar photovoltaic (PV) installed capacity was about 70 GW (gigawatts) in 2012 and was expected to grow by 20% each year, as indicated in the data below. If all installed solar PV cells were operating at capacity, how many gigawatts of power would be produced by solar energy from 2012 through 2020?

Sources: The European Photovoltaic Industry Association; GlobalData

Year	Solar PV Capacity (in gigawatts)
2012	70
2013	84
2014	100.8

68. *Daily Wages.* Suppose you were offered a job for the month of September (30 days) under the following conditions. You will be paid $0.01 for the first day, $0.02 for the second, $0.04 for the third, and so on, doubling your previous day's salary each day. How much would you earn? (Would you take the job? Make a guess before reading further.)

69. *Rebound Distance.* A superball dropped from the top of the Washington Monument (556 ft high) rebounds three-fourths of the distance fallen. How far (up and down) will the ball have traveled when it hits the ground for the 6th time?

70. *Rebound Distance.* Approximate the total distance that the ball of Exercise 69 will have traveled when it comes to rest.

71. *Stacking Paper.* Construction paper is about 0.02 in. thick. Beginning with just one piece, a stack is doubled again and again 10 times. Find the height of the final stack.

72. *Monthly Earnings.* Suppose you accepted a job for the month of February (28 days) under the following conditions. You will be paid $0.01 the first day, $0.02 the second, $0.04 the third, and so on, doubling your previous day's salary each day. How much would you earn?

Aha! 73. Under what circumstances is it possible for the 5th term of a geometric sequence to be greater than the 4th term but less than the 7th term?

74. When r is negative, a series is said to be *alternating*. Why do you suppose this terminology is used?

Skill Review

Solve.

75. $|x - 3| = 11$ [4.3]

76. $|2x + 5| < 6$ [4.3]

77. $|3x - 7| \geq 1$ [4.3]

78. $-5 < 6 - 3x < 7$ [4.2]

79. $x^2 - 5x - 14 < 0$ [8.9]

80. $x \geq \dfrac{1}{x}$ [8.9]

Synthesis

81. Write a problem for a classmate to solve. Devise the problem so that a geometric series is involved and the solution is "The total amount in the bank is $900(1.08)^{40}$, or about $19,550."

82. The infinite series

$$S_\infty = 2 + \frac{1}{2} + \frac{1}{2 \cdot 3} + \frac{1}{2 \cdot 3 \cdot 4} + \frac{1}{2 \cdot 3 \cdot 4 \cdot 5}$$
$$+ \frac{1}{2 \cdot 3 \cdot 4 \cdot 5 \cdot 6} + \cdots$$

is not geometric, but it does have a sum. Using $S_1, S_2, S_3, S_4, S_5,$ and S_6, predict the value of S_∞ and explain your reasoning.

Calculate each of the following sums.

83. $\displaystyle\sum_{k=1}^{\infty} 6(0.9)^k$

84. $\displaystyle\sum_{k=1}^{\infty} 5(-0.7)^k$

85. Find the sum of the first n terms of
$$x^2 - x^3 + x^4 - x^5 + \cdots.$$

86. Find the sum of the first n terms of
$$1 + x + x^2 + x^3 + \cdots.$$

87. The sides of a square are each 16 cm long. A second square is inscribed by joining the midpoints of the sides, successively. In the second square we repeat the process, inscribing a third square. If this process is continued indefinitely, what is the sum of all of the areas of all the squares? (*Hint*: Use an infinite geometric series.)

88. Show that 0.999... is 1.

 89. Using Example 5 and Exercises 43–54, explain how the graph of a geometric sequence can be used to determine whether a geometric series has a limit.

 90. To compare the *graphs* of an arithmetic sequence and a geometric sequence, we plot n on the horizontal axis and a_n on the vertical axis. Graph Example 1(a) of Section 11.2 and Example 1(a) of Section 11.3 on the same set of axes. How do the graphs of geometric sequences differ from the graphs of arithmetic sequences?

 91. *Research.*   How are items such as computers depreciated on an income tax return? Form a sequence listing the value of a computer from the time of its purchase until it is completely depreciated using one of the methods allowed by the IRS. Is the sequence arithmetic, geometric, or neither? Is it realistic?

QUICK QUIZ: SECTIONS 11.1–11.3

1. Find the first 4 terms and a_{20} of the sequence with the general term $a_n = (-1)^n n$.   [11.1]

2. Find S_7 for the sequence 5, 10, 15, 20, ... [11.1]

3. Find the sum of the first 15 terms of the arithmetic series $50 + 47 + 44 + \cdots$. [11.2]

4. Find the sum of the first 10 terms of the geometric series $\frac{1}{2} + 1 + 2 + 4 + \cdots$. [11.3]

5. Find fraction notation for the repeating decimal 1.565656.... [11.3]

PREPARE TO MOVE ON

Multiply.   [5.2]

1. $(x + y)^2$

2. $(x + y)^3$

3. $(x - y)^3$

4. $(x - y)^4$

5. $(2x + y)^3$

6. $(2x - y)^3$

Mid-Chapter Review

A *sequence* is simply an ordered list. A *series* is a sum of consecutive terms in a sequence. Some sequences of numbers have patterns, and a formula can be found for a general term. When every pair of consecutive terms has a common difference, the sequence is *arithmetic*. When every pair of consecutive terms has a common ratio, the sequence is *geometric*.

GUIDED SOLUTIONS

1. Find the 14th term of the arithmetic sequence
$-6, -1, 4, 9, \ldots$ [11.2]

Solution

$$a_n = a_1 + (n - 1)d$$

$$n = \boxed{}, \qquad a_1 = \boxed{}, \qquad d = \boxed{}$$

$$a_{14} = \boxed{} + (\boxed{} - 1)$$

$$a_{14} = \boxed{}$$

2. Find the 7th term of the geometric sequence
$\frac{1}{9}, -\frac{1}{3}, 1, -3, \ldots$ [11.3]

Solution

$$a_n = a_1 r^{n-1}$$

$$n = \boxed{}, \qquad a_1 = \boxed{}, \qquad r = \boxed{}$$

$$a_7 = \boxed{} \cdot (\boxed{})^{\boxed{}-1}$$

$$a_7 = \boxed{}$$

MIXED REVIEW

3. Find a_{20} if $a_n = n^2 - 5n$. [11.1]

4. Write an expression for the general term a_n of the sequence $\frac{1}{2}, \frac{1}{3}, \frac{1}{4}, \frac{1}{5}, \ldots$ [11.1]

5. Find S_{12} for the sequence $1, 2, 3, 4, \ldots$ [11.1]

6. Write out and evaluate the sum

$$\sum_{k=2}^{5} k^2.$$ [11.1]

7. Rewrite using sigma notation:
$1 - 2 + 3 - 4 + 5 - 6$. [11.1]

8. Which term is 22 in the arithmetic sequence
$10, 10.2, 10.4, 10.6, \ldots$? [11.2]

9. For an arithmetic sequence, find a_{25} when $a_1 = 9$ and $d = -2$. [11.2]

10. Find the 12th term of the geometric sequence
$1000, 100, 10, \ldots$ [11.3]

11. Find the *n*th, or general, term for the geometric sequence $2, -2, 2, -2, \ldots$ [11.3]

12. Determine whether the infinite geometric series $100 - 20 + 4 - \cdots$ has a limit. If the limit exists, find it. [11.3]

13. Renata earns $1 on June 1, another $2 on June 2, another $3 on June 3, another $4 on June 4, and so on. How much does she earn during the 30 days of June? [11.2]

14. Dwight earns $1 on June 1, another $2 on June 2, another $4 on June 3, another $8 on June 4, and so on. How much does he earn during the 30 days of June? [11.3]

11.4   The Binomial Theorem

Binomial Expansion Using Pascal's Triangle   ●   Binomial Expansion Using Factorial Notation

The expression $(x + y)^2$ may be regarded as a series: $x^2 + 2xy + y^2$. This sum of terms is the *expansion* of $(x + y)^2$. For powers greater than 2, finding the expansion of $(x + y)^n$ can be time-consuming. In this section, we look at two methods of streamlining binomial expansion.

Binomial Expansion Using Pascal's Triangle

Consider the following expanded powers of $(a + b)^n$.

$$(a + b)^0 = 1$$
$$(a + b)^1 = a + b$$
$$(a + b)^2 = a^2 + 2a^1b^1 + b^2$$
$$(a + b)^3 = a^3 + 3a^2b^1 + 3a^1b^2 + b^3$$
$$(a + b)^4 = a^4 + 4a^3b^1 + 6a^2b^2 + 4a^1b^3 + b^4$$
$$(a + b)^5 = a^5 + 5a^4b^1 + 10a^3b^2 + 10a^2b^3 + 5a^1b^4 + b^5$$

Each expansion is a polynomial. There are some patterns worth noting:

1. There is one more term than the power of the binomial, n. That is, there are $n + 1$ terms in the expansion of $(a + b)^n$.

2. In each term, the sum of the exponents is the power to which the binomial is raised.

3. The exponents of a start with n, the power of the binomial, and decrease to 0 (since $a^0 = 1$, the last term has no factor of a). The first term has no factor of b, so powers of b start with 0 and increase to n.

4. The coefficients start at 1, increase through certain values, and then decrease through these same values back to 1.

Let's study the coefficients further. Suppose we wish to expand $(a + b)^8$. The patterns listed above indicate 9 terms in the expansion:

$$a^8 + c_1a^7b + c_2a^6b^2 + c_3a^5b^3 + c_4a^4b^4 + c_5a^3b^5 + c_6a^2b^6 + c_7ab^7 + b^8.$$

How can we determine the values for the c's? One simple method involves writing down the coefficients in a triangular array as follows. We form what is known as **Pascal's triangle**:

$$
\begin{array}{llccccccccccc}
(a + b)^0\!: & & & & & & 1 & & & & & \\
(a + b)^1\!: & & & & & 1 & & 1 & & & & \\
(a + b)^2\!: & & & & 1 & & 2 & & 1 & & & \\
(a + b)^3\!: & & & 1 & & 3 & & 3 & & 1 & & \\
(a + b)^4\!: & & 1 & & 4 & & 6 & & 4 & & 1 & \\
(a + b)^5\!: & 1 & & 5 & & 10 & & 10 & & 5 & & 1 \\
\end{array}
$$

There are many patterns in the triangle. Find as many as you can.

Perhaps you discovered a way to write the next row of numbers, given the numbers in the row above it. There are always 1's on the outside. Each remaining number is the sum of the two numbers above:

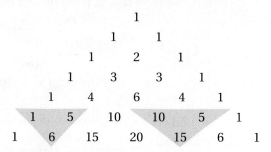

We see that in the bottom (seventh) row

the 1st and last numbers are 1;

the 2nd number is $1 + 5$, or 6;

the 3rd number is $5 + 10$, or 15;

the 4th number is $10 + 10$, or 20;

the 5th number is $10 + 5$, or 15; and

the 6th number is $5 + 1$, or 6.

Thus the expansion of $(a + b)^6$ is

$$(a + b)^6 = 1a^6 + 6a^5b + 15a^4b^2 + 20a^3b^3 + 15a^2b^4 + 6ab^5 + 1b^6.$$

To expand $(a + b)^8$, we complete two more rows of Pascal's triangle:

$$
\begin{array}{ccccccccccccccc}
&&&&&&&& 1 \\
&&&&&&& 1 && 1 \\
&&&&&& 1 && 2 && 1 \\
&&&&& 1 && 3 && 3 && 1 \\
&&&& 1 && 4 && 6 && 4 && 1 \\
&&& 1 && 5 && 10 && 10 && 5 && 1 \\
&& 1 && 6 && 15 && 20 && 15 && 6 && 1 \\
& 1 && 7 && 21 && 35 && 35 && 21 && 7 && 1 \\
1 && 8 && 28 && 56 && 70 && 56 && 28 && 8 && 1
\end{array}
$$

The expansion of $(a + b)^8$ has coefficients found in the 9th row above:

$$(a + b)^8 = 1a^8 + 8a^7b + 28a^6b^2 + 56a^5b^3 + 70a^4b^4 + 56a^3b^5 + 28a^2b^6 + 8ab^7 + 1b^8.$$

We can generalize our results as follows. A proof of this result is outlined in Exercise 70 of Exercise Set 11.4.

THE BINOMIAL THEOREM (FORM 1)

For any binomial $a + b$ and any natural number n,

$$(a + b)^n = c_0a^nb^0 + c_1a^{n-1}b^1 + c_2a^{n-2}b^2 + \cdots + c_{n-1}a^1b^{n-1} + c_na^0b^n,$$

where the numbers $c_0, c_1, c_2, \ldots, c_n$ are from the $(n + 1)$st row of Pascal's triangle.

EXAMPLE 1   Expand: $(u - v)^5$.

SOLUTION   First, note that $(u - v)^5 = (u + (-v))^5$. Using the binomial theorem, we have $a = u$, $b = -v$, and $n = 5$. We use the 6th row of Pascal's triangle: 1   5   10   10   5   1. Thus,

$$(u - v)^5 = 1(u)^5 + 5(u)^4(-v)^1 + 10(u)^3(-v)^2 + 10(u)^2(-v)^3$$
$$+ 5(u)^1(-v)^4 + 1(-v)^5$$
$$= u^5 - 5u^4v + 10u^3v^2 - 10u^2v^3 + 5uv^4 - v^5.$$

Note that the signs of the terms alternate between $+$ and $-$. When $-v$ is raised to an odd power, the sign is $-$; when the power is even, the sign is $+$.

1. Expand: $(u - v)^4$.

YOUR TURN

EXAMPLE 2   Expand: $\left(2t + \dfrac{3}{t}\right)^6$.

SOLUTION   Note that $a = 2t$, $b = 3/t$, and $n = 6$. We use the 7th row of Pascal's triangle: 1   6   15   20   15   6   1. Thus,

$$\left(2t + \frac{3}{t}\right)^6 = 1(2t)^6 + 6(2t)^5\left(\frac{3}{t}\right)^1 + 15(2t)^4\left(\frac{3}{t}\right)^2 + 20(2t)^3\left(\frac{3}{t}\right)^3$$
$$+ 15(2t)^2\left(\frac{3}{t}\right)^4 + 6(2t)^1\left(\frac{3}{t}\right)^5 + 1\left(\frac{3}{t}\right)^6$$
$$= 64t^6 + 6(32t^5)\left(\frac{3}{t}\right) + 15(16t^4)\left(\frac{9}{t^2}\right) + 20(8t^3)\left(\frac{27}{t^3}\right)$$
$$+ 15(4t^2)\left(\frac{81}{t^4}\right) + 6(2t)\left(\frac{243}{t^5}\right) + \frac{729}{t^6}$$
$$= 64t^6 + 576t^4 + 2160t^2 + 4320 + 4860t^{-2} + 2916t^{-4} + 729t^{-6}.$$

2. Expand: $\left(x + \dfrac{4}{x}\right)^5$.

YOUR TURN

Binomial Expansion Using Factorial Notation

The drawback to using Pascal's triangle is that we must compute all the preceding rows in the table to obtain the row we need. The following method avoids this difficulty. It will also enable us to find a specific term—say, the 8th term—without computing all the other terms in the expansion.

To develop the method, we need some new notation. Products of successive natural numbers, such as $6 \cdot 5 \cdot 4 \cdot 3 \cdot 2 \cdot 1$ and $8 \cdot 7 \cdot 6 \cdot 5 \cdot 4 \cdot 3 \cdot 2 \cdot 1$, have a special notation. For the product $6 \cdot 5 \cdot 4 \cdot 3 \cdot 2 \cdot 1$, we write 6!, read "6 factorial."

FACTORIAL NOTATION

For any natural number n,

$$n! = n(n - 1)(n - 2) \cdots (3)(2)(1).$$

Here are some examples:

$$6! = 6 \cdot 5 \cdot 4 \cdot 3 \cdot 2 \cdot 1 = 720,$$
$$5! = 5 \cdot 4 \cdot 3 \cdot 2 \cdot 1 = 120,$$
$$4! = 4 \cdot 3 \cdot 2 \cdot 1 = 24,$$
$$3! = 3 \cdot 2 \cdot 1 = 6,$$
$$2! = 2 \cdot 1 = 2,$$
$$1! = 1 = 1.$$

We also define 0! to be 1 for reasons explained shortly.

To simplify expressions like

$$\frac{8!}{5! \, 3!},$$

note that

$$8! = 8 \cdot 7 \cdot 6 \cdot 5 \cdot 4 \cdot 3 \cdot 2 \cdot 1 = 8 \cdot 7! = 8 \cdot 7 \cdot 6! = 8 \cdot 7 \cdot 6 \cdot 5!$$

and so on.

Student Notes

It is important to recognize factorial notation as representing a product with descending factors. Thus, $7!$, $7 \cdot 6!$, and $7 \cdot 6 \cdot 5!$ all represent the same product.

3. Simplify: $\dfrac{10!}{2! \, 8!}$.

> **CAUTION!** $\dfrac{6!}{3!} \ne 2!$ To see this, note that
>
> $$\frac{6!}{3!} = \frac{6 \cdot 5 \cdot 4 \cdot \cancel{3} \cdot \cancel{2} \cdot \cancel{1}}{\cancel{3} \cdot \cancel{2} \cdot \cancel{1}} = 6 \cdot 5 \cdot 4.$$

EXAMPLE 3 Simplify: $\dfrac{8!}{5! \, 3!}$.

SOLUTION

$$\frac{8!}{5! \, 3!} = \frac{8 \cdot 7 \cdot 6 \cdot 5!}{5! \cdot 3 \cdot 2 \cdot 1} = 8 \cdot 7 \qquad \text{Removing a factor equal to 1: } \frac{6 \cdot 5!}{5! \cdot 3 \cdot 2} = 1$$

$$= 56$$

YOUR TURN

The following notation is used in our second formulation of the binomial theorem.

> $\dbinom{n}{r}$ **NOTATION**
>
> For n and r nonnegative integers with $n \ge r$,
>
> $$\dbinom{n}{r}, \quad \text{read "}n \text{ choose } r\text{,"} \quad \text{means} \quad \frac{n!}{(n-r)! \, r!}.$$
>
> $\dbinom{n}{r}$ can also be written ${}_nC_r$.

Technology Connection

The PRB option of the MATH menu provides access to both factorial calculations and ${}_nC_r$. In both cases, a number must be entered first. To find $\dbinom{7}{2}$, we press ⑦ **MATH**, select PRB and ${}_nC_r$, and press ② **ENTER**.

```
7 nCr 2
                        21
```

1. Find $12!$.

2. Find $\dbinom{8}{3}$ and $\dbinom{12}{5}$.

EXAMPLE 4 Simplify: **(a)** $\dbinom{7}{2}$; **(b)** $\dbinom{6}{6}$.

SOLUTION

a) $\dbinom{7}{2} = \dfrac{7!}{(7-2)! \, 2!}$

$= \dfrac{7!}{5! \, 2!} = \dfrac{7 \cdot 6 \cdot 5!}{5! \cdot 2 \cdot 1} = \dfrac{7 \cdot 6}{2}$ We can write $7!$ as $7 \cdot 6 \cdot 5!$ to aid our simplification.

$= 7 \cdot 3$

$= 21$

b) $\dbinom{6}{6} = \dfrac{6!}{0!\,6!} = \dfrac{6!}{1\cdot 6!}$ Since $0! = 1$

$\qquad\quad = \dfrac{6!}{6!}$

4. Simplify: $\dbinom{9}{6}$.

$\qquad\quad = 1$

YOUR TURN

Now we can restate the binomial theorem using our new notation.

THE BINOMIAL THEOREM (FORM 2)

For any binomial $a + b$ and any natural number n,

$$(a+b)^n = \dbinom{n}{0}a^n + \dbinom{n}{1}a^{n-1}b + \dbinom{n}{2}a^{n-2}b^2 + \cdots + \dbinom{n}{n}b^n.$$

EXAMPLE 5   Expand: $(3x + y)^4$.

SOLUTION   We use the binomial theorem (Form 2) with $a = 3x, b = y$, and $n = 4$:

$$(3x+y)^4 = \dbinom{4}{0}(3x)^4 + \dbinom{4}{1}(3x)^3 y + \dbinom{4}{2}(3x)^2 y^2 + \dbinom{4}{3}(3x)y^3 + \dbinom{4}{4}y^4$$

$$= \dfrac{4!}{4!\,0!}3^4 x^4 + \dfrac{4!}{3!\,1!}3^3 x^3 y + \dfrac{4!}{2!\,2!}3^2 x^2 y^2 + \dfrac{4!}{1!\,3!}3xy^3 + \dfrac{4!}{0!\,4!}y^4$$

$$= 1\cdot 81x^4 + 4\cdot 27x^3 y + 6\cdot 9x^2 y^2 + 4\cdot 3xy^3 + y^4 \quad\Big\} \text{ Simplifying}$$

$$= 81x^4 + 108x^3 y + 54x^2 y^2 + 12xy^3 + y^4.$$

5. Expand: $(a + 2c)^5$.

YOUR TURN

EXAMPLE 6   Expand: $(x^2 - 2y)^5$.

SOLUTION   In this case, $a = x^2, b = -2y$, and $n = 5$:

$$(x^2 - 2y)^5 = \dbinom{5}{0}(x^2)^5 + \dbinom{5}{1}(x^2)^4(-2y) + \dbinom{5}{2}(x^2)^3(-2y)^2$$

$$+ \dbinom{5}{3}(x^2)^2(-2y)^3 + \dbinom{5}{4}(x^2)(-2y)^4 + \dbinom{5}{5}(-2y)^5$$

$$= \dfrac{5!}{5!\,0}x^{10} + \dfrac{5!}{4!\,1!}x^8(-2y) + \dfrac{5!}{3!\,2!}x^6(-2y)^2$$

$$+ \dfrac{5!}{2!\,3!}x^4(-2y)^3 + \dfrac{5!}{1!\,4!}x^2(-2y)^4 + \dfrac{5!}{0!\,5!}(-2y)^5$$

$$= x^{10} - 10x^8 y + 40x^6 y^2 - 80x^4 y^3 + 80x^2 y^4 - 32y^5.$$

6. Expand: $(w^2 - t^2)^4$.

YOUR TURN

Note that in the binomial theorem (Form 2), $\dbinom{n}{0}a^n b^0$ gives us the first term, $\dbinom{n}{1}a^{n-1}b^1$ gives us the second term, $\dbinom{n}{2}a^{n-2}b^2$ gives us the third term, and so on. This can be generalized to give a method for finding a specific term without writing the entire expansion.

FINDING A SPECIFIC TERM

When $(a + b)^n$ is expanded and written in descending powers of a, the $(r + 1)$st term is

$$\binom{n}{r}a^{n-r}b^r.$$

EXAMPLE 7 Find the 5th term in the expansion of $(2x - 3y)^7$.

SOLUTION To find the 5th term, we note that $5 = 4 + 1$. Thus, $r = 4$, $a = 2x$, $b = -3y$, and $n = 7$. Using the above formula, we have

$$\binom{n}{r}a^{n-r}b^r = \binom{7}{4}(2x)^{7-4}(-3y)^4, \quad \text{or} \quad \frac{7!}{3!\,4!}(2x)^3(-3y)^4, \quad \text{or} \quad 22{,}680x^3y^4.$$

7. Find the 6th term in the expansion of $(2a + w)^9$.

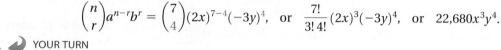 YOUR TURN

It is because of the binomial theorem that $\binom{n}{r}$ is called a *binomial coefficient*.

We can now explain why 0! is defined to be 1. In the binomial theorem,

$\binom{n}{0}$ must equal 1 when using the definition $\binom{n}{r} = \dfrac{n!}{(n-r)!\,r!}$.

Thus we must have

$$\binom{n}{0} = \frac{n!}{(n-0)!\,0!} = \frac{n!}{n!\,0!} = 1.$$

This is satisfied only if 0! is defined to be 1.

11.4 EXERCISE SET

FOR EXTRA HELP

MyMathLab® Math XL
PRACTICE   WATCH   READ   REVIEW

⬥ Vocabulary and Reading Check

Choose the word from the following list that best completes each statement. Words may be used more than once or not at all.

binomial	first
expansion	second
factorial	third

1. The expression $(x + y)^2$ is a(n) _____ squared.

2. The expression $x^2 + 2xy + y^2$ is the _____ of $(x + y)^2$.

3. The _____ number in every row of Pascal's triangle is 1.

4. To use Pascal's triangle to expand $(x + y)^2$, we use the _____ row of coefficients.

5. 8! is an example of _____ notation.

6. $\binom{n}{r}$ represents a(n) _____ coefficient.

⬥ Concept Reinforcement

Complete each of the following statements.

7. The last term in the expansion of $(x + 2)^5$ is _____.

8. The expansion of $(x + y)^7$, when simplified, contains a total of _____ terms.

9. In the expansion of $(a + b)^9$, the sum of the exponents in each term is _____.

10. In the expansion of $(x + y)^9$, the coefficient of y^9 is _____.

Factorial Notation and Binomial Coefficients

Simplify.

11. 4!

12. 9!

13. 10!

14. 12!

15. $\dfrac{10!}{8!}$

16. $\dfrac{12!}{10!}$

17. $\dfrac{9!}{4!\,5!}$

18. $\dfrac{10!}{6!\,4!}$

19. $\dbinom{10}{4}$

20. $\dbinom{8}{5}$

Aha! **21.** $\dbinom{9}{9}$

22. $\dbinom{7}{7}$

23. $\dbinom{30}{2}$

24. $\dbinom{51}{49}$

25. $\dbinom{40}{38}$

26. $\dbinom{35}{2}$

Binomial Expansion

Expand. Use both of the methods shown in this section.

27. $(a - b)^4$

28. $(m + n)^5$

29. $(p + w)^7$

30. $(x - y)^6$

31. $(3c - d)^7$

32. $(x^2 - 3y)^5$

33. $(t^{-2} + 2)^6$

34. $(3c - d)^6$

35. $\left(3s + \dfrac{1}{t}\right)^9$

36. $\left(x + \dfrac{2}{y}\right)^9$

37. $(x^3 - 2y)^5$

38. $(a^2 - b^3)^5$

39. $(\sqrt{5} + t)^6$

40. $(\sqrt{3} - t)^4$

41. $\left(\dfrac{1}{\sqrt{x}} - \sqrt{x}\right)^6$

42. $(x^{-2} + x^2)^4$

Find the indicated term for each binomial expression.

43. 3rd, $(a + b)^6$

44. 6th, $(x + y)^7$

45. 12th, $(a - 3)^{14}$

46. 11th, $(x - 2)^{12}$

47. 5th, $(2x^3 + \sqrt{y})^8$

48. 4th, $\left(\dfrac{1}{b^2} + c\right)^7$

49. Middle, $(2u + 3v^2)^{10}$

50. Middle two, $(\sqrt{x} + \sqrt{3})^5$

Aha! **51.** 9th, $(x - y)^8$

52. 13th, $(a - \sqrt{b})^{12}$

53. Maya claims that she can calculate mentally the first two terms and the last two terms of the expansion of $(a + b)^n$ for any whole number n. How do you think she does this?

54. Without performing any calculations, explain why the expansions of $(x - y)^8$ and $(y - x)^8$ must be equal.

Skill Review

Graph.

55. $y = x^2 - 5$ [8.7]

56. $y = x - 5$ [2.3]

57. $y \geq x - 5$ [4.4]

58. $y = 5^x$ [9.2]

59. $f(x) = \log_5 x$ [9.3]

60. $x^2 + y^2 = 5$ [10.1]

Synthesis

61. Explain how someone can determine the x^2-term of the expansion of $\left(x - \dfrac{3}{x}\right)^{10}$ without calculating any other terms.

62. Devise two problems requiring the use of the binomial theorem. Design the problems so that one is solved more easily using Form 1 and the other is solved more easily using Form 2. Then explain what makes one form easier to use than the other in each case.

63. The notation $\dbinom{n}{r}$ is read "n choose r" because it can be used to calculate the number of ways in which a set of r elements can be chosen from a set containing n elements. Show that there are exactly $\dbinom{5}{3}$ ways of choosing a subset of size 3 from $\{a, b, c, d, e\}$.

64. *Baseball.* During the 2011 season, Michael Young of the Texas Rangers had a batting average of 0.338. In that season, if someone were to randomly select 5 of his "at-bats," the probability of Young's getting exactly 3 hits would be the 3rd term of the binomial expansion of $(0.338 + 0.662)^5$. Find that term and use a calculator to estimate the probability.

Source: www.baseball-reference.com

65. *Widows or Divorcees.* The probability that a woman will be either widowed or divorced is 85%. If 8 women are randomly selected, the probability that exactly 5 of them will be either widowed or divorced is the 6th term of the binomial expansion of $(0.15 + 0.85)^8$. Use a calculator to estimate that probability.

66. *Baseball.* In reference to Exercise 64, the probability that Young will get *at most* 3 hits is found by adding the last 4 terms of the binomial expansion of $(0.338 + 0.662)^5$. Find these terms and use a calculator to estimate the probability.

67. *Widows or Divorcees.* In reference to Exercise 65, the probability that *at least* 6 of the women will be widowed or divorced is found by adding the last three terms of the binomial expansion of $(0.15 + 0.85)^8$. Find these terms and use a calculator to estimate the probability.

68. Find the term of
$$\left(\frac{3x^2}{2} - \frac{1}{3x}\right)^{12}$$
that does not contain x.

69. Prove that
$$\binom{n}{r} = \binom{n}{n-r}.$$
for any whole numbers n and r. Assume $r \le n$.

70. Form 1 of the binomial theorem can be proved using form 2 of the binomial theorem. The key step in that proof is showing that the coefficients inside Pascal's triangle are found by adding the two terms above. Prove this fact by showing that
$$\binom{n}{r} = \binom{n-1}{r-1} + \binom{n-1}{r}.$$

71. Find the middle term of $(x^2 - 6y^{3/2})^6$.

72. Find the ratio of the 4th term of
$$\left(p^2 - \frac{1}{2}p\sqrt[3]{q}\right)^5$$
to the 3rd term.

73. Find the term containing $\dfrac{1}{x^{1/6}}$ of
$$\left(\sqrt[3]{x} - \frac{1}{\sqrt{x}}\right)^7.$$

Aha! 74. Multiply: $(x^2 + 2xy + y^2)(x^2 + 2xy + y^2)^2(x + y)$.

75. What is the degree of $(x^3 + 2)^4$?

QUICK QUIZ: SECTIONS 11.1–11.4

1. Write out and evaluate $\displaystyle\sum_{k=1}^{4}\frac{k+1}{k}$. [11.1]

2. Find the common difference for the arithmetic sequence $2.5, 2.1, 1.7, 1.3, \ldots$. [11.2]

3. Find the common ratio for the geometric sequence $-200, 100, -50, 25, \ldots$. [11.3]

4. Simplify: $\binom{12}{9}$. [11.4]

5. Expand: $(x + w)^4$. [11.4]

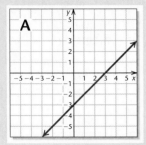

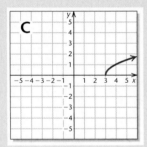

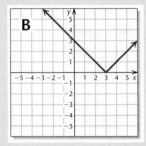

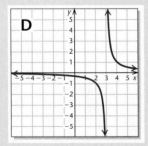

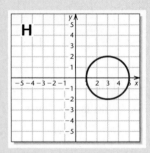

Visualizing for Success

Use after Section 11.3.

Match each equation with its graph.

1. $xy = 2$

2. $y = \log_2 x$

3. $y = x - 3$

4. $(x - 3)^2 + y^2 = 4$

5. $\dfrac{(x - 3)^2}{1} + \dfrac{y^2}{4} = 1$

6. $y = |x - 3|$

7. $y = (x - 3)^2$

8. $y = \dfrac{1}{x - 3}$

9. $y = 2^x$

10. $y = \sqrt{x - 3}$

Answers on page A-64

An additional, animated version of this activity appears in MyMathLab. *To use* MyMathLab, *you need a course ID and a student access code. Contact your instructor for more information.*

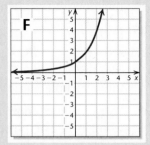

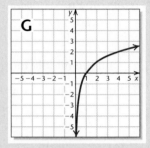

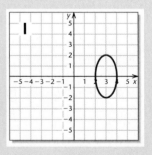

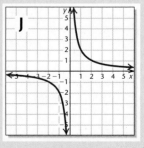

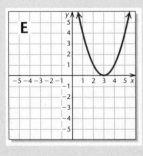

Collaborative Activity   *Bargaining for a Used Car*

Focus: Geometric series
Use after: Section 11.3
Time: 30 minutes
Group size: 2
Materials: Graphing calculators are optional.

Activity[*]

1. One group member ("the seller") is asking $3500 for a car. The second ("the buyer") offers $1500. The seller splits the difference ($3500 − $1500 = $2000, and $2000 ÷ 2 = $1000) and lowers the price to $2500. The buyer splits the difference again ($2500 − $1500 = $1000, and $1000 ÷ 2 = $500) and counters with $2000. Continue in this manner until you are able to agree on the car's selling price to the nearest penny.

2. Check several guesses to find what the buyer's initial offer should be in order to achieve a purchase price of $2000 or less.

---

[*]This activity is based on the article "Bargaining Theory, or Zeno's Used Cars," by James C. Kirby, *The College Mathematics Journal*, **27**(4), September 1996.

3. The seller's price in the bargaining above can be modeled recursively (see Exercises 87, 88, and 96 in Section 11.1) by the sequence

$$a_1 = 3500, \qquad a_n = a_{n-1} - \frac{d}{2^{2n-3}},$$

where d is the difference between the initial price and the first offer. Use this recursively defined sequence to solve parts (1) and (2) above either manually or by using the SEQ MODE and the TABLE feature of a graphing calculator.

4. The first four terms in the sequence in part (3) can be written as

$$a_1, \quad a_1 - \frac{d}{2}, \quad a_1 - \frac{d}{2} - \frac{d}{8}, \quad a_1 - \frac{d}{2} - \frac{d}{8} - \frac{d}{32}.$$

Use the formula for the limit of an infinite geometric series to find a simple algebraic formula for the eventual sale price, P, when the bargaining process from above is followed. Verify the formula by using it to solve parts (1) and (2) above.

Decision Making 𝒮 Connection   *(Use after Section 11.3.)*

Interest. Arithmetic sequences and geometric sequences can be used to compare simple interest and compound interest. For each of the following exercises, assume that you have $1000 to invest at 4% interest at the beginning of next year.

1. Simple interest is calculated on the amount invested. Suppose that interest is calculated annually and that you choose to have the interest sent to you at the end of each year. You collect this interest in an envelope in your desk.

   a) Write an arithmetic sequence with each term representing the sum of the investment and the cumulative interest earned at the beginning of each year. The first term is $1000.
   b) What is the general term of the sequence?
   c) How much money will you have after 20 years (that is, at the beginning of the 21st year)?

2. Compound interest includes interest paid on interest that was previously earned. Suppose that at the end of each year, the interest is added to the principal. This sum becomes the new principal.

   a) Write a geometric sequence with each term representing the amount in the account at the beginning of each year. The first term is $1000.

   b) What is the general term of the sequence?
   c) How much money will you have after 20 years?

3. If interest is calculated n times a year (at evenly spaced intervals), the interest rate for each interval is r/n, where r is the annual interest rate. Suppose that interest is calculated quarterly for your account. Then the interest rate for each quarter is 1%.

   a) Suppose again that interest is sent to you at the end of each quarter and you collect it at home. Write an arithmetic sequence with each term representing the sum of the investment and the cumulative simple interest earnings at the beginning of each quarter. How much money will you have after 20 years?

   b) Suppose instead that the interest is added to the account at the end of each quarter. Write a geometric sequence with each term representing the amount in the account at the beginning of each quarter if interest is compounded quarterly. How much money will you have after 20 years?

4. *Research.* Find the interest rate that a local bank pays on an account and how often interest is calculated. How much will a $1000 investment be worth in 20 years?

Study Summary

KEY TERMS AND CONCEPTS	EXAMPLES	PRACTICE EXERCISES

SECTION 11.1: *Sequences and Series*

The **general term** of a sequence is written a_n.

The sum of the first n terms of a sequence is written S_n.

Find a_{10} if $a_n = 3n - 7$.
$$a_{10} = 3 \cdot 10 - 7$$
$$= 30 - 7 = 23$$

Find S_6 for the sequence $-1, 2, -4, 8, -16, 32, -64$.
$$S_6 = -1 + 2 + (-4) + 8 + (-16) + 32 = 21$$

1. Find a_{12} if $a_n = n^2 - 1$.

2. Find S_5 for the sequence
$$-9, -8, -6, -3, 1, 6, 12.$$

Sigma or **Summation Notation**

k is the **index of summation**

$$\sum_{k=3}^{5}(-1)^k(k^2) = (-1)^3(3^2) \quad \text{For } k = 3$$
$$+ (-1)^4(4^2) \quad \text{For } k = 4$$
$$+ (-1)^5(5^2) \quad \text{For } k = 5$$
$$= -1 \cdot 9 + 1 \cdot 16$$
$$+ (-1) \cdot 25$$
$$= -9 + 16 - 25 = -18$$

3. Write out and evaluate the sum:
$$\sum_{k=0}^{3} 5k.$$

SECTION 11.2: *Arithmetic Sequences and Series*

Arithmetic Sequences and Series

$a_{n+1} = a_n + d$ — d is the **common difference**.

$a_n = a_1 + (n - 1)d$ — The nth term

$S_n = \dfrac{n}{2}(a_1 + a_n)$ — The sum of the first n terms

For the arithmetic sequence 10, 7, 4, 1, ... :
$$d = -3;$$
$$a_7 = 10 + (7 - 1)(-3) = 10 - 18 = -8;$$
$$S_7 = \frac{7}{2}(10 + (-8)) = \frac{7}{2}(2) = 7.$$

4. Find the 20th term of the arithmetic sequence
$$6, 6.5, 7, 7.5, \ldots .$$

5. Find S_{20} for the arithmetic series
$$6 + 6.5 + 7 + 7.5 + \cdots .$$

SECTION 11.3: *Geometric Sequences and Series*

Geometric Sequences and Series

$a_{n+1} = a_n \cdot r$ — r is the **common ratio**.

$a_n = a_1 r^{n-1}$ — The nth term

$S_n = \dfrac{a_1(1 - r^n)}{1 - r}, r \neq 1$ — The sum of the first n terms

$S_\infty = \dfrac{a_1}{1 - r}, |r| < 1$ — Limit of an infinite geometric series

For the geometric sequence 25, -5, 1, $-\frac{1}{5}$, ... :
$$r = -\frac{1}{5};$$
$$a_7 = 25\left(-\frac{1}{5}\right)^{7-1} = 5^2 \cdot \frac{1}{5^6} = \frac{1}{625};$$
$$S_7 = \frac{25(1 - (-\frac{1}{5})^7)}{1 - (-\frac{1}{5})} = \frac{5^2\left(\frac{78,126}{5^7}\right)}{\frac{6}{5}}$$
$$= \frac{13,021}{625};$$
$$S_\infty = \frac{25}{1 - (-\frac{1}{5})} = \frac{125}{6}.$$

6. Find the 8th term of the geometric sequence
$$-5, -10, -20, \ldots .$$

7. Find S_{12} for the geometric series
$$-5 - 10 - 20 - \cdots .$$

8. Find S_∞ for the geometric series
$$20 - 5 + \left(\frac{5}{4}\right) + \cdots .$$

SECTION 11.4: *The Binomial Theorem*

Factorial Notation

$n! = n(n-1)(n-2)\cdots 3 \cdot 2 \cdot 1$

$7! = 7 \cdot 6 \cdot 5 \cdot 4 \cdot 3 \cdot 2 \cdot 1 = 5040$

9. Simplify: $11!$.

Binomial Coefficient

$\binom{n}{r} = {}_nC_r = \dfrac{n!}{(n-r)!\, r!}$

$\binom{10}{3} = {}_{10}C_3 = \dfrac{10!}{7!\,3!} = \dfrac{10 \cdot 9 \cdot 8 \cdot 7!}{7! \cdot 3 \cdot 2 \cdot 1}$

$= 120$

10. Simplify: $\binom{9}{3}$.

Binomial Theorem

$(a+b)^n = \binom{n}{0}a^n + \binom{n}{1}a^{n-1}b$

$\qquad + \cdots + \binom{n}{n}b^n$

$(1-2x)^3 = \binom{3}{0}1^3 + \binom{3}{1}1^2(-2x)$

$\qquad + \binom{3}{2}1(-2x)^2 + \binom{3}{3}(-2x)^3$

$= 1 \cdot 1 + 3 \cdot 1(-2x)$

$\qquad + 3 \cdot 1 \cdot (4x^2) + 1 \cdot (-8x^3)$

$= 1 - 6x + 12x^2 - 8x^3$

11. Expand: $(x^2 - 2)^5$.

$(r+1)$st term of $(a+b)^n$: $\binom{n}{r}a^{n-r}b^r$

3rd term of $(1-2x)^3$:

$\binom{3}{2}(1)^1(-2x)^2 = 12x^2 \qquad r=2$

12. Find the 4th term of $(t+3)^{10}$.

Review Exercises: Chapter 11

Concept Reinforcement

Classify each of the following statements as either true or false.

1. The next term in the arithmetic sequence $10, 15, 20, \ldots$ is 35.   [11.2]

2. The next term in the geometric sequence $2, 6, 18, 54, \ldots$ is 162.   [11.3]

3. $\sum\limits_{k=1}^{3} k^2$ means $1^2 + 2^2 + 3^2$.   [11.1]

4. If $a_n = 3n - 1$, then $a_{17} = 19$.   [11.1]

5. A geometric sequence has a common difference. [11.3]

6. The infinite geometric series $10 - 5 + \frac{5}{2} - \cdots$ has a limit.   [11.3]

7. For any natural number n, $n! = n(n-1)$.   [11.4]

8. When simplified, the expansion of $(x+y)^{17}$ has 19 terms.   [11.4]

Find the first four terms; the 8th term, a_8; and the 12th term, a_{12}.   [11.1]

9. $a_n = 10n - 9$

10. $a_n = \dfrac{n-1}{n^2 + 1}$

Write an expression for the general term of each sequence. Answers may vary.   [11.1]

11. $-5, -10, -15, -20, \ldots$

12. $-1, 3, -5, 7, -9, \ldots$

Write out and evaluate each sum.   [11.1]

13. $\sum\limits_{k=1}^{5} (-2)^k$

14. $\sum\limits_{k=2}^{7} (1 - 2k)$

Rewrite using sigma notation.   [11.1]

15. $7 + 14 + 21 + 28 + 35 + 42$

16. $\dfrac{-1}{2} + \dfrac{1}{4} + \dfrac{-1}{8} + \dfrac{1}{16} + \dfrac{-1}{32}$

17. Find the 14th term of the arithmetic sequence $-3, -7, -11, \ldots$.   [11.2]

18. An arithmetic sequence has $a_1 = 11$ and $a_{16} = 14$. Find the common difference, d.   [11.2]

19. An arithmetic sequence has $a_8 = 20$ and $a_{24} = 100$. Find the first term, a_1, and the common difference, d. [11.2]

20. Find the sum of the first 17 terms of the arithmetic series $-8 + (-11) + (-14) + \cdots$.   [11.2]

21. Find the sum of all the multiples of 5 from 5 to 500, inclusive.   [11.2]

22. Find the 20th term of the geometric sequence $2, 2\sqrt{2}, 4, \ldots$.   [11.3]

23. Find the common ratio of the geometric sequence $40, 30, \frac{45}{2}, \ldots$.   [11.3]

24. Find the nth term of the geometric sequence $-2, 2, -2, \ldots$.   [11.3]

25. Find the nth term of the geometric sequence $3, \frac{3}{4}x, \frac{3}{16}x^2, \ldots$.   [11.3]

26. Find S_6 for the geometric series

$$3 + 15 + 75 + \cdots .$$ [11.3]

27. Find S_{12} for the geometric series

$$3x - 6x + 12x - \cdots .$$ [11.3]

Determine whether each infinite geometric series has a limit. If a limit exists, find it.   [11.3]

28. $6 + 3 + 1.5 + 0.75 + \cdots$

29. $7 - 4 + \frac{16}{7} - \cdots$

30. $-\frac{1}{2} + \frac{1}{2} + \left(-\frac{1}{2}\right) + \frac{1}{2} + \cdots$

31. $0.04 + 0.08 + 0.16 + 0.32 + \cdots$

32. $\$2000 + \$1900 + \$1805 + \$1714.75 + \cdots$

33. Find fraction notation for $0.555555 \ldots$.   [11.3]

34. Find fraction notation for $1.454545 \ldots$.   [11.3]

35. Tyrone begins work in a convenience store at an hourly wage of $11.50. He was promised a raise of 40¢ per hour every 3 months for 8 years. After 8 years, what will be his hourly wage?   [11.2]

36. A stack of poles has 42 poles in the bottom row. There are 41 poles in the second row, 40 poles in the third row, and so on, ending with 1 pole in the top row. How many poles are in the stack?   [11.2]

37. Janine's student loan is for $12,000 at 4%, compounded annually. The total amount is to be paid off in 7 years. How much will she then owe?   [11.3]

38. Find the total rebound distance of a ball, given that it is dropped from a height of 12 m and each rebound is one-third of the height that it falls.   [11.3]

Simplify.   [11.4]

39. $7!$
                                                40. $\begin{pmatrix} 10 \\ 3 \end{pmatrix}$

41. Find the 3rd term of $(a + b)^{20}$.   [11.4]

42. Expand: $(x - 2y)^4$.   [11.4]

Synthesis

43. What happens to a_n in a geometric sequence with $|r| < 1$, as n gets larger? Why?   [11.3]

44. Compare the two forms of the binomial theorem given in the text. Under what circumstances would one be more useful than the other?   [11.4]

45. Find the sum of the first n terms of the geometric series $1 - x + x^2 - x^3 + \cdots$.   [11.3]

46. Expand: $(x^{-3} + x^3)^5$.   [11.4]

Test: Chapter 11

For step-by-step test solutions, access the Chapter Test Prep Videos in MyMathLab®, on YouTube (search "Bittinger Inter Alg CA" and click on "Channels"), or by scanning the code.

1. Find the first five terms and the 12th term of a sequence with general term $a_n = \dfrac{1}{n^2 + 1}$.

2. Write an expression for the general term of the sequence $\frac{4}{3}, \frac{4}{9}, \frac{4}{27}, \ldots$.

3. Write out and evaluate:
$$\sum_{k=2}^{5}(1 - 2^k).$$

4. Rewrite using sigma notation:
$$1 + (-8) + 27 + (-64) + 125.$$

5. Find the 13th term, a_{13}, of the arithmetic sequence $\frac{1}{2}, 1, \frac{3}{2}, 2, \ldots$.

6. Find a_1 and d of an arithmetic sequence when $a_5 = 16$ and $a_{10} = -3$.

7. Find the sum of all the multiples of 12 from 24 to 240, inclusive.

8. Find the 10th term of the geometric sequence $-3, 6, -12, \ldots$.

9. Find the common ratio of the geometric sequence $22\frac{1}{2}, 15, 10, \ldots$.

10. Find the nth term of the geometric sequence $3, 9, 27, \ldots$.

11. Find S_9 for the geometric series
$$11 + 22 + 44 + \cdots .$$

Determine whether each infinite geometric series has a limit. If a limit exists, find it.

12. $0.5 + 0.25 + 0.125 + \cdots$

13. $0.5 + 1 + 2 + 4 + \cdots$

14. $\$1000 + \$80 + \$6.40 + \cdots$

15. Find fraction notation for $0.85858585\ldots$.

16. An auditorium has 31 seats in the first row, 33 seats in the second row, 35 seats in the third row, and so on, for 18 rows. How many seats are in the 17th row?

17. Alyssa's uncle Ken gave her $100 for her first birthday, $200 for her second birthday, $300 for her third birthday, and so on, until her eighteenth birthday. How much did he give her in all?

18. Each week the price of a $10,000 boat will be reduced 5% of the previous week's price. If we assume that it is not sold, what will be the price after 10 weeks?

19. Find the total rebound distance of a ball that is dropped from a height of 18 m, with each rebound two-thirds of the preceding one.

20. Simplify: $\dbinom{12}{9}$.

21. Expand: $(x - 3y)^5$.

22. Find the 4th term in the expansion of $(a + x)^{12}$.

Synthesis

23. Find a formula for the sum of the first n even natural numbers:
$$2 + 4 + 6 + \cdots + 2n.$$

24. Find the sum of the first n terms of
$$1 + \frac{1}{x} + \frac{1}{x^2} + \frac{1}{x^3} + \cdots .$$

Cumulative Review/Final Exam: Chapters 1–11

Simplify.

1. $\left| -\dfrac{2}{3} + \dfrac{1}{5} \right|$ [1.2]

2. $y - [3 - 4(5 - 2y) - 3y]$ [1.3]

3. $(10 \cdot 8 - 9 \cdot 7)^2 - 54 \div 9 - 3$ [1.2]

4. $(2.7 \times 10^{-24})(3.1 \times 10^9)$ [1.7]

Perform the indicated operations to create an equivalent expression. Be sure to simplify your result if possible.

5. $(5a^2 - 3ab - 7b^2) - (2a^2 + 5ab + 8b^2)$ [5.1]

6. $(2a - 1)(2a + 1)$ [5.2]

7. $(3a^2 - 5y)^2$ [5.2]

8. $\dfrac{1}{x - 2} - \dfrac{4}{x^2 - 4} + \dfrac{3}{x + 2}$ [6.2]

9. $\dfrac{3x + 3y}{5x - 5y} \div \dfrac{3x^2 + 3y^2}{5x^3 - 5y^3}$ [6.1]

10. $\dfrac{x - \dfrac{a^2}{x}}{1 + \dfrac{a}{x}}$ [6.3]

11. $\sqrt{12a}\,\sqrt{12a^3b}$ [7.3]

12. $(-9x^2y^5)(3x^8y^{-7})$ [1.6]

13. $(125x^6y^{1/2})^{2/3}$ [7.2]

14. $\dfrac{\sqrt[3]{x^2y^5}}{\sqrt[4]{xy^2}}$ [7.5]

15. $(4 + 6i)(2 - i)$, where $i = \sqrt{-1}$ [7.8]

Factor, if possible, to form an equivalent expression.

16. $4x^2 - 12x + 9$ [5.5]

17. $27a^3 - 8$ [5.6]

18. $12s^4 - 48t^2$ [5.5]

19. $15y^4 + 33y^2 - 36$ [5.7]

20. Divide:

$$(7x^4 - 5x^3 + x^2 - 4) \div (x - 2).$$ [6.6]

Find the domain of each function.

21. $f(x) = \sqrt{2x - 8}$ [7.1]

22. $g(x) = \dfrac{x - 4}{x^2 - 10x + 25}$ [5.8]

23. Find a linear equation whose graph has a y-intercept of $(0, -8)$ and is parallel to the line whose equation is $3x - y = 6$.   [2.5]

24. Write a quadratic equation whose solutions are $5\sqrt{2}$ and $-5\sqrt{2}$.   [8.3]

25. Find the center and the radius of the circle given by
$$x^2 + y^2 - 4x + 6y - 23 = 0.$$ [10.1]

26. Write an equivalent expression that is a single logarithm:
$$\tfrac{2}{3}\log_a x - \tfrac{1}{2}\log_a y + 5\log_a z.$$ [9.4]

27. Write an equivalent exponential equation:
$$\log_a c = 5.$$ [9.3]

Use a calculator to find each of the following. Round to four decimal places.   [9.5]

28. $\log 120$

29. $\log_5 3$

30. Find the distance between the points $(-1, -5)$ and $(2, -1)$.   [7.7]

31. Find the 21st term of the arithmetic sequence $19, 12, 5, \ldots$.   [11.2]

32. Find the sum of the first 25 terms of the arithmetic series $-1 + 2 + 5 + \cdots$.   [11.2]

33. Write an expression for the general term of the geometric sequence $16, 4, 1, \ldots$.   [11.3]

34. Find the 7th term of $(a - 2b)^{10}$.   [11.4]

Solve.

35. $8(x - 1) - 3(x - 2) = 1$ [1.3]

36. $\dfrac{6}{x} + \dfrac{6}{x + 2} = \dfrac{5}{2}$ [6.4]

37. $2x + 1 > 5 \text{ or } x - 7 \le 3$ [4.2]

38. $5x + 6y = -2$,
$3x + 10y = 2$ [3.2]

39. $x + y - z = 0$,
$3x + y + z = 6$,
$x - y + 2z = 5$ [3.4]

40. $3\sqrt{x - 1} = 5 - x$ [7.6]

41. $x^4 - 29x^2 + 100 = 0$ [8.5]

42. $x^2 + y^2 = 8$,
$x^2 - y^2 = 2$ [10.4]

43. $4^x = 12$ [9.6]

44. $\log(x^2 - 25) - \log(x + 5) = 3$ [9.6]

45. $7^{2x+3} = 49$ [9.6]

46. $|2x - 1| \le 5$ [4.3]

47. $15x^2 + 45 = 0$ [8.1]

48. $x^2 + 4x = 3$ [8.2]

49. $y^2 + 3y > 10$ [8.9]

50. Let $f(x) = x^2 - 2x$. Find a such that $f(a) = 80$. [5.8]

51. Solve $I = \dfrac{R}{R + r}$ for R. [6.8]

Graph.

52. $3x - y = 7$ [2.4]

53. $x^2 + y^2 = 100$ [10.1]

54. $\dfrac{x^2}{36} - \dfrac{y^2}{9} = 1$ [10.3]

55. $y = \log_2 x$ [9.3]

56. $f(x) = 2^x - 3$ [9.2]

57. $2x - 3y < -6$ [4.4]

58. Graph: $f(x) = -2(x - 3)^2 + 1$. [8.7]

  a) Label the vertex.

  b) Draw the axis of symmetry.

  c) Find the maximum or minimum value.

Solve.

59. The Brighton recreation department plans to fence in a rectangular park next to a river. (Note that no fence will be needed along the river.) What is the area of the largest region that can be fenced in with 200 ft of fencing? [8.8]

60. The perimeter of a rectangular sign is 34 ft. The length of a diagonal is 13 ft. Find the dimensions of the sign. [10.4]

61. An online movie store offers two types of membership. Limited members pay a fee of $40 per year and can download movies for $2.45 each. Preferred members pay $60 per year and can download movies for $1.65 each. For what numbers of annual movie downloads would it be less expensive to be a preferred member? [4.1]

62. Cosmos Tastes mixes herbs that cost $2.68 per ounce with herbs that cost $4.60 per ounce to create a seasoning that costs $3.80 per ounce. How many ounces of each herb should be mixed together in order to make 24 oz of the seasoning? [3.3]

63. An airplane flies 190 mi with the wind in the same time it takes to fly 160 mi against the wind. The speed of the wind is 30 mph. How fast would the plane fly in still air? [6.5]

64. Jared can tap the sugar maple trees in Southway Farm in 21 hr. Delia can tap the trees in 14 hr. How long would it take them, working together, to tap the trees? [6.5]

65. *National Debt.* The U.S. national debt increased from $9 trillion in 2008 to $15 trillion in 2012.

  Source: CBS News

  a) What was the average rate of change? [2.3]

  b) Find a linear function that fits the data. Let $f(t)$ represent the national debt, in trillions of dollars, t years after 2008. [2.5]

  c) Find an exponential function that fits the data. Let $P(t)$ represent the national debt, in trillions of dollars, t years after 2008. [9.7]

  d) Assume the debt is growing linearly, and use the linear function from part (b) to predict the national debt in 2020. [2.5]

  e) Assume the debt is growing exponentially, and use the exponential function from part (c) to predict the national debt in 2020. [9.7]

  f) How long will it take the national debt to double if it is growing exponentially? [9.7]

Synthesis

Solve.

66. $\dfrac{9}{x} - \dfrac{9}{x + 12} = \dfrac{108}{x^2 + 12x}$ [6.4]

67. $\log_2 (\log_3 x) = 2$ [9.6]

68. Suppose that y varies directly as the cube of x. If x is multiplied by 0.5, what is the effect on y? [6.8]

69. Diaphantos, a famous mathematician, spent $\frac{1}{6}$ of his life as a child, $\frac{1}{12}$ as an adolescent, and $\frac{1}{7}$ as a bachelor. Five years after he was married, he had a son who died 4 years before his father at half his father's final age. How long did Diaphantos live? [3.5]

ANSWERS

CHAPTER 1

Exercise Set 1.1, pp. 8–10

1. Variable **2.** Constant **3.** Value **4.** Base; exponent **5.** Evaluating **6.** Division **7.** Rational **8.** Irrational **9.** Terminating **10.** Repeating
11. Let n represent the number; $n - 5$
13. Let x represent the number; $2x$
15. Let x represent the number; $0.29x$, or $\frac{29}{100}x$
17. Let y represent the number; $\frac{1}{2}y - 6$
19. Let s represent the number; $0.1s + 7$, or $\frac{10}{100}s + 7$
21. Let m and n represent the numbers; $mn - 1$
23. $90 \div 4$, or $\frac{90}{4}$ **25.** 36 sq ft, or 36 ft^2
27. 0.25 sq m, or 0.25 m^2 **29.** 17.5 sq ft, or 17 ft^2
31. 11.2 sq ft, or 11.2 ft^2 **33.** 11 **35.** 39 **37.** 11
39. 35 **41.** 8 **43.** 0 **45.** 5 **47.** 25 **49.** 225
51. 0 **53.** 18 **55.** $\{a, l, g, e, b, r\}$
57. $\{1, 3, 5, 7, \dots\}$ **59.** $\{10, 20, 30, 40, \dots\}$
61. $\{x \mid x$ is an even number between 9 and 99$\}$
63. $\{x \mid x$ is a whole number less than 5$\}$
65. $\{x \mid x$ is an odd number between 10 and 20$\}$
67. (a) $0, 6$; **(b)** $-3, 0, 6$; **(c)** $-8.7, -3, 0, \frac{2}{3}, 6$; **(d)** $\sqrt{7}$;
(e) $-8.7, -3, 0, \frac{2}{3}, \sqrt{7}, 6$ **69. (a)** $0, 8$; **(b)** $-17, 0, 8$;
(c) $-17, -0.01, 0, \frac{5}{4}, 8$; **(d)** $\sqrt{77}$; **(e)** $-17, -0.01, 0, \frac{5}{4}, 8, \sqrt{77}$
71. True **73.** True **75.** False **77.** True
79. True **81.** False **83.**  **85.**
87. Let a and b represent the numbers; $\dfrac{a + b}{a - b}$
89. Let r and s represent the numbers; $\dfrac{1}{2}(r^2 - s^2)$, or $\dfrac{r^2 - s^2}{2}$
91. $\{0\}$ **93.** $\{5, 10, 15, 20, \dots\}$ **95.** $\{1, 3, 5, 7, \dots\}$
97.

[right triangle with legs 3 (bottom) and 2 (right), hypotenuse $\sqrt{13}$]

Technology Connection, p. 13

1. [icon] **2.** 0.97

Exercise Set 1.2, pp. 18–20

1. True **2.** False **3.** True **4.** False **5.** False
6. True **7.** False **8.** True **9.** True **10.** False
11. 10 **13.** 7 **15.** 46.8 **17.** 0 **19.** $1\frac{7}{8}$
21. 4.21 **23.** -5 is less than or equal to -4; true
25. -9 is greater than 1; false **27.** 0 is greater than or equal to -5; true **29.** -8 is less than -3; true

31. -4 is greater than or equal to -4; true **33.** -5 is less than -5; false **35.** 12 **37.** -12 **39.** -2.5
41. $-\frac{11}{35}$ **43.** -9.06 **45.** $\frac{5}{9}$ **47.** -6.25 **49.** 0
51. 3.8 **53.** -2.37 **55.** 56 **57.** 0 **59.** -8
61. $\frac{1}{10}$ **63.** 4.67 **65.** 0 **67.** 6 **69.** -6 **71.** 7
73. -19 **75.** -3.1 **77.** $-\frac{11}{10}$ **79.** 0 **81.** 5.37
83. -24 **85.** 22 **87.** -21 **89.** $-\frac{3}{7}$ **91.** 0
93. $-\frac{1}{2}$ **95.** 4 **97.** -4 **99.** -73 **101.** 0
103. $\frac{1}{8}$ **105.** $-\frac{7}{5}$ **107.** Does not exist **109.** $\frac{7}{10}$
111. $-\frac{6}{5}$ **113.** $\frac{1}{36}$ **115.** 1 **117.** -100 **119.** -9
121. 9 **123.** 25 **125.** $-\frac{6}{11}$ **127.** Undefined
129. $\frac{11}{43}$ **131.** 31 **133.** -3 **135.** $xy + 6$; $6 + yx$
137. $(ab)(-9)$; $-9(ba)$ **139.** $3(xy)$
141. $3y + (4 + 10)$ **143.** $7x + 7$ **145.** $5m - 5n$
147. $-10a - 15b$ **149.** $9ab - 9ac + 9ad$
151. $5(x + 10)$ **153.** $3(3p - 1)$ **155.** $7(x - 3y + 2z)$
157. $17(15 - 2b)$ **159.** $x(y + 1)$ **161.** [icon] **163.** [icon]
165. $(8 - 5)^3 + 9 = 36$ **167.** $5 \cdot 2^3 \div (3 - 4)^4 = 40$
169. 15 **171.** -6.2 **173.** [icon] **175. (a)** Let t represent the temperature at midnight, in °F; $-16 - 5 = t$; $t = -21$. The temperature at midnight was -21°F.
(b) Let x represent the temperature outside Ethan's jet, in °F; $42 - 3.5(20) = x$; $x = -28$. The temperature outside Ethan's jet is -28°F.

Quick Quiz: Sections 1.1–1.2, p. 20

1. Let n represent the number; $2n - 8$ **2.** 120 **3.** 8
4. -6 **5.** 53

Exercise Set 1.3, pp. 26–27

1. Equivalent **2.** Linear **3.** Contradiction
4. Identity **5.** Equivalent expressions
6. Equivalent equations **7.** Equivalent equations
8. Equivalent expressions **9.** Equivalent expressions
10. Equivalent equations **11.** Equivalent
13. Not equivalent **15.** Not equivalent
17. 16.3 **19.** 9 **21.** 45 **23.** -4 **25.** $\frac{17}{2}$
27. $10t^2$ **29.** $15a$ **31.** $-7n$ **33.** $10x$
35. $21p - 4$ **37.** $-5t^2 + 2t + 4t^3$ **39.** $17x - 21$
41. $5a - 5$ **43.** $-5m + 2$ **45.** $5d - 12$
47. $-2x + 22$ **49.** $p - 16$ **51.** $4a - 12$
53. $-310x - 30$ **55.** $14y + 42$ **57.** 7 **59.** -12
61. 6 **63.** 3 **65.** 3 **67.** -3 **69.** 5
71. $\frac{49}{9}$ **73.** $\frac{4}{5}$ **75.** $\frac{19}{5}$ **77.** $-\frac{4}{11}$ **79.** $\frac{23}{8}$
81. $\mathbb{R}$; identity **83.** $\varnothing$; contradiction **85.** $\{0\}$; conditional **87.** $\mathbb{R}$; identity **89.** $\varnothing$; contradiction
91. $\mathbb{R}$; identity **93.** [icon] **95.** [icon]
97. 0.2140224409 **99.** 4 **101.** $\frac{19}{46}$ **103.** [icon]

Quick Quiz: Sections 1.1–1.3, p. 27

1. 6 sq m, or $6\,\text{m}^2$ **2.** 6   **3.** 4   **4.** $\frac{13}{8}$
5. $5(x - 2y + 4)$

Mid-Chapter Review: Chapter 1, p. 28

1. $3x - 2(x - 1) = 3x - 2x + 2$
$\qquad\qquad\qquad = x + 2$
2. $3x - 2(x - 1) = 6x$
$\quad 3x - 2x + 2 = 6x$
$\qquad\quad x + 2 = 6x$
$\qquad\qquad 2 = 5x$
$\qquad\qquad \frac{2}{5} = x$
3. Let n represent the number; $3n - 5$ **4.** 12
5. $\frac{3}{4}\,\text{ft}^2$ **6.** $\frac{5}{6}$ **7.** 40   **8.** 2.52   **9.** $-\frac{3}{25}$ **10.** 11
11. $x + (3 + y)$ **12.** $2x + 7$ **13.** $t + 1$ **14.** $8x + 2$
15. $-2p + 10$ **16.** -11 **17.** 1   **18.** $\mathbb{R}$; identity
19. 0   **20.** $-\frac{13}{4}$

Exercise Set 1.4, pp. 34–36

1. Familiarize., Translate., Carry out., Check., State.
2. Carry out.   **3.** State.   **4.** Translate.   **5.** Familiarize.
6. Check.   **7.** Let x and $x + 9$ represent the numbers;
$x + (x + 9) = 91$ **9.** Let t represent the time, in hours,
that it will take Noah to make the trip; $8 = (4.6 - 2.1)t$
11. Let x, $x + 1$, and $x + 2$ represent the angle measures,
in degrees; $x + (x + 1) + (x + 2) = 180$ **13.** Let t represent the time, in minutes, that it will take Dominik
to reach the top of the escalator; $205 = (100 + 105)t$
15. Let w represent the wholesale price; $w + 0.5w + 1.50 = 22.50$ **17.** Let t represent the number of minutes
spent climbing; $8000 + 3500t = 29{,}000$ **19.** Let
n represent the first odd number; $n + 2(n + 2) + 3(n + 4) = 70$ **21.** Let s represent the length, in centimeters, of a side of the smaller triangle; $3s + 3 \cdot 2s = 90$
23. Let c represent Cody's calls on his next shift;
$\dfrac{5 + 2 + 1 + 3 + c}{5} = 3$ **25.** \$97   **27.** \$750
29. 32 flu shots   **31.** Length: 45 cm; width: 15 cm
33. Length: 52 m; width: 13 m   **35.** 3.2 hr
37. $59°, 60°, 61°$ **39.** \$150   **41.** \$14.00
43. 🖑   **45.** 🖑   **47.** 10 points   **49.** 15.7 million
Americans

Quick Quiz: Sections 1.1–1.4, p. 36

1. $\{c, o, l, e, g\}$ **2.** $\frac{3}{4}$ **3.** $(8 + y) + 3$
4. $\varnothing$; contradiction   **5.** 16 seniors

Connecting the Concepts, p. 39

1. 5   **2.** $x = \dfrac{c + h}{2}$ **3.** $\frac{29}{5}$ **4.** $y = \dfrac{8 + nc}{a + n}$
5. $-\frac{3}{22}$ **6.** $a = \dfrac{3n}{2 - 6x}$

Exercise Set 1.5, pp. 41–45

1. Equation   **2.** Area   **3.** Circumference
4. $A = \frac{1}{2}bh$ **5.** $A = bh$ **6.** Length   **7.** Subscripts
8. Factor   **9.** $A = \dfrac{E}{w}$ **11.** $r = \dfrac{d}{t}$ **13.** $h = \dfrac{V}{lw}$
15. $k = Ld^2$ **17.** $n = \dfrac{G - w}{150}$ **19.** $l = p - 2w - 2h$
21. $y = \dfrac{4 - 2x}{3}$, or $y = -\frac{2}{3}x + \frac{4}{3}$ **23.** $y = \dfrac{C - Ax}{B}$
25. $F = \frac{9}{5}C + 32$ **27.** $r^3 = \dfrac{3V}{4\pi}$ **29.** $n = \dfrac{t}{p + m}$
31. $v = \dfrac{x}{u + w}$ **33.** $n = \dfrac{q_1 + q_2 + q_3}{A}$
35. $t = \dfrac{d_2 - d_1}{v}$ **37.** $d_1 = d_2 - vt$ **39.** $b = \dfrac{c}{d - a}$
41. $w = \dfrac{v}{uv + 1}$ **43.** $m = \dfrac{n}{t^2 + k}$ **45.** 8%
47. 16 cm   **49.** About 239 lb   **51.** About 1504.6 g
53. 9 ft   **55.** 1 year   **57.** 7 times   **59.** 1205
61. 5 ft 7 in.   **63.** 7.22 words per sentence   **65.** \$0.12
67. 512 visits per day   **69.** 34 appointments
71. About 8.5 cm   **73.** 🖑   **75.** 🖑   **77.** About 10.9 g
79. 🖑   **81.** $a = \dfrac{2s - 2v_it}{t^2}$
83. $w = \dfrac{h + p - b(a + p + f)}{b - 1}$ **85.** $b = \dfrac{ac}{1 + c}$
87. $t = \dfrac{1}{s}$

Quick Quiz: Sections 1.1–1.5, p. 45

1. $-\frac{1}{4}$ **2.** -1.875 **3.** 1   **4.** $p = \dfrac{7}{3 - b}$
5. Length: 1 m; width: $\frac{1}{2}$ m

Technology Connection, p. 50

1. Answers may vary; ②[x^y]⑤[(-)][=], ②◠[(-)]
⑤[ENTER], ②[x^y][-x]⑤[ENTER]
2. Compute $1 \div (2 \times 2 \times 2 \times 2 \times 2)$, or
$1 \div 2 \div 2 \div 2 \div 2 \div 2$.

Exercise Set 1.6, pp. 53–55

1. The power rule   **2.** Raising a quotient to a power
3. Raising a product to a power   **4.** The quotient rule
5. The product rule   **6.** The power rule   **7.** Raising a
quotient to a power   **8.** Raising a product to a power
9. The quotient rule   **10.** The product rule
11. 6^{11} **13.** m^8 **15.** $20x^7$ **17.** $24a^8$ **19.** m^8n^3
21. t^5 **23.** $5a^5$ **25.** m^5n^4 **27.** $4x^6y^4$
29. $-4x^8y^6z^6$ **31.** -1 **33.** 1
35. $\dfrac{1}{t^9}$ **37.** $\dfrac{1}{6^2} = \dfrac{1}{36}$ **39.** $\dfrac{1}{(-3)^2} = \dfrac{1}{9}$ **41.** $-\dfrac{1}{3^2} = -\dfrac{1}{9}$
43. $-\dfrac{1}{1^{10}} = -1$ **45.** $10^3 = 1000$ **47.** $\dfrac{6}{x}$ **49.** $\dfrac{3a^8}{b^6}$
51. $\dfrac{2}{x^5z^3}$ **53.** $3y^2z^4$ **55.** $\dfrac{ac}{b}$ **57.** $\dfrac{pv^4}{2q^2r^3u^5}$

59. x^{-3} **61.** $(-10)^{-3}$ **63.** $\dfrac{1}{8^{-10}}$ **65.** $\dfrac{4}{x^{-2}}$

67. $(5y)^{-3}$ **69.** $\dfrac{y^{-4}}{3}$ **71.** 6^{-8}, or $\dfrac{1}{6^8}$ **73.** a^{-7}, or $\dfrac{1}{a^7}$

75. 1   **77.** $-8m^4n^5$ **79.** $35x^{-2}y^3$, or $\dfrac{35y^3}{x^2}$

81. $10a^{-6}b^{-2}$, or $\dfrac{10}{a^6b^2}$ **83.** 10^{-9}, or $\dfrac{1}{10^9}$

85. 2^{-2}, or $\dfrac{1}{2^2}$, or $\dfrac{1}{4}$ **87.** y^9 **89.** $-3ab^2$

91. $\dfrac{3}{2}m^{-5}n^7$, or $\dfrac{3n^7}{2m^5}$ **93.** $\dfrac{1}{4}x^3y^{-2}z^{11}$, or $\dfrac{x^3z^{11}}{4y^2}$ **95.** x^{12}

97. 9^{-12}, or $\dfrac{1}{9^{12}}$ **99.** t^{40} **101.** $25x^2y^2$

103. $(-2)^{-3}a^6b^{-3}$, or $-\dfrac{a^6}{8b^3}$ **105.** $\dfrac{m^6n^{-3}}{64}$, or $\dfrac{m^6}{64n^3}$

107. $32a^{-4}$, or $\dfrac{32}{a^4}$ **109.** 1   **111.** $\dfrac{5a^4b}{2}$ **113.** $\dfrac{8x^9y^3}{27}$

115. 1   **117.** $\dfrac{4}{25}x^{-4}y^{22}$, or $\dfrac{4y^{22}}{25x^4}$ **119.** 🖳   **121.** 🖳

123. $4a^{-x-4}$ **125.** 8^{-2abc} **127.** $-4x^{10}y^8$ **129.** $\dfrac{2}{27}$

131. $\dfrac{a^{-14ac}}{b^{27ac}}$

Quick Quiz: Sections 1.1–1.6, p. 55

1. $\dfrac{3}{7}$ **2.** $7n^3 + 4n^2 + 2n$ **3.** $\dfrac{11}{2}$ **4.** 4 m   **5.** $\dfrac{4w^6}{9x^{14}}$

Exercise Set 1.7, pp. 60–62

1. Is not   **2.** Negative   **3.** Four   **4.** At the very end
5. Positive power of 10   **6.** Negative power of 10
7. Negative power of 10   **8.** Positive power of 10
9. Positive power of 10   **10.** Negative power of 10
11. 6.4×10^{10} **13.** 1.3×10^{-6} **15.** 9×10^{-5}
17. 8.03×10^{11} **19.** 9.04×10^{-7} **21.** 4.317×10^{11}
23. 400,000   **25.** 0.00012   **27.** 0.00000000376
29. 8,056,000,000,000   **31.** 0.00007001   **33.** 8.8×10^7
35. 3.3×10^{-5} **37.** 1.4×10^{11} **39.** 4.6×10^{-11}
41. 6.0   **43.** 2.5×10^{11} **45.** 2.0×10^{-7}
47. 4.0×10^{-16} **49.** 3.00×10^{-22} **51.** 2.00×10^{26}
53. 1.4×10^{21} bacteria   **55.** Approximately 4.5×10^{-16} in³
57. 1.5×10^5 megabytes per person   **59.** 4.50×10^{-3} kg, or 4.50 g   **61.** 1.00×10^5 light years   **63.** 3.08×10^{26} Å
65. 1×10^{22} cu Å, or 1×10^{-8} m³   **67.** 7.9×10^7 bacteria
69. 4.49×10^4 km/h   **71.** 🖳   **73.** 🖳
75. Approximately 5.53 g/cm³   **77.** 6.7×10^4 mph
79. $8 \cdot 10^{-90}$ is larger by 7.1×10^{-90}.   **81.** 8
83. 8×10^{18} grains   **85.** 🖳

Quick Quiz: Sections 1.1–1.7, p. 62

1. 36   **2.** $x + 6$ **3.** $\mathbb{R}$; identity   **4.** $\dfrac{6}{a^5c}$

5. 1.9×10^{-5}

Translating for Success, p. 63

1. F   **2.** D   **3.** I   **4.** C   **5.** E   **6.** J   **7.** O
8. M   **9.** B   **10.** L

Decision Making: Connection, p. 64

1. (a) Tests: 25%; weekly projects: 40%; semester project: 25%; class participation: 5%; **(b)** $g = 0.05q + 0.25t + 0.40w + 0.25s + 0.05p$, where g is the course grade, q is the quiz grade, w is the weekly project grade, s is the semester project grade, and p is the class participation grade; **(c)** tests: 90%; weekly projects: 80%; **(d)** 84.1875%
2. No; the highest grade she can receive is 89.15% if she earns all 500 points on the semester project.   **3.** 🖳

Study Summary: Chapter 1, pp. 65–68

1. Let x and y represent the numbers; $3(x + y)$ **2.** 23
3. 167   **4.** -5 **5.** 14   **6.** 30   **7.** -4
8. $10n + 6$ **9.** $(3a)b$ **10.** $50m + 90n + 10$
11. $13(2x + 1)$ **12.** $3x - 9$ **13.** 6   **14.** $\mathbb{R}$; identity
15. $47\frac{1}{2}$ mi; $72\frac{1}{2}$ mi   **16.** $y = \dfrac{w}{x - 3}$ **17.** 1   **18.** $\dfrac{1}{10}$
19. $\dfrac{y^3}{x}$ **20.** $\dfrac{b}{a}$ **21.** x^{16} **22.** 8^2, or 64   **23.** $\dfrac{1}{t^{20}}$
24. $x^{30}y^{10}$ **25.** $\dfrac{x^{10}}{7^5}$ **26.** 9.04×10^{-4} **27.** 690,000

Review Exercises: Chapter 1, pp. 68–70

1. (e)   **2.** (g)   **3.** (j)   **4.** (a)   **5.** (i)   **6.** (b)
7. (f)   **8.** (c)   **9.** (d)   **10.** (h)   **11.** Let x and y represent the numbers; $\dfrac{x}{y} - 8$ **12.** 22
13. $\{1, 3, 5, 7, 9\}$; $\{x \mid x$ is an odd natural number less than 10$\}$ **14.** 1750 sq cm   **15.** 19   **16.** 0
17. 6.08   **18.** -11 **19.** $\dfrac{1}{20}$ **20.** 4.4   **21.** -3.8
22. 96   **23.** $-\dfrac{5}{12}$ **24.** -9.1 **25.** $-\dfrac{21}{4}$ **26.** 6.28
27. $x + 12$ **28.** $x \cdot 5 + y$, or $y + 5x$ **29.** $4 + (a + b)$
30. $(xy)z$ **31.** $2(6m + 2n - 1)$ **32.** $4x^3 - 6x^2 + 5$
33. $47x - 60$ **34.** $\dfrac{1}{2}$ **35.** $\dfrac{21}{4}$ **36.** $-\dfrac{4}{11}$
37. $\mathbb{R}$; identity   **38.** $\varnothing$; contradiction
39. Let x represent the number; $2x + 15 = 21$ **40.** 48
41. $90°, 30°, 60°$ **42.** $c = \dfrac{xt}{b}$ **43.** $x = \dfrac{c}{m - r}$
44. 14 cm   **45.** $-28m^4n^{10}$ **46.** $4xy^6$ **47.** 1, 64, -64
48. 3^2, or 9   **49.** $8t^{12}$ **50.** $-\dfrac{a^9}{125b^6}$ **51.** $\dfrac{z^8}{x^4y^6}$
52. $\dfrac{n^{12}}{81m^{28}}$ **53.** $\dfrac{3}{7}$ **54.** 0   **55.** 3.07×10^{-4}
56. 3.086×10^{13} **57.** 3.7×10^7 **58.** 2.0×10^{-6}
59. 1.4×10^4 mm³, or 1.4×10^{-5} m³   **60.** 🖳 To write an equation that has no solution, begin with a simple equation that is false for any value of x, such as $x = x + 1$. Then add or multiply by the same quantities on both sides of the equation to construct a more complicated equation with no solution.   **61.** 🖳 **(a)** $-(-x)$ is positive when x is positive; the opposite of the opposite of a number is the number itself; **(b)** $-x^2$ is never positive; x^2 is always

nonnegative, so the opposite of x^2 is always nonpositive; **(c)** $-x^3$ is positive when x is negative; x^3 is negative when x is negative, and the opposite of a negative number is positive; **(d)** $(-x)^2$ is positive when $x \neq 0$; the square of any nonzero number is positive; **(e)** x^{-2} is positive when $x \neq 0$; $x^{-2} = \dfrac{1}{x^2}$, and x^2 is positive when x is nonzero.

62. 0.0000003% **63.** $\frac{25}{24}$ **64.** The 17-in. pizza is a better deal. It costs about 6.6¢ per square inch; the 13-in. pizza costs about 9.0¢ per square inch.   **65.** 729 cm^3

66. $z = y - \dfrac{x}{m}$, or $\dfrac{my - x}{m}$ **67.** $3^{-2a+2b-8ab}$ **68.** -39

69. $-40x$ **70.** $a \cdot 2 + cb + cd + ad = ad + a \cdot 2 + cb + cd = a(d + 2) + c(b + d)$ **71.** $\sqrt{5}/4$; answers may vary

Test: Chapter 1, p. 70

1. [1.1] Let m and n represent the numbers; $mn - 4$
2. [1.1], [1.2] -47 **3.** [1.1] 181.35 sq m   **4.** [1.2] -31
5. [1.2] -3.7 **6.** [1.2] -14.2 **7.** [1.2] -33.92
8. [1.2] $\frac{1}{12}$ **9.** [1.2] $\frac{5}{49}$ **10.** [1.2] 6 **11.** [1.2] $-\frac{4}{3}$
12. [1.2] $-\frac{5}{2}$ **13.** [1.2] $x + 3$ **14.** [1.3] $-3y - 29$
15. [1.3] -2 **16.** [1.3] $\mathbb{R}$; identity   **17.** [1.5] $p = \dfrac{t}{2 - s}$
18. [1.4] 94 **19.** [1.4] $17, 19, 21$ **20.** [1.3] $8x - 11$
21. [1.3] $24b - 9$ **22.** [1.6] $-\dfrac{42}{x^{10}y^6}$
23. [1.6] $-\dfrac{1}{6^2}$, or $-\dfrac{1}{36}$ **24.** [1.6] $-\dfrac{125y^9}{x^3}$ **25.** [1.6] $\dfrac{4y^8}{x^6}$
26. [1.6] 1 **27.** [1.7] 2.01×10^{-7} **28.** [1.7] 3.8×10^2
29. [1.7] 2.0×10^9 neutrinos   **30.** [1.6] $8^c x^{9ac} y^{3bc+3c}$
31. [1.6] $-9a^3$ **32.** [1.3] -2

CHAPTER 2

Technology Connection, p. 77

1. $y = -4x + 3$ $(-1.5, 9), (1, -1)$

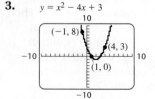

2. $y = 5x - 3$

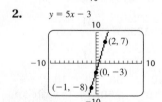

3. $y = x^2 - 4x + 3$

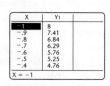

4.

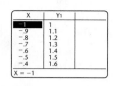

Exercise Set 2.1, pp. 78–80

1. Axes   **2.** Ordered   **3.** Third   **4.** Negative
5. Solutions   **6.** Linear
7. $(5, 3), (-4, 3), (0, 2), (-2, -3), (4, -2),$ and $(-5, 0)$
9.

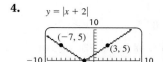

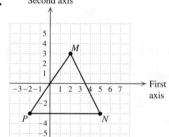

11.

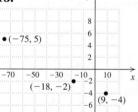

Triangle, 21 units^2

13.

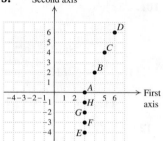

15.

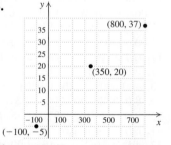

17. IV   **19.** III   **21.** y-axis   **23.** II   **25.** x-axis
27. I   **29.** Yes   **31.** No   **33.** Yes   **35.** Yes
37. No   **39.** No   **41.** Yes
43.

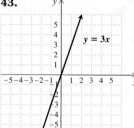

45.

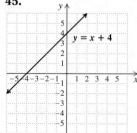

47.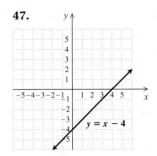

49.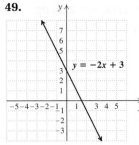

83.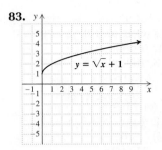

85.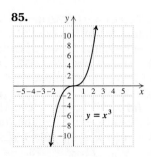

51.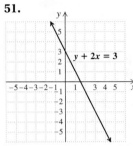

53.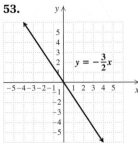

87. **(a)** $y = 0.375x^3$

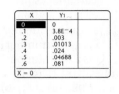

Yscl = 0.1

(b) $y = -3.5x^2 + 6x - 8$

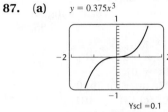

(c) $y = (x - 3.4)^3 + 5.6$

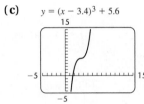

55.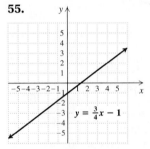

57.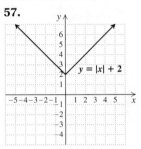

59.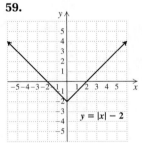

61.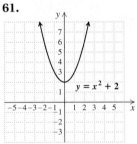

Prepare to Move On, p. 80

1. 43   **2.** 9   **3.** $-\frac{3}{4}$ **4.** 0   **5.** 5   **6.** $-\frac{3}{5}$

63.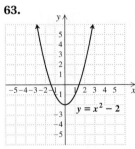

65. 🖉   **67.** -3

68. $15x - 29$ **69.** $4x^{12}y^2$

70. $-\dfrac{12b^{26}}{7a^{12}}$ **71.** 🖉

73. **(a)** III; **(b)** II; **(c)** I; **(d)** IV   **75.** **(a)** III; **(b)** II; **(c)** IV; **(d)** I

77. $(-1, -2)$, $(-19, -2)$, and $(13, 10)$

Exercise Set 2.2, pp. 89–94

1. Correspondence   **2.** Exactly   **3.** Domain
4. Range   **5.** Horizontal   **6.** Vertical   **7.** "f of 3"
8. Vertical   **9.** Yes   **11.** Yes   **13.** No   **15.** Yes
17. Function   **19.** Function   **21.** **(a)** $\{-3, -2, 0, 4\}$;
(b) $\{-10, 3, 5, 9\}$; **(c)** yes   **23.** **(a)** $\{1, 2, 3, 4, 5\}$; **(b)** $\{1\}$;
(c) yes   **25.** **(a)** $\{-2, 3, 4\}$; **(b)** $\{-8, -2, 4, 5\}$; **(c)** no
27. Domain: $\mathbb{R}$; range: $\mathbb{R}$ **29.** Domain: $\mathbb{R}$; range: $\{4\}$
31. Domain: $\mathbb{R}$; range: $\{y \mid y \geq 1\}$ **33.** Domain:
$\{x \mid x \geq 0\}$; range: $\{y \mid y \geq 0\}$ **35.** Yes   **37.** Yes
39. No   **41.** **(a)** 5; **(b)** -3; **(c)** -9; **(d)** 21; **(e)** $2a + 9$;
(f) $2a + 7$ **43.** **(a)** 0; **(b)** 1; **(c)** 57; **(d)** $5t^2 + 4t$;
(e) $20a^2 + 8a$; **(f)** 48   **45.** **(a)** $\frac{3}{5}$; **(b)** $\frac{1}{3}$; **(c)** $\frac{4}{7}$; **(d)** 0;
(e) $\dfrac{x - 1}{2x - 1}$; **(f)** $\dfrac{a + h - 3}{2a + 2h - 5}$ **47.** $4\sqrt{3}\text{ cm}^2 \approx 6.93\text{ cm}^2$

79.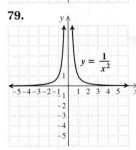

81.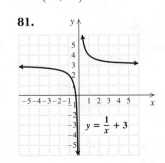

49. 36π in^2 $\approx$ 113.10 in^2 **51.** 164.98 cm
53. (a) -2; (b) $\{x \mid -2 \le x \le 5\}$; (c) 4; (d) $\{y \mid -3 \le y \le 4\}$
55. (a) -2; (b) $\{x \mid -4 \le x \le 2\}$; (c) -2;
(d) $\{y \mid -3 \le y \le 3\}$ **57.** (a) 3; (b) $\{x \mid -4 \le x \le 3\}$;
(c) -3; (d) $\{y \mid -2 \le y \le 5\}$ **59.** (a) 3;
(b) $\{-4, -3, -2, -1, 0, 1, 2\}$; (c) $-2, 0$; (d) $\{1, 2, 3, 4\}$
61. (a) 4; (b) $\{x \mid -3 \le x \le 4\}$; (c) $-1, 3$;
(d) $\{y \mid -4 \le y \le 5\}$ **63.** (a) 2; (b) $\{x \mid -4 \le x \le 4\}$;
(c) $\{x \mid 0 < x \le 2\}$; (d) $\{1, 2, 3, 4\}$ **65.** 11   **67.** 0
69. $-\frac{21}{2}$ **71.** $\frac{25}{6}$ **73.** -3 **75.** -25
77. 75 heart attacks per 10,000 men   **79.** 250 mg/dl
81. $\{x \mid x$ is a real number $and\ x \ne 3\}$ **83.** $\mathbb{R}$ **85.** $\mathbb{R}$
87. $\{x \mid x$ is a real number $and\ x \ne \frac{8}{5}\}$
89. $\{x \mid x$ is a real number $and\ x \ne -1\}$ **91.** $\mathbb{R}$
93. $\{x \mid x$ is a real number $and\ x \ne 0\}$
95. (a) -5; (b) 1; (c) 21   **97.** (a) -15; (b) 0;
(c) -6 **99.** (a) 100; (b) 100; (c) 131   **101.** 🖳
103. Let n represent the number; $n - 7$ **104.** 24
105. $t + 7$ **106.** 4.58×10^7 **107.** 🖳
109. 26; 99   **111.** Worm   **113.** Domain:
$\{x \mid x$ is a real number $and\ x \ne 5\}$; range:
$\{y \mid y$ is a real number $and\ y \ne 2\}$
115. About 2 min 50 sec   **117.** 1 every 3 min
119.

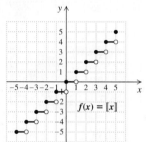

Quick Quiz: Sections 2.1–2.2, p. 94

1. I   **2.** No
3.   **4.** 0

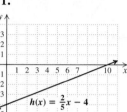

5. $\{x \mid x$ is a real number $and\ x \ne 0\}$

Prepare to Move On, p. 94

1. $-\frac{1}{3}$ **2.** 0   **3.** -1 **4.** $y = 2x - 8$
5. $y = -x + 2$ **6.** $y = \frac{5}{4}x - 2$

Exercise Set 2.3, pp. 101–106

1. (f)   **2.** (c)   **3.** (e)   **4.** (d)   **5.** (a)   **6.** (b)
7.

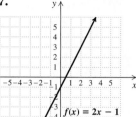

9.

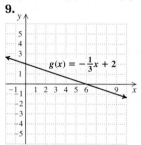

11.

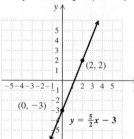

13. $(0, 3)$ **15.** $(0, -1)$
17. $(0, -4.5)$
19. $\left(0, -\frac{1}{4}\right)$ **21.** $(0, 138)$
23. 4   **25.** -2 **27.** $\frac{1}{3}$
29. $-\frac{5}{2}$ **31.** 0

33. Slope: $\frac{2}{3}$; y-intercept: $(0, 4)$ **35.** Slope: 2;
y-intercept: $(0, -3)$ **37.** Slope: 1; y-intercept: $(0, -2)$
39. Slope: $-\frac{4}{5}$; y-intercept: $\left(0, \frac{8}{5}\right)$ **41.** $f(x) = 2x + 5$
43. $f(x) = -\frac{2}{3}x - 2$ **45.** $f(x) = -7x + \frac{1}{3}$
47. Slope: $\frac{5}{2}$;   **49.** Slope: $-\frac{5}{2}$;
y-intercept: $(0, -3)$ y-intercept: $(0, 2)$

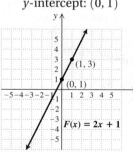

51. Slope: 2;   **53.** Slope: -4;
y-intercept: $(0, 1)$ y-intercept: $(0, 3)$

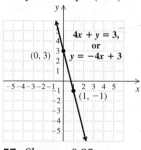

55. Slope: $-\frac{1}{6}$;   **57.** Slope: -0.25;
y-intercept: $(0, 1)$ y-intercept: $(0, 0)$

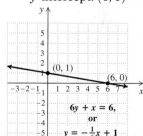

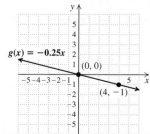

59. Slope: $\frac{4}{5}$;
y-intercept: $(0, -2)$

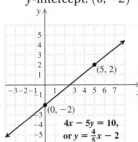

61. Slope: $-\frac{2}{3}$;
y-intercept: $(0, 2)$

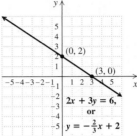

63. Slope: -3;
y-intercept: $(0, 5)$

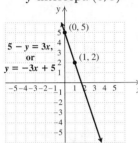

65. Slope: 0;
y-intercept: $(0, 4.5)$

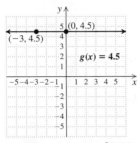

67. The distance from home is increasing at a rate of 0.25 km per minute.   **69.** The distance from the finish line is decreasing at a rate of $6\frac{2}{3}$ m per second.   **71.** The number of bookcases stained is increasing at a rate of $\frac{2}{3}$ bookcase per quart of stain used.   **73.** The average SAT math score is increasing at a rate of 1 point per thousand dollars of family income.   **75.** (a) II; (b) IV; (c) I; (d) III
77. 2.5%/year   **79.** 300 ft/min   **81.** 175,000 hits/year
83. 0.75 signifies that the cost per mile of renting the truck is $0.75; 30 signifies that the minimum cost is $30.
85. $\frac{1}{2}$ signifies that Lauren's hair grows $\frac{1}{2}$ in. per month; 5 signifies that her hair was 5 in. long when cut.
87. $\frac{1}{7}$ signifies that the life expectancy of American women increases $\frac{1}{7}$ year per year, for years after 1970; 75.5 signifies that the life expectancy in 1970 was 75.5 years.
89. 1.02 signifies that the average price of a ticket increases $1.02 per year, for years after 2000; 16.39 signifies that the average price of a ticket was $16.39 in 2000.
91. 2.2 signifies that the sales of organic food increase $2.2 billion per year, for years after 2000; 4.5 signifies that sales of organic food were $4.5 billion in 2000.
93. (a) -5000 signifies that the depreciation is $5000 per year; 90,000 signifies that the original value of the truck was $90,000; (b) 18 years; (c) $\{t \mid 0 \le t \le 18\}$ **95.** (a) -200 signifies that the depreciation is $200 per year; 1800 signifies that the original value of the bike was $1800; (b) after 6 years of use; (c) $\{n \mid 0 \le n \le 9\}$ **97.** 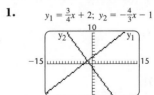   **99.** $-\frac{7}{6}$
100. $-\frac{8}{9}$ **101.** 4.3   **102.** 610.3   **103.**
105. (a) III; (b) IV; (c) I; (d) II   **107.** Sienna to Castellina in Chianti   **109.** Castellina in Chianti

111. Slope: $-\dfrac{r}{r+p}$; y-intercept: $\left(0, \dfrac{s}{r+p}\right)$

113. Since (x_1, y_1) and (x_2, y_2) are two points on the graph of $y = mx + b$, then $y_1 = mx_1 + b$ and $y_2 = mx_2 + b$. Using the definition of slope, we have

$$\begin{aligned}
\text{Slope} &= \frac{y_2 - y_1}{x_2 - x_1} \\
&= \frac{(mx_2 + b) - (mx_1 + b)}{x_2 - x_1} \\
&= \frac{m(x_2 - x_1)}{x_2 - x_1} \\
&= m.
\end{aligned}$$

115. False   **117.** False   **119.** (a) $-\dfrac{5c}{4b}$; (b) undefined;
(c) $\dfrac{a + d}{f}$ **121.**

$y_1 = 1.4x + 2,\ y_2 = 0.6x + 2,$
$y_3 = 1.4x + 5,\ y_4 = 0.6x + 5$

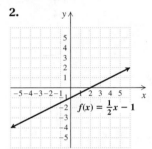

123.

Quick Quiz: Sections 2.1–2.3, p. 106

1.

2.

3. 1   **4.** Domain: $\{x \mid -4 \le x \le 3\}$; range: $\{y \mid -3 \le y \le 1\}$ **5.** Domain: $\{-2, -1, 0, 1, 2\}$; range: $\{-1, 0, 1, 2, 3\}$

Prepare to Move On, p. 106

1. 0   **2.** Undefined   **3.** $-\frac{9}{2}$ **4.** $\frac{3}{4}$ **5.** -7
6. $\frac{7}{2}$

Technology Connection, p. 110

1. $y_1 = \frac{3}{4}x + 2;\ y_2 = -\frac{4}{3}x - 1$

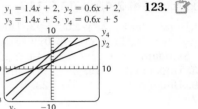

2. $y_1 = -\frac{2}{5}x - 4;\ y_2 = \frac{5}{2}x + 3$

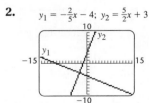

3. $y_1 = \frac{31}{40}x + 2$; $y_2 = -\frac{40}{30}x - 1$ No: $-\frac{40}{30} \neq -\frac{1}{\frac{31}{40}}$

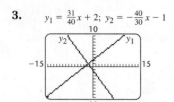

Although the lines appear to be perpendicular, they are not, because the product of their slopes is not -1:

$$\frac{31}{40}\left(-\frac{40}{30}\right) = -\frac{1240}{1200} \neq -1.$$

Exercise Set 2.4, pp. 114–117

1. Horizontal   **2.** y-axis   **3.** Undefined
4. Vertical   **5.** 0; x **6.** 0; y **7.** Intersection
8. Standard   **9.** Linear   **10.** Slope   **11.** 0
13. Undefined   **15.** 0   **17.** Undefined
19. Undefined   **21.** 0   **23.** 0   **25.** Undefined
27. $-\frac{2}{3}$

29.
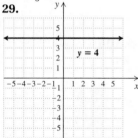

31.
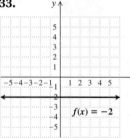

33.
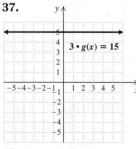

35.
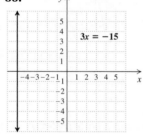

37.
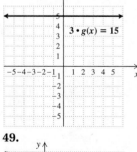

39. Yes   **41.** Yes
43. No   **45.** Yes
47. No

49.
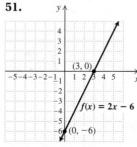

51.

53.
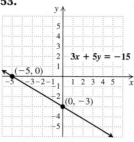

55.
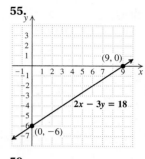

57.
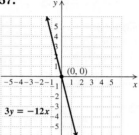

59.
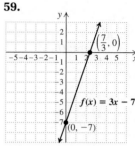

61.
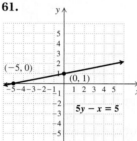

63.
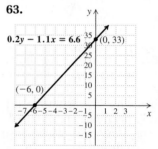

65. 1   **67.** -2 **69.** 4   **71.** $-\frac{3}{2}$ **73.** 2
75. 4 months   **77.** 800 ft^2 **79.** \$16,500 over \$3500
81. 2 hr 15 min   **83.** Linear; $\frac{5}{3}$ **85.** Linear; line is
vertical   **87.** Not linear   **89.** Linear; $\frac{14}{3}$
91. Not linear   **93.** Linear; 3   **95.** Not linear

97. ⬚   **99.** $\frac{17}{3}$ **100.** $-\frac{3}{2}$ **101.** $\mathbb{R}$; identity

102. $\frac{13}{2}$ **103.** $\varnothing$; contradiction   **104.** 0   **105.** ⬚
107. $4x - 5y = 20$ **109.** Linear   **111.** Linear
113. The slope of equation B is $\frac{1}{2}$ the slope of equation A.
115. $a = 7$, $b = -3$
117.
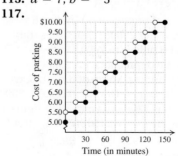

119. $0.\overline{6}$, or $\frac{2}{3}$
121. 2.6, or $\frac{13}{5}$
123. 149 shirts

Quick Quiz: Sections 2.1–2.4, p. 117

1. $\frac{1}{2}$ **2.** Undefined
3.

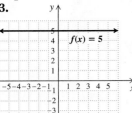

4.

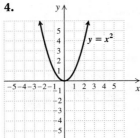

5.

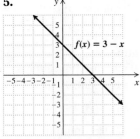

Prepare to Move On, p. 117

1. -1 **2.** -1 **3.** $-10x - 70$ **4.** $\frac{2}{3}x - \frac{2}{3}$
5. $-\frac{3}{2}x - \frac{12}{5}$

Mid-Chapter Review: Chapter 2, p. 118

1. *y-intercept:* $y - 3 \cdot 0 = 6$
　　　　　　　　$y = 6$

　The *y*-intercept is $(0, 6)$.

　x-intercept: $0 - 3x = 6$
　　　　　　　$-3x = 6$
　　　　　　　　　$x = -2$

　The *x*-intercept is $(-2, 0)$.

2. $m = \dfrac{y_2 - y_1}{x_2 - x_1}$

　　$= \dfrac{-1 - 5}{3 - 1}$

　　$= \dfrac{-6}{2}$

　　$= -3$

3. II   **4.** -30 **5.** $\{x \,|\, x \text{ is a real number } and\, x \neq 7\}$
6. $\frac{5}{3}$ **7.** Undefined   **8.** 0   **9.** Undefined
10. Slope: $\frac{1}{3}$; *y*-intercept: $\left(0, -\frac{1}{3}\right)$
11. $f(x) = -3x + 7$ **12.** Perpendicular   **13.** -3
14. Yes
15.

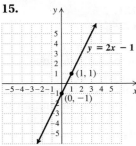

16.

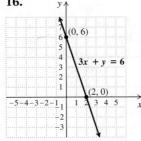

17.

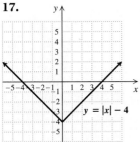

18.

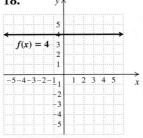

19.

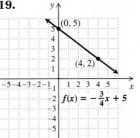

20.

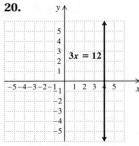

Connecting the Concepts, pp. 124–125

1. Standard form   **2.** Slope–intercept form
3. None of these   **4.** Point–slope form
5. Standard form   **6.** Slope–intercept form
7. $2x - 5y = -5$ **8.** $2x + y = 13$ **9.** $y = \frac{3}{5}x - 2$
10. $y = \frac{1}{2}x - \frac{7}{2}$

Exercise Set 2.5, pp. 125–129

1. False   **2.** True   **3.** False   **4.** True
5. True   **6.** True   **7.** $\frac{1}{4}$; $(5, 3)$ **9.** -7; $(2, -1)$
11. $-\frac{10}{3}$; $(-4, 6)$ **13.** 5; $(0, 0)$
15.

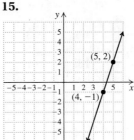

17.

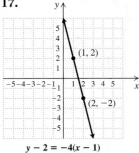

19.

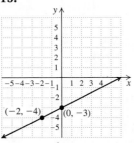

$y - (-4) = \frac{1}{2}(x - (-2))$, or
$y + 4 = \frac{1}{2}(x + 2)$

21.

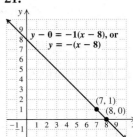

23. $y = 3x + 4$ **25.** $y = \frac{4}{3}x - 12$
27. $y = \frac{2}{3}x + \frac{1}{2}$ **29.** $y = x - 32$
31. $y - 1 = 6(x - 7)$ **33.** $y - 4 = -5(x - 3)$
35. $y - (-5) = \frac{1}{2}(x - (-2))$ **37.** $y - 0 = -1(x - 9)$
39. $f(x) = 2x - 6$ **41.** $f(x) = -\frac{3}{5}x + \frac{28}{5}$

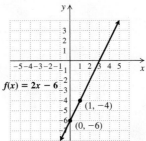

43. $f(x) = -0.6x - 5.8$ **45.** $f(x) = \frac{2}{7}x - 6$

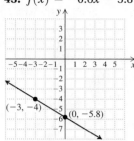

$f(x) = -0.6x - 5.8$

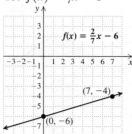

47. $f(x) = \frac{3}{5}x + \frac{42}{5}$

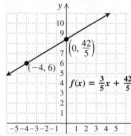

49. $y = \frac{1}{2}x + 4$ **51.** $y = -x - 1$ **53.** $y = -\frac{2}{3}x - \frac{13}{3}$
55. $x = 5$ **57.** $y = -\frac{3}{2}x + \frac{11}{2}$ **59.** $y = x + 6$
61. $y = -\frac{1}{3}x - \frac{8}{3}$ **63.** $y = -\frac{5}{3}x - \frac{41}{3}$ **65.** $x = -3$
67. $f(x) = 4x - 5$ **69.** $f(x) = 4.5x - 9.4$
71. $f(x) = -2x - 1$ **73.** $f(x) = \frac{5}{3}x$ **75.** $y = -6$
77. $x = -10$ **79.** 19 watts; 30 watts

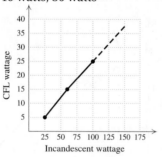

81. 3.5 drinks; 6 drinks

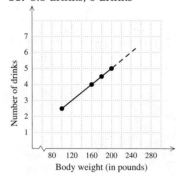

83. \$257,000; \$306,000

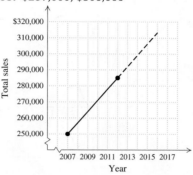

85. **(a)** $N(t) = 1.02t + 74.8$; **(b)** 91.12 million tons
87. **(a)** $E(t) = 0.14t + 79.7$; **(b)** 82.5 years
89. **(a)** $S(t) = \frac{466}{25}t + \frac{2423}{5}$; **(b)** 857.4 tons
91. **(a)** $P(d) = 0.03d + 1$; **(b)** 21.7 atm
93. **(a)** $N(t) = 3.31t + 39.4$; **(b)** 89.05 million households;
(c) about 2036   **95.** 📋   **97.** $-9x$ **98.** $2a^2b$
99. $m = \dfrac{xc}{p}$ **100.** $y = \dfrac{ax}{d - 1}$ **101.** 📋
103. 21.1°C   **105.** \$60   **107.** \$13.80 per pound
109. $\{p \mid p > 4.5\}$ **111.** $-\frac{40}{9}$
113. **(a)** $f(x) = 0.235x - 389.973$; 84.727 years
115. 📋   **117.** Bicycling 14 mph for 1 hr

Quick Quiz: Sections 2.1–2.5, p. 129

1. $f(x) = -5x + 25$ **2.** $y = 3x + 9$ **3.** Slope: 3;
y-intercept: $(0, -6)$ **4.** Yes   **5.** Yes

Prepare to Move On, p. 129

1. $2x^2 + 2x - 5$ **2.** $t + 2$
3. $\{x \mid x \text{ is a real number } and\ x \neq 3\}$ **4.** $\mathbb{R}$

Exercise Set 2.6, pp. 135–138

1. Sum   **2.** Subtract   **3.** Evaluate   **4.** Domains
5. Excluding   **6.** Sum   **7.** 1   **9.** 5   **11.** -7
13. 1   **15.** -5 **17.** $x^2 - 2x - 2$ **19.** $x^2 + 2x - 8$
21. $x^2 - x + 3$ **23.** 5   **25.** 56
27. $\dfrac{x^2 - 2}{5 - x}, x \neq 5$ **29.** $\frac{7}{2}$ **31.** -2
33. $1.2 + 2.9 = 4.1$ million   **35.** $45 - 28 = 17$
37. About 85 million; the number of tons of municipal
solid waste that was composted or recycled in 2009
39. About 240 million; the number of tons of municipal
solid waste in 2000   **41.** About 220 million; the number
of tons of municipal solid waste that was not composted in
2008   **43.** $\mathbb{R}$ **45.** $\{x \mid x \text{ is a real number } and\ x \neq -5\}$

47. $\{x \mid x \text{ is a real number } and \ x \neq 0\}$
49. $\{x \mid x \text{ is a real number } and \ x \neq 1\}$
51. $\{x \mid x \text{ is a real number } and \ x \neq -\frac{9}{2} and \ x \neq 1\}$
53. $\{x \mid x \text{ is a real number } and \ x \neq 3\}$
55. $\{x \mid x \text{ is a real number } and \ x \neq -4\}$
57. $\{x \mid x \text{ is a real number } and \ x \neq 4 \ and \ x \neq 5\}$
59. $\{x \mid x \text{ is a real number } and \ x \neq -1 \ and \ x \neq -\frac{5}{2}\}$
61. $4; 3$ **63.** $5; -1$ **65.** $\{x \mid 0 \leq x \leq 9\};$
$\{x \mid 3 \leq x \leq 10\}; \{x \mid 3 \leq x \leq 9\}; \{x \mid 3 \leq x \leq 9\}$
67.

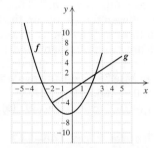

69.   **71.** 82
72. $60°, 30°, 90°$
73. 11 points

74. 2.992×10^{-23} g   **75.**
77. $\{x \mid x \text{ is a real number } and \ x \neq 4 \ and \ x \neq 3 \ and \ x \neq 2 \ and \ x \neq -2\}$
79. Answers may vary.

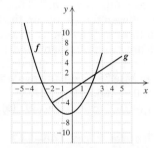

81. $\{x \mid x \text{ is a real number } and -1 < x < 5 \ and \ x \neq \frac{3}{2}\}$
83. Answers may vary. $f(x) = \dfrac{1}{x + 2}, g(x) = \dfrac{1}{x - 5}$
85.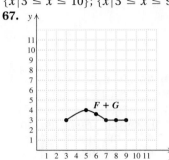

Quick Quiz: Sections 2.1–2.6, p. 138

1. $\frac{3}{16}$% per year   **2.** 1.2 signifies that the number of Americans ages 65 and older increases 1.2 million per year; 40 signifies that there were 40 million Americans ages 65 and older in 2010.   **3.** x-intercept: $(10, 0)$; y-intercept: $(0, -4)$ **4.** Parallel   **5.** $(g - h)(x) = 9x - 13$

Prepare to Move On, p. 138

1. $y = \frac{1}{6}x - \frac{1}{2}$ **2.** $y = -\frac{3}{8}x + \frac{5}{8}$ **3.** Let n represent the number; $2n + 5 = 49$ **4.** Let x represent the number; $\frac{1}{2}x - 3 = 57$ **5.** Let x represent the number; $x + (x + 1) = 145$

Visualizing for Success, p. 139

1. C   **2.** G   **3.** F   **4.** B   **5.** D   **6.** A
7. I   **8.** H   **9.** J   **10.** E

Decision Making: Connection, p. 140

1. $r(t) = 2002.5t + 59{,}710$ **2.** $p(t) = 3550t + 57{,}700$
3. **(a)** Registered nurses; **(b)** speech language pathologists
4. $(p - r)(t) = 1547.5t - 2010$; how much more a speech pathologist makes annually than a registered nurse t years after 2006   **5.** $\$10{,}370$

Study Summary: Chapter 2, pp. 141–143

1.

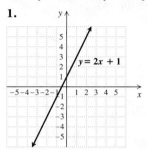

2. 5   **3.** Yes   **4.** Domain: $\mathbb{R}$; range: $\{y \mid y \geq -2\}$;
5. $\mathbb{R}$ **6.** $\frac{1}{10}$ **7.** Slope: -4; y-intercept: $\left(0, \frac{2}{5}\right)$
8.

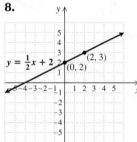

9.

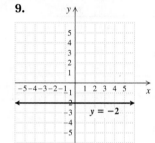

10.

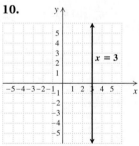

11. x-intercept: $(1, 0)$; y-intercept: $(0, -10)$ **12.** No
13. Yes   **14.** $y - 6 = \frac{1}{4}(x - (-1))$ **15.** $2x - 9$
16. 5   **17.** -6 **18.** $\dfrac{x - 2}{x - 7}, x \neq 7$

Review Exercises: Chapter 2, pp. 144–146

1. True   **2.** False   **3.** True   **4.** False
5. False   **6.** True   **7.** True   **8.** True
9. False   **10.** True   **11.** No   **12.** Yes
13. II   **14.**

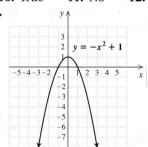

15. The value of the apartment is increasing at a rate of $7500 per year.   **16.** 27,600 homes per month
17. $\frac{4}{7}$ **18.** Undefined   **19.** $-\frac{1}{4}$ **20.** 0
21. Slope: -5; y-intercept: $(0, -11)$ **22.** Slope: $\frac{5}{6}$;
y-intercept: $\left(0, -\frac{5}{3}\right)$ **23.** -2.5 signifies that the percentage of Americans who say they have a great deal of stress decreases by 2.5% per year, for years after 2007; 32 signifies that 32% of Americans said they had a great deal of stress in 2007.   **24.** 0   **25.** Undefined
26. x-intercept: $\left(\frac{8}{3}, 0\right)$; y-intercept: $(0, -4)$

27.

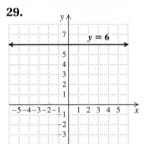

28.

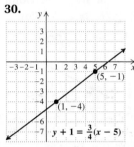

29.

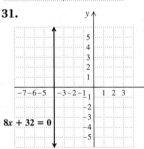

30.

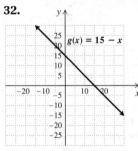

31.

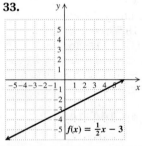

32.

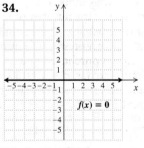

33.

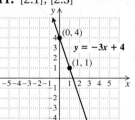

34.

35. -1 **36.** 10 tee shirts   **37.** Perpendicular
38. Parallel   **39.** $f(x) = \frac{2}{9}x - 4$
40. $y - 10 = -5(x - 1)$ **41.** $f(x) = -\frac{1}{4}x + \frac{11}{2}$
42. $y = \frac{3}{5}x - \frac{31}{5}$ **43.** $y = -\frac{5}{3}x - \frac{5}{3}$

44. About 10,000 acres;

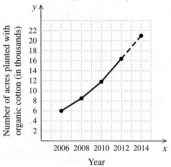

45. About 21,000 acres
46. **(a)** $R(t) = -0.02t + 19.81$; **(b)** about 19.11 sec; about 19.01 sec   **47.** Yes   **48.** Yes   **49.** No   **50.** No
51. **(a)** 3; **(b)** $\{x \mid -2 \le x \le 4\}$; **(c)** -1; **(d)** $\{y \mid 1 \le y \le 5\}$
52. Domain: $\mathbb{R}$; range: $\{y \mid y \ge 0\}$ **53.** Yes   **54.** No
55. -6 **56.** 26   **57.** $3a + 9$ **58.** 102
59. $-\frac{9}{2}$ **60.** $x^2 + 3x - 5$ **61.** $\mathbb{R}$ **62.** $\mathbb{R}$
63. $\{x \mid x$ is a real number $and\ x \ne 2\}$
64. 📋 To find $f(a) + h$, we find the output that corresponds to a and add h. To find $f(a + h)$, we find the output that corresponds to $a + h$.   **65.** 📋 The slope of a line is the rise between two points on the line divided by the run between those points. For a vertical line, there is no run between any two points, and division by 0 is undefined; therefore, the slope is undefined. For a horizontal line, there is no rise between any two points, so the slope is 0/run, or 0   **66.** -9 **67.** $-\frac{9}{2}$
68. $f(x) = 10.94x + 20$ **69.** **(a)** III; **(b)** IV; **(c)** I; **(d)** II

Test: Chapter 2, pp. 146–147

1. [2.1] No   **2.** [2.1], [2.2]

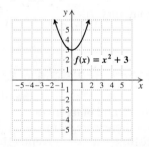

3. [2.3] The number of people on the National Do Not Call Registry is increasing at a rate of 25 million people/year.
4. [2.3] $\frac{5}{8}$ **5.** [2.3] 0   **6.** [2.3] Slope: $-\frac{3}{5}$; y-intercept: $(0, 12)$ **7.** [2.3] Slope: $-\frac{2}{5}$; y-intercept: $\left(0, -\frac{7}{5}\right)$ **8.** [2.4] 0   **9.** [2.4] Undefined
10. [2.4] x-intercept: $(3, 0)$; y-intercept: $(0, -15)$
11. [2.1], [2.3]

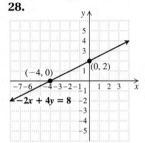

12. [2.5]

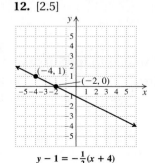

13. [2.4]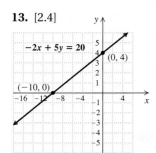

14. [2.4]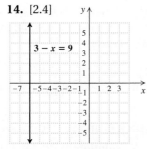

27.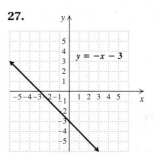

28.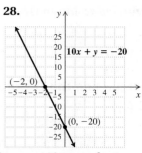

15. [2.4] 3   **16.** [2.4] About 58 million international visitors   **17.** [2.4] (a), (c)   **18.** [2.5] Parallel
19. [2.5] Perpendicular   **20.** [2.3] $f(x) = -5x - 1$
21. [2.5] $y - (-4) = 4(x - (-2))$, or $y + 4 = 4(x + 2)$
22. [2.5] $f(x) = -x + 2$ **23.** [2.5] $y = \frac{2}{5}x + \frac{16}{5}$
24. [2.5] $y = -\frac{5}{2}x - \frac{11}{2}$ **25.** [2.5] (a) $C(m) = 0.3m + 25$;
(b) \$175   **26.** [2.2] (a) 1; (b) $\{x\,|\,{-3} \le x \le 4\}$; (c) 3;
(d) $\{y\,|\,{-1} \le y \le 2\}$ **27.** [2.2] -9
28. [2.6] $\dfrac{1}{x} + 2x + 1$ **29.** [2.2] $\{x\,|\,x$ is a real number
and $x \ne 0\}$ **30.** [2.6] $\{x\,|\,x$ is a real number $and\,x \ne 0\}$
31. [2.6] $\{x\,|\,x$ is a real number $and\,x \ne 0\,and\,x \ne -\frac{1}{2}\}$
32. [2.2], [2.3] (a) 30 mi; (b) 15 mph
33. [2.5] $s = -\dfrac{3}{2}r + \dfrac{27}{2}$, or $\dfrac{27 - 3r}{2}$

34. [2.6] $h(x) = 7x - 2$

Cumulative Review: Chapters 1–2, p. 148

1. 7   **2.** -3 **3.** 9   **4.** $\frac{1}{2}$ **5.** $-2x + 37$
6. $\varnothing$; contradiction   **7.** $\frac{13}{8}$ **8.** $y = \frac{8}{3}x - 4$
9. $\dfrac{y}{3x^2}$ **10.** $-\frac{1}{8}$ **11.** 1   **12.** $\dfrac{x^6}{25y^2}$
13. -5 **14.** 8   **15.** 0   **16.** $f(x) = -x + \frac{1}{5}$
17. $f(x) = 4x + 7$ **18.** $y = -x + 3$ **19.** 99
20. 0   **21.** $\dfrac{x^2 - 1}{x + 5}$, $x \ne -5$
22. $\{x\,|\,x$ is a real number $and\,x \ne -6\}$

23.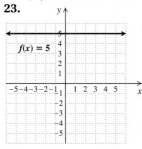

24.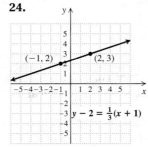

25.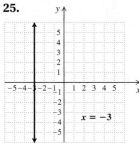

26.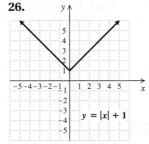

29. Full-time students: 5.6 million; part-time students:
2 million   **30.** (a) $c(t) = 514.75t + 2067$; (b) \$9788.25
31. (a) \$2465.6 million; (b) 245.8 signifies that consumer sales of organic nonfood goods increase \$245.8 million per year, for years after 2005; 745 signifies that consumer sales of organic nonfood goods were \$745 million in 2005.
32. Let x and y represent the numbers; $x^2 - y^2$
33. Let x and y represent the numbers; $(x + y)(x - y)$
34. $f(x) = \frac{1}{2}x + 3$

CHAPTER 3

Technology Connection, p. 153

1. $(1.53, 2.58)$ **2.** $(0.87, -0.32)$

Exercise Set 3.1, pp. 155–158

1. False   **2.** True   **3.** True   **4.** True   **5.** True
6. False   **7.** False   **8.** True   **9.** Yes   **11.** No
13. Yes   **15.** Yes   **17.** $(3, 2)$ **19.** $(2, -1)$
21. $(1, 4)$ **23.** $(-3, -2)$ **25.** $(3, -1)$ **27.** $(3, -7)$
29. $(7, 2)$ **31.** $(4, 0)$ **33.** No solution
35. $\{(x, y)\,|\,y = 3 - x\}$ **37.** All except Exercise 33
39. Exercise 35   **41.** Let x represent the first number and y the second number; $x + y = 10$, $x = \frac{2}{3}y$ **43.** Let s represent the number of e-mails sent and r the number of e-mails received; $s + r = 115, r = 2s + 1$ **45.** Let x and y represent the angles; $x + y = 180, x = 2y - 3$ **47.** Let x represent the number of two-point shots and y the number of foul shots; $x + y = 64, 2x + y = 100$ **49.** Let x represent the number of hats sold and y the number of tee shirts sold; $x + y = 45, 14.50x + 19.50y = 697.50$
51. Let c represent the number of Classic Heart Rate watches purchased and d the number of Drive Plus Heart Rate watches; $c + d = 35, 39.99c + 59.99d = 1639.65$
53. Let l represent the length, in yards, and w the width, in yards; $2l + 2w = 340; l = w + 50$ **55.** ☑   **57.** $-\frac{4}{5}$
58. -0.06 **59.** -29 **60.** $\dfrac{6y^8}{x^5}$ **61.** $-\frac{1}{100}$ **62.** 1
63. ☑   **65.** Answers may vary. (a) $x + y = 6, x - y = 4$;
(b) $x + y = 1, 2x + 2y = 3$; (c) $x + y = 1, 2x + 2y = 2$
67. $A = -\frac{17}{4}, B = -\frac{12}{5}$ **69.** Let x and y represent the number of years that Dell and Juanita have taught at the university, respectively; $x + y = 46, x - 2 = 2.5(y - 2)$
71. Let s and v represent the number of ounces of baking soda and vinegar needed, respectively; $s = 4v, s + v = 16$
73. Mineral oil: 12 oz; vinegar: 4 oz   **75.** $(0, 0), (1, 1)$
77. $(0.07, -7.95)$ **79.** $(0.00, 1.25)$ **81.** 🖥

Prepare to Move On, p. 158

1. $\frac{8}{13}$ **2.** -1 **3.** 11   **4.** $y = 3x - 4$
5. $x = \frac{5}{2}y - \frac{7}{2}$

Connecting the Concepts, pp. 163–164

1. $(1, 1)$ **2.** $(9, 1)$ **3.** $(4, 3)$ **4.** $(5, 7)$
5. $\left(1, -\frac{1}{19}\right)$ **6.** $(3, 1)$ **7.** No solution
8. $\{(x, y) \mid x = 2 - y\}$

Exercise Set 3.2, pp. 164–166

1. Substitution   **2.** Elimination   **3.** Opposites
4. Inconsistent   **5.** (d)   **6.** (e)   **7.** (a)   **8.** (f)
9. (c)   **10.** (b)   **11.** $(2, -1)$ **13.** $(-4, 3)$
15. $(2, -2)$ **17.** $\{(x, y) \mid 2x - 3 = y\}$ **19.** $(-2, 1)$
21. $\left(\frac{1}{2}, \frac{1}{2}\right)$ **23.** $(2, 0)$ **25.** No solution   **27.** $(1, 2)$
29. $(7, -2)$ **31.** $(-1, 2)$ **33.** $\left(\frac{49}{11}, -\frac{12}{11}\right)$ **35.** $(6, 2)$
37. No solution   **39.** $(20, 0)$ **41.** $(3, -1)$
43. $\{(x, y) \mid -4x + 2y = 5\}$ **45.** $\left(2, -\frac{3}{2}\right)$ **47.** $(-2, -9)$
49. $(30, 6)$ **51.** $\{(x, y) \mid 4x - 2y = 2\}$ **53.** No
solution   **55.** $(140, 60)$ **57.** $\left(\frac{1}{3}, -\frac{2}{3}\right)$ **59.** 📋
61. $4 + (m + n)$ **62.** $-2a^2 + 7$ **63.** $47x - 36$
64. -1 **65.** 3.005×10^7 **66.** 0.00061 **67.** 📋
69. $m = -\frac{1}{2}, b = \frac{5}{2}$ **71.** $a = 5, b = 2$ **73.** $\left(-\frac{32}{17}, \frac{38}{17}\right)$
75. $\left(-\frac{1}{5}, \frac{1}{10}\right)$ **77.** Toaster oven: 3 kWh; convection
oven: 12 kWh   **79.** 📋

Quick Quiz: Sections 3.1–3.2, p. 166

1. Yes   **2.** $(3, 1)$ **3.** $(-3, -10)$ **4.** $\left(\frac{7}{5}, -\frac{3}{5}\right)$
5. $\left(\frac{1}{11}, \frac{3}{11}\right)$

Prepare to Move On, p. 166

1. \$105,000   **2.** 90   **3.** 290 mi

Exercise Set 3.3, pp. 175–179

1. Total value   **2.** Principal   **3.** Distance   **4.** Sum
5. 4, 6   **7.** e-mails sent: 38; e-mails received: 77
9. $119°, 61°$ **11.** Two-point shots: 36; foul shots: 28
13. Hats: 36; tee shirts: 9   **15.** Classic Heart Rate
watches: 23; Drive Plus Heart Rate watches: 12
17. Length: 110 yd; width: 60 yd   **19.** Wind: 924 trillion Btu's; solar: 109 trillion Btu's   **21.** 3-credit courses:
37; 4-credit courses: 11   **23.** Regular paper: 16 cases;
recycled paper: 11 cases   **25.** 23-watt bulbs: 60; 42-watt
bulbs: 140   **27.** Youths: 78; adults: 32   **29.** Mexican:
14 lb; Peruvian: 14 lb   **31.** Custom-printed M&Ms:
64 oz; bulk M&Ms: 256 oz   **33.** 50%-chocolate:
7.5 lb; 10%-chocolate: 12.5 lb   **35.** Deep Thought: 12 lb;
Oat Dream: 8 lb   **37.** \$7500 at 6.5%; \$4500 at 7.2%
39. Steady State: 12.5 L; Even Flow: 7.5 L   **41.** 87-octane:
2.5 gal; 95-octane: 7.5 gal   **43.** Whole milk: $169\frac{3}{13}$ lb;
cream: $30\frac{10}{13}$ lb   **45.** 375 km   **47.** 14 km/h
49. About 1489 mi   **51.** Length: 265 ft; width: 165 ft
53. Wii game machines: 92.4 million; PlayStation 3 consoles: 61.6 million   **55.** \$7.99 plans: 190; \$15.98 plans: 90
57. Quarters: 17; fifty-cent pieces: 13   **59.** 📋   **61.** -7
62. 102   **63.** 0   **64.** $x^2 + x - 5$ **65.** $\mathbb{R}$

66. $\{x \mid x$ is a real number $and\ x \neq 7\}$ **67.** 📋
69. 0%: 20 reams; 30%: 40 reams   **71.** $10\frac{2}{3}$ oz
73. 33 boxes   **75.** Brown: 0.8 gal; neutral: 0.2 gal

77. City: 261 mi; highway: 204 mi   **79.** $P(x) = \dfrac{0.1 + x}{1.5}$

(This expresses the percent as a decimal quantity.)

Quick Quiz: Sections 3.1–3.3, p. 179

1. No solution   **2.** $\left(\frac{23}{4}, \frac{25}{16}\right)$ **3.** $(4, 4)$
4. $(3, 0)$ **5.** Large trash bags: 21 rolls; small trash bags:
7 rolls

Prepare to Move On, p. 179

1. 1   **2.** $\frac{1}{2}$ **3.** 17   **4.** 7

Exercise Set 3.4, pp. 185–187

1. True   **2.** False   **3.** False   **4.** True   **5.** True
6. False   **7.** Yes   **9.** $(3, 1, 2)$ **11.** $(1, -2, 2)$
13. $(2, -5, -6)$ **15.** No solution   **17.** $(-2, 0, 5)$
19. $(21, -14, -2)$ **21.** The equations are dependent.
23. $\left(3, \frac{1}{2}, -4\right)$ **25.** $\left(\frac{1}{2}, \frac{1}{3}, \frac{1}{6}\right)$ **27.** $\left(\frac{1}{2}, \frac{2}{3}, -\frac{5}{6}\right)$
29. $(15, 33, 9)$ **31.** $(3, 4, -1)$ **33.** $(10, 23, 50)$
35. No solution   **37.** The equations are dependent.
39. 📋   **41.** Slope: $\frac{1}{3}$; y-intercept: $\left(0, -\frac{7}{3}\right)$ **42.** 0
43. x-intercept: $(10, 0)$; y-intercept: $(0, -4)$ **44.** $\frac{5}{8}$
45. Parallel   **46.** Neither   **47.** 📋   **49.** $(1, -1, 2)$
51. $(1, -2, 4, -1)$ **53.** $\left(-1, \frac{1}{5}, -\frac{1}{2}\right)$ **55.** 14
57. $z = 8 - 2x - 4y$

Quick Quiz: Sections 3.1–3.4, p. 187

1. $\left(\frac{14}{5}, \frac{3}{10}\right)$ **2.** $\{(x, y) \mid 2x - y = 4\}$ **3.** $(5, -6, 9)$
4. 20 mph   **5.** 15% pigment: $1\frac{1}{5}$ gal; 10% pigment: $1\frac{4}{5}$ gal

Prepare to Move On, p. 187

1. Let x represent the first number;
$x + (x + 1) + (x + 2) = 100$ **2.** Let x, y, and z
represent the numbers; $x + y + z = 100$ **3.** Let x, y,
and z represent the numbers; $xy = 5z$ **4.** Let x and y
represent the numbers; $xy = 2(x + y)$

Mid-Chapter Review: Chapter 3, p. 188

1.
$$2x - 3(x - 1) = 5$$
$$2x - 3x + 3 = 5$$
$$-x + 3 = 5$$
$$-x = 2$$
$$x = -2$$

$$y = x - 1$$
$$y = -2 - 1$$
$$y = -3$$

2.
$$2x - 5y = 1$$
$$\underline{x + 5y = 8}$$
$$3x = 9$$
$$x = 3$$

$$x + 5y = 8$$
$$3 + 5y = 8$$
$$5y = 5$$
$$y = 1$$

The solution is $(-2, -3)$.   The solution is $(3, 1)$.

3. $(1, 1)$ **4.** $(9, 1)$ **5.** $(4, 3)$ **6.** $(5, 7)$ **7.** $(5, 10)$
8. $\left(2, \frac{2}{5}\right)$ **9.** No solution   **10.** $\{(x, y) \mid x = 2 - y\}$
11. $(1, 1)$ **12.** $\left(\frac{40}{9}, \frac{10}{3}\right)$ **13.** $(2, -10, -3)$
14. $\left(\frac{1}{2}, -4, \frac{1}{3}\right)$ **15.** No solution   **16.** The equations
are dependent.   **17.** Microsoft: 4%; Yahoo!: 8.3%
18. 5-cent bottles or cans: 336; 10-cent bottles or cans: 94
19. Pecan Morning: 12 lb; Oat Dream: 8 lb   **20.** 18 mph

Exercise Set 3.5, pp. 192–195

1. (a)   **2.** (c)   **3.** (d)   **4.** (b)   **5.** 8, 15, 62
7. 8, 21, −3   **9.** 32°, 96°, 52°   **11.** Reading: 497; mathematics: 514; writing: 489   **13.** Bran muffin: 1.5 g;
banana: 3 g; 1 cup of Wheaties: 3 g   **15.** Base price:
$28,920; tow package: $595; sunroof: $850
17. 12-oz cups: 17; 16-oz cups: 25; 20-oz cups: 13
19. Business-equipment loan: $15,000; small-business loan:
$35,000; home-equity loan: $70,000   **21.** Gold: $55.62/g;
silver: $1.09/g; copper: $0.01/g   **23.** Roast beef:
2 servings; baked potato: 1 serving; broccoli: 2 servings
25. First mezzanine: 8 tickets; main floor: 12 tickets;
second mezzanine: 20 tickets   **27.** Asia: 5.2 billion;
Africa: 2.3 billion; rest of the world: 1.9 billion   **29.** 🖉
31. 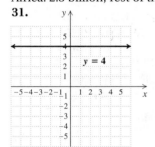   **32.**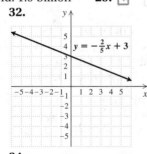

33. 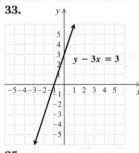   **34.**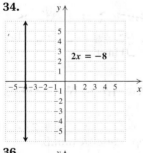

35. 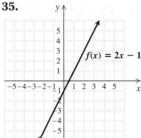   **36.**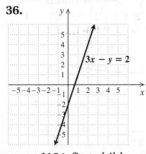

37. 🖉   **39.** Applicant: $150; spouse: $154; first child:
$116; second child: $117   **41.** 20 years   **43.** 35 tickets

Quick Quiz: Sections 3.1–3.5, p. 195

1. $(4, 3)$ **2.** $(1, 1)$ **3.** $(6, -1)$ **4.** $(0, 0)$
5. $(0, 5, -2)$

Prepare to Move On, p. 195

1. $-4x + 6y$ **2.** $7y$ **3.** $-2a + b + 6c$
4. $-12x + 5y - 8z$ **5.** $23x - 13z$

Exercise Set 3.6, pp. 199–200

1. Matrix   **2.** Rows; vertical   **3.** Entry   **4.** Matrices
5. Rows   **6.** First   **7.** $(3, 4)$ **9.** $(-2, 5)$
11. $\left(\frac{3}{2}, \frac{5}{2}\right)$ **13.** $\left(2, \frac{1}{2}, -2\right)$ **15.** $(2, -2, 1)$
17. $\left(4, \frac{1}{2}, -\frac{1}{2}\right)$ **19.** $(1, -3, -2, -1)$ **21.** Dimes: 18;
nickels: 24   **23.** Dried fruit: 9 lb; macadamia nuts: 6 lb
25. $400 at 3%; $500 at 4%; $1600 at 5%   **27.** 🖉
29. 49   **30.** −49   **31.** −49   **32.** 49   **33.** 🖉
35. 1324

Quick Quiz: Sections 3.1–3.6, p. 200

1. $(-1, 5)$ **2.** No solution   **3.** $\left(\frac{1}{2}, \frac{11}{10}, \frac{5}{4}\right)$
4. 4-marker packages: 26; 6-marker packages: 16
5. Drink Fresh: $1\frac{1}{5}$ L; Summer Light: $4\frac{4}{5}$ L

Prepare to Move On, p. 200

1. 17   **2.** −19   **3.** 37   **4.** 422

Exercise Set 3.7, pp. 205–206

1. True   **2.** True   **3.** True   **4.** False   **5.** False
6. False   **7.** 4   **9.** −50   **11.** 27   **13.** −3
15. −5   **17.** $(-3, 2)$ **19.** $\left(\frac{9}{19}, \frac{51}{38}\right)$ **21.** $\left(-1, -\frac{6}{7}, \frac{11}{7}\right)$
23. $(2, -1, 4)$ **25.** $(1, 2, 3)$ **27.** 🖉
29. $f(x) = \frac{1}{2}x - 10$ **30.** $f(x) = 3x - 9$
31. $f(x) = -\frac{8}{5}x + \frac{24}{5}$ **32.** $f(x) = -x + 4$ **33.** 🖉
35. 12   **37.** 10

Quick Quiz: Sections 3.1–3.7, p. 206

1. $(3, 5)$ **2.** $(-8, -23)$ **3.** $(-2, 3)$ **4.** $(1, 0)$
5. $\left(\frac{29}{93}, \frac{28}{93}\right)$

Prepare to Move On, p. 206

1. $70x - 2500$ **2.** 4500   **3.** $\frac{250}{7}$ **4.** $\frac{250}{7}$

Exercise Set 3.8, pp. 210–212

1. (b)   **2.** (f)   **3.** (h)   **4.** (a)   **5.** (e)   **6.** (d)
7. (c)   **8.** (g)   **9.** (a) $P(x) = 20x - 200,000$;
(b) (10,000 units, $550,000)   **11.** (a) $P(x) = 25x - 3100$;
(b) (124 units, $4960)   **13.** (a) $P(x) = 45x - 22,500$;
(b) (500 units, $42,500)   **15.** (a) $P(x) = 16x - 50,000$;
(b) (3125 units, $125,000)   **17.** (a) $P(x) = 50x - 100,000$;
(b) (2000 units, $250,000)   **19.** ($60, 1100)
21. ($22, 474)   **23.** ($50, 6250)   **25.** ($10, 1070)
27. (a) $C(x) = 45,000 + 40x$; (b) $R(x) = 130x$;
(c) $P(x) = 90x - 45,000$; (d) $225,000 profit, $9000 loss
(e) (500 phones, $65,000)   **29.** (a) $C(x) = 10,000 + 30x$;
(b) $R(x) = 80x$; (c) $P(x) = 50x - 10,000$; (d) $90,000
profit, $7500 loss; (e) (200 seats, $16,000)   **31.** 🖉

33. 1×10^{-10} **34.** 5.0×10^{19} **35.** $y = x - \frac{3}{2}C$
36. Yes   **37.** 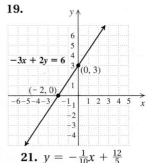   **39.** ($5, 300 yo-yo's)
41. (a) $8.74; **(b)** 24,509 units

Quick Quiz: Sections 3.1–3.8, p. 212

1. Length: 40 ft; width: 30 ft   **2.** Low-fat milk: $9\frac{3}{5}$ oz;
whole milk: $6\frac{2}{5}$ oz   **3.** 90°, 67.5°, 22.5°   **4.** 26
5. −25

Prepare to Move On, p. 212

1. 6   **2.** −2   **3.** $\frac{3}{4}$ **4.** −6   **5.** −5

Visualizing for Success, p. 213

1. C   **2.** H   **3.** J   **4.** G   **5.** D   **6.** I   **7.** A
8. F   **9.** E   **10.** B

Decision Making: Connection, p. 214

1. $28\frac{1}{8}$ years   **2.** $18,750   **3.** $26,250; about 16.4 years
4. $5750   **5.**

Study Summary: Chapter 3, pp. 215–218

1. $(2, -1)$ **2.** $\left(-\frac{1}{2}, \frac{1}{2}\right)$ **3.** $\left(\frac{16}{7}, -\frac{3}{7}\right)$ **4.** Pens:
32 boxes; pencils: 88 boxes   **5.** 40%-acid: 0.8 L;
15%-acid: 1.2 L   **6.** 8 mph   **7.** $\left(2, -\frac{1}{2}, -5\right)$
8. $(2.5, 3.5, 3)$ **9.** $(4, 1)$ **10.** 28   **11.** −25
12. $\left(\frac{11}{4}, -\frac{3}{4}\right)$ **13. (a)** $P(x) = 75x - 9000;$
(b) $(120, $10,800)$ **14.** ($9, 141)

Review Exercises: Chapter 3, pp. 218–220

1. Substitution   **2.** Elimination   **3.** Graphical
4. Dependent   **5.** Inconsistent   **6.** Contradiction
7. Parallel   **8.** Square   **9.** Determinant   **10.** Zero
11. $(4, 1)$ **12.** $(3, -2)$ **13.** $\left(\frac{8}{3}, \frac{14}{3}\right)$
14. No solution   **15.** $\left(\frac{9}{4}, \frac{7}{10}\right)$ **16.** $(-2, -3)$
17. $\left(-\frac{4}{5}, \frac{2}{5}\right)$ **18.** $\left(\frac{76}{17}, -\frac{2}{119}\right)$ **19.** $\{(x, y) \mid 3x + 4y = 6\}$
20. $(4, 3)$ **21.** Private lessons: 7 students; group lessons:
5 students   **22.** 4 hr   **23.** 8% juice: 10 L; 15% juice:
4 L   **24.** $(4, -8, 10)$ **25.** The equations are depen-
dent.   **26.** $(2, 0, 4)$ **27.** $\left(\frac{8}{9}, -\frac{2}{3}, \frac{10}{9}\right)$ **28.** No solution   **29.** A: 90°; B: 67.5°; C: 22.5°   **30.** Man: 1.4;
woman: 5.3; one-year-old child: 50   **31.** $\left(55, -\frac{89}{2}\right)$
32. $(-1, 1, 3)$ **33.** −5   **34.** 9   **35.** $(6, -2)$
36. $(-3, 0, 4)$ **37. (a)** $P(x) = 20x - 15,800;$
(b) $(790 units, $39,500)$ **38.** ($3, 81)
39. (a) $C(x) = 4.75x + 54,000;$ **(b)** $R(x) = 9.25x;$
(c) $P(x) = 4.5x - 54,000;$ **(d)** $31,500 loss, $13,500
profit; **(e)** $(12,000 pints of honey, $111,000)$ **40.** 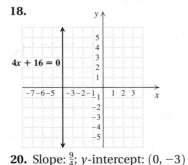 To
solve a problem involving four variables, go through the
Familiarize and *Translate* steps as usual. The resulting
system of equations can be solved using the elimination
method just as for three variables but likely with more
steps.   **41.** A system of equations can be both
dependent and inconsistent if it is equivalent to a system
with fewer equations that has no solution. An example is a
system of three equations in three unknowns in which two

of the equations represent the same plane, and the third
represents a parallel plane.   **42.** 20,000 pints
43. $(0, 2), (1, 3)$

Test: Chapter 3, pp. 220–221

1. [3.1] $(2, 4)$ **2.** [3.2] $\left(3, -\frac{11}{3}\right)$ **3.** [3.2] $(2, -1)$
4. [3.2] No solution   **5.** [3.2] $\{(x, y) \mid x = 2y - 3\}$
6. [3.2] $\left(-\frac{3}{2}, -\frac{3}{2}\right)$ **7.** [3.3] Length: 94 ft; width: 50 ft
8. [3.3] Pepperidge Farm Goldfish: 120 g; Rold Gold
Pretzels: 500 g   **9.** [3.3] 20 mph   **10.** [3.4] The equations are dependent.   **11.** [3.4] $\left(2, -\frac{1}{2}, -1\right)$
12. [3.4] No solution   **13.** [3.4] $(0, 1, 0)$
14. [3.6] $\left(\frac{22}{5}, -\frac{28}{5}\right)$ **15.** [3.6] $(3, 1, -2)$ **16.** [3.7] −14
17. [3.7] −59   **18.** [3.7] $\left(\frac{7}{13}, -\frac{17}{26}\right)$
19. [3.5] Electrician: 3.5 hr; carpenter: 8 hr; plumber: 10 hr
20. [3.8] ($3, 55)   **21.** [3.8] **(a)** $C(x) = 25x + 44,000;$
(b) $R(x) = 80x;$ **(c)** $P(x) = 55x - 44,000;$ **(d)** $27,500 loss,
$5500 profit; **(e)** $(800 hammocks, $64,000)$
22. [2.3], [3.3] $m = 7, b = 10$ **23.** [3.3] $\frac{120}{7}$ lb

Cumulative Review: Chapters 1–3, p. 222

1. x^{11} **2.** $-\dfrac{2a^{11}}{5b^{33}}$ **3.** $\dfrac{81x^{36}}{256y^8}$ **4.** 4.00×10^6

5. 1.12×10^6 **6.** $b = \dfrac{2A}{h} - t$, or $b = \dfrac{2A - ht}{h}$

7. −56   **8.** 6   **9.** −5   **10.** $\frac{10}{9}$ **11.** 1   **12.** $(1, 1)$
13. $\left(-3, \frac{2}{5}\right)$ **14.** $(-3, 2, -4)$ **15.** $(0, -1, 2)$
16. 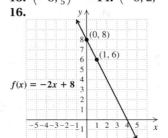   **17.**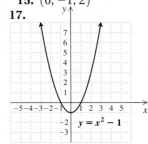

$f(x) = -2x + 8$
$y = x^2 - 1$

18.

$4x + 16 = 0$

19.

$-3x + 2y = 6$

20. Slope: $\frac{9}{4}$; y-intercept: $(0, -3)$ **21.** $y = -\frac{1}{10}x + \frac{12}{5}$
22. Parallel   **23.** $y = -2x + 5$
24. $\{x \mid x \text{ is a real number } and\ x \neq -10\}$ **25.** −31
26. 3   **27.** 7   **28.** $2a^2 + 4a - 4$ **29. (a)** $2.42;
(b) 0.15 signifies that the average fee is increasing at a rate
of $0.15 per year; 1.37 signifies that the average fee was
$1.37 in 2004.   **30. (a)** $B(t) = 3.75t + 570.5;$
(b) 630,500 bank tellers; **(c)** about 2021   **31.** 68 professionals; 82 students   **32.** Sea Spray: 90 oz; Ocean Mist:
30 oz   **33.** 86   **34.** $-12x^{2a}y^{b+y+3}$ **35.** $m = -\frac{5}{9}$,
$b = -\frac{2}{9}$

CHAPTER 4

Exercise Set 4.1, pp. 231–235

1. Solution   **2.** Closed   **3.** Half-open   **4.** Negative
5. Equivalent equations   **6.** Equivalent expressions
7. Equivalent inequalities   **8.** Not equivalent
9. Equivalent equations   **10.** Equivalent inequalities
11. (a) No; **(b)** no; **(c)** yes; **(d)** yes   **13. (a)** Yes; **(b)** no;
(c) yes; **(d)** no
15. $\{y|y < 6\}$, $(-\infty, 6)$
17. $\{x|x \geq -4\}$, $[-4, \infty)$
19. $\{t|t > -3\}$, $(-3, \infty)$
21. $\{x|x \leq -7\}$, $(-\infty, -7]$
23. $\{x|x > -1\}$, or $(-1, \infty)$
25. $\{t|t \leq 10\}$, or $(-\infty, 10]$
27. $\{x|x \geq 1\}$, or $[1, \infty)$
29. $\{t|t < -9\}$, or $(-\infty, -9)$
31. $\{x|x < 50\}$, or $(-\infty, 50)$
33. $\{x|x \leq -0.9\}$, or $(-\infty, -0.9]$
35. $\{y|y \geq -\frac{5}{6}\}$, or $[-\frac{5}{6}, \infty)$
37. $\{x|x < 2\}$, or $(-\infty, 2)$
39. $\{x|x \leq -9\}$, or $(-\infty, -9]$
41. $\{x|x < -26\}$, or $(-\infty, -26)$
43. $\{t|t \geq -\frac{13}{3}\}$, or $[-\frac{13}{3}, \infty)$
45. $\{x|x \geq -3\}$, or $[-3, \infty)$
47. $\{x|x \geq 2\}$, or $[2, \infty)$
49. $\{x|x > \frac{2}{3}\}$, or $(\frac{2}{3}, \infty)$
51. $\{x|x \geq \frac{1}{2}\}$, or $[\frac{1}{2}, \infty)$
53. $\{y|y \leq -\frac{3}{2}\}$, or $(-\infty, -\frac{3}{2}]$ **55.** $\{t|t < \frac{29}{5}\}$, or $(-\infty, \frac{29}{5})$
57. $\{m|m > \frac{7}{3}\}$, or $(\frac{7}{3}, \infty)$ **59.** $\{x|x \geq 2\}$, or $[2, \infty)$
61. $\{y|y < 5\}$, or $(-\infty, 5)$ **63.** $\{x|x \leq \frac{4}{7}\}$, or $(-\infty, \frac{4}{7}]$
65. Let n represent the number; $n < 10$
67. Let t represent the temperature, in °C; $t \leq -3$
69. Let a represent the age of the altar, in years; $a > 1200$
71. Let d represent the distance to Normandale Community
College, in miles; $d \leq 15$ **73.** Let d represent the number
of years of driving experience; $d \geq 5$ **75.** Let c represent
the cost of production, in dollars; $c \leq 12,500$ **77.** Lengths
of time less than $7\frac{1}{2}$ hr   **79.** At least 2.65   **81.** At least
56 questions correct   **83.** Depths less than 437.5 ft
85. Gross sales greater than \$7000   **87.** For more than
17 transactions   **89.** Years after 2015   **91. (a)** Body
densities less than $\frac{99}{95}$ kg/L, or about 1.04 kg/L; **(b)** body
densities less than $\frac{495}{482}$ kg/L, or about 1.03 kg/L

93. Years after 2032   **95. (a)** $\{x|x < 3913\frac{1}{23}\}$, or
$\{x|x \leq 3913\}$; **(b)** $\{x|x > 3913\frac{1}{23}\}$, or $\{x|x \geq 3914\}$
97.   **99.** $-\frac{6}{5}$ **100.** $\frac{15}{8}$ **101.** $(\frac{4}{3}, -\frac{7}{9})$
102. $(\frac{8}{5}, \frac{27}{5})$ **103.** $r = \dfrac{b}{a+c}$ **104.** $n = \dfrac{ty-a}{b}$
105.   **107.** $\{x|x \leq \dfrac{2}{a-1}\}$ **109.** $\{y|y \geq \dfrac{2a+5b}{b(a-2)}\}$
111. $\{x|x > \dfrac{4m-2c}{d-(5c+2m)}\}$ **113.** False; $2 < 3$ and
$4 < 5$, but $2 - 4 = 3 - 5$.   **115.**
117. $\mathbb{R}$
119. $\{x|x$ is a real number *and* $x \neq 0\}$
121. More than 234 mi   **123. (a)** $\{x|x < 4\}$, or $(-\infty, 4)$;
(b) $\{x|x \geq 2\}$, or $[2, \infty)$; **(c)** $\{x|x \geq 3.2\}$, or $[3.2, \infty)$

Prepare to Move On, p. 235

1. $\{x|x$ is a real number *and* $x \neq 0\}$
2. $\{x|x$ is a real number *and* $x \neq \frac{7}{5}\}$ **3.** $\mathbb{R}$
4. $\{x|x$ is a real number *and* $x \neq 0\}$

Exercise Set 4.2, pp. 241–245

1. Intersection   **2.** Union   **3.** Intersection
4. Intersection   **5.** Union   **6.** Union   **7.** (h)
8. (j)   **9.** (f)   **10.** (a)   **11.** (e)   **12.** (d)   **13.** (b)
14. (g)   **15.** (c)   **16.** (i)   **17.** $\{4, 16\}$
19. $\{0, 5, 10, 15, 20\}$ **21.** $\{b, d, f\}$ **23.** $\{u, v, x, y, z\}$
25. $\varnothing$ **27.** $\{1, 3, 5\}$
29. $(1, 3)$
31. $[-6, 0]$
33. $(-\infty, -1) \cup (4, \infty)$
35. $(-\infty, -2] \cup (1, \infty)$
37. $(-2, 4]$
39. $(-2, 4)$
41. $(-\infty, 5) \cup (7, \infty)$
43. $(-\infty, -4] \cup [5, \infty)$
45. $[-3, 7)$
47. $(-7, 0]$
49. $(-\infty, 5)$
51. $\{x|-5 \leq x < 7\}$, or $[-5, 7)$
53. $\{t|4 < t \leq 8\}$, or $(4, 8]$
55. $\{a|-2 \leq a < 2\}$, or $[-2, 2)$
57. $\mathbb{R}$, or $(-\infty, \infty)$
59. $\{x|-3 \leq x \leq 2\}$, or $[-3, 2]$
61. $\{x|7 < x < 23\}$, or $(7, 23)$

63. $\{x|-32 \le x \le 8\}$, or $[-32, 8]$

65. $\{x|1 \le x \le 3\}$, or $[1, 3]$

67. $\left\{x\middle|-\frac{7}{2} < x \le 7\right\}$, or $\left(-\frac{7}{2}, 7\right]$

69. $\{t|t < 0 \ or \ t > 1\}$, or $(-\infty, 0) \cup (1, \infty)$

71. $\left\{a\middle|a < \frac{7}{2}\right\}$, or $\left(-\infty, \frac{7}{2}\right)$

73. $\{a|a < -5\}$, or $(-\infty, -5)$
75. $\varnothing$
77. $\{t|t \le 6\}$, or $(-\infty, 6]$

79. $(-\infty, -6) \cup (-6, \infty)$ **81.** $(-\infty, 0) \cup (0, \infty)$
83. $(-\infty, 4) \cup (4, \infty)$ **85.** $\{x|x \ge 10\}$, or $[10, \infty)$
87. $\{x|x \le 3\}$, or $(-\infty, 3]$ **89.** $\left\{x\middle|x \ge -\frac{7}{2}\right\}$, or $\left[-\frac{7}{2}, \infty\right)$
91. $\{x|x \le 4\}$, or $(-\infty, 4]$ **93.** 🖊   **95.** $\frac{1}{12}$ **96.** 0.35
97. 29   **98.** $-w + 3$ **99.** 🖊
101. $(-1, 6)$ **103.** $0 \text{ ft} \le d \le 198 \text{ ft}$ **105.** Densities between 1.03 kg/L and 1.04 kg/L   **107.** More than 12 trips and fewer than 125 trips
109. $\left\{m\middle|m < \frac{6}{5}\right\}$, or $\left(-\infty, \frac{6}{5}\right)$

111. $\left\{x\middle|-\frac{1}{8} < x < \frac{1}{2}\right\}$, or $\left(-\frac{1}{8}, \frac{1}{2}\right)$

113. False   **115.** True   **117.** $(-\infty, -7) \cup \left(-7, \frac{3}{4}\right]$
119. 〰   **121.** Between 5.24 words and 15.09 words per sentence   **123.** Let w represent the number of ounces in a bottle; $15.9 \le w \le 16.1$; $[15.9, 16.1]$ **125.** 〰
127. 〰   **129.** 💻

Quick Quiz: Sections 4.1–4.2, p. 245

1. $\left\{x\middle|x > \frac{2}{7}\right\}$, or $\left(\frac{2}{7}, \infty\right)$ **2.** $\{x|x \ge 6\}$, or $[6, \infty)$
3. $\{m|m < -18\}$, or $(-\infty, -18)$
4. $\{y|y < 0 \ or \ y > 2\}$, or $(-\infty, 0) \cup (2, \infty)$
5. $\{x|3 < x < 8\}$, or $(3, 8)$

Prepare to Move On, p. 245

1. $\frac{2}{3}$ **2.** 16   **3.** 0   **4.** 7   **5.** 6

Technology Connection, p. 250

1. The x-values on the graph of $y_1 = |4x + 2|$ that are *below* the line $y = 6$ solve the inequality $|4x + 2| < 6$.
2. The x-values on the graph of $y_1 = |3x - 2|$ that are below the line $y = 4$ are in the interval $\left(-\frac{2}{3}, 2\right)$.
3. The graphs of $y_1 = |4x + 2|$ and $y_2 = -6$ do not intersect.

Exercise Set 4.3, pp. 251–253

1. True   **2.** False   **3.** True   **4.** True   **5.** True
6. True   **7.** False   **8.** False   **9.** (g)   **10.** (h)
11. (d)   **12.** (a)   **13.** (a)   **14.** (b)   **15.** $\{-10, 10\}$
17. $\varnothing$ **19.** $\{0\}$ **21.** $\left\{-\frac{1}{2}, \frac{7}{2}\right\}$ **23.** $\varnothing$ **25.** $\{-4, 8\}$
27. $\{6, 8\}$ **29.** $\{-5.5, 5.5\}$ **31.** $\{-8, 8\}$ **33.** $\{-1, 1\}$

35. $\left\{-\frac{11}{2}, \frac{13}{2}\right\}$ **37.** $\{-2, 12\}$ **39.** $\left\{-\frac{1}{3}, 3\right\}$ **41.** $\{-7, 1\}$
43. $\{-8.7, 8.7\}$ **45.** $\left\{-\frac{9}{2}, \frac{11}{2}\right\}$ **47.** $\{-8, 2\}$ **49.** $\left\{-\frac{1}{2}\right\}$
51. $\left\{-\frac{3}{5}, 5\right\}$ **53.** $\mathbb{R}$ **55.** $\left\{\frac{1}{4}\right\}$
57. $\{a|-3 \le a \le 3\}$, or $[-3, 3]$

59. $\{t|t < 0 \ or \ t > 0\}$, or $(-\infty, 0) \cup (0, \infty)$

61. $\{x|-3 < x < 5\}$, or $(-3, 5)$
63. $\{n|-8 \le n \le 4\}$, or $[-8, 4]$

65. $\{x|x < -2 \ or \ x > 8\}$, or $(-\infty, -2) \cup (8, \infty)$

67. $\mathbb{R}$, or $(-\infty, \infty)$

69. $\left\{a\middle|a \le -\frac{10}{3} \ or \ a \ge \frac{2}{3}\right\}$, or $\left(-\infty, -\frac{10}{3}\right] \cup \left[\frac{2}{3}, \infty\right)$

71. $\{y|-9 < y < 15\}$, or $(-9, 15)$

73. $\{x|x \le -8 \ or \ x \ge 0\}$, or $(-\infty, -8] \cup [0, \infty)$

75. $\left\{x\middle|x < -\frac{1}{2} \ or \ x > \frac{7}{2}\right\}$, or $\left(-\infty, -\frac{1}{2}\right) \cup \left(\frac{7}{2}, \infty\right)$

77. $\varnothing$
79. $\left\{x\middle|x < -\frac{43}{24} \ or \ x > \frac{9}{8}\right\}$, or $\left(-\infty, -\frac{43}{24}\right) \cup \left(\frac{9}{8}, \infty\right)$

81. $\{m|-9 \le m \le 3\}$, or $[-9, 3]$

83. $\{a|-6 < a < 0\}$, or $(-6, 0)$

85. $\left\{x\middle|-\frac{1}{2} \le x \le \frac{7}{2}\right\}$, or $\left[-\frac{1}{2}, \frac{7}{2}\right]$

87. $\left\{x\middle|x \le -\frac{7}{3} \ or \ x \ge 5\right\}$, or $\left(-\infty, -\frac{7}{3}\right] \cup [5, \infty)$

89. $\{x|-4 < x < 5\}$, or $(-4, 5)$

91. 🖊   **93.** $f(x) = \frac{1}{3}x - 2$ **94.** $y - 7 = -8(x - 3)$
95. $f(x) = 2x + 5$ **96.** $y = -x - 1$ **97.** 🖊
99. $\left\{\frac{5}{4}, \frac{5}{2}\right\}$ **101.** $\{x|-4 \le x \le -1 \ or \ 3 \le x \le 6\}$, or $[-4, -1] \cup [3, 6]$ **103.** $\left\{t\middle|t \le \frac{5}{2}\right\}$, or $\left(-\infty, \frac{5}{2}\right]$
105. $|x| < 3$ **107.** $|x| \ge 6$ **109.** $|x + 3| > 5$
111. $|x - 7| < 2$, or $|7 - x| < 2$ **113.** $|x - 3| \le 4$
115. $|x + 4| < 3$ **117.** Between 80 ft and 100 ft
119. 🖊, 〰

Quick Quiz: Sections 4.1–4.3, p. 253

1. $\left\{x\middle|x \ge -\frac{13}{2}\right\}$, or $\left[-\frac{13}{2}, \infty\right)$ **2.** $\{2, 3, 5, 7\}$
3. $\{1, 2, 3, 4, 6, 8\}$ **4.** $\{-3, 11\}$ **5.** $\{x|-5 < x < 5\}$, or $(-5, 5)$

Prepare to Move On, p. 253

1.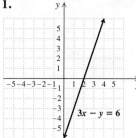

2.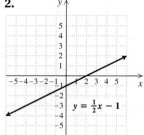

3.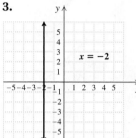

4.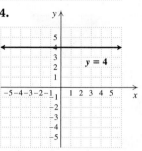

5. $\left(4, -\frac{4}{3}\right)$ **6.** $(5, 1)$

Mid-Chapter Review: Chapter 4, p. 254

1. $2 \le x \le 11$
   The solution is $[2, 11]$.
2. $x - 1 < -9$ or $9 < x - 1$
   $x < -8$ or $10 < x$
   The solution is $(-\infty, -8) \cup (10, \infty)$.
3. $\{-15, 15\}$ **4.** $\{t \mid -10 < t < 10\}$, or $(-10, 10)$
5. $\{p \mid p < -15 \, or \, p > 15\}$, or $(-\infty, -15) \cup (15, \infty)$
6. $\{-4, 3\}$ **7.** $\{x \mid 2 < x < 11\}$, or $(2, 11)$
8. $\{t \mid -4 < t < 4\}$, or $(-4, 4)$
9. $\{x \mid x < -6 \, or \, x > 13\}$, or $(-\infty, -6) \cup (13, \infty)$
10. $\{x \mid -7 \le x \le 3\}$, or $[-7, 3]$ **11.** $\left\{-\frac{8}{3}, \frac{8}{3}\right\}$
12. $\left\{x \mid x < -\frac{13}{2}\right\}$, or $\left(-\infty, -\frac{13}{2}\right)$
13. $\left\{n \mid -9 < n \le \frac{8}{3}\right\}$, or $\left(-9, \frac{8}{3}\right]$
14. $\left\{x \mid x \le -\frac{17}{2} \, or \, x \ge \frac{7}{2}\right\}$, or $\left(-\infty, -\frac{17}{2}\right] \cup \left[\frac{7}{2}, \infty\right)$
15. $\{x \mid x \ge -2\}$, or $[-2, \infty)$ **16.** $\{-42, 38\}$ **17.** $\varnothing$
18. $\{a \mid -7 < a < -5\}$, or $(-7, -5)$ **19.** $\mathbb{R}$, or $(-\infty, \infty)$
20. $\mathbb{R}$, or $(-\infty, \infty)$

Technology Connection, p. 258

1. $y > x + 3.5$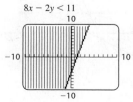

2. $7y \le 2x + 5$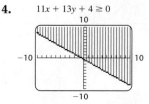

3. $8x - 2y < 11$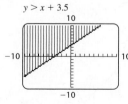

4. $11x + 13y + 4 \ge 0$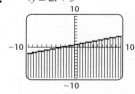

Technology Connection, p. 260

1. $y_1 \le 4 - x, \;\; y_2 > x - 4$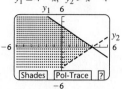

Connecting the Concepts, pp. 261–262

1.

2.

3.

4. 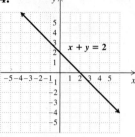 $x + y = 2$

5. 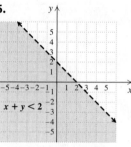 $x + y < 2$

6. 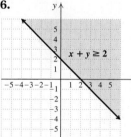 $x + y \ge 2$

7. 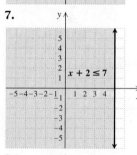 $x + 2 \le 7$

8. 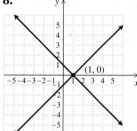 $(1, 0)$

9.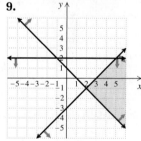

Exercise Set 4.4, pp. 262–264

1. (e)   **2.** (c)   **3.** (d)   **4.** (a)   **5.** (b)   **6.** (f)
7. No   **9.** Yes
11. 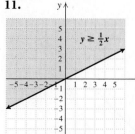 $y \ge \frac{1}{2}x$

13. 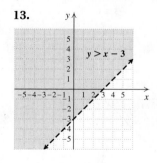 $y > x - 3$

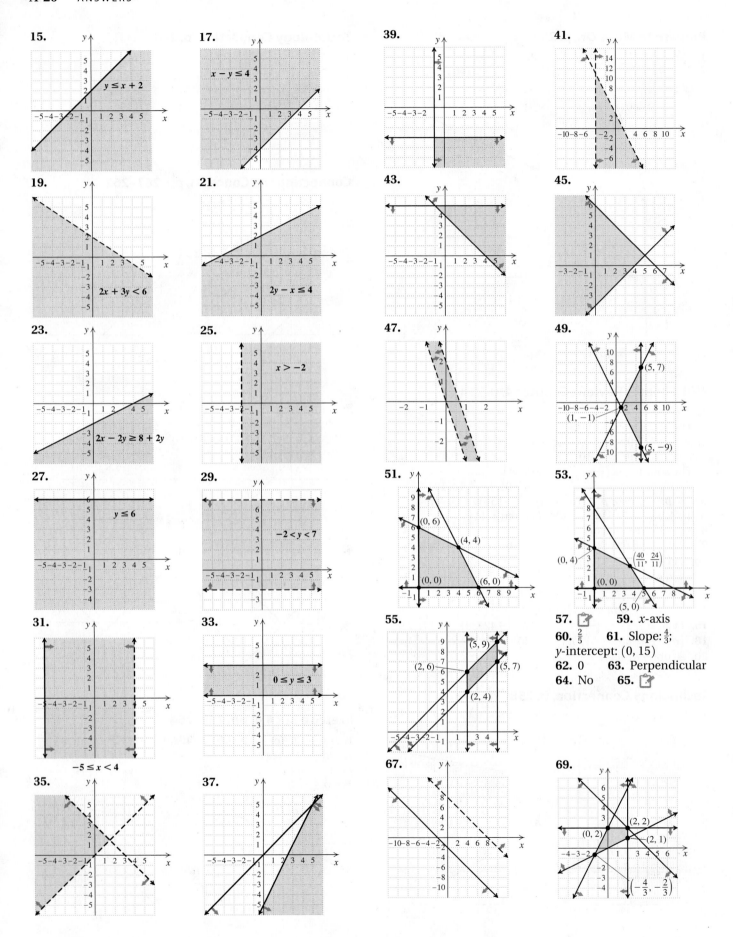

15. $y \le x + 2$

17. $x - y \le 4$

19. $2x + 3y < 6$

21. $2y - x \le 4$

23. $2x - 2y \ge 8 + 2y$

25. $x > -2$

27. $y \le 6$

29. $-2 < y < 7$

31. $-5 \le x < 4$

33. $0 \le y \le 3$

35.

37.

39.

41.

43.

45.

47.

49. (5, 7); (1, −1); (5, −9)

51. (0, 6); (4, 4); (0, 0); (6, 0)

53. (0, 4); $\left(\frac{40}{11}, \frac{24}{11}\right)$; (0, 0); (5, 0)

55. (5, 9); (2, 6); (5, 7); (2, 4)

57. ☑   **59.** x-axis

60. $\frac{2}{5}$ **61.** Slope: $\frac{4}{3}$; y-intercept: (0, 15)

62. 0   **63.** Perpendicular

64. No   **65.** ☑

67.

69. (2, 2); (0, 2); (−2, 1); $\left(-\frac{4}{3}, -\frac{2}{3}\right)$

71.
$$w > 0,$$
$$h > 0,$$
$$w + h + 30 \le 62, \text{ or}$$
$$w + h \le 32,$$
$$2w + 2h + 30 \le 130, \text{ or}$$
$$w + h \le 50$$

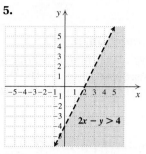

73.
$$q + v \ge 287,$$
$$v \ge 145,$$
$$q \le 170,$$
$$v \le 170$$

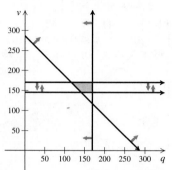

75.
$$35c + 75a > 1000,$$
$$c \ge 0,$$
$$a \ge 0$$

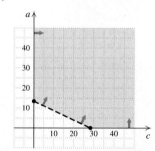

77.
$$h < 2w,$$
$$w \le 1.5h,$$
$$h \le 3200,$$
$$h \ge 0,$$
$$w \ge 0$$

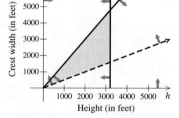

79. (a) $3x + 6y > 2$ **(b)** $x - 5y \le 10$

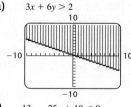
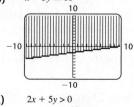

(c) $13x - 25y + 10 \le 0$ **(d)** $2x + 5y > 0$

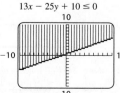
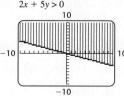

Quick Quiz: Sections 4.1–4.4, p. 264

1. $\{a \mid a \ge -3.6\}$, or $[-3.6, \infty)$

2. $\mathbb{R}$, or $(-\infty, \infty)$

3. $\left\{x \mid x < -\frac{11}{2} \ or \ x > 3\right\}$, or $\left(-\infty, -\frac{11}{2}\right) \cup (3, \infty)$

4. $\varnothing$ **5.**

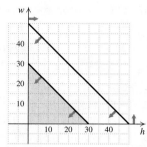

Prepare to Move On, p. 264

1. 3%: $3600; 5%: $6400   **2.** Student tickets: 62; adult tickets: 108   **3.** Corn: 240 acres; soybeans: 160 acres

Exercise Set 4.5, pp. 268–270

1. Objective   **2.** Constraints   **3.** Corner
4. Feasible   **5.** Vertices   **6.** Vertex
7. Maximum 84 when $x = 0$, $y = 6$; minimum 0 when $x = 0$, $y = 0$ **9.** Maximum 76 when $x = 7$, $y = 0$; minimum 16 when $x = 0$, $y = 4$ **11.** Maximum 5 when $x = 3$, $y = 7$; minimum -15 when $x = 3$, $y = -3$
13. Bus: 2 trips; train: 3 trips   **15.** 4-photo pages: 5; 6-photo pages: 15; 110 photos   **17.** Corporate bonds: $22,000; municipal bonds: $18,000; maximum: $1510
19. Short-answer questions: 12; essay questions: 4
21. Merlot: 80 acres; Cabernet: 160 acres   **23.** 2.5 servings of each   **25.** 📝   **27.** $\frac{1}{100}$ **28.** y^{16} **29.** $-2x^{12}$
30. $-\dfrac{8}{a^9 b^{12}}$ **31.** $\dfrac{3d^3}{2c}$ **32.** 1   **33.** 📝
35. T3's: 30; S5's:10

Quick Quiz: Sections 4.1–4.5, p. 270

1. $\left\{x \mid -\frac{4}{5} < x \le -\frac{1}{5}\right\}$, or $\left(-\frac{4}{5}, -\frac{1}{5}\right]$ **2.** $\left\{-\frac{2}{3}, \frac{10}{3}\right\}$
3. $\{x \mid -1 < x < 0\}$, or $(-1, 0)$
4.

5. Maximum 8 when $x = 2$, $y = 0$; minimum -6 when $x = -\frac{3}{2}$, $y = 0$

Prepare to Move On, p. 270

1. 7   **2.** 22   **3.** $21t - 16$ **4.** $22x + 30$ **5.** t
6. $12x - 15$

Visualizing for Success, p. 271

1. B **2.** F **3.** J **4.** A **5.** E **6.** G **7.** C
8. D **9.** I **10.** H

Decision Making: Connection, p. 272

1. Bills greater than $9500 **2.** United Health One:
$1175.76; SmartSense Plus: $2366.04 **3.** The United
Health One plan will save Elisabeth money no matter how
great her bills are.

Study Summary: Chapter 4, pp. 273–275

1. $(-\infty, 0]$ **2.** $\{x|x > 7\}$, or $(7, \infty)$ **3.** $\{x|x \ge -\frac{1}{4}\}$, or
$[-\frac{1}{4}, \infty)$ **4.** Let d represent the distance Luke runs, in
miles; $d \ge 3$ **5.** $\{x|-2 < x \le -\frac{3}{4}\}$, or $(-2, -\frac{3}{4}]$
6. $\{x|x \le 13\ or\ x > 22\}$, or $(-\infty, 13] \cup (22, \infty)$
7. $\{-1, \frac{9}{2}\}$ **8.** $\{x|11 \le x \le 13\}$, or $[11, 13]$
9. $\{x|x < -5\ or\ x > 2\}$, or $(-\infty, -5) \cup (2, \infty)$
10.

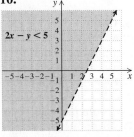

11. Maximum 8 when $x = 4$ and $y = 0$

Review Exercises: Chapter 4, pp. 275–276

1. True **2.** False **3.** True **4.** False **5.** True
6. True **7.** True **8.** True **9.** False **10.** False
11. $\{x|x \le 1\}$, or $(-\infty, 1]$;
12. $\{a|a \le 4\}$, or $(-\infty, 4]$;
13. $\{y|y > -\frac{15}{4}\}$, or $(-\frac{15}{4}, \infty)$;
14. $\{y|y > -30\}$, or $(-30, \infty)$;
15. $\{x|x > -\frac{3}{2}\}$, or $(-\frac{3}{2}, \infty)$;
16. $\{x|x < -3\}$, or $(-\infty, -3)$;
17. $\{y|y > -\frac{220}{23}\}$, or $(-\frac{220}{23}, \infty)$;
18. $\{x|x \le -\frac{5}{2}\}$, or $(-\infty, -\frac{5}{2}]$;
19. $\{x|x \le 2\}$, or $(-\infty, 2]$ **20.** More than 125 hr
21. $3000 **22.** $\{a, c\}$ **23.** $\{a, b, c, d, e, f, g\}$
24. $(-3, 2]$
25. $(-\infty, \infty)$
26. $\{x|-8 < x \le 0\}$, or $(-8, 0]$
27. $\{x|-\frac{5}{4} < x < \frac{5}{2}\}$, or $(-\frac{5}{4}, \frac{5}{2})$
28. $\{x|x < -3\ or\ x > 1\}$, or $(-\infty, -3) \cup (1, \infty)$
29. $\{x|x < -11\ or\ x \ge -6\}$, or $(-\infty, -11) \cup [-6, \infty)$

30. $\{x|x \le -6\ or\ x \ge 8\}$, or $(-\infty, -6] \cup [8, \infty)$
31. $\{x|x < -\frac{2}{5}\ or\ x > \frac{8}{5}\}$, or $(-\infty, -\frac{2}{5}) \cup (\frac{8}{5}, \infty)$
32. $(-\infty, -3) \cup (-3, \infty)$ **33.** $[2, \infty)$ **34.** $(-\infty, \frac{1}{4}]$
35. $\{-11, 11\}$ **36.** $\{t|t \le -21\ or\ t \ge 21\}$, or
$(-\infty, -21] \cup [21, \infty)$ **37.** $\{5, 11\}$
38. $\{a|-\frac{7}{2} < a < 2\}$, or $(-\frac{7}{2}, 2)$
39. $\{x|x \le -\frac{11}{3}\ or\ x \ge \frac{19}{3}\}$, or $(-\infty, -\frac{11}{3}] \cup [\frac{19}{3}, \infty)$
40. $\{-14, \frac{4}{3}\}$ **41.** $\varnothing$ **42.** $\{x|-16 \le x \le 8\}$, or
$[-16, 8]$ **43.** $\{x|x < 0\ or\ x > 10\}$, or $(-\infty, 0) \cup (10, \infty)$
44. $\{x|-6 \le x \le 4\}$, or $[-6, 4]$ **45.** $\varnothing$
46.

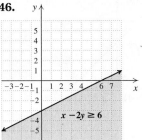

47.

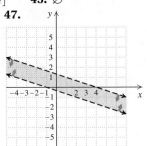

48.

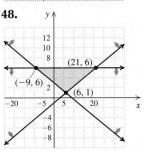

49. Maximum 40 when $x = 7$, $y = 15$; minimum 10 when
$x = 1$, $y = 3$ **50.** East coast: 40 books; West coast:
60 books **51.** ✏️ The equation $|X| = p$ has two solu-
tions when p is positive because X can be either p or $-p$.
The same equation has no solution when p is negative
because no number has a negative absolute value.
52. ✏️ The solution set of a system of inequalities is all
ordered pairs that make *all* the individual inequalities true.
This consists of ordered pairs that are common to all the
individual solution sets, or the intersection of the graphs.
53. $\{x|-\frac{8}{3} \le x \le -2\}$, or $[-\frac{8}{3}, -2]$ **54.** False: $-4 < 3$
is true, but $(-4)^2 < 9$ is false. **55.** $|d - 2.5| \le 0.003$

Test: Chapter 4, p. 277

1. [4.1] $\{x|x < 11\}$, or $(-\infty, 11)$
2. [4.1] $\{t|t > -24\}$, or $(-24, \infty)$
3. [4.1] $\{y|y \le -2\}$, or $(-\infty, -2]$
4. [4.1] $\{a|a \le \frac{11}{5}\}$, or $(-\infty, \frac{11}{5}]$
5. [4.1] $\{x|x > \frac{16}{5}\}$, or $(\frac{16}{5}, \infty)$
6. [4.1] $\{x|x \le \frac{9}{16}\}$, or $(-\infty, \frac{9}{16}]$
7. [4.1] $\{x|x > 1\}$, or $(1, \infty)$ **8.** [4.1] More than $87\frac{1}{2}$ mi

9. [4.1] Less than or equal to 2.5 hr   **10.** [4.2] $\{a, e\}$
11. [4.2] $\{a, b, c, d, e, i, o, u\}$ **12.** [4.2] $(-\infty, 2]$
13. [4.2] $(-\infty, 7) \cup (7, \infty)$
14. [4.2] $\{x | -\frac{3}{2} < x \le \frac{1}{2}\}$, or $\left(-\frac{3}{2}, \frac{1}{2}\right]$

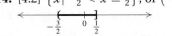

15. [4.2] $\{x | x < 3 \ or \ x > 6\}$, or $(-\infty, 3) \cup (6, \infty)$

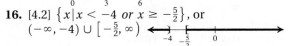

16. [4.2] $\{x | x < -4 \ or \ x \ge -\frac{5}{2}\}$, or
$(-\infty, -4) \cup \left[-\frac{5}{2}, \infty\right)$

17. [4.2] $\{x | -3 \le x \le 1\}$, or $[-3, 1]$

18. [4.3] $\{-15, 15\}$
19. [4.3] $\{a | a < -5 \ or \ a > 5\}$, or $(-\infty, -5) \cup (5, \infty)$

20. [4.3] $\{x | -2 < x < \frac{8}{3}\}$, or $\left(-2, \frac{8}{3}\right)$

21. [4.3] $\{t | t \le -\frac{13}{5} \ or \ t \ge \frac{7}{5}\}$, or $\left(-\infty, -\frac{13}{5}\right] \cup \left[\frac{7}{5}, \infty\right)$

22. [4.3] $\varnothing$
23. [4.2] $\{x | x < \frac{1}{2} \ or \ x > \frac{7}{2}\}$, or $\left(-\infty, \frac{1}{2}\right) \cup \left(\frac{7}{2}, \infty\right)$

24. [4.3] $\left\{-\frac{3}{2}\right\}$
25. [4.4]   **26.** [4.4]

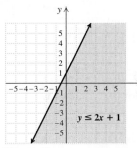

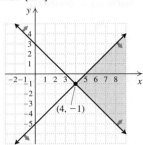

27. [4.4]

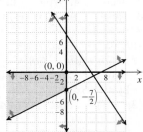

28. [4.5] Maximum 57 when $x = 6$, $y = 9$; minimum 5
when $x = 1$, $y = 0$ **29.** [4.5] Manicures: 35; haircuts: 15;
maximum: $690   **30.** [4.3] $[-1, 0] \cup [4, 6]$
31. [4.2] $\left(\frac{1}{5}, \frac{4}{5}\right)$ **32.** [4.3] $|x + 3| \le 5$

Cumulative Review: Chapters 1–4, p. 278

1. 22   **2.** $c - 6$ **3.** $-\dfrac{6x^4}{y^3}$ **4.** $\dfrac{9a^6}{4b^4}$ **5.** 5

6. $\left\{-\frac{7}{2}, \frac{9}{2}\right\}$ **7.** $\{t | t < -3 \ or \ t > 3\}$, or $(-\infty, -3) \cup (3, \infty)$
8. $\{x | -2 \le x \le \frac{10}{3}\}$, or $\left[-2, \frac{10}{3}\right]$ **9.** $\mathbb{R}$ **10.** $\left(\frac{22}{17}, -\frac{2}{17}\right)$
11. No solution   **12.** $\{x | x < \frac{13}{2}\}$, or $\left(-\infty, \frac{13}{2}\right)$

13.

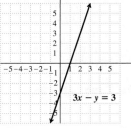

14.

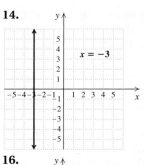

15.

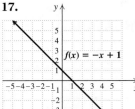

16.

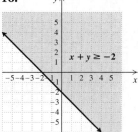

17.

18.

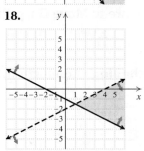

19. Slope: $\frac{4}{9}$; y-intercept: $(0, -2)$ **20.** $f(x) = -7x - 25$
21. $y = \frac{2}{3}x + 4$ **22.** Domain: $\mathbb{R}$; range: $\{y | y \ge -2\}$,
or $[-2, \infty)$ **23.** 22   **24.** $x^2 + 6x - 9$
25.

26. $\{x | x$ is a real number $and \ x \neq 0 \ and \ x \neq \frac{1}{3}\}$

27. $t = \dfrac{c}{a - d}$ **28.** 4.5×10^{10} gal   **29.** Beef: 9300 gal;
wheat: 2300 gal   **30.** **(a)** $f(t) = 0.7t + 0.56$;
(b) $4.76 billion; **(c)** approximately 2018   **31.** $m = \frac{1}{3}$,
$b = \frac{16}{3}$ **32.** $[-4, 0) \cup (0, \infty)$

CHAPTER 5

Technology Connection, p. 286

1. Correct   **2.** Incorrect   **3.** Correct   **4.** Incorrect

Exercise Set 5.1, pp. 286–290

1. (g)   **2.** (d)   **3.** (a)   **4.** (h)   **5.** (b)   **6.** (c)
7. (j)   **8.** (e)   **9.** (f)   **10.** (i)   **11.** $7x^4, x^3, -5x, 8$
13. $-t^6, 7t^3, -3t^2, 6$ **15.** Trinomial   **17.** Polynomial
with no special name   **19.** Binomial   **21.** Monomial
23. 2, 1   **25.** 5, 2, 0   **27.** 3, 7, 4   **29.** 4, 7, -3
31. 1, -1, 4   **33.** 1, -5, 7, 1   **35.** **(a)** 5; **(b)** 6, 4, 3, 1, 0;
(c) 6; **(d)** $-5x^6$; **(e)** -5 **37.** **(a)** 4; **(b)** 4, 5, 3, 0;
(c) 5; **(d)** a^3b^2; **(e)** 1   **39.** 5   **41.** 6
43. $-15t^4 + 2t^3 + 5t^2 - 8t + 4$; $-15t^4$; -15
45. $-x^6 + 6x^5 + 7x^2 + 3x - 5$; $-x^6$; -1
47. $-9 + 4x + 5x^3 - x^6$ **49.** $8y + 5xy^3 + 2x^2y - x^3$
51. -38 **53.** -16 **55.** -13; 11   **57.** 282; -9

59. 6840   **61.** About 250 horsepower   **63.** About 20 W
65. 150   **67.** 14; 55 oranges   **69.** About 260 mg
71. About 340 mg   **73.** 56.5 in^2 **75.** \$18,750
77. \$8375   **79.** $3x^3 - x + 1$ **81.** $-6a^2b - 3b^2$
83. $10x^2 + 2xy + 15y^2$ **85.** $4t^4 - 3t^3 + 6t^2 + t$
87. $-2x^2 + x - 3xy + 2y^2 - 1$ **89.** $6x^2y - 4xy^2 + 5xy$
91. $9r^2 + 9r - 9$ **93.** $-\frac{5}{24}xy - \frac{27}{20}x^3y^2 + 1.4y^3$
95. $-(3t^4 + 8t^2 - 7t - 1),\ -3t^4 - 8t^2 + 7t + 1$
97. $-(-12y^5 + 4ay^4 - 7by^2),\ 12y^5 - 4ay^4 + 7by^2$
99. $-4x^2 - 3x + 13$ **101.** $6a - 6b + 5c$
103. $-2a^2 + 12ab - 7b^2$ **105.** $8a^2b + 16ab + 3ab^2$
107. $x^4 - x^2 - 1$ **109.** $5t^2 + t + 4$ **111.** $13r^2 - 8r - 1$
113. $3x^2 - 9$ **115.** \$9700   **117.** 🖪   **119.** $-\frac{11}{40}$
120. 3.78   **121.** 240   **122.** $-\frac{3}{2}$ **123.** 27   **124.** 6
125. 🖪   **127.** $68x^5 - 81x^4 - 22x^3 + 52x^2 + 2x + 250$
129. $45x^5 - 8x^4 + 208x^3 - 176x^2 + 116x - 25$
131. 494.55 cm^3 **133.** $5x^2 - 8x$
135. $x^{5b} + 4x^{4b} + x^{3b} - 6x^{2b} - 9x^b$ **137.** ◩, 🖪

Prepare to Move On, p. 290

1. x^8 **2.** a^6b^4 **3.** t^8 **4.** $25y^6$ **5.** $4x^{10}y^2$

Technology Connection, p. 295

1.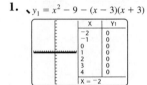
2.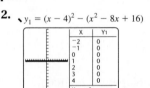

3. 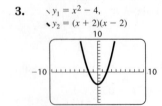   If $y_3 = y_2 - y_1$, the graph of y_3 should be the x-axis.

Exercise Set 5.2, pp. 297–299

1. False   **2.** True   **3.** True   **4.** True   **5.** False
6. False   **7.** True   **8.** True   **9.** $15x^5$ **11.** $-48a^3b^2$
13. $36x^5y^6$ **15.** $21x - 7x^2$ **17.** $20c^3d^2 - 25c^2d^3$
19. $x^2 + 8x + 15$ **21.** $8a^2 + 10a - 3$
23. $x^3 - x^2 - 5x + 2$ **25.** $t^3 - 3t^2 - 13t + 15$
27. $a^4 + 5a^3 - 2a^2 - 9a + 5$ **29.** $x^3 + 27$
31. $a^3 - b^3$ **33.** $t^2 - t - 6$ **35.** $20x^2 + 13xy + 2y^2$
37. $t^2 - \frac{7}{12}t + \frac{1}{12}$ **39.** $3t^2 + 1.5st - 15s^2$
41. $r^3 + 4r^2 + r - 6$ **43.** $x^2 + 10x + 25$
45. $4y^2 - 28y + 49$ **47.** $25c^2 - 20cd + 4d^2$
49. $9a^6 - 60a^3b^2 + 100b^4$ **51.** $x^6y^8 + 10x^3y^4 + 25$
53. $c^2 - 49$ **55.** $1 - 16x^2$
57. $9m^2 - \frac{1}{4}n^2$ **59.** $x^6 - y^2z^2$
61. $-m^2n^2 + 9m^4$, or $9m^4 - m^2n^2$ **63.** $14x + 58$
65. $3m^2 + 4mn - 5n^2$ **67.** $a^2 + 2ab + b^2 - 1$
69. $4x^2 + 12xy + 9y^2 - 16$ **71.** $A = P + 2Pr + Pr^2$

73. $12x^4 - 21x^3 - 17x^2 + 35x - 5$ **75.** $25x^2 - 20x + 4$
77. $4x^2 - \frac{4}{3}x + \frac{1}{9}$ **79. (a)** $t^2 - 2t + 6$; **(b)** $2ah + h^2$;
(c) $2ah - h^2$ **81. (a)** $2a^2$; **(b)** $a^2 + 2ah + h^2 + a + h$;
(c) $2ah + h^2 + h$ **83.** 🖪   **85.** $\frac{87}{4}$
86. $\left\{x \mid x > \frac{8}{5}\right\}$, or $\left(\frac{8}{5}, \infty\right)$ **87.** $\left\{x \mid x < -\frac{2}{3}\ or\ x > \frac{14}{3}\right\}$, or
$\left(-\infty, -\frac{2}{3}\right) \cup \left(\frac{14}{3}, \infty\right)$ **88.** $\{x \mid -5 \le x \le -3\}$, or
$[-5, -3]$ **89.** $\left(\frac{23}{7}, \frac{6}{7}\right)$ **90.** $\left(6, -\frac{7}{2}, \frac{9}{2}\right)$ **91.** 🖪
93. $x^4 - y^{2n}$ **95.** $5x^{n+2}y^3 + 4x^2y^{n+3}$
97. $a^2 + 2ac + c^2 - b^2 - 2bd - d^2$ **99.** $x^4 + x^2 + 25$
101. $x^6 - 1$ **103.** $x^{a^2-b^2}$ **105.** 0
107. $10x^{-5} + 15x^{-4} - 2 - 3x$ **109.** $2a + h$
111.

	$A - B$	B
$A - B$	$(A - B)^2$	$AB - B^2$
B	$AB - B^2$	B^2

with A bracing the left side (rows) and A bracing the bottom, $A - B$ and B labeling the top.

113. (b) and (c) are identities.

Quick Quiz: Sections 5.1–5.2, p. 299

1. 6   **2.** 2   **3.** $y^3 + 4y$ **4.** $25c^6 - 10c^3d + d^2$
5. $a^2 + 6a + 3$

Prepare to Move On, p. 299

1. $5(x + 3y - 1)$ **2.** $7(2a + 5b + 6c)$
3. $x(a + b - c)$ **4.** $b(x + y + 1)$

Technology Connection, p. 301

1. A table should show that $y_3 = 0$ for any value of x.

Exercise Set 5.3, pp. 303–305

1. True   **2.** False   **3.** True   **4.** True   **5.** True
6. False   **7.** True   **8.** False   **9.** $5(2x^2 + 7)$
11. $2y(y - 9)$ **13.** $5(t^3 - 3t + 1)$
15. $a^3(a^3 + 2a - 1)$ **17.** $6x(2x^3 - 5x^2 + 7)$
19. $b(6a^2 - 2a - 9)$ **21.** $5m^3n(3m + 6m^2n + 5n^2)$
23. $3x^2y^4z^2(3xy^2 - 4x^2z^2 + 5yz)$ **25.** $-5(x + 8)$
27. $-16(t^2 - 6)$ **29.** $-2(x^2 - 6x - 20)$
31. $-5(-1 + 2y)$, or $-5(2y - 1)$ **33.** $-4d(-2d + 3c)$,
or $-4d(3c - 2d)$ **35.** $-1(m^3 - 8)$
37. $-1(p^3 + 2p^2 + 5p - 2)$ **39.** $(b - 5)(a + c)$
41. $(x + 7)(2x - 3)$ **43.** $(x - y)(a^2 - 5)$
45. $(y + z)(x + w)$ **47.** $(y - 1)(y^2 + 3)$
49. $(t + 6)(t^2 - 2)$ **51.** $3a^2(4a^2 - 7a - 3)$
53. $(y - 1)(y^7 + 1)$ **55.** $(x - 2)(3 - xy)$, or
$(xy - 3)(2 - x)$ **57. (a)** $h(t) = -8t(2t - 9)$;
(b) $h(1) = 56$ ft   **59.** $\pi r(2h + r)$ **61.** $P(t) = t(t - 5)$
63. $R(x) = 0.4x(700 - x)$ **65.** $P(n) = \frac{1}{2}(n^2 - 3n)$
67. $N(x) = \frac{1}{6}(x^3 + 3x^2 + 2x)$ **69.** 🖪

71.
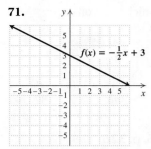

72.
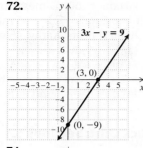

73.
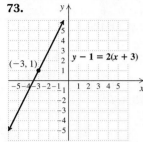

74.
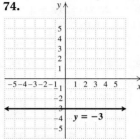

75.
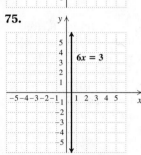

76.
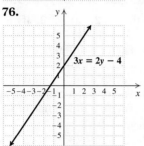

77. 🖮   **79.** $x^5y^4 + x^4y^6 = x^3y(x^2y^3 + xy^5)$
81. $(x^2 - x + 5)(r + s)$
83. $(x^4 + x^2 + 5)(a^4 + a^2 + 5)$ **85.** $x^{-9}(x^3 + 1 + x^6)$
87. $x^{1/3}(1 - 5x^{1/6} + 3x^{5/12})$ **89.** $x^{-5/2}(1 + x)$
91. $x^{-7/5}(x^{3/5} - 1 + x^{16/15})$ **93.** $3a^n(a + 2 - 5a^2)$
95. $y^{a+b}(7y^a - 5 + 3y^b)$ **97.** 🖩

Quick Quiz: Sections 5.1–5.3, p. 305

1. $2xy - y^2 - 3x + 2y$ **2.** $2a^3 - 9$ **3.** $t^2 - \frac{1}{9}$
4. $x(7x^2 - 6y)$ **5.** $(x - 3)(2x^2 + 1)$

Prepare to Move On, p. 305

1. $x^2 + 8x + 15$ **2.** $x^2 - 8x + 15$ **3.** $x^2 + 2x - 15$
4. $x^2 - 2x - 15$ **5.** $2x^2 + 11x + 15$ **6.** $2x^2 + 13x + 15$

Technology Connection, p. 308

1. They should coincide.   **2.** The x-axis
3. Let $y_1 = 2x^2 + x - 15$, $y_2 = (2x + 5)(x - 3)$, and
$y_3 = y_2 - y_1$. The graphs of y_1 and y_2 do not coincide; the
graph of y_3 is not the x-axis.

Exercise Set 5.4, pp. 314–316

1. True   **2.** True   **3.** False   **4.** False   **5.** False
6. True   **7.** True   **8.** False   **9.** $(x + 1)(x + 4)$
11. $(y - 3)(y - 9)$ **13.** $(t - 4)(t + 2)$
15. $(a - 1)(a + 2)$ **17.** $2(x - 6)(x + 9)$
19. $(a + 5)(a + 9)$ **21.** $p(p - 9)(p + 8)$

23. $(a - 4)(a - 7)$ **25.** $(x + 3)(x - 2)$
27. $5(y + 1)(y + 7)$ **29.** $(8 - y)(4 + y)$
31. $x(8 - x)(7 + x)$ **33.** $y^2(y + 12)(y - 7)$
35. Prime   **37.** $(x + 3y)(x + 9y)$
39. $(x - 7y)(x - 7y)$, or $(x - 7y)^2$
41. $n^3(n - 1)(n - 79)$ **43.** $x^4(x - 7)(x + 9)$
45. $(x - 2)(3x + 2)$ **47.** $(2t - 3)(3t + 5)$
49. $2(p - 2)(3p - 4)$ **51.** $(3a + 2)(3a + 4)$
53. $2y(3y - 1)(5y + 3)$ **55.** $6(3x - 4)(x + 1)$
57. $t^6(t + 7)(t - 2)$ **59.** $2x^2(5x - 2)(7x - 4)$
61. $(3y - 4)(6y + 5)$ **63.** $(4x + 1)(4x + 5)$
65. $(x + 4)(5x + 4)$ **67.** $-2(2t - 3)(2t + 5)$
69. $xy(6y + 5)(3y - 2)$ **71.** $(24x + 1)(x - 2)$
73. $3x(7x + 3)(3x + 4)$ **75.** $2x^2(6x + 5)(4x - 3)$
77. $(4a - 3b)(3a - 2b)$ **79.** $(2x - 3y)(x + 2y)$
81. $(2x - 7y)(3x - 4y)$ **83.** $(3x - 5y)(3x - 5y)$, or
$(3x - 5y)^2$ **85.** $(9xy - 4)(xy + 1)$ **87.** 🖮
89. $\dfrac{a^{18}}{8b^3}$ **90.** $-\dfrac{12}{x^7y}$ **91.** $\dfrac{3}{2t^{17}}$ **92.** $\dfrac{9x^{14}}{4y^{10}}$
93. 6.07×10^{-4} **94.** $318{,}750{,}000$ **95.** 🖮
97. $5(4x^4y^3 + 1)(3x^4y^3 + 1)$ **99.** $\left(y + \frac{4}{7}\right)\left(y - \frac{2}{7}\right)$
101. $ab^2(4ab^2 + 1)(5ab^2 - 2)$ **103.** $(x^a + 8)(x^a - 3)$
105. $a(2r + s)(r + s)$ **107.** $(x - 4)(x + 8)$
109. $76, -76, 28, -28, 20, -20$
111. We are given $ax^2 + bx + c = (mx + r)(nx + s)$.
Multiplying the binomials, we have
$ax^2 + bx + c = mnx^2 + msx + rnx + rs =$
$mnx^2 + (ms + rn)x + rs$. Equating coefficients gives
$a = mn$, $b = ms + rn$, and $c = rs$. If $p = ms$ and $q = rn$,
then $p + q = b$ and $pq = ac$.   **113.** 🖩   **115.** 🖮

Quick Quiz: Sections 5.1–5.4, p. 316

1. $-5a^3$; -5 **2.** $x^2y^2 - 9xy + 20$
3. $3x^2y^5(3x + y - 5x^2)$ **4.** $(t - 10)(t + 4)$
5. $n(2n - 5)(3n + 2)$

Prepare to Move On, p. 316

1. $25a^2$ **2.** $9x^8$ **3.** $x^2 + 6x + 9$ **4.** $4t^2 - 20t + 25$
5. $y^2 - 1$ **6.** $16x^4 - 9y^2$

Mid-Chapter Review: Chapter 5, p. 316

1. $(2x - 3)(x + 4) = 2x^2 + 8x - 3x - 12 = 2x^2 + 5x - 12$
2. $3x^3 + 7x^2 + 2x = x(3x^2 + 7x + 2) = x(3x + 1)(x + 2)$
3. $4t^3 + 8t^2 - 13t - 1$ **4.** $12x^3y^2 - 8x^5y + 24x^2y^3$
5. $9n^2 - n$ **6.** $x^2 + 8x + 7$ **7.** $10x^2 - 17x + 3$
8. $\frac{7}{6}x^2 - \frac{1}{6}x - \frac{11}{6}$ **9.** $9m^2 - 60m + 100$
10. $-1.6x^2 - 2.7x - 1.4$ **11.** $a^3 + a^2 - 8a - 12$
12. $c^2 - 81$ **13.** $4x^2y(2y^2z + 3xy - 4z^3)$
14. $(t - 1)(3t^2 + 1)$ **15.** $(x - 10)(x + 9)$
16. $6x(x + 7)(x + 3)$ **17.** $(5x - 3)(x + 2)$
18. $(x + y)(2 + a)$

Exercise Set 5.5, pp. 321–322

1. Difference of two squares   **2.** Perfect-square trinomial
3. Perfect-square trinomial   **4.** Difference of two squares
5. None of these   **6.** Polynomial having a common factor
7. Polynomial having a common factor   **8.** None of these

9. Perfect-square trinomial   **10.** Polynomial having a common factor   **11.** $(x + 10)^2$ **13.** $(t - 1)^2$
15. $4(a - 3)^2$ **17.** $(y + 6)^2$ **19.** $y(y - 9)^2$
21. $2(x - 10)^2$ **23.** $(1 - 4d)^2$ **25.** $-y(y + 4)^2$
27. $(0.5x + 0.3)^2$ **29.** $(p - q)^2$ **31.** $(5a + 3b)^2$
33. $5(a + b)^2$ **35.** $(x + 5)(x - 5)$ **37.** $(m + 8)(m - 8)$
39. $(2a + 9)(2a - 9)$ **41.** $12(c + d)(c - d)$
43. $7x(y^2 + z^2)(y + z)(y - z)$ **45.** $a(2a + 7)(2a - 7)$
47. $3(x^4 + y^4)(x^2 + y^2)(x + y)(x - y)$
49. $(pq + 10)(pq - 10)$ **51.** $a^2(3a + 5b^2)(3a - 5b^2)$
53. $\left(y + \frac{1}{2}\right)\left(y - \frac{1}{2}\right)$ **55.** $\left(\frac{1}{10} + x\right)\left(\frac{1}{10} - x\right)$
57. $(a + b + 6)(a + b - 6)$ **59.** $(x - 3 + y)(x - 3 - y)$
61. $(t + 8)(t + 1)(t - 1)$ **63.** $(r - 3)^2(r + 3)$
65. $(m - n + 5)(m - n - 5)$ **67.** $(9 + x + y)(9 - x - y)$
69. $(r - 1 + 2s)(r - 1 - 2s)$
71. $(4 + a + b)(4 - a - b)$ **73.** $(x + 5)(x + 2)(x - 2)$
75. $(a - 2)(a + b)(a - b)$ **77.** ☑   **79.** 16

80. $w + x$ **81.** $x = \dfrac{y}{a + 3}$ **82.** 9   **83.** $\{1, 3\}$
84. $\{1, 2, 3, 5, 7\}$ **85.** ☑   **87.** $-\frac{1}{54}(4r + 3s)^2$
89. $(0.3x^4 + 0.8)^2$, or $\frac{1}{100}(3x^4 + 8)^2$
91. $(r + s + 1)(r - s - 9)$
93. $(5y^a + x^b - 1)(5y^a - x^b + 1)$ **95.** $3(x + 1 + 2)^2$, or $3(x + 3)^2$ **97.** $(s - 2t + 2)^2$ **99.** $(3x^n - 1)^2$
101. $h(2a + h)$ **103. (a)** $\pi h(R + r)(R - r)$;
(b) $3{,}014{,}400 \text{ cm}^3$ **105.** ☐

Quick Quiz: Sections 5.1–5.5, p. 322

1. $-9x^4y + 2x^3y + 3x^2y^5 + x$ **2.** $p^3 - 5p^2 - 17p - 6$
3. $(c + 2)(a - b)$ **4.** $3(d - 2)(d - 5)$
5. $(xy + z^2)(xy - z^2)$

Prepare to Move On, p. 322

1. $8x^6y^{12}$ **2.** $-1000x^{30}$ **3.** $x^3 + 3x^2 + 3x + 1$
4. $x^3 - 3x^2 + 3x - 1$ **5.** $x^3 + 1$ **6.** $x^3 - 1$

Exercise Set 5.6, pp. 325–326

1. Difference of cubes   **2.** Sum of cubes   **3.** Difference of squares   **4.** None of these   **5.** Sum of cubes
6. Difference of cubes   **7.** None of these
8. Difference of squares   **9.** Difference of cubes
10. None of these   **11.** $(x - 4)(x^2 + 4x + 16)$
13. $(z + 1)(z^2 - z + 1)$ **15.** $(t - 10)(t^2 + 10t + 100)$
17. $(3x + 1)(9x^2 - 3x + 1)$
19. $(4 - 5x)(16 + 20x + 25x^2)$
21. $(x - y)(x^2 + xy + y^2)$ **23.** $\left(a + \frac{1}{2}\right)\left(a^2 - \frac{1}{2}a + \frac{1}{4}\right)$
25. $8(t - 1)(t^2 + t + 1)$
27. $2(3x + 1)(9x^2 - 3x + 1)$
29. $rs(s + 4)(s^2 - 4s + 16)$
31. $5(x - 2z)(x^2 + 2xz + 4z^2)$
33. $\left(y - \frac{1}{10}\right)\left(y^2 + \frac{1}{10}y + \frac{1}{100}\right)$
35. $(x + 0.1)(x^2 - 0.1x + 0.01)$
37. $8(2x^2 - t^2)(4x^4 + 2x^2t^2 + t^4)$
39. $2y(3y - 4)(9y^2 + 12y + 16)$
41. $(z + 1)(z^2 - z + 1)(z - 1)(z^2 + z + 1)$
43. $(t^2 + 4y^2)(t^4 - 4t^2y^2 + 16y^4)$
45. $(x^4 - yz^4)(x^8 + x^4yz^4 + y^2z^8)$ **47.** ☑

49. 20 cm   **50.** Dimes: 3 rolls; nickels: 1 roll; quarters: 6 rolls   **51.** 65 min or more
52. Ken: 54 nests; Kathy: 46 nests   **53.** ☑
55. $(x^{2a} - y^b)(x^{4a} + x^{2a}y^b + y^{2b})$ **57.** $2x(x^2 + 75)$
59. $5\left(xy^2 - \frac{1}{2}\right)\left(x^2y^4 + \frac{1}{2}xy^2 + \frac{1}{4}\right)$
61. $-(3x^{4a} + 3x^{2a} + 1)$ **63.** $(t - 8)(t - 1)(t^2 + t + 1)$
65. $h(2a + h)(a^2 + ah + h^2)(3a^2 + 3ah + h^2)$ **67.** ☐

Quick Quiz: Sections 5.1–5.6, p. 326

1. 6   **2.** $4x + 4y + 10z$ **3.** $36a^2b^2 + 19abx - 7x^2$
4. $(p + w)(p - w)$ **5.** $(p - w)(p^2 + pw + w^2)$

Prepare to Move On, p. 326

1. Common   **2.** Grouping or ac **3.** $(A + B)(A - B)$
4. $(A + B)^2$ **5.** $(A + B)(A^2 - AB + B^2)$

Exercise Set 5.7, pp. 330–331

1. (b)   **2.** (a)   **3.** (f)   **4.** (d)   **5.** (c)   **6.** (e)
7. (a)   **8.** (f)   **9.** Factor a trinomial; $(x + 1)(x - 4)$
10. Factor a difference of cubes; $(x - 1)(x^2 + x + 1)$
11. Factor by grouping; $(2x - 5)(2x^2 - 1)$
12. Factor a perfect-square trinomial; $(t - 10)^2$
13. Factor out a common factor; $8(3a^3 - 2a - 1)$
14. Factor a difference of squares; $(ab + c)(ab - c)$
15. $(x + 9)(x - 9)$ **17.** $9(m^2 + 10)(m^2 - 10)$
19. $2x(x + 2)(x + 4)$ **21.** $(a + 5)^2$
23. $(2y - 3)(y - 4)$ **25.** $3(x + 12)(x - 7)$
27. $(5x + 3y)(5x - 3y)$ **29.** $(t^2 + 1)(t^4 - t^2 + 1)$
31. $(x + y + 3)(x - y + 3)$
33. $2(4a + 5b)(16a^2 - 20ab + 25b^2)$
35. $7x(x + 3)(x - 5)$ **37.** $t^2(4t + 3)(4t - 3)$
39. $(m^3 + 10)(m^3 - 2)$ **41.** $(a + d)(c - b)$
43. $(2c - d)^2$ **45.** $(5x + y)(8x - y)$
47. $(2a - 5)(2 + a^2)$ **49.** $2(x + 3)(x + 2)(x - 2)$
51. $2(3a - 2b)(9a^2 + 6ab + 4b^2)$ **53.** $(6y - 5)(6y + 7)$
55. $4(m^2 + 4n^2)(m + 2n)(m - 2n)$
57. $ab(a^2 + 4b^2)(a + 2b)(a - 2b)$ **59.** $2t(17t^2 - 3)$
61. $2(a - 3)(a + 3)$ **63.** $7a(a^3 - 2a^2 + 3a - 1)$
65. $(9ab + 2)(3ab + 4)$ **67.** $-5t(2t^2 - 3)$
69. $-2x(3x^3 - 4x^2 + 6)$ **71.** $p(1 - 4p)(1 + 4p + 16p^2)$
73. $(a - b - 3)(a + b + 3)$ **75.** ☑   **77.** 98
78. $3a + 3h + 1$ **79.** 39   **80.** $x^2 - 3x - 3$
81. $\mathbb{R}$ **82.** $\{x \mid x \text{ is a real number } and\ x \neq -\frac{1}{3}\}$, or $\left(-\infty, -\frac{1}{3}\right) \cup \left(-\frac{1}{3}, \infty\right)$ **83.** ☑
85. $a(7a - bc)(4a - 3bc)$ **87.** $x(x - 2p)$
89. $y(y - 1)^2(y - 2)$
91. $(2x + y - r + 3s)(2x + y + r - 3s)$
93. $\left(\dfrac{x^9}{10} - 1\right)\left(\dfrac{x^{18}}{100} + \dfrac{x^9}{10} + 1\right)$ **95.** $3(x - 3)(x + 2)$
97. $3(a + 7)^2$ **99.** $2x^{-5}(x^2 + 2)(x^2 - 3)$
101. $a(a^w + 1)^2$

Quick Quiz: Sections 5.1–5.7, p. 331

1. $26 + 9n + 7n^3 - 3n^5$ **2.** $2ah + h^2 + 6h$
3. $10x^3y^2 + 15x^6y^3 - 30x^4y$ **4.** $(3a + c)(2a + 5c)$
5. $2(5c + 2)^2$

Prepare to Move On, p. 331

1. -2 **2.** $\frac{5}{2}$ **3.** 0
4. $\{x \,|\, x \text{ is a real number } and\ x \neq \frac{2}{3}\}$, or $\left(-\infty, \frac{2}{3}\right) \cup \left(\frac{2}{3}, \infty\right)$
5. $\{x \,|\, x \text{ is a real number } and\ x \neq -\frac{1}{2}\}$, or
$\left(-\infty, -\frac{1}{2}\right) \cup \left(-\frac{1}{2}, \infty\right)$

Technology Connection, p. 337

1. The graphs intersect at $(-2, 36)$, $(0, 0)$, and $(5, 225)$, so the solutions are $-2, 0,$ and 5.
2. The zeros are $-2, 0,$ and 5.

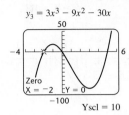

Connecting the Concepts, p. 339

1. $(x + 2)(x + 3)$ **2.** $-3, -2$ **3.** $2, 3$
4. $4x^2 - x - 5$ **5.** $2x^2 - x + 5$ **6.** $(a + 1)(a - 1)$
7. $a^2 - 1$ **8.** $-5, 5$

Exercise Set 5.8, pp. 339–343

1. True   **2.** True   **3.** False   **4.** True   **5.** False
6. True   **7.** $2, 5$ **9.** $-7, -1$ **11.** $-\frac{1}{2}, 0$
13. $0, \frac{4}{5}$ **15.** $-\frac{5}{2}, 7$ **17.** $-3, 6$ **19.** $0, 10$
21. $\frac{1}{3}, \frac{5}{4}$ **23.** $0, \frac{5}{2}$ **25.** $-2, 8$ **27.** $-4, 7$
29. 4 **31.** -10 **33.** $-5, -3$ **35.** $-9, 9$
37. $-7, 0, 9$ **39.** $-5, 5$ **41.** $-6, 6$ **43.** $\frac{1}{3}, \frac{4}{3}$
45. $-\frac{3}{4}, -\frac{1}{2}, 0$ **47.** $-3, 1$ **49.** $-\frac{7}{4}, \frac{4}{3}$
51. $-\frac{1}{10}, \frac{1}{10}$ **53.** $-5, -1, 1, 5$ **55.** $-8, -4$
57. $-4, \frac{3}{2}$ **59.** $-9, -3$ **61.** $\frac{1}{4}, \frac{5}{3}$ **63.** $-5, 0, \frac{3}{2}$
65. $\{x \,|\, x \text{ is a real number } and\ x \neq -1\ and\ x \neq 4\}$, or
$(-\infty, -1) \cup (-1, 4) \cup (4, \infty)$
67. $\{x \,|\, x \text{ is a real number } and\ x \neq -3\ and\ x \neq 3\}$, or
$(-\infty, -3) \cup (-3, 3) \cup (3, \infty)$
69. $\{x \,|\, x \text{ is a real number } and\ x \neq 0\ and\ x \neq \frac{1}{2}\}$, or
$(-\infty, 0) \cup \left(0, \frac{1}{2}\right) \cup \left(\frac{1}{2}, \infty\right)$
71. $\{x \,|\, x \text{ is a real number } and\ x \neq 0\ and\ x \neq 2\ and\ x \neq 5\}$,
or $(-\infty, 0) \cup (0, 2) \cup (2, 5) \cup (5, \infty)$
73. Length: 15 in.; width: 12 in.   **75.** 3 m   **77.** 2 cm
79. 10 in.   **81.** 16, 18, 20   **83.** Height: 30 in.; base: 50 in.
85. 60 ft; 65 ft   **87.** 41 ft   **89.** Length: 100 m; width: 75 m
91. 2 systems   **93.** 2 sec   **95.** 5 sec   **97.** 2010
99. 🖳   **101.** 8   **102.** Slope: $\frac{2}{3}$; y-intercept: $(0, -2)$
103. x-intercept: $(20, 0)$; y-intercept: $(0, -4)$
104. $f(x) = -\frac{1}{2}x + \frac{17}{2}$ **105.** $f(x) = -\frac{5}{2}x + 5$
106. $y = 5x + 7$ **107.** 🖳   **109.** $-\frac{11}{8}, -\frac{1}{4}, \frac{2}{3}$
111. $-3, 1$; $\{x \,|\, -4 \leq x \leq 2\}$, or $[-4, 2]$
113. Answers may vary. $f(x) = 5x^3 - 20x^2 + 5x + 30$
115. Length: 28 cm; width: 14 cm   **117.** About 5.7 sec
119. 🖳   **121.** 6.90   **123.** 3.48   **125.** Since
$n^2 + m^2$ represents the largest number, let $c = n^2 + m^2$.
Then $a = n^2 - m^2$ and $b = 2mn$.

$$
\begin{aligned}
a^2 + b^2 &= (n^2 - m^2)^2 + (2mn)^2 \\
&= n^4 - 2n^2m^2 + m^4 + 4m^2n^2 \\
&= n^4 + 2m^2n^2 + m^4; \\
c^2 &= (n^2 + m^2)^2 \\
&= n^4 + 2m^2n^2 + m^4
\end{aligned}
$$

Since $a^2 + b^2 = c^2$, the expressions satisfy the Pythagorean equation.

Quick Quiz: Sections 5.1–5.8, p. 343

1. 7 **2.** $15x^2 - 8x + 10$ **3.** $p(2p - 3)(4p^2 + 6p + 9)$
4. $0, 2$ **5.** $\{x \,|\, x \text{ is a real number } and\ x \neq -5\ and\ x \neq 4\}$,
or $(-\infty, -5) \cup (-5, 4) \cup (4, \infty)$

Prepare to Move On, p. 343

1. $\frac{3}{2}$ **2.** $-\frac{1}{2}$ **3.** $-\frac{75}{32}$ **4.** $-\frac{2}{27}$ **5.** $\frac{6}{7}$ **6.** $\frac{4}{3}$

Visualizing for Success, p. 344

1. D   **2.** J   **3.** A   **4.** E   **5.** B   **6.** C   **7.** I
8. F   **9.** G   **10.** H

Decision Making: Connection, p. 345

1. (a) About 67 ft^3; **(b)** about 502.5 gal   **2.** 3 coats
3. 5 gal; the area of the windows and door is more than 258 ft^2.   **4.** 10 squares   **5.** 🖳

Study Summary: Chapter 5, pp. 346–348

1. $x^2, -10, 5x, -8x^6$ **2.** 1   **3.** 1   **4.** $-8x^6$ **5.** -8
6. 6   **7.** Trinomial   **8.** $8x^2 + x$ **9.** $10x^2 - 7x$
10. $x^3 - 2x^2 - x + 2$ **11.** $x^2 - 4y^2$
12. $6x(2x^3 - 3x^2 + 5)$ **13.** $(x - 3)(2x^2 - 1)$
14. $(x - 9)(x + 2)$ **15.** $(3x + 2)(2x - 1)$
16. $(10n + 9)^2$ **17.** $(12t + 5)(12t - 5)$
18. $(a - 1)(a^2 + a + 1)$ **19.** $-y(xy + 5)(xy - 2)$
20. $0, \frac{4}{3}$ **21.** 12, 13

Review Exercises: Chapter 5, pp. 349–350

1. (g)   **2.** (b)   **3.** (a)   **4.** (d)   **5.** (e)   **6.** (j)
7. (h)   **8.** (c)   **9.** (i)   **10.** (f)   **11.** 11
12. $-5x^3 + 2x^2 + 3x + 9$; $-5x^3$; -5
13. $-3x^2 + 2x^3 + 8x^6y - 7x^8y^3$ **14.** $0; -6$
15. $2ah + h^2 + 10h$ **16.** $-2a^3 + a^2 - 3a - 4$
17. $-x^2y - 2xy^2$ **18.** $-2x^3 + 2x^2 + 5x + 3$
19. $-2n^3 + 2n^2 - 2n + 11$ **20.** $-5xy^2 - 2xy - 11x^2y$
21. $14x - 7$ **22.** $-2a + 6b$ **23.** $6x^2 - 4xy + 4y^2 + 9y$
24. $-18x^3y^4$ **25.** $x^8 - x^6 + 5x^2 - 3$
26. $8a^2b^2 + 2abc - 3c^2$ **27.** $49t^2 - 1$
28. $9x^2 - 24xy + 16y^2$ **29.** $2x^2 + 5x - 3$
30. $x^4 + 8x^2y^3 + 16y^6$ **31.** $5t^2 - 42t + 16$
32. $x^2 - \frac{1}{2}x + \frac{1}{18}$ **33.** $-3y(y^3 + 3y - 4)$
34. $(a - 9)(a - 3)$ **35.** $(3m + 2)(m - 4)$
36. $(5x + 2)^2$ **37.** $4(y + 2)(y - 2)$
38. $x(x - 2)(x + 7)$ **39.** $(a + 2b)(x - y)$
40. $(y + 2)(3y^2 - 5)$ **41.** $(a^2 + 9)(a + 3)(a - 3)$
42. $4(x^4 + x^2 + 5)$ **43.** $(3x + 2)(9x^2 - 6x + 4)$
44. $\left(\frac{1}{5}b - \frac{1}{2}c^2\right)\left(\frac{1}{25}b^2 + \frac{1}{10}bc^2 + \frac{1}{4}c^4\right)$

45. $(ab^2 + 8)(ab^2 - 8)$ **46.** Prime
47. $(0.1x^2 + 1.2y^3)(0.1x^2 - 1.2y^3)$
48. $4y(x - 5)^2$ **49.** $(3t + p)(2t + 5p)$
50. $(x + 3)(x - 3)(x + 2)$ **51.** $(a - b + 2t)(a - b - 2t)$
52. 6   **53.** $\frac{2}{3}, \frac{3}{2}$ **54.** $0, \frac{7}{4}$ **55.** $-2, 2$
56. $-3, 0, 7$ **57.** $-1, 6$ **58.** $-4, 11$
59. $\{x \mid x \text{ is a real number } and \; x \neq -7 \; and \; x \neq 8\}$, or
$(-\infty, -7) \cup (-7, 8) \cup (8, \infty)$ **60.** Height: 18 m; base: 24 m
61. Length: 8 in.; width: 5 in.   **62.** 15 ft; 17 ft
63. 1990 and 1995   **64.** ☞ When multiplying polynomials, we begin with a product and carry out the multiplication to write a sum of terms. When factoring a polynomial,
we write an equivalent expression that is a product.
65. ☞ The principle of zero products states that if a
product is equal to 0, at least one of the factors must be 0.
If a product is nonzero, we cannot conclude that any one of
the factors is a particular value.
66. $2(2x - y)(4x^2 + 2xy + y^2)(2x + y)(4x^2 - 2xy + y^2)$
67. $-2(3x^2 + 1)$ **68.** $3x^{-6}(1 - 4x^2 + 5x^3)$ **69.** $-1, \frac{1}{2}$
70. No real solution

Test: Chapter 5, p. 351

1. [5.1] 9   **2.** [5.1] $5x^5y^4 - 9x^4y - 14x^2y + 8xy^3$
3. [5.1] $-5a^3$ **4.** [5.1] 4; 2   **5.** [5.2] $2ah + h^2 - 3h$
6. [5.1] $4xy + 3xy^2$ **7.** [5.1] $-y^3 + 6y^2 - 10y - 7$
8. [5.1] $10m^3 - 4m^2n - 3mn^2 + n^2$ **9.** [5.1] $5a - 8b$
10. [5.1] $5y^2 + y^3$ **11.** [5.2] $64x^3y^8$
12. [5.2] $12a^2 - 4ab - 5b^2$ **13.** [5.2] $x^3 - 2x^2y + y^3$
14. [5.2] $16t^2 - 24t + 9$ **15.** [5.2] $25a^6 + 90a^3 + 81$
16. [5.2] $x^2 - 4y^2$ **17.** [5.5] $(x - 5)^2$
18. [5.5] $(y + 5)(y + 2)(y - 2)$
19. [5.4] $(p - 14)(p + 2)$ **20.** [5.3] $t^5(t^2 - 3)$
21. [5.4] $(6m + 1)(2m + 3)$ **22.** [5.5] $(3y + 5)(3y - 5)$
23. [5.6] $3(r - 1)(r^2 + r + 1)$ **24.** [5.5] $5(3x + 2)^2$
25. [5.5] $3(x^2 + 4y^2)(x + 2y)(x - 2y)$
26. [5.5] $(y + 4 + 10t)(y + 4 - 10t)$ **27.** [5.4] Prime
28. [5.5] $5(2a - b)(2a + b)$ **29.** [5.4] $2(4x - 1)(3x - 5)$
30. [5.6] $2ab(2a^2 + 3b^2)(4a^4 - 6a^2b^2 + 9b^4)$
31. [5.8] $-3, 6$ **32.** [5.8] $-5, 5$ **33.** [5.8] $-7, -\frac{3}{2}$
34. [5.8] $-\frac{1}{3}, 0$ **35.** [5.8] 9   **36.** [5.8] 0, 5
37. [5.8] $\{x \mid x \text{ is a real number } and \; x \neq -1\}$, or
$(-\infty, -1) \cup (-1, \infty)$ **38.** [5.8] Length: 8 cm; width: 5 cm
39. [5.8] $4\frac{1}{2}$ sec   **40.** [5.8] 24 ft
41. [5.4] $(a - 4)(a + 8)$ **42.** [5.8] $-\frac{8}{3}, 0, \frac{2}{5}$

Cumulative Review: Chapters 1–5, p. 352

1. -297 **2.** $-8a^2 + 6a^2b + 3b^2$ **3.** $4x^2 - 81$
4. $4x^2 + 36xy + 81y^2$ **5.** $10m^5 - 5m^3n + 2m^2n - n^2$
6. $3(t + 4)(t - 4)$ **7.** $(a - 7)^2$
8. $9x^2y(4xy - 3x^2 + 5y^2)$
9. $(5a + 4b)(25a^2 - 20ab + 16b^2)$
10. $(3y - 2)(4y + 5)$ **11.** $(d + a - b)(d - a + b)$
12. $\varnothing$ **13.** $\left(14, \frac{7}{2}\right)$ **14.** $(-2, -3, 4)$
15. $\{x \mid x \leq -\frac{1}{4}\}$, or $\left(-\infty, -\frac{1}{4}\right]$ **16.** $\{x \mid -8 < x < -6\}$,
or $(-8, -6)$ **17.** $\{x \mid -2 \leq x \leq 3\}$, or $[-2, 3]$
18. $-2, 12$ **19.** $0, 6$ **20.** $b = \dfrac{a - 2}{2}$
21. $f(x) = \frac{1}{3}x - \frac{1}{4}$ **22.** $f(x) = -9x - 13$

23.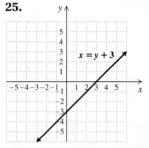
24.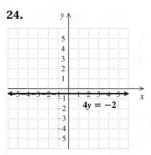

25.

26.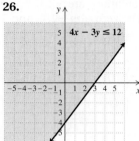

27. $\{x \mid x \text{ is a real number } and \; x \neq 1 \; and \; x \neq 2\}$, or
$(-\infty, 1) \cup (1, 2) \cup (2, \infty)$ **28.** $[-3, \infty)$
29. Between 850 kWh and 950 kWh   **30.** Length: 8 cm;
width: 6 cm   **31.** Length: 13.5 cm; width: 11.5 cm
32. Single-topping pizzas: 29; two-topping pizzas: 19
33. $d = \dfrac{ct - c}{t + 2}$ **34.** $|x + 3| < 4$

CHAPTER 6

Technology Connection, p. 358

1. Let $y_1 = (7x^2 + 21x)/(14x)$, $y_2 = (x + 3)/2$, and
$y_3 = y_1 - y_2$ (or $y_2 - y_1$). A table or the TRACE feature can be
used to show that, except when $x = 0$, y_3 is always 0. As an
alternative, let $y_1 = (7x^2 + 21x)/(14x) - (x + 3)/2$ and
show that, except when $x = 0$, y_1 is always 0.
2. Let $y_1 = (x + 3)/x$, $y_2 = 3$, and $y_3 = y_1 - y_2$ (or $y_2 - y_1$).
Use a table or the TRACE feature to show that y_3 is not
always 0. As an alternative, let $y_1 = (x + 3)/x - 3$ and show
that y_1 is not always 0.

Exercise Set 6.1, pp. 361–364

1. Rational   **2.** Domain   **3.** Factor   **4.** Reciprocal
5. (c)   **6.** (e)   **7.** (f)   **8.** (d)   **9.** (b)   **10.** (a)
11. (a) 5; (b) 1; (c) 5   **13.** (a) $\frac{9}{4}$; (b) does not exist; (c) 0
15. $\frac{30}{11}$ hr, or $2\frac{8}{11}$ hr   **17.** $\dfrac{t^3}{5}$ **19.** $\dfrac{4}{5x^2y^7}$ **21.** $a - 5$
23. $\dfrac{1}{5y - 6}$ **25.** $\dfrac{x - 4}{x + 5}$ **27.** $f(x) = \dfrac{5}{x}, x \neq -6, 0$
29. $g(x) = \dfrac{x - 3}{5}, x \neq -3$ **31.** $h(x) = -\frac{1}{7}, x \neq 2$
33. $f(t) = \dfrac{t + 4}{t - 4}, t \neq 4$ **35.** $g(t) = -\frac{7}{3}, t \neq 3$
37. $h(t) = \dfrac{t + 4}{t - 9}, t \neq -1, 9$ **39.** $f(x) = 3x + 2, x \neq \frac{2}{3}$
41. $g(t) = \dfrac{4 + t}{4 - t}, t \neq 4$ **43.** $\dfrac{6z^3}{7y^3}$ **45.** $\dfrac{8x^2}{25}$

47. $\dfrac{(y+3)(y-3)}{y(y+2)}$ **49.** $-\dfrac{a+1}{2+a}$ **51.** 1 **53.** $c(c-2)$

55. $\dfrac{a^2+ab+b^2}{3(a+2b)}$ **57.** $\dfrac{9a}{b}$ **59.** $\dfrac{5}{x^4}$ **61.** $-\dfrac{5x+2}{x-3}$

63. $-\dfrac{1}{y^3}$ **65.** $\dfrac{(y+6)(y+3)}{3(y-4)}$ **67.** $\dfrac{x^2+4x+16}{(x+4)^2}$

69. $f(t)=\dfrac{t+10}{5}, t\neq -4,10$

71. $g(x)=\dfrac{(x+5)(2x+3)}{7x}, x\neq 0,\dfrac{3}{2},7$

73. $f(x)=\dfrac{(x+2)(x+4)}{x^7}, x\neq -4,0,2$

75. $h(n)=\dfrac{n(n^2+3)}{(n+3)(n-2)}, n\neq -7,-3,2,3$

77. $\dfrac{3(x-3y)}{2(2x-y)(2x-3y)}$ **79.** $\dfrac{(2a-b)(a-1)}{(a-b)(a+1)}$

81. 📋   **83.** $(2n-3)(n-4)$ **84.** $7(y+2)(y-2)$
85. $(2x+5)(4x^2-10x+25)$
86. $(10a+3b)^2$ **87.** $t(t+11)(t-3)$
88. $(z^2+1)(z^4-z^2+1)(z+1)\times$
$(z^2-z+1)(z-1)(z^2+z+1)$ **89.** 📋   **91.** $2a+h$
93.

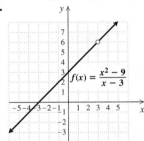

95. $\dfrac{(d-1)(d-4)(d-5)^2}{25d^4(d+5)}$ **97.** $\dfrac{m-t}{m+t+1}$

99. $\dfrac{x^2+xy+y^2+x+y}{x-y}$ **101.** $-\dfrac{2x}{x-1}$

103. (a) $\dfrac{16(x+1)}{(x-1)^2(x^2+x+1)}$; (b) $\dfrac{x^2+x+1}{(x+1)^3}$;

(c) $\dfrac{(x+1)^3}{x^2+x+1}$ **105.** 📋, 〰

Prepare to Move On, p. 364

1. $\dfrac{9}{20}$ **2.** $-\dfrac{27}{100}$ **3.** $4x^2-2$ **4.** $-x^2+29x-4$

Exercise Set 6.2, pp. 370–373

1. True   **2.** True   **3.** False   **4.** False   **5.** False

6. True   **7.** False   **8.** True   **9.** $\dfrac{5}{a}$ **11.** $\dfrac{1}{3m^2n^2}$

13. 2 **15.** $\dfrac{2t+4}{t-4}$ **17.** $\dfrac{1}{x-5}$ **19.** $\dfrac{-1}{a+5}$

21. $f(x)=\dfrac{3x-1}{(x+1)(x+5)}, x\neq -5,-1$

23. $f(x)=\dfrac{-x-5}{(x+1)(x-1)}, x\neq -1,1$ **25.** $24x^5$

27. $(x+3)(x-3)^2$ **29.** $\dfrac{9x+2}{15x^2}$ **31.** $\dfrac{y+3}{2(y-2)}$

33. $\dfrac{x+y}{x-y}$ **35.** $\dfrac{3x^2+7x+14}{(2x-5)(x-1)(x+2)}$

37. $\dfrac{-a^2+7ab-b^2}{(a-b)(a+b)}$ **39.** $\dfrac{x-5}{(x+5)(x+3)}$ **41.** $\dfrac{9}{t}$

43. $-(s+r)$ **45.** $\dfrac{2a^2-a+14}{(a-4)(a+3)}$ **47.** $\dfrac{5x+1}{x+1}$

49. $\dfrac{-x+34}{20(x+2)}$ **51.** $\dfrac{8x+1}{(x+1)(x-1)}$ **53.** $-\dfrac{1}{y+5}$

55. $\dfrac{1}{y^2+9}$ **57.** $\dfrac{1}{r^2+rs+s^2}$ **59.** $\dfrac{y}{(y-2)(y-3)}$

61. $\dfrac{7x+1}{x-y}$ **63.** $\dfrac{-y}{(y+3)(y-1)}$ **65.** $-\dfrac{2y}{2y+1}$

67. $f(x)=\dfrac{3(x+4)}{x+3}, x\neq -3,3$

69. $f(x)=\dfrac{(x-7)(2x-1)}{(x-4)(x-1)(x+3)}, x\neq -4,-3,1,4$

71. $f(x)=\dfrac{-2}{(x+1)(x+2)}, x\neq -3,-2,-1$ **73.** 📋

75. $-2,3$ **76.** $\{x|x\geq -\tfrac{1}{15}\}$, or $\left[-\tfrac{1}{15},\infty\right)$
77. $\{x|-3<x<-\tfrac{1}{2}\}$, or $\left(-3,-\tfrac{1}{2}\right)$
78. $\{x|x\leq -2\,or\,x\geq 6\}$, or $(-\infty,-2]\cup[6,\infty)$
79. $(-2,0)$ **80.** $(4,1)$ **81.** 📋
83. 420 days   **85.** 12 parts
87. $x^4(x^2+1)(x+1)(x-1)(x^2+x+1)(x^2-x+1)$
89. $8a^4, 8a^4b, 8a^4b^2, 8a^4b^3, 8a^4b^4, 8a^4b^5, 8a^4b^6, 8a^4b^7$
91. $\dfrac{x^4+6x^3+2x^2}{(x+2)(x-2)(x+5)}$

93. $\dfrac{x^5}{(x^2-4)(x^2+3x-10)}$ **95.** $\dfrac{2x+1}{x^2}$

97. $\dfrac{9x^2+28x+15}{(x-3)(x+3)^2}$ **99.** $\dfrac{1}{2x(x-5)}$

101. $-4t^4$ **103.** 〰

Quick Quiz: Sections 6.1–6.2, p. 373

1. -1 **2.** $\dfrac{(a-1)(a+1)^2}{3(a-2)(a+2)}$ **3.** $\dfrac{18}{x^3y^3}$

4. $\dfrac{2}{(x+1)(x-1)}$ **5.** $\dfrac{a+3}{a-4}$

Prepare to Move On, p. 373

1. $\dfrac{2}{x}$ **2.** $\dfrac{ab}{(a+b)^2}$ **3.** $9x-6$ **4.** $3ab^3+2a^3$

Technology Connection, p. 378

1. $-1,-\dfrac{1}{2},1$

Exercise Set 6.3, pp. 379–382

1. (b)   **2.** (a)   **3.** (f)   **4.** (c)   **5.** (d)   **6.** (e)
7. 10 **9.** $\dfrac{5}{11}$ **11.** $\dfrac{5x^2}{4(x^2+4)}$ **13.** $\dfrac{(x+1)(x+5)}{(x-3)(x-2)}$

15. $\dfrac{3x^2+2}{x(5x-3)}$ **17.** $\dfrac{6s-r}{2s+3r}$ **19.** $\dfrac{y(3y+2z)}{z(4-yz)}$

21. $\dfrac{a+b}{a}$ **23.** $\dfrac{3}{3x+2}$ **25.** $\dfrac{1}{x+y}$ **27.** $-\dfrac{1}{x(x+h)}$

29. $\dfrac{(a-2)(a-7)}{(a+1)(a-6)}$ **31.** $\dfrac{x+2}{x+3}$ **33.** $\dfrac{1+2y}{1-3y}$

35. $\dfrac{(y+1)(y^2-y+1)}{(y-1)(y^2+y+1)}$ **37.** $\dfrac{3a^2+4b^3}{a^4b^5}$

39. $\dfrac{-x^3y^3}{y^2+xy+x^2}$ **41.** $\dfrac{4x-7}{7x-9}$ **43.** $\dfrac{a^2-3a-6}{a^2-2a-3}$

45. $\dfrac{a+1}{2a+5}$ **47.** $\dfrac{-1-3x}{8-2x}$, or $\dfrac{3x+1}{2x-8}$

49. $y^2+5y+25$ **51.** $-y$ **53.** $\dfrac{6(a^2+5a+10)}{3a^2+2a+4}$

55. $\dfrac{(2x+1)(x+2)}{2x(x-1)}$ **57.** $\dfrac{(2a-3)(a+5)}{2(a-3)(a+2)}$ **59.** -1

61. $\dfrac{t^2+5t+3}{(t+1)^2}$ **63.** ☐ **65.** $-x^3+2x^2-23$

66. $9x^4-y^2$ **67.** $m^2-12m+36$ **68.** 3 **69.** ☐

71. $\dfrac{5(y+x)}{3(y-x)}$ **73.** No; $\dfrac{\frac{a}{b}}{c}=\dfrac{a}{b}\cdot\dfrac{1}{c}=\dfrac{a}{bc};\dfrac{a}{\frac{b}{c}}=a\cdot\dfrac{c}{b}=\dfrac{ac}{b}$

75. \$168.61 **77.** $\dfrac{x^2}{x^4+x^3+x^2+x+1}$

79. $\dfrac{-3}{x(x+h)}$ **81.** $\dfrac{2+a}{3+a}$, $a\neq-2,-3$

83. $\dfrac{x^4}{81}$; $\{x\,|\,x$ is a real number $and\ x\neq3\}$

Quick Quiz: Sections 6.1–6.3, p. 382

1. $f(x)=\dfrac{x-1}{x-2}$, $x\neq-1,2$ **2.** $\dfrac{4(a^2+a+1)}{3a^2}$

3. $\dfrac{(t-4)(t-1)}{(t+3)(t+4)}$ **4.** $\dfrac{2}{n+2}$ **5.** $\dfrac{5(a-b)}{3(a+b)}$

Prepare to Move On, p. 382

1. $\frac{11}{13}$ **2.** $-5,5$ **3.** $2,4$ **4.** $\frac12$

Connecting the Concepts, p. 387

1. $\dfrac{x-2}{x+1}$ **2.** $\dfrac{13t-5}{3t(2t-1)}$ **3.** 6 **4.** $\frac13$

5. $\dfrac{z}{1-z}$ **6.** $-\frac12$

Exercise Set 6.4, pp. 387–389

1. False **2.** True **3.** True **4.** True **5.** Equation
6. Expression **7.** Expression **8.** Equation
9. Expression **10.** Equation **11.** 6
13. $\frac{40}{9}$ **15.** $-7,7$ **17.** -8 **19.** -11
21. $-6,6$ **23.** $\frac83$ **25.** $-\frac23$ **27.** No solution
29. $-4,-1$ **31.** $-2,6$ **33.** $\frac{10}{9}$ **35.** No solution

37. -5 **39.** $-\frac73$ **41.** -1 **43.** $2,3$ **45.** -145
47. -15 **49.** No solution **51.** -1 **53.** $-4,1$
55. 4 **57.** $-2,\frac78$ **59.** No solution **61.** $-\frac32,5$
63. 14 **65.** $\frac34$ **67.** $\frac{5}{14}$ **69.** $\frac35$ **71.** ☐ **73.** -30
74. 3.91×10^8 **75.** $-\frac19$ **76.** 1 **77.** $\dfrac{3}{2a^9c}$

78. $-\dfrac{y^{18}}{125x^6}$ **79.** ☐ **81.** $-8,8$

83. $\{a\,|\,a$ is a real number $and\ a\neq-1\ and\ a\neq1\}$
85. -2 **87.** ☐

Quick Quiz: Sections 6.1–6.4, p. 389

1. $\dfrac{4a}{3(a-1)}$ **2.** $\dfrac{2a}{a-1}$ **3.** $\dfrac{45a^3}{32}$

4. $\dfrac{1}{a-b}$ **5.** $-5,3$

Prepare to Move On, p. 389

1. $\frac43$ hr, or $1\frac13$ hr **2.** 10 m per minute
3. Length: 60 in.; width: 15 in. **4.** -12 and -10, 10 and 12

Mid-Chapter Review: Chapter 6, pp. 390–391

1. $\dfrac{2}{x}+\dfrac{1}{x^2+x}=\dfrac{2}{x}+\dfrac{1}{x(x+1)}$

$=\dfrac{2}{x}\cdot\dfrac{x+1}{x+1}+\dfrac{1}{x(x+1)}$

$=\dfrac{2x+2}{x(x+1)}+\dfrac{1}{x(x+1)}$

$=\dfrac{2x+3}{x(x+1)}$

2. $\dfrac{10}{(t+1)(t-1)}=\dfrac{t+4}{t(t-1)}$

$t(t+1)(t-1)\left(\dfrac{10}{(t+1)(t-1)}\right)=t(t+1)(t-1)\left(\dfrac{t+4}{t(t-1)}\right)$

$10t=(t+1)(t+4)$
$10t=t^2+5t+4$
$0=t^2-5t+4$
$0=(t-1)(t-4)$
$t-1=0\ \ or\ \ t-4=0$
$t=1\ \ or\ \ \ \ \ \ t=4$

The solution is 4.

3. $\dfrac{x-3}{15(x-2)}$ **4.** $\frac13$ **5.** $\dfrac{x^2-2x-2}{(x-1)(x+2)}$

6. $\dfrac{5x+17}{(x+3)(x+4)}$ **7.** $\dfrac{5}{x-4}$ **8.** $\dfrac{x(x+4)}{(x-1)^2}$

9. $\dfrac{x+7}{(x-5)(x+1)}$ **10.** $(t+5)^2$ **11.** $\dfrac{a-1}{(a+2)(a-2)^2}$

12. $\dfrac{3y+2z}{4y-z}$ **13.** $\dfrac{x^2+y}{2(2x+y)}$ **14.** $\dfrac{1}{y+5}$

15. $\dfrac{a^2b^2}{b^2+ab+a^2}$ **16.** $\frac32$ **17.** $-\frac{10}{3}$

18. No solution **19.** $1,6$ **20.** -1

Exercise Set 6.5, pp. 397–400

1. False   **2.** True   **3.** True   **4.** True   **5.** True
6. True   **7.** $\frac{1}{2}$ cake per hour   **8.** $\frac{1}{3}$ cake per hour
9. $\frac{5}{6}$ cake per hour   **10.** 1 lawn per hour
11. $\frac{1}{3}$ lawn per hour   **12.** $\frac{2}{3}$ lawn per hour   **13.** 2
15. $-3, -2$ **17.** -10 and -9, 9 and 10   **19.** $3\frac{3}{5}$ hr
21. $18\frac{6}{13}$ min   **23.** GT-S50: $7\frac{1}{2}$ min; GT-1500: 15 min
25. Anita: 3 days; Tori: 6 days   **27.** Tristan: $\frac{4}{3}$ months;
Sara: 4 months   **29.** Zeno: 35 min; Lia: 15 min
31. 300 min, or 5 hr   **33.** 7 mph   **35.** 5.2 ft/sec
37. Freight: 66 mph; passenger: 80 mph
39. Express: 45 mph; local: 38 mph   **41.** $1\frac{1}{5}$ km/h
43. 40 mph   **45.** 20 mph   **47.** 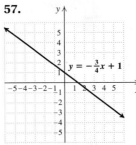   **49.** $228
50. 417 performances per year
51. Oil: $21\frac{1}{3}$ oz; lemon juice: $10\frac{2}{3}$ oz
52. Magic Kingdom: 31; Disneyland: 54; California
Adventure: 22   **53.**   **55.** $49\frac{1}{2}$ hr   **57.** 30 min
59. 2250 people per hour   **61.** $14\frac{7}{8}$ mi   **63.** Page 278
65. $8\frac{2}{11}$ min after 10:30   **67.** $51\frac{3}{7}$ mph

Quick Quiz: Sections 6.1–6.5, p. 400

1. $\dfrac{x - 6}{(x - 8)(x - 3)}$ **2.** $\dfrac{5(n^2 - n + 1)}{3n^2}$ **3.** $-2, 8$
4. $-\frac{11}{4}$ **5.** $3\frac{3}{7}$ hr

Prepare to Move On, p. 400

1. $6x^6y^8$ **2.** $-5a$ **3.** $5x^2 - 4x - 7$ **4.** $-x - 5$

Exercise Set 6.6, pp. 404–406

1. $x - 3$ **2.** $x^2 - x - 1$ **3.** $x + 2$ **4.** 5   **5.** 1
6. 0   **7.** $4x^4 + 2x^3 - 3$ **9.** $-3a^2 - a + \dfrac{3}{7} + \dfrac{2}{a}$
11. $-4z + 2y^2z^3 - 3y^4z^2$ **13.** $8y - \dfrac{9}{2} - \dfrac{4}{y}$
15. $-5x^5 + 7x^2 + 1$ **17.** $1 - ab^2 - a^3b^4$ **19.** $x + 3$
21. $y - 5 + \dfrac{-50}{y - 5}$ **23.** $x - 5 + \dfrac{1}{x - 4}$ **25.** $y - 5$
27. $a^2 - 2a + 4$ **29.** $x - 3 + \dfrac{3}{5x + 1}$
31. $y^2 - 2y - 1 + \dfrac{-8}{y - 2}$ **33.** $2x^2 - x + 1 + \dfrac{-5}{x + 2}$
35. $a^2 - 4a + 6$ **37.** $2y^2 + 2y - 1 + \dfrac{8}{5y - 2}$
39. $3x^2 + x + 1$ **41.** $2x^2 - x - 9 + \dfrac{3x + 12}{x^2 + 2}$
43. $2x - 5, x \neq -\frac{2}{3}$ **45.** $4x^2 + 6x + 9, x \neq \frac{3}{2}$
47. $x^2 + 1, x \neq -5, x \neq 5$
49. $2x^3 - 3x^2 + 5, x \neq -1, x \neq 1$ **51.**

53.

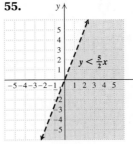
$3x - y = 9$

54.

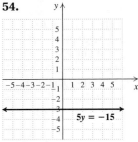
$5y = -15$

55.

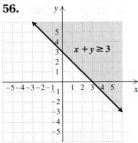
$y < \frac{5}{2}x$

56.

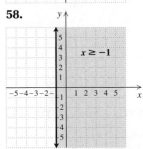
$x + y \geq 3$

57.

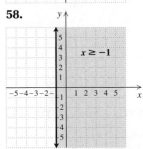
$y = -\frac{3}{4}x + 1$

58.

$x \geq -1$

59. 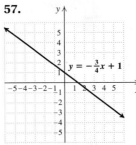   **61.** $a^2 + ab$ **63.** $a^6 - a^5b + a^4b^2 - a^3b^3 + a^2b^4 - ab^5 + b^6$ **65.** $-\frac{3}{2}$ **67.**   **69.**

Quick Quiz: Sections 6.1–6.6, p. 406

1. $\dfrac{8(t + 1)}{(t - 1)(2t - 1)}$ **2.** $\dfrac{(n - 1)(n + 7)}{(n - 2)(n + 1)}$
3. $x + \dfrac{-2x - 3}{x^2 + 1}$ **4.** No solution   **5.** 4, 5

Prepare to Move On, p. 406

1. 2   **2.** 12   **3.** -6 **4.** 2   **5.** -160

Exercise Set 6.7, pp. 410–411

1. True   **2.** True   **3.** False   **4.** False   **5.** True
6. True   **7.** $x^2 - 3x - 5$ **9.** $a + 5 + \dfrac{-4}{a + 3}$
11. $2x^2 - 5x + 3 + \dfrac{8}{x + 2}$ **13.** $a^2 + 2a - 6$
15. $3y^2 + 2y + 6 + \dfrac{-2}{y - 3}$ **17.** $x^4 + 2x^3 + 4x^2 + 8x + 16$
19. $3x^2 + 6x - 3 + \dfrac{2}{x + \frac{1}{3}}$ **21.** 6   **23.** 125   **25.** 0
27.   **29.** $f(x) = 3x - 4$ **30.** $f(x) = \frac{1}{2}x + 6$

31. $f(x) = -\frac{5}{3}x + \frac{26}{3}$ **32.** $f(x) = 2x - 16$
33. $f(x) = -\frac{1}{2}x + 6$ **34.** $f(x) = 5$ **35.** ☞
37. **(a)** The degree of R must be less than 1, the degree of
$x - r$; **(b)** Let $x = r$. Then
$$P(r) = (r - r) \cdot Q(r) + R$$
$$= 0 \cdot Q(r) + R$$
$$= R.$$
39. $0; -3, -\frac{5}{2}, \frac{3}{2}$ **41.** 〰 **43.** 0

Quick Quiz: Sections 6.1–6.7, p. 411

1. $\dfrac{-y(x - y)}{2x^2}$ **2.** $x^2 - x - 5$

3. $x^3 + 3x^2 + 4x + 12 + \dfrac{39}{x - 3}$ **4.** $-5, -1$

5. 9 km/h

Prepare to Move On, p. 411

1. $c = \dfrac{b}{a}$ **2.** $w = \dfrac{x - y}{z}$ **3.** $q = \dfrac{st}{p - r}$

4. $b = \dfrac{d}{a + c}$ **5.** $b = \dfrac{cd + d}{a - 3}$

Exercise Set 6.8, pp. 417–422

1. (d) **2.** (f) **3.** (e) **4.** (b) **5.** (a) **6.** (c)
7. Inverse **8.** Direct **9.** Direct **10.** Inverse
11. Inverse **12.** Direct **13.** $d = \dfrac{L}{f}$

15. $v_1 = \dfrac{2s}{t} - v_2$, or $\dfrac{2s - tv_2}{t}$ **17.** $b = \dfrac{at}{a - t}$

19. $g = \dfrac{Rs}{s - R}$ **21.** $n = \dfrac{IR}{E - Ir}$ **23.** $q = \dfrac{pf}{p - f}$

25. $t_1 = \dfrac{H}{Sm} + t_2$, or $\dfrac{H + Smt_2}{Sm}$ **27.** $r = \dfrac{Re}{E - e}$

29. $r = 1 - \dfrac{a}{S}$, or $\dfrac{S - a}{S}$ **31.** $a + b = \dfrac{f}{c^2}$

33. $r = \dfrac{A}{P} - 1$, or $\dfrac{A - P}{P}$ **35.** $t_1 = t_2 - \dfrac{d_2 - d_1}{v}$, or

$\dfrac{vt_2 - d_2 + d_1}{v}$ **37.** $t = \dfrac{ab}{b + a}$ **39.** $Q = \dfrac{2Tt - 2AT}{A - q}$

41. $w = \dfrac{4.15c - 98.42}{p + 0.082}$ **43.** $k = 6; y = 6x$

45. $k = 1.7; y = 1.7x$ **47.** $k = 10; y = 10x$

49. $k = 100; y = \dfrac{100}{x}$ **51.** $k = 44; y = \dfrac{44}{x}$

53. $k = 9; y = \dfrac{9}{x}$ **55.** $33\frac{1}{3}$ cm **57.** 3.5 hr

59. 12 people **61.** 600 tons **63.** 286 Hz
65. About 21 min **67.** About 33.06 **69.** $y = \frac{1}{2}x^2$

71. $y = \dfrac{5000}{x^2}$ **73.** $y = 1.5xz$ **75.** $y = \dfrac{4wx^2}{z}$

77. 61.3 ft **79.** 64 foot-candles **81.** About 57 mph
83. ☞ **85.** $4a - 7 + h$ **86.** $4a + 4h - 7$
87. $\{x \mid x$ is a real number $and\ x \neq -\frac{1}{2}\}$,
or $\left(-\infty, -\frac{1}{2}\right) \cup \left(-\frac{1}{2}, \infty\right)$ **88.** $\mathbb{R}$

89. $\{x \mid x \geq -4\}$, or $[-4, \infty)$
90. $\{x \mid x$ is a real number $and\ x \neq -1\ and\ x \neq 1\}$,
or $(-\infty, -1) \cup (-1, 1) \cup (1, \infty)$ **91.** ☞ **93.** 567 mi
95. Ratio is $\dfrac{a + 12}{a + 6}$; percent increase is $\dfrac{6}{a + 6} \cdot 100\%$, or
$\dfrac{600}{a + 6}\%$ **97.** $t_1 = t_2 + \dfrac{(d_2 - d_1)(t_4 - t_3)}{a(t_4 - t_2)(t_4 - t_3) + d_3 - d_4}$
99. The intensity is halved. **101.** About 1.7 m
103. $d(s) = \dfrac{28}{s}; 70$ yd

Quick Quiz: Sections 6.1–6.8, p. 422

1. $f(x) = \dfrac{x(x + 1)}{2x - 1}, x \neq 0, \frac{1}{2}, 1$ **2.** $\dfrac{3(x + 2)}{(x - 2)(x + 3)(x + 4)}$
3. -6 **4.** $\frac{3}{2}xy - 2 + x^2$ **5.** 40 kg

Prepare to Move On, p. 422

1. a^{12} **2.** $9t^{10}$ **3.** $x^2 + 4x + 4$ **4.** $9a^2 - 6a + 1$

Visualizing for Success, p. 423

1. A **2.** D **3.** J **4.** H **5.** I **6.** B **7.** E
8. F **9.** C **10.** G

Decision Making: Connection, p. 424

1. 51.2 gal **2.** $2250 **3.** $320; $720; $960 **4.** 🖱

Study Summary: Chapter 6, pp. 425–428

1. $\dfrac{x}{x + 5}, x \neq -5, 5$ **2.** $\dfrac{3(x - 2)^2}{4(x - 1)(2x - 1)}$ **3.** $\dfrac{5(t - 3)}{2(t + 1)}$
4. $\dfrac{9x + 5}{x + 3}$ **5.** $\dfrac{2(2t - 1)}{(t - 1)(t + 1)}$ **6.** $\frac{4}{7}$ **7.** $\frac{7}{2}$
8. $5\frac{1}{7}$ hr **9.** 10 mph **10.** $\frac{16}{3}x^4 + 3x^3 - \frac{9}{2}$
11. $x - 5 + \dfrac{1}{x - 4}$ **12.** 54 **13.** $y = 50x$
14. $y = \dfrac{40}{x}$ **15.** $y = \frac{1}{10}xz$

Review Exercises: Chapter 6, pp. 428–430

1. True **2.** True **3.** False **4.** False **5.** False
6. True **7.** True **8.** False **9.** True **10.** True
11. **(a)** $-\frac{2}{9}$; **(b)** $-\frac{3}{4}$; **(c)** 0 **12.** $120x^3$
13. $(x + 10)(x - 2)(x - 3)$ **14.** $x + 8$ **15.** $\dfrac{20b^2c^6d^2}{3a^5}$
16. $\dfrac{15np + 14m}{18m^2n^4p^2}$ **17.** $\dfrac{(x - 2)(x + 5)}{x - 5}$
18. $\dfrac{(x^2 + 4x + 16)(x - 6)}{(x - 2)^2}$ **19.** $\dfrac{x - 3}{(x + 1)(x + 3)}$
20. $\dfrac{x - y}{x + y}$ **21.** $5(a + b)$ **22.** $\dfrac{-y}{(y + 4)(y - 1)}$
23. $f(x) = \dfrac{1}{x - 1}, x \neq 1, 4$

24. $f(x) = \dfrac{2}{x - 8}, x \neq -8, -5, 8$

25. $f(x) = \dfrac{3x - 1}{x - 3}, x \neq -3, -\frac{1}{3}, 3$ **26.** $\frac{4}{9}$

27. $\dfrac{a^2b^2}{2(b^2 - ba + a^2)}$ **28.** $\dfrac{(y + 11)(y + 5)}{(y - 5)(y + 2)}$

29. $\dfrac{(14 - 3x)(x + 3)}{2x^2 + 16x + 6}$ **30.** 2   **31.** 6

32. No solution   **33.** 0   **34.** $\frac{1}{2}, 5$ **35.** $-1, 4$

36. $5\frac{1}{7}$ hr   **37.** Ben: 30 hr; Jon: 45 hr   **38.** 24 mph

39. Jennifer: 70 mph; Elizabeth: 62 mph

40. $3s^2 + \frac{5}{2}s - 2rs^2$ **41.** $y^2 - 2y + 4$

42. $4x + 3 + \dfrac{-9x - 5}{x^2 + 1}$ **43.** $x^2 + 6x + 20 + \dfrac{54}{x - 3}$

44. 341   **45.** $r = \dfrac{2V - IR}{2I}$, or $\dfrac{V}{I} - \dfrac{R}{2}$

46. $m = \dfrac{H}{S(t_1 - t_2)}$ **47.** $c = \dfrac{b + 3a}{2}$

48. $t_1 = \dfrac{-A}{vT} + t_2$, or $\dfrac{-A + vTt_2}{vT}$ **49.** 20 cm

50. 56 in.   **51.** About 2.9 sec

52. ✏️ The least common denominator was used to add and subtract rational expressions, to simplify complex rational expressions, and to solve rational equations.

53. ✏️ A rational *expression* is a quotient of two polynomials. Expressions can be simplified, multiplied, or added, but they cannot be solved for a variable. A rational *equation* is an equation containing rational expressions. In a rational equation, we often can solve for a variable.

54. All real numbers except 0 and 13

55. 45   **56.** $9\frac{9}{19}$ hr

Test: Chapter 6, p. 431

1. [6.1] $\dfrac{5}{4(t - 1)}$ **2.** [6.1] $\dfrac{x^2 - 3x + 9}{x + 4}$

3. [6.2] $\dfrac{25x + x^3}{x + 5}$ **4.** [6.2] $3(a - b)$

5. [6.2] $\dfrac{a^3 - a^2b + 4ab + ab^2 - b^3}{(a - b)(a + b)}$

6. [6.2] $\dfrac{-2(2x^2 + 5x + 20)}{(x - 4)(x + 4)(x^2 + 4x + 16)}$

7. [6.2] $f(x) = \dfrac{x - 4}{(x + 3)(x - 2)}, x \neq -3, 2$

8. [6.1] $f(x) = \dfrac{(x - 1)^2(x + 1)}{x(x - 2)}, x \neq -2, 0, 1, 2$

9. [6.3] $\dfrac{a(2b + 3a)}{5a + b}$ **10.** [6.3] $\dfrac{(x - 9)(x - 6)}{(x + 6)(x - 3)}$

11. [6.3] $x - 8$ **12.** [6.4] $\frac{8}{3}$ **13.** [6.4] 15

14. [6.4] $-3, 5$ **15.** [6.1] $-5; -\frac{1}{2}$ **16.** [6.4] $\frac{5}{3}$

17. [6.6] $4a^2b^2c - \frac{5}{2}a^3bc^2 + 3bc$

18. [6.6] $y - 14 + \dfrac{-20}{y - 6}$ **19.** [6.6] $6x^2 - 9 + \dfrac{5x + 22}{x^2 + 2}$

20. [6.7] $x^2 + 7x + 18 + \dfrac{29}{x - 2}$ **21.** [6.7] 449

22. [6.8] $s = \dfrac{Rg}{g - R}$ **23.** [6.5] $2\frac{2}{9}$ hr   **24.** [6.5] $3\frac{3}{11}$ mph

25. [6.5] Tyler: 4 hr; Katie: 10 hr   **26.** [6.8] 30 workers

27. [6.8] 637 in²   **28.** [6.4] $-\frac{19}{3}$

29. [6.4] $\{x \mid x$ is a real number *and* $x \neq 0$ *and* $x \neq 15\}$

30. [6.3] a **31.** [6.5] Alex: 56 lawns; Ryan: 42 lawns

Cumulative Review: Chapters 1–6, p. 432

1. Slope: $\frac{7}{4}$; y-intercept: $(0, -3)$ **2.** $y = -2x + 5$

3. (a) 0; (b) $\{x \mid x$ is a real number *and* $x \neq 5$ *and* $x \neq 6\}$

4. $[9, \infty)$

5.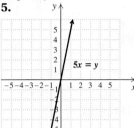

6.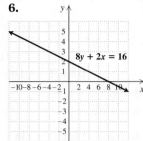

7.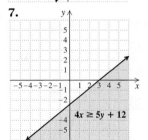

8.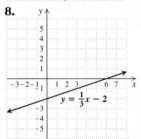

9. $9x^4 + 6x^2y + y^2$ **10.** $4x^4 - 81$ **11.** $\dfrac{y - 6}{2}$

12. $x - 1$ **13.** $\dfrac{a^2 + 7ab + b^2}{(a - b)(a + b)}$ **14.** $\dfrac{y - x}{xy(x + y)}$

15. $9x^2 - 13x + 26 + \dfrac{-50}{x + 2}$ **16.** $(x - 6)(x + 14)$

17. $(4y - 5)(4y + 5)$ **18.** $8(2x + 1)(4x^2 - 2x + 1)$

19. $(t - 8)^2$ **20.** $\left(\frac{1}{2}b - c\right)\left(\frac{1}{4}b^2 + \frac{1}{2}bc + c^2\right)$

21. $(3t - 4)(t + 7)$ **22.** $\frac{1}{4}$ **23.** $-12, 12$

24. $\{x \mid x \geq -1\}$, or $[-1, \infty)$

25. $\{x \mid -5 < x < -1\}$, or $(-5, -1)$

26. $\{x \mid x < -6.4 \text{ or } x > 6.4\}$, or $(-\infty, -6.4) \cup (6.4, \infty)$

27. -1 **28.** $(-3, 4)$ **29.** $(-2, -3, 1)$ **30.** $a = \dfrac{Pb}{4 - P}$

31. Himalayan Diamonds: $6\frac{2}{3}$ lb; Alpine Gold: $13\frac{1}{3}$ lb

32. 72 in.   **33.** IQAir HealthPro: 10 min; Austin Healthmate: 15 min   **34.** $22\frac{1}{2}$ min

35. $\{x \mid -3 \leq x \leq -1 \text{ or } 7 \leq x \leq 9\}$, or $[-3, -1] \cup [7, 9]$

36. All real numbers except 9 and -5 **37.** $-\frac{1}{4}, 0, \frac{1}{4}$

CHAPTER 7

Technology Connection, p. 436

1. False   **2.** True   **3.** False

Exercise Set 7.1, pp. 439–442

1. Two   **2.** Negative   **3.** Positive   **4.** Negative

5. Irrational   **6.** Real   **7.** Nonnegative   **8.** Negative

9. $8, -8$ **11.** $10, -10$ **13.** $20, -20$ **15.** $25, -25$

17. 7   **19.** -4 **21.** $\frac{6}{7}$ **23.** $-\frac{4}{9}$ **25.** 0.2
27. 0.09   **29.** $\sqrt{5}$; 0; does not exist; does not exist
31. -7; does not exist; -1; does not exist
33. 1; $\sqrt{2}$; $\sqrt{101}$ **35.** $10|x|$ **37.** $|-4b|$, or $4|b|$
39. $|8 - t|$ **41.** $|y + 8|$ **43.** $|2x + 7|$ **45.** $|a^{11}|$
47. Cannot be simplified   **49.** -1 **51.** -4 **53.** $5y$
55. $p^2 + 4$; 2   **57.** $\dfrac{x}{y + 4}$; 5   **59.** -4 **61.** $\frac{2}{3}$
63. $|x|$ **65.** t **67.** $6|a|$ **69.** 6   **71.** $|a + b|$
73. $4x$ **75.** $-3t$ **77.** $5b$ **79.** $a + 1$ **81.** $2x$
83. $x - 1$ **85.** t^9 **87.** $(x - 2)^4$ **89.** 2; 3; -2; -4
91. 2; does not exist; does not exist; 3   **93.** $\{x \mid x \geq 6\}$, or
$[6, \infty)$ **95.** $\{t \mid t \geq -8\}$, or $[-8, \infty)$ **97.** $\{x \mid x \leq 5\}$,
or $(-\infty, 5]$ **99.** $\mathbb{R}$ **101.** $\left\{z \mid z \geq -\frac{2}{5}\right\}$, or
$\left[-\frac{2}{5}, \infty\right)$ **103.** $\mathbb{R}$ **105.** 🔲   **107.** 0   **108.** $\mathbb{R}$
109. $\{x \mid x \neq 0\}$, or $(-\infty, 0) \cup (0, \infty)$

110. $(f + g)(x) = 3x - 1 + \dfrac{1}{x}$ **111.** $(fg)(x) = 3 - \dfrac{1}{x}$

112. f **113.** 🔲   **115.** About 840 GPM
117. About 1404 species
119. $\{x \mid x \geq -5\}$, or $[-5, \infty)$ **121.** $\{x \mid x \geq 0\}$, or $[0, \infty)$

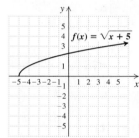

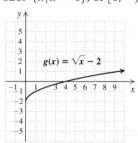

123. $\{x \mid -3 \leq x < 2\}$, or $[-3, 2)$
125. $\{x \mid x < -1 \text{ or } x > 6\}$, or $(-\infty, -1) \cup (6, \infty)$ **127.** 89%

Prepare to Move On, p. 442

1. $15x^3y^9$ **2.** $\dfrac{a^3}{8b^6c^3}$ **3.** $\dfrac{x^3}{2y^6}$ **4.** $\dfrac{y^4z^8}{16x^4}$

Technology Connection, p. 444

1. Without parentheses, the expression entered would be $\dfrac{7^2}{3}$.

2. For $x = 0$ or $x = 1$, $y_1 = y_2 = y_3$; on $(0, 1)$, $y_1 > y_2 > y_3$;
on $(1, \infty)$, $y_1 < y_2 < y_3$.

Technology Connection, p. 446

1. Many graphing calculators do not have keys for radicals
of index 3 or higher. On those graphing calculators that
offer $\sqrt[x]{\ }$ in a MATH menu, rational exponents still require
fewer keystrokes.

Exercise Set 7.2, pp. 446–449

1. Radical   **2.** Subtract   **3.** Equivalent
4. Rational   **5.** (g)   **6.** (c)   **7.** (e)   **8.** (h)
9. (a)   **10.** (d)   **11.** (b)   **12.** (f)   **13.** $\sqrt[3]{y}$
15. 6   **17.** 2   **19.** 8   **21.** $\sqrt{xyz}$ **23.** $\sqrt[5]{a^2b^2}$
25. $\sqrt[6]{t^5}$ **27.** 8   **29.** 625   **31.** $27\sqrt[4]{x^3}$

33. $125x^6$ **35.** $18^{1/3}$ **37.** $30^{1/2}$ **39.** $x^{7/2}$
41. $m^{2/5}$ **43.** $(xy)^{1/4}$ **45.** $(xy^2z)^{1/5}$ **47.** $(3mn)^{3/2}$

49. $(8x^2y)^{5/7}$ **51.** $\dfrac{2x}{z^{2/3}}$ **53.** $\frac{1}{2}$ **55.** $\dfrac{1}{(2rs)^{3/4}}$

57. 8   **59.** $8a^{3/5}c$ **61.** $\dfrac{2a^{3/4}c^{2/3}}{b^{1/2}}$ **63.** $\dfrac{a^3}{3^{5/2}b^{7/3}}$

65. $\left(\dfrac{3c}{2ab}\right)^{5/6}$ **67.** $\dfrac{x}{y^{1/4}}$ **69.** $11^{5/6}$ **71.** $3^{3/4}$

73. $4.3^{1/2}$ **75.** $10^{6/25}$ **77.** $a^{23/12}$ **79.** 64   **81.** $\dfrac{m^{1/3}}{n^{1/8}}$

83. $\sqrt[3]{x}$ **85.** y^5 **87.** $\sqrt{a}$ **89.** x^2y^2 **91.** $\sqrt{7a}$
93. $\sqrt[4]{8x^3}$ **95.** $\sqrt[10]{m}$ **97.** x^3y^3 **99.** a^6b^{12}
101. $\sqrt[12]{xy}$ **103.** 🔲   **105.** -6 **106.** $\left\{y \mid y < \frac{6}{5}\right\}$, or
$\left(-\infty, \frac{6}{5}\right)$ **107.** $\left\{x \mid -2 \leq x \leq \frac{3}{5}\right\}$, or $\left[-2, \frac{3}{5}\right]$
108. $-2, 3$ **109.** $-5, 3$ **110.** $\left(\frac{4}{3}, -\frac{1}{3}\right)$ **111.** 🔲
113. $\sqrt[6]{x^5}$ **115.** $\sqrt[7]{c - d}$, $c \geq d$
117. $2^{7/12} \approx 1.498 \approx 1.5$ **119.** (a) 1.8 m; (b) 3.1 m;
(c) 1.5 m; (d) 5.3 m   **121.** 338 cubic feet
123. Approximately 0.99 m^2 **125.** 🖼, 🔲

Quick Quiz: Sections 7.1–7.2, p. 449

1. -9 **2.** $|x|$ **3.** 2   **4.** 4   **5.** 5

Prepare to Move On, p. 449

1. $x^2 - 25$ **2.** $x^3 - 8$ **3.** $(3a - 4)^2$ **4.** $3(n + 2)^2$

Technology Connection, p. 450

1. The graphs differ in appearance because the domain of
y_1 is the intersection of $[-3, \infty)$ and $[3, \infty)$, or $[3, \infty)$. The
domain of y_2 is $(-\infty, -3] \cup [3, \infty)$.

Exercise Set 7.3, pp. 453–455

1. True   **2.** False   **3.** False   **4.** True   **5.** True
6. True   **7.** $\sqrt{30}$ **9.** $\sqrt[3]{35}$ **11.** $\sqrt[4]{54}$
13. $\sqrt{26xy}$ **15.** $\sqrt[5]{80y^4}$ **17.** $\sqrt{y^2 - b^2}$
19. $\sqrt[3]{0.21y^2}$ **21.** $\sqrt[5]{(x - 2)^3}$ **23.** $\sqrt{\dfrac{6s}{11t}}$
25. $\sqrt[7]{\dfrac{5x - 15}{4x + 8}}$ **27.** $2\sqrt{3}$ **29.** $3\sqrt{5}$ **31.** $2x^4\sqrt{2x}$
33. $2\sqrt{30}$ **35.** $6a^2\sqrt{b}$ **37.** $2x\sqrt[3]{y^2}$ **39.** $-2x^2\sqrt[3]{2}$
41. $f(x) = 2x^2\sqrt[3]{5}$ **43.** $f(x) = |7(x - 3)|$, or $7|x - 3|$
45. $f(x) = |x - 1|\sqrt{5}$ **47.** $a^5b^5\sqrt{b}$ **49.** $xy^2z^3\sqrt[3]{x^2z}$
51. $2xy^2\sqrt[4]{xy^3}$ **53.** $x^2yz^3\sqrt[5]{x^3y^3z^2}$ **55.** $-2a^4\sqrt[3]{10a^2}$
57. $5\sqrt{2}$ **59.** $3\sqrt{22}$ **61.** 3   **63.** $24y^5$ **65.** $a\sqrt[3]{10}$
67. $24x^3\sqrt{5x}$ **69.** $s^2t^3\sqrt[3]{t}$ **71.** $(x - y)^4$
73. $2ab^3\sqrt[4]{5a}$ **75.** $x(y + z)^2\sqrt[5]{x}$ **77.** 🔲
79. $9abx^2$ **80.** $\dfrac{(x - 1)^2}{(x - 2)^2}$ **81.** $\dfrac{x + 1}{2(x + 5)}$
82. $\dfrac{3(4x^2 + 5y^3)}{50xy^2}$ **83.** $\dfrac{b + a}{a^2b^2}$ **84.** $\dfrac{-x - 2}{4x + 3}$ **85.** 🔲
87. 175.6 mi   **89.** (a) $-3.3°$C; (b) $-16.6°$C; (c) $-25.5°$C;
(d) $-54.0°$C   **91.** $25x^5\sqrt[3]{25x}$ **93.** $a^{10}b^{17}\sqrt{ab}$

95. $f(x) = h(x); f(x) \neq g(x)$

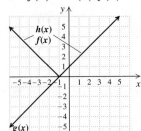

97. $\{x \,|\, x \leq 2 \text{ or } x \geq 4\}$, or $(-\infty, 2] \cup [4, \infty)$ **99.** 6
101. 🖼️ 📋

Quick Quiz: Sections 7.1–7.3, p. 455

1. $(5 - y)^5$ **2.** $2xy^2 \sqrt[3]{5x^2}$ **3.** $\sqrt[5]{c}$ **4.** $3^{9/10}$
5. $2xy$

Prepare to Move On, p. 455

1. $41ab$ **2.** $\dfrac{15m^2}{8n^3}$ **3.** $17xy^4$ **4.** $15x^4$

Exercise Set 7.4, pp. 459–461

1. Quotient rule for radicals   **2.** Multiplying by 1
3. Multiplying by 1   **4.** Quotient rule for radicals
5. (f)   **6.** (a)   **7.** (e)   **8.** (b)   **9.** (c)   **10.** (d)
11. $\frac{7}{10}$ **13.** $\frac{5}{2}$ **15.** $\dfrac{11}{t}$ **17.** $\dfrac{6y\sqrt{y}}{x^2}$ **19.** $\dfrac{3a\sqrt[3]{a}}{2b}$
21. $\dfrac{2a}{bc^2}$ **23.** $\dfrac{ab^2}{c^2}\sqrt[4]{\dfrac{a}{c^2}}$ **25.** $\dfrac{2x}{y^2}\sqrt[5]{\dfrac{x}{y}}$ **27.** $\dfrac{xy}{z^2}\sqrt[6]{\dfrac{y^2}{z^3}}$
29. 3   **31.** $\sqrt[3]{2}$ **33.** $y\sqrt{5y}$ **35.** $2\sqrt[3]{a^2b}$
37. $\sqrt{2ab}$ **39.** $2x^2y^3\sqrt[4]{y^3}$ **41.** $\sqrt[3]{x^2 + xy + y^2}$
43. $\dfrac{\sqrt{10}}{5}$ **45.** $\dfrac{2\sqrt{15}}{21}$ **47.** $\dfrac{\sqrt[3]{10}}{2}$ **49.** $\dfrac{\sqrt[3]{75ac^2}}{5c}$
51. $\dfrac{y\sqrt[4]{45y^2x^3}}{3x}$ **53.** $\dfrac{\sqrt[3]{2xy^2}}{xy}$ **55.** $\dfrac{\sqrt{14a}}{6}$
57. $\dfrac{\sqrt[5]{9y^4}}{2xy}$ **59.** $\dfrac{\sqrt{5b}}{6a}$ **61.** $\dfrac{5}{\sqrt{55}}$ **63.** $\dfrac{12}{5\sqrt{42}}$
65. $\dfrac{2}{\sqrt{6x}}$ **67.** $\dfrac{7}{\sqrt[3]{98}}$ **69.** $\dfrac{7x}{\sqrt{21xy}}$ **71.** $\dfrac{2a^2}{\sqrt[3]{20ab}}$
73. $\dfrac{x^2y}{\sqrt{2xy}}$ **75.** 📋   **77.** $-\frac{1}{3}$ **78.** $\frac{4}{27}$ **79.** -65
80. $9x^3 - x^2 + 9x$ **81.** $12x^2 - 12x + 6 + \dfrac{-14}{x+1}$
82. $49m^2 - 28mn + 4n^2$ **83.** 📋   **85.** (a) 1.62 sec;
(b) 1.99 sec; (c) 2.20 sec   **87.** $9\sqrt[3]{9n^2}$
89. $\dfrac{-3\sqrt{a^2 - 3}}{a^2 - 3}$, or $\dfrac{-3}{\sqrt{a^2 - 3}}$ **91.** Step 1: $\sqrt[n]{a} = a^{1/n}$,
by definition; Step 2: $\left(\dfrac{a}{b}\right)^n = \dfrac{a^n}{b^n}$, raising a quotient to a
power; Step 3: $a^{1/n} = \sqrt[n]{a}$, by definition
93. $(f/g)(x) = 3x$, where x is a real number and
$x > 0$ **95.** $(f/g)(x) = \sqrt{x + 3}$, where x is a real
number and $x > 3$

Quick Quiz: Sections 7.1–7.4, p. 461

1. $\frac{3}{10}$ **2.** $x^{1/3}$ **3.** $6mn^2\sqrt{10mp}$ **4.** $4x\sqrt{x}$
5. $\{x \,|\, x \geq -8\}$, or $[-8, \infty)$

Prepare to Move On, p. 461

1. $-\frac{1}{4}$ **2.** $\frac{11}{18}$ **3.** $a^2 - b^2$ **4.** $a^4 - 4y^2$
5. $15 + 4x - 4x^2$ **6.** $6x^6 + 6x^5$

Connecting the Concepts, p. 464

1. $\dfrac{6\sqrt{7}}{7}$ **2.** $\dfrac{3 + \sqrt{2}}{7}$ **3.** $\dfrac{2\sqrt{xy}}{xy}$ **4.** $\dfrac{5\sqrt{2}}{4}$
5. $\dfrac{\sqrt{10} - \sqrt{6}}{2}$ **6.** $\dfrac{-1 - \sqrt{5}}{2}$ **7.** $\dfrac{\sqrt[3]{xy^2}}{xy}$
8. $\dfrac{\sqrt[4]{ab^2}}{b}$

Exercise Set 7.5, pp. 466–468

1. Radicands; indices   **2.** Indices   **3.** Bases
4. Denominators   **5.** Numerator; conjugate
6. Bases   **7.** $11\sqrt{3}$ **9.** $2\sqrt[3]{4}$ **11.** $10\sqrt[3]{y}$
13. $12\sqrt{2}$ **15.** $13\sqrt[3]{7} + \sqrt{3}$ **17.** $9\sqrt{3}$
19. $-7\sqrt{5}$ **21.** $9\sqrt[3]{2}$ **23.** $(1 + 12a)\sqrt{a}$
25. $(x - 2)\sqrt[3]{6x}$ **27.** $3\sqrt{a - 1}$ **29.** $(x + 3)\sqrt{x - 1}$
31. $5\sqrt{2} + 2$ **33.** $3\sqrt{30} - 3\sqrt{35}$ **35.** $6\sqrt{5} - 4$
37. $3 - 4\sqrt[3]{63}$ **39.** $a + 2a\sqrt[3]{3}$ **41.** $4 + 3\sqrt{6}$
43. $\sqrt{6} - \sqrt{14} + \sqrt{21} - 7$ **45.** 1   **47.** -5
49. $2 - 8\sqrt{35}$ **51.** $23 + 8\sqrt{7}$ **53.** $5 - 2\sqrt{6}$
55. $2t + 5 + 2\sqrt{10t}$ **57.** $14 + x - 6\sqrt{x + 5}$
59. $6\sqrt[4]{63} + 4\sqrt[4]{35} - 3\sqrt[4]{54} - 2\sqrt[4]{30}$
61. $\dfrac{18 + 6\sqrt{2}}{7}$ **63.** $\dfrac{12 - 2\sqrt{3} + 6\sqrt{5} - \sqrt{15}}{33}$
65. $\dfrac{a - \sqrt{ab}}{a - b}$ **67.** -1 **69.** $\dfrac{12 - 3\sqrt{10} - 2\sqrt{14} + \sqrt{35}}{6}$
71. $\dfrac{1}{\sqrt{5} - 1}$ **73.** $\dfrac{2}{14 + 2\sqrt{3} + 3\sqrt{2} + 7\sqrt{6}}$
75. $\dfrac{x - y}{x + 2\sqrt{xy} + y}$ **77.** $\dfrac{1}{\sqrt{a + h} + \sqrt{a}}$ **79.** $\sqrt{a}$
81. $b^2\sqrt[10]{b^3}$ **83.** $xy\sqrt[6]{xy^5}$ **85.** $3a^2b\sqrt[4]{ab}$
87. $a^2b^2c^2\sqrt[6]{a^2bc^2}$ **89.** $\sqrt[12]{a^5}$ **91.** $\sqrt[12]{x^2y^5}$ **93.** $\sqrt[10]{ab^9}$
95. $\sqrt[6]{(7 - y)^5}$ **97.** $\sqrt[12]{5} + 3x$ **99.** $x\sqrt[6]{xy^5} - \sqrt[15]{x^{13}y^{14}}$
101. $2m^2 + m\sqrt[4]{n} + 2m\sqrt[3]{n^2} + \sqrt[12]{n^{11}}$
103. $2\sqrt[4]{x^3} - \sqrt[12]{x^{11}}$ **105.** $x^2 - 7$ **107.** $11 - 6\sqrt{2}$
109. $27 + 6\sqrt{14}$ **111.** 📋   **113.** IV   **114.** 1
115. x-intercept: $(10, 0)$; y-intercept: $(0, -10)$
116. Slope: $-\frac{5}{3}$; y-intercept: $(0, \frac{1}{3})$ **117.** $y = -2x + 12$
118. $y = -2x - 8$ **119.** 📋   **121.** $f(x) = 2x\sqrt{x - 1}$
123. $f(x) = (x + 3x^2)\sqrt[4]{x - 1}$ **125.** $(7x^2 - 2y^2)\sqrt{x + y}$
127. $4x(y + z)^3\sqrt[6]{2x(y + z)}$ **129.** $1 - \sqrt{w}$
131. $(\sqrt{x} + \sqrt{5})(\sqrt{x} - \sqrt{5})$
133. $(\sqrt{x} + \sqrt{a})(\sqrt{x} - \sqrt{a})$ **135.** $2x - 2\sqrt{x^2 - 4}$

137. (a) $(A + B + \sqrt{2AB})(A + B - \sqrt{2AB}) =$
$(A + B)^2 - (\sqrt{2AB})^2 = A^2 + 2AB + B^2 - 2AB = A^2 + B^2$;
(b) $2AB$ must be a perfect square.

Quick Quiz: Sections 7.1–7.5, p. 468

1. $|10t + 1|$ **2.** $\sqrt[8]{(3xy)^7}$, or $(\sqrt[8]{3xy})^7$ **3.** $(17ab)^{1/2}$
4. $\dfrac{\sqrt[3]{10xy^2}}{2x^2y}$ **5.** $\dfrac{1}{2\sqrt{10} - \sqrt{30}}$

Prepare to Move On, p. 468

1. 42   **2.** $-7, 3$ **3.** 11   **4.** 3, 11

Mid-Chapter Review: Chapter 7, p. 469

1. $\sqrt{6x^9} \cdot \sqrt{2xy} = \sqrt{6x^9 \cdot 2xy}$
$= \sqrt{12x^{10}y}$
$= \sqrt{4x^{10} \cdot 3y}$
$= \sqrt{4x^{10}} \cdot \sqrt{3y}$
$= 2x^5\sqrt{3y}$
2. $\sqrt{12} - 3\sqrt{75} + \sqrt{8} = 2\sqrt{3} - 3 \cdot 5\sqrt{3} + 2\sqrt{2}$
$= 2\sqrt{3} - 15\sqrt{3} + 2\sqrt{2}$
$= -13\sqrt{3} + 2\sqrt{2}$
3. 9   **4.** $-\frac{3}{10}$ **5.** $|8t|$, or $8|t|$ **6.** x **7.** -4
8. $\{x \mid x \le 10\}$, or $(-\infty, 10]$ **9.** 4   **10.** $\sqrt[12]{a}$ **11.** y^8
12. $t + 5$ **13.** $-3a^4$ **14.** $3x\sqrt{10}$ **15.** $\frac{2}{3}$
16. $2\sqrt{15} - 3\sqrt{22}$ **17.** $\sqrt[8]{t}$ **18.** $\dfrac{a^2}{2}$ **19.** $-8\sqrt{3}$
20. -4 **21.** $25 + 10\sqrt{6}$ **22.** $2\sqrt{x-1}$
23. $xy\sqrt[10]{x^7y^3}$ **24.** $15\sqrt[3]{5}$ **25.** $6x^3y^2$

Technology Connection, p. 472

1. The x-coordinates of the points of intersection should approximate the solutions of the examples.

Exercise Set 7.6, pp. 474–476

1. Powers   **2.** Radical   **3.** Isolate   **4.** Even
5. True   **6.** True   **7.** False   **8.** True   **9.** 3
11. $\frac{16}{3}$ **13.** 20   **15.** -1 **17.** 5   **19.** 91
21. 0, 36   **23.** 100   **25.** -125 **27.** 16
29. No solution   **31.** $\frac{80}{3}$ **33.** 45   **35.** $-\frac{5}{3}$ **37.** 4
39. 1   **41.** $\frac{106}{27}$ **43.** 3, 7   **45.** $\frac{80}{9}$ **47.** -1
49. No solution   **51.** 2, 6   **53.** 2   **55.** 4
57. ☑   **59.** At least 84%   **60.** $-191{,}800$ permits per year   **61.** 4 mph   **62.** Swiss chocolate: 45 oz; whipping cream: 20 oz   **63.** ☑   **65.** About 68 psi   **67.** About 278 Hz   **69.** 524.8°C   **71.** $t = \dfrac{1}{9}\left(\dfrac{S^2 \cdot 2457}{1087.7^2} - 2617\right)$
73. $r = \dfrac{v^2h}{2gh - v^2}$ **75.** $-\frac{8}{9}$ **77.** $-8, 8$ **79.** 1, 8
81. $\left(\frac{1}{36}, 0\right)$, $(36, 0)$ **83.** ☒

Quick Quiz: Sections 7.1–7.6, p. 476

1. $11n$ **2.** $\sqrt[6]{a}$ **3.** $-\sqrt{3}$ **4.** $x\sqrt[10]{x}$ **5.** 1

Prepare to Move On, p. 476

1. Length: 200 ft; width: 15 ft   **2.** Base: 34 in.; height: 15 in.   **3.** Length: 30 yd; width: 16 yd   **4.** 13 cm

Exercise Set 7.7, pp. 482–486

1. (d)   **2.** (c)   **3.** (e)   **4.** (b)   **5.** (f)   **6.** (a)
7. $\sqrt{34}$; 5.831   **9.** $9\sqrt{2}$; 12.728   **11.** 8   **13.** 4 m
15. $\sqrt{19}$ in.; 4.359 in.   **17.** 1 m   **19.** 250 ft
21. $\sqrt{643{,}600}$ ft $= 20\sqrt{1609}$ ft; 802.247 ft   **23.** 24 in.
25. $(\sqrt{340} + 8)$ ft $= (2\sqrt{85} + 8)$ ft; 26.439 ft
27. $(110 - \sqrt{6500})$ paces $= (110 - 10\sqrt{65})$ paces; 29.377 paces   **29.** Leg $= 5$; hypotenuse $= 5\sqrt{2} \approx 7.071$
31. Shorter leg $= 7$; longer leg $= 7\sqrt{3} \approx 12.124$
33. Leg $= 5\sqrt{3} \approx 8.660$; hypotenuse $= 10\sqrt{3} \approx 17.321$
35. Both legs $= \dfrac{13\sqrt{2}}{2} \approx 9.192$ **37.** Leg $= 14\sqrt{3} \approx 24.249$; hypotenuse $= 28$ **39.** $5\sqrt{3} \approx 8.660$
41. $7\sqrt{2} \approx 9.899$ **43.** $\dfrac{15\sqrt{2}}{2} \approx 10.607$
45. $\sqrt{10{,}561}$ ft ≈ 102.767 ft   **47.** $\dfrac{1089}{4}\sqrt{3}$ ft$^2 \approx 471.551$ ft^2 **49.** $(0, -4)$, $(0, 4)$ **51.** 5
53. $\sqrt{10} \approx 3.162$ **55.** $\sqrt{200} = 10\sqrt{2} \approx 14.142$
57. 17.8   **59.** $\dfrac{\sqrt{13}}{6} \approx 0.601$ **61.** $\sqrt{12} \approx 3.464$
63. $\sqrt{101} \approx 10.050$ **65.** $(3, 4)$ **67.** $\left(\frac{7}{2}, \frac{7}{2}\right)$
69. $(-1, -3)$ **71.** $(0.7, 0)$ **73.** $\left(-\frac{1}{12}, \frac{1}{24}\right)$
75. $\left(\dfrac{\sqrt{2} + \sqrt{3}}{2}, \dfrac{3}{2}\right)$ **77.** ☑
79. 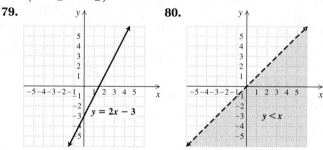   **80.**
81. 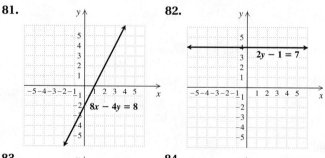   **82.**
83. 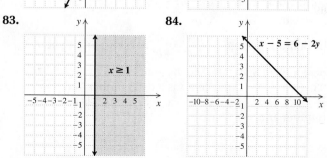   **84.**

85. ☑   **87.** $36\sqrt{3}$ cm²; 62.354 cm²
89. (a) $d = s + s\sqrt{2}$, or $d = s(1 + \sqrt{2})$;
(b) $A = 2s^2 + 2\sqrt{2}s^2$, or $A = 2s^2(1 + \sqrt{2})$
91. 60.28 ft by 60.28 ft   **93.** $\sqrt{75}$ cm

Quick Quiz: Sections 7.1–7.7, p. 486

1. $\dfrac{3x\sqrt[3]{2x}}{5t^2}$ **2.** $x\sqrt[4]{2x}(2 - x^2)$ **3.** $30x^2\sqrt{2}$
4. $\frac{144}{25}$ **5.** $\sqrt{85} \approx 9.220$

Prepare to Move On, p. 486

1. $2 + \sqrt{3}$ **2.** $\sqrt{7} - 6$ **3.** $9x^2 - 4y^2$
4. $25w^2 - 20wx + 4x^2$ **5.** $-8ac + 28c^2$
6. $24a^2 - 14ap - 5p^2$

Exercise Set 7.8, pp. 492–493

1. False   **2.** False   **3.** True   **4.** True   **5.** True
6. True   **7.** False   **8.** True   **9.** $10i$ **11.** $i\sqrt{5}$, or
$\sqrt{5}i$ **13.** $2i\sqrt{2}$, or $2\sqrt{2}i$ **15.** $-i\sqrt{11}$, or $-\sqrt{11}i$
17. $-7i$ **19.** $-10i\sqrt{3}$, or $-10\sqrt{3}i$ **21.** $6 - 2i\sqrt{21}$, or
$6 - 2\sqrt{21}i$ **23.** $(-2\sqrt{19} + 5\sqrt{5})i$ **25.** $(3\sqrt{2} - 8)i$
27. $5 - 3i$ **29.** $7 + 2i$ **31.** $2 - i$ **33.** $-12 - 5i$
35. -40 **37.** -24 **39.** -18 **41.** $-\sqrt{30}$
43. $-3\sqrt{14}$ **45.** $-30 + 10i$ **47.** $28 - 21i$
49. $1 + 5i$ **51.** $38 + 9i$ **53.** $2 - 46i$ **55.** 73
57. 50 **59.** $12 - 16i$ **61.** $-5 + 12i$ **63.** $-5 - 12i$
65. $3 - i$ **67.** $\frac{6}{13} + \frac{4}{13}i$ **69.** $\frac{3}{17} + \frac{5}{17}i$ **71.** $-\frac{5}{6}i$
73. $-\frac{3}{4} - \frac{5}{4}i$ **75.** $1 - 2i$ **77.** $-\frac{23}{58} + \frac{43}{58}i$
79. $\frac{19}{29} - \frac{4}{29}i$ **81.** $\frac{6}{25} - \frac{17}{25}i$ **83.** 1 **85.** $-i$
87. -1 **89.** i **91.** -1 **93.** $-125i$ **95.** 0
97. ☑   **99.** $(x + 10)(x - 10)$
100. $(t + 10)(t^2 - 10t + 100)$ **101.** $(x + 9)(x - 7)$
102. $a(4a - 3)(3a + 1)$ **103.** $(w + 2)(w - 2)(w + 3)$
104. $12x^2y^2(2x - 5y^2 - 1)$ **105.** ☑
107.

109. 5 **111.** $\sqrt{2}$
113. $-9 - 27i$
115. $50 - 120i$
117. $\frac{250}{41} + \frac{200}{41}i$ **119.** 8
121. $\frac{3}{5} + \frac{9}{5}i$ **123.** 1

Quick Quiz: Sections 7.1–7.8, p. 493

1. 3   **2.** $\left\{x \mid x \geq \frac{1}{2}\right\}$, or $\left[\frac{1}{2}, \infty\right)$ **3.** 25
4. $5 - 15\sqrt{6} - \sqrt{2} + 6\sqrt{3}$ **5.** $17 + i$

Prepare to Move On, p. 493

1. $-2, 3$ **2.** 5   **3.** $-5, 5$ **4.** $-\frac{2}{5}, \frac{4}{3}$

Visualizing for Success, p. 494

1. B   **2.** H   **3.** C   **4.** I   **5.** D   **6.** A
7. F   **8.** J   **9.** G   **10.** E

Decision Making: Connection, p. 495

1. Approximately 47.4 mi   **2.** 600 ft
3. (a) Approximately 12.2 mi; **(b)** approximately 24.4 mi;
(c) approximately 32.5 mi

Study Summary: Chapter 7, pp. 496–498

1. -9 **2.** -1 **3.** $|6x|$, or $6|x|$ **4.** x **5.** $\frac{1}{10}$
6. $\sqrt{21xy}$ **7.** $10x^2y^9\sqrt{2x}$ **8.** $\dfrac{2x\sqrt{3x}}{5}$ **9.** $\dfrac{\sqrt{6xy}}{3y}$
10. $-5\sqrt{2}$ **11.** $31 - 19\sqrt{3}$ **12.** $\dfrac{3\sqrt{15} - 5\sqrt{3}}{4}$
13. $x^2\sqrt[6]{x}$ **14.** 3   **15.** $\sqrt{51}$ m ≈ 7.141 m
16. $a = 6$ **17.** $b = 5\sqrt{3} \approx 8.660; c = 10$
18. $\sqrt{185} \approx 13.601$ **19.** $\left(2, -\frac{9}{2}\right)$ **20.** $-3 - 12i$
21. $3 - 2i$ **22.** $-32 - 26i$ **23.** i **24.** $-i$

Review Exercises: Chapter 7, pp. 499–500

1. True   **2.** False   **3.** False   **4.** True   **5.** True
6. True   **7.** True   **8.** False   **9.** $\frac{10}{11}$ **10.** -0.6
11. 5   **12.** $\{x \mid x \geq -10\}$, or $[-10, \infty)$ **13.** $8|t|$
14. $|c + 7|$ **15.** $|2x + 1|$ **16.** -2 **17.** $(5ab)^{4/3}$
18. $\sqrt[5]{3a^4}$ **19.** x^3y^5 **20.** $\sqrt[3]{x^2y}$ **21.** $\dfrac{1}{x^{2/5}}$
22. $7^{1/6}$ **23.** $f(x) = 5|x - 6|$ **24.** $2x^5y^2$
25. $5xy\sqrt{10x}$ **26.** $\sqrt{35ab}$ **27.** $3xb\sqrt[3]{x^2}$
28. $-6x^5y^4\sqrt[3]{2x^2}$ **29.** $y\sqrt[3]{6}$ **30.** $\dfrac{5\sqrt{x}}{2}$ **31.** $\dfrac{2a^2\sqrt[4]{3a^3}}{c^2}$
32. $7\sqrt[3]{4y}$ **33.** $\sqrt{3}$ **34.** $15\sqrt{2}$ **35.** -1
36. $\sqrt{15} + 4\sqrt{6} - 6\sqrt{10} - 48$ **37.** $\sqrt[4]{x^3}$ **38.** $\sqrt[12]{x^5}$
39. $4 - 4\sqrt{a} + a$ **40.** $\dfrac{\sqrt{2xy}}{4y}$ **41.** $-4\sqrt{10} + 4\sqrt{15}$
42. $\dfrac{20}{\sqrt{10} + \sqrt{15}}$ **43.** 19   **44.** -126 **45.** 4
46. 2   **47.** $5\sqrt{2}$ cm; 7.071 cm   **48.** $\sqrt{32}$ ft; 5.657 ft
49. Short leg $= 10$; long leg $= 10\sqrt{3} \approx 17.321$
50. $\sqrt{26} \approx 5.099$ **51.** $\left(-2, -\frac{3}{2}\right)$ **52.** $3i\sqrt{5}$, or $3\sqrt{5}i$
53. $-2 - 9i$ **54.** $6 + i$ **55.** 29   **56.** -1
57. $9 - 12i$ **58.** $\frac{13}{25} - \frac{34}{25}i$ **59.** ☑ A complex number
$a + bi$ is real when $b = 0$. It is imaginary when $b \neq 0$.
60. ☑ An absolute-value sign must be used to simplify
$\sqrt[n]{x^n}$ when n is even, since x may be negative. If x is negative while n is even, the radical expression cannot be
simplified to x, since $\sqrt[n]{x^n}$ represents the principal, or nonnegative, root. When n is odd, there is only one root, and
it will be positive or negative depending on the sign of x.
Thus there is no absolute-value sign when n is odd.
61. $\dfrac{2i}{3i}$; answers may vary   **62.** 3   **63.** $-\frac{2}{5} + \frac{9}{10}i$
64. The isosceles right triangle is larger by about 1.206 ft².

Test: Chapter 7, p. 501

1. [7.3] $5\sqrt{2}$ **2.** [7.4] $-\dfrac{2}{x^2}$ **3.** [7.1] $9|a|$

4. [7.1] $|x-4|$ **5.** [7.2] $(7xy)^{1/2}$ **6.** [7.2] $\sqrt[6]{(4a^3b)^5}$
7. [7.1] $\{x\,|\,x\ge 5\}$, or $[5,\infty)$ **8.** [7.5] $27+10\sqrt{2}$

9. [7.3] $2x^3y^2\sqrt[5]{x}$ **10.** [7.3] $2\sqrt[3]{2wv^2}$ **11.** [7.4] $\dfrac{10a^2}{3b^3}$

12. [7.4] $\sqrt[5]{3x^4y}$ **13.** [7.5] $x\sqrt[4]{x}$ **14.** [7.5] $\sqrt[5]{y^2}$
15. [7.5] $6\sqrt{2}$ **16.** [7.5] $9\sqrt{2xy}$

17. [7.5] $14-19\sqrt{x}-3x$ **18.** [7.4] $\dfrac{\sqrt[3]{2xy^2}}{2y}$

19. [7.6] 4   **20.** [7.6] $-1, 2$ **21.** [7.6] 8
22. [7.7] $\sqrt{10{,}600}$ ft ≈ 102.956 ft   **23.** [7.7] 5 cm;
$5\sqrt{3}$ cm ≈ 8.660 cm   **24.** [7.7] $\sqrt{17}\approx 4.123$
25. [7.7] $\left(\frac{3}{2},-6\right)$ **26.** [7.8] $5i\sqrt{2}$, or $5\sqrt{2}i$
27. [7.8] $12+2i$ **28.** [7.8] $15-8i$
29. [7.8] $-\frac{11}{34}-\frac{7}{34}i$ **30.** [7.8] i **31.** [7.6] 3
32. [7.8] $-\frac{17}{4}i$ **33.** [7.6] 22,500 ft

Cumulative Review: Chapters 1–7, p. 502

1. $-7, 5$ **2.** $\frac{5}{2}$ **3.** -1 **4.** $\frac{1}{5}$
5. $\{x\,|-3\le x\le 7\}$, or $[-3,7]$ **6.** $\mathbb{R}$, or $(-\infty,\infty)$
7. $-3, 2$ **8.** 7   **9.** $(1,-1,-2)$
10.

11.
$y = -x + 5$

12.
$x+y\le 2$

13.
$2x = y$

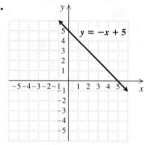

14. $y=7x-11$ **15.** 6   **16.** $4a^2-20ab+25b^2$

17. c^4-9d^2 **18.** $\dfrac{x+13}{(x-2)(x+1)}$ **19.** $\dfrac{(a+2)(2a+1)}{(a-3)(a-1)}$

20. $\dfrac{2x+1}{x^2}$ **21.** 0   **22.** $-1+3\sqrt{5}$ **23.** $\sqrt[15]{y^8}$

24. $(x-7)(x+2)$ **25.** $4y^5(y-1)(y^2+y+1)$
26. $(3t-8)(t+1)$ **27.** $(y-x)(t-z^2)$
28. $\{x\,|\,x$ is a real number *and* $x\ne 3\}$, or $(-\infty,3)\cup(3,\infty)$
29. $\{x\,|\,x\ge \frac{11}{2}\}$, or $[\frac{11}{2},\infty)$ **30.** $6-2\sqrt{5}$
31. $(f+g)(x)=x^2+\sqrt{2x-3}$
32. $2\sqrt{3}$ ft ≈ 3.464 ft   **33. (a)** $m(t)=0.08t+25.02$;
(b) 26.62; **(c)** 2037   **34.** \$36,650   **35.** 5 ft
36. $5x-3y=-15$ **37.** -2 **38.** 9

CHAPTER 8

Technology Connection, p. 511

1. x-intercepts should be approximations of
$-5/2 + \sqrt{37}/2$ and $-5/2 - \sqrt{37}/2$.
2. The graph of $y=x^2-6x+11$ has no x-intercepts.

Exercise Set 8.1, pp. 512–514

1. Quadratic function   **2.** Parabola   **3.** Standard form
4. Zero products   **5.** Square roots   **6.** Complete the
square   **7.** ± 10 **9.** $\pm 5\sqrt{2}$ **11.** $\pm\sqrt{6}$

13. $\pm\frac{7}{3}$ **15.** $\pm\sqrt{\dfrac{5}{6}}$, or $\pm\dfrac{\sqrt{30}}{6}$ **17.** $\pm i$ **19.** $\pm\frac{9}{2}i$

21. $-1, 7$ **23.** $-5\pm 2\sqrt{3}$ **25.** $-1\pm 3i$
27. $-\dfrac{3}{4}\pm\dfrac{\sqrt{17}}{4}$, or $\dfrac{-3\pm\sqrt{17}}{4}$ **29.** $-3, 13$ **31.** $\pm\sqrt{19}$

33. $1, 9$ **35.** $-4\pm\sqrt{13}$ **37.** $-14, 0$
39. $x^2+16x+64=(x+8)^2$
41. $t^2-10t+25=(t-5)^2$
43. $t^2-2t+1=(t-1)^2$ **45.** $x^2+3x+\frac{9}{4}=\left(x+\frac{3}{2}\right)^2$
47. $x^2+\frac{2}{5}x+\frac{1}{25}=\left(x+\frac{1}{5}\right)^2$
49. $t^2-\frac{5}{6}t+\frac{25}{144}=\left(t-\frac{5}{12}\right)^2$ **51.** $-7, 1$ **53.** $5\pm\sqrt{2}$
55. $-8, -4$ **57.** $-4\pm\sqrt{19}$
59. $(-3-\sqrt{2},0), (-3+\sqrt{2},0)$
61. $\left(-\dfrac{9}{2}-\dfrac{\sqrt{181}}{2},0\right), \left(-\dfrac{9}{2}+\dfrac{\sqrt{181}}{2},0\right)$, or
$\left(\dfrac{-9-\sqrt{181}}{2},0\right), \left(\dfrac{-9+\sqrt{181}}{2},0\right)$
63. $(5-\sqrt{47},0), (5+\sqrt{47},0)$ **65.** $-\frac{4}{3}, -\frac{2}{3}$
67. $-\frac{1}{3}, 2$ **69.** $-\dfrac{2}{5}\pm\dfrac{\sqrt{19}}{5}$, or $\dfrac{-2\pm\sqrt{19}}{5}$
71. $\left(-\dfrac{1}{4}-\dfrac{\sqrt{13}}{4},0\right), \left(-\dfrac{1}{4}+\dfrac{\sqrt{13}}{4},0\right)$, or $\left(\dfrac{-1-\sqrt{13}}{4},0\right)$,
$\left(\dfrac{-1+\sqrt{13}}{4},0\right)$ **73.** $\left(\dfrac{3}{4}-\dfrac{\sqrt{17}}{4},0\right), \left(\dfrac{3}{4}+\dfrac{\sqrt{17}}{4},0\right)$,
or $\left(\dfrac{3-\sqrt{17}}{4},0\right), \left(\dfrac{3+\sqrt{17}}{4},0\right)$ **75.** 5%   **77.** 4%

79. About 2.8 sec   **81.** About 15.0 sec   **83.** 🖎
85. $3y(y+10)(y-10)$ **86.** $(t+6)^2$
87. $6(x^2+x+1)$ **88.** $10a^3(a+2)(a-3)$
89. $(4x+3)(5x-2)$
90. $(n+1)(n^2-n+1)(n-1)(n^2+n+1)$
91. 🖎   **93.** ± 18 **95.** $-\frac{7}{2}, -\sqrt{5}, 0, \sqrt{5}, 8$
97. Barge: 8 km/h; fishing boat: 15 km/h   **99.** 〰
101. 🖎, 〰

Prepare to Move On, p. 514

1. 64   **2.** -15 **3.** $10\sqrt{2}$ **4.** $2i$
5. $2i\sqrt{2}$, or $2\sqrt{2}i$

Connecting the Concepts, p. 518

1. $-2, 5$ **2.** ± 11 **3.** $-3 \pm \sqrt{19}$ **4.** $-\frac{1}{2} \pm \frac{\sqrt{13}}{2}$

5. $-1 \pm \sqrt{2}$ **6.** 5 **7.** $1 \pm \sqrt{7}$ **8.** $\pm\frac{\sqrt{11}}{2}$

Exercise Set 8.2, pp. 519–520

1. True **2.** True **3.** False **4.** False **5.** False
6. True **7.** $-\frac{5}{2}, 1$ **9.** $-1 \pm \sqrt{5}$ **11.** $3 \pm \sqrt{6}$
13. $\frac{3}{2} \pm \frac{\sqrt{29}}{2}$ **15.** $-1 \pm \frac{2\sqrt{3}}{3}$ **17.** $-\frac{4}{3} \pm \frac{\sqrt{19}}{3}$
19. $3 \pm i$ **21.** $\frac{1}{2} \pm \frac{\sqrt{3}}{2}i$ **23.** $-2 \pm \sqrt{2}i$ **25.** $-\frac{8}{3}, \frac{5}{4}$
27. $\frac{2}{5}$ **29.** $-\frac{11}{8} \pm \frac{\sqrt{41}}{8}$ **31.** 5, 10 **33.** $\frac{3}{2}, 24$
35. $2 \pm \sqrt{5}i$ **37.** $2, -1 \pm \sqrt{3}i$ **39.** $-\frac{4}{3}, \frac{5}{2}$
41. $5 \pm \sqrt{53}$ **43.** $\frac{3}{2} \pm \frac{\sqrt{5}}{2}$ **45.** $-5.236, -0.764$
47. $0.764, 5.236$ **49.** $-1.266, 2.766$ **51.** ☐
53. 1 **54.** 1000 **55.** $x^{11/12}$ **56.** $\frac{1}{9}$ **57.** $\frac{3a^{10}c^7}{4}$
58. $\frac{9w^8}{4x^{10}}$ **59.** ☐ **61.** $(-2, 0), (1, 0)$
63. $4 - 2\sqrt{2}, 4 + 2\sqrt{2}$ **65.** $-1.179, 0.339$
67. $\frac{-5\sqrt{2} \pm \sqrt{34}}{4}$ **69.** $\frac{1}{2}$ **71.** ☐

Quick Quiz: Sections 8.1–8.2, p. 520

1. $-3, \frac{5}{2}$ **2.** $2 \pm \sqrt{3}$ **3.** $3 \pm \sqrt{5}$ **4.** $\frac{3}{2} \pm \frac{\sqrt{13}}{2}$
5. $-\frac{1}{4} \pm \frac{\sqrt{39}}{4}i$

Prepare to Move On, p. 520

1. $x^2 + 4$ **2.** $x^2 - 180$ **3.** $x^2 - 4x - 3$
4. $x^2 + 6x + 34$

Exercise Set 8.3, pp. 524–525

1. (b) **2.** (a) **3.** (d) **4.** (b) **5.** (c) **6.** (c)
7. Two irrational **9.** Two imaginary **11.** Two irrational
13. Two rational **15.** Two imaginary **17.** One rational
19. Two rational **21.** Two irrational
23. Two imaginary **25.** Two rational
27. Two irrational **29.** $x^2 + x - 20 = 0$
31. $x^2 - 6x + 9 = 0$ **33.** $x^2 + 4x + 3 = 0$
35. $4x^2 - 23x + 15 = 0$ **37.** $8x^2 + 6x + 1 = 0$
39. $x^2 - 2x - 0.96 = 0$ **41.** $x^2 - 3 = 0$
43. $x^2 - 20 = 0$ **45.** $x^2 + 16 = 0$
47. $x^2 - 4x + 53 = 0$ **49.** $x^2 - 6x - 5 = 0$
51. $3x^2 - 6x - 4 = 0$ **53.** $x^3 - 4x^2 - 7x + 10 = 0$

55. $x^3 - 2x^2 - 3x = 0$ **57.** ☐ **59.** $3a^3b^6\sqrt{30a}$
60. $2w^2\sqrt[4]{2w^2}$ **61.** $\sqrt[6]{x^5}$ **62.** $-\sqrt{6}$ **63.** $7 - i$
64. -1 **65.** ☐ **67.** $a = 1, b = 2, c = -3$
69. (a) $-\frac{3}{5}$; (b) $-\frac{1}{3}$ **71.** (a) $9 + 9i$; (b) $3 + 3i$
73. The solutions of $ax^2 + bx + c = 0$ are
$x = \frac{-b \pm \sqrt{b^2 - 4ac}}{2a}$. When there is just one solution,
$b^2 - 4ac$ must be 0, so $x = \frac{-b \pm 0}{2a} = \frac{-b}{2a}$.
75. $a = 8, b = 20, c = -12$ **77.** $x^2 - 2 = 0$
79. $x^4 - 8x^3 + 21x^2 - 2x - 52 = 0$ **81.** ☐, ☐

Quick Quiz: Sections 8.1–8.3, p. 525

1. $-8 \pm \sqrt{3}$ **2.** $\frac{1}{6} \pm \frac{\sqrt{13}}{6}$ **3.** $(-2\sqrt{2}, 0), (2\sqrt{2}, 0)$
4. $-2.414, 0.414$ **5.** Two irrational solutions

Prepare to Move On, p. 525

1. $c = \frac{d^2}{1 - d}$ **2.** $y = \frac{x - 3}{x}$, or $y = 1 - \frac{3}{x}$
3. 10 mph **4.** Homer: 3.5 mph; Gladys: 2 mph

Exercise Set 8.4, pp. 529–532

1. (c) **2.** (d) **3.** (a) **4.** (b)
5. $r = \frac{1}{2}\sqrt{\frac{A}{\pi}}$ **7.** $r = \frac{-\pi h + \sqrt{\pi^2 h^2 + 2\pi A}}{2\pi}$
9. $r = \frac{\sqrt{Gm_1 m_2}}{F}$ **11.** $H = \frac{c^2}{g}$
13. $b = \sqrt{c^2 - a^2}$ **15.** $t = \frac{-v_0 + \sqrt{(v_0)^2 + 2gs}}{g}$
17. $n = \frac{1 + \sqrt{1 + 8N}}{2}$ **19.** $d = \frac{I^2 s}{T^2}$
21. $t = \frac{-b \pm \sqrt{b^2 - 4ac}}{2a}$ **23.** (a) 10.1 sec; (b) 7.49 sec;
(c) 272.5 m **25.** 2.9 sec **27.** 0.872 sec **29.** 2.5 m/sec
31. 4.5% **33.** First part: 60 mph; second part: 50 mph
35. 40 mph **37.** Cessna: 150 mph, Beechcraft: 200 mph;
or Cessna: 200 mph, Beechcraft: 250 mph
39. To Hillsboro: 12 mph; return trip: 9 mph
41. About 14 mph **43.** 12 hr **45.** About 3.24 mph
47. About 11 days **49.** ☐ **51.** 45 **52.** $-\frac{1}{8}$
53. $-\frac{4}{3}, 0$ **54.** $\left(\frac{11}{5}, \frac{2}{5}\right)$ **55.** $\frac{233}{2}$ **56.** $\frac{1}{16}$ **57.** ☐
59. $t = \frac{-10.2 + 6\sqrt{-A^2 + 13A - 39.36}}{A - 6.5}$ **61.** $\pm\sqrt{2}$
63. $l = \frac{w + w\sqrt{5}}{2}$
65. $n = \pm\sqrt{\frac{r^2 \pm \sqrt{r^4 + 4m^4 r^2 p - 4mp}}{2m}}$
67. $A(S) = \frac{\pi S}{6}$

Quick Quiz: Sections 8.1–8.4, p. 532

1. $10 \pm \sqrt{115}$ **2.** $1 \pm \sqrt{26}$ **3.** 2%
4. $x^2 - 12 = 0$ **5.** $d = -1 \pm \sqrt{1 + n}$

Prepare to Move On, p. 532

1. m^{-2}, or $\dfrac{1}{m^2}$ **2.** $y^{1/3}$ **3.** 2   **4.** 81

Exercise Set 8.5, pp. 537–538

1. True   **2.** True   **3.** True   **4.** True   **5.** $\sqrt{p}$
6. $x^{1/4}$ **7.** $x^2 + 3$ **8.** t^{-3} **9.** $(1 + t)^2$ **10.** $w^{1/6}$

11. $\pm 2, \pm 3$ **13.** $\pm\sqrt{3}, \pm 2$ **15.** $\pm\dfrac{\sqrt{5}}{2}, \pm 1$

17. 4   **19.** $\pm 2\sqrt{2}, \pm 3$ **21.** $\pm 2, \pm 3i$ **23.** $\pm i, \pm 2i$
25. $8 + 2\sqrt{7}$ **27.** No solution   **29.** $-\frac{1}{2}, \frac{1}{3}$
31. $-4, 1$ **33.** $-27, 8$ **35.** 729   **37.** 1

39. No solution   **41.** $\dfrac{12}{5}$ **43.** $\left(\dfrac{4}{25}, 0\right)$

45. $\left(\dfrac{3}{2} + \dfrac{\sqrt{33}}{2}, 0\right), \left(\dfrac{3}{2} - \dfrac{\sqrt{33}}{2}, 0\right), (4, 0), (-1, 0)$

47. $(-243, 0), (32, 0)$ **49.** No x-intercepts   **51.**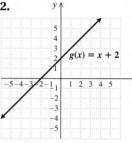
53.

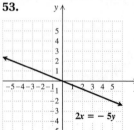
$2x = -5y$

54.

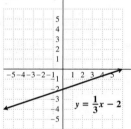
$y = \frac{1}{3}x - 2$

55.

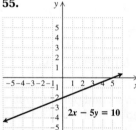
$2x - 5y = 10$

56.

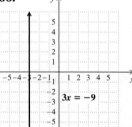
$3x = -9$

57.

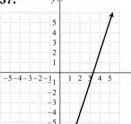

58.

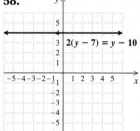
$2(y - 7) = y - 10$

59. 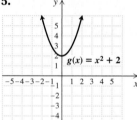   **61.** $\pm\sqrt{\dfrac{-5 \pm \sqrt{37}}{6}}$ **63.** $\dfrac{100}{99}$

65. 9   **67.** $-2, 1, 1 + \sqrt{3}i, 1 - \sqrt{3}i, -\dfrac{1}{2} + \dfrac{\sqrt{3}}{2}i,$
$-\dfrac{1}{2} - \dfrac{\sqrt{3}}{2}i$ **69.** $-3, -1, 1, 4$

Quick Quiz: Sections 8.1–8.5, p. 538

1. $\frac{1}{2}, 1$ **2.** $-\dfrac{1}{2} \pm \dfrac{\sqrt{3}}{2}i$ **3.** $-3, -1, 1, 3$
4. $\dfrac{7}{4} \pm \dfrac{\sqrt{33}}{4}$ **5.** $x = \dfrac{9c^2}{80y}$

Prepare to Move On, p. 538

1.

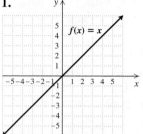
$f(x) = x$

2.

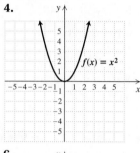
$g(x) = x + 2$

3.

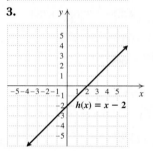
$h(x) = x - 2$

4.

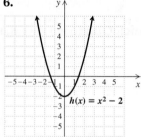
$f(x) = x^2$

5.

$g(x) = x^2 + 2$

6.

$h(x) = x^2 - 2$

Mid-Chapter Review: Chapter 8, p. 539

1. $x - 7 = \pm\sqrt{5}$
   $x = 7 \pm \sqrt{5}$
   The solutions are $7 + \sqrt{5}$ and $7 - \sqrt{5}$.
2. $a = 1, b = -2, c = -1$
   $x = \dfrac{-(-2) \pm \sqrt{(-2)^2 - 4 \cdot 1 \cdot (-1)}}{2 \cdot 1}$
   $x = \dfrac{2 \pm \sqrt{8}}{2}$
   $x = \dfrac{2}{2} \pm \dfrac{2\sqrt{2}}{2}$
   The solutions are $1 + \sqrt{2}$ and $1 - \sqrt{2}$.
3. $-7, 3$ **4.** ± 14 **5.** $1 \pm \sqrt{6}$ **6.** $1 \pm 2i$
7. $\pm 2, \pm 2i$ **8.** $-3 \pm \sqrt{7}$ **9.** $-5, 0$ **10.** $-\frac{5}{6}, 2$
11. $0, \frac{7}{16}$ **12.** $-\dfrac{5}{6} \pm \dfrac{\sqrt{37}}{6}$ **13.** $-6, 5$ **14.** $\pm\sqrt{2}, \pm 2i$
15. Two irrational   **16.** Two rational   **17.** Two imaginary
18. $v = 20\sqrt{\dfrac{F}{A}}$, or $\dfrac{20\sqrt{FA}}{A}$ **19.** $D = d + \sqrt{d^2 + 2hd}$
20. South: 75 mph; north: 45 mph

Technology Connection, p. 541

1. The graphs of y_1, y_2, and y_3 open upward. The graphs of y_4, y_5, and y_6 open downward. The graph of y_1 is wider than the graph of y_2. The graph of y_3 is narrower than the graph of y_2. Similarly, the graph of y_4 is wider than the graph of y_5, and the graph of y_6 is narrower than the graph of y_5.
2. If A is positive, the graph opens upward. If A is negative, the graph opens downward. Compared with the graph of $y = x^2$, the graph of $y = Ax^2$ is wider if $|A| < 1$ and narrower if $|A| > 1$.

Technology Connection, p. 543

1. Compared with the graph of $y = ax^2$, the graph of $y = a(x - h)^2$ is shifted left or right. It is shifted left if h is negative and right if h is positive.   **2.** The value of A makes the graph wider or narrower, and makes the graph open downward if A is negative. The value of B shifts the graph left or right.

Technology Connection, p. 544

1. The graph of y_2 looks like the graph of y_1 shifted up 2 units, and the graph of y_3 looks like the graph of y_1 shifted down 4 units.   **2.** Compared with the graph of $y = a(x - h)^2$, the graph of $y = a(x - h)^2 + k$ is shifted up or down $|k|$ units. It is shifted down if k is negative and up if k is positive.   **3.** The value of A makes the graph wider or narrower, and makes the graph open downward if A is negative. The value of B shifts the graph left or right. The value of C shifts the graph up or down.

Exercise Set 8.6, pp. 546–548

1. False   **2.** True   **3.** True   **4.** False
5.
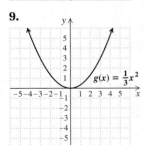
7.
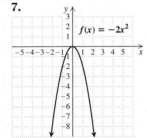
9.
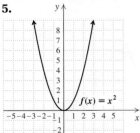
11.
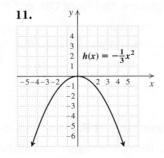

13.
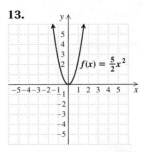

15. Vertex: $(-1, 0)$; axis of symmetry: $x = -1$
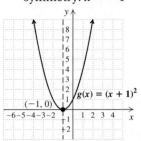

17. Vertex $(2, 0)$; axis of symmetry: $x = 2$
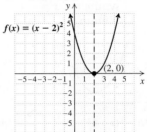

19. Vertex: $(-1, 0)$; axis of symmetry: $x = -1$
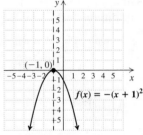

21. Vertex: $(2, 0)$; axis of symmetry: $x = 2$
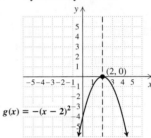

23. Vertex: $(-1, 0)$; axis of symmetry: $x = -1$
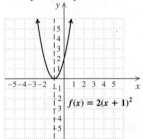

25. Vertex: $(4, 0)$; axis of symmetry: $x = 4$
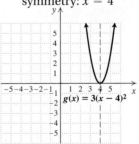

27. Vertex: $(4, 0)$; axis of symmetry: $x = 4$
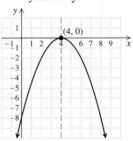

29. Vertex: $(1, 0)$; axis of symmetry: $x = 1$
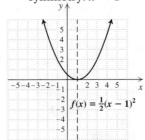

31. Vertex: $(-5, 0)$; axis of symmetry: $x = -5$
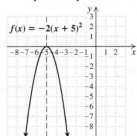

33. Vertex: $\left(\frac{1}{2}, 0\right)$; axis of symmetry: $x = \frac{1}{2}$

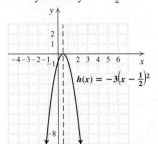

35. Vertex: $(5, 2)$; axis of symmetry: $x = 5$; minimum: 2

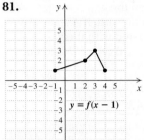

37. Vertex: $(-1, -3)$; axis of symmetry: $x = -1$; minimum: -3

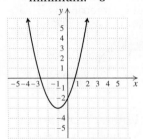

39. Vertex: $(-4, 1)$; axis of symmetry: $x = -4$; minimum: 1

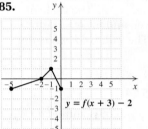

41. Vertex: $(1, -3)$; axis of symmetry: $x = 1$; maximum: -3

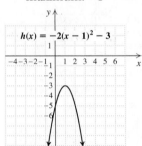

43. Vertex: $(-3, 1)$; axis of symmetry: $x = -3$; minimum: 1

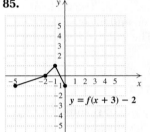

45. Vertex: $(2, 4)$; axis of symmetry: $x = 2$; maximum: 4

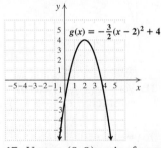

47. Vertex: $(3, 9)$; axis of symmetry: $x = 3$; minimum: 9
49. Vertex: $(-8, 2)$; axis of symmetry: $x = -8$; maximum: 2
51. Vertex: $\left(\frac{7}{2}, -\frac{29}{4}\right)$; axis of symmetry: $x = \frac{7}{2}$; minimum: $-\frac{29}{4}$
53. Vertex: $(-2.25, -\pi)$; axis of symmetry: $x = -2.25$;

maximum: $-\pi$ **55.** ✏   **57.** $\dfrac{x^2 + 3x + 6}{x(x + 2)}$

58. $8\sqrt{2x}$ **59.** $-\sqrt[3]{t} + 5\sqrt{t}$ **60.** $-7a^2 + 3a - 8$

61. $\dfrac{-x^2 + 4x + 1}{(x - 1)(x + 3)}$ **62.** $\dfrac{-3}{x + 4}$ **63.** ✏

65. $f(x) = \frac{3}{5}(x - 1)^2 + 3$ **67.** $f(x) = \frac{3}{5}(x - 4)^2 - 7$
69. $f(x) = \frac{3}{5}(x + 2)^2 - 5$ **71.** $f(x) = 2(x - 2)^2$
73. $g(x) = -2x^2 - 5$ **75.** The graph will move to the right.   **77.** The graph will be reflected across the x-axis.
79. $F(x) = 3(x - 5)^2 + 1$

81.

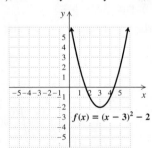

83.

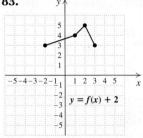

85.

87. ▨   **89.** ✏, ▨

Quick Quiz: Sections 8.1–8.6, p. 548

1. $-7 \pm \sqrt{13}$ **2.** $\frac{3}{2} \pm \frac{\sqrt{3}}{2}i$ **3.** $-\frac{2}{5}, \frac{1}{3}$

4. $z = \pm\sqrt{\dfrac{xy}{3t}}$, or $z = \pm\dfrac{\sqrt{3txy}}{3t}$

5. Vertex: $(3, -2)$; axis of symmetry: $x = 3$; minimum: -2

Prepare to Move On, p. 548

1. x-intercept: $(3, 0)$; y-intercept: $(0, -4)$
2. x-intercept: $\left(\frac{8}{3}, 0\right)$; y-intercept: $(0, 2)$
3. $(-5, 0), (-3, 0)$ **4.** $(-1, 0), \left(\frac{3}{2}, 0\right)$
5. $x^2 - 14x + 49 = (x - 7)^2$
6. $x^2 + 7x + \frac{49}{4} = \left(x + \frac{7}{2}\right)^2$

Exercise Set 8.7, pp. 553–555

1. True   **2.** False   **3.** True   **4.** True   **5.** False
6. True   **7.** False   **8.** True
9. $f(x) = (x - 4)^2 + (-14)$

11. $f(x) = \left(x - \left(-\frac{3}{2}\right)\right)^2 + \left(-\frac{29}{4}\right)$
13. $f(x) = 3(x - (-1))^2 + (-5)$
15. $f(x) = -(x - (-2))^2 + (-3)$
17. $f(x) = 2\left(x - \frac{5}{4}\right)^2 + \frac{55}{8}$
19. (a) Vertex: $(-2, 1)$; axis of symmetry: $x = -2$;
(b)

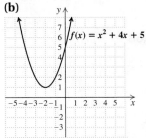

21. (a) Vertex: $(-4, 4)$; axis of symmetry: $x = -4$;
(b)

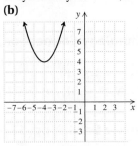

$f(x) = x^2 + 8x + 20$

23. (a) Vertex: $(4, -7)$; axis of symmetry: $x = 4$;
(b)

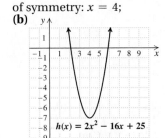

25. (a) Vertex: $(1, 6)$; axis of symmetry: $x = 1$;
(b)

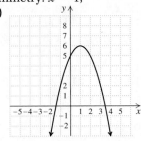

$f(x) = -x^2 + 2x + 5$

27. (a) Vertex: $\left(-\frac{3}{2}, -\frac{49}{4}\right)$; axis of symmetry: $x = -\frac{3}{2}$;
(b)

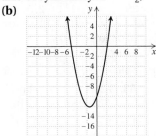

$g(x) = x^2 + 3x - 10$

29. (a) Vertex: $\left(-\frac{7}{2}, -\frac{49}{4}\right)$; axis of symmetry: $x = -\frac{7}{2}$;
(b)

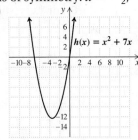

$h(x) = x^2 + 7x$

31. (a) Vertex: $(-1, -4)$; axis of symmetry: $x = -1$;
(b)

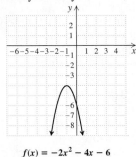

$f(x) = -2x^2 - 4x - 6$

33. (a) Vertex: $(3, 4)$; axis of symmetry: $x = 3$; minimum: 4;
(b)

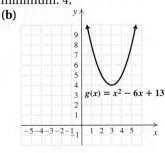

$g(x) = x^2 - 6x + 13$

35. (a) Vertex: $(2, -5)$; axis of symmetry: $x = 2$; minimum: -5;
(b)

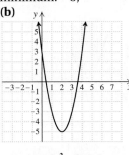

$g(x) = 2x^2 - 8x + 3$

37. (a) Vertex: $(4, 2)$; axis of symmetry: $x = 4$; minimum: 2;
(b)

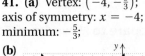

$f(x) = 3x^2 - 24x + 50$

39. (a) Vertex: $\left(\frac{5}{6}, \frac{1}{12}\right)$; axis of symmetry: $x = \frac{5}{6}$; maximum: $\frac{1}{12}$;
(b)

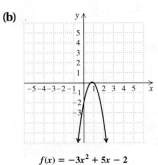

$f(x) = -3x^2 + 5x - 2$

41. (a) Vertex: $\left(-4, -\frac{5}{3}\right)$; axis of symmetry: $x = -4$; minimum: $-\frac{5}{3}$;
(b)

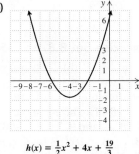

$h(x) = \frac{1}{2}x^2 + 4x + \frac{19}{3}$

43. $(3 - \sqrt{6}, 0), (3 + \sqrt{6}, 0); (0, 3)$ **45.** $(-1, 0), (3, 0)$; $(0, 3)$ **47.** $(0, 0), (9, 0); (0, 0)$ **49.** $(2, 0); (0, -4)$
51. $\left(-\frac{1}{2} - \frac{\sqrt{21}}{2}, 0\right), \left(-\frac{1}{2} + \frac{\sqrt{21}}{2}, 0\right); (0, -5)$
53. No x-intercept; $(0, 6)$ **55.** 🖰   **57.** $x^4 - 4x^2 - 21$
58. $\dfrac{(x - 2)(x + 4)}{x(x - 3)}$ **59.** $3x^2\sqrt[3]{4y^2}$
60. $2x^2 + 2x + 1 + \dfrac{-2}{x - 1}$ **61.** $\dfrac{3a(2a - b)}{b(a - b)}$ **62.** $\dfrac{1}{\sqrt[12]{x^7}}$
63. 🖰   **65. (a)** Minimum: -6.953660714;
(b) $(-1.056433682, 0), (2.413576539, 0); (0, -5.89)$
67. (a) $-2.4, 3.4$; **(b)** $-1.3, 2.3$
69. $f(x) = m\left(x - \dfrac{n}{2m}\right)^2 + \dfrac{4mp - n^2}{4m}$
71. $f(x) = \frac{5}{16}x^2 - \frac{15}{8}x - \frac{35}{16}$, or $f(x) = \frac{5}{16}(x - 3)^2 - 5$
73.

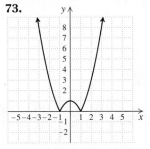

$f(x) = |x^2 - 1|$

75.

$f(x) = |2(x - 3)^2 - 5|$

Quick Quiz: Sections 8.1–8.7, p. 555

1. $\pm\dfrac{\sqrt{15}}{3}$ **2.** $-\dfrac{3}{4}\pm\dfrac{\sqrt{57}}{4}$ **3.** $5x^2+3x-2=0$

4. 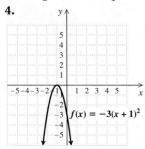   **5.**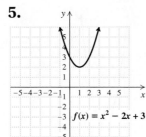

Prepare to Move On, p. 555

1. $(-2,5,1)$ **2.** $(-3,6,-5)$ **3.** $\left(\dfrac{1}{3},\dfrac{1}{6},\dfrac{1}{2}\right)$

Exercise Set 8.8, pp. 559–564

1. (e)   **2.** (b)   **3.** (c)   **4.** (a)   **5.** (d)   **6.** (f)
7. $3\frac{1}{4}$ weeks; 8.3 lb of milk per day   **9.** $120/dulcimer;
350 dulcimers   **11.** 180 ft by 180 ft   **13.** 450 ft²; 15 ft
by 30 ft (The house serves as a 30-ft side.)   **15.** 3.5 in.
17. 81; 9 and 9   **19.** -16; 4 and -4 **21.** 25; -5 and -5
23. $f(x)=ax^2+bx+c,a<0$ **25.** $f(x)=mx+b$
27. Neither quadratic nor linear
29. $f(x)=ax^2+bx+c,a>0$ **31.** $f(x)=mx+b$
33. $f(x)=mx+b$ **35.** $f(x)=2x^2+3x-1$
37. $f(x)=-\frac{1}{4}x^2+3x-5$
39. (a) $A(s)=\frac{3}{16}s^2-\frac{135}{4}s+1750$; **(b)** about 531
accidents for every 200 million km driven
41. $h(d)=-0.0068d^2+0.8571d$ **43.** 🗒
45. $y=-\frac{1}{3}x+16$ **46.** $y=2x+13$
47. $y=-\frac{4}{3}x+\frac{40}{3}$ **48.** $y=\frac{2}{3}x-\frac{23}{3}$ **49.** $y=\frac{1}{2}x-6$
50. $y=-4$ **51.** 🗒   **53.** 158 ft   **55.** $15
57. The radius of the circular portion of the window and the
height of the rectangular portion should each be $\dfrac{24}{\pi+4}$ ft.
59. (a) $a(x)=18.78035714x^2+35.54107143x+165.1107143$; **(b)** 1,651,382 subscriptions   **61.** 🖥

Quick Quiz: Sections 8.1–8.8, p. 564

1. $-\frac{5}{4},\frac{2}{3}$ **2.** 2, 7   **3.** $x^2+25=0$
4. $h=\dfrac{V^2}{12.25}$ **5.**
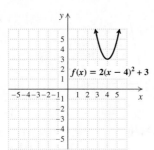

Prepare to Move On, p. 564

1. $\{x\,|\,x\ge-3\}$, or $[-3,\infty)$ **2.** $\left\{x\,|\,-3<x<\frac{5}{2}\right\}$, or
$\left(-3,\frac{5}{2}\right)$ **3.** $f(x)=\dfrac{-4x-23}{x+4},x\ne-4$
4. $f(x)=\dfrac{1}{x-1},x\ne1$ **5.** No solution   **6.** $-6,9$

Technology Connection, p. 568

1. $\{x\,|\,-0.78\le x\le1.59\}$, or $[-0.78,1.59]$
2. $\{x\,|\,x\le-0.21\,or\,x\ge2.47\}$, or $(-\infty,-0.21]\cup[2.47,\infty)$
3. $\{x\,|\,x>-1.37\}$, or $(-1.37,\infty)$

Exercise Set 8.9, pp. 571–573

1. True   **2.** False   **3.** True   **4.** True   **5.** False
6. True   **7.** $\left[-4,\frac{3}{2}\right]$, or $\left\{x\,|\,-4\le x\le\frac{3}{2}\right\}$
9. $(-\infty,-2)\cup(0,2)\cup(3,\infty)$, or
$\{x\,|\,x<-2\,or\,0<x<2\,or\,x>3\}$
11. $\left(-\infty,-\frac{7}{2}\right)\cup(-2,\infty)$, or $\left\{x\,|\,x<-\frac{7}{2}\,or\,x>-2\right\}$
13. $(5,6)$, or $\{x\,|\,5<x<6\}$
15. $(-\infty,-7]\cup[2,\infty)$, or $\{x\,|\,x\le-7\,or\,x\ge2\}$
17. $(-\infty,-1)\cup(2,\infty)$, or $\{x\,|\,x<-1\,or\,x>2\}$
19. $\varnothing$ **21.** $[2-\sqrt{7},2+\sqrt{7}]$, or
$\{x\,|\,2-\sqrt{7}\le x\le2+\sqrt{7}\}$ **23.** $(-\infty,-2)\cup(0,2)$,
or $\{x\,|\,x<-2\,or\,0<x<2\}$ **25.** $[-2,1]\cup[4,\infty)$,
or $\{x\,|\,-2\le x\le1\,or\,x\ge4\}$ **27.** $[-2,2]$, or
$\{x\,|\,-2\le x\le2\}$ **29.** $(-1,2)\cup(3,\infty)$, or
$\{x\,|\,-1<x<2\,or\,x>3\}$ **31.** $(-\infty,0]\cup[2,5]$, or
$\{x\,|\,x\le0\,or\,2\le x\le5\}$ **33.** $(-\infty,5)$, or $\{x\,|\,x<5\}$
35. $(-\infty,-1]\cup(3,\infty)$, or $\{x\,|\,x\le-1\,or\,x>3\}$
37. $(-\infty,-6)$, or $\{x\,|\,x<-6\}$ **39.** $(-\infty,-1]\cup[2,5)$,
or $\{x\,|\,x\le-1\,or\,2\le x<5\}$ **41.** $(-\infty,-3)\cup[0,\infty)$, or
$\{x\,|\,x<-3\,or\,x\ge0\}$ **43.** $(0,\infty)$, or $\{x\,|\,x>0\}$
45. $(-\infty,-4)\cup[1,3)$, or $\{x\,|\,x<-4\,or\,1\le x<3\}$
47. $\left(-\frac{3}{4},\frac{5}{2}\right]$, or $\left\{x\,|\,-\frac{3}{4}<x\le\frac{5}{2}\right\}$ **49.** $(-\infty,2)\cup[3,\infty)$, or
$\{x\,|\,x<2\,or\,x\ge3\}$ **51.** 🗒   **53.** 3.2 hr   **54.** 9 km/h
55. Mileages no greater than 50 mi   **56.** Deanna: 12 hr;
Donna: 6 hr   **57.** 🗒   **59.** $(-1-\sqrt{6},-1+\sqrt{6})$, or
$\{x\,|\,-1-\sqrt{6}<x<-1+\sqrt{6}\}$ **61.** $\{0\}$
63. (a) $(10,200)$, or $\{x\,|\,10<x<200\}$;
(b) $[0,10)\cup(200,\infty)$, or $\{x\,|\,0\le x<10\,or\,x>200\}$
65. $\{n\,|\,n\,is\,an\,integer\,and\,12\le n\le25\}$
67. $f(x)=0$ for $x=-2,1,3$; $f(x)<0$ for
$(-\infty,-2)\cup(1,3)$, or $\{x\,|\,x<-2\,or\,1<x<3\}$; $f(x)>0$
for $(-2,1)\cup(3,\infty)$, or $\{x\,|\,-2<x<1\,or\,x>3\}$
69. $f(x)$ has no zeros; $f(x)<0$ for $(-\infty,0)$,
or $\{x\,|\,x<0\}$; $f(x)>0$ for $(0,\infty)$, or $\{x\,|\,x>0\}$
71. $f(x)=0$ for $x=-1,0$; $f(x)<0$ for
$(-\infty,-3)\cup(-1,0)$, or $\{x\,|\,x<-3\,or\,-1<x<0\}$;
$f(x)>0$ for $(-3,-1)\cup(0,2)\cup(2,\infty)$, or
$\{x\,|\,-3<x<-1\,or\,0<x<2\,or\,x>2\}$
73. $(-\infty,-5]\cup[9,\infty)$, or $\{x\,|\,x\le-5\,or\,x\ge9\}$
75. $(-\infty,-8]\cup[0,\infty)$, or $\{x\,|\,x\le-8\,or\,x\ge0\}$ **77.** 🗒

Quick Quiz: Sections 8.1–8.9, p. 573

1. $\pm\dfrac{\sqrt{3}}{3}i$ **2.** 16   **3.** $\dfrac{1}{10}\pm\dfrac{\sqrt{21}}{10}$
4. $(-\infty,-1]\cup\left[\frac{3}{2},\infty\right)$, or $\left\{x\,|\,x\le -1\ or\ x\ge\frac{3}{2}\right\}$
5. x-intercepts: $(0,0),\left(\frac{3}{4},0\right)$; y-intercept: $(0,0)$

Prepare to Move On, p. 573

1. 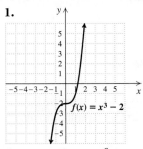   **2.**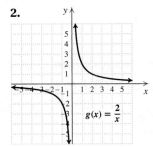

3. $a-8$ **4.** $4a^2+20a+27$

Visualizing for Success, p. 574

1. B   **2.** E   **3.** A   **4.** H   **5.** C   **6.** J
7. F   **8.** G   **9.** I   **10.** D

Decision Making: Connection, p. 575

1. Quadratic   **2.** (a) $p(x)=x-1$; (b) $p(x)=\frac{3}{2}x-6$;
(c) $p(x)=\frac{1}{8}x^2-2x+16.5$; The quadratic function is
the best fit.   **3.** (b)   **4.** Quadratic   **5.** 🖥

Study Summary: Chapter 8, pp. 576–578

1. 1, 11   **2.** $9\pm\sqrt{5}$ **3.** $-21,1$ **4.** $-\frac{3}{2},3$ **5.** Two
imaginary-number solutions   **6.** $n=\pm\sqrt{a-1}$ **7.** 36
8.
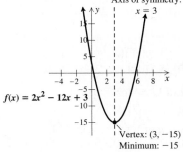

9. 30 ft by 30 ft   **10.** $\{x\,|\,-1<x<12\}$, or $(-1,12)$

Review Exercises: Chapter 8, pp. 578–580

1. False   **2.** True   **3.** True   **4.** True   **5.** False
6. True   **7.** True   **8.** True   **9.** False   **10.** True
11. $\pm\dfrac{\sqrt{2}}{3}$ **12.** $0,-\frac{3}{4}$ **13.** 3, 9   **14.** $2\pm 2i$
15. 3, 5   **16.** $-\dfrac{9}{2}\pm\dfrac{\sqrt{85}}{2}$ **17.** $-0.372, 5.372$

18. $-\frac{1}{4}, 1$ **19.** $x^2-18x+81=(x-9)^2$
20. $x^2+\frac{3}{5}x+\frac{9}{100}=\left(x+\frac{3}{10}\right)^2$ **21.** $3\pm 2\sqrt{2}$ **22.** 4%
23. 5.3 sec   **24.** Two irrational real numbers
25. Two imaginary numbers   **26.** $x^2+9=0$
27. $x^2+10x+25=0$ **28.** About 153 mph   **29.** 6 hr
30. $(-3,0),(-2,0),(2,0),(3,0)$ **31.** $-5,3$
32. $\pm\sqrt{2},\pm\sqrt{7}$
33.
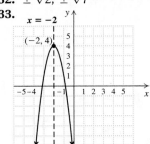
$f(x)=-3(x+2)^2+4$
Maximum: 4
34. (a) Vertex: $(3,5)$; axis of symmetry: $x=3$;
(b)
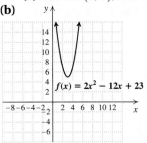

35. $(2,0),(7,0);(0,14)$ **36.** $p=\dfrac{9\pi^2}{N^2}$
37. $T=\dfrac{1\pm\sqrt{1+24A}}{6}$ **38.** Linear   **39.** Quadratic
40. 225 ft²; 15 ft by 15 ft   **41.** (a) $f(x)=\frac{17}{12}x^2-\frac{1}{12}x+7$;
(b) approximately \$109,000   **42.** $(-1,0)\cup(3,\infty)$,
or $\{x\,|\,-1<x<0\ or\ x>3\}$ **43.** $(-3,5]$, or
$\{x\,|\,-3<x\le 5\}$ **44.** 📝 The x-coordinate of the
maximum or minimum point lies halfway between the
x-coordinates of the x-intercepts.   **45.** 📝 The first
coordinate of each x-intercept of the graph of f is a solution of $f(x)=0$. Suppose the first coordinates of the
x-intercepts are a and b. Then $(x-a)$ and $(x-b)$ are
factors of $f(x)$. If the graph of a quadratic function has
one x-intercept $(a,0)$, then $(x-a)$ is a repeated factor of
$f(x)$.   **46.** 📝 Completing the square was used to solve
quadratic equations and to graph quadratic functions by
rewriting the function in the form $f(x)=a(x-h)^2+k$.
47. $f(x)=\frac{7}{15}x^2-\frac{14}{15}x-7$ **48.** $h=60, k=60$
49. 18, 324

Test: Chapter 8, p. 581

1. [8.1] $\pm\dfrac{\sqrt{7}}{5}$ **2.** [8.1] 2, 9   **3.** [8.2] $-1\pm\sqrt{2}i$
4. [8.2] $1\pm\sqrt{6}$ **5.** [8.5] $-2,\frac{2}{3}$
6. [8.2] $-4.193, 1.193$ **7.** [8.2] $-\frac{3}{4},\frac{7}{3}$
8. [8.1] $x^2-20x+100=(x-10)^2$
9. [8.1] $x^2+\frac{2}{7}x+\frac{1}{49}=\left(x+\frac{1}{7}\right)^2$ **10.** [8.1] $-5\pm\sqrt{10}$

11. [8.3] Two imaginary numbers   **12.** [8.3] $x^2 - 11 = 0$
13. [8.4] 16 km/h   **14.** [8.4] 2 hr   **15.** [8.5] $(-4, 0), (4, 0)$
16. [8.6]   **17.** [8.7] **(a)** $(-1, -8), x = -1$;
(b)

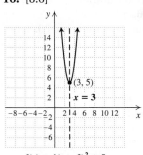

$f(x) = 4(x - 3)^2 + 5$
Minimum: 5

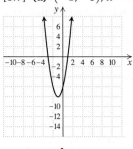

$f(x) = 2x^2 + 4x - 6$

18. [8.7] $(-2, 0), (3, 0); (0, -6)$ **19.** [8.4] $r = \sqrt{\dfrac{3V}{\pi} - R^2}$
20. [8.8] Quadratic   **21.** [8.8] Minimum: \$129/cabinet
when 325 cabinets are built   **22.** [8.8] $f(x) = \frac{1}{5}x^2 - \frac{3}{5}x$
23. [8.9] $(-6, 1)$, or $\{x \mid -6 < x < 1\}$
24. [8.9] $[-1, 0) \cup [1, \infty)$, or $\{x \mid -1 \le x < 0 \ or \ x \ge 1\}$
25. [8.3] $\frac{1}{2}$ **26.** [8.3] $x^4 + x^2 - 12 = 0$; answers may vary
27. [8.5] $\pm \sqrt{\sqrt{5} + 2}, \pm \sqrt{\sqrt{5} - 2}i$

Cumulative Review: Chapters 1-8, p. 582

1. 24   **2.** $14x^2y - 10xy - 9xy^2$ **3.** $81p^4q^2 - 64t^2$
4. $\dfrac{t(t + 1)(t + 5)}{(3t + 4)^3}$ **5.** $13 - \sqrt{2}i$
6. $3(2x^2 + 5y^2)(2x^2 - 5y^2)$ **7.** $x(x - 20)(x - 4)$
8. $100(m + 1)(m^2 - m + 1)(m - 1)(m^2 + m + 1)$
9. $(2t + 9)(3t + 4)$ **10.** 7   **11.** $\{x \mid x < 7\}$, or
$(-\infty, 7)$ **12.** $\left(3, \frac{1}{2}\right)$ **13.** $-6, 11$ **14.** $\frac{1}{2}, 2$ **15.** 4
16. $-5 \pm \sqrt{2}$ **17.** $\dfrac{1}{6} \pm \dfrac{\sqrt{11}}{6}i$
18.

$9x - 2y = 18$

19.

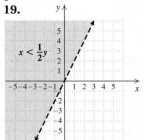

$x < \frac{1}{2}y$

20.

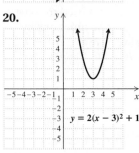

$y = 2(x - 3)^2 + 1$

21.

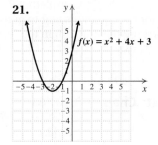

$f(x) = x^2 + 4x + 3$

22. $y = -5x + \frac{1}{2}$ **23.** $-\frac{7}{10}$
24. $(-\infty, 10]$, or $\{x \mid x \le 10\}$
25. $\{x \mid x$ is a real number $and \ x \ne 4\}$, or $(-\infty, 4) \cup (4, \infty)$

26. $a = \dfrac{c}{2b - 1}$ **27.** $t = \dfrac{4r}{3p^2}$
28. **(a)** 1.74 oz; **(b)** \$1200 per ounce; **(c)** 75%
29. **(a)** $f(t) = -t + 52$; **(b)** 18 million Americans; **(c)** 2032
30. Number tiles: 26 sets; alphabet tiles: 10 sets
31. \$125/bunk bed; 400 bunk beds
32. $\dfrac{1}{3} \pm \dfrac{\sqrt{2}}{6}i$ **33.** $\{0\}$
34. $(1 - \sqrt{6}, 16 - 10\sqrt{6}), (1 + \sqrt{6}, 16 + 10\sqrt{6})$

CHAPTER 9

Technology Connection, p. 586

1. To check $(f \circ g)(x)$, we let $y_1 = \sqrt{x}, y_2 = x - 1$,
$y_3 = \sqrt{x - 1}$, and $y_4 = y_1(y_2)$. A table shows that we have
$y_3 = y_4$. The check for $(g \circ f)(x)$ is similar. Graphs can also
be used.

Technology Connection, p. 592

1. Graph each pair of functions in a square window along
with the line $y = x$ and determine whether the first two
functions are reflections of each other across $y = x$. For
further verification, examine a table of values for each pair
of functions.   **2.** Yes; most graphing calculators do not
require that the inverse relation be a function.

Exercise Set 9.1, pp. 593–595

1. True   **2.** True   **3.** False   **4.** False   **5.** False
6. False   **7.** True   **8.** True   **9.** **(a)** $(f \circ g)(1) = 5$;
(b) $(g \circ f)(1) = -1$; **(c)** $(f \circ g)(x) = x^2 - 6x + 10$;
(d) $(g \circ f)(x) = x^2 - 2$ **11.** **(a)** $(f \circ g)(1) = -24$;
(b) $(g \circ f)(1) = 65$; **(c)** $(f \circ g)(x) = 10x^2 - 34$;
(d) $(g \circ f)(x) = 50x^2 + 20x - 5$
13. **(a)** $(f \circ g)(1) = 8$; **(b)** $(g \circ f)(1) = \frac{1}{64}$;
(c) $(f \circ g)(x) = \dfrac{1}{x^2} + 7$; **(d)** $(g \circ f)(x) = \dfrac{1}{(x + 7)^2}$
15. **(a)** $(f \circ g)(1) = 2$; **(b)** $(g \circ f)(1) = 4$;
(c) $(f \circ g)(x) = \sqrt{x + 3}$; **(d)** $(g \circ f)(x) = \sqrt{x} + 3$
17. **(a)** $(f \circ g)(1) = 2$; **(b)** $(g \circ f)(1) = \frac{1}{2}$;
(c) $(f \circ g)(x) = \sqrt{\dfrac{4}{x}}$; **(d)** $(g \circ f)(x) = \dfrac{1}{\sqrt{4x}}$
19. **(a)** $(f \circ g)(1) = 4$; **(b)** $(g \circ f)(1) = 2$;
(c) $(f \circ g)(x) = x + 3$; **(d)** $(g \circ f)(x) = \sqrt{x^2 + 3}$
21. $f(x) = x^4; g(x) = 3x - 5$ **23.** $f(x) = \sqrt{x}$;
$g(x) = 9x + 1$ **25.** $f(x) = \dfrac{6}{x}; g(x) = 5x - 2$ **27.** Yes
29. No   **31.** Yes   **33.** No   **35.** **(a)** Yes;
(b) $f^{-1}(x) = x - 3$ **37.** **(a)** Yes; **(b)** $f^{-1}(x) = \dfrac{x}{2}$
39. **(a)** Yes; **(b)** $g^{-1}(x) = \dfrac{x + 1}{3}$ **41.** **(a)** Yes;
(b) $f^{-1}(x) = 2x - 2$ **43.** **(a)** No   **45.** **(a)** Yes;

(b) $h^{-1}(x) = -10 - x$ **47. (a)** Yes; **(b)** $f^{-1}(x) = \dfrac{1}{x}$

49. (a) No   **51. (a)** Yes; **(b)** $f^{-1}(x) = \dfrac{3x - 1}{2}$

53. (a) Yes; **(b)** $f^{-1}(x) = \sqrt[3]{x - 5}$ **55. (a)** Yes;
(b) $g^{-1}(x) = \sqrt[3]{x} + 2$ **57. (a)** Yes;
(b) $f^{-1}(x) = x^2, x \geq 0$

59. (a) 40, 44, 52, 60; **(b)** Yes; $f^{-1}(x) = (x - 24)/2$, or $\dfrac{x}{2} - 12$

(c) 8, 10, 14, 18

61.

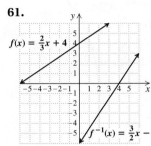

$f(x) = \frac{2}{3}x + 4$

$f^{-1}(x) = \frac{3}{2}x - 6$

63.

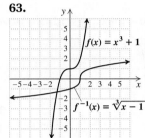

$f(x) = x^3 + 1$

$f^{-1}(x) = \sqrt[3]{x - 1}$

65.

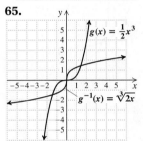

$g(x) = \frac{1}{2}x^3$

$g^{-1}(x) = \sqrt[3]{2x}$

67.

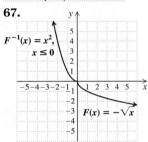

$F^{-1}(x) = x^2,$
$x \leq 0$

$F(x) = -\sqrt{x}$

69.

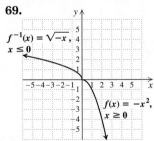

$f^{-1}(x) = \sqrt{-x},$
$x \leq 0$

$f(x) = -x^2,$
$x \geq 0$

71. (1) $(f^{-1} \circ f)(x) = f^{-1}(f(x))$
$= f^{-1}(\sqrt[3]{x - 4}) = (\sqrt[3]{x - 4})^3 + 4$
$= x - 4 + 4 = x;$
(2) $(f \circ f^{-1})(x) = f(f^{-1}(x))$
$= f(x^3 + 4) = \sqrt[3]{x^3 + 4 - 4}$
$= \sqrt[3]{x^3} = x$

73. (1) $(f^{-1} \circ f)(x) = f^{-1}(f(x)) = f^{-1}\left(\dfrac{1 - x}{x}\right)$

$= \dfrac{1}{\left(\dfrac{1 - x}{x}\right) + 1}$

$= \dfrac{1}{\dfrac{1 - x + x}{x}}$

$= x;$

(2) $(f \circ f^{-1})(x) = f(f^{-1}(x)) = f\left(\dfrac{1}{x + 1}\right)$

$= \dfrac{1 - \left(\dfrac{1}{x + 1}\right)}{\left(\dfrac{1}{x + 1}\right)}$

$= \dfrac{\dfrac{x + 1 - 1}{x + 1}}{\dfrac{1}{x + 1}} = x$

75.    **77.** $t^{13/15}$ **78.** $2ab^4 \sqrt[3]{5a^2}$ **79.** $\dfrac{x^{12}}{9y^8}$

80. $-i$ **81.** 25   **82.** 6.3×10^{-15} **83.**

85.

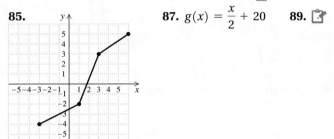

87. $g(x) = \dfrac{x}{2} + 20$ **89.**

91. Suppose that $h(x) = (f \circ g)(x)$. First, note that for
$I(x) = x$, $(f \circ I)(x) = f(I(x)) = f(x)$ for any function f.
(i) $((g^{-1} \circ f^{-1}) \circ h)(x) = ((g^{-1} \circ f^{-1}) \circ (f \circ g))(x)$
$= ((g^{-1} \circ (f^{-1} \circ f)) \circ g)(x)$
$= ((g^{-1} \circ I) \circ g)(x)$
$= (g^{-1} \circ g)(x) = x$
(ii) $(h \circ (g^{-1} \circ f^{-1}))(x) = ((f \circ g) \circ (g^{-1} \circ f^{-1}))(x)$
$= ((f \circ (g \circ g^{-1})) \circ f^{-1})(x)$
$= ((f \circ I) \circ f^{-1})(x)$
$= (f \circ f^{-1})(x) = x.$
Therefore, $(g^{-1} \circ f^{-1})(x) = h^{-1}(x)$.
93. Yes   **95.** No   **97. (1)** C; **(2)** A; **(3)** B; **(4)** D

99.

Prepare to Move On, p. 595

1. $\frac{1}{8}$ **2.** $\frac{1}{25}$ **3.** 32   **4.** Approximately 2.1577
5.

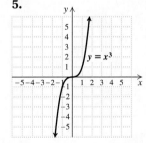

$y = x^3$

6.

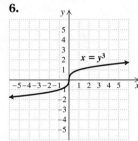

$x = y^3$

Technology Connection, p. 598

1. $y_1 = \left(\dfrac{5}{2}\right)^x;\ y_2 = \left(\dfrac{2}{5}\right)^x$

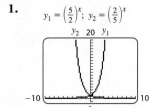

2. $y_1 = 3.2^x;\ y_2 = 3.2^{-x}$

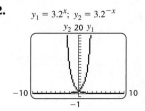

3. $y_1 = \left(\frac{3}{7}\right)^x$; $y_2 = \left(\frac{7}{3}\right)^x$

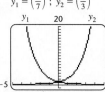

4. $y_1 = 5000(1.08)^x$; $y_2 = 5000(1.08)^{x-3}$

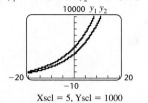

Xscl = 5, Yscl = 1000

Exercise Set 9.2, pp. 600–603

1. True   **2.** True   **3.** True   **4.** False   **5.** False
6. True

7.

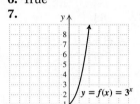

$y = f(x) = 3^x$

9.

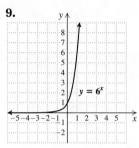

$y = 6^x$

11.

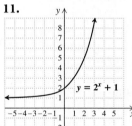

$y = 2^x + 1$

13.

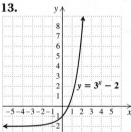

$y = 3^x - 2$

15.

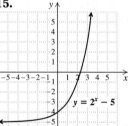

$y = 2^x - 5$

17.

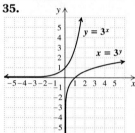

$y = 2^{x-3}$

19.

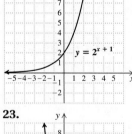

$y = 2^{x+1}$

21.

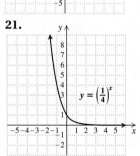

$y = \left(\frac{1}{4}\right)^x$

23.

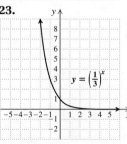

$y = \left(\frac{1}{3}\right)^x$

25.

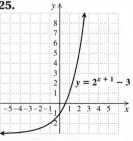

$y = 2^{x+1} - 3$

27.

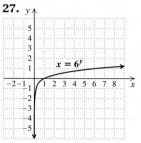

$x = 6^y$

29.

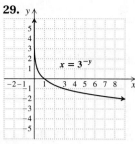

$x = 3^{-y}$

31.

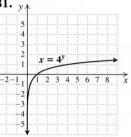

$x = 4^y$

33.

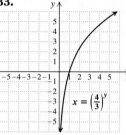

$x = \left(\frac{4}{3}\right)^y$

35.

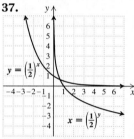

$y = 3^x$

$x = 3^y$

37.

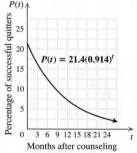

$y = \left(\frac{1}{2}\right)^x$

$x = \left(\frac{1}{2}\right)^y$

39. (a) $3.2 billion; $4.8 billion; $7.3 billion;
(b)

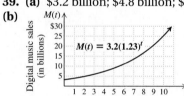

$M(t) = 3.2(1.23)^t$

Digital music sales (in billions)

Years since 2010

41. (a) 19.6%; 16.3%; 7.3%;
(b)

$P(t) = 21.4(0.914)^t$

Percentage of successful quitters

Months after counseling

43. (a) About 44,079 whales; about 12,953 whales;
(b)

$P(t) = 150(0.960)^t$

Humpback whale population (in thousands)

Years since 1900

45. (a) About 8706 whales; about 15,107 whales;
(b)

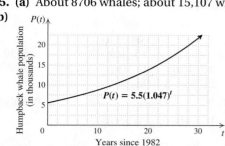

47. (a) About 50 moose; about 1421 moose; about 8470 moose; **(b)**

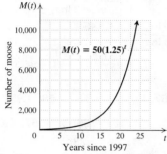

49.
51. $3(x + 4)(x - 4)$ **52.** $(x - 10)^2$
53. $(2x + 3)(3x - 4)$
54. $8(x^2 - 2y^2)(x^4 + 2x^2y^2 + 4y^4)$
55. $(t - y + 1)(t + y - 1)$ **56.** $x(x - 2)(5x^2 - 3)$
57.   **59.** $\pi^{2.4}$
61.

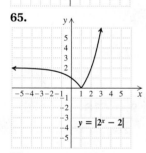

63.

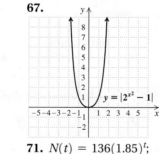

65.

67.

69.

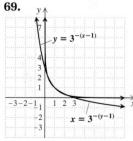

71. $N(t) = 136(1.85)^t$; about 5550 ruffe
73. 📋   **75.** 〰

Quick Quiz: Sections 9.1–9.2, p. 603

1. $(f \circ g)(x) = -3x^2 + x + 9$
2. $f(x) = x^4$; $g(x) = x - 6$ **3.** $g^{-1}(x) = \dfrac{x - 5}{2}$
4.

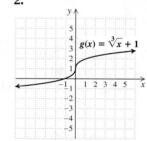

5.

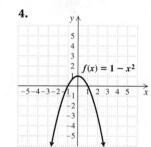

Prepare to Move On, p. 603

1.

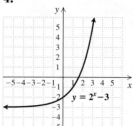

2.

3.

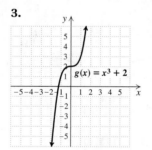

4.

Exercise Set 9.3, pp. 608–610

1. Logarithmic   **2.** Exponent   **3.** Positive   **4.** 1
5. (g)   **6.** (d)   **7.** (a)   **8.** (h)   **9.** (b)   **10.** (c)
11. (e)   **12.** (f)   **13.** 3   **15.** 2   **17.** -2 **19.** -1
21. 4   **23.** 1   **25.** 0   **27.** 5   **29.** -2 **31.** $\frac{1}{2}$
33. $\frac{3}{2}$ **35.** $\frac{2}{3}$ **37.** 29
39.

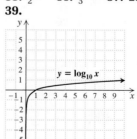

41.

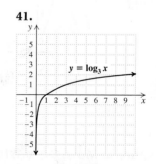

43.

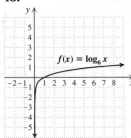

45.

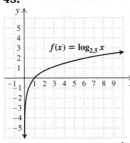

47.

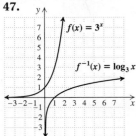

49. $10^x = 8$ **51.** $9^1 = 9$
53. $10^{-1} = 0.1$
55. $10^{0.845} = 7$
57. $c^8 = m$ **59.** $r^t = C$
61. $e^{-1.3863} = 0.25$
63. $r^{-x} = T$
65. $2 = \log_{10} 100$
67. $-3 = \log_5 \frac{1}{125}$
69. $\frac{1}{4} = \log_{16} 2$
71. $0.4771 = \log_{10} 3$
73. $m = \log_z 6$
75. $t = \log_p q$ **77.** $3 = \log_e 20.0855$ **79.** 36
81. 5   **83.** 9   **85.** 49   **87.** $\frac{1}{9}$ **89.** 4   **91.** ✎
93. $30a^2b^4$ **94.** $12 - 2\sqrt{30} + 2\sqrt{15} - 5\sqrt{2}$
95. $3\sqrt{3x}$ **96.** $\sqrt[12]{x}$ **97.** $2y^2\sqrt[3]{y}$ **98.** $\sqrt[10]{a^3y^7}$
99. ✎

101.

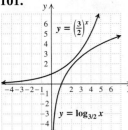

103.

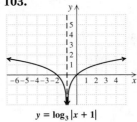

105. 6   **107.** $-25, 4$ **109.** -2 **111.** 0
113. Let $b = 0$, and suppose that $x_1 = 1$ and $x_2 = 2$. Then $0^1 = 0^2$, but $1 \ne 2$. Then let $b = 1$, and suppose that $x_1 = 1$ and $x_2 = 2$. Then $1^1 = 1^2$, but $1 \ne 2$.

Quick Quiz: Sections 9.1–9.3, p. 610

1. No   **2.**

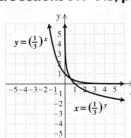

3. 4   **4.** $3^x = t$
5. $\frac{1}{16}$

Prepare to Move On, p. 610

1. c^{16} **2.** x^{30} **3.** a^{12} **4.** $3^{1/2}$ **5.** $t^{2/3}$

Exercise Set 9.4, pp. 615–617

1. (e)   **2.** (f)   **3.** (a)   **4.** (b)   **5.** (c)   **6.** (d)
7. $\log_3 81 + \log_3 27$ **9.** $\log_4 64 + \log_4 16$
11. $\log_c r + \log_c s + \log_c t$ **13.** $\log_a (2 \cdot 10)$, or $\log_a 20$
15. $\log_c (t \cdot y)$ **17.** $8 \log_a r$ **19.** $\frac{1}{3} \log_2 y$
21. $-3 \log_b C$ **23.** $\log_2 5 - \log_2 11$
25. $\log_b m - \log_b n$ **27.** $\log_a \frac{19}{2}$ **29.** $\log_b \frac{36}{4}$, or $\log_b 9$
31. $\log_a \frac{x}{y}$ **33.** $\log_a x + \log_a y + \log_a z$
35. $3 \log_a x + 4 \log_a z$ **37.** $2 \log_a w - 2 \log_a x + \log_a y$
39. $5 \log_a x - 3 \log_a y - \log_a z$
41. $\log_b x + 2 \log_b y - \log_b w - 3 \log_b z$
43. $\frac{1}{2}(7 \log_a x - 5 \log_a y - 8 \log_a z)$
45. $\frac{1}{3}(6 \log_a x + 3 \log_a y - 2 - 7 \log_a z)$ **47.** $\log_a 3x$
49. $\log_a (x^8 z^3)$ **51.** $\log_b \frac{w^2}{zy^4}$ **53.** $\log_a x$
55. $\log_a \frac{y^5}{x^{3/2}}$ **57.** $\log_a (x - 3)$ **59.** 1.953
61. -0.369 **63.** -1.161 **65.** $\frac{3}{2}$ **67.** Cannot be found   **69.** 10   **71.** m **73.** ✎
75. $\left\{ x \middle| -\frac{11}{3} \le x \le -1 \right\}$, or $\left[-\frac{11}{3}, -1 \right]$ **76.** $-\frac{3}{4}, \frac{5}{6}$
77. $-2 \pm i$ **78.** $\frac{1}{2}$ **79.** 16, 256   **80.** $\frac{1}{4}, 9$
81. ✎   **83.** $\log_a (x^6 - x^4y^2 + x^2y^4 - y^6)$
85. $\frac{1}{2}\log_a (1 - s) + \frac{1}{2}\log_a (1 + s)$ **87.** $\frac{10}{3}$
89. -2 **91.** $\frac{2}{5}$ **93.** True

Quick Quiz: Sections 9.1–9.4, p. 617

1. $(g \circ f)(x) = 2x^2 - 20x + 50$ **2.** $\log_m 5 = 10$
3. 5   **4.** $2 \log_a x + 3 \log_a y - \log_a z$
5. $\frac{1}{2} \log_a x + \frac{1}{2} \log_a y + \log_a z$

Prepare to Move On, p. 617

1. $(-\infty, -7) \cup (-7, \infty)$, or $\{x \mid x \text{ is a real number } and\ x \ne -7\}$
2. $(-\infty, -3) \cup (-3, 2) \cup (2, \infty)$, or $\{x \mid x \text{ is a real number } and\ x \ne -3\ and\ x \ne 2\}$
3. $(-\infty, 10]$, or $\{x \mid x \le 10\}$ **4.** $(-\infty, \infty)$, or $\mathbb{R}$

Mid-Chapter Review: Chapter 9, p. 618

1.
$$y = 2x - 5$$
$$x = 2y - 5$$
$$x + 5 = 2y$$
$$\frac{x + 5}{2} = y$$
$$f^{-1}(x) = \frac{x + 5}{2}$$
2.
$$\log_4 x = 1$$
$$4^1 = x$$
$$4 = x$$
3. $x^2 - 10x + 26$ **4.** $f(x) = \sqrt{x}; g(x) = 5x - 3$

5. $g^{-1}(x) = 6 - x$ **6.**

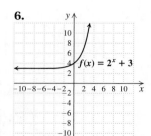

7. 2   **8.** -1 **9.** $\frac{1}{2}$ **10.** 1   **11.** 19   **12.** 0
13. $x^m = 3$ **14.** $2^{10} = 1024$ **15.** $t = \log_e x$
16. $\frac{2}{3} = \log_{64} 16$ **17.** $\log x - \frac{1}{2}\log y - \frac{3}{2}\log z$
18. $\log \dfrac{a}{b^2 c}$ **19.** 4   **20.** $\frac{1}{3}$

Technology Connection, p. 619

1. 〔LOG〕〔7〕〔) 〕〔÷〕〔LOG〕〔3〕〔) 〕〔ENTER〕

Technology Connection, p. 620

1. As x gets larger, the value of y_1 approaches
2.7182818284.... **2.** For large values of x, the graphs of
y_1 and y_2 will be very close or appear to be the same curve,
depending on the window chosen. **3.** Using 〔TRACE〕,
no y-value is given for $x = 0$. Using a table, an error message appears for y_1 when $x = 0$. The domain does not
include 0 because division by 0 is undefined.

Technology Connection, p. 623

1. $y = \log x / \log 7$ **2.** $y = \log (x+2)/\log 5$

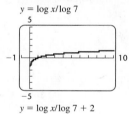   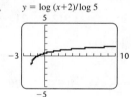

3. $y = \log x / \log 7 + 2$

Exercise Set 9.5, pp. 624–625

1. True   **2.** True   **3.** True   **4.** False   **5.** True
6. True   **7.** True   **8.** True   **9.** True   **10.** True
11. 0.8451   **13.** 1.1367   **15.** 3   **17.** -0.1249
19. 13.0014   **21.** 50.1187   **23.** 0.0011   **25.** 2.1972
27. -5.0832 **29.** 96.7583   **31.** 15.0293   **33.** 0.0305
35. 3.0331   **37.** 6.6439   **39.** 1.1610   **41.** -0.3010
43. -3.3219 **45.** 2.0115

47. Domain: $\mathbb{R}$;
range: $(0, \infty)$

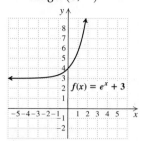

49. Domain: $\mathbb{R}$;
range: $(3, \infty)$

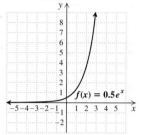

51. Domain: $\mathbb{R}$;
range: $(-2, \infty)$

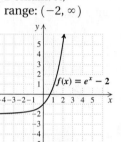

53. Domain: $\mathbb{R}$;
range: $(0, \infty)$

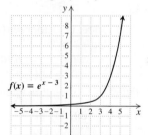

55. Domain: $\mathbb{R}$;
range: $(0, \infty)$

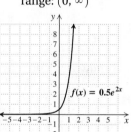

57. Domain: $\mathbb{R}$;
range: $(0, \infty)$

59. Domain: $\mathbb{R}$;
range: $(0, \infty)$

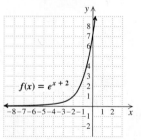

61. Domain: $\mathbb{R}$;
range: $(-\infty, 0)$

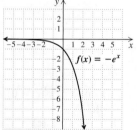

63. Domain: $(0, \infty)$;
range: $\mathbb{R}$

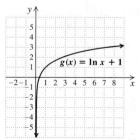

65. Domain: $(0, \infty)$;
range: $\mathbb{R}$

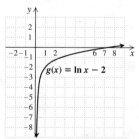

67. Domain: $(0, \infty)$; range: $\mathbb{R}$

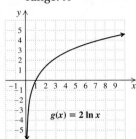

$g(x) = 2 \ln x$

69. Domain: $(0, \infty)$; range: $\mathbb{R}$

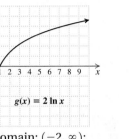

$g(x) = -2 \ln x$

71. Domain: $(-2, \infty)$; range: $\mathbb{R}$

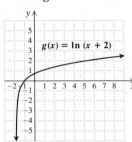

$g(x) = \ln (x + 2)$

73. Domain: $(1, \infty)$; range: $\mathbb{R}$

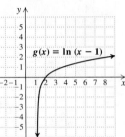

$g(x) = \ln (x - 1)$

75. 🔲   **77.** 1   **78.** $-7\frac{1}{2}$

79. $(g - f)(x) = 5x - 8 - \dfrac{1}{x + 2}$

80. $\{x \mid x$ is a real number $and\ x \neq -2\}$, or $(-\infty, -2) \cup (-2, \infty)$

81. $\{x \mid x$ is a real number $and\ x \neq -2\ and\ x \neq \frac{8}{5}\}$, or $(-\infty, -2) \cup \left(-2, \frac{8}{5}\right) \cup \left(\frac{8}{5}, \infty\right)$

82. $gg(x) = 25x^2 - 80x + 64$ **83.** 🔲   **85.** 2.452

87. 1.442   **89.** $\log M = \dfrac{\ln M}{\ln 10}$ **91.** 1086.5129

93. 4.9855   **95.** (a) Domain: $\{x \mid x > 0\}$, or $(0, \infty)$; range: $\{y \mid y < 0.5135\}$, or $(-\infty, 0.5135)$; **(b)** $[-1, 5, -10, 5]$; **(c)** $y = 3.4 \ln x - 0.25e^x$

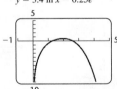

97. (a) Domain: $\{x \mid x > 0\}$, or $(0, \infty)$; range: $\{y \mid y > -0.2453\}$, or $(-0.2453, \infty)$; **(b)** $[-1, 5, -1, 10]$; **(c)** $y = 2x^3 \ln x$

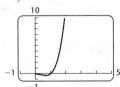

99. 📱🔲

Quick Quiz: Sections 9.1–9.5, p. 625

1. $f(x) = \sqrt{x}$; $g(x) = 3x - 7$ **2.** $\log_a \dfrac{x^2}{y^3}$

3. $\log_a (x^2 - 1)$

4.

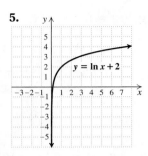

$y = 2^{x-5}$

5.

$y = \ln x + 2$

Prepare to Move On, p. 625

1. $-4, 7$ **2.** $0, \frac{7}{5}$ **3.** $\frac{15}{17}$ **4.** $\frac{5}{6}$ **5.** $\frac{56}{9}$ **6.** 4

Technology Connection, p. 629

1. 0.38   **2.** -1.96 **3.** 0.90   **4.** -1.53
5. 0.13, 8.47   **6.** $-0.75, 0.75$

Connecting the Concepts, p. 630

1. 500   **2.** 8   **3.** $\ln 5 \approx 1.6094$ **4.** $-5, 5$ **5.** 4
6. $\dfrac{\ln 4}{5 \ln 3} \approx 0.2524$ **7.** $-\frac{1}{2}$ **8.** $-\frac{2}{5}$

Exercise Set 9.6, pp. 631–632

1. False   **2.** True   **3.** True   **4.** True   **5.** (d)
6. (a)   **7.** (b)   **8.** (c)   **9.** 2   **11.** $\frac{5}{2}$
13. $\dfrac{\log 10}{\log 2} \approx 3.322$ **15.** -1 **17.** $\dfrac{\log 19}{\log 8} + 3 \approx 4.416$
19. $\ln 50 \approx 3.912$ **21.** $\dfrac{\ln 8}{-0.02} \approx -103.972$
23. $\dfrac{\log 87}{\log 4.9} \approx 2.810$ **25.** $\dfrac{\ln \left(\frac{19}{2}\right)}{4} \approx 0.563$
27. $\dfrac{\ln 2}{-1} \approx -0.693$ **29.** 81   **31.** $\frac{1}{16}$
33. $e^5 \approx 148.413$ **35.** $\dfrac{e^3}{4} \approx 5.021$
37. $10^{1.2} \approx 15.849$ **39.** $\dfrac{e^4 - 1}{2} \approx 26.799$
41. $e \approx 2.718$ **43.** $e^{-3} \approx 0.050$ **45.** -4
47. 10   **49.** No solution   **51.** 2   **53.** $\frac{83}{15}$ **55.** 1
57. 6   **59.** 1   **61.** 5   **63.** $\frac{17}{2}$ **65.** 4   **67.** 🔲
69. $\dfrac{(a + 2)(a - 2)^2}{a^4}$ **70.** $\dfrac{t(2t - 3)(t - 1)}{2t + 3}$
71. $\dfrac{5m - 7}{(m + 1)(m - 5)}$ **72.** $\dfrac{-x^2 + 4x + 2}{x(x - 2)}$
73. $\dfrac{x(3y - 2)}{2y + x}$ **74.** $\dfrac{x + 2}{x + 1}$ **75.** 🔲   **77.** -4
79. 2   **81.** $\pm\sqrt{34}$ **83.** $-3, -1$ **85.** $-625, 625$
87. $\frac{1}{2}, 5000$ **89.** $-3, -1$ **91.** $\frac{1}{100,000}, 100,000$
93. $-\frac{1}{3}$ **95.** 38   **97.** 1

Quick Quiz: Sections 9.1–9.6, p. 632

1. $g^{-1}(x) = x + 6$ **2.** -1 **3.** 11   **4.** 8
5. $10^{2.7} \approx 501.187$

Prepare to Move On, p. 632

1. Length: 9.5 ft; width: 3.5 ft
2. (a) $w(t) = \frac{5900}{11}t + 15{,}200$, where $w(t)$ is the average cost of a wedding t years after 1990; **(b)** about \$28,609
3. $1\frac{1}{5}$ hr

Exercise Set 9.7, pp. 639–644

1. (b)   **2.** (d)   **3.** (c)   **4.** (a)   **5. (a)** 2010;
(b) approximately 1.5 years   **7. (a)** 6.4 years;
(b) 23.4 years   **9. (a)** The 49th key; **(b)** 12 keys
11. (a) 33 months after March 2011, or in December 2013;
(b) approximately 16 months   **13.** 4.9   **15.** 10^{-7}
moles per liter   **17.** 130 dB   **19.** 7.6 W/m²
21. Approximately 42.4 million messages per day
23. (a) $P(t) = P_0 e^{0.025t}$; **(b)** \$5126.58; \$5256.36;
(c) 27.7 years   **25. (a)** $P(t) = 314e^{0.00963t}$, where $P(t)$ is the population, in millions, t years after 2012;
(b) 339 million; **(c)** about 2037   **27.** 21 years
29. (a) About 2044; **(b)** about 2060;
(c)

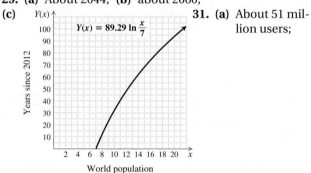

31. (a) About 51 million users;

(b)

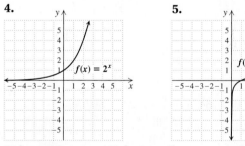

(c) about 2018

33. (a) $k \approx 0.241; P(t) = 4232e^{0.241t}$;
(b) about 2014 **35. (a)** $k \approx 0.341; P(t) = 15.5e^{-0.341t}$;
(b) about 5.6 mcg/mL; **(c)** after about 4 hr; **(d)** 2 hr
37. About 1964 years   **39.** 7.2 days
41. (a) 13.9% per hour; **(b)** 21.6 hr   **43. (a)** $k \approx 0.123$; $V(t) = 9e^{0.123t}$; **(b)** about \$360 million; **(c)** 5.6 years;
(d) 38.3 years   **45.** 🖎   **47.** $f(x) = 18x + \frac{1}{2}$
48. $f(x) = \frac{11}{6}x$ **49.** $f(x) = \frac{2}{3}x + 9$
50. $f(x) = -2x + 8$ **51.** 🖎   **53.** \$14.0 million
55. (a) -26.9; **(b)** 1.58×10^{-17} W/m²
57. Consider an exponential growth function $P(t) = P_0 e^{kt}$. At time T, $P(T) = 2P_0$.
Solve for T:
$$2P_0 = P_0 e^{kT}$$
$$2 = e^{kT}$$
$$\ln 2 = kT$$
$$\frac{\ln 2}{k} = T.$$

Quick Quiz: Sections 9.1–9.7, p. 644

1. Yes   **2.** $\frac{3}{4}$ **3.** 3.9069   **4.** 5
5. Approximately 23.4 years

Prepare to Move On, p. 644

1. $\sqrt{2}$ **2.** $(4, -7)$ **3.** $-4 \pm \sqrt{17}$
4.

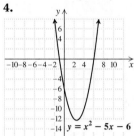

Visualizing for Success, p. 645

1. J   **2.** D   **3.** B   **4.** G   **5.** H   **6.** C   **7.** F
8. I   **9.** E   **10.** A

Decision Making: Connection, p. 646

1. $F(t) = 4430e^{0.062t}$, where t is the number of school years since 2000–2001   **2.** $G(t) = 2050e^{0.103t}$, where t is the number of school years since 2000–2001
3. Answers will vary.   **4.** Answers will vary.   **5.** 🖳

Study Summary: Chapter 9, pp. 647–648

1. $(f \circ g)(x) = 19 - 6x^2$ **2.** Yes   **3.** $f^{-1}(x) = \dfrac{x-1}{5}$
4.   **5.**

6. $\log_5 625 = 4$ **7.** $\log_9 x + \log_9 y$ **8.** $\log_6 7 - \log_6 10$
9. 0   **10.** 19   **11.** 2.3219   **12.** $\frac{4}{3}$
13. $\dfrac{\ln 10}{0.1} \approx 23.0259$ **14. (a)** $P(t) = 15{,}000e^{0.023t}$;
(b) 30.1 years   **15.** 35 days

Review Exercises: Chapter 9, pp. 649–650

1. True   **2.** True   **3.** True   **4.** False   **5.** False
6. True   **7.** False   **8.** False   **9.** True   **10.** False
11. $(f \circ g)(x) = 4x^2 - 12x + 10; (g \circ f)(x) = 2x^2 - 1$
12. $f(x) = \sqrt{x}; g(x) = 3 - x$ **13.** No
14. $f^{-1}(x) = x + 10$ **15.** $g^{-1}(x) = \dfrac{2x-1}{3}$
16. $f^{-1}(x) = \dfrac{\sqrt[3]{x}}{3}$

17.

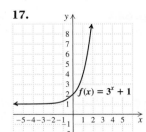

$f(x) = 3^x + 1$

18.

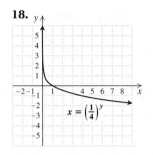

$x = \left(\frac{1}{4}\right)^y$

19.

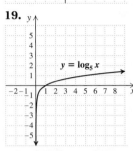

$y = \log_5 x$

20. 2   **21.** -2 **22.** 11
23. $\frac{1}{2}$ **24.** $\log_2 \frac{1}{8} = -3$
25. $\log_{25} 5 = \frac{1}{2}$
26. $16 = 4^x$ **27.** $1 = 8^0$

28. $4\log_a x + 2\log_a y + 3\log_a z$
29. $5\log_a x - (\log_a y + 2\log_a z)$, or
$5\log_a x - \log_a y - 2\log_a z$
30. $\frac{1}{4}(2\log z - 3\log x - \log y)$ **31.** $\log_a (5 \cdot 8)$, or $\log_a 40$
32. $\log_a \frac{48}{12}$, or $\log_a 4$ **33.** $\log \frac{a^{1/2}}{bc^2}$ **34.** $\log_a \sqrt[3]{\frac{x}{y^2}}$
35. 1   **36.** 0   **37.** 17   **38.** 6.93   **39.** -3.2698
40. 8.7601   **41.** 3.2698   **42.** 2.54995   **43.** -3.6602
44. 1.8751   **45.** 61.5177   **46.** -1.2040 **47.** 0.3753
48. 2.4307   **49.** 0.8982
50. Domain: $\mathbb{R}$;   **51.** Domain: $(0, \infty)$;
range: $(-1, \infty)$ range: $\mathbb{R}$

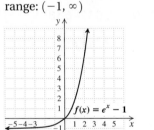

$f(x) = e^x - 1$

$g(x) = 0.6 \ln x$

52. 3   **53.** -1 **54.** $\frac{1}{81}$ **55.** 2   **56.** $\frac{1}{1000}$
57. $e^3 \approx 20.0855$ **58.** $\frac{1}{2}\left(\frac{\log 19}{\log 4} + 5\right) \approx 3.5620$
59. $\frac{\log 12}{\log 2} \approx 3.5850$ **60.** $\frac{\ln 0.03}{-0.1} \approx 35.0656$
61. $e^{-3} \approx 0.0498$ **62.** $\frac{15}{2}$ **63.** 16   **64.** 5
65. (a) 82; (b) 66.8; (c) 35 months
66. (a) 2.3 years; (b) 3.1 years
67. (a) $k \approx 1.565$; $A(t) = 175e^{1.565t}$;
(b) \$91.6 billion; (c) about 2014; (d) 0.44 year
68. (a) $k \approx 0.060$; $F(t) = 2500e^{-0.060t}$;
(b) approximately 3 fatalities; (c) about 2014
69. 11.553% per year   **70.** 16.5 years   **71.** 3463 years
72. 5.1   **73.** About 114 dB   **74.** ☛ Negative numbers
do not have logarithms because logarithm bases are positive, and there is no exponent to which a positive number

can be raised to yield a negative number.   **75.** ☛ If
$f(x) = e^x$, then to find the inverse function, we let $y = e^x$
and interchange x and y: $x = e^y$. If $x = e^y$, then $\log_e x = y$
by the definition of logarithms. Since $\log_e x = \ln x$, we have
$y = \ln x$ or $f^{-1}(x) = \ln x$. Thus, $g(x) = \ln x$ is the inverse
of $f(x) = e^x$. Another approach is to find $(f \circ g)(x)$ and
$(g \circ f)(x)$:
$$(f \circ g)(x) = e^{\ln x} = x, \text{ and}$$
$$(g \circ f)(x) = \ln e^x = x.$$
Thus, g and f are inverse functions.   **76.** e^{e^3}
77. $-3, -1$ **78.** $\left(\frac{8}{3}, -\frac{2}{3}\right)$

Test: Chapter 9, p. 651

1. [9.1] $(f \circ g)(x) = 2 + 6x + 4x^2$;
$(g \circ f)(x) = 2x^2 + 2x + 1$
2. [9.1] $f(x) = \frac{1}{x}$; $g(x) = 2x^2 + 1$ **3.** [9.1] No
4. [9.1] $f^{-1}(x) = \frac{x-4}{3}$ **5.** [9.1] $g^{-1}(x) = \sqrt[3]{x} - 1$
6. [9.2]   **7.** [9.3]

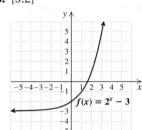

$f(x) = 2^x - 3$

$g(x) = \log_7 x$

8. [9.3] 3   **9.** [9.3] $\frac{1}{2}$ **10.** [9.4] 1   **11.** [9.4] 0
12. [9.3] $\log_5 \frac{1}{625} = -4$ **13.** [9.3] $2^m = \frac{1}{2}$
14. [9.4] $3\log a + \frac{1}{2}\log b - 2\log c$ **15.** [9.4] $\log_a \left(z^2 \sqrt[3]{x}\right)$
16. [9.4] 1.146   **17.** [9.4] 0.477   **18.** [9.4] 1.204
19. [9.5] 1.3979   **20.** [9.5] 0.1585   **21.** [9.5] -0.9163
22. [9.5] 121.5104   **23.** [9.5] 2.4022
24. [9.5]

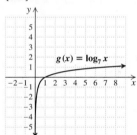

$f(x) = e^x + 3$

Domain: $\mathbb{R}$;
range: $(3, \infty)$

25. [9.5]

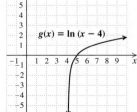

$g(x) = \ln (x - 4)$

Domain: $(4, \infty)$;
range: $\mathbb{R}$

26. [9.6] -5 **27.** [9.6] 2   **28.** [9.6] $\frac{1}{100}$
29. [9.6] $\frac{\log 1.2}{\log 7} \approx 0.0937$ **30.** [9.6] 4

31. [9.7] **(a)** 2.26 ft/sec; **(b)** 1,290,000
32. [9.7] **(a)** $P(t) = 155e^{0.019t}$, where t is the number of years after 2012 and $P(t)$ is in millions; **(b)** 167 million; 319 million; **(c)** about 2025; **(d)** 36.5 years
33. [9.7] **(a)** $k \approx 0.024$; $C(t) = 30,360e^{0.024t}$; **(b)** \$43,516; **(c)** about 2022   **34.** [9.7] 4.3%   **35.** [9.7] 7.0
36. [9.6] $-309, 316$ **37.** [9.4] 2

Cumulative Review: Chapters 1–9, p. 652

1. $\dfrac{y^{12}}{16x^8}$ **2.** $-\dfrac{y^4}{3z^5}$ **3.** $\dfrac{20x^6z^2}{y}$ **4.** 8   **5.** $(3, -1)$

6. $(1, -2, 0)$ **7.** $-7, 10$ **8.** $\dfrac{9}{2}$ **9.** $\dfrac{3}{4}$ **10.** $\pm 4i$

11. $\pm 2, \pm 3$ **12.** 9   **13.** $\dfrac{\log 7}{5 \log 3} \approx 0.3542$

14. $\dfrac{8e}{e - 1} \approx 12.6558$ **15.** $(-\infty, -5) \cup (1, \infty)$, or $\{x \mid x < -5 \text{ or } x > 1\}$ **16.** $-3 \pm 2\sqrt{5}$
17. $\{x \mid x \le -2 \text{ or } x \ge 5\}$, or $(-\infty, -2] \cup [5, \infty)$
18. $\dfrac{a + 2}{6}$ **19.** $\dfrac{7x + 4}{(x + 6)(x - 6)}$ **20.** $\sqrt[10]{(x + 5)^7}$
21. $15 - 4\sqrt{3}i$ **22.** $(3 + 4n)(9 - 12n + 16n^2)$
23. $2(3x - 2y)(x + 2y)$ **24.** $2(m + 3n)^2$
25. $(x - 2y)(x + 2y)(x^2 + 4y^2)$ **26.** $\dfrac{6 + \sqrt{y} - y}{4 - y}$

27. $f^{-1}(x) = \dfrac{x - 9}{-2}$, or $f^{-1}(x) = \dfrac{9 - x}{2}$

28. $f(x) = -10x - 8$
29.
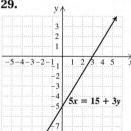
30.
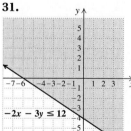
31.
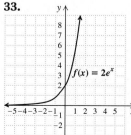
32.
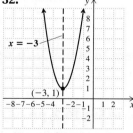
$f(x) = 2x^2 + 12x + 19$
Minimum: 1
33.
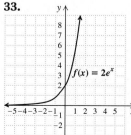
Domain: $\mathbb{R}$; range: $(0, \infty)$

34. 13.5 million acre-feet   **35. (a)** $k \approx 0.076$; $D(t) = 15e^{0.076t}$; **(b)** 108 million m³ per day; **(c)** 2020
36. $5\frac{5}{11}$ min   **37.** Thick and Tasty: 6 oz; Light and Lean: 9 oz
38. $\frac{1}{3}, \frac{10,000}{3}$ **39.** 35 mph

CHAPTER 10

Technology Connection, p. 659

1. $x^2 + y^2 - 16 = 0$
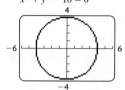
2. $(x - 1)^2 + (y - 2)^2 = 25$
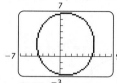

3. $(x + 3)^2 + (y - 5)^2 = 16$
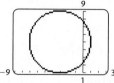

Exercise Set 10.1, pp. 659-663

1. Conic sections   **2.** Circle   **3.** Horizontal
4. Vertex   **5.** Center   **6.** Center   **7.** (f)   **8.** (e)
9. (c)   **10.** (b)   **11.** (d)   **12.** (a)
13.
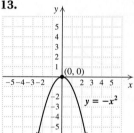
15.
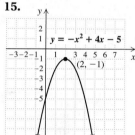
$y = -x^2 + 4x - 5$
$(2, -1)$
$y = -x^2$
17.
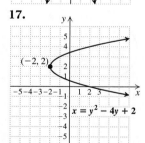
$(-2, 2)$
$x = y^2 - 4y + 2$
19.
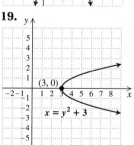
$(3, 0)$
$x = y^2 + 3$
21.
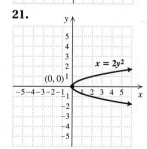
$x = 2y^2$
$(0, 0)$
23.
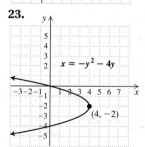
$x = -y^2 - 4y$
$(4, -2)$

25.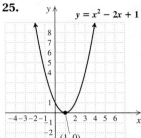

27.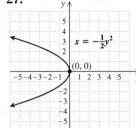

29.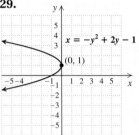

31.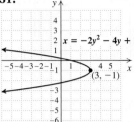

33. $x^2 + y^2 = 64$ **35.** $(x - 7)^2 + (y - 3)^2 = 6$
37. $(x + 4)^2 + (y - 3)^2 = 18$
39. $(x + 5)^2 + (y + 8)^2 = 300$ **41.** $x^2 + y^2 = 25$
43. $(x + 4)^2 + (y - 1)^2 = 20$
45. $(0, 0); 1$ **47.** $(-1, -3); 7$

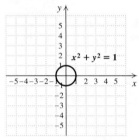   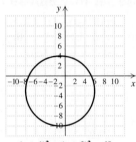

$(x + 1)^2 + (y + 3)^2 = 49$

49. $(4, -3); \sqrt{10}$ **51.** $(0, 0); 2\sqrt{2}$

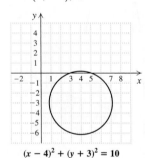   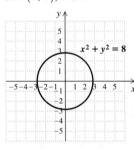

$(x - 4)^2 + (y + 3)^2 = 10$

53. $(5, 0); \frac{1}{2}$ **55.** $(-4, 3); \sqrt{40}$, or $2\sqrt{10}$

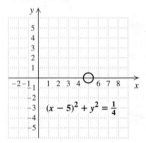   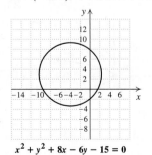

$(x - 5)^2 + y^2 = \frac{1}{4}$ $x^2 + y^2 + 8x - 6y - 15 = 0$

57. $(4, -1); 2$ **59.** $(0, -5); 10$

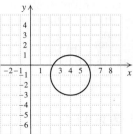   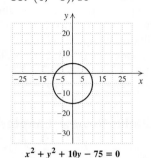

$x^2 + y^2 - 8x + 2y + 13 = 0$ $x^2 + y^2 + 10y - 75 = 0$

61. $\left(-\frac{7}{2}, \frac{3}{2}\right); \sqrt{\frac{98}{4}}$, or $\frac{7\sqrt{2}}{2}$ **63.** $(0, 0); \frac{1}{6}$

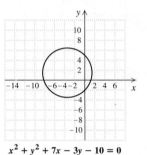   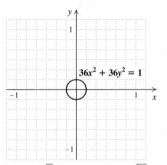

$x^2 + y^2 + 7x - 3y - 10 = 0$

65.   **67.** $2xy^3\sqrt[4]{3x^3}$ **68.** $y\sqrt[6]{y}$ **69.** $10x^2\sqrt{w}$
70. $\sqrt[30]{t^7}$ **71.** $2\sqrt{3}$ **72.** $12\sqrt{3} - 3\sqrt{2} + 4\sqrt{6} - 2$
73. 📋   **75.** $(x - 3)^2 + (y + 5)^2 = 9$
77. $(x - 3)^2 + y^2 = 25$ **79.** $\frac{17}{4}\pi$ m², or
approximately 13.4 m²   **81.** 7169 mm
83. (a) $(0, -3)$; (b) 5 ft   **85.** $x^2 + (y - 30.6)^2 = 590.49$
87. 7 in.   **89.** 📋, 📈

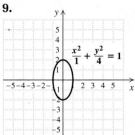

Prepare to Move On, p. 663

1. ± 4 **2.** $\pm a$ **3.** $-4, 6$ **4.** $-3 \pm 3\sqrt{3}$

Exercise Set 10.2, pp. 667–669

1. A   **2.** C   **3.** B   **4.** D   **5.** True   **6.** False
7. True   **8.** True
9.   **11.**

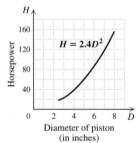   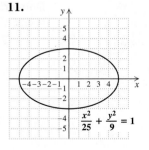

13.

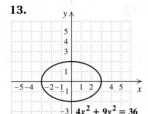

$4x^2 + 9y^2 = 36$

15.

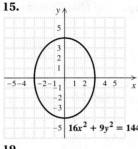

$16x^2 + 9y^2 = 144$

17.

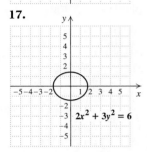

$2x^2 + 3y^2 = 6$

19.

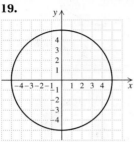

$5x^2 + 5y^2 = 125$

21.

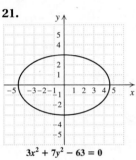

$3x^2 + 7y^2 - 63 = 0$

23.

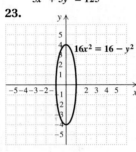

$16x^2 = 16 - y^2$

25.

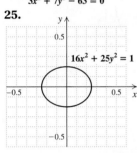

$16x^2 + 25y^2 = 1$

27.

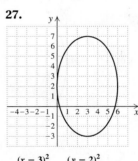

$\dfrac{(x-3)^2}{9} + \dfrac{(y-2)^2}{25} = 1$

29.

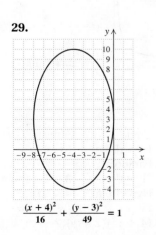

$\dfrac{(x+4)^2}{16} + \dfrac{(y-3)^2}{49} = 1$

31.

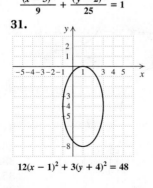

$12(x-1)^2 + 3(y+4)^2 = 48$

33.

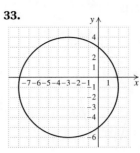

$4(x+3)^2 + 4(y+1)^2 - 10 = 90$

35. 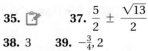   **37.** $\dfrac{5}{2} \pm \dfrac{\sqrt{13}}{2}$

38. 3   **39.** $-\frac{3}{4}, 2$

40. $\frac{5}{2}$ **41.** $-\sqrt{11}, \sqrt{11}$

42. $-10, 6$ **43.**

45. $\dfrac{x^2}{81} + \dfrac{y^2}{121} = 1$

47. $\dfrac{(x-2)^2}{16} + \dfrac{(y+1)^2}{9} = 1$

49. $\dfrac{x^2}{9} + \dfrac{y^2}{25} = 1$

51. (a) Let $F_1 = (-c, 0)$ and $F_2 = (c, 0)$. Then the sum of the distances from the foci to P is $2a$. By the distance formula,

$$\sqrt{(x+c)^2 + y^2} + \sqrt{(x-c)^2 + y^2} = 2a, \text{ or}$$
$$\sqrt{(x+c)^2 + y^2} = 2a - \sqrt{(x-c)^2 + y^2}.$$

Squaring, we get

$$(x+c)^2 + y^2 = 4a^2 - 4a\sqrt{(x-c)^2 + y^2} \\ + (x-c)^2 + y^2,$$

or

$$x^2 + 2cx + c^2 + y^2 = 4a^2 - 4a\sqrt{(x-c)^2 + y^2} \\ + x^2 - 2cx + c^2 + y^2.$$

Thus,

$$-4a^2 + 4cx = -4a\sqrt{(x-c)^2 + y^2} \\ a^2 - cx = a\sqrt{(x-c)^2 + y^2}.$$

Squaring again, we get

$$a^4 - 2a^2cx + c^2x^2 = a^2(x^2 - 2cx + c^2 + y^2) \\ a^4 - 2a^2cx + c^2x^2 = a^2x^2 - 2a^2cx + a^2c^2 + a^2y^2,$$

or

$$x^2(a^2 - c^2) + a^2y^2 = a^2(a^2 - c^2) \\ \dfrac{x^2}{a^2} + \dfrac{y^2}{a^2 - c^2} = 1.$$

(b) When P is at $(0, b)$, it follows that $b^2 = a^2 - c^2$. Substituting, we have

$$\dfrac{x^2}{a^2} + \dfrac{y^2}{b^2} = 1.$$

53. 5.66 ft   **55.**

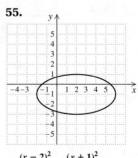

$\dfrac{(x-2)^2}{16} + \dfrac{(y+1)^2}{4} = 1$

57.

Quick Quiz: Sections 10.1–10.2, p. 669

1. $(-5, -1); 2$ **2.** $(x - 9)^2 + (y - (-23))^2 = 200$, or
$(x - 9)^2 + (y + 23)^2 = 200$

3.

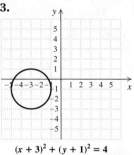

$(x + 3)^2 + (y + 1)^2 = 4$

4.

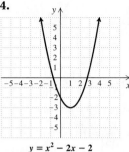

$y = x^2 - 2x - 2$

5.

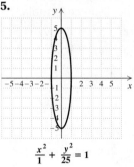

$\dfrac{x^2}{1} + \dfrac{y^2}{25} = 1$

Prepare to Move On, p. 669

1. $y = \dfrac{4}{x}$ **2.** $y = 2x - 4$ **3.** $x^2 - 5x - 7 = 0$

Technology Connection, p. 674

1.
$y_1 = \dfrac{\sqrt{15x^2 - 240}}{2}$;
$y_2 = -\dfrac{\sqrt{15x^2 - 240}}{2}$

2.
$y_1 = \sqrt{\dfrac{16x^2 - 64}{3}}$;
$y_2 = -\sqrt{\dfrac{16x^2 - 64}{3}}$

3.
$y_1 = \dfrac{\sqrt{5x^2 + 320}}{4}$;
$y_2 = -\dfrac{\sqrt{5x^2 + 320}}{4}$

4.
$y_1 = \sqrt{\dfrac{9x^2 + 441}{45}}$;
$y_2 = -\sqrt{\dfrac{9x^2 + 441}{45}}$

Connecting the Concepts, p. 676

1. $(4, 1); x = 4$ **2.** $(2, -1); y = -1$ **3.** $(3, 2)$
4. $(-3, -5)$ **5.** $(-12, 0), (12, 0), (0, -9), (0, 9)$
6. $(-3, 0), (3, 0)$ **7.** $(0, -1), (0, 1)$ **8.** $y = \frac{3}{2}x, y = -\frac{3}{2}x$

Exercise Set 10.3, pp. 677–678

1. B   **2.** E   **3.** A   **4.** D   **5.** F   **6.** C

7.

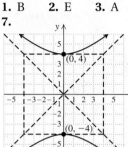

$\dfrac{y^2}{16} - \dfrac{x^2}{16} = 1$

9.

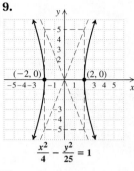

$\dfrac{x^2}{4} - \dfrac{y^2}{25} = 1$

11.

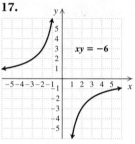

$\dfrac{y^2}{36} - \dfrac{x^2}{9} = 1$

13.

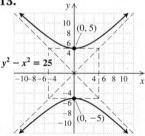

$y^2 - x^2 = 25$

15.

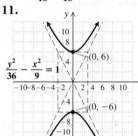

$25x^2 - 16y^2 = 400$

17.

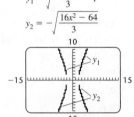

$xy = -6$

19.

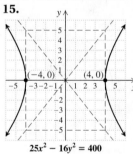

$xy = 4$

21.

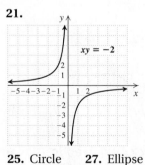

$xy = -2$

23.

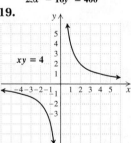

$xy = 1$

25. Circle   **27.** Ellipse
29. Hyperbola   **31.** Circle
33. Parabola
35. Hyperbola   **37.** Parabola
39. Hyperbola   **41.** Circle
43. Ellipse   **45.**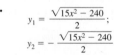
47. $(4 + y^2)(2 + y)(2 - y)$
48. $(3xy - 5)^2$

49. $10c(c - 3)(c - 5)$ **50.** $(x + 1)(x^2 + 3)$
51. $8t(t - 1)(t^2 + t + 1)$ **52.** $(3a - 2b)(2a + 5b)$
53.   **55.** $\dfrac{y^2}{36} - \dfrac{x^2}{4} = 1$
57. $C: (5, 2); V: (-1, 2), (11, 2);$ asymptotes:
$y - 2 = \frac{5}{6}(x - 5), y - 2 = -\frac{5}{6}(x - 5)$

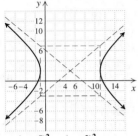

$$\dfrac{(x - 5)^2}{36} - \dfrac{(y - 2)^2}{25} = 1$$

59. $\dfrac{(y + 3)^2}{4} - \dfrac{(x - 4)^2}{16} = 1; C: (4, -3); V: (4, -5), (4, -1);$
asymptotes: $y + 3 = \frac{1}{2}(x - 4), y + 3 = -\frac{1}{2}(x - 4)$

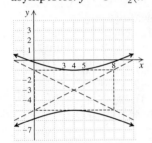

$$8(y + 3)^2 - 2(x - 4)^2 = 32$$

61. $\dfrac{(x + 3)^2}{1} - \dfrac{(y - 2)^2}{4} = 1; C: (-3, 2); V: (-4, 2), (-2, 2);$
asymptotes: $y - 2 = 2(x + 3), y - 2 = -2(x + 3)$

63.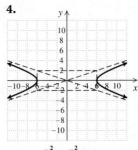

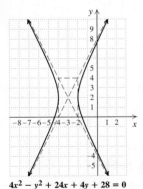

$$4x^2 - y^2 + 24x + 4y + 28 = 0$$

Quick Quiz: Sections 10.1–10.3, p. 678

1.

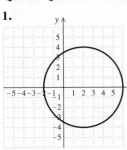

$(x - 2)^2 + y^2 = 16$

2.

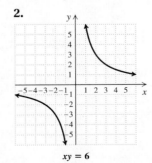

$xy = 6$

3.

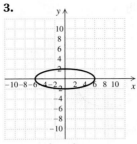

$$\dfrac{x^2}{36} + \dfrac{y^2}{4} = 1$$

4.

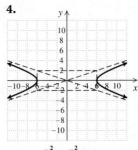

$$\dfrac{x^2}{36} - \dfrac{y^2}{4} = 1$$

5.

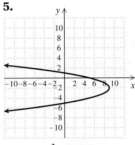

$$x = -y^2 - 4y + 5$$

Prepare to Move On, p. 678

1. $(-3, 6)$ **2.** $(3, 7)$ **3.** $-2, 2$ **4.** $-4, \frac{2}{3}$
5. $\dfrac{3}{2} \pm \dfrac{\sqrt{13}}{2}$ **6.** $\pm 1, \pm 5$

Mid-Chapter Review: Chapter 10, p. 679

1. $(x^2 - 4x) + (y^2 + 2y) = 6$
$(x^2 - 4x + 4) + (y^2 + 2y + 1) = 6 + 4 + 1$
$(x - 2)^2 + (y + 1)^2 = 11$

The center of the circle is $(2, -1)$. The radius is $\sqrt{11}$.
2. (a) Is there both an x^2-term and a y^2-term? Yes;
(b) Do both the x^2-term and the y^2-term have the same sign? No; **(c)** The graph of the equation is a hyperbola.
3. $(x + 4)^2 + (y - 9)^2 = 20$ **4.** Center: $(5, -1)$; radius: 6
5. Circle   **6.** Parabola

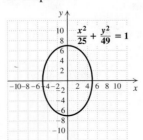

$x^2 + y^2 = 36$

$y = x^2 - 5$

7. Ellipse   **8.** Hyperbola

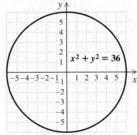

$$\dfrac{x^2}{25} + \dfrac{y^2}{49} = 1$$

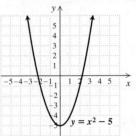

$$\dfrac{x^2}{25} - \dfrac{y^2}{49} = 1$$

9. Parabola

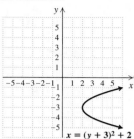

$x = (y + 3)^2 + 2$

10. Ellipse

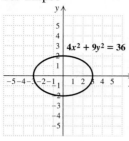

$4x^2 + 9y^2 = 36$

11. Hyperbola

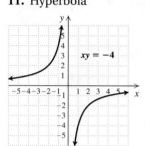

$xy = -4$

12. Circle

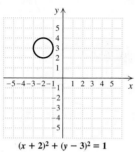

$(x + 2)^2 + (y - 3)^2 = 1$

13. Circle

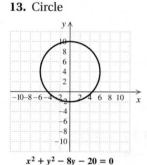

$x^2 + y^2 - 8y - 20 = 0$

14. Parabola

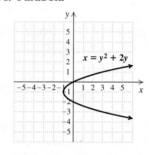

$x = y^2 + 2y$

15. Hyperbola

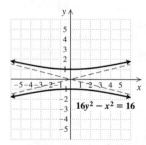

$16y^2 - x^2 = 16$

16. Hyperbola

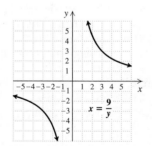

$x = \dfrac{9}{y}$

Technology Connection, p. 682

1. $(-1.50, -1.17); (3.50, 0.50)$
2. $(-2.77, 2.52); (-2.77, -2.52)$

Technology Connection, p. 684

1.
$y_1 = \sqrt{(20 - x^2)/4};\ y_2 = -\sqrt{(20 - x^2)/4};\ y_3 = 4/x$

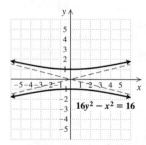

Exercise Set 10.4, pp. 685–687

1. True   **2.** True   **3.** False   **4.** False   **5.** True
6. True   **7.** $(-5, -4), (4, 5)$ **9.** $(0, 2), (3, 0)$
11. $(-2, 1)$
13. $\left(\dfrac{5 + \sqrt{70}}{3}, \dfrac{-1 + \sqrt{70}}{3} \right), \left(\dfrac{5 - \sqrt{70}}{3}, \dfrac{-1 - \sqrt{70}}{30} \right)$
15. $\left(4, \frac{3}{2}\right), (3, 2)$ **17.** $\left(\frac{7}{3}, \frac{1}{3}\right), (1, -1)$ **19.** $\left(\frac{11}{4}, -\frac{5}{4}\right), (1, 4)$
21. $(2, 4), (4, 2)$ **23.** $(3, -5), (-1, 3)$
25. $(-5, -8), (8, 5)$ **27.** $(0, 0), (1, 1),$
$\left(-\dfrac{1}{2} + \dfrac{\sqrt{3}}{2}i, -\dfrac{1}{2} - \dfrac{\sqrt{3}}{2}i \right), \left(-\dfrac{1}{2} - \dfrac{\sqrt{3}}{2}i, -\dfrac{1}{2} + \dfrac{\sqrt{3}}{2}i \right)$
29. $(-4, 0), (4, 0)$ **31.** $(-4, -3), (-3, -4), (3, 4), (4, 3)$
33. $\left(\dfrac{16}{3}, \dfrac{5\sqrt{7}}{3}i \right), \left(\dfrac{16}{3}, -\dfrac{5\sqrt{7}}{3}i \right), \left(-\dfrac{16}{3}, \dfrac{5\sqrt{7}}{3}i \right),$
$\left(-\dfrac{16}{3}, -\dfrac{5\sqrt{7}}{3}i \right)$ **35.** $(-3, -\sqrt{5}), (-3, \sqrt{5}), (3, -\sqrt{5}),$
$(3, \sqrt{5})$ **37.** $(-3, -1), (-1, -3), (1, 3), (3, 1)$
39. $(4, 1), (-4, -1), (2, 2), (-2, -2)$ **41.** $(2, 1), (-2, -1)$
43. $\left(2, -\frac{4}{5}\right), \left(-2, -\frac{4}{5}\right), (5, 2), (-5, 2)$ **45.** $(-\sqrt{2}, \sqrt{2}),$
$(\sqrt{2}, -\sqrt{2})$ **47.** Length: 8 cm; width: 6 cm
49. Length: 2 in.; width: 1 in.   **51.** Length: 12 ft; width: 5 ft
53. 6 and 15; −6 and −15   **55.** Length: 12 in.; width: 7.5 in.
57. Length: $\sqrt{3}$ m; width: 1 m   **59.** 📋   **61.** $\dfrac{9}{2a^3}$
62. $\frac{1}{4}$ **63.** 4   **64.** $-i$ **65.** -253
66. $10\sqrt{5}$ **67.** 📋   **69.** $(-2, 3), (2, -3), (-3, 2),$
$(3, -2)$ **71.** Length: 55 ft; width: 45 ft   **73.** 10 in.
by 7 in. by 5 in.

Quick Quiz: Sections 10.1–10.4, p. 687

1. Circle   **2.** Hyperbola   **3.** Ellipse   **4.** Parabola
5. $(2, -1), (1, -2)$

Prepare to Move On, p. 687

1. -9 **2.** -27 **3.** $\frac{1}{5}$ **4.** 77   **5.** $\frac{21}{2}$

Visualizing for Success, p. 688

1. C   **2.** A   **3.** F   **4.** B   **5.** J   **6.** D   **7.** H
8. I   **9.** G   **10.** E

Decision Making: Connection, p. 689

1. $b \approx 34.9$ in.; $h \approx 19.6$ in.   **2.** $b = 32$ in.; $h = 24$ in.
3. An older TV   **4.** $h = \dfrac{d}{\sqrt{1 + r^2}}$ **5.** 🖥

Study Summary: Chapter 10, pp. 690–691

1.
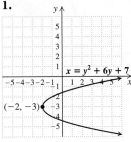

$x = y^2 + 6y + 7$

$(-2, -3)$

2. Center: $(3, 0)$; radius: 2
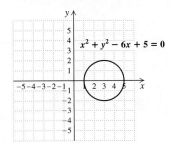

$x^2 + y^2 - 6x + 5 = 0$

3.
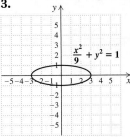

$\dfrac{x^2}{9} + y^2 = 1$

4.
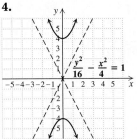

$\dfrac{y^2}{16} - \dfrac{x^2}{4} = 1$

5. $(-5, -4), (4, 5)$

Review Exercises: Chapter 10, pp. 692–693

1. True **2.** False **3.** False **4.** True **5.** True
6. True **7.** False **8.** True **9.** $(-3, 2), 4$
10. $(5, 0), \sqrt{11}$ **11.** $(3, 1), 3$ **12.** $(-4, 3), 3\sqrt{5}$
13. $(x + 4)^2 + (y - 3)^2 = 16$
14. $(x - 7)^2 + (y + 2)^2 = 20$
15. Circle **16.** Ellipse

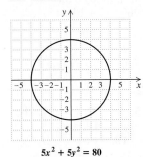

$5x^2 + 5y^2 = 80$

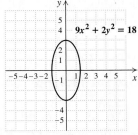

$9x^2 + 2y^2 = 18$

17. Parabola **18.** Hyperbola

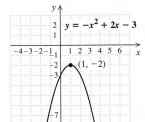

$y = -x^2 + 2x - 3$

$(1, -2)$

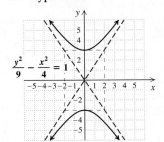

$\dfrac{y^2}{9} - \dfrac{x^2}{4} = 1$

19. Hyperbola
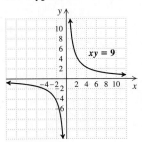

$xy = 9$

20. Parabola
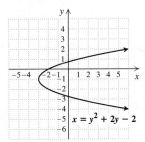

$x = y^2 + 2y - 2$

21. Ellipse
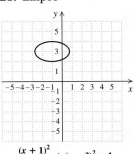

$\dfrac{(x + 1)^2}{3} + (y - 3)^2 = 1$

22. Circle
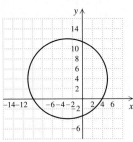

$x^2 + y^2 + 6x - 8y - 39 = 0$

23. $(5, -2)$ **24.** $(2, 2), \left(\frac{32}{9}, -\frac{10}{9}\right)$ **25.** $(0, -5), (2, -1)$
26. $(4, 3), (4, -3), (-4, 3), (-4, -3)$ **27.** $(2, 1), (\sqrt{3}, 0),$
$(-2, 1), (-\sqrt{3}, 0)$ **28.** $(3, -3), \left(-\frac{3}{5}, \frac{21}{5}\right)$ **29.** $(6, 8),$
$(6, -8), (-6, 8), (-6, -8)$ **30.** $(2, 2), (-2, -2),$
$(2\sqrt{2}, \sqrt{2}), (-2\sqrt{2}, -\sqrt{2})$ **31.** Length: 12 m; width: 7 m
32. Length: 12 in.; width: 9 in. **33.** Board: 32 cm; mirror:
20 cm **34.** 3 ft, 11 ft **35.** ✎ The graph of a parabola
has one branch whereas the graph of a hyperbola has two
branches. A hyperbola has asymptotes, but a parabola does
not. **36.** ✎ Function notation rarely appears in this
chapter because many of the relations are not functions.
Function notation could be used for vertical parabolas and
for hyperbolas that have the axes as asymptotes.
37. $(-5, -4\sqrt{2}), (-5, 4\sqrt{2}), (3, -2\sqrt{2}), (3, 2\sqrt{2})$
38. $(x - 2)^2 + (y + 1)^2 = 25$ **39.** $\dfrac{x^2}{100} + \dfrac{y^2}{1} = 1$

Test: Chapter 10, p. 693

1. [10.1] $(x - 3)^2 + (y + 4)^2 = 12$ **2.** [10.1] $(4, -1), \sqrt{5}$
3. [10.1] $(-2, 3), 3$
4. [10.1], [10.3] Parabola **5.** [10.1], [10.3] Circle

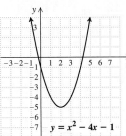

$y = x^2 - 4x - 1$

$x^2 + y^2 + 2x + 6y + 6 = 0$

6. [10.3] Hyperbola

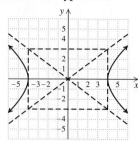

$$\frac{x^2}{16} - \frac{y^2}{9} = 1$$

7. [10.2], [10.3] Ellipse

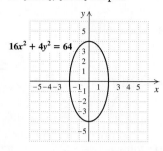

$16x^2 + 4y^2 = 64$

8. [10.3] Hyperbola

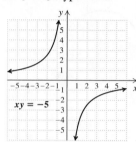

$xy = -5$

9. [10.1], [10.3] Parabola

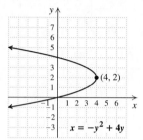

$(4, 2)$
$x = -y^2 + 4y$

10. [10.4] $(0, 6)$, $\left(\frac{144}{25}, \frac{42}{25}\right)$ **11.** [10.4] $(-4, 13)$, $(2, 1)$

12. [10.4] $(3, 2)$, $(-3, -2)$, $\left(-2\sqrt{2}\,i, \frac{3\sqrt{2}}{2}\,i\right)$,
$\left(2\sqrt{2}\,i, -\frac{3\sqrt{2}}{2}\,i\right)$ **13.** [10.4] $(\sqrt{6}, 2)$, $(\sqrt{6}, -2)$,

$(-\sqrt{6}, 2)$, $(-\sqrt{6}, -2)$ **14.** [10.4] 2 by 11

15. [10.4] $\sqrt{5}$ m, $\sqrt{3}$ m

16. [10.4] Length: 32 ft; width: 24 ft

17. [10.4] $1200, 6\%$ **18.** [10.2] $\frac{(x - 6)^2}{25} + \frac{(y - 3)^2}{9} = 1$

19. [10.4] 9   **20.** [10.2] $\frac{x^2}{16} + \frac{y^2}{49} = 1$

Cumulative Review: Chapters 1–10, p. 694

1. $16t^4 - 40t^2s + 25s^2$ **2.** $\frac{4t - 3}{3t(t - 3)}$ **3.** $3t^2\sqrt{10w}$

4. $27a^{1/2}b^{3/16}$ **5.** -4 **6.** 25 **7.** $-\frac{1}{64}$

8. $(10x - 3y)^2$ **9.** $3(m^2 - 2)(m^4 + 2m^2 + 4)$

10. $(x - y)(a - b)$ **11.** $(4x - 3)(8x + 1)$

12. $\left(-\infty, -\frac{25}{3}\right]$, or $\left\{x \mid x \le -\frac{25}{3}\right\}$ **13.** $0, \frac{9}{8}$ **14.** $1, 4$

15. $\pm i$ **16.** 4 **17.** $\frac{\log 1.5}{\log 3} \approx 0.3691$ **18.** 7

19. $(-\sqrt{3}, -1)$, $(-\sqrt{3}, 1)$, $(\sqrt{3}, -1)$, $(\sqrt{3}, 1)$

20.

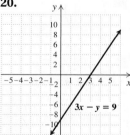

$3x - y = 9$

21.

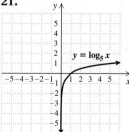

$y = \log_5 x$

22.

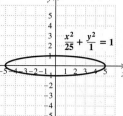

$\frac{x^2}{25} + \frac{y^2}{1} = 1$

23.

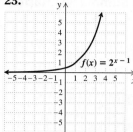

$f(x) = 2^{x-1}$

24.

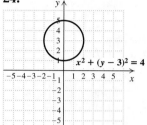

$x^2 + (y - 3)^2 = 4$

25.

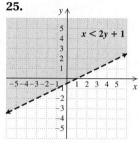

$x < 2y + 1$

26.

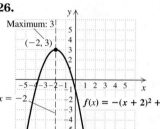

Maximum: 3
$(-2, 3)$
$x = -2$
$f(x) = -(x + 2)^2 + 3$

27. $y = -x + 3$
28. $x^2 - 3 = 0$
29. 2640 mi

30. (a) $g(t) = -687.5t + 48,750$; **(b)** $43,250 \text{ ft}^2$;
(c) about 2019   **31. (a)** $P(t) = 2.19e^{-0.0084t}$;
(b) 1.96 million; **(c)** 83 years   **32.** 8 in. by 8 in.
33. y is divided by 10.   **34.** $(-\infty, 0) \cup (0, 1]$, or
$\{x \mid x < 0 \text{ or } 0 < x \le 1\}$

CHAPTER 11

Exercise Set 11.1, pp. 700–702

1. B   **2.** A   **3.** B   **4.** A   **5.** A   **6.** B   **7.** (f)
8. (a)   **9.** (d)   **10.** (b)   **11.** (c)   **12.** (e)
13. 43   **15.** 364   **17.** -23.5 **19.** -363 **21.** $\frac{441}{400}$
23. 2, 5, 8, 11; 29; 44   **25.** 3, 6, 11, 18; 102; 227
27. $\frac{1}{2}, \frac{2}{3}, \frac{3}{4}, \frac{4}{5}; \frac{10}{11}; \frac{15}{16}$ **29.** $1, -\frac{1}{2}, \frac{1}{4}, -\frac{1}{8}; -\frac{1}{512}; \frac{1}{16,384}$
31. $-1, \frac{1}{2}, -\frac{1}{3}, \frac{1}{4}; \frac{1}{10}; -\frac{1}{15}$ **33.** $0, 7, -26, 63; 999; -3374$
35. $2n$ **37.** $(-1)^n$ **39.** $(-1)^{n+1} \cdot n$ **41.** $2n + 1$
43. $n^2 - 1$, or $(n + 1)(n - 1)$ **45.** $\frac{n}{n + 1}$
47. $(0.1)^n$, or 10^{-n} **49.** $(-1)^n \cdot n^2$ **51.** 5
53. 1.11111, or $1\frac{11,111}{100,000}$ **55.** $\frac{1}{2} + \frac{1}{4} + \frac{1}{6} + \frac{1}{8} + \frac{1}{10} = \frac{137}{120}$
57. $10^0 + 10^1 + 10^2 + 10^3 + 10^4 = 11,111$
59. $2 + \frac{3}{2} + \frac{4}{3} + \frac{5}{4} + \frac{6}{5} + \frac{7}{6} + \frac{8}{7} = \frac{1343}{140}$
61. $(-1)^2 2^1 + (-1)^3 2^2 + (-1)^4 2^3 + (-1)^5 2^4 +$
$(-1)^6 2^5 + (-1)^7 2^6 + (-1)^8 2^7 + (-1)^9 2^8 = -170$
63. $(0^2 - 2 \cdot 0 + 3) + (1^2 - 2 \cdot 1 + 3) +$
$(2^2 - 2 \cdot 2 + 3) + (3^2 - 2 \cdot 3 + 3) +$
$(4^2 - 2 \cdot 4 + 3) + (5^2 - 2 \cdot 5 + 3) = 43$
65. $\frac{(-1)^3}{3 \cdot 4} + \frac{(-1)^4}{4 \cdot 5} + \frac{(-1)^5}{5 \cdot 6} = -\frac{1}{15}$ **67.** $\sum_{k=1}^{5} \frac{k + 1}{k + 2}$

69. $\sum_{k=1}^{6} k^2$ **71.** $\sum_{k=2}^{n}(-1)^k k^2$ **73.** $\sum_{k=1}^{\infty} 6k$

75. $\sum_{k=1}^{\infty} \frac{1}{k(k+1)}$ **77.** [icon] **79.** $t^2 - t + 1$ **80.** $\frac{x}{a}$

81. $\frac{5a-3}{a(a-1)(a+1)}$ **82.** $\frac{2t}{t-1}$ **83.** $\frac{x-4}{4(x+2)}$

84. $\frac{3(y+1)(y-1)}{y}$ **85.** [icon] **87.** 1, 3, 13, 63, 313, 1563

89. \$2500, \$2000, \$1600, \$1280, \$1024, \$819.20, \$655.36, \$524.29, \$419.43, \$335.54 **91.** $S_{100}=0; S_{101}=-1$

93. $i, -1, -i, 1, i; i$ **95.** 11th term

Prepare to Move On, p. 702

1. 98 **2.** −15 **3.** $a_1 + a_n$ **4.** d

Exercise Set 11.2, pp. 708–710

1. Arithmetic series **2.** Common difference
3. First term **4.** Sum **5.** $a_1=8, d=5$
7. $a_1=7, d=-4$ **9.** $a_1=\frac{3}{2}, d=\frac{3}{4}$
11. $a_1=\$8.16, d=\$0.30$ **13.** 154 **15.** −94
17. −\$1628.16 **19.** 26th **21.** 57th **23.** 178
25. 5 **27.** 28 **29.** $a_1=8; d=-3; 8,5,2,-1,-4$
31. $a_1=1; d=1$ **33.** 780 **35.** 31,375 **37.** 2550
39. 918 **41.** 1030 **43.** 35 musicians; 315 musicians
45. 180 stones **47.** \$49.60 **49.** [icon]
51. $y=\frac{1}{3}x+10$ **52.** $y=-4x+11$ **53.** $y=-2x+10$
54. $y=-\frac{4}{3}x-\frac{16}{3}$ **55.** $x^2+y^2=16$
56. $(x+2)^2+(y-1)^2=20$ **57.** [icon]
59. 33 jumps **61.** Let d = the common difference. Since $p, m,$ and q form an arithmetic sequence, $m = p + d$ and

$q = p + 2d$. Then $\dfrac{p+q}{2} = \dfrac{p+(p+2d)}{2} = p+d = m$.

63. 156,375

Quick Quiz: Sections 11.1–11.2, p. 710

1. 19 **2.** $a_n=(n-1)^2$ **3.** $\sum_{k=1}^{\infty} 2k$ **4.** −0.5 **5.** 1

Prepare to Move On, p. 710

1. 315 **2.** 50 **3.** $\frac{2}{3}$

Connecting the Concepts, p. 715

1. −3 **2.** $-\frac{1}{2}$ **3.** 110 **4.** 640 **5.** 4410
6. \$1146.39 **7.** 1 **8.** No

Exercise Set 11.3, pp. 717–720

1. Infinite; geometric; sequence **2.** Finite; geometric; series **3.** Ratio; less; does **4.** 0.22 . . . ; geometric; series **5.** Geometric sequence **6.** Arithmetic sequence
7. Geometric series **8.** Arithmetic series
9. Geometric series **10.** None of these **11.** 2

13. −0.1 **15.** $-\frac{1}{2}$ **17.** $\frac{1}{5}$ **19.** $\frac{6}{m}$ **21.** 1458

23. 243 **25.** 52,488 **27.** \$1423.31 **29.** $a_n = 5^{n-1}$
31. $a_n = (-1)^{n-1}$, or $a_n = (-1)^{n+1}$
33. $a_n = \frac{1}{x^n}$, or $a_n = x^{-n}$ **35.** 3066 **37.** $\frac{547}{18}$
39. $\frac{1-x^8}{1-x}$, or $(1+x)(1+x^2)(1+x^4)$ **41.** \$5134.51
43. 27 **45.** $\frac{49}{4}$ **47.** No **49.** No **51.** $\frac{43}{99}$
53. \$25,000 **55.** $\frac{5}{9}$ **57.** $\frac{343}{99}$ **59.** $\frac{5}{33}$ **61.** $\frac{5}{1024}$ ft
63. 155,797 **65.** 2710 flies **67.** About 1456 GW
69. 3100.35 ft **71.** 20.48 in. **73.** [icon]
75. $\{-8,14\}$ **76.** $\left(x|-\frac{11}{2}<x<\frac{1}{2}\right)$, or $\left(-\frac{11}{2},\frac{1}{2}\right)$
77. $\{x|x \le 2 \text{ or } x \ge \frac{8}{3}\}$, or $(-\infty,2]\cup[\frac{8}{3},\infty)$
78. $\{x|-\frac{1}{3}<x<\frac{11}{3}\}$, or $(-\frac{1}{3},\frac{11}{3})$
79. $\{x|-2<x<7\}$, or $(-2,7)$
80. $\{x|-1 \le x < 0 \text{ or } x \ge 1\}$, or $[-1,0)\cup[1,\infty)$
81. [icon] **83.** 54 **85.** $\frac{x^2[1-(-x)^n]}{1+x}$ **87.** 512 cm²
89. [icon] **91.** [icon]

Quick Quiz: Sections 11.1–11.3, p. 720

1. −1, 2, −3, 4; 20 **2.** 140 **3.** 435 **4.** $511\frac{1}{2}$
5. $\frac{155}{99}$

Prepare to Move On, p. 720

1. $x^2 + 2xy + y^2$ **2.** $x^3 + 3x^2y + 3xy^2 + y^3$
3. $x^3 - 3x^2y + 3xy^2 - y^3$
4. $x^4 - 4x^3y + 6x^2y^2 - 4xy^3 + y^4$
5. $8x^3 + 12x^2y + 6xy^2 + y^3$
6. $8x^3 - 12x^2y + 6xy^2 - y^3$

Mid-Chapter Review: Chapter 11, p. 721

1. $a_n = a_1 + (n-1)d$ **2.** $a_n = a_1 r^{n-1}$
$n = 14, a_1 = -6, d = 5$ $n = 7, a_1 = \frac{1}{9}, r = -3$
$a_{14} = -6 + (14-1)5$ $a_7 = \frac{1}{9} \cdot (-3)^{7-1}$
$a_{14} = 59$ $a_7 = 81$

3. 300 **4.** $\frac{1}{n+1}$ **5.** 78 **6.** $2^2 + 3^2 + 4^2 + 5^2 = 54$

7. $\sum_{k=1}^{6}(-1)^{k+1} \cdot k$ **8.** 61st **9.** −39 **10.** $\frac{1}{10^8}$
11. $2(-1)^{n+1}$ **12.** Yes, a limit exists; $\frac{250}{3}$, or $83\frac{1}{3}$
13. \$465 **14.** \$1,073,741,823

Technology Connection, p. 725

1. 479,001,600 **2.** 56; 792

Exercise Set 11.4, pp. 727–729

1. Binomial **2.** Expansion **3.** First **4.** Third
5. Factorial **6.** Binomial **7.** 2^5, or 32 **8.** 8
9. 9 **10.** 1 **11.** 24 **13.** 3,628,800 **15.** 90
17. 126 **19.** 210 **21.** 1 **23.** 435 **25.** 780
27. $a^4 - 4a^3b + 6a^2b^2 - 4ab^3 + b^4$
29. $p^7 + 7p^6q + 21p^5q^2 + 35p^4q^3 + 35p^3q^4 + 21p^2q^5 + 7pq^6 + q^7$

31. $2187c^7 - 5103c^6d + 5103c^5d^2 - 2835c^4d^3 + 945c^3d^4 - 189c^2d^5 + 21cd^6 - d^7$
33. $t^{-12} + 12t^{-10} + 60t^{-8} + 160t^{-6} + 240t^{-4} + 192t^{-2} + 64$
35. $19{,}683s^9 + \dfrac{59{,}049s^8}{t} + \dfrac{78{,}732s^7}{t^2} + \dfrac{61{,}236s^6}{t^3} +$
$\dfrac{30{,}618s^5}{t^4} + \dfrac{10{,}206s^4}{t^5} + \dfrac{2268s^3}{t^6} + \dfrac{324s^2}{t^7} + \dfrac{27s}{t^8} + \dfrac{1}{t^9}$
37. $x^{15} - 10x^{12}y + 40x^9y^2 - 80x^6y^3 + 80x^3y^4 - 32y^5$
39. $125 + 150\sqrt{5}t + 375t^2 + 100\sqrt{5}t^3 + 75t^4 + 6\sqrt{5}t^5 + t^6$
41. $x^{-3} - 6x^{-2} + 15x^{-1} - 20 + 15x - 6x^2 + x^3$
43. $15a^4b^2$ **45.** $-64{,}481{,}508a^3$ **47.** $1120x^{12}y^2$
49. $1{,}959{,}552u^5v^{10}$ **51.** y^8 **53.** ☐

55.
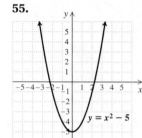

56.
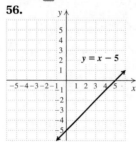

57.
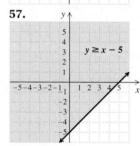

58.
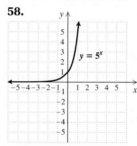

59.
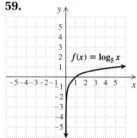

60.
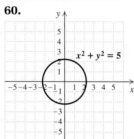

61. ☐   **63.** List all the subsets of size 3: $\{a, b, c\}$, $\{a, b, d\}$, $\{a, b, e\}$, $\{a, c, d\}$, $\{a, c, e\}$, $\{a, d, e\}$, $\{b, c, d\}$, $\{b, c, e\}$, $\{b, d, e\}$, $\{c, d, e\}$. There are exactly 10 subsets of size 3 and $\binom{5}{3} = 10$, so there are exactly $\binom{5}{3}$ ways of forming a subset of size 3 from $\{a, b, c, d, e\}$.
65. $\binom{8}{5}(0.15)^3(0.85)^5 \approx 0.084$ **67.** $\binom{8}{6}(0.15)^2(0.85)^6 + \binom{8}{7}(0.15)(0.85)^7 + \binom{8}{8}(0.85)^8 \approx 0.89$
69. $\binom{n}{n-r} = \dfrac{n!}{[n-(n-r)]!(n-r)!}$
$= \dfrac{n!}{r!(n-r)!} = \binom{n}{r}$

71. $-4320x^6y^{9/2}$ **73.** $-\dfrac{35}{x^{1/6}}$ **75.** 12

Quick Quiz: Sections 11.1–11.4, p. 729

1. $\dfrac{1+1}{1} + \dfrac{2+1}{2} + \dfrac{3+1}{3} + \dfrac{4+1}{4} = \dfrac{73}{12}$ **2.** -0.4
3. $-\frac{1}{2}$ **4.** 220   **5.** $x^4 + 4x^3w + 6x^2w^2 + 4xw^3 + w^5$

Visualizing for Success, p. 730

1. J   **2.** G   **3.** A   **4.** H   **5.** I   **6.** B   **7.** E
8. D   **9.** F   **10.** C

Decision Making: Connection, p. 731

1. (a) $1000, 1040, 1080, 1120, \ldots$;
(b) $a_n = 1000 + 40(n-1)$; **(c)** $1800
2. (a) $1000, 1040, 1081.60, 1124.87, \ldots$;
(b) $a_n = 1000(1.04)^{n-1}$; **(c)** $2191.12
3. (a) $1000, 1010, 1020, 1030, \ldots$; or $a_n = 1000 + 10(n-1)$; $1800
(b) $1000, 1010, 1020.10, 1030.30, \ldots$; or $a_n = 1000(1.01)^{n-1}$; $2216.72   **4.** ☐

Study Summary: Chapter 11, pp. 732–733

1. 143   **2.** -25 **3.** $5 \cdot 0 + 5 \cdot 1 + 5 \cdot 2 + 5 \cdot 3 = 30$
4. 15.5   **5.** 215   **6.** -640 **7.** $-20{,}475$
8. 16   **9.** 39,916,800   **10.** 84
11. $x^{10} - 10x^8 + 40x^6 - 80x^4 + 80x^2 - 32$ **12.** $3240t^7$

Review Exercises: Chapter 11, pp. 733–734

1. False   **2.** True   **3.** True   **4.** False   **5.** False
6. True   **7.** False   **8.** False   **9.** $1, 11, 21, 31$; 71; 111
10. $0, \frac{1}{5}, \frac{1}{5}, \frac{3}{17}; \frac{7}{65}; \frac{11}{145}$ **11.** $a_n = -5n$
12. $a_n = (-1)^n(2n-1)$
13. $-2 + 4 + (-8) + 16 + (-32) = -22$
14. $-3 + (-5) + (-7) + (-9) + (-11) + (-13) = -48$
15. $\sum\limits_{k=1}^{6} 7k$ **16.** $\sum\limits_{k=1}^{5} \dfrac{1}{(-2)^k}$ **17.** -55 **18.** $\frac{1}{5}$
19. $a_1 = -15$, $d = 5$ **20.** -544 **21.** 25,250
22. $1024\sqrt{2}$ **23.** $\frac{3}{4}$ **24.** $a_n = 2(-1)^n$
25. $a_n = 3\left(\dfrac{x}{4}\right)^{n-1}$ **26.** 11,718   **27.** $-4095x$ **28.** 12
29. $\frac{49}{11}$ **30.** No   **31.** No   **32.** $40,000   **33.** $\frac{5}{9}$
34. $\frac{16}{11}$ **35.** $24.30   **36.** 903 poles   **37.** $15,791.18
38. 6 m   **39.** 5040   **40.** 120   **41.** $190a^{18}b^2$
42. $x^4 - 8x^3y + 24x^2y^2 - 32xy^3 + 16y^4$

43. ☐ For a geometric sequence with $|r| < 1$, as n gets larger, the absolute value of the terms gets smaller, since $|r^n|$ gets smaller.   **44.** ☐ The first form of the binomial theorem draws the coefficients from Pascal's triangle; the second form uses factorial notation. The second form avoids the need to compute all preceding rows of Pascal's triangle, and is generally easier to use when only one term of an expansion is needed. When several terms of an expansion are needed and n is not large (say, $n \leq 8$), it is often easier to use Pascal's triangle.   **45.** $\dfrac{1 - (-x)^n}{x + 1}$

46. $x^{-15} + 5x^{-9} + 10x^{-3} + 10x^3 + 5x^9 + x^{15}$

Test: Chapter 11, p. 735

1. [11.1] $\frac{1}{2}, \frac{1}{5}, \frac{1}{10}, \frac{1}{17}, \frac{1}{26}; \frac{1}{145}$ **2.** [11.1] $a_n = 4\left(\frac{1}{3}\right)^n$

3. [11.1] $-3 + (-7) + (-15) + (-31) = -56$

4. [11.1] $\displaystyle\sum_{k=1}^{5}(-1)^{k+1}k^3$ **5.** [11.2] $\frac{13}{2}$

6. [11.2] $a_1 = 31.2; d = -3.8$ **7.** [11.2] 2508

8. [11.3] 1536   **9.** [11.3] $\frac{2}{3}$ **10.** [11.3] 3^n

11. [11.3] 5621   **12.** [11.3] 1   **13.** [11.3] No

14. [11.3] $\frac{\$25,000}{23} \approx \$1086.96$ **15.** [11.3] $\frac{85}{99}$

16. [11.2] 63 seats   **17.** [11.2] \$17,100

18. [11.3] \$5987.37   **19.** [11.3] 36 m   **20.** [11.4] 220

21. [11.4] $x^5 - 15x^4y + 90x^3y^2 - 270x^2y^3 + 405xy^4 - 243y^5$ **22.** [11.4] $220a^9x^3$

23. [11.2] $n(n + 1)$ **24.** [11.3] $\dfrac{1 - \left(\frac{1}{x}\right)^n}{1 - \frac{1}{x}}$, or $\dfrac{x^n - 1}{x^{n-1}(x - 1)}$

Cumulative Review/Final Exam: Chapters 1–11, pp. 736–737

1. $\frac{7}{15}$ **2.** $-4y + 17$ **3.** 280   **4.** 8.4×10^{-15}

5. $3a^2 - 8ab - 15b^2$ **6.** $4a^2 - 1$

7. $9a^4 - 30a^2y + 25y^2$ **8.** $\dfrac{4}{x + 2}$

9. $\dfrac{(x + y)(x^2 + xy + y^2)}{x^2 + y^2}$ **10.** $x - a$ **11.** $12a^2\sqrt{b}$

12. $-27x^{10}y^{-2}$, or $-\dfrac{27x^{10}}{y^2}$ **13.** $25x^4y^{1/3}$

14. $y\sqrt[12]{x^5y^2}, y \geq 0$ **15.** $14 + 8i$

16. $(2x - 3)^2$ **17.** $(3a - 2)(9a^2 + 6a + 4)$

18. $12(s^2 + 2t)(s^2 - 2t)$ **19.** $3(y^2 + 3)(5y^2 - 4)$

20. $7x^3 + 9x^2 + 19x + 38 + \dfrac{72}{x - 2}$

21. $[4, \infty)$, or $\{x \mid x \geq 4\}$ **22.** $(-\infty, 5) \cup (5, \infty)$, or $\{x \mid x < 5 \text{ or } x > 5\}$ **23.** $y = 3x - 8$

24. $x^2 - 50 = 0$ **25.** $(2, -3); 6$

26. $\log_a \dfrac{\sqrt[3]{x^2} \cdot z^5}{\sqrt{y}}$ **27.** $a^5 = c$ **28.** 2.0792

29. 0.6826   **30.** 5   **31.** -121 **32.** 875

33. $16\left(\frac{1}{4}\right)^{n-1}$ **34.** $13,440a^4b^6$ **35.** $\frac{3}{5}$ **36.** $-\frac{6}{5}, 4$

37. $\mathbb{R}$, or $(-\infty, \infty)$ **38.** $\left(-1, \frac{1}{2}\right)$ **39.** $(2, -1, 1)$

40. 2   **41.** $\pm 2, \pm 5$ **42.** $(\sqrt{5}, \sqrt{3}), (\sqrt{5}, -\sqrt{3}), (-\sqrt{5}, \sqrt{3}), (-\sqrt{5}, -\sqrt{3})$ **43.** 1.7925   **44.** 1005

45. $-\frac{1}{2}$ **46.** $\{x \mid -2 \leq x \leq 3\}$, or $[-2, 3]$

47. $\pm i\sqrt{3}$ **48.** $-2 \pm \sqrt{7}$ **49.** $\{y \mid y < -5 \text{ or } y > 2\}$, or $(-\infty, -5) \cup (2, \infty)$ **50.** $-8, 10$ **51.** $R = \dfrac{Ir}{1 - I}$

52.
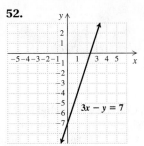
$3x - y = 7$

53.
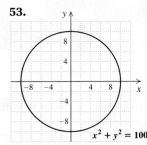
$x^2 + y^2 = 100$

54.
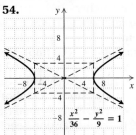
$\dfrac{x^2}{36} - \dfrac{y^2}{9} = 1$

55.
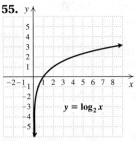
$y = \log_2 x$

56.
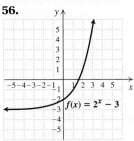
$f(x) = 2^x - 3$

57.
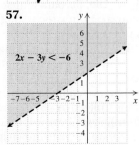
$2x - 3y < -6$

58.
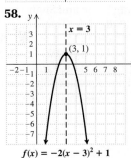
$f(x) = -2(x - 3)^2 + 1$
Maximum: 1

59. 5000 ft^2 **60.** 5 ft by 12 ft

61. More than 25 downloads

62. \$2.68 herb: 10 oz; \$4.60 herb: 14 oz   **63.** 350 mph

64. $8\frac{2}{5}$ hr, or 8 hr 24 min

65. **(a)** \$1.5 trillion per year;
(b) $f(t) = 1.5t + 9$;
(c) $P(t) = 9e^{0.128t}$;
(d) \$27 trillion;
(e) 41.8 trillion;
(f) about 5.4 years

66. All real numbers except 0 and -12 **67.** 81

68. y gets divided by 8   **69.** 84 years

PHOTO CREDITS

GLOSSARY

Abscissa [2.1] The first coordinate in an ordered pair

Absolute value [1.2, 4.3] The distance that a number is from zero on the number line

***ac*-method [5.4]** A method for factoring a trinomial that uses factoring by grouping; also called *grouping method*

Additive identity [1.2] The number 0

Additive inverse [1.2] A number's opposite; two numbers are additive inverses of each other if their sum is 0.

Algebraic expression [1.1] An expression consisting of variables and/or numerals, often with operation signs and grouping symbols

Arithmetic sequence (or **progression**) **[11.2]** A sequence in which the difference between any two successive terms is constant

Arithmetic series [11.2] A series for which the associated sequence is arithmetic

Ascending order [5.1] A polynomial in one variable written with the terms arranged according to degree, from least to greatest

Associative law for addition [1.2] The statement that when three numbers are added, regrouping the addends gives the same sum

Associative law for multiplication [1.2] The statement that when three numbers are multiplied, regrouping the factors gives the same product

Asymptote [10.3] A line that a graph approaches more and more closely as x increases or as x decreases

Axes (singular, **Axis**) **[2.1]** Two perpendicular number lines used to identify points in a plane

Axis of symmetry [8.6] A line that can be drawn through a graph such that the part of the graph on one side of the line is an exact reflection of the part on the opposite side

Base [1.1] In exponential notation, the number being raised to an exponent

Binomial [5.1] A polynomial with two terms

Branches [10.3] The two curves that comprise a hyperbola

Break-even point [3.8] In business, the point of intersection of the revenue function and the cost function

Cartesian coordinate system (*x, y*-**coordinate system**) **[2.1]** A coordinate system in which two perpendicular number lines are used to identify points in a plane; the variable x is usually represented on the horizontal axis and the variable y on the vertical axis

Circle [10.1] A set of points in a plane that are a fixed distance r, called the radius, from a fixed point (h, k), called the center

Closed interval [a, b] [4.1] The set of all numbers x for which $a \leq x \leq b$; thus, $[a, b] = \{x \mid a \leq x \leq b\}$

Coefficient [5.1] The part of a term that is a constant factor

Combined variation [6.8] A mathematical relationship in which a variable varies directly and/or inversely, at the same time, with more than one other variable

Common logarithm [9.5] A logarithm with base 10

Commutative law for addition [1.2] The statement that when two real numbers are added, the order in which the numbers are written does not affect the result

Commutative law for multiplication [1.2] The statement that when two real numbers are multiplied, the order in which the numbers are written does not affect the result

Completing the square [8.1] A method of adding a particular constant to an expression so that the resulting sum is a perfect square

Complex number [7.8] Any number that can be written in the form $a + bi$, where a and b are real numbers and $i = \sqrt{-1}$

Complex-number system [7.8] A number system that contains the real-number system and is designed so that negative numbers have defined square roots

Complex rational expression [6.3] A rational expression that contains rational expressions within its numerator and/or its denominator

Composite function [9.1] A function in which some quantity depends on a variable that, in turn, depends on another variable

Compound inequality [4.2] A statement in which two or more inequalities are joined by the word "and" or the word "or"

Compound interest [8.1] Interest earned on both the initial investment and the interest from the first time period

Conditional equation [1.3] An equation that is sometimes true and sometimes false, depending on the replacement for the variable

Conic section [10.1] A curve formed by the intersection of a plane and a cone

Conjugates [7.5] Pairs of radical expressions like $\sqrt{a} + \sqrt{b}$ and $\sqrt{a} - \sqrt{b}$

Conjugate of a complex number [7.8] The *conjugate* of a complex number $a + bi$ is $a - bi$, and the *conjugate* of $a - bi$ is $a + bi$.

Conjunction [4.2] A sentence in which two or more statements are joined by the word *and*

Consistent system of equations [3.1, 3.4] A system of equations that has at least one solution

Constant [1.1] A particular number that never changes

Constant function [2.4] A function given by an equation of the form $f(x) = b$, where b is a real number

Constant of proportionality [6.8] The constant, k, in an equation of direct or inverse variation; also called *variation constant*

Contradiction [1.3] An equation that is never true

Coordinates [2.1] The numbers in an ordered pair

Counting numbers [1.1] The set of numbers used for counting: $\{1, 2, 3, 4, 5, \dots\}$; also called *natural numbers*

Cube root [7.1] The number c is the *cube root* of a if $c^3 = a$.

Cubic function [5.1] A polynomial function in one variable of degree 3

Decay rate [9.7] The rate of decay of a population or other quantity at any instant in time

Degree of a polynomial [5.1] The degree of the term of highest degree in a polynomial

Degree of a term [5.1] The number of variable factors in a term

Demand function [3.8] A function modeling the relationship between the price of a good and the quantity of that good demanded

Dependent equations [3.1, 3.4] Equations in a system from which one equation can be removed without changing the solution set

Descending order [5.1] A polynomial in one variable written with the terms arranged in order according to degree, from greatest to least

Determinant [3.7] A descriptor of a square matrix; the determinant of a two-by-two matrix $\begin{bmatrix} a & c \\ b & d \end{bmatrix}$ is denoted by $\begin{vmatrix} a & c \\ b & d \end{vmatrix}$ and is defined as $ad - bc$.

Difference [1.1] The result when two numbers are subtracted

Difference of two cubes [5.6] An expression that can be written in the form $A^3 - B^3$

Difference of two squares [5.2, 5.5] An expression that can be written in the form $A^2 - B^2$

Direct variation [6.8] A situation that can be modeled by a linear function of the form $f(x) = kx$, or $y = kx$, where k is a nonzero constant

Discriminant [8.3] The radicand $b^2 - 4ac$ from the quadratic formula

Disjunction [4.2] A sentence in which two or more statements are joined by the word *or*

Distance formula [7.7] The formula $d = \sqrt{(x_2 - x_1)^2 + (y_2 - y_1)^2}$, where d is the distance between any two points (x_1, y_1) and (x_2, y_2)

Distributive law [1.2] The statement that multiplying a factor by the sum of two numbers gives the same result as multiplying the factor by each of the two numbers and then adding

Domain [2.2] The set of all first coordinates of the ordered pairs in a function

Doubling time [9.7] The amount of time necessary for a population to double in size

Element [1.1] Each object belonging to a set; also called *member*

Elements of a matrix [3.6] The individual entries in a matrix

Elimination method [3.2] An algebraic method that uses the addition principle to solve a system of equations

Ellipse [10.2] The set of all points in a plane for which the sum of the distances from two fixed points F_1 and F_2 is constant

Empty set [1.3] The set containing no elements, denoted $\varnothing$ or $\{\,\}$

Equation [1.1] A number sentence formed by placing an equals sign between two expressions

Equilibrium point [3.8] The point of intersection between the demand function and the supply function

Equivalent equations [1.3] Equations that have the same solution(s)

Equivalent expressions [1.2] Expressions that have the same value for all possible replacements

Evaluate [1.1] To substitute a number for each variable in an expression and calculate the result

Even root [7.1] A root with an even index

Exponent [1.1] In an expression of the form a^n, the number n is the exponent.

Exponential decay model [9.7] A decrease in quantity over time that can be modeled by an exponential function of the form $P(t) = P_0 e^{-kt}$, $k > 0$, where P_0 is the quantity present at time 0, $P(t)$ is the amount present at time t, and k is the exponential decay rate

Exponential equation [9.6] An equation with variables in exponents

Exponential function [9.2] The function $f(x) = a^x$, where a is a positive constant, $a \neq 1$, and x is any real number

Exponential growth model [9.7] An increase in quantity over time that can be modeled by an exponential function of the form $P(t) = P_0 e^{kt}$, $k > 0$, where P_0 is the quantity present at time 0, $P(t)$ is the amount present at time t, and k is the exponential growth rate

Exponential notation [1.1] A representation of a number using a base raised to an exponent

Extrapolation [2.4] The process of estimating a value that goes beyond the given data

Factor [1.2] *Verb:* to write an equivalent expression that is a product; *noun:* a multiplier

Factoring [1.2] The process of rewriting an expression as a product

Finite sequence [11.1] A function having for its domain a set of natural numbers: $\{1, 2, 3, 4, 5, \ldots, n\}$, for some natural number n

Finite series [11.1] The sum of the first n terms of a sequence: $a_1 + a_2 + a_3 + \cdots + a_n$; also called *partial sum*

Fixed costs [3.8] In business, costs that must be paid regardless of how many items are produced

Foci (singular, **Focus**) **[10.2]** Two fixed points that determine the points of an ellipse

FOIL method [5.2] To multiply two binomials $A + B$ and $C + D$, multiply the First terms AC, the Outer terms AD, the Inner terms BC, and then the Last terms BD. Then combine like terms if possible.

Formula [1.5] An equation using numbers and/or letters to represent a relationship between two or more quantities

Function [2.2] A correspondence between a first set, called the *domain*, and a second set, called the *range*, such that each member of the domain corresponds to *exactly one* member of the range

General term of a sequence [11.1] The nth term, denoted a_n

Geometric sequence [11.3] A sequence in which the ratio of every pair of successive terms is constant

Geometric series [11.3] A series for which the associated sequence is geometric

Graph [2.1, 4.1] A picture or a diagram of the data in a table; a set of points forming a line, a curve, or a plane that represent all the solutions of an equation or an inequality

Greatest common factor [5.3] The greatest common factor of a polynomial is the largest common factor of the coefficients times the greatest common factor of the variable(s) in the terms; also called *largest common factor.*

Grouping method [5.4] A method for factoring a trinomial that uses factoring by grouping; also called *ac-method*

Growth rate [9.7] The rate of growth of a population or other quantity at any instant in time

Half-life [9.7] The amount of time necessary for half of a quantity to decay

Half-open intervals $(a, b]$ and $[a, b)$ [4.1] An interval that contains one endpoint and not the other; thus, $(a, b] = \{x | a < x \leq b\}$ and $[a, b) = \{x | a \leq x < b\}$

Horizontal line [2.4] The graph of any function of the form $f(x) = b$ or $y = b$ is a horizontal line that crosses the y-axis at $(0, b)$; the slope of any horizontal line is 0.

Horizontal-line test [9.1] If it is impossible to draw a horizontal line that intersects a function's graph more than once, then the function is one-to-one.

Hyperbola [10.3] The set of all points P in the plane such that the difference of the distance from P to two fixed points is constant

Hypotenuse [5.8, 7.7] In a right triangle, the side opposite the 90° angle

***i* [7.8]** The square root of -1; that is, $i = \sqrt{-1}$ and $i^2 = -1$

Identity [1.3] An equation that is true for all replacements

Imaginary number [7.8] A number that can be written in the form $a + bi$, where a and b are real numbers and $b \neq 0$

Inconsistent system of equations [3.1, 3.4] A system of equations for which there is no solution

Independent equations [3.1, 3.4] Equations that are not dependent

Index (plural, Indices) [7.1] In the radical, $\sqrt[n]{a}$, the number n is called the index.

Inequality [1.2, 4.1] Formed when an inequality symbol, $(<, >, \leq, \geq, \text{or} \neq)$ is placed between two algebraic expressions

Infinite geometric series [11.3] The sum of the terms of an infinite geometric sequence

Infinite sequence [11.1] A function having for its domain the set of natural numbers: $\{1, 2, 3, 4, 5, \dots\}$

Infinite series [11.1] Given the infinite sequence $a_1, a_2, a_3, a_4, \dots, a_n, \dots$, the sum of the terms $a_1 + a_2 + a_3 + a_4 + \cdots + a_n + \cdots$ is called an infinite series.

Input [2.2] An element of the domain of a function

Integers [1.1] The set of all whole numbers and their opposites: $\{\dots, -4, -3, -2, -1, 0, 1, 2, 3, 4, \dots\}$

Interpolation [2.5] The process of estimating a value between given values

Intersection of A and B [4.2] The set of all elements that are common to both A and B; denoted $A \cap B$

Interval notation [4.1] The use of a pair of numbers inside parentheses or brackets to represent the set of all numbers between and sometimes including those two numbers; see also *closed interval* and *open interval*

Inverse relation [9.1] The relation formed by interchanging the members of the domain and the range of a relation

Inverse variation [6.8] A situation that can be modeled by a rational function of the form $f(x) = k/x$, or $y = k/x$, where k is a nonzero constant

Irrational number [1.1, 7.1] A real number that, when written as a decimal, neither terminates nor repeats

Isosceles right triangle [7.7] A right triangle in which both legs have the same length

Joint variation [6.8] A situation that can be modeled by an equation of the form $y = kxz$, where k is a nonzero constant

Largest common factor [5.3] The largest common factor of a polynomial is the greatest common factor of the coefficients times the greatest common factor of the variable(s) in the terms; also called *greatest common factor.*

Leading coefficient [5.1] The coefficient of the term of highest degree in a polynomial

Leading term [5.1] The term of highest degree in a polynomial

Least common denominator (LCD) [6.2] The least common multiple of the denominators of two or more fractions

Least common multiple (LCM) [6.2] The smallest number that is a multiple of two or more numbers

Legs [5.8, 7.7] In a right triangle, the two sides that form the 90° angle

Like radicals [7.5] Radical expressions that have the same indices and radicands

Like terms [1.3, 5.1] Terms that have exactly the same variable(s) raised to the same power(s); also called *similar terms*

Linear equation [1.3, 2.1] In two variables, any equation whose graph is a straight line, and can be written in the form $y = mx + b$, or $Ax + By = C$, where x and y are variables, and m, b, A, B, and C are constants

Linear equation in three variables [3.4] An equation equivalent to one of the form $Ax + By + Cz = D$, where A, B, C, and D are real numbers

Linear function [2.3, 5.1] A function whose graph is a straight line and can be written in the form $f(x) = mx + b$, where m and b are constants

Linear inequality [4.4] An inequality whose related equation is a linear equation

Linear regression [2.5] A method for finding an equation for a line that best fits a set of data

Logarithmic equation [9.6] An equation containing a logarithmic expression

Logarithmic function, base a [9.3] The inverse of an exponential function $f(x) = a^x$, written $f^{-1}(x) = \log_a x$

Mathematical model [1.5] A mathematical representation of a real-world situation

Matrix (plural, Matrices) [3.6] A rectangular array of numbers

Maximum value [8.6] The greatest function value (output) achieved by a function

Member [1.1]   Each object belonging to a set; also called *element*

Midpoint formula [7.7]   The formula $\left(\dfrac{x_1 + x_2}{2}, \dfrac{y_1 + y_2}{2}\right)$, which represents the midpoint of the line segment with endpoints (x_1, y_1) and (x_2, y_2)

Minimum value [8.6]   The least function value (output) achieved by a function

Monomial [5.1]   A constant, a variable, or a product of a constant and one or more variables

Motion problem [3.3, 6.5]   A problem that deals with distance, speed (rate), and time

Multiplicative identity [1.2]   The number 1

Multiplicative inverses [1.2]   *Reciprocals*; two numbers whose product is 1

Natural logarithm [9.5]   A logarithm with base e; also called Napierian logarithm

Natural numbers [1.1]   The numbers used for counting: $\{1, 2, 3, 4, 5, \ldots\}$; also called *counting numbers*

Nonlinear equation [2.1]   An equation whose graph is not a straight line

***n*th root [7.1]**   A number c is called the *n*th root of a if $c^n = a$.

Odd root [7.1]   A root with an odd index

One-to-one function [9.1]   A function for which different inputs have different outputs

Open interval (a, b) [4.1]   The set of all numbers x for which $a < x < b$; thus, $(a, b) = \{x \mid a < x < b\}$

Opposites [1.2]   Two expressions whose sum is 0; additive inverses

Opposite of a polynomial [5.1]   The opposite of polynomial P can be written as $-P$ or, equivalently, by replacing each term in P with its opposite

Ordered pair [2.1]   A pair of numbers of the form (x, y) for which the order in which the numbers are listed is important

Ordinate [2.1]   The second coordinate in an ordered pair

Origin [2.1]   The point on a coordinate plane where the two axes intersect

Output [2.2]   An element of the range of a function

Parabola [8.1, 8.6, 10.1]   A graph of a quadratic function

Parallel lines [2.4]   Lines in the same plane that never intersect; two lines are parallel if they have the same slope or if both lines are vertical

Partial sum [11.1]   The sum of the first n terms of a sequence: $a_1 + a_2 + a_3 + \cdots + a_n$; also called *finite series*

Pascal's triangle [11.4]   A triangular array of coefficients of the expansion $(a + b)^n$ for $n = 0, 1, 2, \ldots$

Perfect-square trinomial [5.5]   A trinomial that is the square of a binomial

Perpendicular lines [2.4]   Two lines that intersect to form a right angle; two lines are perpendicular if the product of their slopes is -1 or if one line is vertical and the other line is horizontal

Piecewise function [2.2]   A function defined by different equations for various parts of their domains

Point–slope form [2.5]   Any equation of the form $y - y_1 = m(x - x_1)$, where the slope of the line is m and the line passes through (x_1, y_1)

Polynomial [5.1]   A monomial or a sum of monomials

Polynomial equation [5.8]   An equation in which two polynomials are set equal to each other

Polynomial function [5.1]   A function in which outputs are determined hy evaluating a polynomial

Polynomial inequality [8.9]   An inequality that is equivalent to an inequality with a polynomial as one side and 0 as the other

Prime polynomial [5.3]   A polynomial that cannot be factored

Principal square root [7.1]   The nonnegative square root of a number

Product [1.1]   The result when two numbers are multiplied

Progression [11.1]   A function for which the domain is a set of counting numbers beginning with 1; also called *sequence*

Pure imaginary number [7.8]   A complex number of the form $a + bi$, in which $a = 0$ and $b \neq 0$

Pythagorean theorem [5.8, 7.7]   In any right triangle, if a and b are the lengths of the legs and c is the length of the hypotenuse, then $a^2 + b^2 = c^2$.

Quadrants [2.1]   The four regions into which the horizontal axis and the vertical axis divide a plane

Quadratic equation [5.8, 8.1]   An equation equivalent to one of the form $ax^2 + bx + c = 0$, where $a \neq 0$

Quadratic formula [8.2]   The formula $x = (-b \pm \sqrt{b^2 - 4ac})/(2a)$, which gives the solutions of $ax^2 + bx + c = 0$, $a \neq 0$

Quadratic function [5.1, 8.2]   A second-degree polynomial function in one variable

Quadratic inequality [8.9]   A second-degree polynomial inequality in one variable

Quartic function [5.1]   A polynomial function in one variable of degree 4

Quotient [1.1]   The result when two numbers are divided

Radical equation [7.6]   An equation in which a variable appears in a radicand

Radical expression [7.1]   Any expression in which a radical sign appears

Radical sign [7.1]   The symbol $\sqrt{}$

Radical term [7.5]   A term in which a radical sign appears

Radicand [7.1]   The expression under the radical sign

Radius (plural, **Radii**) **[10.1]**   The distance from the center of a circle to a point on the circle; a segment connecting a point on the circle to the center of the circle

Range [2.2]   The set of all second coordinates of the ordered pairs in a function

Rational equation [6.4]   An equation that contains one or more rational expressions

Rational expression [6.1]   An expression that consists of a polynomial divided by a nonzero polynomial

Rational function [6.1]   A function described by a rational expression

Rational inequality [8.9]   An inequality involving a rational expression

Rational numbers [1.1]   Numbers that can be expressed as an integer divided by a nonzero integer

Rationalizing the denominator [7.4]   A procedure for finding an equivalent expression without a radical expression in the denominator

Rationalizing the numerator [7.4] A procedure for finding an equivalent expression without a radical expression in the numerator

Real numbers [1.1] Numbers that are either rational or irrational

Reciprocals [1.2] Two numbers whose product is 1; *multiplicative inverses*

Reflection [9.2] The mirror image of a graph

Relation [2.2] A correspondence between a first set, called the *domain,* and a second set, called the *range,* such that each member of the domain corresponds to at least one member of the range

Remainder theorem [6.7] The remainder obtained by dividing the polynomial $P(x)$ by $x - r$ is $P(r)$.

Repeating decimal [1.1] A decimal in which a block of digits repeats indefinitely

Right triangle [5.8] A triangle that has a 90° angle

Roster notation [1.1] Set notation in which the elements of a set are listed within { }

Row-equivalent operations [3.6] Operations used to produce equivalent systems of equations

Scientific notation [1.7] An expression of the form $N \times 10^m$, where N is in decimal notation, $1 \leq N < 10$, and m is an integer

Sequence [11.1] A function for which the domain is a set of consecutive counting numbers beginning with 1; also called *progression*

Series [11.1] The sum of specified terms in a sequence

Set-builder notation [1.1] The naming of a set by specifying conditions under which an element is in the set

Sigma notation [11.1] The naming of a sum using the Greek letter Σ (sigma) as part of an abbreviated form; also called *summation notation*

Similar terms [1.3, 5.1] Terms that have exactly the same variable(s) raised to the same power(s); also called *like terms*

Slope [2.3] The ratio of the rise to the run (or the vertical change to the horizontal change) for any two points on a line

Slope–intercept form [2.3] Any equation of the form $y = mx + b$, where the slope of the line is m, and the y-intercept is $(0, b)$

Solution [1.1, 4.1] A replacement or substitution that makes an equation or inequality true

Solution set [1.3, 4.1] The set of all solutions of an equation, an inequality, or a system of equations or inequalities

Solve [1.1], [3.1], [4.1] To find all solutions of an equation, an inequality, or a system of equations or inequalities

Square matrix [3.7] A matrix with the same number of rows and columns

Square root [7.1] The number c is a *square root* of a if $c^2 = a$.

Standard form of a linear equation [2.4] Any equation of the form $Ax + By = C$, where A, B, and C are real numbers and A and B are not both 0

Subset [1.1] When every member of one set is a member of a second set, the first set is a *subset* of the second set.

Substitute [1.1] To replace a variable with a number or an expression

Substitution method [3.2] An algebraic method for solving a system of equations

Sum [1.1] The result when two numbers are added

Sum of two cubes [5.6] An expression that can be written in the form $A^3 + B^3$

Summation notation [11.1] The naming of a sum using the Greek letter Σ (sigma) as part of an abbreviated form; also called *sigma notation*

Supply function [3.8] A function modeling the relationship between the price of a good and the quantity of that good supplied

Synthetic division [6.7] A method used to divide a polynomial by a binomial of the type $x - a$

System of equations [3.1] A set of two or more equations, in two or more variables, for which a common solution is sought

Term [1.3, 5.1] A number, a variable, or a product of numbers and/or variables, or a quotient of numbers and/or variables

Terminating decimal [1.1] A decimal that can be written using a finite number of decimal places

Total cost [3.8] The amount spent to produce a product

Total profit [3.8] The money taken in less the money spent, or total revenue minus total cost

Total revenue [3.8] The amount taken in from the sale of a product

Trinomial [5.1] A polynomial with three terms

Undefined [1.2] An expression that has no meaning attached to it

Union of A and B [4.2] The collection of elements belonging to A and/or B; denoted $A \cup B$

Value [1.1] The numerical result after a number has been substituted for a variable in an expression and calculations have been carried out

Variable [1.1] A letter that represents an unknown number

Variable costs [3.8] In business, costs that vary according to the quantity being produced

Variation constant [6.8] The constant k in an equation of direct or inverse variation; also called *constant of proportionality*

Vertex [8.6] The "turning point" of the graph of a quadratic equation

Vertical line [2.4] The graph of any equation of the form $x = a$ is a vertical line that crosses the x-axis at $(a, 0)$.

Vertical-line test [2.2] The statement that if it is possible for a vertical line to cross a graph more than once, then the graph is not the graph of a function

Whole numbers [1.1] The set of natural numbers with 0 included: $\{0, 1, 2, 3, 4, 5, \ldots\}$

x-intercept [2.4] The point at which a line crosses the x-axis

x, y-coordinate system (Cartesian coordinate system) [2.1] A coordinate system in which two perpendicular number lines are used to identify points in a plane; the variable x is usually represented on the horizontal axis and the variable y on the vertical axis

y-intercept [2.4] The point at which a line crosses the y-axis

Zeros [8.9] The x-values for which a function $f(x)$ is 0

INDEX

Multiplication. *See also* Product(s)
associative law for, 17, 66
of binomials, 291
commutative law for, 16–17, 66
of complex rational expressions,
by one, 374–376, 426
of monomials, 290–291
of polynomials. *See* Multiplying
polynomials
of radical expressions, 449–452,
462–463, 496, 497
of rational expressions, 359, 425
of real numbers, 14–15, 65
significant digits and, 56–57
Multiplication principle, 21–22, 66
for inequalities, 227–228, 273
Multiplicative identity, 14
Multiplicative inverses.
See Reciprocals
Multiplying polynomials, 290–297,
346–347
FOIL method and, 292–293
function notation and, 296–297
multiplying monomials and,
290–291
multiplying monomials and
binomials and, 291
products of sums and differences
and, 294–296
special products and, 292–293, 347
squares of binomials and, 293–294

Napier, John, 620
Napierian logarithms. *See* Natural
logarithms
Natural logarithms, 619
on a calculator, 620–621
Natural numbers, 5
Negative exponents, factors and, 49
Negative integers as exponents, 48–50
Nonlinear equations, 76–77
systems of. *See* Nonlinear systems
of equations
Nonlinear systems of equations,
680–685, 691
involving one equation, 680–682
problem solving with, 684–685
of two equations, 682–683
Notation
absolute-value, 11
decimal, 3
exponential, 3
factorial. *See* Factorial notation
function, for polynomials, 296–297
interval (set-builder). *See* Interval
notation
roster, 5
scientific. *See* Scientific notation
set-builder, 6
sigma (summation), 698–699
*n*th roots, 438–439, 496
*n*th term of arithmetic sequence,
finding, 703–704

Number(s), 3
complex. *See* Complex numbers
fraction, 3
imaginary. *See* Imaginary numbers
irrational, 7, 435
natural (counting), 5
rational, 3
real. *See* Real numbers
relationships among kinds of, 7
repeating, 3
terminating, 3
whole, 5
Number *e*, 620
Number *i*, 487, 498
powers of, 491
Numerators
with one term, rationalizing, 457–458
with two terms, rationalizing, 463–464

Objective function, 265, 275
Odd roots, 438
One
multiplication of complex rational
expressions by, 374–376, 426
as multiplicative identity, 14
One-to-one functions, 587, 647
Open interval, 225
Opposites, 12
law of, 12
of a polynomial, 284–285, 285, 346
Order of operations, 4
Ordered pairs, 72–73
Ordinate, 72n
Origin
equation of a hyperbola centered
at, 670
of a graph, 72

Pairs, ordered, 72–73
Parabolas, 504, 540, 654–657, 690.
See also Quadratic equations;
Quadratic functions
axes of symmetry of, 540, 541, 577
equation of, 654, 690
vertices of, 540, 541, 551, 577
Parallel lines, 108–109, 143
Pascal's triangle, binomial expansion
using, 722–723
Perfect-square trinomials, factoring,
317–318
Perpendicular lines, 109, 143
Piecewise-defined functions, 88
Point–slope form of a linear equation,
119–120, 143
Polynomial(s), 280–281, 346
addition of, 284, 346
ascending order of, 282
coefficient of a term of, 281, 346
degree of, 281, 346
degree of a term of, 281, 346
descending order of, 282
division by a monomial,
400–401, 427

division by a polynomial,
401–404, 427
factored completely, 301
greater common factor of, 300, 347
leading term of, 281, 346
multiplication of. *See* Multiplying
polynomials
opposite of, 285, 346
prime, 301
subtraction of, 284–285, 346
synthetic division and, 406–408, 428
terms of. *See* Terms of a polynomial
Polynomial equations, 332
principle of zero products and,
332–334, 348
second-degree, in one variable.
See Quadratic equations
Polynomial functions, 279, 282–283
division of, remainder theorem
and, 409, 428
Polynomial inequalities, 565–568, 578
Power(s). *See also* Exponent(s)
of *i*, 491
principle of, 470–472, 497
raising a product to, 51–53, 68
raising a quotient to, 52, 68
Power rule, 51, 68
for logarithms, 611–612, 648
Price, cost versus, 207
Prime polynomials, 301
Principal square root, 434–435, 496
Principles
absolute-value, 246
addition. *See* Addition principle
Corner, 266–268, 275
of exponential equality, 607, 626,
648
of logarithmic equality, 626, 648
multiplication. *See* Multiplication
principle
of powers, 470–472, 497
of square roots, 505–507
of zero products, 332–334, 348
Problem solving, 29–34
of absolute-value problems,
principles of, 249
five-step strategy for, 29–32, 67
with geometric sequences and
series, 716–717
with inequalities, 229–231, 274
with nonlinear systems of
equations, 684–685
with quadratic equations, 335–338,
510–511, 527–529
with quadratic functions, 555–558,
578
scientific notation in, 58–59
tips for, 170, 175
Product(s). *See also* Multiplication
with different indices, simplifying,
465–466
of functions, 130–131, 143
raising to a power, 51–53, 68